TALL BUILDING DESIGN

Steel, Concrete, and Composite Systems

Bungale S. Taranath, Ph.D., P.E., S.E.

TALL BUILDING DESIGN

Steel, Concrete, and Composite Systems

CRC Press
Taylor & Francis Group
Boca Raton London New York

CRC Press is an imprint of the
Taylor & Francis Group, an **informa** business

CRC Press
Taylor & Francis Group
6000 Broken Sound Parkway NW, Suite 300
Boca Raton, FL 33487-2742

Printed on acid-free paper
Version Date: 20160610

International Standard Book Number-13: 978-1-4665-5620-1 (Hardback)

Library of Congress Cataloging-in-Publication Data

Taranath, Bungale S.
Tall building design : steel, concrete, and composite systems / Bungale S. Taranath.
pages cm
Includes bibliographical references and index.
ISBN 978-1-4665-5620-1
1. Building, Iron and steel. 2. Concrete construction. 3. Composite construction. 4. Tall buildings--Design and construction. I. Title.

TH1611.T373 2015
720'.483--dc23 2013047690

Visit the Taylor & Francis Web site at
http://www.taylorandfrancis.com

and the CRC Press Web site at
http://www.crcpress.com

Printed and bound in the United States of America by
Edwards Brothers Malloy on sustainably sourced paper

This book is dedicated to my wife, Saroja.

My dearest and best friend, whose considerable expertise and interest in all aspects of life has influenced my entire existence including my book-writing avocation.

Contents

Preface

Tall buildings have a unique appeal, even an air of romance and mystery associated with their design. The adoration that super- and ultratall buildings command lies in their apparent freedom from gravity loads—they do not just stand tall, they seem to do so effortlessly resisting gravity as well as laterally directed force generated by wind gusts and seismic ground motions.

Tall buildings have fascinated humans from the beginning of civilization—the primary motivation was to create monuments rather than human habitats. Today's structures, on the other hand, are human habitats—not allowed by economics and design to be nearly as simple, heavy, stiff, and robust as their relatively recent counterparts such as the Empire State Building of the 1930s.

Although tall buildings are unique from certain aspects such as consideration of lateral deflection, their design, in a manner of speaking, is similar to the design of their lower brethren. Thus the material presented in this book applies equally to not-so-tall buildings as well.

This book is an outgrowth of my previous publications. It attempts to maintain the same basic approach: first to establish a firm understanding of the behavior of structural members and systems and then to develop proficiency in the methods used in current design practice with particular reference to the provisions of the following publications:

- Minimum Design Loads for Buildings and Other Structures, ASCE/SEI 7-10 Specifications
- Specifications for Structural Steel Buildings, ANSI/ASCE 360-10
- Seismic Provisions of Structural Steel Buildings, ANS1/AISC 341-10
- Pre-qualified Connections for Special and Intermediate Steel Moment Connections for Seismic Applications, ANSI/AISC 358-10
- Building Code Requirements for Structural Concrete, ACI 318-11
- Seismic Rehabilitation of Existing Buildings, ASCE/SEI 41-06

Much of the present-day design is carried out using commercially available computer software or spreadsheets written by individuals for their particular needs. It is generally recognized that mere proficiency in navigating through computer software is inadequate, and often dangerous, for successful professional practice. Moreover, code provisions and procedures are subject to change periodically, oftentimes too frequent for the comfort of design professionals. To understand and keep abreast of these rapid developments is no small task. To do so successfully, the engineer needs a thorough grounding in the behavior of structural components and systems. Familiarity in the present-day methodology is essential to design structures that comply with legally adapted standards and to do so safely, economically, and efficiently.

The fundamental laws governing the static and dynamic analysis of structures subjected to the forces of nature are over 150 years old. Therefore, anyone who claims that they have invented a new fundamental principle is a victim of their own knowledge gap. The real challenge in writing a text in the structural engineering field, then, is to describe in physical and practical terms the underlying theory and how it relates to the modern world, where structural analysis and even the interpretation of analysis results are typically done by the computer.

Thus, the foremost objectives of this book are as follows:

- To promote a better understanding of the structural behavior of steel, concrete, and composite members and systems.
- To develop a cohesive wind- and earthquake-resistant design procedure for tall building structures and their lower brethren.

- To bridge the gap between two design approaches, one based on skill and experience and the other that relies upon computer skills, to imagine the design possibilities when that wonderful ability—the intuition we humans have—marries unfathomable precision and numerical accuracy.
- To cultivate imaginative approaches by presenting examples, and where appropriate relate these specific examples to building codes and standards that are essential and mandatory tools of the trade.
- To address the question frequently proposed to the designer by architects: "Can we do this?" In tackling this seemingly simple question, we need to acknowledge that in the fast paced world we live in, the time frame for answering such questions is measured in days and even in hours. What is needed at this juncture is the proverbial back-of-the-envelope analysis that confirms the applicability and efficiency of a concept, which would then also serve as a check of computer solutions.
- To promote the idea that design is a creative process as opposed to a mere execution of framing proposals.
- To reiterate the adage that computers assist us in the analysis phase, but it is the designer who harmonizes system components so as to optimize both cost and behavior.

Utilizing the aforementioned goals as a guide, I have set for myself a challenge to prepare a comprehensive text that will explore the world of steel, concrete, and composite materials as applied to the construction of buildings, particularly those that are super- and ultratall.

Using conceptual thinking and basic strength of material concepts as foundations, I have ventured to show how to use imperfect information to estimate the answer to much larger and complex design problems. To do so requires a certain intuitive feel for numbers as well as an appreciation of the fact that the "right answer" in this context is only of an order of magnitude of a more precise computer solution, but good enough to put us on the right track. The whole idea is to break seemingly intractable problems down to more manageable pieces that can be quickly approximated. Thus, I attempt to base the entire text on that wonderful ability of intuition we humans have developed in visualizing and realizing economical structural systems.

Developments in the last decade have produced many slender high-rise buildings, demanding that particular attention be paid to their complex behavior under lateral loads. Economic considerations routinely call for leaner and sparser designs that increasingly challenge the design professional to come up with safe and economical structural solutions.

In today's engineering practice, it is obligatory to prepare several schematic options before a final scheme is selected. Even experienced engineers find it difficult to readily come up with diversified structural schemes because, other than their own library of experience, very little reference material is available. This book attempts to alleviate this problem by providing a systematic basis for arriving at preliminary structural schemes.

The trend in building design today is for the architect to define the building shape while the structural engineer, as a facilitator, comes up with a structural system that fulfils the architect's dream within the owner's budget requirements. This trend has resulted in innovative and daring structural schemes. Fortunately for the layperson, the result has been an interesting, varied, and flamboyant architecture that adds to the variety and interest of the skyline in urban cities.

Therefore, there is a need today for the structural engineer to be familiar with the run-of-the-mill design as well as with the less usual structural solutions. To this end, emphasis is placed in this book on the state-of-the-art solutions that have evolved as a natural extension of the proven systems.

Structural steel, as we know today, has been with us for well over a hundred years. It was in the year 1894 that the first specification for structural steel was published, and an examination of test results of that era suggests that the properties of this early steel were not very different from the A36 steel of the 1950s and 1960s. The first design specifications for steel buildings published by the

American Institute of Steel Construction (AISC) in the 1920s firmly established steel as a building material, and ever since its growth has been phenomenal in the construction of buildings and bridges.

Reinforced concrete has been known to humans for over two hundred years. However, its recognition as a viable product for seismic areas and loads is relatively recent. In fact, it was at an American Concrete Institute (ACI) convention held in San Francisco in 1980 that reinforced concrete was presented as a modern, earthquake-resistant material capable of being at once strong and ductile. Since then, we have witnessed a phenomenal increase in its load-resisting and ductile properties.

At first glance, composite construction may appear to be a new, emerging technology, but in reality it has been with us also for over a hundred years. However, only recently has its use been officially formalized by the AISC. We can now design, with equal assurance, composite buildings in areas of high seismic risk.

In today's world of high expectations, we seem to place less emphasis on learning the fundamentals of conceptual thinking. If we were to retain these skills as a profession, we engineers would be more adept at identifying what is critical for capturing essential behavior of the structural system instead of addressing every component of design independently. Computer analysis, then, works to solidify and extend the creative idea or concept that might have started out as a sketch on the proverbial back of the envelope. Our unique gift as engineers is our critical thinking, and we risk shortchanging ourselves and our field, in general, if we remain convinced that the output of voluminous calculations of every structural member is proof of good design.

When designing buildings in accordance with the ASCE 7-10 Minimum Loads Standards, considerations of wind- and seismic-resistant design is required for most building structures in the United States. The use of these documents can be daunting, particularly for those engineers that have little formal training in seismology, seismic hazard analysis, structural dynamics, and inelastic behavior. Given this perspective, this book has been designed to provide guidance on how to use code-based procedures while at the same time providing sufficient technical background to explain why the provisions are written the way they are. Where possible, the technical background is presented simultaneously with the explanation of the building code provisions. In many cases, such explanations are presented as part of a series of detailed numerical examples that are presented throughout the book. Information is provided on the wind and seismic detailing requirements of structural steel, reinforced concrete, and composite structures in the context of building system selection and behavior.

The first three introductory chapters present a discussion of various loads and load combinations typically used in building design. Tall buildings, like their lower brethren, are utilitarian creations. Out of all concerns related to structural design, that of safety is paramount as it is directly related to the loads. If the earth did not pull, the wind did not blow, the earth's surface did not sink or shake, and the air temperature and humidity did not change, loads would not exist, and a formal structural design would be unnecessary. We would all be out of work. This, however, is not the case.

In Chapter 1, we discuss basic dead and live loads, the two types of loads that exert gravitational loads on buildings. Dead loads are the self-weight components that make up the building, while live loads determined on the basis of statistic probabilities include all the loads that are variable within the operation cycle of a building.

Chapter 2 discusses the wind forces that must be accounted for in a properly engineered lateral force resisting system, regardless of building size or magnitude of load. Special emphasis is on the technical background to explain why the code provisions are written the way they are. Methods for assessing wind loads to examine building performance in severe windstorms are also discussed.

In Chapter 3, we discuss the ground and building characteristics so essential to give designers a *feel* for how their building will react to ground shaking. The chapter emphasizes the fact that in spite of the complexity of the interactions between the building and the ground during the first few seconds of shaking, there is ample evidence from extensive observations of buildings in earthquakes worldwide as to how different building types will perform under different shaking conditions.

Chapter 4 deals with methods for determining design wind loads using the provisions of ASCE 7-10. Wind tunnel procedures are discussed, including analytical methods for determining along-wind and across-wind response.

Conceptual seismic design, defined here as the avoidance or minimization of problems created by the effects of seismic excitation, is discussed in Chapter 5. From the analysis of general equations for predicting earthquake response, it becomes clear that to overcome the detrimental effects of many of the uncertainties in the predictions, one needs to apply a two-pronged approach: (1) control or decrease the demand as much as possible and (2) be generous in the supply of capacity, particularly by providing large ductility with stable hysteretic behavior, also called toughness. Using this philosophy as a basis, the first part of this chapter translates the complex field of structural dynamics into a simplified language that will be comprehensible to anyone concerned with the seismic design of buildings. The primary emphasis is on visual and descriptive analysis. The engineering mechanics is kept to a basic level and the mathematics to slide rule accuracy. Design requirements of ASCE 7-10 that implicitly provide for acceptable performance beyond elastic range are discussed using static, dynamic, and time-history procedures.

In Chapter 6, we introduce the concept of performance-based design (PBD). Although not revolutionary, it represents an evolution in design thinking that is in tune with the increasing complexity of today's buildings and also takes advantage of development and innovations in building technology. PBD suggests that rather than relying on the building code for protection against seismic hazards, a more systematic investigation is conducted to ensure that the specific concerns of building owners and occupants are addressed. Building codes focus on providing life safety, and property protection is secondary; PBD provides additional levels of protection that cover property damage and avoidance of functional interruption within a financially feasible context.

PBD has become the high-end, cutting-edge technology in building design. In lieu of prescriptive provisions that tend to discourage innovation required of ever more complex buildings, PBD provides analytical tools to assist in the earthquake design assurance process. It is expected that the profession will be able to avail itself of PBD techniques within this decade. This is so because owners like them for they are likely to cost less if designed only for traditional code compliance, architects love them because it offers more design freedom, and engineers being thrifty go for it because it can result in higher quality structures with the least amount of material.

Chapter 7 presents preliminary analysis and design techniques. Approximate methods are developed using fundamental principles of mechanics because it is only through sound understanding of these principles that engineers can successfully perform preliminary designs without resorting to full-blown computer analysis. The chapter concludes with a discussion of preliminary methods for determining axial shortening of tall steel building columns, and graphical aids for estimating unit quantity of structural materials for the purpose of conceptual estimates.

Chapter 8 is devoted to the structural rehabilitation of seismically vulnerable steel and concrete buildings. Design differences between a code-sponsored approach and the concept of ductility trade-off for strength are discussed, including seismic deficiencies and common upgrade methods. The ASCE standard, Seismic Rehabilitation of Buildings, ASCE/SEI 41-06, forms the basis of this chapter.

In Chapter 9, we address a number of topics, including serviceability considerations, prediction of tall building motions, damping devices, seismic isolation, blast-resistant design, and progressive collapse. The structural systems for selected tall buildings are also described.

Chapter 10 covers warping torsion, as it applies to open-section shear walls and wide flange sections. It includes worked examples to give the readers a feel of the magnitude of axial stresses resulting from warping torsion.

Chapter 11 is somewhat unique in that we attempt to capture the essence of seismic design using only illustrations with elaborate captions where necessary.

Finally, Chapters 12 and 13 are dedicated to explaining gravity and lateral systems for steel, and composite buildings, respectively. Also discussed in these chapters is the nonquantifiable, nonautomatic phase of design that engineers call their art—the art of connection design.

It is of interest to recognize that the debate over the perceived inadequacies of structural engineering graduates has reached fever pitch. Few deny that engineers today are faced with more information than they did even five years ago. Building information modeling and sustainability are just a few of the new design paradigms, and globalization, intelligent technology, digital fabrication, and self-consolidating concrete are just a few of the new industry standards.

Additionally, the codified laws by which we create structures have also expanded and sharpened. The bureaucratic, legalistic, rule-fixated viewpoint of our society has given rise to building codes and design guidelines that are voluminous and complex without precedent. Gone are the old days when an entire code book was no more than an inch thick, while the rest of the design process was left to the engineer's specific principles and experience.

To be sure, today's flamboyant architecture does not allow—by design and economics—structures that are simple, heavy, stiff, and robust as were the buildings of the 1930s and 1940s. However, even in today's computer age, the same timeless principles of engineering judgment apply as much now as ever before, demanding that we perform *back-of-the-envelope* decision-making calculations based on intuition and engineering judgment.

No one really starts with intuition, but cultivates it slowly over time. Computers can in fact help the engineer develop understanding because it challenges one's conventional thinking. The trick is to establish a link between those who *have* knowledge and those young engineers who simply run analytical models. Thus a business office, as I see, also becomes a place of continuing education between masters and apprentices.

What else can we do to prepare tomorrow's engineers to design safe, cost-effective projects, accounting for greater complexity and uncertainty with less formal education? The answer is by motivating them to cultivate engineering judgment and intuition with a constant objective of educating oneself. Every moment of every workday can be a learning experience practically regardless of the actual task: every drawing glanced at, or an engineering conversation overheard, can be another bit of experience gained, with the right attitude.

Another avenue is to have available—out of books, notes, and individual experience—all the *rules of thumb* and *reality checks* engineers have acquired over the years. No matter how complicated an analysis becomes, it is practically guaranteed that at some point in the process you will need to *prove* your design succinctly, in the space of a single page or two, to someone who has a stake in the project, but, above all, to your own conscience. When faced with these challenges, one learns what cannot be taught.

The very magnitude of efforts required to achieve the said goals begs for a communal effort on a national scale. The work presented in this book is but a modest attempt by a single author.

Design specifications for steel, concrete, and composite construction get more and more complicated with each edition, and there seems to be no let up in the drive of code writing agencies to increase the complication. Every expert in the field wants to incorporate what he or she considers to be the proper structural action, typically resulting in long and barely understandable formulas. It has gotten so that in many cases it is not possible to understand the rationality behind these equations.

How far should we go to increase the complications? What have the super specifications accomplished? Do we have better structures? Are there fewer failures? Have we balanced the complications against the need to maintain simplicity so that we will always understand the structure?

We need to stop and take a hard look at what the so-called increased precision has accomplished. If we feel that the specifications are not accomplishing their purpose, then we should make our opinions known. It seems there is no real input from practicing engineers to the decision of code writing authorities.

One answer to this problem is perhaps for someone to write a simple specification that will satisfy the intent of the code, equations. In some cases, it may be necessary to be much more specific, particularly in areas that are fundamental to the stability of structures. Some of the ideas in the commentary to the specifications could be incorporated with much more discussion.

No committee could do this. If done by an individual, such a document would not have the voice of authority, but if it was well done it would be used with confidence by practicing engineers.

The most important duty of engineers is to understand the structure they are designing. If this is not accomplished, then there is a risk that there will be mistakes that will cause problems. Specifications ought to help rather than hinder this process.

Tall Building Design: Steel, Concrete, and Composite Systems addresses the foregoing anxieties while integrating the design aspects of building structures within a single text. It is my hope that a commonsense approach for the modern world presented in this book will serve as a comprehensive design guide and reference for practicing engineers and educators, and more importantly, as a welcome mat for recent graduates entering the structural engineering profession by assuring them that they have discovered an exciting world of challenges and opportunity.

Bungale S. Taranath PhD, PE, SE
Structural Consultant
Chino Hills, California

Acknowledgments

I wish to express my sincere gratitude to Samantha Roy, for typing the manuscript and for her help in organizing the entire manuscript package.

My sincere appreciation and thanks to Arun Kumar Aranganathan, project manager, SPi Global, Pondicherry, India, for the elegant layout and copyediting of this book.

Thanks are also due to Jennifer Ahringer, production coordinator, and Joseph Clements, acquisitions editor, CRC Press, Taylor & Francis Group, for their cooperation in the production of this book.

Thanks in no small measure are due to my friend of many years, M.V. Ravindra, CEO, LeMessuirier Consultants, Boston, Massachusetts, for his valuable advice during the preparation of this book.

Everlasting thanks to my daughter, Dr. Anupama Taranath; son, Abhi Taranath; son-in-law, Dr. Rajesh Rao; daughter-in-law, Kristin Taranath; grandsons, Vijay and Kavi; and granddaughters, Anika and Maya; for their love throughout the writing of this book that stole valuable time from my family life.

Special thanks to my daughter Anu for taking the time to work with me on various sections of this manuscript. My sincere thanks to my son, Abhiman B. Taranath, for typing sections of the manuscript.

The Publisher wishes to thank Syed Mehdi Ashraf, Nate Roy, Mike Mota, and Hans William Hagen for their help with preparing the final manuscript for publication.

Special Acknowledgment

In my previous five books, the Acknowledgment section highlighted the extraordinary support my wife Saroja has offered me as I prepared each manuscript. The process of preparing this particular manuscript, however, took a different and unexpected turn when I became sick. In the midst of all the health challenges we faced, the camaraderie and good humor Saroja showered on me daily literally helped birth this sixth book into being. She is in many ways my coauthor on this book and, in life, my copilot. For her incredible care, patience, and above all her love, I thank her deeply. Without her, my humble contributions to structural engineering would not have been possible.

Bungale S. Taranath
Chino Hills, California

Author

Dr. Bungale S. Taranath, PhD, PE, SE, was a structural consultant based in Chino Hills, California. He had extensive experience in the design of concrete, steel, and composite tall buildings and served as principal-in-charge for many notable high-rise buildings. He held positions as a senior project engineer in Chicago, Illinois, and as vice president and principal-in-charge with two consulting firms in Houston, Texas. He also served as senior project manager with a consulting firm in Los Angeles, California. Dr. Taranath was a member of the American Society of Civil Engineers and the Concrete Institute and a registered structural and professional engineer in several states. He conducted research on the behavior of tall buildings and shear wall structures and authored a number of published papers on torsion analysis and multistory construction projects. He has published five other books: *Structural Analysis and Design of Tall Buildings*; *Steel, Concrete, and Composite Design of Tall Buildings*; *Wind and Earthquake Resistant Buildings: Structural Analysis and Design*; *Reinforced Concrete Design of Tall Buildings*; and *Structural Analysis and Design of Tall Buildings: Steel and Composite Construction.* Three of his books were translated into Chinese and Korean and are widely referenced throughout Asia. Dr. Taranath conducted seminars on tall building design in the United States, China, Hong Kong, Singapore, Mexico, India, and England. He was awarded a bronze medal in recognition of a paper presented in London, when he was a fellow of the Institution of Structural Engineers, London, England. Taranath's passion for tall buildings never waned. His greatest joy was sharing that enthusiasm with owners, architects, and fellow structural engineers to develop imaginative solutions for seemingly impossible structures. Dr. Taranath passed away as this book was being produced.

1 Loads on Building Structures

PREVIEW

Loads acting on a structure are generated either directly by the forces of nature or by humans themselves. Thus, there are two basic sources of building loads: geophysical and human made.

The geophysical forces, being the result of continuous changes in nature, may be further subdivided into gravitational, meteorological, and seismological forces. As a result of gravity, the weight of a building itself produces on the structures forces called dead load, and this load remains constant throughout the building's life span. The ever-changing occupancy of a building is also subject to gravitational effects producing a variation of loads over a period of time. Meteorological loads vary with time and location and appear in the form of wind, temperature, humidity, rain, snow, and ice. Seismological forces result from the erratic motion of the ground.

The human-made sources of loading may be the variations of shocks generated by cars, elevators, machines, and so on, or they may be the movement of people and equipment or the result of blast and impact. Furthermore, forces may be locked into the structures during the manufacturing and construction processes. The stability of the building may require prestressing, which induces forces.

Geophysical and human-made sources for building loads are often mutually dependent. The mass, size, shape, and materials of a building influence the geophysical force action. For instance, if building elements are restrained from responding to temperature and humidity changes, forces are induced into the building.

Relative to the gravitational forces to which a building is subjected, loads can be classified into two distinct categories: static and dynamic. Static loads are always a permanent part of the structure. Dynamic loads are temporary: they change as time and season change or as a function of spaces within or on a structure.

Dead loads may be defined as the static forces caused by the weight of every element within the structures. The forces resulting in dead load consist of the weights of the load-bearing elements of the building, floor, and ceiling finishes, permanent partitioning walls, facade cladding, storage tanks, mechanical distribution systems, and so on. The combination weights of all these elements make up the dead load of a building.

It appears to be a simple matter to determine the weights of materials, thus the dead load of a structure. However, the estimate of dead loads may be in error by 15%–20% or more because of various problems in making an accurate analysis of the loads. At an early design stage, it is impossible for the analyst to predict accurately the weight of building materials not yet selected. Specific nonstructural materials to be chosen include facade panels, light fixtures, ceiling systems, pipes, ducts, electrical lines, and components of special interior requirements. The weight of stiffening elements and joinery systems for steel structures is estimated only on a percentage basis. The unit weights of materials given by the suppliers are not always consistent with those of the final manufactured product. The nominal sizes of building elements may differ from the actual sizes; the formwork for cast-in-place concrete may have inaccuracies of ½ in. or even more.

These few examples indicate that in the absence of precise information, it makes sense to make an allowance for imprecision in calculating dead loads.

Dead Loads

Dead loads can be defined as vertical loads that are fixed in position and are produced by the weight of the elements of the structure or the whole structure with all its permanent components. Although these loads are known quite accurately once the design of the structure is completed, at the beginning of an analysis, a dead load has to be assumed as close as possible. Previous design experience, available data, and weight tables, as well as some empirical formulas, are helpful in this stage of the design process.

Weights of building materials and of types of built-up roofs and building floors can be found in handbooks or manuals.

Occupancy Loads on Buildings

All loads other than dead loads are live loads. The live loading on buildings is highly variable, depending upon the use of the building. Minimum values are usually specified by local or national building codes. Some types of live loads may be practically permanent in nature, although subject to removal or relocation. Movable partitions, hung ceilings, and building equipment fall in this category.

To produce a safe design, occupancy loads are taken conservatively, derived more from experience and current practice than from accurately computed values from statistical data based on the probability of their occurrence. The ASCE 7 gives the minimum values for such loads. To make sure these loads are more realistic, allowance is made for some percentage reduction from the full loading.

Snow Loads on Buildings

During the winter of 1979, more than 200 roofs collapsed in the northern counties of Illinois and Wisconsin, following more than 1300 collapses in the northern United States in the winter of 1978. Subsequent investigations focused on two problems most common in heavy snow areas: unpredictable amounts of snow and the nonuniform distribution of it. Roof failure caused by snow usually does not occur as a result of a uniform load but from a localized drift or ponding load. Wind drifting of snow has been the root cause of many failures.

The current ASCE 7 gives a basic snow load P_f, which is then multiplied by the appropriate coefficients C_e, C_t, and I. The basic snow load corresponds to the ground load in psf for 50 years mean recurrence interval (MRI). These loads are used for all permanent structures except those that are judged to represent an unusually high degree of hazard to life and property in case of failure. For those, a 100-year MRI must be used. If the risk to human life is negligible, a 25-year MRI may be used. ASCE 7-10 gives a map of the United States showing isolines of ground snow. See ASCE 7-10, Figure 7.1.

The snow-load coefficient depends on the wind speed and direction, the geometry of the structure, and the temperature gradient between the inside of the structure and the outside. The basic slope factor coefficient, C_s, may be decreased to reflect slide-off of snow where sloped roofs qualify, based on their roof surface roughness and whether they are considered to be warm or cold, and must be increased to reflect nonuniform accumulation on pitched or curved roofs as well as in valleys formed by multiple series roofs.

For simplicity, it is usually assumed that forces acting on building structures can be reduced to static (unchanging) loads, in pounds per square foot (psf or kg/m^2). In fact, these loads are not always static. Sometimes they are dynamic, changing over a small interval of time. Live loads, seismic disturbances, gusting of wind, movement of machinery, or any other source of fairly rapid load variations will produce dynamic loads. At other times, they are produced by the strain and movement in the structure caused by temperature and shrinkage. Additional strains, and thus forces, may also be produced by the uneven settling of foundations, even though dead-load conditions are static.

Nonetheless, it is generally possible to express the effects of these more or less changing forces in terms of equivalent static load in psf.

An even more truly dynamic source of additional live load can result from the sudden application (or movement) of live load in the form of lifting equipment, oscillating machinery, cars in a garage, and so on. These produce impact forces that act in addition to the gravity weight of the basic live load itself. But because the use of a building can be anticipated, these impact loads can also be expressed by an *impact factor* that yields an additional statistical load as a percentage of the basic live load.

Internal forces may also be produced in a building as a result of a temperature differential between various parts of the building. In this situation, one part of a structure will tend to resist the expansion or contraction movement of another part. Relative shrinkage of the materials, or uneven settling of foundations, produces internal forces similar to those of self-stressing produced by a temperature gradient within the structure itself. Each condition can cause unequal movement across a structure that can produce significant forces in various parts of a building.

In this opening chapter, we study the effect of various types of loads, except those caused by wind and seismic. Their effects are considered in subsequent chapters.

1.1 DEAD LOADS

The engineer's first job is to determine which loads will act on a structure and how strong they might be in extreme cases. Structural engineering would be unnecessary and we would all be out of work, if the earth did not pull, the wind did not blow, the earth's surface did not shake or sink, and the air temperature and humidity did not change. But in the real world, we must concern ourselves with all the loads that act unavoidably on buildings.

A structure consists of elements like columns, beams, floors, arches, or domes that must, first of all, support their own weight, the so-called dead load. And here lies the paradox of structural design. To determine the weight of a structure, once the dimensions of its elements are established and the material chosen, one has only to compute the volume of the elements and multiply it by the weight of a unit volume of the material. The trouble is that, for example, in order to make sure that a beam will carry its own weight and other loads on it, we must first know its dimensions, but these in turn depend on the beam's weight. Thus, structural design, the determination of the shape and dimensions of structural elements, can only be learned by experience.

The dead load is a load *permanently there*. In some structures built of masonry or concrete, it is often the heaviest load to be supported by the structure. By the way, any other load permanently residing on the structure is always included in the dead load—the weights of the flooring, ceiling, and insulation materials, for example. Similarly, the weight of permanent partitions—the walls dividing one space from another that may be changed or shifted in rearranging the plan of a building but will always be there—must also be included in the dead load.

Dead loads consist of the weight of all materials of construction incorporated into the building including, but not limited to, walls, floors, roofs, ceiling, stairways, building partitions, finishes, cladding, and other similarly incorporated architectural and structural items and fixed service equipment including the weight of cranes.

To establish uniform practice among designers, the ASCE 7 Standard, Minimum Design Loads for Buildings and Other Structures, in its commentary, Table C3-1, presents a list of materials generally used in building construction, together with their unit weights. A condensed version of the table is given here in Table 1.1.

For ease of computation, most values are given in terms of pounds per square foot (psf) (kN/m^2). Pounds per cubic foot (lb/ft^3) (kN/m^3) values, consistent with the pounds per square foot (kilonewtons per square meter) values, are also presented in some cases. Some constructions for which a single value is given actually have a considerable range in weight. The average value given is suitable for general use, but when there is reason to suspect a considerable deviation from this, the actual weight should be determined.

Although engineers cannot be responsible for circumstances beyond their control, experience has shown that conditions are encountered, which, if not considered in design, may reduce the

TABLE 1.1
Weights of Building Materials

Materials	Weight (psf)
Ceilings	
Channel suspended system	1
Lathing and plastering	See partitions
Acoustical fiber tile	1
Floors	
Steel deck	See manufacturer
Concrete—reinforced 1 in.	
Stone	12½
Slag	11½
Lightweight	6–10
Concrete—plain 1 in.	
Stone	12
Slag	11
Lightweight	3–9
Fills 1 in.	
Gypsum	6
Sand	8
Cinders	4
Finishes	
Terrazzo 1 in.	13
Ceramic or quarry tile ¾ in.	13
Linoleum ¼ in.	1
Mastic ¾ in.	9
Hardwood ⅞ in.	4
Softwood ¾ in.	2½
Roofs	
Copper or tin	1
3-ply ready roofing	1
3-ply felt and gravel	5½
5-ply felt and gravel	6
Shingles	
Wood	2
Asphalt	3
Clay tile	9–14
Slate ¼	10
Sheathing	
Wood ¾ in.	3
Gypsum 1 in.	4
Insulation 1 in.	
Loose	½
Poured in place	2
Rigid	1½
Partitions	
Clay tile	
3 in.	17
4 in.	18
6 in.	28
8 in.	34
10 in.	40

(*Continued*)

TABLE 1.1 (*Continued*)
Weights of Building Materials

Materials	Weight (psf)
Gypsum block	
2 in.	9½
3 in.	10½
4 in.	12½
5 in.	14
6 in.	18½
Wood studs 2 × 4 12–16 in. o.c.	2
Steel partitions	4
Plaster 1 in.	
Cement	10
Gypsum	5
Lathing	
Metal	½
Gypsum board ½ in.	2
Walls	
Brick	
4 in.	40
8 in.	80
12 in.	120
Hollow concrete block (heavy aggregate)	
4 in.	30
6 in.	43
8 in.	55
12½ in.	80
Hollow concrete block (light aggregate)	
4 in.	21
6 in.	30
8 in.	38
12 in.	55
Clay tile (load bearing)	
4 in.	25
6 in.	30
8 in.	33
12 in.	45
Stone 4 in.	55
Glass block 4 in.	18
Windows, glass, frame, and sash	8
Curtain walls	See manufacturer
Structural glass 1 in.	15
Corrugated cement asbestos ¼ in.	3

Source: ASCE/SEI 7-10. American Society of Civil Engineers Minimum Design Loads for Buildings and Other Structures, 2010.

future utility of a building or reduce its margin of safety. For example, there have been numerous instances in which the actual widths of concrete members and construction materials have exceeded the values used in design. Allowances should therefore be made for such factors as the influence of formwork and support deflections on the actual thickness of a concrete slab of prescribed nominal thickness. Also, allowance should be made for the weight of future wearing or protective surfaces where there is a good possibility that such may be applied. Special consideration should be given to the likely types and position of partitions, as insufficient provision for partitioning may reduce the future utility of the building.

When considering a building as a whole at the preliminary design stage, it is best and easiest to estimate the total weight of the building in terms of average psf of floor area. As an approximation for steel buildings, the gross dead-load average will be on the order of 50–80 psf. For ordinary reinforced concrete buildings, it will likely be between 100 and 150 psf. For prestressed concrete buildings, a value of 70%–80% of that suggested for ordinary reinforced concrete building may be used.

Although expressed in terms of floor area, these overall dead-load values attempt to include the weight of the structural floor itself, the roof, walls, shafts, columns, and possibly the floor surfacing, ceilings, and partitions. But it should be clear that they are only rough approximations. They can vary greatly depending upon construction, the basic structural system, the floor construction, the extent of walls, and the type of building façade.

Observe, as shown in Figure 1.1, that the weight of the structural framing members for steel buildings may be only a small part of the total dead load and will vary from as low as 10 psf for low-rise buildings to 40 psf or more for high-rise buildings with long spans. Therefore, the rough starting values suggested earlier should eventually be verified or modified when a preliminary design is being refined.

However, despite the roughness, the ability to assume a reasonable, if approximate, dead load for a building structure is useful in schematic design of major subsystems. Provision for the overall interaction of walls, shafts, and columns as major structural subsystems can be facilitated by approximation of their dead-load values to provide a basis for initiating and quickly comparing alternate overall structural schemes. Such overall estimates can be particularly useful when dealing with the lower stories of a building where the total structure and the foundation interface and the overall layout and component dimensions are critical. The overall estimates enable the designer to consider the overturning effect of wind or earthquake forces, since the gross weight of a building will have to act to help stabilize the building. Approximate knowledge of the distribution of the dead-load values can also be used to indicate whether the structural proportioning of the building is tending to be effective insofar as lateral force resistance is concerned.

When one is analyzing the dead load acting on a specific floor or roof, loading assumptions based on the overall average of the dead load may not be accurate enough. But greater accuracy calls for an estimate of a particular type of floor structural subsystem to get a start. This will involve making preliminary decisions about the design of particular floor or roof framing and covering systems in order to compute their actual dead load. Assumptions will have to be made regarding the likely thickness of materials involved to obtain a reasonable picture of the load condition. Then the unit weight of the material can be used to estimate the actual dead load of the subsystem itself.

Walls and partitions can be expressed in psf of their own projected area and applied to the floor along their plan lines. However, when walls are more or less distributed, it is more convenient to reduce such dead load to pounds psf distributed over the entire floor area. This means that one will need to have an idea of the ratio of the wall area per floor compared to the floor area. Note that the minimum allowance for movable partitions according to ASCE 7-10 provisions is 15 psf and shall be considered a live load.

However, in calculating the dead load for a given exterior column, it is preferable to include the weight of a curtain wall as a line item expressed in plf of curtain wall tributary to the columns.

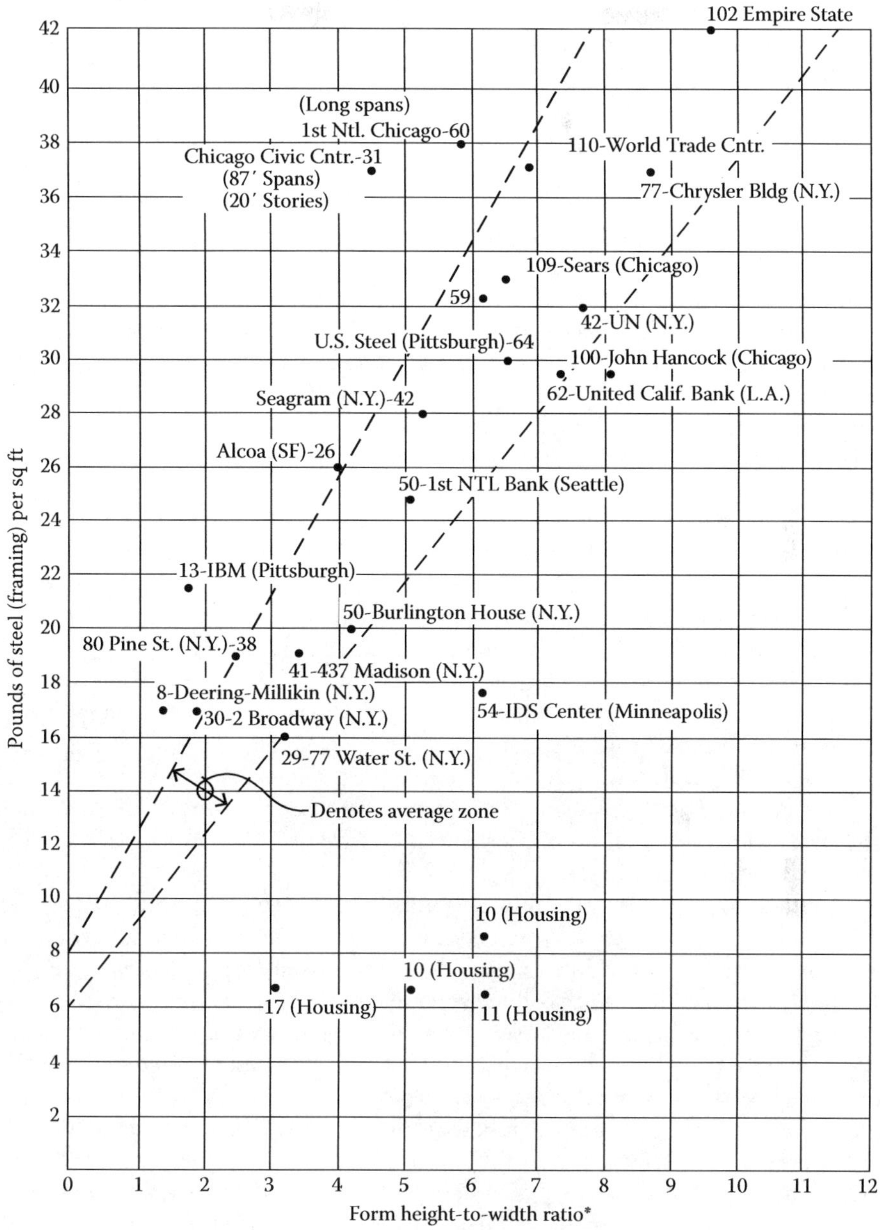

FIGURE 1.1 Unit weight of structural steel frame versus building height-to-width ratio. *The number beside each entry indicates building height in stories.

1.2 LIVE LOADS

In addition to its dead load, a structure must support a variety of other weights—people, furniture, equipment, and stored goods. These impermanent or live loads may be shifted around and they may change in value. It is obvious that we can never know exactly what the live load is and how it is going to be distributed. Therefore, concern for public safety suggests that live loads must be established on the basis of the worst loading conditions one may expect during the entire life of the structure. These are contained in building codes, which are published by cities, counties, and states. In the United States, the International Building Code (IBC) has gained general acceptance and most local codes are based on its prescriptions.

The values of the code loads are conventional. They assume that the worst effect of the varying and shifting live loads may be represented by a uniform load, that is, a load evenly spread over the surface of the floors.

On the other hand, the chances are absolutely minimal that each square foot of each space at each floor of a building will be loaded at the same time by the full code allowance. Hence, the codes allow a live load reduction, which may reach 60% for a high-rise building column.

The value of the live load varies with the type of building, its location, and its importance. The floors of a warehouse must be expected to carry a much greater live load than those of an apartment complex. The public areas of a building, its corridors and halls, which at times may be jammed with people, must be designed for larger live loads than a private room.

Live-load calculations are not very demanding on the imagination and intelligence of the engineer. Computer programs do most of the evaluations now, saving engineering time and increasing both speed of calculation and accuracy of results.

Live loads differ from dead loads in their character: they are variable and unpredictable. Changes in live loads occur not only over time but also as a function of location. The change may be a short- or long-term one, thus making it almost impossible to predict live loads in statical terms.

Loads caused by the contents of objects within or on a building are called occupancy loads. These loads include allowance for the weights of people, furniture, movable partitions, safes, books, filing cabinets, fixtures, electronic equipment (e.g., computers and copying machines), automobile, industrial equipment, and all other semipermanent or temporary loads that act on a building system but are not part of the structure and are not considered under dead loads.

Given the potential versatility of buildings, it is nearly impossible to predict the possible live-load conditions to which a structure will be subjected. Through experience, survey analysis, and practice, however, recommended load values for various occupancies have been developed. Load tables are given in the building codes and include built in empirical safety factors to account for maximum possible loading conditions. See ASCE 7-10, Table 4.1, for a complete list of minimum uniformly distributed live loads. A partial list of the loads is given here in Table 1.2. Also given in the tables are minimum concentrated live loads.

The load values take the form of equivalent uniform loads and prescribed concentrated loads. Equivalent uniform loads reflect the varying, actual occupancy loading conditions. The values, established by approximation of actual loads, appear to be rather conservative. A survey taken on the actual occupancy load in various office buildings showed a maximum load much less than the recommended design value of 50 psf. A load survey on apartments noted that the maximum load intensity measured in a 10-year period was about 26 psf; the usual value for design, however, is 40 psf.

Concentrated loads indicate possible single load action at critical locations such as on stair treads, accessible ceilings, parking garages (e.g., jack for changing a tire), and other vulnerable areas that are subject to high concentrated stresses.

Although it may appear that the regulations are too conservative, there is always the unpredictable element to consider. The minimum regulated live loads are warranted by such uncontrollable and extraordinary situations as people crowding because of ceremonies, parties, and fire drills or overloading of parts of the building due to change of occupancy or furniture and wall rearrangements that will exert more load on a specific area.

As stated previously, the likelihood of having a full occupancy load simultaneously on every square foot of every floor is very slim. The actual loading consists of different areas with different loading conditions. Generally, the smaller the area, the larger the potential load intensity. The actual occupancy loads on floors are never uniform. Building codes take this into account by allowing the use of live-load reduction factors. In determining the live loads, the possibility of temporary change in the use of building needs particular attention as in the case of clearing a dormitory for a dance or other recreational purpose.

Codes do not take into account that live loads on a building element are reduced because of the ability of a continuous building structure to redistribute loading as it deforms. On the other hand, the load capacity of buildings is likely to be reduced, since they are subject to fatigue brought about

TABLE 1.2
Minimum Uniformly Distributed Live Loads, L_o, and Minimum Concentrated Live Loads

Occupancy or Use	Uniform psf (kN/m²)	Conc. lb (kN)
Apartments (see residential)		
Access floor systems		
Office use	50 (2.4)	2000 (8.9)
Computer use	100 (4.79)	2000 (8.9)
Armories and drill rooms	150 (7.18)	
Assembly areas and theaters		
Fixed seats (fastened to floor)	60 (2.87)	
Lobbies	100 (4.79)	
Movable seats	100 (4.79)	
Platforms (assembly)	100 (4.79)	
Stage floors	150 (7.18)	
Balconies (exterior)	100 (4.79)	
On one- and two-family residences only and not exceeding 100 ft² (9.3 m²)	60 (2.87)	
Bowling alleys, poolrooms, and similar recreational areas	75 (3.59)	
Catwalks for maintenance access	40 (1.92)	300 (1.33)
Corridors		
First floor	100 (4.79)	
Other floors, same as occupancy served except as indicated	100 (4.79)	
Dance halls and ballrooms	100 (4.79)	
Garages (passenger vehicles only)	40 (1.92)[a,b,c]	
Trucks and buses		
Grandstands (see *stadium and arena bleachers*)		
Gymnasiums—main floors and balconies	100 (4.79)	
Hospitals		
Operating rooms, laboratories	60 (2.87)	1000 (4.45)
Private rooms	40 (1.92)	1000 (4.45)
Corridors above first floor	80 (3.83)	1000 (4.45)
Hotels (see residential)		
Libraries		
Reading rooms	60 (2.87)	1000 (4.45)
Stack rooms	150 (7.18)[a,b]	1000 (4.45)
Corridors above first floor	80 (3.83)	1000 (4.45)
Manufacturing		
Light	125 (6.00)	2000 (8.90)
Heavy	250 (11.97)	3000 (13.40)
Office buildings		
File and computer rooms shall be designed for heavier loads based on anticipated occupancy		
Lobbies and first floor corridors	100 (4.79)	2000 (8.90)
Offices	50 (2.40)	2000 (8.90)
Corridors above first floor	80 (3.83)	2000 (8.90)
Penal institutions		
Cell blocks	40 (1.92)	
Corridors	100 (4.79)	

(*Continued*)

TABLE 1.2 (*Continued*)
Minimum Uniformly Distributed Live Loads, L_o, and Minimum Concentrated Live Loads

Occupancy or Use	Uniform psf (kN/m²)	Conc. lb (kN)
Residential		
Dwellings (one- and two-family)		
Uninhabitable attics without storage	10 (0.48)	
Uninhabitable attics with storage	20 (0.96)	
Habitable attics and sleeping areas	30 (1.44)	
All other areas except stairs and balconies	40 (1.92)	
Hotels and multifamily houses		
Private rooms and corridors serving them	40 (1.92)	
Public rooms and corridors serving them	100 (4.79)	
Reviewing stands, grandstands, and bleachers	100 (4.79)[d]	

Source: ASCE/SEI 7-10. American Society of Civil Engineers Minimum Design Loads for Buildings and Other Structures, 2010.

[a] Live load reduction for this use is not permitted by Section 4.7 unless specific exceptions apply.

[b] Floors in garages or portions of a building used for the storage of motor vehicles shall be designed for the uniformly distributed live loads of Table 4-1 or the following concentrated load: (1) for garages restricted to passenger vehicles accommodating not more than nine passengers, 3,000 lb (13.35 kN) acting on an area of 4.5 in. by 4.5 in. (114 mm by 114 mm); and (2) for mechanical parking structures without slab or deck that are used for storing passenger vehicles only, 2,250 lb (10 kN) per wheel.

[c] Design for trucks and buses shall be per AASHTO LRFD Bridge Design Specifications; however, provisions for fatigue and dynamic load allowance are not required to be applied.

[d] In addition to the vertical live loads, the design shall include horizontal swaying forces applied to each row of the seats as follows: 24 lb per linear ft of seat applied in a direction parallel to each row of seats and 10 lb per linear ft of seat applied in a direction perpendicular to each row of seats. The parallel and perpendicular horizontal swaying forces need not be applied simultaneously.

by years of combating wind loads, vibrations, temperature changes, settlement, and the continuous change of environmental forces. However, concrete and masonry materials have the advantage of gaining strength with age, therefore increasing their loading capacity.

From a structural standpoint, the choice of an appropriate structural system depends on the knowledge of three factors:

1. The loads to be carried.
2. The property of the construction materials.
3. The structural action by which the load forces are transferred through the members into the ground. The structural engineer has some control over the last two factors but none on the mandated live loads. Future research will perhaps make possible a more accurate prediction of actual live-load conditions.

1.2.1 Live Load Reduction

The concept of determining live load reduction as a function of the tributary area supported by a member is nearly 50 years old. Since then, there have been a few changes in the format to the formula for determining live load reduction. The most recent ASCE 7-10 has the following formula for determining the reduced live load:

$$L = L_0\left(0.25 + \frac{15}{\sqrt{K_{LL}A_T}}\right)$$

where

L is the reduced design live load per ft^2 of area supported by the member

L_0 is the unreduced design live load per ft^2 of area supported by the member

K_{LL} is the live load element factor that varies from 4 (for an interior column) to 1 (for a member such as one-way and two-way slabs) (see ASCE 7-10, Table 4.2)

A_T is the tributary area in ft^2

L shall not be less than $0.50L_0$ for members supporting one floor and L shall not be less than $0.40L_0$ for members supporting two or more floors

It should be noted that live-load reduction is permitted only when the value of $K_{LL}A_T$ is equal to or greater than 400 ft^2. Other restrictions are as follows:

- Live loads that exceed 100 psf shall not be reduced except that members supporting two or more floors shall be permitted to be reduced by 20%. Similarly, the live loads shall not be reduced in passenger vehicle garages, except the live loads for members supporting two or more floors shall be permitted to be reduced by 20%.
- The tributary area, A_T, for one-way slabs shall not exceed an area defined by the slab span times a width normal to the span of 1.5 times the slab span.

1.3 CONSTRUCTION LOADS

Structural members are generally designed for dead and live loads; however, a member may be subject to loads larger by far than the design loads during construction of a building. These loads, called construction loads, constitute an important consideration in the design of structural elements.

Although engineers may design a building to suit a particular construction system, they may not know the individual practices of the contractor. Contractors commonly stockpile heavy equipment and materials on a small area of the structure. This causes concentrated loads that are much larger than the assumed live loads for which the structure was designed. Structural failures have resulted from such conditions.

A major problem in concrete construction results when the contractor fails to allow sufficient curing time before removal of shoring and formwork. Concrete increases its strength with time, but since time is money to the contractor, the contractor may remove the forms before the concrete has reached its minimum design strength, whereupon the structural element may be subject to loads it is unable to support, and failure or unacceptable large permanent deflections may result.

Construction loads also must be considered for a beam designed to act compositely with a concrete slab, assuming that no temporary shoring is used during the construction process. In this case, the beam needs to be verified with respect to carrying construction loads in noncomposite action.

For precast concrete, the most critical period is at the time of lifting of a heavy panel element from its form. The number of lifting points and their placement must be known. Also, since the element has to be designed for any possible position it may encounter during handling and erection, impact and stress at that time must be considered.

It is quite likely that plaza areas surrounding the footprint of a building may be subjected to vehicular traffic during and after construction of the building. The loading, for example, due to fire trucks and emergency vehicles particularly on framed plazas must be accounted for in the design. There appears to be no guidance regarding this type of loading. Perhaps one should use the recommended highway traffic loads.

Highway traffic is made up of four principal kinds of vehicles: the tractor-trailer, the truck, the bus, and the passenger car. These units vary widely in weight, and their average weights are considerably less than the weights of the heaviest units. The heaviest truck, which is used to any considerable extent at present, weighs about 50,000 lb when fully loaded, and since it is about 40 ft long, it represents an average load of, say, 1,250 plf on traffic lane. An average passenger car weighs perhaps 4,000 lb fully loaded, while the heaviest car may weigh something less than 10,000 lb, with corresponding intensities of 200–500 plf. The distance between vehicles in the same traffic lane is obviously important in its effect on load intensity. This distance may range from 25 to 50 ft center to center at speeds of 10 mph.

Because of widely variable character and distribution of highway traffic, it is expedient to adopt a conventionalized loading. Nearly all highway bridges in the United States are designed for one of the four classes of load recommended by the American Association of State Highway and Transportation Officials (AASHTO). These loads consist of a system of concentrated loads to represent a truck or of a load distributed uniformly along the traffic lane, together with a concentrated load, to represent a long line of medium-weight traffic with a heavy vehicle somewhere in the line.

The heaviest loading of the AASHTO specifications is the HS20-44, pictured in Figure 1.2.

One of the most common machines for steel erection is the crawler crane (Figure 1.3). Self-propelled, such cranes are mounted on a mobile base having tracks or crawlers for propulsion. The base of the crane contains a turntable that allows 360° rotation. Crawlers come with booms up to 540 ft high and capacities up to 1000 tons. Self-contained counterweights move the center of gravity of the loaded crane to the rear to increase the lift capacity of the crane. Crawler cranes can also be fitted with counterweights on attached mobile carriages or ring attachments to increase their capacity.

Truck cranes (Figure 1.4) are similar in many respects to crawler cranes. The principal difference is that truck cranes are mounted on rubber tires and are therefore much more mobile on hard surfaces. Truck cranes can be used up to 500 ft long and have capacities up to 750 tons. Typically, they have outriggers to provide stability.

Use of any erecting equipment that loads the structure requires verification that such loads can be adequately supported by the structure. If not, installation of additional bracing or temporary supports may be necessary.

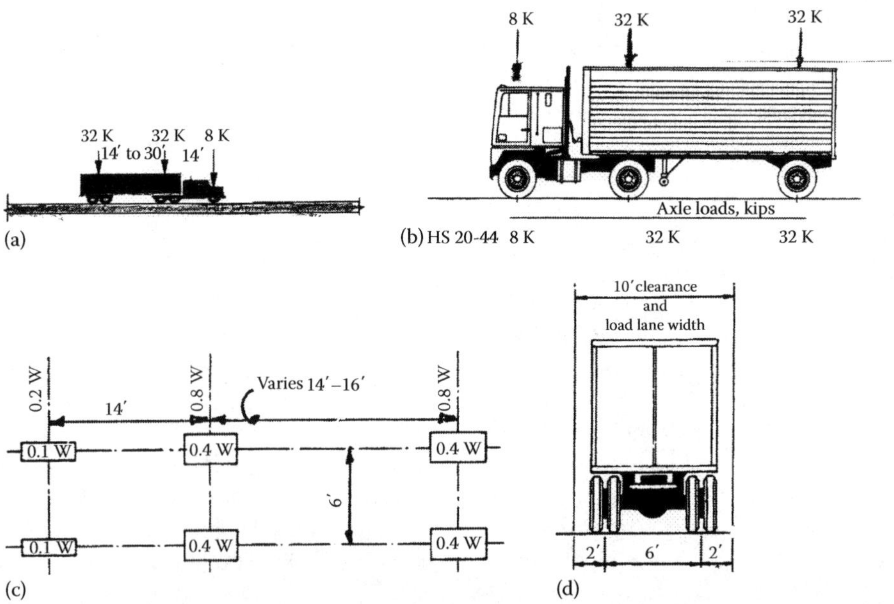

FIGURE 1.2 (a–d) Standard HS20-44 truck loading for highway bridges.

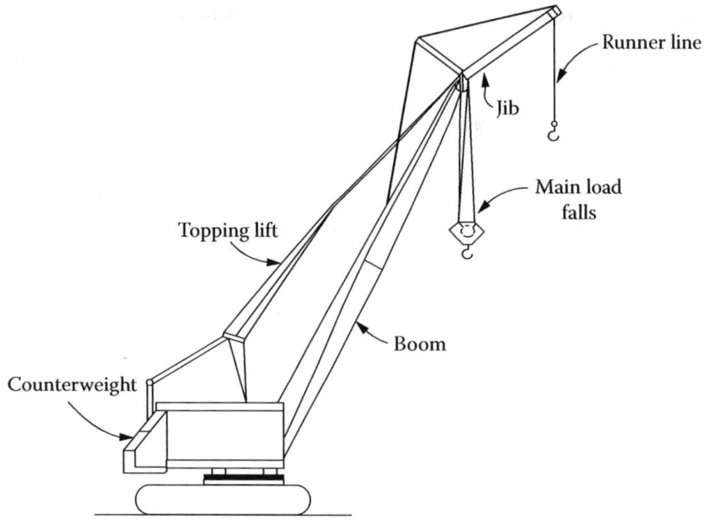

FIGURE 1.3 Crawler crane: one of the most common machines for steel erection with booms up to 540 ft high and capacities up to 1000 tons.

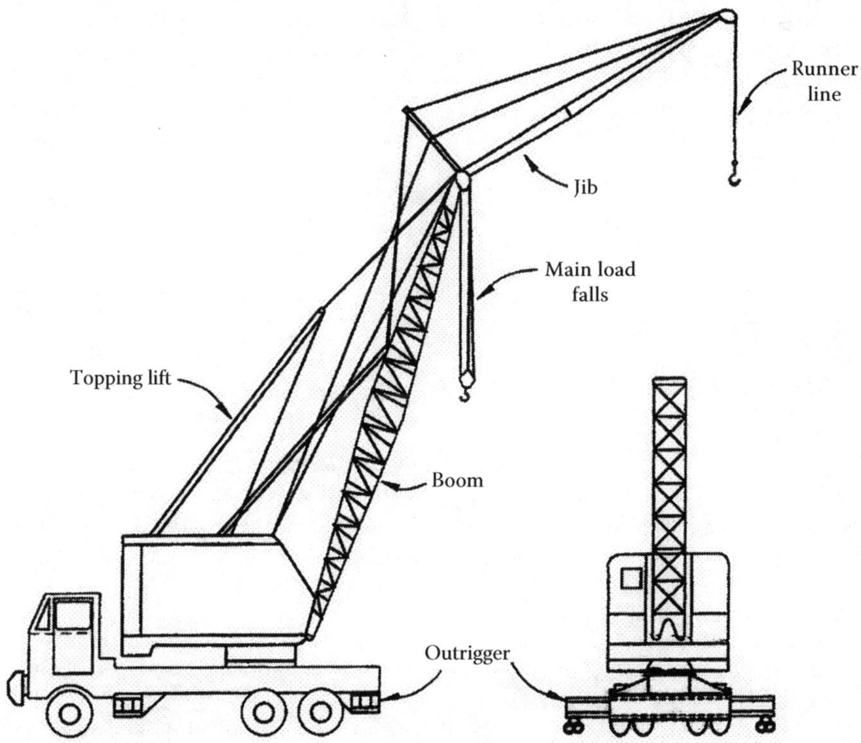

FIGURE 1.4 Truck crane mounted on rubber tires: booms up to 500 ft long and capacities up to 750 tons.

1.4 LATERAL SOIL LOAD

In the design of structures below grade, provision shall be made for the lateral pressure of adjacent soil. Minimum lateral pressures varying anywhere from 35 per foot of depth to as much as 100 psf are specified in ASCE 7-10 Table 3.2-1 depending upon the soil classification. An abbreviated version of the table is given in Table 1.3.

Due allowance shall be made for possible surcharge from fixed or movable loads, and if the adjacent soil is below a free-water surface, the weight of the soil may be reduced for buoyancy but full hydrostatic pressure must be included in calculating the lateral pressure.

The lateral pressure shall be increased if soils with expansion potential are present at the site as determined by geotechnical investigations.

Expansive soils exist in many regions of the United States and may cause serious damage to basement walls unless special design considerations are provided. Expansive soils should not be used as backfill because they can exert very high pressures against walls. Special soil testing is required to determine the magnitude of these pressures. It is preferable to excavate expansive soils and backfill with nonexpansive freely draining sands or gravels. The excavated back slope adjacent to the wall should be no steeper than 45° from the horizontal to minimize the transmission

TABLE 1.3
Design Lateral Soil Load

Description of Backfill Material	Unified Soil Classification	Design Lateral Soil Load[a] psf per Foot of Depth (kN/m² per Meter of Depth)
Well-graded, clean gravels; gravel–sand mixes	GW	35 (5.50)[b]
Poorly graded clean gravels; gravel–sand mixes	GP	35 (5.50)[b]
Silty gravels, poorly graded gravel–sand mixes	GM	35 (5.50)[b]
Clayey gravels, poorly graded gravel–clay mixes	GC	45 (7.07)[b]
Well-graded, clean sands; gravel–sand mixes	SW	35 (5.50)[b]
Poorly graded clean sands; sand–gravel mixes	SP	35 (5.50)[b]
Silty sands, poorly graded sand–silt mixes	SM	45 (7.07)[b]
Sand–silt–clay mix with plastic fines	SM-SC	85 (13.35)[c]
Clayey sands, poorly graded sand–clay mixes	SC	85 (13.35)[c]
Inorganic silts and clayey silts	ML	85 (13.35)[c]
Mixture of inorganic silt and clay	ML-CL	85 (13.35)[c]
Inorganic clays of low to medium plasticity	CL	100 (15.71)
Organic silts and silt–clays, low plasticity	OL	[d]
Inorganic clayey silts, elastic silts	MH	[d]
Inorganic clays of high plasticity	CH	[d]
Organic clays and silty clays	OH	[d]

Source: ASCE/SEI 7-10. American Society of Civil Engineers Minimum Design Loads for Buildings and Other Structures, 2010.

[a] Design lateral soil loads are given for moist conditions for the specified soils at their optimum densities. Actual field conditions shall govern. Submerged or saturated soil pressures shall include the weight of the buoyant soil plus the hydrostatic loads.

[b] Unsuitable as backfill material.

[c] For relatively rigid walls, as when braced by floors, the design lateral soil load shall be increased for sand and gravel type soils to 60 psf (9.43 kN/m²) per foot (meter) of depth. Basement walls extending not more than 8 ft (2.44 m) below grade and supporting light floor systems are not considered as being relatively rigid walls.

[d] For relatively rigid walls, as when braced by floors, the design lateral load shall be increased for silt and clay-types oils to 100 psf (15.71 kN/m²) per foot (meter) of depth. Basement walls extending not more than 8 ft (2.44 m) below grade and supporting light floor systems are not considered as being relatively rigid walls.

of swelling pressure from the expansive soil through the new backfill. Other special details are recommended, such as a cap of nonpervious soil on top of the backfill and provision of foundation drains.

If expansive soils are present under floors or footings, large pressures can be exerted and must be resisted by special design. Alternatively, the expansive soil can be removed and replaced with nonexpansive material. A geotechnical engineer should make recommendations in these situations.

1.5 SNOW, RAIN, AND ICE LOADS

Observation of the depth and density of snowfalls over many years has resulted in a reasonable prediction of maximum snow loads. The ground snow-load map given in ASCE 7-10, Figure 7-1 indicates the minimum snow loads for various regions, in pounds per square foot, ranging from zero psf in the south to as much as 100 psf in the northeastern United States. One inch of snow weighs approximately 0.5–0.7 psf, depending on its density.

Snow loads need to be considered only for roofs and other areas of a building that may gather snow, such as elevated courtyards, balconies, and sundecks. The snow loads, as established by codes, are based on the maximum snow on the ground. In general, these loads tend to be higher than the snow loads acting on a roof, since the wind blows the loose snow off the roof or the snow melts and evaporates because of heat loss through the roof skin. Codes generally allow a percentage reduction of the load value on pitched roofs, since the snow can easily slide off the roof. However, certain roof conditions may influence the behavior of the wind, resulting in high snow-load accumulations locally.

The procedure established in ASCE 7-10 for determining design snow loads is as follows:

1. Determining the ground snow load for the geographic location
2. Generating a flat roof snow load from the ground load considering (a) roof exposure, (b) roof thermal condition, and (c) occupancy and function of the structure
3. Considering roof slope
4. Considering partial loading
5. Considering unbalanced loads
6. Considering snow drifts: (a) on lower roofs and (b) at projections such as parapets and rooftop equipment
7. Considering sliding snow
8. Considering extra loads from rain on snow
9. Considering ponding loads
10. Considering existing roofs
11. Considering the consequences of loads in excess of the design value; for example, if a roof is deflected at the design snow load so that slope to drain is eliminated, *excess* snow load might cause ponding and perhaps progressive failure

The snow-load/dead-load ratio of a roof structure is an important consideration when assessing the implications of *excess* loads. If the design snow load is exceeded, the percentage increase in total load would be greater for a lightweight structure than for a heavy structure.

Water, though not often thought of when calculating live load, should certainly be kept in mind when designing roofs. Rain loads in general are less than snow loads, but it should be remembered that the accumulation of water, weighing 62.4 lb/ft^3, will result in appreciable loads. Heavy loads can occur on flat roofs because of clogged drains. As water accumulates, the roof deflects, allowing more water to collect and resulting in more deflection. This process is called ponding and may cause the eventual collapse of the roof.

Ice will collect on protruding elements, especially on exterior ornamental elements that otherwise receive no load other than their own. It is therefore necessary to design and secure such elements to withstand heavy loads of icicles. Furthermore, the ice formation on open-trussed structures will increase the exposed and projected area of the members, as well as their weight, resulting in large wind pressure.

1.6 THERMAL AND SETTLEMENT LOADS

In many high-rise buildings built in the sixties and seventies, the structural frame is set outside the curtain wall, rather than hidden inside it. We are told that the reason for this architectural innovation is to emphasize the importance of structure and a feeling for its aesthetic value. While there may be aesthetic value to some of these exposed structures, it must be noticed that they create problems for the engineer. The interior of these buildings is air-conditioned and maintained at a constant temperature of 65°F–72°F, while the exposed structure is subjected to air temperature changes. In summer, the exterior columns may reach a temperature of up to 120°F and become 2 or 3 in. longer than the interior columns, while in winter, they may become that much shorter when their temperature goes down to 20°F. These variations in length of columns bend the beams connecting the outer to the inner columns particularly at the higher floors; see Figure 1.5. These would be damaged if they were not properly designed, either by reinforcing them or by allowing their ends to rotate, that is, by hinging them to the columns (Figure 1.6).

Bending of the beams connecting outer to inner columns may also occur if the soil under a building settles unevenly (Figure 1.7).

It must be emphasized that most damage to buildings is caused by foundation problems. Soil mechanics, the study of soil behavior, has moved to a science from an art only during the last 50 years. The island of Manhattan is blessed by a rocky soil, which permitted in 1913 the erection of the first high-rise building (the Woolworth Building). Mexico City, on the other hand, is built on a mixture of sand and water. Such soils settle when heavy buildings are erected, squeezing the water out of the sand.

Invariably the design of buildings is performed according to the applicable building code requirements. The codes, however, are set up as minimum requirements and serve as a guide for the design. They are, therefore, at best a set of empirical approximations.

Nonetheless, it must be acknowledged that the great majority of code-designed buildings generally behave quite well. Modern building codes favor the *load factor* method of approach, also known

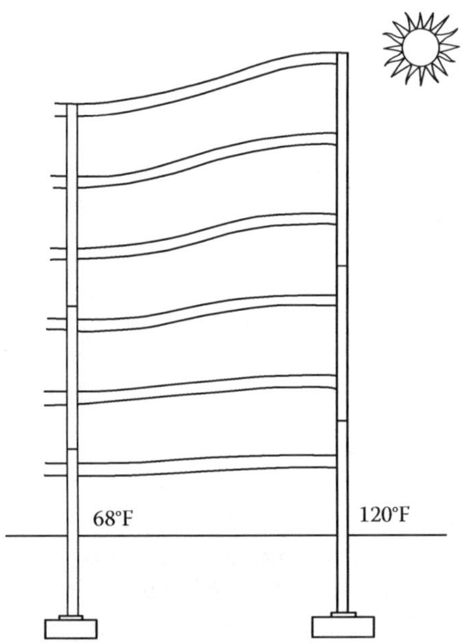

FIGURE 1.5 Thermal bending on fixed-end beam.

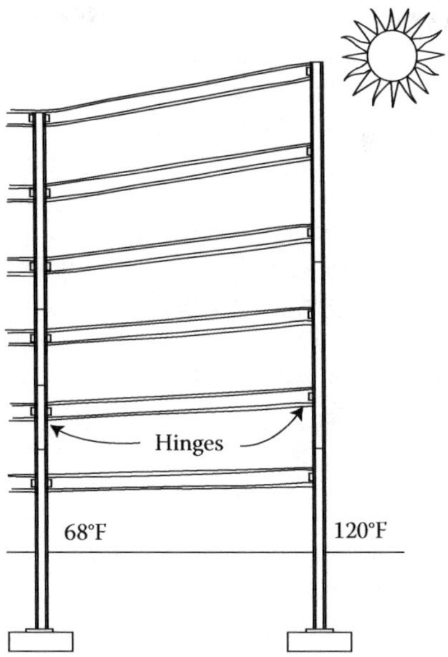

FIGURE 1.6 Thermal rotation of hinged beam.

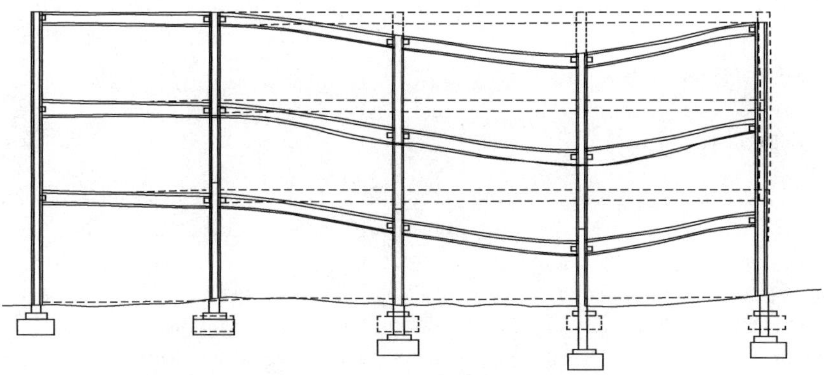

FIGURE 1.7 Uneven settlement of building foundations.

as *ultimate strength* design. Thus, a code-specified load is multiplied by a factor to be equated to the ultimate strength of the structure. And to provide an additional safety margin, the computed ultimate strength of a structure should be reduced by some 10%–20% to take care of other unknown elements, such as variation in material properties, dimensions, workmanship, and computation discrepancies. For component design, this is a more realistic approach. However, the problem still remains that it is quite difficult to predict the actual load capacity of a building that acts as a total space structure. That is, all the various components will work together in a rather complicated manner and not as independent elements designed out of the total system context. Nevertheless, this ultimate strength method is now *accepted* as a more rational approach.

When using the ultimate strength design method, one must check to be certain that under normal loadings, such as dead load and/or live load (and in combination with wind and earthquake load), the structure should have no excessive deflections, vertical or lateral, and no excessive vibrations, cracks, and other undesirable movements.

Although all buildings vibrate to a certain extent under moving live, wind, or earthquake load, it is rather a difficult problem to determine how much vibration is undesirable. Studies have been made of human sensitivity to vibrations as well as the possibility of structural damage produced by vibrations. Note that the higher the frequency, or the larger the amplitude, the more uncomfortable humans would feel and the higher are the chances for possible damage. In the actual design of a building structure, one does not often have the knowledge to decide whether the vibration is objectionable. Hence, the usual approach is to limit the depth–span ratio of structural components in a floor. For ordinary structures, comparison of the ratio to existing ones that have behaved well would give a fair guidance. For unusual structures, special studies should be made. Therefore, a building designer should use building codes only as a reference and as a guide. It is only through a thorough understanding of the strength and behavior of structures as total systems and of the requirement for subsystem and component interaction that one can design a safe and economical structure to fit the various functional requirements and environmental conditions found in modern building structures.

1.7 SELF-STRAINING FORCES

Constrained structures that experience dimensional changes develop self-straining forces. Examples include moments in rigid frames that undergo differential foundation settlements and shears in bearing walls that support concrete slabs that shrink. Unless provisions are made for self-straining forces, stresses in structural elements, either alone or in combination with stresses from external loads, can be high enough to cause structural distress.

Generally, the magnitude of self-straining forces can be anticipated by analyses of expected shrinkage, temperature fluctuations, foundation movement, and so forth. However, it is not always practical to calculate the magnitude of self-straining forces. Therefore, it is better to provide for self-straining forces by specifying relief joints, suitable framing systems, or other details to minimize the effects of self-straining forces.

1.8 DYNAMIC LOADS

The dead load is permanent and unchanging and the live loads are tacitly assumed to change slowly, if at all. Together, these unchanging or slowly changing loads are called *static loads*, loads that stay.

But other loads change value rapidly and even abruptly, like the pressure of a wind gust or the action of an object dropped on the floor. Such loads are called *dynamic* and often have a much greater effect than the same loads applied slowly. Similarly, impact loads such as the dynamic pressure of a slap, compared to the static pressure of a gentle touch, would be many times its static equivalent.

The effects of a dynamic load depend on how fast the load varies. This raises a question: For example, should the pressure on a building created by a wind gust, first increasing and then decreasing, be considered a static or a dynamic load? The answer is that no varying load is ever static or dynamic in *itself*. Its effects can be static or dynamic depending on the structure to which it is applied. To illustrate this, let us consider a tall building acted upon by a wind gust.

Under the wind pressure, the building bends slightly and its top moves. Its back and forth movement also referred to as oscillations may be small enough not to be seen by the naked eye, or even sensed, but since structural materials are never totally rigid, all buildings do sway in the wind.

It is easy to visualize these oscillations by considering the building as an upside-down pendulum, which also swings back and forth when displaced from its lowest position. The time it takes a pendulum to complete a full swing from extreme right to extreme left and back is called the period of the pendulum. Similarly, the time it takes a building to swing through a complete oscillation is called its period. For example, the period of the oscillations of the now nonexistent steel towers of the World Trade Center in New York City, which were 1350 ft high, was 10 s, while the period of a low-rise stiff building such as a 10-story brick building may be as short as half a second.

The action of the gust depends not only on how long it takes for the gust to reach its maximum value and decrease again but on the period of the building on which it acts. If the wind load increases to its maximum value and vanishes in a time much shorter than the period of the building, its effects are dynamic. It is static if the load grows and vanishes in a time much larger than the period of the building. For example, a wind gust increasing to its strongest pressure and decreasing in 2 s is a dynamic load for a building with a period greater than seconds, but the same 2 s gust is a static load for the 10-story brick building with a period of only half a second. In a sense, a force the building can slowly absorb is static; an unexpected one is dynamic. Similarly, the explosion of a weapon that reaches its maximum effect and decreases so rapidly (less than a thousandth of a second) is a dynamic load.

Interestingly enough, there are loads that, though not increasing rapidly, do have dynamic effects, not instantaneously, but progressively in time. This phenomenon, called resonance, is one of the most dangerous a structure may be subjected to. To understand resonance, let us consider how a heavy church bell, which swings like a pendulum, is made to ring by the relatively small but sudden pulls of a single person on its rope. If the bell weighs a few tons—often the case—the ringer might try in vain to move it with a single pull. But if the ringer starts pulling the rope with a small pull of, say, a few pounds and, before yanking it again, waits for the bell to go through its first tiny swing, then keeps yanking in rhythm with the bell's oscillations, eventually the bell swings widely and rings. The strategy here consists in pulling the rope at the beginning of each new oscillation, that is, at time intervals equal to the period of the bell, so the applied pulls will add up.

When a force is rhythmically applied to a structure with the same period as that of the structure, the force is said to be in resonance with it. Resonant forces do not produce large effects immediately, as impact forces do, but their effects increase steadily with time and may become catastrophic if they last long enough. If a long series of wind gusts, growing and waning in pressure with a relatively slow period of, say, eight seconds, were to hit a building with a fundamental period close to eight seconds, the sway of the building would slowly increase until it might sway so widely as to collapse.

1.9 ABNORMAL LOADS

Through accident, misuse, or sabotage, properly designed structures may be subjected to conditions that could lead to either general or local collapse. It is usually impractical for a structure to be designed to resist general collapse caused by gross misuse of a large part of the system or severe abnormal loads acting directly on a large portion of it. However, precautions can be taken in the designs of structures to limit the effects of local collapse and to prevent or minimize progressive collapse. Progressive collapse is defined as the spread of an initial local failure from element to element resulting, eventually, in the collapse of an entire structure or a disproportionately large part of it.

Because accidents, misuse, and sabotage are normally unforeseeable events, they cannot be defined precisely. Likewise, general structural integrity is a quality that cannot be stated in simple terms.

In addition to unintentional or willful misuse, some of the incidents that may cause local collapse are explosions due to ignition of gas or industrial liquids, boiler failures, vehicle impact, impact of falling objects, effects of adjacent excavations, gross construction errors, very high winds such as tornadoes, and sabotage. Generally, such abnormal events would not be a part of normal design considerations.

1.9.1 Explosion Effects

Explosions occur when a solid or concentrated gas is transformed into a large volume of hot gases in a fraction of a second. In the case of high-explosives detonation, conversion of energy occurs at a very high rate (as high as 4 mi/s), while low explosives such as gunpowder undergo rapid burning at the rate of about 900 ft/s. The resulting rapid release of energy consists of sound (bang), heat and light (fireball), and a shock wave that propagates radially outward from the source at subsonic speeds for most low explosives and supersonic speeds for high explosives. It is the shock wave, consisting of highly compressed particles of air, that causes most of the damage to structures. When natural gas explosions occur within structures, gas pressures can build up within confined spaces causing extensive damage. In all explosions, large, weak, and/or lightly attached wall, floor, and roof surfaces may be blown away. The columns and beams may survive a blast, but their stability may be compromised by the removal of their bracing elements such as floor diaphragms. In large explosions, concrete slabs, walls, and even columns may be blown away, leading to conditions that will produce progressive collapse. In 1967, a progressive collapse started when a natural gas explosion caused the collapse of an exterior wall on the 18th floor of a 22-story building in the United Kingdom. The force of falling debris from floors 19 to the roof then caused the remaining floors to collapse in that section of the building. In the case of an exterior explosion such as from a bomb, the shock wave is initially reflected and amplified by the building face and then penetrates through openings, subjecting floor and wall surfaces to great pressure. Diffraction occurs as the shock propagates around corners, creating areas of amplification and reduction in pressure. Finally, the entire building is engulfed by the shock wave, subjecting all building surfaces to overpressure. A secondary effect of an air blast is a very high-velocity wind that propels the debris, which become deadly missiles. In very large explosions at close proximity to building surfaces, the effect can be so severe that the structure is locally disintegrated and separated away from the main structure.

1.9.2 Floods

Forces are generated on buildings due to hydrostatic lateral and lifting pressure, hydrodynamic forces, and debris impacts. Hydrostatic pressures can overload foundation and basement walls and lift up structures, when water level is not equalized between exterior and interior spaces. River and ocean currents may load frontal and side walls that are submerged and ocean waves can produce pressures as high as 1000 psf. Debris varying in size from floating wood pieces to floating structures can impact a building causing anything from broken windows to a total collapse.

1.9.3 Vehicle Impact Loads

There are many examples where structures have been severely damaged and set on fire by vehicle impacts. The most hazardous configurations include soft (high, open) first stories and open-front buildings typical of retail one- and two-story structures.

1.10 CLASSIFICATION OF BUILDINGS, RISK CATEGORIES, AND IMPORTANCE FACTORS

ASCE 7-10 classifies buildings into four risk categories, I, II, III, and IV, based on the risk to human life, health, and welfare associated with their damage or failure by nature of their occupancy and use for the purpose of applying wind, earthquake, flood, snow, and ice provisions. Each building is assigned to the highest applicable risk category for purposes of incorporating an importance factor. Importance factor is a factor that accounts for the degree of risk to human life, health, and welfare associated with damage to property or loss of use of functionality.

The risk categories, referred to as occupancy categories in previous editions of ASCE 7, are used to relate the criteria for maximum environmental loads to the consequence of the load being exceeded for the structure and its occupants. The specific importance factors differ according to the statistical characteristics of the environmental loads and the manner in which the structure responds to the loads.

See Tables 1.4 and 1.5 for classification of building risk categories and corresponding importance factors.

TABLE 1.4
Risk Category of Buildings for Flood, Wind, Snow, Earthquake, and Ice Loads

Use or Occupancy of Buildings and Structures	Risk Category
Buildings and other structures that represent a low risk to human life in the event of failure	I
All buildings and other structures except those listed in Risk Categories I, III, and IV	II
Buildings and other structures, the failure of which could pose a substantial risk to human life	III
Buildings and other structures, not included in Risk Category IV, with potential to cause a substantial economic impact and/or mass disruption of day-to-day civilian life in the event of failure	
Buildings and other structures not included in Risk Category IV (including, but not limited to, facilities that manufacture, process, handle, store, use, or dispose of such substances as hazardous fuels, hazardous chemicals, hazardous waste, or explosives) containing toxic or explosive substances where their quantity exceeds a threshold quantity established by the authority having jurisdiction and is sufficient to pose a threat to the public if released	
Buildings and other structures designated as essential facilities	IV
Buildings and other structures, the failure of which could pose a substantial hazard to the community	
Buildings and other structures (including, but not limited to, facilities that manufacture, process, handle, store, use, or dispose of such substances as hazardous fuels, hazardous chemicals, or hazardous waste) containing sufficient quantities of highly toxic substances where the quantity exceeds a threshold quantity established by the authority having jurisdiction to be dangerous to the public if released and is sufficient to pose a threat to the public if released[a]	
Buildings and other structures required to maintain the functionality of other Risk Category IV structures	

Source: ASCE/SEI 7-10. American Society of Civil Engineers Minimum Design Loads for Buildings and Other Structures, 2010.

[a] Buildings and other structures containing toxic, highly toxic, or explosive substances shall be eligible for classification to a lower risk category if it can be demonstrated to the satisfaction of the authority having jurisdiction by a hazard assessment as described in Section 1.5.2 that a release of the substances is commensurate with the risk associated with that risk category.

TABLE 1.5
Importance Factors by Risk Category for Snow, Ice, and Earthquake Loads

Risk Category	Snow Importance Factor, I_s	Ice Importance Factor—Thickness, I_i	Ice Importance Factor—Wind, I_w	Seismic Importance Factor, I_e
I	0.80	0.80	1.00	1.00
II	1.00	1.00	1.00	1.00
III	1.10	1.25	1.00	1.25
IV	1.20	1.25	1.00	1.50

Source: ASCE/SEI 7-10. American Society of Civil Engineers Minimum Design Loads for Buildings and Other Structures, 2010.

2 Wind Loads

PREVIEW

A well-designed, constructed, and maintained building may be damaged by a wind event that is much stronger than what the building was designed for; however, except for tornado damage, this scenario is a very rare occurrence. Rather, most damage occurs because various building elements have limited wind resistance due to inadequate design or material deterioration. Wind with sufficient speed to cause damage to weak buildings can occur anywhere in the United States and its possessions. Although the magnitude and frequency of strong windstorms vary by locale, all buildings should and can be designed, constructed, and maintained to avoid wind damage (other than that associated with tornadoes). In tornado-prone regions, considerations should be given to designing and constructing portions of buildings to provide occupant protection.

There are fundamental differences between design methods for wind and earthquake loading. Wind-loading design is concerned with safety, but occupant comfort and serviceability is a dominant concern. Wind loading does not require any greater understanding of structural behavior beyond that required for gravity and other similar loadings, although it is noted that complex, large, or aerodynamically sensitive structures frequently require wind tunnel testing or more sophisticated dynamic analysis to assure occupant comfort during wind storms. However, design of buildings for wind forces does require a greater understanding of the load mechanism than many other aspects of structural design. To fulfill this need, the primary emphasis in this chapter is on wind loading and its distribution over the entire surface of the building.

2.1 DESCRIPTION OF WIND FORCES

One of the basic questions to be resolved before designing a building is: "What is the strongest wind to be expected at its site?" To answer it, wind measurements have been taken and cataloged in most parts of the world and maps are plotted, like that in Figure 2.1, which gives at a glance the maximum wind speeds to be expected for the United States. It is noted that the wind speed map is taken from ASCE 7-05; hence, they represent allowable stress design (ASD) values for wind design.

Besides depending on wind speed and building height, wind forces vary with the shape of the building. The wind exerts a pressure on the windward surface of a rectangular building because the movement of the air particles is stopped by this face. The air particles, forced from their original direction, go around the building in order to continue their flow and get together again behind the building as shown in Figure 2.2. In so doing, the air particles suck on the leeward face of the building and a negative pressure or suction is exerted on it. The total wind force, thus, is the sum of the windward pressure and the leeward suction, but each of these two forces has its own local effects.

In designing for wind, a building cannot be considered independent of its surroundings. The influence of nearby buildings and of the land configuration can be substantial.

The lateral deflection also referred to as sway at the top of a building due to wind may not be seen by the passerby, but it may feel substantial to those who occupy the top stories of a high building. Under a strong wind, the top of a 1500 ft tall building may swing left and right of its vertical position by as much as 3 ft, and a hurricane can produce swings of 6–7 ft on each side of the vertical. These horizontal swings may not be structurally dangerous, but they may be inconvenient for those who work at such great heights: occupants of the top floor of flexible buildings may sometimes

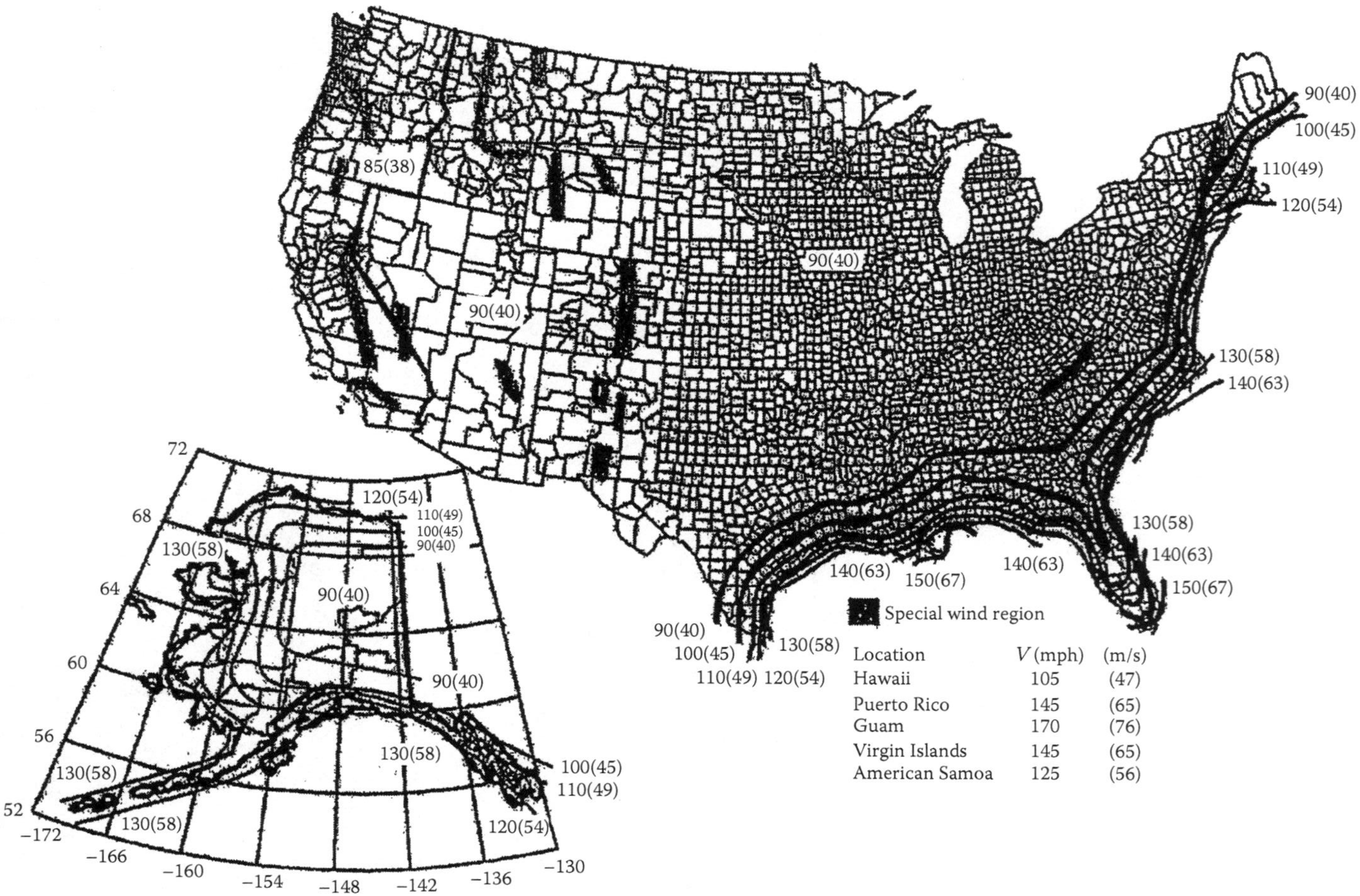

FIGURE 2.1 Wind speed map for United States and Alaska. (From ASCE 7-05.) *Notes*: 1. Values are nominal design 3-s gust wind speeds in miles per hour (m/s) at 33 ft (10 m) above ground for Exposure C category. 2. Linear interpolation between wind contours is permitted. 3. Islands and coastal areas outside the last contour shall use the last wind speed contour of the coastal area. 4. Mountainous terrain, gorges, ocean promotories, and special wind regions shall be examined for unusual wind conditions.

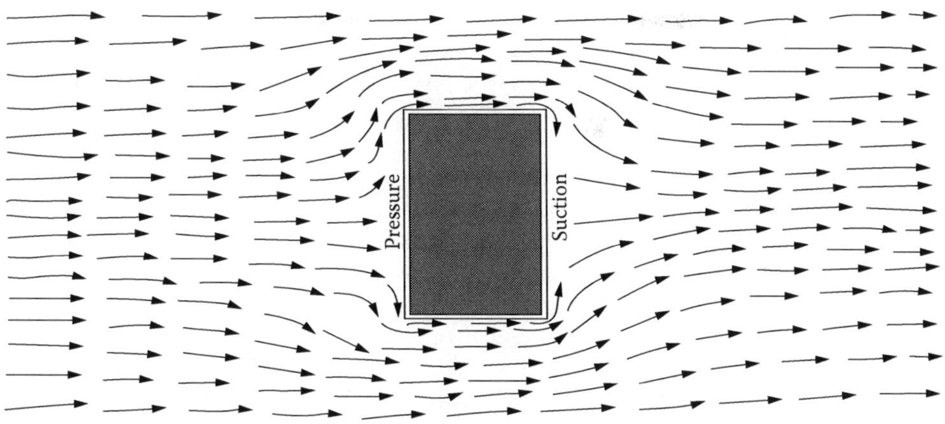

FIGURE 2.2 Plan showing wind flow past a building.

become airsick. Recent research indicates that the airsickness is induced by wind motion in tall buildings in a resonance phenomenon. It occurs when the period of the building more or less coincides with the period of the up-and-down oscillations of our own insides. This explains why some, but not all, of the people in a building may feel queasy. To avoid excessive wind defections, buildings should be stiffened so that their tops will never swing more than, say, 1/500 of their height. Thus, a 3 ft wind drift is perhaps acceptable in a 1500 ft building. However, it is the accelerations of top floors that come into play in determining the serviceability of a building.

How does one prevent the resonant oscillations in a building? The basic method consists in changing its period by reinforcing its structure to make it stiffer. The stiffer the structures, the shorter the period but this is a costly remedy. One method used for years in tall buildings is to use a device called tuned mass damper (TMD), which consists of a heavy mass attached to the top of the building by means of lateral springs (see Figure 2.3). This heavily springed mass is designed to have the same period as the building. When the building oscillates with its own period, the tuned damper after a short while also tends to oscillate with the same period, but in the opposite direction. One could say that the damper moves in antiresonance with the building. When this happens, the oscillations of the building are damped out by the counteraction of the damper. The damper's resonant oscillations do not grow because they are controlled by large shock absorbers that brake its motion.

The first skyscrapers were not vulnerable to the complex consequences of lateral action caused by wind. The enormous weight of the masonry bearing wall building was such that wind action could not overcome the locked-in-gravity forces. Even when the bearing wall system was replaced by the

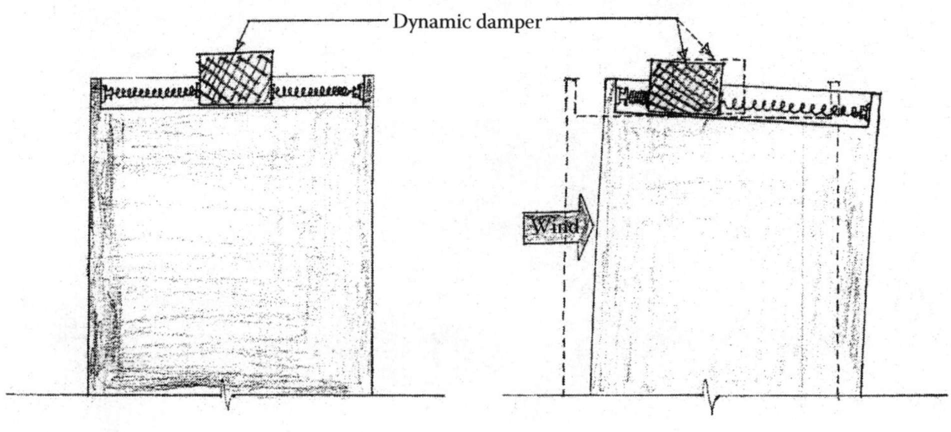

FIGURE 2.3 TMD.

rigid frame structures in the late 1800s, gravity remained the prime determining factor. Heavy stone facades with small openings, closely spaced columns, massive built-up frame members, and heavy partition walls still generated so much weight that wind action was not a major problem.

The glass-walled skyscraper of the 1950s with its optimum interior open space and relatively small weight was first to make us aware of the complexity of wind forces. With the introduction of the steel and lightweight concrete frame, weight was no longer a factor limiting the potential height of buildings. The era of the high-rise building, however, had brought with it new problems. To reduce dead weight and create larger, more flexible spaces, longer spanning beams, movable non-load-bearing interior partitions, and non-load-carrying curtain walls were routinely used. All these innovations have taken away from the overall rigidity of the structure; now the lateral stiffness of a building may be a more important consideration than its strength.

To understand the wind and predict its behavior in precise scientific terms may be impossible. Wind action on a building is dynamic and is influenced by such environmental factors as large-scale roughness and form of terrain, the shape, slenderness, and facade texture of the structure itself, and the arrangement of adjacent buildings.

Wind pressure on a building surface depends primarily on its velocity, the shape and surface structure of the building, the protection from wind offered by surrounding natural terrain or man-made structures, and, to a smaller degree, the density of air, which decreases with altitude and temperature. All other factors remaining the same, the pressure due to wind is proportionate to the square of the velocity:

$$p = 0.00256V^2$$

where

p is the pressure, in psf
V is the velocity of wind, in mph

The multiplier 0.00256 is a constant that reflects the mass density of air for the standard atmosphere and dimensions associated with wind speed in mph, discussed later in this chapter.
During storms, velocities for a 3 s gust wind may reach values up to or greater than 150 mph, which corresponds to dynamic pressure of about 90 psf at a height of 500 ft (153 m). Pressure as high as this is exceptional, and in general, values of 40–50 psf are common.

In an engineered structure, wind loads have long been a factor in the design of the basic lateral-force-resisting system, with added significance as the height of the building increased. For many decades, the cladding systems of high-rise buildings, particularly around corners of buildings, have been scrutinized for the effects of wind on building enclosure. Glass and curtain wall systems are regularly developed and tested to resist cladding pressures and suctions induced by the postulated wind event.

As wind hits the structure and flows around it, several effects are possible as illustrated in Figures 2.4 and 2.5. Pressure on the windward face and suction on the leeward face creates *drag forces.*

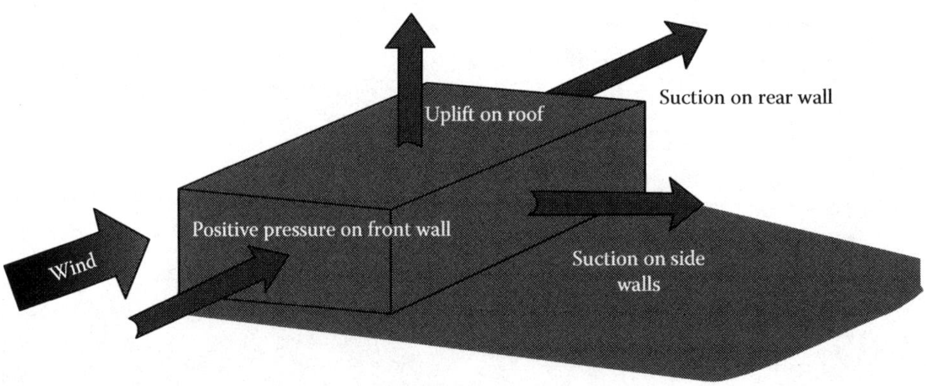

FIGURE 2.4 Schematics of wind-induced pressures.

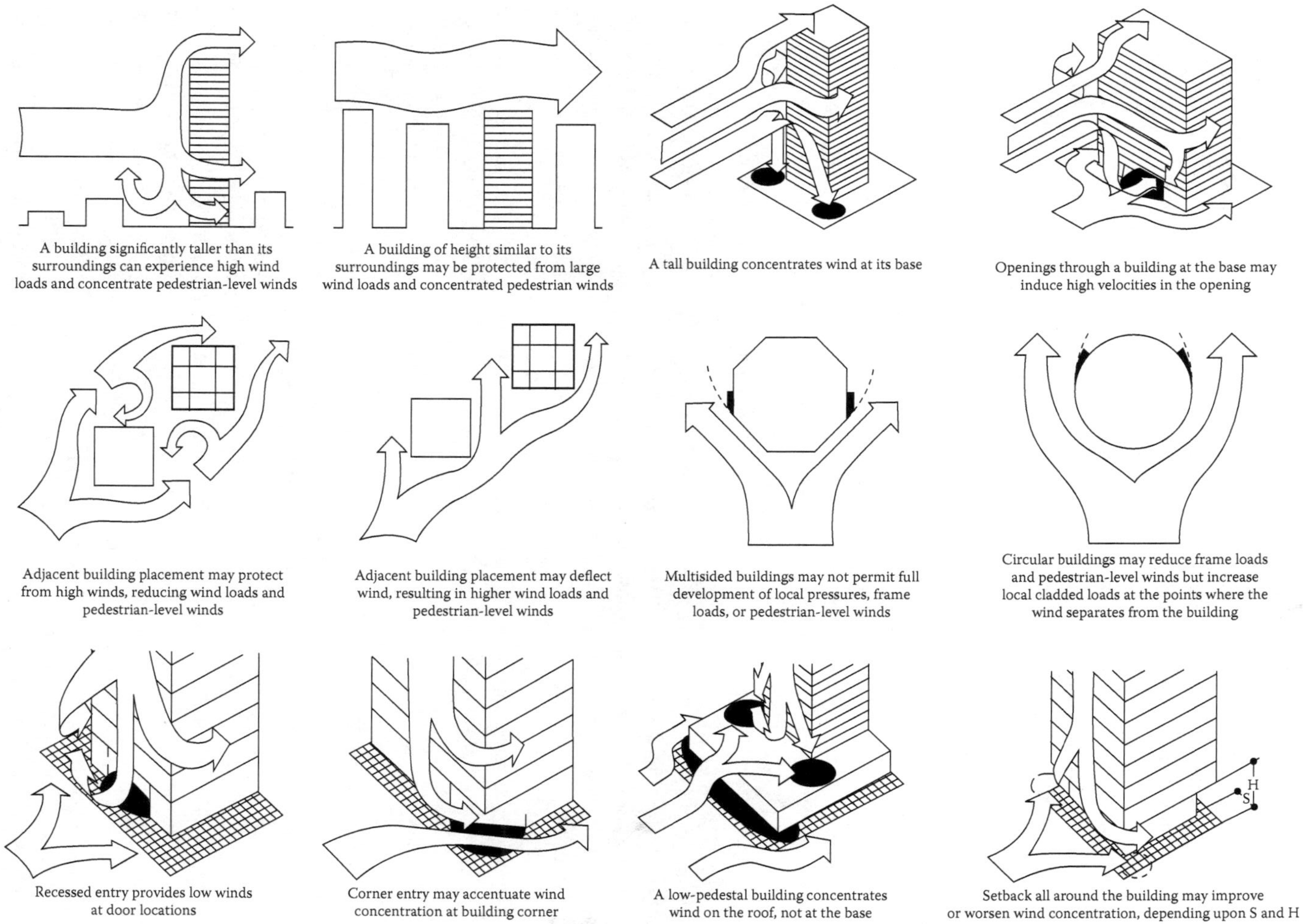

FIGURE 2.5 Wind flow around buildings. (From Taranath, B.S., *Reinforced Concrete Design of Tall Buildings*, CRC Press, Boca Raton, FL, 2009.)

Analogous to flow around an airplane wing, unsymmetrical flow around the structure can create *lift forces.* Air turbulence around the leeward corners and edges can create *vortices*, which are high-velocity air currents that create circular updrafts and suction streams adjacent to the building. Periodic shedding of vertices creates building oscillation transverse to the direction of the wind and may result in unacceptable accelerations at the upper floors of tall buildings. The effects of downdrafts shown schematically in Figure 2.5 must also be considered: Downdrafts have been known to completely strip trees in plaza areas and to buffet pedestrians dangerously. Some tall buildings that extend into high-wind-velocity regions have swayed excessively in strong wings. High suction forces have blown off improperly anchored lightweight roofs.

To be sure, all buildings sway during windstorms, but the motion in earlier tall buildings with heavy full-height partitions has usually been imperceptible and, therefore, has not been a cause for concern. Structural innovations coupled with lightweight construction have reduced the stiffness, mass, and damping characteristics of modern buildings. In these buildings, objects may vibrate, doors and chandeliers may swing, pictures may lean, and books may fall off shelves. Additionally, if the building has a twisting action, its occupants may get an illusory sense that the world outside is moving, creating symptoms of vertigo and disorientation. In more violent storms, windows may break, creating safety problems for pedestrians below. Sometimes, strange and frightening noises may be heard by occupants as the wind shakes elevators, strains floors and walls, and whistles around the building sides.

It is generally agreed that acceleration response that includes the effects of torsion at the top floors of a tall building is the best standard for evaluation of motion perception. A commonly used criterion is to limit accelerations of the building upper floors to no more than 2.0% of gravity (20 milli-g) for a 10-year wind. Other commonly applied guidelines include those published by the council on tall buildings and urban habitat and the international organization for standardization (ISO 6899-1984).

Although one cannot see wind, we know by experience, its flow is quite random and turbulent. Imagine taking a walk on a windy day. You will no doubt experience a constant flow of wind, but intermittently you may also experience sudden gusts of rushing wind. This sudden variation in wind speed, called gustiness or *turbulence*, is an important factor in determining dynamic response of tall buildings.

Unlike steady flow of wind, which for design purposes is considered static, turbulent wind loads associated with gustiness cannot be treated in the same manner. This is because gusty wind velocities change rapidly and even abruptly, creating effects much larger than if the same loads were static. Wind loads, therefore, need to be studied as if they were dynamic, somewhat similar to seismic loads. The intensity of dynamic load depends on how fast the velocity varies and also on the response of the structure itself. Therefore, whether pressures on a building due to wind gust are dynamic or static entirely depends on the gustiness of wind and the dynamic properties of the building to which it is applied.

Consider, for example, the lateral movement of an 800 ft tall building designed for a drift index of *H*/400, subjected to a wind gust. Under wind loads, the building bends slightly as its top moves. It first moves in the direction of wind, with a magnitude of, say, 2 ft (0.61 m) and then starts oscillating back and forth. After moving in the direction of wind, the top goes through its neutral position, then moves approximately 2 ft (0.61 m) in the opposite direction, and continues oscillating back and forth progressively with smaller drifts, until it eventually stops. The time it takes a building to cycle through a complete oscillation is known as *period* of the building. The period of oscillation for a tall steel building in the height range of 700–1400 ft (214–427 m) normally is in the range of 8–12 s, whereas for a 10-story concrete or masonry building, it may be in the range of 0.5–1 s. The action of a wind gust depends not only on how long it takes for the gust to reach its maximum intensity and decrease again, but on the period of the building itself. If the wind gust reaches its maximum value and vanishes in a time much shorter than the period of the building, its effects are dynamic. On the other hand, the gusts can be considered as static loads if the wind load increases and vanishes in a time much longer than the period of the building. For example, a wind gust that develops to its strongest intensity and decreases to zero in 2 s is a dynamic load for a tall building with a period of considerably larger than 2 s, but the same 2 s gust is a static load for a low-rise building with a period much less than 2 s.

Every structure has a natural frequency of vibration. Should dynamic loading occur at or near its natural frequency, structural damage out of all proportion to size of load may result. It is well known, for example, that bridges capable of carrying far greater loads than the weight of a company of soldiers may oscillate dangerously and may even break down under dynamic loading of soldiers marching over them in step. Similarly, certain periodic gust within the wide spectrum of gustiness in wind may find resonance with natural vibration frequency of a building, and although the total force caused by that particular gust frequency would be much less than the static design load for the building, dangerous oscillations may be set up. This applies not only to the structure as a whole but also to components such as curtain wall panels and sheets of glass.

A second dynamic effect is caused by instability of flow around certain structures. Long narrow structures such as smoke stacks, light standards, and suspension bridges are particularly susceptible to this sort of loading, causing an alternating pattern of eddies to form in its wake. A side thrust is thus exerted on the structure similar to the lift on an aerofoil, and since this thrust alternates in direction, the structure responds in a direction perpendicular to the wind direction.

Wind is not constant either with height or time, is not uniform over the windward side of the building, and does not always cause positive pressure. In fact, wind is a complicated phenomenon; it is air in turbulent flow, which means that motion of individual particles is so erratic that in studying wind, one ought to be concerned with statistical distributions of speeds and directions rather than with simple averages.

Wind is the term used for air in motion and is usually applied to the natural horizontal motion of the atmosphere. Motion in a vertical or nearly vertical direction is called a *current*. Movement of air near the surface of the earth is 3D, with horizontal motion much greater than the vertical motion. Vertical air motion is of importance in meteorology but is of less importance near the ground surface. On the other hand, the horizontal motion of air, particularly the gradual retardation of wind speed and high turbulence that occur near the ground surface, is of importance in building engineering. In urban areas, this zone of wind turbulence often referred to as the *surface boundary layer* extends to a height of approximately one-quarter of a mile aboveground. Above this layer, the horizontal airflow is no longer influenced by the retarding effect of the ground surface. The wind speed at this height is called *gradient wind speed*, and it is precisely within this boundary layer where most construction activity occurs. Therefore, how wind effects are felt within this zone is of concern in building design.

Air flowing over the earth's surface is slowed down and made turbulent by the roughness of the surface. As the distance from the surface increase, these friction effects are felt less and less until a height is reached where the influence of the surface roughness is negligible. This height as mentioned earlier is referred to as the gradient height, and the layer of air below this, where the wind is turbulent and its speed slowly increases with height, is referred to as the boundary layer. During periods of neutral atmospheric stability, typical of strong wind conditions, the gradient height or depth of the earth's boundary layer is determined largely by the terrain roughness and typically varies from 900 ft (270 m) over open country to about 1660 ft (500 m) over built-up urban areas.

Wind tunnel testing provides information regarding the dependence of particular response parameters on wind speed and direction. In order to make the most rational use of this aerodynamic information, it is necessary to synthesize it with the actual wind climate characteristics at the site. The characteristics necessary to define are those governing wind speed and direction at a suitable height above ground level at the site. The joint probability distribution of wind speed and direction at a suitable height and wind climate is determined statistically, whereas all of the aerodynamic information, which includes sensitivities to building orientation and to its surroundings, is obtained from the wind tunnel data. For most wind engineering applications, natural wind over a particular terrain can be simulated by a turbulent boundary layer flow developed in a wind tunnel over a long fetch of appropriate model terrain roughness.

Winds are produced by differences in atmospheric pressure, which are primarily attributable to differences in temperature. These differences are caused largely by the unequal distribution of heat from the sun and the difference in thermal properties of land and ocean surfaces. When temperatures of adjacent regions become unequal, the warmer, lighter air rises and flows over the colder, heavier air. Winds initiated in this way are modified by rotation of the earth.

In describing global circulation of wind, modern meteorology relies on wind phrases used by early long-distance sailors. For example, terms like trade winds and westerlies were used by sailors who recognized the occurrence of steady winds blowing for long periods of time in the same direction.

Near the equator, the lower atmosphere is warmed by the sun's heat. The warm air rises, depositing much precipitation and creating a uniform low-pressure area. Into this low-pressure area, air is drawn from the relatively cold high-pressure regions from northern and southern hemispheres, giving rise to trade winds between the latitudes of 30° from the equator. The air going aloft flows counter to the trade winds to descend into these latitudes, creating a region of high pressure. Flowing northward and southward from these latitudes in the northern and southern hemispheres, respectively, are the prevailing westerlies, which meet the cold dense air flowing away from the poles in a low-pressure region characterized by stormy variable winds. It is this interface between cold, dense air and warm, moist air that is of main interest to the television meteorologists of northern Europe and North America.

As the air above hot earth expands and rises, air from cooler areas such as the oceans floats in to take its place. The process produces two types of wind circulation:

1. General global circulation extending around the earth
2. Smaller secondary circulations producing local wind conditions

Figure 2.6 shows a model of global circulation of prevailing winds that result from the general movement of air around the earth. Observe that there are no prevailing winds within the equatorial belt,

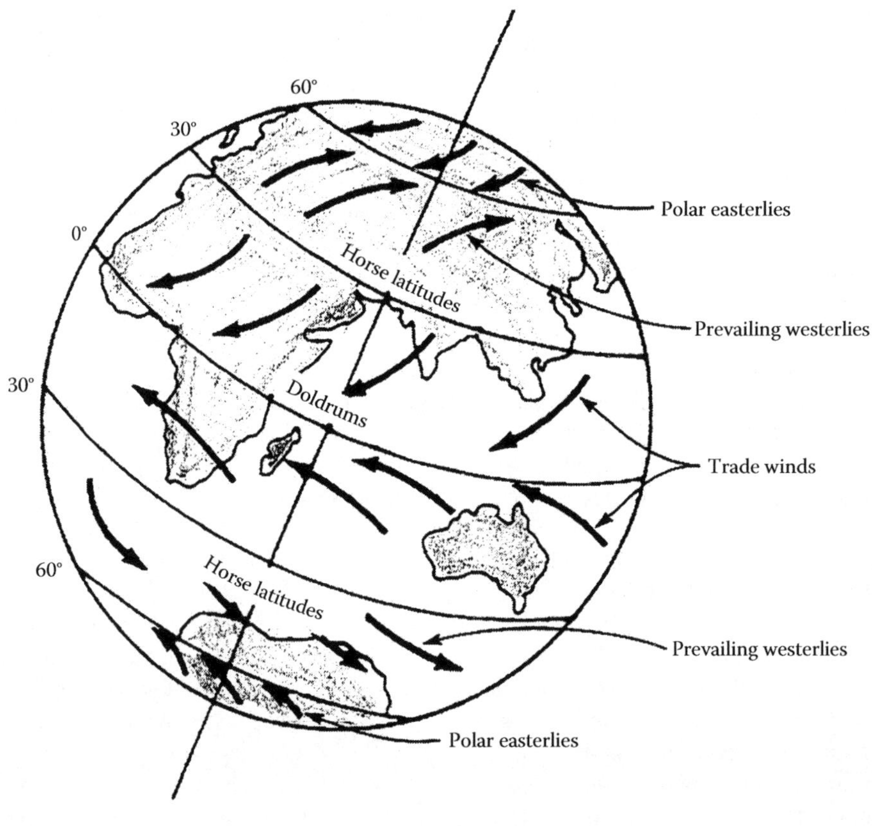

FIGURE 2.6 Global circulation of winds.

which lies roughly between latitudes 10°S and 10°N. Therefore, near the equator and up to about 700 mi (1127 km) on either side of it, there exists a region of relative calm called the doldrums. In both hemispheres, some of the air that has risen at the equator returns to the earth's surface at about 30′ latitude, producing little or no wind. These high-pressure areas are called horse latitudes, possibly because many horses died on the sailing ships that got stalled because of lack of wind. The winds that blow between the horse latitudes and the doldrums are called trade winds because sailors relied on them in sailing ships. The direction of trade winds is greatly modified by the rotation of the earth as they blow from east to west. Two other kinds of winds that result from the general circulation of the atmosphere are called the prevailing winds and the polar easterlies. The prevailing winds blow in to the belts bounded by the horse latitudes and 60° north and south of the equator. Thus, the moving surface air produces six belts of winds around the earth as shown in Figure 2.6.

Wind forces occur on the exterior envelope of the building and are a function of its height, local ground surface roughness (hills, trees, and other buildings). Their magnitude is directly proportional to square of the wind velocity. The weight of the building, unlike in earthquake design, has no effect on wind forces, but is helpful in resisting uplift. Unless the structure has large openings, all wind forces are applied to the building exterior. This is in contrast to earthquake forces where both exterior and interior elements are loaded proportionally to their weight. Wind pressures act inward only on the windward face and outward on the most sides and roof surfaces.

2.2 TYPES OF WIND STORMS

A variety of wind storm types occur in different part of the United States that impose loads on buildings many times more severe than those calculated in their design. The characteristics of the type of storms that can impact the building site should be considered in the design. The primary storm types are as follows:

2.2.1 Straight-Line Wind

This type of wind event is the most common and is considered, in general, to blow in a straight line. Straight-line wind speeds range from very low to very high. High winds associated with intense low pressure can last for upward of a day at a given location. Straight-line winds occur throughout the United States and its possessions.

2.2.2 Down-Slope Wind

Wind flowing down the slope of mountains is referred to as down-slope wind. Down-slope winds with very high wind speeds frequently occur in Alaska and Colorado. In the continental United States, mountainous areas are referred to as *special wind regions*. These are shown in Figure 2.7 along with hurricane-prone regions. Neither ASCE 7 nor IBC provide values for wind speeds in special wind regions. Instead, they advise that if the local building department has not established the basic speed, use of regional climatic data and consultation with a wind engineer or meteorologist for establishing design wind speed.

2.2.3 Downburst

Also known as microburst, it is a powerful downdraft associated with thunderstorm. When the downdraft reaches the ground, it spreads out horizontally and may form one or more horizontal vortex rings around the downdraft. The outflow is typically 6,000–12,000 ft across and the vortex ring may rise 2,000 ft above the ground. The life cycle of a downburst is usually between 15 and 20 min. Observations suggest that approximately 5% of all thunderstorms produce a downburst, which can result in significant damage in a localized area.

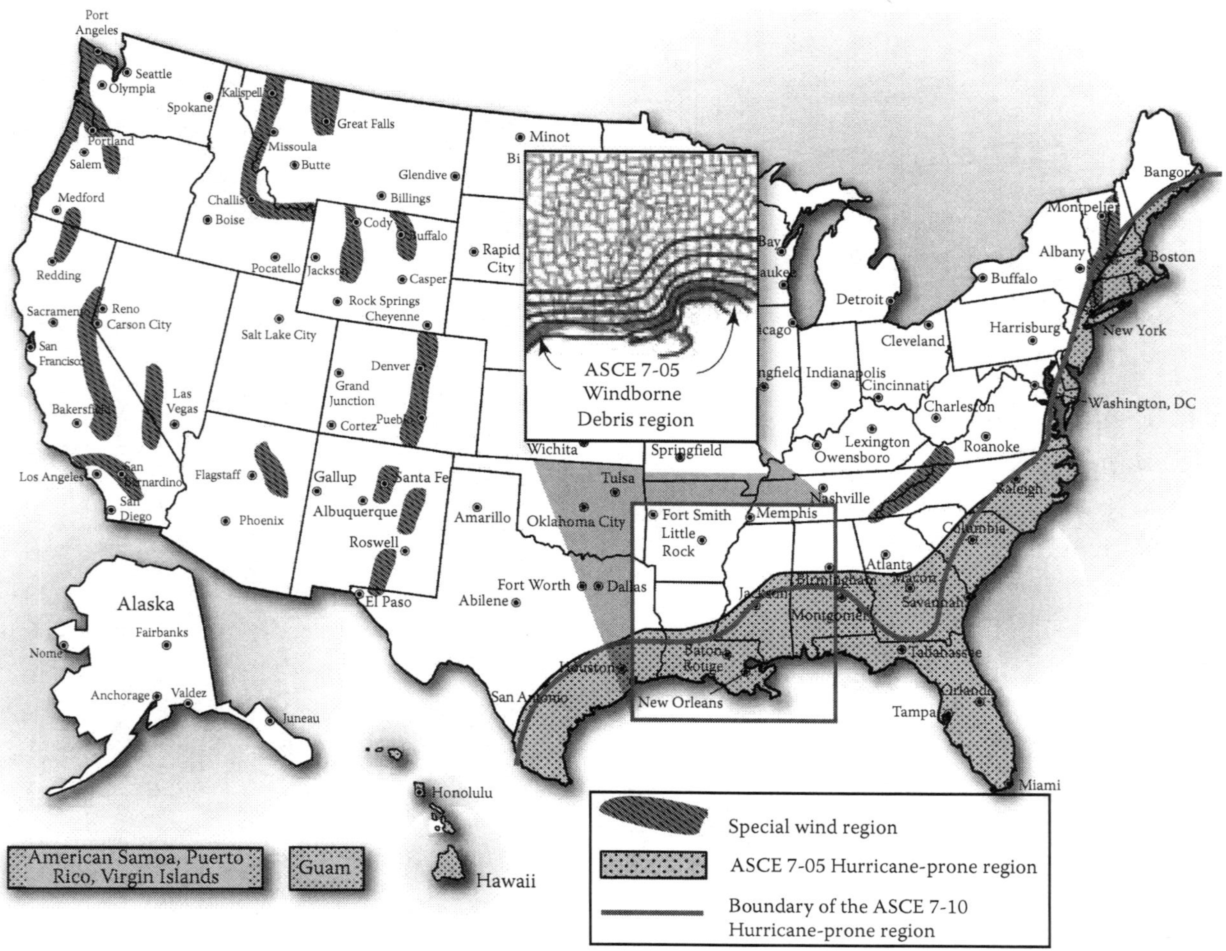

FIGURE 2.7 Hurricane-prone regions and special wind regions. *Note*: Hurricane/typhoon-prone regions also include American Samoa, Guam, Puerto Rico, and the U.S. Virgin Islands. (From FEMA P-804. *Wind Retrofit Guide for Residential Buildings*, Washington, DC: Federal Emergency Management Association, December 2010.)

2.2.4 Northeastern Winds

This type of storm is cold and violent and occurs along the northeastern coast of the United States. These storms blow in from the northeast and may last for several days.

2.2.5 Thunderstorm

This type of storm can rapidly form and produce high wind speeds. Approximately 10,000 severe thunderstorms occur in the United States each year, typically in the spring and summer. They are most common in the southeast and Midwest. Besides producing high winds, they often create heavy rain. Hail and tornadoes are also sometimes produced. Thunderstorms commonly move through an area quite rapidly, often causing high winds for only a few minutes at a given location. However, thunderstorms can also stall and become virtually stationary.

Thunderstorms are one of the most familiar features of temperate summer weather, characterized by long hot spells punctuated by release of torrential rain. The essential conditions for the development of thunderstorms are warm, moist air in the lower atmosphere and cold, dense air at higher altitudes. Under these conditions, warm air at ground level rises, and once it has started rising, it continues to rise faster and faster, building storm clouds in the upper atmosphere. Thunder and lightning accompany downpours, creating gusty winds that sometimes blow violently at great speeds. Wind speed of 50–70 mph are typically reached in a thunderstorm and are often accompanied with swirling wind action exerting high suction forces on roofing and cladding elements.

2.2.6 Hurricane

This is a system of spiraling winds converging with increasing speed toward the storm's center (the eye of the hurricane). Hurricanes form over warm oceans. The diameter of the storm varies between 50 and 600 mi. A hurricane's forward movement (translational speed) can vary between approximately 10 and 25 mph. Besides being capable of delivering extremely strong winds for several hour, many hurricanes also bring very heavy rainfall. Hurricanes also occasionally spawn tornadoes. The Saffir–Simpson Hurricane Scale rates the intensity of hurricanes. The five-step scale ranges from Category I (the weakest) to Category V (the strongest).

Of all the storm types, hurricanes have the greatest potential for devastating a very large geographical area and, hence, affect great numbers of people. The terms *hurricanes, tropical cyclones, and typhoons* are synonymous for the same type of storm. See Figure 2.7 for hurricane-prone regions.

Hurricanes are severe atmospheric disturbances that originate in the tropical regions of the Atlantic Ocean or Caribbean Sea. They travel north, northwest, or northeast from their point of origin and usually cause heavy rains. They originate in the doldrums and consist of high-velocity winds blowing circularly around a low-pressure center known as the eye of the storm. The low-pressure center develops when the warm saturated air prevalent in the doldrums interacts with the cooler air. From the edge of the storm toward its center, the atmospheric pressure drops sharply and the wind velocity rises. In a fully developed hurricane, winds reach speeds up to 70–80 mph (31–36 m/s), and in severe hurricanes, it can attain velocities as high as 200 mph (90 m/s). Within the eye of the storm, the winds cease abruptly, the storm clouds lift, and the seas become exceptionally violent.

The maximum basic wind velocity (3 s gust) for any area of the United States specified in ASCE 7-10 for buildings in Risk Categories I, II and III, and IV are 170, 180, and 200 mph, respectively. Observe these values are much less than the highest wind speeds in hurricanes. Except in rare instances, such as defense installations, a structure is not normally designed for full hurricane wind speeds.

Hurricanes are one of the most spectacular forms of terrestrial disturbances and produce the heaviest rains known on earth. They have two basic components, warmth and moisture, and consequently they develop only in the tropics. Almost invariably they move in a westerly direction at first and then swing away from the equator, either striking land with devastating results or moving out over the oceans until they encounter cool surface water and die out naturally. The region of greatest storm frequency is the northwestern Pacific, where the storms are called typhoons—a name of Chinese origin meaning *wind which strikes*. The storms that occur in the Bay of Bengal and the seas of north Australia are called cyclones. Although there are some general characteristics common to all hurricanes, no two are exactly alike. However, a typical hurricane can be considered to have a 375-mi (600 km) diameter, with its circulating winds spiraling in toward the center at speeds up to 112 mph (50 m/s). The size of the eye can vary in diameter from as little as 3.7–25 mi (6–40 km). However, the typhoon that roared past the island of Guam in 1979 had a very large diameter of 1400 mi (2252 km) with the highest wind reaching 190 mph (85 m/s). Storms of such violence have been known to drive a plank of wood right through the trunk of a tree and blow straws end-on through a steel deck. Fortunately, storms of such magnitude are not common.

2.2.7 Tornado

This is a violently rotating column of air extending from the base of a thunderstorm to the ground. The Fujita scale categorizes tornado severity based on observed damage. The six-step scale ranges from F0 (light damage) to F5 (incredible damage). Weak tornadoes (F0–F1) are most common, but strong tornadoes (F2 and F3) frequently occur. Violent tornadoes (F4–F5) are rare. See Figure 2.8 for recorded tornado occurrence historical data.

Tornado path widths are typically less than 1000 ft; however, widths of approximately 1 mi have been reported. Wind speed rapidly decreases with increased distance from the center of a tornado. A building on the periphery of a strong or violent tornado could be subjected to moderate to high wind speeds, depending upon the distance from the core of the tornado. However, even though the wind speed might not be great, a building on a periphery could still be impacted by many large pieces on wind-borne debris. Tornadoes are responsible for the greatest number of wind-related deaths each year in the United States.

Tornadoes develop within severe thunderstorms and occasionally in hurricanes. They consist of a rotating column of air usually accompanied by a funnel-shaped downward extension of a dense cloud having a vortex of several hundred feet, typically 200–800 ft (61–244 m) in diameter whirling destructively at speeds up to 300 mph (134 m/s). A tornado contains the most destructive of all wind forces, usually destroying everything along its path of approximately 10 mi (16 km) long and directed predominately toward the northeast. Tornadoes form when a cold storm front runs over warm, moist surface air. The warm air rises through the overlaying cold storm clouds and is intercepted by the high-altitude winds, which are even colder and are rapidly moving above the clouds. Warm air collides with the cooler air and begins to whirl. The pressure at the center of the spinning column of air is reduced because of the centrifugal force. This reduction in pressure causes more warm air to be sucked into it, creating a violent outlet for the warm air trapped under the storm. As the velocity increases, more warm air is drawn in to the low-pressure area created in the center of the vortex. As the vortex gains strength, the funnel begins to extend toward the ground, eventually touching it. Funnels usually form close to the leading edge of the storm. Larger tornadoes may have several vortices within a single funnel. If the bottom of the funnel can be seen, it usually means that the tornado has touched down and begun to pick up visible debris from the ground. A typical tornado travels 20–30 mph (9–14 m/s), touches ground for 5–6 mi (8–10 km), and has a funnel 300–500 ft (92–152 m) wide. Distance from the ground to the cloud averages about 2000 ft (610 m). Although it is impractical to design buildings to sustain a direct hit from a tornado, it behooves the engineer to pay extra attention to anchorage of roof decks and curtain walls for buildings in areas of high tornado frequency. Rolling plains and flat country make a natural home for tornadoes. Statistically, flat plains get more tornadoes than other parts of the country. In North America, communities in Kansas, Nebraska, and Texas have many tornadoes and are classified as *tornado belt* areas. No accurate measurement of the inner speed of a tornado has been made because tornadoes destroy standard measuring instruments. However, photographs of tornadoes suggest the wind

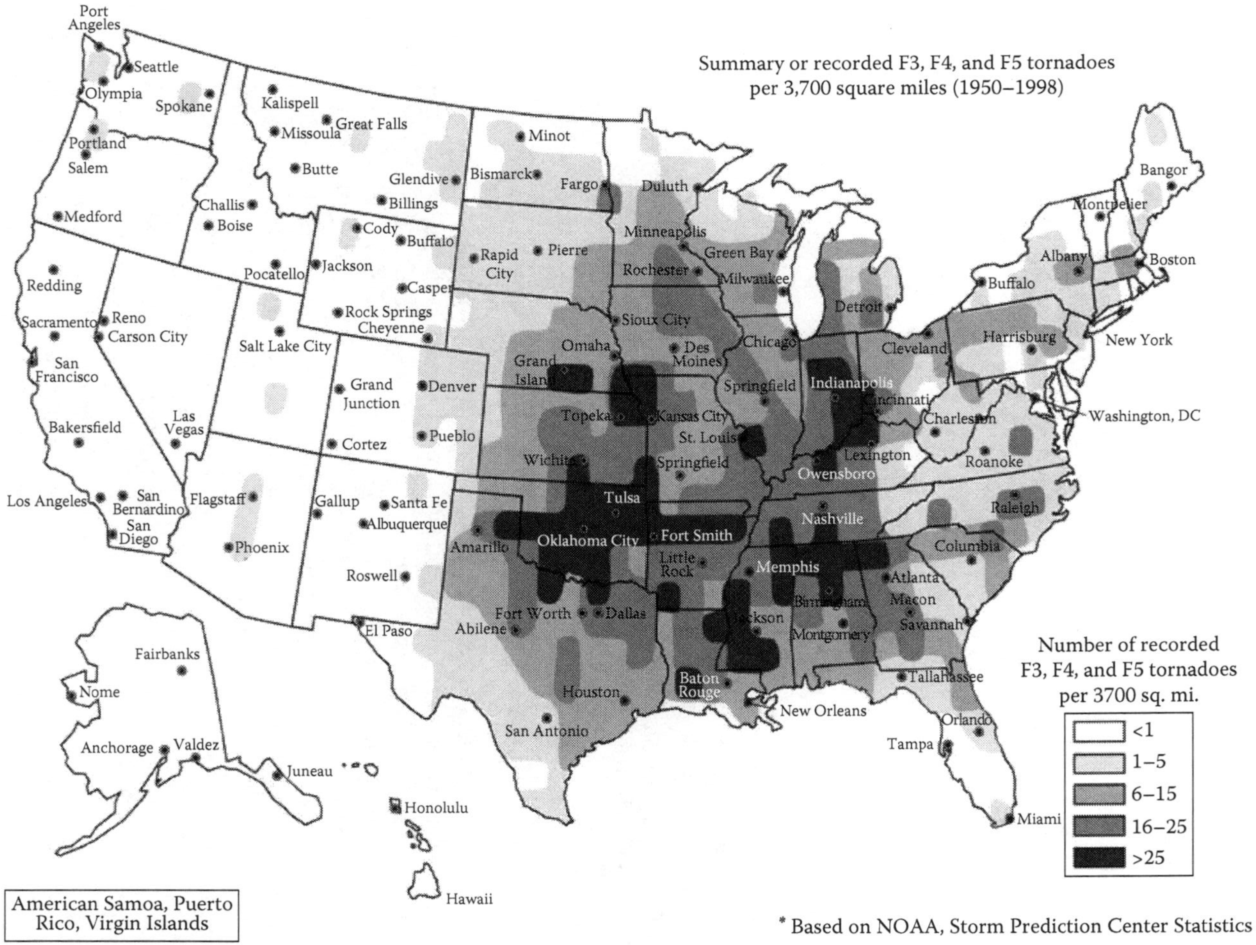

FIGURE 2.8 Tornado occurrence in the United States based on historical data. (From FEMA 361. Design and construction guidance for community shelters, Washington, DC: Federal Emergency Management Association, July 2000.)

speeds are of the order of 167–224 mph (75–100 m/s). Although there definitely are tornado seasons, tornadoes can occur at any time. Like a hurricane, the tornado consists of a mass of unstable air rotating furiously and rising rapidly around the center of an area with low atmospheric pressure. The similarity ends here, because whereas the hurricane is generally of the order of 300–400 mi (483–644 km) in diameter, a large tornado is unlikely to be more than 1500 ft (458 m) across. However, in terms of destructive violence, no other atmospheric disturbance comes even close to that caused by tornadoes.

Although the probability of any one particular building being hit by a tornado is very small, tornadoes account for the greatest incidence of death and serious injury of building occupants due to structural failure and cause considerable economic loss. With some exceptions, such as nuclear power plants, it is generally not economical to design buildings for tornadoes. It is, however, important to provide key construction details for the safety of building occupants. Investigations of tornado-damaged areas have shown that the buildings in which well over 90% of the occupants were killed or seriously injured by tornadoes did not satisfy the following two key details of building construction:

1. The anchorage of house floors into the foundation or ground (the floor takes off with the occupant in it)
2. The anchorage of roofs to concrete block walls (the roof takes off and the unsupported block wall collapses onto the occupants)

Deficiency of the second construction detail is especially serious for open assembly occupancies because there is nothing inside, such as stored goods, to protect the occupants from wall collapse. For such buildings in tornado-prone areas, it is recommended that the block walls contain vertical reinforcing linking the roof to the foundation.

For tornado protection, key details such as those indicated earlier should be designed on the basis of design wind speeds shown in Figure 2.9.

2.2.8 Statistical Likelihood of Natural Hazards

The warning times for natural hazards such as wind and earthquakes hazards vary. Earthquakes are unique among the natural hazards because there is no warning at all, although new sensing devices can now give a few seconds warning to locations far from the epicenter. Floods (except flash floods) can be predicted so as to give hours or days of warning. Hurricanes can be tracked for days and give several hours of warning before hitting a specific location. Tornadoes are more localized and, though visible, may hit a specific location almost without notice.

Although the tornado gives warning and its approach is visible during daylight, its winds are often so strong that damage or destruction in its immediate vicinity is common. Hurricanes are tracked by the national hurricane tracking system and their movement is carefully and thoroughly reported. The hurricane's movement along its path is slower and its size is much larger than a tornado, yet even then its precise route and timing cannot be predicted until a few hours before making landfall.

In earthquake-prone areas that experience frequent events, such as California and Alaska, there is a continuous generalized prediction, but earthquake always strikes totally without warning. Therefore, earthquakes are regarded as the most random events within a general envelope of probability.

For all hazards, the regional probabilities are much higher than the local ones, and the extreme events are relatively rare for a given site.

Values for high winds are commonly expressed in return periods, much shorter than earthquakes because their incidence is much more frequent. To the public, these return periods seem very long (i.e., why would a business owner confronting small crises every day and large ones every month on

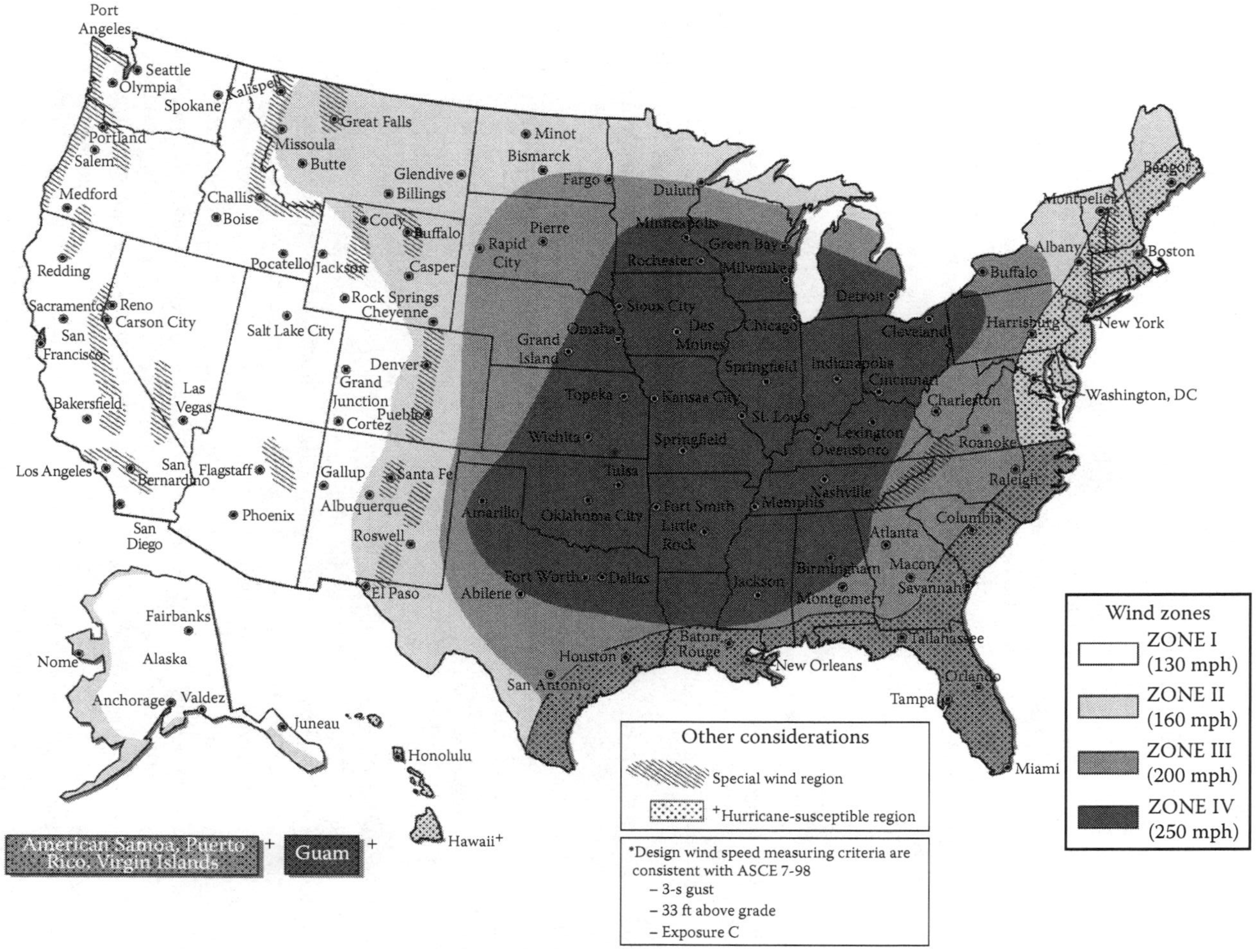

FIGURE 2.9 Design wind speeds for community tornado shelters. (From FEMA 361. Design and construction guidance for community shelters, Washington, DC: Federal Emergency Management Association, July 2000.)

so be worried about an event that might not occur for 500 years—let alone 2000 years?). And if the return period for California is 500 years, would it not be another 400 plus years before something of the magnitude of the 1906 San Francisco earthquake occurs?

The problem is that these figures represent mean or average return periods over a very long period of time, with the result that the return period is often quite inaccurate in relation to the shorter time periods in which most of us are interested (i.e., the next year or the next 10 years). Because floods and high winds are relatively frequent, the discrepancy between the actual return period and the mean return period used in the codes is much more noticeable than the corresponding probabilities for earthquakes.

Currently, these statements of probability are the best we can do. Because they express mean values over long periods of time, they tell little about what will really happen this year or next year, but they may give a hint as to what will happen in our lifetime.

For example, a succession of hurricanes roam the Atlantic seaboard every year, bringing the risk of extreme winds and storm surge. Traditionally, residents in tornado-prone areas retreated to their basements, but engineered safe rooms are now being constructed in homes, schools, and other buildings. Earthquakes are perhaps the most difficult to deal with, because of their complete lack of warning, their rarity, and their possible extreme consequences. Although an earthquake of a given magnitude is still, in practical terms, unpredictable, its probability of occurrence can safely be predicted as far higher in California or Alaska than in, for example, Massachusetts or Tennessee. Even in California, the rarity of a large earthquake is such that many people will not experience one in their lifetime. In less seismic parts of the country, one must go back several generations, or to folklore, for earthquake stories, but even then there is a probability of an event.

Because natural hazards are only broadly predictable, the incidence of future events can only be expressed as probabilities. This presents a problem because what may be perfectly rational and useful to a mathematician may be confusing or even counterproductive to the public and their decision makers. The probability of occurrence of earthquakes, floods, and high winds is commonly expressed by use of the term *return period* or *mean recurrence interval* (*MRI*). This is defined as the average or mean time in years between the expected occurrence of an event of specified intensity. For example, earthquake codes use a level of shaking (an acceleration value) that correspond to a 1% probability of exceedance in 50 years (or a probability that it would be exceeded one time in approximately 500 years, a 500-year return period).

Winds that are of interest in the design of buildings can be classified into three major types: prevailing winds, seasonal winds, and local winds.

1. *Prevailing winds*: Surface air moving toward the low-pressure equatorial belt is called prevailing wind or trade wind. In northern hemisphere, the northerly wind blowing toward the equator is deflected by the rotation of the earth to a northeasterly direction and hence commonly known as the northeast trade wind. The corresponding wind in the southern hemisphere is the southeast trade wind.
2. *Seasonal winds*: Air over the land is warmer in summer and colder in winter than the air adjacent to oceans during the same seasons. During summer, the continents become seats of low pressure, with wind blowing in from the colder oceans. In winter, the continents experience high pressure with winds directed toward the warmer oceans. These movements of air caused by variations in pressure difference are called seasonal winds. The monsoons of the China Sea and the Indian Ocean are examples.
3. *Local winds*: These are associated with the regional weather patterns and include whirlwinds and thunderstorms. They are caused by daily changes in temperature and pressure, generating local effects in winds. The daily variations in temperature and pressure may occur over irregular terrain, causing valley and mountain breezes.

All three types of wind are of importance in building design. However, for the purpose of determining wind loads, the characteristics of prevailing and seasonal winds are grouped together, whereas those of local winds are studied separately. This grouping is to distinguish between the widely differing scales of fluctuations of the winds; prevailing and seasonal winds fluctuate over a period of several months, whereas local winds may vary every few seconds. The variations in the mean velocity of prevailing and seasonal winds are referred to as *fluctuations*, whereas the variations in local winds occurring over a very short period of time are referred to as *gusts*.

Flow of wind unlike that of other fluids fluctuates in a random fashion. Because of this, wind loads imposed on buildings are studied statistically.

2.2.9 Probabilistic Approach in Wind Engineering

In many engineering sciences, the intensity of certain events is considered as a function of the duration recurrence interval (return period). For instance, in hydrology, the intensity of rainfall expected in a region is considered in terms of a return period. For example, it is self-evident that the rainfall expected once in 10 years is less than the one expected, say, once every 50 years. Similarly, in wind engineering, the speed of wind is considered to vary with return periods. For example, the nominal design 3 s gust speed for most of nonhurricane areas of the United States at 33 ft (10 m) aboveground, corresponding to a MRI of 700 years, 115 mph (51 m/s), compared 120 mph (54 m/s) for an MRI of 1700 years. See Figures 26, 5-IA and IB of ASCE 7-10, for basic wind speed maps.

A 700-year return-period wind of 115 mph (51 m/s) means that on the average, these areas will experience a wind faster than 115 mph within a period of 700 years. A return period of 700 years corresponds to a probability of occurrence of $1/700 = 0.00143 \sim 0.14\%$. Thus, the chance that a wind exceeding 115 mph (51 m/s) will occur in these regions within a given year is 0.14%.

Suppose a building is designed for a 100-year lifetime using a design wind speed of 115 mph. What is the probability that this wind will exceed the design speed within the lifetime of the structure? The probability that this wind speed will not be exceeded in any year is 49/50. The probability that this speed will not be exceeded 100 years in a row is $(49/50)^{100}$. Therefore, the probability that this wind speed will be exceeded at least once in 100 years is

$$1-\left(\frac{49}{50}\right)^{100} = 0.87 = 87\%$$

This signifies that although a wind with low annual probability of occurrence is used to design structures, there still exists a high probability of that wind being exceeded within the lifetime of the structure. However, in structural engineering practice, it is believed that the actual probability of overstressing a structure is much less because of the factors of safety and the generally conservative values used in design.

It is important to understand the notion of probability of occurrence of design wind speeds during the service life of buildings. The general expression for probability P that a design wind speed will be exceeded at least once during the exposed period of n years is given by

$$P = 1 - (1 - P_a)^n$$

where

P_a is the annual probability of being exceeded (reciprocal of the MRI)
n is the exposure period in years

Consider a building designed for a 50-year service life instead of 100 years. The probability of exceeding the design wind speed at least once during the 50-year lifetime of the building is

$$P = 1 - (1 - 0.02)^{50} = 1 - 0.36 = 0.64 = 64\%$$

TABLE 2.1
Probability of Exceeding Design Wind Speed during Design Life of a Building

Annual Probability P_a	Mean Recurrence Interval ($1/P_a$) Years	Exposure Period (Design Life), *n* (Years) 1	5	10	25	50	100
0.1	10	0.1	0.41	0.15	0.93	0.994	0.999
0.04	25	0.04	0.18	0.34	0.64	0.87	0.98
0.034	30	0.034	0.15	0.29	0.58	0.82	0.97
0.02	50	0.02	0.10	0.18	0.40	0.64	0.87
0.013	75	0.013	0.06	0.12	0.28	0.49	0.73
0.01	100	0.01	0.05	0.10	0.22	0.40	0.64
0.0067	150	0.0067	0.03	0.06	0.15	0.28	0.49
0.005	200	0.005	0.02	0.05	0.10	0.22	0.39

Source: Taranath, B.S., *Reinforced Concrete Design of Tall Buildings*, CRC Press, Boca Raton, FL, 2009.

The probability that wind speeds of a given magnitude will be exceeded increases with a longer exposure period of the building and the MRI used in the design. Values of P for a given MRI and a given exposure period are shown in Table 2.1.

2.3 WIND/BUILDING INTERACTIONS

When wind interacts with a building, both positive and negative (i.e., suction) pressures occur simultaneously (see Figures 2.2, 2.4, and 2.5). (*Note*: Negative pressures are less than ambient pressure.) A building must have sufficient strength to resist the applied loads in order to prevent wind-induced building failure. The magnitude of the pressures is a function of the following primary factors.

2.3.1 Exposure Categories

The characteristics of the ground roughness and surface irregularities in the vicinity of a building influence the wind loading. ASCE 7-10, like its predecessor ASCE 7-05, defines three exposure categories, Exposures B, C, and D. Exposure B is the roughest terrain and Exposure D is the smoothest. Exposure B includes urban, suburban, and wooded areas. Exposure C includes flat open terrain with scattered obstructions and areas adjacent to water surfaces in hurricane-prone regions (which are defined in the succeeding text under *basic wind speed*). Exposure D includes areas adjacent to water surfaces, mud flats, salt flats, and unbroken ice. See Figure 2.10 for wind velocity profiles corresponding to Exposures B, C, and D.

Note that recent findings have shown that wind speed retardation in hurricane coastal regions is not as pronounced as was estimated earlier. Therefore, Exposure D is back in ASCE 7-10 for hurricane regions.

The smoother the terrain, the greater the wind load; therefore, buildings (subject to the same basic wind speed) located in Exposure D would receive higher wind loads than those located in Exposure C.

2.3.2 Basic Wind Speed

ASCE 7-10 defines basic wind speeds as nominal 3 s gust speeds measured at 33 ft above grade in Exposure C (flat open terrain). If the building is located in Exposure B or D, rather than C, an adjustment for the actual exposure is made in the ASCE 7 calculation procedure.

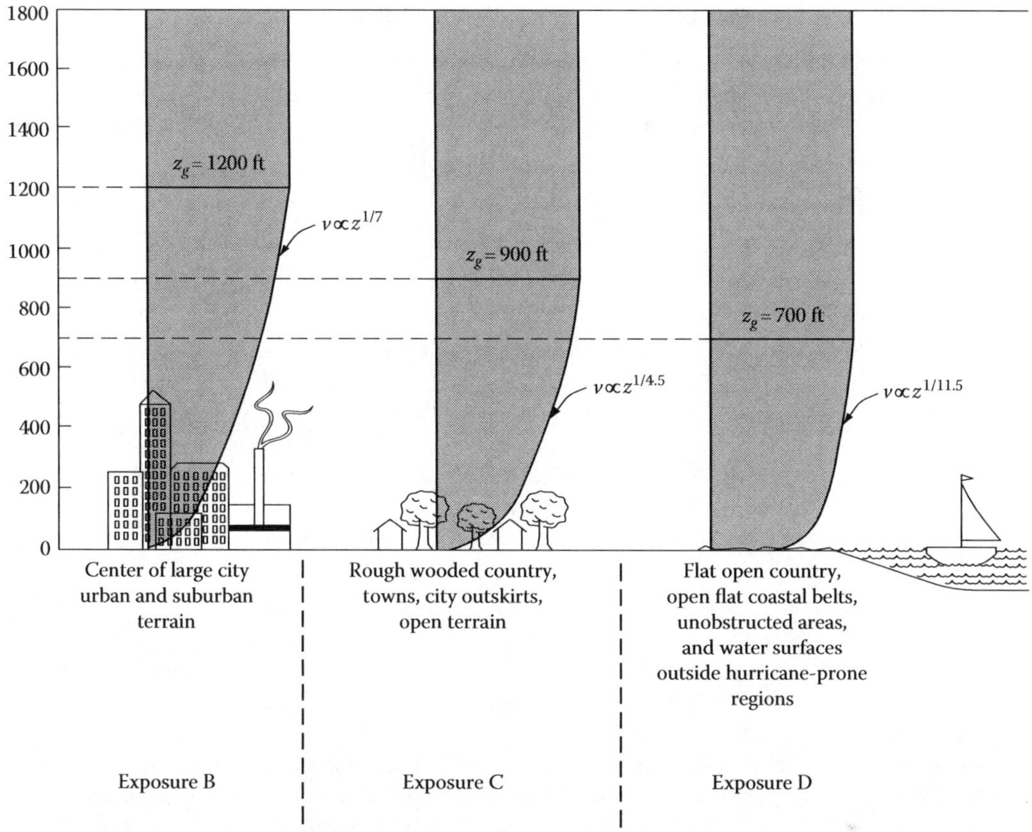

FIGURE 2.10 Wind velocity profile corresponding to Exposures B, C, and D. (From Taranath, B.S., *Reinforced Concrete Design of Tall Buildings*, CRC Press, Boca Raton, FL, 2009.)

Since the 1995 edition of ASCE 7, the basic wind speed has been a peak 3 s gust speed. Prior to that time, the basic wind speed was a fastest-mile speed (i.e., the speed averaged over the time required for a mile-long column of air to pass a fixed point). Because the measuring time for peak gust versus fastest mile is different, peak gust speeds are typically about 20 mph faster than fastest-mile speeds (e.g., a 90 mph wind speed is equivalent to a 70 mph fastest-mile wind speed).

In determining wind pressures, the basic wind speed is squared; therefore, as the velocity is increased, the pressures are exponentially increased. For example, let us assume the uplift load on the high roof covering at a corner area of a given high-rise building is 50 pounds per square foot (psf) with a basic wind speed of 110 mph. If the speed is increased to 150 mph, the roof corner load increases by a factor of $(150/110)^2 = 1.86$.

2.3.3 Topography

Abrupt changes in topography, such as isolated hills, ridges, and escarpments, cause wind speedup; therefore, a building located near a ridge would receive higher wind loads than a building located on relatively flat land. ASCE 7 provides a procedure to account for topographic influences.

2.3.4 Building Height

Wind speed increases with height above the ground. Therefore, the taller the building the greater the speed and, hence, the grater the wind loads. ASCE 7 provides a procedure to account for building height.

2.3.5 Internal Pressure

Wind striking a building can cause either an increase in the pressure within the building (i.e., positive pressure) or a decrease in the pressure (i.e., negative pressure). Internal pressure changes occur because of the porosity of the building envelope. Porosity is caused by openings around doors and window frames and by air infiltration through walls that are not absolutely airtight. A door or window left in the open position also contributes to porosity.

Wind striking an exterior wall exerts a positive pressure on the wall, which forces air through openings and into the interior of the building (this is analogous to blowing up a balloon). At the same time, the windward wall is receiving positive pressure and the side and rear walls are receiving negative (suction) pressure; therefore, air within the building is being pulled out at openings in these other walls. As a result, if the porosity of the windward wall is greater than the combined porosity of the side and rear walls, the interior of the building is pressurized. But if the porosity of the windward wall is less than the combined porosity of the side and rear walls, the interior of the building is depressurized (this is analogous to letting air out of a balloon).

When a building is pressurized, the internal pressure pushes up on the roof. This push from below the roof is combined with the suction above the roof, resulting in an increased wind load on the roof. The internal pressure also pushes on the side and rear walls. This outward push is combined with the suction on the exterior side of these walls. Therefore, a pressurized building increases the wind load on the side and rear walls as well as on the roof, as shown schematically in Figure 2.11.

When a building is depressurized, the internal pressure pulls the roof down, which reduces the amount of uplift exerted on the roof. The decreased internal pressure also pulls inward on the windward wall, which increases the wind load on that wall. See Figure 2.12 for schematics.

When a building becomes fully pressurized (e.g., due to window breakage), the loads applied to the exterior walls and roof are significantly increased. The buildup of high internal pressure can also blow down interior partitions and blow ceiling boards out of their support grid. The breaching of a small window is typically sufficient to cause full pressurization of the building's interior.

ASCE 7 provides a design procedure to assess the influence of internal pressure on the wall and roof loads, and it provides positive and negative internal pressure coefficients for use in load calculations. Buildings that can be fully pressurized are referred to as partially enclosed buildings. Buildings that have limited internal pressurization capability are referred to as enclosed buildings.

2.3.6 Aerodynamic Pressure

Because of building aerodynamics (i.e., the interaction between the wind and the building), the highest uplift loads occur at roof corners. The roof perimeter has a somewhat lower load. Exterior walls typically have lower loads than the field at the roof. The ends of walls have higher suction loads than the portion of wall between the ends. However, when the wall is loaded with positive pressure, the entire wall is uniformly loaded. The negative values indicate suction pressure acting upward from the roof surface and outward from the wall surface. Positive values indicate pressures acting inward on the wall surface.

Aerodynamic influences are accounted for by use of external pressure coefficients, which are used in load calculations. The magnitude of the coefficient is a function of the location on the building (e.g., roof corner) and building shape as discussed in the succeeding text.

Building shape affects the magnitude of pressure coefficients and, therefore, the loads applied to the various building surfaces. For example, the uplift loads on a low-slope roof are larger than the loads on a gable or hip roof. The steeper the slope, the lower the uplift load. Pressure coefficients for monoslope roofs, sawtooth roofs, and domes are all different from those for low-slope and gable/hip roofs.

Building irregularities such as bay window projections, a stair tower projecting out from the main walls, dormers, chimneys, etc., can cause localized turbulence. Turbulence causes wind speedup, which increases the wind loads in the vicinity of the building irregularity.

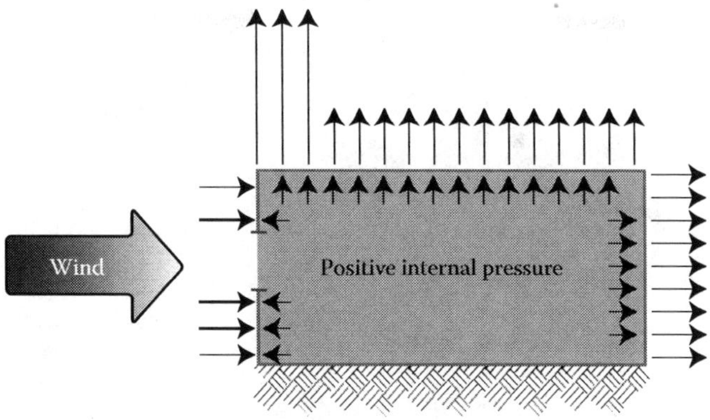

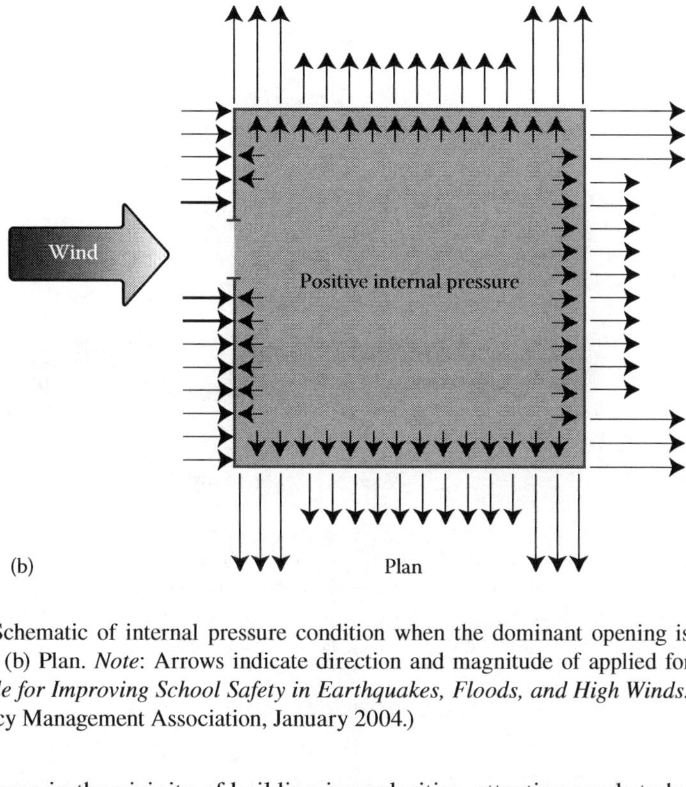

FIGURE 2.11 Schematic of internal pressure condition when the dominant opening is in the windward wall. (a) Section. (b) Plan. *Note*: Arrows indicate direction and magnitude of applied force. (From FEMA 424. *Design Guide for Improving School Safety in Earthquakes, Floods, and High Winds*. Washington, DC: Federal Emergency Management Association, January 2004.)

To avoid damage in the vicinity of building irregularities, attention needs to be given to attachment of building elements located in turbulent flow areas.

2.3.7 Probability of Occurrence

Most buildings are typically assigned a Risk Category II (Occupancy Category II) and are designed for a 700-year MRI wind event (7% probability of exceedance in 50 years). A 50-year storm would be expected to happen once every 50 years on an average but may not occur within any given interval. On the other hand, two 50-year storms could occur within a year.

ASCE 7-10 requires buildings assigned to higher risk categories (Occupancy Categories III and IV) to be designed for a 1700-year MRI event (3% probability of exceedance in 50 years). Therefore, these buildings are designed to resist stronger, rarer storms than Risk Category I and II buildings.

When designing building wind resistance, the following types of wind should be considered.

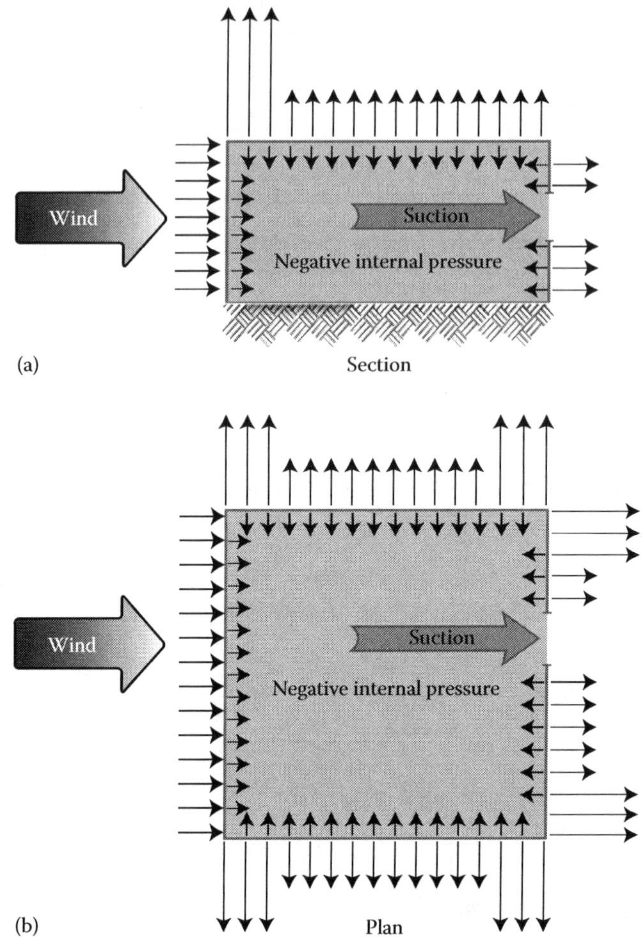

FIGURE 2.12 Schematic of internal pressure condition when the dominant opening is the leeward walls. (a) Section. (b) Plan. *Note*: Arrows indicate direction and magnitude of applied force. (From FEMA 424. *Design Guide for Improving School Safety in Earthquakes, Floods, and High Winds*. Washington, DC: Federal Emergency Management Association, January 2004.)

2.3.7.1 Routine Winds

In many locations, winds with low to moderate speeds occur daily. Damage is not expected to occur during these events.

2.3.7.2 Stronger Winds

At a given site, stronger winds (e.g., winds with a basic wind speed in the range of 70–80 mph peak gust) may occur from several times a year to only once a year or less frequently. Seventy to eighty miles per hour is the threshold at which damage normally begins to occur to building elements that have limited wind resistance due to problems associated with inadequate design, strength, application, or material deterioration.

2.3.7.3 Design Level Winds

Buildings that experience design level events and events that are somewhat in excess of design level should experience little, if any damage; however, design level storms frequently cause extensive building envelope damage. Structural damage also occurs, but less often. Damage experienced with design level events is typically associated with inadequate design, or material deterioration. The exceptions are wind-driven water infiltration and wind-borne debris (missiles) damage.

2.3.7.4 Tornadoes

Although more than 1200 tornadoes typically occur each year in the United States, the probability of a tornado occurring at any given location is quite small. Only a few areas of the United States frequently experience tornadoes, and tornadoes are very rare in the west. The Oklahoma City area is the most active location in the United States, with 106 recorded tornadoes between the years 1890 and 2000. Except for window breakage, well-designed, constructed, and maintained buildings should experience little if any damage from weak tornadoes. However, weak tornadoes often cause building envelope damage. Most buildings experience significant damage if they are in the path of a strong or violent tornado.

Missile damage is very common during hurricanes and tornadoes. Missiles can puncture roof coverings, many types of exterior walls, and glazing. The IBC does not address missile-induced damage, except for glazing in wind-borne debris regions. (Wind-borne debris regions are limited to portions of hurricane-prone regions.) In hurricane-prone regions, significant damage should be expected even during design level hurricane events, unless special enhancements are incorporated.

2.4 BEHAVIOR OF TALL BUILDINGS SUBJECTED TO WIND

The traditional approach to wind loading that has been used for the design of tall buildings since the 1960s is one in which the wind pressure is assumed to act statically. This is convenient in that it has enabled the appropriate coefficients of pressure to be estimated from wind tunnel tests carried out in a uniform steady velocity in a wind tunnel. The velocity used in design has been variously determined from maximum gust speeds or average velocities measured over a minute of a *mile of wind*, depending on the nature of the routine meteorological measurements. There is usually an adjustment for the variation of velocity with height, which is sometimes based on the maximum gust variation with height and sometimes on the mean speed variation with height. The resulting pressures usually are in the range 15–50 psf and are applied to various elements of the structure such as the mainframe and the glass.

Although convenient, this quasistatic approach to wind loading is unrealistic in several respects. Furthermore, there is a need for broadening the basis of design against wind to include more explicitly such factors as allowable deflections, comfort of occupants, and strength. The diversity of structures now being built makes the problem of formulating wind loads using simplified umbrella loadings more and more difficult if at the same time they are to be economical and satisfactory from a performance standpoint.

The use of static wind loads in the design of tall buildings, although convenient, can at times lead to false impression as to the real behavior of a tall building—or any structure for that matter—in the wind. In the natural wind over a city, a tall building is constantly buffeted by gusts and other aerodynamic forces, and although it does tend to deflect toward a mean position, it is continuously swaying with an amplitude that may be at least as large as the steady deflection. This continuous swaying motion is illustrated in Figure 2.13, which is taken from actual observations on two skyscrapers in New York. Stresses measured in a column of Empire State Building shown in Figure 2.14 also confirm sway motions of tall buildings.

In describing the observation on the Empire State Building, Rathbun (1940) states that the "building tended to vibrate continuously like the tines of a turning fork." All the traces show that the sway motion occurs primarily in what turns out to be the fundamental frequency of the building with little evidence of the higher harmonics.

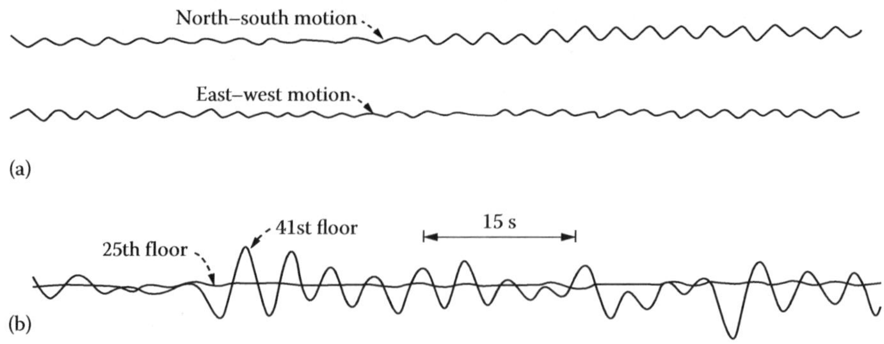

FIGURE 2.13 Observed deflection of two New York skyscrapers (circa 1940). (a) Stiff building (51st floor). (b) Comparatively flexible building east–west motion.

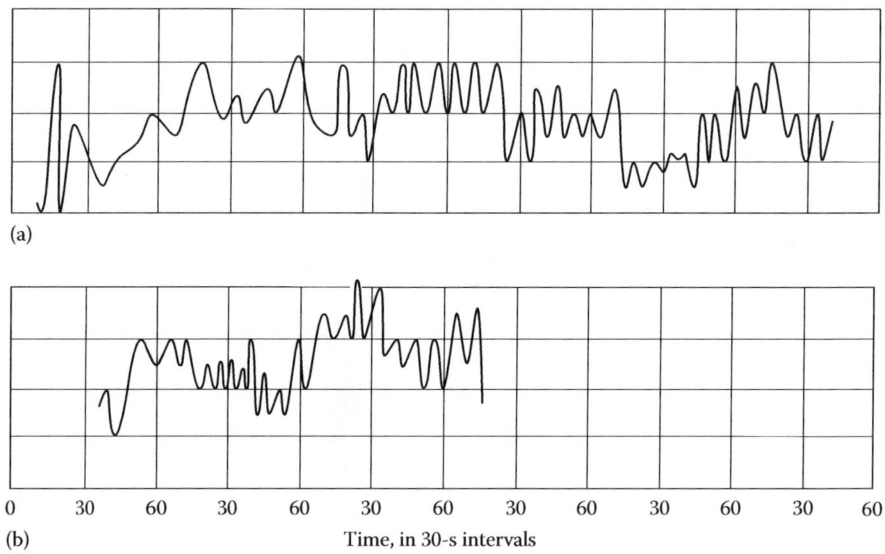

FIGURE 2.14 Measured stress in column of the Empire State Building. (a) 17 cycles in 140 s. (b) 17 cycles in 136 s.

The sway motion is in fact probably the most significant aspect of the behavior determining whether a structure performs satisfactorily or not in service. The sway acceleration directly determines the comfort of occupants and a large proportion of the total deflection, which governs the cracking of walls is due to sway. The fact that the dynamic response is influenced by factors other than stiffness alone, such as the mass and damping, and the fact that the criterion of performance cannot be based on stress and strength only but also on other factors such as deflection and acceleration considerations makes design for the dynamic effects of wind more subtle.

In order to develop an approach for the design of tall buildings against the wind, it is first necessary to understand in general terms the character of the wind and the important factors that influence its action on buildings. While it cannot be pretended that the understanding is yet complete, research has clarified a great deal and provided a framework in which to fit ideas.

The motion of the atmosphere, as it is manifested by the wind, is comprised of air movements of a very wide range of scales. On a very large scale, there are seasonal fluctuations in the wind. On a scale comparable with the weather maps seen in the press or used on airlines, there are large-scale fluctuations identified by the patterns of isobars moving across the country. On a still smaller scale are fluctuations that are best observed on high-speed anemometer records. Such a record is shown in Figure 2.5, which was obtained from three instruments mounted on a tall mast.

It is convenient for analytical purposes to separate the widely different scales of fluctuations into two categories. The large-scale fluctuations, down to the scale of weather map fluctuations, are referred to as fluctuations in the mean velocity. Fluctuations of a much smaller character such as those appearing in the anemometer record are referred to as gusts.

It turns out that mean wind speeds averaged over a 10 min to 1 h period are reasonably stable quantities and are unaffected by slight shifts in the time origin. This is evident from the constancy of the mean velocity measured in anemometer records. This is less true of shorter averaging periods, and speeds averaged over a minute may differ radically from minute to minute. A 10 min or one 1 h average, therefore, represents a suitable unit to define the mean velocity while the characteristics of the gusts are described separately for an average period of a few seconds.

2.4.1 Properties of the Mean Wind Loads

Reference to Figure 2.15 suggests that the mean wind speed at each of the three heights remain more or less constant throughout the period of record and gust fluctuations take place about this mean. A further feature revealed by this record is that the magnitude of the mean velocity increases with height. This increase of the mean velocity with height is a fairly well-understood phenomenon. The rate of increase is much influenced by the roughness of the terrain.

This variation of velocity with height is best considered as a gradual retardation of the wind nearer the ground due to surface friction at heights great enough for the wind to be virtually independent of surface friction the wind moves freely under the influence of the pressure gradient and attains the so-called gradient velocity. The height at which this occurs is the gradient height. Figure 2.16 suggests typical mean velocity profiles for a nominal gradient wind speed of 100. This shows that the wind speed at 100 ft in a city is approximately one-half of that in open country. This reduction in velocity

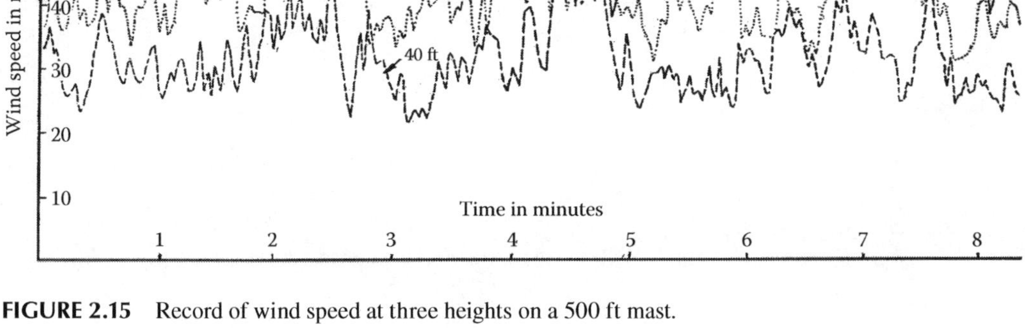

FIGURE 2.15 Record of wind speed at three heights on a 500 ft mast.

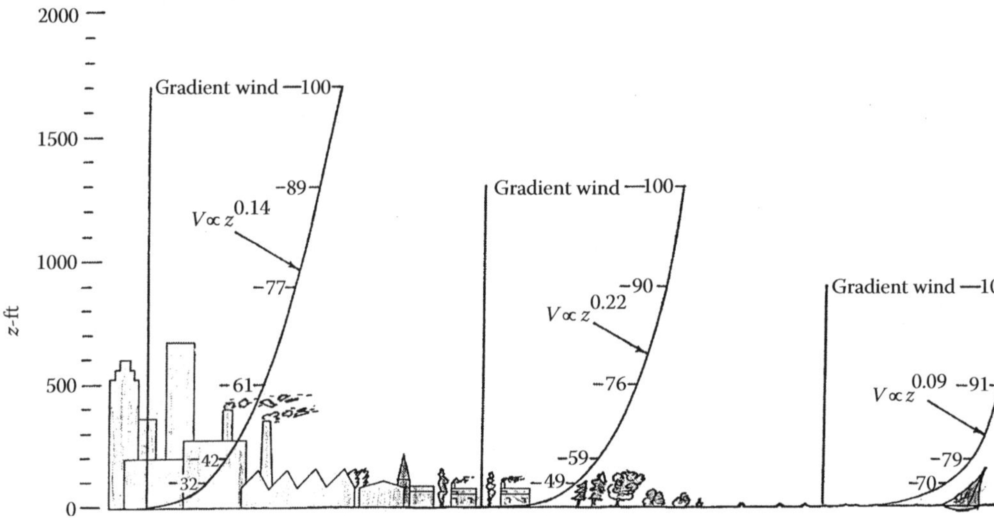

FIGURE 2.16 Profiles of mean wind velocity over level terrains of differing roughness.

in built-up cities has long been confirmed by the measurements shown in Figure 2.17, which compares the wind speed in cities and their local airports. Although anemometers in the city are generally mounted higher aboveground than at the airport, the mean wind speed is seen to be invariably lower.

2.4.1.1 Variation of Wind Velocity with Height

The roughness of the earth's surface due to friction causes drag on wind flow. Therefore, wind speed at the surface is much less than speeds at higher levels. The effect of frictional drag gradually decreases with height, and at gradient level (around 1000–2000 ft), frictional drag effect becomes negligible. For strong winds, the shape of the vertical profile of wind speed depends mainly on the degree of surface roughness, including the overall drag effects of buildings, trees, and any other projections that impede flow of wind at the surface.

The height at which the drag effects cease to exist is called gradient height, and the corresponding velocity, *gradient velocity*. This characteristic increase of wind velocity with height is a well-understood phenomenon, as evidenced by higher design pressures specified at higher elevations in most building codes.

At heights of approximately 1200 ft (366 m) aboveground, wind speed is virtually unaffected by surface friction. Its movement at and above this level is solely a function of seasonal and local wind effects. The height through which the wind speed is affected by topography is called the *atmospheric boundary layer.*

The wind speed profile within the atmospheric boundary layer is given by

$$V_z = V_g \left(\frac{z}{z_g} \right)^{1/\alpha} \tag{2.1}$$

where

V_z is the mean wind speed at height Z aboveground
V_g is the gradient wind speed assumed constant above the boundary layer
z is the height aboveground
z_g is the nominal height of boundary layer also referred to as gradient height
α is the power law coefficient

See Figure 4.1 for schematics of wind profile and values of z_g and α.

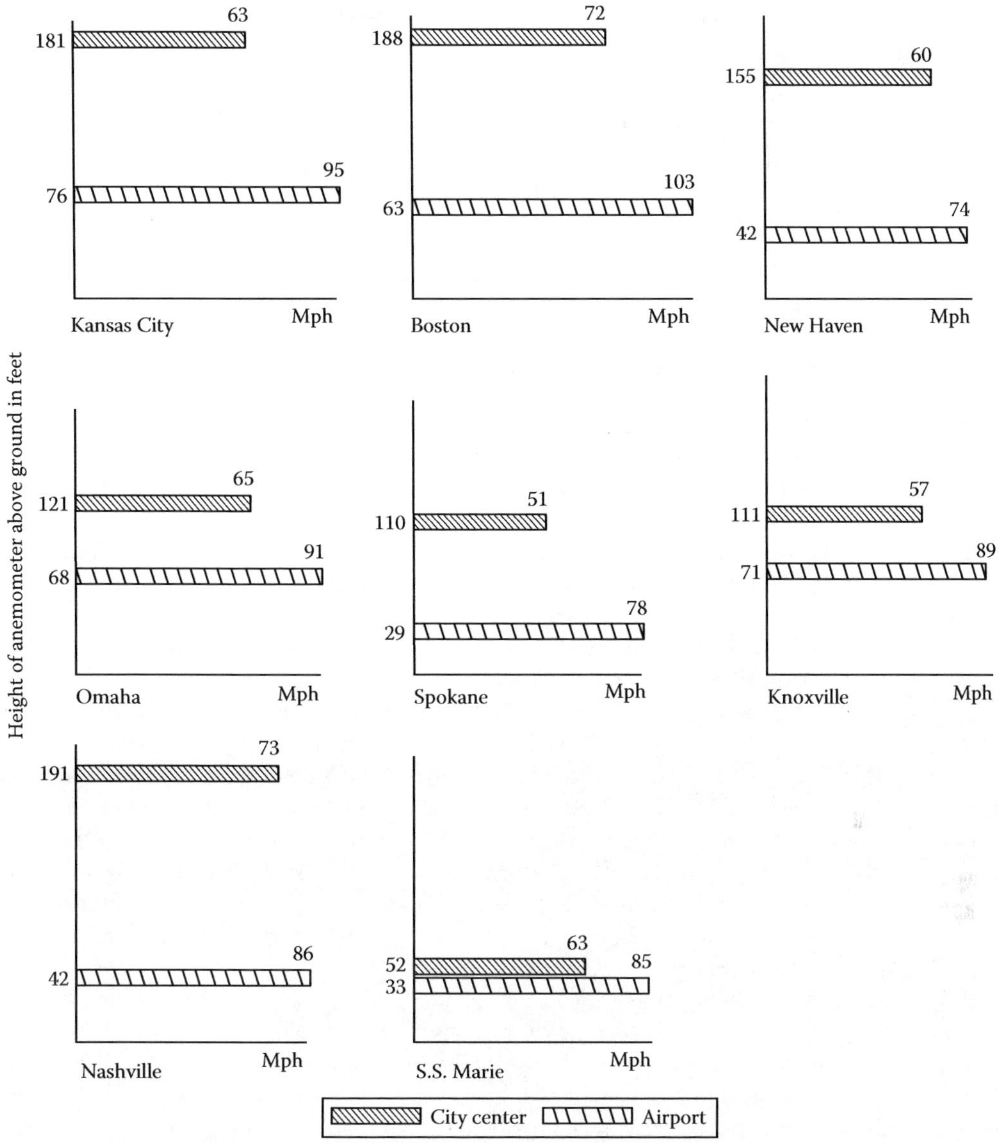

FIGURE 2.17 Comparison of year wind speeds at airport and meteorological stations in the United States (Circa 1960).

With known values of mean wind speed at gradient height and exponent α, wind speeds at height z are calculated by using Equation 2.1. The exponent $1/\alpha$ and the depth of boundary layer z_g vary with terrain roughness and the averaging time used in calculating wind speed. The values of a range from a low of 0.087 for open country and 0.20 for built-up urban areas, signifying that wind speed reaches its maximum value over a greater height in an urban terrain than in the open country.

2.4.1.2 Wind Turbulence

Motion of wind is turbulent. A concise mathematical definition of turbulence is difficult to give, except to state that it occurs in wind flow because air has a very low viscosity—about one-sixteenth that of water. Any movement of air at speeds greater than 2–3 mph (0.9–1.3 m/s) is turbulent, causing particles of air to move randomly in all directions. This is in contrast to the laminar flow of particles of heavy fluids, which move predominantly parallel to the direction of flow.

FIGURE 2.18 Wind turbulence showing gusts and lulls.

The velocity profiles shown in Figures 2.10 and 2.16 describe only one aspect of wind at lower levels. Superimposed on mean speed are gusts and lulls, which are deviations above and below the mean values. Theses gusts and lulls have a random distribution over a wide range of frequencies and amplitudes over the entire height of the boundary layer as shown schematically in Figure 2.18. Gusts are frequently the result of the introduction of fast-moving parcels of air from higher levels into slower moving air strata. This mixing produces turbulence due to surface roughness and thermal instability.

When this occurs, dynamic instability of flow may result with eddies separating first from one side and then form again. Turbulence caused by surface roughness is similar to the turbulent boundary layer flow at the walls of pipes. Flow near surface encounters small obstacles that change wind speed and introduce random vertical and horizontal velocity components at right angles to main direction of flow. Turbulence generated by obstacles may persist downwind from projections as much as 100 times their height. Large-scale topographical features are not included in the aforementioned surface roughness. They can influence the flow, so they are given special consideration in design by using a topographic factor K_{zt}. For instance, wind is usually much stronger over the brow of a hill or ridge. This is because, to pass the same quantity of air over the obstructing feature, a higher speed is required. Large valleys often have a strong funneling effect that increases wind speed along the axis of the valley.

Every structure has a natural frequency of vibration. Should dynamic loading occur at or near its natural frequency, structural damage out of all proportion to size of load may result. It is well known, for example, that bridges capable of carrying far greater loads than the weight of a company of soldiers may oscillate dangerously and may even break down under dynamic loading of soldiers marching over them in step. Similarly, certain periodic gust within the wide spectrum of gustiness in wind may find resonance with natural vibration frequency of a building, and although the total force caused by that particular gust frequency would be much less than the static design load for the building, dangerous oscillations may be set up. This applies not only to the structure as a whole but also to components such as curtain wall panels and sheets of glass. A second dynamic effect is caused by instability of flow around certain structures. Long narrow structures such as smoke stacks, light standards, and suspension bridges are particularly susceptible to this sort of loading, causing an alternating pattern of eddies to form in its wake. A side thrust is thus exerted on the object similar to the lift on an aerofoil, and since this thrust alternates in direction, a vibration may result. Side-to-side wobbling effect of a straight stick pulled through water is an example of this phenomenon.

2.4.2 Action of Wind on Tall Buildings

The mechanism whereby pressures are induced on a bluff body by a flow field is still incompletely understood. This is particularly true of the fluctuating forces generated by gusts in the wind flow.

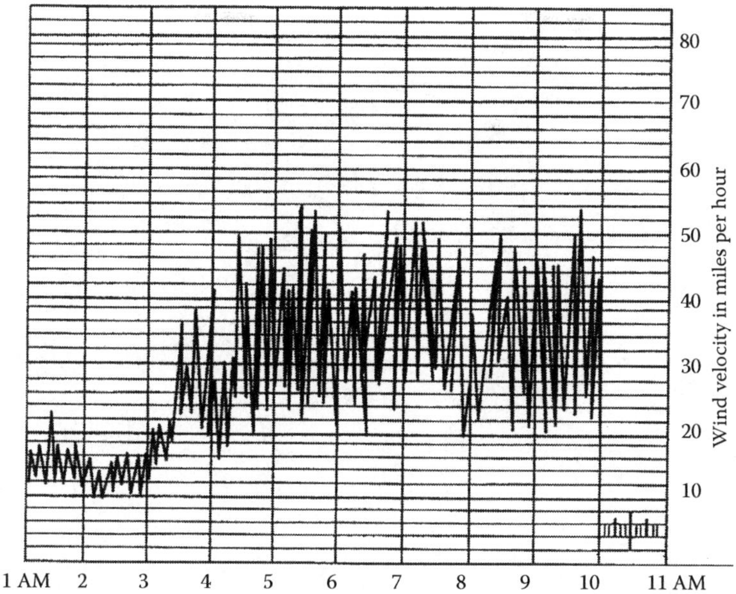

FIGURE 2.19 Wind speed as measured by an anemometer.

An impression of the nature of gust wind pressures (which are proportional to the square of velocity) can be obtained from the wind speed measurements shown in Figure 2.19. It is apparent from this figure that the fluctuating pressures are of the same magnitude as the mean pressures. The frequency of the more significant pressure fluctuations turns out to be of the same order as the natural frequency of the building.

2.4.3 Dynamic Action of Wind

The dynamic action of the wind on tall buildings can be associated with the following influences:

1. Buffeting by guts
2. Buffeting by turbulence and vortices shed by the structure itself
3. Buffeting by the wake from another structure
4. Aerodynamic damping

2.4.4 Buffeting due to Vortex Shedding

In general, wind buffeting against a bluff body is diverted in three mutually perpendicular directions, giving rise to these sets of forces and moments. In aeronautical engineering (see Figure 2.20), all six components are significant. However, in building engineering, the force and moment corresponding to the vertical axis (lift and yawing moment) are of little significance. Therefore, aside from the effects of uplift forces on roof areas, the flow of wind is considered 2D, as shown in Figure 2.21, consisting of *along wind* and *transverse wind*. If uplift forces are considered, the wind flow would be 3D (Figure 2.22). Observe that the more streamlined the wind flow is, the less the buffeting force on the building (Figure 2.23).

The term along wind—or simply wind—is used to describe windward and leeward forces, while transverse wind is the term used to describe crosswind. Generally, in tall building design, the crosswind motion perpendicular to the direction of wind is often more critical than along-wind motion.

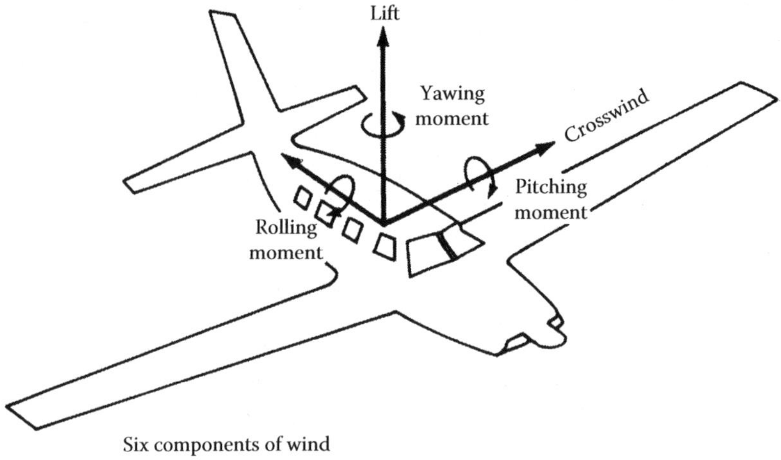

FIGURE 2.20 Critical components of wind in aeronautical engineering.

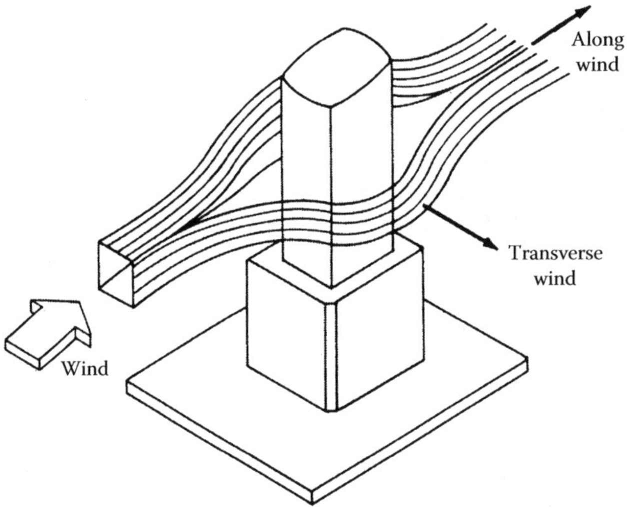

FIGURE 2.21 Simplified 2D wind flow consisting of along wind and across wind.

Consider a prismatic building subjected to a smooth wind flow. The originally parallel upwind streamlines are displaced on either side of the building, as illustrated in Figure 2.24. This results in spiral vortices being shed periodically from the sides into the downstream flow of wind. At relatively low wind speeds of, say, 50–60 mph (22.3–26.8 m/s), the vortices are shed symmetrically in pairs, one from each side. When the vortices are shed, that is, break away from the surface of the building, an impulse is applied in the transverse direction.

At low wind speeds, since the shedding occurs at the same instant on either side of the building, there is no tendency for the building to vibrate in the transverse direction. Therefore, the building experiences only along-wind oscillations parallel to wind direction. However, at higher speeds, vortices are shed alternately, first from one and then from the other side. When this occurs, there is an impulse in the along-wind direction as before, but in addition, there is an

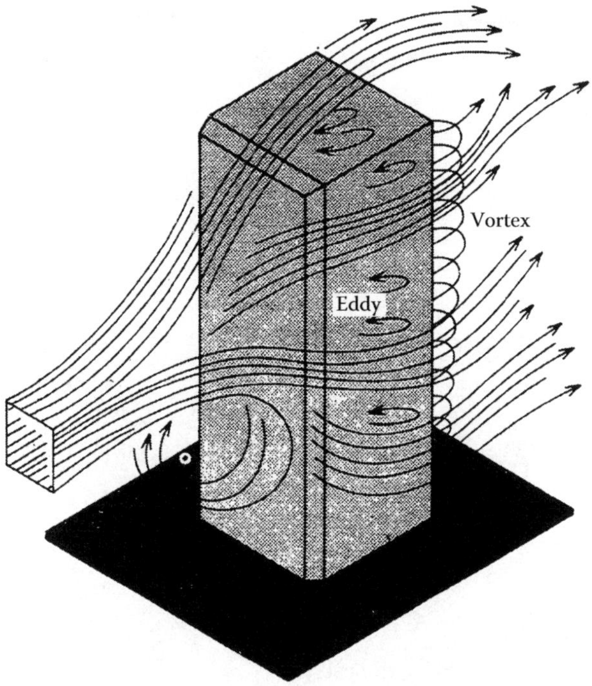

FIGURE 2.22 3D flow of wind.

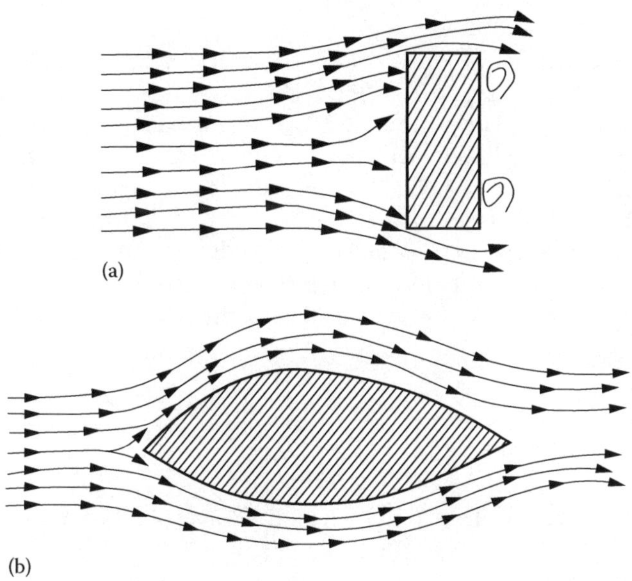

FIGURE 2.23 Wind flow around bluff and streamlined buildings: (a) bluff (rectangular) building (b) streamlined building.

impulse in the transverse direction. However, the transverse impulse occurs alternately on opposite sides of the building with a frequency that is precisely half that of the along-wind impulse. This impulse due to transverse shedding gives rise to vibrations in the transverse direction. The phenomenon is called *vortex shedding or Karman vortex street*, terms well known in the field of fluid mechanics.

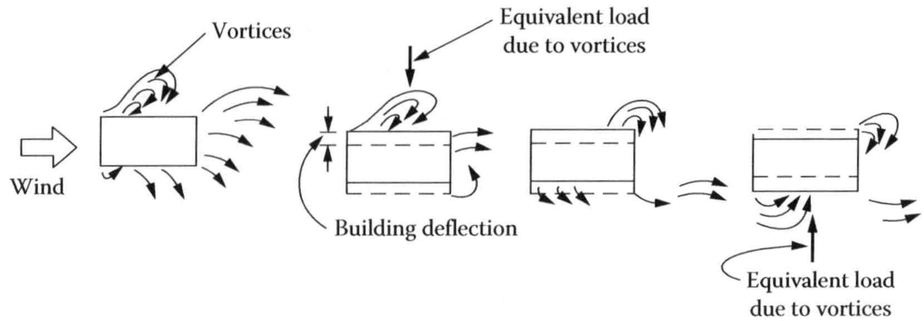

FIGURE 2.24 Vortex shedding: periodic shedding of vertices generates building vibrations transverse to wind direction.

There is a simple formula to calculate the frequency of the transverse pulsating forces caused by vortex shedding:

$$f = \frac{V \times S}{D} \tag{2.2}$$

where

f is the frequency of vortex shedding in hertz
V is the mean wind speed at the top of the building
S is a dimensionless parameter called the Strouhal number for the shape
D is the diameter of the building

In Equation 2.2, the parameters V and D are expressed in consistent units such as ft/s and ft, respectively.

The Strouhal number is not a constant but varies irregularly with wind velocity. At low air velocities, S is low and increases with the velocity up to a limit of 0.21 for a smooth cylinder. This limit is reached for a velocity of about 50 mph (22.4 m/s) and remains almost a constant at 0.20 for wind velocities between 50 and 115 mph (22.4 and 51 m/s).

Consider, for illustration purposes, a circular prismatic-shaped high-rise building having a diameter equal to 110 ft (33.5 m) and a height-to-width ratio of 6 with a natural frequency of vibration equal to 0.16 Hz. Assuming a wind velocity of 60 mph (27 m/s), the vortex-shedding frequency is given by

$$f = \frac{V \times 0.2}{110} = 0.16 \text{ Hz} \tag{2.3}$$

where V is in ft/s.

If the wind velocity increases from 0 to 60 mph (27.0 m/s), the frequency of vortex excitation will rise from 0 to a maximum of 0.16 Hz. Since this frequency happens to be very close to the natural frequency of the building, and assuming very little damping, the structure would vibrate as if its stiffness were zero at a wind speed somewhere around 60 mph (27 m/s). Note the similarity of this phenomenon to the ringing of church bells or the shaking of a tall lamppost whereby a small impulse added to the moving mass at each end of the cycle greatly increases the kinetic energy of the system. Similarly, during vortex shedding, an increase in deflection occurs at the end of each swing. If the damping characteristics are small, the vortex shedding can cause building displacements far beyond those predicted on the basis of static analysis.

When the wind speed is such that the shedding frequency becomes approximately the same as the natural frequency of the building, a resonance condition is created. After the structure has begun to resonate, further increases in wind speed by a few percent will not change the shedding frequency, because the shedding is now controlled by the natural frequency of the structure. The vortex-shedding frequency has, so to speak, locked in with the natural frequency. When the wind speed increases significantly above that causing the lock-in phenomenon, the frequency of shedding is again controlled by the speed of the wind. The structure vibrates with the resonant frequency only in the lock-in range. For wind speeds either below or above this range, the vortex shedding will not be critical.

Vortex shedding occurs for many building shapes. The value of S for different shapes is determined in wind tunnel tests by measuring the frequency of shedding for a range of wind velocities. However, one does not have to know the value of S very precisely because the lock-in phenomenon occurs within a range of about 10% of the exact frequency of the structure.

Unlike steady flow of wind, which for design purposes is considered static, turbulent wind loads associated with gustiness cannot be treated in the same manner. This is because gusty wind velocities change rapidly and even abruptly, creating effects much larger than if the same loads were static. Wind loads, therefore, need to be studied as if they were dynamic, somewhat similar to seismic loads. The intensity of dynamic component of wind load depends on how fast the velocity varies and also on the response of the structure itself. Therefore, whether pressures on a building due to a wind gust is dynamic or static entirely depends on the gustiness of wind and the dynamic properties of the building to which it is applied.

Buffeting is a comparatively well-recognized phenomenon that a bluff cylinder exposed to a uniform flow of air sheds eddies or vortices from either side of structure at a frequency n, dependent on the diameter of the structure the D and the velocity of the flow V. The frequency can be expressed in terms of the dimensionless Strouhal number S

where

$$S = \frac{n_0 D}{V} \tag{2.4}$$

For square cylinders $S = 0.14$.

Associated with this shedding of vortices are lateral-force fluctuations having a frequency n_e and longitudinal force fluctuations having a frequency $2n_e$. In uniform steady flow, these forces are powerful and well organized at the Strouhal frequency. As a consequence, in steady flow, large amplitude oscillations can be set up on lightly damped structures with the Strouhal frequency that is resonant with the natural frequency. At other frequencies, the body is generally quiescent owing to the absence of dynamic magnification.

Uniform steady flow, however, is not particularly representative of the natural wind impinging on a tall building. In this case, the flow, being usually in a city, is highly turbulent and the velocity varies with height. Both these factors affect the phenomenon considerably in the following manner:

1. The effective vortex-shedding frequency varies down the length of the structure owing to the mean velocity gradient.
2. The regularity of the shedding is disturbed by the turbulence in the flow.

As a consequence, the forces associated with the vortex shedding become random and less effective. The *end effect* of a tall building also tends to inhibit the vortex shedding compared to that on an infinite cylinder.

2.4.5 Aerodynamic Damping

A factor that can be a considerable influence in the dynamic response is the aerodynamic damping. A damping force is generally a force induced by the motion of a body; it is usually dissipative in character, but not under all circumstances.

The aerodynamic damping forces are proportional to both the ratio of air density to average building density and the frequency.

The aerodynamic damping associated with shapes typical of tall buildings shows several features of importance. The aerodynamic damping for a square tower of height-to-width ratio of $6^1/_2$:1 is positive. Certain exceptions occur in which the lateral damping is negative and comparatively large. In the case of positive aerodynamic damping, the forces generated by motion through the airstream tend to oppose the motion. Negative damping tends to reinforce the motion.

The nature of the damping forces shows that for a square shape subjected to winds normal to a face, the longitudinal oscillation induces forces opposing the motion, while transverse motion induces forces that reinforce the motion. The latter is due to the property of square shapes in which a small change in relative wind direction from the normal produces an into-wind component of transverse force.

The phenomenon of negative damping has for a long time been recognized as the source of instability of the so-called galloping variety that occurs with ice-covered transmission cables and suspension bridges. Instability occurs when the net damping available in the system, including both the mechanical and aerodynamic, approaches zero. Under these circumstances, the amplitude of oscillations grows progressively until usual event nonlinearities in the system restrain the motion. This restraint, however, is unlikely to occur until unacceptably large amplitudes have been reached.

It would appear that aerodynamic instability is not likely to arise with the stiffness mass and damping parameters of representative modern structures of conventional shapes and not exposed to wind speeds much in excess of 100 mph. There is, however, the likelihood that with structures of lighter weight, lower damping, lower flexibility, and unconventional shape, values of negative aerodynamic damping may be encountered, which reduce the overall damping available so significantly that sway amplitudes will result, which, if not large enough to cause structural failure, may cause disquieting sway amplitudes and serious secondary damage to walls and partitions. There are, at present, trends toward all of the aforementioned characteristics brought about by the increasing application of lightweight structural systems, higher strength materials, more monolithic forms of construction, and the wide range of architectural forms. Desirable at these trends may be from other viewpoints, and it is clear that the possibility of large negative aerodynamic damping should be kept under careful surveillance.

2.4.6 Design Criteria for Wind

Satisfactory performance in the wind can be related to the following major factors:

1. Damage to the fabric of the structure, such as cracking of the partitions, elevator shafts, and opening of joints in the exterior fabric, caused by excessive deflections of the main structure
2. Damage to glass and cladding due to high local loading
3. Discomfort of occupants due to relatively frequent occurrence of high sway accelerations
4. Partial or complete collapse of the structure due to dynamic and static buckling or plastic collapse

   In the context of present-day buildings, it would appear the first three of these factors are as significant as or of greater significance than the last. That is to say, if satisfactory deflection and acceleration characteristics are achieved, the chances of collapse are probably remote. The converse, however, is not necessarily true.

Tolerable amplitudes of acceleration are related to the number of times such amplitudes are experienced in a year. Moderate damage to walls and partitions may be accepted once or twice during the life time of a structure but not much more frequently. Complete collapse must, of course, be associated with an extremely small probability.

Research into human response to acceleration at building frequencies indicates that the threshold lies in the range 1–10 mg (0.032–0.32 ft/s^2). For a 40-story building with a 4 s period of vibration, the corresponding sway amplitude is of the order of 1.0–1.6 in. Cracking of the partitions may occur if story deflections greater than about 3/8 in. are exceeded.

The actual damping in structures at small amplitudes appears to be of the order of 1% in steel structures and 2% in concrete structures. At larger amplitudes, the damping undoubtedly increases. This fact is perhaps the salvation of many taller buildings.

The influence of wind loading on the economics of the design of tall buildings varies greatly with the slenderness and height of the building. It follows that the sophistication required in the design varies a great deal. A common practice, however, is to design buildings of unusual shape and height greater than say 400 ft using wind tunnel testing in realistic flow conditions and meteorological study of wind conditions pertaining to the site. Design criteria based on evaluation of human comfort and maximum deflections must be given particular attention.

2.4.7 Building Sway

The questions regarding building sway may be asked in many different ways but, in essence, boils down to this: "How much can we allow a building to sway in the wind?" It seems no one proposes to answer the question—for the most part because a reasonably simple answer is exceedingly difficult to formulate. We will, however, examine several of its facets.

Let us start by considering how the engineers start their wind design. First, they must establish the nature of the wind that is the environment of the particular building in question. For a very tall building, the engineer would like the characteristics of the wind to be described in relatively great detail. Second, the engineer must determine the response of the structure to the turbulent wind environment.

In spite of the advances in computer field, it is not yet possible to select a structure that accepts the wind environment and meets the acceptable deflection and acceleration criteria. We must still resort to a trial-and-error procedure in which the structure response is calculated, compared with the acceptance criteria and modified as required. The procedure is repeated over and over until the acceptance criteria are just met. At the moment, we lack sufficient background and knowledge to go about a direct design.

Great steps forward are being made both in describing the nature of the wind and the effects of that wind on a structural system. Much less is known as to what levels of structural response can be tolerated. Deflection, as such, and particularly total deflection or sway appears to have little value in an engineering sense. It does not really matter if a building sways 10 or 20 in. so long as its performance is satisfactory. The three things that do matter are the integrity of the structural system, the integrity of the architectural finishes, and the comfort of the building occupants. All are related to the dynamic and to the static response of the structure to its wind environment.

The disposition of the natural wind—particularly over large built-up areas—is exceedingly complex to quantify. Tall buildings in these terrains are buffeted by gusts and other aerodynamic forces and are set into oscillation. In addition, the building tends toward a steady deflection. The sway amplitudes can equal or exceed the steady deflection—this is almost always true for very tall buildings. It follows, then, that the occupants of upper levels of tall buildings are constantly subjected to motion. The design challenge therefore is to only limit motion to the extent that it should not adversely affect the integrity of the structure nor cause discomfort to occupants due to high sway accelerations.

In elevators, people are subject to speeds in excess of 20 mph and accelerations in excess of 150 milli-gs. Passengers in airplanes are occasionally subject to accelerations far greater than 1*g*; the speeds, of course, are very high. The human species apparently has no sensory apparatus that is known to detect absolute displacement or velocity and can detect only relative quantities or each; clearly, we can fly at jet speeds with no sense of displacement or velocity. The second derivative of displacement is *acceleration* and can be detected by all normal persons. The third derivative can be called *jerk* and is also detectable. In fact, it appears that we are more *sensitive* to the rate of change of acceleration than we are to acceleration itself. Even so, in examining the reaction of humans to building response, we can normally neglect the third derivative except as it is manifested in the relationship between the period of vibration and the level of acceleration.

To determine acceptable levels of acceleration, we must consider the extremes—panic levels, so to speak—as well as various working and perception levels. It would be convenient if we could establish a simple number as an acceptance level for acceleration and then limit building response to that level. At first glance, it would seem to be the simplest way, but in reality, it is not.

In attempting to evaluate this swaying motion of a structure, the engineer needs to know how much and how often the swaying motion occurs.

Knowing the periods of vibration of the structure, the engineer can readily translate sway amplitudes into accelerations. As with amplitude, knowledge of modes shapes will allow an evaluation of acceleration to be made throughout the height of the structure. An increase in structural damping will decrease structures response and decrease accelerations. It is interesting to recognize that a change in structure stiffness will affect response but will have little effect on acceleration levels. Now, should the engineer have knowledge of acceptable levels of acceleration, he or she can adjust the stiffness of his structure and the damping contained therein until the response of the structure is within the desired limits. At this stage in the technology, it remains to the individual engineer to set his own criteria for swaying motion. Guidance, however, is available in several published papers and standards.

2.5 SCOPE, EFFECTIVENESS, AND LIMITATIONS OF BUILDING CODES

2.5.1 Scope

With respect to wind performance, the scopes of the building codes have greatly expanded since the mid-1980s. Significant improvements include the following:

- *Recognition of increased uplift loads at the roof perimeter and corners*: Prior to 1982, wind provisions did not account for the increased uplift at the roof perimeter and corners. Therefore, buildings designed in accordance with earlier standards are likely to be susceptible to blowoff of the roof deck and/or roof covering.
- *Adoption of ASCE 7 for wind design loads*: Although the Uniform Building Code (UBC) permitted use of ASCE 7, the 2000 edition of the IBC was the first model code to require ASCE 7 for determining wind design loads. ASCE 7 has been more reflective of the current state of the knowledge than the model codes, and use of this procedure has typically resulted in higher design loads.
- *Roof coverings*: Several performance and prescriptive requirements pertaining to wind resistance of roof coverings have been incorporated. The majority of these additional provisions were added after Hurricanes Hugo (1989) and Andrew (1992). Poor performance of roof coverings was widespread in both of those storms. Prior to the 1991 edition of the UBC, wind provisions were essentially silent on roof covering wind loads and test methods for determining uplift resistance.
- *Glazing protection*: The 2000 edition of the IBC was the first model code to address wind-borne debris requirements for buildings located in the wind-borne debris regions of hurricane-prone regions (via reference to the 1998 edition of ASCE 7). (The 1995 edition of ASCE 7 was the first edition to address wind-borne debris requirements.)
- *Parapets and rooftop equipment*: The 2003 edition of the IBC was the first model code to address wind loads on parapets and rooftop equipment (via reference to the 2002 edition of ASCE 7, which was the first edition of ASCE 7 to address these elements).

2.5.2 Effectiveness

Except for hurricanes and tornadoes, the 2009 edition of the IBC is believed to be a relatively effective code, provided that it is properly followed and enforced. This code is also believed to be an effective code for hurricanes, except that it does not account for water infiltration due to puncture of the roof membrane by missiles nor does it adequately address the vulnerabilities of brittle roof coverings (such as tile) to missile-induced damage and subsequent progressive cascading failure.

The IBC relies on several referenced standards and test methods developed or updated in recent years. Most of these standards and test methods have not been validated by actual building performance during design level wind events. Therefore, the actual performance of buildings designed and constructed to the minimum provisions of the IBC remains to be determined. Future poststorm building performance evaluations may or may not show the need for further enhancements.

2.5.3 Limitations

- Adoption of the current model code does not always ensure good wind performance. Rather, the code is a minimum tool that should be used by knowledgeable design professionals in conjunction with their professional judgment. To achieve good wind performance, in addition to good design, the construction work must be effectively executed and the building must be adequately maintained and repaired.
- Specific limitations of the model codes include lack of provisions pertaining to blowoff of aggregate from built-up and sprayed polyurethane foam roofs and limitations of some of the test methods used to assess wind and wind-driven rain resistance of building envelope components. In addition, the codes do not address protection of occupants in buildings located in tornado-prone regions.
- The current wind-load provisions do not address the need for continuity, redundancy, or energy-dissipating capability (ductility) to limit the effects of local collapse and to prevent or minimize progressive collapse in the event of the loss of one or two primary structural members such as a column. However, ASCE 7-10, like its predecessors, does address general structural integrity and provides some guidance on this issue.

2.6 ASCE 7-10 WIND LOAD PROVISIONS, OVERVIEW

Wind engineering as a separate discipline can be traced to the United Kingdom in the 1960s, when informal meetings were held at the National Physical Laboratory, the Building Research Establishment, and elsewhere. Since then, we have come to understand that the magnitude of the wind-induced pressure or suction depends upon the complex interrelationship of wind velocity; air mass density; structural geometry including dimensions, stiffness, orientation, and location; and the surrounding ground surface conditions.

A conceptual understanding of the behavior of buildings buffeted by wind is not that complicated, but predicting their behavior in precise scientific terms is another matter. Therefore, building standards such as ASCE 7-10 have given us relatively simple procedures by introducing certain adjustment factors that take into account the complex 3D dynamic nature of wind flow. These are examined later in Chapter 10, but for now, we content ourselves with an overview.

2.6.1 Design Wind Loads for Main Wind-Force-Resisting Systems

The more robust design procedure for determining design wind loads for the MWFRS of buildings is the directional procedure. Also called analytical method, it applies to rigid building structures of any height and shape. It can even design *flexible* structures with some limitations. Most practicing engineers use this method.

A significant benefit of this method is its applicability for a large variety of structures. It provides pressure coefficients for myriad shapes and heights of any type of regular-shaped buildings. In this method, internal pressure coefficients related to enclosure classification are determined and then combined algebraically with external coefficients, both of which can be positive or negative relative to the exterior and interior building surfaces. However, in determining wind forces for the design of the main wind-force-resisting systems (MWFRS) of enclosed buildings, we may ignore the effects of internal pressures. This is because the internal pressure pushes or pulls equally on all surfaces and therefore cancels out. However, roofs are directly affected by internal pressure and it must be accounted for. And all partially enclosed buildings always need to account for internal pressures on all surfaces.

The design equation for determining the wind pressure, P_z, is based on the well-known equation

$$P_z = \frac{1}{2}\rho(V_{33})^2 \times K_z \tag{2.5}$$

where

P_z is the pressure induced on a particular surface at height z above mean ground level, psf
V_{33} is the basic 3 s gust speed in airport exposure (Exposure Category C) at reference height of 33 ft (10 m) above mean ground level, mph
ρ is the mass density of air = 0.07651 lb/ft^3
K_z is the exposure coefficient to account for variation in velocity with height z above mean ground level as influenced by terrain exposure

This equation that converts the chaotic nature of wind into a more manageable formulation is attributed to the scientist Daniel Bernoulli (1700–1792). Since then, our understanding of the effects of wind has improved most notably due to studies conducted on scaled building models in wind tunnels. The tendency has been to introduce refinements (and attendant complexities) with an increasing degree of sophistication considered proper for codifying wind loads. The resulting relationship as defined in ASCE 7-10 is summarized as follows:

$$\begin{bmatrix}\text{Velocity}\\\text{Pressure}\end{bmatrix} = \begin{bmatrix}\text{Reference}\\\text{Velocity}\\\text{Pressure}\end{bmatrix} \cdot \begin{bmatrix}\text{Exposure}\\\text{Height}\\\text{Factor}\end{bmatrix} \cdot \begin{bmatrix}\text{Topographic}\\\text{Factor}\end{bmatrix} \cdot \begin{bmatrix}\text{Directionality}\\\text{Factor}\end{bmatrix} \cdot \begin{bmatrix}\text{Importance}\\\text{Factor}\end{bmatrix}$$

$$q_z = \frac{1}{2}p\left(V_{33}\right)^2 \cdot K_z \cdot K_{zt} \cdot K_d$$

$$[\text{wind pressure}]_{\text{windward face}} = q_z \cdot G_h \cdot (C_p - C_{pi})$$

$$[\text{wind pressure}]_{\text{non-windward face}} = q_h \cdot G_h \cdot (C_p - C_{pi})$$

The internal pressure coefficient, C_{pi}, is of no consequence for enclosed buildings. Therefore,

$$[\text{wind pressure}]_{\text{windward face}} = q_z \cdot G_h \cdot C_p$$

$$[\text{wind pressure}]_{\text{non-windward face}} = q_h \cdot G_h \cdot C_p$$

In ASCE 7-10, q_z is referred to as velocity pressure and is given by

$$q_z = 0.00256 K_z K_{zt} K_d V^2 \quad (q_z \text{ in psf, } V \text{ in mph})$$

where
q_z is the velocity pressure also referred to as dynamic pressure, at height z aboveground level
K_z is the velocity exposure coefficient
K_{zt} is the topographic factor
K_d is the directionality factor
V is the wind speed at an elevation 33 ft (10 m) aboveground in flat open country (Exposure C)

The constant 0.00256 reflects the mass density of air for the standard atmosphere, that is, temperature of 59°F and sea level pressure of 29.92 in. of mercury and dimensions associated with wind speed in mph. The constant is obtained as follows:

$$\begin{aligned}\text{Constant} &= \frac{1}{2}\left[\frac{0.0765\ \text{lb/ft}^3}{32.2\ \text{ft/s}^2}\right] \\ &\times \Big[(\text{mi/h})(5280\ \text{ft/mi}) \\ &\times (1\text{h}/3600\ \text{s})\Big]^2 \\ &= 0.00256\end{aligned}$$

The velocity pressure exposure coefficient K_z adjusts the basic wind speed V to height z and terrain roughness (i.e., exposure category). Three exposure categories—B, C, and D—are defined. Exposure A, meant for heavily built-up city centers, was deleted in the 2002 edition of ASCE 7.

Exposure B corresponds to surface roughness typical of urban and suburban areas, while Exposure C represents surface roughness in flat open country. Exposure D is representative of flat unobstructed area and water surface outside hurricane-prone regions. Exposure C, the default category, applies to all cases where Exposures B and D do not. Interpolation between exposure categories permitted for the first time in the ASCE 7-02 is retained in the ASCE 7-10 with specific procedures added for interpolating the values of K_z in-between exposure categories. This, however, is rarely done in practice.

The basic wind speed for a given geographical area specified on the ASCE map is based on flat, open terrain. The same wind passing through a wooded area or city center neighborhood for your building will be effected by numerous obstructers of various size and shape. Wind speed also varies with height. The degree to which wind speed varies with height depends upon the terrain and obstructions that exist at ground level. In order to assign a value to K_z to our building, we need to ask ourselves questions such as what is the height of our building and what obstructions, if any, are surrounding it? The design provisions provide answers to these questions by classifying the surround setting and furnishing adjustment factors.

Then there are the special adjustment factors that need to be addressed. Is the building located on a hill? Is it a 2D hill like a ridge or is it a 3D hill like a knoll? The reason for asking these questions is to make sure we take into account the increase in wind speed uphills. The term K_z is used to capture this effect and is given by

$$K_{zt} = (1 - K_1 K_2 K_3)^2$$

Next, the wind directionality factor, K_d, with a value equal to 0.85, accounts for the directionality of wind. It refers to the fact that wind rarely, if ever, strikes along the most critical direction of a building. Additionally, wind direction changes from one instant to the next. Thus, wind can be only instantaneous along the most critical direction because at the very next instant, it will not be blowing from the same direction.

Next, we need to know what the building is used for. The basic wind speed V given in the ASCE 7-10 maps accounts for the occupancy category of the building. The higher occupancy category, the higher the wind speed. For example, the wind speed in Los Angeles, California, is 100 mph for Occupancy Category I buildings, 110 for Category IV, and 115 for Categories III and IV.

Examples of lowest Occupancy Category I are agricultural and minor storage facilities, and those at the high-end Category IV are hospitals and buildings critical to national defense functions.

Internal pressures and suctions do not affect wind-load calculations for the MWFRS of enclosed buildings. Therefore, external pressures and suctions that we use for calculating wind loads on the MWFRS of typical buildings may be calculated using the following simplified equations:

$P_z = q_z G_f C_p$ (positive pressures, varies up the height on the windward wall)

$P_z = q_h G_f C_p$ (negative pressures calculated at roof height, remains constant for the entire building height on all walls except the windward wall)

The overall wind load is the summation of positive and negative pressures on the windward and leeward surfaces, respectively. Using typical value of $G_f = 0.85$ permitted for rigid buildings, and $C_p = 0.8$ and 0.5 for the windward and leeward walls, the overall wind pressure at height z for a typical squarish building is given by

$$p_z = 0.85\ (0.8q_z + 0.5q_h)$$

2.6.2 Design Wind Pressures for Components and Cladding

It is worth reiterating that during exposure to wind, the surface areas of a building are subjected to variations in pressure that are constantly changing. At any given instant, the pressure on the surface of a building, if visible, would look like a mountain range with peaks and valleys. At a given location, the pressure can be extreme at one moment and seconds later be practically nothing. To address this, pressure coefficients that involve time and area averaging techniques are used to provide design pressures for varying area size on the building surface. It is also known that pressures are significantly greater where airflow is disrupted; such as corners and roof overhangs. Wind pressures at the corner of a building can be two times the pressures in the mid-areas of the wall. To account for this, the surface area is mapped into zones with different sets of coefficients derived for each zone.

Let us say we want the localized pressures and suctions for the design of a given window. You need to know the size of the window (wind area) and its location on the building (height and distance to corner or roof eave). An examination of the related pressure coefficients reveals that a larger window will experience a lower design pressure than a smaller window at the same location. This makes sense if we remember that the actual pressures and suctions are peaks and valleys throughout the surface area. The larger the area, the lower the average pressure on that area. Conversely, smaller windows are subjected to greater average loads.

Pressure coefficients as stated previously are derived for both positive and negative pressures. Positive pressures, as one would expect, act toward the surface. This is easy to comprehend—wind blows against the building. Negative pressure (acting away from the surface—suction) is a little more difficult for some to accept. When wind blows against a building, the windward side feels positive pressure. The wind flowing around the sides and collapsing around the back (leeward) side develops negative surface pressures. Because wind can come from any direction, we must consider both positive and negative wind pressures. The coefficients provided in the design provisions are developed from measurements of pressure from wind in all directions.

Openings in the building envelope create ports for airflow resulting in the development of internal pressures. The magnitude of internal pressure depends upon the ratio of open area on one side to the open area on the remainder of the building envelope. As with external pressure, internal pressure can be positive or negative depending on the location of the opening relative to the wind direction. The design provisions provide criteria to classify the potential for internal pressure based on openness and assigns pressure coefficients accordingly.

The design pressure for a given component (door, mullion, etc.) can now be determined. The basic velocity pressure for the building is established considering its geographical location, exposure (height and surrounding setting), and use (school, hospital, home, shed, etc.). The exterior pressure coefficients are established based on the component size and location on the building. The internal pressure coefficients are determined based on the potential for wind being blown into or sucked out of the building.

The product of all these factors gives the positive and negative design pressures for the component.

In the case of a large opening due to operable windows or breach of the building envelope, large internal pressures may develop. Typically, the external pressure at the opening will be transmitted into the building interior volume. Building envelope at other locations within the building volume will experience both the external pressures at those external locations as well as the large internal pressure transmitted from the opening.

The importance of determining internal pressures is clear, but it is not a quantity that can be determined exactly. In fact, internal pressures are influenced by many factors that are uncertain in themselves, such as the character of the leakage paths or whether windows or other exterior openings will be left open or will be broken during wind storms. The complex distribution of exterior pressures and the influence of the dynamic exterior pressures on the internal pressures must also be taken into account. In spite of these difficulties, reasonable estimates of the internal pressure can be made expressing the uncertainties in statistical terms.

Internal pressures induced by wind do not include stack effects nor any effects of interior partitions and other restriction of interior flows, which could lead to considerably higher load in special cases.

Although failure of exterior cladding resulting in broken glass may be of less consequence than collapse of a building, the expense of replacement and hazards posed to pedestrians is of major concern. Cladding breakage in a windstorm is an erratic occurrence, as witnessed in Hurricane Alicia, which hit Galveston and downtown Houston, Texas, on August 18, 1983, causing breakage of glass in several buildings. It is now known that glass breakage is also influenced by other factors, such as solar radiation, mullion and sealant details, tempering of the glass, double or single glazing, and fatigue of glass. It is also known with certainty that glass failure starts at nicks and scratches that may have occurred during manufacture and handling operations.

There appears to be no analytical approach available for rational design of curtain walls that come in all shapes and sizes. Although most codes identify regions of high wind loads such as building corners, the modern architectural trend with nonprismatic and curvilinear shapes combined with unique topography of each site has made wind tunnel determination of design loads a common and necessary practice.

In the past two decades, curtain wall has developed into an ornamental item and has emerged as a significant architectural statement. Sizes of window panes have increased considerably, requiring that glass panes be designed for various combinations of forces due to wind, shadow effects, and temperature movement. Glass in curtain walls must not only resist large wind forces, particularly in tall buildings, but must also be designed to accommodate various distortions of the total building structure.

2.6.2.1 Distribution of Pressures and Suctions

Shown in Figure 2.25 are schematics of relative roof uplift pressures and wall positive and negative pressures for a low-rise building of 30 ft roof height. It is assumed that the building is enclosed and is subjected to a basic wind speed of 90 mph, Exposure B. Observe that corner uplift pressures for hip or gable roof (Figure 2.25c) is 46 psf as compared to a relatively low positive pressure of 15 psf on the windward surface (Figure 2.25a).

Winds flowing around edges of a building result in higher pressures and suctions at corners than those at the center of elevation. This has been evidenced by damage caused to corner windows,

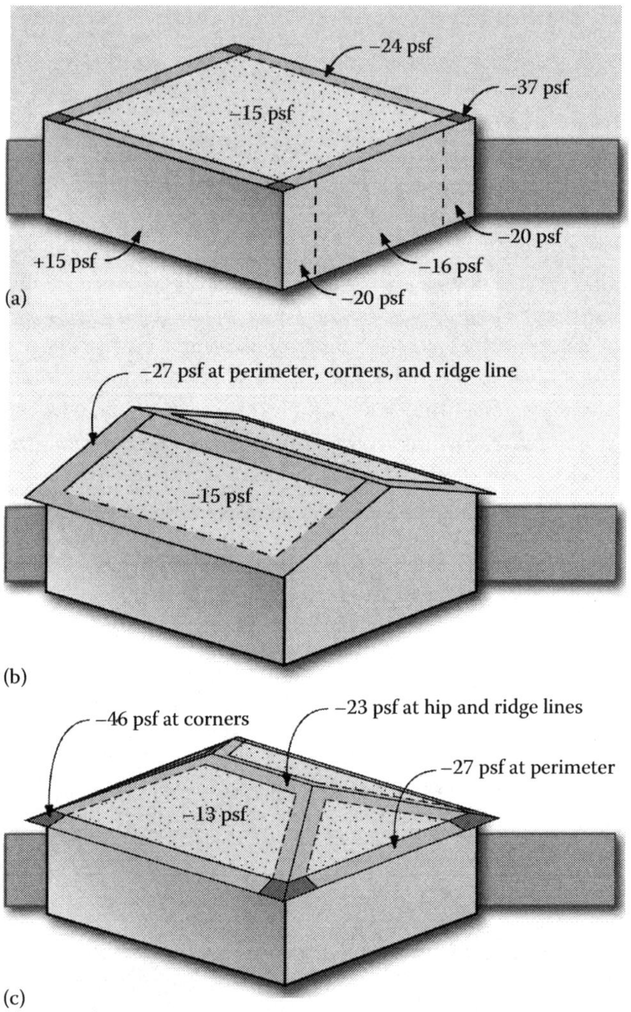

FIGURE 2.25 Relative roof uplift pressures as a function of roof geometry, roof slope, and location on roof, and relative positive and negative wall pressures as a function of location along the wall: (a) flat or gable, up to 7° roof slope, no overhang, and (b) gable, 27°–45° roof slope, overhang all sides. (c) Hip or gable roof, 7° to 7° roof slope, overhang all sides. *Note*: Design pressures all assume an enclosed building with the same basic wind speed of 90 mph, Exposure B, and 30 ft roof height. (From FEMA 424. *Design Guide for Improving School Safety in Earthquakes, Floods, and High Winds.* Washington, DC: Federal Emergency Management Association, January 2004.)

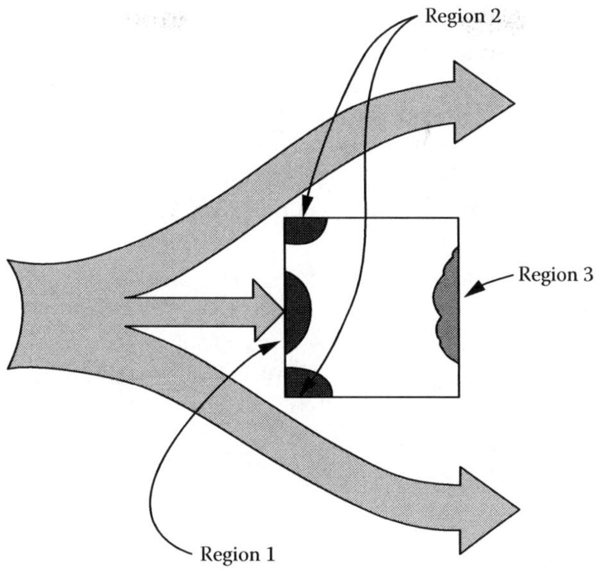

FIGURE 2.26 Wind pressures and suctions around building, plan.

eave and ridge tiles, etc., in windstorms. Wind tunnel studies on scale models of buildings have confirmed that three distinct high-pressure and suction areas develop around buildings, as shown schematically in Figure 2.26:

1. Positive-pressure zone on the upstream face (Region 1)
2. Negative-pressure zones at the upstream corners (Regions 2)
3. Negative-pressure zone on the downstream face (Region 3)

The highest negative pressures are generated in the upstream corners designated as Regions 2 in Figure 2.26. Wind pressures on a building's surface are not constant, but fluctuate continuously. The positive pressure on the upstream or the windward face fluctuates more than the negative pressure on the downstream or the leeward face. The negative-pressure region remains relatively steady as compared to the positive-pressure zone. The fluctuation of pressure is random and varies from point to point on the building surface. Therefore, the design of the cladding is strongly influenced by local pressures. As mentioned earlier, the design pressure can be thought of as a combination of the mean and the fluctuating velocity. As in the design of buildings, whether or not the pressure component arising from the fluctuating velocity of wind is treated as a dynamic or as a pseudostatic load is a function of the period of the cladding. The period of cladding on a building is usually on the order of 0.2–0.02 s, which is much shorter than the period it takes for wind to fluctuate from a gust velocity to a mean velocity. Therefore, it is sufficiently accurate to consider both the static and the gust components of winds as equivalent static loads in the design of cladding.

The strength of glass, and indeed of any other cladding material, is not known with the same certainty as the strength of other construction materials such as steel or concrete. For example, it is not possible to buy glass based on yield strength criteria as with steel. Therefore, the selection, testing, and acceptance criteria for glass are based on statistical probabilities rather than on absolute strength. The glass industry has addressed this problem and commonly uses 8 failures per 1000 lights (panes) of glass as an acceptable probability of failure.

2.6.2.2 Local Cladding Loads and Overall Design Loads

The design wind loads that we use in lateral analysis are an overall combination of positive and negative pressures occurring simultaneously around the building. The local wind loads that act on specific areas of the building are not required for overall building design but are vital for the design of exterior cladding elements and their connections to building. The methodology of determining these two types of loads (1) overall building design load and (2) local components and cladding design load, differ significantly. Important differences are the following:

1. Local winds are more influenced by the configuration of the building than the overall loading.
2. Local load is the maximum load that may occur at any location at any instant of time on any wall surface, whereas the overall load is the summation of positive and negative pressures occurring simultaneously over the entire building surface.
3. Intensity and character of local loading for any given wind direction and velocity differ substantially on various parts of the building surface, whereas the overall load is considered to have a specific intensity and direction.
4. Local loading is sensitive to the momentary nature of wind, but in determining critical overall loading, only gusts of about 2 s or more are significant.
5. Generally, maximum local negative pressures, also referred to as suctions, are of greater intensity than overall load.
6. Internal pressures caused by leakage of air through cladding systems have a significant effect on local cladding loads but are of no consequence in determining overall load on typical fully enclosed buildings.

The relative importance of designing for these two types of wind loading is quite obvious. Although proper assessment of overall wind load is important, very few, if any, buildings have been toppled by winds. There are no classic examples of building failures comparable to the Tacoma bridge disaster. On the other hand, local failures of roofs, windows, and wall cladding are not uncommon.

The analytical determination of wind pressure or suction at a specific surface of a building under varying wind direction and velocity is a complex problem. Contributing to the complexity are the vagaries of wind action as influenced both by adjacent surroundings and the configuration of the wall surface itself. Much research is needed on the micro effects of common architectural features such as projecting mullions, column covers, and deep window reveals. In the meantime, model testing of building in wind tunnels is perhaps the only recourse.

Probably the most important fact established by tests is that the negative or outward-acting wind loads on wall surfaces are greater and more critical than had formerly been assumed. They may be as much as twice the magnitude of positive loading. In most instances of local cladding failures, glass panels have blown off of the building, not into it, and the majority of such failures have occurred in areas near building corners. Therefore, it is important to give careful attention to the design of both anchorage and glazing details to resist outward-acting forces.

Another feature that has come to light for quite some time from model testing is that wind loads, both positive and negative, do not vary in proportion to height aboveground (see Figure 2.27). Typically, the positive-pressure contours follow a concentric pattern as illustrated in Figure 2.27b, with the highest pressure near the lower center of the facade, and pressures at the very top somewhat less than those a few stories below the roof. Figure 2.27e shows a pressure diagram for the design of cladding of a high-rise building measured in wind tunnel tests. The block pressure diagram shown in Figure 2.27f gives zones of design pressures based on the building grid system, to assist in the cladding design.

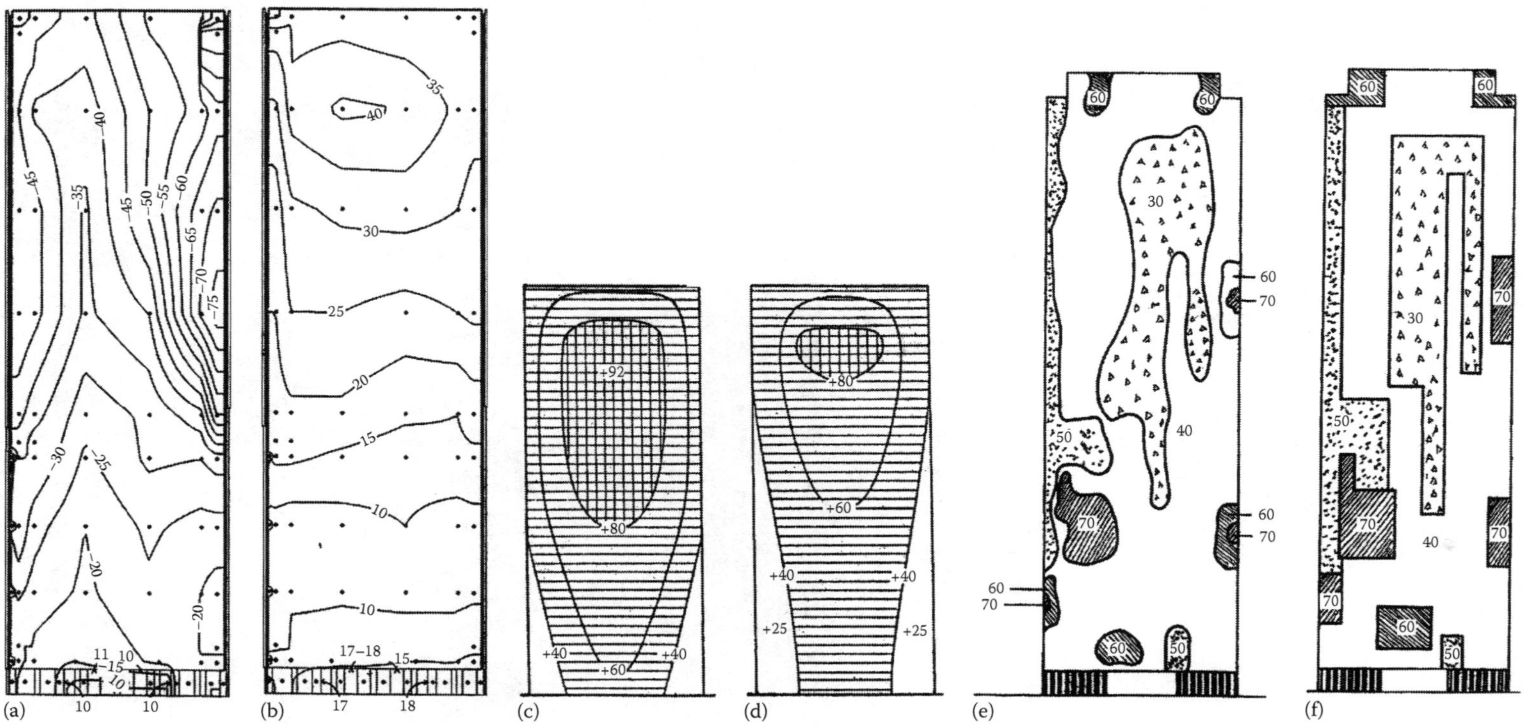

FIGURE 2.27 Pressure contours for cladding design: (a) peak negative pressures; (b) peak positive pressures; (c) peak positive pressures, building in open country; (d) peak positive pressures, building in city; (e) pressures measured in wind tunnel; and (f) simplified pressure diagram tied to building grid. *Note*: Pressures shown are in psf.

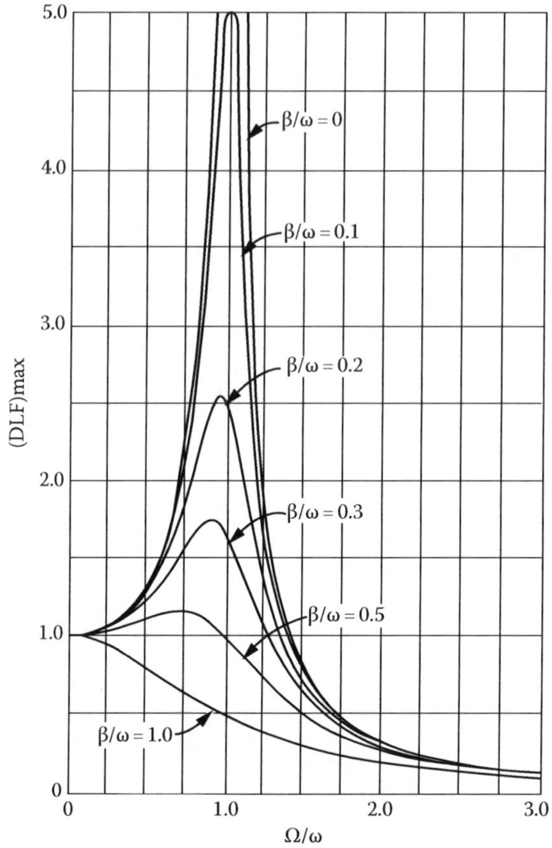

FIGURE 2.28 Maximum dynamic load factor for sinusoidal load, $F_1 \sin \Omega t$, damped systems.

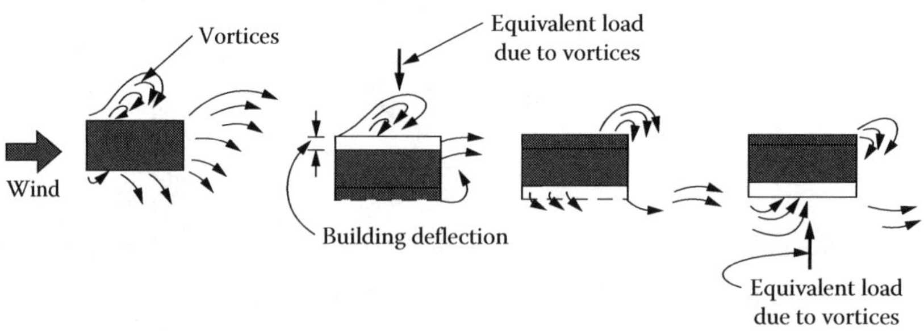

FIGURE 2.29 Vortex shedding: periodic shedding of vertices generates building vibrations in the transverse direction.

2.6.3 Comments on ASCE 7-10 Wind Provisions

It is perhaps evident from the brief discussion given earlier that the wind-load provisions of the ASCE 7-10 are quite tedious if not migranious. Could the provisions be simplified? Yes, certainly they can be but at a cost—the cost resulting from adding a level of conservatism that would cover all special conditions. However, the current trend, particularly with the availability of computers, is to perform rigorous calculations that result in the most economical and efficient use of our resources, so don't expect the ASCE wind provisions to get easier any time soon (Figures 2.28 and 2.29).

3 Earthquake Effects on Buildings

PREVIEW

Earthquakes have wreaked destruction since oldest antiquity, and it is only in the last 50 years that our knowledge of earthquakes and of their impact on buildings has resulted in the design of earthquake-resistant structures. These are built with particularly strong lateral bracing systems capable of resisting the jerking forces of an earthquake. Even so, the number of quake victims is still high all over the world. When 27,000 people died in the Guatemala earthquake of 1967, we thought we had seen the worst, but when 242,000 people died in an earthquake later in the same year in the region north of Peking. The Earth's crust floats over a core of molten rock and some of its parts have a tendency to move with respect to one another. This movement creates stresses in the crust, which may break out along fractures called *faults*. The break occurs through a sudden sliding motion in the direction of the fault and jerks the buildings in the area. Since the dynamic impact forces due to this jerky motion are mostly horizontal, they can be resisted by the same kind of bracing used against wind.

Earthquake strengths are evaluated on scales like the Richter scale, which measures the magnitude of the energy in the earthquake. For example, an earthquake measuring 4 or 5 on the Richter scale does little damage to well-built buildings, while one measuring 8 or above collapses buildings and may cause many deaths. Not all parts of the earth are subjected to earthquakes, but there are two wide zones on the Earth's surface where the worst earthquakes take place. One follows a line through the Mediterranean, Asia Minor, the Himalayas, and the East Indies, and the other the western, northern, and eastern shores of the Pacific.

Earthquakes are catastrophic events that occur mostly at the boundaries of portions of the Earth's crust called tectonic plates. When movement occurs in these regions, along faults, waves are generated at the Earth's surface that can produce very destructive effects.

Aftershocks are smaller quakes that occur after all large earthquakes. They are usually most intense in size and number within the first week of the original quake. They can cause very significant reshaking of damaged structures, which makes earthquake-induced disasters more hazardous. A number of moderate quakes (6+ magnitude) have had aftershocks that were very similar in size to the original quake. Aftershocks diminish in intensity and number with time. They generally follow a pattern of there being at least 1 large (within 1 Richter magnitude) aftershock, at least 10 lesser (within 2 Richter magnitude) aftershocks, 100 within 3, and so on. The Loma Prieta earthquake had many aftershocks, but the largest was only magnitude 5.0 with the original quake being magnitude 7.1.

Some of the most destructive effects caused by earthquake shaking are those that produce lateral loads in a structure. The input shaking causes the foundation of a building to oscillate back and forth in a more or less horizontal plane. The building mass has inertia and wants to remain where it is, and therefore, lateral forces are exerted on the mass in order to bring it along with the foundation. For analysis purposes, this dynamic action is simplified as a group of horizontal forces that are applied to the structure in proportion to its mass and to the height of the mass above the ground. In multistory buildings with floors of equal weight, the loading is further simplified as a group of loads, each being applied at a floor line and each being greater than the one below in a triangular distribution (see Figure 3.1). Seismically resistant structures are designed to resist these lateral forces through

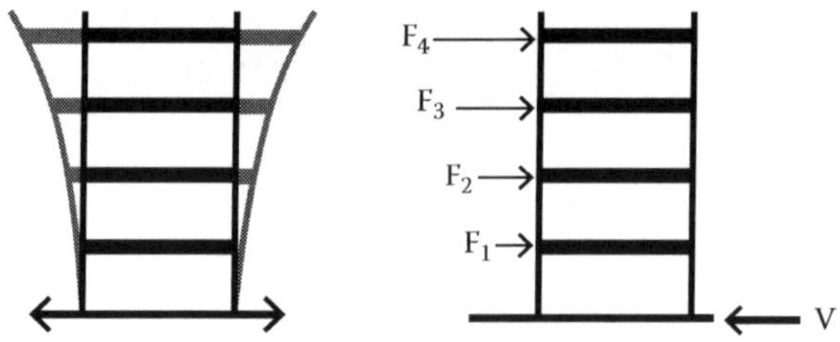

FIGURE 3.1 Lateral loads due to earthquakes.

inelastic and primarily for economic reason and must, therefore, be detailed accordingly. These loads are often expressed in terms of a percent of gravity weight of the building and can vary from a few percent to near 50% of gravity weight. There are also vertical loads generated in a structure by earthquake shaking, but these forces rarely overload the vertical-load-resisting system. However, earthquake-induced vertical forces have caused damage to structures with high dead load compared to design live load. These vertical forces also increase the chance of collapse due to either increased or decreased compression forces in the columns. Increased compression overloads columns, while decreased compression reduces column bending strength.

In earthquake engineering, we deal with random variables, and therefore, the design must be treated differently from the orthodox design. The orthodox viewpoint maintains that the objective of design is to prevent failure; it idealizes variables as deterministic. This simple approach is still valid, applied to design under only mild uncertainty. But when confronted with the effects of earthquakes, this orthodox viewpoint seems so over trustful as to be worthless. In dealing with earthquakes, we must contend with appreciable probabilities that failure will occur in the near future. Otherwise, all the wealth of this world would prove insufficient to fill our needs: the most modest structures would be fortresses. We must also face uncertainty on a large scale while designing engineering systems—whose pertinent properties are still debated to resist future earthquakes—about whose characteristics we know even less.

Although over the years, experience and research have diminished our uncertainties and concerns regarding the characteristics of earthquake motions and manifestations, it is unlikely, though, that there will be such a change in the nature of knowledge to relieve us of the necessity of dealing openly with random variables. In a way, earthquake engineering is a parody of other branches of engineering. Earthquake effects on structures systematically bring out the mistakes made in design and construction, even the minutest mistakes. Add to this the undeniable dynamic nature of disturbances, the importance of soil–structure interaction, and the extremely random nature of it all; in a manner of speaking, earthquake engineering is to the rest of the engineering disciplines what psychiatry is to other branches of medicine. This aspect of earthquake engineering makes it challenging and fascinating and gives it an educational value beyond its immediate objectives. If structural engineers are to acquire fruitful experience in a brief span of time, expose them to the concepts of earthquake engineering, even if their interest in earthquake-resistant design is indirect. Sooner or later, they will learn that the difficulties encountered in seismic design are technically intriguing and begin to exercise that nebulous trait called engineering judgment to make allowance for these unknown factors.

To understand the seismic behavior of buildings, it is helpful to study strong-motion seismograms (also called *time histories*). The familiar *wiggly line* graphic records shown in Figure 3.2 are not the actual motion of the ground but have been filtered in some way by both the recording instrument and by the agency providing the data. In most cases, however, for practical applications, the engineer need not be concerned about the difference.

Modern instruments capable of recording large motions strategically placed in structures provide information on the structural response. In this case, it is evident that there is amplification of both short-period and long-period motions in the upper floors. This effect is reflected in seismic design by applying larger loads up the building height.

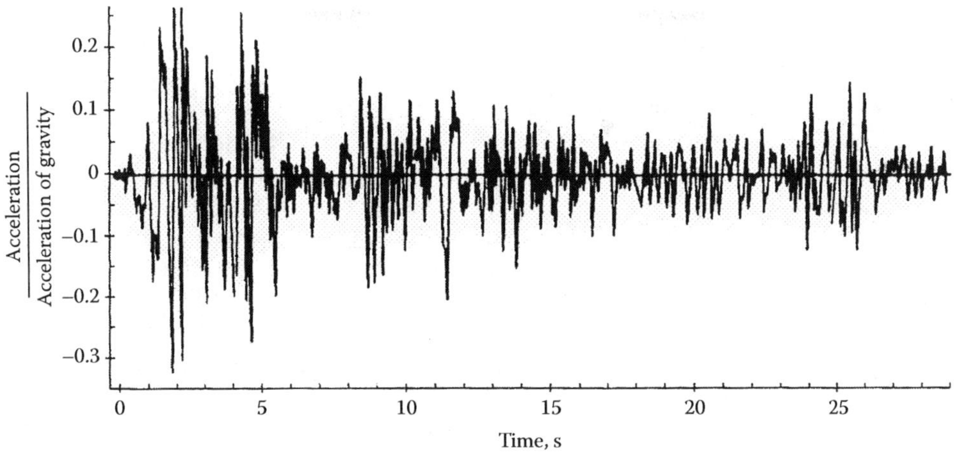

FIGURE 3.2 Strong-motion seismogram accelegram from El Centro earthquake, May 18, 1940 (NS Component).

Until the 1990s, seismic building codes used a single map of the United States that divided the country into numbered seismic zones (0, 1, 2, 3, 4) in which each zone was assigned a single acceleration value in %*g*, which was used to determine seismic loads on the structure.

Starting in the 1970s, new hazard maps began to be developed on a probabilistic basis. In the 1994 National Earthquake Hazards Reduction Program (NEHRP) Recommended Provisions, two maps of the United States were provided, showing effective peak acceleration coefficients and effective peak velocity–related coefficients by use of contour lines that designate regions of equal value. The ground motions were based on estimated probabilities of 10% of exceedance in various exposure times (50, 100, and 250 years).

The probabilistic analysis is typically represented in maps in the form of a percentage probability of exceedance in a specified number of years. For example, commonly used probabilities are a 10% probability of exceedance in 50 years (a return period of about 475 years) and 2% probability of exceedance in 50 years (a return period of about 2,500 years). These maps show ground motions that may be equaled but are not expected to be exceeded in the next 50 years: the odds that they will not be exceeded are 90% and 98%, respectively.

Seismic hazard probability maps are produced by the United States Geological Survey (USGS). The latest sets of USGS of maps provide a variety of maps for peak ground acceleration and spectral acceleration, with explanatory material, and are available on the USGS website.

The USGS map is a probabilistic representation of hazard for the contiguous United States. This shows the spectral acceleration in %*g* with a 2% probability of exceedance in 50 years: this degree of probability is the basis of the maps used in the building codes.

The return period of 1 in 2,500 years may seem very infrequent, but this is a statistical value, not a prediction, so some earthquakes will occur much sooner and some much later. The design dilemma is that if a more frequent earthquake—for example, the return period of 475 years—was used in the lower seismic regions, the difference between the high- and low-probability earthquakes is a ratio of between 2 and 5. Design for the high-probability earthquake would be largely ineffective when the low-probability event occurred.

In practical terms, the building designer must assume that the large earthquake may occur at any time. Thus, use of the 2,500 return period earthquakes in the lower seismic regions ensures protection against rare earthquakes, such as the recurrence of the 1811–1812 earthquake sequence in New Madrid, Missouri, or the 1898 Charleston, South Carolina, earthquake. The selection of 2% in 50-year likelihood as the maximum considered earthquake (MCE) ground motion is believed to result in acceptable levels of seismic safety for the nation.

The acceleration experienced by a building will vary depending on the period of the building, and in general, short-period buildings will experience more accelerations than long-period buildings. The USGS maps recognize this phenomenon by providing acceleration values for periods of 0.2 s (short) and 1.0 s (long). These are referred to as spectral acceleration, and the values are approximately what are experienced by a building (as distinct from the peak acceleration that is experienced at the ground). The spectral acceleration is usually considerably more than the peak ground accelerations.

The USGS maps are based on MCE ground motion—the most severe earthquake considered in the US seismic standards. They are based on a 2% probability of occurrence in 50 years.

These USGS probability maps provide the basis for the maps used in building codes that provide design values for spectral acceleration used by structural engineers to calculate the seismic forces on a structure. These design value maps differ by use of an MCE for the regions. For most regions of the country, the MCE is defined as ground motion with a uniform likelihood of exceedance of 2% in 50 years (a return period of about 2,500 years) and is identical to the USGS probability maps. However, in regions of high seismicity, such as coastal California, the seismic hazard is typically controlled by large-magnitude events occurring on a limited number of well-defined fault systems. For these regions, rather than using the 2% in 50-year likelihood, it is considered more appropriate to directly determine the MCE ground motions based on the characteristic earthquakes of those defined faults.

It is to be noted that the acceleration values shown on the maps are not used directly for design. Instead, they are reduced by two-thirds of this value to determine the design earthquake (DE) and are the values used by engineers for design. The reason for this is that it is believed by engineers that the design provisions contain at least a margin of 1.5 against structural failure. MCE is inferred to provide collapse prevention level, while the actual design is done using the DE, which is 2/3 MCE for code-level, life-safety protection level. This belief is the result of the study of the performance of many types of buildings in earthquakes, mostly in California.

The building response to earthquake shaking occurs over the time of a few seconds. During this time, several types of seismic waves are combining to shake the building in ways that are different in detail for each earthquake. In addition, as the result of variations in fault slippage, differing rock through which the waves pass, and the different geological nature of each site, the resultant shaking at each site is different. The characteristics of each building are different, whether in size, configuration, material, structural system, age, or quality of construction: each of these characteristics affects the building response.

In spite of the complexity of the interactions between the building and the ground during the few seconds of shaking, there is broad understanding of how different building types will perform under different shaking conditions. This understanding comes mainly from extensive observation of buildings in earthquakes all over the world and to a lesser extent from analytical and experimental research.

Understanding the ground and building characteristics discussed in this chapter is essential to give designers a *feel* for how their building will react to shaking, which is necessary to guide the conceptual design of their building.

In this chapter, we

- Provide an introduction to some of the key issues involved in seismic design, including a summary of the effect of earthquakes on building structures
- Outline the characteristics of earthquake that are important for building design
- Explain the basic ways in which earthquake-induced ground motion affects buildings
- Discuss how the building becomes more prone to failure and less predictable as the building becomes more complex in its configuration

3.1 INERTIAL FORCES AND ACCELERATION

The seismic waves create internal forces within the building. Inertial forces are generated when an outside force tries to make the building move if it is at rest or change its rate or direction of motion if it is moving.

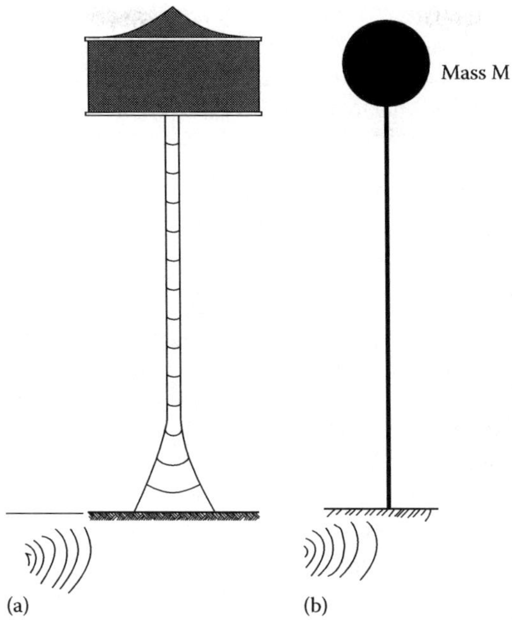

FIGURE 3.3 Analytical model for a freestanding water tower. (a) Water tower and (b) equivalent SDOF cantilever.

Consider a freestanding water tower shown in Figure 3.3a subjected to earthquake ground motions. A simplified analytical model for the tower may be represented by a cantilever column with a concentrated mass M at top, as shown in Figure 3.3b. When the base of the cantilever is subjected to sudden ground motion, the initial tendency for the water tower, that is, the mass *M*, is to stay put. The shifting of the ground is too rapid for the tower to keep up.

After a moment, the tower accelerates laterally to catch up with the movement of the ground. From Newton's second law of motion, we can surmise that the equivalent lateral force (ELF) *F* at the top is equal to the mass *M* of the tower and the acceleration at the base. Thus,

$$F = Ma \tag{3.1}$$

where

F is an inertial force

M is the mass (equal to building weight divided by acceleration due to gravity, g)

a is the acceleration

This part of Newton's law explains why light buildings, such as wood-frame houses, tend to perform better in earthquakes than large heavy ones.

The acceleration, or the rate of change of the velocity of the seismic waves setting the building in motion, determines the percentage of the building mass or weight that must be dealt with as an equivalent horizontal force.

Acceleration is measured in terms of the acceleration due to gravity or *g*. One *g* is the rate of change of velocity of a free-falling body in space. This is an additive velocity of 32 ft/s. Thus, at the end of the first second, the velocity is 32 ft/s; a second later, it is 64 ft/s, and so on. When parachutists or bungee jumpers are in the free fall, they are experiencing an acceleration of 1*g*. While the roller-coaster riders reach as much as 4*g*. The aerobatic pilots are undergoing about 9*g*. The human body is very sensitive and can feel accelerations as small as 0.001*g*, such as when you shake hands with another person. A building in an earthquake experiences for a fraction of a second very high forces in one direction before they abruptly change direction. Poorly constructed buildings begin to suffer damage at about 10% *g* (or 0.1*g*). In a moderate earthquake, vibration may last for a few seconds, and accelerations may be approximately 0.2*g*. Short accelerations may, for a fraction of a second, exceed 1.0*g*. In the Northridge earthquake in 1994, a recording station in Tarzana, 5 miles from the epicenter, recorded 1.92*g*.

3.2 DURATION, VELOCITY, AND DISPLACEMENT

Acceleration is a key factor in determining the forces on a building, but a more significant measure is that of acceleration combined with duration, which takes into account the impact of earthquake forces over time. In general, a number of cycles of moderate acceleration, sustained over time, can be much more difficult for a building to withstand than a single much larger peak. Continued shaking weakens a building structure and reduces its resistance to earthquake damage.

The duration of strong motion, termed the *bracketed duration*, is measured above a certain threshold acceleration value, commonly taken as 0.05*g*, and is defined as the time between the first and last peaks of motion that exceeds this threshold value. In the San Fernando earthquake of 1971, the bracketed duration was only about 6 s. In both the Loma Prieta and the Northridge earthquakes, the strong motion lasted a little over 10 s yet caused much destruction. In the 1906 San Francisco earthquake, the severe shaking lasted 45 s, while in Alaska, in 1964, the severe motion lasted for over 3 min.

Two other measures of wave motion are directly related to acceleration and can be mathematically derived from it. Velocity, which is measured in inches or centimeters per second, refers to the rate of motion of the seismic waves as they travel through the earth. This is very fast. Typically the *P wave* travels at between 3 and 8 km/s or 7,000–18,000 mph. The *S wave* is slower, traveling at between 2 and 5 km/s or 4,000–11,000 mph.

Displacement refers to the distance that points on the ground are moved from their initial locations by the seismic waves. These distances, except immediately adjacent to or over the fault rupture, are quite small and are measured in inches or centimeters. For example, in the Northridge earthquake, parking structures at Burbank, about 18 miles (29 km) from the epicenter, recorded displacements at the roof of 1.6 in. (4.0 cm) at an acceleration of 0.47*g*. In the same earthquake, the Olive View hospital in Sylmar, about 7.5 miles (12 km) from the epicenter, recorded a roof displacement of 13.5 in. (34 cm) at an acceleration of 1.5*g*.

The velocity of motion on the ground caused by seismic waves is quite slow—huge quantities of earth and rock are being moved. The velocity varies from about 2 cm/s in a small earthquake to about 60 cm/s in a major shake. Thus, typical building motion is slow and the displacements are small, but because thousands of tons of steel and concrete are wrenched in all directions several times a second, building failure or severe damage is likely to occur.

In earthquakes, the values of ground displacement, velocity, and acceleration (DVA) vary a great deal in relation to the frequency of the wave motion. High-frequency waves (higher than 10 Hz) tend to have high amplitudes of acceleration but small amplitudes of displacement, compared to low-frequency waves, which have small accelerations and relatively large velocities and displacements.

3.3 ACCELERATION AMPLIFICATION DUE TO SOFT SOIL

Earthquake shaking is initiated by a fault slippage in the underlying rock. As the shaking propagates to the surface, it may be amplified, depending on the intensity of shaking, the nature of the rock, and the surface soil type and depth.

A layer of soft soil, measuring from a few feet to a hundred feet or so, may result in an amplification factor ranging from 1.5 to 6 over the rock shaking. This amplification is most pronounced at longer periods and may not be so significant at short periods. The amplification also tends to decrease as the level of shaking increases.

As a result, earthquake damage tends to be more severe in areas of soft ground. This characteristic became very clear when the effects of 1906 San Francisco earthquake were studied. Also, inspection of records from soft clay sites during the 1989 Loma Prieta earthquake indicated a maximum amplification of long-period shaking of three to six times. In this earthquake, extensive damage was caused to buildings in San Francisco's Marina District, which was largely built on filled ground, some of it rubble deposited after the 1906 earthquake.

Because of the possibility of considerable shaking amplification related to the nature of the ground, seismic codes have some very specific requirements that relate to the characteristics of the site. These require the structure to be designed for higher force levels if it is located on poor soil. Specially designed foundations may also be necessary.

3.4 NATURAL PERIODS

All buildings have a natural or fundamental period; this is the rate at which they will move back and forth if they are given a horizontal push (Figure 3.4). In fact, without pulling and pushing it back and forth, it is not possible to make an object vibrate at anything other than its natural period.

Another characteristic of importance in seismic design is the period of frequency of earthquake waves. Whether the waves are quick and abrupt or slow and rolling is particularly important for determining building seismic forces.

When a building's motion is started with a seismic push, to be effective, this shove must be as close as possible to the natural period of the building. When earthquake motion starts a building vibrating, it will tend to sway back and forth at its natural period.

Periods are the time in seconds (or fractions of a second) that is needed to complete one cycle of a seismic wave. Frequency is the inverse of this—the number of cycles that will occur in a second—and is measured in *hertz*. One hertz is one cycle per second.

Natural periods vary from about 0.05 s for a piece of equipment, such as a filing cabinet, to about 0.1 s for a one-story building. Period is the inverse of frequency, so the cabinet will vibrate at 1 divided by 0.05 = 20 cycles a second or 20 Hz.

A four-story building will sway at about 0.5 s period, and taller buildings between about 10 and 20 stories will swing at periods of about 1–2 s.

A rule of thumb for preliminary design is that the building period equals the number of stories divided by 10; therefore, period is primarily a function of building height. The 60-story Citicorp office building in New York has a measured period of 7 s; give it a push, and it wall sway slowly

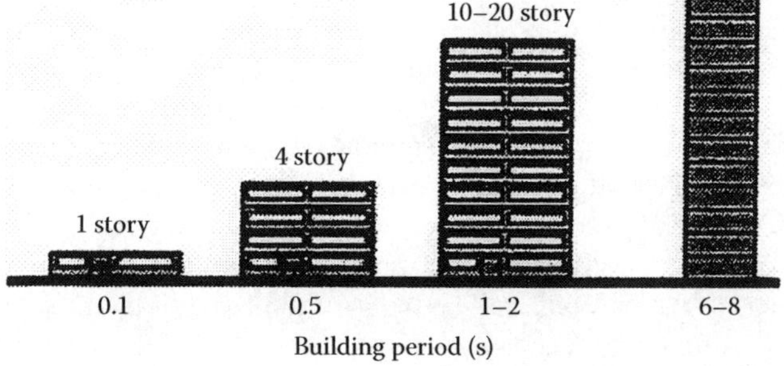

FIGURE 3.4 Effect of building height on period.

back and forth completing a cycle every 7 s. Other factors, such as the building's structural system, its construction materials, its content, and its geometric proportions, also affect the period, but height is the most important consideration (Figure 3.4).

The building's period may also be changed by earthquake damage, which has the effect of increasing the structure's period of vibration: the structure is *softening*. This may result in the structure's period approaching that of the ground and experiencing resonance, which may prove fatal to an already weakened structure.

3.5 BUILDING RESONANCE

In structural dynamics, resonance is defined as the tendency of the system to oscillate with larger amplitude at some frequencies than at others. It should be noted that frequency, typically denoted as ω, is the reciprocal of the period T. Thus,

$$\omega = \frac{1.0}{T} \tag{3.2}$$

At resonant frequencies, even small periodic driving forces can produce large amplitude oscillations.

A familiar example of resonance is a playground swing, which acts as a pendulum. Pushing a person in a swing in time with the natural interval of the swing (its resonant frequency) will make the swing go higher and higher (maximum amplitude), while attempts to push the swing at a faster or slower tempo will result in smaller arcs. This is because the energy the swings absorb is maximized when the pushes are in phase with the swing oscillations, while some of the swing's energy is actually extracted by the opposing force of the pushes when they are not.

Avoiding resonance disasters is a major concern in every building, tower, and bridge construction project. As a countermeasure, shock mounts can be installed to absorb resonant frequencies and thus dissipate the absorbed energy. The Taipei 101 building relies on a 730-ton pendulum—a tuned mass damper—to cancel resonance. Buildings in seismic zones are often constructed to take into account the oscillating frequencies of expected ground motion.

It is of interest to note that resonance was first recognized by Galileo Galilei with his investigations of pendulums and musical strings beginning in 1602.

It is reassuring to know that in practice, exact resonance does not really occur, because buildings never completely respond as linear elastic systems. As distortions become large, the characteristics of the building change because of nonlinear response due to plastic deformation. Therefore, the concern as to whether building displacements become infinite is of course of academic interest only. Furthermore, the maximum amplitude is attained not after just a few cycles but many cycles of vibration. The important engineering conclusion is that, at or near resonance, the deflection of the building is likely to become very large and hence problematic.

When a vibrating building is given further pushes that are also at its natural period, its vibrations increase dramatically in response to even rather small pushes, and in fact, its accelerations may increase as much as four to five times.

Perhaps it is hard to imagine, but the ground obeys the same physical law and also vibrates at its natural period, if set in motion by an earthquake. The natural period of ground varies from about 0.4 to 2 s, depending on the nature of the ground. Hard ground or rock will experience short-period vibration. Very soft ground may have a period of up to 2 s, but unlike a structure, it cannot sustain longer-period motions except under certain unusual conditions. Since this range is well within the range of common building periods, it is quite possible that the pushes that earthquake ground motion impacts to the building will be the natural period of the building. This may create resonance, causing the structure to encounter accelerations of perhaps 1g when the ground is only vibrating with accelerations of 0.2g. Because of this, buildings suffer the greatest damage from ground motion at a frequency close or equal to their own natural frequency.

FIGURE 3.5 Vibration modes.

The terrible destruction in Mexico City in the earthquake of 1985 was primarily the result of response amplification caused by coincidence of building and ground motion periods. Mexico City was some 250 miles from the earthquake focus, and the earthquake caused the soft ground in margins of the old lake bed under the downtown buildings to vibrate for over 90 s at its long natural period of around 2 s. This caused buildings that were between about 6 and 20 stories in height to resonate at a similar period, greatly increasing the accelerations within them. Taller buildings suffered little damage. This amplification in building vibration is very undesirable. The possibility of it happening can be reduced although not always, by trying to ensure that the building period will not coincide with that of the ground. Thus, on soft (long-period) ground, it would be best to design a short, stiff (short-period) building.

Taller buildings also will undergo several modes of vibration so that the building will wiggle back and forth like a snake (Figure 3.5).

However, higher modes of vibration are generally less critical than the natural period, although they may be significant in a high-rise building. For low-rise buildings, the natural period (which, for common structures, will always be relatively short) is the most significant. Note, however, that the low-period, low-to-midrise building is more likely to experience resonance from the more common short-period ground motion.

3.6 SITE RESPONSE SPECTRUM

From the preceding discussion, it is evident that buildings with different periods (or frequency responses) will respond in widely differing ways to the same earthquake ground motion. Conversely, any given building will act differently during different earthquakes, so for design purposes, it is necessary to represent the building's range of response to ground motion of different frequency content. Such a representation is termed a site response spectrum. A site response spectrum is a graph that plots the maximum response values of DVA against period (and frequency).

Figure 3.6 shows a simplified version of a response spectrum. These spectra show, on the vertical ordinate, the DVA that may be expected at varying periods (the horizontal ordinate). Thus, the response spectrum illustrated shows a maximum acceleration response at a period of about 0.3 s—the fundamental period of a midrise building. This shows how building response varies with building period: as the periods lengthen, accelerations decrease and displacement increases. On the other hand, one- or two-story buildings with short periods undergo higher accelerations but smaller displacements.

In general, a taller, more flexible building may be expected to experience proportionately lesser accelerations than a stiffer low-rise building. A glance at a response spectrum will show why this is so: as the period of the building lengthens (moving toward the right of the horizontal axis of the spectrum), the accelerations reduce. However, there's a tradeoff, in that the lower accelerations in taller buildings come at the expense of more displacement. This increased displacement may be such that the building may suffer considerable damage to its nonstructural components, such as ceilings and partitions, in even a modest earthquake.

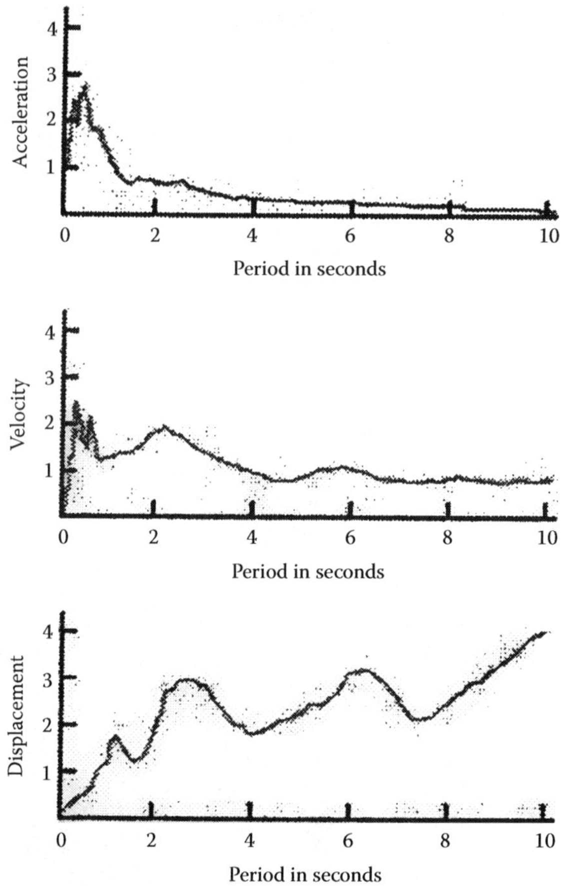

FIGURE 3.6 Simplified response spectra for acceleration, velocity, and displacement.

The response spectrum enables the engineer to identify the resonant frequencies at which the building will undergo peak accelerations. Based on this knowledge, the building design might be adjusted to ensure that the building period does not coincide with the site period of maximum response. This is perhaps easier said than done for a building but is quite possible for a nonbuilding structure such as a flagpole shown in Figure 3.7.

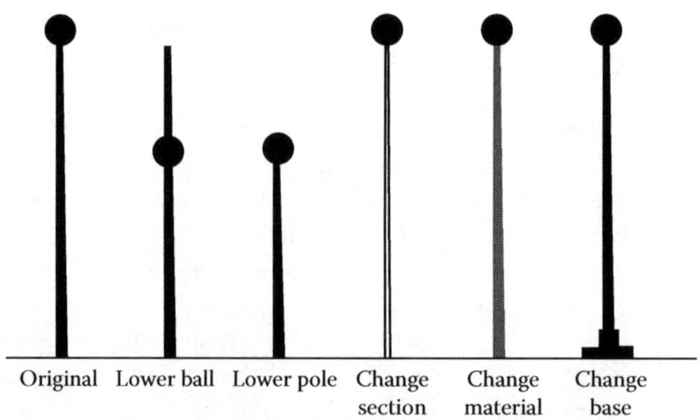

FIGURE 3.7 Turning response of a flagpole: these changes shorten its period.

3.7 DAMPING

If a building is made to vibrate, the amplitude of the vibration will decay over time and eventually cease. Damping is a measure of this decay in amplitude, and it is principally due to internal friction and absorbed energy. The nature of the structure and its connections affects the damping; a heavy concrete structure will provide more damping than a light steel frame.

Architectural features such as partitions and exterior façade construction also contribute to the damping.

Damping is measured by reference to a theoretical damping level termed critical damping. This is the least amount of damping that will allow the structure to return to its original position without any continued vibration. For most structures, the amount of damping in the system will vary between 3% and 10% of critical. The higher values would apply to older buildings (such as offices and government buildings) that employed a structure of steel columns and beams encased in concrete together with some structural walls, which also had many heavy fixed partitions (often concrete block or hollow tiles), and would have high damping values. The lower values would apply to a modern office building with a steel moment frame, a light metal and glass exterior envelope, and open office layouts with a minimum of fixed partitions.

The main significance of damping is that accelerations generated by ground motion decrease rapidly as the damping value increases. The response spectra in Figure 3.8 show that the peak acceleration is about 3.2*g* for a damping value of 0%, 0.8*g* for a damping value of 5%, and 0.65*g* for a damping value of 10%.

Response spectra generally show acceleration values for 5% damping. A damping value of zero might be used in the design of a simple vibrator, such as a flagpole or a water tank supported on a single cantilever column. For typical structures, engineers generally use a value of 5% critical.

Damping used to be regarded as a fixed attribute of buildings, but in recent years, a number of devices have been produced that enable the engineer to increase the damping and reduce the building response. This greatly increases the designer's ability to provide a *tuned* response to the ground motion.

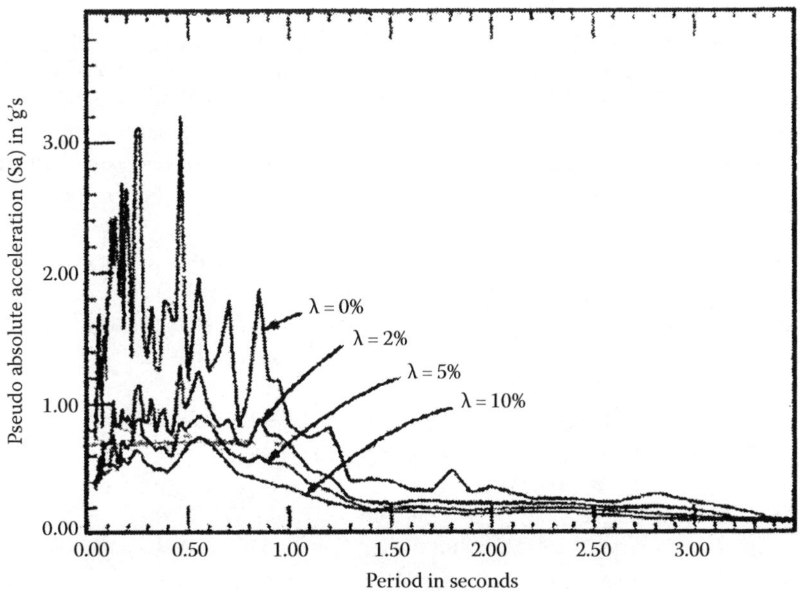

FIGURE 3.8 Response spectra for a number of damping values.

3.8 DUCTILITY

This is the property of certain materials to fail only after considerable inelastic deformation has taken place, meaning that the material does not return to its original shape after distortion. This deformation, or distortion, dissipates earthquake energy.

This is why it is much more difficult to break a metal spoon by bending it than one made of plastic. The metal spoon will remain intact, through distorted, after successive bending to and fro, while the plastic spoon will snap suddenly after a few bends. The metal is far more ductile than the plastic.

The plastic deformation of the metal absorbs energy and defers absolute failure of the structure. The material bends but does not break and so continues to resist forces and support loads, although with diminished effectiveness. The effect of earthquake motion on a building is rather like that of bending a spoon rapidly back and forth: the heavy structures are pushed back and forth in a similar way several times a second (depending on its period of vibration).

Brittle materials, such as unreinforced masonry or inadequately reinforced concrete, fail suddenly, with a minimum of prior distortion. However, the steel bars embedded in reinforced concrete with heavier and more closely spaced ties and special detailing of their placement can give this material considerable ductility equal to that of structural steel.

Ductility and reserve capacity are closely related: past the elastic limit (the point at which forces cause permanent deformation); ductile materials can take further loading before complete failure.

In addition, the member proportions, end conditions, and connection details will also affect ductility. Reserve capacity is the ability of a structure to resist overload and is dependent on the ductility of its individual members and connections. The only reason for not requiring ductility is to provide so much resistance that members would never be pushed beyond elastic limits.

Thus, buildings are designed in such a way that in the rare case when they are subjected to forces higher than those required by code, the materials and connections will distort but not snap. In so doing, they will safely absorb the energy of the earthquake vibrations, and the building, although distorted and possibly beyond repair, is at least still standing.

3.9 EARTHQUAKES AND OTHER GEOLOGIC HAZARDS

Earthquakes have long been feared as one of nature's most terrifying phenomena. Early in human history, the sudden shaking of the earth and the death and destruction that resulted were seen as mysterious and uncontrollable. We now understand the origin of earthquakes and know that they must be accepted as a natural environmental process. Scientific explanations, however, have not lessened the terrifying nature of the earthquake experience. Earthquakes continue to remind us that nature can, without warning, in a few seconds create a level of death and destruction that can only be equaled by the most extreme weapons of war.

This uncertainly, together with the terrifying sensation of earth movement, creates our fundamental fear of earthquakes. Beyond the threat to life is the possibility of the destruction of public and private property. Jobs, services, and business revenues can disappear instantly, and for many, homelessness can suddenly be very real. The aftermath of a great earthquake can endure for years or even decades.

Other types of phenomena sometimes accompany earthquake-caused ground shaking and are generally identified as geologic hazards. These are the following:

- *Liquefaction* that occurs when loose granular soils and sand in the presence of water change temporarily from a solid to a liquid state when subjected to ground shaking. This condition occurs mainly at sites located near rivers, lakes, and bays.
- *Landslides*, which involve the slipping of soil and rock on sloping ground. This can be triggered by earthquake ground motion.

- *Tsunamis* that are earthquake-caused wave movements in the ocean. They travel at high speeds and may result in large coastal waves of 30 ft or more. They are sometimes, and incorrectly, called tidal waves.
- *Seiches* that are similar to tsunamis but take the form of sloshing in closed lakes or bays. They have the potential to cause serious damage, although such occurrences have been very rare.

For all the aforementioned geologic hazards, the only truly effective defense is the application of good land-use practices that limit development in hazard-prone locations. Seismic design and construction is aimed at reducing the consequences of earthquake-caused ground shaking, which is by far the main cause of damage and casualties.

Earthquakes in the United States are a national problem. Most people now know that earthquakes are not restricted to just a few areas in the United States, most notably California and Alaska. Structural engineers recognize that two of the greatest earthquakes known occurred not in California, but near New Madrid, MO, in 1811 and 1812. As can be seen on a map of earthquake probability in the United States, more than 40 of the 50 states are at risk from earthquake-caused damage, life loss, injuries, and economic impacts (see ASCE 7-10 for seismic maps). Certainly the likelihood of a damaging earthquake occurring west of the Rocky Mountains, and particularly in California, the states of Oregon and Washington, and Salt Lake City is much greater than it is in the East, Midwest, or South. However, the New Madrid, Missouri, and Charleston, South Carolina, regions are subject to the possibility of severe earthquakes, although with a lesser probability than the Western United States.

3.10 EARTHQUAKE MEASUREMENTS

There are several common measures of earthquakes. Perhaps the most familiar is the Richter magnitude, devised by Professor Charles Richter of the California Institute of Technology in 1935. Richter scale is based on the maximum amplitude of certain seismic waves recorded on a standard seismograph at a distance of 100 km from the earthquake epicenter. Because the instruments are unlikely to be exactly 100 km from the source, Richter devised a method to allow for the diminishing of wave amplitude with increased distance. The Richter scale is logarithmic, and each unit of magnitude indicates a 10-fold increase in wave amplitude. The energy increase represented by each unit of scale is approximately 31 times. The scale is open ended, but a magnitude of about 9.5 represents the largest possible earthquake.

Table 3.1 shows significant earthquakes (magnitude 6 or over) that occurred in 47 of the 50 US states between 1568 and 1989.

Records show that some seismic zones in the United States experience moderate-to-major earthquakes approximately every 50–70 years, while other areas have *recurrence intervals* for the same size earthquake of about 200–400 years. These frequencies of occurrence are simply statistical probabilities and one or several earthquakes could occur in a much shorter than average period. With current knowledge, there is no practical alternative to assume that a large earthquake is likely to occur at any time and to take appropriate action.

Moderate and even very large earthquakes are inevitable, although very infrequent, in areas of normally low seismicity. Consequently, in these regions, buildings are very seldom designed to deal with an earthquake threat; therefore, they are extremely vulnerable. In other places, however, the earthquake threat is quite familiar. Buildings in many areas of California and Alaska will be shaken by an earthquake perhaps two or three times a year, and some level of *earthquake-resistant* design has been accepted as a way of life since the early twentieth century.

Although, on a national basis, the areas where earthquakes are likely to occur and the potential size or *magnitude* of these earthquakes are well identified and scientists have a broad statistical knowledge of the likelihood of their occurrence, it is not yet possible to predict the near-term occurrence of

TABLE 3.1
Numerical Integration Results

I, s	$1/2F(t)$, ft/s^2	$1000y$, ft/s^2	y, Eq. ft/s^2	$y\,(\Delta t)^2$, ft	y, ft
0.0	25	0	25.0	0.0100	0
0.02	30	5.0	25.0	0.0100	0.0050*
0.04	35	20.0	15.0	0.0060	0.0200
0.06	40	41.0	−1.0	−0.0004	0.0410
0.08	45	61.6	−16.6	−0.0066	0.0616
0.10	50	75.6	−25.6	−0.0102	0.0756
0.12	37.5	79.4	−41.9	−0.0168	0.0794
0.14	25	66.4	−41.4	−0.0166	0.0664
0.16	25	36.8	−11.8	−0.0047	0.0368
0.18	25	2.5	22.5	0.0090	0.0004
0.20	25	−22.8	47.8	0.0191	−0.0228
0.22	25	−29.0	54.0	0.0216	−0.0290
0.24	25	−13.6	38.6	0.0154	−0.0136
0.26	25	17.2	7.8	0.0031	0.0172
0.28	25	51.1	−26.1	−0.0104	0.0511
0.30	25	74.6	−49.6	−0.0198	0.0746
0.32	25	78.3	−53.3	−0.0123	0.0783
0.34	—	—	—	—	0.0607

a damaging earthquake. Therefore, lacking useful predictions, it makes sense in any seismic region to take at least the minimum affordable prudent actions directed at saving lives. Because most lives are lost in earthquakes when buildings collapse, US seismic building code provisions focus on requiring that the minimum measures necessary to prevent building collapse are taken.

3.11 DETERMINATION OF LOCAL EARTHQUAKE HAZARDS

Until quite recently, the United States was divided into a number of seismic zones, which were shown on the maps in the model codes. Zones ranged from Zone 0 (indicating no seismicity) to Zones 1, 2A, 2B, 3, and 4. Zone 4 indicates the highest level of seismicity (see Figure 3.9). Each zone was allocated a factor, or coefficient, from 0.075 to 0.40; this value was a multiplier representing the acceleration value for which the building was to be designed. These values indicate a fourfold range in acceleration values between Zones 1 and 4. Within a zone, all buildings must be designed to the same acceleration value.

Current seismic standards, such as the ASCE 7-10, define site seismicity in a different way. The United States is still divided into zones by contour lines, but their areas are much smaller. Numerical values are also shown on the maps and also represent the acceleration value to be used for design, but they are calculated in a different way, and many more values are shown that reflect greater precision of knowledge. Also, acceleration values for both long- and short-period buildings are shown in a separate series of maps. The simplicity of the old seismic zones is lost, but the design information is much more detailed.

Seismic performance in current codes is specified by selecting a maximum tolerable damage level for a given earthquake-shaking intensity. The shaking intensity can be specified *probabilistically*, derived by considering all future potential shaking at the site regardless of the causative fault, or *deterministically*, giving the expected shaking at the site for a given sized earthquake on a given fault.

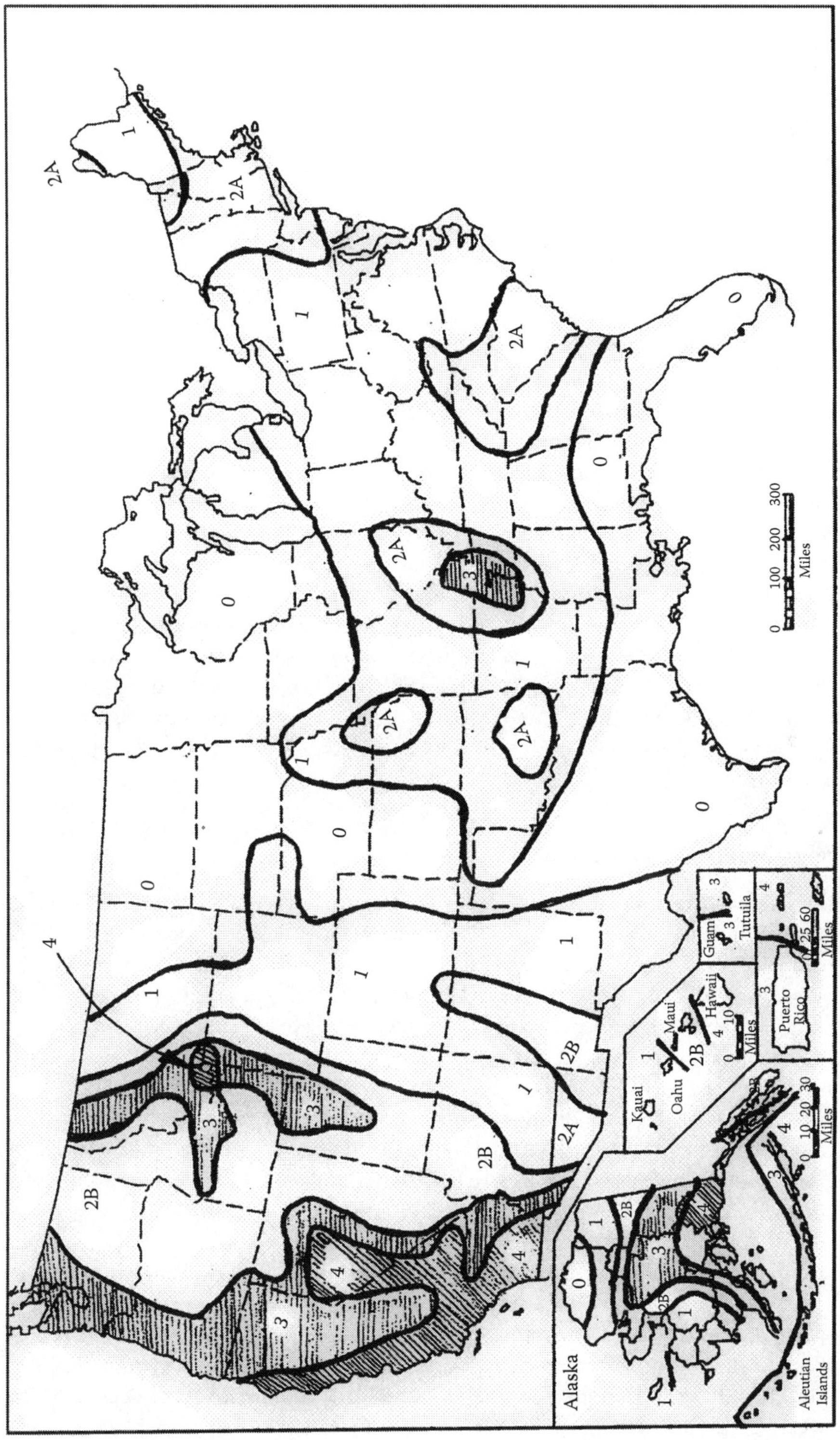

FIGURE 3.9 1997 UBC seismic zone map of the United States. The map is based on 10% probability of exceedance in 50 years.

For some time, the earthquake shaking used by building codes for a new building has been described probabilistically, as shaking with a 10% chance of being exceeded in a 50-year time period (50 years being judged as the average life of buildings). This can also be specified, similar to methods used with storms or floods, as the shaking with a return period of 475 years. (Actually, for ease of use, the return period is often rounded to 500 years, and since actual earthquake events are more understandable than probabilistic shaking, the most common term, although slightly inaccurate, is *the 500-year event.*)

Nationally applicable building codes were therefore based on the level of shaking intensity expected at any site once every 500 years (on average). However, engineers in several areas of the country, most notably Salt Lake City, Utah; Charleston, South Carolina; and Memphis, Tennessee, felt that this standard was not providing sufficient safety in their regions because very rare, exceptionally large earthquakes could occur in those areas, producing shaking intensities several times that of the 500-year event. Should such a rare earthquake occur, the building code design would not provide the same level of protection provided in areas of high seismicity, particularly California, because rare, exceptionally large shaking in California is estimated to be only marginally larger (about 1.5 times) than the 500-year shaking. It was therefore decided to determine the national mapping parameters on a much longer return period—one that would capture the rare events in the regions at issue, and a 2.500-year event was chosen (known as the MCE). Finally, it was judged unnecessary, and in fact undesirable, to significantly change seismic design practices in California, so the MCE was multiplied by 2/3 to make California design shaking levels about the same as before.

If the new shaking level—about 1.5 times the old—were multiplied by 2/3, the final design parameter would not change. However, in a region where the MCE is three times the previously used 500-year event, the new parameter of 2/3 MCE would result in a shaking level twice that previously used—providing the sought-after additional level of safety in those regions. Currently, national standards such as the ASCE define the level of shaking to be considered for evaluation of existing buildings to be 2/3 MCE, which, as previously explained, is about the same as the 500-year event for much of California.

3.11.1 Probabilistic Seismic Hazard Analysis

Probabilistic seismic hazard analysis (PSHA) provides an estimate of the likelihood of hazard from earthquakes based on geological and seismological studies. It is probabilistic in the sense that the analysis takes into consideration the uncertainties in the size and location of earthquakes and the resulting ground motions that could affect a particular site. Seismic hazard is commonly described as the probability of occurrence of some particular earthquake characteristic (such as peak ground acceleration). For statistical reasons, these probabilities cover a range of values, and because risk involves values being greater than expected, the word *exceedance* has been coined in their usage.

The effects of ground shaking on building response are well known and extensively documented. It is also acknowledged that severe ground shaking can significantly damage buildings designed in accordance with seismic codes and can even cause partial collapse of buildings with inadequate seismic resistance.

Seismic shaking is typically quantified using a parameter of motion, such as DVA. In current seismic codes, seismic design forces are defined in terms that relate to acceleration in the horizontal direction.

The earthquake ground-shaking hazard for a given region or site can be determined in two ways: deterministically or probabilistically. A deterministic hazard assessment estimates the level of shaking, including the uncertainty in the assessment, at the building site for a selected earthquake scenario. Typically, the earthquake is selected as the maximum-magnitude earthquake considered to be capable of occurring on an identified active earthquake fault; this maximum-magnitude earthquake is termed a characteristic earthquake. A deterministic analysis is often made when there is a well-defined active fault for which there is a sufficiently high probability of a characteristic earthquake occurring during the life of the building. The known past occurrence of such an earthquake, or geologic evidence of the periodic occurrence of such earthquakes in the past, is often considered to be indicative of a high probability for a future repeat occurrence of the event.

Probabilistic hazard assessment expresses the level of ground shaking with a specific, low probability of being exceeded in a selected time period, for example, 10% probability of being exceeded in 50 years or 2% probability of being exceeded in 50 years, where 50 years is commonly chosen as the building design life. The seismic loading criteria in current US building codes define design force levels based on ground motions specified in probabilistic seismic hazard maps. Such maps show expected peak ground acceleration response at different building periods of vibration S_s and S_1 defined elsewhere in this text.

These maps indicate that, although the level of earthquake activity is high in California, most parts of the United States are also exposed to a significant earthquake ground-shaking hazard. In fact, large historic earthquakes in the United States have occurred outside California, in Missouri, Arkansas, South Carolina, Nevada, Idaho, Montana, Washington, Alaska, and Hawaii. Furthermore, current geologic studies have shown increasing evidence for large earthquake potential in areas that are popularly believed to be relatively quiet. Examples include the now-recognized subduction zones in Oregon and Washington, the Wasatch fault zone in Utah, and the Wabash Valley seismic zone in Illinois and Indiana.

Decisions to mitigate seismic risk require a logical and consistent approach to evaluate the effects of future earthquakes on people and structures. One method to achieve this logic and consistency is the PSHA, which gives a probabilistic description (a frequency of exceedance) of earthquake characteristics such as ground motion amplitudes and fault displacement.

There is a distinction between earthquake damage and loss: damage refers to physical effects, such as the effect on structures. Loss is the associated monetary or social consequence.

Structural engineers can estimate damage, but estimating loss may involve considering additional factors like inflated costs of labor and materials after a major earthquake (which is called *demand surge*) or insurance deductibles. The destruction of a building's contents and the interruption of business are additional losses that might occur. Buildings suffer damage during an earthquake, but the owner (or perhaps the insurer) incurs the loss. Confusion occurs between these terms because both are often quantified as a percentage of the replacement value of the structure.

The term *exposure time* is often used in estimating natural hazards. It is used, for example, in defining the ground motion with a 2.475-year return period as the motion "that will be exceeded with a 2% probability during an exposure time of 50 years, "with 50 years being the nominal lifetime of major civil structures.

A 10% probability of exceedance in 50 years corresponds to a 475-year return period, and the question sometimes arises, "What is special about the 475-year return period?" This period is derived by assuming a Poisson process for ground motion occurrences, wherein the probability of an event, P, is related to the annual frequency of exceedance of the ground motion Y and the exposure time t through

$$P = 1 - \exp(-yt) \tag{3.3}$$

Rearranging this gives

$$Y = \frac{-[\ln(1-P)]}{t} \tag{3.4}$$

Substituting a probability $P = 0.1$ and an exposure time $t = 50$ years gives $y = 0.002107$ per year, which is 1/475 years.

Whereby using PSHA, a geotechnical engineer determines and integrates contributions to the probability of exceedance of a ground motion level from all earthquake faults and magnitudes that could produce potentially damaging ground shaking at the site. Using the results from PSHA, ground motions can be readily obtained for any selected probability of exceedance and building design life.

For applications in performance-based design discussed in Chapter 14, both a probabilistic approach and a deterministic approach for the ground-shaking hazard assessment may be used. Using a probabilistic approach, the seismic hazard can be integrated with the building resistance characteristics to estimate the probability of exceeding some level of damage during a time period of significance such as the anticipated building life. Using a deterministic approach, the probability of exceeding a specified damage level may be assessed for an earthquake likely to occur during the anticipated building life.

3.11.2 Range of Earthquake Performance Criteria

These are several performance descriptions that are currently available. However, it is believed that those given in ASCE 7-10 cover the full range of performance criteria typically used in seismic evaluation. These are the following:

- *Operational*: Buildings meeting this performance level are expected to sustain minimal or no damage to their structural and nonstructural components. The building will be suitable for its normal occupancy and use, although possibly in a slightly impaired mode, with power, water, and other required utilities provided from emergency sources. The risk to life safety is extremely low.
- *Immediate occupancy (IO)*: Buildings meeting this performance level are expected to sustain minimal or no damage to their structural elements and only minor damage to their nonstructural components. Although immediate reoccupancy of the building will be possible, it may be necessary to perform some cleanup and repair and await the restoration of utility service to function in a normal mode. The risk to life safety is very low.
- *Life safety*: Buildings meeting this performance level may experience extensive damage to structural and nonstructural components. Structural repair may be required before reoccupancy, and the combination of structural and nonstructural repairs may be deemed economically impractical. The risk to life safety is low.
- *Collapse prevention*: Buildings meeting this performance level will not suffer complete or partial collapse nor drop massive portions of their structural or cladding on to the adjacent property. Internal damage may be severe, including local structural and nonstructural damage that poses risk to life safety. However, because the building itself does not collapse, gross loss of life is avoided. Many buildings in this damage state will be a complete economic loss.

3.12 NONSTRUCTURAL COMPONENTS

For many decades, seismic building codes focused exclusively on the structure of the building, that is, the system of columns, beams, walls, and diaphragms that provides resistance against earthquake forces. Although this focus remains dominant for obvious reasons, experience in more recent earthquakes has shown that damage to nonstructural components is also of great concern. In most modern buildings, the nonstructural components account for 60%–80% of the value of the building. Most nonstructural components are fragile (compared to the building structure), easily damage, and costly to repair or replace.

The distinction between structural and nonstructural components and system is, in many instances, artificial. The engineer labels as nonstructural all those components that are not designed as part of the seismic lateral-force-resisting system (LFRS). Nature, however, makes no such distinction and tests the whole building. Many nonstructural components may be called upon to resist forces even though not designed to do so.

The nonstructural components or system may modify the structural response in ways detrimental to the safety of the building. Examples are the placing of heavy nonstructural partitions in locations that result in severe torsion and stress concentration or the placement on nonstructural partitions between columns in such a way as to produce a short-column condition, as described later in this text. This can lead to column failure, distortion, and further nonstructural damage. Failure of the fire protection system, because of damage to the sprinkler system, may leave the building vulnerable to postearthquake fires caused by electrical or gas system damage.

While distance does not always guarantee safety (San Francisco was approximately 60 miles from the focus of the Loma Prieta earthquake), in general, being a substantial distance from the earthquake will lessen the effects of the earthquake on the building and its nonstructural components. Nonstructural failures are commonly seen at greater distances than structural failures.

Historically, the model earthquake codes paid little attention to the vertical component of the shaking generated by earthquakes. As a rule of thumb, the maximum vertical ground motion is generally 60%–70% of the maximum horizontal ground motion. While it may be unnecessary to consider the vertical motions of the structure as a whole, this is often not the case with nonstructural design. The model codes have little reference to vertical acceleration design requirements for nonstructural components. The building, usually due to its configuration, can act as an amplifier for both horizontal and vertical motions. Therefore, even though the code most often does not require vertical design resistance, the designer must be cognizant of the implications of vertical motions during an earthquake and their potential effects.

It is quite evident for some time that seismic design of regular and prismatic buildings and their nonstructural components is at once easy to visualize and execute. The current vogue for complex shapes in architectural design has, however, increased the complexity of the nonstructural systems. This increase in complexity decreases our ability to visualize how systems and components will respond and interact.

Since many nonstructural issues involve the intermixing of several engineering disciplines, perhaps the best way to address the concern is to sit down with all the engineering professionals early and often to discuss earthquake performance objectives of the facility. This should help in visualizing potential interactions between building systems and components.

Nonstructural components may also, however, influence structural performance in response to ground shaking. Structural analysis assumes a bare structure. Nonstructural components that are attached to the structure and heavy contents, depending on their location, may introduce torsional forces. Characteristic examples of structural/nonstructural interaction are as follows:

- Heavy masonry partitions rigidly attached to columns and floor slabs can, if asymmetrically located, introduce localized stiffness and create stress concentrations and torsional forces. A particular form of this condition that has caused significant structural damage is when short-column conditions are created by the insertion of partial masonry walls between columns. The addition of such partial walls after the building completion is often treated as a minor remodel that is not seen to require engineering analysis. The result is that the shortened column with high relative stiffness attracts a large percentage of the earthquake forces often resulting in failures.
- In smaller buildings, stairs can act as bracing members between floors, introducing torsion. The solution is to detach the stair from the floor slab at one end to allow free structural movement.
- In storage areas or library stacks, heavy nonsymmetric loading can introduce torsion into a structure.

3.12.1 Response of Elements Attached to Buildings

Elements attached to the floors of buildings (e.g., mechanical equipment, ornamentation, piping, nonstructural partitions) respond to floor motion in much the same manner as the building responds to ground motion. However, the floor motion may vary substantially from the ground motion. The high-frequency components of the ground motion that correspond to the natural periods of vibrations of the building tend to be magnified. If the elements are rigid and are rigidly attached to the structure, the forces on the elements will be in the same proportion to the mass as the forces on the structure. But elements that are flexible and have periods of vibration close to any of the predominant modes of the building vibration will experience forces substantially greater than the forces on the structure.

When individual elements of the structure are analyzed separately, it becomes necessary to consider seismic effects differently than for lateral-load-resisting systems including diaphragms. The reason for this is that certain elements that are attached to the structure respond dynamically to the motion of the structure rather than to the motion of the ground. Resonance between the structure and the attached elements may occur.

3.13 SEISMIC ANALYSIS PROCEDURES

3.13.1 Equivalent Lateral Force Procedure

The ELF procedure simplifies the dynamic effects of earthquakes by using a static model. Historically, the procedure was used for the design of all structures, but the current codes restrict its application to small buildings of regular configuration and larger buildings of limited height constructed with flexible diaphragms that are not considered to be essential or hazardous to the public.

$$V = C_S W \tag{3.5}$$

where

- V is the seismic base factor
- W is the building weight including permanent and long-term contents
- $C_s = S_{DS}/(R/I)$
- S_{DS} is an attenuation parameter that varies according to soil conditions and the structures fundamental period
- R is a response modification factor that reflects the structural behavior of the seismic-force-resisting system
- I is an importance factor based on building use

In an earthquake, buildings experience ground motions that cause high accelerations and proportionately large internal forces in the building structure for short durations. In the ELF procedure, static loads with a lesser magnitude than the actual earthquake forces are applied. This relies on the anticipated ability of structures to withstand larger forces for short periods of time and allows for a less conservative, more affordable seismic design. The seismic base shear V is specified as a given percentage of the building weight. The value is determined by combining factors representing properties of the structure, soil, and use of the building.

The tendency for the building to sway from side to side in response to ground motion produces greater accelerations in the upper parts of the building. This back-and-forth motion, called the fundamental mode, dominates the response of most regular building structures. To model this effect statically, the ELF procedure redistributes the load applied to the building floors by considering for their distance from the base of the building.

3.13.2 Linear Dynamic Analysis

A linear dynamic analysis is useful for evaluating irregular or dynamically complex buildings. An irregular building is defined as having a distribution of mass or stiffness that is nonuniform and is often created in buildings that have complex space planning requirements or asymmetrical configurations. Dynamic complexity is common in flexible structural systems. Flexibility is greatly influenced by the selection of structural system and building height. Flexible buildings tend to have a significant response to higher mode shapes. Mode shapes are movement patterns that occur naturally in structures that have been set in motion by ground shaking. Schematics of the mode shapes of a four-story building are shown in Figure 3.5.

Designers use linear dynamic analysis to determine the degree of influence each mode shape will have on a structure's performance. The importance of higher modes depends on the relationship between the fundamental mode of the structure and the dynamic ground-shaking characteristics of the site. Designers express mode shape influence in terms of the percent of building mass assigned to a particular mode. If the building mass vibrates primarily in the first or fundamental mode, a static analysis is permitted by the code. Although linear dynamic analysis methods are becoming routine in engineering practice, they are more complicated because they require detailed information about ground motion. When linear dynamic analysis is used, the structure–ground shaking interaction is usually modeled using a response spectrum. The ASCE 7-10 includes a procedure for developing a design response spectrum, as explained elsewhere in this text.

An alternative and significantly more complex method for modeling ground shaking, called a time-history analysis, examines modal response using actual ground motion data. The seismic standards, such as ASCE 7-10, require that time-history analyses consider several different ground motion records to insure that the structure response is sufficiently representative to account for future unknown ground motion patterns.

3.14 SYSTEM SELECTION

Selecting a good structure requires engineering common sense. Common sense requires understanding the earthquake motion and its demands and understanding the structural behavior of the individual systems available. There are differences of scale between elastic and inelastic behavior and dynamic responses and seismic energy dissipation. Structural and architectural configurations (such as regular versus irregular forms) are also significant in the performance. The many variables often make it difficult to select an appropriate system. The ASCE 7-10 lists numerous structural systems, but it does not provide guidance in the selection of a system, and the many systems are not equal in performance.

Key performance issues are elastic behavior, inelastic behavior, and the related cyclic behavior resulting from pushing a structure back and forth. The cyclic behavior, often referred to as hysteretic behavior, should be stable, nondegrading, predictable, and capable of dissipating a large amount of seismic energy.

3.14.1 Elastic Behavior

The ELF approach to seismic design requires diminishing an acceleration spectra plot by use of an R value. Elastic design is expressed by an R value referred to as seismic response modification coefficient. This is a simple but frequently questionable method. It does not consider performance, nonlinear cyclic behavior, or—most importantly—energy dissipation.

3.14.2 Postelastic Behavior

Inelastic design is a better indication of realistic lateral drift or deflection that results from actual earthquake motions. Nonlinear drift impacts structural and nonstructural behavior. For significant seismic energy dissipation, the drift should be large, but for favorable nonstructural behavior, this drift should be small. A building with a large but unstable structural drift is likely to collapse. A building with a limited or small structural drift generally will not dissipate significant seismic energy without substantial damage.

3.14.3 Cyclic Behavior

A good measure of seismic performance is stable cyclic hysteretic behavior. The plot of load versus deformation of a member, for motion in both directions, represents cyclic behavior (Figure 3.10). If the load curves are full, undiminished, without *necking down*, they represent a stable system that is ductile and has sufficient capacity to deliver a constant level of energy dissipation during the shaking imposed by an earthquake. Degrading cyclic systems may, however, be acceptable if they degrade slowly and in a predictable manner.

The aforementioned attributes show that a range of possibilities exist for selecting a structural system. The favorable systems will do the following:

- Possess stable cyclic behavior
- Control lateral drift
- Dissipate seismic energy without failure
- Create a low postearthquake repair cost

The design reduction value *R*, discussed at length elsewhere in this work, does not necessarily correlate with performance. The *R* value was a consensus value developed for conventional elastic design. With the advent of performance design based on nonlinear evaluation, the *R* value serves only as a *rough estimate* of system behavior, but not a realistic estimate of performance.

We have learned from detailed investigations of past earthquake damages that problems occur because of

- Inappropriate building or structural configuration
- Brittle, nonductile structural systems
- Buildings not able to dissipate sufficient seismic energy
- Excessive loads caused by resonance between the ground shaking and building dynamic response

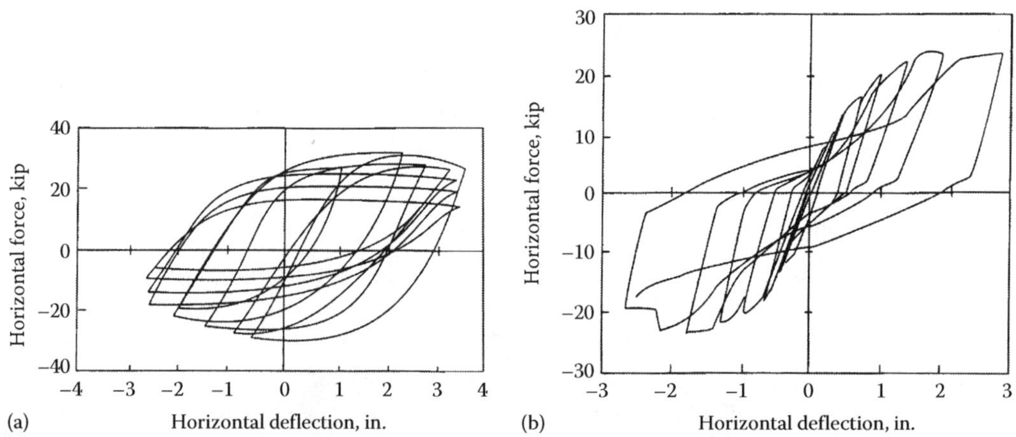

FIGURE 3.10 Hysteretic behavior: (a) curve representing large energy dissipation and (b) curve representing limited energy dissipation. (From Taranath, B.S., *Structural Analysis and Design of Tall Buildings*, CRC Press, Boca Raton, FL, 2011.)

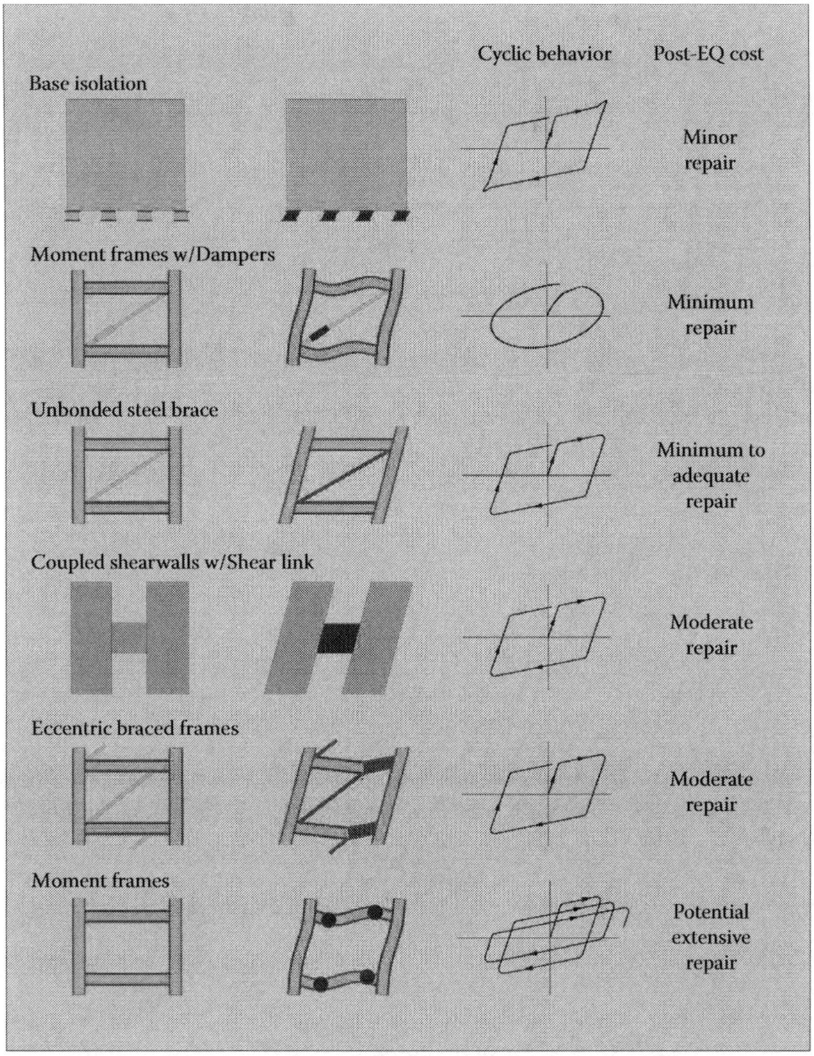

FIGURE 3.11 Structural systems with excellent to acceptable seismic performance. (From FEMA-454, *Designing for Earthquakes: A Manual for Architects*. Washington, DC: Federal Emergency Management Agency, 2006.)

Why, with all our accumulated knowledge, does all this failure continue? It is because buildings tend to be constructed essentially in the same manner, even after an earthquake. It takes a significant effort to change habits, styles, and techniques of construction.

Sometimes bad seismic ideas get passed on without too much investigation and modification.

And finally as a guide to engineers are, shown in Figure 3.11, six schematic structures that have exhibited excellent to acceptable performance in previous earthquakes.

3.15 SEISMIC ISSUES DUE TO CONFIGURATION IRREGULARITIES

A building's structural system is directly related to its architectural configuration, which largely determines the size and location of structural elements such as walls, columns, horizontal beams, floors, and roof structure. In today's flamboyant architecture, it is more than likely that engineers are faced with seismic design of building structures that are at once unique and at the same time are of irregular configuration.

In the following sections, the effects of irregular configurations on seismic performance are explained using the two main conditions created by vertical and plan irregularities. A number of deviations from regular characteristics are identified as problematical from a seismic viewpoint. Four of these deviations are then discussed in more detail, and conceptual solutions are provided for reducing or eliminating the detrimental effects.

3.15.1 Vertical Lateral-Load-Resisting Systems

Seismic designers have the choice of three basic alternative types of vertical lateral-load-resisting systems (VLLRSs) as illustrated in Figure 3.12.

These basic systems have a number of variations, mainly related to the structural materials used and the ways in which the members are connected. See Table 3.2 for a summary of seismic performance of structural systems in previous earthquakes.

Shown in Figures 3.13 and 3.14 are some examples of structural systems suitable for different site conditions and occupancy types.

3.15.1.1 Shear Walls

Shear walls are designed to receive lateral forces from diaphragms and transmit them to the ground. The forces in these walls are predominately shear forces in which material fibers within the wall try to slide past one another. To be effective, shear walls must run from the top of the building to the foundation with no offsets and a minimum of openings.

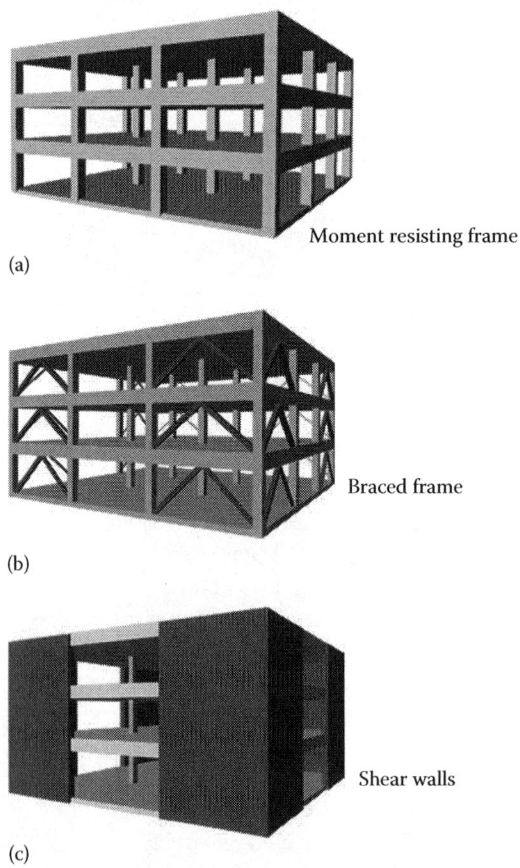

FIGURE 3.12 The three basic vertical seismic systems. (From FEMA-454, *Designing for Earthquakes: A Manual for Architects*. Washington, DC: Federal Emergency Management Agency, 2006.)

TABLE 3.2
Comparative Seismic Performance of Selected Structural Systems

Structural System	Earthquake Performance	Specific Building Performance and Energy Absorption	General Comments
Reinforced Concrete Wall	San Francisco, 1957 Alaska, 1964 Japan, 1966 Los Angeles, 1994 Variable to Poor	• Buildings in Alaska, San Francisco and Japan performed poorly with spandrel and pier failure • Brittle system	• Proportion of spandrel and piers is critical, detail for ductility and shear.
Steel Brace	San Francisco, 1906 Taft, 1952 Los Angeles, 1994 Variable	• Major braced systems performed well. • Minor bracing and tension braces performed poorly.	• Details and proportions are critical.
Steel Moment Frame	Los Angeles, 1971 Japan, 1978 Los Angeles, 1994 ? Good	• Los Angeles and Japanese buildings 1971/78 performed well. • Energy absorption is excellent. • Los Angeles 1994, mixed performance.	• Both conventional and ductile frame have performed well if designed for drift.
Concrete Shear Wall	Caracas, 1965 Alaska, 1964 Los Angeles, 1971 Algeria, 1980 Variable	• Poor performance with discontinuous walls. • Uneven energy absorption.	• *Configuration is critical*; soft story or L-shape with torsion have produced failures.
Reinforced Concrete Ductile Moment Frame	Los Angeles, 1971 ? Good	• Good performance in 1971, Los Angeles • System will crack • Energy absorption is good. • Mixed performance in 1994 Los Angeles	• Details *critical*

Source: Federal Emergency Management Agency. *Primer for Design Professionals: Communicating with Owners and Managers of New Buildings on Earthquake Risk.* FEMA-389. Washington, DC: FEMA, 2004.

3.15.1.2 Braced Frames

Braced frames act in the same way as shear walls; however, they generally provide less resistance but better ductility depending on their detailed design. They provide more architectural design freedom than shear walls.

There are two general types of braced frames: conventional concentric and eccentric. In the concentric frame, the center lines of the bracing members meet the horizontal beam at a single point.

In the eccentric-braced frame, the braces are deliberately designed to meet the beam some distance apart from one another: the short piece of beam between the ends of the braces is called a link beam. The purpose of the link beam is to provide ductility to the system: under heavy seismic forces, the link beam will distort and dissipate the energy of the earthquake in a controlled way, thus protecting the remainder of the structure.

3.15.1.3 Moment-Resistant Frames

A moment-resistant frame is the engineering term for a frame structure with no diagonal bracing in which the lateral forces are resisted primarily by bending in the beams and columns

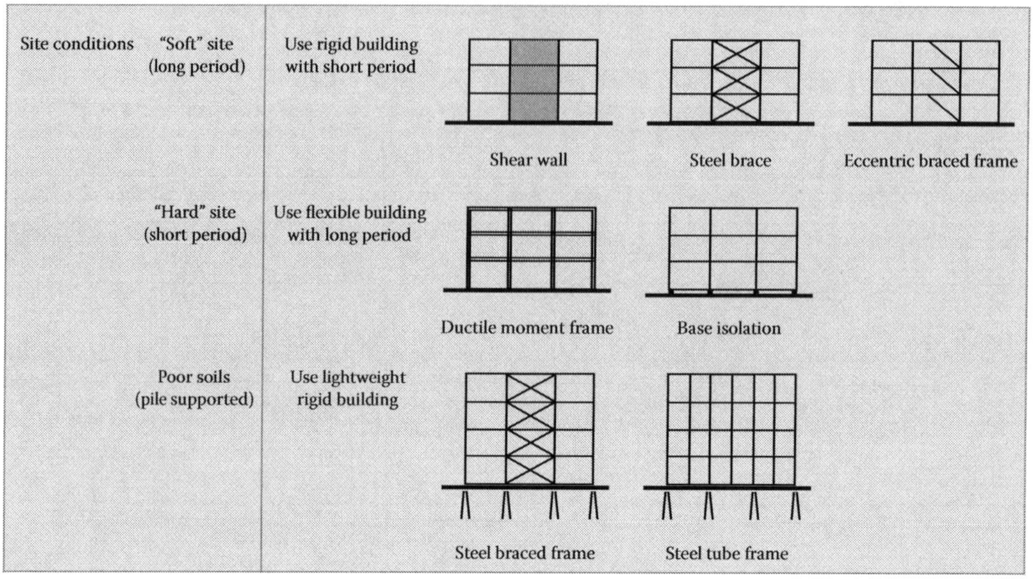

FIGURE 3.13 Example structural systems for site conditions. (From FEMA-389, Primer for Design Professionals: Communicating with Owners and Managers of New Buildings on Earthquake Risk. Washington, DC: Federal Emergency Management Agency, 2004.)

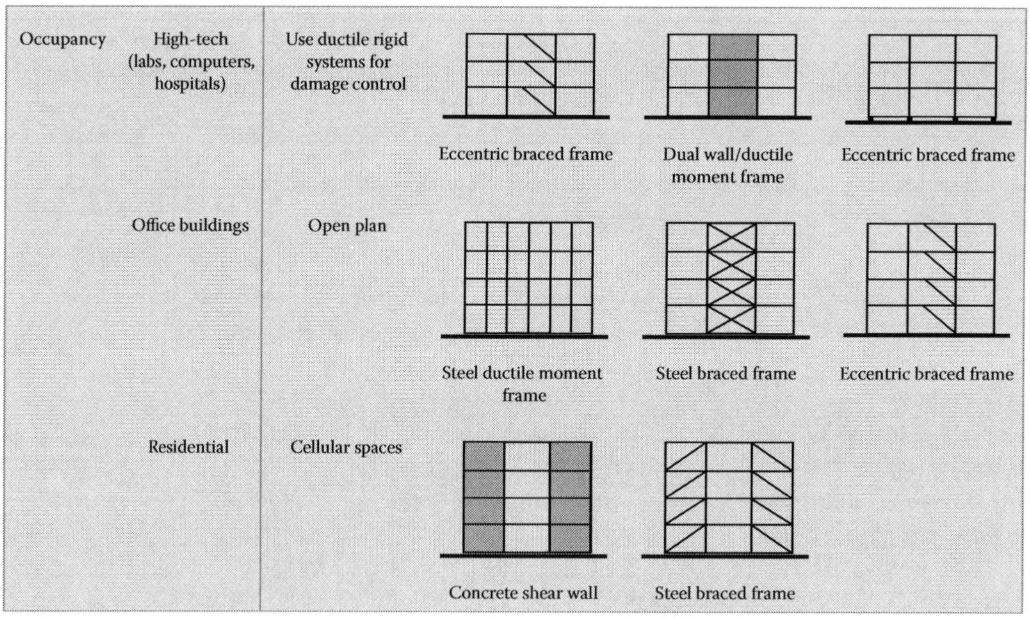

FIGURE 3.14 Structural systems for occupancy types. (From FEMA-389, Primer for Design Professionals: Communicating with Owners and Managers of New Buildings on Earthquake Risk. Washington, DC: Federal Emergency Management Agency, 2004.)

mobilized by strong joints between columns and beams. Moment-resistant frames provide the most architectural design freedom.

These systems are, to some extent, alternatives, although designers oftentimes mix systems, using one type in one direction and another type in the other or combining them in a given direction. This must be done with care, however, mainly because the different systems are of varying stiffness (shear-wall systems are much stiffer than moment-resisting frame systems, and braced systems fall in between), and it is difficult to obtain balanced resistance when they are mixed. However, for high-performance structures, there is now increasing use of dual systems. Examples of effective mixed systems are the use of shear-wall core together with a perimeter moment-resistant frame or a perimeter steel moment frame with interior eccentric-braced frames. Another variation is the use of shear walls combined with a moment-resistant frame in which the frames are designed to act as a fail-safe backup case of shear-wall failure.

The framing system must be chosen at an early stage in the design because the seismic system plays the major role in determining the seismic performance of the building and more importantly has considerable influence on the architecture of design. For example, if shear walls are chosen as the seismic-force-resisting system, the building planning must be able to accept a pattern of permanent structural walls with limited openings that run uninterrupted through every floor from roof to foundation.

3.15.2 Diaphragms

The term *diaphragm* is used to identify horizontal resistance members that transfer lateral forces between vertical resisting elements such as shear walls or frames. The diaphragm action is generally provided by the floor and roof systems of the building; sometimes, however, horizontal bracing systems independent of the roof or floor structure serve as diaphragms.

The diaphragm can be visualized as a wide horizontal beam with components at its edges, termed chords, designed to resist tension and compression: chords are similar to the flanges or a vertical beam (Figure 3.15).

A diaphragm that forms part of a resistant system may act either in a *flexible* or *rigid* manner, depending partly on its size (the area between enclosing resistance elements) and also on its material. The flexibility of the diaphragm, relative to that of the vertical lateral-load-resisting (VLLR) elements such as shear walls whose forces it is transmitting, also has a major influence on the nature and magnitude of those forces. With flexible diaphragms made of steel decking without concrete, walls take loads according to tributary areas (if mass is evenly distributed). With rigid diaphragms (usually concrete slabs and steel deck with concrete topping), walls share the loads in proportion to their stiffness.

Perhaps, now it is as good a time as any to discuss behavior of diaphragms with particular emphasis on the following:

- Collectors
- Role of diaphragms
- Types of diaphragms
- Diaphragm design procedures
- Shear transfer from diaphragm to the VLLRS
- Modeling of rigid diaphragms

See Figure 3.16 for diaphragm design terminology.

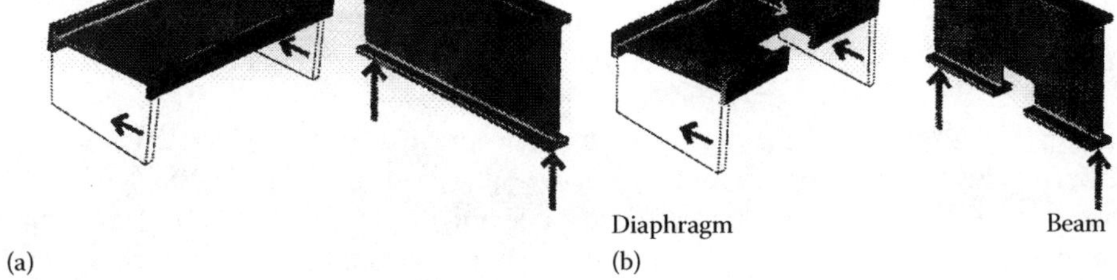

FIGURE 3.15 Deep beam action of diaphragm.

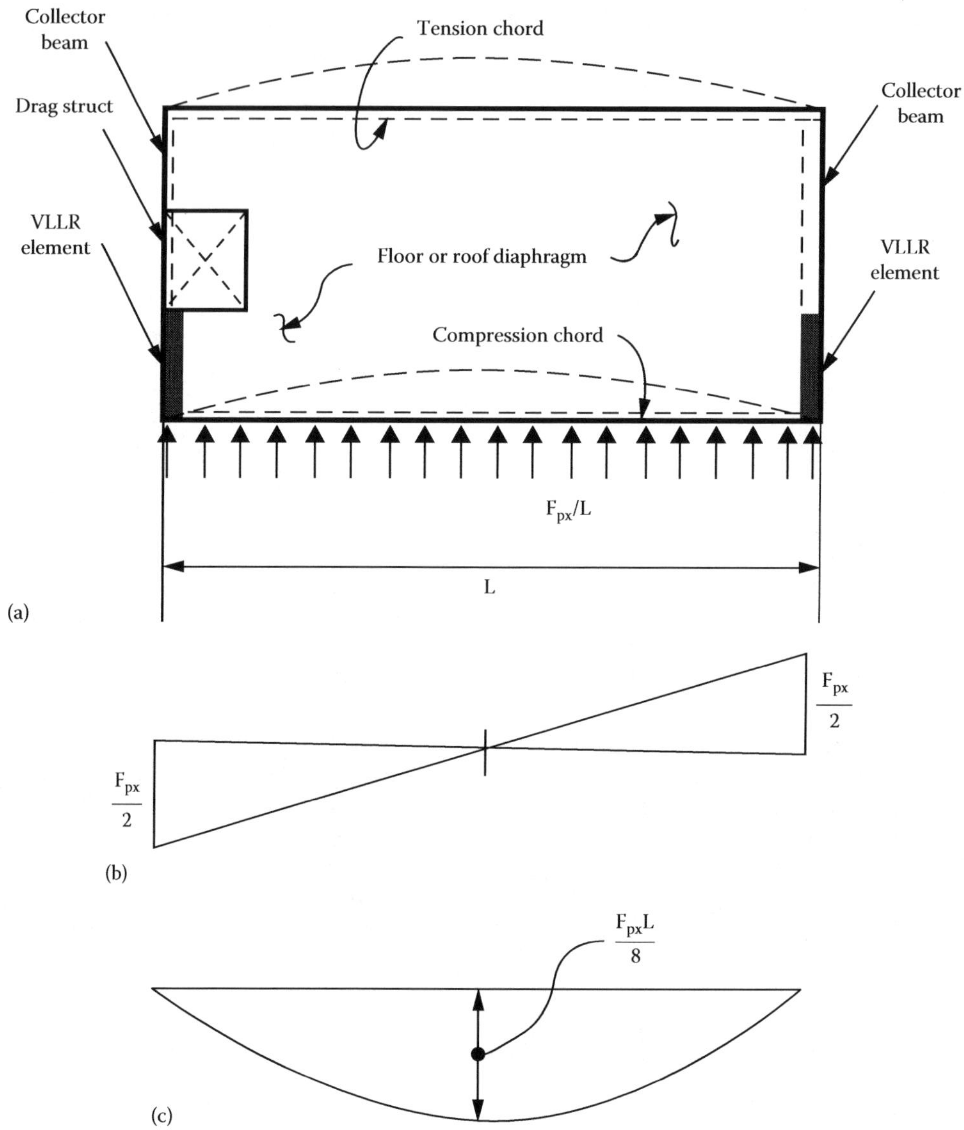

FIGURE 3.16 Diaphragm design terminology: (a) plan, (b) shear force diagram, and (c) bending moment diagram. *Note*: F_{px} diaphragm design force (see ASCE 7-10 Section 12.10.1.1).

3.15.2.1 Collectors

Collectors, also called drag struts or ties, are diaphragm framing members that *collect* or *drag* diaphragm shear forces from laterally unsupported areas to vertical resisting elements (Figure 3.16).

Floors and roofs, more often than not, have to be penetrated by openings required for certain features such as staircases, elevators and duct shafts, skylights, and atria. The size and location of these penetrations are critical to the effectiveness of the diaphragm. The reason for this is not hard to see when the diaphragm is visualized as a beam. For example, it can be seen that openings cut in the tension flange of a beam will seriously weaken its load-carrying capacity. In a vertical-load-bearing situation, a penetration through a beam flange would occur in either a tensile or compressive region.

In a lateral-load system, the opening would be in a region of both tension and compression, since the seismic loading alternates rapidly in direction.

3.15.2.2 Role of Diaphragms

Earthquake-resistant design requires that components of the structure be connected or tied together in such a manner that they behave as a unit. Diaphragms are an important structural element in achieving this interconnection. Diaphragms are horizontally spanning members, analogous to deep beams that distribute the seismic loads from their origin to the vertically oriented lateral-force-resisting frames (braced frames, moment frames, etc.). Diaphragms are commonly analyzed as simple-span or continuously spanning deep beams and hence is subject to shear, moment, and axial forces (for truss diaphragms and collectors) as well as the associated deformations. Figure 3.16 shows typical loading, shear and moment diagrams for the analysis and design of diaphragms.

In steel buildings, the floor or roof deck is usually designed as the shear-resistant member (which is analogous to the web of a beam), and the beams or supplemental deck reinforcing at the boundaries of the diaphragms is designed as the flexural-resistant member or chord (which is analogous to the flanges of a beam).

Diaphragms are classified into one of three categories: rigid, flexible, or semirigid. Rigid diaphragms are those that possess the strength and stiffness to distribute the lateral forces to the lateral-force-resisting frames in proportion to the relative stiffness of the individual frames, without significant deformation in the diaphragm. On the other hand, the distribution of the lateral forces through a flexible diaphragm is independent of the relative stiffness of the lateral-force-resisting frames.

A semirigid diaphragm, as the name implies, distributes lateral forces in proportion to the stiffness of the diaphragm *and* the relative stiffness of the lateral-force-resisting frames. Semirigid diaphragms are often analyzed using the analogy of a beam on elastic supports, where the beam represents the stiffness of the diaphragm and the elastic supports represent the stiffness of the lateral-force-resisting frames.

Since many buildings have lateral-force-resisting frames that are not uniformly spaced and continuous around the diaphragm boundaries, collector elements are utilized. Collector elements are tension and compression members that serve to deliver the diaphragm forces to the lateral-force-resisting frames. A redistribution of collector forces can occur as yield mechanisms form in the lateral-force-resisting frames.

The purpose of diaphragm as stated earlier is to distribute lateral forces to the elements of the VLLRS. In doing so, it

- Ties the building together as a unit
- Behaves as a horizontal continuous beam spanning between and supported by the VLLRS
- Acts as web of a continuous beam
- Causes members at floor edges to act as flanges/chords of the continuous beam
- Provides for stability of structure

3.15.2.3 Types of Diaphragms

There are many types of materials and system for use as floor and roof diaphragms such as

- Concrete slab
- Composite steel deck with concrete topping
- Precast elements with or without concrete topping slab
- Untopped steel deck (roof deck)
- Plywood sheathing

In merits noting, however, that standing seam roof panels are *not* considered as diaphragms.

As stated previously, diaphragms are typically classified into three categories: rigid, flexible, and semirigid. A diaphragm is considered rigid if it exhibits the following characteristics:

- Distributes horizontal forces to VLLR elements in direct proportion to relative rigidities of VLLR elements
- Diaphragm deflection insignificant compared to that of VLLR elements

Examples of rigid diaphragm are

- Composite steel deck slabs and concrete slabs (under most conditions)

A diaphragm is considered flexible if

- It distributes horizontal forces to VLLR elements independent of relative rigidities of VLLR elements
- It distributes horizontal forces to VLLR elements based on tributary areas
- Diaphragm deflection is significantly large compared to that of VLLR elements

As an example, an untopped steel deck, under certain conditions, may be considered as a flexible diaphragm.

A semirigid diaphragm, as the name implies, is one whose behavior is in between those of rigid and flexible diaphragms. It should be observed, however, that the classification is only for analytical purpose because diaphragms in building design are neither completely rigid nor completely flexible.

The behavior of a semirigid diaphragm may be considered analogous to that of a beam on elastic foundation. This is because deflection of VLLR elements and diaphragm under horizontal forces are of the same order of magnitude. Therefore, the design must account for relative rigidities of VLLR elements and diaphragm. Most seismic standards mandate that a diaphragm be considered flexible when the maximum lateral deformation of the diaphragm is more than two times the average story drift of the associated story. It is quite evident that in order to classify a diaphragm, one must compare the stiffness of the diaphragm to that of the VLLRS. The Steel Deck Institute (SDI), however, gives us the following guidelines for diaphragm classifications.

SDI diaphragm classifications

Shear Rigidity G′	Classification	Deck Type
6.67–14.3	Flexible	Bare steel deck
14.3–100	Semiflexible	Bare steel deck
100–1000	Semirigid	Concrete-filled steel deck
>1000	Rigid	Concrete-filled steel deck

Refer to SDI manual for definition of shear rigidity, G′.

It is perhaps self-evident that the in-plane deflection of the diaphragm shall be limited to that which will permit the attached element to maintain its structural integrity under the applied lateral loads and continue to support self-weight and vertical load if applicable.

It should be noted that the generally accepted drift ratio of h/400 is for the in-plane deflection experienced by the cladding and the out-of-plane drift limits are considerably less stringent (see Figure 3.17 for schematic explanation of this concept).

Another aspect equally important in diaphragm classification is the geometry (span-to-depth ratio) of the diaphragm itself. For example, shown in Figure 3.18a is a diaphragm that has a span-to-depth ratio of 62.5/50 = 1.25, which, for all practical purposes, would allow us to judge the diaphragm as rigid.

Next, let us consider the same diaphragm with the interior VLLR elements removed from the system as shown in Figure 3.18b. The diaphragm now has to span the entire width of 250 ft resulting

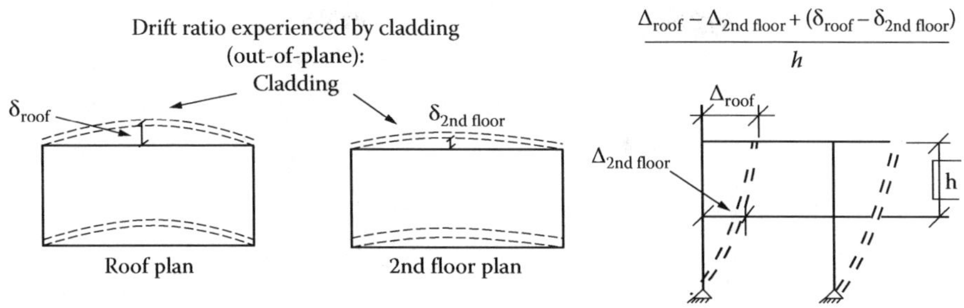

FIGURE 3.17 Out-of-plane drift. *Note*: The ratio h/400 limit is generally for the in-plane deflections. The out-of-plane drift limits are generally less stringent.

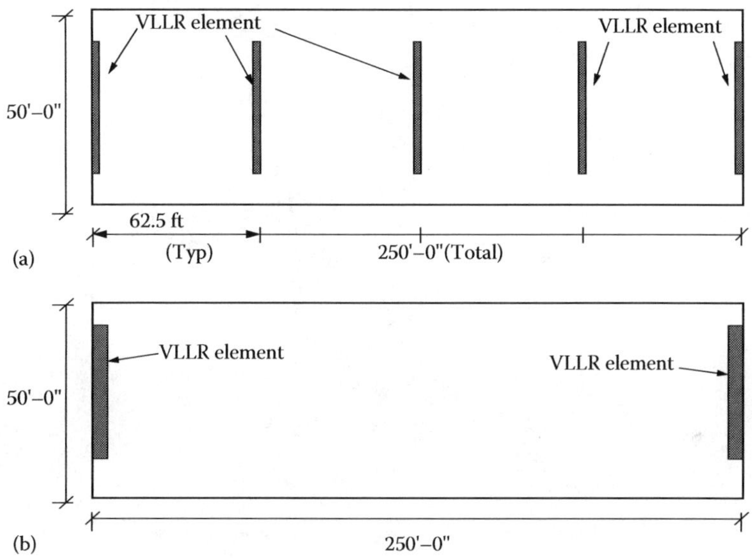

FIGURE 3.18 Rigid and flexible diaphragms: (a) rigid diaphragm with a span-to-depth ratio of 62.5/50 = 1.25 and (b) flexible diaphragm with a span-to-depth ratio of 250/50 = 5.

in a span-to-width ratio = 250/50 = 5. This would generally require the diaphragm to be considered as flexible or semiflexible.

3.15.2.4 Diaphragm Design Procedures

To understand diaphragm behavior and design procedures, it is convenient to define certain terms unique to diaphragm design (see Figure 3.16).

Shown therein are the schematic bending moment and shear force diaphragms resulting from beam action of the diaphragm spanning between the VLLR elements. The term F_{px} is the diaphragm design force occurring at this level due to inertial force. To this, we must add any shear force resulting from discontinuity of VLLR element such as a shear wall, as shown in Figure 3.19.

3.15.2.5 Shear Transfer from Diaphragm to VLLRS

Basically, there are two design approaches. In the first method, the entire diaphragm shear is transferred to the VLLRS without using collector beams or drag struts.

The shear transfer is assumed to occur only over the length of the VLLRS (see Figure 3.20).

In the second approach, the diaphragm shear is transferred to collector beams and then to the VLLRS (see Figure 3.21). Observe that this approach is required when there is no direct

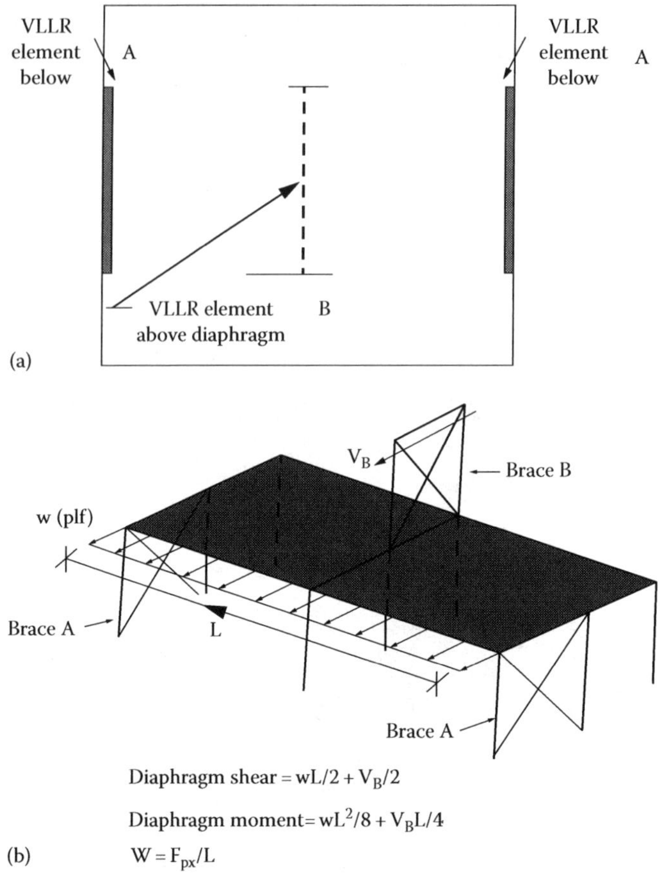

FIGURE 3.19 Transfer of VLLR element: (a) plan and (b) schematic elevation.

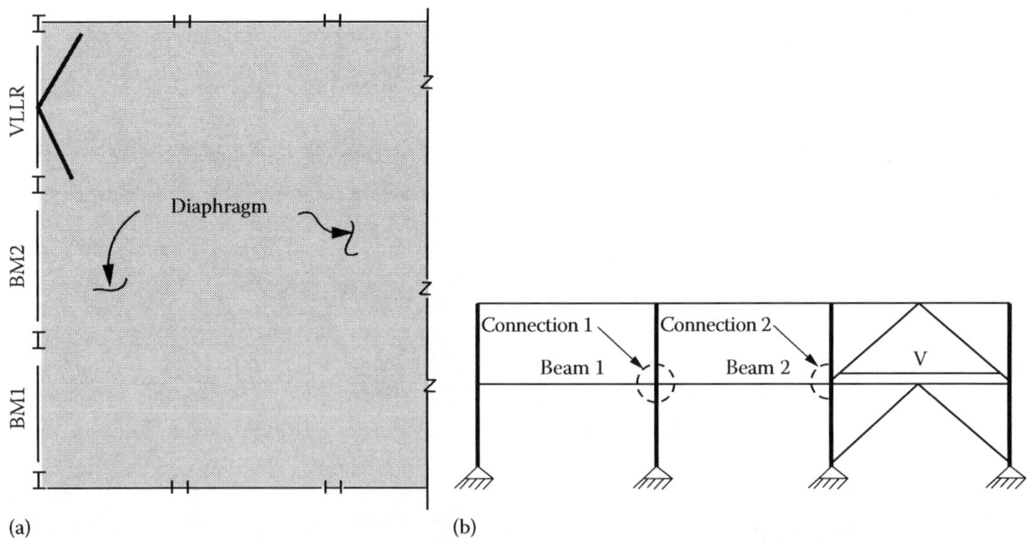

FIGURE 3.20 Diaphragm design procedures, approach 1: (a) partial plan of diaphragm and (b) elevation. *Note*: In this method, the entire diaphragm shear is transferred directly to the VLLRS. Connections 1 and 2 or beams 1 and 2 need not be designed for diaphragm forces.

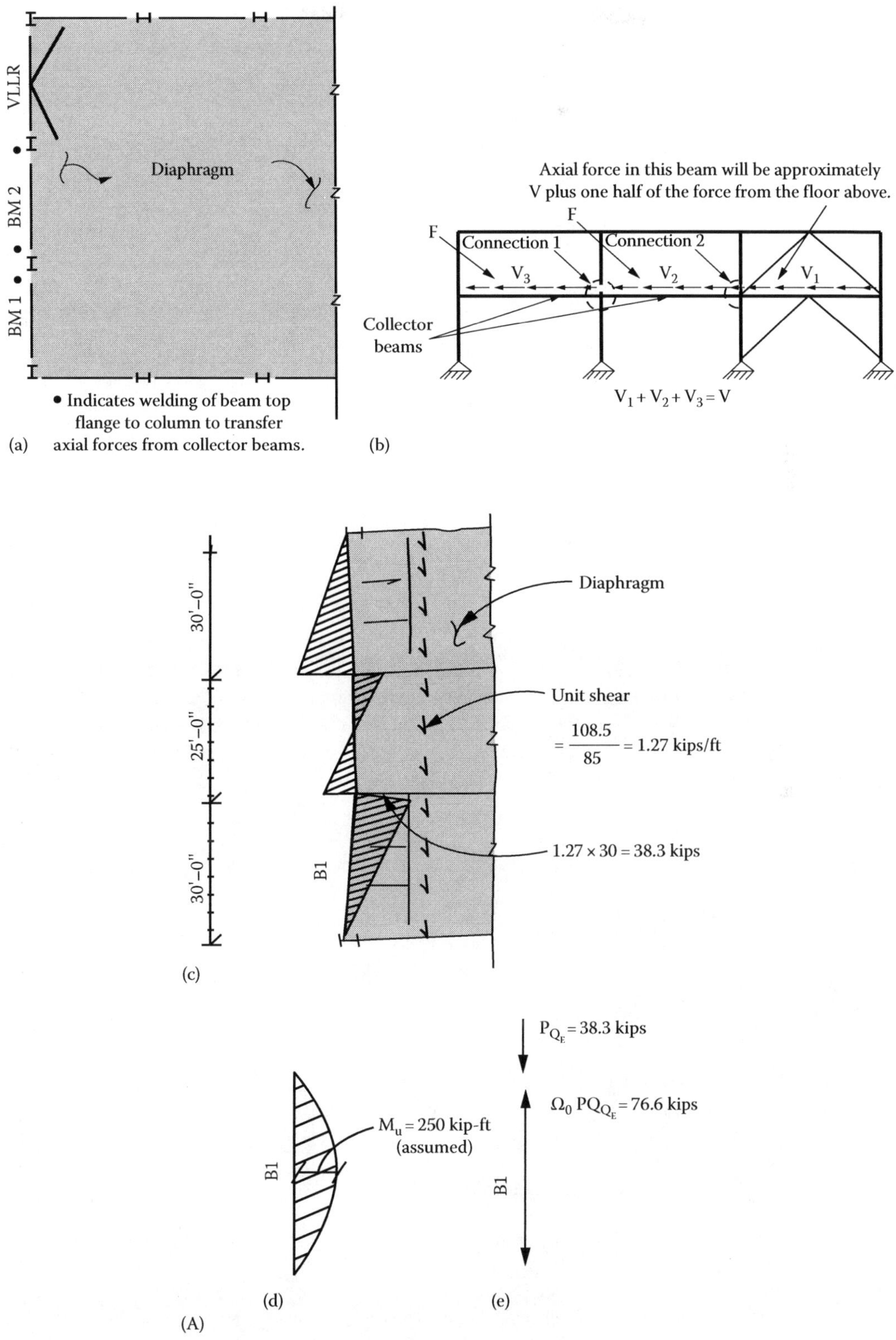

FIGURE 3.21 Diaphragm design procedure, approach 2: (A) (a) partial plan of diaphragm; (b) elevation; (c) axial load in beam B1 due to diaphragm shear; (d) gravity moment Mu; and (e) axial load Ω_0 P_{QE}. *(Continued)*

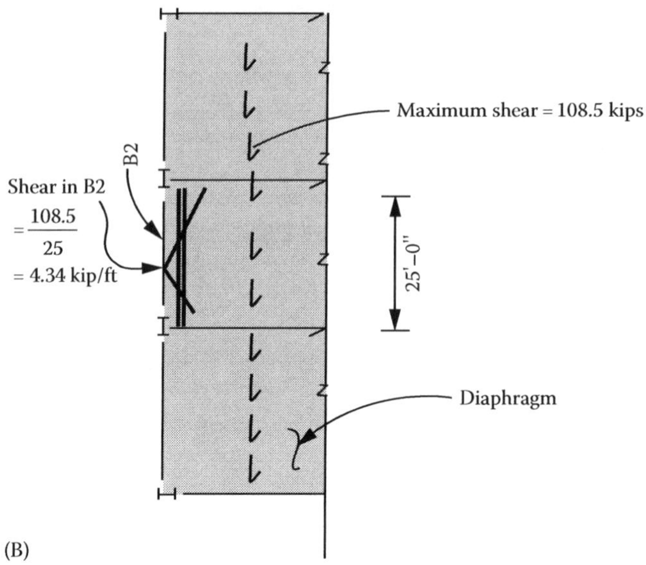

FIGURE 3.21 (*Continued*) Diaphragm design procedure, approach 2: (B) schematic diaphragm, shear transfer to VLLR without using collectors. *Note*: In this approach, diaphragm shear not directly transferred to the VLLR system is through collector beams.

connection of the diaphragm to the VLLR element or when the length of VLLR element is insufficient to transfer the diaphragm shear. As an example of the former, shown in Figure 3.22 is a diaphragm that has an opening adjacent to the VLLRS. Because there is no direct connection between the diaphragm and beam BM2, there is no direct transfer of diaphragm shear force from the diaphragm to this beam. It does, however, carry the axial force V_2 resulting from the shear accumulated by the beam BM1 (see Figure 3.22).

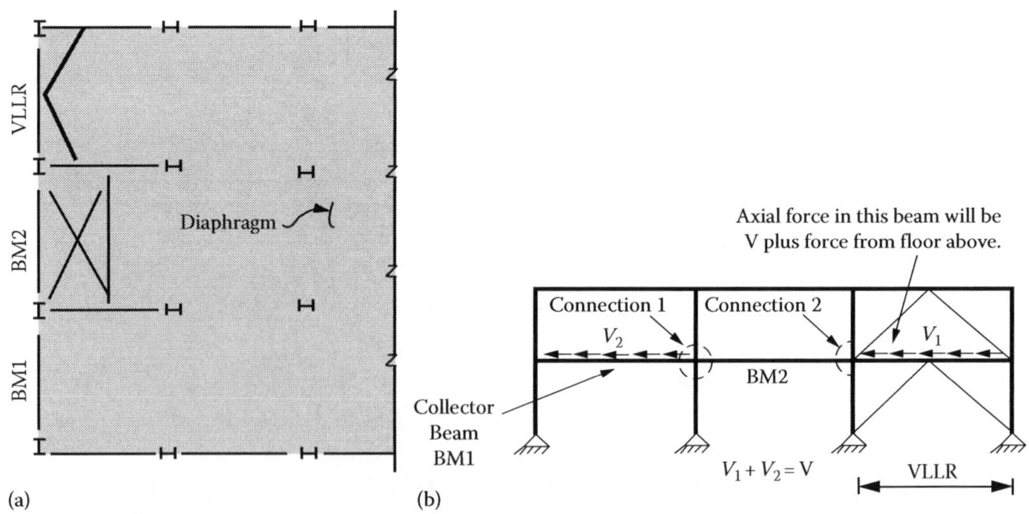

FIGURE 3.22 Diaphragm with opening along drag beam, approach 2: (a) plan and (b) elevation. *Note*: (1) Beam BM2 does not collect diaphragm shear forces but must be designed to carry the axial force V_2, (2) provide connection between diaphragm and collector beam BM1 to transfer diaphragm force, and design collector beam for this force. (3) Connections 1 and 2 must be designed to carry the axial force V_2 through the beam–column connection.

Cast-in-place concrete diaphragm

The strength of cast-in-place concrete diaphragms discussed in Chapter 21 of ACI 318-11 is governed by Equation 3.6:

$$V_n = A_{cv}\left(2\lambda\sqrt{f_c'} + \rho_t f_y\right) \tag{3.6}$$

and shall not exceed

$$V_{n\,\max} = A_{cv}\left(8\sqrt{f_c'}\right) \tag{3.7}$$

Observe that

- Design of concrete diaphragms is based on ultimate strength
- F_y shall not exceed 60 ksi

$$\begin{aligned} \Phi V_n &> V_u \\ \Phi &= 0.75 \end{aligned} \tag{3.8}$$

3.15.2.6 Modeling of Rigid Diaphragms

By definition, a rigid diaphragm has no relative in-plane displacement of joints within the diaphragm. Beams that are part of VLLRSs with both end joints connected to a rigid diaphragm do not experience, in an analytical sense, any axial force. And none is reported in a typical computer analysis output. Therefore, any design that uses post processors without proper adjustment of axial loads in the beams would be wrong. How can we overcome this stumbling block?

One method, often attempted in practice, is to release the beam ends strategically from the rigid diaphragm so that axial force and deformation are captured in the analysis. It should, however, be kept in mind that release of excessive joints may create instability. To overcome this problem, it seems that we have no choice but to revert back to separate manual or computer analysis of each frame for proper determination of axial force in beams that are part of VLLRSs.

Observe that in braced frames, the building stiffness is likely to be overestimated by as much as 15% if axial deformation of beams is not considered. Keep in mind that computer analysis results do not tell the engineer if there is a problem in the analytical model.

A similar problem is likely to occur when modeling story-deep transfer trusses subjected to gravity and lateral loads (see Figure 3.23). If the diaphragms attached to the top and bottom chords of the truss are assumed rigid, then the computer thinks that the areas of chords are infinitely large.

Thus, $A_{\text{top chord}} = A_{\text{bot.chord}} = \infty$.

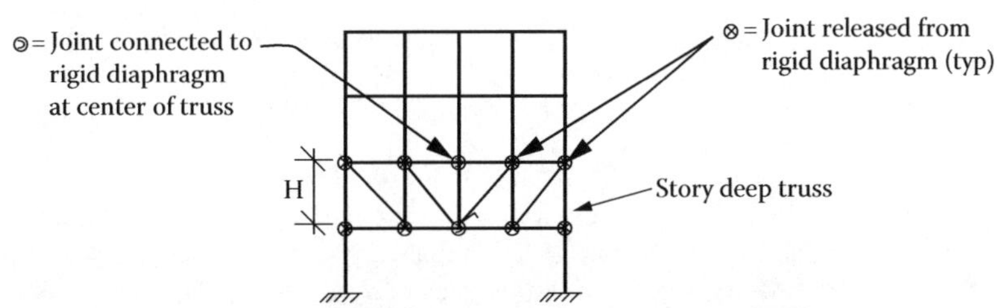

FIGURE 3.23 Top and bottom chords of story-deep truss attached to rigid diaphragms. *Note*: To capture correct axial forces in the truss top and bottom chords, release all but those at center of truss.

Since the moment of inertia of the truss (neglecting the flexibility of diagonals) is equal to

$$I_{Trus} = \sum A\left(\frac{H}{2}\right)^2 \tag{3.9}$$

any analysis based on this assumption of rigid diaphragms is obviously incorrect.

One method of circumventing this problem is to release all but the center joint from the rigid diaphragm (see Figure 3.23).

The design of the diaphragm, at least in theory, is no more complicated than that of a deep beam. However, there are many reasons why you may have diaphragm-related problems. Chief among them are as follows:

- Diaphragm shear capacity is not checked.
- Connections are not designed to transfer chord and collector forces.
- Force transfer from the diaphragm to collector beams/VLLRS is not considered (load path is not clearly defined).
- Chord and collector beams are not designed properly.
- Diaphragms are not modeled properly in the analytical model.
- Axial force in beams that are part of the VLLRSs is incorrect.
- Boundary elements for large openings in the floors and roof are not given proper attention.

Schematics of diaphragm conditions that require particular attention are summarized in Figures 3.24 through 3.32.

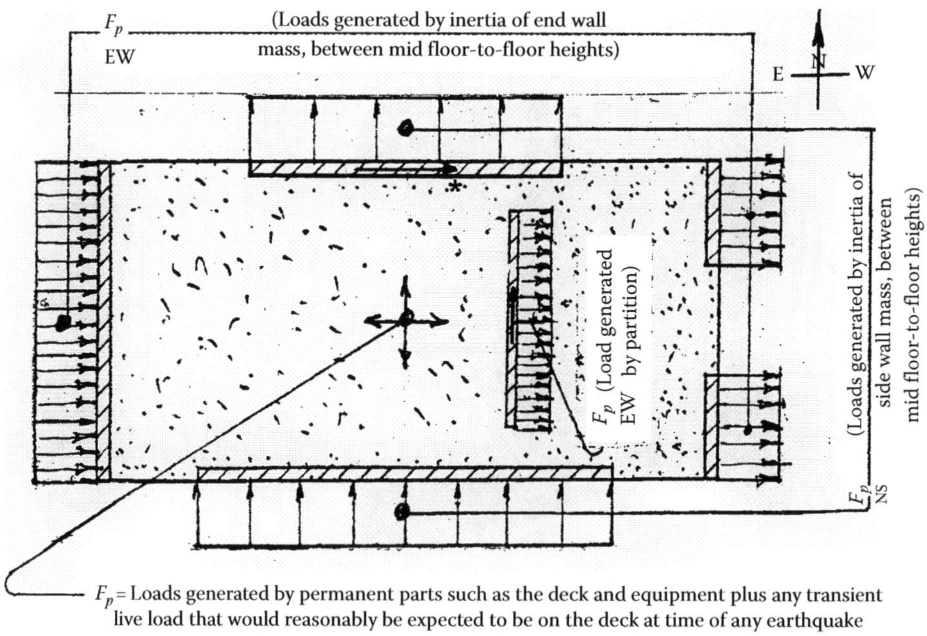

FIGURE 3.24 Diaphragm design loads. *Notes*: (1) Floors and roofs used as diaphragms are designed for a lateral force of F_p acting in any direction. Generally, it is assumed that the in-plane mass of a shear wall does not contribute to the diaphragm loading unless the shear wall is interrupted at the specific level. In case a shear wall does not extend below the floor level, both its horizontal and vertical loads must be distributed to the remaining walls with due considerations to major differences in rigidities. (2) The total shear F_p at any level will be distributed to the various vertical elements of the lateral-force-resisting system in proportion to their rigidities considering the rigidity of the diaphragm.

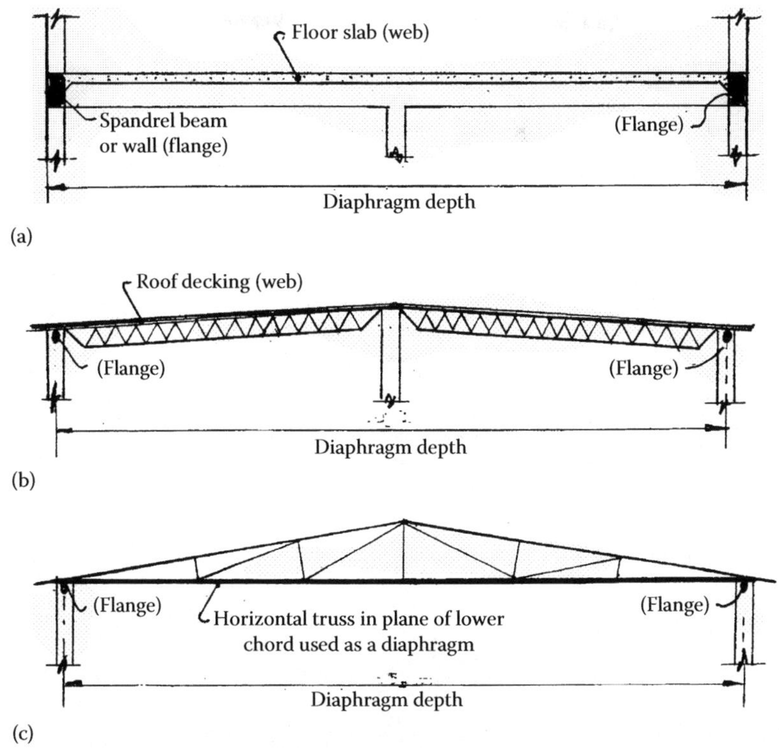

FIGURE 3.25 Various types of diaphragm: (a) floor slab diaphragms, (b) roof deck diaphragm, and (c) truss diaphragm. *Note*: A diaphragm may be considered analogous to a plate girder laid in a horizontal plane or inclined in the roof plane where the floor or roof deck performs the function of the plate girder web, the joints or beams function as web stiffeners, and the peripheral beams or integral reinforcement functions as flanges.

3.15.3 Optimizing Structural Configuration

For near-optimum seismic performance, the following characteristics are desirable:

- Continuous load path

  Uniform loading of structural elements and no stress concentrations
- Equal floor heights

  Equalizes column or wall stiffness, no stress concentrations
- Symmetrical plan shape

  Minimizes torsion
- Identical resistance on both axes

  Eliminates eccentricity between the centers of mass and resistance and provides balanced resistance in all directions, thus minimizing torsion
- Identical vertical resistance

  No concentrations of strength or weakness
- Uniform section and elevations

  Minimizes stress concentrations
- Seismic-force-resisting elements at perimeter

  Maximum torsional resistance

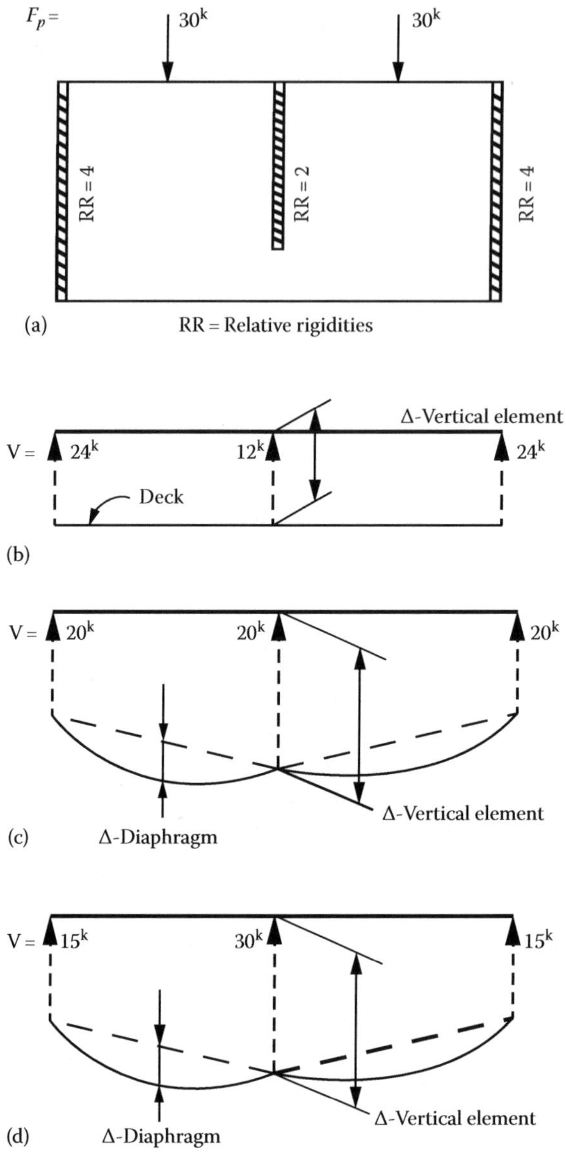

FIGURE 3.26 Relative effects of diaphragm stiffness. (a) Schematic plan, (b) rigid diaphragm, (c) semi-rigid diaphragm, and (d) flexible diaphragm.

- Short spans

  Multiple columns provide redundancy. Loads can be redistributed if some columns are lost.
- No cantilevers

  Reduced vulnerability to vertical accelerations
- No openings in diaphragms (floors and roofs)

  Ensures direct transfer of lateral forces to the lateral force resisting system

In seismic terms, engineers refer to a building that has the desirable characteristics as a regular building. As the building characteristics deviate from this model, the building becomes increasingly irregular. It is these irregularities that affect the building's seismic performance.

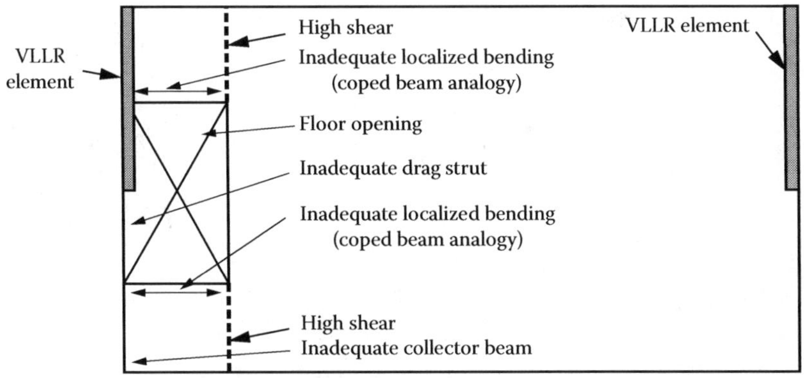

FIGURE 3.27 Diaphragm with limited connection to VLLRS.

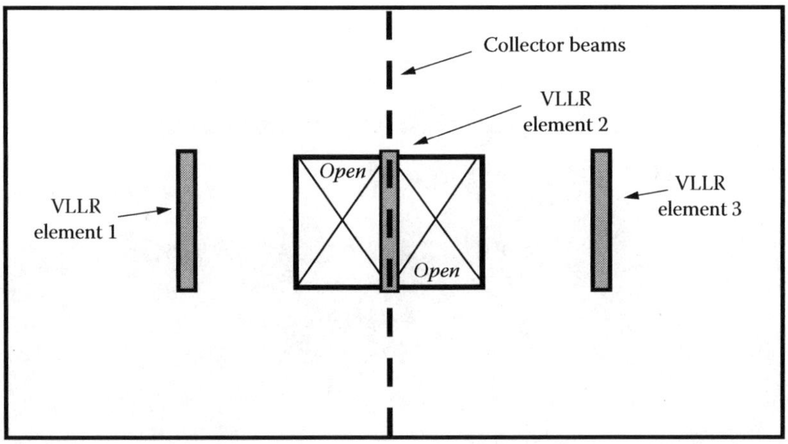

FIGURE 3.28 Isolate VLLR element with no direct connection to diaphragm: *Note*: Entire force for VLLR 2 must come through the collector beam. Beam and connections must be designed for these forces.

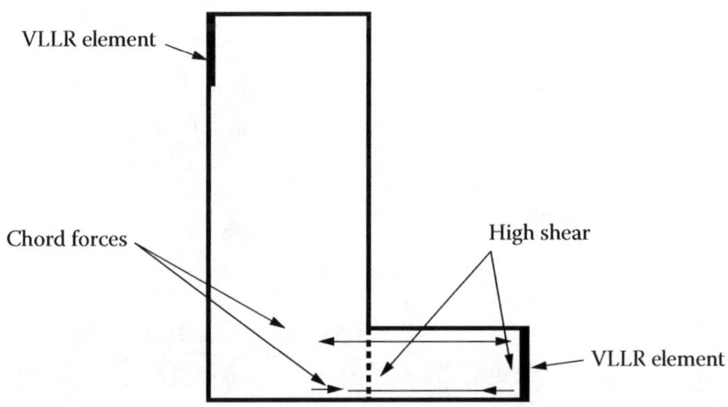

FIGURE 3.29 Narrow diaphragm with a VLLR element results in high shears.

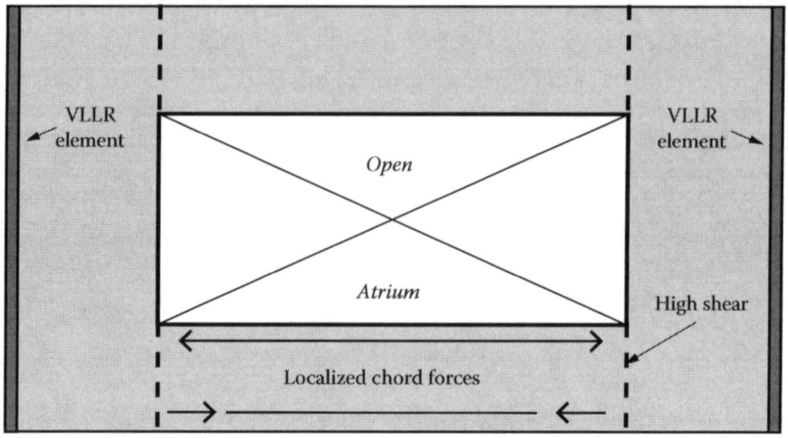

FIGURE 3.30 Large opening in diaphragm results in high shear and/or bending stress in diaphragm.

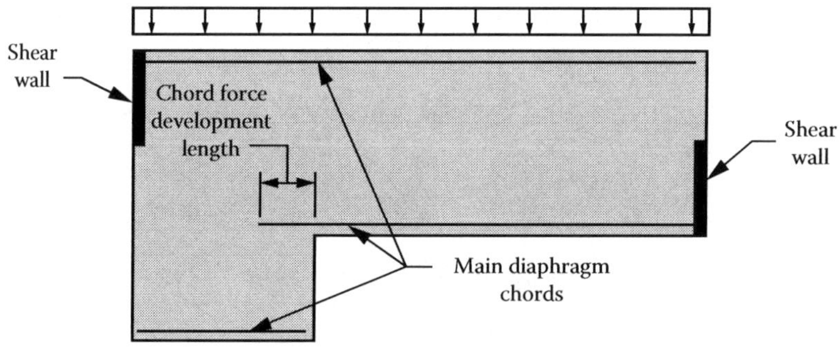

FIGURE 3.31 Diaphragm with a reentrant corner.

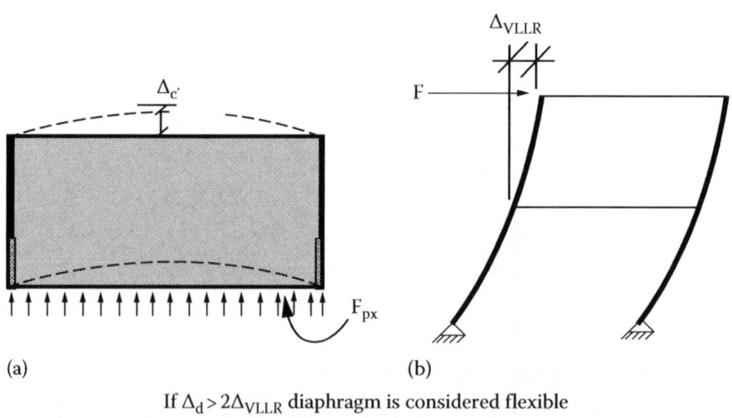

FIGURE 3.32 Criteria for diaphragm classification. (a) Plan and (b) elevation.

3.15.4 Effects of Configuration Irregularity

Configuration irregularity is largely responsible for two undesirable conditions: stress concentrations and torsion. These conditions often occur concurrently.

3.15.4.1 Stress Concentrations

Irregularities tend to create abrupt changes in strength or stiffness that may concentrate forces in an undesirable way resulting in stress concentration.

Stress concentration occurs when large forces are concentrated at one or a few elements of the building, such as a particular set of beams, columns, or walls. These few members may fail and, by a chain reaction, damage or even bring down the whole building.

Stress concentrations can be created by both horizontal and vertical stiffness irregularities. The short-column phenomenon discussed elsewhere in this book is an example of stress concentration created by vertical dimensional irregularity in the building design. In plan, a configuration that is most likely to produce stress concentrations is the *reentrant corner* condition as exemplified in buildings with plan forms such as an *L* or a *T*. Other reentrant corner configurations are shown in Figure 3.43.

The vertical irregularity of the *soft- or weak-story* types can produce dangerous stress concentrations along the plane of discontinuity. Soft and weak stories are discussed in more detail elsewhere in this book.

3.15.4.2 Torsion

Configuration irregularities in plan may cause torsional forces to develop, which contribute a significant component of uncertainty to an analysis of building resistance, and are perhaps the most frequent cause of structural failure. Torsional forces are created in a building eccentricity between the center of mass and the center of resistance. This eccentricity originates either in the lack of symmetry in the arrangement of the perimeter-resistant elements or in the plan configuration of the building, as in the reentrant corner forms discussed earlier.

Torsional moment is generated whenever the cg of the lateral forces fails to coincide with the center of rigidity (cr) of the vertical resisting elements, provided that the diaphragm is sufficiently rigid to transfer torsion. The magnitude of the torsional moment that is required to be distributed to the vertical resisting elements by a diaphragm is determined by the sum of the moments created by the physical eccentricity of the translational forces at the level of the diaphragm from the cr of the resisting elements ($M_T = F_p e$, where e = distance between cg and cr) and the *accidental* torsion of 5%. The *accidental* torsion is an arbitrary code requirement equivalent to the story shear acting with an eccentricity of not less than 5% of the maximum building dimension at that level. The torsional distribution by the more rigid diaphragms to the resisting elements is assumed to be in proportion to the stiffness of the elements and its distance from the cr. Negative torsional shears are neglected. Flexible diaphragms are not used for torsional distribution. Cantilever diaphragms on the other hand will distribute translational forces to vertical resisting elements, even if the diaphragm is flexible. In this case, the diaphragm and its chord act as a flexural beam on supports (vertical resisting elements) whose resistance is in the same direction as the forces.

Diaphragm deflections: A diaphragm is designed to provide such stiffness and strength so that walls and other vertical elements laterally supported by the diaphragm can safely sustain the stresses induced by the response to seismic motion. The total computed deflection (Δ_d) of diaphragms under seismic forces consists of the sum of two components. The first component is the flexural deflection (Δ_f) of the diaphragm, which is determined in the same manner as the deflection of beams. The assumption that flexural stresses on the diaphragm web are neglected is used except for reinforced concrete slabs. For such slabs, the proportional flexural stresses also may be assumed to be carried by the web. The second component is the web deflection (Δ_w) of the diaphragm. The specific nature of the web deflection will vary depending on the type of diaphragm. The deflection of the diaphragm under seismic forces is used as the criteria for the adequacy of the stiffness of a diaphragm.

3.15.5 Configuration Irregularities in Seismic Standards

Many of the configuration conditions that present seismic problems were identified by observers early in the twentieth century. However, the configuration problem was first defined for code purposes in the 1975 *Commentary to the Structural Engineers Association of California (SEAOC) Recommended Lateral Force Requirements* (commonly called the SEAOC Blue Book). In this commentary, over 20 specific types of *irregular structures or framing systems* were noted as examples of designs that should involve further analysis including dynamic consideration, rather than the use of the simple equivalent static force method in unmodified form. These irregularities vary in importance in their effect, and their influence also varies in degree, depending on which particular irregularity is present. Thus, while in an extreme form, the reentrant corner is a serious plan irregularity; in a lesser form, it may have little or no significance. The determination of the point at which a given irregularity becomes serious was left up to the judgment of the engineer.

Because of the belief that this approach was ineffective, in the 1988 codes, a list of six horizontal (plan) and six vertical (section and elevation) irregularities was provided that, with minor changes, is still in today's codes. This list also stipulated dimensional or other characteristics that established whether the irregularity was serious enough to require regulation and also provided the provisions that must be met in order to meet the code.

The seismic provisions of ASCE 7-10 provide descriptions of 10 irregularities—5 for horizontal and 5 for vertical irregularities as shown in Figures 3.33 and 3.34. Observe that the ASCE 7-10 provides only descriptions of these conditions; the diagrams are added in this text to illustrate each condition by showing how it would modify our optimized configuration and to also illustrate the failure pattern that is created by the irregularity.

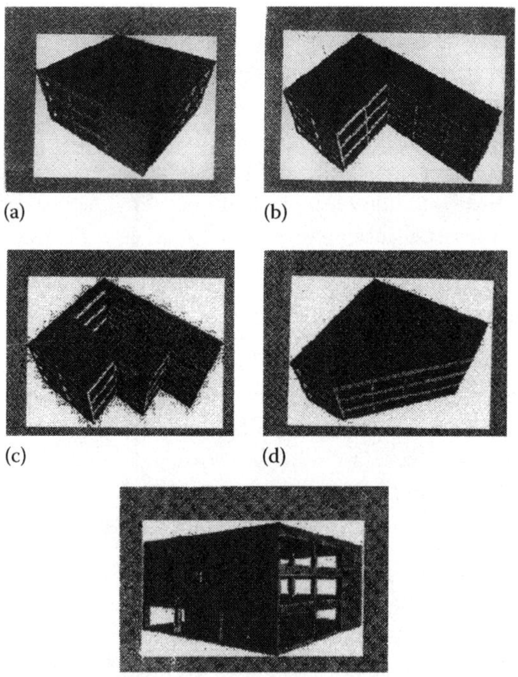

FIGURE 3.33 Plan irregularities. (a) Torsional irregularity may cause localized damage and collapse in extreme cases. (b) Reentrant corners may cause load damage to diaphragm and attached elements. Collapse may occur in extreme conditions. (c) Diaphragm eccentricity and large cutouts may cause localized damage. (d) Nonparallel lateral-load-resisting system may lead to torsional instability and localized damage. (e) Out-of-plane offsets of VLLRS may cause collapse mechanism in severe offsets.

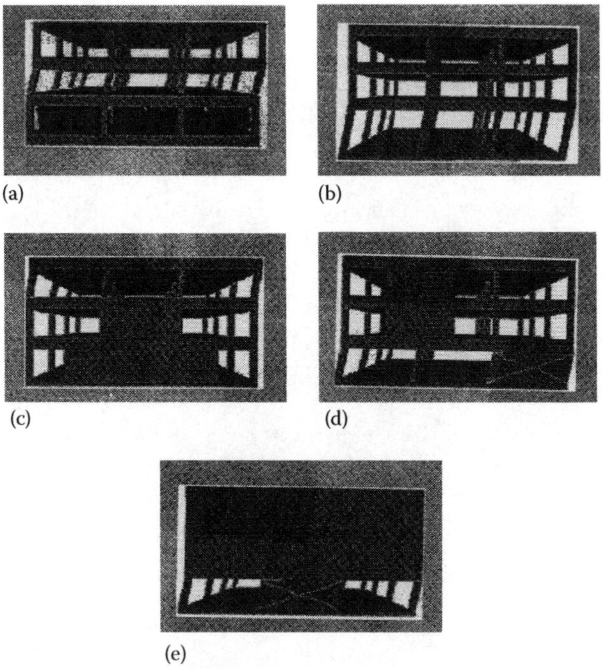

FIGURE 3.34 Vertical irregularities. (a) Stiffness irregularity, soft story (common collapse mechanism, deaths, and much damage in 1994 Northridge earthquake). (b) Weight/mass irregularity (collapse mechanism in extreme instances). (c) Vertical geometric irregularity (localized structure failure). (d) In-plane irregularity in vertical LFRS (localized structural failure). (e) Capacity discontinuity, weak story (collapse mechanism).

For the most part, code provisions seek to discourage irregularity in design by imposing penalties, which are of three types:

- Requiring increased design forces
- Requiring a more advanced analysis procedure
- Disallowing extreme soft stories and extreme torsional imbalance in high seismic design categories

It should be noted that the code provisions treat the symptom of irregularity rather than the cause. The irregularity is still allowed to exist; the hope is that the penalties will be sufficient to cause the designers to eliminate the irregularities. Increasing the design forces or improving the analysis to provide better information does not, in itself, solve the problem. The problem must be solved by design.

The code-defined irregularities serve as a checklist for ascertaining the possibility of configuration problems. Four of the more serious configuration conditions are described in more detail in the following sections along with some conceptual suggestions for their solution.

3.15.6 Four Serious Configuration Conditions

Four configuration conditions (two vertical and two in plan) that have the potential to seriously impact seismic performance are

1. Soft and weak stories
2. Discontinuous shear wall
3. Variations in perimeter strength and stiffness
4. Reentrant corners

3.15.6.1 Soft and Weak Stories

The term weak story has commonly been applied to buildings whose ground-level story is less stiff than those above. The building code distinguishes between *soft* and *weak* stories. Soft stories are less stiff, or more flexible, than the story above; weak stories have less strength. A soft or weak story at any height creates a problem, but since the cumulative loads are greatest toward the base of the building, a discontinuity between the first and second floor tends to result in the most serious condition.

The way in which severe stress concentration is caused at the top of the first floor is shown in the diagram sequence in Figure 3.35. Normal drift under earthquake forces that is distributed equally among the upper floors is shown in Figure 3.35a. With a soft story, almost all the drift occurs in the first floor, and stress concentrates at the second-floor connections. This concentration overstresses the joints along the second-floor line, leading to distortion or collapse (see Figure 3.35c).

Three typical conditions create a soft first story (Figure 3.36). The first condition (Figure 3.36a) is where the vertical structure between the first and second floor is significantly more flexible than that of the upper floors. (The seismic code provides numerical values to evaluate whether a soft-story condition exists.) This discontinuity most commonly occurs in a frame structure in which the first-floor height is significantly taller than those above, resulting in a large discrepancy in stiffness.

The second form of soft story (Figure 3.36b) is created by a common design concept in which some of the vertical framing elements do not continue to the foundation but rather are terminated at the second floor to increase the openness at ground level. This condition creates a discontinuous load path that results in an abrupt change in stiffness and strength at the plane of change.

Finally, the soft story may be created by an open first floor that supports heavy structural or non-structural walls above (Figure 3.36c). This situation is most serious when the walls above are shear walls acting as major lateral-force-resisting elements.

The best solution to the soft- and weak-story problem is to avoid the discontinuity through architectural design. There may, however, be good programmatic reasons why the first floor should be

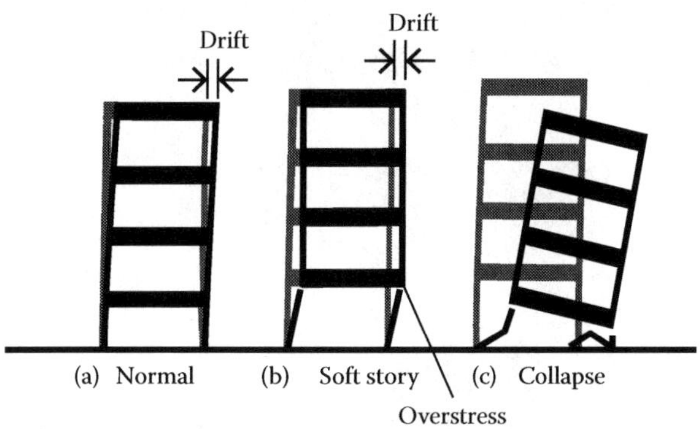

FIGURE 3.35 Soft and weak stories: (a) building drift without soft story, (b) drift in soft story, and (c) collapse due to soft story.

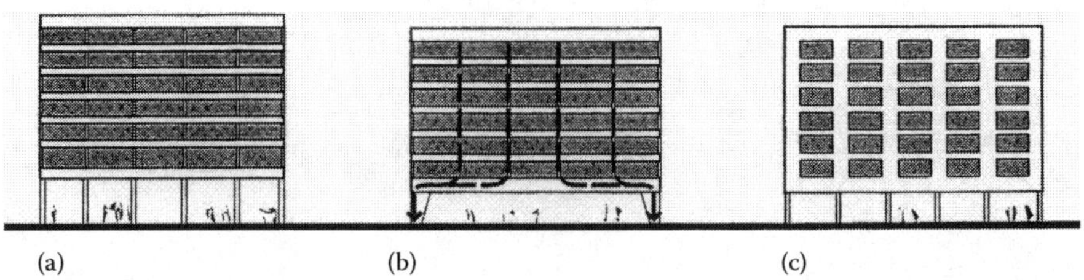

FIGURE 3.36 Three types of soft first story: (a) flexible first floor, (b) discontinuity—indirect load path—and (c) heavy superstructure.

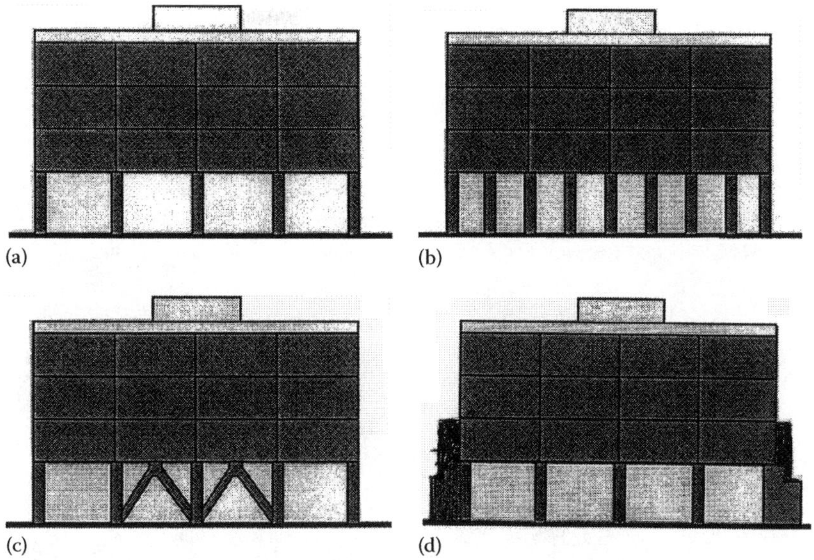

FIGURE 3.37 Design solutions for soft-story condition: (a) soft-story condition, (b) add columns, (c) add bracing, and (d) add external buttresses.

more open or higher than the upper floors. In these cases, careful design must be employed to reduce the discontinuity. Some conceptual methods for doing this are shown in Figure 3.37.

Not all buildings that show slender columns and high first floors are soft stories. For a soft story to exist, the flexible columns must be the main lateral-force-resistant system.

3.15.6.2 Discontinuous Shear Walls

When shear walls form the main lateral-force-resistant elements of a structure, and there is no continuous load path through the walls from roof to foundation, the result can be serious, overstressing at the points of discontinuity. This discontinuous shear wall condition represents a special, but common, case of the *soft*-first-story problem.

The discontinuous shear wall is a fundamental design contradiction: the purpose of the shear wall is to collect diaphragm loads at each floor and transmit them as directly and efficiently as possible to the foundation. To interrupt this load path is undesirable; to interrupt it at its base, where the shear forces are greatest, is a major error. Thus, the discontinuous shear wall that terminates at the second floor represents perhaps a *worst case* of the soft-first-floor condition. A discontinuity in vertical stiffness and strength leads to a concentration of stresses, and the story that must hold up all the rest of the stories in a building should be the last, rather than the first, element to be sacrificed.

Olive View hospital, which was severely damaged in the 1971 San Fernando, California, earthquake, represents an extreme form of the discontinuous shear wall problem. The general vertical configuration of the main building was a *soft*-two-story layer of rigid frames on which was supported by a four-story (five, counting penthouse), stiff, shear-wall-plus-frame structure (Figures 3.38 through 3.40). The second

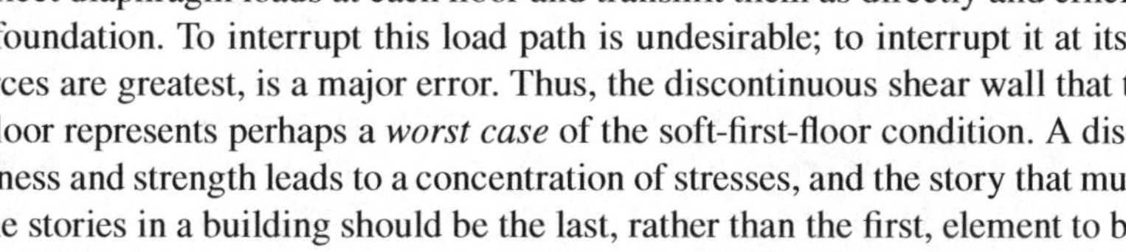

FIGURE 3.38 Olive View Hospital, cross section. *Note*: The shear walls stop at first floor. (From FEMA-454, *Designing for Earthquakes: A Manual for Architects*. Washington, DC: Federal Emergency Management Agency, 2006.)

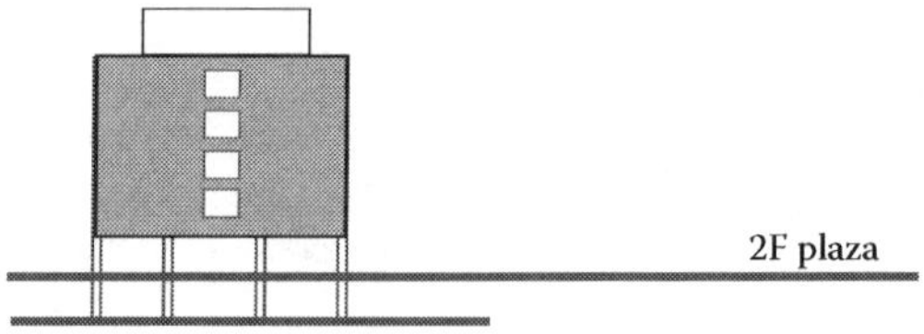

FIGURE 3.39 Olive View Hospital, second-floor plaza and the discontinuous shear wall. (From FEMA-454, *Designing for Earthquakes: A Manual for Architects*. Washington, DC: Federal Emergency Management Agency, 2006.)

FIGURE 3.40 Olive View Hospital, San Fernando earthquake, 1971. Note the extreme deformation of the columns above the plaza level. (From FEMA-454, *Designing for Earthquakes: A Manual for Architects*. Washington, DC: Federal Emergency Management Agency, 2006.)

floor extends out to form a large plaza. Sever damage occurred in the soft-story portion. The upper stories moved as a unit and moved so much that the column at ground level could not accommodate such a high displacement between their bases and tops and hence failed. The largest amount by which a column was left permanently out of plumb was 2 ft 6 in. (Figure 3.40). The building did not collapse, but two occupants in intensive care and maintenance person working outside the building were killed.

The solution to the problem of the discontinuous shear walls is unequivocally to eliminate the condition. To do this may create architectural problems of planning or circulation or image. If this is so, it indicates that the decision to use shear walls as resistant elements was wrong from the inception of the design. If the decision is made to use shear walls, then their presence must be recognized from the beginning of schematic design, and their size and location made the subject of careful architectural and engineering coordination early.

3.15.6.3 Variations in Perimeter Strength and Stiffness

This problem may occur in buildings whose configuration is geometrically regular and symmetrical, but nonetheless irregular for seismic design purposes.

A building's seismic behavior is strongly influenced by the nature of the perimeter design. If there is wide variation in strength and stiffness around the perimeter, the center of mass will not coincide with the center of resistance, and torsional forces will tend to cause the building to rotate around the center of resistance (see Figure 3.41).

A common instance of an unbalanced perimeter is that of open-front design in buildings, such as fire stations and motor maintenance shops in which it is necessary to provide large doors for the

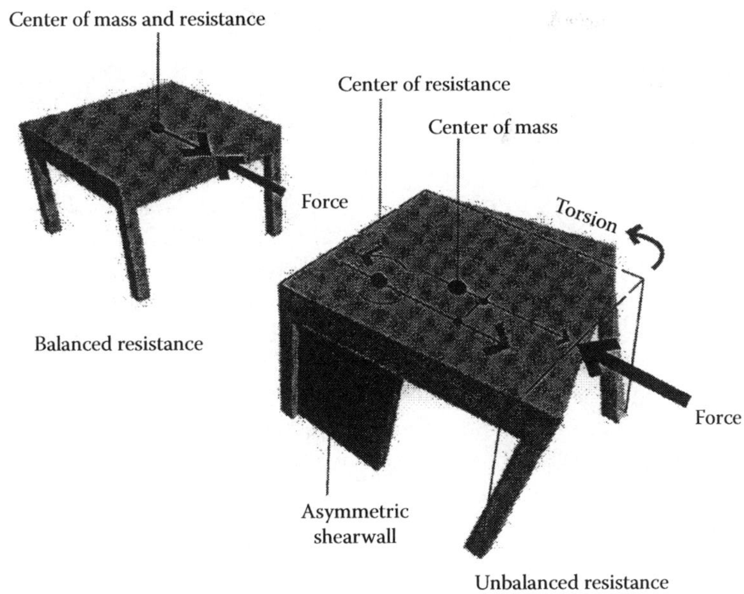

FIGURE 3.41 Torsion due to wide variation in strength and stiffness around the building perimeter.

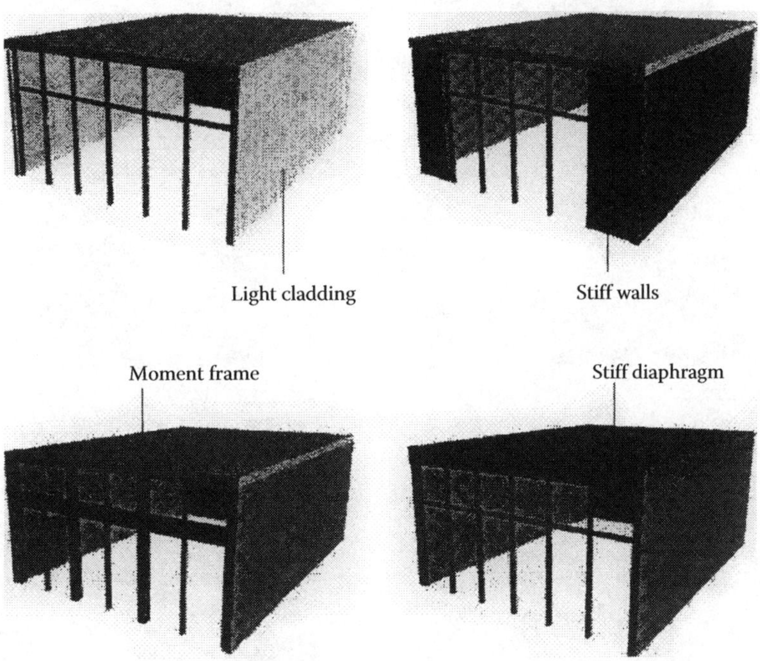

FIGURE 3.42 Some solutions to store-front-type unbalanced-perimeter-resistance conditions.

passage of vehicles. Stories, individually or as a group in a shopping mall, are often designed as boxes with three solid sides and an open glazed front.

The solution to this problem is to reduce the possibility of torsion by balancing the resistance around the perimeter. The example shown is that of the store front. A number of alternative design strategies can be employed that could also be used for the other building type conditions noted (Figure 3.42).

The strategy is to design a frame structure of approximately equal strength and stiffness for the entire perimeter. The opaque portion of the perimeter can be constructed of nonstructural cladding, designed so that it does not affect the seismic performance of the frame. This can be done either by using lightweight cladding or by ensuring that heavy materials, such as concrete or masonry, are isolated from the frame.

A second approach would be to increase the stiffness of the open facades by adding sufficient shear walls, at or near the open face, designed to approach the resistance provided by the other walls. A third solution would be to use a strong moment-resisting or braced frame at the open front, which approaches the solid wall in stiffness. The ability to do this will depend on the size of the facades. However, it should be noted that a long steel frame can never approach a long concrete wall in stiffness.

The possibility of torsion may be accepted, and the structure designed to have the capacity to resist it, through a combination of moment frames, braced frames, and shear walls.

3.15.6.4 Reentrant Corners

The reentrant corner is the common characteristic of building forms that, in plan, assume the shape of L, T, H, etc., or a combination of these shapes (Figures 3.43 and 3.44).

Invariably these forms result in torsion due to eccentricity of center of mass and center of resistance shown schematically in Figure 3.45.

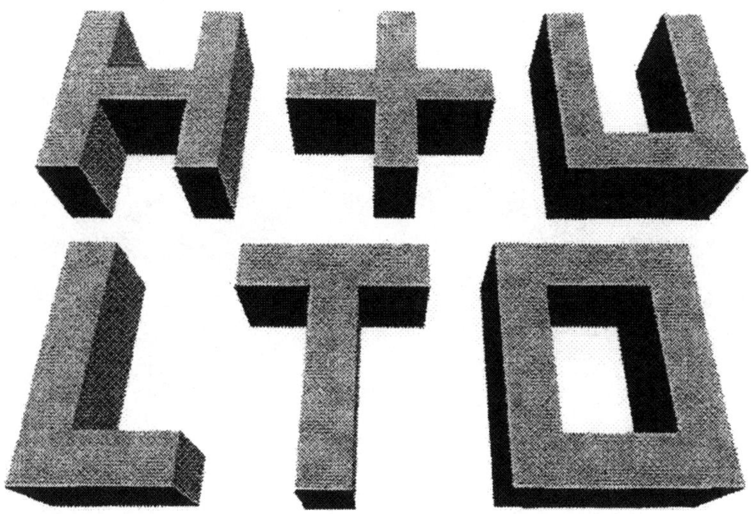

FIGURE 3.43 Reentrant corner plan forms.

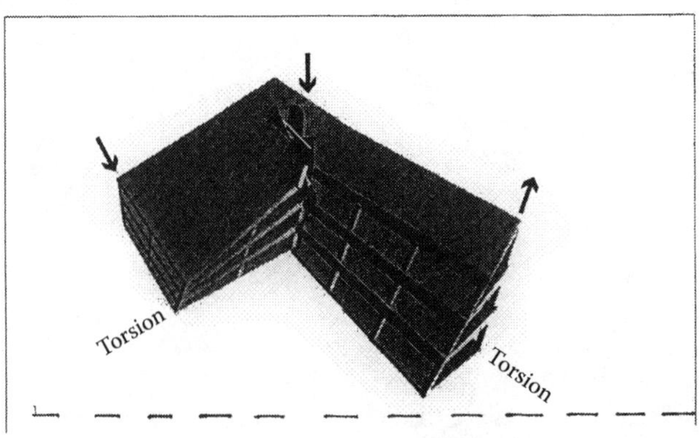

FIGURE 3.44 Stress concentrations at reentrant corner.

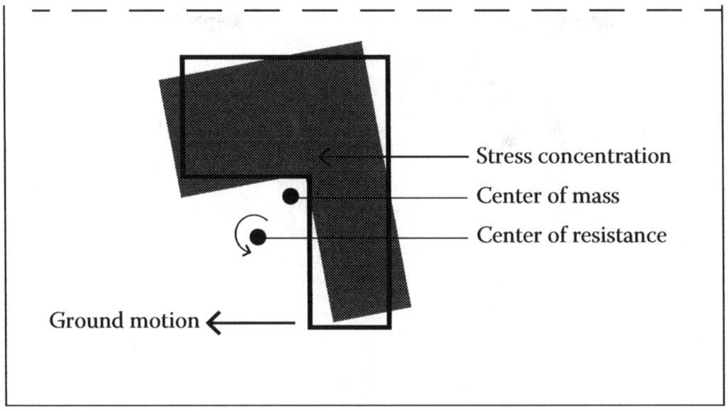

FIGURE 3.45 Torsion due to eccentricity of center of mass and center of resistance.

There are two problems created by these shapes. The first is that they tend to produce differential motions between different wings of the building that, because of stiff elements that tend to be located in this region, result in local stress concentrations at the reentrant corner or *notch*.

The second problem of this form is torsion caused because the center of mass and the cr cannot geometrically coincide for all possible earthquake directions. The result is rotation. The resulting forces are very difficult to analyze and predict. Observe that the stress concentration at the *notch* and the torsional effects are interrelated. The magnitude of the forces and the severity of the problems will depend on

- The characteristics of the ground motion
- The mass of the building
- The type of structural systems
- The length of the wings and their aspect ratios (length to width proportion)
- The height of the wings and their height/depth ratios

Reentrant corner plan forms are, however, the most useful set of building shapes for urban sites, particularly for residential apartments and hotels. This is because large plan areas may be accommodated in relatively compact form yet still provides a high percentage of perimeter rooms with access to air and light.

There are two basic alternative approaches to the problem of reentrant corner forms: structurally to separate the building into simpler shapes or to tie the building together more strongly with elements positioned to provide a more balanced resistance (Figure 3.46).

Once the decision is made to use separation joints, the structurally separated entities of a building must be fully capable of resisting vertical and lateral forces on their own, and their individual configuration must be balanced horizontally and vertically.

To design a separation joint, the maximum drift of the two units must be calculated. The worst case is when the two individual structures would lean toward each other simultaneously. Therefore, the dimension of the separation space must allow for the statistical sum of the building deflections.

Several considerations arise if it is decided to dispense with the separation joint and tie the building together. Collectors at the intersection can transfer forces across the intersection area, but only if the design allows for these beam-like members to extend straight across without interruption. Since the portion of the wing that typically distorts the most is the free end, it is desirable to place stiffening elements at that location.

The use of splayed rather than right-angle reentrant corners lessens the stress concentration at the notch (Figure 3.47). This is analogous to the way a rounded hole in a steel plate creates less stress concentration than a rectangular hole or the way a tapered beam is structurally more desirable than an abruptly notched one.

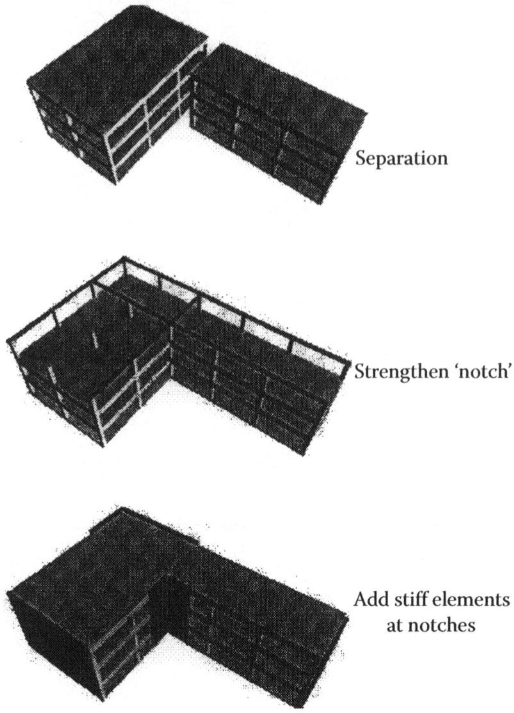

FIGURE 3.46 Solutions for the reentrant corner condition.

FIGURE 3.47 Relieving the stress on a reentrant corner by using a splay.

3.15.7 Other Seismic Issues

3.15.7.1 *P*-Delta Effect

When flexible structures are subjected to lateral forces, the resulting horizontal displacements lead to additional overturning moments because the gravity load is also displaced. Thus, in the simple cantilever model of Figure 3.48a, the total base moment is

$$M_{ub} = V_u H + P_u \Delta \tag{3.10}$$

Therefore, in addition to the overturning moments produced by lateral force, V_u, the secondary moment $P_u\Delta$ must also be resisted. This moment increment in turn will produce additional lateral displacement, and hence Δ will increase further. In very flexible structures, instability, resulting in collapse, may occur. However, analyses of typical building frames have indicated that *P*-delta effects are small when maximum interstory drift is less than 1%.

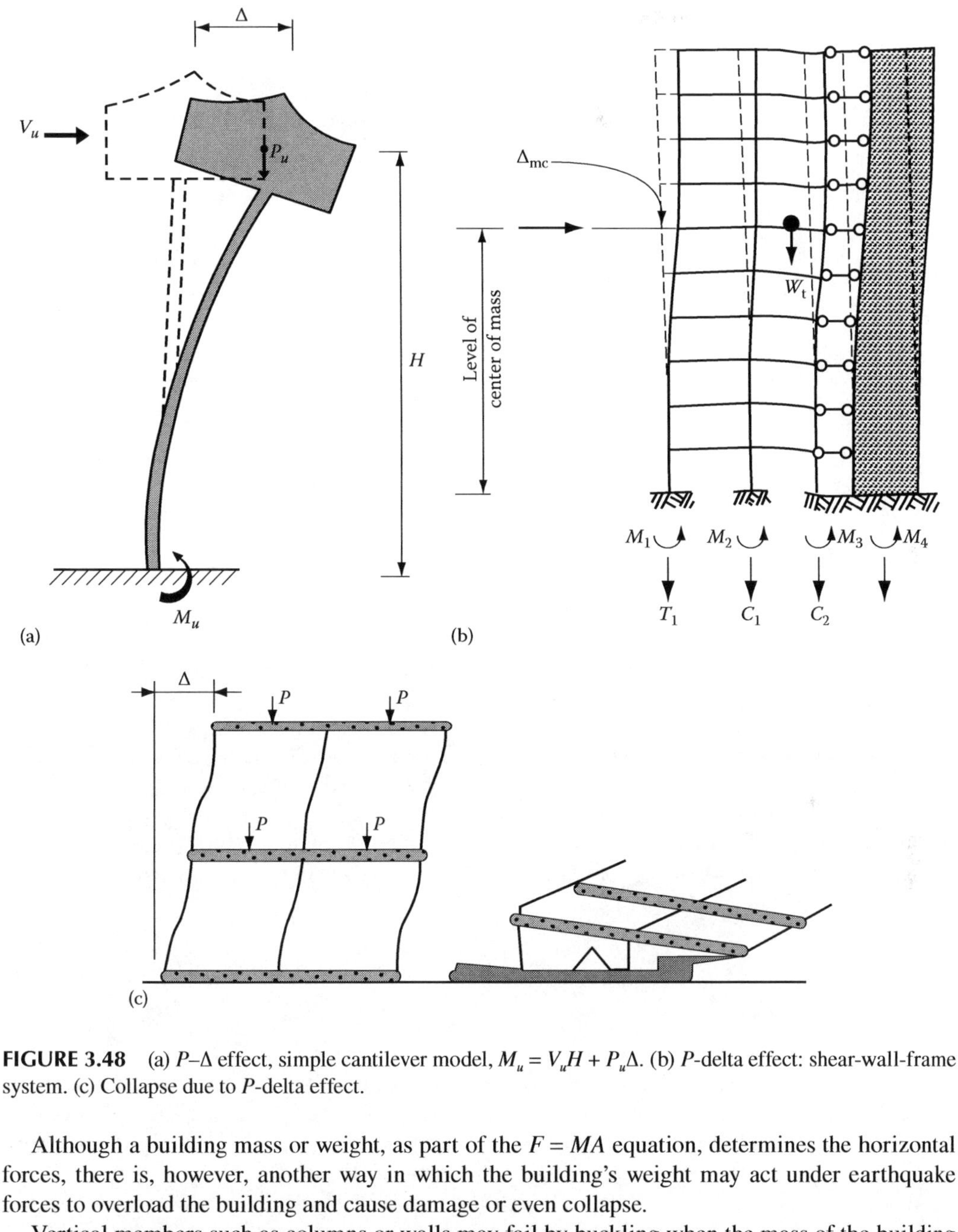

FIGURE 3.48 (a) P–Δ effect, simple cantilever model, $M_u = V_uH + P_u\Delta$. (b) P-delta effect: shear-wall-frame system. (c) Collapse due to P-delta effect.

Although a building mass or weight, as part of the $F = MA$ equation, determines the horizontal forces, there is, however, another way in which the building's weight may act under earthquake forces to overload the building and cause damage or even collapse.

Vertical members such as columns or walls may fail by buckling when the mass of the building exerts its gravity force on a member distorted or moved out of plumb by the lateral forces. This phenomenon as stated previously is the P-delta effect, where P is the gravity force or weight and *delta* is the eccentricity or the extent to which the force is offset (see Figure 3.48b and c for schematics of P-delta effects).

The geometrical proportions of the building also may have a great influence on whether the P-delta effect will pose a problem. A tall, slender building is much more likely to be subject to overturning forces than a low, squat one. It should be noted, however, that if the lateral resistance is provided by shear walls of braced frames, it is their proportions that are significant rather than those of the building as a whole.

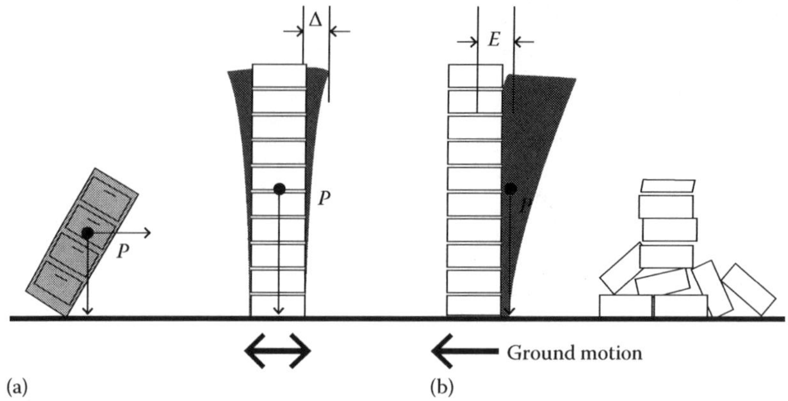

FIGURE 3.49 Why buildings generally fall down, not over: (a) filing cabinet and (b) building structure.

However, in earthquakes, buildings seldom overturn, because structures are not homogenous but are composed of many elements connected together. The earthquake forces will pull the components apart, and the building will fall down, not over. Strong homogeneous structures such as filing cabinets, however, will fall over (see Figure 3.49 for schematics).

3.15.7.2 Strong Beam, Weak Column

Structures are commonly designed so that under sever shaking, the beams will fail before the columns. This reduces the possibility of complete collapse. The short-column effect, discussed elsewhere in this text, is analogous to a weak-column, strong-beam condition, which is sometimes produced inadvertently when strong or stiff nonstructural members are inserted between columns.

3.15.7.3 Setbacks and Planes of Weakness

Vertical setbacks can introduce discontinuities, particularly if columns or walls are offset at the plane of the setback. A horizontal plane of weakness can be created by the placement of windows or other openings that may lead to failure.

3.15.8 Earthquake Collapse Patterns

We typically accept higher risks of damage under seismic design loads than under other comparable extreme loads, such as maximum live load or wind loads. The reason being that the seismic forces generated during severe ground motions are too high to be resisted within the elastic range of material response. Common practice therefore is to design for forces that are a fraction of those corresponding to elastic response. We then expect the structures to survive strong earthquakes by large inelastic deformations and energy dissipation characteristics.

Although earthquake shaking causes damage to a structure, it is the gravity load that causes collapse. Redundancy for the load path and ductile behavior of critical members can prevent or reduce the extent of collapse. On the other hand, brittle behavior enhances possibility and increases extent of collapse.

With increased awareness that excessive strength is not essential or even necessarily desirable, the emphasis in seismic design has shifted from the resistance of large seismic forces to the *evasion* of these forces. Inelastic structural response has become an essential reality in structural design for earthquake forces. Inelastic deformations that provide ductility are considered essential for preventing building collapse while the structure is subjected to back-and-forth motions during severe ground shaking.

Seismic design encourages structural forms that are more likely to exhibit ductility than those that do not. Thus, for concrete structures, the shear strength provided in a member must exceed its actual flexural strength.

As stated earlier, one of the most common causes of building failures during earthquakes is the *soft-story mechanism* that may develop when one level, typically the lowest, is weaker than upper levels. This condition results from a functional desire to open up lower levels for circulation. When subjected to strong ground motions, the columns develop high compression strains due to the combined effects of axial force and bending moment. Unless adequate, closely spaced, well-detailed transverse reinforcement is placed in the potential plastic hinge region, spalling of concrete followed by instability of the compression reinforcement will follow. It must be recognized that even with a weak beam/strong column design in which seismic energy dissipation is primarily in well-confined beam plastic hinges, a column plastic hinge may still form at the base of the column, resulting in partial or total collapse.

While there is something new to be learned from each earthquake, it may be said that the majority of structural lessons have been learned. However, there is still plenty to learn about the unpredictable and unquantifiable effects of earthquakes.

Well-established techniques, used for design of structures for various static loads, including wind forces, cannot simply be extended and applied to conditions that arise during earthquakes. In earthquake design, it is imperative that we consider forces corresponding to the largest seismic displacement.

3.15.8.1 Unintended Addition of Stiffness

A source of major damage, particularly in columns, repeatedly observed in earthquakes, is the interference with the deformations of members by rigid nonstructural elements, such as infill walls. As Figure 3.50a and b shows, the top edge of a brick wall will reduce the effective length of one of the columns, thereby increasing its lateral stiffness. Since seismic forces are proportional to the stiffness, the braced column will attract larger horizontal shear forces than it would otherwise. The failure of such gravity-load-carrying members may lead to collapse of the entire building. Therefore, it is important to ensure that the inelastic column deformations can take place without interference from nonstructural construction.

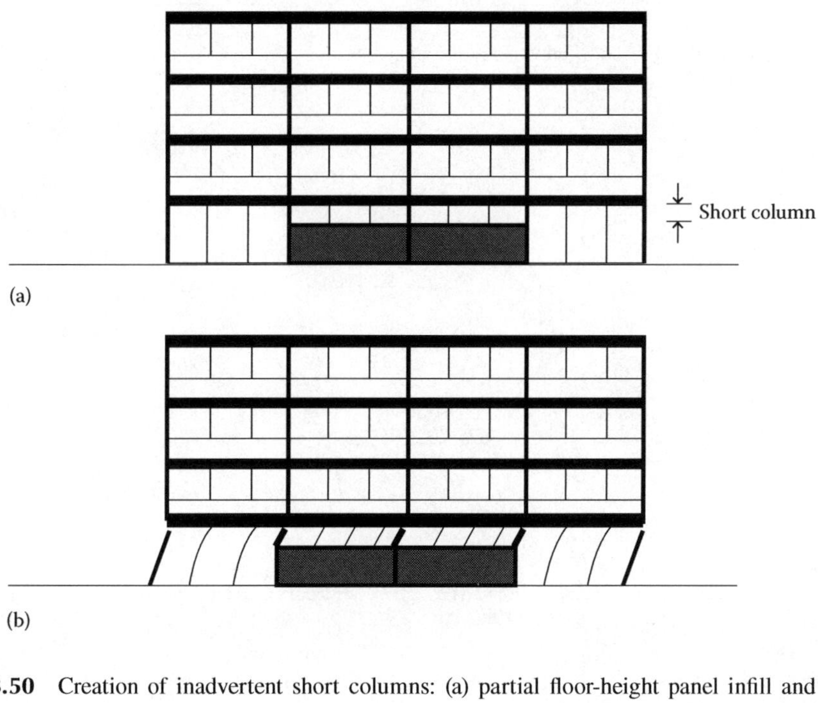

FIGURE 3.50 Creation of inadvertent short columns: (a) partial floor-height panel infill and (b) failure pattern.

3.15.8.2 Inadequate Beam–Column Joint Strength

Earthquake damage or even failures can occur in buildings that have poorly detailed joints. The back-and-forth cycling of the structure during severe ground shaking may cause moment-resistant joints to unravel. The gravity load can no longer be supported by these columns. Consequently, the structure is driven earthward until it stops on the ground or lower floors that have sufficient strength to stop the falling mass as shown schematically in Figure 3.51. The resulting collapse may be a pancaked group of slabs held apart by broken columns and building contents or a condition where columns are left standing, punched through the slabs.

3.15.8.3 Tension/Compression Failures

These types of failures usually occur in taller structures (see Figure 3.52a, b, and c). The tension that is concentrated at the edges of a concrete frame or shear wall can produce very rapid loss of stability. In walls, if the reinforcing steel is inadequately proportioned or poorly embedded, it can fail in tension and result in rapid collapse of the wall by overturning. A more common condition occurs, when the tension causes the joints in a concrete moment frame to lose bending and shear strength. A rapid degradation of the structure can result in a partial or complete pancaking as in a beam/column failure.

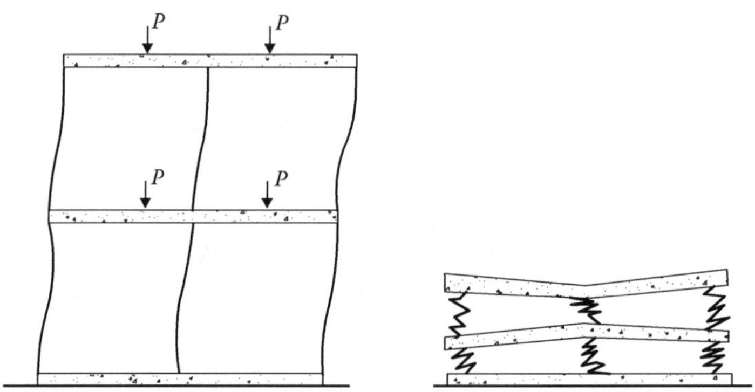

FIGURE 3.51 Breaking up of poorly detailed joints may result in failure of columns.

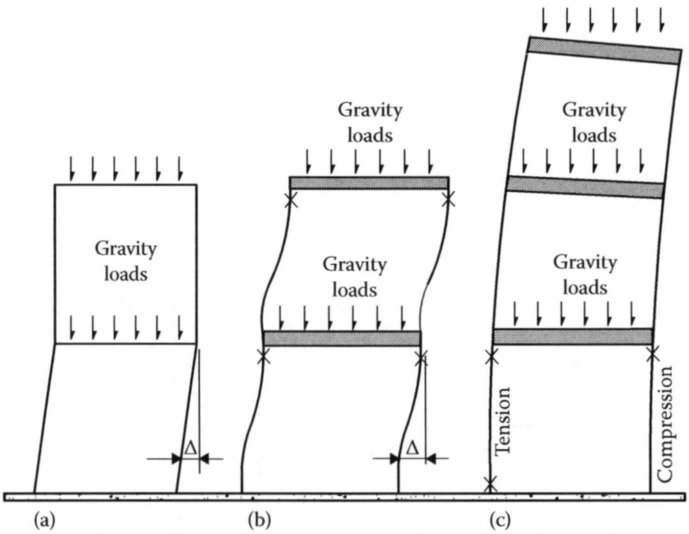

FIGURE 3.52 Collapse patterns: (a) inadequate shear strength, (b) inadequate beam/column strength, and (c) tension compression failure due to overturning.

3.15.8.4 Wall-to-Roof Connection Failure

In this case, stability is lost for both the roof system and the wall. The vertical support of the roof is lost, as well as the horizontal out-of-plane support of the wall.

3.15.8.5 Local Column Failure

This can lead to loss of stability and/or progressive collapse in part of a structure as shown schematically in Figure 3.53. Observe that in most collapses, the driving force is the gravity load acting on a structure that has become unstable due to horizontal offset or insufficient vertical capacity. In addition, subsequent lateral loads from aftershocks can increase the offset, exaggerating the instability. The structure is often disorderly as it collapses. Some parts may remain supported by uncollapsed adjacent bays.

3.15.8.6 Heavy Floor Collapse

Schematically shown in Figure 3.54, this type of collapse can be partial or complete. It is usually caused when columns or walls, weakened by earthquake motion, are unable to support the heavy floors. Tall, moment frame structures may overturn, but more often, they collapse within their plan boundaries due to high gravity forces. Many partially collapsed concrete frame structures will contain parts of slabs and/or walls that are hanging off an uncollapsed area. This has been observed in corner buildings when only the street-front bays collapse due to torsion effects.

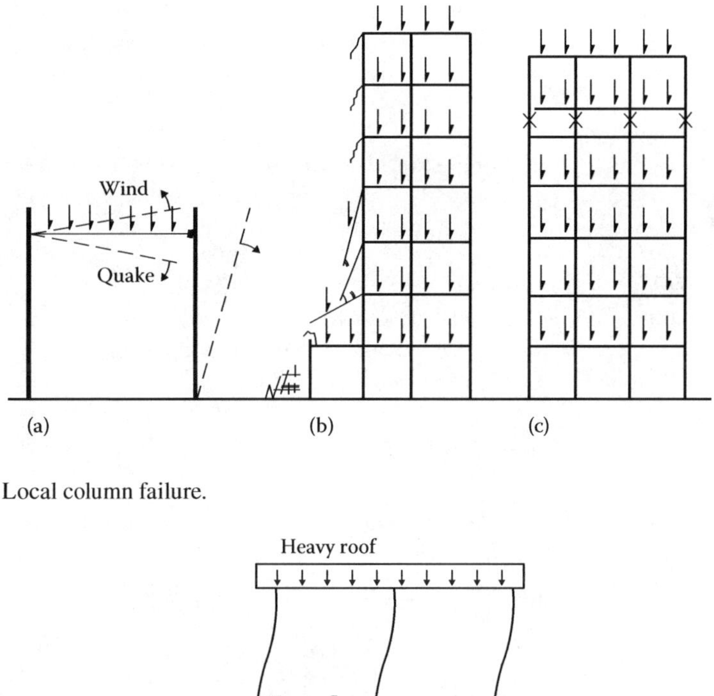

FIGURE 3.53 Local column failure.

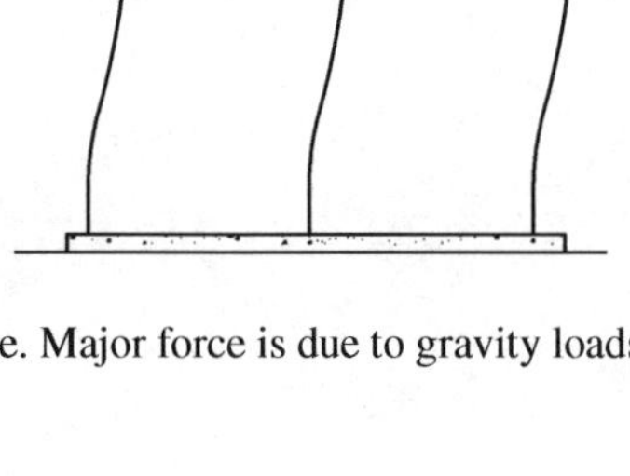

FIGURE 3.54 Heavy floor collapse. Major force is due to gravity loads due to inertial force at each floor. If column fails, heavy floors collapse.

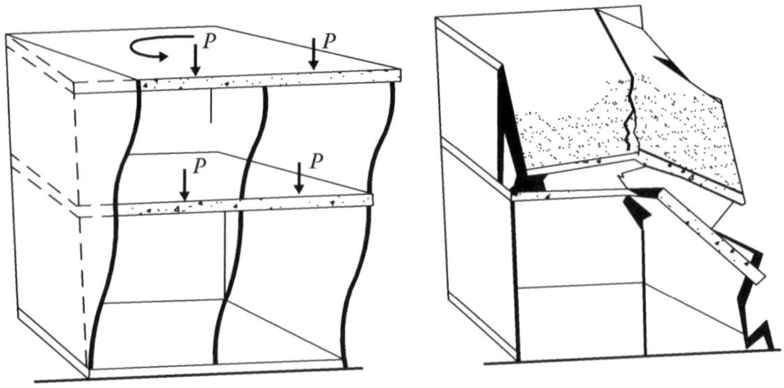

FIGURE 3.55 Torsion effect: unsymmetrical placement of walls can lead to collapse.

3.15.8.7 Torsion Effects

Torsion may occur in frame structures when an infill wall is placed in between columns. These infilled bays become stiffer than other parts of the building and cause an eccentric condition that can lead to collapse (see Figure 3.55).

3.15.8.8 Soft-Story Collapse

This occurs in buildings that are configured such that they have significantly less stiffness because much fewer or no walls are provided in the first story than in the stories above (see Figure 3.56). The collapse is often limited to the one story only, as the building becomes one story shorter.

3.15.8.9 Midstory Collapse

This can occur when a midstory is configured with much different stiffness than the stories above and below. Examples are when a story has no walls and the ones above and below have significant walls, or when a story has stiff, short columns and the ones above and below have longer, more limber columns.

3.15.8.10 Pounding

Damage due to pounding normally occurs when two adjacent buildings have floors that are at different elevations. The very stiff/strong edge of a floor in one building may cause damage to or even collapse of the adjacent building's column when they collide.

FIGURE 3.56 Soft-story collapse: lower story that is weakened by too many openings becomes racked resulting in failure of first-story columns.

3.15.9 Conclusions

Earthquakes are catastrophic events that occur mostly at the boundaries of portions of the Earth's crust called tectonic plates. When movement occurs in these regions, along faults, waves are generated at the Earth's surface that can produce very destructive effects.

Aftershocks are smaller quakes that occur after all large earthquakes. They are usually most intense in size and number within the first week of the original quake. They can cause very significant reshaking of damaged structures, which makes earthquake-induced disasters more hazardous. A number of moderate quakes (6+ magnitude) have had aftershocks that were very similar in size to the original quake. Aftershocks diminish in intensity and number with time.

They generally follow a pattern of there being at least one large (within one Richter magnitude) aftershock, at least 10 lesser (within two Richter magnitude) aftershocks, 100 within 3, and so on. The Loma Prieta earthquake had many aftershocks, but the largest was only magnitude 5.0 with the original quake being magnitude 7.1.

Some of the most destructive effects caused by earthquake shaking are those that produce lateral loads in a structure. The intense ground shaking causes the foundation of a building to oscillate back and forth in a more or less horizontal plane. The building mass has inertia and wants to remain where it is, and therefore, lateral forces are exerted on the mass in order to bring it along with the foundation. For analysis purposes, this dynamic action is simplified as a group of horizontal forces that are applied to the structure in proportion to its mass and to the height of the mass above the ground. In multistory buildings with floors of equal weight, the loading is further simplified as a group of loads, each being applied at a floor line and each being greater than the one below, thus resulting in a triangular distribution. Seismically resistant structures are designed to resist these lateral forces through inelastic action and must, therefore, be detailed accordingly. Earthquake loads are often expressed in terms of a percent of gravity weight of the building and can vary from a few percent to near 50% of gravity weight. There are also vertical loads generated in a structure by earthquake shaking, but these forces rarely overload the vertical-load-resisting system. However, earthquake-induced vertical forces have caused damage to structures with high dead load compared to design live load. These vertical forces also increase the chance of collapse due to either increased or decreased compression forces in the columns. Increased compression may overload columns while decreased compression reduces column bending strength (Taranath, Book 6, Chapter 3).

In earthquake engineering, we deal with random variables, and therefore, the design must be treated differently from the orthodox design. The orthodox viewpoint maintains that the objective of design is to prevent failure; it idealizes variables as deterministic. This simple approach is still valid when applied to design under only mild uncertainty. But when confronted with the effects of earthquakes, this orthodox viewpoint is no longer applicable. This is because in dealing with earthquakes, we must contend with appreciable probabilities that failure will occur in the near future. Otherwise, all the wealth of this world would prove insufficient to design and construct structures that are *earthquake safe*—the most modest structures would be fortresses. We must also face considerable uncertainty while designing engineering systems—whose pertinent properties are still debated to resist future earthquakes—about whose characteristics we know even less.

Although over the years, experience and research have diminished our uncertainties and concerns regarding the characteristics of earthquake motions and manifestations, it is unlikely, though, that there will be such a change in the nature of knowledge to relieve us of the necessity of dealing openly with random variables. In a way, earthquake engineering is a parody of other branches of engineering. Earthquake effects on structures systematically bring out the mistakes made in design and construction, even the minutest mistakes. Add to this the undeniable dynamic nature of disturbances, the importance of soil–structure interaction, and the extremely random nature of it all; in a manner of speaking, earthquake engineering is to the rest of the engineering disciplines what psychiatry is to other branches of medicine. This aspect of earthquake engineering makes it at once challenging and fascinating and gives it an educational value beyond its immediate objectives. If structural engineers are to acquire fruitful experience (Taranath, Book 6, Chapter 3) in a brief span of time, they should be exposed to the concepts of earthquake engineering, even if their interest in earthquake-resistant design is indirect. Sooner or later, they will learn that the difficulties encountered in seismic design are technically intriguing and begin to exercise that nebulous trait called engineering judgment to make allowance for these unknown factors.

Regardless of building type, size, or function, it is clear that the attempt to encourage or enforce the use of regular configurations is frequently not going to succeed; the architect's search for original forms is very powerful.

The seismic standards such as those in ASCE 7-10 are oriented toward *everyday* economical building and go a modest route of imposing limited penalties on the use of irregular configurations in the form of increased design forces and, for larger buildings, the use of more advanced analytical methods; both these measures translate into cost penalties that may be acceptable by the building owners. Only two irregularities are banned outright: extreme soft stories and extreme torsion in essential buildings in high seismic zones (i.e., buildings assigned to Seismic Design Category [SDC] E or F). This suggests a strategy that exploits the benefits of the *ideal* configuration but permits the architect to use irregular forms when they suit the design intentions.

Extreme irregularities may require extreme engineering solutions; these may be costly, but it is likely that a building with these conditions will be unusual and important enough to justify additional costs in materials, finishes, and systems.

A soft or weak story should perhaps never be used: this does not mean that high stories or varied story heights cannot be used, but rather that appropriate structural measures be taken to ensure balanced resistance.

In looking at architectural design through a seismic *filter*, it appears that many useful and common architectural forms are in conflict with seismic design needs.

The ultimate solution to these conflicts depends on the architect and engineer working together on building design from the outset of the project and engaging in knowledgeable negotiation. It is unfair to expect the engineer to convince the architect of some of the conventional virtues of seismic design, such as simplicity, symmetry, and regularity.

Such discussions on building configurations are valid only for projects in which economy is the paramount objectives. When the architect and the client are looking for high-style design, the forms will probably be irregular, unsymmetrical, and fragmented. The successful engineer will enjoy the challenges. New methods of analysis will help, but engineers must also continue to develop their own innate feeling on how buildings perform and be able to visualize the complex interaction of building elements that result from many influences, both functional and aesthetic.

3.16 STRUCTURAL DYNAMIC

In one sense, the static loading condition may be considered merely as a special form of dynamic loading. It is convenient, however, to analytically distinguish between the static and the dynamic components of the applied loading, to evaluate the response to each type of loading separately, and then to superpose the two response components to obtain their total effect. When treated thusly, the static and dynamic methods of analysis become fundamentally different in character.

The term *dynamic* may be defined simply as time varying; thus, a dynamic load is any load of which its magnitude, direction, and/or position varies with time. Similarly, the structural response to a dynamic load, that is, the resulting stresses and deflections, is also time varying or dynamic.

Even though seismic ground motions are highly oscillatory and irregular in character, the resulting loads may be considered *as deterministic dynamic loads*, as opposed to *random dynamic loads*. Therefore, the emphasis in earthquake engineering here is on development of methods of deterministic dynamic analysis, rather than on random or nondeterministic analysis that requires statistical information about dynamic response quantities.

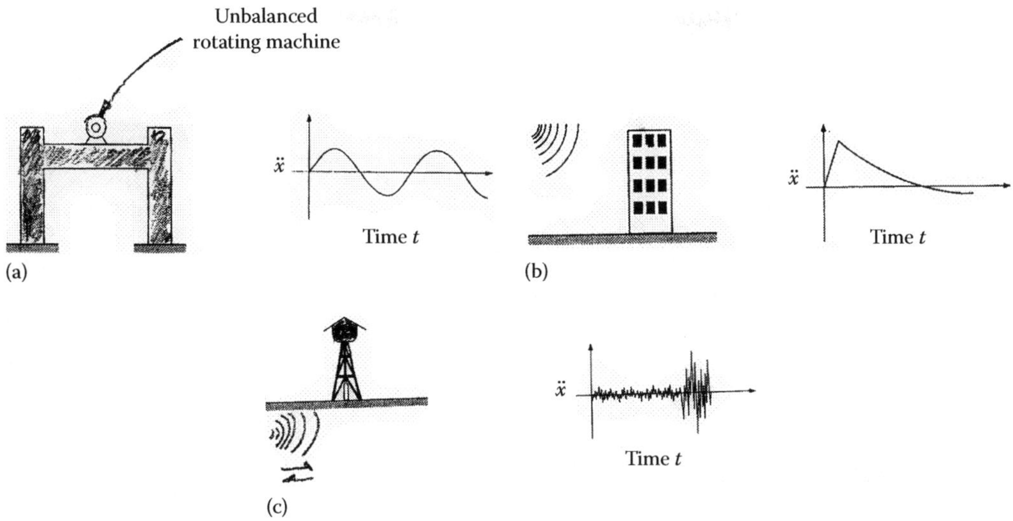

FIGURE 3.57 Characteristics of dynamic loading: (a) simple harmonic, (b) impulsive, and (c) long duration. *Note*: $\ddot{x}$ denotes acceleration.

In general, structural response to any dynamic loading is expressed basically in terms of the displacements of the structure. Thus, a deterministic analysis leads directly to displacement time histories corresponding to the prescribed loading history; other related response quantities, such as stresses, strains, and internal forces, are usually obtained as a secondary phase of the analysis. On the other hand, a nondeterministic analysis provides only statistical information about the displacements resulting from the statistically defined loading; corresponding information on the related response quantities are then generated using independent nondeterministic analysis procedures.

From an analytical point of view, it is convenient to divide deterministic dynamic loadings into two distinct categories, periodic and nonperiodic. A periodic loading exhibits the same time rotating machinery variation successively for a large number of cycles as typified by the unbalanced rotating machinery.

Nonperiodic loadings, on the other hand, may be either short-duration impulsive loading or relatively long-duration general loading. A blast or explosion is a typical source of impulsive load; for such short-duration loads, special simplified forms of dynamic analysis may be employed. On the other hand, a general, relatively long-duration loading suck as might from an earthquake can be treated only by a completely general dynamic analysis procedure (see Figure 3.57 for characteristics of different types of dynamic loading described earlier).

3.16.1 Dynamic Loads

A good number of loads that occur in buildings may be considered stationary requiring only static analysis. Although almost all loads except dead loads are transient, meaning they change with time, customarily in structural design, they are treated as static loads. For example, lateral loads due to transitory wind gusts that are often dynamic are usually treated as static loads. To realize acceptable results, the dynamic characteristics due to the sudden variation of wind velocity are taken into account by including a gust factor in the determination of wind loads. Therefore, the analysis reduces to a static case requiring but one unique solution.

Let us consider a building that, instead of being buffeted by wind, is subjected to ground motions due to an earthquake. The seismic shock causes the foundation of the building to oscillate back and forth, principally in a horizontal plane. The building would follow the movement of the ground without experiencing lateral loads if the ground oscillation took place very slowly over a long period of time. The building would merrily ride to the new displaced position as if no load was ever applied to it. On the other hand, when the ground moves suddenly as in an earthquake, the building mass, which has inertia, attempts to stay in its preearthquake position, thus resisting the accelerations of the structure. The resulting initial forces in a building with several floors may be visualized as a group of horizontal forces applied at various floors in proportion to the floor mass and its height above the ground.

These earthquake forces are considered dynamic, because they vary with time. Since the load is time varying, the response of the structure, including deflections, axial and shear forces, and bending moments, is also time dependent. Therefore, instead of a single solution, a separate solution is required to capture the response of the building at each instant of time for the entire duration of an earthquake. Because the resulting inertial forces are a function of building accelerations, which are themselves related to the inertial forces, it is necessary to formulate the dynamic problem in terms of differential equations.

3.16.2 Concept of Dynamic Load Factor

A special feature of earthquake excitation of structures is that it is applied in the form of support motions that are neither harmonic nor periodic. Defining such excitations that vary arbitrarily with time is perhaps the most difficult and uncertain phase of predicting structural response to earthquake-induced ground motions. However, when these motions have been established, the calculation of response of a given structure is a standard structural dynamics problem.

Simply stated, the dynamic load factor (DLF) is the ratio of the dynamic response such as deflection of an elastic system to the corresponding static deflection that results from the static application of a time–load function *Ft*. In a static problem, time variation of the load has no effect on the response since it is assumed that the load is applied in a gradual manner over a long period of time. In dynamic problems, however, the time it takes to apply the load has a large influence on the structural behavior and thus must be given consideration in the analysis. Shown in Figure 3.58 are some

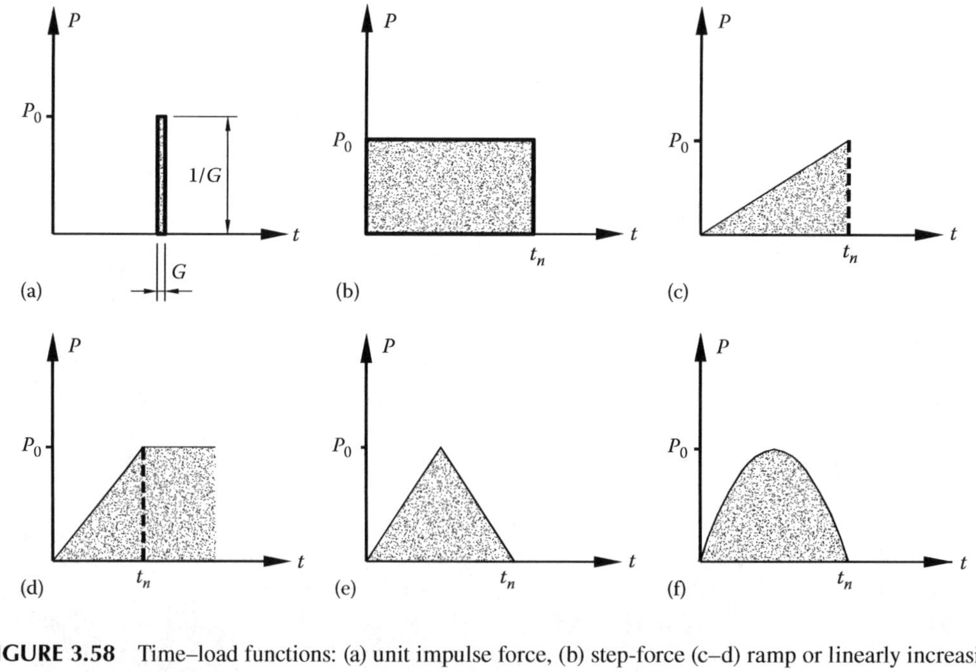

FIGURE 3.58 Time–load functions: (a) unit impulse force, (b) step-force (c–d) ramp or linearly increasing force, (e) triangular pulse force, and (f) half-cycle sine pulse force.

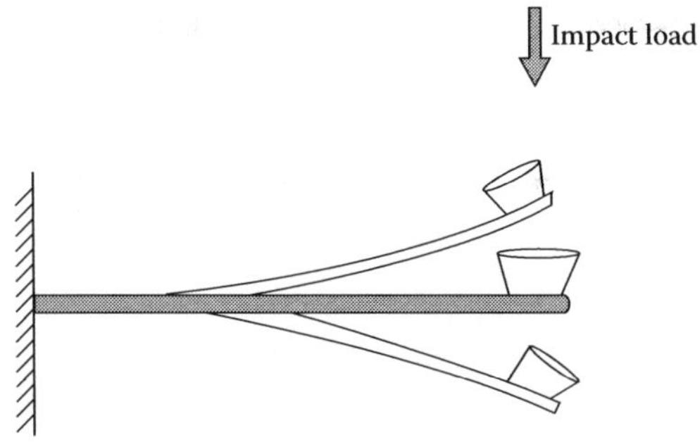

FIGURE 3.59 Dynamic response of cantilever. *Note*: DLF = 2 for suddenly applied point load.

examples of time–load function, *Ft*, in which load variation is graphed with respect to time *t*. The graphs are for unit impulse force, step force, ramp or linearly increasing force, and step force with finite rise time.

Consider the cantilever beam shown in Figure 3.59 with an empty container of weight W_D at the free end. Ignoring the self-weight of the beam and using common notations, the deflection of the cantilever due W_D is given by

$$\Delta_c = \frac{W_D L^3}{3EI} \tag{3.11}$$

Let the container be filled with water of weight W_w, flowing gradually, say, drop by drop into the container. The additional deflection due to the weight of water W_w is

$$\Delta w = \frac{W_w L^3}{3EI} \tag{3.12}$$

Instead of a drop-by-drop loading, if all water is suddenly gushed into the container instantly, the cantilever deflection will no longer be the same as the static deflection Δ_w. We know, by intuition, it will be larger than Δ_w (and in our case, it can be shown that it is two times the static deflection Δ_w).

Hence, the DLF = 2 for this suddenly applied load case. Observe that in addition to experiencing a larger deflection due to dynamic load, the cantilever also springs up and then down and continues to do so about its static position. In other words, the cantilever, when subject to a sudden load, exhibits a dynamic response by vibrating typically in a sinusoidal manner. But for the effects of damping, it would continue to do so indefinitely as shown schematically in Figure 3.60a, as opposed to damped free vibrations shown in Figure 3.60b.

Earthquake ground motions are erratic; they are neither harmonic nor periodic and vary arbitrarily with time and last no more than a few seconds. The longest recorded earthquake lasted only about 90 s although it may have seemed like an eternity for those experiencing the trauma. Although ground motions are unrhythmical in an earthquake, we may, for analytical purposes, consider their effects as equivalent to the summation of the responses to appropriately scaled forces such as those shown in Figure 3.58.

Thus, the response spectrum of single-degree-of-freedom (SDOF) systems to ground acceleration, as explained presently in this chapter, can be plotted from the DLFs determined for a chosen set of time–load functions such as those shown in Figure 3.58.

It should be noted that the DLF also varies with time (see Figure 3.61). However, in structural engineering, we are interested in the maximum design values (such as displacements, stresses) that occur during an earthquake. Therefore, the maximum response, irrespective when it occurs during

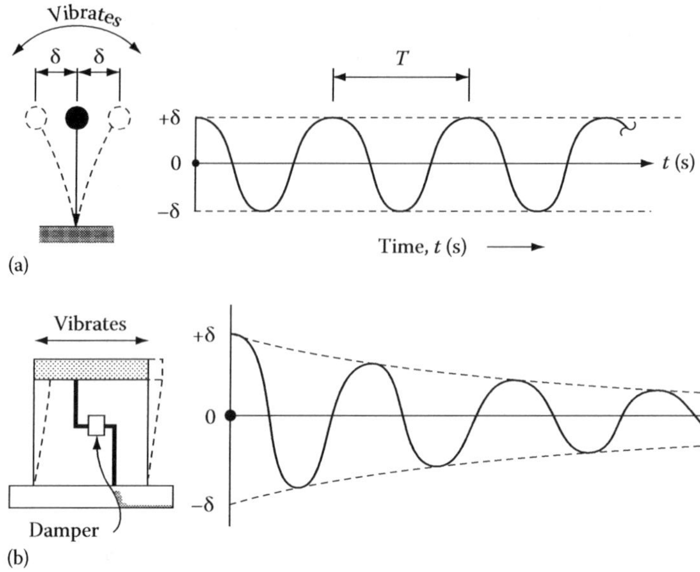

FIGURE 3.60 Vibrations of one-DOF systems: (a) undamped and (b) damped.

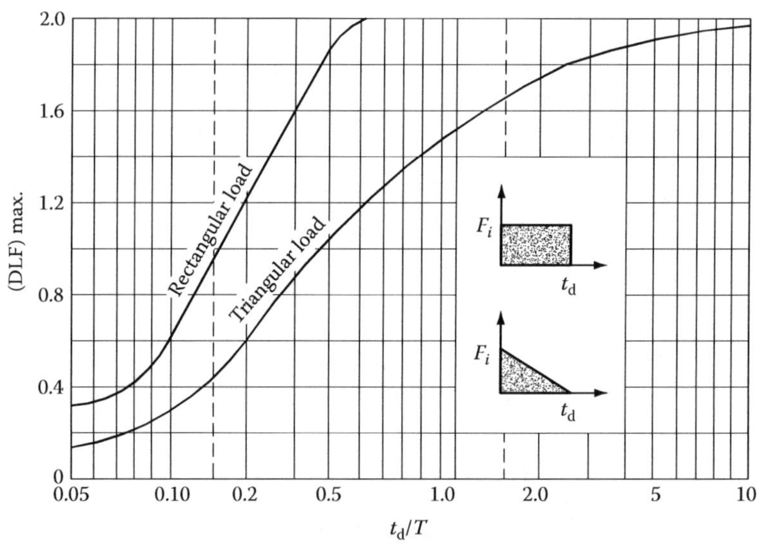

FIGURE 3.61 Maximum response of one-DOF undamped systems to rectangular and triangular pulse having zero rise time. *Note*: T = system period and t_d = pulse duration.

an earthquake, is of concern. It is worthwhile to note again that when we use response spectra to calculate design values for SDOF systems, it is not necessary to amplify the design values by DLF, because the computationally intensive dynamic analysis has been completed in generating the response spectra.

3.16.3 Difference between Static and Dynamic Analyses

In our day-to-day structural engineering practice, we are quite comfortable with the idea that if a member is overstressed, say in flexure, we simply increase the member size, and thus its section modulus, to decrease the flexural stresses. This concept is as old as engineering itself and has served us quite well in our design of structures subject to static loads. However, this thought of automatically reducing the stresses by increasing member sizes may not always hold together when designing structures subject to seismic ground motions.

In the case of earthquake excitation, the increase in member size resulting in higher stiffness shortens the natural vibration period that may have the effect of increasing the equivalent static force. Whether the corresponding stress decreases or increases by increasing the size of the member depends on the increase in section modules, S, and the increase in the equivalent static force that, in turn, depends on the response spectrum.

Example

Given

A 15 ft tall cantilever, an 8 in. nominal diameter standard steel pipe supporting a 40 kip weight at the top as shown in Figure 3.62. The properties of the column are

$$I = 106 \text{ in.}^4$$

$$S = 24.5 \text{ in.}^3$$

$$\text{Weight } W = 43.39 \text{ lb/ft}$$

$$E = 29{,}000 \text{ ksi}$$

Required

1. Determine the peak deformation and bending stress in the cantilever column using the seismic ground motion, acceleration response spectrum, for a damping $\beta = 5\%$, shown in Figure 3.63.
2. Repeat the calculations using a bigger section, a 12 in. diameter standard pipe. Discuss the pros and cons of using the bigger pipe.

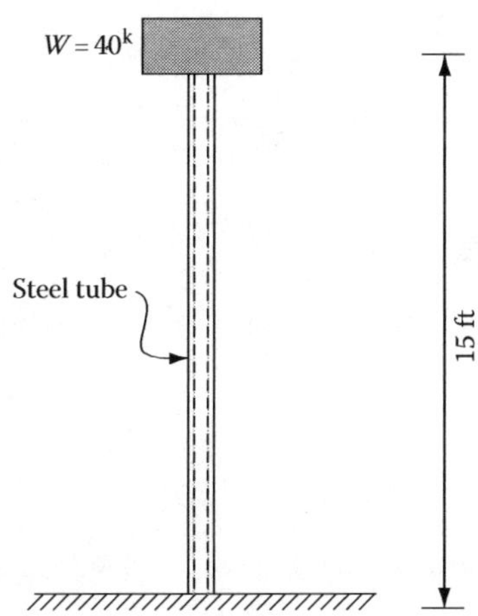

FIGURE 3.62 Cantilever columns with weight W at top.

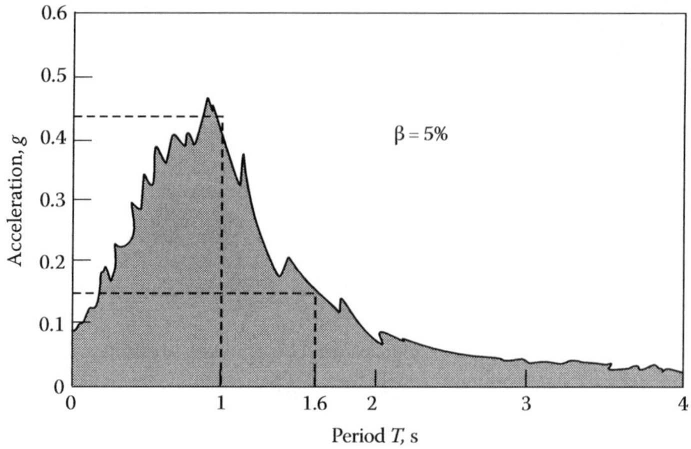

FIGURE 3.63 Acceleration response spectrums.

Solution

1. The weight of pipe at 43.39 plf = 15 × 43.39 = 650 lbs. Compared to the load of 40 kip at the top of the cantilever, the self-weight of the column is small and therefore can be ignored. The lateral stiffness of the column is

$$k = \frac{3EI}{h^3} = \frac{3 \times 29{,}000 \times 106}{(15 \times 12)^3} = 1.584 \text{ kip/in.}$$

$$\text{Mass at top} = \frac{W}{g} = \frac{40}{386} = 0.104 \text{ kip-s}^2/\text{in.}$$

The natural vibration frequency is given by

$$\omega_n = \sqrt{\frac{k}{m}} = \sqrt{\frac{1.584}{0.104}} = 3.91 \text{ rad}$$

$$\text{Period } T_a = \frac{2\pi}{\omega_a} = \frac{2 \times 3.14}{3.91} = 1.6 \text{ s}$$

From the acceleration response spectrum curve (Figure 3.63) for T_a = 1.6 s, acceleration $\alpha = 0.15\,g$.

The peak value of the equivalent static force is

$$v = \frac{a}{g}\omega = 0.15 \times 40 = 6 \text{ kip}$$

The bending moment at the base of the column is

$$6 \times 15 = 90 \text{ kip-ft}$$

$$\text{Bending stress } f_b = \frac{90 \times 12}{24.5} = 44.08 \text{ ksi}$$

2. Because the bending stress is relatively high, the designer elects to try a bigger pipe, a 12 in. diameter standard pipe. Let us verify if the resulting bending stress situation gets any better.

   The properties of the 12 in. column are

$$I = 279 \text{ in.}^4$$

$$S = 43.8 \text{ in.}^3$$

$$\text{Weight} = 49.56 \text{ lb/ft}$$

   As before, we ignore the self-weight of the column.

   The lateral stiffness of the new column is

$$k = \frac{3EI}{h^3} = \frac{3 \times 29{,}000 \times 279}{(15 \times 12)^3} = 4.17 \text{ kip/in}$$

   Mass at top = 0.104 kip-s^2/in., as before

   The natural vibration frequency is

$$w_n = \sqrt{\frac{k}{m}} = \sqrt{\frac{4.17}{0.104}} = 6.33 \text{ rad}$$

$$\text{Period } T_n = \frac{2\pi}{\omega_n} = \frac{2 \times 3.14}{6.33} = 0.992 \text{ s} = \sim 1.0 \text{ s}$$

   From the response spectrum curve (Figure 3.63) for $T_n = 1$ s, acceleration $a = 0.43\ g$.

   The peak value of the equivalent static force is

$$v = \frac{a}{g}\omega = 0.43 \times 40 = 17.2 \text{ kip}$$

   The bending moment at the base of the column is

$$17.5 \times 15 = 258 \text{ kip-ft}$$

$$\text{Bending stress } f_b = \frac{258 \times 12}{43.8} = 70.68 \text{ ksi}$$

   that is about 60% more than the calculated value for the 8 in. diameter pipe.

This example (admittedly cooked up to make a point) would make you think twice before you add stiffness to seismic structural systems. The concept of designing with sufficient strength has been viewed traditionally as a key to more effectively control the behavior of buildings. However, in certain instances, the benefits associated with an increase in strength may be small or even have a negative impact. The doctored-up example given here is not intended to discount strength as an important design consideration but rather to allude to the possible negative impacts. Keep in mind the excessive strength of the yielding element will impose more demand on the brittle components along the lateral load path. A better design strategy would be to increase ductility; treat ductility as a wealthy person treats money—you cannot have too much of it.

3.16.4 Dynamic Effects due to Wind Gusts

Perhaps the reader will recall that in Chapter 2, it was stated that the load on a building due to a wind gust becomes a dynamic load if the rise time of the gust is considerably shorter than the fundamental period of the building. Conversely, if the rise time is the same as or higher than the building's period, it may be considered as a static load.

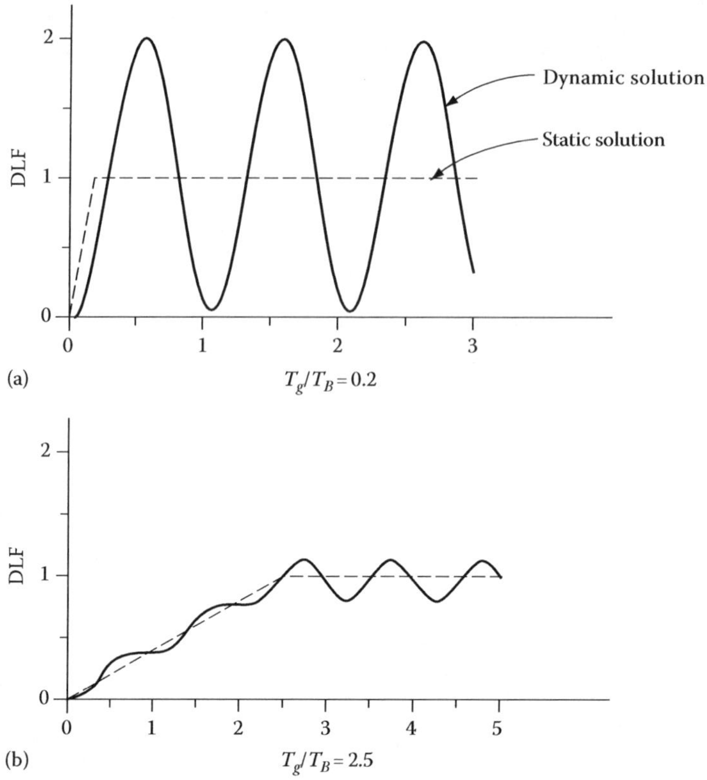

FIGURE 3.64 DLF for undamped SDOF system: (a) $T_g/T_B = 0.2$, (b) $T_g/T_B = 2.5$. *Note*: DLF = 2 for l s gust, and 1.0 for 12.5 s gust.

This characteristic that depends entirely on rise time of wind gust, T_g, is shown in Figure 3.64, T for two values of (T_g/T_B), the ratio of rise time T_g of the gust to the fundamental period, T_B, of the building. Observe that for $(T_g/T_B) = 2.5$ (which signifies a wind gust of relatively long rise time), DLF is nearly equal to 1.

To understand the significance of these figures, consider a tall building with a period, T_B, of 5 s (frequency of 0.2 H_z) subject to a constant wind pressure of 30 psf and then suddenly to a 1 s gust of 15 psf. A 1 s gust simply means that wind pressure has increased to 30 + 15 = 45 psf in a time period of 1 s. In other words, the 30 psf of constant pressure has increased to 45 psf, with the increase taking place in a short time interval of 1 s, and remains at 45 psf for the time duration of interest.

Referring to Figure 3.50a, for $(T_g/T_B) = (1/5) = 0.20$, the DLF is equal to 2.0 signifying that the dynamic effect of wind gust is as though it is an impulse load culminating with a maximum DLF equal to 2.0.

Next, consider the same increase in the wind load but assume that the increase occurs over a relatively long time interval of say, 12.5 s. As can be seen in Figure 3.64b, for the ratio $(T_g/T_B) = (12.5/5) = 2.5$, the corresponding DLF is close to unity, signifying that the dynamic effect can, in fact, be ignored for this case.

The concept of DLF presented earlier for wind gust is equally applicable to seismic design, except the load–time functions are referenced to support accelerations.

3.16.5 Characteristics of a Dynamic Problem

A structural dynamic problem differs from its static loading counterpart in two important respects. The first difference as implied by the definition is the time-varying nature of the dynamic problem.

Because both loading and response vary with time, it is evident that a dynamic problem does not have a single solution, as static problem does; instead, the engineer must establish a succession of solutions corresponding to all times of interest in the response history. Thus, a dynamic analysis is clearly more complex and time consuming than a static analysis.

The second and more fundamental distinction between static and dynamic problems is illustrated in Figure 3.65. If a single-bay, single-story frame is subjected to a static load *F*, as shown in Figure 3.65, the internal moments and shears in the columns and deflected shape of the frame depend only upon this load, and they can be computed by established principles of force equilibrium. On the other hand, if the *F*(*t*) is applied dynamically, as shown in Figure 3.66, the resulting displacements of the frame depend not only upon this load but also upon inertial forces, which oppose the accelerations producing them. Thus, the corresponding internal moments and shears in the columns must equilibrate not only the externally applied force *F*(*t*) but also the inertial forces resulting from the accelerations of the beam.

Inertial forces, which resist accelerations of the structure in this way, are the most important distinguishing characteristic of a structural dynamics problem. In general, if the inertial forces represent a significant portion of the total load equilibrated by the internal elastic forces of the structure, then the dynamic character of the problem must be accounted for in its solution. On the other hand, if the motions are so slow that the inertial forces are negligibly small, the analysis of response for any desired instant of time may be made by static structural analysis procedures even though the load and response may be time varying.

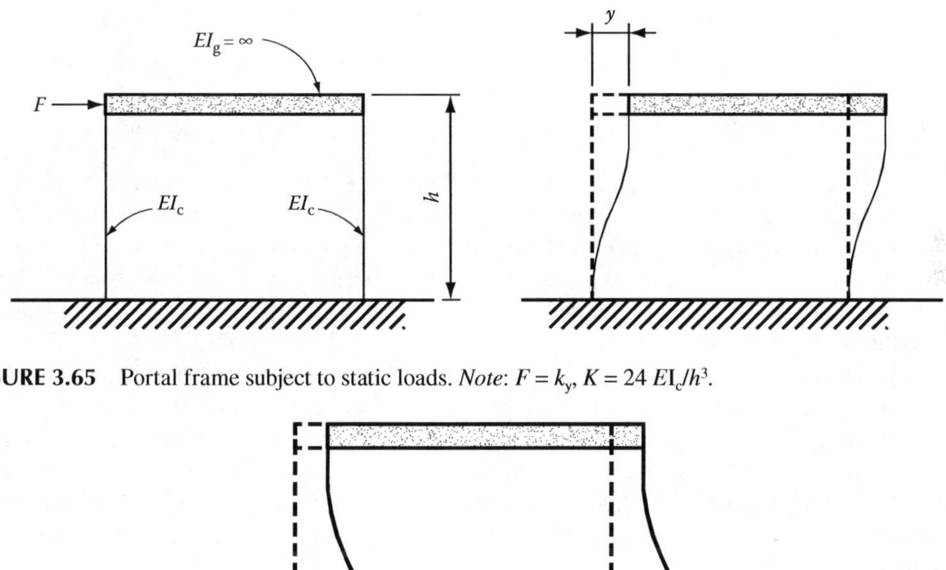

FIGURE 3.65 Portal frame subject to static loads. *Note*: $F = k_y$, $K = 24\ EI_c/h^3$.

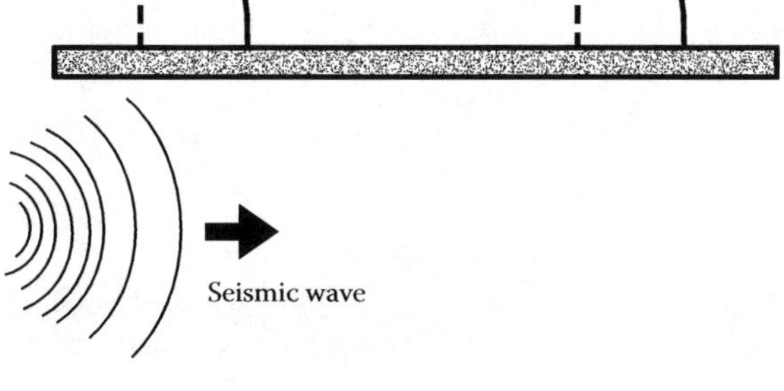

FIGURE 3.66 Portal frame subject to earthquake ground motions.

3.16.6 Multiple Strategy of Seismic Design

The design of structures to perform satisfactorily under expected seismic conditions requires that realistic earthquake loadings be specified and that the structural components be proportioned to resist these and other combined loads within the limits of certain design requirements. In regions of high seismicity, earthquake loading is often critical among the types of loading that must be considered because a large earthquake will usually cause greater stresses and deformations in the various critical components of a structure than will all other loadings combined, yet the probability of such an earthquake occurring within the life of the structure is very low. In order to deal effectively with this combination of extreme loading and low probability of occurrence, a strategy based on the following dual criteria has usually been adopted:

1. A moderate earthquake, which reasonably may be expected to occur once at the site of a structure during its lifetime, is taken as the basis of design. The structure should be proportioned to resist the intensity of ground motion produced by this earthquake without significant damage to the basic system.
2. The most severe earthquake, which could possibly ever be expected to occur at the site, is applied as a test of structural safety. Because this earthquake is very unlikely to occur within the life of the structure, the designer (in concert with the building owners) is perhaps economically justified in accepting significant structural damage; however, collapse and serious personal injury or loss of life must be avoided.

Currently, the trend is to strengthen the second of these criteria for critical and expensive structures by calling for limited repairable damage, thus focusing not only on life safety but on protection of financial investment as well.

A special feature of earthquake excitation of structures, compared with most other forms of dynamic excitation, is that it is applied in the form of support motions rather than as external loads; thus, the effective seismic loadings must be established in terms of these motions. Defining the support motions is the most difficult and uncertain phase of the problem of predicting structural response to earthquakes. When these input motions have been established, however, the calculation of the corresponding stresses and deflections in any given structure is a standard problem in structural dynamics.

The earthquake excitation considered to act on a structure is the free-field ground motion at support points. Inherent in this treatment is the assumption that the same free-field ground motion acts simultaneously at all support points of the structure. This assumption is equivalent to considering the foundation soil or rock to be rigid. This hypothesis clearly is not consistent with the concept of earthquake waves propagating through the Earth's crust from the source of energy release; however, if the base dimensions of the structure are small relative to the predominant wavelengths, the assumption is acceptable.

When specifying input motions at the base of a structure, it should be recognized that the actual structure base motions during an earthquake may be significantly different from the corresponding free-field motions that would have occurred without the structure being present. This *soil–structure interaction* effect will be of slight importance if the foundation is relatively stiff and the structure is relatively flexible; in this case, the structure transmits little energy into the foundation and the free-field motions are adequate measures of the actual foundation displacements. On the other hand, if a stiff structure is supported on a deep, relatively soft layer of soil, considerable energy will be transferred from the structure to the soil and the base motions will differ from those experienced by the soil under free-field conditions. This soil–structure interaction mechanism is independent of, and in addition to, the effect of local soil conditions on the free-field motions.

3.16.7 Example of Portal Frame Subject to Ground Motions

Given

A moment frame with bay length $L = 16$ ft, height $h = 12$ ft, elastic modulus $E = 29{,}000$ ksi. The columns are $W8 \times 67$ with $I_c = 272$ in.^{4} and are fixed at the base (see Figure 3.67a).

The beams are relatively heavy, $W36 \times 330$ with $I_b = 23{,}300$ in.^{4}

Required

Assuming the beam is infinitely rigid as compared to the column, compute the maximum displacement and maximum base shear produced in this frame by an earthquake having the acceleration response spectrum shown in Figure 3.67b. Assume that the effective weight including the weight of the beam is equal to 596 kips.

Solution

The first step is to determine the vibration period of the structure. The circular frequency of the structure is given by

$$\omega = \sqrt{(k/m)}$$

where

ω is the circular frequency
k is the stiffness of the rigid frame
m is mass

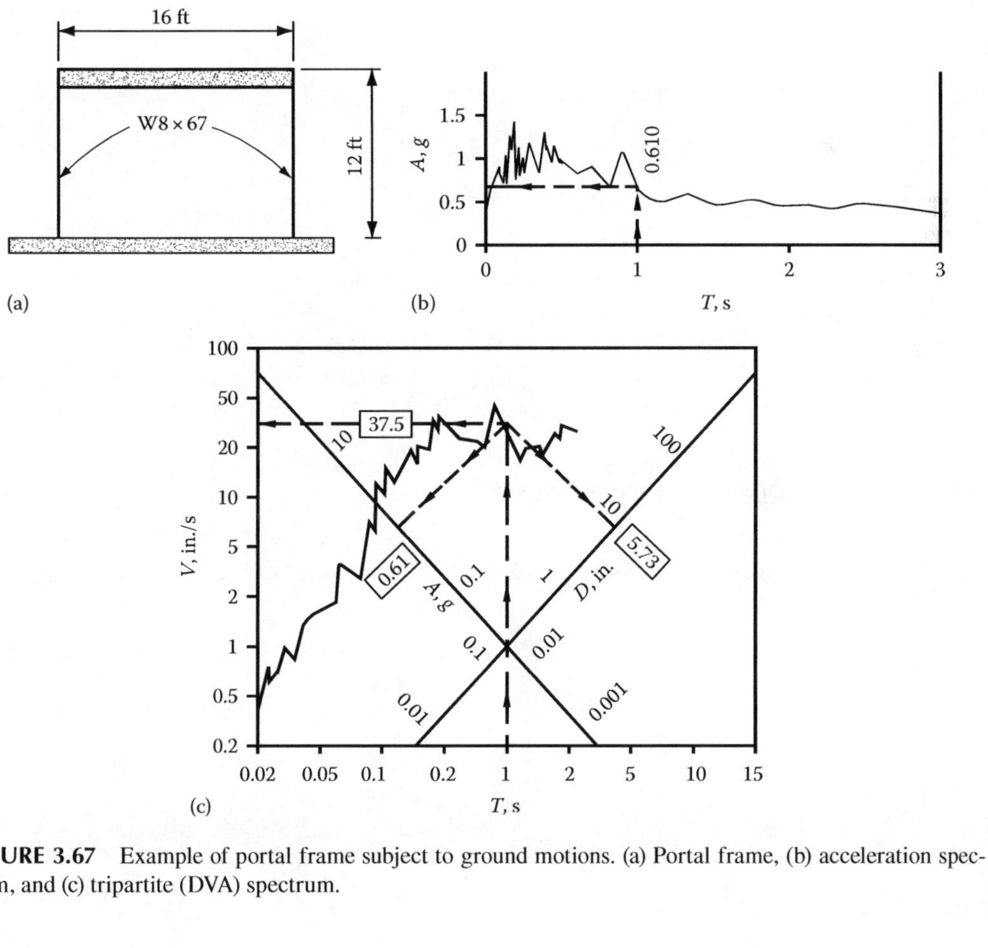

FIGURE 3.67 Example of portal frame subject to ground motions. (a) Portal frame, (b) acceleration spectrum, and (c) tripartite (DVA) spectrum.

Because the beam is considered rigid, that is, its flexural stiffness is large as compared to the column stiffness, the lateral stiffness, k, of the frame is

$$k = \sum \frac{12EI_c}{h^3}$$

$$= \frac{2 \times 12EI_c}{h^3} = \frac{24 \times 29{,}000 \times 272}{(144)^3} = 63.4 \text{ kips/in.}$$

The mass

$$m = \frac{W}{g} = \frac{596}{386.4} = 1.54 \text{ kips} \cdot \text{s}^2/\text{in.}$$

$$\omega = \sqrt{\frac{k}{m}} = \sqrt{\frac{63.4}{1.54}} = 6.41 \text{ rad/s}$$

$$\text{Period } T = \frac{2\pi}{\omega} = \frac{2 \times 3.1416}{6.41} = 0.98 \text{ s} \cong 1 \text{ s}$$

From Figure 3.67b, the spectral acceleration for this period is 0.61g. Hence, the maximum base shear

$$V = 0.61 \times 596$$

$$= 363.6 \text{ kips}$$

$$\text{The displacement at top} = \frac{V}{K}$$

$$= \frac{363.6}{63.4} = 5.73 \text{ in.}$$

Shown in Figure 3.67 is a tripartite response spectrum of the same earthquake shown in Figure 3.63. The procedure for developing tripartite response spectrum is explained presently. Suffice here to note that the displacement for the example portal frame can be read directly from the tripartite response spectrum without performing the calculations.

3.16.8 Concept of Dynamic Equilibrium

We begin our study of dynamic equilibrium in structural dynamics with a simple single-story portal frame shown in Figure 3.68, where the mass is distributed along the girder and only horizontal

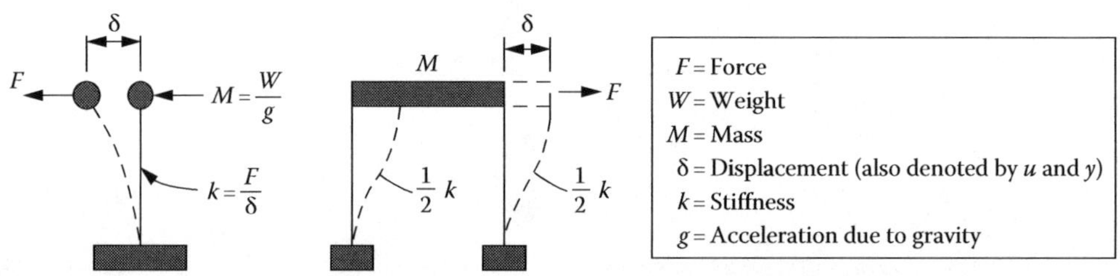

FIGURE 3.68 Idealized SDOF system.

motions are considered. The supporting structure consisting of the beams and columns is assumed massless. The rigid frame is defined as a one-degree system in which only one type of motion is possible, or in other words, the position of the mass at any instant can be defined in terms of a single coordinate, which, in our case, is the horizontal displacement u. Thus, the mass can move in a horizontal direction only. As an example of dynamic analysis, let us determine the motion of this mass resulting from the application of a time-varying horizontal force.

A convenient method of deriving the equation of motion is by the use of D'Alembert's principle of *dynamic equilibrium*. Having been trained to think in terms of equilibrium of forces, structural engineers may find this method particularly appealing. This principle states that with *inertial forces* included, a system is in dynamic equilibrium at each time instant. The inertial force, a force equal to the product of mass times its acceleration and acting in a direction opposite to the acceleration, is a fictitious force that allows us to treat structural dynamics in exactly the same manner as a problem in static equilibrium. From Figure 3.69, the equilibrium equation is

$$F_t - ku - M\ddot{u} = 0 \tag{3.13}$$

where

F_t is the time-varying horizontal force
k is the lateral stiffness of the portal frame
u is the horizontal displacement of the mass
M is the mass of the system = W/g
$\ddot{u}$ is the time-dependent acceleration of the mass M

Throughout this text, $\dot{u}$ and $\ddot{u}$ will be used to the first and second derivatives of displacement u with respect to time t. In other words, $\dot{u} = (du/dt)$ is the velocity and $\ddot{u} = (du^2/d^2t)$ is the acceleration.

The rearranged and expanded form of Equation 3.13:

$$\frac{\delta^2 u}{dt^2} + \frac{k}{M}u = F_t \tag{3.14}$$

This is a differential equation of second order, which may be solved by using the principles of calculus. However, the solution of the differential equations will not be discussed herein since it is assumed that the reader is familiar with such procedures. The deliberate omission of mathematical techniques in this text enables us to concentrate on the physical phenomena involved and helps us to develop a physical feel and intuition for dynamic response, which is necessary for successful analysis of more complicated dynamic problems.

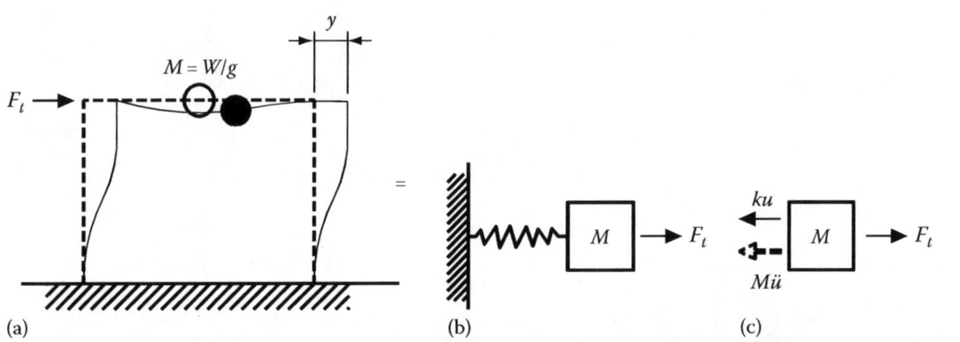

FIGURE 3.69 D'Alembert's principle of dynamic equilibrium: (a) portal frame subject to dynamic force F_t, (b) equivalent dynamic model, and (c) free-body diagram.

3.16.9 Free Vibrations

Consider, first, the elementary case where F_t equals zero, that is, *there is no external dynamic excitation. Such a structure is said to be undergoing free vibration* when it is disturbed from its static equilibrium position and then allowed to vibrate. Motion occurs only if the system is given an initial disturbance, such as an initial displacement u_0. Imagine that the mass m shown in Figure 3.70 is pulled horizontally by a rope and then suddenly released at $t = 0$. The resulting motion unaffected by any external force is the free vibration. The differential equation of motion for this case is given by

$$\ddot{u} + \frac{k}{M}u = 0 \tag{3.15}$$

The solution for this equation in terms of displacement u is given by

$$u = \frac{\dot{u}_0}{w}\sin \omega t + u_0 \cos \omega t \tag{3.16}$$

A plot of displacement for initial displacement u_0 is given in Figure 3.71. The parameter ω in the equation is called the natural frequency of the system and is given by the relation

$$\omega = \sqrt{\frac{k}{M}} \text{ rad/s} \tag{3.17}$$

As stated previously, D'Alembert's principle, named after its discoverer, the French mathematician Jean le Rond d'Alembert, permits the reduction of a problem in dynamics to one in statics. This is accomplished by intruding a fictitious force equal in magnitude to the product of the mass of the body and its acceleration and directed opposite to the acceleration. While D'Alembert's principle is merely another way of writing Newton's second law of motion, $F = Ma$, it has the advantage of changing a problem in engineering dynamics into a problem in statics.

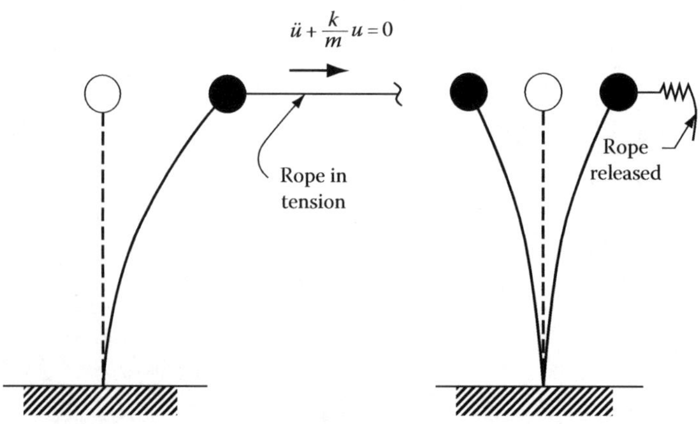

FIGURE 3.70 Concept of free vibrations.

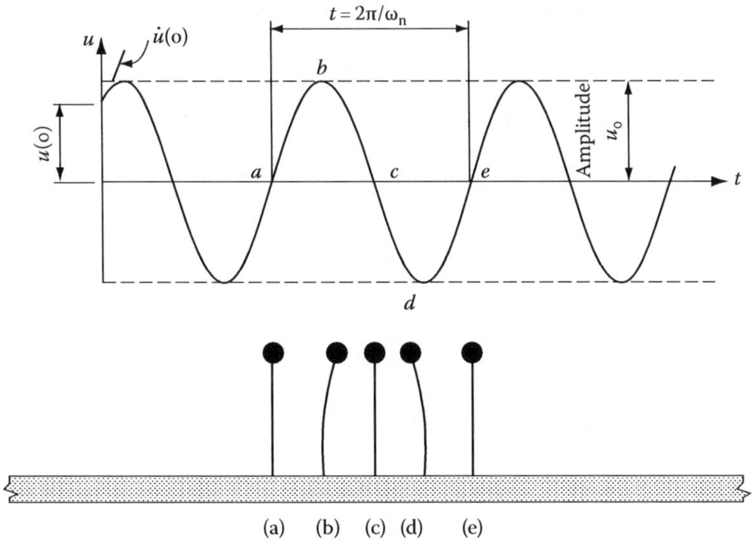

FIGURE 3.71 Displacement plots of free vibrations with damping.

3.16.10 Earthquake Excitation

The difficulties encountered with the design of structures to withstand earthquake-induced ground motions of the base structure are technically intriguing. The major problem lies in the prediction of the character and intensity of the earthquakes to which a structure might be subjected to during its life. Another difficulty lies in the fact that a realistic analysis for earthquakes should account for inelastic behavior of the structure because very few structures could withstand a strong earthquake without some plastic deformation. In this section, attention will be restricted to a few basic concepts, which will provide a foundation for understanding seismic design principles.

3.16.11 Single-Degree-of-Freedom Systems

Consider a portal frame subject to a displacement of the ground denoted by u_g. The total displacement u^t of the mass at each instant of time is given by

$$u^t(t) = u(t) + u_g(t) \tag{3.18}$$

The rigid body component of the displacement of the structure produces no internal forces; only the relative motion u between the base and the mass produces elastic forces.

Thus, the relative displacement $u(t)$ of the structure due to ground acceleration $\ddot{u}(t)$ will be identical to the displacement $u(t)$ of the structure if its base were stationary and if it were subjected to an external force = $m\ddot{u}_g(t)$ acting in a direction opposite to the ground motion. The ground acceleration can therefore be replaced by an *effective earthquake force*

$$F_{eff}(t) = -m\ddot{u}_g(t) \tag{3.19}$$

This force is equal to mass times the ground acceleration acting opposite to its acceleration.

It was seen earlier that the deflection of the cantilever due to sudden application of the load is twice as much as the static deflection. If we knew this information, beforehand, we could have obtained the maximum dynamic response without going through the dynamic analysis procedure. However, the procedure, in a manner of speaking, is still a dynamic analysis, because it uses the vibration properties of the structure and the dynamic characteristics of ground motion. It is just that the engineers do not have to carry out any excruciating dynamic analysis. Somebody has done these in telling us that DLF = 2 for this particular load function.

Observe that this argument also holds well when we use acceleration values from a response spectrum. There is no need to multiply the acceleration value by a DLF because this has already been done prior to plotting the response spectrum.

The beam-to-column stiffness ratio p indicates how much the system may be expected to behave as a frame. For $p = 0$, the beams impose no restraint on joint rotations behaving as if they were pin connected to the columns. The portal frame then behaves as if it consists of two independent cantilever columns fixed at the base. For $p = \infty$, the beam restrains completely the joint rotations, and the portal frame behaves as a shear beam with double-curvature bending of the columns in each story. It is thus evident that an intermediate value of p represents a combination of both the cantilever bending mode of columns and the shear mode deformations due to beam bending. Observe that strong column–weak beam design is what we try to realize in seismic design of moment frames.

3.16.12 Numerical Integration Technique

To understand the need for numerical integration in structural dynamics, perhaps it is instructive to recall the definition of differential equations and their application in engineering.

A differential equation in a broad sense is a mathematical formulation for the determination of an unknown variable that itself relates to the derivatives of the variable orders.

Take, for example, the use of differential equations in determining the velocity of a ball falling through the air. Consider only gravity and air resistance. The ball's acceleration toward the ground is the acceleration due to gravity minus the deceleration due to air resistance. Gravity is constant but air resistance is proportional to the ball's velocity. Therefore, the ball's acceleration, which is a derivative of its velocity, depends on the velocity. Finding the velocity as a function of time involves solving a differential equation.

Very few simple differential equations can be solved by using explicit formulas. If an explicit solution is not available, which usually is the case in seismic design, the solution is approximated using computers. In earthquake engineering, we call these methods time-history or numerical methods.

Anyone who tries to explain the concept of numerical integration in less than 30 pages is asking for trouble. Here, we take the risk and go for relatively fewer pages by describing only the dynamic response of SDOF systems. The reader is referred to the vast body of literature that exists in several textbooks for further study.

Structural dynamics is too often taught as a course in advanced mathematics, making this approach unnecessarily difficult for practicing engineers. Many engineers may find the mathematical manipulation so intriguing that they fail to develop physical understanding so essential to good design. In this text, we avoid mathematical complexities, which, although are useful in understanding advanced dynamic topics, are not necessary for developing an insight into practical seismic designs.

Today, in the year 2013, the use of closed-form solutions for analyzing practical dynamic problems, very rarely if ever, is undertaken. As in other fields, most often numerical analysis techniques are used to execute solutions to dynamic problems. Therefore, the author believes that summarizing a numerical analysis executed by hand develops a physical feel for dynamic behavior much more rapidly than does the closed-form solution using differential equations.

Consider an SDOF system, such as the portal frame shown in Figure 3.72. In terms of dynamic behavior, the system is equivalent to a spring connected to a mass, a model most often illustrated in structural dynamics textbooks. Let us recognize at the outset that this simple system is not only just an ideational model but in fact represents practical single-bay, single-story portal frames.

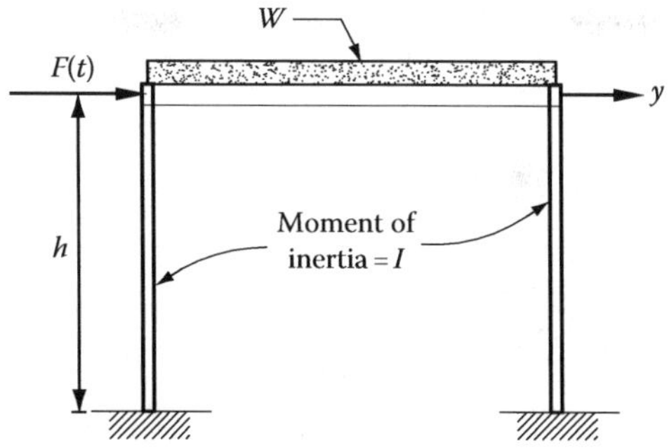

FIGURE 3.72 Portal frame subject to ground motions.

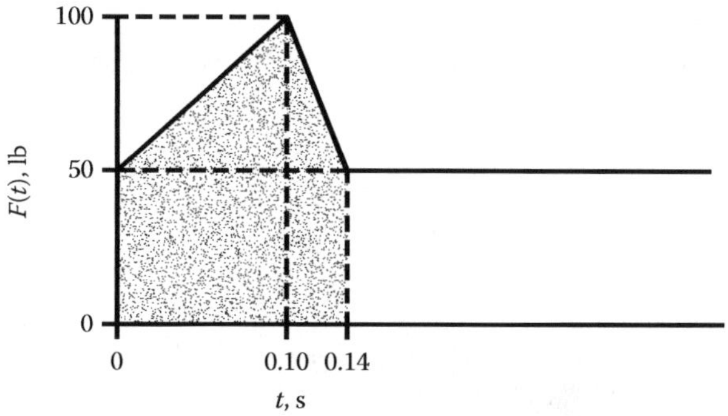

FIGURE 3.73 Equivalent load–time function due to base motions.

To demonstrate the numerical integration procedure, suppose that the base of the portal frame is subjected to a given set of ground accelerations. Let us assume further that the accelerations result in an equivalent dynamic load at the top, as shown in Figure 3.73 by the load–time function. To simplify the numerical solution, we will make two assumptions: first, the weights of the columns are negligible, and second, the girder is sufficiently rigid to prevent significant rotation at the tops of the columns. These assumptions are not necessary for demonstrating the analysis but will serve to simplify the problem and, in fact, are essentially correct for many actual frames of this type.

In the load–time function shown in Figure 3.73, the dynamic load is equal to 50 lbs at the time $t =$ 0 s, rising linearly to 100 lbs at $t = 0.10$ s and dropping back to 50 lb at time $t = 0.14$ s. The purpose here is to determine the variation of the horizontal displacement of the beam, with time, starting with the system at rest $t = 0$ all the way to $t = 0.14$ s.

The example portal frame is an SDOF system defined as one in which only one type of motion is possible. Or, in other words, the position of the system at any instant can be defined in terms of a single coordinate. In our case, the mass m can move in a horizontal direction only.

Before considering a specific example, we shall discuss the process of numerical integration in general terms. This is a procedure by which the differential equation of motion is solved step by step, starting at zero time, when the displacement and velocity are presumably known. The time scale is divided into discrete intervals, and one progresses by successively extrapolating the displacement from one time station to the next.

There are many such methods available. One of the simpler versions presented by John M. Biggs is called the *constant-velocity or lumped-impulse procedure*. The reader is referred to Biggs' textbook for further explanation of this procedure.

To illustrate the numerical integration procedures, consider again the portal frame subjected to the load–time function shown in Figure 3.73. It is desired to determine the variation of displacement with time, starting with the system at rest at $t = 0$. Substituting into Equation 3.13 and being careful to keep the units consistent, the equation of motion is written as follows:

$$F(t) - 2000u = \frac{64.4}{32.2}\ddot{u}$$

or

$$\ddot{u} = \frac{1}{2}F(t) - 1000u \tag{3.20}$$

Thus, knowing the load and the displacement at any time enables one to compute acceleration at the time.

The next step is to select a time interval for the numerical integration. As mentioned earlier, this should not be greater than one-tenth of the natural period of the system. The *natural period* T of a one-degree system is given by

$$T = 2\pi\sqrt{\frac{W}{kg}} \tag{3.21}$$

This, in this example, is 0.198 s. The *natural frequency* f is the inverse of the natural period or 5.04 cps. The *natural circular frequency* ω is $2\pi f$ or 31.6 rad/s.

One-tenth of the natural period is approximately 0.02 s, and this value will be used in the computation. However, a second criterion must be considered when selecting the time interval: the interval should be small enough to represent properly the variation of load with time. In this example, it will be noted that the stations (i.e., $\Delta = 0.02$) occur at the sudden breaks in the load function and, furthermore, that the interval is small enough so that points at this spacing accurately represent the load function (Figure 3.73). Therefore, the time interval selected is satisfactory.

The result of the complete calculation is plotted in Figure 3.74 where displacement versus time is shown. The ordinate also represents the time variation of the spring force if the displacements are multiplied by the spring constant k. To assist in the interpretation of the result, the hypothetical displacement corresponding to the static application of the load at any instant is also plotted. The maximum displacement, which occurs at 0.12 s, corresponds to a spring force of 159 lb, which is 1.59 times the maximum external load. Subsequent peaks are somewhat smaller and correspond to a spring force of 157 lbs. The latter peak value would remain constant indefinitely since we have not included damping in the present example. The time interval between successive positive (or negative) peaks is exactly equal to the natural period of the system. After the load becomes constant at 0.14 s, the spring force carries, in a sinusoidal fashion, the positive and negative peaks being equidistant above and below the value of the external load.

The parameters of the idealized system needed in performing the numerical integration are as follows:

$$W = 64.4 \text{ lb}$$

$$k = \frac{12E(21)}{h^3} = 2000 \text{ lb/ft}$$

The spring constant k is simply equal to the inverse of the deflection at the top of the frame due to a unit horizontal load, and the equation employed may be easily verified by simple elastic analysis.

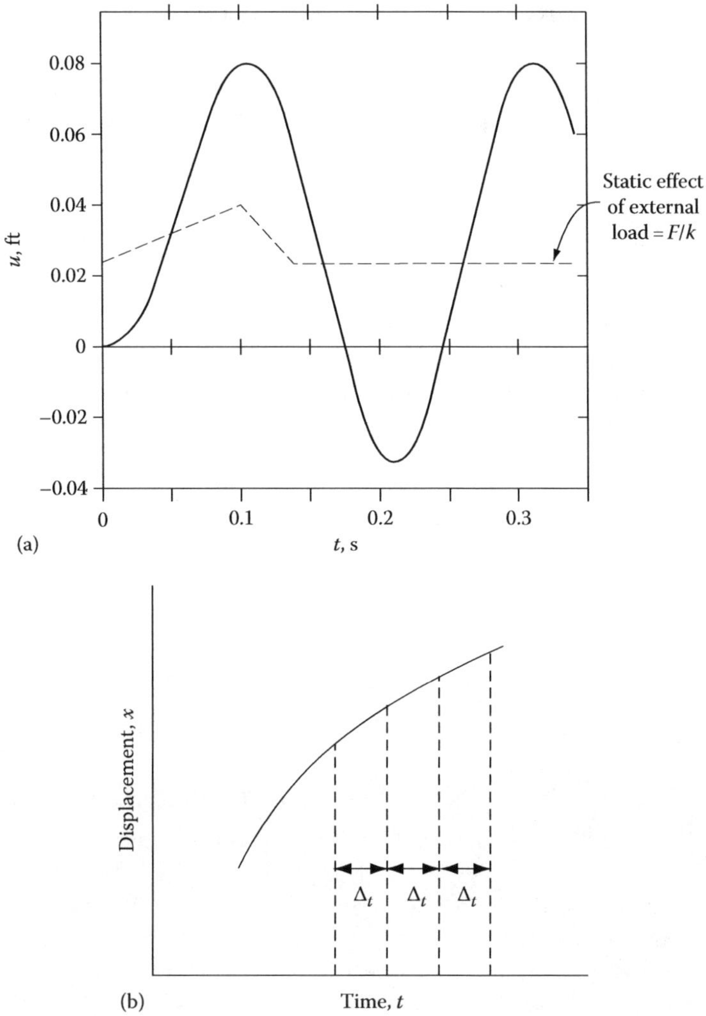

FIGURE 3.74 (a) Dynamic response of portal frame showing plot of displacement versus time. (b) Numerical integration, lumped-impulse procedure.

Having these parameters, we would then proceed to determine the dynamic displacements for the given load function. These displacements would be equal to the actual horizontal deflections at the top of the frame. The spring force computed for the ideal system would at all times be equal to the total shear in the two columns, and the maximum column bending moment would be given by $6EIu/h^2$. Thus, the dynamic stresses at any time could be easily determined.

3.16.13 Summary of Numerical Integration Technique

The process of numerical integration, also referred to as time-history analysis, is a method in which the differential equation of motion is solved step by step, starting at zero time, when the displacement and velocity are presumably known. The time duration of earthquake of interest is divided into discrete intervals, Δt, and we progress by successively extrapolating the displacement from one time station to the next.

There are many methods available for successive extrapolation of displacements, but only one of the more simple versions called the constant-velocity technique is discussed here.

Suppose that the curve shown in Figure 3.74b represents the displacement–time variation for a dynamic system. Suppose, further, that a successive numerical analysis for the determination of displacement is in progress with the displacements x_t at time station t and x_{t-1} at time station $(t - 1)$ previously known. Our task is to determine the next displacement at station $(t + 1)$ by extrapolation. This could be done by the following self-evident formula:

$$x_{(t+1)} = x_t + \dot{x}_{ay}\Delta t \tag{3.22}$$

where

$\dot{x}_{ay}$ is the average velocity between time stations t and $(t + 1)$
Δt is the time interval between the stations
$\dot{x}_t$ is the average velocity between the time intervals between the stations
$\dot{x}_{ay}$ is given by the relation

$$\dot{x}_{ay} = \frac{x_t - x_{(t-1)}}{\Delta t} + \dot{x}_t\Delta t \tag{3.23}$$

Substituting Equation 3.23 into Equation 3.22, the following recurrence formula is obtained:

$$x_{(t+1)} = 2x_t - x_{t-1} + \dot{x}_t(\Delta t)^2 \tag{3.24}$$

The acceleration $\ddot{x}_t$ at time station t can be determined by the use of D'Alembert's principle of dynamic equilibrium:

$$F_t - kx_t = M\ddot{u}_t \tag{3.25}$$

In keeping with the policy stated at the beginning of this section, we will not go further into explaining the process of numerical integration. Suffice it to note that a recurrence formula such as the one given in Equation 3.24 plays a pivotal role in time-history analysis of dynamic systems. The reader is referred to standard textbooks on structural dynamics for an in-depth discussion of numerical integration techniques.

3.16.14 Summary of Structural Dynamics

1. A very convenient way of writing the equation of dynamic motion is by the use of D'Alembert's principle of dynamic equilibrium. In this method, an imaginary force, referred to as inertial force and equal to the product of mass m and acceleration a, is added to the static equilibrium condition in a direction opposite to positive displacement.
2. The natural period T of a one-degree system is given by

$$T = 2\pi\sqrt{\frac{W}{kg}} \tag{3.26}$$

3. Motion that occurs only if the system is given an initial disturbance such as an initial displacement and is unaffected by any external force is called free vibration.
4. Motion that occurs as a result of an applied force is called forced vibration.
5. DLF, for a given load F_1, is defined as the ratio of dynamic deflection to the deflection that would have resulted from the static application of the load F_1. DLF varies as the ratio

(t_x/T), where t_x is the duration of load and T is the fundamental period of the system. The time at which the DLF becomes a maximum is also a variant but is of little consequence in seismic design because we are interested only in the maximum values such as deflection and stresses and not in the time when they occur.

6. Damping is conventionally indicated by a dash pot. It produces a force $c\dot{y}$ that opposes the motion and dissipates some of the energy of the system. Damping is the process by which vibration steadily diminishes in amplitude. In damping, the kinetic energy and strain energy of the vibrating system are dissipated by various mechanisms and should be included in the structural idealization in order to incorporate the feature of decaying motion. Damping has the effect of lengthening the period of a system. However, these effects are negligible for damping ratios below 20%. Observe that in seismic design, the damping ratios used for most structures are in the range of 5%–10%.
7. The determination of building response by numerical procedures involves the time-wise sequential determination of system's velocities and displacements. This type of numerical procedures in structural engineering is often referred to as a time-history analysis. The prediction of response by time-history analysis allows the designer to develop a physical feeling as to why a building responds as it does when subjected to ground motion.

In order to develop a physical feeling for dynamic behavior, consider what happens to our portal frame that has columns hinged to the ground, when the ground moves. Assume that you are sitting at the top of the beam and an earthquake occurs. The ground moves, but you do not, at least for the first small interval of time. You will, of course, move along with ground motions right after the first shock.

The initial differential displacement, d_o, between the mass (that is you) and the ground causes strain energy to be stored in the columns. The displacement may be, for analytical purposes, considered as though it is due to an equivalent force F_o applied at top.

F_o is given by

$$F_o = Kd_o = \frac{2\times 3EI}{h^3}(d_o) \tag{3.27}$$

This force, F_o, which is in dynamic equilibrium, is given by Newton's second law of motion. Force equals mass times acceleration:

$$F_o = M\ddot{x}_o = \frac{W}{g}\ddot{x}_o \tag{3.28}$$

The acceleration you feel during the ride is given by

$$\ddot{x}_o = \frac{Kd_o}{M} \tag{3.29}$$

The inertial force imparted on the mass is given by $F = k(d - x)$. Using subscripts to identify sequential increments, the acceleration, $\ddot{x}$, experienced by the mass is $\ddot{x}_i = (K/M)(d_i - x_i)$. Once we know the acceleration experienced by the mass, we can determine the corresponding values for the velocity and displacement. The velocity $(\dot{x})$ and displacement (x) of the mass can be calculated through the use of integral calculus if the acceleration can be described in terms of a continuous function. If not, we have no choice but to use numerical techniques.

Earthquake motions are erratic. It is not always possible to describe ground accelerations as continuous functions. Therefore, numerical integration is a more appropriate means of predicting successive velocities and displacements. The change in velocity experienced by the mass during a time interval of Δt is then determined by multiplying the average acceleration, which occurs during this time interval by the interval (Δt). The change in displacement is determined by multiplying the average velocity experienced by the mass during the time interval (Δt). Several numerical methods can be used to predict incremental changes in velocity and displacement.

It should be noted that given a small enough time interval, these methods will describe the characteristics of the dynamic response and maximum values sought in structural design, to a degree of accuracy consistent with our ability to predict ground motions.

In order to perform response-history analysis, it is necessary to have a digitized ground motion acceleration record. In linear response-history (LRH) analysis, the stiffness of the structure K is assumed to be independent of the prior displacement history. In nonlinear response-history (NRH) analysis, the structure's stiffness at a given instant of time t is dependent on the displacement history up to that point in time and varies to account for yielding, buckling, and other behaviors that may have occurred earlier in the structure's response.

Response-history analysis is useful because it allows solutions of the deflected shape and force state of the structure at each instant of time during an earthquake. Since each earthquake record has different characteristics, the results obtained from response-history analysis are valid only for the particular earthquake record analyzed. Therefore, when performing response-history analysis to determine forces and displacements for use in design, it is necessary to run a suite of analyses, each using different ground motion records as input. Present building codes require a minimum of three records. If three records are used, the maximum forces and displacements obtained from any of the analyses must be used for design purposes. If seven or more records are used, the code permits use of the mean forces and displacements obtained from the suite of analyses.

In design practice, LRH analysis is seldom used. This is because for design purposes, one is usually interested only in the maximum values of the response quantities (forces and displacements), and these quantities can more easily be approximated by an alternative form of analysis known as response spectrum analysis. However, NRH analysis is an essential part of the design of structures using seismic isolation or energy dissipation technologies. It is also routinely used in high-end performance-based design approaches.

3.17 RESPONSE SPECTRUM METHOD

The word spectrum in seismic engineering conveys the idea that the response of buildings having a broad range of periods is summarized in a single graph. For a given earthquake motion and a percentage of critical damping, a typical response spectrum gives a plot of earthquake-related responses such as acceleration, velocity, and deflection for a complete range, or spectrum, of building periods. An understanding of the concept of response spectrum is pivotal to performing seismic design.

A response spectrum (Figures 3.75 and 3.76) may be visualized as a graphical representation of the dynamic response of a series of progressively long cantilever pendulums with increasing natural periods subjected to a common lateral seismic motion of the base. Imagine that the fixed base of the cantilevers is moved rapidly back and forth in the horizontal direction, its motion corresponding to that occurring in a given earthquake. A plot of maximum dynamic response, such as accelerations versus the periods of the pendulums, gives us an acceleration response spectrum as shown in Figure 3.75, for the given earthquake motion. In this figure, the absolute value of the peak acceleration occurring during the excitation for each pendulum is represented by a point on the acceleration spectrum curve. As an example, an acceleration response spectrum from the 1940 EI Centro earthquake is illustrated in Figure 3.77. Using ground acceleration as an input, a family of response spectrum curves can be generated for various levels of damping, where higher values of damping result in lower spectral response.

To establish the concept of how a response spectrum is used to evaluate seismic lateral forces, consider two SDOF structures: (1) an elevated water tank supported on columns and (2) a revolving

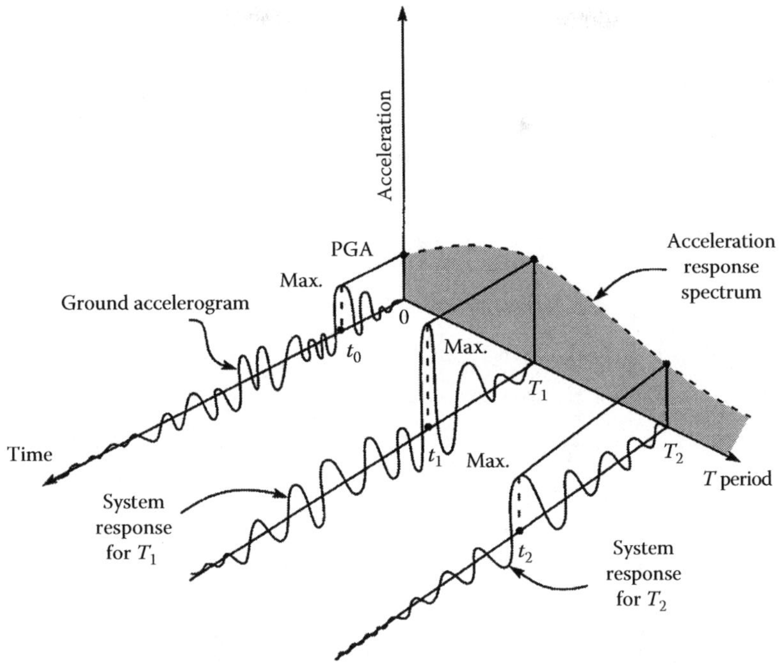

FIGURE 3.75 Graphical description of response spectrum.

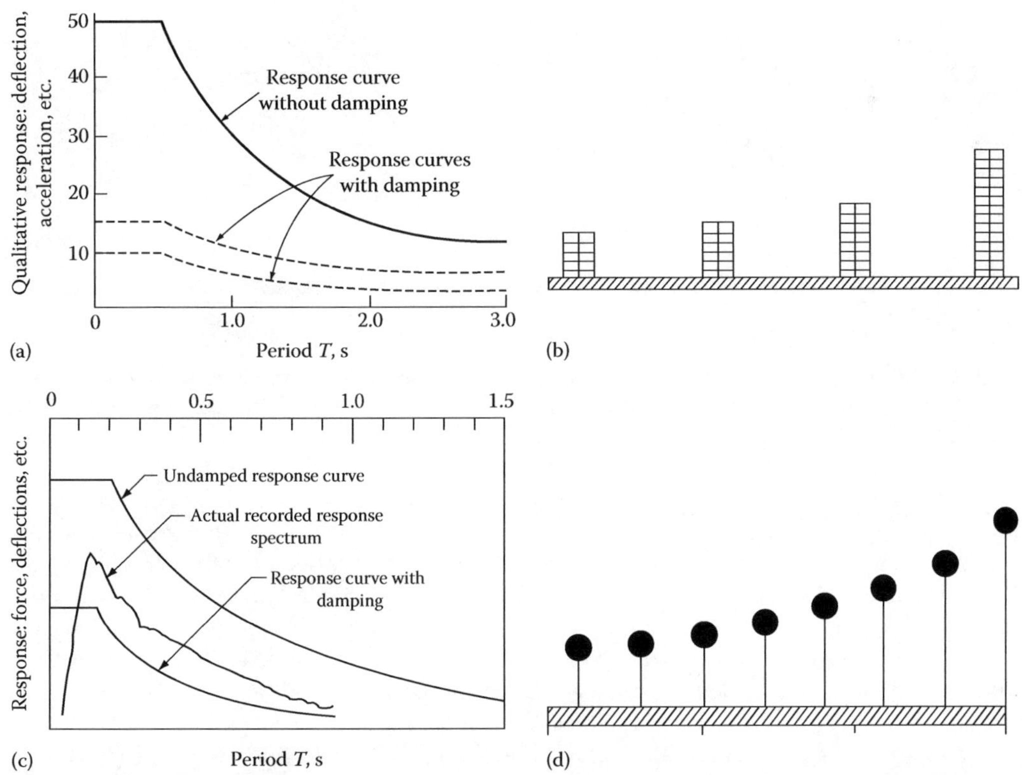

FIGURE 3.76 (a and c) Concept of response spectrum (b and d) pendulums of varying heights representing progressively taller building.

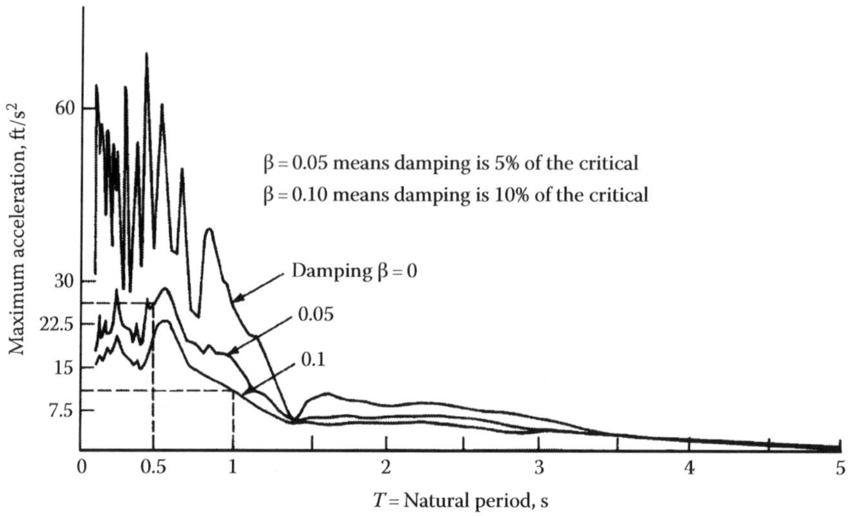

FIGURE 3.77 Acceleration response spectrum: El Centro earthquake.

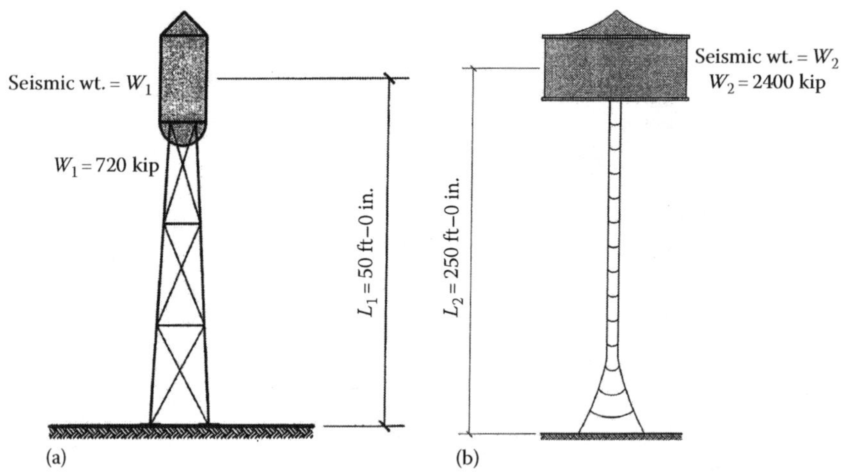

FIGURE 3.78 Examples of SDOF systems: (a) elevated water tank and (b) restaurant atop tall concrete core. *Note*: The acceleration for the water tank = 26.25 ft/s^2 for T = 0.5 s and β = 0.05 and the acceleration for the water tank = 11.25 ft/s^2 for T = 1.00 s and β = 0.10.

restaurant supported at the top of a tall concrete core (see Figure 3.78). We will neglect the mass of the columns supporting the tank and consider only the mass m_1 of the tank in the dynamic analysis. Similarly, the mass m_2 assigned to the restaurant is the only mass considered in the second structure. Given the simplified models, let us examine how we can calculate the lateral loads for both these structures resulting from an earthquake, for example, one that has the same ground motion characteristics as the 1940 El Centro earthquake shown in Figure 3.79. To evaluate the seismic lateral loads, we shall use the recorded ground acceleration for the first 30 s. Observe that the maximum acceleration recorded is 0.33g occurring about 2 s after the start of the record.

As a first step, the base of the two structures is analytically subjected to the same acceleration as the El Centro recorded acceleration. The purpose is to calculate the maximum dynamic response experienced by the two masses during the first 30 s of the earthquake. The maximum response such as DVA response of an SDOF system such as the two examples considered here may be obtained by

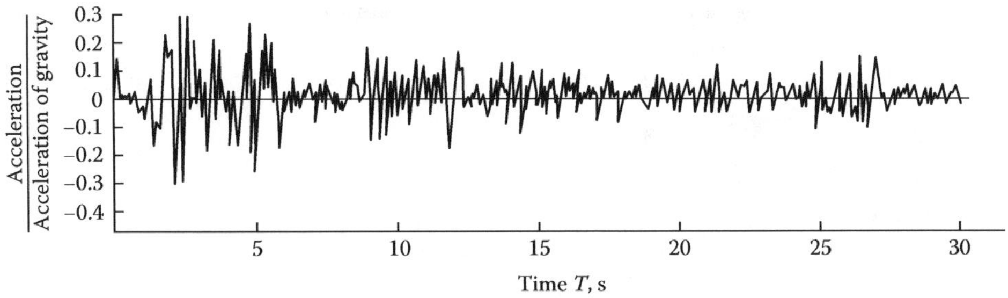

FIGURE 3.79 Recorded ground acceleration: El Centro earthquake.

considering the earthquake effects as a series of impulsive loads and then integrating the effect of individual impulses over the duration of the earthquake. This procedure, known as Duhamel integration method, requires considerable analytical effort. However, it is not necessary for us to carry out this procedure because maximum response for postulated earthquakes at a given site is already established. The spectral acceleration responses for the north–south component of the El Centro earthquake, shown in Figure 3.79, are one such example.

To determine the seismic lateral loads, assume that the tank and restaurant structures weigh 720 and 2400 kips, with corresponding periods of vibration of 0.5 and 1 s, respectively. Since the response of a structure is strongly influenced by damping, it is necessary to estimate the damping factors for the two structures. Let us assume that the percentage of critical damping £ for the tank and restaurant are 5% and 10% of the critical damping, respectively. From Figure 3.77, the acceleration for the tank structure is 26.25 ft/s^2, and the horizontal force in kips would be equal to the mass at the top times the acceleration.

Thus, the horizontal force, F, for the water tank is equal to

$$F = \frac{720}{32.2} \times 26.25 = 587 \text{ kip} \tag{3.30}$$

Similarly, for the restaurant structure, the force at top is equal to

$$F = \frac{2400}{32.2} \times 11.25 = 838.5 \text{ kip} \tag{3.31}$$

The two structures can then be designed by applying the seismic loads at the top and determining the associated forces, moments, and deflections. The lateral load, evaluated by multiplying the response spectrum acceleration by the effective mass of the system, is referred to as base shear, and its evaluation forms one of the major tasks in earthquake analysis.

In the example, SDOF structures were chosen to illustrate the concept of spectrum analysis. A multistory building, however, is not typically modeled as an SDOF system because it will have as many modes of vibration as its degrees of freedom (DOFs). For practical purposes, the distributed mass of a building is lumped at discrete levels to reduce the DOFs to a manageable number. In multistory buildings, the masses are typically lumped at each floor level.

Thus, in the 2D analysis of a building, the number of modes of vibration corresponds to the number of levels, with each mode having its own characteristic frequency. The actual motion of a building is a linear combination of its natural modes of vibration. During vibrations, the masses vibrate in phase with the displacements as measured from their initial positions, always having the same relationship to each other. Therefore, all masses participating in a given mode pass the equilibrium position at the same time and reach their extreme positions at the same instant.

Using certain simplifying assumptions, it can be shown that each mode of vibration behaves as an independent SDOF system with a characteristic frequency. This method, called the modal superposition method, consists of evaluating the total response of a building by statistically combining the response of a finite number of modes of vibration.

A building, in general, vibrates with as many mode shapes and corresponding periods as its DOFs. Each mode contributes to the base shear, and for elastic analysis, this contribution can be determined by multiplying a percentage of the total mass, called effective mass, by an acceleration corresponding to that modal period. The acceleration is typically read from the response spectrum modified for damping associated with the structural system. Therefore, the procedure for determining the contribution of the base shear for each mode of a multidegree-of-freedom (MDOF) structure is the same as that for determining the base shear for an SDOF structure, except that an effective mass is used instead of the total mass. The effective mass is a function of the lumped mass and deflection at each floor with the largest value for the fundamental mode, becoming progressively less for higher modes. The mode shape must therefore be known in order to compute the effective mass.

Because the actual deflected shape of a building consists of a linear combination of its modal shapes, higher modes of vibration also contribute, although to a lesser degree, to the structural responses. These can be taken into account through use of the concept of a participation factor. The base shear for each mode is determined as the summation of products of effective mass and spectral acceleration at each level. The force at each level for each mode is then determined by distributing the base shear in proportion to the product of the floor weight and displacement. The design values are then computed using modal combination methods, such as the complete quadratic combination (CQC) or the square root of the sum of the squares (SRSS), the preferred method being the former.

3.17.1 Earthquake Response Spectrum

Response spectrum analysis is a means of using acceleration response spectra to determine the maximum forces and displacements in a structure that remains elastic when it responds to ground shaking. For SDOF structures, the maximum elastic structural displacement is given by

$$\Delta = \frac{T^2}{4\pi^2} S_a \tag{3.32}$$

In this equation, T is the structure's period and S_a is the spectral acceleration obtained from the response spectrum plot at period T. The maximum force demand on the structure is given by

$$F = \frac{W}{g} S_a = K\Delta \tag{3.33}$$

For MDOF structures, the response of the structure can be determined by combining the response quantities for a series of SDOF structures having the same period and mass as each of the structure's modes. For mode i, the maximum inertial force produced in the structure by the earthquake, which is also termed the modal base shear, V_i, is given by

$$V_i = M_i S_a \tag{3.34}$$

In this equation, M_i is the modal mass for mode i and S_a is the spectral acceleration obtained from the response spectrum at natural period T_i.

The results of the analyses conducted for the various modes must be combined in order to obtain an estimate of the structure's actual behavior. Since it is unlikely that peak structural response in all modes will occur simultaneously, statistical combination rules are used to combine the modal results in a manner that more realistically assesses the probable combined effect of these modes. One such combination method takes the combined value as the SRSS of the peak response quantities in each mode.

When several modes have similar periods, the SRSS method does not adequately account for modal interaction. In this case, the CQC technique is used. Most structural analysis software used in design offices today provides the capability to perform these computations automatically.

For SDOF structures, the response spectrum analysis gives exact results, as long as the response spectrum that is used to represent the loading accurately represents the ground motion. However, the response spectra contained in building codes only approximate the ground motion from real earthquakes, and therefore, analysis using these spectra will be approximate. For MDF structures, response spectrum analysis is always approximate because the way that the peak displacements and forces from the various modes are combined does not accurately represent the way these quantities will actually combine in a real structure subjected to ground shaking. Although the results of response spectrum analysis are approximate, it is universally accepted as a basis for earthquake-resistant design, when properly performed.

An acceleration response spectrum as stated previously is a plot of the maximum acceleration that SDOF structures having different periods, T, would experience when subjected to a specific earthquake ground motion. This plot is constructed by performing response-history analyses for a series of structures, each having a different period, T, obtaining the maximum acceleration of each structure from the analysis, and plotting this as a function of T. Linear acceleration response spectra are most common and are obtained by performing LRH analysis.

Although the response spectra obtained from each earthquake record will be different, spectra obtained from earthquakes having similar magnitudes on sites with similar characteristics tend to have common characteristics. This has permitted the building codes to adopt standard response spectra that incorporate these characteristics and envelop spectra that would be anticipated at a building site during a DE. The response spectra contained in the building code are called smoothed design spectra because the peaks and valleys that are common in the spectrum obtained from any single record are averaged out to form smooth functional forms that generally envelope the real spectra.

Earthquake response spectrum gives engineers a practical means of characterizing ground motions and their effects on structures. Introduced in 1932, it is now a central concept in earthquake engineering that provides a convenient means to summarize the peak response of all possible linear SDOF systems to a particular ground motion. It also provides a practical approach to apply the knowledge of structural dynamics to the design of structures and development of lateral force requirements in building codes.

A plot of the peak value of response quantity as a function of the natural vibration period T_n of the system (or a related parameter such as circular frequency ω_n or cyclic frequency f_n) is called the response spectrum for that quantity. Each such plot is for SDOF systems having a fixed damping ratio ξ. Oftentimes, several such plots for different values of ξ are included to cover the range of damping values encountered in actual structures. Whether the peak response is plotted against f_n or T_n is a matter of personal preference. In this text, we use the latter because engineers are more comfortable in using natural period rather than natural frequency because the period of vibration is a more familiar concept and one that is intuitively appealing. Although a variety of response spectra can be defined depending on the chosen response quantity, it is almost always the acceleration response spectrum, a plot of pseudoacceleration, against the period T_n for a fixed damping, ξ, that is most often used in the practice of earthquake engineering.

A similar plot of displacement u is referred to as the deformation spectrum, while that of velocity $\dot{u}$ is called a velocity spectrum.

It is worthwhile to note that only the deformation $u(t)$ is needed to compute internal forces. Obviously, then, the deformation spectrum provides all the information necessary to compute the peak values of deformation and internal forces. The pseudovelocity and pseudoacceleration response spectra are important, however, because they are useful in studying the characteristics of response spectra, constructing design spectra, and relating structural dynamics results to building codes.

3.17.2 Deformation Response Spectrum

To explain the procedure for determining the deformation response spectrum, we start with the spectrum developed for El Centro ground motion, which has been studied extensively in textbooks.

The acceleration recorded for the first 30 s of ground motion is shown in Figure 3.80a. The corresponding deformations in the three SDOF systems of varying periods are presented in Figure 3.80b.

The peak deformations are

$u_0 = 2.67$ in. for a system with natural period $T_n = 0.5$ s and damping ratio $\xi = 2\%$
$u_0 = 5.97$ in. for a system with $T_n = 1$ s and $\xi = 2\%$
$u_0 = 7.47$ in. for a system with $T_n = 2$ s and $\xi = 2\%$

The u_0 value so determined for each system provides one point on the deformation response spectrum. Repeating such computations for a range of values of T_n while keeping c constant at 2% provides the deformation response spectrum shown in Figure 3.81. The spectrum shown is for a single damping value, $\xi = 2\%$. However, a complete response spectrum would include such spectrum curves for several values of damping.

3.17.3 Pseudovelocity Response Spectrum

The pseudovelocity response spectrum is a plot of V as a function of the natural vibration period T_n or natural vibration frequency f_n of the system. For a given ground motion, the peak pseudovelocity

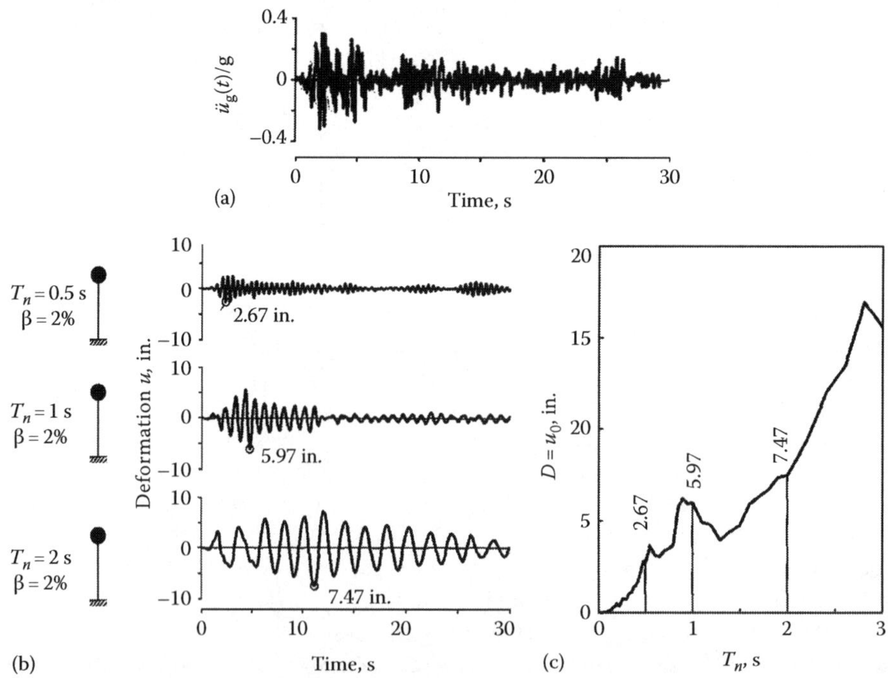

FIGURE 3.80 (a) Ground acceleration; (b) deformation response of three SDOF systems with β = 2% and $T_n = 0.5$, 1, and 2 s; and (c) deformation response spectrum for β = 2%.

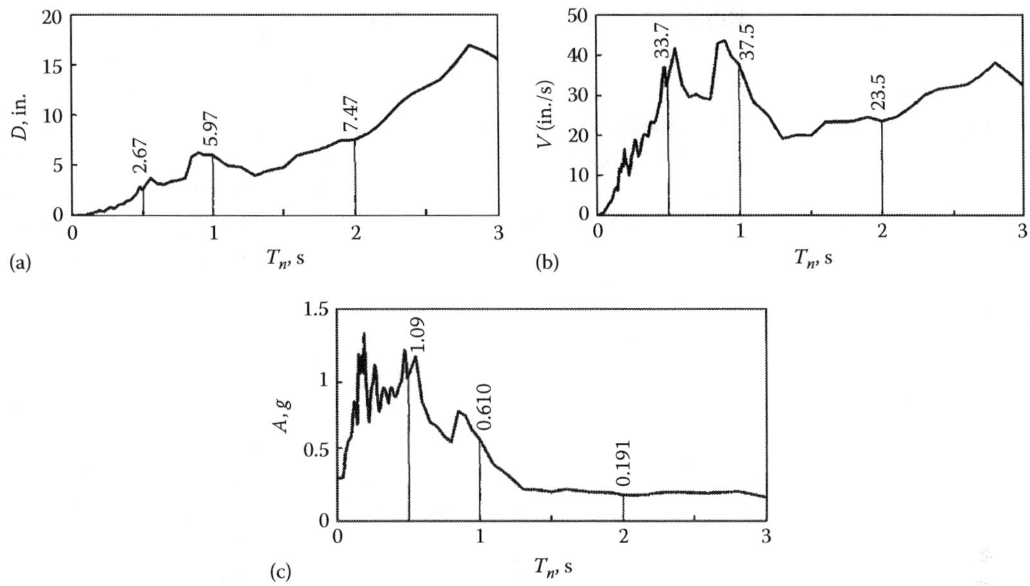

FIGURE 3.81 Response spectra ($\beta = 2\%$) for El Centro ground motion: (a) deformation response spectrum, (b) pseudovelocity response spectrum, and (c) pseudoacceleration.

V for a system with natural period T_n can be determined from the following equation using the deformation $\xi = 2\%$ of the same system from the response spectrum of Figure 3.81:

$$V = \omega_a D = \frac{2\pi}{T_n} D \tag{3.35}$$

As an example, for a system with $T_n = 0.5$ s and $\xi = 2\%$, $D = 2.67$ in.:

$$V = \omega_a D = \frac{2\pi}{T_n} D = \left(\frac{2\pi}{0.5}\right) 2.67 = 33.7 \text{ in./s} \tag{3.36}$$

Similarly, for $T_n = 1.0$ s and $\xi = 2\%$, $D = 5.97$ in.:

$$V = \frac{2\pi}{1} 5.97 = 37.5 \text{ in./s} \tag{3.37}$$

And for $T_n = 2.0$ s and the same damping $\xi = 2\%$, $D = 7.47$ in.:

$$V = \left(\frac{2\pi}{2}\right) 7.47 = 23.5 \text{ in./s} \tag{3.38}$$

These three values of peak pseudovelocity V are identified in Figure 3.81b. Repeating such computations for a range of values of T_n while keeping ξ constant at 2% provides the pseudovelocity spectrum shown in Figure 3.81b. The prefix *pseudo* is used for V because it is not equal to the peak velocity, although it has the same units for velocity.

3.17.4 Pseudoacceleration Response Spectrum

It has been stated many times in this chapter that the base shear is equal to the inertial force associated with the mass *m* undergoing acceleration *A*. This acceleration *A* is generally different from the peak acceleration of the system. It is for this reason that *A* is called the peak pseudoacceleration; the prefix *pseudo* is used to avoid possible confusion with the true peak acceleration, just as we did for velocity *V*. The pseudoacceleration response spectrum is a plot of acceleration *A* as a function of the natural vibration period T_n or natural vibration frequency f_n of the system. For a given ground motion, peak pseudoacceleration *A* for a system with natural period T_n and damping ratio ξ can be determined from the following equation using the peak deformation *D* of the system from the response spectrum:

$$A = \omega_n^2 D = \left(\frac{2\pi}{T_n}\right)^2 D \tag{3.39}$$

As an example, for a system with $T_n = 0.5$ s and $\xi = 2\%$, $D = 2.67$ in.:

$$A = \omega_n^2 D = \left(\frac{2\pi}{T_n}\right)^2 D = \left(\frac{2\pi}{0.5}\right)^2 2.67 = 1.09\,g \tag{3.40}$$

where $g = 386$ in./s^2.

Similarly, for a system with $T_n = 1$ s and $\xi = 2\%$, $D = 5.97$ in.:

$$A = \left(\frac{2\pi}{T_n}\right)^2 D = \left(\frac{2\pi}{1}\right)^2 5.97 = 0.610\,g \tag{3.41}$$

And for a system with $T_n = 2$ s and the same damping $\xi = 2\%$, $D = 7.47$ in.:

$$A = \left(\frac{2\pi}{2}\right)^2 7.47 = 0.191\,g \tag{3.42}$$

The three values T_n of pseudoacceleration, *A*, are shown in Figure 3.82. Repeating similar computations for a range of T_n values, while keeping β constant at 2%, yields the pseudoacceleration spectrum shown in Figure 3.81c.

3.17.5 Tripartite Response Spectrum: Combined Displacement–Velocity–Acceleration Spectrum

It was shown in the previous section that each of the deformation, pseudovelocity, and pseudoacceleration response spectra for a given ground motion contains the same information, no more and no less. The three spectra are simply distinct ways of displaying the same information on structural response. With knowledge of one of the spectra, the other two can be derived by algebraic operations using the procedure given in the previous section.

If each of the spectra contains the same information, why do we need three spectra? There are two reasons. One is that each spectrum directly provides a physically meaningful quantity: The deformation spectrum provides the peak deformation of a system, the pseudovelocity spectrum gives the peak strain energy stored in the system during the earthquake, and pseudoacceleration spectrum yields

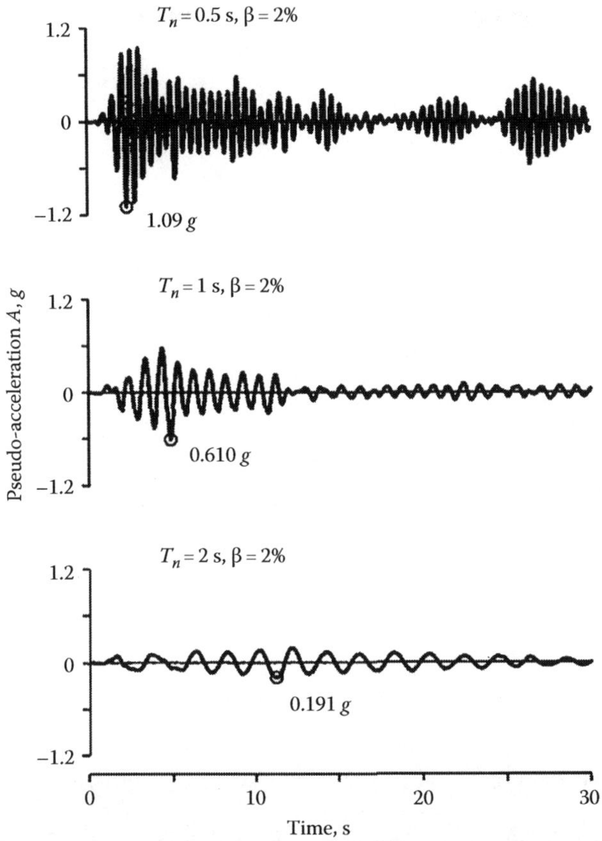

FIGURE 3.82 Pseudoacceleration responses of SDOF systems to El Centro ground motion.

directly the peak value of the equivalent static force and base shear. The second reason lies in the fact that the shape of the spectrum can be approximated more readily for design purposes with the aid of all three spectral quantities rather than any one of them alone. For this purpose, a combined plot showing all three of the spectral quantities is especially useful. This type of plot was developed for earthquake response spectra for the first time by A.S. Veletsos and N.M. Newmark in 1960.

In an integrated DVA spectrum, the vertical and horizontal scales for V and T_n are standard logarithmic scales. The two scales for D and A sloping at +45° and −45°, respectively, to the T_n-axis are also logarithmic scales but not identical to the vertical scale. The pairs of numerical data for V and T_n that were plotted in Figure 3.81 on linear scales are replotted in Figure 3.83 on logarithmic scales. For a given natural period T_n, the D and A values can be read from the diagonal scales. As an example, for $T_n = 2$ s in Figure 3.83, it gives $D = 7.47$ in. and $A = 0.191g$. The four-way plot is a compact presentation of the three—deformation, pseudovelocity, and pseudoacceleration—response spectra, for a single plot of this form replaces the three plots.

The benefit of the response spectrum in earthquake engineering may be recognized by the fact that spectra for virtually all ground motions strong enough to be of engineering interest are now computed and published soon after they are recorded. From these, we can get a reasonable idea of the kind of motion that is likely to occur in future earthquakes. It should be noted that for a given ground motion response spectrum, the peak value of deformation, pseudovelocity, and base shear in any linear SDOF can be readily read from the spectra without resorting to dynamic analyses. This is because the computationally intense dynamic analysis has been completed in generating the response spectrum.

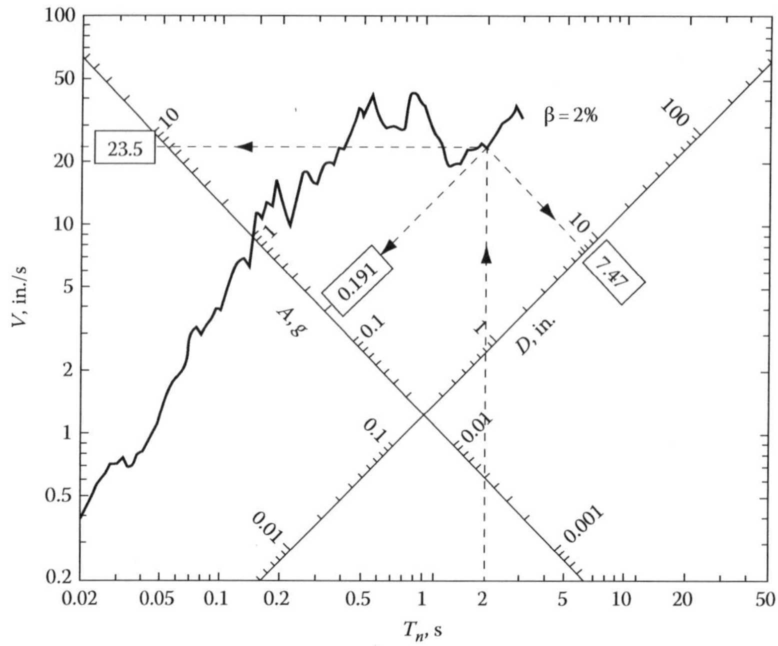

FIGURE 3.83 Combined DVA response for El Centro ground motion, β = 2%.

Given these advantages, it makes good sense to have geotechnical engineers provide tripartite response spectrum rather than just an acceleration spectrum, when site-specific studies are commissioned.

An example of tripartite response spectrum is shown in Figure 3.84. The spectrum is for maximum capable earthquake (MCE) of magnitude 8.5 occurring at San Andreas fault. The project site is in downtown Los Angeles located at a distance of 34 miles from San Andreas fault.

The response spectrum tells us that the forces experienced by buildings during an earthquake are not just a function of the quake but are also their dynamic response characteristics to the quake. The response primarily depends on the period of the building being studied. A great deal of single-mode information can be read directly from the response spectrum. Referring to Figure 3.85, the horizontal axis of the response spectrum expresses the period of the building being affected by the quake. The vertical axis shows the velocity attained by this building during the quake. The diagonal axis running up toward the left-hand corner reads the maximum accelerations to which the building is subjected. The axis at right angles to this will read the displacement of the building in relation to the support. Superimposed on these tripartite scales are the response curves for an assumed 5% damping. Now let us study how buildings with different periods respond to the earthquake described by these curves.

If the building is to be studied had a natural period of 1 s, we would start at the bottom of the chart at $T = 1$ s and reference vertically until we intersect the response curve. From this intersection, point A, we travel to the extreme right and read a velocity of 16 in./s. Following a displacement line diagonally down to the right, we find a displacement of 2.5 in. Similarly following the acceleration line down to the left, we see that the building will experience an acceleration of 0.25*g*. If we then move to the 2 s period, point B, in the same sequence, we find that we will have the same velocity of 16 in./s, a displacement of 4 in., and a maximum acceleration of 0.10*g*. If we then move to 4 s, point C, we see a velocity of 16 in./s, a displacement of 10 in., and an acceleration of 0.06*g*. If we run all out to 10 s, point D, we find a velocity of 7 in./s, a displacement of 10 in. the same as for point C, and an acceleration of 0.0l*g*. Notice that the values vary widely, as started earlier, depending on the period of building exposed to this particular quake.

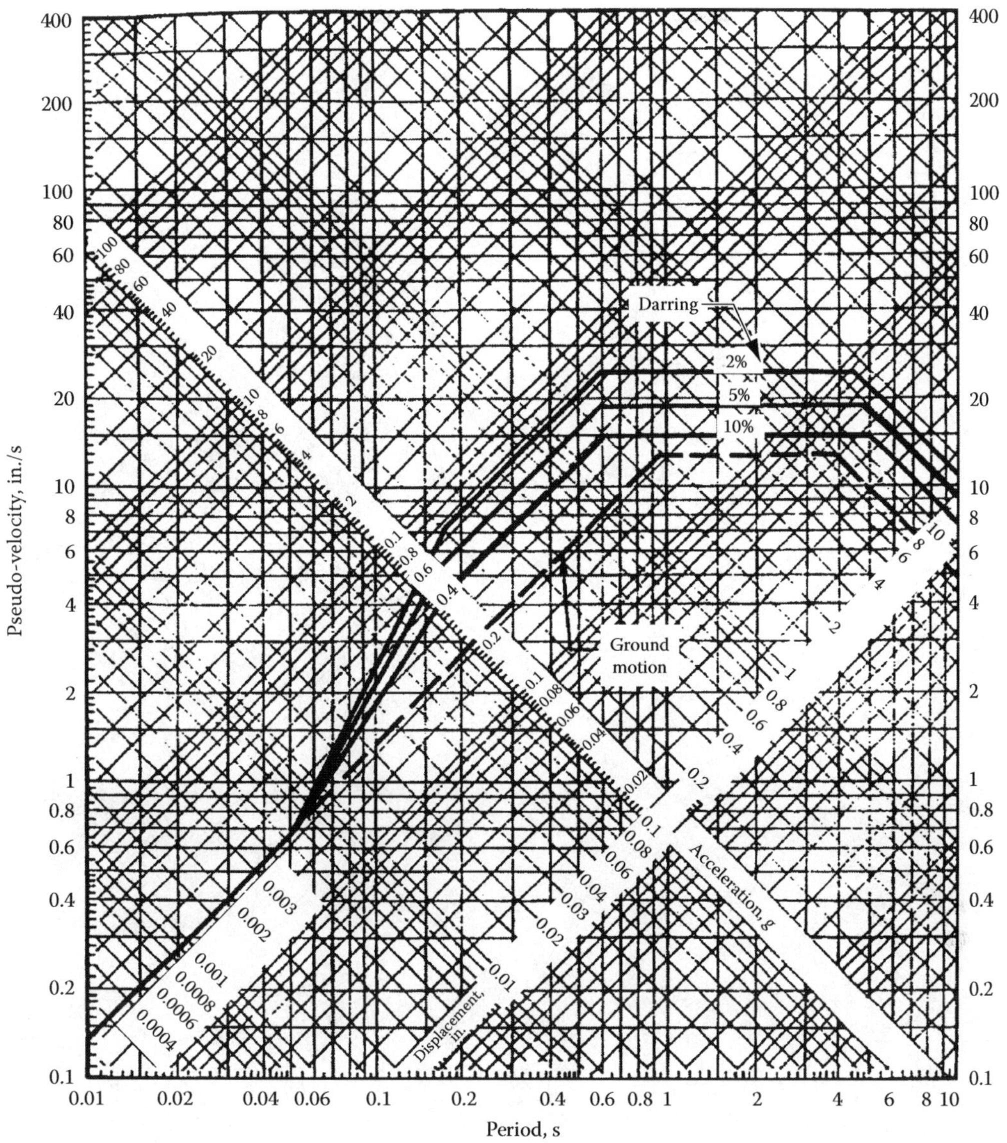

FIGURE 3.84 Tripartite response spectrum.

3.17.6 Characteristics of Response Spectrum

We now study the important properties of earthquake response spectra. For this purpose, we use once again an idealized response spectrum for El Centro ground motion shown in Figure 3.85. The damping, ξ, associated with the spectrum is 5%. The period T_n plotted on a logarithmic scale covers a wide range, $T_n = 0.01–10$ s.

Consider a system with a very short period, say 0.03 s. For this system, the pseudoacceleration A approaches the ground acceleration, while the displacement D is very small. There is a physical reasoning for this trend. For purposes of dynamic analysis, a very-short-period system is extremely stiff and may be considered essentially rigid. Such a system would move rigidly with the ground as if it is a part of the ground itself. Thus, its peak acceleration would be approximately equal to the ground acceleration as shown in Figure 3.86.

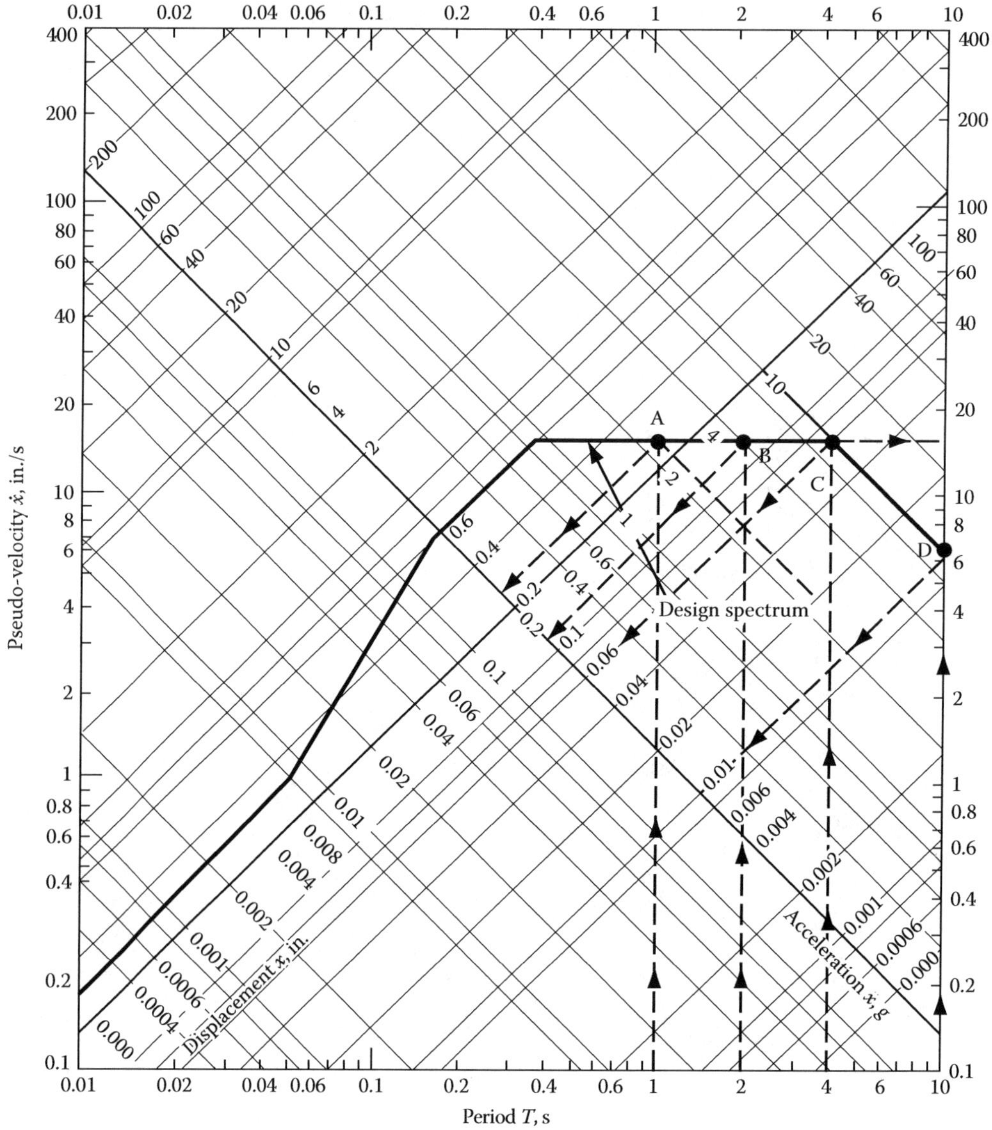

FIGURE 3.85 Velocity, displacement, and acceleration readout from response spectra.

Next, we examine a system with a very long period, say $T_n = 10$ s. The acceleration A, and thus the force in the structure, which is related to mA, would be small. Again, there is a physical reasoning for this trend: A very-long-period system is extremely flexible. Therefore, the mass at top would remain stationary while the base moves with the ground below (see Figure 3.87).

Based on these two observations, and those in between the two periods (not examined here), it is logical to divide the spectrum into three period ranges: (1) the long-period region to the right of point D, called the displacement-sensitive region because structural response is most directly related to ground displacement; (2) the short-period region to the left of point C, called the acceleration-sensitive region because structural response is most directly related to ground acceleration; and (3) the intermediate-period region between points C and D, called the velocity-sensitive region because structural response appears to be better related to ground velocity than to other ground motion parameters.

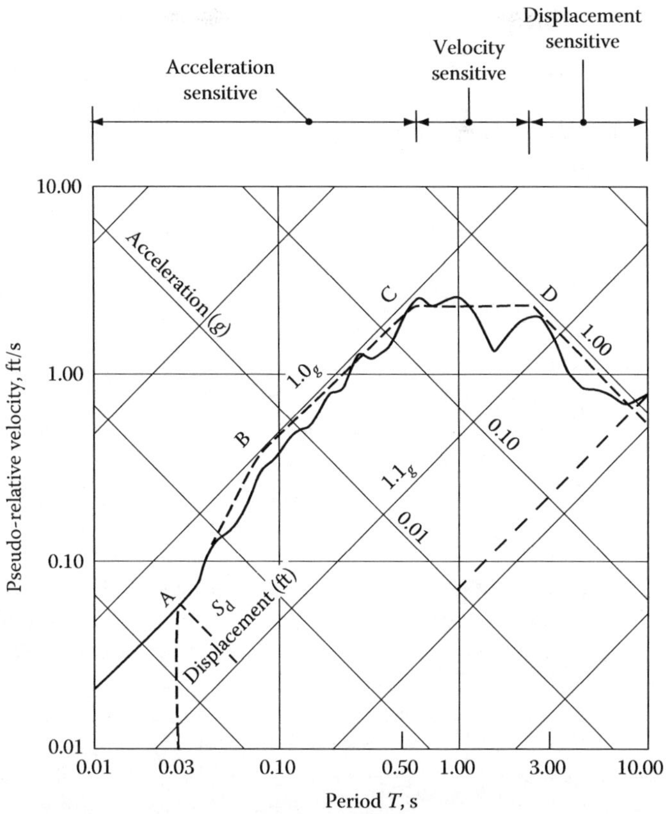

FIGURE 3.86 Idealized response spectrum for El Centro ground motion.

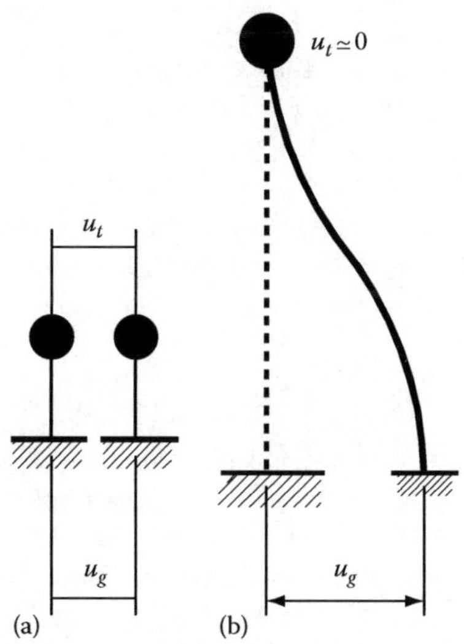

FIGURE 3.87 Schematic response of rigid and flexible systems: (a) rigid system, acceleration at top is nearly equal to the ground acceleration; (b) flexible system, structural response is most directly related to ground displacement.

The preceding discussion should be helpful in recognizing the usefulness of four-way logarithmic plot of the combined deformation, pseudovelocity, and pseudoacceleration response spectra. These observations would be difficult to discover from the three individual spectra.

We now turn our attention to damping that has significant influence on the earthquake response of buildings. It reduces the response of a structure, as expected. However, the reduction achieved with a given amount of damping is different in the three spectral regions. In the limit as $T_n \to \infty$, damping does not affect the response because the structural mass stays still while the ground underneath moves. Among the three period regions, the effect of damping tends to be greatest in the velocity-sensitive region of the spectrum. In this spectral region, the effect of damping depends on the ground motion characteristics. If the ground motion is harmonic over many cycles as it was in the Mexico City earthquake of 1985, the effect of damping would be especially large for systems near resonance.

The motion of structure and the associated forces could be reduced by increasing the effective damping of the structure. The addition of dampers achieves this goal without significantly changing the natural vibration period of the structure. Viscoelastic dampers have been used in many structures; for example, 10,000 dampers were installed throughout the height of each tower of the now nonexistent World Trade Center in New York City to reduce wind-induced motion to within a comfortable range for the occupants. In recent years, there is a growing interest in developing dampers suitable for structures in earthquake-prone regions. Because the inherent damping in most structures is small, their earthquake response can be reduced significantly by the addition of dampers. These can be especially useful in improving the seismic safety of existing structures.

As stated previously, the response of an SDOF system at the extremes of the fundamental periods is intuitively obvious. For example, a system with a very short period, that is, a very stiff system, experiences peak acceleration approaching that of the ground. The distortion of the system, however, is negligible since the motion of the mass is the same as that of the ground. On the other hand, for systems with a very long period, that is, flexible buildings, the mass remains stationary while the ground moves beneath it. The relative displacement between the two is equal to the ground displacement. However, the acceleration of the mass is zero or nearly so. Thus, the force in the structure related to *ma* would also be small. This concept is illustrated in Figure 3.88.

3.17.7 Difference between Design and Actual Response Spectra

The response spectra for a given earthquake are typically a jagged plot representing the peak response of SDOF systems. It is a unique plot representative of the particular ground motion. The design spectrum is, however, a smooth plot specifying the level of seismic force as a function of the systems' period and damping ratio. This is the first difference. The second difference is the design spectrum is an envelope of two or more different elastic spectra that could affect ground motions at a given site. After determining the design spectrum for each of the postulated earthquakes, the design spectrum for the site is developed by enveloping the design spectra for all earthquakes considered for the site, as shown schematically in Figure 3.88a.

3.17.8 Summary of Response Spectrum Analysis

This is a well-devised method of determining peak response of a system directly from the response spectrum for the ground motion without having to carry a time-history analysis. For SDOF systems, the method is quite accurate. For MDOF systems, in a manner of speaking, the result is not exact. However, the result is accepted in practice, as being accurate enough for seismic design applications.

In general, modal responses attain their peaks at different time instants. Therefore, approximations must be used in combining the peak responses determined from response spectrum analysis because no information is available as to when these peak modal values occur. Summing up the absolute values of the maximum values would certainly give us an upper bound solution but this is not used in building design practice because the results tend to be too conservative.

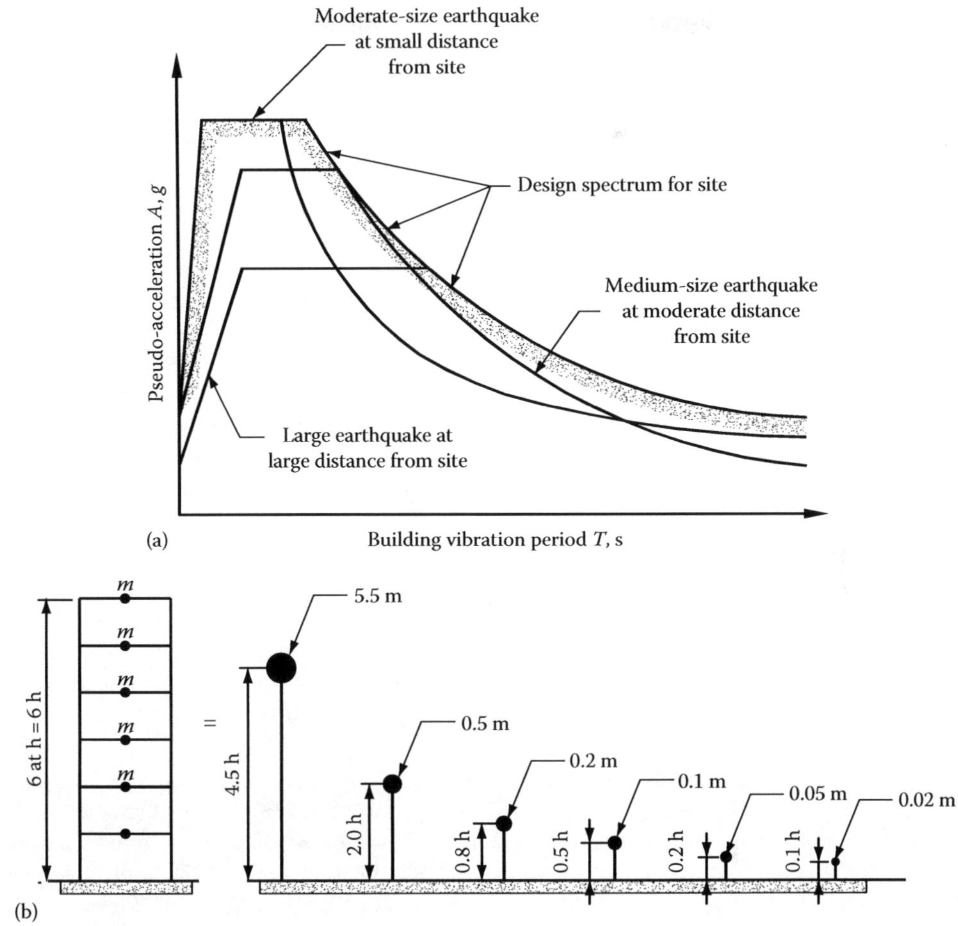

FIGURE 3.88 (a) Comparison of design and actual response spectra and (b) concept of effective modal masses and effective modal heights.

In practice, two other modal combination rules, the SRSS and CQC, both of which are based on random vibration theory, are popular. The latter is the preferred method applicable to a wider class of structures as it is also applicable to structures with not-so-well-separated periods.

The response spectrum analysis is a procedure for dynamic analysis of a structure subjected to earthquake ground motions. But it reduces to a series of static analysis. For each mode considered, static analysis of the structure subjected to corresponding modal shears provides the peak modal static response. This procedure avoids the dynamic analysis of SDOF systems necessary for time-history analysis. Thus, the response spectrum analysis may be considered as a pseudo-dynamic analysis procedure, because it indirectly uses the vibration properties of the structure and the dynamic characteristics of the ground motion. The intensive time-history calculations are not required because in developing smooth design spectra, somebody has already done these by converting the dynamic loads into equivalent static loads. Conceptually equivalent loads may be thought of as static loads multiplied by their corresponding deflection load factors.

Response spectrum analysis is a procedure to compute the peak response of a structure during an earthquake directly from the earthquake design spectrum without the need for response-history analysis of the structure. The procedure is not an exact predictor of peak response, but it provides an estimate that is considered sufficiently accurate for structural design applications.

To assist us in developing an understanding of the response behavior of multistory buildings, a conceptual explanation of modal analysis is given in Figure 3.88b.

It should be obvious by now that the effective modal masses and effective modal heights for a given building depend on the characteristics of the response spectrum and the dynamic properties of the building itself. Therefore, the values shown for a six-story example building in Figure 3.88b are for conceptual purposes only and are not based on any given design response spectrum or building properties.

3.17.9 Hysteresis Loop

The ability of structural elements to withstand deformations in the elastic range is called *ductility*. In a major earthquake, a structure will not remain *elastic* but will be forced into the *inelastic* range. Inelastic action absorbs significantly more energy from the system. Therefore, if a structure is properly detailed and constructed so that it can perform in a ductile manner (i.e., deform in the inelastic range), it can be designed for considerably smaller lateral forces. A good way to visualize the concept of ductility is perhaps to follow through the load–deformation characteristics of a member subject to an applied cyclic loading.

If dynamic loads in a seismic event deform a structure beyond the elastic range of the material, the resulting motion is called inelastic response. Such excursions beyond the elastic range are not usually permitted in normal operating conditions such as under gravity and wind loads. However, the behavior of and the resulting damage in structures subject to extreme load conditions are quite important in structural design. For example, a building subjected to blast or severe earthquake loading will probably be deformed inelastically, and therefore, it is of interest to evaluate its stability to assure that the building can sustain gravity loads without collapse.

Let us consider an idealized 2D steel frame shown in Figure 3.89 subjected to a lateral load P applied at the top. If the flexural rigidities of the columns are less than that of the beam and the load is increased infinitely, at some point in the loading history, so-called plastic hinges will form at the ends of the column. A plot of the load P against the displacement x is linear up to the value of P_{y1} (see the line labeled 1 in Figure 3.90), where yielding of the material begins. Subsequently, it curves (see the curve labeled 2), due to softening of columns at the base and top. Upon unloading, the material rebounds elastically, as indicated by part 3 of the plot in Figure 3.90.

If a reverse loading is then applied, the parts of Figure 3.90 labeled 4 and 5 result, and a subsequent unloading produces line 6. If the maximum positive and negative forces P_{m1} and $-P_{m2}$ (the ordinates of points B and E on the diagram) are numerically equal, the *hysteresis loop* formed by cyclic loading is symmetric with respect to the origin.

The curved portions of Figure 3.90 are often replaced by straight lines approximating the true behavior. Figure 3.91 illustrates such a simplified load–displacement diagram, called a *bilinear inelastic restoring force*. It consists of two parallel lines (labeled 2 and 4 in Figure 3.91)

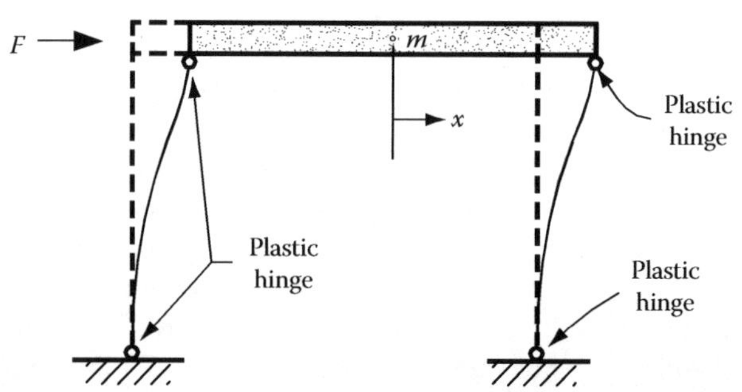

FIGURE 3.89 Idealized steel portal frame subject to load P at top.

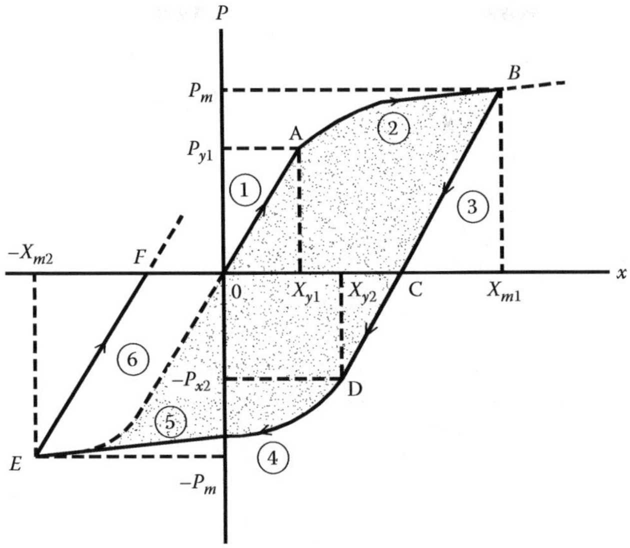

FIGURE 3.90 Displacement plot of load P versus displacement x.

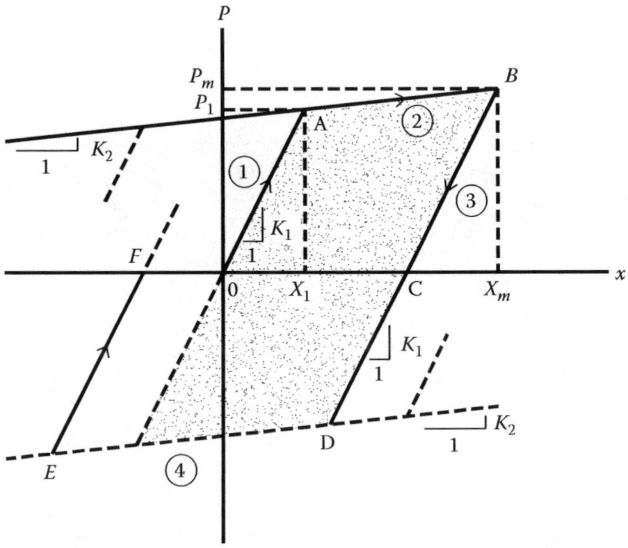

FIGURE 3.91 Plot of idealized displacement x versus load P: observe the curved portion is replaced by straight lines depicting bilinear inelastic behavior.

for inelastic behavior and a family of parallel lines (of which those labeled 1 and 3 are representative) for elastic behavior. If the slopes of lines 2 and 4 are zero, as in Figure 3.92, the diagram represents an *elastoplastic restoring force*. That is, the plot of P against x consists of straight-line segments, where the behavior is assumed to be either perfectly elastic or perfectly plastic. For example, let us reconsider the frame in Figure 3.89 and suppose that the load P increases to the value P_m (see point A in Figure 3.92). If the plastic hinges in Figure 3.92 are assumed to form instantaneously, the displacement increases without a corresponding increase of the load, as shown by the horizontal line from A to B in Figure 3.92. A decrease in the load causes a decrease of displacement in accordance with line (3) in Figure 3.93 and so on.

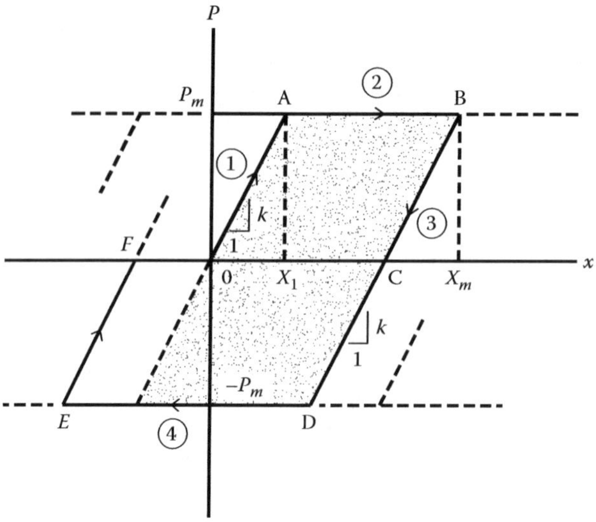

FIGURE 3.92 Elastoplastic hysteresis loop.

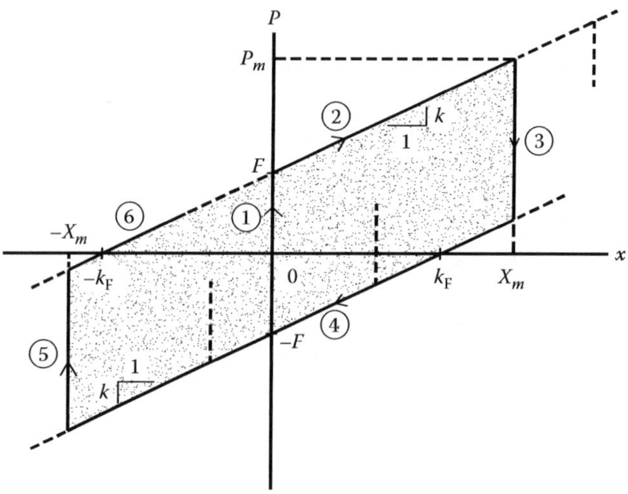

FIGURE 3.93 Rigid-plastic hysteresis loop.

The hysteresis loop inherent in elastoplastic analyses represents a discretized form of structural damping. In this case, all of the dissipated energy is tacitly assumed to be absorbed at the plastic hinges. This dissipative mechanism is often referred to as *elastoplastic damping*, which represents a particular case of *hysteretic damping*.

It is interesting to note that if the slopes of inclined lines 1 and 2 are taken to be infinite (meaning that the deformation of the portal in its elastic range is assumed to be small in comparison to the plastic deformation), the hysteresis loops simplify to a rigid-plastic dissipative mechanism as shown in Figure 3.93.

The importance of understanding the use of hysteresis loops in seismic design may be explained as follows:

Inelastic response occurs when the amplitude of earthquake shaking is strong enough to cause forces in a structure that exceed the strength of any of the structure's elements or connections. When this occurs, the structure may experience a variety of behaviors. If the elements that are strained beyond their elastic strength limit are brittle, they will tend to break and lose the ability to resist any further load. This type of behavior is typified by a steel tension member that is stretched such that the force in the brace exceeds the ultimate strength of its end connections or by an unreinforced concrete element that is strained beyond its cracking strength. If the element is ductile, it may exhibit plastic behavior, being able to maintain its yield strength as it is strained beyond its elastic limit. This type of behavior is typified by properly braced, compact section beams in moment frames; by the cores of buckling-restrained braces; and by the shear links in eccentrically braced frames. Even elements that are ductile and capable of exhibiting significant postyielding deformation without failure will eventually break and lose load-carrying capacity due to low-cycle fatigue if plastically strained over a number of cycles.

Modern structural analysis software provides the capability to analyze structures at deformation levels that exceed their elastic limit. In order to do this, these programs require input on the hysteretic (nonlinear force vs. deformation) properties of the deforming elements. Hence, knowledge of how these loops are generated is of importance in seismic design.

3.17.10 Seismology

The essential background for practice in the field of earthquake engineering is knowledge about the earthquake itself. The detailed study of earthquakes and earthquake mechanisms lies in the province of seismology, but in their studies, the earthquake engineers must take a different point of view than the seismologist. Seismologists have focused their attention primarily on the global effects of earthquakes and are also concerned with small-amplitude ground motions that induce no significant structural response. Engineers, on the other hand, are concerned mainly with the local effects of large earthquakes where the ground motions are intense enough to cause structural damage. Nevertheless, even though the objectives of earthquake engineers differ from those of seismologist, there are many topics in seismology that are of immediate engineering interest.

The seismicity of a region determines the extent to which earthquake loadings may control the design of a structure planned for that location and the principal indicator of the degree of seismicity in the historical record of earthquakes that have occurred in the region. Because major earthquakes often have had disastrous consequences, they have been noted in chronicles dating back to the beginnings of civilization. In China, records have been kept that are thought to include every major destructive seismic event for a time span of nearly 3000 years.

More recently, beginning from the 1970s, earthquake occurrence information has been compiled from strong seismograph motion, which indicates the location and magnitude of all earthquakes. The most obvious conclusion to be drawn from these maps is that earthquake occurrences are not distributed uniformly over the surface of the Earth. Instead, they tend to be concentrated along well-defined lines, which are known to be associated with the boundaries of segments or *plates* of the Earth's crust.

Within the structure of the Earth, the mantle is considered to consist of two distinct layers. The upper mantle together with the crust forms a rigid layer called the lithosphere. Below that, a zone called the asthenosphere is thought to be partially molten rock consisting of solid particles incorporated within a liquid component. Because of its highly plastic character, the lithosphere can act as if it is floating on a liquid and thus can be subjected to large crustal deformations. The lithosphere does not move as a single unit; however, instead it is divided into a pattern of plates of various sizes, and it is the relative movements along the plate boundaries that cause the earthquake occurrence patterns. The detailed description of the motions of these plates is subjectedly called plate tectonics; development of understanding of this subject is one of the great advances of geology and seismology during the present century.

A special feature of earthquake excitation of structures, compared with most other forms of dynamic excitation, is that it is applied in the form of support motions rather than by external loads; thus, the effective seismic loadings must be established in terms of these motions. Defining the support motions is the most difficult and uncertain phase of the problem of predicating structural response to earthquakes. When these input motions have been established, however, the calculation of the corresponding stresses and deflections in any given structure is a standard problem of structural dynamics.

The most important aspect of an earthquake's ground motions is the effect they will have on structures, that is, the stresses and deformations or the amount of damage they would produce. This damage potential is, of course, least partly dependent on the *size* of the earthquake, and a number of measures of size are used for different purposes. The most important measure of size from a seismological point of view is the amount of strain energy released at the source, and this is indicated quantitatively as the magnitude, measured in micrometers (10^{-6}), of the earthquake record obtained by a Wood–Anderson seismograph, corrected to a distance of 100 km. This magnitude rating has been related empirically to the amount of earthquake energy released by the formula

$$\log E = 11.8 + 1.5\,M \tag{3.43}$$

in which M is the magnitude. By this formula, the energy increases by a factor of 12 for each unit increase of magnitude. More important to engineers, however, is the empirical observation that earthquakes of magnitude less than 5 are not expected to cause structural damage, whereas for magnitudes greater than 5, potentially damaging ground motions will be produced.

The magnitude of an earthquake by itself is not sufficient to indicate whether structural damage can be expected. This is a measure of the size of the earthquake at its source, but the distance of the structure from the source has an equally important effect on the amplitude of its response. The severity of the ground motions observed at any point is called the earthquake intensity; it diminishes generally with distance from the source, although anomalies due to local geological conditions are not uncommon. The oldest measures of intensity are based on observations of the effects of the ground motions on natural and man-made objects. In the United States, the standard measure of intensity for many years has been the Modified Mercalli (MM) scale. This is a 12-point scale ranging from I (not felt by anyone) to XII (total destruction). Results of earthquake-intensity observations are typically complied in the form of isoseismic maps. Although such subjective intensity ratings are very valuable in the absence of any instrumented records of an earthquake, deficiencies in providing criteria for the design of earthquake-resistant structures are obvious.

Basic information on the characteristics of earthquakes comes from strong-motion-recording accelerographs. Although their installations are radically increasing in the seismic region of the United States and various other parts of the world, the distribution of instruments is quite limited. Consequently, basic data concerning the influence of such factors as magnitude, distance, and local soil conditions of the characteristics of earthquake motions are still very scarce.

The three components of ground motion recorded by a strong-motion accelerograph provide a complete description of the earthquake, which would act upon any structure at that site. However, the most important features of the record obtained in each component from the standpoint of its effectiveness in producing structural response are the amplitude, the frequency content, and the duration. The amplitude generally is characterized by the peak value of acceleration or sometimes by the number of acceleration peaks exceeding a specified level. The frequency content can be represented roughly by the number of zero crossings per second in the accelerogram and the duration by the length of time between the first and last peaks exceeding a given threshold level. It is evident, however, that all these quantitative measures taken together provide only a very limited description of the ground motion and certainly do not quantify its damage-producing potential adequately.

3.18 SEISMIC DESIGN CONSIDERATIONS

Although structural design for seismic loading is primarily concerned with structural safety during major earthquakes, serviceability and the potential for economic loss are also of concern.

As such, seismic design requires an understanding of the structural behavior under large inelastic, cyclic deformations. Behavior under this loading is fundamentally different from wind or gravity loading. It requires a more detailed analysis and application of a number of stringent detailing requirements to assure acceptable seismic performance beyond the elastic range. Some structural damage can be expected when the building experiences design ground motions.

The seismic analysis and design of buildings has traditionally focused on reducing the risk of loss of life in the largest expected earthquake. Building codes base their provisions on the historic performance of buildings and their deficiencies and have developed provisions around life-safety concerns, by focusing their attention to prevent collapse under the most intense earthquake expected at a site during the life of a structure. These provisions are based on the concept that the successful performance of buildings in areas of high seismicity depends on a combination of strength, ductility manifested in the details of construction, and the presence of a fully interconnected, balanced, and complete LFRS. In regions of low seismicity, the need for ductility reduces substantially. And in fact, strength may even substitute for a lack of ductility. Very brittle LFRSs can be excellent performers as long as they are never pushed beyond their elastic strength.

Seismic provisions specify criteria for the design and construction of new structures subjected to earthquake ground motions with three goals:

1. Minimize the hazard to life from all structures
2. Increase the expected performance of structures having a substantial public hazard due to occupancy or use
3. Improve the capability of essential facilities to function after an earthquake

Some structural damage can be expected as a result of design ground motion because the codes allow inelastic energy dissipation in the structural system. For ground motions in excess of the design levels, the intent for the codes is for structures to have a low likelihood of collapse.

In most structures that are subjected to moderate-to-strong earthquakes, economical earthquake resistance is achieved by allowing yielding to take place in some structural members at predetermined locations. It is generally impractical to design a structure to respond in the elastic range when subjected to the maximum expected earthquake ground motions. Therefore, in seismic design, yielding is permitted in predetermined structural members or locations, with the provision that the vertical-load-carrying capacity of the structure is maintained even after strong earthquakes. However, for certain types of structures such as nuclear facilities, yield cannot be tolerated, and as such, the design needs to be elastic.

Structures that contain facilities critical to postearthquake operations—such as hospitals, fire stations, power plants, and communication centers—must not only survive without collapse but must also remain operational during and after an earthquake. Therefore, in addition to life safety, damage control is an important design consideration for structures deemed vital to postearthquake functions.

An idea of the behavior of a building during an earthquake may be grasped by considering the simplified response shown in Figure 3.94. As the ground on which the building rests is displaced, the base of the building moves with it. However, the building above the base is reluctant to move with it because the inertia of the building mass resists motion and causes the building to distort. This distortion wave travels along the height of the structure and, with continued shaking of the base, causes the building to undergo a complex series of oscillations.

Although both wind and seismic forces are essentially dynamic, there is a fundamental difference in the manner in which they are induced in a structure. Wind loads, applied as external loads, are characteristically proportional to the exposed surface of a structure, while the earthquake forces are principally internal forces resulting from the distortion produced by the inertial resistance of the structure to earthquake motions. Whereas in wind design, one would feel greater assurance about the safety of a structure made up of heavy sections, in seismic design, this does not necessarily produce a safer design.

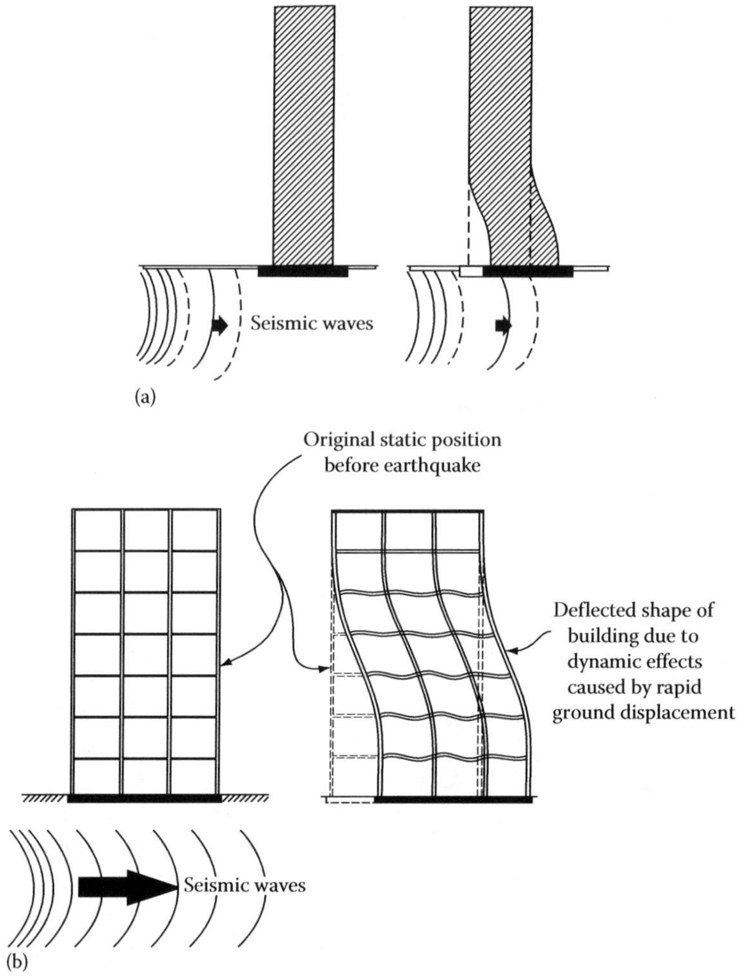

FIGURE 3.94 (a and b) Building behavior during earthquakes.

3.18.1 Seismic Response of Buildings

Earthquakes can cause structural collapse in several different ways. First, if the structure is not adequately connected and *tied together*, the motions induced in the structure by earthquake shaking can allow the building components to pull apart and, if one member is supported by another, to eventually collapse. This type of collapse is observed in bridges and other long structures that incorporate expansion joints.

Another way that earthquakes can cause structures to collapse is by overstressing gravity-load-bearing elements such that they lose their load-carrying capacity. As an example, if the overturning loads on the columns in a braced frame exceed the buckling capacity of the columns, these columns could buckle and lose their ability to continue to support the structure above.

The third way that earthquakes cause collapse is by inducing sufficient lateral displacement into a building to allow $p\Delta$ effects to induce lateral sideway collapse of the frame. Sideway collapse can occur in a single story or can involve multiple stories. Often, it is difficult to distinguish these collapses from the local failures of elements because the large displacements associated with sideway collapse can often trigger concurrent local collapse.

If the base of a structure is suddenly moved, as in a seismic event, the upper part of the structure will not respond instantaneously but will lag because of the inertial resistance and flexibility of the structure. The resulting stresses and distortions in the building are the same as if the base of the structure were to remain stationary while time-varying horizontal forces are applied to the upper part of the building. These forces, called inertial forces, are equal to the product of the mass of the structure times acceleration, that is, $F = ma$ (the mass m is equal to weight divided by the acceleration of gravity, i.e., $m = w/g$). Because earthquake ground motion is 3D (one vertical and two horizontal), the structure, in general, deforms in a highly complex manner. Generally, the inertial forces generated by the horizontal components of ground motion require greater consideration for seismic design since adequate resistance to vertical seismic loads is usually provided by the member capacities required for gravity-load design.

The behavior of a building during an earthquake is thus a vibration problem. The seismic motions of the ground do not damage a building by impact, as does a wrecker's ball, or by externally applied pressure such as wind but by internally generated inertial forces caused by vibration of the building mass. An increase in mass, in this context, has two undesirable effects on the earthquake design. First, it results in an increase in the force, and second, it can cause buckling or crushing of columns and walls when the mass pushes down on a member bent or moved out of plumb by the lateral forces. This effect is known as the $p\Delta$ effect, and the greater the vertical forces, the greater the movement due to $p\Delta$. It is almost always the vertical load that causes buildings to collapse; in earthquakes, buildings very rarely fall over—they fall down.

In general, tall buildings respond to seismic motions quite differently than low-rise buildings as shown schematically in Figure 3.95. They are invariably more flexible than low-rise buildings and, in general, experience much lower accelerations. But a tall building subjected to ground motions for a prolonged period may experience much larger forces if its natural period is near that of the ground waves. Thus, the magnitude of lateral forces is not a function of the acceleration of the ground alone but is influenced to a great extent by the type of response of the structure itself and its foundation as well. This interrelationship of building behavior and seismic ground motion also depends on the building period as formulated in the so-called response spectrum, explained in this chapter.

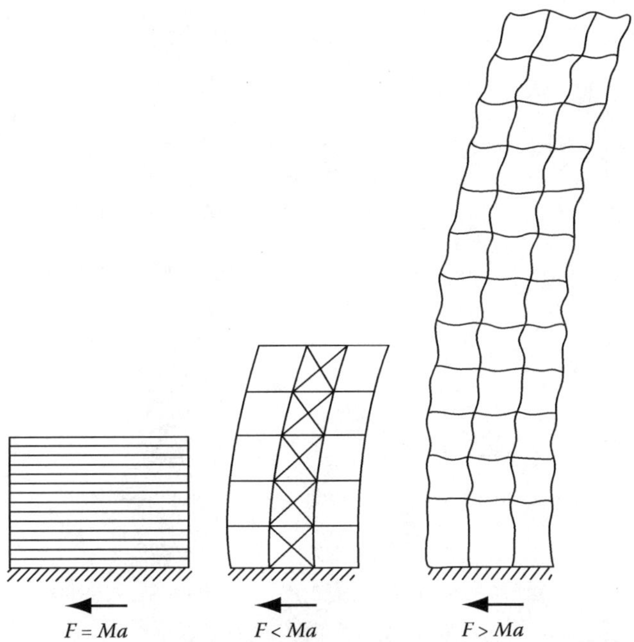

FIGURE 3.95 Schematic magnitude of seismic force. *Note*: The magnitude of seismic force depends on the building mass M, ground acceleration a, and the dynamic response of the building itself.

An effective seismic design generally includes the following:

1. Selecting an overall structural concept including layout of an LFRS that is appropriate to the anticipated level of ground shaking. This includes providing a redundant and continuous load path to ensure that the building responds as a unit when subject to ground motion.
2. Determining forces and deformations generated by the ground motion and distributing the same vertically to the LFRS with due consideration to the structural system, configuration, and site characteristics.
3. Analysis of the building for the combined effects of gravity and seismic loads to verify that adequate vertical and lateral strength and stiffness are achieved to satisfy the structural performance and acceptable deformation levels prescribed in the governing building code.
4. Providing details that allow for expected structural movements without damage to nonstructural elements such items as piping, window glass, plaster, veneer, and partitions. To minimize this type of damage, special care in detailing, either to isolate these elements or to accommodate the expected movement, is required. Breakage of glass windows can be minimized by providing adequate clearance at edges to allow for frame distortions. Damage to rigid nonstructural partitions can be largely eliminated by providing a detail, which will permit relative movement between the partitions and the adjacent structural elements.
5. In piping installations, use of expansion loops and flexible joints to accommodate relative seismic deflections between adjacent equipment items and the building floors.
6. Fasten freestanding shelving to walls to prevent toppling.
7. Stairways often suffer seismic damage because they tend to prevent drift between connected floors. This can be avoided by providing a slip joint at the lower end of each stairway to eliminate their bracing effect or by tying stairways to stairway shear walls.

3.18.2 Building Motions and Deflections

Earthquake-induced motions, even when they are more violent than those induced by wind, evoke a totally different human response—first, because earthquakes occur much less frequently than windstorms, and second, because the duration of motion caused by an earthquake is generally short. People who experience earthquakes are grateful that they have survived the trauma and are less inclined to be critical of the building motion. Earthquake-induced motions are, therefore, a safety rather than a human discomfort issue.

Lateral deflections that occur during earthquakes should be limited to prevent distress in structural members and architectural components. Non-load-bearing in-fills, external wall panels, and window glazing should be designed with sufficient clearance or with flexible supports to accommodate the anticipated movements.

3.18.3 Building Drift and Separation

Drift is generally defined as the lateral displacement of one floor relative to the floor below. Drift control is necessary to limit damage to interior partitions, elevator and stair enclosures, and glass and cladding systems. Stress or strength limitations in ductile materials do not always provide adequate drift control, especially for tall buildings with relatively flexible moment-resisting frames.

Total building drift is the absolute displacement of any point relative to the base. Adjoining buildings or adjoining sections of the same building may not have identical modes of response and therefore may have a tendency to pound against one another. Building separations or joints must be provided to permit adjoining buildings to respond independently to earthquake ground motion.

3.18.4 Adjacent Buildings

Buildings are often built right up to property lines in order to make maximum use of space. Historically, buildings have been built as if the adjacent structures do not exist. As a result, the buildings may pound during an earthquake. Building pounding can alter the dynamic response of both buildings and impart additional inertial loads to them.

Buildings that are the same height and have matching floors are likely to exhibit similar dynamic behavior. If the buildings pound, floors will impact other floors, so damage usually will be limited to nonstructural components. When floors of adjacent buildings are at different elevations, the floors on one building will impact the columns of the adjacent building, causing structural damage. When buildings are of different heights, the shorter building may act as a buttress for the taller neighbor. The shorter building receives an unexpected load, while the taller building suffers from a major discontinuity that alters its dynamic response. Since neither is designed to weather such conditions, there is potential for extensive damage and possible collapse.

One of the basic goals in seismic design is to distribute yielding throughout the structure. Distributed yielding dissipates more energy and helps prevent the premature failure of any one element or group of elements. For example, in moment frames, it is desirable to have strong columns relative to the beams to help distribute the formation of plastic hinges in the beams throughout the building and prevent a story collapse mechanism.

3.18.5 Continuous Load Path

A continuous load path, or preferably more than one path, with adequate strength and stiffness should be provided from the origin of the load to the final lateral-load-resisting elements. The general path for load transfer is in reverse to the direction in which seismic loads are delivered to the structural elements. Thus, the path for load transfer is as follows: Inertial forces generated in an element, such as a segment of exterior curtain wall, are delivered through structural connections to a horizontal diaphragm (i.e., floor slab or roof); the diaphragms distribute these forces to vertical components such as moment frames and braced frames; and finally, the vertical LFRS transfers the forces into the foundations. While providing a continuous load path is an obvious requirement, examples of common flaws in load paths include a missing collector, or a discontinuous chord because of an opening in the floor diaphragm, or a connection that is inadequate to deliver diaphragm shear to LFRF.

The horizontal elements such as floor and roof slabs distribute lateral forces to the LFRS acting as horizontal diaphragms. In special situations, horizontal bracing may be required in the plane of diaphragms to transfer large shears from discontinuous walls or braces.

A complete load path is a basic requirement. There must be a complete gravity and LFRS that forms to a continuous load path between the foundation and all portions of the building. If there is a discontinuity in the load path, the building is unable to resist seismic forces regardless of the strength of the elements. Interconnecting the elements needed to complete the load path is necessary to achieve the required seismic performance. Examples of gaps in the load path in addition to those stated earlier would include a shear wall that does not extend to the foundation, a missing shear transfer connection between a diaphragm and vertical elements, a discontinuous chord at a diaphragm's notch, or a missing collector.

A good way to remember this important design strategy is to ask yourself the question, How does the inertial load get from here (meaning the point at which it originates) to there (meaning the shear base of the structure, typically the foundations)?

3.18.6 Building Configuration

A building with an irregular configuration may be designed to meet all code requirements, but it will not perform as well as a building with a regular configuration. If the building has an odd shape that is not properly considered in the design, good details and construction are of a secondary value.

Two types of structural irregularities, vertical and plan irregularities, are typically defined in most seismic standards. These irregularities result in building responses significantly different from those assumed in the equivalent static force procedure and, to a lesser extent, from the dynamic analysis procedure. Although seismic provisions give certain recommendations for assessing the degree of irregularity and corresponding penalties and restrictions, it is important to understand that these recommendations are not an endorsement of their design; rather, the intent is to make the designer aware of the potential detrimental effects of irregularities.

Consider, for example, a reentrant corner, resulting from an irregularity characteristic of a building's plan. If the plan configuration has an inside corner, as shown in Figure 3.96, then it has a reentrant corner. It is, however, unavoidable in buildings of L, H, T, and X plan shapes.

Two problems related to seismic performance are created by these shapes: (1) differential lateral deformation modes between different wings of the building may result in a local stress concentration at the reentrant corner, and (2) torsion may result because the eccentricity between the center of rigidity and center of mass.

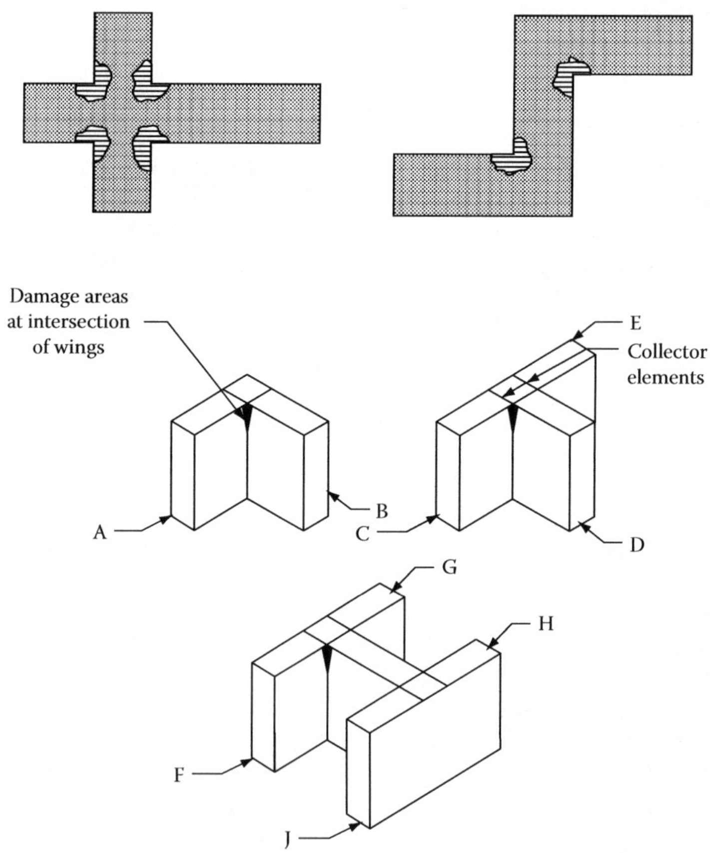

FIGURE 3.96 Reentrant corners in L-, T-, and H-shaped buildings (as a solution, add collector elements and/or stiffen end walls A, B, C, D, E, F, G, and J).

There are two alternative solutions to this problem: Tie the building together at lines of stress concentration and locate seismic-force-resisting elements at the extremity of the wings to reduce torsion or separate the building into simple shapes. The width of the separation joint must allow for the estimated inelastic deflections for adjacent wings. The purpose of the separation is to allow adjoining portions of buildings to respond to earthquake ground motions independently without pounding on each other. If it is decided to dispense with the separation joints, collectors at the intersection must be added to transfer forces across the intersection areas. Since the free ends of the wings tend to distort most, it is beneficial to place seismic-force-resisting members at these locations.

The seismic design of regular buildings is based on two concepts. First, the linearly varying lateral force distribution is a reasonable and conservative representation of the actual response distribution due to earthquake ground motions. Second, the cyclic inelastic deformation demands are reasonably uniform in all of the seismic-force-resisting elements. However, when a structure has irregularities, these concepts may not be valid, requiring corrective factors and procedures to meet the design objectives.

The impact of irregular parameters in estimating seismic force levels, first introduced into the UBC in 1973, long remained a matter of engineering judgment. Beginning in 1988, however, some configuration parameters were quantified to establish the condition of irregularity. Additionally, specific analytical treatments and/or corrective measures have been mandated to address these flaws.

Typical building configuration deficiencies include an irregular geometry, a weakness in a story, a concentration of mass, or a discontinuity in the LFRS. Vertical irregularities are defined in terms of strength, stiffness, geometry, and mass. Although these are evaluated separately, they are related to one another and may occur simultaneously. For example, a building that has a tall first story can be irregular because of a soft story, a weak story, or both, depending on the stiffness and strength of this story relative to those above.

Those who have studied the performance of buildings in earthquakes generally agree that the building's form has a major influence on performance. This is because the shape and proportions of the building have a major effect on the distribution of earthquake forces as they work their way through the building. Geometric configuration, type of structural members, details of connections, and materials of construction all have a profound effect on the structural dynamic response of a building. When a building has irregular features, such as asymmetry in plan or vertical discontinuity, the assumptions used in developing seismic criteria for buildings with regular features may not apply. Therefore, it is best to avoid creating buildings with irregular features. For example, omitting exterior walls in the first story of a building to permit an open ground floor leaves the columns at the ground floor level as the only elements available to resist lateral forces, thus causing an abrupt change in rigidity at that level. When irregular features are unavoidable, special design considerations are required to account for the unusual dynamic characteristics and the load transfer and stress concentrations that occur at abrupt changes in structural resistance. Examples of plan and elevation irregularities are illustrated in Figures 3.97 and 3.98. Note that plan irregularities are also referred to as horizontal irregularities.

The ASCE 7-10 qualifies irregularity by defining geometrically or by use of dimensional ratios, the points at which a specific irregularity becomes an issue requiring remedial measures. These issues are discussed later in this chapter. It is worth noting that no corrective course is required for certain irregularities other than performing modal analysis for determining design seismic forces.

Irregularities are divided into two broad categories: (1) vertical and (2) plan irregularities. Vertical irregularities include soft or weak stories, large changes in mass from floor to floor, and discontinuities in the dimensions or in-plane locations of lateral-load-resisting elements. Buildings with plan irregularities include those that experience substantial torsion when subjected to seismic

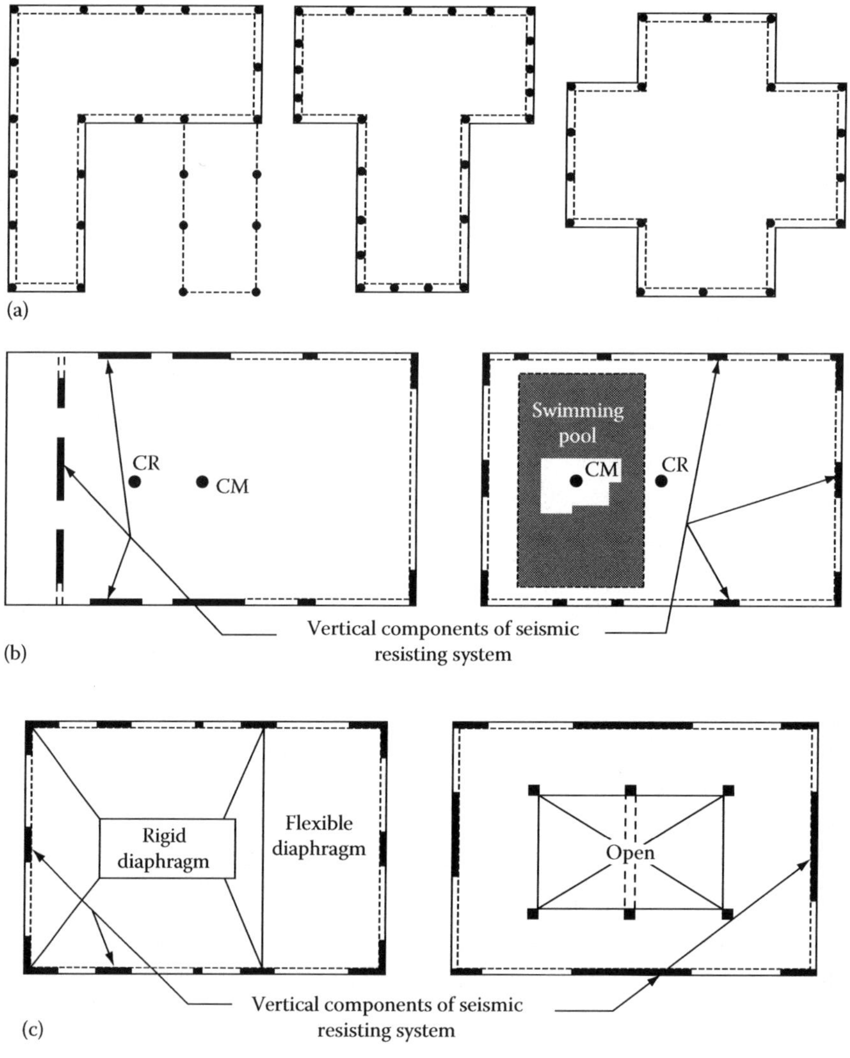

FIGURE 3.97 Plan irregularities: (a) geometric irregularities, (b) irregularity due to mass-resistance eccentricity, and (c) irregularity due to discontinuity in diaphragm stiffness. *Note*: CR, center of resistance; CM, center of mass.

loads, or have reentrant corners and discontinuities in floor diaphragms or in the lateral force path, or have lateral-load-resisting elements that are not parallel to each other or to the principal axes of the building.

3.18.7 Influence of Soil

The intensity of ground motion reduces with the distance from the epicenter of the earthquake. The reduction, called attenuation, occurs at a faster rate for high-frequency (short-period) components than for lower-frequency (long-period) components. The cause of the change in attenuation rate is not understood, but its existence is certain. This is a significant factor in the design of tall buildings, because a tall building, although situated farther from a causative fault than a low-rise building, may

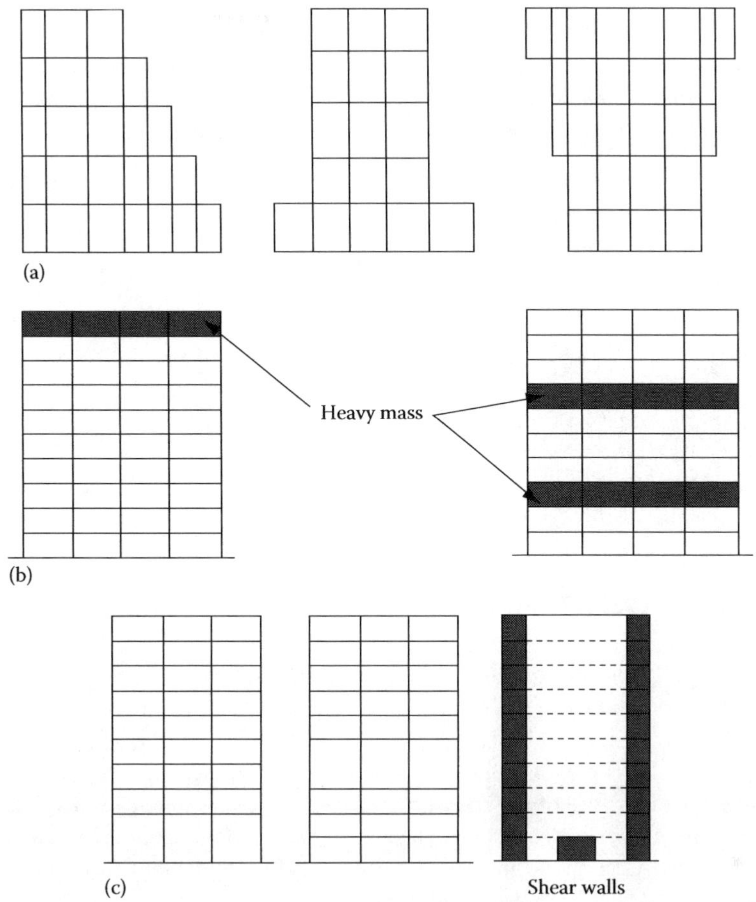

FIGURE 3.98 Elevation irregularities: (a) abrupt change in geometry, (b) large difference in floor masses, and (c) large difference in story stiffnesses.

experience greater seismic loads because long-period components are not attenuated as fast as the short-period components. Therefore, the area influenced by ground shaking potentially damaging to, say, a 50-story building is much greater than for a 1-story building.

As a building vibrates due to ground motion, its acceleration will be amplified if the fundamental period of the building coincides with that of the soil it rests upon. It is worth noting that natural periods of soil are typically in the range of 0.5–1.0 s. Therefore, it is entirely possible for a building and ground it rests upon to have the same fundamental period. This was the case for many 5- to 10-story buildings that were damaged in the September 1985 earthquake in Mexico City.

Experience in several other earthquakes has confirmed that local soil conditions can have a significant effect on earthquake response. It is perhaps more challenging to picture, but the soil layers beneath a structure have a period of vibration T_{soil} similar to the period of vibration of a building T. Greater structural damage is likely to occur when the period of the underlying soil is close to the fundamental period of the structure. In these cases, a partial resonance effect may develop between the structure and the underlying soil. These conditions are addressed in ASCE 7-10 by classifying soil profiles, into Site Class A through F.

3.18.8 Ductility

Ductility is the property exhibited by certain structural elements and structures composed of such elements that enable them to sustain loads when strained beyond their elastic limit. For structures that have well-defined yield and ultimate deformation capacities, ductility, μ, is defined by

$$\mu = \frac{\delta_u}{\delta_y} \tag{3.44}$$

In this equation, δ_u and δ_y are the displacements at which failure and yielding, respectively, initiate.

Ductility is an important parameter in computing seismic resistance. It enables structures that do not have adequate elastic strength to survive strong ground motions through inelastic response. The principal benefit of ductile response is that it makes it possible to place ductile elements at key locations in the seismic-load-resisting system to protect other nonductile elements from being overstressed. This is a key strategy in design of structures for seismic resistance.

As stated earlier, in seismic design, structures are designed for forces much smaller than those the design ground motion would produce in a structure with completely linear elastic response. The reduction is possible for a number of reasons. As the structure begins to yield and deform inelastically, the effective period of response of the structure tends to lengthen, resulting in a reduction in strength demand. Furthermore, the inelastic action results in a significant amount of energy dissipation through hysteretic damping.

The energy dissipation resulting from hysteretic behavior can be measured as the area enclosed by the force–deformation curve of the structure as it experiences several cycles of excitation. Some structures have far more energy dissipation capacity than do others. The extent of energy dissipation capacity is largely dependent on the amount of stiffness and strength degradation of the structure as it experiences repeated cycles of inelastic deformation.

Let us consider the load–deformation curves for a beam–column assembly earlier in Figure 3.10. Hysteretic curve (a) in the figure is representative of the behavior of substructures that have been detailed for ductile behavior. The substructure can maintain nearly all of its strength and stiffness over a number of large cycles of inelastic deformation. The resulting force–deformation *loops* are quite wide and open, resulting in a large amount of energy dissipation capacity. Hysteretic curve (b) represents the behavior of a substructure that has not been detailed for ductile behavior. It rapidly loses stiffness under inelastic deformation and the resulting hysteretic loops are quite pinched. The energy dissipation capacity of such a substructure is much lower than that for the substructure (a). Hence, structural systems with large energy dissipation capacity are assigned higher *R* values, resulting in design for lower forces, than systems with relatively limited energy dissipation capacity.

In providing for ductility, it should be kept in mind that severe penalties are imposed by seismic provisions on structure with nonuniform ductility.

Some examples of nonuniform ductility due to vertical discontinuities are shown in Figure 3.99. Avoid them, if you can.

3.18.9 Redundancy

Redundancy is a fundamental characteristic for good seismic performance. It provides a building with a redundant system such that failure of a single connection or component does not adversely affect the entire lateral stability of the structure

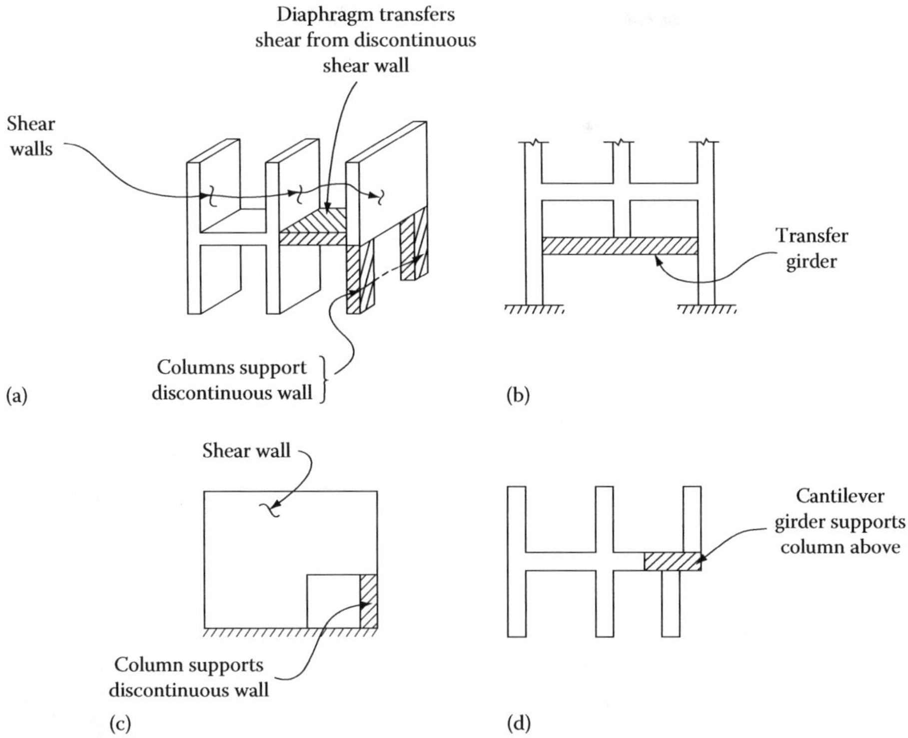

FIGURE 3.99 Examples of nonuniform ductility in structural systems due to vertical discontinuities.

3.18.10 Damping

Buildings do not resonate with the purity of a tuning fork because they are damped; the extent of damping depends upon the construction materials, type of connections, and the influence of nonstructural elements on the stiffness characteristics of the building. Damping is measured as a percentage of critical damping. In a dynamic system, critical damping is defined as the minimum amount of damping necessary to prevent oscillation altogether. To visualize critical damping, imagine a tensioned string immersed in water. When the string is plucked, it oscillates about its rest position several times before stopping. If we replace water with a liquid of high viscosity, the string will oscillate but (Taranath, Book 6, Chapter 3) certainly not as many times as when it was in water. By progressively increasing the viscosity of the liquid, it is easy to visualize that a state can be reached where the string, once plucked, will return to its neutral position without ever crossing it. The minimum viscosity of the liquid that prevents the vibration of the string altogether can be considered equivalent to the critical damping.

The damping of structures is influenced by a number of external and internal sources. Chief among them are the following:

1. External viscous damping caused by air surrounding the building. Since the viscosity of air is low, this effect is negligible in comparison to other types of damping.
2. Internal viscous damping associated with the material viscosity. This is proportional to velocity and increases in proportion to the natural frequency of the structure.
3. Friction damping, also called Coulomb damping, occurring at connections and support points of the structure. It is a constant, irrespective of the velocity or amount of displacement.
4. Hysteretic damping, which contributes to a major portion of the energy absorbed in ductile structures.

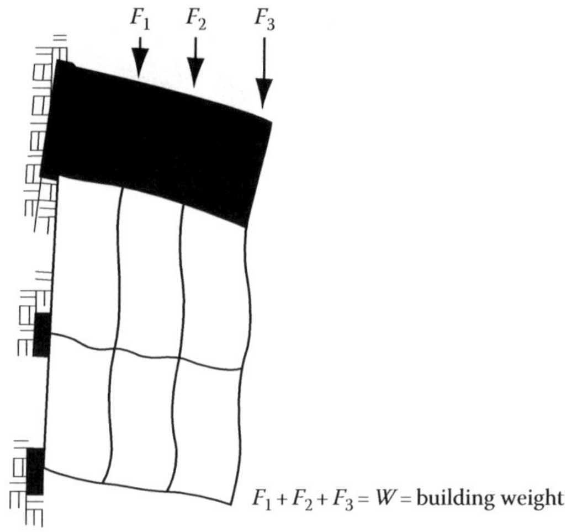

FIGURE 3.100 Concept of 100% *g*. A building subjected to 1*g* acceleration conceptually behaves as if it cantilevers horizontally from a vertical surface. *Note*: This is not a misprint!

For analytical purposes, it is a common practice to lump different sources of damping into a single viscous damping. For non-base-isolated buildings, analyzed for code-prescribed loads, the damping ratios used in practice vary anywhere from 1% to 10% of critical. The low-end values are for wind, while those for the upper end are for seismic design. The damping ratio used in the analysis of seismic base-isolated buildings is rather large compared to values used for nonisolated buildings and varies from about 0.20 to 0.35 (20%–35% of critical damping).

Base isolation, discussed elsewhere in this book, consists of mounting a building on an isolation system to prevent horizontal seismic ground motions from entering the building. This strategy results in significant reductions in interstory drifts and floor accelerations, thereby protecting the building and its contents from earthquake damage.

A level of ground acceleration on the order 0.1*g*, where *g* is the acceleration due to gravity, is often sufficient to produce some damage to weak construction. An acceleration of 1.0*g*, or 100% of gravity, is analytically equivalent, in the static sense, to a building that cantilevers horizontally from a vertical surface as shown in Figure 3.100.

As stated previously, the process by which free vibration steadily diminishes in amplitude is called damping. In damping, the energy of the vibrating system is dissipated by various mechanisms, and often, more than one mechanism may be present at the same time. In simple laboratory models, most of the energy dissipation arises from the thermal effect of repeated elastic straining of the material and from the internal friction. In actual structures, however, many other mechanisms also contribute to the energy dissipation. In a concrete building, these include opening and closing of microcracks in concrete and friction between the structure itself and nonstructural elements such as partition walls. Invariably, it is impossible to identify or describe mathematically each of these energy-dissipating mechanisms in an actual building.

Therefore, the damping in actual structures is usually represented in a highly idealized manner. For many purposes, the actual damping in structures can be idealized by a linear viscous damper or dashpot. The damping coefficient is selected so that the vibrational energy that dissipates is equivalent to the energy dissipated in all the damping mechanisms. This idealization is called equivalent viscous damping.

Figure 3.101 shows a linear viscous damper subjected to a force f_D. The damping forces f_D are related to the velocity $\dot{u}$ across the linear viscous damper by

$$f_D = c\dot{u} \tag{3.45}$$

where the constant *c* is the viscous damping coefficient; it has units of force × time/length.

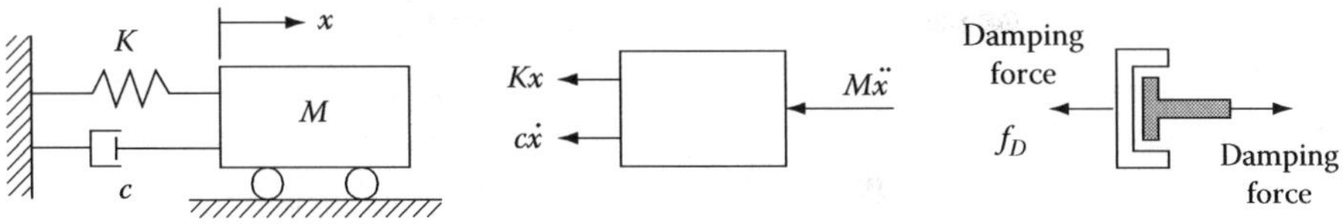

FIGURE 3.101 Linear viscous damper. Damping is defined as a force that resists dynamic motion. A simple and yet realistic damping model for analysis purposes is to assume that the damping force, f_D, is proportional to viscous friction of a fluid in a dash pot, and therefore, it is called *viscous damping.*

Unlike the stiffness of a structure, the damping coefficient cannot be calculated from the dimensions of the structure and the sizes of the structural elements. This is understandable because it is not feasible to identify all the mechanisms that dissipate vibrational energy of actual structures. Thus, vibration experiments on actual structures provide the data for evaluating the damping coefficient. These may be free-vibration experiments that lead to measured rate at which motion decays in free vibration. The damping property may also be determined from forced vibration experiments.

The equivalent viscous damper is intended to model the energy dissipation at deformation amplitudes within the linear elastic limit of the overall structure. Over this range of deformations, the damping coefficient c determined from experiments may vary with the deformation amplitude. This nonlinearity of the damping property is usually not considered explicitly in dynamic analyses. It may be handled indirectly by selecting a value for the damping coefficient that is appropriate for the expected deformation amplitude, usually taken as the deformation associated with the linearly elastic limit of the structure. Additional energy is dissipated due to inelastic behavior of the structure at larger deformations. Under cyclic forces or deformations, this behavior implies formation of a force–deformation hysteresis loop as shown in Figure 3.102. The damping energy dissipated during one deformation cycle between deformation limits $\pm u_0$ is given by the area within the hysteresis loop abcda (Figure 3.102).

This energy dissipation is usually not modeled by a viscous damper, especially if the excitation is earthquake ground motion. Instead, the most common and direct approach to account for the energy dissipation through inelastic behavior is to recognize the inelastic relationship between resisting force and deformation. Such force–deformation relationships are obtained from experiments on structures or structural components at slow rates of deformation, thus excluding any energy dissipation arising from rate-dependent effects.

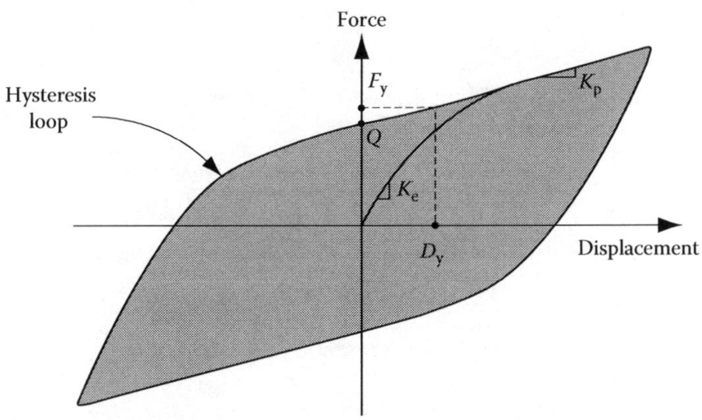

FIGURE 3.102 Force–displacement hysteresis loop: the area inside of the loop is a measure of energy dissipation due to nonelastic behavior. *Note*: K_e = initial elastic stiffness, K_p = stiffness in the plastic range, F_y = stress at yield, D_y = deformation at yield.

3.18.11 Diaphragms

Buildings are composed of vertical and horizontal structural members that resist lateral forces caused by wind and seismic. The primary purpose of a diaphragm, which consists of the roof and floors of a building, is to transfer the horizontal forces to the vertical elements resisting the lateral loads.

It is customary to consider a diaphragm analogous to a deep plate girder laid in a horizontal plane. In a composite floor system, the steel deck and concrete topping perform the function of the plate girder web. Just as the plate girder flanges carry flexural forces, so must the diaphragm elements located at the perimeter. These edge members are commonly referred to as *chords* using the analogy of a truss spanning between supports. The resulting forces in the chords are either tension or compression acting in a direction perpendicular to the direction of lateral loading width consideration. The diaphragm chord forces at any point along the diaphragm boundary member are equal to the diaphragm moment at that point divided by the depth of the diaphragm at that location. In the absence of such chord members to take the moment couple, the diaphragm must act as a deep plate resisting both bending and shear forces.

Drag force is the tension or compression force in the diaphragm boundary member parallel to the direction of loading under consideration. Where reentrant corner irregularities are present, drag struts (also called *collectors*) are required to prevent localized tearing of the diaphragm. The collector essentially collects the diaphragm shear and delivers it to the vertical elements resisting the lateral loads.

The stiffness of the diaphragm has an important effect on the proportionate distribution of lateral loads to various components of the lateral support system. This effect is displayed in Figure 3.103, which

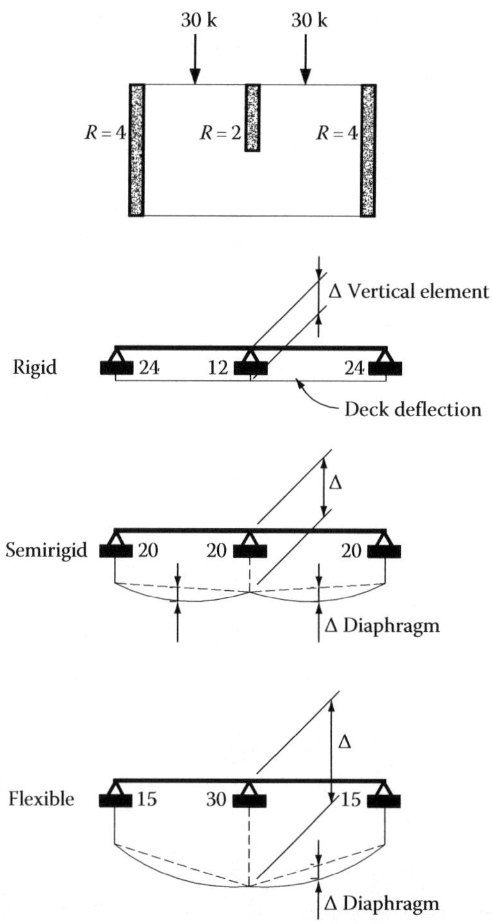

FIGURE 3.103 Relative effects of diaphragm stiffness.

shows how the relative rigidities of diaphragm and lateral support elements influence load distribution. At one extreme is the *rigid diaphragm* that distributes the lateral loads to the supports in proportion to *just* their rigidities. At the other extreme is the flexible diaphragm, which delivers lateral loads based on tributary area. The diaphragm, which falls in between the two extremes, is the *semirigid diaphragm* in which the load distribution involves both aspects of behavior. All three types of diaphragm behavior require effective transfer of bending and shear forces in the plane of the diaphragm, necessitating careful design detailing of connections between diaphragm elements and the lateral support system.

In steel buildings, floor diaphragms most commonly consist of composite steel deck with concrete topping. Of concern are the design of the diaphragm itself and the transfer of diaphragm forces into the lateral-load-resisting system.

Depending upon the magnitude of lateral load to be transferred to the vertical system, the designer has two choices in detailing the force transfer: assigns the transfer to occur either along the entire width frame of the line or to a selected segment of the frame width. A combination of the two is, of course, quite logical.

As stated previously, tension and compression chord forces are developed at the perimeter of the diaphragm due to lateral loads. The topping concrete slab over steel deck can typically resist the compression chord forces. Tension chord forces can be resisted by spandrel steel beams, continuous steel closure plates, or by mild steel reinforcement placed within the concrete topping.

Earthquake loads at any given level of a building are distributed to the lateral-load-resisting vertical elements through the floor and roof slabs. For analytical purpose, the diaphragms are assumed to behave as deep beams. The slab is the web of the deep beam carrying shear, and the perimeter spandrel or shear wall, if any, is the flange of the beam resisting bending (see Figure 3.104a and b). In the absence of perimeter members, the slab is analyzed as a plate subjected to in-plate bending.

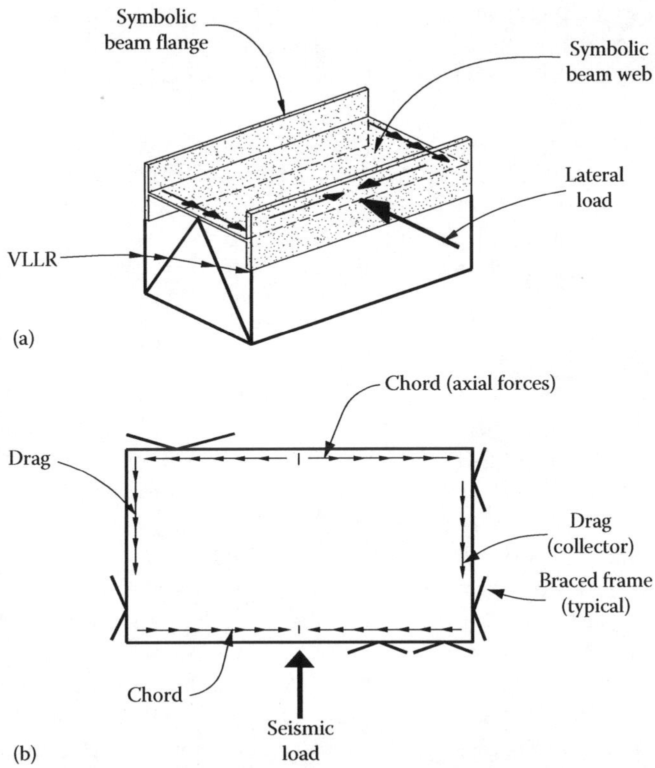

FIGURE 3.104 (a) Diaphragm action of floor or roof system. *Note*: VLLRS. (b) Schematic drag and chord for north–south seismic loads.

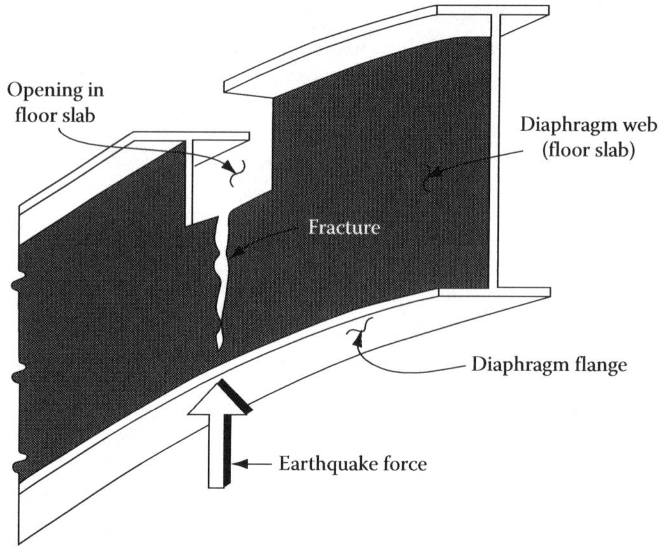

FIGURE 3.105 Diaphragm web failure due to large opening.

Three factors are important in diaphragm design:

1. The diaphragm must be adequate to resist both the bending and shear stresses and be tied together to act as one.
2. The collectors and drag members (see Figure 3.104) must be adequate to transfer loads from the diaphragm into the lateral-load-resisting vertical elements.
3. Openings or reentrant corners in the diaphragm must be property placed and adequately reinforced.

Inappropriate location or large-size openings such as for stairs or elevator cores, atriums, and skylights create problems similar to those related to cutting of beam flanges and holes in the beam web of a beam adjacent to the flange. This reduces the ability of the diaphragm to transfer the chord forces and, if not designed properly, may cause rupture in the web (see Figure 3.105).

3.18.12 Strategies to Reduce Seismic Hazards

- Locate the building in a region of lower seismicity, where earthquakes occur less frequently or with typically smaller intensities. This option is generally the most effective strategy solely in terms of reducing the potential for earthquake damage to a facility, whether it is caused by ground shaking, fault rupture, liquefaction, landslide, or inundation. Locating a building in Dallas, Texas, for example, will almost guarantee that it will never be damaged by an earthquake. Of course, this option isn't possible for many building owners.

  It is, however, fairly common for high-technology manufacturing plants to be located far from their headquarter locations, at sites with low seismicity such as Texas, Massachusetts, or Idaho. While it would be very rare for a retail establishment to make a siting decision based on seismic risk over the demographics of the market, moving a facility even a few miles in some cases can make a measurable difference in seismic hazard, for example, moving a proposed building location from within a mile of a major fault to 5 miles away.
- Locate the building on a soil profile that reduces the hazard. Local soil profiles can be highly variable, especially near water, on sloped surfaces, or close to faults. In an extreme case, siting on poor soils can lead to liquefaction, land sliding, or lateral spreading.

Often, as was the case in the 1989 Loma Prieta earthquake near San Francisco, similar structures located less than a mile apart each performed in dramatically different ways because of differing soil conditions.

Even when soil-related hazards are not present, the amplitude, duration, and frequency content of earthquake motions that have to travel through softer soils can be significantly different from those traveling through firm soils or rock.

- Engineer the soil profile to increase building performance and reduce vulnerability. If relocating to a region of lower seismicity or to an area with a better natural soil profile is not a cost-effective option, the soil at the designated site can often be reengineered to reduce the hazard. On a liquefiable site, for instance, the soil can be grouted or otherwise treated to reduce the likelihood or liquefaction occurring. Soft soils can be excavated and replaced or combined with foreign materials to make them stiffer. The building foundation itself can be modified to account for the potential effects of the soil, reducing the building's susceptibility to damage even if liquefaction or limited land sliding does occur.

3.18.13 Strategies to Improve Building Seismic Performance

Seismic vulnerability may be reduced by increasing the performance of the building, thereby reducing the damage expected in earthquakes. There are two methods by which this is typically accomplished:

Reduce the response of the building to earthquake shaking. An earthquake generates inertial forces in a building that are a function of the structure's mass, stiffness, and damping, and of the acceleration and frequency of the earthquake motion. While the actual mass of the building (the weight of the structure, contents, and people) typically cannot be significantly altered, the effective mass can be changed by providing special devices, such as passive or active mass dampers, that can effectively reduce the overall mass that is accelerated by the earthquake. Stiffness can be altered by modifying the structural system. The building's fundamental period, which is an important parameter in determining building response, can be significantly increased (and resulting forces reduced) by providing seismic isolating devices at the building foundation.

3.19 LESSONS FROM PAST EARTHQUAKES

The seismic design and construction requirements contained in US building codes have been developed over many years mostly by observing the way actual structures perform in earthquakes. Following each major earthquake that causes damage to modern engineered construction, engineers investigate the behavior of typical buildings. When these engineers observe that certain design and construction practices lead to unacceptable types of damage in buildings, they develop building code provisions to discourage the continued used of these practices in future design and construction. This process has been underway in the United States and worldwide for more than 100 years and, to be sure, will continue forever. This section summarizes some of the more significant lessons that have been learned from past earthquakes and how they are implemented in the US codes.

3.19.1 1906 San Francisco Earthquake

The great M7.9 San Francisco earthquake of April 18, 1906, remains one of the worst natural disasters to affect the United States and had great significance with regard to the development of US building codes.

Primary lessons learned from this earthquake included the importance of soil type on structural behavior and the benefits of having a complete vertical-load-carrying steel frame for seismic resistance.

Most of the urban center of San Francisco was constructed on land that was reclaimed from the surround San Francisco Bay. The fill soils used to reclaim this land were of mixed characteristics, consisting of debris from building construction, excavation spoils from building basements on the adjacent dry land, and even the rotting hulls of ships, abandoned in the bay.

Observers noted that buildings constructed on this *made* or *infirm* ground performed far worse than buildings that had been constructed on the natural ground that defined San Francisco prior to 1849. This observation was included in early building code provisions for seismic resistance, developed in the 1920s, which required higher design forces for buildings sited on infirm sites. As time progressed, memory of these effects faded, and the building codes of the 1940s no longer included this factor. It was to be rediscovered in the 1970s and 1980s and reinstituted into present building codes.

It was also observed that the only buildings in the commercial center of San Francisco that remained standing were those constructed with complete vertical-load-carrying steel frames and infill masonry walls. This observation led to the eventual requirement contained in building codes that tall buildings have complete vertical-load-carrying space frames and that buildings taller than 240 ft have moment-resisting space frames as part of their seismic-load-resisting system.

3.19.2 1933 Long Beach Earthquake

On March 10, 1933, a moderate M6.3 earthquake struck near Long Beach, California, causing extensive damage to many unreinforced masonry buildings. The massive damage to these buildings prompted California to adopt legislation prohibiting the further construction of unreinforced masonry buildings in the state and also empowering the Office of the State Architect to adopt occupancy-specific design, construction, and quality assurance requirements for the construction of schools. This legislation marked the beginning of the recognition that some buildings are more important than others and, therefore, should be designed and constructed with greater precautions to protect the safety of the public.

3.19.3 1940 Imperial Valley and 1952 Kern County Earthquakes

The 1940 Imperial Valley earthquake, with a magnitude of nearly M7.0, occurred in the California desert, east of San Diego. This earthquake affected few buildings in this sparsely populated region. However, the USGS did obtain a high-quality, strong ground motion recording from this event. Much of the work performed by researchers in developing response spectra contained in building codes was based on these recorded motions.

The 1952 Kern County earthquake was a large event, with a magnitude of nearly M7.4. Located in the arid area east of Bakersfield, California, this earthquake also caused little building damage of significance, but it did result in extensive damage to oil field and refining facilities in the region. Following this earthquake, ASCE and SEAOC formed a joint committee to make recommendations for seismic design provisions in building codes. The recommendations of the committee eventually resulted in adoption of the building codes requirements to determine seismic design forces for building based on spectral response analysis concepts.

3.19.4 1971 San Fernando Earthquake

The M6.6 San Fernando earthquake of February 9, 1971, though not of great magnitude, was one of the most significant events with regard to its effect on building codes. Prior to the 1971 earthquake, the seismic provisions in building codes were largely limited to specification of minimum design lateral forces and contained few requirements related to structural detailing.

Because a large-magnitude earthquake had not affected California in nearly 20 years prior to this event, engineers felt confident that the building codes in effect at the time were capable of providing reliable protection of buildings in earthquakes. However, the San Fernando earthquake severely damaged many modern code-conforming buildings. Among the most famous of these damaged buildings was the Olive View hospital, a large multibuilding complex located near the epicenter of this earthquake. This complex consisted of a series of reinforced concrete frame buildings. Although the 1967 edition of the UBC contained provisions for ductile reinforced concrete frame design, these requirements were not mandatory and had not been included in the hospital's design. One of the buildings, which housed ambulances and emergency vehicles, collapsed. Stair towers separated and fell away from the main hospital building, and a single-story mechanism formed in the columns at the first level of the main structure, resulting in large permanent drift and leaving the building unrepairable.

In response to this damage, a number of major revisions were introduced into the building codes in the following years. Perhaps the most important of these changes was the recognition of the importance of ductile detailing to seismic performance. The voluntary provisions for ductile concrete moment frames were made mandatory in regions of high seismicity, and similar provisions began to be introduced into the code for other structural systems, essentially resulting in the precedent for the special, intermediate, and ordinary classifications of seismic-load-resisting systems contained in building codes today.

Another important change related to more formal consideration of building occupancy when determining the seismic design requirements. Following this earthquake, the concept of occupancy categories was introduced into the building code, with higher design forces required for the design of hospitals and other buildings deemed to be essential to the public safety.

Additionally, following the San Fernando event, engineers once again recognized that the types of soil present at a site had great significance with regard to the intensity and character of ground motions experienced by buildings. This resulted in the formal introduction of site classes into the determination of seismic design forces.

3.19.5 1979 Imperial Valley Earthquake

The M6.4 Imperial Valley earthquake of October 15, 1979, affected relatively few buildings due to the sparse population of the affected region, near the California–Mexico border. However, one building, the six-story Imperial County Services Building, did experience noteworthy damage. This six-story concrete shear wall building had an out-of-plane offset between the shear walls and frames above the first story and those below, in order to accommodate an arcade feature at the ground level. Overturning forces from the shear walls above the first story crushed the first-story columns immediately below the walls and frames.

Research into the behavior of this building led to the present code requirement to design columns beneath discontinuous walls and frames for the amplified forces that consider the overstrength of the structure above. Later, the building code applied this same requirement to other irregularities.

3.19.6 1985 Mexico City Earthquake

Prior to the great M8.1 Mexico City earthquake of September 19, 1985, there had been little record of any significant damage to steel frame buildings. However, in this earthquake, two high-rise steel frame buildings, one 22-story tall and other 16-story tall, collapsed. These buildings were braced steel frame structures that utilized built-up box section columns. Investigation of damage suggested that overturning forces imposed on the columns resulted in local buckling of plate sections in the built-up box section columns, which then led to failure of the seam welds in the boxes.

This observation led to the design requirement to proportion the columns in steel seismic-load-resisting systems with adequate strength to resist the maximum axial forces that can be delivered to the columns, considering the overstrength of the structural system supported above, whether or not the structural system is irregular.

3.19.7 1987 Whittier Narrows Earthquake

The M5.9 Whittier Narrows earthquake of October 17, 1987, was a relatively modest event, both in size and effect. However, it did cause damage to some modern buildings prompting several code changes. One of the most significant of these changes was based on observation of damage sustained by the California Federal Savings company's data processing center. This braced steel frame building employed chevron-pattern braces. The braces buckled in compression and the floor beams at the apex of the chevrons were bent downward, causing damage to the floor systems. Observations of this damage led to the introduction of provisions requiring design of beams at the apex of chevron-pattern braces for the unbalanced forces that result following buckling of one of these braces.

3.19.8 1989 Loma Prieta Earthquake

The 1989 M7.1 Loma Prieta earthquake, occurring approximately 70 miles south of San Francisco, caused relatively little damage to modern structures designed to recent editions of the building code. However, it caused extensive damage to older buildings and structures.

The large array of ground motion obtained in this earthquake was key to the determination that the ground shaking in the region within the near field—that is, within a few kilometers of the zone of fault rupture—is not only stronger but also significantly different from the shaking experienced farther from the rupture zone. Recordings obtained in this event, together with the limited near-fault recordings available from other earthquakes (such as the 1971 San Fernando and 1992 Landers and Big Bear events), allowed seismologists to identify the pulse-like characteristics of near-field motions. However, it was only after the 1994 Northridge earthquake when these effects were again noted, and the building code was actually modified to account for these effects.

3.19.9 1994 Northridge Earthquake

The M6.7 earthquake that struck Northridge, California, on January 17, 1994, was one of the most significant earthquakes of the past century with regard to the wealth of engineering data that were obtained and analyzed and subsequently implemented into the building codes. This is because, like the San Francisco Bay area in 1989, the affected region had many strong ground motion instruments present, but also, unlike the Loma Prieta earthquake, this event damaged many modern code-conforming buildings.

The Northridge earthquake provided valuable earthquake experience data on the performance of four types of structures: concrete tilt-up buildings with wood roofs, precast concrete parking structures, braced steel frames, and moment-resisting steel frames. As a result of the observations of damage that occurred in this earthquake, extensive revisions were made to the 1997 editions of the UBC, as well as the NEHRP provisions, which form the basis for seismic design requirements of SEI/ASCE 7.

With regard to braced steel frames, the observation of fractures in hollow structural section (HSS) braces, following buckling, resulted in significant restrictions on the permissible width–thickness ratios for braced elements. It also resulted in severe restrictions of the use of ordinary concentric braced frame (OCBF).

Perhaps the most significant lesions learned were associated with the unanticipated discovery of fractures in the welded joints of modern steel moment frames. These fractures were attributed to a variety of factors, including connection geometries that resulted in stress concentrations and high restraint, limiting ductility; wide variations in the yield and tensile strengths of the common ASTM A36 and A572 grades of material used in building construction at that time; the low toughness of weld filler metals commonly used in steel construction; and poor adherence to the requirements of the American Welding Society (AWS) welding code when making welded joints. Another important contributing factor to these unexpected failures was that the connection practices commonly in use prior to the earthquake had been validated years earlier by testing of specimens that were much lighter than those present in the damaged buildings. Over the years since the initial research, design practice had deviated away from the use of highly redundant frames with relatively small members, relatively few participating members, and use of very large sections. The earlier testing was not applicable to these heavier frames, but this was not understood until after the earthquake and discovery of the damage.

These observations resulted in a number of changes in the building codes and design practice as well as a major rewrite of AISC 341 and the introduction of the AWS D1.8 seismic supplement to AWS Dl.l. Significant changes included the introduction of the new ASTM A992 grade of structural steel, with controlled yield strengths and yield to tensile ratios; requirements to demonstrate (through laboratory testing of full-scale specimens) that moment connections are capable of attaining minimum inelastic deformation demands; requirements to use weld filler metals with minimum rated notch toughness in seismic-load-resisting systems; and requirements to remove weld backing bars and weld tabs from critical joints.

Another major feature introduced into the building codes following the Northridge earthquake was the requirement to quantify the redundancy inherent in a structural design to adjust the design seismic forces and permissible drifts for the structure, based on this redundancy. This was based on the observation that many modern structures that had been severely damaged in the earthquake were less redundant than earlier structures that performed better.

3.20 SEISMIC DESIGN WRAP-UP

Current seismic building codes and accepted engineering practice do not assure that all structures designed and built in strict compliance with the codes will not be severely damaged in a moderate-to-strong earthquake shaking. This basic purpose of a building code is to provide public safety. Codes are not concerned with high quality, serviceability, or limited repairable damage; the owner who is interested in these has to establish his or her own criteria. He or she may prescribe more stringent criteria so the owner can minimize property damage as well as protecting the occupants. The engineer may desire to apply more stringent requirements for any number of reasons, such as to lower insurance rates, to reduce expensive repair costs by lowering earthquake damage risk, to provide better quality for improved service and longer-life structures, or to reduce maintenance and operating costs. Performance can be improved at little, if any, increase in cost of proper structural concepts, and detailing techniques are selected.

The current ASCE 7-10 seismic standard is an advance in the art and standard that a building conforming to the seismic won't suffer damage in an earthquake; it is not all-inclusive; there are areas where good judgment and experience of the earthquake-conscious engineer will have to prevail in order to produce a sound, safe, serviceable, and economical structure. It provides improved definitions and more specific guidance in several instances, but because of the great number of variables and complexity of the problem, the structural engineer should realize that the requirements of the code are minimum that emphasis must be placed on the structural concepts and detailing techniques rather than on refinement of stress calculations.

There are three basic steps in the development of adequate earthquake-resistant design: (1) architectural concept, (2) stress analysis, and (3) proper details. The seismic design begins with and is dominated by the architectural concept of the building. The materials of construction, the nonstructural component configuration (both horizontal and vertical), fenestration, and the nature of the structural elements all have a profound influence upon the success of the seismic design. Code-prescribed seismic coefficients are applicable only to a few simplified, stereotyped structures. Consequently, the structural engineer must visualize the response of the structure to earthquake ground motions and adapt the design to the distortions and stress paths, which will occur in the building. Assumptions used in the calculations are consistent with the as-built structure. Details for earthquake-resistant structures differ in basic concept from details for wind or vertical forces. Foremost among requirements vital to earthquake-resistant design of all types of buildings is the necessity of tying the various components together so that they act as a unit or the isolation of certain components so that the structure will respond as assumed in the calculations. Generally, these requirements cannot be stated in detail by a building code but have to be left to the understanding and judgment of the responsible engineer. By proper coordination in the development of basic concepts of the particular facility, the design team can secure adequate resistance, frequently without significant premium cost, just by better planning, simple mathematics, and the execution of better details.

Field observations made following several earthquake emphasized that improperly designed and constructed joints between vertical and horizontal resisting elements, as well as joints between precast and prestressed components, performed poorly under the abnormal earthquake intensity. Further, observations and computer dynamic analysis of a typical building show that a moderately strong earthquake will cause deflections and forces that are considerably greater than those that would result from application of the static seismic loads required by the code. Thus, during any reasonably severe earthquake, it can be expected that significant inelastic (nonlinear) deformations will be produced in a typical building. Hence, ductility that will increase the energy-absorbing capacity is essential if structures are to tolerate deformations beyond the range of elastic (linear) behavior.

Principal elements that influence damage to buildings are (1) intensity of the earthquake waves, (2) duration of earthquake motion, (3) response of the structure to ground motions, and (4) design of the building components and their connections. The code substitutes static design forces for a dynamic response of a structure to a complex earthquake ground motion. The code approach is to apply a horizontal force, in each direction alternately, to each principal mass, such as at each floor. Such static forces cause the structure to deflect in a rocking motion. However, in an actual earthquake with oscillating ground motion, the response of the structure may result in a floor moving in one direction, while at the same time, an adjacent floor may be moving in an opposite direction. When this happens, a building does not tilt just back and forth, but it deflects in a weaving, snakelike fashion. Using dynamic loading and a computer analysis, one can more nearly predict how a proposed building will act and deform under ground motions of a specific earthquake. It will be found that this response may sometimes cause deflections, joint rotations, and stresses in components quite different than that determined from the static loadings.

3.20.1 Determination of Earthquake Lateral Forces

An earthquake causes random erratic vibratory ground motions at the base of a structure and the structure actively responds to these motions. Seismic design involves two distinct steps: (1) the prediction of ground motions at the base of the structure based on the seismicity of the site and (2) the selection of a structure that will deform without catastrophic collapse when responding to these ground motions.

Destructive earthquakes are all due to tectonic motions. According to the generally accepted elastic rebound theory, defaulting is the cause of earthquakes, not a consequence of them. The primary mechanism is the fracturing of the rock in a zone of large deformation along a fault. The consequent sudden displacement as the deformation and strain energy are released initiates vibrations in the Earth's crust (bedrock). Vibrations of such intensity as to cause significant damage to structures have been found only in the immediate environs (up to 150+ miles) of the fault movement. As the response of buildings depends on the characteristic of the ground motion, it is highly desirable that a quantitative description of the ground motion be available at the site of the building. Large amplitude vibrations near faults are too great to be recorded by ordinary seismographs, and special *strong-motion seismographs* have been developed, which directly record the three components of ground accelerations as functions of time. Unfortunately, there are absolutely no ground motion records available at the location of destructive damage or maximum ground motion for the truly major earthquakes such as Alaska (1964) or San Francisco (1906). Seismologists classify earthquakes according to the Richter scale of magnitude (M), which characterizes the total energy released during an earthquake. From the viewpoint of effect on a particular structure, the engineers are interested in the intensity of the vibrations of the Earth's surface at the site. Such intensity of vibrations is a function of (1) amount of energy released (M), (2) distance from the fault to the structure, and (3) the character and thickness of the foundation material. Seismological analysis allows prediction on a regional basis of the magnitude of future earthquakes. Mathematical theories have been developed for predicting the effect of distance from the fault movement for various soil conditions underlying a site, using assumed bedrock vibrations. But as mentioned previously, ground motion records of destructive earthquakes are not sufficiently available to substantiate these theories. Each new earthquake appears to differ from the previous one. Hence, there are too many unknowns to be able to predict with any firm degree of certainty (quantitatively) the time-varying foundation vibration input for some unknown future earthquake. On the other hand, these historical records indicate the general trend, give the best estimate available, and should be used where appropriate. Such appropriate analytical procedures provide the capability of obtaining a semiquantitative guide as well as a qualitative guide to site response during future earthquakes. Qualitatively, the following are apparent:

1. Ground shaking will be strongest in the general vicinity of the causative fault and the intensity diminishes with distance from the fault.
2. The period of the ground motion increases with an increase in the distance from the causative fault.
3. Deep deposits of stiff soils result in ground motions having predominately short-period characteristics.
4. Shallow deposits of stiff soils result in ground motions having predominantly short-period characteristics.
5. The soil amplification varies with the frequency and intensity of the bedrock motions. The amplification is very sensitive to the intensity of the bedrock motions. For small intensity, the amplification may be 4 or more. However, for large bedrock motions, the amplification is considerably smaller. Although the problem cannot be considered as solved, it is estimated that for strong motions, the amplification factor in case of soft sediment might generally be taken in the order of 1.5–2.

3.20.2 Structural Response

The loads or forces that a structure is called upon to sustain due to earthquake motions result directly from the distortions induced in the structure by the motion of the ground on which it rests. Base motion is characterized by displacements, velocities, and accelerations, which are erratic in direction, magnitude, duration, and sequence. Earthquake loads are essentially inertial forces related to the mass, stiffness, and damping characteristics of the structure. The problem is completely defined by these physical properties of the structural system and by the time-varying displacements applied at its foundation. Thus, the evaluation of these structural properties and the selection of an appropriate earthquake input are the most critical factors in the earthquake response analysis. Assuming that the structural and earthquake characteristics are known, the problem of the determination of the vibratory response of the structure can be formulated in theory and solved in very general terms, even for situations involving plastic (inelastic) deformations. In the formulation of any dynamic response analysis, it must be recognized that the structure generally will be subjected to static loadings (e.g., gravitational forces) in addition to the dynamic excitation. If the structure is linearly elastic, so that the principle of superposition is applicable, it is convenient to consider separately the static and dynamic loadings; then the total structural response is obtained by adding the static stresses and deflections to the results of the dynamic analysis. However, if the structure yields or is subject to some other nonlinear behavior during the dynamic loading, superposition is not valid. Since the mechanics and mathematics of dynamic analysis are given in standard textbooks, it will suffice here to point out that the application of such analysis has certain limitations. The structural properties and the earthquake input must be specified, but these properties are simplified and idealized for the model study as an assembly of masses interconnected by springs and damping elements. Since the mechanics of dynamic analysis requires that a separate solution must be obtained for each instant of time during the entire history of interest, computations using computers are necessary. The specific object of this theory is to predict the stresses and deflections that will be developed in any given structural system. Unlike the stiffness and mass properties of a structure, damping capacity cannot be calculated. Therefore, in formulating a mathematical model, it is important that some results on the damping capacity of structures similar to the one being modeled are available. Our knowledge of the response of buildings to earthquake motions is far from exact or complete. However, *qualitatively*, the following are apparent:

1. If the earthquake motion is severe, some portions of most structures will exceed their yield capacity.
2. It is economically impractical to design buildings to resist the maximum expected earthquake forces within the elastic range of stress.

3.20.3 Equivalent Lateral Load Procedure

In this analysis, we substitute assumed equivalent design static lateral forces for the true response to ground motions. This approach attempts to recognize our limited recorded experience and the qualitative dynamic analysis of simplified structures. The ASCE formulas for base shear V may be rewritten in a compact form as follows:

$$V = C_s W \tag{3.46}$$

where
C_s = the seismic response coefficient
W = the effective seismic weight

The seismic response coefficient, C_s,

$$C_s = \frac{S_{DS}}{\left(\frac{R}{I_e}\right)}$$

The value of C_s computed need not exceed the following:

$$C_s = \frac{S_{D1}}{T\left(\frac{R}{I_e}\right)} \quad \text{for } T \le T_L$$

$$C_s = \frac{S_{D1}T_L}{T^2\left(\frac{R}{I_e}\right)} \quad \text{for } T > T_L$$

C_s shall not be less than

$$C_s = 0.044 S_{DS} I_e \ge 0.01$$

In addition, for strutures located where S_1 is equal to or greater than 0.6g, C_s shall not be less than

$$C_s = 0.5S_1/(R/I_e)$$

where

S_{DS} is the short-period design spectral response accelerations
S_{D1} is the 1 s design spectral response acceleration
R is the response modification factor
I is the importance factor
T is the building period
S_1 is the mapped 1 s spectral acceleration
T_L is the transition period

1. The value of the coefficients S_{D1} and S_{DS} in the base shear formula may be considered analogous to the relative intensity of the ground motion at the site of the structure.
2. The value of the coefficient R is determined empirically in a crude attempt to recognize the effect of ductility and energy absorption qualities of certain types of construction, which historically have shown varying degrees of earthquake resistance.
3. The value of the period coefficient T has been established empirically in an attempt to recognize the effect of the period of the structure in response to the ground motions. In the absence of more refined methods, the ASCE structural empirically determines the period based on shape and/or type of structure.
4. Further, the distribution of the base shear vertically along the height of a structure is determined by the general formula:

$$F_x = C_{vx} V$$

$$C_{vx} = \frac{w_x h_z^k}{\sum_{iwl}^{n} W_i h_i^k}$$

These terms in these equations are defined in ASCE 7-10 (and elsewhere in this book) and need no further explanation except to indicate that the shear diagram corresponds roughly to the shear envelope obtainable from a dynamic analysis (for a uniform section and weight). Invariably, the design forces specified by the ASCE are smaller than the corresponding maximum values that would be indicated by an elastic–dynamic analysis, using moderate-intensity input. Therefore, their application must be used with sound engineering judgment. The type of structure must be selected, detailed, and constructed to fit the intent of the ASCE and the particular site conditions. Even though the assumed equivalent static forces specified in the ASCE are only approximations, they are applicable to a few types of simple structures and under restricted conditions. The assignment of lateral load design coefficients to inappropriate type of structures or conditions could be disastrous in case of a strong earthquake. When unusual structures are involved, a dynamic analysis should be made to aid in making decisions. It is incumbent upon the architect, in cooperation with the structural engineer, to develop a concept that will avoid problems.

3.20.4 Architectural Implications

As previously indicated, the seismic forces specified are quite small relative to the actual forces expected at least once in the life cycle of any building. Therefore, they are meant to be used with certain limitations. These limitations and principles are important since they constitute the basic justification for designing with lateral earthquake forces of 3%–13% gravity (G) as compared with spectrum analysis requirements of over 50% G.

Accelerations derived from actual earthquakes are surprisingly high as compared to the ASCE forces used in ordinary design. Any design that merely comes up to the code is apt to be marginal. The manipulation of numbers in accordance with code requirements will certainly not assure adequate earthquake resistance in case of a major earthquake. Important basic assumptions are necessary before design enters the mathematical stage. Generally, these requirements cannot be stated in detail by a building code but must be left to the judgment of the responsible engineer. The most sensible approach is to design on some reasonable basis, to recognize the uncertain nature of the demands, and to provide all the reserve capacity that can be accomplished at little or no extra cost in initial construction or at only a slight sacrifice in architectural features. Of course, unusual risks should have additional design considerations and refinements. If potential elements of distress can be eliminated before calculations are made, the problems are much simpler. The first step is to develop an architectural concept that will avoid as many problems as feasible. The shape of the building (both horizontal and vertical configuration and fenestrations), the materials of construction, and detailing all have a profound effect on the response to lateral forces. Architects and engineers must realize the following:

1. A great deal of a building's inherent resistance to lateral forces is determined by its basic play layout. Desirable objectives in this regard are symmetry about both axes, not only of the building itself but of the wall openings, columns, shear walls, etc. It is most desirable to consider the effect of the lateral forces from the start of the layout since this may save considerable money without detracting particularly from the function nor appearance of the building.
2. In many cases, two or more schemes must be considered. Reentrant corners of L-, T-, or U-shaped buildings are points of great stress and must be avoided or reinforced accordingly.
3. As related to tall structures especially, another very important factor in approach to earthquake design is the concept of making the building as tough as feasible.
4. To neglect in analysis or detailing the effects of *nonstructural* partitions and filler walls creates fictitious structures on paper. In reality, the filler walls and partitions cause a substantial change in the magnitude and distribution of earthquake forces, causing a shear wall-like response with considerably higher lateral forces and overturning moments.
5. Frequently, partitions and exterior walls are omitted in the first story of a building to permit an open floor because of usage or for architectural fenestration. This leaves the columns on the ground level as the only elements available to resist lateral forces. Omission of walls and partitions in the ground stories (freestanding columns) causes an abrupt change in rigidities at that level. This fact must be considered in analysis, and proper details provided to accommodate larger distortions due to increased flexibility of the first story. It is advisable to carry all shear walls down to the foundation.

3.20.5 Structural Concept

The objective is to reduce project costs to a minimum without compromising its function, quality, or reliability. Final selection of materials and systems will be made with due considerations to cost of construction, architectural requirements, fire and other safety hazards, low maintenance and operating costs over the life cycle of the facility, and fund limitations. Of equal importance along with costs is the necessity to make certain that the most efficient systems, methods, and materials are employed and that the physical appearance is a credit to the design agency.

Usually, the major structural–architectural components of a building that mostly affect the cost of construction are the exterior walls, partitions, floor and roof decks, and the structural framing system. In some instances, the selection of a type of foundation may be a major factor in a cost study. Skillful planning, simple detailing, and arrangement of space to be compatible with repetitive construction all add greatly to reducing total building costs; however, such saving may not be easily measured. Including contractor's options will create competition and thereby lower costs. On the other hand, the use of exotic, unconventional, sophisticated, and/or proprietary materials and methods of construction is usually not well suited to open competitive bidding requirements, and the reliability of the earthquake-resistance performance may be difficult to determine.

Participation of all disciplines of the design team in the conceptual planning in a selection of basic construction materials, at an early stage, will pay big dividends in ensuring the most efficient design at lowest construction cost and minimize the total design effort. Procedures in the approach to develop a concept will vary depending upon the type of facility and the experience of the individuals on the design team.

3.20.6 Damage Control Features

The design of a structure in accordance with the seismic provisions will not ensure against earthquake damage, since the horizontal deformations from design loads are lower than can be expected during a major earthquake. However, a number of things can be done without increasing construction cost to limit damage, which otherwise would be expensive to repair following strong earthquake. An important factor to keep in mind is the nature and geometry of the building when it responds to earthquake motions. As a rough guide, it should be assumed that deflections (story drift) may be four times that resulting from the required lateral forces. A list of features, which can aid in avoiding excessive damage, follows:

1. Provide details that allow structural movement without damage to nonstructural elements. Damage to such items as piping, glass, plaster, and partitions may constitute a major financial loss even though the damage to structural elements is minor. Special care in detailing is required to minimize this type of damage.
2. Breakage of glass windows can be minimized by providing extra clearance at edges to allow for frame distortions.
3. Damage to rigid or nonstructural partitions can be largely eliminated by locating them away from columns and by providing a detail at the top that will permit relative motions between the partitions and the floor above.
4. In piping installations, the expansion loops and bellows joints used to accommodate temperature movement are often adaptable to handling the relative seismic deflections between adjacent equipment items attached to floors.
5. Fasten shelving to walls to prevent tipping over.
6. Concrete stairways often suffer seismic damage due to inhibition to drift between connected floors. This can be avoided by providing open joints at each floor to eliminate the bracing effect of the stairway.
7. If only cosmetic paint and plaster repairs are undertaken without regard to structural rehabilitation after damage from an earthquake, the structure may be left vulnerable to further damage and possible collapse in the event of a second strong earthquake.

3.20.7 Techniques of Seismic Design

For gravity loads, it has been a long-standing practice to design for strength and deflections within the elastic limits of the members. However, to control design within elastic behavior for likely horizontal seismic forces is impractical. Hence, designers need to resort to other techniques to meet acceptable performance. A number of structural features contributing to seismic resistance under earthquake conditions are enumerated and discussed as follows:

1. *Adequate foundations*. Differential movement of foundations that is due to seismic motions is an important cause of structural damage, especially to heavy, rigid structures that cannot accommodate these movements. Adequate design must minimize the possibility of relative displacement between the various parts of the foundation and between the foundation and superstructure.
2. *Lightweight mass*. With other things being equal, the greater the mass of the structure, the greater are the seismic forces. Lightweight construction materials minimize seismic forces but do not necessarily result in lower costs.

3. *Structural symmetry.* Past experience has shown that buildings, which are unsymmetrical in plan, have greater susceptibility to earthquake damage than symmetrical structures. The effect of dissymmetry is to induce torsion (out-of-phase) oscillations of the structure. Dissymmetry in plan can be eliminated or improved by separating L-, T-, and U-shaped buildings into distinct units by use of seismic joints at junctions of the individual wings. In regular structures, dissymmetry can also be caused by eccentric structural elements. Such a condition can exist, for example, in a building with a flexible front as a result of large openings and an essentially stiff (solid) rear wall. Buildings with this type of structural dissymmetry can usually be avoided by better conceptual planning, or may be improved by modifying the stiffness of the rear wall, or by the judicial insertion of rigid structural partitions so as to make the cr of the vertical resisting elements more nearly coincide with the center of the lateral forces.
4. *Damping.* The damping characteristics of the structure have a major effect on its response to ground motions because a small amount of damping significantly reduces the maximum deflections due to resonant response. In this connection, reinforced concrete has a higher degree of damping than structural steel. However, damping in itself is not a complete index of the antiseismic value of a material or system.
5. *Ductility.* Ductile materials are highly desirable for earthquake-resistant design. Brittle material such as concrete and unit masonry shall not be used to resist seismic forces unless properly reinforced. Under the combined effect of compression (overturning of the structure as a whole) and flexure, a common mode of failure for concrete columns is by buckling of the main steel and spalling of the concrete cover near the floor levels. Columns with spiral reinforcing or hooping have greater reserve strength and are less vulnerable to this type of failure.
6. *Diaphragms.* In floor and roof slabs used as diaphragms, it is customary to provide for tensile stresses by means of flange steel reinforcement concentrated at the edges of the slabs (or in steel spandrel beams). Too frequently, it is forgotten that these flanges must be made continuous at columns.
7. *Shear walls.* Horizontal forces at any floor or roof level may be transferred to the ground (foundation) by using the strength and rigidity of the walls (and partitions) as shear walls. The strength of shear walls is usually governed by flexure and not by shear, except for very low and long walls. A shear wall may be considered analogous to a cantilever plate girder standing on end in a vertical plane where the wall performs the function of a plate girder web, the pilaster or floor diaphragms function as web stiffeners, and the integral reinforcement of the vertical boundaries functions as flanges.
   a. Walls may be subjected to both vertical (gravity) and horizontal (wind or earthquake) forces. A wall carrying a vertical load other than its own weight is called a bearing wall. The horizontal forces acting on a wall may be either normal to the wall or parallel to the wall. Where a wall resists horizontal forces parallel to the wall, it is called a shear wall. This is one of the several types of vertical resisting elements. Any wall or partition that carries a vertical load other than its own weight or that resists a horizontal force parallel to the wall is classified as a structural wall. The combined effect of horizontal forces and vertical loads on a wall must be considered. Wall and partitions shall be designed to withstand all vertical loads and horizontal forces, both parallel to the normal and to the flat surface, with due allowance for the effect of any eccentric loading or overturning forces generated. The tensile forces resulting from combination of vertical loads and overturning moments due to lateral loads must be resisted by anchorage into the foundation medium unless they can be overcome by gravity loads mobilized from neighboring elements. Any wall, which is isolated so as not to resist external loads or forces, is classified as nonstructural. Nonisolated walls will obviously participate in shear resistance to horizontal forces parallel to the wall, since they tend to deflect and be stressed when the framework or horizontal diaphragms deform under lateral forces.

   b. Tensile stress due to bending moments in shear walls is also provided for by steel concentrated at vertical edges of the wall. Since the walls are acting as cantilever beams, the tensile stress in the boundary steel must be anchored into a foundation that is capable of transferring the forces to the ground.

8. *Connections*. Past experience has shown the connections between floor and roof diaphragms and the shear walls to be the most vulnerable spot for getting a box-type building into trouble during strong earthquakes. In harder to develop latent capacity of the structural elements, the manual requires that the design forces for connections between lateral-force-resisting elements be larger than the calculated shear from prescribed seismic forces.
9. *Future expansion*. If future expansion of a building is contemplated, generally, it is far better to plan for horizontal rather than vertical growth, since (1) there will be greater freedom in planning any future increment, (2) be less interruption of existing operations when additions are made, and (3) the first increment will not have to bear a large share of cost of the second increment. When designing for vertical expansion, the foundation, floor/roof system, and the structural frame shall be proportioned for both the initial and future loadings and seismic forces. When designing for horizontal expansion, provide for complete structural separation between the two phases; otherwise, the first increment must be designed for both conditions—before and after expansion.
10. *Design parameters*. Experience shows that the cause of earthquake damage is about equally divided between design and construction deficiencies. Design must start with conceptual planning and be carried through all phases of the design and construction program. The major check points include site investigations and collaboration of the architect and engineers (structural, mechanical, and electrical) to establish a plan, systems, and the selection of materials of construction; establish design criteria for the specific facility; identify and locate primary structural elements; determine and distribute lateral and seismic forces; prepare design calculations; detail connections; detail nonstructural parts for damage control; check shop drawings; perform quality control inspection; and maintain surveillance over changed condition during the entire construction period.

3.21 DYNAMIC ANALYSIS, THEORY

Earthquake forces are dynamic because they vary with time. Since the load is time varying, the response of the structure, including deflections, axial and shear forces, and bending moments, is also time dependent. Therefore, instead of a single solution, a separate solution is required to capture the response of the building at each instant of time for the entire duration of an earthquake. Because the resulting inertial forces are a function of building accelerations, which are themselves related to the inertial forces, it is necessary to formulate the dynamic problem in terms of differential equations.

3.21.1 Single-Degree-of-Freedom Systems

Consider a portal frame, shown in Figure 3.106, consisting of an infinitely stiff beam supported by flexible columns that have negligible mass as compared to that of the beam. For horizontal motions, the structure can be visualized as a spring-supported mass, as shown in Figure 3.107a, or as a weight W suspended from a spring, as shown in Figure 3.107b. Under the action of a force W, the spring will extend by a certain amount x. If the spring is very stiff, x is small, and vice versa. The extension x can be related to the stiffness of the spring k by the relation

$$x = \frac{W}{k} \tag{3.47}$$

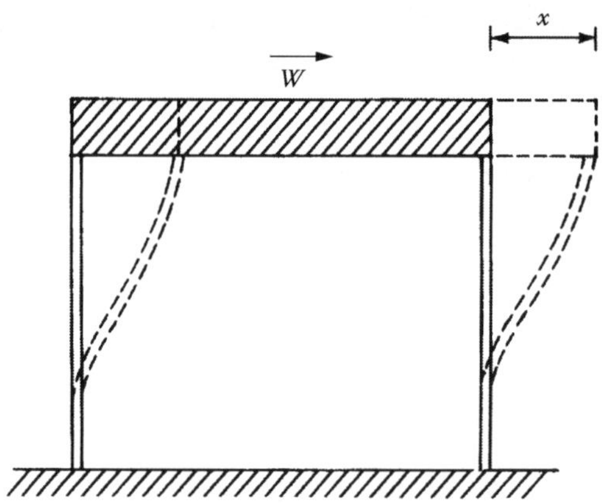

FIGURE 3.106 Single-bay, single-story postal frame.

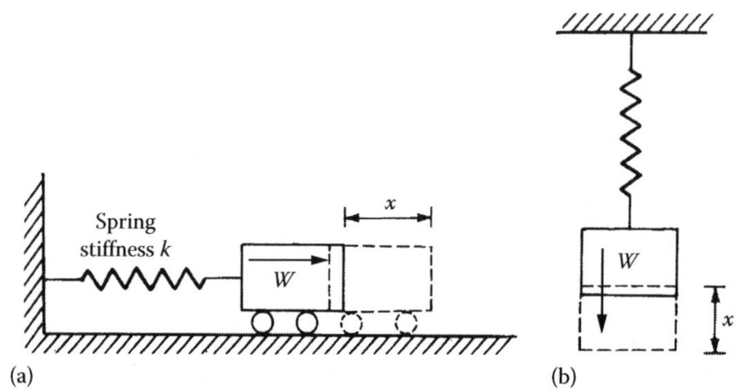

FIGURE 3.107 Analytical models for SDOF stem: (a) model in horizontal position and (b) model in vertical position.

The spring constant or spring stiffness k denotes the load required to produce the unit extension of the spring. If W is measured in kip and the extension in inches, the spring stiffness will have a dimension of kip per inch. The weight W comes to rest after the spring has extended by the length x. Equation 3.47 expresses the familiar static equilibrium condition between the internal force in the spring and the externally applied force W.

If a force is applied or removed suddenly, vibrations of the system are produced. Such vibrations, maintained by the elastic force in the spring along, are called free or natural vibrations. The weight moves up and down and therefore is subjected to an acceleration x given by the second derivative of displacement x, with respect to time t. At any instant t, there are three forces acting on the body: the dynamic force equal to the product of the body mass and its acceleration, the force W acting downward, and the force in the spring equal to $W + Kx$ for the position of weight shown in Figure 3.108. These are in a state of dynamic equilibrium for an undamped system given by the relation

$$\frac{W}{g}\ddot{x} = W - \left(W + kx\right) = -kx \tag{3.48}$$

The preceding equation of motion is called Newton's law of motion and is governed by the equilibrium of inertial force that is a product of the mass W/g and acceleration x and the resisting forces that are a function of the stiffness of the spring.

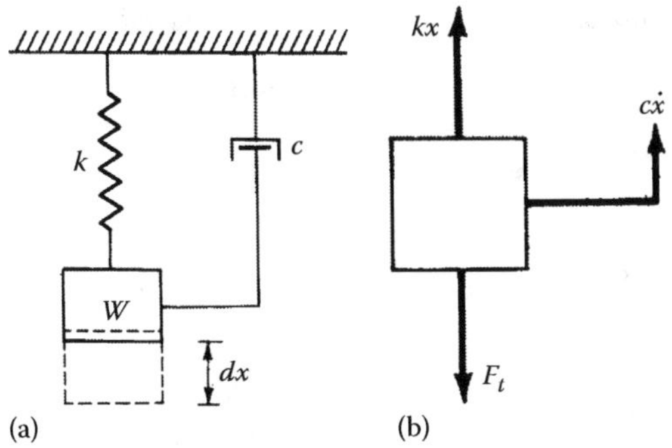

FIGURE 3.108 Damped oscillator: (a) analytical model and (b) forces in equilibrium.

The principle of virtual work can be used as an alternative to derive Newton's law of motion. Although the method was first developed for static problems, it can be readily applied to dynamic problems by using D'Alembert's principle. The method establishes dynamic equilibrium by including inertial forces in the system.

The principle of virtual work can be stated as follows: For a system in equilibrium, the work done by all the forces during a virtual displacement is equal to 0. Consider a damped oscillator subjected to a time-dependent force F_t, as shown in Figure 3.108. The free-body diagram of the oscillator subjected to various forces is shown in Figure 3.108b.

Let δ_x be the virtual displacement. The total work done by the system is zero and is given by

$$\begin{gathered} m\ddot{x}\delta x + c\dot{x}\delta x + kx\delta x - F_t\delta x = 0 \\ (m\ddot{x} + c\dot{x} + kx - F_t)\,\delta x = 0 \end{gathered} \tag{3.49}$$

Since δx is arbitrarily selected,

$$m\ddot{x} + c\dot{x} + kx - F_t = 0 \tag{3.50}$$

This is differential equation of motion of the damped oscillator.

The equation of motion for an undamped system can also be obtained from the principle of conservation of energy. It states that if no external forces are acting on the system, and there is no dissipation of energy due to damping, then the total energy of the system must remain constant during motion, and consequently, its derivative with respect to time must be equal to zero.

Consider again the oscillator shown in Figure 3.108 without the damper. The two energies associated with this system are the kinetic energy of the mass and the potential energy of the spring.

The kinetic energy of the spring

$$T = \frac{1}{2}m\dot{x}^2 \tag{3.51}$$

where $\dot{x}$ is the instantaneous velocity of the mass.

The force in the spring is kx; work done by the spring is $kx\,\delta x$. The potential energy is the work done by this force and is given by

$$V = \frac{1}{2}kx^2 \tag{3.52}$$

The total energy in the system is a constant. Thus,

$$\frac{1}{2}m\dot{x}^2 + \frac{1}{2}kx^2 = \text{constant} \tag{3.53}$$

Differentiating with respect to x, we get

$$m\dot{x}\ddot{x} + kx\dot{x} = 0 \tag{3.54}$$

Since $\dot{x}$ cannot be zero for all values of t, we get

$$m\ddot{x} + kx = 0 \tag{3.55}$$

This differential equation has a solution of the form

$$\begin{aligned} x &= A \sin(\omega t + a) \\ \dot{x} &= \omega A \cos(\omega t + a) \end{aligned} \tag{3.56}$$

where
A is the maximum displacement
ωA is the maximum velocity
a is the phase of the wave

Maximum kinetic energy is given by

$$T_{max} = \frac{1}{2} m(\omega A)^2 \tag{3.57}$$

Maximum potential energy is

$$V_{max} = \frac{1}{2} kA^2 \tag{3.58}$$

Since $T = V$,

$$\frac{1}{2} m(\omega A)^2 = \frac{1}{2} kA^2 \tag{3.59}$$

or

$$\omega = \sqrt{\frac{k}{m}} \tag{3.60}$$

This is the natural frequency of the simple oscillator. This method, in which the natural frequency is obtained by equating maximum kinetic energy and maximum potential energy, is known as Rayleigh's method.

3.21.2 Multidegree-of-Freedom Systems

In these systems, the displacement configuration is determined by a finite number of displacement coordinates. The true response of multidegree system can be determined only by evaluating the inertial effects at each mass particle because structures are continuous systems with an infinite number of DOF's. Although analytical methods are available to describe the behavior of such systems, these are limited to structures with uniform material properties and regular geometry. The methods are complex, requiring the formulation of partial differential equations. However, the analysis is greatly simplified by replacing the entire displacement of the structure by a limited number of displacement components and assuming the entire mass of the structure is concentrated in a number of discrete points.

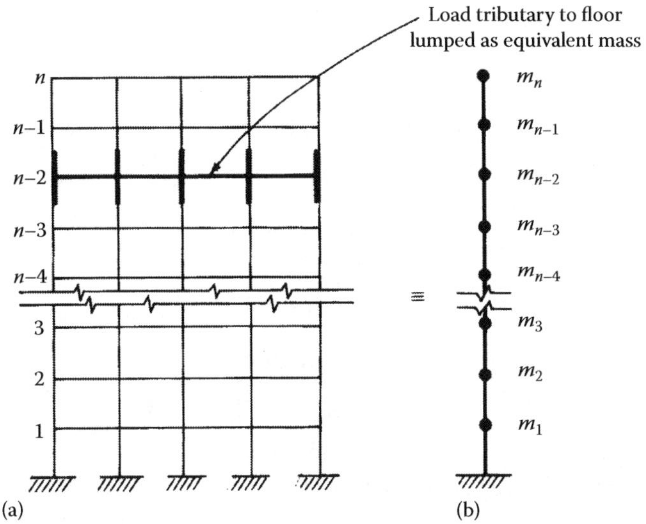

FIGURE 3.109 Multistory analytical model with lumped masses.

Consider a multistory building with n DOFs, as shown in Figure 3.109. The dynamic equilibrium equations for undamped free vibration can be written in the general form.

Writing the equations in matrix form

$$[M]\{\ddot{x}\} + [K]\{x\} = 0 \tag{3.61a}$$

where

[M] is the mass or inertia matrix
$\{\ddot{x}\}$ is the column vector of accelerations
[K] is the structure stiffness matrix
{x} is the column vector of displacements of the structure

If the effect of damping is included, the equations of motion would be in the form

$$[M]\{\ddot{x}\} + [C]\{\dot{x}\} = [K]\{x\} = \{P\} \tag{3.61b}$$

where

[C] is the damping matrix
$\{\dot{x}\}$ is the column vector of velocity
{P} is the column vector of external forces

General methods of solutions of these equations are available but tend to be cumbersome. Therefore, in solving seismic problems, simplified methods are used; the problem first is solved by neglecting damping. The absence of precise data on damping does not usually justify a more rigorous treatment. Neglecting damping results in dropping the second term, and limiting the problem to free vibrations results in dropping the right-hand side of Equation 3.61b. The resulting equations of motion will become identical to Equation 3.61a. During free vibrations, the motions of the system are simple harmonic, which means that the system oscillates about the stationary position in a sinusoidal manner; all masses follow the same harmonic function, having similar angular frequency, ω. Thus,

$$x_1 = a_1 \sin \omega_1 t$$
$$x_2 = a_2 \sin \omega_2 t$$
$$\vdots$$
$$x_n = a_n \sin \omega_n t$$

or in matrix notation,

$$\{x\} = \{a_n\} \sin \omega_n t$$

where

$\{a_n\}$ represents the column vector of modal amplitudes for the nth mode
ω_n is the corresponding frequency

Substituting for {x} and its second derivative $\{\ddot{x}\}$ in Equation 3.61a results in a set of algebraic expressions:

$$-\omega_n^2[M]\{a_n\}+[K]\{a_n\}=0 \tag{3.61c}$$

Using a procedure known as Cramer's rule, the preceding expressions can be solved for determining the frequencies of vibrations and relative values of amplitudes of motion a_{11}, a_{12}, …, a_n. The rule states that nontrivial values of amplitudes exist only if the determinant of the coefficients of a is equal to 0 because the equations are homogeneous, meaning that the right-hand side of the equation is 0. Setting the determinant of the equation equal to 0, we see that the values for all the stiffness coefficients k_{11}, k_{12}, etc., and the masses m_1, m_2, etc., are known, the determinant of the equation can be expanded, leading to an ω^2 polynomial expression. The solution of the polynomial gives one real root for each mode vibration. Hence, for a system with n DOFs, n natural frequencies are obtained. The smallest of the values obtained is called the fundamental frequency and the corresponding mode, the fundamental or first mode.

In mathematical terms, the vibration problem is similar to those encountered in stability analyses: The determination of frequency of vibrations can be considered similar to the determination of critical loads, while the modes of vibration can be likened to the evaluation of buckling modes. Such types of problems are known as eigenvalue, or characteristic value, problems. The quantities ω^2, which are analogous to critical loads, are called eigenvalues, or characteristic values, and in a broad sense can be looked upon as unique properties of the structure similar to geometric properties such as area or moment of inertia of individual elements.

Unique values for characteristic shapes, on the other hand, cannot be determined because the substitution of ω^2 for a particular mode into the dynamic equilibrium equation results in exactly n unknowns for the characteristic amplitudes x_1, …, x_n for that mode. However, it is possible to obtain relative values for all amplitudes in terms of any particular amplitude. We are, therefore, able to obtain the pattern or the shape of the vibrating mode, but not its absolute magnitude. The set of modal amplitudes that describe the vibrating pattern is called eigenvector or characteristic vector.

3.21.3 Modal Superposition

In this method, the equations of motions are transformed from a set of n simultaneous differential equations to a set of n independent equations by the use of normal coordinates. The equations are solved for the response of each mode, and the total response of the system is obtained by superposing individual solutions. Two concepts are necessary for the understanding of the modal superposition method: (1) the normal coordinates and (2) the property of orthogonality.

3.21.4 Normal Coordinates

In a static analysis, it is common to represent structural displacements by a Cartesian system of coordinates. For example, in a planar system, coordinates x and y and rotation q are used to describe the position of a displaced structure with respect to its static position. If the structure is restrained to move only in the horizontal direction and if rotations are of no consequence, only one coordinate x is sufficient to describe the displacement. The displacements can also be identified by using any other independent system of coordinates. The only stipulation is that a sufficient number of coordinates are included to capture the deflected shape of the structure. These coordinates are commonly referred to as generalized coordinates and their number equals the number of DOFs of the system. In dynamic analysis, however, it is advantageous to use free-vibration mode shapes known as normal modes to represent the displacements. While a mathematical description of normal modes and their properties may be intriguing, there is nothing complicated about their concept. Let us indulge in some analogies to bring home the idea. For example, normal modes may be considered as being similar to the primary colors red, blue, and yellow. None of these primary colors can be obtained as a combination of the others, but any secondary color such as green, pink, or orange can be created by combining the primary colors, each with a distinct proportion of the primary colors. The proportions of the primary colors can be looked upon as scale factors, while the primary colors themselves can be considered similar to normal modes. To further reinforce the concept of generalized coordinates, recall beam bending problems in which the deflection curve of beam is represented in the form of trigonometric series. Considering the case of a simply supported beam subjected to vertical loads, as shown in Figure 3.110, the deflection y, at any point, can be represented by the following series:

$$y = a_1 \frac{\sin \pi x}{1} + a_2 \frac{\sin 2\pi x}{1} a_3 \frac{\sin 3\pi x}{1} \tag{3.62}$$

Geometrically, this means that the deflection curve can be obtained by superposing the simple sinusoidal curves shown in Figure 3.110.

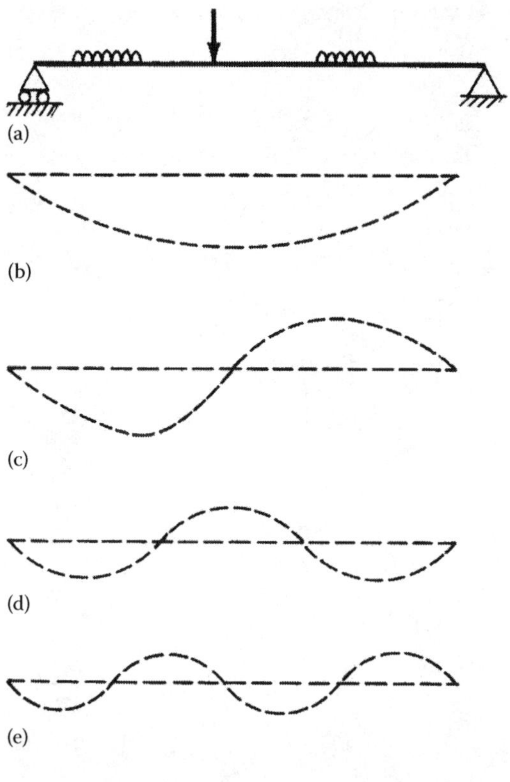

FIGURE 3.110 Generalized displacement of a simply supported beam (a) loading, (b) full-sine curve, (c) half-sine curve, (d) one-third-sine curve, and (e) one-fourth-sine curve.

The first term in Equations 3.62 represents the full-sine curve, the second term, the half-sine, etc. The coefficients a_1, a_2, a_3, etc., represent the maximum ordinates of the curves, while the numbers 1, 2, 3, etc., the number of waves or mode shapes. By determining the coefficients a_1, a_2, a_3, etc., the series can represent the deflection curve to any desired degree of accuracy, depending on the number of terms considered in the series.

3.21.5 Orthogonality

This force–displacement relationship is rarely used in static problems but is of great significance in structural dynamics. This is best explained with an example shown in Figure 3.111.

Consider a two-story, lumped-mass system subjected to free vibrations. The system's two modes of vibrations can be considered as elastic displacements due to two different loading conditions, as shown in Figure 3.111. We will use a theorem known as Betti's reciprocal theorem to demonstrate the derivation of orthogonality conditions. This theorem states that the work done by one set of loads on the deflections due to a second set of loads is equal to the work done by the second set of loads acting on the deflections due to the first. Using this theorem with reference to Figure 3.111, we get

$$\omega_1^2 m_1 x_{1b} + \omega_1^2 m_2 x_{2b} = \omega_2^2 m_1 x_{1a} + \omega_1^2 m_2 x_{2a} \tag{3.63}$$

This can be written in matrix form
or

$$(\omega_1^2 - \omega_2^2)\{x_b\}^{\mathrm{T}}[\mathrm{M}]\{x_a\} = 0 \tag{3.64}$$

If the two frequencies are not the same, that is, $\omega_1 \neq \omega_2$, we get

$$\{x_b\}^{\mathrm{T}}[\mathrm{M}]\{x_a\} = 0 \tag{3.65}$$

This condition is called the orthogonality condition, and the vibrating shapes, $\{x_a\}$ and $\{x_b\}$, are said to be orthogonal with respect to the mass matrix [M]. By using a similar procedure, it can be shown that

$$\{x_a\}^{\mathrm{T}}[\mathrm{k}]\{x_b\} = 0 \tag{3.66}$$

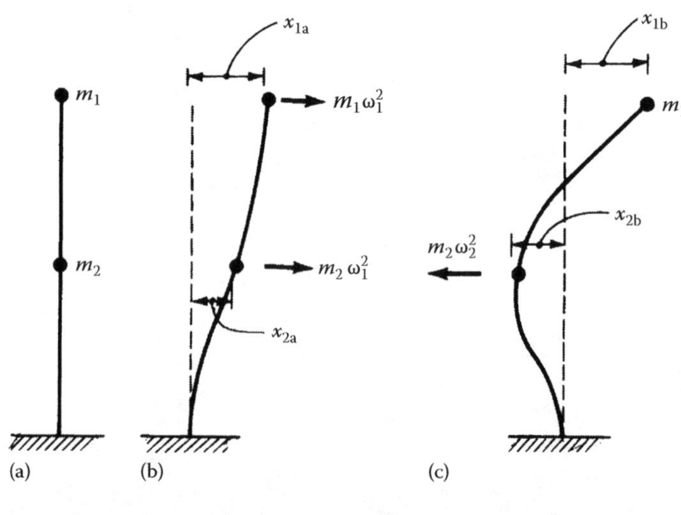

FIGURE 3.111 Two-story lumped-mass system illustrating Betti's reciprocal theorem: (a) lumped model (b) forces during the first mode of vibration and (c) forces acting during second mode of vibration.

The vibrating shapes are therefore orthogonal with respect to the stiffness matrix as they are with respect to the mass matrix. In the general case of the structures with damping, it is necessary to make a further assumption in the modal analysis that the orthogonality condition also applies for the damping matrix. This is for mathematical convenience only and has no theoretical basis. Therefore, in addition to the two orthogonality conditions mentioned previously, a third orthogonality condition of the form

$$\{x_a\}^{\mathrm{T}} c\{x_b\} = 0 \tag{3.67}$$

is used in the modal analysis.

To bring out the essentials of the normal mode method, it is convenient to consider the dynamic analysis of a two-DOF system. We will first analyze the system by a direct method and then show how the analysis can be simplified by the modal superposition method.

Consider a two-story dynamic model of a shear building shown in Figure 3.112a through c, subject to free vibrations. The masses m_1 and m_2 at levels 1 and 2 can be considered connected to each other and to the ground by two springs having stiffnesses k_1 and k_2. The stiffness coefficients are mathematically equivalent to the forces required at levels 1 and 2 to produce unit horizontal displacements relative to each level.

It is assumed that the floors and, therefore, the masses m_1 and m_2 are restrained to move in the direction x and that there is no damping in the system. Using Newton's second law of motion, the equations of dynamic equilibrium for masses m_1 and m_2 are given by

$$m_1\ddot{x}_1 = -k_1 x + k_2 (x_2 - x_1) \tag{3.68}$$

$$m_2\ddot{x}_2 = -k_2 x + k_2 (x_2 - x_1) \tag{3.69}$$

Rearranging terms in these equations gives

$$m_1\ddot{x}_1 + (k_1 + k_2)x_1 - k_2 x_2 = 0 \tag{3.70}$$

$$m_2\ddot{x}_2 - k_2 x_1 + k_2 x_2 = 0 \tag{3.71}$$

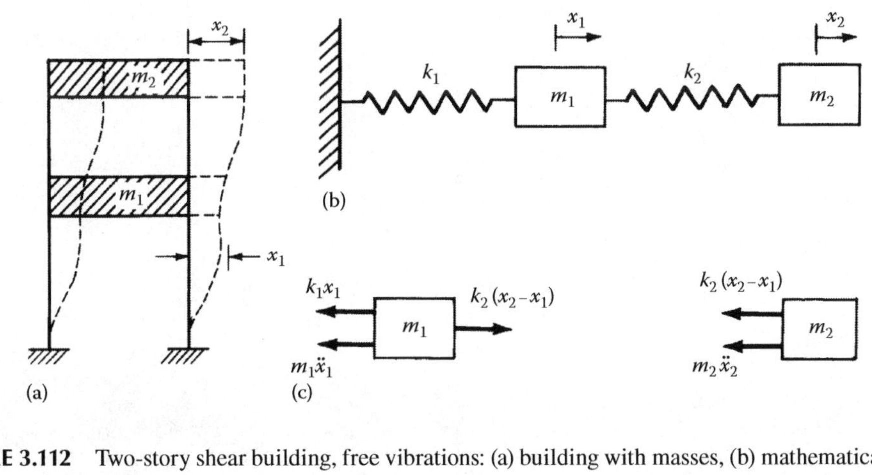

FIGURE 3.112 Two-story shear building, free vibrations: (a) building with masses, (b) mathematical model, and (c) free-body diagram with masses.

The solutions for the displacements x_1 and x_2 can be assumed to be of the form

$$x_1 = A \sin(\omega t + a) \tag{3.72}$$

$$x_2 = B \sin(\omega t + a) \tag{3.73}$$

where
- ω represents the angular frequency
- a represents the phase angle of the harmonic motion of the two masses
- A and B represent the maximum amplitudes of the vibratory motion

The substitution of Equations 3.72 and 3.73 into Equations 3.70 and 3.71 gives the following equations:

$$(k_1 + k_2 - \omega^2 m_1)A - k_2 B = 0 \tag{3.74}$$

$$k_2 A + (k_2 - \omega^2 m_2)B = 0 \tag{3.75}$$

To obtain the solution for the nontrivial case of A and $B \neq 0$, the determinant of the coefficients of A and B must be equal to 0. Thus,

$$\begin{bmatrix} (k_1 - k_2 - \omega^2 m_1) & -k_2 \\ -k_2 & (k_2 - \omega^2 m_2) \end{bmatrix} = 0 \tag{3.76}$$

The expansion of the determinant gives the relation

$$(k_1 + k_2 - \omega^2 m_1)(k_2 - \omega^2 m_2) - k_2^2 = 0 \tag{3.77}$$

or

$$m_1 m_2 \omega^4 - m_1 k_2 + m_2(k_1 + k_2)\omega_2 + k_1 k_2 = 0 \tag{3.78}$$

The solution of this quadratic equation yields two values for ω^2 of the form

$$\omega_1^2 = -b + \sqrt{b^2} - \frac{4ac}{2a} \tag{3.79}$$

$$\omega_2^2 = -b + \sqrt{b^2} - \frac{4ac}{2a} \tag{3.80}$$

where
- $a = m_1 m_2$
- $b = -[m_1 k_2 + m_2(k_1 + k_2)]$
- $c = k_1 k_2$

As mentioned previously, the two frequencies ω_1 and ω_2, which can be considered intrinsic properties of the system, are uniquely determined.

The magnitudes of the amplitudes A and B cannot be determined uniquely but can be obtained in terms of ratios $r_1 = A_1/B_1$ and $r_2 = A_2/B_2$ corresponding to ω_1^2 and ω_2^2, respectively. Thus,

$$r_1 = \frac{A_1}{B_1} = \frac{k_2}{k_1} + k_2 - \omega_1^2 m_1 \tag{3.81}$$

$$r^2 = \frac{A_2}{B_2} = \frac{k_2}{k_1} + k_2 - \omega_1^2 m_1 \tag{3.82}$$

The ratios r_1 and r_2 are called the amplitude ratios and represent the shapes of the two natural modes of vibration of the system.

Substituting the larger angular frequency ω_1 and the corresponding ratio r_1 in Equations 3.72 and 3.73, we get

$$x_1' = r_1 B_1 \sin(\omega_1 t + a_1) \tag{3.83}$$

$$x_2' = B_1 \sin(\omega_1 t + a_1) \tag{3.84}$$

These expressions describe the first mode of vibration, also called the fundamental mode. Substituting the larger angular frequency ω_2 and the corresponding ratio r_2 in Equations 3.72 and 3.73, we get

$$x_1'' = r_2 B_2 \sin(\omega_2 t + a_2) \tag{3.85}$$

$$x_2'' = B_2 \sin(\omega_2 t + a_2) \tag{3.86}$$

The displacements x_1' and x_2'' describe the second mode of vibration. The general displacement of the system is obtained by summing the modal displacements:

$$x_1 = x_1' + x_1''$$

$$x_2 = x_2' + x_2''$$

Thus, for systems having two DOFs, we are able to determine the frequencies and mode shapes without undue mathematical difficulties. Although the equations of motions for multidegree systems have similar mathematical form, solutions for modal amplitudes in terms of geometrical coordinates become unwieldy. The use of orthogonal properties of modes shapes makes this laborious process unnecessary. We will demonstrate how the analysis can be simplified by using the modal superposition method. Consider again the equations of motion for the idealized two-story building in the previous section. As previously stated, damping is neglected, but instead of free vibrations, we will consider the analysis of the system subject to time-varying force functions F_1 and F_2 at levels 1 and 2. The dynamic equilibrium for masses m_1 and m_2 is given by

$$m_1 \ddot{x}_1 + (k_1 + k_2)\, x_1 - k_2 x_2 = F_1 \tag{3.87}$$

$$m_2 \ddot{x}_2 - k_2 x_1 + k_2 x_2 = F_2 \tag{3.88}$$

These two equations are interdependent because they contain both the unknowns x_1 and x_2. These can be solved simultaneously to get the response of the system, which was indeed the method used in the previous section to obtain the values for frequencies and mode shapes. The modal superposition method offers an alternate procedure for solving such problems. Instead of requiring the simultaneous solution of equations, we seek to transform the system of interdependent or coupled equations into a system of independent or uncoupled equations. Since the resulting equations contain only one unknown function of time, solutions are greatly simplified. Let us assume that the solution for the preceding dynamic equations is of the form

$$x_1 = a_{11}z_1 + a_{12}z_{12} \tag{3.89}$$

$$x_2 = a_2z_1 + a_{22}z_2 \tag{3.90}$$

What we have done in the preceding equations is to express displacements x_1 and x_2 at levels 1 and 2 as a linear combination of properly scaled values of two independent modes. For example, a_{11} and a_{12}, which are the mode shapes at level 1, are combined linearly to give the displacement x_1; z_1 and z_2 can be looked upon as scaling functions. Substituting for x_1 and x_2 and their derivatives x_1 and x_2 in equilibrium (Equations 3.87 and 3.88), we get

$$m_1a_{11}\ddot{z}_1 + (k_1 + k_2)\,a_{11}z_1 - k_2a_{21}z_1 - m_1a_{12}\ddot{z}_2 + (k_1 + k_2)\,a_{12}z_2 - k_2a_{22}\ddot{z}_2 = F_1 \tag{3.91}$$

$$m_2a_{21}\ddot{z}_1 - k_2a_{11}z_1 + k_2a_{21}z_1 - m_2a_{24}\ddot{z}_2 - k_2a_{12}z_2 + k_2a_{22}z_2 = F_2 \tag{3.92}$$

We seek to uncouple Equations 3.87 and 3.88 by using the orthogonality conditions. Multiplying Equation 3.91 by a_{11} and Equation 3.92 by a_{21}, we get

$$m_1a_{11}^2\ddot{z}_1 + \left(k_1 + k_2\right)a_{11}^2z_1 - k_2a_{11}a_{21}z_1 + m_1a_{11}a_{12}z_2 + \left(k_1 + k_2\right)a_{11}a_{12}z_2 - k_2a_{11}a_{22}\ddot{z}_2 = a_{11}F_1 \tag{3.93}$$

$$m_1a_{21}^2\ddot{z}_1 - k_2a_{11}a_{21}z_1 + k_2a_{21}^2z_1 + m_2a_{21}a_{22}\ddot{z}_2 - (k_2a_{12}a_{21}z_2 + k_2a_{21}a_{22}z_2) = a_{21}F_2$$

Adding the preceding two equations, we get

$$\left(m_1a_{11}^2 + m_2a_{21}^2\right)\ddot{z}_1 + \omega_1^2\left(m_1a_1^2 + m_2a_{21}^2\right)z_1 = a_{11}F_1 + a_{21}F_2 \tag{3.94}$$

Similarly, multiplying Equations 5.48 and 5.49 by a_{12} and a_{22} and adding, we obtain

$$\left(m_1a_{12}^2 + m_2a_{22}^2\right)\ddot{z}_2 + \omega_2^2\left(m_1a_{12}^2 + m_2a_{22}^2\right)z_2 = a_{12}F_1 - a_{22}F_2 \tag{3.95}$$

Equations 3.91 and 3.92 are independent of each other and are the uncoupled form of the original system of coupled differential equations. These can be further written in a simplified form by making use of the following abbreviations:

$$\begin{aligned} M_1 &= m_1a_{11}^2 + m_2a_{21}^2 \\ M_2 &= m_1a_{12}^2 + m_2a_{22}^2 \end{aligned} \tag{3.96}$$

$$\begin{aligned} K_1 &= \omega_1^2M_1 \\ K_2 &= \omega_2^2M_2 \end{aligned} \tag{3.97}$$

$$\begin{aligned} P_1 &= a_{11}F_1 + a_{21}F_2 \\ P_2 &= a_{12}F_1 + a_{22}F_2 \end{aligned} \tag{3.98}$$

where

M_1 and M_2 are called the generalized masses
K_1 and K_2 are the generalized stiffnesses
P_1 and P_2 are the generalized forces

Using these notations, each of Equations 3.94 and 3.95 takes the form similar to the equations of motion of an SDOF system:

$$M_1\ddot{z}_1 + k_1 z_1 = P_1 \tag{3.99}$$

$$M_2\ddot{z}_2 + k_2 z_2 = P_2 \tag{3.100}$$

The solution of these uncoupled differential equations can be found by any of the standard procedures given in textbooks on vibration analysis. In particular, Duhame's integral provides a general method of solving these equations irrespective of the complexity of the loading function. However, in seismic analysis, usually a response spectrum is used instead of a forcing function to obtain the maximum values of the response corresponding to each modal equation. The direct superposition of modal maximum would, however, give only an upper limit for the total system that, in many engineering problems, would be too conservative. To alleviate this problem, approximations based on probability considerations are generally employed. One method employs the so-called root-mean-square (RMS) procedure, also called the SRSS method. As the name implies, a probable maximum value is obtained by evaluating the SRSS of the modal quantities. Although this method is simple and widely used, it is not always a conservative predictor of earthquake response because more combinations of modal quantities can occur, for example, when two modes have nearly the same natural period. In such cases, it is more appropriate to use the CQC procedure.

The aim of this section is to bring out the essentials of structural dynamics as related to seismic design of buildings. A certain amount of mathematical presentation has been unavoidable. Lest the reader lose the physical meaning of the various steps, it is worthwhile to summarize the essential features of dynamic analysis.

The dynamic analysis of buildings is performed by idealizing them as MDOF systems. The dead load of the building together with a percentage of live load (estimated to be present during an earthquake) is considered as lumped masses at each floor level. In a planar analysis, each mass has one DOF corresponding to lateral displacement in the direction under consideration, while in a 3D analysis, it has three DOF corresponding to two translational and one torsional displacements. Free vibrations of the buildings are evaluated, without including the effect of damping. The damping is taken into account by modifying the design response spectrum. The dynamic model representing a building has the number of mode shapes equal to the number of DOFs of the model. Mode shapes have the property of orthogonality, which means that no given mode shaped can be constructed as a combination of others, yet any deformation of the dynamic model can be described as a combination of its mode shapes, each multiplied by a scale factor. Each mode shape has a natural frequency of vibration. The mode shapes and frequencies are determined by solving for the eigenvalues. The total response of the building to a given response spectrum is obtained by statistically summing a predetermined number of modal responses. The number of modes required to adequately determine the design forces is a function of the dynamic characteristics of the building. Generally, for regular buildings, 6–10 modes in each direction are considered sufficient. Since each mass responds to earthquakes in more than one mode, it is necessary to evaluate effective modal mass values. These values indicate the percentage of the total mass that is mobilized in each mode. The acceleration experienced by each mass undergoing various modal deformations is determined from the response spectrum, which has been adjusted for damping. The product of the acceleration for a particular mode, multiplied by the effective modal mass for that mode, gives the static equivalent for forces at each discrete level. Since these forces do not reach their maximum values simultaneously, statistical methods such as SRSS or CQC are used for the combinations. The resulting forces are used as design static forces.

3.21.6 Design Example: Dynamic Displacement

Given: A weight W when attached to the end of a rubber band produces in it a static elongation $\delta_{st} = 5$ in. If the weight is raised until the tension in the band is zero and then released without initial velocity, what maximum elongation will be produced in the band due to this sudden application of load and with what frequency will the suspended weight W oscillate?

Solution: Taking the position of static equilibrium of the suspended weight as the origin and considering downward displacement as positive, we conclude that at the moment of release, the weight has the initial displacement

$$y_o = -\delta_{st}$$

and that the initial velocity is 0. Hence, from the equation of simple harmonic motion, the displacement of weight W from the position of static equilibrium at any instant t is

$$u(t) = W/k\,(1 - \cos W_n\, t)$$

When the angle $pt = \pi$, $\cos pt = -1$ and the displacement y have its maximum positive value,

$$\gamma_{\max} + \delta_{st}$$

We conclude that the total elongation produced in the band by the sudden application of the load W is twice that produced by the same load when gradually applied. The frequency of vibration is given by

$$F = \frac{l}{2\pi}\sqrt{\frac{g}{\delta_{st}}} = \frac{1}{2\pi}\sqrt{\frac{386}{5}} = 1.4 \text{ oscillations/s}$$

3.22 ANATOMY OF COMPUTER RESPONSE SPECTRUM ANALYSES

Now that we have learned the fundamentals of dynamic analysis in excruciating detail, perhaps it is instructive to scrutinize how dynamic analysis is performed internally in computer programs. This will enhance our understanding of the modal superposition techniques used in computer programs.

Two examples presented illustrate the modal analysis method. In the first part of each example, the analysis is performed to determine the base shear for each mode using given building characteristics and ground motion spectra. In the second part, the story forces, accelerations, and displacements are calculated for each mode and are combined statistically using the SRSS combination.

The base shear is determined from

$$V_m = a_m S_{am} W \tag{3.101}$$

where

V_m is the base shear contributed by the mth mode
a_m is the modal base shear participation factor for the mth mode
S_{am} is the spectral acceleration for the mth mode determined from the response spectrum
W is the total weight of the building including dead loads and applied portions of other loads

The modal base shear participation factor, a_m, for the mth mode is determined from

$$\alpha_m = \frac{\left(\sum_{i=1}^{n} \frac{w_i}{g}\phi_{im}\right)^2}{\sum_{i=1}^{n} \frac{w_i}{g} \sum_{i=1}^{n} \frac{w_i}{g}\phi_{im}^2}$$

The story modal participation, PF_{xm}, for the mth mode is determined from

$$PF_{xm} = \left(\frac{\sum_{i=1}^{n} \frac{w_i}{g} \phi_{im}}{\sum_{i=1}^{n} \frac{w_i}{g} \phi_{im}^2} \right) \phi_{xm}$$

where

PF_{xm} is the modal participation factor at level x for the mth mode
w_i/g is the mass assigned to level i
ϕ_{im} is the amplitude of the mth model at level i
ϕ_{xm} is the amplitude of the mth mode at level x
n is the level n under consideration

The modal story lateral displacement, δ_{xm}, is determined from

$$\delta_{xm} = PF_{xn} S_{am} \tag{3.102}$$

where

δ_{xm} is the lateral displacement at level x for the mth mode
S_{am} is the spectral acceleration for the mth mode determined from the response spectrum
T_m is the period of vibration of the mth mode

3.22.1 Example 1: Three-Story Building

Given: The example is illustrated in Figure 3.113.

Weights and masses

$$W_R = 187 \text{ kip}$$

$$M_R = 187/32.2 = 5.81 \text{ kip-s}^2\text{/ft}$$

$$W_2 = W_3 = 236 \text{ kip}$$

$$m_2 = m_3 = 236/32.2 = 7.33 \text{ kip-s}^2\text{/ft.}$$

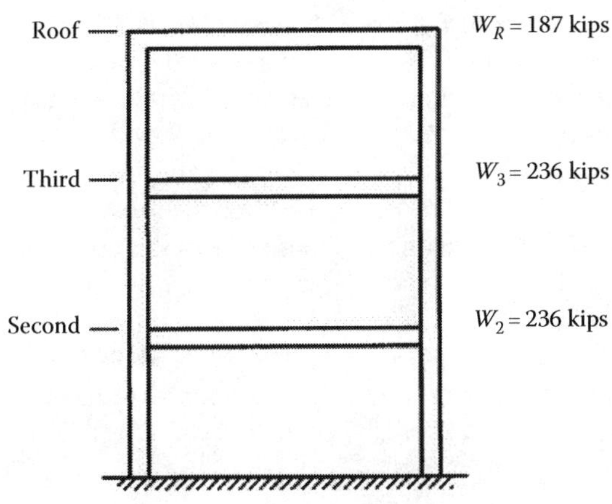

FIGURE 3.113 Three-story building example: dynamic analysis.

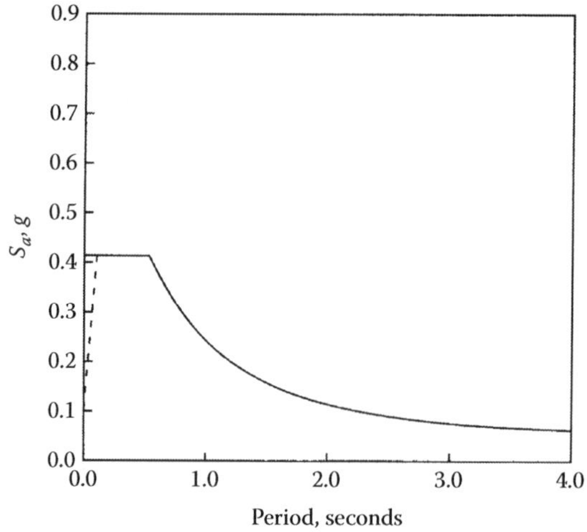

	Period							
	0.0	.586	.80	1.0	1.5	2.0	3.0	4.0
S_a, g	.14	.41	.300	.240	.160	.120	.080	.060

FIGURE 3.114 Three-story building; response spectrum.

Periods

Spectral acceleration: From the response spectrum of Figure 3.114, the spectral accelerations are

$$S_{a1} = 0.25g \text{ for mode 1}$$

$$S_{a2} = 0.41g \text{ for mode 2}$$

$$S_{a3} = 0.25\ lg \text{ for mode 3}$$

Required

1. Modal analysis to determine base shears
2. Story forces, overturning moments, accelerations, and displacements for each mode
3. SRSS combinations

Solution: The results of the modal analysis are shown in Tables 3.3 through 3.5. It should be noted that higher modes of response become increasingly important for taller or irregular buildings. For the regular three-story building, the first mode dominates the lateral response as shown in the comparison of the modal story shears and the SRSS story shears in Table 3.5. For example, if only the first-mode shears had been used for analysis, we would have obtained 89% of the SRSS shear at the roof, 99% at the third floor, and 95% at the second floor. While the second-mode shear at the roof is 50% of the first-mode shear, when combined on SRSS basis, the first mode accounts for 79% of the SRSS response, with 20% for the second mode and 0.6% for the third mode. These percentages are 91%, 8%, and 1% at the base. The effective modal weight factor, a_m, also shows the relative importance of each mode. In this example, with $a_1 = 0.804$, $a_2 = 0.149$, and $a_3 = 0.048$, this indicate that 80.4% of the building mass participation is in the first mode, 14.9% in the second, and 4.8% in the third.

3.22.2 Example 2: Seven-Story Building

Given: See the seven-story building illustration in Figure 3.115.

TABLE 3.3
Three-Story Building: Modal Analysis to Determine Base Shears

T_m, s		0.964			0.356			0.182		

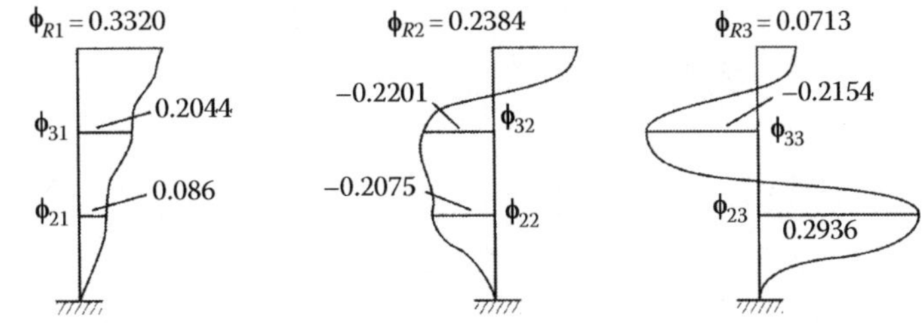

		Mode 1			Mode 2			Mode 3		
Level	Mass $\left(\frac{k\cdot s^2}{ft}\right)$	ϕ_{x1}	$m_x\phi_{x1}$	$m_x\phi_{x1}^2$	ϕ_{x2}	$m_x\phi_{x2}$	$m_x\phi_{x2}^2$	ϕ_{x3}	$m_x\phi_{x3}$	$m_x\phi_{xs}^2$
R	5.81	0.3320	1.929	0.640	0.2384	1.385	0.330	0.0713	0.4143	0.030
3	7.32	0.2044	1.496	0.306	−0.2201	−1.611	0.355	−0.2154	−1.577	0.340
2	7.32	0.0860	0.630	0.054	−0.2075	−1.519	0.315	0.2936	2.149	0.631
Σ	20.45		4.055	1.000[b]		−1.745	1.000		0.9863	1.001
PF_{Rm}^{a}		$\frac{\Sigma m\phi}{\Sigma m\phi^2}\phi_{R1} = 1.346$				−0.416			0.070	
PF_{3m}			0.829			0.384			−0.212	
PF_{2m}			0.349			0.362			0.289	
α_m		$\frac{(\Sigma m\phi)^2}{\Sigma m(\Sigma m\phi^2)} = 0.8040$				0.149			0.048	
S_a			0.251 g			0.41 g			0.41 g	
$v = \alpha_m S_a W$			132.7 kips			40.2 kips			13.0 kips	

[a] Note that the sum of the modal participation factors $\sum_{m=1}^{3} PF_{xm} = 1.0$ and the sum of modal base shear participation factors $\sum_{m=1}^{3} \alpha_m = 1.0$.

[b] The mode shapes have been normalized by the computer program so that $\sum m\phi^2 = 1.0$.

Weights and masses

$$W_R = 1410 \text{ kip}$$

$$m_R = W_R/g = 1410/32.2 = 43.79 \text{ kips-s}^2\text{/ft.}$$

$$W_7 = W_6 = W_5 = W_4 = W_3 = 1460 \text{ kip}$$

$$m_7 = m_6 = m_5 = m_4 = m_3 = 1460/32.2 = 45.34 \text{ kip-s}^2\text{/ft.}$$

$$W_2 = 1830 \text{ kip}$$

$$m_2 = 1830/32.2 = 56.83 \text{ kip-s}^2\text{/ft.}$$

Periods:
$T_1 = 0.880$ s
$T_2 = 0.288$ s
$T_3 = 0.164$ s

TABLE 3.4
Three-Story Building: Modal Analysis to Determine Story Forces, Accelerations, and Displacements

Level	PF_{xm}	$\frac{m_x\phi_{xm}}{\sum m_x\phi_{xm}}$	F_{xm} (k)	V_{xm} (k)	ΔOTM_{xm} (ft·k)	OTM_{xm} (ft·k)	$a_{xm} = \frac{F_{xm}}{w_x}$	δ_{xm} (in.)	Δ_{xm} (in.)
(a) Mode 1									
R	1.346	0.476	63.2	63.2	772	0	0.337	3.065	1.182
3	0.829	0.369	48.9	112.1	1233	772	0.208	1.892	1.101
2	0.349	0.155	20.6	132.7	1416	2005	0.087	0.791	0.791
		1.000				3421			
(b) Mode 2									
R	−0.416	−0.793	−31.9	−31.9	−389	0	−0.171	−0.212	0.407
3	0.384	0.923	37.1	5.2	57	−389	−0.157	0.195	0.011
2	0.362	0.870	35.0	40.2	429	−332	−0.148	0.184	0.184
		1.000				97			
(c) Mode 3									
R	0.070	0.420	5.5	5.5	67	0	−0.029	0.0094	0.037
3	−0.212	−1.599	−20.8	−15.3	−168	67	−0.087	−0.028	0.066
2	0.289	2.179	28.3	13.0	139	−101	0.118	0.038	0.038
		1.000				38			
(d) SRSS combination									
R			71.0	71.0	867	0	0.379	3.072	1.251
3			64.8	113.3	1246	867	0.275	1.893	1.094
2			49.5	139.3	1486	2035	0.208	0.812	0.813
						3423			

TABLE 3.5
Three-Story Building: Comparison of Modal Story Shears and the SRSS Story

		Mode 1			Mode 2		Mode 3	
Level	V_{SRSS}	V_1	V_1/V_{SRSS}	$(V_1/V_{SRSS})^2$	V_2	$(V_2/V_{SRSS})^2$	V_1	$(V_3/V_{SRSS})^2$
R	71.0	63.2	0.89	0.79	−31.9	0.202	5.5	0.006
3	119.3	112.1	0.989	0.98	5.2	0.002	−15.3	0.018
2	139.3	132.7	0.953	0.91	40.2	0.083	13.0	0.009

Spectral accelerations: From the response spectrum of Figure 3.116a through c, the spectral accelerations are

$S_{a1} = 0.276g$
$S_{a2} = 0.500g$
$S_{a3} = 0.500g$

Observe that all three parts of Figure 3.116 contain the same information related to the acceleration response, S_a. Only the format is different. Figure 3.116a shows the building periods and spectral accelerations in a format similar to that in 1997 UBC and IBC-03. Figure 3.116b is a tripartite

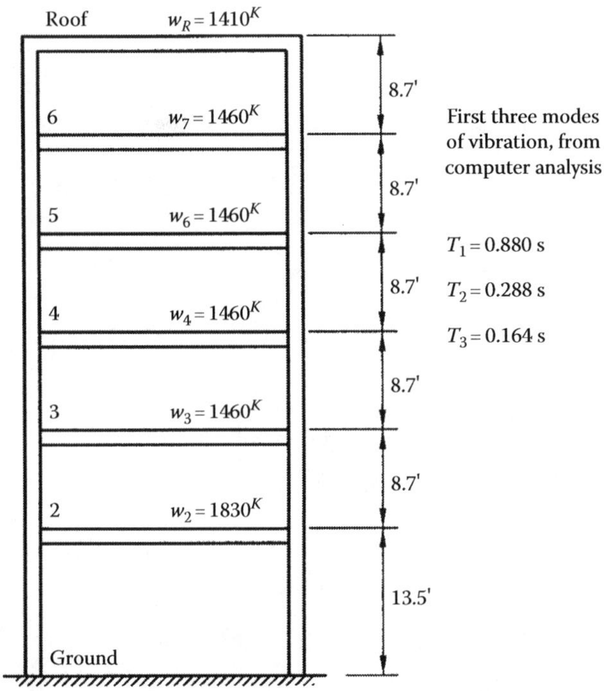

FIGURE 3.115 Seven-story building example: dynamic analysis.

response spectrum with additional values for displacements and velocities. Figure 3.116c shows the building periods and response accelerations in tabular format.

It should be noted that in the computer program used for the calculation of the eigenvalues, each mode is normalized for a value of $\sum_g^w \Phi_2 = 1.0$. In some programs, Φ is normalized to 1.0 at the uppermost level.

Required

1. Modal analysis to determine base shears
2. First-, second-, and third-mode force and displacements
3. Modal analysis summary

Solution: From the modal analysis results shown in Table 3.6, the sum of the participation factors, PF_{xm}, and a_m adds up to 1.08 and 0.986, respectively. These values being close to 1.0 indicate that most of the modal participation is included in the three modes considered in the example. The story accelerations and the base shears are combined by the SRSS. The modal base shears are 2408, 632, and 200 kip for the first, second, and third modes, respectively. These are used in Figure 3.117 to determine story forces. The SRSS base shear is 2498 kip.

Story forces, accelerations, and displacements: Tables 3.6 through 3.9 are set up in a manner similar to the static design procedure described previously. In the static lateral procedure, $Wh/\Sigma Wh$ is used to distribute the force on the assumption of a straight-line mode shape. In the dynamic analysis, the more representative $W\Phi/SW\Phi$ distribution is used to distribute the forces. The story shears and overturning moments are determined in the same manner for each method. Modal story accelerations are determined by dividing the story force by the story

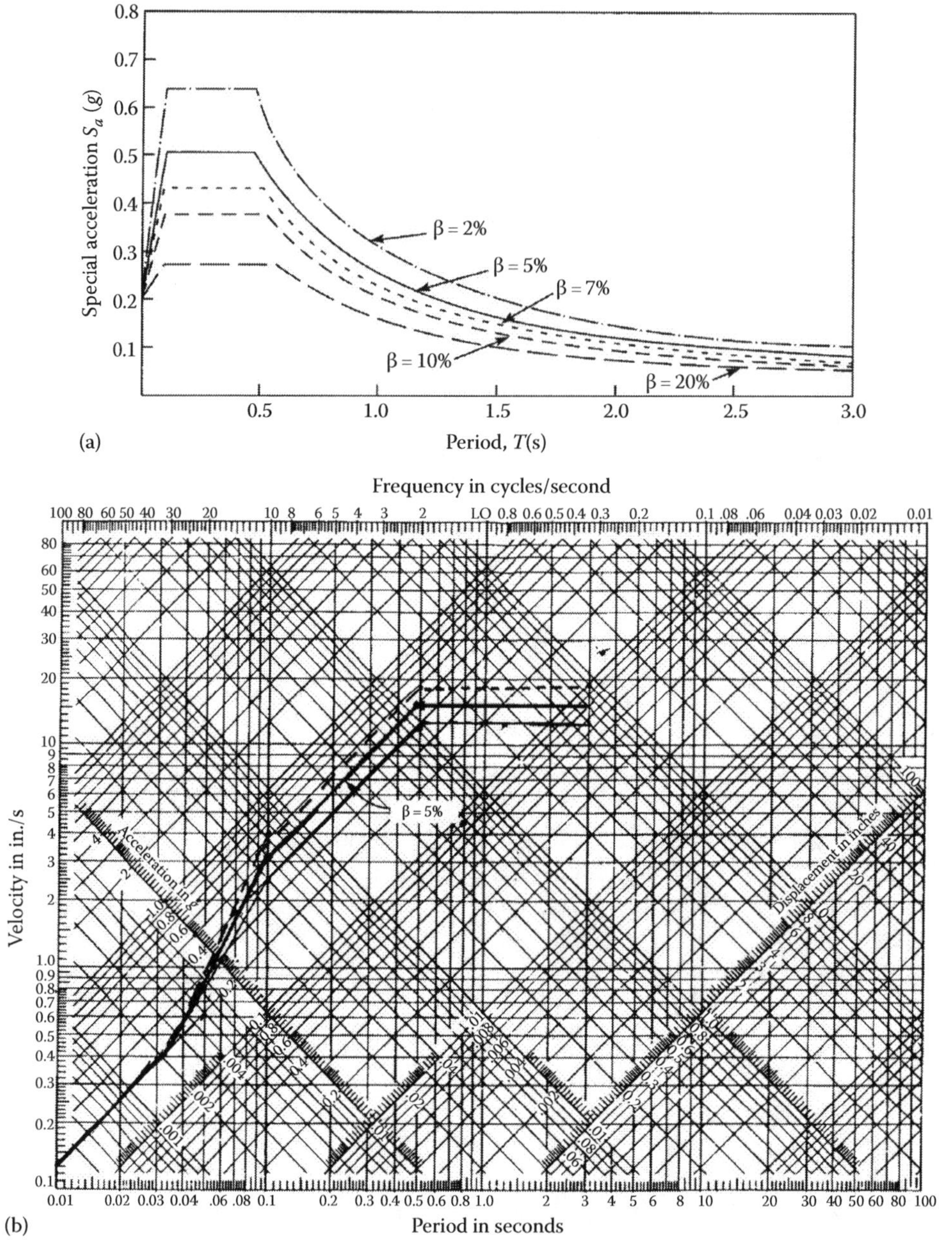

	Spectral Acceleration, $S_\alpha(g)$											
β \ T	**0.1**	**0.48**	**0.50**	**0.80**	**1.00**	**1.75**	**1.50**	**1.75**	**2.00**	**2.25**	**2.50**	**3.00**
2%	0.64	0.64	0.59	0.37	0.30	0.24	0.20	0.17	0.15	0.13	0.17	0.10
5%	0.50	0.50	0.48	0.30	0.24	0.192	0.16	0.137	0.12	0.107	0.096	0.08
7%	0.44	0.44	0.44	0.28	0.22	0.18	0.15	0.13	0.11	0.10	0.09	0.07
10%	0.38	0.38	0.38	0.25	0.20	0.16	0.13	0.11	0.10	0.09	0.08	0.066
20%	0.27	0.27	0.27	0.20	0.16	0.12	0.10	0.09	0.08	0.07	0.06	0.05

(c)

FIGURE 3.116 Response spectrum for three-story building example: (a) acceleration spectrum, (b) tripartite diagram, and (c) response spectra numerical representation.

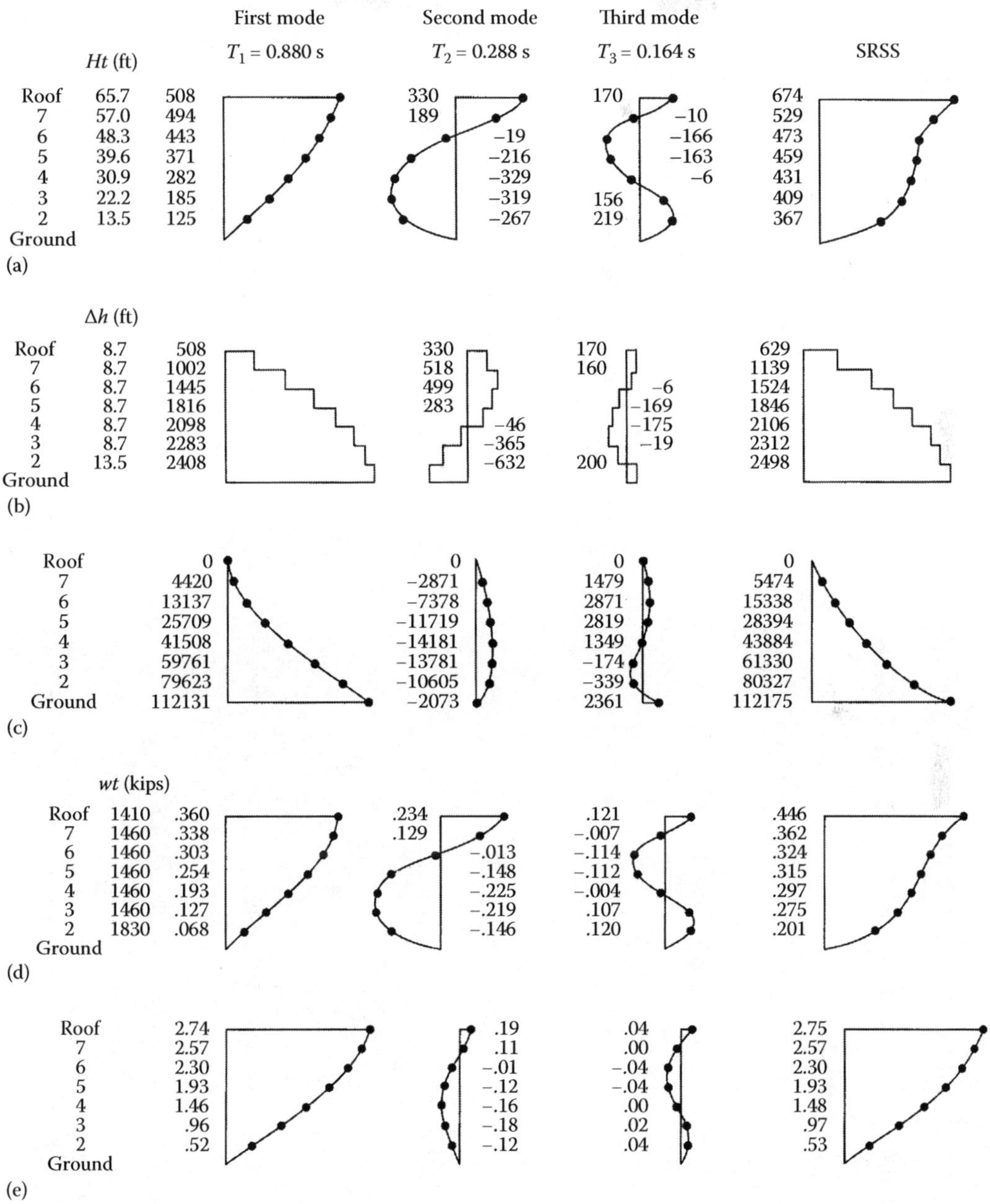

FIGURE 3.117 Seven-story building, modal analysis summary: (a) modal story forces, kip; (b) modal story shears, kip; (c) modal story overturning moments, kip-ft; (d) modal story accelerations, g; and (e) modal lateral displacement, inches.

TABLE 3.6
Seven-Story Building: Modal Analysis to Determine Base Shears

		Mode 1				Mode 2				Mode 3				SRSS
Level	$\frac{w}{g}\left(\frac{k-s^2}{ft}\right)$	ϕ_1	$\frac{w}{g}\phi_1$	$\frac{w}{g}\phi_1^2$	a_1 (g)	ϕ_2	$\frac{w}{g}\phi_2$	$\frac{w}{g}\phi_2^2$	a_2 (g)	ϕ_3	$\frac{w}{g}\phi_3$	$\frac{w}{g}\phi_3^2$	a_3 (g)	a_x (g)
Roof	43.78	0.0794	3.48	0.276	0.362	0.0747	3.27	0.744	−0.235	0.0684	2.99	0.205	0.120	0.448
7	45.34	0.0745	3.38	0.252	0.340	0.0411	1.86	0.076	−0.129	−0.0040	−0.18	0.001	−0.007	0.364
6	45.34	0.0666	3.02	0.201	0.304	−0.0042	−0.19	0.001	0.013	−0.0644	−2.92	0.188	−0.113	0.325
5	45.34	0.0558	2.53	0.141	0.254	−0.0471	−2.14	0.101	0.148	−0.0630	−2.86	0.180	−0.111	0.314
4	45.34	0.0425	1.93	0.082	0.194	−0.0718	−3.26	0.234	0.226	−0.0023	−0.10	0.000	−0.004	0.298
3	45.34	0.0279	1.27	0.035	0.127	−0.0697	−3.16	0.220	0.219	0.0604	2.74	0.166	0.106	0.275
2	56.83	0.0149	0.85	0.013	0.068	−0.0467	−2.65	0.124	0.147	0.0677	3.85	0.261	0.119	0.201
1	—	0	0	0	0	0		0	0	0	0	0	0	0
Σ	327.31		16.46	1.000			−6.27	1.000			3.52	1.001		
PF_{roof}	Eq. (3.8)	$\frac{16.46}{1.000}(0.0794) = 1.31$				$\frac{-6.37}{1.000}(0.0747) = -0.47$				$\frac{3.52}{1.001}(0.0684) = 0.24$				$\Sigma = 1.08$
α	Eq. (3.9)	$\frac{(16.46)^2}{(327.31)(1.000)} = 0.828$				$\frac{(-6.27)^2}{(327.31)(1.000)} = 0.120$				$\frac{(3.52)^2}{(927.31)(1.001)} = 0.038$				$\Sigma = 0.986$
T		0.880 sec				0.288 sec				0.164 sec				
S_a		0.276 g				0.500 g				0.500 g				
a_{roof}	Eq. (3.10)	(1.31)(0.276) = 0.362 g				(−0.47)(0.500) = −0.235 g				(0.24)(0.500) = 0.120 g				0.448
V	Eq. (3.11)	(0.828)(0.276)(10,539) = 2408 kips				(0.12)(0.500)(10,539) = 632 kips				(0.038)(0.500)(10,539) = 200 kips				2498 kips (SRSS)
V/W		0.229				0.060				0.019				0.237

$W = \sum\left(\frac{w}{g}\right) \times g = 327.31 \times 32.2 = 10{,}539$ kips = Building weight.

$A_G = 0.20\ g$ Site PGA.

$\beta = 0.05$ Damping factor.

TABLE 3.7
Seven-Story Building: First-Mode Forces and Displacements

$T_1 = 0.880$ s

Modal Base Shear $V_1 = 2408$ kips

(1)	(2)	(3)	(4)	(5)	(6)	(7)	(8)	(9)	(10)	(11)		
Story	ϕ	h ft	Δh ft	w kips	$\frac{w\phi}{\sum w\phi}$	F kips $(V_1) \times (6)$	V kips $\sum(7)$	ΔOTM K-ft (4)–(8)	OTM K-ft $\sum(9)$	Accel. g (7) ÷ (5)	δ[a] ft	$\Delta\delta$ ft
Roof	0.0794	65.7	8.7	1410	0.211	508	508	4420	0	0.360	0.228	0.014
7	0.7450	57.0	8.7	1460	0.205	494	1002	8717	4420	0.338	0.214	0.022
6	0.0666	48.3	8.7	1460	0.184	443	1445	12,572	13,137	0.303	0.192	0.031
5	0.0558	59.6	8.7	1460	0.154	371	1816	15,799	25,709	0.254	0.161	0.039
4	0.0425	30.9	8.7	1460	0.117	282	2098	10,253	41,508	0.193	0.122	0.042
3	0.0279	22.2	8.7	1460	0.077	185	2283	19,862	59,761	0.127	0.080	0.057
2	0.0149	13.5	13.5	1830	0.052	125	2408	32,508	79,623	0.068	0.043	0.043
Grd.	0	0		0	0	0			112,131	0	0	
				Σ	1.000	2408		112,191				

[a] Displacement $\delta_{x1} = \frac{g}{4\pi^2} \times T_1^2 \times \frac{F}{w}$

$= \frac{32}{4\pi^2} \times 0.88^2 \times \text{acceleration}$

$= 0.632 \times \text{acceleration}.$

TABLE 3.8
Seven-Story Building: Second-Mode Forces and Displacements

$T_2 = 0.288$ s

Modal Base Shear $V_2 = 632$ kips

(1)	(2)	(3)	(4)	(5)	(6)	(7)	(8)	(9)	(10)	(11)		
Story	ϕ	h ft	Δh ft	w kips	$\frac{w\phi}{\sum w\phi}$	F kips $(V_2) \times (6)$	V kips $\sum(7)$	ΔOTM K-ft (4) × (8)	OTM k-ft $Z(9)$	Accel. g (7) ÷ (5)	δ[a] ft	$\Delta\delta$ ft
Roof	0.0747	65.7	8.7	1410	0.522	−330	−330	−2871	0	−0.234	−0.016	0.007
7	0.0411	57.0	8.7	1460	0.297	−188	−518	−4507	−2871	−0.129	−0.009	0.010
6	−0.0042	48.3	8.7	1460	0.030	19	−499	−4341	−7378	0.013	0.001	0.009
5	−0.0471	39.6	8.7	1460	0.341	216	−283	−2462	−11,719	0.148	0.010	0.005
4	−0.0718	30.9	8.7	1460	0.520	329	46	400	−14,181	0.225	0.015	0.000
3	−0.0697	22.2	8.7	1460	0.504	319	365	3176	−13,781	0.219	0.015	0.005
2	−0.0467	13.5	13.5	1830	0.423	267	632	8532	−10,605	0.146	0.010	0.010
Grd.	0	0							−2073	0	0	
				Σ	0.999	632		−2073				

[a] Displacement $\delta_{x2} = \frac{g}{4\pi^2} \times T_2^2 \times \frac{F}{w}$

$= \frac{32}{4\pi^2} \times 0.288^2 \times \text{acceleration}$

$= 0.068 \times \text{acceleration}.$

TABLE 3.9
Seven-Story Building: Third-Mode Forces and Displacements

$T_3 = 0.164$ s

Modal Base Shear $V_3 = 200$ kips

(1)	(2)	(3)	(4)	(5)	(6)	(7)	(8)	(9)	(10)	(11)		
Story	ϕ	h ft	Δh ft	w kips	$\frac{w\phi}{\sum w\phi}$	F kips $(V_3) \times (6)$	V kips $\sum$ (7)	ΔOTM K-ft (4) × (8)	OTM K-ft $\sum$ (9)	Accel. g (7) ÷ (5)	δ[a] ft	$\Delta\delta$ ft
Roof	0.0684	65.7	8.7	1410	0.849	170	170	1479	0	0.121	0.003	0.003
7	−0.0040	57.0	8.7	1460	−0.051	−10	160	1392	1479	−0.007	0.000	0.003
6	−0.0644	48.3	8.7	1460	−0.830	−166	−6	−52	2871	−0.114	−0.003	0.000
5	−0.0630	39.6	8.7	1460	−0.813	−163	−169	−1470	2819	−0.112	−0.003	0.003
4	−0.0023	30.9	8.7	1460	−0.028	−6	−175	−1523	1349	−0.004	0.000	0.002
3	0.0604	22.2	8.7	1460	0.778	156	−19	−165	−174	0.107	0.002	0.001
2	0.0677	13.5	13.5	1830	1.094	219	200	2700	−339	0.120	0.003	0.003
Grd.	0	0							2361	0	0	
				$\sum$	0.999	200		2361				

[a] Displacement $\delta_{x3} = \frac{g}{4\pi^2} \times T_3^2 \times \frac{F}{W}$

$= \frac{32}{4\pi^2} \times 0.64^2 \times \text{acceleration}$

$= 0.022 \times \text{acceleration}.$

weight. Modal story displacements are calculated from the accelerations and the period by using the following equations:

$$\delta_{xm} = \text{PF}_{xm} S_{am} \left(\frac{T_m}{2\Phi\pi} \right)^2 g$$

where

δ_{xm} is the lateral displacement at level x for mode m
S_{am} is the spectral displacement for mode m calculated from response spectrum
T_m is the modal period of vibration

Modal interstory drifts are calculated by taking the difference between the values of adjacent stories. The values shown in Tables 3.6 through 3.9 are summarized in Figure 3.117.

The fundamental period of vibration as determined from a computer analysis is 0.88 s. The periods of the second and third modes of vibration are 0.288 and 0.164 s, respectively. From Tables 3.7 through 3.9 using a response curve with 5% of critical damping ($\beta = 0.05$), it is determined that the second- and third-mode spectral accelerations are 0.276g. On the basis of mode shapes and modal participation factors, modal story forces, shears, overturning moments, accelerations, and displacements are determined.

Figure 3.117 shows story forces obtained by multiplying the story acceleration by the story mass. The shapes of story force curves are quite similar to the shapes of the acceleration curves because the building mass is essentially uniform.

Figure 3.117b shows story shears that are a summation of the modal story forces in Figure 3.117. The higher modes become less significant in relation to the first mode because the forces tend to cancel each other due to the reversal of direction. The SRSS values do not differ substantially from the first-mode values.

Figure 3.117c shows the building overturning moments. Again, the higher modes become somewhat less significant because of the reversal of force direction. The SRSS curve is essentially equal to the first-mode curve.

Figure 3.117d shows story accelerations. Observe that the second and third modes do play a significant role in the structure's maximum response. While the shape of an individual mode is the same for displacements and accelerations, accelerations are proportional to displacements divided by the squared value of the modal period, which accounts for the greater accelerations in the higher modes. The shape of the SRSS combination of the accelerations is substantially different from the shapes of any of the individual modes because it accounts for the predominance of the various modes at different story levels.

Figure 3.117e shows the modal displacements. Observe that the fundamental mode predominates, while the second- and third-mode displacements are relatively insignificant. The SRSS combination does not differ greatly from the fundamental mode. It should be noted, however, that for taller and irregular buildings, the influence of the higher modes will become larger.

4 Wind Load Analysis of Buildings

PREVIEW

This chapter addresses wind effects on building structures, provides guidelines for assessing design wind loads for buildings, and offers a discussion on the advantages of wind tunnel testing. The prevailing standard for the calculation for wind forces in the United States is the American Society of Civil Engineers (ASCE) publication, *Minimum Design Loads for Buildings and Other Structures* (ASCE 7-10). The fundamental goals of this chapter are outlined in the following.

- To present engineering understanding of wind, structural dynamics, and wind effects on buildings
- To describe how ASCE 7-10 interprets and incorporates the fundamentals of wind engineering
- To apply the provisions of ASCE 7-10 with correct interpretation to assess wind loads on buildings
- To gain an in-depth understanding of ASCE 7-10 wind load provisions and wind loads on buildings
- To discuss interpretations and limitations of key provisions of ASCE 7-10

The chapter has several references to the sections, figures, tables, and equations of the ASCE 7-10. It is advised that the reader keeps a copy of the ASCE 7-10 while reading this chapter and frequently refers to it. An abbreviation of "F" for figures and "T" for tables is used at several locations in this chapter. Where the term ASCE is used along with equations, figures, tables, sections it refers to ASCE 7-10 otherwise it is a reference to the section, figure or table of this book. In several sections of this chapter tables are provided to explain concepts.

The chapter discusses hurricanes, history of codes, and standards pertaining to wind loads and general wind concepts in the first three sections. In order to explain the use of ASCE 7-10, Section 4.4 is dedicated to the organization of the chapters of ASCE 7-10 regarding information about wind. Sections 4.5 and 4.6 deal with the general requirements in determining the wind loads for various cases, and the concept of velocity pressure, respectively. Subsequent Sections 4.7 through 4.10 deal with Chapters 27 through 30 of the ASCE 7-10. Section 4.11 explains the major differences between ASCE 7-05 and ASCE 7-10. Some examples of the calculations of various factors of ASCE 7-10 are provided in Section 4.12. Detailed discussions regarding the wind tunnel procedures are provided in Section 4.13.

4.1 MAJOR CAUSES OF WIND FORCES

Lateral forces caused by winds are the major factor in the design of tall buildings. Even in locations belonging to low wind zones, tall buildings are designed for the wind effects. The greatest wind effect is caused by hurricanes and tornadoes. Hurricane is the highest wind storm on the earth and few natural disasters can pose as much calamity as a hurricane can do. Hurricane can make a landfall with sustained winds greater than 155 mph. During their life time, they can expend as much energy as 10,000 nuclear bombs. They are called by different names in different parts

TABLE 4.1
How Is a Hurricane Formed?

Number	Description of Event
1	Warm, moist air moves over the ocean.
2	Water vapor rises into the atmosphere.
3	As the water vapor rises, it cools and condenses into liquid droplets.
4	Condensation releases heat into the atmosphere, making the air lighter.
5	The warmed air continues to rise, with moist air from the ocean taking its place and creating more wind.

of the world. They are called "typhoons" in the western Pacific and China Sea area. In Australia, Bangladesh, India, and Pakistan, they are called "cyclones," and they are called "baguios" in the Philippines. Their scientific name is "Tropical Storm." They are the storm systems consisting of a large low-pressure center and numerous thunderstorms that produce strong winds and heavy rain (See Table 4.1). When saturated air rises, water evaporated from the ocean is released and the storms and water vapor contained in the moist air condenses. At any height in the atmosphere, the center of a tropical cyclone will be warmer than its surroundings. In general, it is a large system of spinning air that rotates around a point of low pressure.

The first sign of hurricane formation is the appearance of a cluster of thunderstorms over the tropical oceans. It is called tropical disturbance. When winds converge, the collision forces the air to rise, initiating thunderstorms. These convergences take place either at the meeting point of the northern and southern hemispheres at the eastern side of the equator or along the boundary between masses of warm and cold water. Thunderstorms created get organized into a more unified storm system and results in the fall of surface air pressures in the area around them. Winds begin to spin. Water vapors condense in rising air and release energy, which increases the buoyancy of air and makes it rise. To compensate for this rising air, the surrounding air sinks and is compressed by the air above it and warms. The pressure rises at the top of the layer of warming air, pushing air at the top of the layer outward. Now there is less air in the layer making the pressure of the ocean surface to drop, drawing more air at the surface, which converges near the center of the storm to form more clouds. This becomes a chain reaction and the storm gets intensified. The lower the surface pressure, the more rapidly the air flows into the storm at the surface, increasing the wind and causing more thunderstorms. Stronger winds are triggered. When the wind speed is about 25 mph, it is called a tropical depression and at about 40 mph, it is called a tropical storm and at 75 mph, it is a hurricane. However, if the atmospheric condition 3–6 miles above the surface is not favorable, the storm withers away. Hurricanes can diminish when the storm moves over cooler water that can't supply warm air or moves over land or move into an area where strong winds high in the atmosphere disperse latent heat reducing the warm temperatures aloft and raising the surface pressure.

Tropical cyclones produce extremely powerful winds and torrential rain and also high waves and damaging storm surges as well as tornadoes. Once they make landfall, they lose their strength due to increased surface friction with the ground and loss of the warm ocean as an energy source. The coastal regions receive significant damage from a tropical cyclone, while inland regions are winds with lesser velocity. Heavy rains produce significant flooding inland and storm surges flooding up to 25 miles from the coastline. Storm surge is the most destructive weapon accompanying hurricanes, a rise in the ocean levels of up to about 33 ft.

A major hurricane is a category 3, 4, or 5 hurricane on Saffir/Simpson scale, capable of inflicting great damage and loss of life. The Saffir/Simpson hurricane wind scale provides specific wind values for each hurricane category. The original Saffir/Simpson hurricane scale category assignment of U.S. hurricanes was based on a combination of wind, central pressure, and storm surge values. It consists of five categories (one being the weakest and five being the strongest).

Scale 1 hurricane has a wind velocity of 75–95 mph. Storm surges generally 4–5 ft above normal. There is no real damage to building structures. The damage is primarily to unanchored mobile homes, shrubbery, and trees. There is some damage to poorly constructed signs. Also, some coastal road flooding and minor pier can get damaged.

Scale 2 hurricane has a wind velocity of 96–110 mph. Storm surges generally 6–8 ft above normal. There is some roofing material, door, and window damage of buildings. Considerable damage is done to shrubbery and trees, with some trees blown down. There is also considerable damage to mobile homes, poorly constructed signs, and piers. Coastal and low-lying escape routes flood 2–4 h before arrival of the eye of the hurricane. Small craft in unprotected anchorages break the moorings.

Scale 3 hurricane has a wind velocity of 111–130 mph. Storm surges generally 9–12 ft above normal. There is some structural damage to small residences and utility buildings, with a minor amount of curtain wall (non-load-bearing exterior wall) failures. Damage to shrubbery and trees is experienced, with foliage blown off trees and large trees blown down. Mobile homes and poorly constructed signs are destroyed. Low-lying escape routes are cut by rising water 3–5 h before arrival of the center of the hurricane. Flooding near the coast destroys smaller structures, with larger structures damaged by battering from floating debris.

Scale 4 hurricane has a wind velocity of 131–155 mph. Storm surges generally 13–18 ft above normal. More extensive curtain wall failures with some complete roof structure failures on small residences. Shrubs, trees, and all signs are blown down. Mobile homes are completely destroyed. Extensive damage to doors and windows is experienced. Low-lying escape routes may be cut by rising water 3–5 h before arrival of the center of the hurricane. There is major damage to lower floors of structures near the shore. Terrain lower than 10 ft above sea level may be flooded.

Scale 5 hurricane has a wind velocity of greater than 155 mph. Storm surges generally greater than 18 ft above normal. Roof could completely fail in single family residences and industrial buildings. Complete failure of one- or two-storied buildings could occur. All shrubs, trees, and signs are blown down. There is severe and extensive window and door damage. Low-lying escape routes are cut by rising water 3–5 h before arrival of the center of the hurricane. Major damage to lower floors of all structures located less than 15 ft above sea level and within 500 yd of the shoreline.

The National Hurricane Center has recorded historic information about hurricanes that made land falls in the United States. Florida, Texas, Louisiana, and North Carolina are the states with more frequencies of hurricane land falls. Katrina, Andrew, Ike, Wilma, Ivan, Charley, Hugo, Rita, Agnes, and Betsey are among the most damage causing hurricanes.

When it comes to hurricanes, wind speeds do not tell the whole story. Hurricanes produce storm surges, tornadoes, and often the most deadly of all—inland flooding. While storm surge is always a potential threat, more people have died from inland flooding. Intense rainfall is not directly related to the wind speed of tropical cyclones. In fact, some of the greatest rainfall amounts occur from weaker storms that drift slowly or stall over an area. Inland flooding can be a major threat to communities hundreds of miles from the coast as intense rain falls from these huge tropical air masses. Persistent high wind and changes in air pressure push water toward the shore, causing a storm surge, which can be several feet high. Waves can be highly destructive as they move inland, battering structures in their path. On open coasts, the magnitude varies with the tides. An increase in the level of the ocean during the high tide will flood larger areas than a storm that strikes during the low tide. Major coastal storms can significantly change the shape of shoreline landforms, making sandy coastal floodplains unstable places for development. Wind and waves shape sand dunes, bluff, and barrier islands. The preservation of the landforms is important for the internal development because they form a protection from the effects of the storm.

4.2 BUILDING CODES ADDRESSING WIND LOADS AND FLOODS

In the 21st century, Hurricane Katrina has been the most devastating natural event in the United States with a life loss of more than 1200 and a financial damage of $108 billion in Florida, Louisiana, and Mississippi. A category 4 hurricane hit Galveston in Texas at a sustained wind of more than 140 mph killing more than 8000 people in 1900. There were no preparations made to resist the impact of that hurricane. The hurricane covered the buildings like an ocean and had cost several million dollars to recuperate. The year 1992 experienced the most devastating hurricane in the modern times when in August, Hurricane Andrew hit South Florida. People in Miami and Homestead were unprepared as Andrew changed its route. Building codes were rewritten due to Andrew. There was a major change to both the Miami-Dade and Broward edition of the *South Florida Building Code*, which was later incorporated in the *Florida Building Code* for the High-Velocity Hurricane Zone (HVHZ) portion of the code.

The two standards published by the American Society of Civil Engineers—ASCE 7-10 *Minimum Design Loads for Buildings and Other Structures* and ASCE 24-05 *Flood Resistant Design and Construction* address the loading requirements for wind and flood. Wind and flood along with earthquake, fire, and snow are the main reasons for major building damage in the United States.

The first code related to wind loads was published as ANSI A58.1-1972 by the American National Standard Institute (ANSI) in 1972. A subsequent edition of the ANSI A58.1 was published in 1982. Then the American Society started publishing the ASCE 7. Editions of the standard were published in 1988, 1993, 1995, 1998, 2002, 2005, and 2010. The ASCE 7-10 has been significantly reorganized in comparison with the previous codes with six chapters dedicated to wind. In the 1995 edition, the wind speed was changed from the fastest wind speed to the 3-second gust. Each revision made changes to several different factors like the importance factor, terrain factor, directionality factor, gust-effect factor, and the pressure/force coefficients.

Flood Resistant Design and Construction—ASCE 24 was published in 1998 and subsequently in 2005. ASCE 24 deals with the minimum requirements and expected performance for the design and construction of buildings and structures in flood hazard areas. It is not a restatement of all of the National Flood Insurance Program NFIP regulations but offers additional specificity, some additional requirements, and some limitations. The parts 59, 60, 65, and 70 of the Chapter 44 of the *Code of Federal Regulations* (CFR) deal with the National Flood Insurance Program (NFIP). These parts of the CFR 44 describe the program, flood plain management criteria, identification, and mapping of special hazard areas and procedures of map corrections.

The *International Building Code* (2009) uses ASCE 7-05 for wind loads and ASCE 24-05 for flood loads. The *Florida Building Code* (2007) uses ASCE 7-05 for the wind loads. However, the freeboard of ASCE 24-05 was not adopted in the *Florida Building Code* (2007). The *Florida Building Code* (2011) has adopted the ASCE 7-10 for wind loads and ASCE 24-05 for the flood loads.

In general, the "Building Envelope" is the physical separator between the interior and the exterior environments of a building. Another emerging term is "Building Enclosure". It serves as the outer shell to help maintain the indoor environment (together with the mechanical conditioning systems) and facilitate its climate control. Building envelope design is a specialized area of architectural and engineering practice that draws from all areas of building science and indoor climate control. The building envelope provides air barrier system in buildings, blast safety, seismic safety, wind safety, CBR safety, flood resistance, and provides indoor air quality and mold prevention, sustainability and HVAC integration. In terms of wind engineering, building envelope provides protection against strong wind actions to the interior environment and the occupants. The building envelope generally consists of the cladding, roofing, exterior walls, glazing, door assemblies, window assemblies, skylight assemblies, and other components enclosing the building.

Various states and International Code Council have their product evaluation agencies that review and approve products for the building envelopes. Like in Florida, Miami-Dade County Product

Control Division and the Florida Department of Business and Professional Regulation are the two agencies, which approve products to be used in the envelope of the structure to resist wind. International Code Council Evaluation Services (ICC-ES) is the agency that gives approvals of envelope products in accordance with the *International Building Code*.

The state of Florida has its own code based upon the *International Building Code* with local amendments. There are two portions of the code the aforementioned HVHZ and the rest of Florida. The HVHZ consists of Miami-Dade and Broward counties. *Florida Building Code* has requirements for product approvals that include approved testing laboratories, testing standards, evaluation criteria, and quality assurance verification for the building envelope elements. There are several Roofing Application Standards (RAS) and several Testing Application Standards (TAS) supplementing the *Florida Building Code* for envelope product approvals.

4.3 BASIC WIND ENGINEERING CONCEPTS

There are static, dynamic, and aerodynamic effects of wind on structures. In general, static effects of wind are sufficient in the design of low-rise buildings. In tall buildings, the dynamic and aerodynamic effects along with the static effects are required to be analyzed. Flexible slender structures and structural elements like tall buildings are subjected to wind induced along and across the direction of wind. The along-wind effects are mainly due to buffeting effects caused by turbulence and across-wind effects are mainly due to alternate-side vortex shedding. The across-wind effect can sometimes become primary because it could exceed along-wind accelerations if the building is slender about both axes.

Galloping and flutter are two important wind-induced motions. Galloping is transverse oscillations of structures due to the development of aerodynamic forces, which are in phase with the motion. It is demonstrated by the progressively increasing amplitude of transverse vibration with the increase of wind speed. The structural elements that are not circular are more prone to galloping. Flutter is unstable oscillatory motion of a structure due to coupling between aerodynamic force and elastic deformation of the structure. Combined bending and torsion are among the most common form of oscillatory motion.

Gust, vortex shedding, and buffeting are three major dynamic components of wind which cause the oscillation of structures.

During a hurricane, the wind velocity is not constant. There is a steady component of wind and there are effects of gusts which last for a few seconds at a time. The wind velocity in the American standards is based upon a 3-second gust (explained later). Gust gives a more realistic assessment of wind load. The intensity of gusts is related to the duration of gusts that affect the structures. Larger structures are affected by larger duration gusts and are subjected to smaller pressure compared to smaller structures. The gust-effect factor accounts for additional dynamic amplification of loading in the along-wind direction due to wind turbulence and structure interaction. It does not include allowances for cross wind loading effects, vortex shedding, and instability due to galloping or flutter, or dynamic torsional effects. Where cross wind loading effects, vortex shedding, galloping, flutter, and dynamic torsion are anticipated, wind tunnels are used to determine wind pressures on buildings, which take into consideration random wind gusts acting for short durations over entire or part of structure, fluctuating pressures induced in the wake of a structure, including vortex shedding forces and fluctuating forces induced by the motion of a structure.

When wind acts on a building, forces and moments in three mutually perpendicular directions are generated (three translations and three rotations). Since the weight of a building is high as compared to wind pressure in the upward direction, only the along-wind response and across transverse wind responses are considered. Only on the roof elements, the uplift due to wind is considered. The across-wind response causing motion in a plane perpendicular to the direction of wind typically dominates over the along-wind response for tall buildings. In a building subjected to a smooth wind flow, the originally parallel upwind streamlines are displaced on either side of the building due to boundary layer separation. This results in spiral vortices being shed periodically from the sides into the downstream flow of wind creating a low pressure zone due to shedding of eddies called the "wake." When the vortices shed, cross-wind components are generated in the transverse direction. At low wind speeds, since the shedding occurs at the same instant on either side of the building, there is no tendency for the building to vibrate in the transverse direction. It is therefore subject to along-wind oscillations parallel to the wind direction. At higher speeds, the vortices are shed alternately, first from one and then from the other side. When this occurs, there is a force in the along-wind direction as before, but in addition, there is a force in the transverse direction. This type of shedding, which gives rise to structural vibrations in the flow direction as well as in the transverse direction, is called vortex shedding. The frequency of shedding depends mainly on shape and size of the structure, velocity of flow and to a lesser degree on surface roughness, and turbulence of flow. Changing the cross-sectional shape of the building over its height can ensure that vortices are broken up and cannot be shed coherently over the entire height of the building, thus reducing across-wind loading. The Willis Tower in Chicago and the Burj Khalifa in Dubai use this technique to great effect.

Large buildings affect the wind loading of the nearby low buildings. There are significant adverse effects for particular building proximity configurations. These effects are called buffeting. A downwind structure could oscillate due to vortex shedding of adjacent large structure.

4.4 ORGANIZATION OF ASCE 7-10 FOR WIND LOAD CALCULATIONS

There is a significant difference in the organization of ASCE 7. In ASCE 7-05, the entire wind load information is in Chapter 6 and in ASCE 7-10 the wind load information is in Chapters 26 through 31. Table 4.2 familiarizes a reader with the chapters of ASCE 7-10 related to wind loads.

4.5 GENERAL REQUIREMENTS OF WIND LOAD CALCULATIONS

The building codes in the United States have adopted ASCE 7-10, *Minimum Design Loads for Building and Other Structures* to design the Main Wind Force Resisting System (MWFRS) and Components and Cladding (C & C) of buildings and other structures to resist wind loads.

4.5.1 MWRFS and C & C

The MWFRS in accordance with Chapter 26 of the ASCE 7-10 is defined as an assemblage of structural systems assigned to provide support and stability for the overall structure. The system generally receives wind loading from more than one surface. MWFRS consists of entire assembly that is used to transfer the wind loads to ground. The elements of the building envelope that do not qualify as part of the MWFRS are the C & C. They transfer the load to the MWFRS. Claddings receive wind loads directly and Components receive wind loads either directly or from cladding. Components can also be part of the MWFRS when they act as roof diaphragm or shear walls (See Table 4.3 for examples of components and cladding).

4.5.2 General Requirements

In order to calculate the wind loads for any case, Chapter 26 of the ASCE 7-10 is used to determine the basic parameters for both MWFRS and C & C. The important change made to the 2010 version of the ASCE 7 is the elimination of the importance factor. For all occupancy categories, the importance factor is 1.0 and it is not used in the calculations.

A risk category is assigned to a building occupancy in Table 1.5-1 of the ASCE 7-10 to determine the basic wind speed. The basic wind speed can be determined from the ASCE maps in

TABLE 4.2
Organization of ASCE 7-10 (in Relation to Wind Pressures Determination)

Chapter	Title and Intent	Content with the Section, Figure and Table Numbers of ASCE 7-10
1	General MWFRS/C & C	Definitions (Section 1.2) Risk Categories (Table 1.5-1)
2	Combination of Loads MWFRS/C & C	Strength Design (Section 2.3) Allowable Stress Design (Section 2.4)
26	Wind Loads: General Requirements MWFRS/C & C	Definitions (Section 26.2), Wind Hazard Maps (Figure 26.5-1A, 1B and 1C), Directionality Factor (Section 26.6), Exposure Category (Section 26.7), Topographic Factor (Section 26.8), Gust-Effect Factor (Section 26.9), Enclosure Classifications (Section 26.10), Internal Pressure Coefficient (Section 26.11)
27	Wind Loads on Building—Directional Procedure MWFRS	Part (1): Enclosed, Partially Enclosed and Open Low-Rise Buildings of all heights Part (2): Enclosed Simple Diaphragm Buildings
28	Wind Loads on Buildings—Envelope Procedure MWFRS	Part (1): Enclosed, Partially Enclosed and Open Low-Rise Buildings Part (2): Enclosed Simple Diaphragm Low-Rise Buildings
29	Wind Loads on Other Structures and Building Appurtenances MWFRS	Solid Free Standing Walls and Attached Sign (Section 29.4) Rooftop Structures and Equipment (Section 29.5) Parapets (Section 29.6) Roof Overhangs (Section 29.7)
30	Wind Loads C & C	Part (1): Low-Rise Buildings Part (2): Low-Rise Buildings (Simplified) Part (3): Buildings with height > 60′ Part (4): Buildings with height ≤ 160′ Part (5): Open Buildings Part (6): Building Appurtenances and Rooftop Structures and Equipment
31	Wind Tunnel Procedures MWFRS/C & C	Wind Tunnel Procedure

TABLE 4.3
Examples of Components and Claddings

Element	Example
Components	Fasteners, Purlins, Girts, Studs, Roof decking, Roof trusses
Claddings	Wall coverings, Curtain walls, Roof coverings, Exterior doors and windows

Figures 26.-5-1A (Risk Category II), 26.5-1B (Risk Category III and IV) and 26.5-1C (Risk Category I). These are maps with isotachs (lines of equal pressure) representing a 3 s gust speed at 33 ft above the ground. The maps are for Occupancy Category I, II and III buildings and are standardized for 300, 700, and 1700-year recurrence intervals respectively for exposure C topography (flat, open, country and grasslands with open terrain and scattered obstructions generally less than

TABLE 4.4
General Wind Load Parameters

Factor	Notation	ASCE 7-10 Reference
Wind Directionality	K_d	Section 26.6
Exposure Category		Section 26.7
Topographic Factor	K_{zt}	Section 26.8
Gust-Effect Factor	G or GC_p, GC_p and GC_{pf}	Section 26.9
Exposure Classification		Section 26.10
Internal Pressure Coefficient	GC_{pi}	Section 26.11

30 ft in height). The minimum wind speed provided in the standard is 100 miles per hour (mph) for a mean recurrence interval (MRI) of 300 years. Increasing the minimum wind speed for special topographies such as mountain terrain, gorges, and ocean fronts is recommended. The abandonment of the fastest-mile speed in favor of a 3-second gust speed first took place in the ASCE 7-1995 edition primarily due to the following reasons:

1. Modern weather stations no longer measure wind speeds using the fastest-mile method.
2. The 3-second gust speed is closer to the sensational wind speeds often quoted by news media.
3. It matches closely the wind speeds experienced by small buildings and components of all buildings.

The wind directionality factor, exposure categories, topographic factor, gust-effect factor, exposure classification, and internal pressure coefficient in the ASCE 7-10 have no significant changes from ASCE 7-05. The references to the ASCE 7-10 to determine these factors are provided in Table 4.4. These factors are discussed in the subsequent Sections 4.5.3 through 4.5.8.

4.5.3 Wind Directionality Factor (K_D)

Wind directionality factor (K_d) accounts for the reduced probability of maximum winds coming from any given direction and for the reduced probability of maximum pressure coefficient occurring for any given wind direction. The factor K_d accounts for the directionality of wind. Directionality refers to the fact that wind rarely, if ever, strikes along the most critical direction of a building. Wind direction changes from one instance to the next. Wind can be only instantaneous along the most critical direction because at the very next instant, it will not be from the same direction. K_d can only be used with load combinations of ASCE Sections 2.3 and 2.4.

4.5.4 Exposure Category

There are three exposure categories B, C and D defined in accordance with three categories of surface roughness (B, C and D). There is an additional exposure category A which is used in the wind tunnel testing. The exposure category of a building or other structure should be very carefully selected as the velocity pressure coefficient (K_h or K_z) depends upon the exposure category. The velocity pressure is directly proportional to velocity pressure coefficient. There is a significant numerical difference between the velocity coefficients for different exposures. The exposure categories are explained in Table 4.5 and the surface roughness categories are explained in Table 4.6 in accordance with the ASCE 7-10.

TABLE 4.5
Exposure Categories

Exposure Category	Description
B	Mean Roof Height ≤ 30′, if surface roughness B prevails > 1500′ length in the upwind direction
	Mean Roof Height > 30′, if surface roughness B prevails > 2600′ length in the upwind direction
D	If surface roughness D prevails > 5000′ length or 20 times the height of building in the upwind direction
	If surface roughness D prevails > 600′ or 20 times the height of building and surface roughness is B or C is immediately upwind of the site
C	Where exposure categories B and D do not apply

TABLE 4.6
Surface Roughness

Surface Roughness	Description
B	Urban and suburban areas, wooded areas, or other terrain with numerous closely spaced obstruction having the size of single-family dwellings or larger
C	Open terrain with scattered obstructions having heights generally < 30′. Includes flat open country and grasslands
D	Flat unobstructed areas and water surfaces. Includes smooth mud flats, salt flats and unbroken ice

TABLE 4.7
Range of z_0

Exposure Category	Range of Z_0 (in Feet)
A	>2.3
B	0.5–2.3
C	0.033–0.5
D	<0.033

The ground surface roughness is measured in terms of a roughness length parameter called Z_0 and can be estimated by the following equation (See Table 4.7 and Example 4.12.1).

$$Z_0 = 0.5H_{ob}\frac{S_{ob}}{A_{ob}} \qquad \text{ASCE Equation C26.7-1}$$

where

H_{ob} is the average height of roughness in the upwind stream
S_{ob} is the average vertical frontal area per obstruction presented to the wind
A_{ob} is the average area of ground occupied by each obstruction, including the open area surrounding

4.5.5 Topographic Factor (K_{ZT})

Topographic factor (K_{zt}) is used to include the wind speed-up effect in the calculations of the design wind loads. Wind speed-up effects at isolated hills, ridges, and escarpments with abrupt changes in topography. Escarpment is defined as cliff or steep slope generally separating two levels or gently sloping areas. Topographic effects are considered if all of the following ASCE 7-10 conditions listed in the ASCE Section 26.8.1 are met. It is not the intent of ASCE Section 26.8 to address the general case of wind flow over hilly or complex terrain for which engineering judgment, expert advice or the wind tunnel procedure may be required.

1. The hill, ridge, or escarpment is isolated and unobstructed upwind by other similar topographic features of comparable height for 100 times the height of the topographic feature (100H) or 2 mi, whichever is less. This distance shall be measured horizontally from the point at which the height H of the hill, ridge, or escarpment is determined.
2. The hill, ridge, or escarpment protrudes above the height of upwind terrain features within a 2-mile radius in any quadrant by a factor of two or more.
3. The structure is located as shown in ASCE Figure 26.8-1 in the upper one-half of a hill or ridge or near the crest of an escarpment.

Factors K_1, K_2 and K_3 are calculated from Figure 26.8-1 and the topographic factor K_{zt} is calculated using the following equation. If the site conditions do not meet all of the above conditions then K_{zt} is 1.0

$$K_{zt} = (1 + K_1K_2K_3)^2 \qquad \text{ASCE Equation 26.8-1}$$

4.5.6 Gust Effects

Gust is a sudden, brief increase in speed of the wind. According to U.S. weather observing practice, gusts are reported when the peak wind speed reaches at least 17.5 mph, and the variation in wind speed between the peaks and lulls is at least 10 mph. The duration of a gust is usually less than 20 s. The ASCE 7-10 considers 3-second gusts. "Gust-effect factor" is an increasing function of speed. In addition, most structures will experience yielding as a "pushover" loading is increased, resulting in a reduced natural frequency and therefore even higher load factor.

To determine whether a building is rigid or flexible, the fundamental natural frequency, n_a, shall be established using the structural properties and deformational characteristics of the resisting elements in a properly substantiated analysis. Low-rise buildings are permitted to be considered rigid.

Rigid structures are structures with the fundamental frequency $\geq$ 1 Hz. If the fundamental frequency <1 Hz then the structure is flexible. ASCE Section 26.9.3 can be used to determine the approximate natural frequency (n_a) if:

Building height $\leq$ 300′ (and)
Building height < 4 × effective length (L_{eff})

The effective length is determined using the ASCE Equation 26.9-1; the approximate natural frequencies for structural steel moment-resisting frame buildings, concrete moment-resisting frame buildings, structural steel and concrete buildings with other lateral-force-resisting systems and concrete, or masonry shear wall buildings are determined using ASCE Equations 26.9-2 through 26.9-5, respectively (See Examples 4.12.2 through 4.12.5).

The gust-effect factor of rigid buildings can be assumed as 0.85 or can be calculated using ASCE Equations 26.9-6 through 26.9-9. The gust-effect factor of flexible buildings can be calculated using ASCE Equations 26.9-10 through 26.9-16 (See Example 4.12.6).

As stated in the previous paragraph, ASCE 7 contains a single gust-effect factor of 0.85 for rigid buildings. As an option, the designer can incorporate specific features of the wind environment and building size to more accurately calculate a gust-effect factor. One such procedure is specified in the standard. A procedure is also included for calculating the gust-effect factor for flexible structures. The rigid structure gust factor is 0%–10% lower than the simple, but conservative, value of 0.85 permitted in the standard without calculation. The procedures for both rigid and flexible structures (1) provide a superior model for flexible structures that displays the peak factors g_Q and g_R and (2) cause the flexible structure value to match the rigid structure as resonance is removed. A designer is free to use any other rational procedure in the approved literature.

The gust-effect factor accounts for the loading effects in the along-wind direction due to wind turbulence-structure interaction and also along-wind loading effects due to dynamic amplification for flexible buildings. It does not include allowances for cross-wind loading effects, vortex shedding, and instability due to galloping or flutter, or dynamic torsional effects. For structures susceptible to loading effects that are not accounted for in the gust-effect factor, information should be obtained from wind tunnel tests.

4.5.7 Enclosure Classifications

A structure can be classified as "open," "partially enclosed," and "enclosed" in accordance with ASCE Section 26.2. A building having each wall at least 80% open is an "open structure." A "partially enclosed structure" complies with both of the following conditions:

1. The total area of openings in a wall that receives positive external pressure exceeds the sum of the areas of openings in the balance of the building envelope (walls and roof) by more than 10%.
2. The total area of openings in a wall that receives positive external pressure exceeds 4 ft^2 or 1% of the area of that wall, whichever is smaller, and the percentage of openings in the balance of the building envelope does not exceed 20%.

If a structure is not an "open" or "partially open," then it is an "enclosed" structure. This concept does not work properly for some structures such as a 10′ × 10′ building with a roof and two walls without openings. Since there are no openings in two walls, it cannot be an "open" structure. If the two walls without openings are receiving the positive external pressure, the total area of the openings on these walls is zero and not exceeding 4 ft^2 or 1% of the area of the wall. Hence it cannot be a "partially open" structure. So, is the building "enclosed?" In the commentary of the ASCE Section 26.2, this type of building is classified as "enclosed."

Hurricane-prone regions are areas vulnerable to hurricanes defined in ASCE Section 26.2 as the U.S. Atlantic Ocean and Gulf of Mexico coasts, where the basic wind speed for risk category II buildings is greater than 115 mph, and Hawaii, Puerto Rico, Guam, Virgin Islands, and American Samoa.

Wind-borne debris regions as defined in ASCE Section 26.10.3.1 are located within one mile of the coastal mean high water line, where the basic wind speed is equal to or greater than 130 mph, or in areas where the basic wind speed is equal to or greater than 140 mph. The expanded risk category III wind-borne debris region in ASCE Figure 26.5-1B is applied only to health-care facilities and not to any other type of building of risk category III. However, the wind-borne debris region in the ASCE Figure 26.5-1B is applied to entire risk category IV buildings.

Glazed opening shall be protected with impact-resistant glass or shutters, tested in accordance with ASTM E1886 to comply with ASTM E1996. Glazing shall be protected up to the height of 60′ above the ground and up to the height of 30′ above an aggregate-surface roof located within 1500′ of the building.

4.5.8 Internal Pressure Coefficient (GC_{pi})

The magnitude and sense of internal pressure is dependent upon the magnitude and location of openings around the building envelope with respect to a given wind direction. In accordance with Table 26.11-1, the internal pressure coefficients (GC_{pi}) are as follows:

Open buildings	0.0
Partially enclosed buildings	±0.55
Enclosed buildings	±0.18

(The value of GC_{pi} shall be used with two cases of positive and negative, with q_z or q_h to determine the design wind pressures)

4.5.9 Structural Damping

Structural damping is a measure of energy dissipation in a vibrating structure that results in bringing the structure to a quiescent state. The damping is defined as the ratio of the energy dissipated in one oscillation cycle to the maximum amount of energy in the structure in that cycle. There are as many structural damping mechanisms as there are modes of converting mechanical energy into heat. The most important mechanisms are material damping and interfacial damping.

In engineering practice, the damping mechanism is often approximated as viscous damping because it leads to a linear equation of motion. This damping measure, in terms of the damping ratio, is usually assigned based on the construction material. The calculation of dynamic load effects requires damping ratio as an input. In wind applications, damping ratios of 1%–2% are typically used in the United States for steel and concrete buildings at serviceability levels, respectively. Damping values for steel support structures for signs, chimneys, and towers may be much lower than buildings and may fall in the range of 0.15%–0.5%. Damping values of special structures like steel stacks can be as low as 0.2%–0.6% and 0.3%–1.0% for unlined and lined steel chimneys, respectively. These values may provide some guidance for design. Damping levels used in wind load applications are smaller than the 5% damping ratios common in seismic applications because buildings subjected to wind loads respond essentially elastically whereas buildings subjected to design level earthquakes respond inelastically at higher damping levels. Because the level of structural response in the serviceability and survivability states is different, the damping values associated with these states may differ.

In addition to structural damping, aerodynamic damping may be experienced by a structure oscillating in air. In general, the aerodynamic damping contribution is quite small compared to the structural damping, and it is positive in low-to-moderate wind speeds. Depending on the structural shape, at some wind velocities, the aerodynamic damping may become negative, which can lead to unstable oscillations. In these cases, reference should be made to a wind tunnel study.

4.6 WIND VELOCITY PRESSURE

Wind velocity pressure can be calculated at a height "z" above the ground (q_z) or at the mean roof height (q_h). The basic wind speed (V) in mph is converted to a velocity pressure in psf using the basic wind parameters (velocity pressure coefficient, topographic factor, directionality factor, and basic wind speed).

$$Q_z = 0.00256 K_z K_{zt} K_d V^2$$ ASCE Equations 27.3-1, 28.3-1, 29.3-1, 30.3-1

(Refer to Section 4.5 for the notation of terms)

In the foot-pound system, the constant 0.00256 reflects the mass density of air for the standard atmosphere, that is, temperature of 59°F and sea level pressure of 29.92 in. of mercury, and dimensions associated with wind speed in mile/h. The constant is obtained as follows:

$$\text{Constant} = 1/2[(0.0765\ \text{lb/ft}^3)/(32.2\ \text{ft/s}^2)] \times [(\text{mile/h})(5280\ \text{ft/mi}) \times (1\ \text{h}/3600\ \text{s})]^2 = 0.00256$$

0.0765	lb/ft^3	Average ambient air density
32.2	ft/s^2	Acceleration due to gravity
5280		Used to convert miles to feet
3600		Used to convert hours to second

The constant 0.00256 takes into account the equation, pressure (p) = ½ × ρ × V^2 and the conversion of mph to ft/s^2.

Where ρ is mass density and V is velocity in mph. The basic wind velocity is multiplied by the constant 0.00256, velocity pressure coefficient, topographic factor, and directionality factor to obtain the wind velocity pressure.

4.7 DIRECTIONAL PROCEDURE (CHAPTER 27, ASCE 7-10)

The directional procedure of the ASCE 7-10 is used for the determination of wind pressures of the MWFRS on enclosed, partially enclosed, and open buildings of all heights. It is the analytical method of the ASCE 7-05. A simplified method is added as Part (2) for special buildings less than 160′ high. There are two parts in Chapter 27 of ASCE 7-10. Part (1) deals with buildings of all heights where it is necessary to separate applied wind loads onto the windward, leeward, and side walls and part (2) deals with special buildings designated as enclosed simple diaphragm with $h \leq 160'$. Both parts are explained in the following. The direction procedure prescribes minimum design load of 16 psf for the walls and 8 psf of the roof for enclosed and partially enclosed MWFRS buildings and 16 psf for open buildings. The tables and figures used in the part (1) of Chapter 27 are described in Table 4.8. The conditions and limitations of the directional procedure are listed as follows:

- For regular shaped buildings
- For buildings not subject to across-wind loading, vortex shedding, or instability due to galloping or flutter

TABLE 4.8
Explanation of Tables and Figures of Part (1) of Chapter 27

Table/Figure	Description
T27.2-1	Steps to determine the wind loads
T27.3-1	Velocity pressure exposure coefficients (K_z or K_{h+}) for B, C and D till height of 500′
F27.4-1	External pressure coefficient (C_p) for walls and gable/hip/monoslope/mansard roof (Enclosed and partially enclosed structures)
F27.4-2	External pressure coefficient (C_p) for doomed roof (Enclosed and partially enclosed structures)
F27.4-3	External pressure coefficient (C_p) for arched roof (Enclosed and partially enclosed structures)
F27.4-4	External pressure coefficient (C_p) for monoslope free roofs (Open structures)
F27.4-5	External pressure coefficient (C_p) for pitched free roofs (Open structures)
F27.4-6	External pressure coefficient (C_p) for troughed free roofs (Open structures)
F27.4-7	External pressure coefficient (C_p) for free roofs (Open structures)
F27.4-8	Design wind load cases 1, 2, 3 and 4 with equations for eccentricities and moments due to torsion

- Building not located at a site for which channeling effects or buffeting in the wake of upwind obstructions warrant special consideration
- Load magnification effect caused by gusts in resonance with along-wind vibrations of flexible buildings considered
- No reductions in velocity pressure due to apparent shielding afforded by buildings and other structures or terrain features

Part (1): Enclosed, Partially Enclosed and Open Buildings of All Heights

1. Determine V, K_d, exposure category, K_{zt}, G, enclosure classification and GC_{pi}.
2. Determine velocity pressure coefficient (K_z or K_h) from ASCE Table 27.3-1.
3. Determine C_p from ASCE Figures 27.4-1 through 27.4-3 for the cases of wall and flat, gable, hip, monoslope, or mansard roof; domed roof, arched roof, monoslope roof/open building, pitched roof/open building, troughed roof/open building, and along-ridge/valley wind load for monoslope, pitched or troughed roof of open building
4. Calculate velocity pressure (q_z)

$$q_z = 0.00256K_zK_{zt}K_dV^2 \qquad \text{ASCE Equation: 27.3-1}$$

5. Calculate wind force as shown in the following for each case
   a. For enclosed and partially enclosed rigid buildings

$$p = qGC_p - q_i(GC_{pi}) \qquad \text{ASCE Equation: 27.4-1}$$

   b. For enclosed and partially enclosed flexible buildings

$$p = qG_fC_p - q_i(GC_{pi}) \qquad \text{ASCE Equation: 27.4-2}$$

   c. Open buildings with monoslope, pitched, troughed free roofs

$$p = q_hGC_N \qquad \text{ASCE Equation: 27.4-3}$$

   (q_h evaluated at mean roof height and C_N is net pressure coefficient determined from ASCE Figures 27.4-4 through 27.4-7)
   d. Roof overhangs
   The positive external pressure on the bottom surface of the windward roof overhangs shall be determined using $C_p = 0.8$ and combined with the top surface pressures determined using ASCE Figure 27.4-1
   e. Parapets

$$p_p = q_p(GC_{pn}) \qquad \text{ASCE Equation: 27.4-4}$$

   (q_p evaluated at the top of the parapet and GC_{pn} is 1.5 for windward parapet and (–) 1.0 for leeward parapet)

   Where
   $q = q_z$ for windward walls at height z
   $q = q_h$ for leeward walls, side walls, and roof at mean roof height
   $q_i = q_h$ for windward walls, side walls, leeward walls, and roofs of enclosed building and negative internal pressure evaluation in partially enclosed buildings
   $q_i = q_z$ for positive internal pressure evaluation in partially enclosed building where height z is the height of the highest opening in the building that could affect the positive internal pressure

TABLE 4.9
Design Wind Load Cases

Case	Action	Value
1	Separately on each plane	Direct Forces P_{WX}, P_{WY}
2	Separately on each plane	Direct Forces $0.75P_{WX}$, $0.75P_{WY}$
		Torsion
		$M_{TX} = 0.75(P_{WX} + P_{LX})B_Xe_X$
		$e_X = \pm 0.15B_X$
		$M_{TY} = 0.75(P_{WY} + P_{LY})B_Ye_Y$
		$e_Y = \pm 0.15B_Y$
3	Simultaneously on both planes	Direct Forces $0.75P_{WX} + 0.75P_{WY}$
4	Simultaneously on both planes	Direct Forces $0.563P_{WX} + 0.563P_{WY}$
		Torsion
		$M_{TX} = 0.75(P_{WX} + P_{LX})B_Xe_X + 0.75(P_{WY} + P_{LY})B_Ye_Y$
		$e_X = \pm 0.15B_X$, $e_Y = \pm 0.15B_Y$

P_{WX}, P_{WY}, windward face design pressure acting in the x, y principal axis, respectively; P_{LX}, P_{LY}, leeward face design pressure acting in the x, y principal axis, respectively; $e(e_X, e_Y)$, eccentricity for the x, y principal axis of the structure, respectively; M_T, torsional moment per unit height acting about a vertical axis of the building.

There are four design load cases in the directional procedure for buildings of all heights where it is necessary to separate applied wind loads onto the windward, leeward, and side walls. Case (1) is the full design wind pressure acting on the project area perpendicular to each principal axis of the structure along the principal plane. Case (2) is 75% of the design wind pressure acting on the project area perpendicular to each principal axis of the structure in conjunction with torsional moment along the principal plane. Case (3) is 75% of the design wind pressure acting on the project area perpendicular to each principal axis of the structure along the principal plane, and Case (4) is 56.3% of the design wind pressure acting on the project area perpendicular to each principal axis of the structure in conjunction with torsional moment along the principal plane. Forces obtained in each of the cases are shown in Table 4.9.

For flexible structures the eccentricities shall be modified using the following equation.

$$e = \frac{e_Q + 1.7I_z\sqrt{\left((g_Q Q e_Q)^2 + (g_R R e_R)^2\right)}}{1 + 1.7I_z\sqrt{\left((g_Q Q)^2 + (g_R R)^2\right)}} \qquad \text{ASCE Equation: 27.4-5}$$

where

e_Q is the eccentricity e for rigid structure as defined in the before table

e_R is the distance between the elastic shear center and center of mass of each floor

I_z, g_Q, Q, g_R and R determined from ASCE Section 26.9 dealing with the gust-effect factor.

The ASCE Appendix D lists buildings exempted from torsional wind cases. They include one and two story buildings with flexible diaphragm or light-frame construction, seismic controlled buildings, torsionally regular buildings, flexible diaphragm buildings that are designed for increased wind loadings and class 1 and 2 simple diaphragm buildings. Refer to ASCE Appendix D for criteria of exemption for torsional wind.

TABLE 4.10
Explanation of Tables and Figures of Part (2) of Chapter 27

Table/Figure	Description
T27.5-1	Steps to determine wind loads
F27.5-1	Geometry requirements of class 1 and 2 enclosed simple diaphragm buildings
F27.6-1	Application of wind pressures for enclosed simple diaphragm buildings
F27.6-2	Application of parapet wind loads for enclosed simple diaphragm buildings
F27.6-3	Application of roof overhang wind loads for enclosed simple diaphragm buildings
T27.6-1	Application of wall pressures for enclosed simple diaphragm buildings MWFRS Wind loads—Walls—Exposure B, C and D
T27.6-2	Application of roof pressures for enclosed simple diaphragm buildings MWFRS Wind loads—Walls—Exposure C Exposure adjustment factors for exposures B and D Description of flat/gable/hip/monoslope/mansard roofs

Part (2): Enclosed Simple Diaphragm Buildings with Heights ≤ 160′

In accordance with ASCE Chapter 26, Simple diaphragm building is a building in which both windward and leeward wind loads are transmitted by roof and vertically spanning wall assemblies, through continuous floor and roof diaphragms, to the MWFRS. The tables and figures used in the part (2) of ASCE Chapter 27 are described in Table 4.10.

1. Determine V, K_d, exposure category, K_{zt}, G, enclosure classification and GC_{pi}
2. From ASCE Table 27.6-1, determine net pressures on walls at top and base (p_h, $p0$, respectively) of building.
3. From ASCE Table 27.6-2, determine net roof pressures (p_z)
4. Apply topographic factors to the wall and roof pressures (if applicable)
5. Apply loads to walls and roof simultaneously.
6. Where two load cases are shown in the table of roof pressures, effects of each load case shall be dealt separately. The MWFRS shall be designed for the four wind load cases of Figure 27.4-8 with the exceptions of Appendix D.

Building shall meet the class 1 or class 2 requirements of the code.

CLASS (1)—Simple diaphragm building, $h \leq 60'$, range of L/B is (0.2–5.0), $K_{zt} = 1.0$

For $L/B < 0.5$, use tabulated value of $L/B = 0.5$

For $L/B > 2.0$, use tabulated value of $L/B = 2.0$

CLASS (2)—Simple diaphragm building,

$60' < h \leq 160'$, range of L/B is (0.5–2.0),

Fundamental natural frequency $(f) \geq 75/h$,

$K_{zt} = 1.0$

Parapets: Wind pressure = 2.25 × wind pressure for the wall with $L/B = 1.0$ applied simultaneously with wall and roof pressures. The height to determine the parapet wind pressures is the height of the building at the top of the parapet.

Roof Overhangs: Positive wind pressure on the underside of the roof overhang is 75% of the roof edge pressure for the applicable zone applied on the windward roof overhang.

4.8 ENVELOPE PROCEDURE (CHAPTER 28, ASCE 7-10)

The envelope procedure is used to determine the MWFRS wind loads on low-rise buildings. There are two parts of this procedure. Part I is the former "low-rise buildings" provision of method 2 of ASCE 7-05. Part 2 is derived from the MWFRS provisions of method 1 for simple diaphragm buildings up to 60′ in height. There are two parts in Chapter 28 of ASCE 7-10. Part (1) deals with low-rise buildings where it is necessary to separate applied wind loads onto the windward, leeward, and side walls of the building to properly assess the internal forces in the MWFRS members and part (2) deals with special class of low-rise buildings designated as enclosed simple diaphragm building. The tables and figures used in the part (1) of Chapter 28 are described in Table 4.11. The conditions and limitations of the envelope procedure are as follows:

- For regular shaped buildings
- For buildings not subject to across-wind loading, vortex shedding, or instability due to galloping or flutter
- Building not located at a site for which channeling effects or buffeting in the wake of upwind obstructions warrant special consideration
- Load magnification effect caused by gusts in resonance with along-wind vibrations of flexible buildings considered
- No reduction in velocity pressure due to apparent shielding afforded by buildings and other structures or terrain features.

Part (1): Enclosed and Partially Enclosed Low-Rise Buildings

1. Determine V, K_d, exposure category, K_{zt}, enclosure classification and GC_{pi}.
2. Determine velocity pressure coefficient (K_z or K_h) from ASCE Table 28.3-1.
3. Calculate velocity pressure (q_z)

$$q_z = 0.00256 K_z K_{zt} K_d V^2 \qquad \text{ASCE Equation: 28.3-1}$$

4. Calculate wind pressure (p) for each case as shown down.
   a. For low-rise buildings

$$p = q_h[(GC_{pf+}) - (GC_{pi+})] \qquad \text{ASCE Equation: 28.4-1}$$

   q_h is the velocity pressure evaluated at mean roof height
   GC_{pf} is the external pressure coefficient (ASCE Figure: 28.4-1)

   b. Parapets

$$P_P = q_p(GC_{pn}) \qquad \text{ASCE Equation: 28.4-2}$$

   q_p is velocity pressure evaluated at top of the parapet and GC_{pn} is 1.5 for windward parapet and (–) 1.0 for leeward parapet

TABLE 4.11
Explanation of Tables and Figures of Part (1) of Chapter 28

Table/Figure	Description
T28.2-1	Steps to determine the wind loads
T28.3-1	Velocity pressure exposure coefficients (K_z or K_{h+}) for B, C and D till height of 60′
F28.4-1	External pressure coefficient (GC_{pf}) for low-rise walls and roof (Enclosed and partially enclosed structures)

TABLE 4.12
Explanation of Tables and Figures of Part (2) of Chapter 28

Table/Figure	Description
T28.5-1	Steps to determine the wind loads
F28.6-1	Design wind pressures for walls and roof of enclosed buildings
	Tables for wind pressures
	Table for adjustment factor for height and exposure

c. Roof overhangs
The positive external pressure on the bottom surface of the windward roof overhangs shall be determined using $C_p = 0.7$ and combined with the top surface pressures determined using Figure 28.4-1

The minimum design load in the design of MWFRS for an enclosed or partially enclosed building shall not be less than 16 psf.

Part (2): Enclosed Simple Diaphragm Low-Rise Buildings

In accordance with ASCE Chapter 26, Simple diaphragm building is a building in which both windward and leeward wind loads are transmitted by roof and vertically spanning wall assemblies, through continuous floor and roof diaphragms, to the MWFRS. The tables and figures used in the part (2) of ASCE Chapter 28 are described in Table 4.12.

1. Determine V, exposure category, K_{zt} and enclosure classification
2. From Figure 28.6-1, determine wind pressure (p_{s30}) for $h = 30'$ and exposure B
3. From Figure 28.6-1, determine the adjustment factor (λ) for height and exposure.
4. Determine adjusted wind pressure

$$p_s = \lambda K_{zt} p_{s30} \qquad \text{ASCE Equation: 28.6-1}$$

Conditions:

- Simple diaphragm, low-rise, enclosed, regular-shaped, rigid and conforms to the wind-borne debris provisions of ASCE Section 26.10.3.
- For buildings not subject to across-wind loading, vortex shedding or instability due to galloping or flutter. Building not located at a site for which channeling effects or buffeting in the wake of upwind obstructions warrant special consideration.
- The building has an approximately symmetrical cross-section in each direction with either a flat roof or a gable or hip roof with $\theta \leq 45°$.
- The building is exempted from torsional load cases as indicated in Note 5 of Figure 28.4-1, or the torsional load cases defined in Note 5 do not control the design of any of the MWFRS of the building.

4.9 OTHER STRUCTURES AND BUILDING APPURTENANCES (CHAPTER 29, ASCE 7-10)

Building appurtenances generally consists of rooftop structures and rooftop equipment. Other structures generally consist of solid freestanding walls, freestanding sold signs, chimneys, tanks, open signs, lattice frameworks, and trusses towers. Equations for determining the wind forces for solid free-standing walls, solid signs, other structures, and lateral and vertical forces acting on rooftop

TABLE 4.13
Explanation of Tables and Figures of Chapter 29

Table/Figure	Description
T29.1-1	Steps to determine the wind loads
T29.3-1	Velocity pressure exposure coefficients (K_z or K_{h+}) for B, C and D till height of 500′
F29.4-1	Force coefficients (C_f) for solid freestanding walls and signs
F29.5-1	Force coefficients (C_f) for chimneys, tanks, rooftop equipment and similar structure
F29.5-2	Force coefficients (C_f) for open signs and lattice frameworks
F29.5-3	Force coefficients (C_f) for trussed towers

structures are provided in the ASCE Chapter 29. The tables and figures used in ASCE Chapter 29 are described in Table 4.13. The conditions and limitations of ASCE Chapter 29 are listed in the following:

- For regular shaped buildings
- For buildings not subject to across-wind loading, vortex shedding, or instability due to galloping or flutter
- Building not located at a site for which channeling effects or buffeting in the wake of upwind obstructions warrant special consideration
- Load magnification effect caused by gusts in resonance with along-wind vibrations of flexible buildings considered
- No reduction in velocity pressure due to apparent shielding afforded by buildings and other structures or terrain features.

Procedure:

1. Determine V, K_d, G, exposure category, K_{zt} and enclosure classification.
2. Determine velocity pressure coefficient (K_z or K_h) from Table 29.3-1.
3. Calculate velocity pressure (q_z)

$$q_z = 0.00256K_zK_{zt}K_dV^2 \qquad \text{ASCE Equation: 28.3-1}$$

4. Calculate wind pressure (p) as shown in the following for each case.
   a. Solid Freestanding Walls and Solid Signs

$$F = q_hGC_fA_s \qquad \text{ASCE Equation: 29.4-1}$$

   C_f is net force coefficient from ASCE Figure 29.4-1 and A_s is the gross area of the structure

   b. Other Structures

$$F = q_zGC_fA_s \qquad \text{ASCE Equation: 29.5-1}$$

   C_f is obtained from the ASCE Figures 29.5-1 (for chimneys, tanks, rooftop equipment, and similar structures); 29.5-2 (for open signs and lattice work) and 29.5-3 (for trussed towers)

   A_f is the projected area normal to wind except where C_f is specified for actual surface area

c. Rooftop Structures and Equipment for $H \leq 60'$ (Lateral Force)

$$F_h = q_h(GC_r)A_f \qquad \text{ASCE Equation: 29.5-2}$$

(GC_r) is 1.9 for rooftop structures and equipment with $A_f < (0.1\text{Bh})$ and (GC_r) reduced from 1.9 to 1.0 if A_f increased from (0.1Bh) to (Bh)
A_f is the vertical projected area normal to wind

d. Rooftop Structures and Equipment for $H \leq 60'$ (Vertical Uplift Force)

$$F_v = q_h(GC_r)A_f \qquad \text{ASCE Equation: 29.5-3}$$

(GC_r) is 1.5 for rooftop structures and equipment with $A_f < (0.1\text{Bh})$ and (GC_r) reduced from 1.5 to 1.0 if A_f increased from (0.1Bh) to (Bh)
A_f is the horizontal projected area normal to wind

Use the methods of directional and envelope procedures to calculate wind loads for parapets and roof overhangs. The minimum design wind force shall be not less than 16 psf multiplied by the area A_f.

4.10 COMPONENTS AND CLADDING (CHAPTER 30, ASCE 7-10)

Conditions and Limitations:

- For regular shaped buildings
- For buildings not subject to across-wind loading, vortex shedding, or instability due to galloping or flutter
- Building not located at a site for which channeling effects or buffeting in the wake of upwind obstructions warrant special consideration
- Load magnification effect caused by gusts in resonance with along-wind vibrations of flexible buildings considered
- No reduction in velocity pressure due to apparent shielding afforded by buildings and other structures or terrain features.
- Used for air permeable cladding unless lower loads are demonstrated by test or research
- Elements with tributary area greater than 700 sft can be designed as MWFRS

General Requirements:

1. Determine V, K_d, exposure category, K_{zt}, G, enclosure classification and GC_{pi}.
2. Determine velocity pressure coefficient (K_z or K_h) from Table 30.3-1.
3. Calculate velocity pressure (q_z)

$$q_z = 0.00256 K_z K_{zt} K_d V^2 \qquad \text{ASCE Equation: 30.3-1}$$

4. Determine design wind pressure (p) for each case as defined in Table 4.14

Part (1): Low-Rise Buildings

$$p = q_h[(GC_p) - (GC_{pi})] \qquad \text{ASCE Equation: 30.4-1}$$

(GC_p) is the external pressure coefficients given in ASCE Figure 30.4-1 for walls; ASCE Figures 30.4-2A to 30.4-2C for flat, gable roof, and hip roofs; ASCE Figure 30.4-3 for stepped roofs; ASCE Figure 30.4-4 for multi-span gable roofs; ASCE Figures 30.4-5A and 30.4-5B for monoslope roofs;

TABLE 4.14
Arrangement of Chapter 30

Part	Height	Applicability	Description
1	≤60′	Enclosed or partially enclosed	Flat, gable, multi-span gable, hip, monoslope, stepped or saw-tooth roof
2	≤60′	Enclosed	Flat, gable or hip roof
3	>60′	Enclosed or partially enclosed	Flat, pitched, gable, hip, mansard, arched or domed roof
4	≤60′	Enclosed	Flat, gable, hip, monoslope or mansard roof
5	All	Open	Pitched free, monoslope free or trough free roof
6	All		Building appurtenances such as roof overhangs and parapets and rooftop equipment

ASCE Figure 30.4-6 for saw-tooth roofs; ASCE Figure 30.4-7 for domed roofs and ASCE Figure 27.4-3, footnote 4 for arched roofs. q_h is evaluated at mean roof height

Explanation of Tables and Figures of Part (1) of Chapter 30:

Table/Figure	Description
T30.4-1	Steps to determine the wind loads
F30.4-1	External pressure coefficient (GC_p) for walls with $h \leq 60'$ (Enclosed and partially enclosed structures)
F30.4-2A	External pressure coefficient (GC_p) for gable roof ($\theta \leq 7°$) with $h \leq 60'$ (Enclosed and partially enclosed structures)
F30.4-2B	External pressure coefficient (GC_p) for gable/hipped roof ($7° < \theta \leq 27°$) with $h \leq 60'$ (Enclosed and partially enclosed structures)
F30.4-2C	External pressure coefficient (GC_p) for gable/hipped roof ($27° < \theta \leq 45°$) with $h \leq 60'$ (Enclosed and partially enclosed structures)
F30.4-3	External pressure coefficient (GC_p) for stepped roof with $h \leq 60'$ (Enclosed and partially enclosed structures)
F30.4-4	External pressure coefficient (GC_p) for multispan gable roof with $h \leq 60'$ (Enclosed and partially enclosed structures)
F30.4-5A	External pressure coefficient (GC_p) for mono-slope roof($3° < \theta \leq 10°$) with $h \leq 60'$ (Enclosed and partially enclosed structures)
F30.4-5B	External pressure coefficient (GC_p) for mono-slope roof ($10° < \theta \leq 30°$) with $h \leq 60'$ (Enclosed and partially enclosed structures)
F30.4-6	External pressure coefficient (GC_p) for sawtooth with $h \leq 60'$ (Enclosed and partially enclosed structures)
F30.4-7	External pressure coefficient (GC_p) for domed roof for all heights (Enclosed and partially enclosed structures)

Part (2): Low-Rise Building

$$p_{\text{net}} = \lambda K_{zt}\, p_{\text{net30}} \qquad \text{ASCE Equation: 30.5-1}$$

λ is the adjustment factor for building height and exposure from ASCE Figure 30.5-1 and p_{net30} is the net design wind pressure for Exposure B, at $h = 30$ ft, from ASCE Figure 30.5-1. K_{zt} is evaluated at 0.33 h.

Explanation of Tables and Figures of Part (2) of Chapter 30:

Table/Figure	Description
T30.5-1	Steps to determine the wind loads
F30.5-1	Design wind pressures of walls and roof of enclosed structures with $h \leq 60'$
	Tables include for
	Net design wind pressures for walls and roofs
	Flat roof
	Hip roof ($7° < \Theta \leq 27°$)
	Gable roof ($\Theta \leq 7°$)
	Gable roof ($7° < \Theta \leq 45°$)
	Net design wind pressures for roof overhang
	Adjustment factor for building height and exposure

Part (3): Buildings with h > 60′

$$p = q_h[(GC_p) - (GC_{pi})] \qquad \text{ASCE Equation: 30.6-1}$$

$q = q_z$ for windward walls at height z
$q = q_h$ for leeward walls, side walls, and roof at mean roof height
$q_i = q_h$ for windward walls, side walls, leeward walls of enclosed building and negative internal pressure evaluation in partially enclosed buildings
$q_i = q_z$ for positive internal pressure evaluation in partially enclosed building where height z is the height of the highest opening in the building that could affect the positive internal pressure.
(GC_p) is the external pressure coefficients given in ASCE Figure 30.6-1 for walls and flat roofs; 27.4-3, footnote 4, for arched roofs; 30.4-7 for domed roofs;—Note 6 of 30.6-1 for other roof angles and geometries

Explanation of Tables and Figures of Part (3) of Chapter 30:

Table/Figure	Description
T30.6-1	Steps to determine the wind loads
F30.6-1	External pressure coefficient (GC_p) for walls and roof with $h > 60'$ (Enclosed and partially enclosed structures)

Part (4): Buildings with h ≤ 160′

$$p = p_{\text{table}}(EAF)(RF)K_{zt} \qquad \text{ASCE Equation: 30.7-1}$$

RF is the effective area reduction factor from Table 30.7-2 and *EAF* is the exposure adjustment factor from Table 30.7-2

Windward Parapet (Load Case A)

Windward parapet pressure is determined using the positive wall pressure zones 4 or 5 from ASCE Table 30.7-2 and ASCE Figure 30.7-1.

Leeward parapet pressure is determined using the negative roof pressure zones 2 or 3 from ASCE Table 30.7-2 and ASCE Figure 30.7-1.

Leeward Parapet (Load Case B)

Windward parapet pressure is determined using the positive wall pressure zones 4 or 5 from ASCE Table 30.7-2 and ASCE Figure 30.7-1.

Leeward parapet pressure is determined using the negative wall pressure zones 4 or 5 from ASCE Table 30.7-2 and ASCE Figure 30.7-1.

Height shall be height of the top of the parapet. A wind effective area of 10 ft^2 is used in the calculations of wind pressures of ASCE Table 30.7-2 and the reduction factor (RF) can be used to calculate the wind pressure for higher tributary area.

Roof Overhang

The net outward pressure of the roof overhang is the edge zone pressure of zones 1 and 2 read from ASCE Table 30.7-2. For zone 3, the wind pressure read from the ASCE Table 30.7-2 shall be increased by 15%.

Explanation of Tables and Figures of Part (4) of Chapter 30:

Table/Figure	Description
T30.7-1	Steps to determine the wind loads
T30.7-2	Wall and roof pressures for enclosed buildings with $h \leq 160'$ for flat, gable, mono-slope, hip and mansard roofs
	Tables include for
	C & C exposure adjustment factor
	Effective wind area for reduction factors
	C & C wind pressures for different heights and velocities
F30.7-1	Parapet wind loads for enclosed simple diaphragm buildings with $h \leq 160'$
F30.7-2	Roof overhang wind loads for enclosed simple diaphragm buildings with $h \leq 160'$

Part (5): Open Buildings

$$p = q_h G C_N \qquad \text{ASCE Equation: 30.8-1}$$

C_N is evaluated from ASCE Figure 30.8-1 for mono-sloped roof, ASCE Figure 30.8-2 for pitched roof and ASCE Figure 30.8-3 for troughed roof.

Explanation of Tables and Figures of Part (5) of Chapter 30:

Table/Figure	Description
T30.8-1	Steps to determine the wind loads
F30.8-1	Net pressure coefficient C_N for open buildings mono-slope free roof ($\Theta \leq 45°$) ($0.25 \leq h/L \leq 1.0$)
F30.8-2	Net pressure coefficient C_N for open buildings pitched free roof ($\Theta \leq 45°$) ($0.25 \leq h/L \leq 1.0$)
F30.8-3	Net pressure coefficient C_N for open buildings troughed free roof ($\Theta \leq 45°$)($0.25 \leq h/L \leq 1.0$)

Part (6)(a): Parapets

$$p = q_P[(GC_p) - (GC_{pi})] \qquad \text{ASCE Equation: 30.9-1}$$

q_P is the velocity pressure evaluated at the top of the parapet.

(GC_p) is the external pressure coefficients given in ASCE Figures 30.4-1 for walls with $h \leq 60'$; 30.4-2A to 30.4-2C for flat, gable roof, and hip roofs; 30.4-3 for stepped roofs; 30.4-4 for multi-span gable roofs; 30.4-5A and 30.4-5B for monoslope roofs; 30.4-6 for saw-tooth roofs; 30.4-7 for domed roofs of all heights; 30.6-1 for walls and flat roofs with $h > 60'$ and 27.4-3, footnote 4 for arched roofs. Loads cases of ASCE Figure 30.9-1 apply.

Load Case (A)—Windward Parapet: Apply positive wall pressure from ASCE Figure 30.4-1 ($h \leq 60'$) or from ASCE Figure 30.6-1 ($h > 60'$) to the windward surface of the parapet. Apply negative edge or corner zone roof pressure from ASCE Figure 30.4-2 (A, B or C), 30.4-3, 30.4-4, 30.4-5 (A or B), 30.4-6, 30.4-7, ASCE Figure 27.4-3 footnote 4, or Figure 30.6-1 ($h > 60$ ft) to the leeward surface of the parapet.

Load Case (B) and Leeward Parapet: Apply positive wall pressure from ASCE Figure 30.4-1 ($h \leq 60'$) or from ASCE Figure 30.6-1 ($h > 60'$) to the windward surface of the parapet. Apply the negative wall pressure from ASCE Figure 30.4-1 ($h \leq 60$ ft) or ASCE Figure 30.6-1 ($h > 60$ ft) to the leeward surface.

Part (6)(b): Roof Overhang

$$p = q_h[(GC_p) - (GC_{pi})] \qquad \text{ASCE Equation: 30.10-1}$$

GC_p is the external pressure co-efficient for overhangs from ASCE Figure 30.4-2A to 30.4-2C (flat roofs, gable roofs, and hip roofs), including contributions from top and bottom surfaces of overhang. The external pressure co-efficient for the covering on the underside of the roof overhang is the same as the external pressure co-efficient on the adjacent wall surface, adjusted for effective wind area, determined from ASCE Figure 30.4-1 or 30.6-1. GC_{pi} is the internal pressure coefficient from ASCE Table 26.11-1.

Part (6)(c): Rooftop Structures and Equipment for Buildings with $h \leq 60'$
Wall pressures of the rooftop structures are calculated in accordance with ASCE Section 29.5.1 and divided by respective wall surface area. They shall be considered to act inward or outward.

Roof pressures of the rooftop structures are calculated in accordance with ASCE Section 29.5.1 and divided by the horizontal projected area of the roof. They shall be considered to act upwards.

Explanation of Tables and Figures of Part (6) of Chapter 30:

Table/Figure	Description
T30.9-1	Steps to determine the wind loads for parapets
T30.10-1	Steps to determine the wind loads for roof overhang
F30.9-1	Parapet wind loads for all buildings types
F30.10-1	Roof overhang wind loads for all buildings types

4.11 SIGNIFICANT CHANGES IN ASCE 7-10 AS COMPARED TO ASCE 7-05

There have been significant changes between the 2005 and 2010 versions of ASCE 7. In ASCE 7-05, the entire wind load information is in Chapter 6 and in ASCE 7-10 the wind load information is in Chapters 26 through 31. The changes are demonstrated in Tables 4.15 through 4.24.

In both versions, tornadoes have not been considered in developing the basic wind speed distributions.

Some Similarities between ASCE 7-05 and ASCE 7-10

1. Wind directionality factor
2. Definitions of surface roughness
3. Information about topographic effects
4. Definitions of the enclosure classifications
5. The values of internal pressure coefficient for the various enclosure classifications and the equation for the calculation for the reduction factor for large volume buildings

TABLE 4.15
Arrangement of the Code

ASCE 7-05	ASCE 7-10	
Only Chapter 6 –"Wind Loads" deal with the complete information	Chapters 26–31 have been introduced.	
	Chapter	Title
	26	Wind Loads—General Requirements
	27	Wind Loads on Buildings—MWFRS (Directional Procedure)
	28	Wind Loads on Buildings—MWFRS (Envelope Procedure)
	29	Wind Loads on Other Structures and Building Appurtenances—MWFRS
	30	Wind Loads—Components and Cladding
	31	Wind Tunnel Procedure

TABLE 4.16
Definitions

ASCE 7-05	ASCE 7-10
In the ASCE 7-10, the definitions for Building, Torsionally Regular under Wind Load; Diaphragm; Directional Procedure, Envelope Procedure and Wind Tunnel Procedure were added	
Hurricane Prone region is US Atlantic Ocean and Gulf of Mexico coasts where the basic wind speed is greater than 90 mph.	Hurricane Prone region is US Atlantic Ocean and Gulf of Mexico coasts where the basic wind speed for Risk II category is greater than 115 mph.
Wind-borne debris region areas within hurricane prone regions located:	Wind-borne debris region areas within hurricane prone regions located:
1. Within 1 mile of coastal mean high water line where the basic wind speed ≥ 110 mph and in Hawaii, or	1. Within 1 mile of coastal mean high water line where the basic wind speed ≥ 130 mph or
2. In areas where the basic wind speed ≥ 120 mph	2. In areas where the basic wind speed ≥ 140 mph

TABLE 4.17
Basic Wind Speed

ASCE 7-05	ASCE 7-10	
Basic wind speed is determined using Figure 6.1 for all exposure categories with exception of Special Wind Regions and estimation of basic wind speeds from Regional Climatic Data.	Basic wind speed is determined using Figure 26.5-1 according to exposure categories with exception of Special Wind Regions and estimation of basic wind speeds from Regional Climatic Data.	
	Figure	Risk category
	26.5-1A	II
	26.5-1B	III and IV
	26.5-1C	I

TABLE 4.18
Permitted Procedures

ASCE 7-05		ASCE 7-10		
For Both MWFRS and C & C, the following Procedures are Allowed		Directional, Envelope and Wind Tunnel Procedures have been Used in Five Different Chapters		
Method	**Description**	**Chapter**	**Component**	**Description**
1	Simplified Procedure	27	MWFRS—Buildings of all height	Directional Procedure
2	Analytical Procedure	28	MWFRS—low-rise buildings	Envelope Procedure
3	Wind Tunnel Procedure	29	Building appurtenances and Other structures	Directional Procedure
		30	C & C	Envelope Procedure (Parts 1 and 2) Directional Procedure (Parts 3, 4 and 5) Building Appurtenances (Part 6)
		31	MWFRS/C & C	Wind Tunnel Procedure

TABLE 4.19
Importance Factor

ASCE 7-05	ASCE 7-10
Based upon the building categories listed in the Table 1.1, importance factors differ.	The importance factor is the same for building categories. It is not used in the calculations of the wind velocity pressure.

TABLE 4.20
Exposure

ASCE 7-05	ASCE 7-10
Exposure B is applied to buildings of all heights where Surface Roughness B prevails in the upwind direction for a distance of at least 2600′ or 20 times the height of the building, whichever is greater.	For buildings of mean roof height ≤ 30′, exposure B is applied where Surface Roughness B prevails in the upwind direction for a distance of at least 1500′. For buildings of mean roof height ≥ 30′, Exposure B is applied where Surface Roughness B prevails in the upwind direction for a distance of at least 2600′ or 20 times the height of the building, whichever is greater.

TABLE 4.21
Gust Effects

ASCE 7-05	ASCE 7-10
	The calculations of the gust effect remaining the same, calculations for approximate natural frequency in Sections 26.9.2 and 26.9.3 are added.

TABLE 4.22
Velocity Pressure

ASCE 7-05	ASCE 7-10
$q_z = 0.00256K_zK_{zt}K_dIV^2$ (lb/ft²)	$q_z = 0.00256K_zK_{zt}K_dV^2$ (lb/ft²) Since importance factor $I = 1$ for all building classifications, it is not used in the calculations of the velocity pressure

TABLE 4.23
Analytical Procedure of ASCE 7-05 and Directional and Envelope Procedures of ASCE 7-10

ASCE 7-05 Method II (Analytical Procedure)	ASCE 7-10 Directional and Envelope Procedures
MWFRS Rigid buildings of all heights	MWFRS Enclosed and Partially enclosed rigid buildings (Chapter 27/Part 1)
MWFRS Low-Rise buildings	MWFRS Enclosed and Partially enclosed Low-Rise buildings (Chapter 28/Part 1)
MWFRS Flexible buildings	MWFRS Enclosed and Partially enclosed flexible buildings (Chapter 27/Part 1)
MWFRS Parapets	MWFRS Parapets (Chapter 27/Part 1)
C & C Low-Rise Buildings, $h \leq 60'$	C & C Enclosed and Partially enclosed Low-Rise Buildings, $h \leq 60'$ (Chapter 30, Part 1)
C & C Buildings, $h > 60'$	C & C Enclosed and Partially enclosed Buildings, $h > 60'$ (Chapter 30, Part 3)
MWFRS for monoslope, pitched or troughed roof Open buildings	MWFRS Open buildings with monoslope, pitched or troughed roof (Chapter 27/Part 1)
C & C for monoslope, pitched or troughed roof Open buildings	C & C Open buildings with monoslope, pitched or troughed roof (Chapter 30, Part 5)
Solid Freestanding walls and solid signs	Solid Freestanding walls and solid signs (Chapter 29)
Other Structures	Other Structures (Chapter 29)
Rooftop structures and equipment of buildings with $h \leq 60'$ uses Equation 6.28 in accordance with the note of Section 6.5.15.1	The lateral force (F_h) and the vertical uplift force (F_v) are defined by Equations 29.5-2 and 29.5-3 with the gust factor GC_r varying in accordance with the vertical and horizontal project areas (Chapter 29)

4.12 SOLVED EXAMPLES

Example (4.12.1)—Roughness Length Parameter

Problem Statement: A building 50′ wide and 20′ high is located on an open lot of 10,000 ft². Calculate the surface roughness parameter and determine the exposure category for the wind load calculations.

Solution

Height of the building = 20′; Width of the building = 50′; Ground area of building = 10,000 ft²
Hence, $H_{ob} = 20'$; $S_{ob} = 20 \times 50 = 1000$ ft², $A_{ob} = 10{,}000$ ft²

$$\text{Roughness length parameter } (z_0) = \frac{(0.5)(20')(1000)}{10{,}000} = 1.0$$

Hence building falls under exposure "B"

TABLE 4.24
Simplified Procedure of ASCE 7-05 and Directional and Envelope Procedures of ASCE 7-10

ASCE 7-05	ASCE 7-10
Method I (Simplified Procedure)	**Directional and Envelope Procedures**
Simplified method is used only for enclosed simple diaphragm buildings < 60′ high with either a flat or gable end roof with Θ ≤ 45° or a hip roof with Θ ≤ 45° for MWFRS and Θ ≤ 27° for C & C. Wind pressure values tabulated for walls and roof at 30′, 85–170 mph wind velocity and exposure B with the related adjustment factors.	For MWFRS of enclosed simple diaphragm buildings using the Directional Procedure, the height has been extended to 160′. The buildings have been classified as Class 1 and Class 2 buildings. Class 1 buildings have mean roof height ≤ 60′ while Class 2 buildings have 60′ < h ≤ 160′. Separate tables for roofs (110–200 mph wind velocity at exposure B) and walls (110–200 mph wind velocity at exposure C) are provided with related adjustment factors (Chapter 27/Part 2). For MWFRS of enclosed simple diaphragm low-rise buildings for height ≤ 60′, using the Envelope Procedure, tables (at h = 30′ and exposure B) are provided to determine wind pressures for basic wind speeds 110–200 mph. Adjustment factors are used for height and exposure (Chapter 28/Part 2). For C & C of low-rise enclosed buildings with height ≤ 60′, simplified procedure is used to tabulate wind pressures for roof and wall evaluated at 30′ height and exposure B. The values are tabulated for basic wind speed 110–200 mph and for roofs (0°–7°, 7°–27° and 27°–45°) and walls for effective wind areas 10–100 ft^2. Adjustment factors are used for height and exposure (Chapter 30/Part 2). For C & C of low-rise enclosed buildings with height ≤ 160′, simplified procedure is used to tabulate wind pressures for roof and wall for exposure C. The values are tabulated for basic wind speed 110–200 mph and for flat, gable, mansard, hip and monoslope roofs. Reduction factors for Effective Wind Area and Adjustment factors for height and exposure are provided (Chapter 30/Part 4).

Example (4.12.2)—Approximate Natural Frequency—Steel Moment Resisting Frames

Problem Statement: Calculate the natural frequency of a steel moment resisting frame building 300′ tall. Determine the type of building.

Solution

Natural frequency $\eta_a = \dfrac{22.2}{h^{0.8}}$ ASCE Equation 26.9-2

Hence flexible $= \dfrac{22.2}{(300)^{0.8}} = 0.232\,\text{Hz} < 1$ ASCE Section 26.2

Example (4.12.3)—Approximate Natural Frequency—Concrete Moment Resisting Frames

Problem Statement: Calculate the natural frequency of a concrete moment resisting frame building 300′ tall. Determine the type of building.

Solution

$$\text{Natural frequency } \eta_a = \frac{43.5}{h^{0.8}} \quad \text{ASCE Equation 26.9-3}$$

$$\text{Hence flexible} \quad = \frac{43.5}{(300)^{0.8}} = 0.256\ \text{Hz} < 1 \quad \text{ASCE Section 26.2}$$

Example (4.12.4)—Approximate Natural Frequency—Steel and Concrete Building with Other Lateral-Force Resisting Systems

Problem Statement: Calculate the natural frequency of a steel and concrete building with masonry lateral resisting system. The building is 300′ tall. Determine the type of building.

Solution

$$\text{Natural frequency } \eta_a = \frac{75}{h} \quad \text{Equation 26.9-4}$$

$$\text{Hence flexible} \quad = \frac{75}{300} = 0.25\ \text{Hz} < 1 \quad \text{Section 26.2}$$

Example (4.12.5)—Approximate Natural Frequency—Concrete Shear Wall Building

Problem Statement: A 300′ × 100′ building has seven concrete shear walls in the shorter direction as shown in the figure. The lengths of shear walls are marked on the figure. Calculate the natural frequency and determine the type of building.

Shear Wall	Length (Feet)	Cross Section Area (sq. ft)	Height (Feet)	$\frac{(h)^2}{(h_i)}\frac{A_i}{\left[1+0.83(h_i/D_i)^2\right]}$ (Equation 26.9-5) for C_w
SW-1	80	53.6	300	4.23
SW-2	70	53.6	200	3.09
SW-3	70	53.6	200	3.09
SW-4	60	53.6	300	2.46
SW-5	70	53.6	200	3.09
SW-6	70	53.6	200	3.09
SW-7	80	53.6	300	4.23

$$\text{Natural frequency } \eta_a = \frac{385(C_w)^{0.5}}{h} \quad \text{ASCE Equation 26.9-5}$$

$$\text{Where } C_w = \frac{100}{A_B}\sum_{i=1}^{n}\left(\frac{h}{h_i}\right)^2 \frac{A_i}{\left[1+0.83\left(\frac{h_i}{D_i}\right)^2\right]}$$

$$\text{Hence } C_w = \frac{100}{100\times 300}[4.23+3.09+3.09+2.46+3.09+3.09+4.23] = 0.0776$$

$$\eta_a = \frac{385(0.0776)^{0.5}}{300} = 0.36$$

Hence, it is a flexible building.

Example (4.12.6)—Gust-Effect Factor

Problem Statement: A 300′ × 300′ building located in exposure "C" is made of concrete moment resisting frame and is 300′ tall. The basic wind velocity in the region is 150 mph. Find the gust-effect factor for concrete moment resisting frame.

Solution:

Approximate Natural Frequency $(n_a) = 43.5/h^{0.9}$ ASCE Equation 26.9-3

Hence, it is a flexible building. Use ASCE Section 26.9.5

$$g_R = \sqrt{2\ln(3600n_1)} + \frac{0.577}{\sqrt{2\ln(3600n_1)}}$$ ASCE Equation 26.9-11

$$= \sqrt{2\ln(3600\times 0.256)} + \frac{0.577}{\sqrt{2\ln(3600\times 0.256)}} = 3.85$$ ASCE Section 26.9-4

$$\bar{z} = 0.6h = 0.6(300') = 180',\ l = 500',\ \varepsilon = 1/5.0$$ ASCE Table 26.9-1

$$L_{\bar{z}} = l\left(\frac{\bar{z}}{33}\right)^{\varepsilon} = 500\left(\frac{180}{33}\right)^{1/5.0} = 702$$ ASCE Equation 26.9-9

$$\bar{b} = 0.65,\ \bar{\alpha} = \frac{1}{6.5}$$

$$\overline{V_z} = \bar{b}\left(\frac{\bar{z}}{33}\right)^{\bar{\alpha}}\left(\frac{88}{60}\right)\mathrm{Y}$$ ASCE Equation 26.9-16

$$= 0.65\left(\frac{180}{33}\right)^{1/6.5}\left(\frac{88}{60}\right)150 = 185.6$$

$$N_1 = \frac{n_1 L_{\bar{z}}}{\overline{V_z}} = 0.256\frac{(702)}{(185.6)} = 0.968$$ ASCE Equation 26.9-14

$$R_n = \frac{7.47N_1}{(1+10.3N_1)^{5/3}} = \frac{7.47(0.968)}{(1+10.3(0.968))^{5/3}} = 0.153$$ ASCE Equation 26.9-13

$$\text{Where } R_l = R_h,\ \eta = \frac{4.6n_1 h}{\overline{V_z}} = \frac{(4.6)(0.256)(300)}{185.6} = 1.90$$

$$\text{Where } R_l = R_B,\ \eta = \frac{4.6n_1 B}{\overline{V_z}} = \frac{(4.6)(0.256)(300)}{185.6} = 1.90$$

$$\text{Where } R_l = R_L,\ \eta = \frac{15.4n_1 L}{\overline{V_z}} = \frac{(15.4)(0.256)(300)}{185.6} = 6.37$$

In general,

$$R_\ell = \frac{1}{\eta} - \frac{1}{2\eta^2}(1-e^{-2n}) \text{ for } \eta > 0 \qquad \text{ASCE Equation 26.9-15a}$$

$$R_h = \frac{1}{1.9} - \frac{1}{2(1.9)^2}(1-e^{-2(1.9)}) = 0.390$$

$$R_B = \frac{1}{1.9} - \frac{1}{2(1.9)^2}(1-e^{-2(1.9)}) = 0.390$$

$$R_L = \frac{1}{6.37} - \frac{1}{2(6.37)^2}(1-e^{-2(6.37)}) = 0.144$$

$$R = \sqrt{\frac{1}{\beta}R_nR_hR_B(0.53+0.47R_L)} \qquad \text{ASCE Equation 26.9-12}$$

Use 5% damping ratio, β = 0.05

$$R = \sqrt{\left(\frac{1}{0.05}\right)(0.153)(0.39)(0.39)(0.53+0.47\times 0.144)} = 0.36$$

$$Q = \sqrt{\frac{1}{1+0.63\left(\frac{B+h}{L_z}\right)^{0.63}}} \qquad \text{ASCE Equation 26.9-8}$$

$$= \sqrt{\frac{1}{1+0.63\left(\frac{600}{702}\right)^{0.63}}} = 0.798$$

$$I_{\bar{z}} = C\left(\frac{33}{z}\right)^{1/6} \quad C = 0.20 \qquad \text{ASCE Equation 26.9-7}$$

ASCE Table 26.9-1

$$I_{\bar{z}} = 0.2\left(\frac{33}{180}\right)^{1/6} = 0.151$$

$$G_f = 0.925\left[\frac{1+1.7I_{\bar{z}}\sqrt{g_Q{}^2Q^2+g_R{}^2R^2}}{1+1.7g_rI_{\bar{z}}}\right] \qquad \text{ASCE Equation 26.9-10}$$

$$= 0.925\left[\frac{1+1.7(0.151)\sqrt{(3.4)^2(0.798)^2+(3.85)^2(0.36)^2}}{1+1.7(3.4)(0.151)}\right]$$

$$= 0.952$$

With a 2% damping, $R = 0.678$ and $G_f = 1$

Now, assume the same building as a rigid building,

$$G = 0.925\left(\frac{1+1.7g_Q I_{\bar{z}} Q}{1+1.7g_r I_{\bar{z}}}\right)$$

$$= 0.925\left(\frac{1+(1.7)(3.4)(0.151)(0.798)}{1+(1.7)(3.4)(0.151)}\right) = 0.84$$

Example (4.12.7)—Topographic Factor

Problem Statement: In a building shown with the profile in the diagram in the following, determine the topographic factor (K_{zt}).

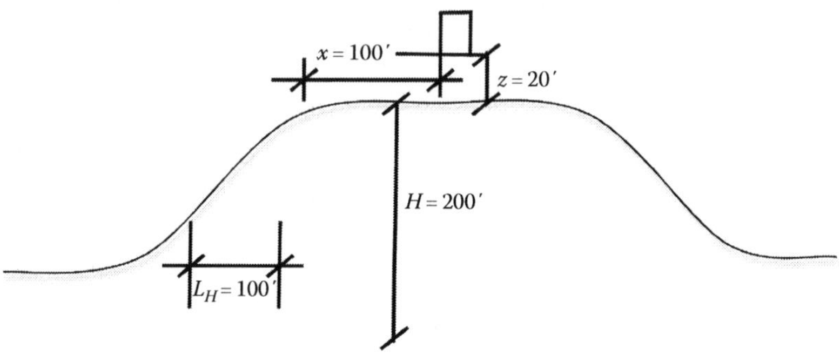

Solution:

H = Height of hill
L_H = Distance upwind of crest to where the difference in ground elevation is half the height of the hill.
x = Distance from the crest to the building site
z = Height above the ground surface at building site
Assume: $H = 200'$, $L_H = 100'$, $x = 100'$, $z = 20'$

$$\frac{H}{L_n} = \frac{200}{100} = 2.0,\ K_1 = 0.43$$

$$\frac{x}{L_n} = \frac{100}{100} = 1.0,\ K_2 = 0.75$$ Figure 26.8-1

$$\frac{z}{L_n} = \frac{30}{100} = 0.3,\ K_3 = 0.47$$

$$K_{zt} = (1 + K_1K_2K_3)^2$$

$$= \left[1+(0.43)(0.75)(0.47)\right]^2 = 1.33$$

Example (4.12.8)—MWFRS and Window Pressures

Problem Statement: A residential building is 300′ by 300′ in plan and 300′ high located at the flat coast line of Miami. Assume a gust-effect factor of 0.85. It has 30 floors of equal height. The building has impact resistant windows and exterior doors.

Calculate the windward wall, leeward wall, and side walls wind pressures for each floor.

Calculate the wind pressure on the window (5′ high and three equal panels of 3′ width) located in the corner at the 30th floor.

Solution:

Since the building is residential and located at the Miami coast line and the terrain is flat,

Basic wind velocity for Occupancy Category II	= 175 mph	ASCE Figure 26.5-1A
Exposure	D	ASCE Section 26.7.3
Topographic factor (K_{zt})	= 1.0	ASCE Section 26.8.2
Directionality factor (K_d)	= 0.85	ASCE Table 26.6-1
Velocity pressure (q_z)	$= (0.00256)(K_z)(K_{zt})(K_d)V^2$	ASCE Equation 27.3-1
	$= (0.00256)(K_z)(1)(0.85)$ $(175)^2 = 66.64\ K_z$	
Gust-effect factor = 0.85 (assumed rigid building)		
External pressure coefficient (C_p)		= 0.85
$L/B = 300/300 = 1$		
External pressure coefficient (C_p) from		ASCE Figure 27.4-1
For windward wall, 0.8		
For leeward wall, −0.5		
For side walls, −0.7		
Internal pressure coefficient (GC_{pi})	= ±0.18	ASCE Table 26.11-1
For mean roof height = 300′, $K_z = 1.73$		ASCE Table 27.3-1
Hence, $q_h = (66.64)(1.73) = 115.3$ psf		
Wind pressure (p) $= qGC_p - q_i(GC_{pi})$		ASCE Equation 27.4-1

$$\text{For windward wall, } p = q_z(0.85)(0.8) - (115.3)(0.85)(\pm 0.18) = 0.68\ q_z \pm 17.64$$

$$\text{For leeward wall, } p = (115.3)(0.85)(-0.5) - (115.3)(0.85)(\pm 0.18) = -49.0 \pm 17.64 = -66.64 \text{ psf}$$

$$\text{For side walls, } p = (115.3)(0.85)(-0.7) - (115.3)(0.85)(\pm 0.18) = -68.64 \pm 17.64 = -86.24 \text{ psf}$$

In order to get the worst case, −ve sign for the windward wall and +ve sign for the leeward and side walls are ignored.

The wind pressures for the windward wall are tabulated in Table 4.25 for each floor of equal height. K_z is estimated in intervals of 10′ and for the wall spanning 10′, an average value of K_z is taken.

From ASCE Tables 27.3.1 and 26.9.1, $z_g = 700$ and $\alpha = 7$ and

$$K_z = 2.01(z/z_g)^{2/\alpha} \text{ for } z > 15',\ K_z = 2.01(z/700)^{2/7}$$

$$K_z = 2.01(15/z_g)^{2/\alpha} \text{ for } z \le 15',\ K_z = 2.01(z/700)^{2/7}$$

In order to calculate the wind pressure of the window located in the corner (Zone 5), the effective wind area is taken as area of one panel and not the entire window, as explained in the commentary of the Chapter 6 of ASCE 7-05. Hence the effective wind area of the window = (5)(3) = 15 SF.

Wind pressure on the windows $(p) = qGC_p - q_i(GC_{pi})$ ASCE Equation 30.6-1

Even though windows are located at a lower level than the roof, because an average K_z was calculated in Table 4.12.1, $q = 115.25$ is adopted.

$GC_p = +0.9$ and -1.8 ASCE Figure 30.6-1

$$\text{Hence } p(+) = (115.25)(0.9) - (115.25)(-0.18) = +124.74 \text{ psf}$$

$$P(-) = (115.25)(-1.8) - (115.25)(0.18) = -228.2 \text{ psf}$$

TABLE 4.25
Wind Pressures MWFRS

Height (h) (in Feet)	K_z	K_z Used for the "p" Calculation	Velocity Pressure (q_z)	Pressure (p) (psf)
10	1.03	1.03	68.64	64.31
20	1.08	1.06	70.42	65.52
30	1.16	1.12	74.81	68.51
40	1.22	1.19	79.44	71.66
50	1.27	1.25	83.03	74.10
60	1.31	1.29	86.01	76.13
70	1.35	1.33	88.56	77.86
80	1.38	1.36	90.80	79.38
90	1.41	1.39	92.81	80.75
100	1.43	1.42	94.62	81.98
110	1.46	1.44	96.29	83.12
120	1.48	1.47	97.83	84.16
130	1.50	1.49	99.26	85.13
140	1.52	1.51	100.60	86.05
150	1.54	1.53	101.86	86.90
160	1.55	1.55	103.04	87.71
170	1.57	1.56	104.17	88.48
180	1.59	1.58	105.24	89.21
190	1.60	1.59	106.27	89.90
200	1.62	1.61	107.25	90.57
210	1.63	1.62	108.18	91.20
220	1.64	1.64	109.08	91.82
230	1.66	1.65	109.95	92.41
240	1.67	1.66	110.78	92.97
250	1.68	1.67	111.59	93.52
260	1.69	1.69	112.37	94.05
270	1.70	1.70	113.12	94.56
280	1.71	1.71	113.86	95.06
290	1.72	1.72	114.56	95.54
300	1.73	1.73	115.25	96.01

4.13 WIND TUNNEL PROCEDURE

Wind tunnel procedures may be used for determining wind pressures for the main wind force resisting system (MWFRS) and/or for the components and cladding (C & C) of any building irrespective of its shape, height, dynamic characteristics, and wind environment. This method given in Chapter 31 of ASCE 7-10 is considered to be the most robust of the procedures specified there in.

4.13.1 Test Requirements

Wind tunnel tests, used for the determination of design wind loads shall meet all of the following conditions:

1. The natural atmospheric boundary layer has been modeled to account for the variation of wind speed with height.
2. The relevant macro-(integral) length and micro-length scales of the longitudinal component of atmospheric turbulence are modeled to approximately the same scale as that used to model the building or structure.

3. The modeled building and surrounding structures and topography are geometrically similar to their full-scale counterparts.
4. The projected area of the modeled building and surrounding is less than 8% of the test section cross-sectional area unless correction is made for blockage.
5. The longitudinal pressure gradient in the wind tunnel test section is accounted for.
6. Reynolds number effects on pressures and forces are minimized.
7. Response characteristics of the wind tunnel instrumentation are consistent with the required measurements.

The structural model and associated analysis used for the purpose of determining the dynamic response of a building shall account for mass distribution, stiffness, and damping.

4.13.2 Load Effects and Limitations

Mean recurrence intervals of load effects—the load effect required for strength design shall be determined for the same mean recurrence interval as for the analytical method.

Loads for the main wind force resisting system determined by wind tunnel testing shall not be less than 80% of those that would be obtained from Part 1 of Chapter 27 or Part 1 of Chapter 28 of ASCE 7-10. The overall principal load shall be based on the overturning moment for flexible buildings and the base shear for other buildings.

Pressures for components and cladding determined by wind tunnel testing shall not be less than 80% of those calculated for Zone 4 for walls and Zone 1 for roofs using the procedure of Chapter 30 of ASCE 7-10.

The limiting values of 80% may be reduced to 50% for the main wind force resisting system and 65% for components and cladding if either of the following conditions applies:

1. There were no specific influential buildings or objects within the detailed proximity model.
2. Loads and pressures from supplemental tests for all significant wind directions in which specific influential buildings or objects are replaced by the roughness representative of the adjacent roughness condition, but not rougher than exposure B, are included in the test results.

4.13.3 Types of Wind Tunnel Tests

Wind tunnel testing accounts for shielding or channeling effects and can more accurately determine wind loads for a complex building shape than the directional procedure, envelope procedure, or analytical procedure for components and cladding.

It is common practice to resort to wind tunnel tests when design data are required for the following wind-induced loads:

1. Curtain wall pressures resulting from irregular geometry
2. Across-wind and/or torsional loads
3. Periodic loads caused by vortex shedding
4. Loads resulting from instabilities, such as flutter or galloping

Boundary-layer wind tunnels capable of developing flows that meet the conditions stipulated earlier typically have test-section dimensions in the following ranges: width of 6–12 ft (2–4 m), height of 6–10 ft (2–3 m), and length of 50–100 ft (15–30 m). Maximum wind speeds are ordinarily in the range of 25–100 mile/h (10–45 m/s). The wind tunnel may be either an open-circuit or closed-circuit type.

Three basic types of wind-tunnel tests models are commonly used. These are designated as follows:

1. Rigid pressure model (PM)
2. Rigid high-frequency base balance model (H-FBBM)
3. Aero-elastic model (AM)

One or more of the models may be employed to obtain design loads for a particular building or structure. The PM provides local peak pressures for design of elements, such as cladding and mean pressures, for the determination of overall mean loads. The H-FBBM measures overall fluctuating loads for the determination of dynamic responses. When motion of a building influences the wind loading, the AM is employed for direct measurement of overall loads, deflections, and accelerations. Each of these models together with a model of the surrounding (proximity model) can provide information other than wind loads, such as snow loads on complex roofs, wind data to evaluate environmental impact on pedestrians, and concentrations of air-pollutant emission for environmental impact determinations.

The pressure model provides local peak pressures for design of cladding elements and mean pressures for the determination of overall mean loads. The high-frequency base balance model, H-FBBM, measures overall fluctuating loads for the determination of dynamic responses. The aero-elastic model, AM, is used for direct measurement of overall loads, deflections, and accelerations and deemed necessary, when the lateral motions of a building are considered to have a large influence on wind loading.

Many techniques are used in aeronautical tunnels to generate turbulence and atmospheric boundary layer by using devices such as screens, spires, and grids. IN special wind tunnels with long test sections, turbulent boundary layer is generated by installing appropriate roughness elements in the upstream flow. Another approach is to use counter jet technique. In every case there is some question whether the natural wind turbulence characteristic is appropriately modeled and proper gust simulation is included. The degree of scaling required to appropriately account for these may yield a very extreme scale for the building, on the order of 1:500 or even more for urban environment studies.

4.13.3.1 Rigid Pressure Model

Although the primary purpose of the rigid-model test is for obtaining cladding design pressures, the data acquired from the wind-tunnel tests may be extrapolated to provide floor-by-floor shear forces for design of the overall main wind-force-resisting system.

Most commonly, pressure study models are made from methyl methacrylate commonly known as Plexiglas, Lucite, and Perspex. This material has several advantages over wooden or aluminum alloy models because it can be easily and accurately machined and drilled and is transparent, facilitating observation of the instrumentation inside the model. It can also be formed into curved shapes by heating the material to about 200°C. Model panels can either be cemented together or joined, using flush-mount screws.

A scale model of the prototype typically in a 1:300 to 1:500 range is constructed using architectural drawings of the proposed project. In rigid models, building features of significance to wind flow, such as building profile, protruding mullions, and overhangs are simulated. Wind measurements provide mean and fluctuating pressures acting on the building.

The model is typically instrumented with as many as 500–700 pressure taps. It includes detailed topography of nearby surroundings within a radius of 1500 ft (457 m). Flexible, transparent vinyl, or polyethylene tubing of about 1/16 in (1.5 mm) internal diameter is used as pressure tapping. Pressure tap locations are generally more concentrated in regions of high pressure gradients such as around corners.

The wind tunnel test is run for a duration of about 60 s, which corresponds to approximately 1 h in real time. Sufficient numbers of readings are gathered from each port to offset the effects of time-dependent fluctuations. From the measured values, the mean and the root-mean-square values of pressure fluctuations are evaluated.

The boundary-layer wind tunnel, BLWT, by virtue of having a long working section with roughened floor and turbulence generators at the upwind end, is believed to correctly simulate the mean wind profile and turbulence. The model mounted on a turntable allows measurement of pressures for any wind direction. Near-field characteristics around the building are duplicated, typically using polystyrene foam models.

The aerodynamic damping, which is building-motion dependent, cannot be measured using the rigid model method. Rather, it can be measured with conventional aero-elastic models or by using a technique in which the simple force balance model is oscillated during the test. Nevertheless, there are indications that the value of the aerodynamic damping is small and positive for typical buildings in the drag direction, and positive too in the lift direction up to the velocity where vortex shedding has a major contribution to the response. This velocity is usually higher than typical design wind speeds.

4.13.3.1.1 Component and Cladding Pressures

Measurements are taken for representative wind directions, generally spaced about 10°–20° apart. From the data acquired, full-scale peak exterior pressures and suctions at each tap location are derived by combining the wind-tunnel data with a statistical model of windstorms expected at the building site. The results are typically given for 25-, 50-, and 100-year return periods.

In evaluating peak wind loads on the exterior of the prototype, the effects of internal pressures arising from air leakage, mechanical equipment, and stack effect should be included. The possibility of window breakage caused by roof gravel scoured from roofs of adjacent buildings and other flying debris during a windstorm should also be included. As a rough guide, the resulting internal pressure can be considered to be in the range of ±5 psf (25 kg/m) at the base, to as much as ±20 psf (100 kg/m) at the roof of a 50-story building.

In the design of glass, a 1 min loading is commonly used. The duration of measured peak pressure in wind tunnels is quite different from the 1 min interval used in design. Usually it corresponds to 5–10 s. Therefore, it is necessary to reduce the peak loads measured in wind-tunnel tests. Empirical reduction factors of 0.80, 0.94, and 0.97 are given in glass manufacturers' recommendations for three different types of glass—annealed float glass, heat-strengthened glass, and tempered glass.

4.13.3.1.2 Overall Loads for Design of MWFRS

Although rigid-model test results are used to predict wind loads for design of glass and other cladding elements, they can nevertheless be extrapolated to provide lateral loads for the design of the main wind-force resisting system. The procedure entails introducing a gust factor for converting the mean wind load to gust loads. An appropriate gust factor estimation should take into account:

- Averaging period of the mean wind load
- Terrain roughness in relation to the building height
- Peak gust factor, which depends on the natural frequency of the building
- Effect of turbulence
- Critical damping ratio of the building

In spite of the fact that rigid-model wind study does not take into account all of the preceding factors, it is still considered adequate to provide design data for buildings with height-to-width ratios of less than 5.

The development of solid state pressure scanners, which permit the simultaneous measurement of pressures at many points on the surface of a building, allows the determination of instantaneous overall wind forces from the local pressure measurements. The advantages of this technique are that a single model used in a single testing session can produce both overall structural loads and cladding loads. The testing parameters would be extended to ensure that the local pressure data taken is also sufficient for the analysis of structural loads.

A disadvantage of this technique is that the cladding pressure test model typically includes more instrumentation and takes longer time to construct than force balance test model. Also, as in the force balance technique, it does not include any effects of the building's motion through the air, such as aerodynamic damping; however, neglecting these effects is usually slightly conservative. Proper integration of local pressures to obtain overall wind forces requires that all buildings surfaces are properly represented in the model instrumentation and subsequent calculations. This may not be possible for buildings or structures with complex geometry.

An approach to the determination of total design loads from the simultaneously measured external point pressures is as follows: the instantaneous generalized forces are determined from the pressure measurements and are then used in a standard random vibration analysis to provide estimates of the total dynamic loads and responses of the structure. These loads and responses are then combined with the statistics of the full-scale wind climate at the site, to provide predictions of loads and responses for various return periods.

4.13.3.2 High-Frequency Base Force Balance Model (H-FBBM)

The effect of wind load on a flexible building can be considered as an integrated response resulting from three distinct sources. The first is the mean wind load that bends and twists the building. The second is the fluctuating load from the unsteady wind that results in oscillation of the building about a steady deflected shape. The third contribution comes from the inertia forces similar to the lateral forces induced during earthquakes. However, for design purposes, the inertial effects can be considered as an additional equivalent wind load.

A rigid model as mentioned previously is convenient for measuring local wind positive and negative pressures distributed uniquely around a building. These local pressures are integrated to derive net lateral forces in two perpendicular directions and a torsional moment about a vertical axis, at each level. The cumulative shear and the overturning and torsional moments at each floor are obtained from simple statics as are the base shear and overturning moments. These values derived from the mean measurements would have been sufficient for the design of building lateral system, except for the drawback that they ignore the influence of gust factor. Therefore, when using rigid-model pressure studies, it is necessary to assume a conservative gust factor to increase the mean values. An alternative and better approach is to take the guesswork out of gust factor by experimentally determining it.

This is precisely what the aerostatic study does; it provides comprehensive information on dynamic loads and motions because the essential structural features such as flexibility, mass, and damping of the prototype are simulated in the model. However, an aero-elastic model is quite complex to design and build, and may take 10–12 weeks to complete the required tests.

A high-frequency base force balance model provides an alternative, more economical, and time-efficient method of furnishing the same design information as provided by aero-elastic models.

Two basic types of force balance models are in vogue. In the first, the outer shell of the model is connected to a flexible metal cantilever bar. Accelerometers and strain gauges are fitted into the model, and the aerodynamic forces are derived from the acceleration and strain measurements. In the second type, a simple foam model of the building is mounted on a five-component, high-sensitivity force balance that measures bending moments and shear forces in two orthogonal directions and torsion about a vertical axis. In both of models, the resulting overall fluctuating loads are determined, and by making certain simplifying assumptions, the information of interest to the structural engineer—floor-by-floor lateral loads—and the expected acceleration at top floors is deduced. A conceptual description of each of the two types of models is given in the following.

4.13.3.2.1 Rigid Model Connected to Flexible Bar

It consists of a lightweight rigid model mounted on a high-frequency-response force balance. Design lateral loads and expected building motions are computed from the test results. The method is suitable when building motion does not, itself, affect the aerodynamic forces, and when torsional effects are not of prime concern. In practice, this method is applicable to many tall buildings.

The high-frequency force balance model is typically constructed to a scale on the order of 1:500. The model itself is constructed of a lightweight material such as balsa wood and is mounted on top of a torsion spring through a relatively rigid plate. Strain gauges attached to the bar measure the instantaneous overturning and torsional moments at the base.

From the measured bending and twisting moments and known frequency and mass distribution of the prototype, wind forces at each floor and the expected peak acceleration are derived.

4.13.3.2.2 Five-Component High-Frequency Base Force Balance Model

In this model, prototype building is represented as a rigid model. Made of lightweight material such as polystyrene foam, the model is attached to a measuring device consisting of a set of five highly sensitive load cells attached to a three-legged miniature frame and an interconnecting rigid beam. A typical configuration is shown in Figure 4.1, in which the load cells are schematically represented as extension springs. Horizontal forces acting in the x direction produce extension of the vertical spring at 1, which can be related to the base overturning moment M_y, with the known extension of the spring and the pivotal distance P_x. Similarly, the base-overturning moment M_y can be calculated from a knowledge of extension of the spring at 2 and the pivotal distance P_y. The horizontal spring at 3 measures the shear force in the x direction, while those at 4 and 5 measure the shear force in the y direction. The difference in the measurements of springs at 4 and 5 serves to compute the torsional moment at the base about

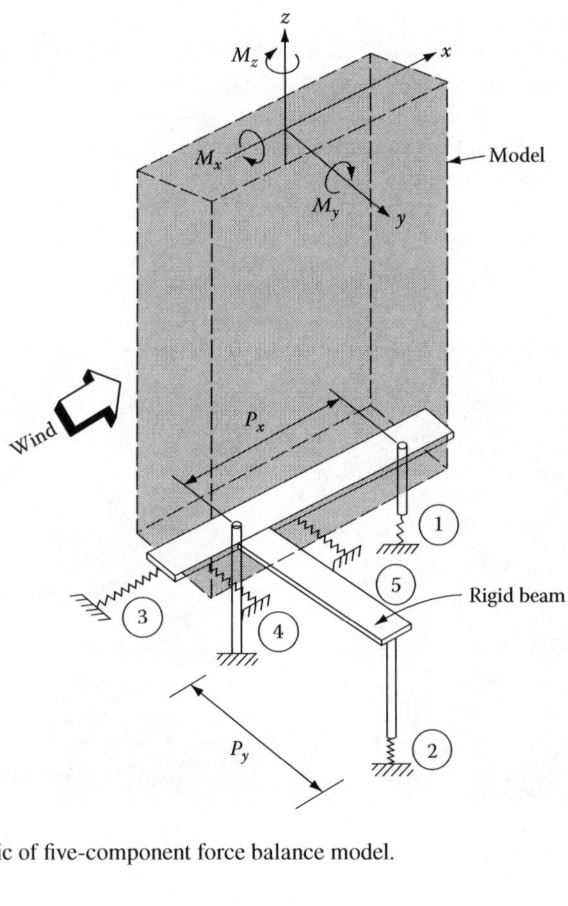

FIGURE 4.1 Schematic of five-component force balance model.

the z-axis. It should be noted, however, that the test results for torsion are an approximation of the true response because the model does not account for the relative twist present in the prototype.

This technique requires scaling the dynamic properties (mass, stiffness, and period) of the building in the fundamental sway models and measuring the response to wind loads directly. The building is modeled as a rigid body, pivoted near the base, with the elasticity provided by appropriately selected springs. Implicit in this technique is the assumption that the sway modes do not include any coupling and can be approximated as linear, and that torsion is unimportant.

These provide to be reasonable assumptions for a large range of buildings.

The advantage of this technique are (1) the measurements will include effects of aerodynamic damping that are not included when using the force balance technique. (2) It is also simpler, less expensive technique than a multi-degree-of-freedom aero-elastic test. The disadvantages of the technique are (1) it is limited by the assumptions noted earlier, and it is more complicated and expensive than the force balance technique, (2) it is less accommodating of changes to the dynamic properties of the building after the test, (3) its advantage over the force balance technique, namely the inclusion of aerodynamic damping effects, rarely proves to be necessary since the aerodynamic damping is usually small, and (4) it cannot capture the negative effects of vortex shedding plays an important role in the dynamic response.

As with other types of tests, once the aerodynamic data has been measured for a full range of wind directions, it is combined with the statistics of the full scale wind climate at the site, to provide predictions of loads and responses for various return periods.

The main objective of this study is to determine a more accurate design wind load and more importantly to predict building oscillations to assess occupant sensitivity to building motions.

Rigid-model study is based on the assumption that the fundamental displacement mode of a tall building varies linearly along the height. In terms of aerodynamic modeling, it is not necessary to achieve the correct density distribution along the building height as long as the mass moment of inertia about a chosen pivot point is the same as that of the correct density distribution. It should be noted that the pivot point is chosen to obtain a mode shape that provides the best agreement with the calculated fundamental mode shapes of the prototype. For example, modal calculations for a tall building with a relatively stiff podium may show that the pivot point is located at the intersection of podium and the tower and not at the ground level. Therefore the pivot point for the model should be at a location corresponding to this intersection point rather than at the base of the building.

Figure 4.2 shows a rigid aero-elastic model mounted on gimbals. The purpose of springs located near the gimbals is to achieve the frequency correlation in the two fundamental sway modes. An electromagnet or oil dashpot provides the model with a damping corresponding to that of the full-scale building.

FIGURE 4.2 Rigid AM mounted on gimbals.

FIGURE 4.3 Rigid AM mounted as a flexible steel bar.

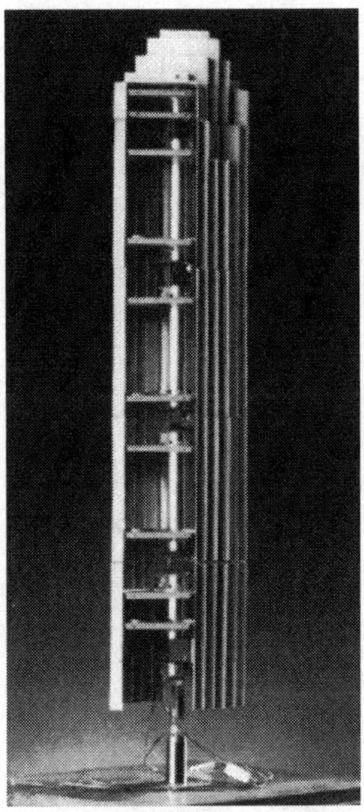

FIGURE 4.4 AM: cutaway view.

An alternative method is to mount the model on a flexible steel bar attached to a vibration-free table. The width and thickness of the bar are chosen to simulate the building stiffness in two horizontal directions. Damping is simulated by using dashpots. Figure 4.3 shows schematics of a rigid aero-elastic model mounted on a flexible steel bar. Shown in Figure 4.4 is a photograph of an aero-elastic model of a 62-story building. In either type, torsional modes are not simulated because the model effectively rotates as a rigid body about the vertical axis.

4.13.3.2.3 Rigid Model Simulating Torsion

Torsion is a consequence of eccentricity between the building mass and its shear center as in seismic design, or it may occur because of eccentric disposition of lateral loads with respect to the center of stiffness of the building as in wind design. Centrally supported concrete-core buildings often use open section shear walls with their shear centers located at considerable distance from the geometrical center of the core. Unless additional lateral resisting elements such as moment frames, braces, or shear walls are used on the building perimeter, torsional may play an important role in its design. When such characteristics are present, it is necessary to simulate not only the bending characteristics of the building but also its torsional behavior. This is achieved by introducing torsional springs in the aero-elastic model at appropriate locations along the height. To allow one section of the model to rotate relative to the next, the model shell is cut around the periphery at the spring locations. Figure 4.5 shows schematics of a model with two cuts. The resulting model with three vertical segments behaves as a three-degrees-of-freedom system in torsion and can therefore capture the dynamic behavior of the three lowest torsional modes.

4.13.3.3 Aero-Elastic Model

Aero-elastic model (AM) tests attempt to take the guesswork out of the gust factor. Instead of guessing a gust factor for extrapolating overall design loads from rigid model tests, here we directly determine the value of gust by testing flexible models. A variety of models ranging from very simple rigid models mounted on flexible supports to models that mimic multimode vibration characteristics of tall buildings are used for this purpose. The more common types can be broadly classified into two categories: (1) stick models, and (2) multi degree-of-freedom models.

In addition to the similarity of the exterior geometry between the prototype and the model, the aero-elastic study requires similarity in inertia, stiffness, and damping characteristics of the building. Although a building in reality responds dynamically to wind loads in a multimode configuration enough evidence exists to show that the dynamic response occurs primarily in the

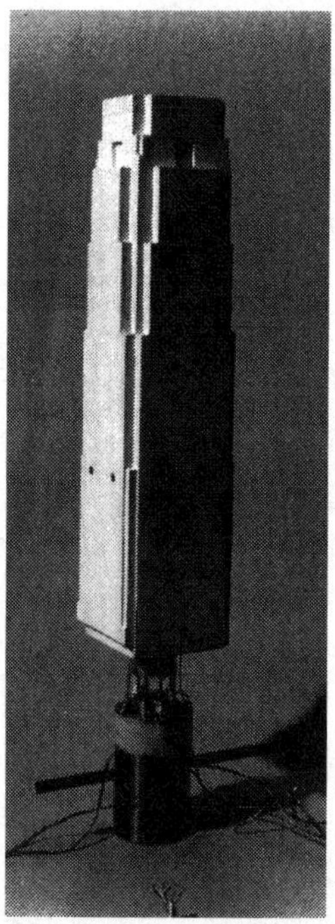

FIGURE 4.5 High-frequency force balance model.

lower modes of vibration. Therefore, it is possible to study dynamic behavior of buildings by using simple dynamic models.

Aero-elastic study basically examines the wind-induced sway response, in addition to providing information on the overall wind-induced mean and dynamic loads. These tests are important for slender, flexible, and dynamically sensitive structures where aero-elastic or body-motion-induced effects are of significance. When a tall building sways and twists under wind action, the resulting accelerations generate inertial loads, causing fluctuating stresses. At any given instant, the amplitude of twisting and swaying motion is not just a function of the magnitude of wind load at the instant but also depends on the integrated effect of the wind over the several previous minutes. Therefore, for slender buildings it is important to consider dynamic response when predicting design wind loads. In addition to providing more reliable loads for structural design, aero-elastic model is believed as one of the most reliable approaches to predicting building response to wind. The results may be used by the designer to ensure that the predicted motion will not cause discomfort to building occupants.

Typically, aero-elastic measurements are carried out at several wind speeds because engineers need information on both relatively common events, such as 10-year wind loads for assessing serviceability and occupant comfort, and relatively rare events, such as 50- and 100-year winds to determine loads for strength design. As mentioned earlier, the modeling of dynamic properties requires simulation of inertial, stiffness, and damping characteristics. It is necessary, however, to simulate these properties for only those modes of vibration that are susceptible to wind excitation. It is often difficult to determine quantitatively when aero-elastic study is required. The following factors may be used as a guide in making a decision:

1. The building height-to-width ratio is greater than about 5; that is, the building is relatively slender.
2. Approximate calculations show that there is a likelihood of vortex shedding phenomenon.
3. The structure is light in density on the order of 8–10 lb/ft^3 (1.25–1.57 kN/m^3).
4. The structure has very little inherent damping, such as a building with welded steel construction.
5. The structural stiffness is concentrated in the interior of the building, making it torsionally flexible. A building with a central concrete core is one example.
6. The calculated period of oscillation of the building is long, in excess of 4 or 5 s.
7. Existence of nearby buildings that could create unusual channeling of wind, resulting in torsional loads and strong buffeting action.
8. The building is sited such that predominant winds occur from a direction most sensitive to the building oscillations.
9. The building is a high-rise apartment, condominium, or hotel, Occupants in these buildings are more likely to experience discomfort from building oscillations. This is because residents in these buildings are likely to remain longer in a given location than they would in a typical office setting.

For a building that is uniform for the entire height, it is reasonable to assume that sway modes of vibration vary linearly along the height. However, for buildings of complex shapes with step backs or major variations in stiffness, this assumption may not yield satisfactory results because fundamental modes shape may not be linear, and more importantly higher modes could contribute significantly to the dynamic behavior.

In such cases, it is essential to simulate the multimode behavior of the building. This is achieved using a model with several lumped masses interconnected with elastic columns.

Schematics of such a model are shown in Figure 4.6, in which the building is divided into three zones, with the mass of each zone located at the center. The masses are concentrated in diaphragms representing the floor system that are interconnected by flexible columns. A lightweight shell simulating the building shape encloses the assembly of the floor system, masses, and columns. The outer

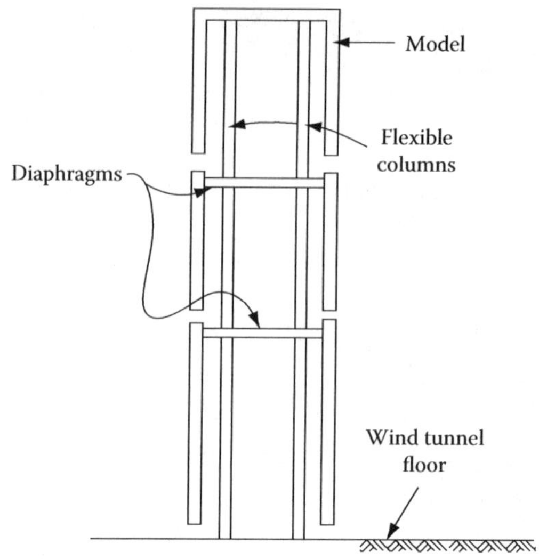

FIGURE 4.6 AM: schematic section.

shell is cut at the diaphragm levels to allow for relative movements between the masses. Similarity between the elastic properties of the prototype and the model is achieved to varying degrees depending upon the predominant characteristics of the building.

This technique requires scaling the dynamic properties (mass, stiffness, periods, and mode shapes) of the building in the fundamental sway modes and the fundamental torsion mode, including any coupling within modes. Some higher modes of vibration are also modeled. The responses to wind loads are then measured directly. Typically, a building is modeled as a series of lumped masses joined by appropriately sized columns.

The advantage of this technique is that the measurements will include effects of aerodynamic damping, vortex shedding, coupling within modes, and some higher modes that are not fully dealt with when using the force balance technique. For most buildings, however, it can be argued that the aerodynamic damping effects are likely to be small, higher modes can be neglected and that the force balance adequately handles coupled modes analytically. Nevertheless, for more complicated structures, the additional reassurance of an aero-elastic test may be justified.

The disadvantage of the technique are that the model is time consuming and expensive to build. The model is also designed for a single set of building dynamic properties and approximations must be made if these change. The force balance technique on the other hand, yields results equally applicable to any set of building dynamic properties.

As with other types of tests, once the aerodynamic data has been measured for a full range of wind directions, it is combined with the statistics of the full scale wind climate at the site, to provide predictions of loads and responses for various return periods.

In addition to reproducing the exterior geometry, an aero-elastic model of a tall building simulates its dynamic properties in the lower modes of vibration, which are likely to be excited by wind action. In a typical study, the building is represented by a lumped mass system in which each mass has two sway and one torsional degree of freedom. For illustration, Figure 4.7 shows a schematic representation of a multi-degree-of-freedom model for a tall building. In the model, each mass consists of a stiff plate or "floor" made of either aluminum or magnesium. The floors are connected by clamped aluminum columns, which provide the required stiffness in sway and in torsion. The required structural damping is achieved by adding strips of foam tape connecting the floors. These strips add negligible stiffness while dissipating vibrational energy. The geometric modeling is completed with sections of balsa wood skins, which are attached to each floor so as not to make contact with each other.

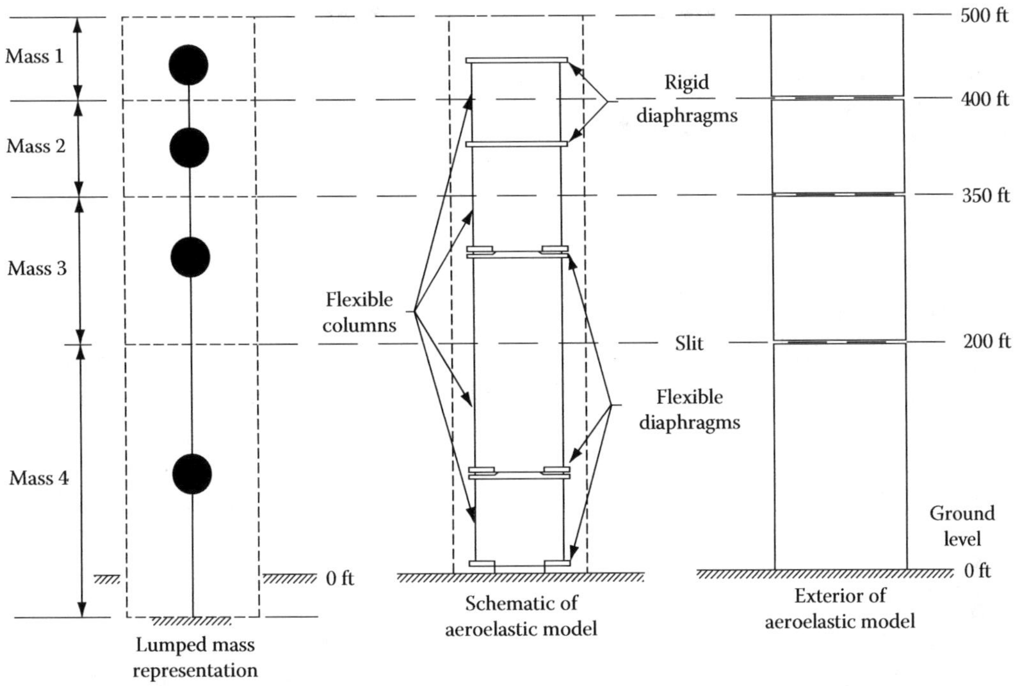

FIGURE 4.7 AM with provisions for simulating torsion.

The completed model is attached to a strain-gauged balance at its base. The balance measures overall base bending moments and torsion. Three accelerometers are normally located at the top mass to provide measures of the translational acceleration and the rotational accelerations at the height. Once accelerometer is mounted to measure acceleration in one direction, say the x direction, while the other two are mounted to measure accelerations in the orthogonal direction.

These two accelerometers are mounted symmetrically about the axis of twist so that their difference yields values of the torsional acceleration.

The choice of the aero-elastic modeling approach for a tall building depends largely on the complexity of the structural system and the exterior geometry. For many tall buildings, the structural system and geometry are such that the response is primarily in the fundamental sway modes of vibration. However, for some buildings the contribution of the fundamental torsional mode becomes significant. As well, for some structural systems or geometries there is strong coupling between directions within modes such that the modes are 3-dimensional in character. This latter detail in particular makes it necessary to design an aero-elastic model, which simulates these complicated modes while maintaining consistent scaling parameters in both sway and torsion directions. Such a model is known as a multi-degree-of-freedom aero-elastic model. The scaling requirements for aero-elastic modeling are based on established similarity theory.

4.13.4 Lower Limits on Wind Tunnel Test Results

Wind tunnel tests frequently measure wind loads that are significantly lower than required by other methods specified in ASCE 7-10. This is due to the shape of the building, the likelihood that the highest wind speeds occur at directions where the building's shape or pressure coefficients are less than their maximum values. Specific buildings included in a detailed proximity model that may provide shielding in excess of that implied by exposure categories and necessary conservatism in enveloping load coefficients is specified in the ASCE 7-10, Chapters 28 and 30. In some cases, adjacent structures may shield the structure sufficiently that removal of one or two structures could significantly increase wind loads. Additional wind tunnel testing without specific nearby buildings (or with additional buildings if they might cause increased loads through channeling or buffeting) is an effective method for determining the influence of adjacent buildings.

For this reason, ASCE 7-10 limits the reduction that can be expected from wind tunnel tests to 80% of the results obtained from the other methods. If supplemental testing is performed to quantity the shielding effect of influential buildings, it is permissible to justify a limit lower than 80%.

However, the absolute minimum reduction permitted is 65% of the baseline result for components and cladding, and 50% for the main wind force resisting system. A higher reduction is permitted for MWFRS, because components and cladding loads are more subject to changes due to local channeling effects when surroundings change and can be easily dramatically increased when a new adjacent building is constructed. It is also recognized that cladding failures are much more common than failures of the MWFRS.

4.13.5 Structural Properties Required for Wind Tunnel Data Analysis

For a rigorous interpretation of wind tunnel test results, certain dynamic properties of a structure are required. These are furnished by the structural engineer and consist of

1. Natural frequencies and mode shapes
2. Mass distribution, mass moments of inertia, and centroid location for each floor.
3. Damping ratio
4. Miscellaneous information such as origin and orientation of the global coordinate system, floor heights, reference elevation for "base" overturning moments

4.13.5.1 Natural Frequencies and Mode Shapes

The natural frequencies (reciprocal of building periods) of buildings are typically determined by performing a 3D analysis using computer programs that can provide the entire spectrum of frequencies and mode shapes. Generally, however, only the first few modes that correspond to say, the three fundamental modes in each of the sway (x and y), and torsional (z) directions. It should be noted that if the structure or mass distribution is unsymmetrical, then at least two of these components will be coupled together in some modes. Normally, the higher modes (four through six) are required only to insure that all of the fundamental directions have been included.

Each mode of vibration is described by both natural frequency and shape. The mode shapes consist of a tabulated values of the x, y, and z deformations of each degree-of-freedom in the structure. For wind–tunnel purposes, the floor diaphragm is typically considered rigid, and a single set of x, y, and z deformations is established for each floor. It should be noted that today's buildings of structural complexity commonly feature three-dimensional mode shapes coupling the three degrees of freedom, two translational, and one torsional, at each floor.

Mode shapes have no units. They are of intermediate magnitude and can be scaled to any desired size. However, it should be remembered that when multidimensional mode shapes include both translational and rotational (twist) components, the same sealing factor must be applied to all components.

Another aspect of mode shapes concerns the reference system used in conveying the mode shapes. Most commercial programs specify the components with respect to the center of mass of each floor. If the shape consists of coupled twist and displacement, then the displacement magnitude is dependent on the location of the reference origin. If the centers of mass do not align on a straight vertical axis—as in setbacks of shear wall drop-offs—then the displacements will contain offsets or "kinks." It is essential, therefore, that the wind engineer knows the reference system used in the modal data received from the structure engineer.

4.13.5.2 Mass Distribution

The mass and moment of inertia are required at each floor, which typically includes the structure's dead weight. The moment of inertia is taken about a vertical axis through the centroid (center of mass) of each floor. The location of the centroid is also needed.

All of the mass in the structure should be included since it will affect the natural frequency, which in turn, will influence the loads determined from the wind-tunnel tests. As a crude rule of thumb, an x percent change in the natural frequency may cause the loads to change by $0.5x$ to $2x$ percent.

The mass distribution is needed for two reasons. First, the mass and mode shape are used together with the natural frequency, to determine the generalized stiffness of each mode. The stiffness is combined with the generalized load (measured on the wind-tunnel model) to determine the fluctuating displacement response in each mode. This is needed to evaluate the acceleration at the top of the building.

Second, the static-equivalent loads from the wind-tunnel analysis consist of mean, background, and resonant contributions. The resonant contributions (which in many cases are the single largest contributor) are applied to the structure as concentrated forces at each concentrated mass, and are in proportion to the mass (and also to the modal displacement of that mass). Thus, the accuracy of the wind-tunnel load distribution (i.e., the floor-by-floor forces) is dependent on the relative mass throughout the structure.

4.13.5.3 Damping Ratio

This parameter is the most difficult one to obtain since there is currently no known method to compute it and assumptions are made based on limited field data. Customary practice in many parts of the world is to use a value of 0.01 for a steel frame structure, and 0.02 for a concrete structure for the prediction of 50–100 year loads. For special structures, for example, a mixed or composite frame, or those with extreme aspect ratios, other values may be appropriate. It should be noted that wind design is based on different principles from earthquake design, for which very high values of damping, usually 0.05, are considered. This value comes from those schooled in seismic design, based on ultimate conditions that don't apply to wind design. Therefore, wind engineers strongly encourage structural engineers to consider a lower value. As stated earlier, a service-level damping ratio of 0.01 or 0.02 is still the most used to determine the service load effects. An extreme damping level (say, 0.03) in combination with wind speeds that have (See Chapter 4, page 99) recurrence interval of approximately 500–1000 years is also used. For the prediction of low return period accelerations such as a 10 year return period, it may be appropriate to use a lower damping value.

4.13.6 Building Drift

Drift (lateral deflection) in a building is a serviceability issue primarily from the effects of wind. Drift limits are imposed on buildings to minimize damage to cladding and to nonstructural walls and partitions. Lateral frame deflection is evaluated for the building as a whole, where the applicable parameter is the total building drift (defined as the lateral frame deflection at the top of the most occupied floor divided by the height of the building to that level, Δ/H). For each floor, the applicable parameter is *inter-story drift* [defined as the lateral deflection of a floor relative to that of the floor immediately below, divided by the distance between floors, $(\delta - \delta_{n-1})/h$].

Typical drift limits most widely used values are H (or h)/400 to H (or h)/500. An absolute limit on inter-story drift is sometimes imposed in light of evidence that damage nonstructural partitions; cladding and glazing may occur if the inter-story drift exceeds about 3/8 in. (10 mm), unless special details are provided to accommodate larger movements. It should be noted that many components can accept deformations that are significantly larger.

It is important to recognize that frame racking or shear distortion is the real cause of damage to building elements such as cladding and partitions, and that drift control limits by themselves in wind-sensitive buildings do not provide comfort of the occupants under wind load.

4.14 HUMAN RESPONSE TO WIND-INDUCED BUILDING MOTIONS

Designers of wind-sensitive buildings have long recognized that perception of building motion under the action of wind may be described by various physical quantities including maximum displacement, velocity, acceleration, and rate of change of acceleration (sometimes called "jerk"). Acceleration has become the standard for evaluation because it is readily measured in the field and can be calculated analytically. Human response to building motion is a complex phenomenon involving many psychological arid physiological factors. Perception and tolerance thresholds of acceleration as a measure of building motion are known to depend on factors such as frequency of the building, occupant gender, age, body posture (sitting, standing, or reclining), body orientation, expectation of motion, body movement, visual cues, acoustic clues, and the type of motion (translational or torsional). Different thresholds and tolerance levels exist for different people and responses can be very subjective. It is known that some people will become accustomed to building motion and tolerate higher levels than others. Limited research exists on this subject but certain standards have been applied for design as discussed in the following.

Acceleration in wind-sensitive buildings maybe expressed as either root mean square (RMS) or peak acceleration. Both measures are used in practice and there is no clear agreement as to which is the more appropriate measure of motion perception. Target peak accelerations of 21 milli-g (0.021 times the acceleration of gravity) for commercial buildings (occupied mostly during daylight hours) and 15 milli-g for residential buildings (occupied mostly during the entire day) under a 10 year mean recurrence interval wind storm have been successfully used in practice for many tail building designs. The target is generally more strict for residential buildings because of the continuous occupancy, the perception that people are less sensitive and more tolerant at work than at home, the fact that there is more turnover in commercial buildings, and the fact that commercial buildings are more easily evacuated for peak wind events.

It is important to recognize that perception to building motion is strongly influenced by building mass and available damping as well as stiffness. For this reason, building drift limits by themselves should not be used as the sole measure of controlling building motion. Damping levels for use in building motion under wind events are generally taken as approximately 1% of critical damping for steel buildings, 1.5% for composite buildings, and 2.0 for reinforced concrete buildings.

Every building must satisfy a strength criterion typically requiring each member be sized to carry its design load without buckling, yielding, or fracture. It should also satisfy the intended function (serviceability) without excessive deflection and vibration. While strength requirements are traditionally specified, serviceability limit states are generally not included in building codes. The reasons for not codifying the serviceability requirements are several: Failure to meet serviceability limits is generally no catastrophic, is a matter of judgment as to the requirements' application, and entails the perceptions and expectations of the user or owner, and because the benefits themselves are often subjective and difficult to quantify. However, the fact that serviceability limits are not codified should not diminish their importance. A building that is designed for code loads may nonetheless be too flexible for its occupants, due to lack of deflection criteria. Excessive building drifts can cause safety-related frame stability problems because of large $P\Delta$ effects. It can also cause portions of building cladding to fall, potentially injuring pedestrians below.

Perception of building motion under the action of wind is a serviceability issue. In locations where buildings are close together, the relative motion of an adjacent building may make occupants of the other buildings more sensitive to an otherwise imperceptible motion. Human response to building motions is a complex phenomenon encompassing many physiological and psychological factors.

Some people are more sensitive than others to building motions. Although building motion can be described by various physical quantities, including maximum values of velocity, acceleration, and rate of change of acceleration—sometimes called jerk—it is generally agreed that acceleration, especially when associated with torsional rotations, is the best standard for evaluation of motion perception in tall buildings. A commonly used criterion as noted previously is to limit the acceleration of a building's upper floors to no more than say, 2.0% of gravity (20 milli-g) for a 10-year return period. The building motions associated with this acceleration are believed to not seriously affect the comfort and equanimity of the building's occupants.

The design of tall, slender buildings can be strongly influenced by the need to keep the wind-induced motions within levels that are acceptable to the occupants. Perception of building motions under the action of wind can be described by various physical quantities including maximum values of velocity, acceleration, and the rate of change of acceleration sometimes referred to as jerk. Human response to motion in buildings is a complex phenomenon involving several psychological and physiological factors. It is unlikely that human beings are directly sensitive to velocity if they are isolated from visual effects because, once in motion at constant velocity, no forces act on the human body to keep it in such motion. Acceleration, on the other hand, requires a force to act that stimulates various body organs and senses. This changing acceleration is an important component of motion perception in tall, slender buildings. Consequently it has become the widely used parameter for the evaluation of motion perception in buildings. It is the standard for comparison and establishment of motion perception guidelines of various countries and international organizations.

Certain criteria for wind-induced accelerations in tall buildings are defined as limits that should not be exceeded more than once in a particular return period. The Council on Tall Buildings and Urban Habitat (CTBUH) recommends 10-year peak resultant accelerations of 10–15 milli-g for residential buildings, 15–20 milli-g for hotels, and 20–25 milli-g for office buildings. Generally, more stringent requirements are suggested for residential buildings, which would have continuous occupancy in comparison to office buildings usually occupied only part of the time and whose occupants have the option of leaving the building in advance of a wind storm.

The target criterion commonly used for residential buildings is for the 10-year return period peak acceleration not to exceed 15 milli-g. However, on some of the extremely slender towers this proves difficult to achieve structurally. Even after doing all that it is practically possible in terms of adding stiffness and mass the results of wind tunnel tests may come in as high as 18 milli-g. It seems the remaining measure that can be taken is to install a supplementary damping system.

One of the basic reasons for conducting aero-elastic study is to evaluate the effect of building motions on the comfort of its occupants. It is generally known that quantitative prediction of human discomfort is difficult if not impossible to den in absolute terms because perception of motion and associated discomfort are subjective by their very nature. However, in practice certain thresholds of comfort have been established by relating acceleration due to building motion at top floors to the frequency of windstorms. One such criterion is to limit accelerations of top floors to 20-milli-g (2% of acceleration due to gravity) in a 10-year wind storm.

In wind-tunnel tests, accelerations are measured directly by accelerometers. Two are typically used to measure components in the x and y directions, while a third records the torsional component. Peak acceleration is evaluated from the expression

$$a = G_p a_x^2 + a_y^2 + a_z^2$$

where

a is the peak acceleration

G_p is a peak factor for acceleration, usually in the range of 3.0–3.5

a_x and a_y is the accelerations due to the sway components in the jc and **y** directions

a_z is the acceleration due to torsional component

The peak accelerations measured for a series of wind directions and speeds are combined with the meteorological data to predict frequency of occurrence of human discomfort, for various levels of accelerations. A commonly accepted criterion is that for human comfort, the maximum acceleration in upper floors should not exceed 2.0% of gravitational acceleration for a 10-year return period storm.

4.15 BUILDING PERIODS

Many simple formulas have been proposed over the years to estimate a tall building's fundamental periods, for preliminary design purposes, until a rigorous calculation can be made. Sometimes this rough approximation is the only calculation of fundamental period that can be made when the wind-tunnel engineer is asked to perform a "desktop" prediction of the eventual loads. The formulas

$$(1)\ T_1 = H/100 \quad (2)\ T_1 = H/75 \quad (3)\ T_1 = H/150 \quad (4)\ T_1 = H/200,\ T_1 = H/164$$

Where
T_1 is the fundamental period
H is the building height in feet

Wind-tunnel engineers are typically hesitant to "out-guess" the design engineer or substitute their own estimate of the structure's period. They are most likely to produce loads consistent with the modal properties provided by the engineer. So, this is an issue worthy of further research. Until then, it is appropriate for discussion between the wind-tunnel engineer and design engineer.

Another consideration that goes hand-in-hand with the determination of building periods is the value of damping for the structure. Damping for buildings is any effect that reduces its amplitude of vibrations. It results from many conditions ranging from the presence of interior partition walls, to concrete cracking, to deliberately engineered damping devices. While for seismic design, five percent of critical damping is typically assumed for systems without engineered damping devices, the corresponding values for wind design are much lower as buildings subject to wind loads generally respond within the elastic range as opposed to inelastic range for seismic loading. The additional damping is assumed to come from severe concrete cracking and plastic hinging.

The ASCE 7-10 commentary suggests a damping value of one percent for steel buildings and two percent for concrete buildings. These wind damping values are typically associated with determining wind loads for serviceability check. Without recommending specific values, the commentary implicitly suggests that higher values may be appropriate for checking the survivability states.

So, what values are design engineers supposed to use for ultimate level wind loads? Several resources are available that provide values of damping for service and ultimate loads. The values vary greatly depending upon which reference cited in the commentary is used. Depending upon the type of lateral force resisting system the damping value may vary from a low of 0.5% to a high of 16% or more.

Although the level of damping has only a minor effect on the overall base shear for wind design for a large majority of low and mid-rise buildings, where serviceability criteria govern, such as accelerations for tall buildings, a more in-depth study of damping criteria is typically warranted.

The computation of the fundamental building period is essential for calculating lateral load effects due to both seismic and wind forces. Code prescribed empirical equations for calculating approximate building periods are now provided for both seismic and wind design. The following paragraphs explain the background and assumptions behind these equations.

While the use of the fundamental building period for seismic design calculations is well established, the parameters used for wind design have not been as clear. For wind design, the building period is only relevant for those buildings designated as "flexible" (having a fundamental building period exceeding 1 s). When a building is designated as flexible, the natural frequency (inverse of the building's fundamental period) is introduced into the gust-effect factor, G_f.

Prior to ASCE 7-05, designers typically used either the approximate equations within the seismic section or the values provided by a computer eigenvalue analysis. The first can actually be un-conservative because the approximate seismic equations are intentionally skewed toward shorter building periods. Thus for wind design, where longer periods equate to higher base shears, their use can provide potentially un-conservative results. Also, the results of an eigenvalue analysis can yield building periods much longer than those observed in actual tests, thus providing potentially overly conservative results.

We now have new formulas in the ASCE 7-10 commentary, which are based solely on the building height. Some of these equations, according to some wind engineers, are quite un-conservative by as much as a factor of 2, and therefore, not appropriate for wind studies.

4.16 PEDESTRIAN WIND STUDIES

A sheet of air moving over the earth's surface is reluctant to rise when it meets an obstacle such as a tall building. If the topography permits, it prefers to flow around the building rather than over it. There are good physical reasons for this tendency, the predominant one being that wind, if it has to pass an obstacle, will find the path of least resistance, that is, a path that requires minimum expenditure of energy. As a rule, it requires less energy for wind to flow around an obstacle at the same level than for it to rise. Also, if wind has to go up or down, additional energy is required to compress the column of air above or below it. Generally, wind will try to seek a gap at the same level. However, during high winds when the air stream is blocked by the broadside of a tall, flat building, its tendency is to drift in a vertical direction rather than to go around the building at the same level; the circuitous path around the building would require expenditure of more energy. Thus, wind is driven in two directions. Some of it will be deflected upward, but most of it will spiral to the ground, creating a so-called standing vortex or mini tornado at sidewalk level.

Buildings and their smooth walls are not the only victims of wind buffeting. Pedestrians who walk past tall, smooth-skinned skyscrapers may be subjected to what is called the Mary Poppins syndrome, referring to the tendency of the wind to lift the pedestrian literally off his or her feet. Another effect, known humorously as the Marilyn Monroe effect, refers to the billowing action of women's skirts in the turbulence of wind around and in the vicinity of a building. The point is that during windy days, even a simple activity such as crossing a plaza or taking an afternoon stroll becomes an extremely unpleasant experience to pedestrians, especially during winter months in cold climates. Walking may become irregular, and the only way to keep walking in the direction of the wind is to bend the upper body windward (See Figure 4.8a through d).

Although one can get some idea of wind flow patterns from the preceding examples, analytically it is impossible to estimate pedestrian-level wind conditions in the outdoor areas of building complexes. This is because there are innumerable variations in building location, orientation, shape, and topography, making it impossible to formulate an analytical solution. Based on actual field experience and results of wind-tunnel studies, it is, however, possible to qualitatively recognize situations that adversely affect pedestrian comfort within a building complex.

Model studies can provide reliable estimates of pedestrian-level wind conditions based on considerations of both safety and comfort. From pedestrian-level wind speed measurements taken at specific locations of the model, acceptance criteria can be established in terms of how often wind speed occurrence is permitted to occur for various levels of activity. The criterion is given for both summer and winter seasons, with the acceptance criteria being more severe during the winter months. For example, the occurrence once a week of a mean speed of 15 mph (6.7 m/s) is considered acceptable for walking during the summer, whereas only 10 mph (4.47 m/s) is considered acceptable during winter months.

The pedestrian level wind speed test is usually performed using the same model that was used for the cladding loads test, and may include some landscaping details. The model is instrumented with omni-directional wind speed sensors at various locations around the development where measurements of the mean and fluctuating wind speed are made for a full range of wind angles, usually at 10° intervals.

(a) (b)

(c) (d)

FIGURE 4.8 (a through d) Pedestrian reactions.

The scaling involved is the same as that of the modeled wind flow. Thus, the ratio of wind speed near the ground to a reference wind speed near the top of the building is assumed to be the same in model and full scale, and to be invariant with both test speed and prototype speed. Since the thermal effects in the full scale wind are neglected, strictly speaking, the results are only applicable to neutrally-stable flows, which are usually associated with stronger wind speeds. However, near tall buildings, local acceleration effects due to the local geometry are usually dominant over thermal effects and are also the most important for design considerations.

The measure aerodynamic data is combined with the statistics of the full scale wind climate at the site, using the methodology outlined in Appendix D, to provide predictions of wind speeds at the site. Two types of predictions are typically provided:

1. Wind speeds exceeded for various percentages of the time on an annual basis. Wind speeds exceeded 5% of the time can be compared to comfort criteria for various levels of activity. Very roughly, this is equivalent to a storm even of several hours duration occurring about once per week.
2. Predictions of wind speeds exceed during events or storms with different frequencies of occurrence. Wind speeds exceeded once per year can be compared to criteria for pedestrian safety.

Other, nonquantitative techniques are also available to determine levels of windiness over a project site. One of these techniques is a scour technique in which a granular material is spread uniformly over the area of interest. The wind speed is then slowly increased in increments. The areas where the granular material is scoured away first are the windiest areas, while areas that are scoured later as the wind speed increases represent progressively less windy areas. Photographs of the scour patterns at increasing wind speeds can be superimposed using image processing technology to develop contour diagrams of windiness. This information can be used to determine locations for quantitative measurements, or simply to identify problem area where remedial measures are necessary. Testing several configurations can provide comparative information for use in evaluating the effects of various architectural or landscaping details. The advantage of the scour technique is that it can provide continuous information on windiness over a broad area, as opposed to the quantitative techniques, which provide wind speeds as discrete points.

An even more qualitative technique is to introduce smoke to visualize flow paths and accelerations at arbitrary places. This can be a useful exploratory technique by which to understand the flow mechanisms and how best to alter them.

Pedestrian comfort depends largely on the magnitude of the ground level wind speed regardless of the local wind direction. As a result, quantitative evaluation of the pedestrian level wind environment at the wind tunnel laboratory is normally restricted to measurements of the magnitude of ground level wind speeds unless information on local wind direction is of special interest. Measurements are made of coefficients of wind speed (pedestrian level wind speed as a fraction of an upper level reference speed) for a full range of wind directions at various locations near the site, and in some cases, at a location well away from the building to provide a form of calibration with existing experience.

These wind speed coefficients are subsequently combined with the design probability distribution of gradient wind speed and direction for the area to provide predictions of the full-scale pedestrian level wind environment.

5 Seismic Design with Particular Reference to ASCE 7-10 Seismic Provisions

PREVIEW

Earthquakes in the United States are regional in their occurrence. While California is famous for its earthquakes, other states, such as Texas, have much less concern for the threat of seismic activity. However, structural engineering practice is becoming increasingly national and global, and the engineers in Texas may find that their next project is in California. Thus, it has become necessary for the professional engineer to have some knowledge of the earthquake problem and how design seeks to control it.

Earthquakes have long been feared as one of nature's most terrifying phenomena. Early in human history, the sudden shaking of the earth and the death and destruction that resulted were seen as mysterious and uncontrollable.

We now understand the origin of earthquakes and know that they must be accepted as a natural environmental process, one of the periodic adjustments that the earth makes in its evolution. Arriving without warning, the earthquake can, in a few seconds, create a level of death and destruction that can only be equaled by the most extreme weapons of war. This uncertainty, combined with the terrifying sensation of earth movement, creates our fundamental fear of earthquakes.

Earthquakes are catastrophic events that occur mostly at the boundaries of portions of the earth's crust called tectonic plates. When movement occurs in these regions, along faults, waves are generated at the earth's surface that can produce very destructive effects.

Aftershocks are smaller quakes that occur after all large earthquakes. They are usually most intense in size and number within the first week of the original quake. They can cause very significant reshaking of damaged structures, which makes earthquake-induced disasters more hazardous. A number of moderate quakes (6+ magnitude on the Richter scale) have had aftershocks that were very similar in size to the original quake. Aftershocks diminish in intensity and number with time. They generally follow a pattern of being at least 1 large (within magnitude 2 on the Richter scale) aftershocks, 100 within magnitude 3 on the Richter scale, and so on. The Loma Prieta earthquake had many aftershocks, but the largest was only magnitude 5.0, with the original quake being magnitude 7.1.

Some of the most destructive effects caused by shaking as a result of the earthquake are those that produce lateral loads in a structure. The input shaking causes the foundation of a building to oscillate back and forth in a more or less horizontal plane. The building mass has inertia and wants to remain where it is, and therefore, lateral forces are exerted on the mass in order to bring it along with the foundation. For analysis purposes, this dynamic action is simplified as a group of horizontal forces that are applied to the structure in proportion to its mass and to the height of the mass above the ground. In multistory buildings with floors of equal weight, the loading is further simplified as a group of loads, each being applied at a floor line and each being greater than the one below in a triangular distribution. Seismically resistant structures are designed to resist these lateral forces through inelastic action and must, therefore, be detailed accordingly. These loads are often expressed in terms of a percent of gravity weight of the building and can vary from a few percent to near 50% of gravity weight. There are also vertical loads generated in a structure by earthquake shaking, but these forces rarely overload the vertical load-resisting system. However, earthquake-induced vertical forces have caused damage to structures with high dead load compared to design live load. These vertical forces also increase the chance of collapse due to either increased or decreased compression forces in the columns. Increased compression overloads columns, while decreased compression reduces column bending strength.

In earthquake engineering, we deal with random variables and therefore the design must be treated differently from the orthodox design. The orthodox viewpoint idealizes variables as deterministic. This simple approach is valid when applied to design under only mild uncertainty. But when confronted with the effects of earthquake design, we must contend with appreciable probabilities, particularly so when dealing with building failures that may occur in the near future. Otherwise, all the wealth of this world would prove insufficient to fill out needs; the most modest structures would be fortresses. We must also face uncertainty on a large scale while designing engineering systems—whose pertinent properties are still debated, about whose characteristics we know even less.

Although over the years experience and research have diminished our uncertainties and concerns regarding the characteristics of earthquake motions and its manifestations, it is unlikely, though, that there will be such a change in the nature of knowledge to relieve us of the necessity of dealing openly with random variable. In a way, earthquake engineering is a parody of other branches of engineering. Earthquake effects on structures systemically bring out the mistakes made in design and construction, even the minutest mistakes. Add to this the undeniable dynamic nature of disturbances, the importance of soil structure interaction, and the extremely random nature of it all; it could be said that earthquake engineering is to the rest of the engineering disciplines what psychiatry is to other branches of medicine. This aspect of earthquake engineering makes it challenging and fascinating and gives it an educational value beyond its immediate objectives. If structural engineers are to acquire fruitful experience in a brief span of time, expose them to the concepts of earthquake engineering, even if their interest in earthquake-resistant design is indirect. Sooner or later, they will learn that the difficulties encountered in seismic design are technically intriguing and begin to exercise that nebulous trait called engineering judgment to make allowance for these unknown factors.

Seismic design entrails determination of maximum response of a structure subjected to ground motions that vary in magnitude and direction. Design of structures to withstand earthquakes is technically invigorating because major challenges lie in the prediction of future earthquakes, their character and energy content imparted to the structure at the base, and the dynamic behavior of the structure itself.

5.1 ASCE 7-10, CHAPTER 11, SEISMIC DESIGN CRITERIA

Chapter 11 of ASCE 7-10 includes introductory material required for establishing seismic design requirement for structures assigned to SDC, A through F. It defines how to construct a general design response spectrum by using acceleration parameters S_s and S_1 and explains the procedures for establishing the seismic importance factor, I_E, and the SDCs, A through F.

The limitation of citing SDC E and F buildings is given in the final sections of Chapter 11 followed by the requirements for the investigation of the building site for potential geologic and seismic hazards.

Tucked in between the introductory material and the requirements of geotechnical investigation are the seismic design requirements for SDC A buildings. If the building being designed is assigned to SDC A, the designer is not required to comply with the requirements of other seismic chapters.

The definition for the MCE ground motion has been changed to risk-targeted maximum considered earthquake (MCE_R) ground motion. MCE_R is the most severe earthquake effects that result in maximum response to horizontal ground motion *with* adjustment for targeted risk.

A new definition of MCE_G has been added. MCE_G is the most severe earthquake effects for geometric peak ground acceleration and *without* adjustment for targeted risk and is used for evaluation of liquefaction, lateral spreading, seismic settlements, and other soil-related issues.

When prescribed wind loading governs the stress or drift design, the resisting system still must conform to the special requirements for seismic-force-resisting systems. This is required in order to resist, in a ductile manner, potential seismic loads in excess of the prescribed wind loads. A proper, continuous load path is an obvious design requirement, but experience has shown that it often is overlooked and that significant damage and collapse can result. The basis for this design requirement is twofold:

1. To ensure that the design has fully identified the seismic-force-resisting system and its appropriate design level
2. To ensure that the design basis is fully identified for the purpose of future modifications or changes in the structure

5.1.1 Alternate Materials and Alternate Means and Methods of Construction

ASCE 7 Seismic Provisions do not provide criteria for the use of all possible materials and their combinations and methods of construction, either existing or anticipated. While not citing specific materials or methods of construction currently available that require approval, ASCE 7 serves to emphasize that the evaluation and approval of alternate materials and methods require a recognized and accepted approval system. The requirements for materials and methods of construction contained within the seismic provisions represent the judgment of the best use of the materials and methods based on well-established expertise and historical seismic performance. It is important that any replacement or substitute can be evaluated with an understanding of all the ramifications of performance, strength, and durability implied by the standard.

5.1.2 Seismic Ground Motion Values, S_s and S_1

Two parameters, S_s and S_1, play a key role in the determination of ground motion values used in seismic design. These are derived using the maps given in the ASCE provisions or by accessing web-based information. The seismicity maps showing contours of 5% damped 0.2 and 1 s spectral acceleration values for the MCE ground motions are based on USGS probabilistic maps. These maps incorporate improved earthquake data in terms of updated fault parameters (such as slip rates, recurrence time, and magnitude), and additional attenuation parameters. The interpolated ground motion for the continuous 48 states by latitude and longitude are calculated using a closer grid spacing in areas of known fault regions.

S_s is the mapped value of the 5% damped MCE spectral response acceleration, for short-period structures on Class B, firm rock, sites. The short-period acceleration has been determined at a period of 0.2 s. This is because 0.2 s is a reasonable representation of the shortest effective period of buildings that are designed by the ASCE provisions considering the effects of soil compliance, foundation rocking, and other factors typically neglected in structural analysis.

Similarly, S_1 is the mapped value of the 5% damped MCE spectral response acceleration at a period of 1 s on site class B. The spectral response acceleration at periods other than 1 s can typically be derived from the acceleration at 1 s. Consequently, these two response acceleration parameters, S_s and S_1, are sufficient to define an entire response spectrum for the period range of importance for most buildings and structures. See Figures 5.1 and 5.2 for mapped ground motion acceleration values S_s and S_1.

In order to obtain acceleration response parameters that are appropriate for sites other than site class B, it is necessary to modify the S_s and S_1 values. This modification is performed using *two coefficients, F_a and F_v. The MCE spectral response accelerations adjusted for site class are designated S_{MS} and S_{M1}, respectively, for short-period and 1 s period response.*

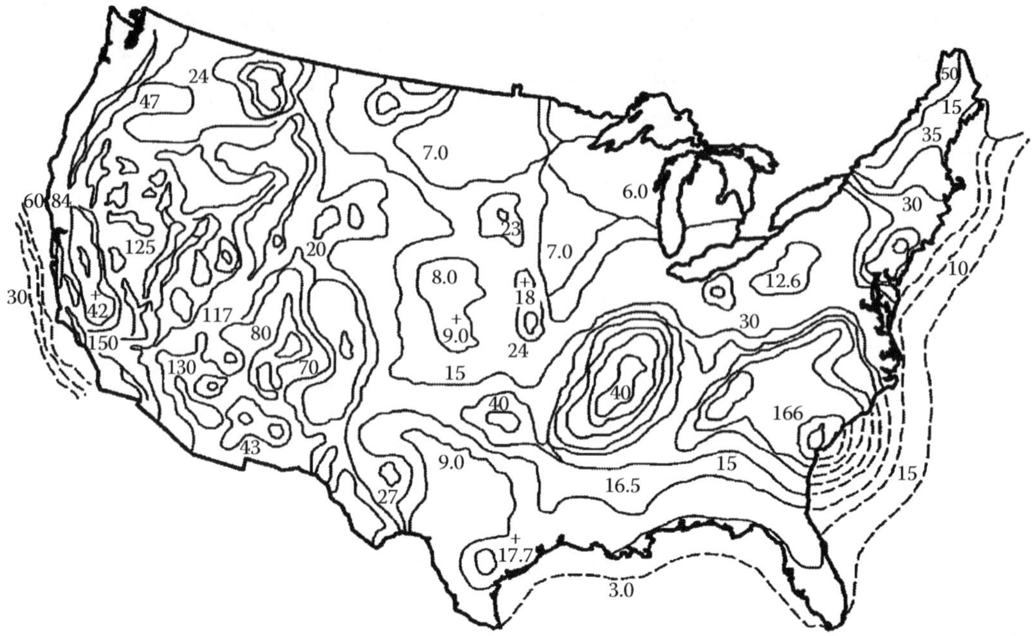

FIGURE 5.1 MCE ground motion for 0.2 s period. Spectral response acceleration, S_s, as a percent of gravity, site class B, 5% critical damping. (From Taranath, B.S., *Structural Analysis and Design of Tall Buildings*, CRC Press, Boca Raton, FL, 2011.)

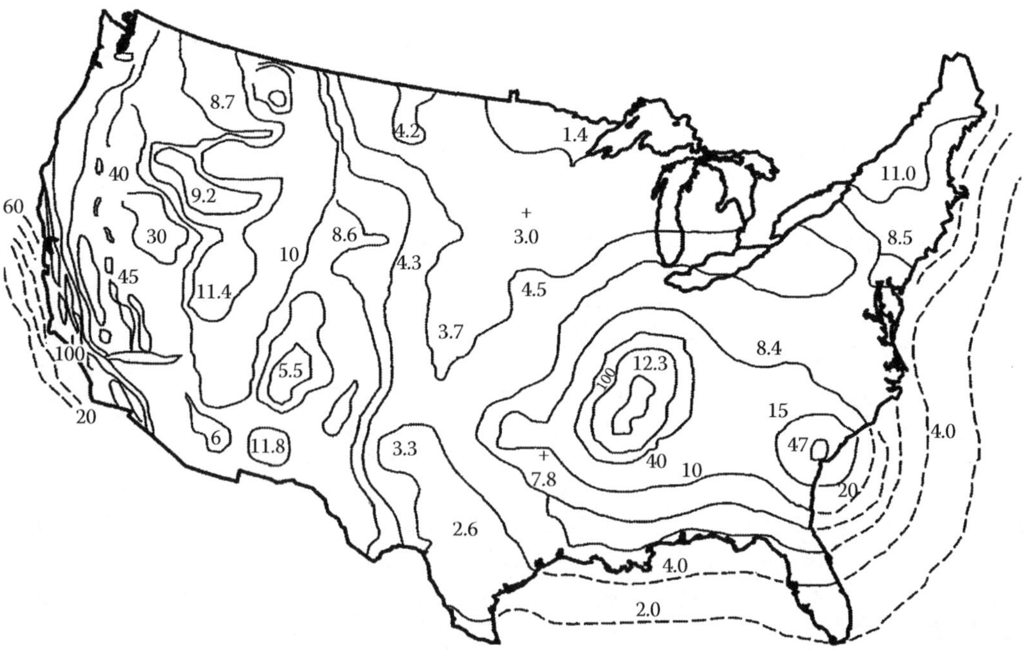

FIGURE 5.2 MCE ground motion for 1.0 s period. Spectral response acceleration, S_1, as a percent of gravity, site class B, 5% critical damping. (From Taranath, B.S., *Structural Analysis and Design of Tall Buildings*, CRC Press, Boca Raton, FL, 2011.)

The approach adopted in ASCE 7-10, Section 11 is intended to provide for a uniform margin against collapse at the design ground motion. In order to accomplish this objective, ground motion hazards are defined in terms of MCE ground motions, which are based on a set of rules that depend on the seismic hazard of a region. Design ground motions are based on a lower-bound estimate of the margin against collapse inherent in structures designed to the seismic provisions in the standard. This lower bound is based on experience to correspond to a factor of about 1.5 in ground motion. Consequently, the DE ground motion is selected at a ground-shaking level that is 1/1.5 (or 2/3) of the MCE ground motion.

For most regions of the nation, the MCE ground motion is defined with a uniform probability of exceedance of 2% in 50 years (return period of about 2500 years). While stronger shaking than this could occur, it would be economically impractical to design for such very rare ground motions, and therefore, the selection of the 2% probability of exceedance in 50 years as the MCE ground motion would result in acceptable levels of seismic safety.

In regions of high seismicity, such as in many areas of California, the seismic hazard is typically controlled by large-magnitude events occurring on a limited number of well-defined fault systems. Probabilistic ground motions calculated at a 2% probability of exceedance in 50 years can be much larger than deterministic ground motions computed based on characteristic magnitudes of earthquakes on these known active faults. These probabilistic motions are greater if these major active faults produce characteristic earthquakes every few hundred years. For these regions, it is considered more appropriate to determine MCE ground motions directly by deterministic methods based on the characteristic earthquake of these defined faults. In order to provide an appropriate level of conservatism in the design process when the deterministic approach is used to calculate MCE ground motion, the median ground motion estimate for the characteristic event is multiplied by 1.5.

Probabilistic estimates of ground shaking at a given site can also be determined from a probabilistic ground-shaking analysis for the site (often termed a *probabilistic seismic hazard analysis* or PSHA), whereby a geotechnical engineer determines and integrates contributions to the probability of exceedance of a ground motion level from all earthquake faults and magnitudes that could produce potentially damaging ground shaking at the site. Using the results from PSHA, ground motions can be readily obtained for any selected probability of exceedance and building design life.

For applications in PBD presented in Chapter 6, probabilistic approach and a deterministic approach for the ground-shaking hazard assessment may be used. Using a probabilistic approach, the seismic hazard can be integrated with the building resistance characteristics to estimate the probability of exceeding some pre-agreed-upon level of damage, during a time period of significance such as the anticipated building life. Using a deterministic approach, the probability of exceeding a specific damage level may be assessed for an earthquake likely to occur during the anticipated building life.

5.1.2.1 Mapped Acceleration Parameters

In the general procedure, these are computed from mapped values of the spectral response acceleration at short periods, S_s and a 1 s S_1, for Class B sites. Although S_s and S_1 values may be obtained directly from Figures 22-1 through 22-14 (in Chapter 221) in ASCE 7-10, it is best to get the values from the USGS website: http://eaithquake.usgs.gove/desigrimaps.

S_s is the mapped value of the 5% damped MCE spectral response acceleration for short-period structures founded on site class B (firm rock) sites. The short-period acceleration has been determined at a period of 0.2 s because 0.2 s is reasonably representative of the shortest effective period of buildings designed using the standard, considering the effects of soil compliance, foundation rocking, and other factors typically neglected in structural analysis.

Similarly, S_1 is the mapped value of the 5% damped MCE spectral response acceleration at a period of 1 s on site class B. The spectral response acceleration at periods other than 1 s typically can be derived from the acceleration at 1 s. Consequently, for MCE ground shaking on site class B sites, these two response acceleration parameters, S_s and S_1, are sufficient to define an entire response spectrum for the period range of importance for most buildings and structures.

5.1.3 Site Coefficients and Adjusted Acceleration Parameters

To obtain acceleration response parameters that are appropriate for sites with classification other than site class B, the S_s and S_1 values must be modified. This modification is performed using two coefficients, F_a and F_v, that respectively scale the S_s and S_1 values determined for site class B to values appropriate for other site classes. The MCE spectral response accelerations adjusted for site class designated S_{MS} and S_{M1}, respectively, for short-period and 1 s period response. As described earlier, structural design in ASCE/SEI 7-10 is performed for earthquake demands that are 2/3 of the MCE response spectra. Therefore, two additional parameters, S_{DS} and S_{D1}, are used to define the acceleration response spectrum for this design-level event. These parameters are 2/3 of the respective S_{MS} and S_{M1} values and define a design response spectrum for sites of any characteristics and for natural periods of vibration less than the transition period, T_L. Values of S_{MS}, S_{M1}, S_{DS}, and S_{D1} can also be obtained from the USGS website cited earlier.

The site coefficients, F_a and F_v, present, respectively, in ASCE 7-10 Tables 11.4-1 and 11.4-2 (the latter of which is given in Table 5.1 below) for the various site classes are based on the results of empirical analyses of strong-motion data and analytical studies of site response.

The amount of ground motion amplification by a soil deposit relative to bedrock depends on the wave-propagation characteristics of the soil, which can be estimated from measurements of inferences of shear wave velocity and in turn the shear modulus for the materials as a function of the level of shaking. In general, softer soils exhibit greater amplifications than stiffer soils. Increased levels of ground shaking result in increased damping, which, in general, reduces the amplification, especially for shorter periods. Furthermore, for soil deposits of sufficient thickness, soil amplification is generally greater at longer periods than at shorter periods.

5.1.3.1 Site Class S_A, S_B, S_C, S_D, S_E, and S_F

A set of six site classifications, S_A through S_F, based on the average properties of the upper 100 ft of soil profile are defined in ASCE-7-10 Table 20.3-1. Since, in practice, geotechnical investigations are seldom conducted to such depths, ASCE-7 allows the geotechnical engineers to determine site class based on site-specific data and professional judgment.

The site class should reflect the soil conditions that will affect the ground motion input to the structure or a significant portion of the structure. For structures receiving substantial ground motion input from shallow soils (e.g., structures with shallow spread footings, laterally flexible piles or structures with basements where it is judged that substantial ground motion input to the structure may come through the side walls), it is reasonable to classify the site on the basis of the

TABLE 5.1
Site Coefficient, F_v

	Mapped MCE Spectral Response Acceleration Parameter at 1 s Period				
Site Class	$S_1 \le 0.1$	$S_1 = 0.2$	$S_1 = 0.3$	$S_1 = 0.4$	$S_1 \ge 0.5$
A	0.8	0.8	0.8	0.8	0.8
B	1.0	1.0	1.0	1.0	1.0
C	1.7	1.6	1.5	1.4	1.3
D	2.4	2.0	1.8	1.6	1.5
E	3.5	3.2	2.8	2.4	2.4
F	See Section 11.4.7, ASCE 7-10				

Source: ASCE 7-10. With permission.
Note: Use straight-line interpolation for intermediate values of S_1.

top 100 ft (30 m) of soils below the ground surface. Conversely, for structures with basements supported on firm soils or rock below soft soils, it is reasonable to classify the site on the basis of the soil or rock below the mat, if it can be justified that the soils contribute very little to the response of the structure.

Buildings on sloping bedrock sites and/or having highly variable soil deposits across the building area require careful study since the input motion may vary across the building (e.g., if a portion of the building is on the rock and the rest is over weak soils). Site-specific studies including 2D or 3D modeling of the soil may be appropriate in such cases to evaluate the subsurface conditions and site and superstructure response. Other conditions that may warrant site-specific evaluation include the presence of low shear wave velocity soils below a depth of 100 ft (30 m), location of the site near the edge of a filled-in basin, or other subsurface or topographic conditions with strong 2D and 3D site response effects.

5.1.4 Design Response Spectrum

The design response spectrum (Figure 5.3) consists of several segments. The constant-acceleration segment covers the period band from T_0 to T_S; response accelerations in this band are constant and equal to S_{DS}. The constant-velocity segment covers the period band from T_S to T_L, and the response accelerations in this band are proportional to $1/T$ with the response acceleration at 1 s period equal to S_{D1}. The long-period portion of the design response spectrum is defined on the basis of the parameter, T_L, the period that marks the transition from the constant-velocity segment to the constant-displacement segment of the design response spectrum. Response accelerations in the constant-displacement segment, where $T \geq T_L$, are proportional to $1/T^2$. Values of T_L are provided on maps in Figures 22-12 through 22-16 of ASCE 7-10.

It should be pointed out that a number of earthquake ground acceleration records are available, and each record could be used to generate a unique response spectrum for a given damping level and soil condition. However, ASCE 7-10 simplifies this process by providing a smoothed design response spectrum based on an assumed damping coefficient and enveloping results from a large

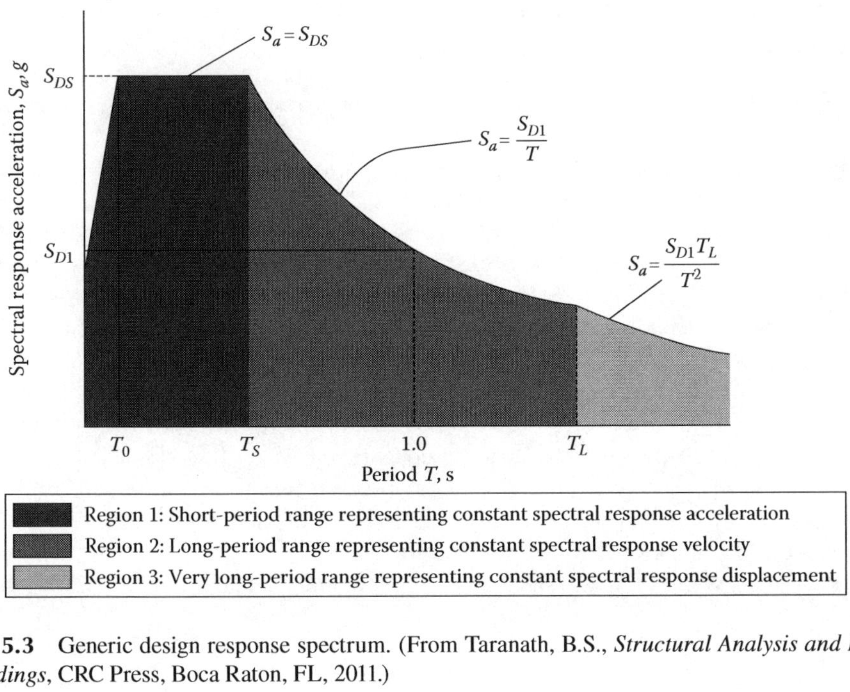

FIGURE 5.3 Generic design response spectrum. (From Taranath, B.S., *Structural Analysis and Design of Tall Buildings*, CRC Press, Boca Raton, FL, 2011.)

number of individual ground motion records. The spectrum is specific to the mapped spectral accelerations, S_s and S_1, and the particular site class. The development of the ASCE 7 design spectra includes a modest amount of damping that is applicable to all building types except base-isolated and artificially damped buildings.

5.1.4.1 Design Base Shear

Design base shear is the total lateral force or shear at the base of the building. It is also equal to the sum of the seismic design forces at each level of a building. The symbol V is used to represent the base shear. For the purpose of calculating the base shear, the base of the building is the level at which the earthquake forces are considered to be imparted to the structure or the level at which the structure, as a dynamic vibrator, is supported.

The total horizontal base shear, V, is calculated from an expression that is essentially of the form

$$F = Ma = \left(\frac{W}{g}\right)a = W\left(\frac{a}{g}\right)$$

where

F is the inertia force equal to the base shear V
M is the mass
a is the acceleration
g is the acceleration of gravity

In ASCE 7-10, the base shear equation is somewhat modified. The (a/g) term is replaced by a *seismic base shear coefficient*, C_s. Thus,

$$V = C_s W$$

where

V is the *base shear* equal to the *strength level* horizontal seismic force acting at the base of the structure
C_s is the seismic base shear coefficient
W is the *weight of structure* that is assumed to contribute to the development of seismic forces

See Figure 5.4 for schematics of tributary weights contributing to the seismic dead-load calculation. For most structures, this weight is simply taken as the dead load. However, in structures where a large percentage of the live load is likely to be present at any given time, it is reasonable to include at least a portion of this live load in the value of W. ASCE 7-10 lists four specific live-load items that are to be included in the weight of the structure, W. For example, in storage warehouse, W is to include at least 25% of the floor live load. In offices and other buildings where the locations of partitions are subject to relocation, ASCE 7 requires that floors be designed for a live load of not less than 15 psf. However, this 15 psf value is to account for localized partition loads, and it is intended to be used only for gravity-load design. For seismic design, it is recognized that the 15 psf loading does not occur at all locations at the same time. Consequently, an average floor load of 10 psf may be used for the weight of partitions in determining W for seismic design. Roof live loads need not be included in the calculation of W, but 20% of the snow load must be included if it exceeds 30 psf.

The ASCE 7-10 formulas for base shear V may be rewritten in a compact form as follows:

$$V = C_s W$$

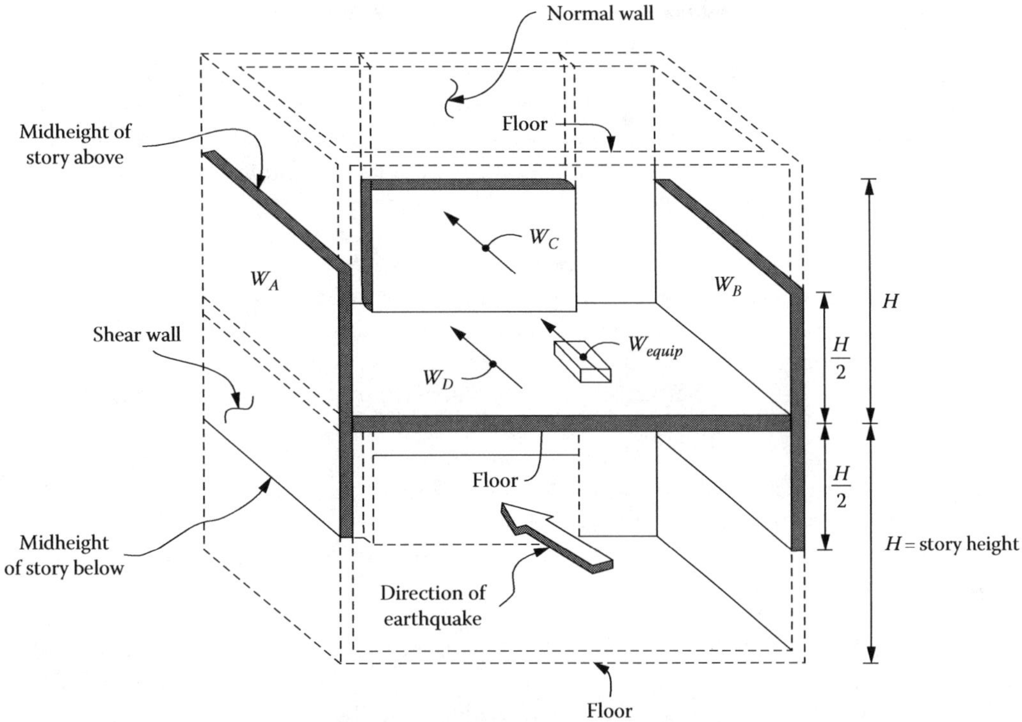

Story weight for calculation of lateral forces

W_x = walls + floor + equipment

$= W_A + W_B + W_C + W_D + W_{equip}$

Weight for design of diaphragm W_{Px} = normal walls + floor + equipment

$= W_C + W_D + W_{equip}$

Note: Floor weight W_D includes floor structure, suspended ceiling, mechanical equipment, and an allowance for partitions.

FIGURE 5.4 Tributary weights for seismic dead-load calculations. (From Taranath, B.S., *Structural Analysis and Design of Tall Buildings*, CRC Press, Boca Raton, FL, 2011.)

$$C_s = \frac{S_{DS}I}{R} = \left(\begin{array}{lll} \leq \dfrac{S_{D1}I}{RT} & \text{for} & T \leq T_L \\ \leq \dfrac{S_{D1}T_L I}{RT^2} & \text{for} & T > T_L \\ = 0.044 \vert S_{DS} & & I \geq 0.01 \\ \geq \dfrac{0.5 S_1 I}{R} & \text{for} & S_1 \geq 0.6\,g \end{array} \right)$$

where

S_{DS} is the short-period design spectral acceleration
S_{D1} is the 1 s design spectral response acceleration
R is the response modification factor
I is the importance factor
T is the building period
S_1 is the mapped 1 s spectral acceleration
T_L is the transition period assigned to regions

It should be noted that ASCE 7 permits the maximum value of S_s used in calculating S_{DS} to be 1.5 g for regular structures of five stories or less and having a period of 0.5 s or less.

Experience in several earthquakes has shown that local soil conditions can have a significant effect on earthquake response. The 1985 Mexico earthquake is a prime example of earthquake ground motions being amplified by local soil conditions. It is perhaps more difficult to visualize, but the soil layers beneath a structure also have a period of vibration T_{soil} similar to the period of vibration of a building, T. Greater structural damage is likely to occur when the fundamental period of the structure is close to the period of the underlying soil. In these cases, a resonance effect between the structure and the underlying soil develops.

Soil–structure resonance is the term used to refer to the amplification of earthquake effects caused by local soil conditions. The soil characteristics associated with a given building site (site specific) are incorporated into the definition of site coefficients F_a and F_v.

As noted previously, ASCE 7 establishes six *site classes* (classes A through F). Assigned to these are different values of site coefficients F_a and F_v are assigned to each class for each tabulated range of mapped spectral acceleration. If a structure is supported directly on hard rock (site class A), then F_a and F_v are 0.8 for all mapped spectral accelerations. However, if the structure rests on softer soil, the earthquake ground motion originating in the bedrock may be amplified. In the absence of a geotechnical evaluation, site class D is the default site class normally permitted for use in determining the site coefficients F_a and F_v.

Shown in Figures 5.5 through 5.7 are response spectrum curves made for specific locations in Los Angeles, CA; Boston, MA; and Seattle, WA. Default site class D is used in determining the site coefficients F_a and F_v.

Figures 5.5 through 5.7 show the spectral accelerations for buildings of varying periods. The level plateau, defined by S_{DS}, can be considered to apply to stiffer buildings. For periods above T_S, the downward trend of the curve shows that as buildings become more flexible, they tend to experience lower seismic forces. On the other hand, the more flexible buildings will experience greater deformations, and therefore damage to finishes and contents could become a problem.

5.1.5 Site-Specific Ground Motion Procedures

The objective of a site-specific ground motion analysis is to determine ground motions for local seismic and site conditions with higher confidence than is possible using the general procedure

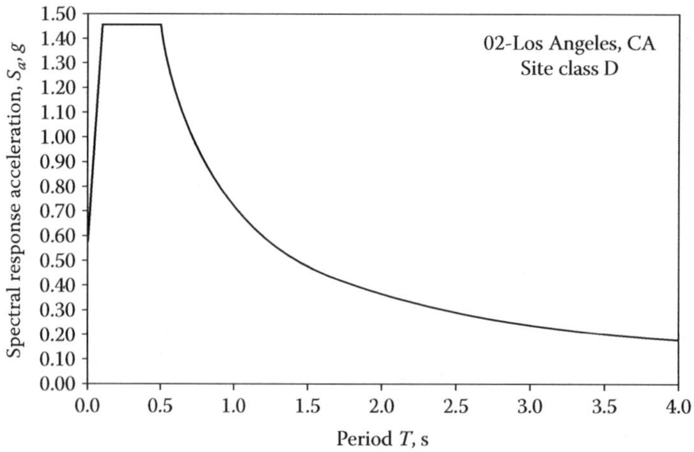

FIGURE 5.5 Response spectrum for a specific site in Los Angeles, CA, latitude 34°3′N, longitude 118°14′W, site class D. (From Taranath, B.S., *Structural Analysis and Design of Tall Buildings*, CRC Press, Boca Raton, FL, 2011.)

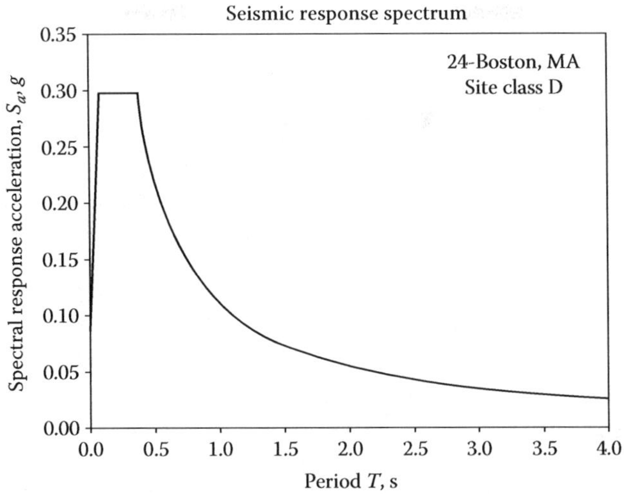

FIGURE 5.6 Response spectrum for a specific site in Boston, MA, latitude 42°22′N, longitude 71°2′W, site class D. (From Taranath, B.S., *Structural Analysis and Design of Tall Buildings*, CRC Press, Boca Raton, FL, 2011.)

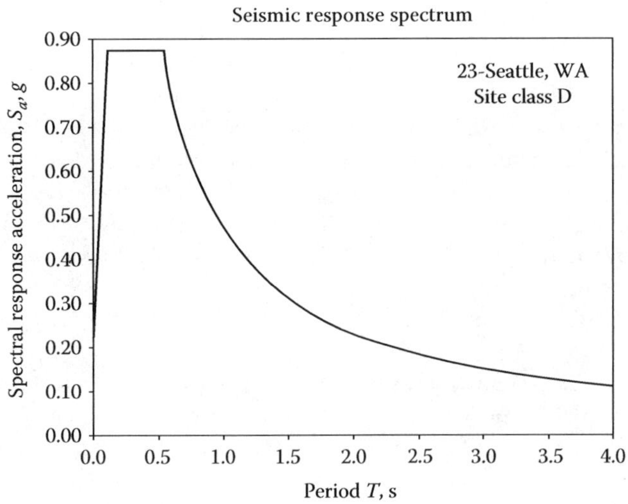

FIGURE 5.7 Response spectrum for a specific site in Seattle, WA, latitude 47°39′N, longitude 122°18′W, site class D.

of the ASCE 7-10 provisions. Accordingly, such studies must be comprehensive and incorporate current scientific interpretations. Because there is typically more than one scientifically credible alternative for models and parameter values used to characterize seismic sources and ground motions, it is important to formally incorporate these uncertainties in a site-specific probabilistic analysis. For example, uncertainties may exist in seismic source location, extent, and geometry; maximum earthquake recurrence rate; choices for ground motion attenuation relationships; and local site conditions including soil layering and dynamic soil properties as well as possible 2D or 3D wave-propagation effects. The use of a peer review for a site-specific ground motion analysis is encouraged.

Near-fault effects on horizontal response spectra include (1) directivity effects that increase ground motions for periods of vibration greater than approximately 0.5 s for fault rupture propagating toward the site and (2) directionality effects that increase ground motions for periods greater than approximately 0.5 s in the direction normal (perpendicular) to the strike of the fault.

ASCE 7-10 requires that site-specific geotechnical investigations and dynamic site response analysis be performed for sites having site class F soils.

For purposes of obtaining data to conduct a site response analysis, site-specific geotechnical investigations should include borings with sampling, standard penetration tests (SPTs) for sandy soils, cone penetrometer tests (CPTs), and/or other subsurface investigative techniques and laboratory soil testing to establish the soil types, properties, and layering and the depth to rock or rocklike material. For very deep soil sites, the depth of investigation need not necessarily extend to bedrock but to a depth that may serve as the location of input motion for a dynamic site response analysis. It is desirable to measure shear wave velocities in all soil layers. Alternatively, shear wave velocities may be estimated based on shear wave velocity date available for similar soils in the local area or through correlations with soil types and properties.

The development of a site-specific design response spectrum requires reviewing seismological and geological data and performing engineering analyses. ASCE 7-10 allows development of a site-specific design spectrum using probabilistic seismic hazard and site response analyses. However, it should be noted that the design spectrum for structural analysis may not be less than 80% of the general response spectrum developed using Section 11.4 of ASCE 7-10.

5.1.6 Importance Factor and Occupancy Category

Large earthquakes are rare events that will include severe ground motions. Such events are expected to result in damage to structures even if they were designed and built in accordance with the minimum requirements of the standard. The consequence of structural damage or failure is not the same for the various types of structures located within a given community. Serious damage to certain classes of structures, such as critical facilities (e.g., hospitals), will disproportionally affect a community. The fundamental purpose of importance factor and occupancy category is to improve the ability of a community to recover from a damaging earthquake by tailoring the seismic protection requirements to the relative importance of that structure. The purpose is achieved by requiring better performance of those structures that

1. Are necessary to response and recovery efforts immediately following an earthquake
2. Present the potential for catastrophic loss in the event of an earthquake
3. House a very large number of occupants or occupants less able to care for themselves than the average

The first basis for seismic design in the standard is that structures will have a suitably low likelihood of collapse in the very rare event defined as the MCE ground motion. A second basis is that life-threatening damage, primarily from failure of nonstructural elements in and on structures, will be unlikely in an unusual but less rate earthquake ground motion, which is given as the DE ground motion (defined as two-thirds of the MCE). Given the occurrence of ground motion equivalent to the MCE, a population of structures built to meet these design objectives will probably still experience substantial damaging in many structures, rendering these structures unfit for occupancy or use.

Experience in past earthquakes around the world has demonstrated that there will be an immediate need to treat injured people, to extinguish fires and prevent conflagration, to rescue people from severely damaged or collapsed structures, and to provide sustenance to a population deprived of its normal means. Experience also has shown that these needs are best met when structures essential to response and recovery activities remain functional.

ASCE 7-10 addresses these objectives by requiring that each structure be assigned to one of the four occupancy categories and by assigning an importance factor to the structure based upon that occupancy category. The occupancy category is then used as one of the two components in determining the SDC and is a primary factor in setting drift limits for building structures under the DE ground motion.

The expected performance of structures is controlled by assignment of each structure to one of four occupancy categories. The ASCE provisions specify progressively more conservative strength, drift control, system selection, and detailing requirements for structures assigned to higher occupancy groups. The purpose is to attain minimum levels of earthquake performance deemed suitable to the individual occupancies. In terms of postearthquake recovery, certain types of occupancies are vital to public needs. These special occupancies are identified in ASCE 7-10 and given specific recognition. In terms of disaster preparedness, regional communication centers identified as critical emergency services are placed in a higher classification than retail stores, office buildings, and factories.

5.1.6.1 Importance Factor I_E

The purpose of this factor, I_E (referred to as I without the subscript E in ASCE 7-10), is to improve the capability of certain types of buildings such as essential facilities and structures containing substantial quantities of hazardous materials to function during and after DEs. This is achieved by introducing an importance factor of 1.25 for occupancy category III structures and 1.5 for category IV structures. This factor is intended to reduce the ductility demands and result in less damage. When combined with the more stringent drift limits, the result is improved performance of such facilities.

ASCE 7-05, Table 1.1 gives the seismic importance factors I_E. Also shown therein are the wind importance factors, I_{w1}, for wind in nonhurricane regions and I_{w2} for wind in hurricane-prone regions.

A value of I_E greater than unity has the effect of reducing the ductility expected of a structure. However, added strength due to higher design forces by itself is not sufficient to ensure superior seismic performance. Connection details that assure ductility, quality assurance procedures, and limitations on building deformation are also important to improve the functionality and safety in critical facilities and those with high-density occupancy. Consequently, the reduction in the damage potential of critical facilities is also addressed by using more conservative drift controls and by providing special design and detailing requirements and material limitations. The assignment of the importance factor, I_E, is not the sole responsibility of the structural engineer. It is a decision to be made concurrently with the building owners, the architect, and the building official.

Specific consideration is given to essential facilities required for postearthquake recovery. Also included are structures that contain substances deemed to be hazardous to the public. It is at the discretion of the authority having jurisdiction to identify which structures are required for post-earthquake response and recovery.

Although the ASCE provisions explicitly require design for only a single level of ground motion, it is expected that structures designed and constructed in accordance with the provisions will generally be able to meet a number of performance criteria, when subjected to earthquake ground motions of differing severity. Occupancy category I, II, or III structures located where the mapped spectral response acceleration parameter at 1 s period, S_1, is greater than or equal to 0.75 shall be assigned to SDC F. Occupancy category IV structures located where the mapped spectral response acceleration parameter at 1 s period, is greater than or equal to 0.75 shall be assigned to SDC F. All other structures shall be assigned to an SDC based on their occupancy category and the design spectral response acceleration parameters, S_{DS} and S_m. Each building and structure shall be assigned to the more adverse SDC, irrespective of the fundamental period of vibration of the structure, T.

The importance factor, I, is used to quantify criteria for strength. In most of those quantitative criteria, the importance factor is shown as a divisor on the factor R or R_p in order to send a message to designers that the objective is to reduce damage from important structures in addition to preventing collapse in larger ground motions. The R and R_p factors adjust the computed linear elastic response to a value appropriate for design.

5.1.6.2 Protected Access for Category IV Structures

Those structures *considered* essential facilities for response and recovery efforts must be accessible to carry out their purpose. For example, if the collapse of a simple canopy at a hospital could lock ambulances from the emergency room admittance area, the canopy must meet the same structural standard as the hospital. This requirement must be considered in the siting of essential facilities in densely built urban areas.

Where operational access to an occupancy category IV structure is required through an adjacent structure, the adjacent structure shall conform to the requirements for occupancy category IV structures. Where operational access is less than 10 ft from an interior lot line or another structure on the same lot, protection from potential falling debris from adjacent structures shall be provided by the owner of the occupancy category IV structure.

The value for occupancy category III buildings, which includes buildings with an occupancy greater than 5000 and college buildings with a capacity greater than 500 students, is now 1.25, while in previous codes, it has been 1.0. There is new emphasis in attempting to control the amount of ductility demand for these occupancies.

5.1.7 Seismic Design Categories

The earthquake limit state is based upon system performance, and considerable energy dissipation through repeated cycles of inelastic straining is assumed. The reason for dependence on inelastic excursions is entirely due to economics. The associated cost of providing enough strength to maintain linear elastic response in ordinary buildings is simply too much. However, structures that include facilities critical to postearthquake operations—such as hospitals, fire stations, and communication centers—must not only survive without collapse but must also remain operational after an earthquake. Therefore, in addition to life safety, damage control is also a design consideration for structures deemed vital to postearthquake function. The current requirements of achieving this goal are to increase the magnitude of design forces by a factor of 1.25 or 1.5 depending upon the nature of occupancy. For certain structures such as nuclear facilities, yielding cannot be tolerated, and as such, the design needs to be based on forces determined by elastic analysis.

ASCE-7 establishes five SDCs that are key for establishing design requirements for any building or structure. The SDC A, B, C, D, E, or F is established, using the short-period and 1 s period response parameters, S_{DS} and S_{D1}, and the occupancy category.

SDC A represents structures in regions where anticipated ground motions are minor, even for very long return periods. For such structures, ASCE-7-10 requires only that a complete LFRS be provided and that all elements of the structure be tied together. A nominal design base shear equal to 1% of the weight of the structure is used to proportion the lateral system.

SDC B includes structures in regions of seismicity where only moderately destructive ground shaking is anticipated. In addition to the requirements for SDC A, structures in SDC B must be designed for forces determined using ASCE 7-10 seismic maps.

SDC C includes buildings in regions where moderately destructive ground shaking may occur. The use of some structural systems is limited and some nonstructural components must be specifically designed for seismic resistance.

SDC D includes structures located in regions expected to experience destructive ground shaking, but not located very near major active faults. In SDC D, severe limits are placed on the use of some structural systems, and irregular structures must be subjected to dynamic analysis techniques as part of the design process.

SDC E includes structures in regions located very close to major active faults and SDC F includes occupancy category IV structures in these locations. Very severe limitations on systems, irregularities, and design methods are specified for SDC E and F. For the purpose of determining if a structure is located in a region that is very close to a major active fault, ASCE-10 uses a trigger of mapped MCE spectral response acceleration at 1 s periods of 0.75 g or more regardless of the structure's fundamental period. The mapped short-period acceleration, S_s, is not used for this purpose because short-period response accelerations do not tend to be affected by near-source conditions as strongly as do response accelerations at long periods.

Refer to ASCE 7-10, Tables 11.6-1 and 11.6-2 and Table 5.2 below for SDCs. It should be noted that each building shall be assigned to the more severe SDC in accordance with these tables, irrespective of the fundamental period of vibration, T, of the building.

SDCs provide a means to step progressively from simple, easily performed design and construction procedures and minimums to more sophisticated, detailed, and costly requirements as both the level of seismic hazard and the consequence of failure escalate.

In this regard, SDCs perform one of the functions of the seismic zones used in earlier US building codes and still in use throughout much of the world. However, SDCs also are dependent on a building's occupancy and, therefore, its desired performance. Further, unlike the traditional implementation of seismic design, the ground motions used to define the SDCs include the effects of individual site conditions on probable ground-shaking intensity.

Except for the lowest level of hazard (SDC A), the SDC also depends on the occupancy categories. For a given level of ground motion, the SDC is one category higher for occupancy category IV structures than for lower-risk structures. This has the effect of increasing the confidence that the design and construction requirements will deliver the intended performance in the extreme event.

In sum, at the MCE level, SDC A structures should not see motions that are normally destructive to structural systems, whereas the MCE level motions for SDC D structures can destroy vulnerable structures. The grouping of step function requirements by SDC is such that there are a few basic structural integrity requirements imposed at SDC A graduating to a more stringent requirements at SDC D based upon observed performance in past earthquakes analysis.

TABLE 5.2
SDC Based on S_{D1} and S_1

			S_{D1}				S_1
Nature of Occupancy (Typical Examples)	**Occupancy Category**	**Importance Factors**	$S_{D1} <$ 0.067	$0.067 \le S_{D1}$ <0.133	$0.133 \le S_{D1}$ <0.20	$0.20 \le S_{D1}$	$S_1 \ge$ 0.75
Agricultural, temporary facilities, and minor storage facilities	I	$I_w = 0.83$ or 0.77 $I_E = 1.0$	A	B	C	D	E
Typical residential and office buildings	II	$I_w = 1.0$ $I_E = 1.0$	A	B	C	D	E
Schools, colleges, fire and police stations, detention facilities, and buildings containing toxic substances	II	$I_w = 1.15$ $I_E = 1.25$	A	B	C	D	E
Hospitals, health-care facilities, and designated emergency shelters	IV	$I_w = 1.15$ $I_E = 1.50$	A	C	D	D	E

Source: Taranath, B.S., *Structural Analysis and Design of Tall Buildings*. CRC Press, Boca Raton, FL, 2011.

The nature of ground motions within a few kilometers of a fault can be very different from more distant motions. For example, some near-fault motions tend to be highly destructive to irregular structures even if they are well detailed. For ordinary occupancies, the boundary between SDCs D and E is set to define sites likely to be close enough to a fault that these unusual ground motions may be present. Note that this boundary is defined in terms of mapped bedrock outcrop motions affecting response at 1 s, not site-adjusted values, in order to better discriminate between sites near and far from faults. Short-period response is not normally affected as the longer-period response. The additional design criteria imposed on structures in SDCs E and F specifically are intended to provide acceptable performance under these very tense near-fault ground motions.

For most buildings, the SDC is determined without consideration of the building's period. Structures are assigned to an SDC based on the more severe condition determined from 1 s acceleration and short-period acceleration. This is done for several reasons. Perhaps the most important of these is that it is often difficult to estimate precisely the period of a structure. In order to avoid misclassifying a building's SDC by inaccurately estimating the structural period, the standard generally requires that the more severe SDC determined on the basis of a short- and long-period shaking be used.

Another reason for this requirement is a desire to simplify building regulation by requiring all buildings on a given soil profile in a particular region to be assigned to the same SDC regardless of the structural type. This has the advantage of permitting uniform regulation of structural system selection, seismic design requirements for nonstructural components, regulated on the basis of SDC, within a community.

Note that classification of a building as an SDC C instead of B or D can have significant impact on the cost of construction. Therefore, the seismic provisions include an exception permitting reliable classification of buildings as having short structural periods on the basis of short-period shaking alone.

5.1.8 Design Requirements for Seismic Design Category A

SDC A is assigned when the MCE_G ground motions are well known to be below those normally associated with hazardous damage. Damaging earthquakes are not unknown or impossible in such regions, however, and ground motions close to such events may be large enough to produce serious damage. Providing a minimum level of resistance reduces both the radius over which the ground motion exceeds structural capacities and resulting damage in such rare events. These are reasons beyond seismic risk for minimum levels of structural integrity.

The requirements for SDC A are all minimum strengths at force levels appropriate for direct use in the strength design load combinations. The two fundamental requirements are a minimum strength for a structural system to resist lateral forces and a minimum strength for connections of structural members.

SDC A represents structures in regions where anticipated ground motions are minor, even for very long return periods. For such structures, the ASCE *provisions* require only that a complete seismic-force-resisting system be provided and that all elements of the structure be tied together. A nominal design force equal to 1% of the weight of the structure is used to proportion the lateral system. The requirements are given in Section 1.4, General Integrity of ASCE 7-10.

It is not considered necessary to specify seismic-resistant design on the basis of an MCE ground motion for SDC A structures because the ground motion computed for the areas where these structures are located is determined more by the rarity of the event with respect to the chosen level of probability than by the level of motion that would occur if a small but close earthquake actually did occur. However, it is desirable to provide some protection against earthquakes and many other types of unanticipated loadings. Thus, the requirements for SDC A provide a nominal amount of structural integrity that will improve the performance of buildings in the event of a possible but rare earthquake even though it is possible that the ground motions could be large enough to cause serious damage or even collapse. The result of design to SDC A requirements is that fewer buildings would collapse in the vicinity of such an earthquake.

Structural integrity is provided by a combination of requirements. First, a complete load path for lateral forces must be designed for a lateral force based on a 1% acceleration of the mass. The minimum connection forces specified for SDC A also must be satisfied.

The 1% value has been used in other countries as a minimum value for structural integrity. For many structures, design for the wind loadings specified in the local building codes normally will control the lateral force design when compared to the minimum integrity force on the structure. However, many low-rise, heavy structures or structures with significant dead loads resulting from heavy equipment may be controlled by the nominal 1% acceleration. Also, minimum connection forces may exceed structural forces due to wind in some structures.

The minimum lateral force for SDC A structures is a structural integrity issue related to the load path. It is intended to specify design forces in excess of wind loads in heavy low-rise construction. The design calculation is simple and easily done to ascertain if it governs or the wind load governs. The ASCE 7-10 provision requires a nominal lateral force, F_x, equal to 1% of the gravity load assigned to a story to assure general structural integrity.

The complete list of the seismic design activities required are as follows:

- Analyze the structure for the effect of static lateral forces, F_x, applied independently in each of the two orthogonal directions.
- Apply the static forces, F_x, at all levels simultaneously.
- Determine the force, F_x, at each level by using the following equation:

$$F_x = 0.01\ W_x$$

where

F_x is the design lateral force applied at level x

W_x is the portion of the dead load, D, located or assigned to level x

- Connect all parts of the structure to form a continuous load path to the LFRS.
- Tie any smaller portion of the structure to the remainder of the structure with connections capable of transmitting at least 5% of the portion's weight.
- Provide a positive connection for resisting a horizontal force acting parallel to each horizontal member such as slab, beam, girder, or truss. The connection shall be adequate to transfer a minimum horizontal force equal to 5% of the dead plus live-load vertical reaction.
- Anchor concrete walls at the roof and all floors. Design the anchors for a horizontal force equal to 5% of the wall weight, but not less than 280 lb per linear foot.

Because of the very low seismicity associated with sites with S_{DS} less than 0.25 and S_{D1} less than 0.10, it is considered appropriate for category A buildings to require only a complete seismic-force-resisting system, good quality of construction materials, and adequate ties and anchorage as specified in this section. Category A buildings will be constructed in a large portion of the United States that is generally subject to strong winds but low earthquake risk. Those promulgating construction regulations for these areas may wish to consider many of the low-level seismic requirements as being suitable to reduce the windstorm risk. Since the seismic provisions consider only earthquakes, no other requirements are prescribed for Category A buildings. Only a complete seismic-force-resisting system, ties, and wall anchorage are required.

Construction qualifying under Category A may be built with no special detailing requirements for earthquake resistance. Special details for ductility and toughness are not required in Category A.

5.1.8.1 Lateral Forces

SDC A buildings are not designed for resistance to any specific level of earthquake ground shaking as the probability that they would ever experience shaking of sufficient intensity to cause life-threatening damage is very low so long as the structures are designed with basic levels of structural integrity. Minimum levels of structural integrity are achieved in a structure by assuring that all elements in the structure are tied together so that the structure can respond to shaking demands in an integral manner and also by providing the structure with a complete seismic-force-resisting system. It is believed that structures having this level of integrity would be able to resist, without collapse, the very infrequent earthquake ground shaking that could affect them. In addition, requirements to provide such integrity provide collateral benefit with regard to the ability of the structure to survive other hazards such as high wind storms, tornadoes, and hurricanes.

The procedure outlined for Category A buildings is intended to be a simple approach to ensuring both that a building has a complete seismic-force-resisting system and that it is capable of sustaining at least a minimum level of lateral force. In this analysis procedure, a series of static lateral forces equal to 1% of the weight at each level of the structure is applied to the structure independently in each of two orthogonal directions. The structural elements of the seismic-force-resisting system then are designed to resist the resulting forces in combination with other loads under the load combinations specified by the building code.

The selection of 1% of the building weight as the design force for SDC A structures is somewhat arbitrary. This level of design lateral force is consistent with prudent requirements for lateral bracing of structures to prevent inadvertent buckling under gravity loads. It is also sufficiently small as to not present an undue burden on the design of structures in zones of very low seismic activity.

The seismic weight W is the total weight of the building and that part of the service load that might reasonably be expected to be attached to the building at the time of an earthquake. It includes permanent and movable partitions and permanent equipment such as mechanical and electrical equipment, piping, and ceilings. The normal human live load is taken to be negligibly small in its contribution to the seismic lateral forces. Buildings designed for storage or warehouse usage should have at least 25% of the design floor live load included in the weight, W. Snow loads up to 30 psf are not considered. Freshly fallen snow would have little effect on the lateral force in an earthquake; however, ice loading would be more or less firmly attached to the roof of the building and would contribute significantly to the inertia force. For this reason, the effective snow load is taken as the full snow load for those regions where the snow load exceeds 30 psf with the proviso that the local authority having jurisdiction may allow the snow load to be reduced up to 80%. How much snow load should be included in W depends upon the expected ice buildup or snow entrapment for a given roof configuration. This is best left to the discretion of the local plan approving authority.

5.1.8.2 Anchorage of Concrete or Masonry Walls

The intent is to ensure that out-of-plane inertia forces generated within a concrete or masonry wall can be transferred to the adjacent roof or floor construction. The transfer can be accomplished only by reinforcement or anchors.

For many buildings, the strength of the LFRS is controlled by wind force but for low-rise buildings of heavy construction with large plan aspect ratio, it is controlled by the minimum seismic force. Note that the requirement is for strength and not for toughness, energy dissipation capacity, or some measure of ductility. The force level is not tied to any postulated seismic ground motion. The boundary between SDCs A and B is based on a spectral response acceleration of 25% of gravity (MCE level) for short-period structures; clearly, the 1% acceleration level is far smaller. For ground motions below the A/B boundary, the spectral displacements generally are on the order of a few inches or less depending on the period. Experience has shown that even a minimal strength is beneficial in providing resistance to small ground motions, and it is an easy provision to implement in design. The low probability of motions greater than the MCE is a factor in taking the simple approach without requiring details that would produce a ductile response. Another factor is that larger design forces are specified for connections between main elements of the lateral force load path.

The minimum connection force is specified in three ways: a general minimum horizontal capacity for all connections; a special minimum for horizontal restraint of beams and trusses in line, which also includes the live load on the member; and a special minimum for horizontal restraint of concrete and masonry walls perpendicular to their plane. The 5% coefficient used for the first two is a simple and convenient value that provides some margin over the minim strength of the system as a whole. The value for anchorage of concrete and masonry walls is simply scaled upward from the value of 200 lb per linear foot traditionally used in past building codes for ASD.

5.1.9 Site Limitations for Seismic Design Categories E and F

Because of the difficulty of designing a structure for the direct shearing displacement of fault rupture and the relatively high seismic activity of SDCs E and F, locating a structure on an active fault having the potential to cause rupture of the ground surface at the structure is prohibited.

5.1.10 Additional Geotechnical Investigation Report Requirements for Seismic Design Categories D, E, and F

The dynamic lateral earth pressure on basement and retaining walls during earthquake ground shaking is considered to be an earthquake load, E, for use in design load combinations. This dynamic earth pressure is superimposed on the preexisting static lateral earth pressure during ground shaking. The preexisting static lateral earth pressure is considered to be an H load.

Liquefaction potential should be evaluated for the DE ground motions consistent with peak ground accelerations of $S_{DS}/2.5$. The occurrence and consequence of geologic hazards for MCE_G peak ground accelerations also should be considered when evaluating structural stability and other pertinent performance criteria.

The geotechnical report shall include the following:

1. The determination of lateral pressures of basement and retaining walls due to earthquake motions.
2. The potential for liquefaction and soil strength loss evaluated for site peak ground accelerations, magnitudes, and source characteristics consistent with the DE ground motions. Peak ground acceleration is permitted to be determined based on a site-specific study taking into account soil amplification effects, or in the absence of such a study, peak ground accelerations shall be assumed equal to $S_s/2.5$.
3. The assessment of potential consequences of liquefaction and soil strength loss, including estimation of differential settlement, lateral movement, lateral loads on foundations, reduction in foundation soil-bearing capacity, increases in lateral pressures on retaining walls, and flotation of buried structures.
4. The discussion of mitigation measures such as, but not limited to, ground stabilization, selection of appropriate foundation type and depths, selection of appropriate structural systems to accommodate anticipated displacements and forces, or any combination of these measures and how they shall be considered in the design of the structure.

5.2 ASCE 7-10, CHAPTER 12, SEISMIC DESIGN REQUIREMENTS FOR BUILDING STRUCTURES*

Structures designed in accordance with the ASCE 7-10 standard are likely to have a low probability of collapse but may suffer serious damage if subjected to the MCE or stronger ground motion. The uncertainty in performance results from variability of both ground motion and structural characteristics.

Earthquakes load structures indirectly. As the ground displaces, a structure follows and vibrates. The vibration produces structural deformations with associated strains and stresses. Computation of

* This section is heavily excerpted from FEMA P-750, *NEHRP Recommended Seismic Provisions for New Buildings and Other Structures:* Part 2, Commentary to ASCE/SEI 7-05. National Institute of Building Sciences, Washington, D.C., 2009. http://www.fema.gov/media-library-data/20130726-1730-25045-5820/femap_750_commentary.pdf.

dynamic response to earthquake ground shaking is complex. The basic methods of analysis employ the common simplification of a response spectrum. A response spectrum for a specific earthquake ground motion approximates the maximum value of response spectrum specified in the provision, which is a smoothed and normalized approximation for many different ground motions.

Although the seismic requirements are stated in terms of forces and loads, there are no external forces applied to the aboveground portion of a structure during an earthquake. The design forces are intended only as approximations to generate internal forces suitable for proportioning the strength of structural elements and for estimating the deformations (when multiplied by the deflection amplification factor, C_d) that would occur in the same structure in the event of design-level (not MCE) ground motion.

5.2.1 Basic Requirements

The basic steps in structural design for acceptable seismic resistance are as follows:

1. Select gravity and seismic-force-resisting systems appropriate to the anticipated intensity of ground shaking, depending on the SDC.
2. Lay out these systems to produce a continuous, regular, and redundant load path so that the structures act as integral units in responding to ground shaking.
3. Analyze a mathematical model of the structure subjected to lateral seismic motions and gravity forces.
4. Proportion members and connections to have adequate lateral and vertical strength and stiffness.

The baseline seismic forces for proportioning structural elements (individual members, connections, and supports) are static horizontal forces derived from a linear elastic response spectrum procedure. A basic requirement is that horizontal motion can come from any direction. For most structures, the effect of vertical ground motions is not analyzed specifically; it is included in an approximate fashion by adjusting the load factors for dead load up and down. Certain conditions requiring more detailed analysis of vertical response are defined for nonstructural components.

Higher levels of seismic analysis are permitted (and encouraged) for any structure and are required for some structures, but lower limits based on the ELF procedures apply. The basic procedure uses response spectra that are representative of, but substantially reduced from, the anticipated ground motions. As a result, at the MCE level of ground shaking, structural elements are expected to yield, buckle, or otherwise behave inelastically.

This approach has substantial historical precedent. In past earthquakes, structures with appropriately ductile, regular, continuous systems designed for reduced forces have performed acceptably. In the standard, such design forces are computed by dividing the forces that would be generated in a structure behaving linearly when subjected to the design ground motion by the response modification coefficient, R, and the design ground motion is taken as two-thirds of the MCE ground motion.

The elastic deformations calculated under these reduced design forces are multiplied by the deflection amplification factor, C_d, to estimate the deformations likely to result from the design ground motion. The amplified deformations are used to assess story drifts and to determine seismic demands on elements of the structure that are not part of the seismic-force-resisting system and on nonstructural components within structures. Where C_d is substantially less than R, the system is considered to have damping greater than the nominal 5% of critical damping.

The seismic-force-resisting system is expected to reach significant yield for forces in excess of the design forces. Significant yield is the point where complete plastification of the most critical region of the structure (e.g., formation of a first plastic hinge in the structure) occurs, not the point where first yield occurs in any member. Figure 5.8 shows the lateral force versus deformation relation for a typical structure. Significant yield is shown as the lowest yield hinge on the force-deformation diagram. With increased lateral loading, additional plastic hinges form, and the resistance increases (following the solid curve) until a maximum is reached. The maximum resistance developed along the curve is substantially higher than that at first significant yield, and the margin is referred to as the overstrength capacity.

ASCE 7-10 provisions contemplate a seismic-force-resisting system with redundant characteristics wherein significant structural overstrength above the level of significant yield can be obtained by plastification at other points in the structure prior to the formation of a complete mechanism. The overstrength obtained by this inelastic action provides the reserve strength necessary for the structure to resist the extreme motions of the actual seismic forces that may be generated by the design ground motion.

The structural overstrength described earlier results from the development of sequential plastic hinging in a properly designed, redundant structure. Several other sources will further increase structural overstrength. First, material overstrength (i.e., actual material strengths higher than the nominal material strengths specified in the design) may increase the structural overstrength significantly. For example, it is known that the mean yield strength of A36 steel is about 30%–40% higher than the minimum specified strength used in design calculations. Second, member design strengths usually incorporate a strength reduction (or resistance) factor, Φ, to produce a low probability of failure under design loading. Third, designers themselves introduce additional overstrength by selecting sections or specifying reinforcing patterns that exceed those required by the computation. Similar situations occur where prescriptive minimums control the design. Finally, the design of many flexible structural systems (e.g., moment-resisting frames) often is controlled by the drift rather than strength limitations of the standard with sections selected to control lateral deformations rather than to provide the specified strength.

The result is that structures typically have a much higher lateral resistance than that specified as a minimum by the provisions and first significant yielding of structures may occur at lateral-load levels that are 30%–100% higher than the prescribed design seismic forces. If provided with adequate ductile detailing, redundancy, and regularity, full yielding of structures may occur at load levels that are two to four times the prescribed design force levels.

Most structural systems have some components that cannot provide reliable inelastic response or energy dissipation. Such components must be designed considering that the actual forces in the structure will be larger than those at first significant yield. Therefore, an overstrength factor is specified to amplify the prescribed forces for use in design of such components. This specified overstrength factor is neither an upper nor a lower bound; it is simply an approximation specified to provide a nominal degree of protection against undesirable behavior.

Figure 5.8 illustrates the significance of design parameters contained including the response modification coefficient, R, the deflection amplification, C_d, and the system overstrength factor, Ω_o. These design values, provided in Table 12.2-1 of ASCE 7 as well as the criteria for story drift and P-delta effects, have been established considering the characteristics of typical properly designed structures. The actual structural overstrength, Ω, often will be less than the tabulated factor, Ω_o. This means that the required ductility, R_d, usually will exceed R/Ω_o. If excessive *optimization* of a structural design is performed with lateral resistance provided by only a few elements, the successive yield hinge behavior depicted in Figure 5.8 will not be able to form. The actual overstrength (Ω) will be small, and use of the design parameters in the standard may not provide the intended seismic performance.

The response modification coefficient, R, represents the ratio of the forces that would develop under the specified ground motion if the structure had entirely linear elastic response to the prescribed design forces (see Figure 5.8). The structure must be designed so that the level of significant yield exceeds the prescribed design force. The ratio, R, expressed as $R = V_E/V_s$, is always larger than 1.0; thus, all structures are designed for forces smaller than those the design ground motion

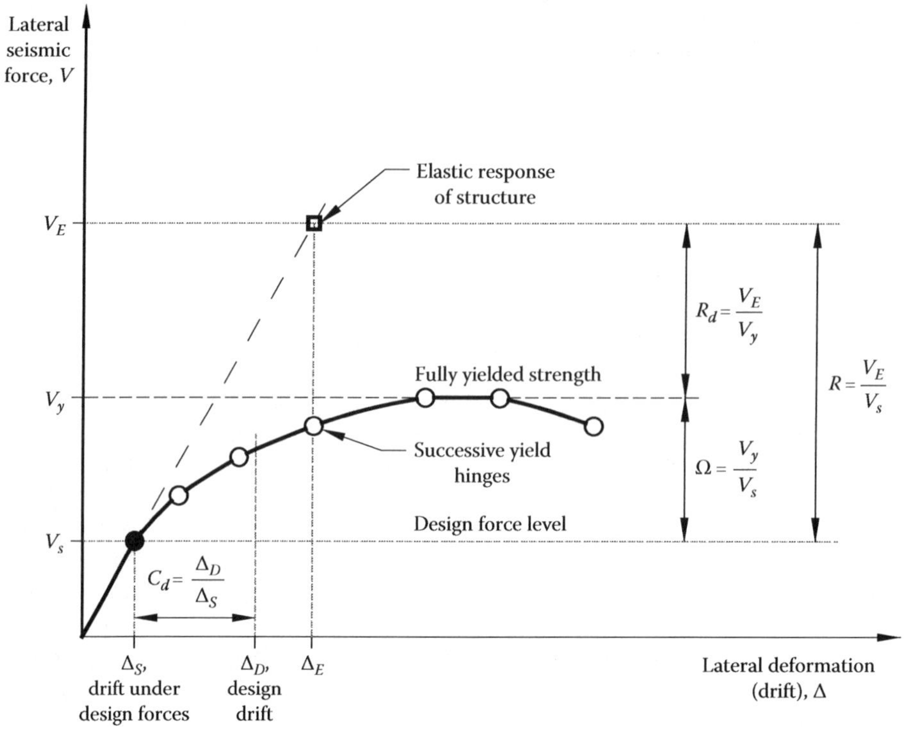

FIGURE 5.8 Inelastic force-deformation curve.

would produce in a structure with completely linear elastic response. This reduction is possible for a number of reasons. As the structure begins to yield and deform inelastically, the effective period of response of the structure lengthens, which, for most structures, results in a reduction in strength demand. Furthermore, the inelastic action results in a significant amount of energy dissipation (hysteretic damping) in addition to other sources of damping present below significant yield. The combined effect, which is also known as the ductility reduction, explains why a properly designed structure with a fully yield strength (V_y in Figure 5.8) that is significantly lower than the elastic seismic-force demand (V_E in Figure 5.8) can be capable of providing satisfactory performance under the design ground motion excitations.

The energy dissipation resulting from hysteretic behavior can be measured as the area enclosed by the force-deformation curve of the structure as it experiences several cycles of excitation. Some structures have far more energy dissipation capacity than others. The extent of energy dissipation capacity available depends largely on the amount of stiffness and strength degradation the structure undergoes as it experiences repeated cycles of inelastic deformation.

Figure 5.8 shows representative load deformation curves for two simple substructures such as a beam–column assembly in a frame. The hysteretic curve in Figure 5.9a in the figure is representative of the behavior of substructures that have been detailed for ductile behavior. The substructure can maintain nearly all of its strength and stiffness over several large cycles of inelastic deformation. The resulting force-deformation *loops* are quite wide and open, resulting in a large amount of energy dissipation. The hysteretic curve in Figure 5.9b represents the behavior of a substructure that has not been detailed for ductile behavior. It loses stiffness rapidly under inelastic deformation, and the resulting hysteretic loops are quite pinched. Such a substructure has much less energy dissipation than that for the structure (Figure 5.9a) but has a greater change in response period. The structural response is determined by a combination of energy dissipation and period modification.

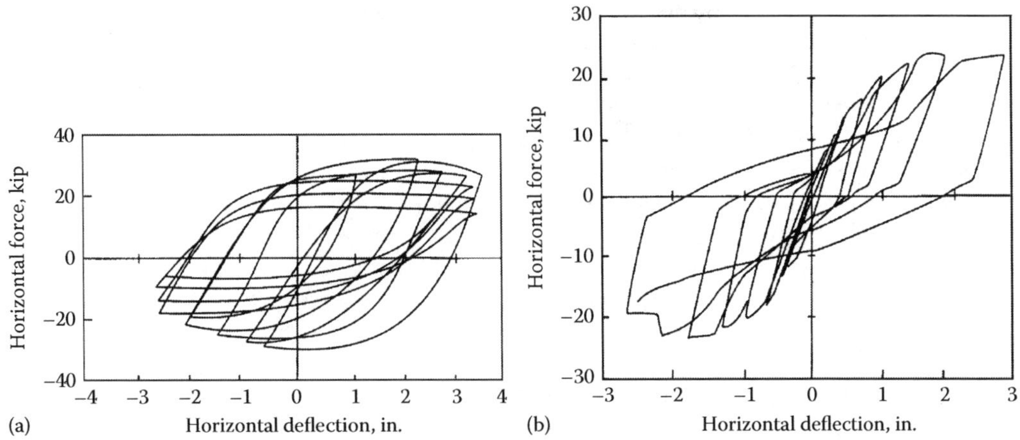

FIGURE 5.9 Hysteretic behavior: (a) curve representing large energy dissipation and (b) curve representing limited energy dissipation. (From Taranath, B.S., *Structural Analysis and Design of Tall Buildings*, CRC Press, Boca Raton, FL, 2011.)

The *R* values in the standard are based largely on engineering judgment of the performance of the various materials and system in past earthquakes. The *R* factor for a specific project should be chosen and used with care. For example, lower values should be used for structures possessing a low degree of redundancy wherein all the plastic hinges required for the formation of a mechanism may be formed essentially simultaneously and at a force level close to the specified design strength. This situation can result in considerably more detrimental *P*-delta effects. Since it is difficult for individual designers to judge the extent to which *R* factors should be adjusted based on the inherent redundancy of their designs, Section 12.3.4 of ASCE 7-10 provides a coefficient, ρ, that is calculated based on the removal of individual seismic-force-resisting elements.

5.2.2 Member Design, Connection Design, and Deformation Limit

Given that key elements of the seismic-force-resisting system will likely yield in response to ground motions, it might be expected that structural connections would be required to develop the strength of the connected members. Although that is a logical procedure, it is not a general requirement. The actual requirement varies by system and design of the various structural materials. Good seismic design requires careful consideration of this issue.

5.2.3 Continuous Load Path and Interconnection

In effect, seismic design needs to be complete and in accordance with the principles of structural mechanics. The loads must be transferred rationally from their point of origin to the final point of resistance. This should be obvious, but it often is overlooked by those inexperienced in earthquake engineering. Design consideration should be given to potentially adverse effects where there is a lack of redundancy. Given the many unknowns and uncertainties in the magnitude and characteristics of earthquake loading, in the materials and systems of construction for resisting earthquake loadings and in the methods of analysis, good earthquake engineering practice has been to provide as much redundancy as possible in the seismic-force-resisting system of buildings. Redundancy plays an important role in determining the ability of the building to resist earthquake forces. In a structural system without redundant components, every component must remain operative to preserve the integrity of the building structure. On the other hand, in a highly redundant system, one or more redundant components may fail and still leave a structural system that retains its integrity and can continue to resist lateral forces, albeit with diminished effectiveness.

The overall system redundancy can be improved by making all joints of the vertical-load-carrying frame moment resisting and incorporating them into the seismic-force-resisting system. These multiple points of resistance can prevent a catastrophic collapse due to distress or failure of a member or joint. The designer should be particularly aware of the proper selection of R when using only one- or two-bay rigid frames in one direction for resisting seismic loads. A single, one-bay frame or a pair of such frames provides little redundancy so the designer may wish to consider a reduced R to account for a lack of redundancy if the calculated redundancy is considered to be too low. As more one-bay frames are added to the system, however, overall system redundancy increases. The increase in redundancy is a function of frame placement and total number of frames.

The response modification factor, R, reduces the design seismic forces as a function of the ductility and overstrength of the LFRS. This is based on the experience in past earthquakes that have demonstrated that buildings designed to a significantly lower base shear can adequately resist seismic forces without collapse. This experience provides the basis for use of the R factor. The reason for adequate performance at a lower base shear is thought to be the result of both the extra/reserve strength in the structural system and nonstructural elements and finishes and the stable inelastic behavior of the structural elements. Based on reserve strength and inelastic behavior, the code allows the use of R values significantly greater than 1.0 (as much as 8), resulting in correspondingly lower seismic base shears.

Some elements of properly detailed structures are not capable of safely resisting ground-shaking demands through inelastic behavior. To ensure safety, these elements must be designed with sufficient strength to remain elastic. The Ω_o coefficient approximates the inherent overstrength in typical structures having different seismic-force-resisting systems. The special seismic loads, factored by the Ω_o coefficient, are an approximation of the maximum force these elements are ever likely to experience. ASCE 7-10 permits the special seismic loads to be taken as less than the amount computed by applying the Ω_o coefficient to the design seismic forces when it can be shown that yielding of other elements in the structure will limit the amount of load that can be delivered to the element. A case in point is the axial load induced in a column of a moment-resisting frame from the shear forces in the beams that connect to this column. The axial loads due to lateral seismic action need never be taken greater than the sum of the shears in these beams at the development of a full structural mechanism, considering the probable strength of the materials and strain-hardening effect. For frames controlled by beam hinge-type mechanisms, this would typically be $2M_p/L$, where for steel frames, M_p is the expected plastic moment capacity of the beam as defined in the AISC Seismic Specifications. For concrete frames, M_p is the probable flexural strength of the beams. L is the clear-span length for both steel and concrete beams. In the context of seismic design, the term capacity means the expected or median anticipated strength of the element, considering potential variation in material yield strength- and strain-hardening effects. When calculating the capacity of elements for this purpose, material strengths should not be reduced by capacity or resistance factors.

The seismic load effect with overstrength is intended to address those situations where failure of an isolated, individual, brittle element can result in the loss of a complete seismic-force-resisting system or in instability and collapse.

This somewhat arbitrary factor attempts to quantify the maximum force that can be delivered to sensitive elements based on historic observation that the real force that could develop in a structure may be three to four times the design levels. The use of the coefficient provides an estimate of the maximum forces likely to be experienced by an element.

Most structures designed with a given seismic-force-resisting system will fall within a range of overstrength values. Since the purpose of the Ω_o factor to estimate the maximum force that can be delivered to a component that is sensitive to overstress, the values of this factor are intended to be representative of the larger values in this range for each system.

While overstrength can be quite beneficial in permitting structures to resist actual seismic demands that are larger than those for which they have been specifically designed, it is not always beneficial. Some elements incorporated in structures behave in a brittle manner and can fail in an abrupt manner if substantially overloaded. The existence of structural overstrength results in a condition where such overloads are likely to occur, unless they are specifically accounted for in the design process.

One case where structural overstrength should specifically be considered is in the design of column elements beneath discontinuous braced frames and shear walls, such as that occurs at vertical in-plane and out-of-plane irregularities. Overstrength in shear walls could cause buckling failure of such columns with resulting structural collapse. Columns subjected to tensile loading in which splices are made are another example of a case where the seismic effect with overstrength should be used.

Although the most common cases in which structural overstrength can lead to an undesirable failure mode are identified in the ASCE provisions, not all such conditions are noted. Therefore, designers should be alert to conditions where the isolated independent failure of any element can lead to a condition of instability or collapse and should use the seismic effect with overstrength for the design of such elements. Other conditions, which may warrant such a design approach, include the design of transfer structures beneath discontinuous lateral-force-resisting elements and the design of diaphragm force collectors to shear walls and braced frames, when these are the only method of transferring force to these elements at a diaphragm level.

5.2.3.1 R, C_d, and Ω_o Values for Vertical Combinations

Consider a building with a vertical combination consisting of steel special moment-frame system for the upper levels, supported on a steel special concentrically braced frame system for the lower levels. ASCE 7-10 permits such a system to be designed using the seismic design coefficients R, C_d, and Ω_o applicable to the corresponding systems. In other words, we design the upper level using

$$R = 8\ \Omega = 3 \quad \text{and} \quad C_d = 5\tfrac{1}{2}$$

and the lower levels using

$$R = 6\ \Omega = 2 \quad \text{and} \quad C_d = 5$$

However, the forces transmitted from the upper system shall be increased by multiplying the ratio of higher response modification coefficient to the lower response modification coefficient.

In our case, the ratio is equal to 8/6 = 1.33.

On the other hand, if our building consists of the braced frame for the upper levels supported on the moment-frame system, the design procedure is quite different. ASCE 7-10 mandates that the seismic design coefficient for the upper level be used for the design of both systems. In our case, we use $R = 6$, $\Omega_o = 2$, and $C_d = 5$ not only for the upper system but also for the lower system. See Figure 5.8 for schematic explanation of the requirement.

For the past several decades, building codes have allowed two-stage static analysis for certain structures with a vertical combination of dynamically uncoupled systems. While this approach may be used for any structure that meets the requirements, it is most often used for the design of light-framed construction built on a rigid concrete base. The design process requires that the *flexible* upper structure and *rigid* lower structure be designed separately with the reactions from the upper portion amplified by the ratio of respective R/p values. This ratio, which must be taken as no less than 1, produces demands for the *rigid* lower portion that are commensurate with its inelastic capability.

5.2.3.2 R, C_d, and Ω_o Values for Horizontal Combinations

For nearly all conditions, the least value of R of different structural systems in the same direction must be used in design. This requirement reflects the expectation that the entire system will undergo the same deformation with its behavior controlled by the least ductile system. However, where the listed conditions are met, the R value for each independent line of resistance can be used. This exceptional condition is consistent with light-frame construction that utilizes the ground for parking with residential use above.

5.2.4 Dual System

The moment frame of a dual system must be capable of resisting at least 25% of the design seismic forces; this percentage is based on judgment. The purpose of the 25% frame is to provide a secondary lateral system with higher degrees of redundancy and ductility in order to improve the ability of the building to support the effect of gravity loads after strong earthquake shaking. The primary system (walls or bracing) acting together with the moment frame must be capable of resisting all of the design seismic forces. The following analyses are required for dual systems:

1. The moment frame and shear walls or braced frames must resist the design seismic forces considering fully the force and deformation interaction of the walls or braced frames and the moment frames as a single system. This analysis must be made in accordance with the principles of structural mechanics considering the relative rigidities of the elements and torsion in the system. Deformations imposed upon members of the moment frame by their interaction with the shear walls or braced frames must be considered in this analysis.
2. The moment frame must be designed with sufficient strength to resist at least 25% of the design seismic forces including torsional effects.

5.2.5 Irregular and Regular Classification

Prior to the 1988 UBC, building codes published a list of irregularities defining the conditions, but provided no quantitative basis for determining the relative significance of a given irregularity. However, starting in 1988, seismic codes have attempted to quantify irregularities by establishing geometrically, or by use of building dimensions, the points at which the specific irregularity becomes an issue as to require extra analysis and design considerations over and above those of the equivalent lateral procedure. The code requirements for determining the presence of plan and vertical irregularity, and the required methods to compensate for it, have now become complex, as can be seen in ASCE 7-05, Tables 12.3-1 and 12.3-2. Observe that the remedial measures range from a simple requirement of a dynamic distribution of lateral forces (e.g., mass irregularity) to special load combination of gravity and seismic forces (e.g., out-of-plane offset irregularity).

5.2.5.1 Plan (Horizontal) Irregularity

ASCE 7-05, Table 12.3-1 indicates under what circumstances a building must be designed as having a plan irregularity. A building may have a symmetrical geometric shape without reentrant corners or wings but still be classified as irregular in plan because of distribution of mass or vertical seismic-force-resisting elements. Torsional effects in earthquakes can occur even when the static centers of mass and resistance coincide. For example, ground motion waves acting with a skew at respect to the building axis can cause torsion. Cracking or yielding in a nonsymmetrical fashion also can cause torsion. These effects also can magnify the torsion due to eccentricity between the static centers. For this reason, buildings having an eccentricity dimension perpendicular to the direction of the seismic force should be classified as irregular. The vertical resisting components may be arranged so that the static centers of mass and resistance are within the limitations given earlier and still be unsymmetrically arranged so the prescribed torsional forces would be unequally distributed to the various components. Torsional irregularities are subdivided into two categories, with a category of extreme irregularity. Extreme torsional irregularities are prohibited for structures located very close to major active faults and should be avoided when possible, in all structures.

There is a second type of distribution of vertical resisting components that, while not being classified as irregular, does not perform well in earthquakes. This arrangement is termed a core-type building with the vertical components of the seismic-force-resisting system concentrated near the center of the building. Better performance has been observed when the vertical components are distributed near the perimeter of the building. In recognition of the problems leading to torsional instability, a torsional amplification factor is prescribed in the ASCE provisions.

A building having a regular configuration can be square, rectangular, or circular. A square or rectangular building with minor reentrant corners would still be considered regular, but large reentrant corners creating a crucifix form would be classified as an irregular configuration. The response of the wings of this type of building is generally different from the response of the building as a whole, and this produces higher local forces than would be determined by analysis. Other plan configurations such as H-shapes that have a geometrical symmetry also would be classified as irregular because of the response of the wings.

Significant differences in stiffness between portions of a diaphragm at a level are classified as irregularities since they may cause a change in the distribution of seismic forces to the vertical components and create torsional forces not accounted for in the normal distribution considered for a regular building.

Where there are discontinuities in the path of lateral force resistance, the structure can no longer be considered *regular*. The most critical of the discontinuities is the out-of-plane offset of vertical elements of the seismic-force-resisting elements. Such offsets impose vertical and lateral-load effects on horizontal elements that are at the least difficult to provide for adequately.

Where vertical elements of the LFRS are not parallel to or symmetric about major orthogonal axes, the static lateral force procedures cannot be applied, and thus, the structure must be considered to be *irregular*.

5.2.5.2 Vertical Irregularity

ASCE 7-05, Table 12.3-2 indicates under what circumstances a structure must be considered to have a vertical irregularity. Vertical irregularities affect structural response and induce loads at the irregularity levels that are significantly different from the distribution assumed in the ELF procedure.

A building would be classified as irregular if the ratio of mass to stiffness in adjoining stories differs significantly. This might occur when a heavy mass, such as a swimming pool, is placed at one level while the floors above or below have typical floor loads. A comparative stiffness ratio between stories is given at a benchmark to exempt structures from being designed as having a vertical irregularity.

Another type of vertical irregularity is created by unsymmetrical geometry with respect to the vertical axis of the building. The building may have a geometry that is symmetrical about the vertical axis and still be classified as irregular because of significant horizontal offset in the vertical elements of the LFRS at one or more levels. An offset is considered to be significant if the ratio of the larger dimension to the small dimension is more than 130%. The building also would be considered irregular if the smaller dimension was below the larger dimension, thereby creating an inverted pyramid effect.

Weak-story irregularities occur whenever the strength of a story to resist lateral demands is significantly less than that of the story above. This is because buildings with this configuration tend to develop all of their inelastic behavior at the weak story. This can result in a significant change in the deformation pattern of the building, with most earthquake-induced displacement occurring within the weak story. This can result in extensive damage within the weak story and even instability and collapse. Note that an exception has been provided in where there is considerable overstrength of the *weak* story.

The soft-story irregularity is subdivided into two categories with an extreme soft-story category. Like weak stories, soft stories can lead to instability and collapse. Buildings with extreme soft stories are prohibited on sites located very close to major active faults.

The configuration of a structure can significantly affect its performance during a strong earthquake producing the ground motion contemplated in the standard. Configuration can be divided into two aspects: horizontal and vertical. Most seismic design provisions were derived for buildings having regular configurations, but earthquakes have shown repeatedly that buildings having irregular configuration suffer greater damage. This situation prevails even with good design and construction. There are several reasons for this poor behavior of irregular structures. In a regular structure, the inelastic response produced by strong ground shaking, including energy dissipation and damage, tends to be well distributed throughout the structure. However, in irregular structures, inelastic behavior can be concentrated by irregularities and result in rapid failure of structural elements in these areas. In addition, some irregularities introduce unanticipated demands into the structure, which designers frequently overlook when detailing the structural system. Finally, the elastic analysis methods typically employed in the design of structures often cannot predict the distribution of earthquake demands in an irregular structure very well, leading to inadequate design in the areas associated with the irregularity. For these reasons, the standard encourages regular configurations and prohibits gross irregularity in buildings located on sites close to major active faults where very strong ground motion and extreme inelastic demands are anticipated.

Vertical configuration irregularities affect the responses at the various levels and induce loads at these levels that differ significantly from the distribution assumed in the ELF procedure. A moment-resisting frame building might be classified as having a vertical irregularity if one story is much taller than the adjoining stories and the design did not compensate for the resulting decrease in stiffness that normally would occur. ASCE 7-05, Table 12.3-2 illustrates vertical irregularities.

A building is classified as irregular where the ratio of mass to stiffness in adjacent stories differs significantly. This might occur where a heavy mass (e.g., an interstitial mechanical floor) is placed at one level. Irregularity Type 3 applies regardless of whether the larger dimension is above or below the smaller one. Buildings with a weak-story irregularity tend to develop all of their inelastic behavior and consequent damage at the weak story, possibly leading to collapse. ASCE 7-10, Section 12.3.3.2 provides an exception for SDC B or C structures where essentially elastic response of the weak story is expected.

5.2.5.3 Prohibited Horizontal and Vertical Irregularities in Seismic Design Categories D through F

The irregularity prohibitions stem from poor performance in past earthquakes and the potential to concentrate large inelastic demands in certain portions of the structure. Even when such irregularities are permitted, they should be avoided whenever possible in all structures. Since extreme weak-story irregularities are prohibited for buildings located in SDCs D, E, and F, the extreme weak-story limitations and exceptions apply only to buildings assigned to SDC B or C.

5.2.5.4 Elements Supporting Discontinuous Walls or Frames

The purpose of this requirement is to protect the supporting elements from overload caused by overstrength of a discontinued seismic-force-resisting element. Columns, beams, slabs, or trusses may be subject to such failure so all are included in the design requirement. Overload may result from forces in either the downward or upward direction; therefore, both possibilities must be considered. Such load reversals may be especially problematic for reinforced concrete beams, weaker top laminations of glulam beams, unbraced flanges of steel beams, and steel trusses.

The connection between the discontinuous element and the supporting member must be adequate to transmit the forces for which the discontinuous element is designed. For example, where the discontinuous element must be designed using the load combinations of ASCE 7.10, Sections 12.4.3, as in the case for a steel column in a braced frame or moment frame, its connection to the supporting member must be designed using the same load combinations. Since concrete shear walls are not required to be designed using the load combinations of Sections 12.4.3, the connection between a discontinuous shear wall and the supporting member may be designed using the loads associated with the shear wall and not the load combinations with overstrength factor.

5.2.6 Increase in Forces due to Irregularities for Seismic Design Categories D through F

The irregularities listed may result in loads that are distributed differently than assumed in the ELF procedure.

The 25% increase in force is intended to account for this difference. Where the load combinations with overstrength apply, no further increase is warranted.

5.2.7 Redundancy

The desirability of redundancy, or multiple lateral-force-resisting load paths, has long been recognized. The redundancy provisions of this section reflect the belief that an excessive loss of story shear strength or development of an extreme torsional irregularity may lead to structural failure. The redundancy factor determined for each direction may differ.

5.2.7.1 Redundancy Factor, p, for Seismic Design Category D through F

There are two approaches to establishing a redundancy factor of 1.0. Where neither condition is satisfied, p is taken equal to 1.3. It is permitted to take p equal to 1.3 without checking either condition.

The first approach is a check of the elements outlined in ASCE 7-05, Table 12.3-3, for cases where the story shear exceeds 35% of the base shear. Parametric studies (conducted by Building Seismic Council Technical Subcommittee 2 but unpublished) were used to select the 35% value. Those studies indicated that stories with at least 35% of the base shear include all stories of low-rise buildings (buildings up to 5 stories) and about 87% of the stories of tall buildings. The intent of this limit is to exclude penthouses and the uppermost stories from the redundancy requirements.

This approach requires the removal (or loss of moment resistance) of an individual lateral-force-resisting element to determine its effect on the remaining structure. If the removal of elements, one-by-one, does not result in more than a 33% reduction in story strength or an extreme torsional irregularity, p may be taken as 1.0. For this evaluation, the determination of story strength requires an in-depth calculation. The intent of the check is to use a simple measure (elastic or plastic) to determine whether an individual member has a significant effect on the overall system. If the original structure has an extreme torsional irregularity to begin with, the resulting p is 1.3.

As indicated in the table, braced frame, moment frame, shear wall, and cantilever column systems must conform to redundancy requirements. Dual systems also are included but, in most cases, are inherently redundant. Shear walls or wall piers with a height-to-length aspect ratio greater than 1.0 within any story have been included; however, the required design of collector elements and their connections for Ω_o times the design force may address the key issues. In order to satisfy the collector force requirements, a reasonable number of shear walls usually is required. Regardless, shear wall systems are addressed in this section so that either an adequate number of wall elements are included or the proper redundancy factor is applied. For wall piers, the height is taken as the height of the adjacent opening and generally is less than the story height.

The second approach is a deemed-to-comply condition wherein the structure is regular and has a specified arrangement of seismic-force-resisting elements to qualify for p of 1.0. As part of the parametric study, simplified braced frame and moment-frame systems were investigated to determine their sensitivity to the analytical redundancy criteria. This simple deemed-to-comply condition is consistent with the results of the study.

Lack of redundancy exists when the failure of a building component results in the failure of the entire structure. A conceptual explanation of a highly redundant and not-so-highly redundant building is quite easy as shown in Figure 5.10, but quantifying redundancy with numerical values is another matter.

Redundancy provisions were first introduced into building codes and standards via the 1997 UBC. Since their original inception, the redundancy provisions created much controversy with respect to the interpretation and implementation. The debate centered mainly on the following issues:

- A sliding redundancy value as was proposed originally, based on the force in only one of the elements of the system, was too precise and not technically justified.
- A better approach to determining redundancy is to base it on whether the loss or removal of an important component within the system would result in more than a 33% reduction in story strength, or would the resulting system have an extreme torsional irregularity (horizontal structural irregularity Type lb, ASCE 7-10, Table 12.3-1).
- Checking redundancy throughout the entire building height is not necessary. Only those stories resisting more than 35% of the base shear in the direction of interest need to be checked.
- Well-distributed perimeter systems should automatically qualify for a redundancy value of 1.0—specifically, those structures that are regular in plan at all levels.

Now it seems that the issue is settled by permitting the redundancy factor to be taken as 1.0 for

1. Structures assigned to SDC B or C
2. Drift calculation and $P\Delta$ effects
3. Design of nonstructural components
4. Design of nonbuilding structures that are not similar to buildings

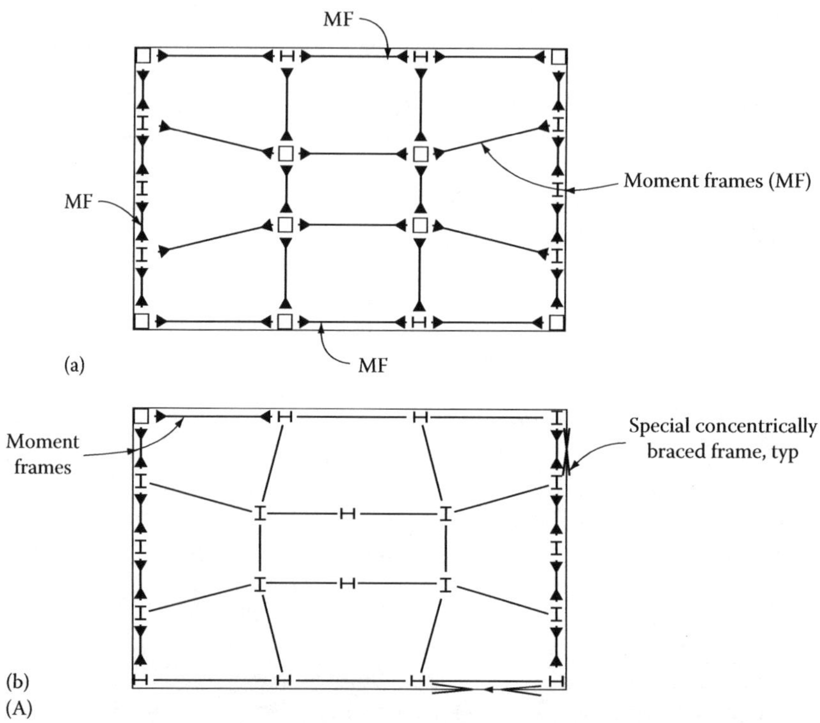

FIGURE 5.10 (A) (a) Highly redundant building and (b) not-so-redundant building. *(Continued)*

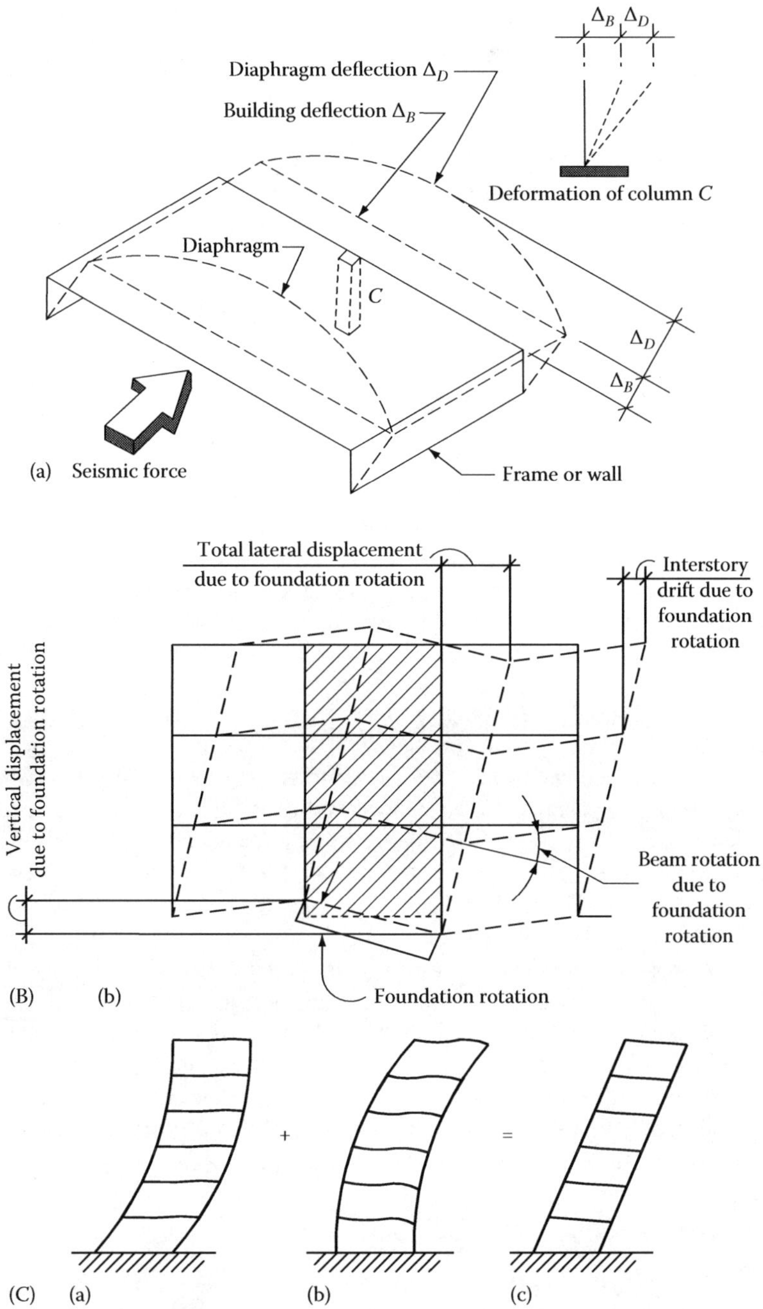

FIGURE 5.10 (*Continued*) (B) Column deformation for use in compatibility considerations. (a) Deformation of column = building deflection Δ_B + diaphragm deflection Δ_D. (b) Foundation flexibility for deformation compatibility considerations. (C) Characteristic shape of fundamental mode: (a) shear distortion; (b) bending deformation; (c) overall combined deformation. (From Taranath, B.S., *Structural Analysis and Design of Tall Buildings*, CRC Press, Boca Raton, FL, 2011.)

5. Design of collector elements, splices, and their connections for which the load combinations with overstrength factors are used
6. Design of members or connections where the load combinations with overstrength are required for design
7. Diaphragm loads determined using Equation 12.10-1 of ASCE 7-10
8. Structures with damping systems

A logical way to determine the lack of redundancy is to check whether a component's failure results in an unacceptable amount of the loss of story strength or in the development of extreme torsional irregularity.

In the ASCE 7-10, $p = 1.0$ or 1.3, depending on whether or not an individual element can be removed from the LFRS without

1. Causing the remaining structure to suffer a reduction of story strength of more than 33%
2. Creating an extreme torsional irregularity

Additionally, the redundancy factor, p, is 1 for the following cases:

1. *Moment frames*: The loss of moment resistance at the beam-to-column connections at both ends of a single beam would not result in more than a 33% reduction in story strength, nor does the resulting system have an extreme torsional irregularity.
2. *Shear wall or wall pier with a height-to-length ratio of greater than 1.0*: The removal of a shear wall or wall pier with a height-to-length ratio of greater than 1.0 within any story, or collector connections thereto, would not result in more than a 33% reduction in story strength, nor does the resulting system have an extreme torsional irregularity.

The redundancy factor p is used to encourage designers to provide a reasonable number and distribution of vertical LFRS elements. As stated previously, ASCE 7 contains a list of eight circumstances when p is permitted to be set equal to 1.0. Item one permits p to be 1.0 for all structures assigned to SDCA A, B, or C. Item two permits p to be set to 1.0 for drift calculation and determining P-delta effects. The balance of the items applies to components and elements rather than the primary LFRS.

ASCE 7 requires that p be 1.3 for all structures in SDC D, E, or F, unless the structure is qualified for a p of 1.0 using one of two possible methods. Both methods require further evaluation of each story that resists more than 35% of the base shear. In a three-story building, it would be anticipated that the bottom two stories require evaluation, but possibly not the top story.

In the first method, the user is asked to remove vertical resisting elements one at a time and check to see if (1) the story strength is reduced by more than 33% or (2) if an extreme torsional irregularity is created with the element removed. If either of these conditions exist, a p of 1.3 must be used; if it does not, p may be taken as 1.0.

ASCE 7-05, Table 12.3-3 lists evaluation requirements for determining seismic vulnerability of each story resisting more than 35% of base shear, for purposes of establishing $p = 1.0$ or 1.3.

5.2.8 Seismic Load Effect and Combinations

Structural elements designed as part of the seismic-force-resisting system typically are designed directly for seismic load effects. None of the seismic forces associated with the design base shear are formally assigned to structural elements that are not designed as part of the seismic-force-resisting system, but such elements must be designed using the load conditions of Section 5.4 and must accommodate the deformations resulting from application of seismic loads.

The load combinations are taken from the basic load combinations of Chapter 2 of the standard with further elaboration of the seismic load effect, E. The seismic load effect includes horizontal and vertical components. For strength design, the effect of vertical seismic forces, E_V, is based on an assumed effective vertical acceleration of 0.2 S_{DS} times gravity.

It may be helpful to recognize that the quantities E_h and E_V are the effects of loads, not the loads themselves. They can be tension or compression axial forces, shear, bending moments, or torsional moments. For a one-story shear wall, application of the horizontal seismic forces from V causes overturning moment and shear in the wall, both of which are E_h effects. The factor 0.2 S_{DS} times gravity dead load corresponds to an E_V load effect that increases or decreases the axial force in the wall. In this simple example, an E_h force or moment is never added directly to an E_V force or moment because the former affects only moment and shear, while the latter affects only axial force.

While the shear and moment are independent of the axial force, the capacity check of the wall may need to include all three terms (or certainly moment and axial force) simultaneously.

For a diagonal brace that carries earthquake and gravity load, application of the horizontal seismic force from V causes a brace force that has both horizontal and vertical components, and the factor 0.2 S_{DS} times dead load produces a load effect that also affects both the horizontal and vertical components of axial force. In this case, the brace force is based on $E_h \pm E_V$.

The 0.2 S_{DS} vertical acceleration effect is required to be considered in the design of all members of a structure—even though they are not part of the seismic-force-resisting system. For example, design of a gravity load-resisting prestressed concrete girder may be governed by the dead and earthquake condition, where 0.2 $S_{DS}D$ is subtracted from the dead load. This could be the controlling condition for tension at the top of the girder.

Certain structural elements or actions, such as collectors in SDCs C through F or columns supporting discontinuous walls, are required to be designed for seismic load combinations with overstrength. In such cases, the seismic load effect, E_m, has its horizontal component multiplied by the overstrength factor Ω_o.

In SDCs D, E, and F, horizontal cantilevers are designed for an upward force that results from an effective vertical acceleration of 1.2 times gravity. This is to provide some minimum strength in the upward direction and to account for possible dynamic amplification of vertical ground motions resulting from the vertical flexibility of the cantilever. The requirement is not applied to downward forces on cantilevers, for which the typical load combinations are used.

5.2.9 Direction of Loading

Seismic forces are delivered to a building through ground accelerations that may approach from any direction relative to the orthogonal directions of the building; therefore, seismic effects are expected to develop in both directions simultaneously. The standard requires structures to be designed for the most critical loading effects from seismic forces applied in any direction, and the procedures outlined in this section are deemed to satisfy that requirement.

The orthogonal combination procedure combines the effects from 100% of the seismic load applied in one direction with 30% of the seismic load applied in the perpendicular direction.

For horizontal structural elements such as beams and slabs, orthogonal effects may be minimal; however, for vertical elements of the seismic-force-resisting system that participate in both orthogonal directions, the design likely will be governed by these combinations.

The maximum effect of seismic forces, Q_E, from orthogonal load combinations must be modified by the redundancy factor, ρ, or the overstrength factor, Ω_o, and consider the effects of vertical seismic forces, E_V, to obtain the seismic load effect E.

5.2.10 Analysis Procedure

The procedure given in ASCE 7-10, Table 12.6.1 applies only to buildings without seismic isolation or passive energy devices. The procedures addressed are ELF analysis, modal response spectrum (MRS) analysis, LRH analysis, and NRH analysis.

The value of T_S (= S_{DI}/S_{DS}) depends on the site class because S_{DS} and S_{DI} include such effects. Where ELF is not allowed, analysis must be performed using MRS or response-history analysis.

ELF is not allowed for buildings with the listed irregularities because it assumes a gradually varying distribution of mass and stiffness along the height and negligible torsional response. The $3.5T_S$ limit recognizes that higher modes are more significant in taller buildings such that the ELF method may underestimate the design base shear and may not predict correctly the vertical distribution of seismic forces.

5.2.11 Foundation Modeling Criteria

Structural systems consist of three interacting components: the structural framing, the foundation, and the supporting soil. The ground motion that a structure experiences, as well as the response to that ground motion, depends on the complex interaction between these components.

Those aspects of ground motion that are affected by site characteristics are assumed to be independent of the structure foundation system as these effects would occur in the free-field in the absence of the structure. Hence, site effects are considered separately.

Given a site-specific ground motion or response spectrum, the dynamic response of the structure will depend on the foundation system and on the characteristics of the soil that support the system. The dependence of the response on the structure–foundation–soil system is referred to as soil–structure interaction. Such interactions will usually but not always result in a reduction of base shear. This reduction in shear is due to the flexibility of the foundation–soil system and an associated lengthening of the period of vibration of the structure. In addition, the soil system may provide an additional source of damping. However, that total displacement typically increases with soil–structure interaction.

If the foundation is considered to be rigid, the computed base shears usually will be conservative, and it is for this reason that rigid foundation analysis is allowed. The designer may ignore soil–structure interaction or may consider it explicitly or implicitly.

It should be noted that in the ASCE 7-10 provisions, like in its predecessor, there are no modification factors to account for the effects of higher mode except at the soil–foundation interface. A 25% reduction is allowed in overturning values if the analysis is performed by using ELF procedure and a 10% reduction if the analysis is by modal response method. However, no reduction is permitted for cantilevered column-type structures.

5.2.12 Effective Seismic Weight

During an earthquake, the structure accelerates laterally, and these accelerations of the structural mass produce inertial forces. These inertial forces, accumulated over the height of the structure, produce the design base shear.

When a building vibrates during an earthquake, only that portion of the mass or weight that is physically tied to the structure needs to be considered as effective. Hence, live loads (e.g., loose furniture, loose equipment, and human occupants) need not be included. However, certain types of live loads such as storage loads may develop inertial forces, particularly where they are densely packed.

Also considered as effective weight is all permanently attached equipment (e.g., air conditioners, elevator equipment, and mechanical systems), movable partitions (a minimum of 10 psf is required), and 20% of significant roof snow load. The full snow load need not be considered because maximum snow load and maximum earthquake load are unlikely to occur simultaneously and loose snow does not move with the roof.

5.2.13 Structural Modeling

The development of a mathematical model of a structure is always required because the story drifts and the design forces in the structure cannot be computed without such a model. In some cases, the mathematical model can be as simple as a free-body diagram as long as that model can appropriately capture the strength and stiffness of the structure.

The most realistic analytical model is 3D, includes all sources of stiffness (and flexibility) of the structure and the soil–foundation system as well as *P*-delta effects, and allows for nonlinear inelastic behavior in all parts of the structure–foundation–soil system. The development of such an analytical model is very time consuming, and such analysis is rarely warranted for typical building designs. Instead of performing a nonlinear analysis, inelastic effects are accounted for indirectly in the linear analysis methods by means of the response modification factor, R, and the deflection amplification factor, C_d.

Using modern software, it often is more difficult to decompose a structure into planar models than it is to develop a full 3D model so 3D models now are commonplace. Increased computational efficiency has reduced the motivation to model rigid diaphragms, allowing for easy and efficient modeling of diaphragm flexibility. Three-dimensional models are required where the structure has torsional irregularities, out-of-plane offset irregularities, or nonparallel system irregularities.

In general, the same 3D model may be utilized for ELF, MRS, and LRH analysis. The response spectrum and LRH models require a realistic modeling of structural mass, and the response-history method also requires an explicit representation of inherent damping. Five percent critical damping is automatically included in the MRS approach.

It is well known that deformations in the panel zones of the beam–column joints of steel moment frames are a significant source of flexibility. For elastic structures, centerline analysis provides reasonable, but not always conservative, estimate of frame flexibility. Fully rigid end zones should not be used, as this will always result in an over estimation of lateral stiffness in steel moment-resisting frames. Partially rigid end zones may be justified in certain cases such as where double plates are used to reinforce the panel zone.

Including the effect of composite slabs on the stiffness of beams and girders may be warranted in some circumstances. Where composite behavior is included, due consideration should be paid to the reduction in effective composite stiffness for portions of the slab in tension.

For reinforced concrete buildings, it is important to address the effects of axial, flexural, and shear cracking in modeling the effective stiffness of the structural components. Determining appropriate effective stiffness of the structural components would take into consideration the anticipated demands on the components, their geometry, and the complexity of the model.

5.2.14 Interaction Effects

The interaction requirements are intended to prevent unexpected failures in members of moment-resisting frames, a typical situation where masonry infill is used, and this masonry is fitted tightly against reinforced concrete columns. Since the masonry is much stiffer than the columns, column hinges form at the top of the column at the top of the masonry rather than at the top and bottom of the columns. If the column flexural capacity is M_p, the shear in the columns increases by the factor H/h, and this may cause an unexpected nonductile shear failure in the columns. Many building collapses have been attributed to this effect.

5.2.15 Equivalent Lateral Force Procedure

The ELF procedure provides a simple way to incorporate the effects of inelastic dynamic response into a linear static analysis. This procedure is useful in preliminary design of all structures and is allowed for final design of the vast majority of structures. The procedure is valid only for structures without significant discontinuities in mass and stiffness along the height, where the dominant response to ground motions is in the horizontal direction without significant torsion.

The ELF procedure has three basic steps:

1. Determining the seismic base shear
2. Distributing the shear vertically along the height of the structure
3. Distributing the shear horizontally across the width and breadth of the structure

5.2.16 Seismic Base Shear

Equation 12.8-1 of the ASCE 7-10 provisions simply expresses the base shear as the product of the effective seismic weight, W, and a response coefficient, C_S. The response coefficient is a spectral pseudoacceleration, in g units, which has been modified by R and I to account for inelastic behavior and to provide for improved performance for high occupancy or essential structures.

There are five equations for determining the response coefficient C_S; the first three are plotted in Figure 5.11.

Equation 12.8-2, representing the constant-acceleration part of the spectrum, controls where $0/0 < T < T_S$. As shown in Table 12.6-1 (which provides values of $3.5T_S$), T_S is a function of seismicity and site. It may be as low as 0.2 s for hazard regions on site class B or as high as 0.9 s in high hazard regions on site class E.

The true pseudoacceleration response spectrum transitions to the peak ground acceleration as the period approaches zero. This transition is not used in the ELF method. One reason is that simple reduction of the response spectrum by $(1/R)$ in the very short-period region would exaggerate inelastic effects.

Equation 12.8-3, representing the constant-velocity part of the spectrum, controls where $T_S < T < T_L$. In the region, the seismic response coefficient is inversely proportional to period, and the pseudovelocity (pseudoacceleration divided by circular frequency, ω) is constant. T_L, the long period, is provided in Figures 22-15 through 22-20. T_L ranges from 4 s in the north-central conterminous states and western Hawaii to 16 s in the Pacific Northwest and in western Alaska.

Equation 12.8-4, representing the constant-displacement part of the spectrum, controls where $T > T_L$. Given the current mapped values of T_L, this equation only affects tall and flexible structures.

Equation 12.8-5 is the minimum base shear and provides (working stress) strength of approximately 3% of the weight of the structure. This minimum base shear was originally enacted in 1933 by the state of California's Riley Act.

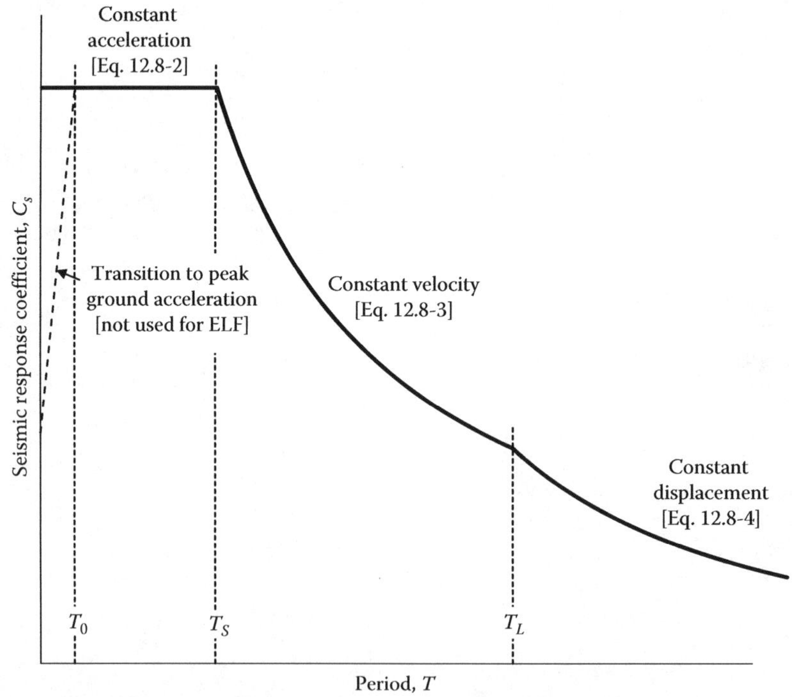

FIGURE 5.11 Seismic response coefficient versus period. (From FEMA P-750, National Institute of Buliding Sciences, Washington, D.C., 2009.)

Equation 12.8-6 applies to sites near major active faults (as reflected by values of S_1) where pulse effects can increase long-period demands.

5.2.17 Period Determination

The fundamental period of the structure, T, is used to determine the design base shear as well as the exponent k that establishes the distribution of the shear along the height of the structure. Equation 12.8-7 ASCE 7-10 is an empirical relationship determined through statistical analysis of the measured response of buildings in California.

Since the empirical expression is based on the lower bound of the data, it produces a lower bound for the period of a building of given height. This lower-bound period, used in Equations 12.8-3 and 12.8-4, provides a conservative estimate of base shear.

The fundamental period determined from a rational analysis may be used in design unless it exceeds the approximate period times the coefficient provided in Table 12.8-1 ASCE 7-10. This period limit prevents the use of unusually low ELF base shear for design of buildings that are overly flexible. The coefficients in the table have two effects. First, the conservatism of lower-bound empirical formulas for T_a is removed. Second, the period is increased in regions of lower seismicity as buildings in such areas generally are more flexible (and, hence, have longer periods) than buildings in regions of higher seismicity.

In the denominator of base shear equations, T is the fundamental period of vibration of the building. It is preferable to determine this using the structural properties and deformational characteristics of the building, by using a dynamic analysis. However, a dynamic analysis can calculate the period only after the building has been designed. Therefore, an approximate method is necessary to estimate building period, with minimal information available on the building characteristics. Hence, the ASCE 7-10 provides simple formulas that involve only a general description of the building type (such as shear wall building, concrete moment frame) and the overall height or number of stories.

Building periods, computed even with the use of very sophisticated software, are only as good as the modeling assumptions used in the analysis and, to a great extent, are dependent on stiffness assumptions. The smaller the assumed stiffness, the longer the computed period, which translates directly into a lower design base shear. The computed period is thus open to possible abuse. Therefore, certain limits are imposed on the computed period. For design purposes, it may not be taken any larger than a coefficient C_u times the approximate period calculated. Reasonable mathematical rules should be followed such that the increase in period allowed by the C_u coefficient is not taken advantage of when the structure does not merit it. *Note that for purposes of drift analysis only, the upper bound limitation on the computed fundamental period T of the building does not apply.* It may be noted from ASCE 7-10, Table 12.8.1 that larger values of C_u are permitted as the design spectral response acceleration parameter, at 1 s, S_{D1}, of a location decreases. This is because buildings in areas with lower lateral-force requirements are thought likely to be more flexible. High values of C_u for lower values of S_{D1} also result in less dramatic changes from prior practice in seismic risk areas. It is generally accepted that the equations for T_a are tailored to fit the types of construction common in areas with high lateral-force requirements. It is unlikely that buildings in lower seismic risk areas would be designed to produce as high a drift level as allowed by ASCE 7-10, due to stability (PA) considerations and wind requirements.

Every building has a set of periods in which it *wants* to oscillate when set in motion by seismic ground motions or wind guests. The period of vibrations are based on building mass and stiffness characteristics. The longest period of the system, and more specifically the inverse of the fundamental period, is the natural frequency.

In seismic design, the closer the frequency of an earthquake is to the natural frequency of a building, the more energy is introduced into the building structure. Buildings with shorter fundamental periods attract higher forces as the code-based or site-specific response spectrum exhibits higher accelerations at shorter periods.

Conversely, taller buildings because of long fundamental periods tend to attract lower seismic forces. For wind design, the opposite behavior is observed. Longer fundamental periods are indicative of buildings that are more susceptible to dynamic amplification effects from wind gusts and result in higher design forces. In order to investigate the magnitudes of these wind and seismic effects, the fundamental period of the building must first be determined.

Most designers are familiar with the use of the fundamental period of the structure, T, in conjunction with calculating the seismic response coefficient, C_s, for base shear determination using the ELF procedure. The most straightforward method for determining the building period is to use the empirical formulas for the calculation of the approximate building period, T_a, presented in Chapter 12 of ASCE 7-10.

The equations are based on data from several instrumental buildings subjected to ground motion during seismic events. The formulas are intentionally skewed to represent a conservative (short) estimation of the fundamental building period. Shorter building periods result in higher and more conservative base shears.

If desired, the approximate period, T_a, may be used for the strength design and checking the drift limits of the building. However, this practice typically results in significantly overly conservative results.

Therefore, ASCE 7-10 allows the use of a *properly sustained analysis*. In today's practice, this means use of software programs to perform an eigenvalue analysis to determine the mode shapes and periods of a building. It is important to note that the periods determined using an eigenvalue analysis are significantly longer than those determined using the approximate equations.

For strength design, ASCE 7-10 caps the maximum building period to the approximate building period T_a, multiplied by the factor, C_u, from ASCE 7-10 Table 12.8-1. The cap is intended to prevent the so-called sharp pencil effects resulting from erroneous assumptions used in the *properly substantiated analysis*. However, for the determination of seismic drift, ASCE 7-10 removes the cap and allows the engineer to use the building period resulting from analysis without restriction.

The first and most basic dynamic property of a structure is its fundamental period of vibration. ASCE 7 provides the following formula for the approximate period of vibration:

$$T_a = C_t h_n^x$$

where

h_n is the height of the highest (nth) level above the base, ft
x is the exponent dependent on structure type
= 0.80 for moment-resisting systems of steel
= 0.75 for eccentrically braced steel frames
= 0.75 for all other structures
C is the coefficient dependent on structure type
= 0.028 for moment-resisting systems of steel
= 0.030 for eccentrically braced steel frames
= 0.020 for all other structures

ASCE 7-10 provides optional alternative definitions for the approximate period T_a in structures with steel moment frames.

5.2.17.1 Vertical Distribution of Seismic Force

It is interesting to note that seismic forces are calculated and distributed throughout the structure in the reverse order used for evaluating wind forces. For wind design, forces at each level are calculated first from the external pressure and suctions, and then the total shear (referred to as base shear in seismic terminology) is determined by summing the forces in the horizontal direction.

For determining seismic forces, the process is just the reverse. The seismic shear at the base, the base shear, is calculated as a product of the applicable seismic coefficient, C_s, and the total mass M of the structure equal to W/g, where W is the total seismic weight of the building and g is the acceleration due to gravity.

The next step is the vertical distribution of V as story forces F_x at each story.

The formula for F_x will produce a triangular distribution of horizontal story forces if the masses (tributary weights) assigned to the various story levels are all equal. If the weights are not equal, some variation from the straight-line distribution will result, but the trend will follow the first-mode shape. Accelerations and, correspondingly, inertia forces ($F = Ma$) increase with increasing height above the base.

F_x distribution is the same for all SDCs and is given by

$$F_x = C_{vx}V \quad \text{(ASCE 7-10 Eq. 12.8-12)}$$

and

$$C_{vx} = \frac{w_x h_x^k}{\sum_{i=1}^{n} w_i h_i^k} \quad \text{(ASCE 7-10 Eq. 12.8-12)}$$

where

C_{vx} is the vertical distribution factor
V is the total base shear
w_i and w_x are the tributary weights assigned to level i or x
h_i and h_x are the height from the base of structure to level i or x
k is an exponent related to the structural period
= 1 for structures having a period of 0.5 s or less

Equation 12.8-12 ASCE 7-10 is based on the simplified first-mode shape shown in Figure 12.8-3 ASCE 7-10. In the Figure, F_x is the inertial force at level x, which is simply the total acceleration at level x times the mass at level x. The base shear is the sum of these inertial forces, and Equation 12.8 simply gives the ratio of the force at level x to the total base shear.

The deformed shape of the structure is a function of the exponent k, which is related to the fundamental period of vibration of the structure. The exponent k is intended to approximate the effect of higher modes, which are generally more dominant in structures with longer fundamental period of vibration. Although the actual first-mode shape for a structure is also a function of the type of seismic-force-resisting system, that effect is not reflected in these equations.

The horizontal forces computed using Equation 12.8-12 do not reflect the actual inertial forces impacted on a structure at any particular time. Instead, they are intended to provide design story shears that are consistent with enveloped results from more accurate analysis.

As stated previously, the distribution of base shear over the building height is based on the assumption that the characteristic shape of the fundamental mode is a straight line from foundation to the top of the building. Although this is an approximation, the supposition is quite reasonable for typical buildings. The justification for the assumption may be understood if it is recognized that the total distortion of a typical building is the sum of two effects:

1. The shear distortion of the frame
2. The change in length of columns due to overall bending of the building

The former tends to produce a deflected shape that is concave to the left, and the latter a shape concave to the right. The combined effect results in a shape that approaches a straight line.

5.2.17.2 Seismic Loads due to Vertical Ground Motions

During an earthquake, vertical ground motion creates vertical forces in addition to the horizontal seismic forces. The vertical forces generated by earthquake ground motions are invariably much smaller than the horizontal forces. ASCE 7 directly incorporates vertical ground motions into the seismic-force equations. The result of including the vertical component of ground motion in the equations is an increase in net uplift and downward forces when considering overturning.

The method used to calculate the horizontal story forces involves three parts. The first part is calculating the *base shear* (the horizontal force acting at the base of the building, V). The second part is assigning the appropriate percentages of this force to the various story levels throughout the height of the structure (story forces). The third part is to determine the forces on particular elements as a result of the story forces (element forces). However, there are several multiplying factors required to convert these *element forces* into design seismic forces for the elements at a story.

The story forces are given the symbol F_x (the force at level x). It should be clear that the sum of the F_x forces must equal the base shear V. It should be noted that the story forces increase with increasing height above the base of the building. The magnitudes of the story forces depend on the mass distribution throughout the height of the structure. The vertical distribution provisions have an exponent of between one and two on the height term in the vertical distribution. This exponent accounts for the increased top-story forces that can occur in buildings with long periods. The exponent is taken as unity for structures with a calculated period of 0.5 s or less. With the exponent taken as one, a triangular distribution of story forces occurs.

The reason for this distribution is that ASCE 7 bases its forces on the fundamental mode of vibration of the structure. The fundamental mode is also known as the *first mode* of vibration, and it is the significant mode for almost all structures.

A *mode shape* is a simple displacement pattern that occurs as a linear elastic structure moves when subjected to dynamic forces.

The first-mode shape is defined as the displacement pattern where all lumped masses are on one side of the reference axis. Higher mode shapes will show masses on both sides of the vertical reference axis. In a dynamic analysis, the complex motion of the complete structure is described by adding together the appropriate percentages of all of the modes of vibration.

The notation E is used for seismic forces. Its magnitude on an element of a structure is defined as

$$E = \rho Q_E \pm 0.2 S_{DS} D$$

in which ρ is a factor representing redundancy and reliability, Q_E is the horizontal seismic-force component, S_{DS} is the design spectral response acceleration at short periods, and D is the deal load. The first term represents horizontal forces, while the second term represents forces acting vertically, reducing dead load for overturning resistance and increasing downward vertical reactions.

5.2.18 Horizontal Distribution of Forces

Within the context of an elastic ELF analysis, the distribution of lateral forces to various seismic-force-resisting elements depends on the type, geometric arrangement, and vertical extents of the resisting elements and on the shape and flexibility of the floor diaphragms. Because seismic-force-resisting elements are expected to respond inelastically to design ground motions, the distribution of forces to the various elements also depends on the strength of the elements and their sequence of yielding. Clearly, such effects cannot be captured accurately by a linear elastic static analysis. Nonlinear dynamic analysis is too cumbersome to be applied to the design of most buildings so other approximate methods are used.

Of particular concern is the torsional response of the structure during the earthquake. This response has been observed in structures that are designed to be nearly symmetric in plan and layout of seismic-force-resisting systems. This torsional response is due to a variety of *accidental* eccentricities that exist due to uncertainties in quantifying the mass and stiffness distribution of the structure, as well as torsional components of ground motion that are not included explicitly in code-based designs.

5.2.19 Inherent and Accidental Torsion

When lateral forces in a particular direction are applied statically at each story of a building with rigid diaphragms, torsional displacement (twisting about the vertical axis) occurs if the centers of stiffness and mass of each story are not perfectly coincident in plan. When 3D analysis is used, this inherent torsion is included automatically. When planar analysis is used, the centers of mass and rigidity for each story must be determined explicitly. Unfortunately, it is difficult to determine the cr for a multistory building to compute the inherent torsion; the cr for a particular story depends on the configuration of the seismic-force-resisting elements above and below that story and may be load dependent.

For buildings with fully flexible diaphragms, vertical elements are assumed to resist inertial forces from the mass that is tributary to the elements, but with no explicitly computed torsion. No diaphragm is perfectly flexible so some torsional forces always develop even when they are ignored.

Even for perfectly symmetric buildings, the true locations of the centers of mass and rigidity are uncertain. Other effects also may produce torsion. The requirement to consider accidental torsion is intended to address this concern.

Accidental and inherent torsions result in forces that must be combined with those obtained from the application of the lateral story forces; all components must be designed for the maximum effects determined considering positive accidental torsion, negative accidental torsion, and no accidental torsion.

In calculating the torsional amplification factor, A_x, the applied loads include inherent and accidental torsion, but with no further amplification; the calculation is not iterative.

The torsional moment to be considered in the design elements in a story consists of two parts:

1. M_t, the moment due to eccentricity between centers of mass and resistance for that story, which is computed as the story shear times the eccentricity perpendicular to the direction of applied earthquake forces
2. M_{ta}, commonly referred to as *accidental torsion*, which is computed as the story shear times the *accidental eccentricity*, equal to 5% of the dimension of the structure (in the story under consideration) perpendicular to the applied earthquake forces

Computation of M_{ta} in this manner implies that the dimension of the structure is the dimension in the story where the torsional moment is being computed and that all the masses above that story should be assumed to be displaced in the same direction at one time (e.g., first, all of them to the left and then, to the right).

Dynamic analyses assuming linear behavior indicate that the torsional moment due to eccentricity between centers of mass and resistance may significantly exceed M_t. However, such dynamic magnification is not included in the ASCE 7-10 partly because its significance is not well understood for structures designed to deform well beyond the range of linear behavior.

Accidental torsion is intended to cover the effects of several factors that have not been explicitly considered in the design. These factors include the rotational component of ground motion about a vertical axis; unforeseeable differences between computed and actual values of stiffness yield strengths and dead-load masses and unforeseeable unfavorable distributions of dead- and live-load masses.

The way in which the story shears and the effects of torsional moments are distributed to the vertical elements of the seismic-force-resisting system depends on the stiffness of the diaphragms relative to vertical elements of the system.

Where the diaphragm stiffness in its own plane is sufficiently high relative to the stiffness of the vertical components of the system, the diaphragm may be assumed to be indefinitely rigid for purposes of analysis. Then, in accordance with compatibility and equilibrium requirements, the shear in any story is to be distributed among the vertical components in proportion to their contributions to the lateral stiffness of the story while the story torsional moment produces additional shears in these components that are proportional to their contributions to the torsional stiffness of the story about its center of resistance. This contribution of any component is the product of its lateral stiffness and the square of its distance to the center of resistance of the story. Alternatively, the story shears and torsional moments may be distributed on the basis of a 3D analysis of the structure, consistent with the assumption of linear behavior.

Where the diaphragm in its own plane is very flexible relative to the vertical components, each vertical component acts nearly independently of the rest. The story shear should be distributed to the vertical components considering these to be rigid supports. Analysis of the diaphragm acting as a continuous horizontal beam or truss on rigid supports leads to the distribution of shears. Because the properties of the beam or truss may not be accurately computed, the shears in vertical elements should not be taken to be less than those on *tributary areas*. Accidental torsion may be accounted for by adjusting the position of the horizontal force with respect to the supporting vertical elements.

5.2.20 Story Drift Determination

Equation 12.8-15 is used to estimate inelastic deflections, which are then used to calculate design story drifts. These story drifts must be less than the allowable story drifts of ASCE 7-10, Table 12.12-1. For buildings without torsional irregularity, computations are performed using deflections at the centers of mass of adjacent stories. For SDC C, D, E, or F structures that are torsionally irregular, Section 12.12.1 requires that drifts be computed along the edges of the structure.

$$\delta_x = \frac{C_d \delta_{xe}}{I} \qquad \text{(ASCE 7-10 Eq. 12.8-15)}$$

where

C_d is the deflection amplification factor
δ_{xe} is the delfections determined by an elastic analysis
I is the importance factor

The term C_d in Equation 12.8-15 amplifies the displacements from elastic analysis at design-level forces, which are reduced by R.

There is a relationship between elastic response, response to reduced design-level forces, and the expected inelastic response. If the structure remained elastic during an earthquake, the force would be V_E, and the corresponding displacement would be e. Note that V_E does not include the reduction factor, R, which accounts primarily for ductility and overstrength. According to the equal displacement *rule* of seismic design, the maximum displacement of an inelastic system is approximately equal to that of an elastic system with the same initial stiffness. This condition has been observed for structures idealized with bilinear inelastic response and a fundamental period greater than T_S. For shorter-period structures, peak displacement of an inelastic system tends to exceed that of the corresponding elastic system. Since the forces used for design include the response modification coefficient, R, the resulting displacements are too small and must be amplified.

Because of overstrength and associated stiffness increases, the actual inelastic response differs from the idealized inelastic response; the actual displacement of the system may be less than *R* times. The standard accounts for this difference by multiplying the fictitious (design-level) elastic displacements by the factor C_d, which is usually less than *R*.

The design forces used to compute include the importance factor, *I*, so Equation 12.8-15 includes *I* in the denominator. This is appropriate since the allowable story drifts (except for masonry shear wall structures) are more stringent for higher occupancy categories.

5.2.21 Period for Computing Drift

Where the response spectrum of Section 11.4.5 or the corresponding equations of Section 12.8.1 are used and the structural period is less than T_L, displacement increases with increasing period (even through forces may decrease). Section 12.8.2 applies a period limit so that design forces are not too low, but if the lateral forces used to compute drifts are inconsistent with the forces corresponding to the computed period, displacements will be overestimated. Therefore, the standard allows the determination of drift using forces that are consistent with the computed period of vibration of the structure.

Computed periods greater than C_uT_a are common, particularly for moment frames. In such cases, the seismic design forces used to proportion strength may produce displacements that violate drift limits, whereas displacement based on the computed period will satisfy drift limits.

The more flexible the structure, the more likely it is that *P*-delta effects will ultimately control the design. Computed periods that are significantly greater than (perhaps more than 1.5 times) C_uT_a may indicate a modeling error.

5.2.22 *P*-Delta Effects

P-delta (also noted as *P*Δ) effects influence both stiffness and strength of structures. The intent is to determine whether *P*-delta effects are significant and if so, to modify the strength and stiffness of the structure to account for such effects. Also, maximum permitted values are established.

$$\theta = \frac{P_x\Delta}{V_x h_{sx} C_d} \qquad \text{(ASCE 7-10 Eq. 12.8-16)}$$

where

P_x is the total vertical design load at and above Level *x* (kip or kN); where computing P_x, no individual load factor need exceed 1.0

Δ is the design story drift simultaneously with V_x (in. or mm)

V_x is the seismic shear force acting between Levels *x* and *x*–1 (kip or KN)

h_{sx} is the story height below level *x* (in. or mm)

C_d is the deflection amplification factor

Equation 12.8-16 is used to determine the stability coefficient of each story of a structure. Where the stability coefficient exceeds 0.1, *P*-delta effects must be considered using one of two approaches. Displacements and member forces are either multiplied by $1/(1-\theta)$ to reflect the conditions in accordance with the equal displacement rule or determined by rational analysis. Two types of rational analysis are envisioned. First, a nonlinear static (pushover) analysis could be performed to show that the post-yield slope of the pushover curve is continuously positive up to the target displacement. Second, a nonlinear dynamic response-history analysis could be repeated with and without *P*-delta effects to determine if the behavior including *P*-delta meets all performance criteria.

Although the *P*-delta procedure given in ASCE 7-10 is a simple static idealization, the real issue is one of dynamic stability. For that reason, NRH analysis is appealing. Such analysis should reflect variability of ground motions and system properties, including initial stiffness, strain-hardening stiffness, initial strength hysteretic behavior, and magnitude of gravity load. Unfortunately, the dynamic response of structures is highly sensitive to such parameters, causing considerable dispersion to appear in the results. This dispersion, which increases dramatically with stability coefficient θ, is due primarily to the incrementally increasing residual deformations that occur during the response. Residual deformations may be controlled by increasing either the initial strength or the secondary stiffness.

$$\theta_{max} = \frac{0.5}{\beta C_d} \leq 0.25 \qquad \text{(ASCE 7-10 Eq. 12.8-17)}$$

where β is the ratio of shear demand to shear capacity for the story between Levels x and $x - 1$. This ratio is permitted to be conservatively taken as 1.0.

Equation 12.8-17 establishes the maximum stability coefficient permitted. The intent of these requirements is to protect structures from the possibility of stability failures triggered by postearthquake residual deformation.

The *P*-delta effects in a given story are due to the eccentricity of the gravity load above that story. If the story drift due to the lateral forces were Δ, the bending moments in the story would be augmented by an amount equal to Δ times the gravity load above the story. The ratio of the *P*-delta moment to the lateral force story moment is designated as a stability coefficient, θ. If the stability coefficient is less than 0.10 for every story, the *P*-delta effects on story shears and moments and member forces may be ignored. If, however, the stability coefficient θ exceeds 0.10 for any story, the *P*-delta effects on story drifts, shears, member forces, etc. for the entire structure must be determined by a rational analysis.

The *P*-delta procedure effectively checks the static stability of a structure based on its initial stiffness.

There is justification for using the *P*-delta amplifier based on elastic stiffness because of the following:

1. Many structures display strength well above the strength implied by code-level forces. This overstrength likely protects structures from stability-related failures.
2. The likelihood of a failure due to instability decreases with increased intensity of expected ground shaking. This is due to the fact that the stiffness of most structures designed for extreme ground motion is significantly greater than the stiffness of the same structures designed for lower-intensity shaking or for wind. Since damaging, low-intensity earthquakes are somewhat rare, there would be little observable damage.

5.2.23 Analysis Procedures with Particular Emphasis on Response Spectrum Analysis

A flow chart of permitted analysis procedures, developed from ASCE 7-10, Table 12.6-1, is shown in Figure 5.12.

The modal superposition method is a general procedure for linear analysis of the dynamic response of structures. In various forms, modal analysis has been widely used in the earthquake-resistant design of special structures such as very tall buildings, offshore drilling platforms, dams, and nuclear power plants for a number of years; however, its use has become more common for ordinary structures as well because of the advent of high-speed, desktop computers and the availability of relatively inexpensive structural analysis software capable of performing 3D modal analyses.

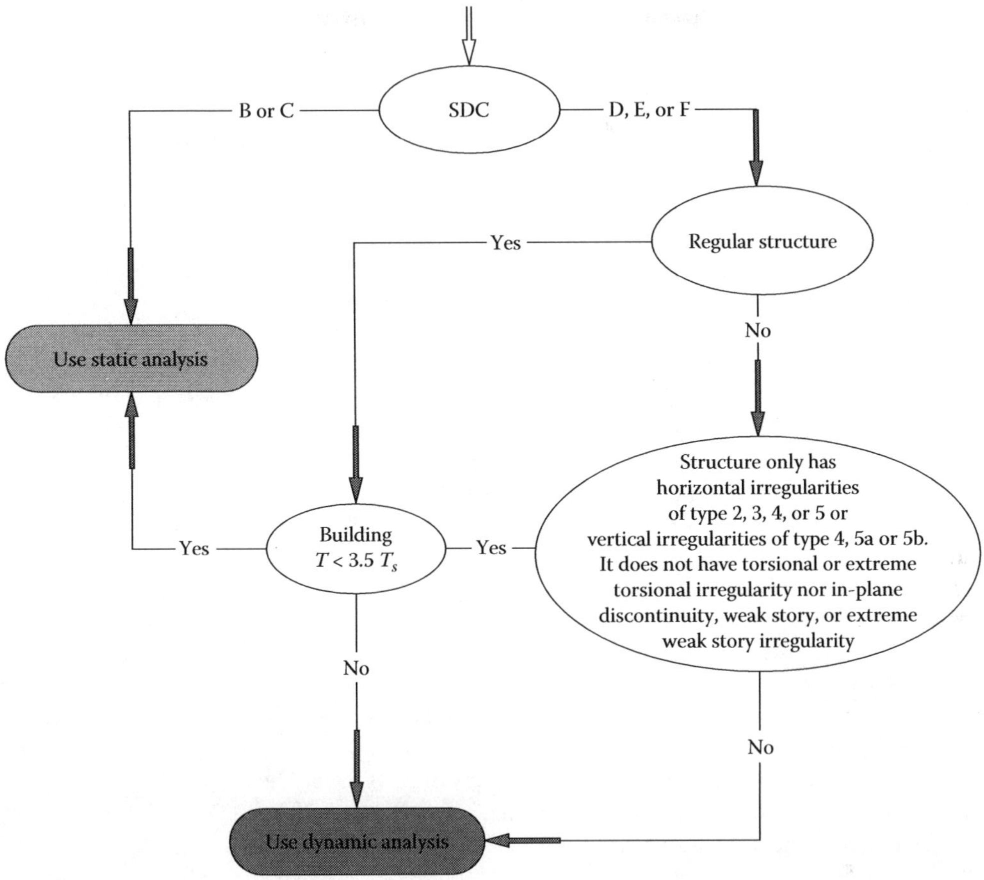

FIGURE 5.12 Permitted seismic analysis procedures. (Based on ASCE 7-10, Table 12.6-1.)

When modal analysis is specified by the ASCE provisions, a 3D analysis generally is required except in the case of highly regular structures or structures with flexible diaphragms.

The ELF procedure and the response spectrum procedure are both based on the approximation that the effects of yielding can be adequately accounted for by the linear analysis of the seismic-force-resisting system for the design spectrum, which is the elastic acceleration response spectrum reduced by the response modification factor, R. The effects of the horizontal component of ground motion perpendicular to the direction under consideration in the analysis, the vertical component of ground motion, and torsional motions of the structure are all considered in the same simplified approaches in the two procedures. The main difference between the two procedures lies in the distribution of the seismic lateral forces over the height of the building. In the modal analysis procedure, the distribution is based on properties of the natural vibration modes, which are determined from the mass and stiffness distribution. In the ELF procedure, the distribution is based on simplified formulas that are appropriate for regular structures. Otherwise, the two procedures are subject to the same limitations.

The simplifications inherent in the ELF procedure result in approximations that are likely to be inadequate if the lateral motions in two orthogonal directions and the torsional motion are strongly coupled. Such would be the case if the buildings were irregular in its plan configuration or if it had a regular plan, but its lower natural frequencies were nearly equal. The modal analysis method includes a general model that is more appropriate for the analysis of such structures. It requires at least three DOFs per floor: two for translational motion and one for torsional motion.

The methods of modal analysis can be generalized further to model the effect of diaphragm flexibility, soil–structure interaction, etc. In the most general form, the idealization would take the form of a large number of mass points, each with six DOFs (three translational and three rotational) connected by generalized stiffness elements.

The ELF procedure and the response spectrum procedure are all likely to be systematically on the unsafe side if story strengths are distributed irregularly over height. This feature is likely to lead to the concentration of ductility demands in a few stories of the building. The nonlinear static (or the so-called pushover) procedure is a method to more accurately account for irregular strength distribution. However, it also has limitations and is not particularly applicable to tall structures or structures with relatively long fundamental periods of vibration.

The actual strength properties of the various components of a structure can be explicitly considered only by a nonlinear analysis of dynamic response by the direct integration of the coupled equations of motion. This method has been used extensively in earthquake research studies of inelastic structural response. If the two lateral motions and the torsional motion are expected to be essentially uncoupled, it would be sufficient to include only one DOF per floor, for motion in the direction along which the structure is being analyzed; otherwise, at least three DOFs per floor, two translational and one torsional, should be included. It should be recognized that the results of an NRH analysis of such mathematical structural models are only as good as the models chosen to represent the structure vibrating at amplitudes of motion large enough to cause significant yielding during strong ground motions. Furthermore, reliable results can be achieved only by calculating the response to several ground motion-recorded accelerograms and/or simulated motions and examining the statistics of response.

It is possible with presently available computer programs to perform 2D and 3D inelastic analyses of reasonably simple structures. The intent of such analyses could be to estimate the sequence in which components become inelastic and to indicate those components requiring strength adjustments so as to remain within the required ductility limits. It should be emphasized that with the present state of the art in analysis, there is not one method that can be applied to all types of structures. Further, the reliability of the analytical results is sensitive to

1. The number and appropriateness of the input motion records
2. The practical limitations of mathematical modeling including interacting effects of inelastic elements
3. The nonlinear solution algorithms
4. The assumed hysteretic behavior of members

Because of these sensitivities and limitations, the maximum base shear produced in an inelastic analysis should not be less than 80% of that required by ELF.

The least rigorous analytical procedure that may be used in determining the design seismic forces and deformations in structures depends on the SDC and the structural characteristics (in particular, regularity). Except for structures assigned to SDC A, the ELF procedure is the minimum level of analysis except that a more rigorous procedure is required for some SDCs D, E, and F structures. The modal analysis procedure adequately addresses vertical irregularities of stiffness, mass, or geometry:

1. Structures with irregular mass and stiffness properties in which case the simple equations for the vertical distribution of lateral forces may lead to erroneous results
2. Structures (regular or irregular) in which the lateral motions in two orthogonal directions and the torsional motion are strongly coupled
3. Structures with the irregular distribution of story strengths leading to the possible concentration of ductility demand in a few stories of the building

In such cases, a more rigorous procedure that considers the dynamic behavior of the structure should be employed.

Many of the standard procedures for the analysis of forces and deformations in structures subjected to earthquake ground motion are listed in the following in order of increasing rigor and expected accuracy:

1. ELF procedure
2. Response spectrum (modal analysis) procedure
3. LRH procedure
4. Nonlinear static procedure, involving the incremental application of a pattern of lateral forces and the adjustment of the structural model to account for progressive yielding under load applications (pushover analysis)
5. NRH procedure, involving the step-by-step integration of the coupled equations of motion

Each procedure becomes more rigorous if effects of soil–structure interaction are considered. In the model response spectrum analysis method, the structure is decomposed into a number of SDOF systems, each having its own mode shape and natural period of vibration. The number of modes available is equal to the number of mass DOFs of the structure, so the number of modes can be reduced by eliminating mass DOFs. For example, rigid diaphragm constraints may be used to reduce the number of mass DOFs to one per story for planar models and to three per story (two translations and rotation above the vertical axis) for 3D structures. However, where the vertical elements of the seismic-force-resisting system have significant differences in lateral stiffness, rigid diaphragm models should be used with caution as relatively small in-plane diaphragm deformations can have a significant effect on the distribution of forces.

For a given direction of loading, the displacement in each mode is determined from the corresponding spectral acceleration, modal participation, and mode shape. Because the sign (positive or negative) and the time of occurrence of the maximum acceleration are lost in creating a response spectrum, there is no way to recombine modal responses exactly. However, statistical combination of modal responses produces reasonably accurate estimates of displacement and component forces. The loss of signs for computed quantities leads to problems in interpreting force results where seismic effects are combined with gravity effects, produces forces that are not in equilibrium, and makes it impossible to plot deflected shapes of the structure.

5.2.23.1 Number of Modes

The key motivation to perform MRS analysis is to determine how the actual distribution of mass and stiffness of a structure affects the elastic displacements and component forces. Where at least 90% of the model mass participates in the response, the distribution of forces and displacements is sufficient for design. The scaling required by Section 12.9.4 ASCE 7-10 controls the overall magnitude of design values so that incomplete mass participation does not produce unconservative results.

The number of modes required to achieve 90% modal mass participation is usually a small fraction of the total number of modes.

5.2.23.2 Modal Response Parameters

The design response spectrum whether the general spectrum from Section 11.4.5 or a site-specific spectrum is representative of linear elastic structures. Division of the spectral ordinates by R accounts for inelastic behavior, and multiplication of spectral ordinates by I provides the additional strength needed to improve the performance of important structures. The displacements that are computed using the response spectrum that has been modified by R and I (for strength) must be amplified by C_d and reduced by I to produce the expected inelastic displacements.

Most computer programs provide either the SRSS or the CQC method of modal combination. The two methods are identical when applied to planar structures, or when zero damping is specified for the computation of the cross-modal coefficients in the CQC method. The modal damping specified in each mode for the CQC method should be equal to the damping level that was used in the development of the response spectrum.

The SRSS and CQC method is applied to loading in one direction at a time. Where explicit consideration of orthogonal loading effects is required, the results from one direction of loading may be added to 30% of the results from the loading in an orthogonal direction. A more accurate approach is to use the SRSS method to combine 100% of the results from each of two orthogonal directions where the individual directional results have been combined by SRSS and CQC, as appropriate.

5.2.23.3 Scaling Design Values of Combined Response

The modal base shear, V_t, may be less than the ELF base shear, V, because of the following: (1) the calculated fundamental period may be longer than that used in computing V, (2) the response is not characterized by a single mode, and (3) the ELF base shear assumes 100% mass participation in the first mode, which is always an overestimate. The scaling required by the provisions provides, in effect, a minimum base shear for design. This minimum base shear is provided because the computed period of vibration may be the result of an overly flexible provisions analytical model. The possible 15% reduction in design base shear may be considered as an incentive for using an MRS analysis in lieu of the ELF procedure.

Displacements from the MRS are not scaled because the use of an overly flexible model will result in conservative estimates of displacement that need not be further scaled.

5.2.23.4 Horizontal Shear Distribution

Accidental torsion must be included in the analysis. For modal analysis, there are two basic approaches to include accidental torsion.

The first approach is to perform static analyses with accidental torsions applied at each level of the structure and then add these results to those obtained from the MRS analysis. Where this approach is used, torsional amplification is required.

The second approach, which applies only to 3D analysis, is to offset the centers of mass of each story 5% in each direction, thus requiring four separate models. The advantage of this method is that the effects of direct loading and accidental torsion are combined automatically. A practical disadvantage is the increased bookkeeping for multiple analyses.

When this approach is used, further amplification of accidental torsion is not required because repositioning the center of mass in a dynamic analysis changes the natural mode shapes and frequencies, producing torsions larger than the static accidental torsion.

5.2.23.5 Deflection Amplification due to *P*-Delta Effects

Amplification of displacements and member forces as a result of *P*-delta effects may be accomplished through the use of the geometric stiffness. For the purpose of dynamic analysis, the linearized geometric stiffness, which includes the story-wise *P*-delta effect, is usually sufficient. Using the consistent geometric stiffness (*P*-delta effect), which is associated with the deflected shape of the individual elements of the structure, slightly improves accuracy. Including *P*-delta effects directly in dynamic analysis lengthens the periods of vibration of each mode of response and increases lateral displacements.

The forces for designing the vertical elements are given the symbol F_x, and the forces applied to the design of horizontal diaphragms are given the symbol F_{px}. Both F_x and F_{px} are horizontal *story forces* applied to level x in the structure. Thus, the horizontal forces are assumed to be concentrated at the story levels in as much the same manner as the masses tributary to a level are *lumped* or assigned to a particular story height.

The rationale behind two different F_t and F_{ps} distributions has to do with the fact that the forces occurring during an earthquake change rapidly with time. Because of these rapidly changing forces and the different modes of vibration, it is likely that maximum force on an individual horizontal diaphragm will not occur at the same instant in time as the maximum force on another horizontal diaphragm. Hence, the loading given by F_{px} is to account for the possible larger *instantaneous* forces that will occur on individual horizontal diaphragms. Therefore, F_{px} story force is to be used in the design of individual horizontal diaphragms, diaphragm collectors (drag struts), and related connections.

Floor and roof assemblies obligatory for resisting gravity loads also must provide the necessary strength and stiffness to effectuate load path continuity. This linkage between the vertical and horizontal structural elements is necessary to distribute lateral loads due to wind and seismic, to the lateral support subsystems. The subsystems typically consisting of vertical bracing systems such as moment frames, braced frames, and shear walls are generally referred to as VLLRSs. However, to keep the verbiage easy-flowing, we will refer to these simply as vertical systems, it being understood that these systems are the designated lateral-load-resisting systems. Their primary function, in addition to carrying vertical loads, is to resist lateral loads due to seismic and wind loads.

In this role, the floor and roof surfaces, which are most typically horizontal, act as beams or diaphragms spanning between the vertical subsystems. They may effectively engage beam elements at the perimeter (transverse to the direction of the load) as top and bottom *chords* or *flanges*, in which case the bending moment can be resolved into a tension and compression couple and considered resisted by the beam elements (with shear resisted by the diaphragm surface). In the absence of such flange elements to take the moment couple, the floor or roof must act as a deep plate taking both bending and shear forces. Either type of diaphragm behavior requires effective transfer of bending and shear forces in the plane of the floor or roof, necessitating careful detailing of connections between the diaphragm and the lateral support subsystem.

The stiffness of the diaphragm has an important effect on the proportionate distribution of lateral loads to the various components of the lateral support system. This effect is indicated in Figure 3.27, which shows how diaphragm and lateral support element relative rigidities influence load distribution. At one extreme is the *rigid diaphragm,* which distributes the lateral loads to the lateral support elements in proportion to just *their* rigidities. At the other extreme is the *flexible diaphragm,* which transmits lateral loads based on tributary area. The *semirigid diaphragm* falls somewhere in between these two extremes and involves both aspects of behavior.

No new concepts are involved in the analysis and design of floor and roof assemblies for gravity loads. The slab, beam, or truss elements are designed using the well-accepted assumption that gravity loads are proportional to the areas tributary to these elements. However, the role of a diaphragm in seismic design is best understood by asking ourselves a question: *How does the inertia load generated in a unit area of a given diagram get from there to the foundation?* Designing a logical load path for the said question will typically result in an acceptable diaphragm design.

When the floor and roof assembly must also act as a diaphragm, the first step is to determine bending moments and shears due to the lateral loads. Considering the most elementary case of a simple span diaphragm spanning between two walls of equal stiffness, we may idealize that diaphragm as a simple span beam on its side. We note the following formulas apply for our diaphragm.

For the diaphragm shear

$$S = \frac{wl}{2b}$$

where

S is the maximum web shear, lb/ft
w is the uniform lateral load, lb/ft
l is the span of diaphragm between lateral support elements, ft
b is the width of diaphragm, parallel to the direction of lateral load, ft

For diaphragm bending stress

$$F_b = \frac{Mc}{I}$$

where

F_b is the maximum bending stress, ksi
$M = wl^2/8$ (ft-kips)
$c = b/2$ (ft)
I is the moment of inertia in the plane of bending, ft^4

If the diaphragm has flanges, the moment can be resisted by a tension and compression couple computed as follows:

$$C = T = \frac{M}{b}$$

The deflection of diaphragms can be evaluated with the stiffness factor (G') and the following expression, which is simply the superposition of bending and shear deflections:

$$\Delta d = \Delta f + Aw$$

$$\Delta w = \frac{q_{ave} L_1}{G'}, \quad \Delta f = \frac{5wl^4}{384EI}$$

where

q_{ave} is the average web shear over length L_1, kips/ft
L_1 is the distance between vertical resisting element and the point to which the deflection is to be determined, ft
Δw is the web component of total deflection, Δd
Δf is the flexural component of total deflection, Δd
G' is the diaphragm stiffness, kips/in.
w is the uniform lateral load
l is the span of diaphragm
E is the modulus of elasticity, diaphragm material
I is the moment of inertia of diaphragm

ASCE 7 specifies the distribution of F_{px} diaphragm story forces. As stated previously, these forces require a second vertical distribution that recognizes the higher instantaneous forces that can act on one story at a time. In practice, the F_x story forces must be determined first because they are then used to evaluate the F_{px} story forces. The F_x story forces are to be applied simultaneously to all levels in the primary LFRS for designing the vertical elements in the system. In contrast, the F_{px} story forces are applied individually to each level x in the primary LFRS for designing the horizontal diaphragms.

The story shear, F_{px}, for diaphragm design is given by

$$F_{px} = \frac{\sum_{i=x}^{n} F_i}{\sum_{i=x}^{n} w_i} w_{px}$$

and

$$0.2S_{DS}Iw_{px} \le F_{px} \le 0.4S_{DS}Iw_{px}$$

where

F_{px} is the horizontal force on primary LFRS at story level x for designing horizontal elements

F_i is the lateral force applied to level i (this is story force determined in accordance with formula for F_x)

w_{px} is the weight of diaphragm and elements tributary to diaphragm at level x

Other terms are as defined for F_x.

In the formulas for distributing the seismic force over the height of the structure, the superscript k is to account for whip action in tall, slender buildings and to allow for the effects of the higher modes of vibration. When the period of vibration is less than 0.5 s, there is no whipping effect.

The formulas for F_x and F_{px} can be simplified to a form that is similar to the base shear expression. In other words, the earthquake force can be written as the mass of the structure multiplied by a seismic coefficient. For example,

$$V = (\text{seismic coefficient})\ W$$

The seismic coefficient in the formula for V is known as the *base shear coefficient.* When all of the terms in the formulas for the story forces (F_x and F_{px}) are evaluated except the dead load w, seismic story *coefficients* are obtained. Obviously, since there are two formulas for story forces, there are two sets of seismic story coefficients.

The story coefficients used to define forces for designing LFRS is referred to as the F_x *story coefficient.* It is obtained by factoring out the story weight from the formula for F_x:

$$F_x = C_{vx}V = \frac{w_x h_x^k}{\sum_{i=1}^{n} w_i h_i^k} V$$

$$F_x = \left[\frac{V h_x^k}{\sum_{i=1}^{n} w_i h_i^k}\right] w_x$$

$$= (F_x \text{ story coefficient}) w_x$$

Likewise, the formula for F_{px} for use in diaphragm design can be viewed in terms of an F_{px} story coefficient. The formula for F_{px} is initially expressed in the format

$$F_{px} = \left[\frac{\sum_{i=1}^{n} F_i}{\sum_{i=1}^{n} w_i}\right] w_{px}$$

$$= \left(F_{px} \text{ story coefficient}\right) w_{px}$$

5.2.23.6 Seismic-Force Distribution for Diaphragm Design

Having explained the concept behind the formula for F_{px} for use in diaphragm design, we now turn our attention to the distribution of this force in the plane of the diaphragm. The theory behind this distribution is based on the idea of inertial forces generated in the diaphragm. If it is visualized that the mass tributary to each square foot of diaphragm has a corresponding inertial force generated by an earthquake force, then the loading shown in Figure 3.27A through C becomes clear. If each square foot of area has the same weight (and hence the same mass), the distributed seismic force is in proportion to the length of the diaphragm that is parallel to the direction of seismic force. Hence, the magnitude of the distributed force is large where the dimension of the diaphragm parallel to the force is larger, and it is small where this dimension is small. The same argument holds true if the distributed unit mass of the diaphragm is not the same for the entire diaphragm.

5.2.23.6.1 General Procedure for Diaphragm Design

In its role as a distributor of seismic loads to vertical lateral supports, the floor or roof surface acts as a diaphragm essentially responding as a horizontal beam, spanning between lateral support points. It may engage beam elements at the perimeter transverse to the direction of load as top and bottom *chords* or flanges. In this case, the bending moment can be resolved into a tension and compression couple and considered resisted by the beam elements to resist the moment couple; the floor or roof must act as a deep plate resisting both bending and shear forces. Either types of diaphragm behavior requires effective transfer of bending and shear forces in the plane of the roof or floor, necessitating careful detailing of connections between the diaphragm and the lateral support system.

Collectors are elements of the floor or roof structures that serve to transmit lateral forces from their location of origin to the vertical seismic-force-resisting elements (e.g., walls or moment frames) of the building. Typically, collectors transfer earthquake forces in axial tension or compression. When a collector is a part of the gravity-force-resisting system, it is designed for seismic axial forces along with the bending moment and shear force from the applicable gravity loads acting simultaneously with seismic forces.

When subjected to lateral forces corresponding to a DE, most buildings are expected to undergo inelastic, nonlinear behavior. Typically, the structural elements of a building that are intended to perform in the nonlinear range are the vertical elements of the seismic-force-resisting system, such as structural walls or moment frames. For the intended seismic response to occur, other parts of the seismic-force-resisting elements should have the strength to remain essentially elastic during an earthquake. It is for this reason ACI 318-08, Section 21.11 mandates design of collectors and their connections in buildings assigned to SDC D and higher, for seismic forces amplified by a factor, Ω_o. The amplification factor ranges between 2 and 3 depending upon the type of seismic system.

The intent of the Ω_o amplification factor, as stated previously, is to allow for the likely overstrength of the vertical seismic-force-resisting elements so that major yielding does not occur in collectors and their connections prior to yielding and inelastic response at the vertical elements of the building's seismic-force-resisting system.

Two values of compressive stress index calculated for the factored forces on gross section of the structural diaphragm are used to determine whether confining reinforcement is required. A compressive stress index of $0.2f_c'$ is used. Because the integrity of the entire structure depends on the ability of the collector to resist substantial compressive force under severe cyclic loading, transverse reinforcement is required to provide confinement for the concrete and reinforcement.

Diaphragms are generally treated as horizontal deep beams or trusses and distribute lateral forces to the vertical elements of the seismic-force-resisting system. As deep beams, diaphragms must be designed to resist the resultant shear and bending stresses. Diaphragms are commonly compared to girders, with the roof or floor deck analogous to the girder web in resisting shear and the boundary elements (chords) analogous to the flanges of the girder in resisting flexural tension and compression. As in girder design, the chord members (flanges) must be sufficiently connected to the body of the diaphragm (web) to prevent separation and to force the diaphragm to work as single unit.

Diaphragm openings may require additional localized reinforcement (subchords and collectors) to resist the subdiaphragm chord forces above and below the opening and to collect shear forces where the diaphragm depth is reduced. Collectors on each side of the opening drag shear into the subdiaphragms above and below the opening. The subchord and collector reinforcement must extend far enough into the adjacent diaphragm to develop the axial force through shear transfer. The required development length is determined by dividing the axial force in the subchord by the shear capacity (in force/unit length) of the main diaphragm.

Chord reinforcement at reentrant corners must extend far enough into the main diaphragm to develop the chord force through shear transfer. Continuity of the chord members also must be considered where the depth of the diaphragm is not constant.

Inertial forces are those seismic forces that originate at the specified diaphragm level, while the transfer forces originate above the specified diaphragm level. The redundancy factor, *p*, used for the design of the seismic-force-resisting elements also applies to diaphragm transfer forces, thus completing the load path.

The requirements of using the same overstrength factor, as for the seismic-force-resisting system, is intended to keep inelastic behavior in the ductile elements of the seismic-force- resisting system (consistent with the R factor) rather than in collector elements.

Diaphragm design summary for buildings assigned to SDC C, D, E, and F are as follows:

Step 1

- Evaluate the diaphragm inertial force F_{px} at the floor and roof levels by the formula

$$F_{px} = \frac{\sum_{i=x}^{n} F_i}{\sum_{i=x}^{n} W_i} W_{px} \qquad \text{ASCE 7-10 Eq. (12.10.1)}$$

where

F_{px} is the diaphragm design force
F_i is the design force applied to level i
W_i is the weight tributary to level i
W_{px} is the weight tributary to the diaphragm at level x
F_{px} need not exceed $0.4S_{DS}Iw_{px}$ but shall not be less than $0.2S_{DS}IW_{px}$

- Observe that additional shear forces resulting from transfer of vertical seismic elements or changes in their relative stiffness must be added to F_{px}. These additional forces shall be multiplied by the redundancy factor, ρ, equal to that used in the design of the structure.
- Observe that F_{px} computed from this equation is typically larger than force F_x determined by

$$F_x = C_{vx}V \qquad \text{ASCE 7-10 Eq. (12.8-11)}$$

where

$$C_{vx} = \text{vertical distribution factor} = \frac{W_x h_x^k}{\sum_{i=1}^{n} W_i h_i^k} \qquad \text{ASCE 7-10 Eq. (12.8-12)}$$

V is the total design lateral force often referred to as base shear (kip)
W_i and W_x are the portion of the total gravity load of the structure (W) located or assigned to level i or x

h_i and h_x are the height (ft or m) from the base to level i or x
k is an exponent related to the structure period as follows:

For structures having a period of 0.5 s or less, $k = 1$
For structures having a period of 2.5 s or more, $k = 2$
For structures having a period between 0.5 and 2.5 s, k shall be 2 or shall be determined by linear interpolation between 1 and 2

The formula for F_{px} allows for a higher-mode participation that can result in larger forces at individual diaphragm levels than predicted by the equation for F_x.

Step 2

- Perform a 3D lateral-load analysis of the building by applying F_{px} at floor and roof levels. Include torsion but ignore its effect if it reduces shear in the VLLR elements.

Step 3

- Determine the net shear in the vertical elements due to F_{px} equal to the difference in shears resisted by the vertical elements immediately above and below the level of the diaphragm being designed. Conceptually, the shear forces may be considered as reactions to the inertial forces of the diaphragm at that level.

Step 4

- Determine a set of equivalent loads at the diaphragm level that is in equilibrium with the shear forces determined in step 3. Use both force and moment equilibrium conditions. The equivalent loads may be determined as a combination of primary action due to F_{px} and a secondary action due to torsional effects. Refer to the numerical example given in the succeeding text.

Step 5

- Using the equivalent loads, determine the shear and bending moment at critical sections of the diaphragm.
- Compute the shear per unit length to check the shear capacity of the diaphragm. Provide collectors, also referred to as drag beams, to carry the shear that is in excess of force transferred directly into the vertical elements.

Step 6

- Calculate the ultimate shear capacity of the diaphragm as follows:

$$
\begin{aligned}
V_u &= \phi(V_c + V_s) \\
&= \phi A_{cv}\left(2\sqrt{f_c'} + \rho_n f_y\right)
\end{aligned}
$$

Note that the strength reduction factor for shear, ϕ, in diaphragm designs must not exceed the value used for the shear design of vertical elements of LFRSs.

Step 7

- Check perimeter beams (or equivalent widths of slab assumed to act as beams) and their connections for diaphragm chord forces.
- Extend chords at reentrant corners, if any, to develop the forces calculated at the critical sections.

5.2.24 Drift and Deformation

Deflection is the absolute lateral displacement of any point in a structure relative to its base, and story drift is the difference in deflection across a story (i.e., the deflection of a floor relative to that of the floor below).

The drifts and deflections are checked for the DE ground motion, which is two-thirds of the MCE_R ground motion.

There are many reasons to control drift; the most significant are to address the structural performance concerns of member inelastic strain and system stability and to limit damage of nonstructural components, which can be life-threatening. Drifts provide a direct but imprecise measure of member strain and structural stability. Under small lateral deformations, secondary stresses due to the P-delta effect are normally within tolerable limits. The drift limits provide indirect control of structural performance.

Buildings subjected to earthquakes need drift control to restrict damage of partitions, shaft and stair enclosures, glass, and other fragile nonstructural elements. The drift limits have been established without regard to economic considerations such as a comparison of present worth of future repairs with additional structural costs to limit drift. These are matters for building owners and designers to address.

The drift limits of Table 12.12-1 ASCE 7-10 reflect consensus judgment taking into account life safety and damage control objectives described earlier. Since the displacements induced in a structure include inelastic effects, structural damage in the design-level earthquake is likely. This may be seen from the seismic drift limits stated in ASCE 7-10, Table 12.12-1. For ordinary structures (occupancy category I or II), the drift limit is $0.02h_{sx}$, which is about 10 times the drift ordinarily allowed under wind loads. If deformations well in excess of the seismic drift limits were to occur repeatedly, structural components could lose so much stiffness or strength that they compromise the safety and stability of the structure.

To provide better performance for occupancy category IV essential facilities, their drift limits generally are more stringent than those for occupancy categories II and III. However, those limits are still greater than the damage thresholds for most nonstructural components. Therefore, while the performance of occupancy category IV buildings should be better than that of lower occupancy category buildings, there still can be considerable damage in the DE.

The limits set forth in are for story drifts and apply to each and every story. For some structures, satisfying strength requirements may produce a system with adequate drift control. However, the design of moment-resisting frames and of tall, narrow shear walls or braced frames often is governed by drift considerations. Where design spectral response accelerations are large, seismic drift considerations are expected to control the design of mid-rise buildings. Where design spectral response accelerations are small or the building is very tall, design for wind generally will control.

In the ASCE 7-10, for the first time, allowable drifts are based on building risk category. The higher the risk category, the more restrictive the allowable story drift. For occupancy category IV buildings (essential facilities), the allowable story drift has been reduced by a factor of two relative to recent versions of the UBC. The current requirement specifies an allowable story drift of 0.01, while the UBC value was 0.02 depending on the selected building system; this can have a significant effect on the seismic design.

The design story drift is the difference of the deflections, δ_x, at the top and bottom of the story under consideration. The deflections, δ_x, are determined by multiplying the deflections, δ_x (determined from an elastic analysis), by the deflection amplification factor, C_d. The elastic analysis is to be made for the seismic-force-resisting system using the prescribed seismic design forces and considering the structure to be fixed at the base. Stiffness other than those of the seismic-force-resisting system should not be included since they may not be reliable at higher inelastic strain levels.

The deflections are to be determined by combining the effects of joint rotation of members, shear deformations between floors, the axial deformations of the overall lateral resisting elements, and the shear and flexural deformations of shear walls. Centerline dimensions between the frame elements often are used for analysis, but clear-span dimensions with consideration of joint panel zone deformation also may be used.

The term *drift* has two connotations: story drift and absolute displacement.

1. *Story drift* is the maximum lateral displacement with a story (i.e., the displacement of one floor relative to the floor below caused by the effects of seismic loads).
2. *Absolute displacement* is the lateral deflection of any point in the structure relative to the base. This is not *story drift* and is not to be used for drift control or stability considerations since it may give a false impression of the effects in critical stories. However, it is important when considering seismic separation requirements.

There are many reasons for controlling drift, one is to control the member inelastic strain and the other stems from stability considerations. The stability of members under elastic and inelastic deformation is a direct function of both axial loading and bending of members. A stability problem is resolved by limiting the drift on the vertical-load-carrying elements and the resulting secondary moment from this axial load and deflection, that is, the *P*-delta effect. Under small lateral deformations, secondary stresses are normally within tolerable limits. However, large deformations with heavy vertical loads can lead to significant secondary moments from the *P*-delta effects in the design. The drift limits indirectly provide upper bounds for these effects.

Another reason for drift control is to restrict damage to partitions, elevator shafts, stair enclosures, glass, and other fragile nonstructural elements. The design of some nonstructural components that span vertically in the structure can be complicated when supports from the element do not occur at horizontal diaphragms. Therefore, story drift must be accommodated in the elements that will actually distort. For example, a glazing system supported by precast concrete spandrels must be designed to accommodate the full story drift, even though the height of the glazing system is only a fraction of the floor-to-floor height. The condition arises because the precast spandrels will behave as rigid bodies relative to the glazing system, and therefore, all the drift must be accommodated by the joint between the precise spandrel and the glazing unit.

Determination of design story drift involves the following steps:

1. Determine the lateral deflections at the various floor levels by an elastic analysis of the building under the design base shear. The lateral deflection at floor level x, obtained from this analysis, is termed δ_{xe}. The subscript e stands for elastic analysis.
2. Amplify d_{xe} by the deflection amplification factor, C_d. The resulting quantity, $C_d\delta_{xe}$, is an estimated DE displacement at floor level x. ASCE 710 requires this quantity to be divided by the importance factor, I_E, because the forces under which the δ_{xe} displacement is computed are already amplified by I_E. Since drift limits are tighter for buildings in higher occupancy categories, this division by I_E is important. Without it, there would be a double tightening of drift limitations for buildings with seismic importance factors greater than one. The quantity $C_d\delta_{xc}/I_E$ at floor level x is δ_x, the adjusted DE displacement.
3. Calculate the design story drift δ_x for story x (the story below floor level x) by deducting the adjusted DE displacement at the bottom of story x (floor level $x - 1$) from the adjusted DE displacement at the top of story x.

$$\Delta_x = \delta_x - \delta_x - 1$$

The Δ_x values must be kept within limits, as given in ASCE 7-10, Table 12.12-1.

Three items are worth noting:

The design story drift must be computed under the strength level DE forces irrespective of whether member design is done using the strength design or the ASD load combinations. Note that this comment does not apply to reinforce concrete structures that are designed by ultimate strength method.

The redundancy coefficient, p, is equal to 1.0 for the computation of the design story drift.

For determining compliance with the story drift limitations, the deflections, δ_x, may be calculated as indicated previously using design forces corresponding to the fundamental period of the structure, T, calculated without the limit, $T < C_u T_a$. The same model of the seismic-force-resisting system used in determining the deflections must be used for determining T. The waiver does not pertain to the calculation of drifts for determining $P\Delta$ effects on member forces, overturning moments, etc. If $P\Delta$ effects are significant, the design story drift must be increased by the resulting incremental factor.

The $P\Delta$ effects in a given story are due to the eccentricity of the gravity load above the story. If the design story drift due to the lateral forces is Δ, the bending moments in the story are augmented by an amount equal to Δ times the gravity load above the story. The ratio of the $P\Delta$ moment to the lateral-force story moment is designated as the stability coefficient. If the stability coefficient, q, is less than 0.10 for every story, then the $P\Delta$ effects on story shears and moments and member forces, etc., must be determined by a rational analysis. However, with the availability of computer programs that take into consideration $P\Delta$ effects automatically within the analysis, hand calculations of q for determining whether $P\Delta$ is significant are rarely necessary.

$P\Delta$ effects are much more significant in buildings assigned to low SDCs than in buildings assigned to high SDCs. This is because lateral stiffness of buildings is typically greater for higher SDCs.

The design story drift limits reflect consensus judgment taking into account the goals of drift control outlined earlier. In terms of life safety and damage control objectives, the drift limits should yield a substantial, though not absolute, measure of safety for well-detailed and constructed brittle elements.

To provide a higher performance standard, the drift limit for structures contained in the four occupancy categories is progressively more stringent in order to attain minimum levels of earthquake performance suitable to the individual occupancies.

It should be emphasized that the drift limits, Δ_a, are story drifts and, therefore, are applicable to each story (i.e., they must not exceed in any story even though the drift in other stories may be well below the limit).

Stress or strength limitations imposed by design-level forces may provide adequate drift control for low-rise buildings. However, it is expected that the design of moment-resisting frames and tall, narrow shear wall buildings will be governed at least in part by drift considerations. In areas having large design spectral response accelerations, S_{DS} and S_{D1}, it is expected that seismic drift considerations will predominate for buildings of medium height. In areas having low design spectral response accelerations and for very tall buildings in areas with large design spectral response accelerations, wind considerations generally will control. However, as stated many times in this chapter, the detailing of members must comply with the governing seismic provisions.

Due to probable first-mode drift contributions, the ELF procedure may be too conservative for drift design of very tall moment-frame buildings. It is suggested for these buildings, where the first mode would be responding in the constant-displacement region of a response spectrum (where displacement would be essentially independent of stiffness), that the response spectrum procedure be used for design even when not required, the reason being economy.

5.2.25 Building Separation

The intent of this section is to address separations (also called seismic joints) between adjacent structures or portions of the same structure for the purpose of permitting independent response to earthquake ground motion. For irregular structures that cannot be expected to act reliably as a unit, seismic joints should be used to produce separate units whose independent response to earthquake ground motion can be predicted.

ASCE 7-10 does not give a precise formulation for the separations, but it does require that the distance be *sufficient to avoid damaging contact under total deflection*. It is recommended that the distance be no less than the square root of the sum of the squares of the lateral deflections, which represent the anticipated maximum inelastic deformations including torsion, of the two units assumed to deflect toward each other (thus increasing with height). If the effects of impact can be shown not to be detrimental, these distances can be reduced. For very rigid shear wall structures with rigid diaphragms whose lateral deflections cannot be reasonably estimated, it is suggested that older code requirements for structural separation of at least 1 in. (25 mm) plus ½ in. (13 mm) for each 10 ft (3 m) of height above 20 ft (6 m) be followed.

Building separations are necessary between two adjoining buildings or parts of the same building, with or without frangible closures, for the purpose of permitting the adjoining buildings or parts to respond independently to earthquake ground motion. Unless all portions of the structure have been designed and constructed to act as a unit, they must be separated by seismic joints. For irregular structures that cannot be expected to act reliably as a unit, seismic joints should be used to separate the building into units whose independent response to earthquake ground motion can be predicted.

The separation should be sufficient to avoid damaging contact under total deflection to prevent interference and possible destructive hammering between buildings. It is recommended that the distance be equal to the statistical sum of the lateral deflections, δ_x, of the two units. This involves increasing separations with height. If the effects of hammering can be shown not to be detrimental, these distances can be reduced.

Adjacent structures on the same property shall be separated at least by a distance S_{MT} determined as the statistical sum of S_{M1} and S_{M2} as follows:

$$S_{MT} = \sqrt{(S_{M1})^2 + (S_{M2})^2}$$

In dense urban settings, there exists the potential for closely spaced buildings to pound against each other. Pounding occurs when buildings with different dynamic response characteristics, which are governed by building stiffness (period of vibration), floor height, and number of stories, vibrate or sway out of sync when subjected to ground shaking. The potential for pounding is most acute when the story heights of adjacent buildings are dissimilar. Pounding has caused severe damage and even collapse in urban earthquakes, such as during the magnitude 8.1 earthquake that affected Mexico City in 1985. Although building codes call attention to this problem, building designers are often reluctant to provide the necessary space between buildings to eliminate the problem, principally because the required space would reduce available square footage in the building being developed. Consequently, adequate seismic gaps between buildings are seldom implemented in densely populated urban areas of seismically hazardous regions of the United States. In Japan, even with the acute shortage of space in its largest cities, the problem is taken seriously, with new buildings seldom built closer than a meter or so from adjacent structures. In suburban or campus-type site planning in which building sites tend to be much larger, the problem seldom arises.

Closely spaced buildings in dense urban environments are also subjected to the failure of hazardous adjacent buildings or building components. The problem is most acute if there is an older adjacent building, built to less stringent seismic codes, that is taller than the new building being constructed. Designers should carefully assess neighboring structures and design against possible falling objects from them (e.g., unreinforced parapets, walls, or chimneys). In the 1989 Loma Prieta earthquake, several fatalities were caused when a large portion of an unreinforced masonry building collapsed onto the roof of a lower adjoining building.

5.2.26 Deformation Compatibility for Seismic Design Categories D, E, and F

The purpose is to require that the seismic-force-resisting system provide adequate deformation control to protect elements of the structure that are not part of the seismic-force-resisting system. In regions of high seismicity, many designers apply ductile detailing requirements to elements that are intended to resist seismic forces but neglect such practices in nonstructural elements or elements intended to resist only gravity forces. Even where elements of the structure are not intended to resist seismic forces and are not detailed for such resistance, they can participate in the response and suffer severe damage as a result.

In the 1994 Northridge earthquake, such participation was a cause of several failures. For example, the several partial collapses in precast concrete parking structures seem to have been precipitated by damage to the gravity-load system. Possibly, the combination of large lateral deformation and vertical load caused crushing in poorly confined columns that were not detailed to be part of the lateral-load-resisting system. Punching shear failures were observed in some structures at slab-to-column connections. The primary lateral-load-resisting system was a perimeter ductile frame that performed quite well. However, the interior slab–column system was incapable of undergoing the same lateral deflections and experienced punching failures.

Rather than relying on designers to assume appropriate levels of stiffness, ASCE 7-10 Provisions explicitly require that the stiffening effects of adjoining rigid structural and nonstructural elements be considered and that a rational value of member and restraint stiffness be used for the design of components that are not part of the seismic-force-resisting system. Also included is a requirement to address shears that can be induced in structural components that are not part of the seismic-force-resisting system, since sudden shear failures have been catastrophic in past earthquakes.

The provisions encourage the use of intermediate or special detailing in beams and columns that are not part of the seismic-force-resisting system. In return for better detailing, such beams and column are permitted to be designed to resist moments and shears from unamplified deflections. This reflects observations and experimental evidence that well-detailed components can accommodate large drifts by responding inelastically without losing significant vertical-load-carrying capacity.

Noteworthy features of deformation compatibility requirements are the following:

1. Expected deformations must be the greater of the maximum inelastic response displacement, δ_x, considering $P\Delta$ effects and deformation induced by a story drift of 0.0025 times the story height.
2. When computing expected deformations, stiffening effects of those elements not part of the LFRS must be neglected.
3. Forces induced by expected deformations may be considered factored forces.
4. In computing forces, restraining effect of adjoining rigid structures and nonstructural elements must be considered.
5. For concrete elements that are not part of the LFRS, assigned flexural and shear stiffness properties must not exceed one-half of gross section properties, unless a rational cracked section analysis is performed.
6. Additional deformations that may result from foundation flexibility and diaphragm deflection must be considered.

The deformation compatibility requirements for D, E, or F buildings provide a means of protecting elements of the structure that are not part of the seismic-force-resisting system. The fact that many elements of the structure are not intended to resist seismic forces and are not detailed for such resistance does not prevent them from actually providing this resistance and become severely damaged, hence, the compatibility requirements.

It should be noted that diaphragm and foundation deformations must be included in checking compatibility requirements (see Figure 5.9b and c).

5.2.27 Catalog of Seismic Design Requirements for Buildings Assigned to SDC A, B, C, D, E, or F

The seismic design requirements given in the ASCE 7-10 are cascading, meaning that what applies to a lower SDC building also applies to buildings in higher SDCs. For example, if analyses using bidirectional lateral loads are required for an SDC B building, they are also a requirement for SDC C, D, E, and F buildings. In other words, what applies to SDC A applies to SDC B through F, what applies to SDC C applies to SDC D through F, and so on.

5.2.27.1 Buildings in SDC A

1. Determine a pseudo seismic lateral force at each level by using the following equation:

$$F_x = 0.01W_x$$

   where
   F_x is the design lateral force applied at story x
   W_x is the portion of the total dead load of the structure, D, located at or assigned to level x

   This minimum base shear equal to 1% of the seismic weight of the structure applies to all buildings irrespective of SDC. The term W_x applies only to the dead load of the floor. It is a revision to ASCE 7-02 provision in which W_x included a list of other loads such as 25% of the floor live load in a storage structure, partition load, weight of permanent equipment, and 20% of the flat roof snow load. The load F_x does not represent a seismic related calculation, but rather meant to provide a certain level of strength relative to the mass of the structure to ensure structural integrity.
2. Apply the static lateral force F_x independently in two orthogonal directions. The lateral forces in each direction shall be applied at all levels simultaneously. Orthogonal combination procedure of applying 100% of the forces in one direction plus 30% in the perpendicular direction is not required. Similarly, simultaneous applications of orthogonal ground motions are not warranted.
3. Interconnect all parts of the structure to provide a continuous load path. Tie any smaller portion of the structure to the remainder of the structure with connections having a lateral design strength not less than 5% of the portion's weight.
4. Provide connections at supports for each beam, girder, or truss for resisting a horizontal force acting parallel to the member. The connections shall have a minimum design strength of 5% of the dead plus live-load reaction.
5. Anchor concrete walls to the roof and all floors to provide lateral support for the wall. The connection shall be capable of resisting a horizontal force equal to 5% of the wall weight but not less than 280 lb/linear ft.

   In the aforementioned equations, the term E refers to the effect of earthquake loads as determined by applying a horizontal load of F_x equal to 1% of the building dead load at each level. Vertical effects, E_v, of the seismic loads, design coefficients, and seismic factors such as response modification coefficient R, overstrength factors Ω_o, and deflection amplification factor, C_d, are not of concern in the design of structures assigned to SDC A.
6. Just about any structural system that has a continuous load path is permitted for buildings assigned to SDC A, you are not restricted to the list given in ASCE 7-10, Table 12.2.1. The structural members and connections need only to be designed for the forces determined from the prescribed analysis. The nonductile detailing requirements given in the first 20 chapters of ACI 318-05/08 are presumed to be sufficient to provide the necessary strength and ductility for buildings assigned to SDC A.
7. There is no requirement to increase the design loads due to horizontal and vertical structural irregularities. Extreme irregularities are not prohibited. This is not the case for buildings assigned to SDC B through F.
8. There is no requirement to multiply the torsional moment by a torsional amplification factor.

9. There is no requirement to compute the story drift, Δ, along the building edges.
10. Elements supporting discontinuous walls or fames need not be designed for load combinations along with overstrength factor Ω_o.
11. There is no height limit for buildings exhibiting extreme weak-story irregularity.

5.2.27.2 SDC B Buildings

It is well worth remembering that Category B and Category C buildings will be constructed in the largest portion of the United States. Earthquake-resistant requirements are increased appreciably over Category A requirements, but they still are quite simple compared to requirements in areas of high seismicity or for buildings assigned to SDC D and higher. The category B requirements specifically recognize the need to design diaphragms, provide collectors, and reinforce around openings. These requirements may seem elementary and obvious, but because they are not specifically covered in many codes, some engineers totally neglect them.

SDC B includes occupancy category I, II, and III structures in regions of moderate seismicity. Structures in this category must be designed for the calculated forces in addition to the requirements of SDC A. The design requirements are the following:

1. Instead of lateral force $F_x = 0.01\ W_x$ at each level, we now use a rational procedure to determine the total design seismic force and its distribution over the height of the building. Typically, the ELF procedure or a dynamic procedure is used for this purpose.
2. The building structure shall include complete lateral and vertical force-resisting systems capable of providing adequate strength stiffness and *energy dissipating capacity* to withstand the design ground motions without exceeding prescribed limits of deformation and strength demand.
3. The directions of application of seismic forces used in the design shall be those which will produce the most critical effects. However, application of seismic forces independently in two orthogonal directions (as for SDC A buildings) is deemed sufficient. Similarly, orthogonal effects are permitted to be neglected.
4. The connections shall develop the strength of the connected members or the forces determined from seismic analysis. Adequate strength shall be provided in individual members to resist the shears, moment, and axial forces determined from seismic analysis.

   It is always good to remember that analysis of a structure for a given design ground motion alone does not make a structure earthquake resistant; additional details are necessary to provide adequate earthquake resistance in structures. Experienced seismic designers normally fill these requirements, but because some are not formally specified, they often are overlooked by inexperienced engineers.

   Probably, the most important single requirement of an earthquake-resistant structure is that it be tied together to act as a unit. This not only is important in earthquake-resistant design but also is an indispensable strategy in resisting high wind, flood, explosion forces, and provide protection against, progressive failure, and detrimental foundation settlement. Any part of the structure must be tied to the rest to resist a force of $0.133S_{DS}$ (but not less than 0.05) times the weight of the smaller. In addition, beams must be tied to their supports or columns and columns to footings for a minimum of 5% of the dead- and live-load reaction.

   The connections shall be capable of transmitting the seismic force, F_p, induced by the parts being connected. Any small portion of the structure shall be tied to the main structure with connections capable of delivering a seismic force equal to $0.33S_{DS}$, times the weight of the smaller portion, or 5% of the portion weight, whichever is greater.
5. Extreme weak-story irregularity is permitted for SDC B buildings not over two stories or 30 ft in height. The height limit does not apply where the extreme weak story is capable of resisting a total seismic force equal to Ω_0 times the design force.

6. There is no requirement to increase the diaphragm forces due to horizontal or vertical irregularities.
7. Torsional moment due to accidental torsion need not be amplified.
8. Performing a 3D dynamic analysis is a requirement for structures exhibiting torsional irregularity, extreme torsional irregularity, out-of-plane irregularity, or nonparallel systems irregularity.
9. There is no requirement to compute the story drift, Δ, along the building edges.
10. Extreme torsional irregularity, extreme soft-story irregularity, or extreme weak-story irregularity are not prohibited.
11. Designing the elements supporting discontinuous walls or frames for load combinations using overstrength factor Ω_o is a requirement.
12. Performing a 3D dynamic analysis with due consideration for diaphragm stiffness, the use of cracked section properties for concrete elements, and inclusion of $P\Delta$ effects, is a requirement.
13. There is no need to multiply, M_{ta}, the torsional moment due to accidental torsion by a torsional amplification $A_x = (\delta_{max}/1.2\delta_{ave})^2 \leq 3.0$.

5.2.27.3 SDC C Buildings

The requirements for Category C are more restrictive than those for Categories A and B. Also, a nominal interconnection between pile caps and caissons is required.

SDC C includes occupancy category I, II, and III structures in regions of severe seismicity. The design requirements are the following:

1. All requirements of SDC A and B also apply to SDC C.
2. A geotechnical investigation shall be conducted, and a report shall be submitted that includes an evaluation of the following potential geologic and seismic hazards:
   a. Slope instability
   b. Liquefaction
   c. Differential settlement
   d. Surface displacement due to faulting or lateral spreading

   The report shall contain recommendations for appropriate foundation designs or other measures to mitigate the effects of the previously mentioned hazards. Where deemed appropriate by the authority having jurisdiction, a site-specific geotechnical report is not required where prior evaluations of nearby sites with similar soil conditions provide sufficient direction relative to the proposed construction.
3. Foundations

   Pole-type structures: Where construction employing posts or poles as columns embedded in earth or embedded concrete footings in the earth is used to resist lateral loads, depth of embedment required for posts or poles to resist seismic forces shall be determined by means of the design criteria as established in the foundation investigation report.

   Foundation ties: Individual pile caps, drilled piers, or caissons shall be interconnected by ties. All ties shall have a design strength in tension or compression at least equal to a force equal to 10% of S_{DS} times the larger pile cap or column factored dead plus factored live load unless it is demonstrated that equivalent restraint will be provided by reinforced concrete beams within slabs on grade or reinforced concrete slabs on grade or confinement by competent rock, hard cohesive soils, very dense granular soils, or other approved means.

   Pile anchorage requirements: Where required for resistance to uplift forces, anchorage of piles to the pile cap shall be made by means other than concrete bond to the base steel section.

Exception: Anchorage of concrete-filled steel pipe piles is permitted to be accomplished using deformed bars developed into the concrete portion of the pile.

4. Increase design forces determined by static procedure by 25% for connection of diaphragms to vertical elements and to collectors, and for connections of collectors to the vertical elements. Collectors and their connections also shall be designed for these increased forces unless they are designed for load combinations with overstrength factor Ω_o.
5. Multiply, M_{ta}, the torsional moment due to accidental torsion by a torsional amplification

$$A_x = \frac{(\delta_{max})^2}{1.2\delta_{ave}} \leq 3.0.$$

6. Compute story drift, Δ, as the largest difference of the deflections along any of the edges.
7. Use $100x + 30y$, if you are using ELF or modal analysis. Use simultaneous application of load, if you are analyzing the structure using a linear or NRH procedure.

5.2.27.4 SDC D Buildings

Category D requirements compare roughly to present design practice in California seismic areas for buildings. Interaction effects between structural and nonstructural element must be investigated. Foundation interaction requirements are increased.

SDC D includes occupancy category I, II, III, and IV structures in regions of high seismicity, but not located close to a major fault, as well as occupancy category IV structures in regions of somewhat less severe seismicity. The use of some structural systems is restricted in this design category, and dynamic analysis must be used for design of irregular structures. The design requirements are as follows:

1. All requirements of SDC A, B, and C also apply to SDC D.
2. Additional geotechnical investigation report requirements shall include the following:
   a. The determination of lateral pressures on basement and retaining walls due to earthquake motions.
   b. The potential for liquefaction and soil strength loss evaluated for site peak ground accelerations, magnitudes, and source characteristics consistent with the DE ground motions. Peak ground acceleration is permitted to be determined based on a site-specific study taking into account soil amplification effects or, in the absence of such a study, peak ground accelerations shall be assumed equal to $S_S/2.5$.
   c. Assessment of potential consequences of liquefaction and soil strength loss, including estimation of differential settlement, lateral movement, lateral loads on foundations, reduction in foundation soil-bearing capacity, increases in lateral pressures on retaining walls, and flotation of buried structures.
   d. Discussion of mitigation measures such as, but not limited to, ground stabilization, selection of appropriate foundation type and depths, selection of appropriate structural systems to accommodate anticipated displacements and forces, or any combination of these measures and how they shall be considered in the design of the structure.
3. A special moment frame that is used but not required by ASCE 7-10, Table 12.2-1, shall not be discontinued and supported by a more rigid system with a lower response coefficient R, unless the building height is limited to two stories, and a 25% increase in loads is taken in the design of diaphragm elements.
4. Vertical irregularity Type 5b of ASCE 7-10 shall not be permitted.

5. For structures having a horizontal structural irregularity of Type 1a, 1b, 2, 3, or 4 in ASCE 7-10 Table 12.3-1 or a vertical structural irregularity of Type 4 in Table 12.3-2, the design forces shall be increased 25% for connections of diaphragms to vertical elements and to collectors and for connections of collectors to the vertical elements. Collectors and their connections also shall be designed for these increased forces unless they are designed for the load combinations with overstrength factor.
6. Horizontal structural components shall be designed for a net upward force of 0.2 times the dead load in additional to the other applicable load combinations.
7. Redundancy factor, ρ, shall equal 1.3 unless one of the two conditions is met, whereby ρ is permitted to be taken as 1.0:
   a. Each story resisting more than 35% of the base shear in the direction of interest shall comply with ASCE 7-10, Table 12.3-3.
   b. Structures that are regular in plan at all levels provided that the seismic-force- resisting systems consist of at least two bays of seismic-force-resisting perimeter framing on each side of the structure in each orthogonal direction at each story resisting more than 35% of the base shear. The number of bays for a shear wall shall be calculated as the length of shear wall divided by the story height or two times the length of shear wall divided by the story height for light-framed construction.
8. *Foundations*

Anchorage of the pile into the pile cap should be conservatively designed to allow energy dissipating mechanisms, such as rocking, to occur in the soil without structural failure of the pile. Precast prestressed concrete piles are exempt from the concrete special moment-frame column confinement requirements since these requirements were never intended for slender, precast prestressed concrete elements and will result in unbuildable piles. These piles have been proven through cyclic testing to have adequate performance with substantially less confinement reinforcing than required by ACI 318. Therefore, a transverse steel ratio reduced from that required in frame columns is permitted in concrete piles. It should be noted that confinement provided by the soil improves the behavior of concrete piles.

Design and construction of concrete foundation components shall confirm to the requirements of ACI 318, Section 21.8, except as modified here in:

Pole-type structures: Where construction employing posts or poles as columns embedded in earth or embedded concrete footings in the earth is used to resist lateral loads, depth of embedment required for posts or poles to resist seismic forces shall be determined by means of the design criteria as established in the foundation investigation report.

Foundation ties: Individual pile caps, drilled piers, or caissons shall be interconnected by ties. In addition, individual spread footings founded on soil defined in Chapter 20 as site class E or F shall be interconnected by ties. All ties shall have a design strength in tension or compression at least equal to a force equal to 10% of S_{DS} times the larger pile cap or column factored dead plus factored live load unless it is demonstrated that equivalent restraining will be provided by reinforced concrete beams within slabs on grade or reinforced concrete slabs on grade or confinement by competent rock, hard cohesive soils, very dense granular soils, or other approved means.

General pile design requirement: Piling shall be designed and constructed to withstand deformations from earthquake ground motions and structure response. Deformations shall include both free-field soils strains (without the structure) and deformations induced by lateral pile resistance to structure seismic forces, all as modified by soil-pile interaction.

Batter piles: Batter piles and their connections shall be capable of resisting forces and moments from the load combinations with overstrength factor of Section 12.4.3.2 or 12.14.3.2.2. Where vertical and batter piles act jointly to resist foundation forces as a group, these factors shall be distributed to the individual piles in accordance with the relative horizontal and vertical rigidities and the geometric distribution of the piles within the group.

Batter pile systems that are partially embedded have historically performed poorly under strong ground motions. Difficulties in examining fully embedded batter piles have led to uncertainties as to the extent of damage for this type of foundation. Batter piles are considered as limited ductile systems and should be designed using the special seismic load combinations.

5.2.27.5 SDC E Buildings

SDC E includes occupancy category *I*, *II*, and *III* structures located close to major active fault that is defined as a region with $S_1 \geq 0.75g$. Severe restrictions are placed on the use of some structural systems, irregular structures and analysis methods. The design requirements are the following:

1. All requirements of SDC A, B, C, and D also apply to SDC E.
2. Sitting of a structure is prohibited where there is a known potential for an active fault to cause rupture of the ground surface at the structure.

5.2.27.6 SDC F Buildings

SDC F includes occupancy category IV structures located close to major active faults (regions where $S_1 \geq 0.75g$). As in SDC E structures, severe restrictions are placed on the use of some structural systems, irregular structures, and analysis methods. The design requirements are the following:

1. All requirements of SDC A, B, C, D, and E also apply to SDC F.
2. Single-story steel ordinary moment frames and intermediate moment frames are permitted to a height of 65 ft where the deal load supported by and tributary to roof does not exceed 20 psf. In addition, the dead load of the exterior walls tributary to the moment frame shall not exceed 20 psf.
3. Steel intermediate moment frames are permitted in light-frame construction.

6 Performance-Based Design

PREVIEW

Performance-based seismic design (PBSD), more commonly denoted as PBD, is a relatively new concept that reflects a natural evolution in engineering design practice. It is based on investigations of building performance in past earthquakes and is enabled by improvements in analytical tools and computational capabilities. PBSD concepts have been made possible by the collective intellect of an interested professional and financial resources provided by the federally funded National Earthquake Hazards Reduction Program.

Interestingly enough, currently applied concepts in PBSD were developed for the rehabilitation of existing buildings, as opposed to the design of new buildings. These concepts, however, apply equally well to new buildings, and model codes for new building seismic design are beginning to adopt and adapt the performance-based concepts created for seismic rehabilitation of existing buildings.

PBD seeks to augment current code approaches rather than replacing them. In the natural hazards area, PBD is well developed for seismic design. However, prescriptive approaches are still typical for high winds. A sound multihazard design approach should provide an impetus to adopt a performance-based philosophy for design against high wind and flood risk.

Building codes establish minimum requirements for life safety through the specification of prescriptive criteria that regulate acceptable materials of construction, identify approved structural and nonstructural systems, specify required minimum levels of strength and stiffness, and control the details of how a building is to be put together. Although these prescriptive criteria are intended to result in buildings capable of providing certain levels of performance, the actual performance of individual building designs is not assessed as part of the traditional code design process. As a result, the performance capability of buildings designed to these prescriptive criteria can be better or worse than the minimum standards anticipated by the code.

PBSD explicitly evaluates how a building is likely to perform, given the potential hazard it is likely to experience, considering uncertainties inherent in the quantification of potential hazard and uncertainties in assessment of the actual building response. It permits design of new buildings or upgrade of existing buildings with a realistic understanding of the risk of occupancy interruption and economic loss that may occur as a result of future earthquakes.

It also established a vocabulary that facilitates meaningful discussion between stakeholders and design professionals on the development and selection of design options. It provides a framework for determining what level of safety and what level of property protection are acceptable to building owners, tenants, lenders, insurers, regulators, and other decision makers based upon the specific needs of a project.

In contrast to prescriptive design approaches, PBD provides a systematic methodology for assessing the performance capability of a building, system, or component. It can be used to verify the equivalent performance of alternatives, possibly deliver standard performance at a reduced cost, or confirm higher performance need for critical facilities.

First-generation procedures introduced the concept of performance in terms of discretely defined performance objectives: collapse prevention, life safety, immediate occupancy, and operational performance. Also introduced is the concept of performance related to damage of both structural and nonstructural components. Although intended for new buildings, these procedures were first developed for use in seismic vulnerability study and retrofit design of existing buildings.

To introduce the subject, we begin with a description of expected performance of structural and nonstructural components designed to current seismic standards and how it can be improved to reduce seismic risk. This is followed by an outline of current specifications of PBD that improves our ability, through analytical means, to achieve a building design that will reliably perform in a prescribed manner under postulated seismic hazards scenarios.

6.1 DEFINITIONS OF PERFORMANCE-BASED DESIGN

PBD is an evolving concept. The term as currently used has multiple definitions and three are presented in the following:

- A design approach that meets the life safety and building performance intents of the traditional code while providing designers and building officials with a more systematic way to evaluate alternative design options currently available in codes. In this regard, PBD facilitates innovation and makes it easier for designers to propose new building systems not covered by existing code provisions.
- A design approach that identifies and selects a performance objective from several performance options. These performance options are conceived to achieve higher-than-code-minimum design requirements.
- A design approach that provides designers with tools to achieve specific performance objectives such that the performance of a structure can be reliably predicted.

6.2 PRESCRIPTIVE APPROACH TO CODES

The traditional approach used in building codes in the United States has been that of prescriptive-based codes. Prescriptive-based codes are quantitative and rely on fixed values that are prescribed by the codes and intended to achieve a reasonable levels of safety from hazards such as earthquakes. Prescriptive requirements are based on broad classifications of buildings and occupancies and are typically stated in terms of fixed values.

Prescriptive codes provide limited rules for addressing various designs, and construction issues and buildings designed and built under the prescriptive-based codes are presumed safe, but it is important to understand that the requirements in the prescriptive-based codes are only the minimum necessary to safeguard the public health, safety, and general welfare. In some instances, it may be desirable, appropriate, or even necessary to raise the level of safety above the prescribed minimums.

Under the normal prescriptive code approach, all buildings assigned to a particular SDC are essentially treated alike. This approach does not address the building features and systems necessary for satisfying performance objectives, other than life safety, as may be desired by owners of high-end facilities.

How can the issues such as these and others be addressed? An innovative procedure that is becoming increasingly adopted is the use of a performance-based approach to improve or supplement the prescriptive requirements.

6.3 PERFORMANCE-BASED APPROACH

Although having detailed requirements for *performance* is relatively new to the building codes used in the United States, the concept is not. The various *prescriptive* life-safety codes contain provisions for what is known as *alternative methods and materials* or *equivalencies*. These code provisions allow for the use of methods, equipment, or materials not specified or prescribed in the code provided the alternative is approved by the code official. It is under these provisions the traditional codes that the PBD approach can be undertaken.

Under the concept of an alternative method, material, or equivalency, the code official must approve the alternative or equivalency if it can be shown to be equivalent in quality, strength, durability, and safety. The proponent of the alternative method or equivalency is responsible for providing all necessary documentation to the code official. Based on the ability of the code official to permit alternate methods and materials in the existing prescriptive codes, performance-based codes simply offer the code official a system with which to accept alternative designs based on performance. In other words, this is nothing new to the code official; it is just a more formal way to review designs.

As mentioned previously, taking a *performance* approach is not new to building design because decisions based upon performance occur in almost every project. However, PBD provides a formal and structured way of making decisions that is particularly applicable to the issue of life safety and damage reduction from natural and man-made hazards. From a designer's standpoint, the performance-based codes provide a more formalized system to develop, document, and submit alternative materials, methods, and equivalencies.

Unlike relying solely on a prescriptive code, PBD addresses an individual building's unique aspects or uses and specified and *stakeholder* needs. *Stakeholders* include everyone who as interest in the successful completion of a project (i.e., members of the design team, the builders, the owners, and the code enforcement officials). The design team is a subgroup of the *stakeholders*, which includes individuals such as representatives of the architect and other pertinent consultants.

It is critical to the proper development, approval, and implementation of any PBD for all the stakeholders to be actively involved in the process. Because the stakeholders establish the acceptable level of risk, it is crucial that all stakeholders be involved in the project from the earliest stages.

The performance-based procedure provides the basis for the development and selection of design options, based upon the needs of the specific project, to augment the broad occupancy classification requirements. The approach structures a comparison of performance objectives provided by various alternative designs and also provides a mechanism for determining what level of performance, at what cost, is acceptable to the stakeholders. PBD aims at property protection and life-safety strategies in which the systems are integrated rather than designed in isolation.

6.3.1 Performance-Based Design for Natural Hazards

As noted earlier, a performance-based approach to building design is not new, because decisions based on performance occur frequently in almost any project. What is new is the attempt to formalize a decision-making process related to expected performance and ultimately to develop performance-based codes to regulate building design and construction.

In the natural hazards area, *performance* is used to signify a level of damage or load. This, in itself, represents a major change in perception, because the building owner or occupancy generally believe that adherence to building codes provides a safe environment and anticipated degrees of damage are not normal source of conversation between an architect and an owner or even an architect and his engineer. Earthquake experience in recent years has forced recognition that damage (sometimes severe) will occur in a building designed in accord with the code.

The theory and practice of PBD currently is most advanced in seismic design. Advanced seismic engineering practitioners have, for some time, recognized several performance objectives in relation to owner's needs and have used them as a basis for establishing design parameters. These objectives, or performance levels, can be simply stated as follows:

- **Level 1**: The building is essentially undamaged and can be immediately operational (operational level).
- **Level 2**: The building is damaged, and needs some repairs, but can remain occupied and be functional after minor repairs (of a nonstructural nature) are complete (immediate occupancy level).
- **Level 3**: The building is both structurally and nonstructurally damaged, but the threat to life is minimal and occupancy injuries should be minor and few (life-safety level).
- **Level 4**: The building is severely damaged and will probably have to be demolished; it has not collapsed, although there is some likelihood of occupancy injury (near collapse level).

In this spectrum, the code-conforming building is fairly far down the scale (at level 3) and many private and public owners are prepared to pay more to achieve a higher level of performance. A hospital should achieve at least level 2, and preferably level 1. A high-tech manufacturing plant might desire to achieve the same level, because of the high value of its contents and the business losses if the plant must shut down production. The owner of a warehouse that houses a modest and easily replaced commercial inventory, with very few occasional occupants, might opt for the economies of level 4.

In the last decade or so, this informal pragmatic approach to PBSD has become formalized; the performance levels have been named and carefully defined. Detailed observation of damaged buildings, together with advances in materials science, experimental research, and analytical methods, has led to much more sophisticated understanding of building response and have enabled engineers predict more reliably how a structure will behave under various levels of shaking. This prediction is still far from a guarantee, but it has a scientific and engineering basis that was nonexistent even two decades ago. Meanwhile, extensive studies of all aspects of PBSD are underway around the country, particularly in California.

PBD is not proposed as an immediate substitute for design to traditional codes. Rather, it is seen as an opportunity for enhancement and the tailoring of the design to match the objectives of building owners. Design to the code remains as the minimum baseline to ensure safety.

To achieve a building code that regulates performance rather than easily inspected design construction methods will not be easy, but ultimately one can expect to see a rational mix of performance and prescription in the regulatory mix. That shift took place in advanced industries (e.g., airplane design) a few decades ago, and airplanes are now habitually designed to stringent performance requirements, specified by the military or the airline companies.

Designers and owners of buildings in flood and high seismic areas need to begin to think in terms of few basic objectives:

- Can the real probabilities and frequencies of events during the useful life of the building be defined with a useful degree of accuracy?
- Can the extent and kinds of damage (if any) that can be tolerated be defined?
- Are there ways (if any) in which this acceptable level can be achieved?
- Are there alternative levels of performance that can be achieved and how much do they cost over the lifetime/ownership of the building?
- Are these levels below, at, or above design to code enforced criteria?

Serious thought about these basic issues by all the stakeholders is the beginning of design for performance.

6.3.2 Performance-Based Seismic Design

As discussed previously, procedures for the application of PBSD are well advanced. However, the procedures are still evolving and issues such as terminology, analytical methods, and achieving reliable performance prediction are still subject of much research and development. This section outlines the general approaches that are current in PBSD.

6.3.2.1 Determining Acceptable Risk

The PBD procedure starts with the definition of acceptable risk. Prior to inception of design work for a new or retrofitted building, discussion should be initiated between the design team and the owner's representatives to explain the level of seismic performance that will be achieved by conformance to the code and other possible performance options that may be available. In this discussion, *seismic performance* refers to the extent of damage and loss that is likely to occur in earthquakes of differing magnitudes. This discussion focuses on ensuring that all parties understand that *earthquake* or damage-free performance is not possible and compromises must be made between seismic performance and cost. *Acceptable risk* refers to the extent and types of damage and loss that the building owners can tolerate. Clearly, avoidance of casualties is of the highest priority, but what are the priorities for issues such as damage to the building's structure, nonstructural components, and systems and contents?

The discussion of acceptable risk begins with determining the answer to the following question: If the building is designed strictly to the minimum code requirements, are the damage and loss that might occur in the design-level earthquake acceptable? If the answer to this question is positive, an implicit level of acceptable risk has been set and design can proceed. If the answer is negative or undecided, the following should be addressed:

- What lesser extent and types of damage can be accepted?
- What are the implications for long-term costs and benefits over the life of the building?
- Is the desired performance level affordable within the first cost of the owners (minimum code requirements must always be provided)?

Issues of uncertainty must also be made clear. It should be noted that the degree of uncertainty in predicting performance will be dependent on the existing design in addition to the application of code requirements. The design team for a new building has control over this issue; however, for a retrofit, some of the existing building characteristics may be less than desirable.

A new design in which key parameters of good seismic design are provided (i.e., continuous load path, structural redundancy, symmetry in plan and section, short spans, and well-designed nonstructural connections and bracing) will be more economical and more predicable in performance than a design in which these characteristics are not present.

The discussion of these issues should lead to a formal conclusion on performance objectives that then serve as a target for the designers, but it is the owner's representative who must make the final performance objective decision. The implications of this decision must be fully understood and it is the responsibility of the design team to provide necessary information, to the extent that is available.

Traditionally, the architect has been the source of all design information for the building owners, but due to the technical sophistication of PBD, the structural engineer will invariably be consulted. On large projects, the key consultants such as peer reviewers may be involved in early meetings. In these instances, the peer reviewers may be expected to be able to discuss the project on equal terms with the design team. Whether all parties are familiar with the language of PBSD may have significant impact on the extent to which seismic performance issue can be subject for useful discussion and decision making.

If community representatives or committees, whose technical expertise may be more limited, are involved, the design team should try to ensure that the issues are understood.

For most building owners, the discussion of acceptable risk will be an entirely new kind of discussion and the language of seismic performance may be unfamiliar. Historically, it has not been common practice to initiate a discussion of damage tolerance for a new project. In the initial meetings, some owners consider it a bad omen. However, agreement on design goals and expectations can help achieve a desired level of performance and limit later surprises due to unexpected earthquake damage. Such performance objectives statements might properly be part of a project's building program and serve as the basis for a PBD procedure.

ASCE/SEI 41-06 contains tables that show expected damage to vertical and horizontal structural elements: architectural, mechanical, electrical, and plumbing. These expectations refer to a building designed using the appropriate analytical tools, which provides the necessary methods of analysis and detailing to achieve these performance levels for high, moderate, and low earthquake intensity regions.

6.3.3 Expected Performance When Designing to Current Codes

Current seismic design codes are essentially aimed at the preservation of life and safety for the benefit of the community. The recommended provisions express expectations and provide no guarantees; they assume that there may be damage to a building as a result of an earthquake. A general set of performance statements to qualify the nature of expected damage may be summarized as follows:

Structures designed in accordance with current seismic recommendations should, in general, be able to

- Resist a minor level of earthquake ground motion without damage
- Resist a moderate level of earthquake ground motion without structural damage, but possibly experience some nonstructural damage
- Resist a major level of earthquake ground motion having an intensity equal to either the strongest experience or forecast for the building site without collapse, but possible with some structural as well as nonstructural damage.

It is expected that structural damage, even in a major design-level earthquake, will be limited to a repairable level for most structures that meet these requirements. In some instances, damage may not be economical to repair. The level of damage depends upon a number of factors, including the intensity and duration of ground shaking, structure configuration, type of lateral-force-resisting systems, materials used in the construction, and construction workmanship.

Designers use codes as a resource, as they provide minimum acceptable consensus standards. Codes provide no guidance on the selection of materials and systems, rather only criteria for their use once selected. Codes also do not provide the designer with the difference in performance between systems, for example, the difference between the stiffness of shear walls and frames and the importance of this characteristic for the overall seismic performance of the building. Lastly, codes do not discuss that the use of some structural systems will result in more nonstructural damage than others, even though the structural systems perform equally well in resisting the earthquake forces. The following subsections describe the expected performance of structural and nonstructural components, respectively.

6.3.4 Expected Performance of Structural Components

As mentioned earlier, current seismic design provisions for nonessential facilities are intended to provide life safety, that is, no damage in a minor earthquake, limited structural damage in a moderate earthquake, and resistance to collapse in a major earthquake (typically the design ground motion). Resistance to collapse means that the structure may have lost a substantial amount of its original lateral stiffness and strength, but the gravity-load-bearing elements still function and provide some margin of safety against collapse. The structure may have permanent lateral offset, and some elements of the seismic-force-resisting systems may exhibit substantial cracking, spalling, yielding, buckling, and localized failure. Following a major earthquake, the structure is not safe for continued occupancy until repairs are done. Shaking associated with strong aftershocks could threaten the stability of the structure. Repair to a structure in this state is expected to be feasible; however, it may not be economically attractive to do so.

6.3.5 Expected Performance of Nonstructural Components

While current seismic design provisions provide minimum structural performance standards in terms of resistance to collapse, they typically do not address performance of nonstructural components such as room partitions, filing cabinets and book cases, hung lighting and ceiling, entryway canopies, and stairwells; they also do not address performance of mechanical, electrical, or plumbing systems including fire sprinklers, heating and air conditioning equipment or ductwork electoral panels, or transformers. The vast majority of damage and resulting loss of building functionality during recent damaging earthquakes has been the result of damage to nonstructural components and systems. Many building owners have been surprised when a building withstands the effects of a moderate earthquake from a structural perspective but is still rendered inoperable from a nonstructural standpoint.

Current seismic design provisions typically require that nonstructural components be secured so as to not present a falling hazard; however, these components can still be severely damaged such that they cannot function. The loss of electric power, breaks in water supply and sewer outflow lines, or nonfunctioning hearing or air conditioning will render a building unusable by tenants. Breaks in fire sprinklers will cause flooding within all or part of a building, soaked carpets and walls, inundated files and records, and electrical shorts or failures in electrical equipment and computers. Other examples of nonstructural damage that can be expected in a code-compliant building subjected to strong ground shaking include extensive cracking in cladding, glazing, partitions, and chimneys; broken light fixtures; racked doors; and dropped ceiling tiles.

6.4 IMPROVING PERFORMANCE TO REDUCE SEISMIC RISK

Improving performance to reduce seismic risk is a multifaceted issue that requires consideration of a broad range of factors. These include understanding of the concept of seismic risk management and two of the fundamental factors affecting improved seismic performance: consideration of the seismic hazards affecting the site and consideration of the desired seismic performance of structural and nonstructural components for the range of earthquakes of concern.

In this section, we address seismic design issues that are fundamentally important to improved seismic performance, regardless of the occupancy type:

- Selection of the structural materials and systems
- Selection of the architectural/structural configuration
- Consideration of the expected performance of nonstructural components, including ceiling, partitions, heating, ventilation, and air conditioning equipment, piping and other utility systems, and cladding

6.4.1 Selection of Structural Materials and Systems

An earthquake has no knowledge of building function but uncovers weaknesses in the building that are the result of errors or deficiencies its design and construction. However, variations in design and construction will affect its response, perhaps significantly, and to the extent that these variation are determined by the occupancy, then each building type tends to have some unique seismic design determinants. A building that uses a moment-frame structure will have a different ground motion response than a building that uses shear walls; the frame structure is more flexible, so it will experience lower earthquake forces, but it will deflect more than the shear wall structure, and this increased motion may cause more damage to nonstructural components such as partitions and ceiling. The shear wall building will be stiffer but this will attract more force: the building will deflect less but will experience higher accelerations and this will affect acceleration-sensitive components such as air conditioning equipment and heavy tanks.

These structural and nonstructural system characteristics can be deduced from the information in the seismic provisions, but the provisions are not a design guide and give no direct guidance on the different performance characteristics of available systems or how to select an appropriate structural system for a specified site or building type.

6.4.2 Selection of the Architectural Configuration

The architectural configuration—the building's size, proportions, and 3D form—plays a large role in determining seismic performance. This is because the configuration largely determines the distribution of earthquake forces, that is, the relative size and nature of the forces as they work their way through the building. A good configuration will provide for a balanced force distribution, both in plan and section, so that the earthquake forces are carried directly and easily back to the foundations. A poor configuration results in stress concentrations and torsion, which at their worst are dangerous.

Configuration problems have long been identified, primarily as the result of extensive observation of building performance in earthquakes. However, many of the problem configurations arise because they are useful and efficient in supporting the functional needs of the building or accommodating site constraints. The design task is to create configuration alternatives that satisfy both the architectural needs and provide for structural safety and economy. This requires that the architect and engineer must cooperate from the outset of the design process: first to arrive at an appropriate structural system to satisfy building needs and then to negotiate detailed design alternatives that avoids, or reduces, the impact of potential problem configurations.

Seismic codes now have provisions intended to deal with configuration problems. However, the code approach is to accept the problems and attempt to solve them by either increasing design forces or requiring a more sophisticated analysis. Neither of these approaches is satisfactory, for they do not remove the problem. The problem can only be solved by design and not by a prescriptive code.

Design solutions for a soft first story condition that the architect and engineer might explore together include

- The architectural implications of eliminating it (which solves the structural problem)
- Alternative framing designs, such as increasing the number of columns or increasing the system stiffness by changing the design, to alleviate the stiffness discrepancy between the first and adjacent floors
- Adding bracing at the end of line of columns (if the site constraints permit this)

A more general problem is the increasing unpredictability of building response as the architectural/structural configuration increasingly deviates from an ideal symmetrical form. This has serious implications for PBD, which depends for its effectiveness on the ability of the engineer to predict structural performance.

6.4.3 Consideration of Nonstructural Component Performance

We know by now that the majority of the damage that has resulted in building closure following recent US earthquakes has been the result of damage to nonstructural components and systems. A building designed to current seismic regulations may perform well structurally in a moderate earthquake, but be rendered nonfunctional due to nonstructural damage.

Nonstructural components may also, however, influence structural performance in response to ground shaking. Structural analysis assumes a bare structure. Nonstructural components that are attached to the structure and heavy contents, depending on their location, may introduce torsional forces. Characteristic examples of structural/nonstructural interaction are as follows:

- Heavy masonry partitions that are rigidly attached to columns and under floor slabs can, if asymmetrically located, introduce localized stiffness and create stress concentrations and torsional forces. A particular form of this condition, which has caused significant structural damage, is when short column conditions are created by the insertion of partial masonry walls between columns. The addition of such partial walls after the building completion is often treated as a minor remodel that is not seen to require engineering analysis. The result is that the shortened columns have high relative stiffness, attract a large percentage of the earthquake forces, and fail.

- In smaller buildings, stairs can act as bracing members between floors, introducing torsion; the solution is to detach the stair from the floor slab at one end to allow free structural movement.
- In storage areas or library stacks, heavy storage items can introduce torsion into a structure. The structure may have been calculated to accommodate the maximum dead load, but consideration may be lacking for the effect of nonsymmetric loading over time as, for example, when library books are acquired.

6.5 DESIGN AND PERFORMANCE ISSUES RELATING TO COMMERCIAL OFFICE BUILDINGS

Commercial office buildings represent a large building segment and house the core of American business operations. Corporate headquarters, banks, law firms, consulting firms, accountants, insurance companies, nonprofit organizations—the list is almost endless—use office space buildings around the country to house their operations. As these companies make decisions about the buildings that they construct or office space that they lease, seismic considerations can easily be factored into the decision process.

The following are some unique issues associated with commercial office buildings that should be kept in mind during the design and construction phase of new facilities:

- Protection of building occupants is a very high priority.
- Occupants are predominately workforce, with high daytime *8 am to 5 pm* occupancy.
- Most office building occupants are generally familiar with the characteristics of their building; a small percentage of occupants may be disabled to some degree and visitors will generally not be familiar with the building.
- Office buildings change their interior layouts frequently, to respond to tenant needs, fluctuations in workforce, or organizational changes.
- Ensuring the survival of business records, whether in electronic or written form, is essential for continued business operation.
- Closure of the building for any length of time represents a serious business problem.

Commercial buildings may be owner operated, particularly if owned by national or global corporations, but many are developer owned (at least initially) housing tenant (lease holder) operations. In many instances, the developer and building designers provide an empty *shell*, which is fitted out according to the tenants' planning, spatial, and environmental needs; design and construction is generally undertaken by the tenant's consultants and contractors. This tends to split the responsibility for interior nonstructural and other risk reduction design and construction measures between the building designers and contractor and a multiplicity of tenant designers and contractors.

Financing for these facilities is typically through private loans. The effective life of an office building is 20–30 years, after which major renovation and updating is normally necessary. Interior renovation is usually on a much shorter interval, particularly for rental office structures.

6.5.1 Performance of Office Buildings in Past Earthquakes

The seismic performance of modern office buildings designed to codes adopted since the late 1970s has been good as far as providing life safety. However, the recognition by building owners that satisfactory life-safety code-level performance may still encompass considerable damage, along with repair costs and possible business interruption of the building for weeks or even months, even in moderate earthquake, suggests that some PBD strategies may be useful.

Where server structural damage has occurred in commercial office buildings, it has generally been to older buildings, often the result of configuration irregularities.

Newer office buildings have also been damage, most notably the more than 100 welded steel moment-frame buildings (healthcare and residential structures as well as commercial, higher education, and industrial buildings) that failed during the 1994 Northridge earthquake. The damage occurred primarily at welded beam-to-column connections, which had been designed to act in a ductile manner and to be capable of withstanding repeated cycles of large inelastic deformation.

While no casualties or collapses occurred as a result of these failures, the incidence of damage was sufficiently high in regions of strong motion to cause widespread concern by structural engineers and building officials. Initial investigations showed that in some cases, 50% of the connections were broken and very occasionally the beam or column was totally fractured. Possible causes focused on incorrect connection design, incorrect fabrication, poor welding techniques and materials, and the impact of the need for economy on design strategies and construction techniques.

As a result, new guidelines for these types of structures have been developed, but remedial measures have resulted in more costly designs and extended approval procedures, with the result that many engineers have avoided welded steel moment-resistant frames in recent projects.

6.5.2 Performance Expectations and Requirements

The following guidelines are suggested as seismic performance objectives for commercial office buildings:

- Persons within and immediately outside the building must be protected to at least a life-safety performance level during design-level earthquake ground motions.
- Persons should be able to evacuate the building quickly and safely after the occurrence of design-level earthquake ground motions.
- Emergency systems in the facility should remain operational after design-level earthquake ground motions.
- Emergency workers should be able to enter the building immediately after the occurrence of design-level earthquake ground motions, encountering minimum interference and danger.

6.5.3 Seismic Hazard and Site Issues

Unusual site conditions, such as a near-source location, poor soil characteristics, or other seismic hazards, may lead to lower performance than expected by the code design. If any of this other suspected conditions are geologic hazards, a geotechnical engineering consultant should conduct a site-specific study. If defects are encountered, an alternative site should be considered (if possible), or appropriate soil stabilization, foundation, and structural design approaches should be employed to reduce consequence of ground motion beyond code design values or costly damage caused by geologic or other seismic hazards (see Chapter 3 for additional information). If possible, avoid sites that lack redundant access and are vulnerable to bridge or highway closure.

6.5.4 Structural System Issues

Office buildings are typically low to midrise in suburban locations and occasionally high rise in downtown locations of larger cities or in satellite suburban office complexes. Office buildings are intrinsically simple and often are of simple rectangular configuration, not least because economy is usually a prime concern for commercial structures. Thus, their seismic design can be economical and use simple equivalent lateral force analysis procedures with a good probability of meeting performance expectations as far as life safety is concerned. The protection of nonstructural components requires structural design to a higher performance level. Configuration irregularities may be introduced for image reasons or site constraints in odd-shaped urban lots, and the structural design may become more complex and expensive. To assist the protection of nonstructural components, special attention should be paid to drift control.

The need for planning flexibility requires minimization of fixed interior structural elements and a preference for column-free space. The need for flexibility in power and electronic servicing has resulted in increasing use of underfloor servicing to work cubicles, and structural systems have been developed to provide this.

Office buildings typically employ steel or reinforced concrete frames to permit maximum planning flexibility. Steel or reinforced concrete moment frames provide maximum flexibility but tend to be expensive in high and moderate seismic zones. New guidelines for the design of post-Northridge, welded moment-frame connections have increased the cost of these types of structural systems, increasing the already common use of steel braced frames. Elevator cores duct shafts and toilet rooms, being permanent, can be used as shear walls if or suitable size and locations. Since these elements are much stiffer than a surrounding frame, they may be a source of stress concentration and torsion, if asymmetrically located.

6.5.5 Nonstructural System Issues

The extensive use of frame structures for commercial office buildings, together with the tendency for them to be designed to minimum code standards, has resulted in structures that are subject to considerable drift and motion (sway). The result has been a high level of nonstructural damage, particularly to partitions, ceilings, and lighting. This kind of damage is costly and its repair is disruptive.

In addition, storage units, free standing work stations, and filing cabinets are subject to upset. Excessive drift and motion may also lead to damage to rooftop equipment and localized damage to water system and fire suppression piping and sprinklers; thus, the likelihood of water damage is greater.

Design and performance issues related to other types of buildings such as hospitals, schools, and retail commercial properties may be found in Federal Emergency Management Agency (FEMA) publication 389.

6.6 CURRENT SPECIFICATIONS FOR PERFORMANCE-BASED SEISMIC DESIGN

As described earlier, an important yet emerging concept in the successful implantation of seismic risk management strategies is the application of PBD approaches. The primary function of PBD is the ability to achieve, through analytical means, a building design that will reliably perform in a prescribed manner under one or more seismic hazard conditions. The fact that alternative levels of building performance are being defined and can be chosen as performance objectives is a relatively new development in seismic design. Some of its origins lie in studies of building performance during recent earthquakes, in which owners of buildings that suffered hundreds of thousands of dollars in damage were surprised to learn that the buildings met the intent of life-safety provisions of the seismic code under which they were designed, since no one was killed or seriously injured. Out of these experiences came the realization that design professionals need to be more explicit about what *design to code* represents and what seismic design in general can and cannot accomplish. At the same time, studies of damaged buildings have led to a much more sophisticated understanding of building response under the range of earthquake ground motion that can be expected to occur.

6.6.1 Building Performance Objectives

A fundamental concept behind the implementation of PBSD is the development of a consensus set of performance objectives. The performance objectives described the intended performance of the building (e.g., in terms of life safety, levels of acceptable damage, and post-earthquake functionality) when subjected to an earthquake hazard of defined intensity (e.g., a maximum credible event or an event with a certain return period). As earthquake intensity increases, building performance generally decreases. The goal of specifying performance objective is to achieve a reliable estimate of performance under one or more earthquake scenarios.

6.6.2 Building Performance Levels

Building performance can be described qualitatively in terms of the

- Safety afforded by building occupants, during and after an earthquake
- Cost and feasibility of restoring the building to pre-earthquake conditions
- Length of time the building is removed from service to conduct repairs
- Economic, architectural, or historic impacts on the community at large

These performance characteristics will be directly related to the extent of damage sustained by the building during a damaging earthquake. In general, four performance levels may be defined as follows.

6.6.2.1 Operational Level

This is the lowest level of overall damage to the building. The structure will retain nearly all of its pre-earthquake strength and stiffness. Expected damage includes minor cracking of facades, partitions, and ceilings, as well as structural elements. All mechanical, electrical, plumbing, and other systems necessary for normal operation of the buildings are expected to be functional, possibly from standby sources. Negligible damage to nonstructural components is expected. Under very low levels of earthquake ground motion, most buildings should be able to meet or exceed this performance level. Typically, however, it will not be economically practical to design for this level of performance under severe levels of ground shaking, except for buildings that house essential services.

6.6.2.2 Immediate Occupancy Level

Overall damage to the building is light. Damage to the structural systems is similar to the operational performance level; however, somewhat more damage to nonstructural systems is expected. Nonstructural components such as cladding and ceilings and mechanical and electrical components remain secured; however, repair and cleanup may be needed. It is expected that utilities necessary for normal function of all systems will not be available, although those necessary for life-safety systems would be provided. Many building owners may wish to achieve this level of performance when the building is subjected to moderate levels of earthquake ground motion. In addition, some owners may desire such performance for very important buildings, under severe levels of earthquake ground shaking. This level provides most of the protection obtained under the operational building performance level, without the associated cost of providing standby utilities and performing rigorous seismic qualification to validate equipment performance.

6.6.2.3 Life-Safety Level

Structural and nonstructural damage is significant. The building may lose a substantial amount of its pre-earthquake lateral strength and stiffness, but the gravity-load-bearing elements function. Out-of-plane wall failures and tipping of parapets are not expected, but there will be some permanent drift and select elements of the lateral-force-resisting system that have substantial cracking, spalling, yielding, and buckling. Nonstructural components are secured and not presenting a falling hazard, but many architectural, mechanical, and electrical systems are damaged. The building may not be safe for continued occupancy until repairs are done. Repair of the structure is feasible, but it may not be economically attractive to do so. This performance level is generally the basis for the intent of code compliance.

6.6.2.4 Collapse Prevention Level or Near-Collapse Level

The structure sustains severe damage. The lateral-force-resisting system loses most of its pre-earthquake strength and stiffness. Load-bearing columns and walls function, but the building is near collapse. Substantial degradation of structural elements occurs, including extensive cracking and spalling of masonry and concrete elements and buckling and fracture of steel elements. Infills and unbraced parapets may fail and exits may be blocked. The building has large permanent drifts. Nonstructural components experience substantial damage and may be falling hazards. The building is unsafe for occupancy. Repair and restoration is probably not practically achievable. This building performance level has been selected as the basis for mandatory seismic rehabilitation ordinances enacted by some municipalities, as it results in mitigation of the most severe life-safety hazards at relatively low cost.

6.6.2.5 Alternative Design Criteria: 2008 LATBSDC

The acronym LATBSDC stands for Los Angeles Tall Buildings Structural Design Council. The council has put forth a document *Alternate Design Criteria* in which a performance-based approach for seismic analysis and design of tall buildings is provided.

For the purpose of this document, tall buildings are defined as those with height greater than 160 ft above average adjacent ground surface, roughly corresponding to a 12-story office building or an 18-story apartment building.

In preparing this document, the council has invoked the use of alternate analysis and design methods traditionally permitted in most building standards.

The design methodology is based on capacity design principles followed by a series of PBD evaluations. Thus, to qualify under this procedure, a structure must necessarily have a suitable ductile yielding mechanism capable of experiencing nonlinear lateral deformations. For example, a special moment-resisting frame (SMRF) must have inelastic zones typified by (1) flexural yielding of beam ends, (2) shear yielding in column–beam panel zones, and (3) yielding at column bases.

The adequacy of the design is then demonstrated by analytically verifying the building performance using two distinct levels of earthquake ground motions due to the following:

1. *Extremely rare earthquake* defined as an event having a 2% probability of exceedance in 50 years (approximated to a 2500 return period). This earthquake is defined as the MCE in ASCE 7-05/10. The purpose of this evaluation is to demonstrate that collapse does not occur when the building is subjected to ground motions corresponding to extremely rare seismic events. Performance of all structural members including those not considered as part of lateral-load-resisting systems is evaluated. Cladding and their connections to the structure must accommodate displacements corresponding to MCE without failure.
2. *Frequent earthquakes* defined as an event having a **50%** probability of exceedance in **30** years (**43**-year return period). The purpose of this evaluation is to verify that the structural and nonstructural components of the building remain essentially elastic, thus retaining their functionality during and after the postulated seismic event. Repairs, if necessary, are expected to be minor and could be carried out without unduly affecting the normal functionality of the building. This criterion is deemed satisfied if the analysis shows only minor yielding of ductile yielding of primary structural system.

Buildings designed according to the Alternate Design Criteria of LATBSDC shall be designed for a minimum base shear equal to 3% of the building's seismic weight W:

$$V_{\text{min}} = 0.030W$$

The base shear may be distributed along the height of the building according to the equivalent lateral procedure of ASCE. As an alternate, an elastic response spectrum analysis may be performed using a scaled design spectrum and a CQC procedure for modal response combination. The requirement of design for a minimum base shear of 3% of the building weight is not a PBD provisions but has been retained to continue design tradition of Los Angeles tall buildings.

The analysis shall be evaluated for the following load combinations:

$$1.0D + L_{exp} + 1.0E$$

where

D is the service dead load
L_{exp} is the expected service live load
E is the earthquake load

The provisions of ASCE 41 and ASCE 7-05, Section 16.1.3, may be used for verifying the acceptability criteria for both collapse prevention (strength limit) state and serviceability limit state.

Independent project-specific peer review is a requirement under the provisions of the alternative procedure. A document of interest for designers of tall buildings in seismically active regions anywhere in the world is the *Recommendations for the Seismic Design of High-Rose Buildings*, 2008, published by the CTBHU.

6.6.3 Recommended Administrative Bulletin on the Seismic Design and Review of Tall Buildings Using Nonprescriptive Procedures AB-083

This document known as AB-083 for short was published in 2007 by the Structural Engineers Association of Northern California (SEAONC) and adapted by the San Francisco Department of Building Inspection (SFDB1). While both AB-083 and 2008 LATBSDC share the same philosophy of PBD for the seismic analysis and design of tall buildings, their approach is somewhat different.

For example, AB-083 permits code approaches with certain exceptions, while 2008 LATBSD bases its provisions on capacity design procedures only. However, it should be noted that the 2008 LATBC requirement of 3% of the building weight as the minimum base shear is in fact a prescriptive equivalent lateral force procedures permitted in most seismic codes.

The reader is referred to publication AB-083 for further details.

6.7 CLOSING COMMENTS

The purpose of PBD is to provide a realistic assessment of how a structure will perform when subjected to a preselected set of earthquake ground motions. While code-sponsored design provides capacity to resist a prescribed lateral force, this force level is substantially less than what the building may experience during a postulated major earthquake. It is assumed, however, that the building will be able to withstand the corresponding earthquake ground motions by yielding of the components into the inelastic range. However, this well-accepted design philosophy does not provide a direct method for determining how the structure will actually perform under such conditions. Fulfilling this void is the role of PBD.

PBD in a broad sense is not entirely new. Historically, its use can be traced back to the 1960s, when the now nonexisting New York World Trade Center Towers, Chicago's John Hancock Building, and Sears Towers (now Willis Tower) were designed using PBD principals to wind engineering.

When these buildings were designed, wind engineering was in its early childhood. The definition of appropriate demand levels and corresponding acceptance criteria for wind design were nonexistent and were developed from scratch by leading industry experts at the time. Definitions of recurrence intervals of wind events for strength and serviceability limits and commensurate acceptance criteria for thresholds of occupant comfort were developed for the first time. These have been refined over the years and are now well established and accepted by practicing professionals. The basic framework for PBSD may thus be traced back to the 1960s. However, the introduction of PBD to seismic community is relatively new. It was introduced in FEMA 273/274, published in October 1997.

The reasons for the emergence of PBDs (also noted in some publications as PBSD) are manyfold. Based on experience gained from about 20 tall buildings built recently in San Francisco, it seems owners like them because they cost less. Architects go for it because they offer more design freedom; contractors love them because they are easier to build. Engineers being thrifty like them because PBD can result in higher-quality structures with the least amount of materials and labor.

PBD can be used for a variety of purposes such as design verification for new construction, evaluation of an existing structure to identify damage states, and correlation of damage scenarios to various amplitudes of ground motion. Simply stated, the procedure compares the capacity of the structure, typically determined using a pushover analysis. It should be noted that in order to account for nonlinear behavior of the structural system, effective viscous damping is sometimes applied to linear elastic spectra.

It is of interest to watch as groups of researchers hyperbolize PBD as an exact solution and compare other *approximate* solutions to this *exact* solution. In their opinion, a nonlinear dynamic analysis is often deemed to be the exact solution. It is, however, difficult to understand how such a claim can be made because every computer analysis requires development of a computer model of the building's structural system. In the computer, the model is then subjected to a digitized version of each of the ground motions. At best, the computer model is an idealized mathematical model that is based on a mixed bag of assumptions regarding strength, behavior, and configuration of the component structural materials and assemblages. Therefore, the computer analysis is, at best, able to generate an exact numerical solution to a reasonable but inexact set of assumptions.

The insight we gain by a nonlinear time-history evaluation of how a building has actually performed in a past earthquake may not be all that is required to give us total confidence in its use. This is because there is always a nagging doubt that such an *after-the-fact* analysis is more than likely to benefit from prior knowledge of the building's actual performance and more often than not liberally tweaked to match the analysis to the actual observed performance. Therefore, it may not be prudent to rely entirely on the results of an after-the-fact analysis of an existing building to predict equally well *before-the-fact* seismic performance of a building to a yet-to-occur earthquake with unknown characteristics.

To accurately predict how a building will perform, one must have accurate data on both demand and capacity; however, the accuracy of the answers cannot be greater than the accuracy of the input data. This is not to say that nonlinear time-history analysis does not provide important information about structural response; but it is not *exact* as claimed by its promoters.

Over the years, PBD procedures have evolved significantly from their beginnings; however, there has been a push to develop increasingly complex, codified, PBD procedures. In the author's opinion, by codifying PBD, the very essence of PBD, that is, the focus on the attributes and behavior of an individual building, is destroyed. Engineers must be given sufficient latitude to arrive at the best estimate of a building's capacity.

PBD is a useful tool for design and to estimate the performance characteristics of buildings subjected to strong earthquake ground motion. It is not a *magic predict all* procedure. It takes a combination of analytical procedures, data evaluation, judgment, and experience to get a credible approximation of how a building works in the inelastic range of lateral motion.

Since the publication of the ASCE 31 and ASCE 41 standards, the PBD process that underlies these standards has been widely used. As popular as these first-generation procedures have become, they are of unknown reliability, present difficulties with regard to definition of performance intent, and may create significant liability for the design professional using them. Since 2001, FEMA has been sponsoring the ATC-58 project to develop next-generation PBD criteria. This new methodology is expected to permit engineers to characterize performance directly in terms of probable repair costs, occupancy interruption time, and casualties associated with building response to earthquakes. Slated for completion in 2011, the new procedures are anticipated to revolutionize the practice of performance-based earthquake engineering.

The hype surrounding has been picking up a lot of steam lately. While current design codes explicitly require life-safety design for only a single level of ground motion, it is expected that figure design codes will provide engineers with the necessary guideline to design and construct buildings that meet a number of performance criteria when subjected to earthquake ground motion of differing severity.

It is clear that performance-based strategies will be included in future seismic design codes. Regardless of when this actually occurs, it is now time for design professionals to get the hang of this new concept to provide owners and manager with a much clearer picture of what may be expected in terms of damage, downtime, and occupant safety for a given building under various intensities of ground motion.

When communicating with building owner representatives during the development of seismic performance criteria for a new building, it would be useful to

1. Explain the concepts of PBSD using the concepts and materials currently available
2. Help the owner to determine if a performance level higher than life safety is needed for the design earthquake; if so, assist the owner in developing a design that would be accepted by the governing regulatory agency

It is of interest to recognize that PBD reflects the evolution of design practice as it takes place in changing technical and political contexts. As the shift to a performance-based approach takes place, designers are raising questions about possible impacts on design practice. How will a performance-based seismic code affect professional liability? How will it affect the cost of professional design services? Will it be particularly difficult or expensive for small firms or inexperienced owners to implement? Will it really help designers manage uncertainty? Without real data on real buildings in real earthquakes to confirm our performance predictions, how confident can we be that our PBD methods work?

As the dialog surrounding building code development continues, new insights are emerging, particularly the recognition that this process needs the attention of all stakeholders concerned with the building environment. For architects, the new codes and the performance-based concepts behind them will require greater involvement in seismic design decisions. As architects help owners investigate the feasibility of proposed building projects and lead their clients through the design process, they will need to be aware of the interaction between design decisions and seismic design regulations.

Well-informed engineers fear that PBD will turn structural engineering over to lawyers instead of keeping it within the structural engineering profession where it belongs.

7 Preliminary Calculations to Ensure Validity of Computer Analysis

PREVIEW

The advent of computer-based approaches to structural analysis and design over the last three decades has only accentuated the need for structural engineers to recognize that we are dealing with models of structures, and not with the actual structures. Further, as tempting as it is to run innumerable computer simulations, closed-form estimates and approximate calculations can be effectively used to both guide and check numerical results, as well as to confirm in clear terms physical insights and intuitions. What is truly remarkable is that the way of thinking about structures and their models that we propose is rooted in classic elementary elasticity: it depends less on advanced mathematical techniques and far more on thinking about the dimensions and magnitudes of the underlying physics.

The author has based this chapter in particular, and the entire book in general, on the premise that it is now even more important to understand basic structural modeling, with strong emphasis on understanding behavior and interpreting results in terms of the limitations of the models being applied. In fact, one could argue that the generation of numerical analyses for particular cases is, in the *real world*, increasingly a task performed by technicians or entry-level engineers, rather than by seasoned professional engineers. As numerical analysis becomes both more common and significantly easier, those structural analysts and designers, who know which calculations to perform, how to validate and interpret those calculations, and what the subsequent results mean, will be the most highly regarded engineers. The knowledge needed to do these tasks can often be encapsulated and illustrated with the ability to obtain and properly use analytical, closed-form estimates, or, in other words, the ability to obtain and properly use *back-of-the-envelope* models and formulas.

Structural engineering, in spite of its roots in scientific principles, is vastly driven today by the use of the computer and solutions of complicated equations that are totally devoid of physical feeling.

Physical intuition is really important for successful modeling of structural behavior, and yet good intuition often develops with repeated application of structural models. Design today, however, seems to be going in a degenerating direction. We seldom create but are experts in executing a given design, confidently so because of the immense computer power at our disposal. Humans have the amazing ability to subconsciously call upon a wealth of what is largely *unconsciously gained knowledge* often referred to as experience. The waiting period to gain this experience, often estimated as being perhaps 20 years, can be considerably shortened by the student or professional who chooses to cultivate a basic understanding of fundamental principles and system behavior.

The cultivation of a fundamental knowledge foundation is considerably complicated by the codification process. Current codes give the appearance of a meticulous technology, a technology not particularly suited to subliminal recall. The engineer who does not clearly understand the basis of a code criterion cannot be expected to intelligently apply the criterion. However, this development of an in-depth understanding of system behavior will not be difficult if focus is maintained on the fundamental issues.

Careful examination of the properties of most materials indicates that they are not isotropic or homogenous; nonetheless, it is common practice to use isotropic properties for most analysis.

It should be remembered that the result obtained from a computer model is our estimation of the behavior of the real structure. The actual behavior is dictated by the fundamental laws of physics and is not required to satisfy the building code or the computer program's user manual. We should always remember that *equilibrium is essential—compatibility is optional.* In other words, you can get away with a design that does not satisfy compatibility—the resulting structure may display distress such as unacceptable cracks and yielding of steel—but if you have not taken care of equilibrium provision, you are going to reach judiciary courts at computer speeds.

It is my hope that the reader will find this chapter more fun than sweat, for the objective here is to put to good use the creativity and inventiveness of the designer. A variety of design problems are appropriately treated at a conceptual level, to allow the reader to explore the design process and understand how to create sound designs. This superabundant overview of design solutions strives to stimulate creativity by instilling basic structural behavior concepts.

The practicing engineer who takes time to mediate on the concepts presented here should find the material useful for exploiting better and more economical designs.

7.1 CHARACTERIZING STRUCTURAL BEHAVIOR

Before the student or practicing professional can effectively design a complex building system, an understanding of the system behavior must be acquired. An independent treatment of building components such as beams, columns, frames, and panel zones allows the designers to assess how each component affects the behavior of the entire system and thus empowers them to allocate material and optimize system efficiency.

Structural analysis is the process by which the structural engineer determines the response of a structure to specified loads or actions. This response is usually measured by establishing the forces and deformations throughout the structure. A given method of structural analysis is based on information gained through the application of engineering mechanics theory, laboratory research, model and field experimentation, experience, and engineering judgment.

The earliest demands for sophisticated analysis, coupled with some serious limitations on computational capability, led to a host of special techniques for solving a corresponding set of special problems. These so-called classical methods incorporated some ingenious innovations and served the needs of the structural engineer very well for many years. However, the advent and subsequent development of the digital computer increased computational capabilities by several orders of magnitude and thus obviated the need for special techniques. The ingenious specializations of the classical methods were replaced by the sweeping generalities of the modern matrix methods.

The transition from the classical methods to the modern methods has triggered some revolutionary changes in structural engineering and in the education of structural engineers. Although the matrix methods have become the foundation of modern structural analysis as it is employed in the practice of structural engineering, is it not completely clear what the role of the classical methods will be especially in the education process.

By either classical or matrix methods, the analysis process can be part of preliminary design; however, for reasons explained in the introduction, the author has selected the classical method to introduce the fundamental concepts of preliminary design.

However, it is important to note that structural analysis plays a limited role in the structural design process and even smaller role in the overall design process. Furthermore, the role that it plays is entirely supportive of the design process. That is, structural analysis is not an end in itself. It is particularly important to understand the supportive nature of structural analysis as one studies the subject. It is easy for an engineering student to become enamored of this intriguing subject to the point of aspiring to become a structural analyst. However, such a goal is unrealistic.

Good structural engineers are necessarily good structural analysts, and they will use their analytic ability intelligently as they fulfill their primary responsibility as structural engineers.

Simply defined, structural analysis is a mathematical process by which the engineer verifies the adequacy of the structure with respect to its strength and stiffness. It is not always possible to obtain rigorous mathematical solutions for building engineering problems. In fact, rigorous analytical solutions can be obtained only for certain simplified cases. High-rise structural problems, like most other practical engineering problems, involve complex material property, loading, and boundary conditions. The engineers introduce assumptions and idealizations deemed necessary to make the problem mathematically manageable, but still capable of providing sufficiently accurate solutions and satisfactory results from the point of view of safety and economy. They establish a link between the real physical system and the mathematically feasible solution by providing an analytical model that is the symbolic designation for the substitute idealized system, including all the assumptions imposed on physical problems. Modeling techniques, therefore, can be defined as a way to reduce, synthesize, and properly represent the structural system.

The basic principles and mathematical relationships used in the design and analysis of tall buildings are not unique to this type of construction. Virtually all of the fundamental relationships are based on the normal assumptions of elastic design, which form the backbone of the study of the strength of materials. Although the form of certain relationships is somewhat modified, their application to the analysis and design of high-rise buildings will not impose undue difficulty.

Two major types of problems are encountered by the engineer engaged in the design of tall buildings: (1) review of a set of completed working drawings and (2) the actual design, starting from the preliminary stages. The review of a completed design consists of the determination of the stresses and deflections under appropriate conditions of loading in order to confirm their compliance with the design criteria and applicable codes. The strength of a member under all loading conditions, bending, shear, torsion, axial, and bond must be determined to equal or exceed the minimum strength requirements. It should be apparent that in order to review the design, the dimensions and material properties of all structural elements that are used in the makeup of the tall building, together with knowledge of the loads to which the structure is subjected, must be known.

The task of checking the work of another engineer is done to ensure that the design satisfies the safety requirements as specified in the applicable codes. Although there is no uniform procedure for carrying out this work, a balance must be maintained between checking for safety compliance and the avoidance of malicious damage to the reputation of another engineer. The check should be carried out in a climate of mutual understanding. The checker must recognize that he or she has no duty to comment on the choice of design, only on its validity and its satisfactory compliance with applicable codes. It is a mistake to concentrate on the minute accuracy of the calculations when time can be saved by assessing the soundness of the structure.

The design of a building, on the other hand, consists of selecting and proportioning member sizes in which the stresses do not exceed the permissible values under any combination of loads. The design also includes the study of the deformation characteristics to ensure that the building meets applicable serviceability criteria. In common with other types of design, member sizes in a tall building are arrived at from a trial-and-error procedure. In the design of a member, several adjustments of the trial section are normally required before a satisfactory solution is found. Of course, it is just as important to adjust members that are found to be excessively conservative.

At the schematic stages of architectural design, overall options associated with different space forms of the building are thought through with due consideration given to the basic relationship of the building to the available site, environmental conditions, intended use of the building, and other performance criteria. The architectural task at this stage is to organize and orient various space components such as service corps, stairs, elevator cores, and mechanical rooms. These are arranged around the typical floor plan configuration, with the understanding that it works around a typical floor; it is relatively easy to force the arrangement to work at other nontypical levels. The various components are organized around the typical floor to achieve maximum efficiency, measured in terms of the gross to net leasable floor space. A structural appraisal is made of the general geometry of the building, especially the height-to-width ratio, function of the structure, whether it is a single-use or multiuse project, whether there is a basement, parking, or other requirements that may necessitate transfer of large vertical elements, limitations on layout and sizes of structural members, head room, and span requirements. The process of preparing structural system alternatives starts simultaneously with due regard to choice of construction materials, availability of building materials, and local workmanship. Generally speaking, the economy of a structure depends to a great extent upon the design criteria and framing layout adopted but to a far lesser extent upon the detailed design of structural members. Although the decision on the framing may be somewhat subjective, it should have the backing of at least some preliminary economic comparison. While the analytical phase of structural engineering is based on physical sciences, designing remains essentially an art for which knowledge of structural analysis, imagination, judgment, and experience are prerequisites.

It is difficult, if not impossible, to outline the thought process that would go through a structural engineer's mind when he or she conceptualizes the structure at the schematic level. It is very likely that more than one solution may present itself at this stage, and proper guidance to the architect may require knowledge of relative cost and construction procedures on the part of the structural engineer. The point is that one does not really make a lot of calculations or hardline drawings at this stage but guides the architect with enough confidence obtained through experience. What is needed is a thorough knowledge of the structural systems to augment the creative thinking of the architect by concentrating on the design evolution of the whole rather than becoming entangled in premature consideration of consequential details of the parts. The key to successful application of structural design ideas at the schematic stages of architectural design rests in being able to look at the big picture first without getting bogged down by the details.

Although, for optimum results, a comprehensive interactive approach between architectural and engineering fields is necessary, there is a growing tendency for the architect to initiate a space form and then to have the engineer find a way to technically implement the form. This idea appears to stem from the premise that building forms, like pure art, need to be developed without the restriction imposed by engineering disciplines. The results have often been very daring and interesting building forms.

Even in today's high-tech, computer-oriented world with all its sophisticated design capability, there still is a need to perform approximate analysis of structures. First, it provides a basis for selecting preliminary member sizes because the design of a structure, no matter how simple or complex, begins with a tentative selection of members. With the preliminary sizes, an analysis is made to determine if design criteria are met. If not, an analysis of the modified structure is made to improve its agreement with the requirements, and the process is continued until a design is obtained within the limits of acceptability. Starting the process with the best possible selections of members results in a rapid convergence of the iterative process to the desired solution.

Second, because of the ever-increasing cost of labor and building materials, it is almost mandatory for the structural engineer to compare several designs before choosing the one most likely to be the best from the points of view of structural economy and how well it minimizes the premium required by the mechanical, electrical, and curtain wall systems. Of the myriad structural systems, which present themselves as possibilities, only two or three schemes may be worthy of further refinement requiring full-blown computer solutions. Approximate methods are all that may be required to logically arrive at cost figures and to sort out the few final contenders from among the innumerable possibilities. It is very time-consuming, costly, and indeed unnecessary to undertake a complete sophisticated analysis for all the possible schemes. Preliminary designs are therefore very useful in weeding out the weak solutions.

Sophisticated computer analyses are indispensable in reducing the number of inaccuracies caused by hand analysis techniques and are being used routinely in everyday engineering practice. Although such computer analyses may intimidate the structural engineer by virtue of their unbelievable amount of documentation and output, the prudent engineer will always verify the reasonableness of the computer analysis by using approximate hand-calculated values for forces, moments, and deflections. Approximate analysis is, therefore, a powerful tool in providing the engineer with (1) a basis for preliminary sizing of members, (2) an orderly method for evaluating several schemes to select the most likely for further study, and (3) methods for obtaining approximate values of forces, moments, and deflections to check on the validity of the computer solutions.

Having established the need for preliminary analysis techniques, what then are the techniques available for the structural engineer? The techniques are many and range from sophisticated solutions satisfying both compatibility and equilibrium conditions requiring lengthy calculations to simple ones based on the considerations of equilibrium alone. In this chapter, we examine building behavior by considering only equilibrium aspects for equilibrium is essential while compatibility is optional.

Conceptual design is unquestionably more artistry than a systematic mathematical pursuit. It requires a degree of cultivated talent. Like science, there exist procedures that, if followed, will produce logical solutions. The *art* or acquired talent portion of the design process involves, in part, the development of a feeling or understanding for the appropriate level of detail required to produce a meaningful preliminary design. As there are many artistic approaches, so are there many methods that may be used by designers to develop a preliminary design. The procedure described in this section is but one way to design.

The goal is selecting the appropriate type of framing system and visualizing suitable connections. These decisions, however, are not independent. The art then comes in arranging the decisions in a logical fashion while at the same time being alert to those later decisions, which may have an impact on the decision at hand. Today, the computer is used invariably as a design aid. If used inappropriately, the computer can obscure the objective by introducing an insurmountable level of usually meaningless detail while still failing to provide a meaningful insight.

This section will endeavor to show how a designer might organize and implement the decision-making process—just enough detail to preserve its focus. The designer must nevertheless understand the importance of and need for each decision in the design process, as well as the basic principles that allow focus and promote conclusions.

It is my hope that this chapter would motivate even those hardcore computerized designers who refuse to push a pencil. There is ample reward and a lot more fun in mulling over the art of design.

7.1.1 History of Structural Engineering

Humans are builders of structures; but more than that, they are envisioners and designers. If we postulate that the first structure was a tree that conveniently fell across a chasm and was subsequently used as a bridge, then since that meager and accidental beginning, humans have indeed advanced in their ability to design and build structures. When the structures of humans began to reflect their ability to conceive and design them as well as to construct them, structural engineering was born, and it has grown in sophistication as it has endeavored to meet the demands of humanity.

The evolution of structural engineering to its present form involved the development of several individual areas of endeavor. These were the development of the theories of mechanics of materials and structural analysis, the formulation of the computational techniques necessary to solve the governing equations of these theories, the introduction of new building materials, the application of the theories and materials to the creation of new structural forms, and the inventive development of construction techniques. Some of these areas required the analytic talents of mathematician, scientist, or engineer, whereas others required the daring and artistic skills of the entrepreneur or builder. Although each of these areas has its own historical chronology, their juxtaposition shows how the individual areas are nested—how a development in one area sparked a need and, therefore, a subsequent development in another area.

A beginning point for structural engineering as we presently view it would be at about 500 BC From that point until the time of Christ, the Greeks primarily used stone to build post and lintel structures, that is, structures whose columns supported short beams. An example of this form is the Parthenon in Athens. Even though experience and empirical rules formed the basis of their structural activity, Aristotle (384–322 BC) and Archimedes (287–212 BC) were establishing the beginning of the principles of statics. Although some metals and wood were introduced, stone and masonry continued as the primary building material of the Romans until about AD 500. They introduced new structural forms such as the arch, vault, dome, and even the wooden truss. Some of these structures remain with us today as monuments of that era. However, the Romans were not analytic in their approach but rather were builders who concentrated on certain structural forms.

During the Middle Ages (500–1500), much of what the Greeks and Romans had developed was lost. The only major structural accomplishment during this time was achieved by the nine Gothic builders; witness their splendid cathedrals, characterized by pointed arches stabilized by *flying buttresses*.

Following the inactivity of the Middle Ages, the Renaissance saw new impetus in many areas, including structural engineering. In the early part of the period, Leonardo da Vinci (1452–1519) formulated the beginning of structural theory. However, Galileo (1564–1642), who published *Two New Sciences*, is generally credited with originating the mechanics of materials. He studied the failure of a cantilever beam, and even though his writings were not wholly correct, they did establish an important beginning. Europe was rife with the activity of rebirth, which spawned more than a few significant analytic accomplishments. The most important of these were those of A. Pallidio (1518–1580), who introduced the modern truss; R. Hooke (1635–1703), who established the law governing the linear behavior of materials; Johann Bernoulli (1667–1748), who stated the principle of virtual displacements; Daniel Bernoulli (1700–1782), who contributed to the understanding of elastic curves and the strain energy of flexure; Leonard Euler (1707–1783), who examined column buckling and energy methods; and Louis Navier (1785–1836), who followed up the earlier work of C.A. de Coulomb (1736–1836) and published a book on strength of materials, which dealt with the elastic analysis of beam flexure.

These accomplishments were paralleled by new developments in building construction materials. Timber was used by German and Swiss engineers to construct bridges up to 300 ft long. And iron arrived with revolutionary impact. As a material, it exhibited elastic properties much better than those of wood or stone, and thus, the new theories could be applied to enable more daring structural forms to be used with confidence. A whole host of *firsts* in both form and dimension followed—cast iron arch bridges, iron trusses, suspension bridges, etc.

However, the golden age of structural engineering is considered to be 1800–1900. During this period, most of the present-day theories of mechanics of materials and structural analysis were developed. This age also saw new materials on the scene. Portland cement appeared early in the 1800s, and the first reinforced concrete bridge was constructed before the end of that century. Iron rolling mills made iron more usable, and quantity steel production was introduced by H. Bessemer. These developments led to new structural forms. In fact, the theoretical developments of the mid-1800s paved the way for the analysis of continuous beams and frames, and these forms grew popularity by the early 1900s.

The twentieth century brought in some modest advancement in structural theory and some significant developments in solution techniques. A few are as follows: G. Maney (1888–1947) introduced the slope deflection method, which was the forerunner of modem displacement methods; H. Cross (1885–1959) contributed the moment distribution method, and R. Southwell (1888–1970) presented the more general relaxation methods (these two developments allowing systematic solution of statistically indeterminate structures and serving as the cornerstone of frame analysis for a quarter of a century); and several analysts contributed to the margining of matrix algebra and fame and continuum analysis to form the modern matrix and finite element methods of analysis. At the same time, the areas of inelastic analysis and strength methods were introduced.

The 1900s also saw a host of new materials, techniques, and structural forms introduced. Material developments brought forth aluminum, high-strength steels and concretes, special cements, plastics, laminated timber, and composites. Developments in technique include the introduction of experimental research, the use of electric welding and prestressed concrete, and the development of improved construction methods; however, the event that had the greatest impact was the introduction of electronic computation in the 1950s. New advances in structural form included the perfection of long-span bridges of many configuration, record-breaking heights in buildings, and newer forms such as shells, panels, and stress-skin structures.

7.1.2 Design Process

The *engineering design process* encompasses much more than structural design. Although the primary role of the structural engineer is in structural design, he or she is necessarily enmeshed in the entire design process. This is illustrated by the following breakdown of the engineering process as it is relative to a typical civil engineering project in which a structural engineer is involved.

7.1.2.1 Conceptual Stage

Any engineering project must be directed toward the satisfaction of a unique set of objectives. During the conceptual or planning stage, the specific needs are identified, and the objectives are carefully articulated to meet these needs. These objectives must be consistent with the desires of the client and the interests of other involved parties.

This stage requires input from client, architects, planners, the public as represented by elected officials, governmental regulatory agencies or civic organizations, and the engineer. During this stage, the engineer frequently serves as a resource person regarding the engineering feasibility and the economic soundness of the various alternatives under consideration.

The conceptual stage should bring forth a plan that maximizes the satisfaction of the stated objectives while minimizing any objectionable features of the project.

7.1.2.2 Preliminary Design Stage

The plan that emerges from the conceptual stage frequently includes several alternatives that are to be investigated through the preparation of individual preliminary designs.

The preliminary designs are of vital importance, and the structural engineer plays a central role during this stage. Key decisions regarding the positioning of the structure, the structural form to be used, and the manner in which the structural components are to be connected will have a bearing on the final design. It is during this stage that the creative talent of the engineer is vitally important as he or she considers the options brought forward from the conceptual stage. Yet, the engineer must keep in mind that the structure he or she designs has to be built, and thus, the construction and fabrication aspects must be carefully considered. In many cases, the most severe loading conditions occur during construction. The loading on these partially erected structures is vastly different from the loading that the final structures will support.

Each preliminary design involves a thorough consideration of the loads and actions that the structure will have to support, including the conditions that will occur during fabrication. For each case, a structural analysis is necessary; that is, the forces and deformations throughout the structure must be determined. It is this area, the structural analysis of the system, that is examined in detail in this textbook.

Frequently, the preliminary designs are based on approximate theories of structural analysis in order to minimize the time and effort invested in the preliminary phase. This phase must produce sufficient detail so that intelligent decisions can be made in the final selection of one of the alternatives that was proposed in the conceptual stage.

7.1.2.3 Selection Stage

Once the preliminary designs are completed, a selection must be made. At this point, the parties involved in the conceptual stage are reconvened so that they may participate in the selection process. The prime consideration centers on how each alternative satisfies the original objectives. Again, due consideration is given to any objectionable features of the alternatives.

The structural engineer is concerned at this stage with the relative economies of the alternatives, the impact that any unique features of each alternative might have on the structural behavior or practicality of construction, and any other areas related to the decision.

The result of this stage is usually a decision to proceed with one of the alternatives for which a preliminary design has been prepared.

7.1.2.4 Final Design Stage

The results of the preliminary design stage constitute a starting point for the final design stage; however, the structural engineer must proceed with greater care from this point. Here, the loads are determined with greater accuracy than was necessary during preliminary design, and all plausible loading conditions and combinations must be considered. The structural analysis that is required for this stage must be carried out with great precision, and the approximations of the preliminary design stage must be eliminated. Each member is proportioned, and the connections are detailed to ensure that the structure will behave in accordance with the assumptions made in the structural analysis.

The results of the final design stage are presented in a set of complete design drawings, which give a graphic portrayal of the details of the entire system. These are generally accompanied by written specifications that stipulate the materials to be used, the quality of workmanship, the pertinent codes to be employed, and many other items.

7.1.2.5 Construction Stage

The goal of this stage is to bring into existence that which was described in the final design stage. The complete documents of the final design stage serve as the basis for bidding by the prospective building contractors. The successful bidder frequently prepares additional drawings related to the fabrication of the structure.

7.1.3 Basic Principles of Structural Analysis

The purpose of reviewing, here, the basic principles of, and approaches to, structural analysis is to bring out the general features of analysis common to all structures. Such a treatment would give the student as well as the seasoned professional the confidence of being able to identify the essence of computer-assisted analysis routinely used in the design office. A structure is an assembly of a number of small units generally called members or elements. A building frame—both from the point of view of construction and analysis—consists of beams, columns, floors, wall panels, and foundations. Similarly, the components of a bridge are longitudinal girders, cross girders, decking slab, braces, etc. We generally call such components as members or elements and their assembly as a structure. A structure while serving its functional purpose is subjected to forces, and the essential task of the designer is to proportion the different elements such that the resulting structure is strong and stiff enough to fulfill its intended function with due considerations of economy.

A load-resisting structure body may be conveniently classified from the point of view of analysis as (1) framed structures, (2) surface structures, and (3) solids.

A framed structure is a network of beam (line) elements. A beam element is a unidimensional element with its cross-sectional dimensions very small compared with its length. In framed structures straight beam elements are the simplest class of structural elements. *Surface structures* are built up of two-dimensional elements such as thin slabs, membranes, and shells. Their thickness is very small compared with their lateral dimensions. Massive structures consist of *solid elements*. We can classify a dam, or a rock mass, or a thick container like the modem prestressed concrete pressure vessel into this class. A solid element is truly three-dimensional from the point of view of analysis.

In a practical structure, these different kinds of elements almost always occur together. Thus in buildings, unidimensional (line) elements such as beams and columns interact with two-dimensional (plane) elements, namely, floors and wall panels. Similar situation occurs in a bridge with edge beams and deck slab elements. In the modern context where digital computers are invariably used, computer analysis procedures do not distinguish between such elements: the entire structure is treated as an assembly of structural elements of different types.

7.1.3.1 Requirements of Structural Analysis

The essential requirement of a structure is that it should resist the loads without undue deformations in doing so; all materials develop internal resistance and undergo deformations. The first requirement is that the structural member is strong enough to develop internal resistance equal to the resultant action of the loads at any location. This is the *requirement of equilibrium* between internal forces given and external loads and must always hold good for both static and dynamic behavior. Since the material is not rigid, deformations (distortions) of the members occur, such as axial straining, bending, twisting, etc. The second aspect of structural analysis is the load deformation characteristics of the structural member, which are related to the loads through stiffness parameters of the member, which depend on the strength of the material and the shape of the member. The study of *stiffness properties* of structural members is, therefore, the second aspect of structural analysis. The *third* aspect of concern is the one of obtaining assessing displacements such as deflections and rotations at certain locations, such as center of a bridge or top of a building. The deformations of structural members are, in general, different for different members. They do not occur freely since the members are connected to one another to form the structure. It deforms the whole according to the constraints exerted by the connectivity of the members and their supports. For example, the deflection of a node point of a pin-connected truss can only be determined by considering simultaneously the elongations and constraints of all members connected. Similarly the change in the sag of a suspended cable due to temperature changes, or the shortening of the column due to an axial load applied at the top are examples which are purely matters of geometry. This is referred to as *kinematics* or compatibility requirements that depend on the structural form. Kinematics relates the deformations of structural members to the displacements at certain chosen locations.

In the analysis of any structure, the three aspects, namely, equilibrium, stiffness properties of members, and kinematics of the structure, are always of concern.

7.1.3.2 Equilibrium Requirements

The requirements of equilibrium between loads and internal forces are an absolutely necessary condition and are satisfied in any structure under any kind of equilibrium, be it static, dynamic, or neutral. The conditions of statics are to be satisfied by the structures as a whole and also element of the structure. In case of a structure vibration, the conditions of equilibrium are changing with respect to time, but at any given instant, the internal forces are in equilibrium with the inertia and external force.

Statically determinate structures are those in which the internal forces are directly determinable from the conditions of statics applied to different parts of the structure. If the displacements are small, the changes in the geometrical configuration of the structure due to loads can be ignored. This leads to equations of statics that will be linear in load terms, enabling the use of superposition principle. Determinate structures form an important class of structures possessing certain advantages and are widely used in practice. They also occupy an important position in structural analysis because statically indeterminate structures are often reduced to equivalent determinate structures as an intermediate stage of final analysis. In the context of present-day design, they offer an excellent method for determining preliminary sizes of structural members.

The consideration of equilibrium between the internal forces and the loads in a *redundant* structure is rather involved because there is no unique force distribution for a given set of loads. In other words, more than one set of force distribution is possible because the internal force distribution would change with the changes in the stiffness of the structural elements.

A redundant structure also has the feature that its members can experience forces without apparent external loads; such loads are temperature, creep, lack of fit members, and yielding of supports.

There would be similar self-equilibrating internal forces when some of the members have residual deformations such as when loaded beyond the yield limit.

It is pertinent to emphasize here the significance of statics in structural analysis relative to that of other requirements, especially from the practical viewpoint of the safety of a structure. Although the criteria of deformations are important from the point of view of functional efficiency of the structure, such as limitations on deflections, crack widths, etc. If the conditions of statics are not ensured, there is no other aspect of structural behavior that can come to our rescue except perhaps the redistribution of forces due to plasticity of the material.

7.2 ADVANTAGES AND DISADVANTAGES OF INDETERMINATE STRUCTURES

One might legitimately ask, however, why indeterminate structures are used. They are obviously more difficult to analyze than determinate ones, so why aren't statically determinate structures used exclusively? And if there are specific advantages in statically indeterminate structures, why are determinate structures used in certain cases? As it turns out, specific advantages and disadvantages are associated with each type of structure.

The major advantages of statically indeterminate structures are manifested in three ways. First, a statically indeterminate structure displays greater stiffness in resisting load than does a comparable statically determinate structure. For example, consider the two structures shown in Figures 7.1a and 7.1b, where the individual elements of each structure have the same cross-sectional dimensions and lengths. The post and lintel arrangement shows a beam member (lintel) that is supported atop

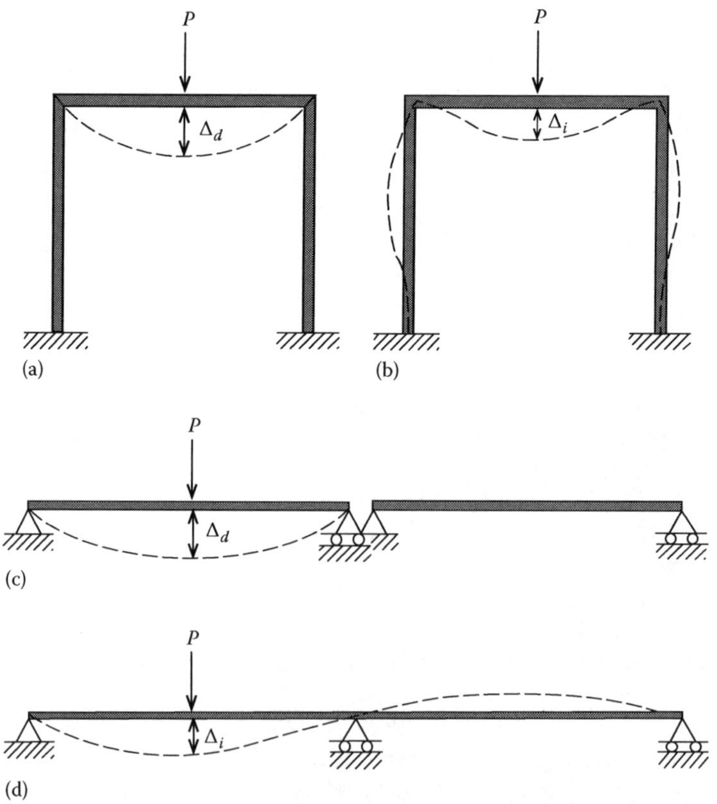

FIGURE 7.1 Comparative responses for statically determinate and indeterminate structures (a) post and lintel, (b) portal frame, (c) two simply supported beams, and (d) two-span continuous beam.

two vertical struts (posts). When the load P is applied at midspan of the beam, the beam deflects with a vertical displacement of Δd as the load point. If the beam is integrally connected to the column, a portal frame is formed. In this case, the same load P will cause a vertical displacement of $\Delta i < \Delta d$ as the load point. Computing the stiffness of the each case, that is, the load per unit displacement, we see $P/\Delta i > P/\Delta d$. That is, the stiffness of the indeterminate portal frame is greater than that of the determinate post and lintel system. The increased stiffness of the portal frame stems from the fact that the ends of the beam are restrained by the columns. Thus, as the beam deflects, the columns assist in resisting the load. This behavior differs from the post and lintel construction, where the beam simply deflects, and the struts are merely passive supports for the beam.

A similar situation exists when the two simply supported beams of Figure 7.1c are compared with the two-span continuous beam of Figure 7.1d. Again, if the individual elements in each system have the same lengths and cross-sectional dimensions, then a $\Delta i < \Delta d$, and thus, $P/\Delta i > P/\Delta d$. Again, the statically determinate case reduces to a simply supported beam, with the loaded span carrying the entire load. However, the statically indeterminate beam provides an end restraint for the loaded span, and the unloaded span participates in resisting the load.

7.2.1 Free-Body Diagrams

One of the most useful tools at the structural analyst's disposal is the *free-body diagram*, which is a sketch of a structural component with all of the appropriate forces acting on it. It may be of a whole structure or part of one.

For instance, consider the structural body shown in Figure 7.2. Figure 7.2a shows a free-body diagram of the entire structure along with the applied forces and the reaction forces. If the body is cut along the line a–b shown, a free-body diagram of either section must show the internal stresses (force intensities) that act on the cut face. These internal stresses act in an equal and opposite fashion on the two free-body diagrams of Figure 7.2b.

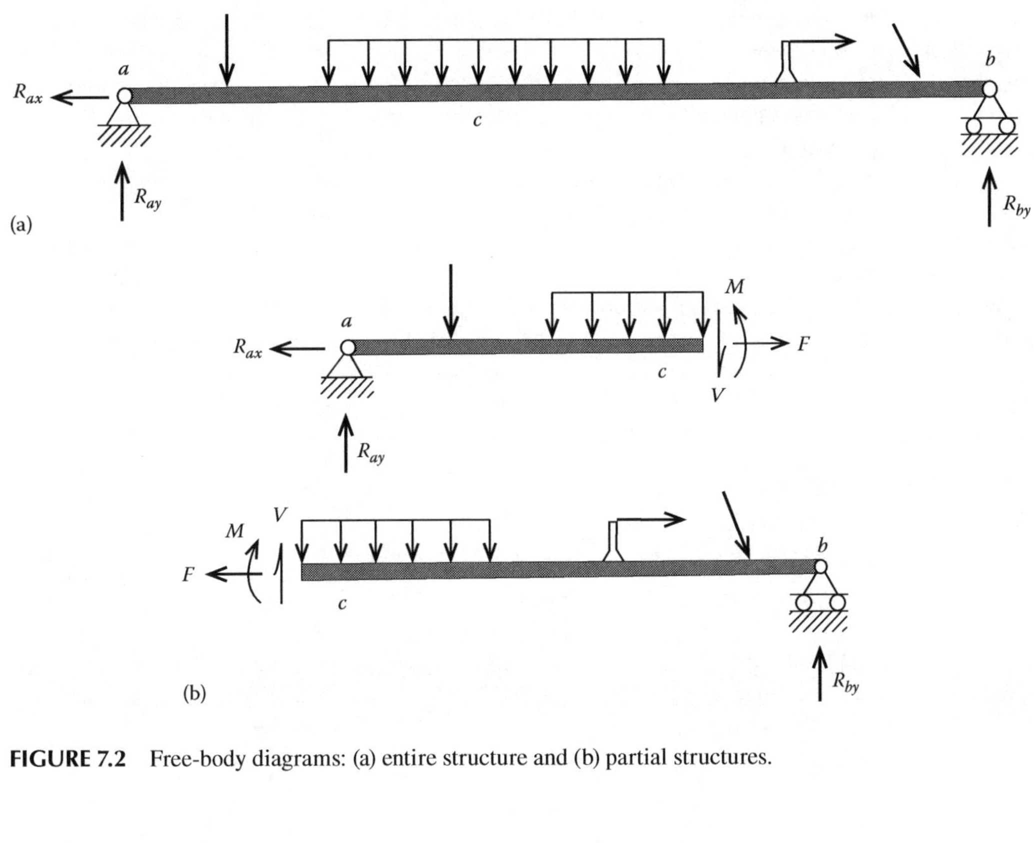

FIGURE 7.2 Free-body diagrams: (a) entire structure and (b) partial structures.

A free-body diagram of an entire structure leads to relationships between the external forces and reactions, whereas free-body diagrams of subsections of the structure will lead to relationships between the external and internal forces. In any case, each free-body diagram represents a structural analysis. The judicious selection of free-body diagrams and the subsequent analysis of each structural component are fundamental to the field of structural analysis.

7.2.2 Stiffness Requirements

The relationship between the deformation of a member and the corresponding force can be expressed in the form of either flexibility or stiffness of the member. For small deformations, these are constants depending only on the geometrical properties of the member and the material property. The *stiffness or flexibility* relations can be derived either by experiments or deduced from known material properties and appropriate assumptions regarding stress and strain distributions.

In a structure, the calculated internal force and the corresponding deformation should be such that they satisfy the stiffness (or flexibility) relations. This is the characteristic relation between the resisting force and deformation of a member.

In the analysis of determinate structures, the internal forces are determinable without the consideration of the stiffness properties of the members. However, in a redundant structure, the internal force distributions directly depend on the stiffness characteristics of the members. Thus, there is more than one possible distribution of internal forces depending on the stiffness properties of a redundant structure.

7.2.3 Kinematic Requirements

Kinematics is the term applied to the requirement of geometric relationships between deformations of structural members and nodal displacements. In a pin-jointed frame, the deformation of any given member is the relative elongation (or shortening) that can be calculated in terms of the displacements of the node to which the member is connected. Strictly speaking, kinematic relationships are nonlinear. However, this is rendered linear by assuming that the deformations and the displacements are both small when compared. This is the essential assumption of linear structural analysis, consistent with neglecting the effect to the geometry of displacements in establishing the equations of equilibrium.

Different types of kinematic considerations are made in structural analysis depending on the restrictive conditions of approximations. For example, in pin-connected frameworks, bending moments in the members are absent by virtue of pin connections between the members, and the problem of kinematics is one of obtaining only the member elongations in terms of joint displacements, while, if the members are monolithic with one another at their junctions, in addition to elongations, rotational deformations are also required to be evaluated in terms of the joint displacements.

It is relatively easy to visualize the kinematics of a source than the internal force distributions. Thus, it is easier to incorporate approximations more confidently in the kinematic behavior of structure than in the internal force distribution.

7.2.4 Analysis Summary

As stated earlier, the general requirements of structural analysis are as follows: equilibrium, member stiffness characteristics, and kinematics. Any method of structural analysis must therefore necessarily ensure that (1) the conditions of equilibrium between the internal and external forces are satisfied, (2) the internal forces of the members and the deformations of the member are consistent with their stiffness properties, and (3) the deformation of members obey the kinematical characteristics and restraints of the structure.

In a determinate structure, the previously discussed three aspects are separate and are handled independently. Laws of statics give the internal forces whatever may be the stiffness properties of the members. When the internal forces are thus determined, each element can be isolated with the forces acting on it and can be analyzed separately for the resulting deformations. The displacements of the structure are then computed from these known member deformations and the kinematic conditions of the structural assembly. However, in a redundant structure, all the three aspects are coupled and are required to be handled simultaneously.

7.2.5 Braced Frames as Beams

In this section, we will examine a cantilevered braced frame under a tip load, F. The purpose is to show how its behavior is remarkably like the beam models. In particular, we will see how tall braces behave like classical Euler–Bernoulli beams, and that in shorter frames, shear displacement becomes dominant.

Consider the braced frame shown in Figure 7.3, which is about as simple a braced frame as we can develop by adding a single diagonal in each story. All members are assumed to carry only axial loads that do not bend or support shear.

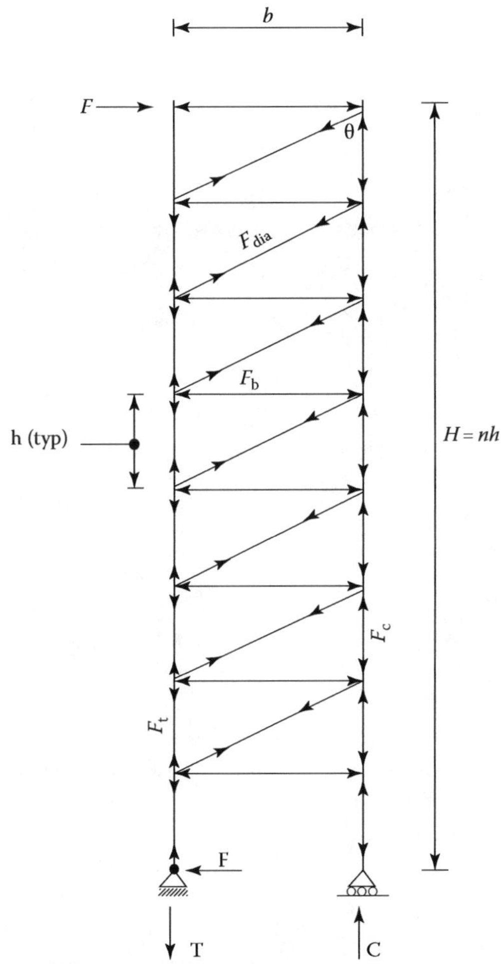

FIGURE 7.3 Braced frame.

The braced frame is n stories tall, each of height h for a total height of $H = nh$. The width of the frame is equal to b. We assume, for simplicity, that all the members of the braced frame, columns, beams, and braces, are made of the same material and have a common area A.

The structure is a statically determinate vertical truss that is supported by two vertical and one horizontal reaction. The frame is also statically determinate internally, meaning that we can calculate all of the forces in the members directly from the equations of equilibrium.

By taking a horizontal section through any of the panels, we can immediately see that from horizontal equilibrium, all of the forces in the diagonal members are the same, that is,

$$F\text{dia} = F/\sin(\theta)$$

Similarly, a section through any adjoining pair of stories shows that each of the $(n - 1)$ beams carries the same compressive load F_b:

$$F_b = F$$

By summing the forces at each joint along the frame's left and right columns, it is easily seen that the column forces

$$F_c = F_t \text{ or}$$

$$F_c = F_t = F \cot \theta$$

Note that columns on the right are in tension while those on the left are in compression.

To illustrate the similarity of the braced-frame behavior with that of beam, we calculated the deflection Δt of the frame at top subjected to a load F at top. It can be shown, by using Castiglione's second theorem, that the deflection

$$\Delta t = FH^3/3\varepsilon I[2n^2 + 1/2n^2 + 3\ (n - 1)/2n^3(b/n)^3 + 3/2n^2(1 + (b/n)^2)^{3/2}]$$

where

I is the second moment of area of the frame about the vertical axis $= Ab^2/2$
H is the height of braced frame $= nh$
ε is the modulus of elasticity of the material
n is the member of stories
b is the frame width
h is the story height

The first term in the brackets results from the extension and compression of the columns. The term $(n - 1)/n^3$ derives from the compression of $(n - 1)$ beams, while the $(1 + (b/h)^2)^3/^2$ term stems from the tension in the diagonals.

It is evident that since the braced-frame height is proportional to the number of stories n, the frame will get taller and more slender as the number of stories increases. Consequently, the effect of the horizontal beams and diagonals decreases with the square of the number of panels, n.

Using the previous equation for Δt, it can be shown that even for modestly tall braced frames with as few as five stories, the tip deflection is virtually identical to the corresponding Euler–Bernoulli beam theory result. Conversely, for short frames, the contribution of the beams and diagonals become much more important, with the diagonals that support the shear becoming the most influential structural elements. Thus, a modestly tall braced frame is basically little more than an articulated beam.

7.2.6 Preliminary Analysis of Rigid Frames

Years ago, before computers developed such a strong mental on the engineering office, engineers were still able to design complex indeterminate bracing systems. Simplifying assumptions were used to reduce these indeterminate problems to manageable determinate forms. The engineer was, in essence, forced to understand how the components behaved. Design objectives were simpler then, for strength was the almost exclusive concern. Drift was not of much concern, for steel frames were encased in concrete or stiffened by masonry facades. The flexible steel structures, which comprise the bulk of today's tall buildings, require that drift be considered in the design process. Unfortunately, today, few engineers understand how system configuration and component characteristics affect system strength or drift. As a consequence, the design process is reduced to an *iterative analysis*, in which the engineer presumes that the design process is complete when an almost randomly generated bracing system is analyzed many times over on a computer until both the strength and drift criteria are satisfied.

A moment frame, and for that matter any other type of bracing system, is typically a highly redundant combination of elements. It is now taken for granted that the assessment of its behavior is a task that only a computer can accomplish.

However, the configuration and type of bracing system proposed must be conveyed to the computer along with member sizes and connection characteristics. The computer will then advise where limit states have been exceeded, and the designer will then attempt to adjust the original design to correct the identified deficiencies. The process will be repeated until an acceptable system is produced. This iterative approach may attain the design objectives and be accomplished quickly, but success will depend entirely on the experience or understanding of bracing system behavior acquired by the designers. A skilled or experienced designer of bracing systems will understand how to control the behavior of a bracing system in a cost-effective manner.

Understanding how building systems behave is one of the keys to becoming a good designer. The goal of this book, as stated earlier is statically determinate to develop an understanding of how building systems behave from the perspective of both strength and stiffness so that the appropriate system can be selected. Complex systems will be reduced to simple systems so that the designer can understand how system behavior can be cost-effectively controlled.

After determining wind and seismic effects, the structure must be then analyzed to determine forces and moments for the design of the members and connections. Member and connection design proceeds quite normally for wind load design after these internal forces are determined, but seismic design is also subject to the detailed ductility considerations. Today, building frames are nearly always designed with the aid of a computer analysis to consider the frame stiffness, deformation, and distribution of forces. However, approximate analysis methods are desirable for preliminary analysis and initial member sizing needed prior to development of computer models. Two such methods for frames subject to lateral loads are the portal and cantilever methods.

The portal method is used for buildings of intermediate or shorter height. In this method, a bent is treated as if it were composed of a series of two-column rigid frames, or portals. Each portal shares one column with an adjoining portal. Thus, an interior column serves as both the windward column of one portal and the leeward column of the adjoining portal. Horizontal shear in each story is distributed in equal amounts to interior columns, while each exterior column is assigned half the shear for an interior column, since exterior columns do not share the loads of adjacent portals. If the bays are unequal, shear may be apportioned to each column in proportion to the lengths of the girders and ports. When bays are equal, the axial load in interior columns due to lateral load is zero.

Inflection points (points of zero moment) are placed at midheight of the columns and midspan beams. This approximates the deflected shapes and moment diagrams of those members under the loads. The location of their inflection points may be adjusted for special cases, such as fixed or pin base columns, or roof beams and top-story columns, or other special situations. On the basis of the preceding assumptions, member forces and bending moments can be determined entirely from equilibrium condition.

The cantilever method is used for tall buildings. It is based on the recognition that axial shortening and elongation of the columns contributes too much of the lateral deflections of such buildings. In this method, the floors are assumed to remain rigid, and the axial force in each column is assumed proportional to the distance of the column from the centroid of the columns. Inflection points are assumed to occur at midheight of the columns and midspan of the beams. The internal moments and forces are determined from the equations of equilibrium, as with the portal method.

Analysis of dual systems. Approximate analysis of braced frames can be performed as if the bracing were a truss. However, many braced structures are dual systems that combine resisting-frame behavior with braced-frame behavior. Under these conditions, an approximate analysis can be performed by first distributing the lateral forces between the brace frame and moment-resisting frame portions of the structure in proportion to their relative stiffness. Braced frames are commonly very stiff and normally would carry the largest portion of the loads. Once this distribution is completed, the moment-resisting frame can be approximately analyzed by the portal or cantilever method, and the braced frame can be analyzed as a truss.

Even with the availability of computers, it is desirable to make a rough check of results, using approximate means, to detect gross errors. For this reason, many engineers at some stage in the design process estimate the values of moment, shears, and thrusts at critical locations, using approximate sketches of the structure deflected by its loads.

Provided the points of inflection (locations in members at which the bending moment is zero and there is a reversal of curvature of the elastic curve) fan be located accurately, the stress resultants for a framed structure can usually be found on the basis of static equilibrium alone. Each portion of the structure must be in equilibrium under the application of its external loads and the internal stress resultants.

For the fixed-end beam of Figure 7.4a, for example, the points of inflection under uniformly distributed load are known to be located 0.211/from the ends of the span. Since the moment at these points is zero, imaginary hinges can be placed there without modifying the member behavior. The individual segments between hinges can be analyzed by statics, as shown in Figure 7.4b. Starting with the center segment, shears equal to $0.29wl$ must act at the hinges. These, together with the transverse load (Figure 7.4c), produce a midspan moment of $0.0417wl^2$. Proceeding next to the outer segments, a downward load is applied at the hinge representing the shear from the center segment.

This, together with the applied load, produces support moments of $0.0833wl^2$. Note that, for this example, since the correct position of the inflection points was known at the start, the resulting moment diagram agrees exactly with the true moment diagram for a fixed-end beam. In more practical cases, inflection points must be estimated, and the results obtained will only approximate the true values.

The use of approximate analysis in determining stress resultants in frames is illustrated by Figure 7.5. Figure 7.5a shows the geometry and loading of a two-member rigid frame. In Figure 7.5b, an exaggerated sketch of the probable deflected shape is given, together with the estimated location of points of inflection. On this basis, the central portion of the girder is analyzed by statics, to obtain girder shears at the inflection points of 14 kips, acting with a thrust P (still not determined). Similarly, the requirements of statics applied to the outer segments of the girder give vertical shears of 22 and 26 kips at B and C, respectively, and end moments of 36 and 60 ft kips at the same locations. Proceeding then to the upper segment of the column, shown in Figure 7.5f with the known thrust of 22 kips and top moment of 36 ft kips acting, a horizontal shear of 9 kips

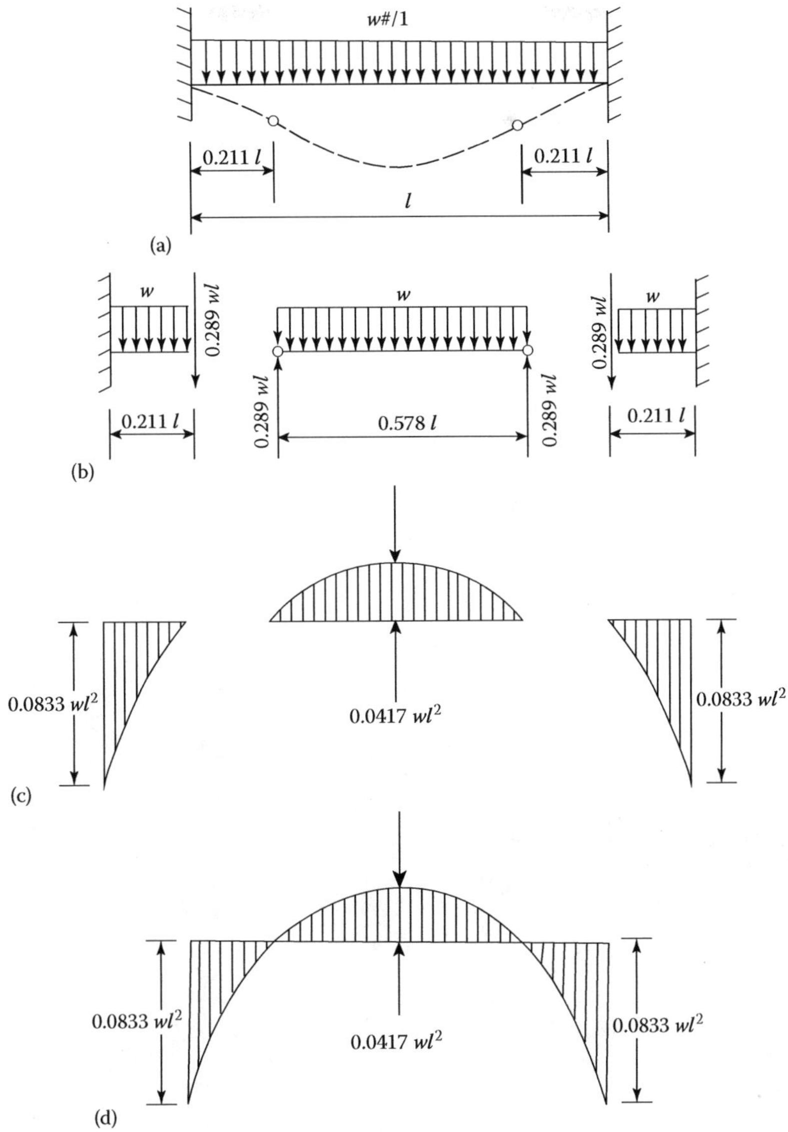

FIGURE 7.4 (a–d) Approximate analysis of a fixed beam.

at the inflection point is required for equilibrium. Static analysis of the lower part of the column indicates a requirement of 18 ft kips moment at A, as shown in Figure 7.5g. The value of P equal to 9 kips is obtained by summing horizontal forces at joint B.

The moment diagram resulting from approximate analysis is shown in Figure 7.5h. For comparison, an exact analysis of the frame indicates member end moments of 18 ft kips at A, 36 ft kips at B, and 60 ft kips at C. The results of the approximate analysis would be a valuable check on the magnitude of results.

Two specializations of the approximate method described, known as the *portal method* and *cantilever method,* are commonly used to estimate the effects of sideways due to lateral forces acting on multistory building frames. For such frames, it is usual to assume that horizontal loads are applied at the joints only.

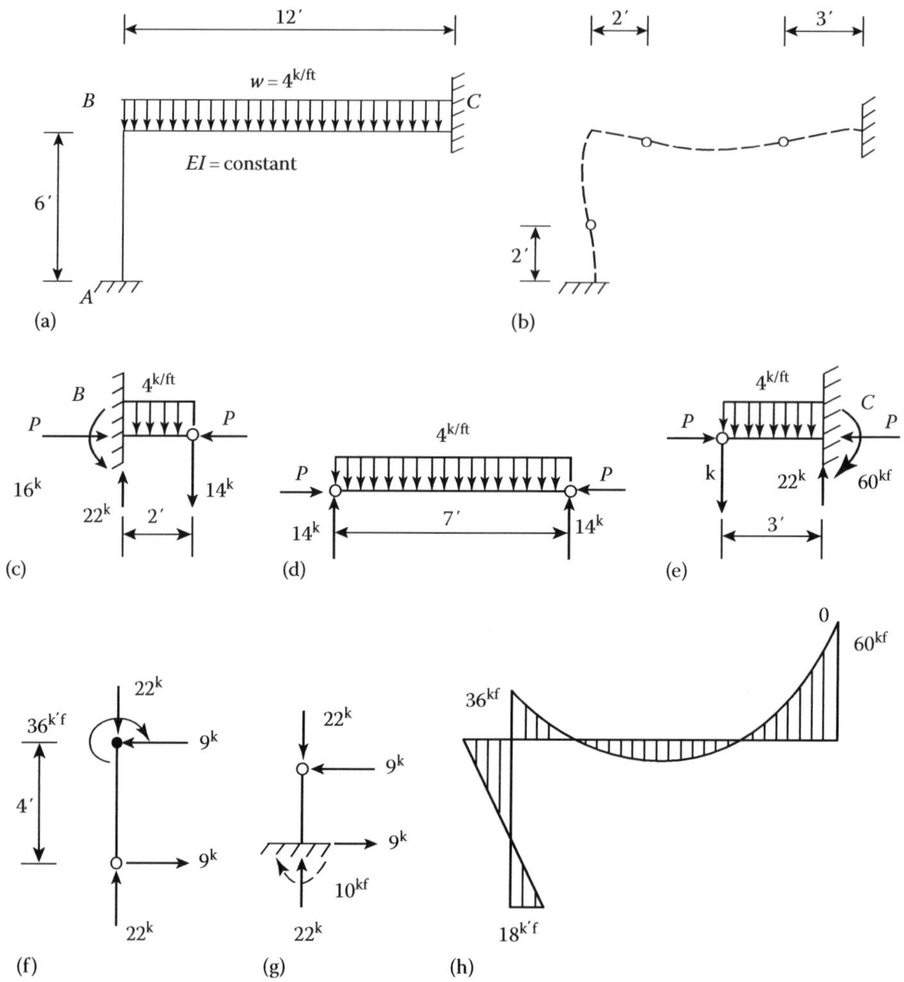

FIGURE 7.5 (a–h) Approximate analysis of half-frame.

7.2.6.1 Portal Method

This method is best explained using a single-story single-bay portal frame shown in Figure 7.6. The portal is statically indeterminate to the third degree. Therefore, three assumptions are needed to make an analysis by statics alone. If we assume (1) the magnitude of one of the horizontal reactions, (2) the location of the point of inflection on the left-hand column, and (3) the location of the point of inflection of the right-hand column, we can solve for the remaining horizontal reaction, the vertical reactions, and the end moments. Figure 7.6a shows the reactions; Figure 7.6b, the deflected shapes of the portal; and Figure 7.6c, the bending-moment diagrams, assuming points of inflection to lie at the midpoint of each column and the horizontal reactions to be equal.

The portal method is easy to use. Once the shears in the columns of each story are known, moments at the top and bottom of each column in a story are found by multiplying the shear by the half-length of the column. This follows from the assumption that the point of inflection is at midheight. Next, the end moment in a girder connecting to an outside column is equal to the sum of the column moments at that joint. But if there is a point of inflection at the midpoint of the girder, the end moments on the girder are numerically equal so that the moment at the interior end of the girder is also known. Next, the end moment for the first interior girder is found from the condition

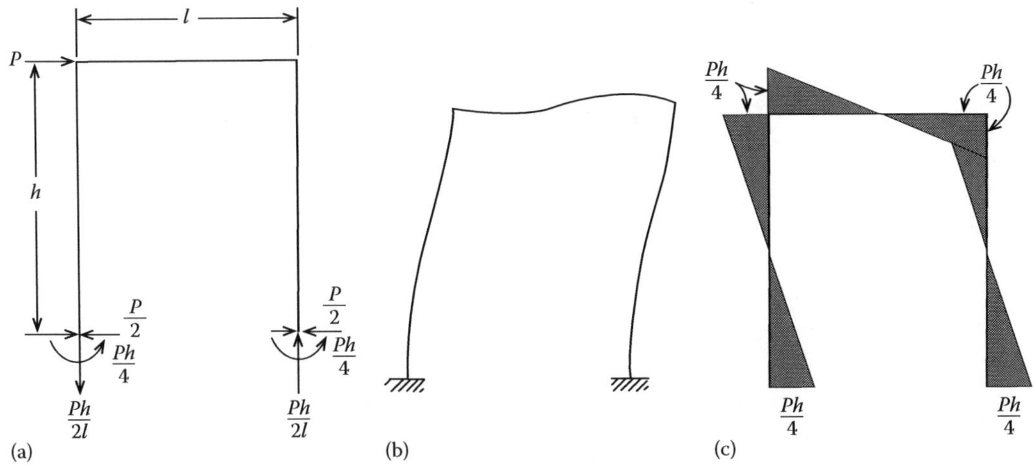

FIGURE 7.6 Single-bay, single-story portal frame: (a) reactions, (b) deflected shape, and (c) bending moments.

that the sum of the girder moments must equal the sum of the column moments at an interior joint. The shear in any girder is found from the condition that the sum of the end moments on the girder by the span. Evidently, the axial force in any column of any story is equal to the sum of the shears on all the girders connecting to it above the story in question.

The portal method is considered to be generally satisfactory for buildings of moderate height–width ratio and not over about 25 stories high. Story heights and girder spans should be approximately equal and the configuration reasonably symmetrical.

In this method, it is assumed that (1) points of contraflexure are located at midpoints of girders and columns and (2) the shear in columns is distributed in a rational manner. Under the second category, some engineers assume that the shear in exterior columns equals one-half the shear in an interior column, while some assume that the shear is distributed in proportion to the tributary bay width. For unequal bays, the former assumption results in direct stresses in the interior columns equal to the difference in girder shears on either side of the column. The latter assumption keeps the interior columns free from direct stresses.

We will now consider the application of the portal method to a 30-story frame shown in shown in Figure 7.7, consisting of two equal exterior bays and a smaller interior bay. Table 7.1 lists the lateral loads assumed in the analysis. The procedure is as follows: Distribute the accumulated story shears to each column in proportion to the aisle widths to keep the interior columns free of direct stresses. Calculate the moments in the top and bottom of each column as a product of the known shear in the column and one-half the story height. Next, starting at the upper left corner of the frame, write the girder moments where the column and girder moments are the same. Since the points of contraflexure are assumed at the center of girder, the moments at each end are equal but opposite in sign. Determine the girder shears by the relation that shear multiplied by half of span lengths equals girder end moment. Next, the direct stresses at the exterior columns are written directly from girder shears. The results for the example frame are shown in Figure 7.7.

In the cantilever method of analysis, points of inflection of all the columns and girders of a building frame are assumed to lie at midpoint, just as in the portal method. However, instead of an assumption about the distribution of the shear among the columns of a given story, in the cantilever method, an assumption is made about the distribution of the axial stresses in the columns of a given story. It is assumed that the bent acts as a cantilever beam and that in resisting the bending moments produced by the lateral forces, it develops in its columns axial forces distributed linearly about the vertical neutral axis of the frame.

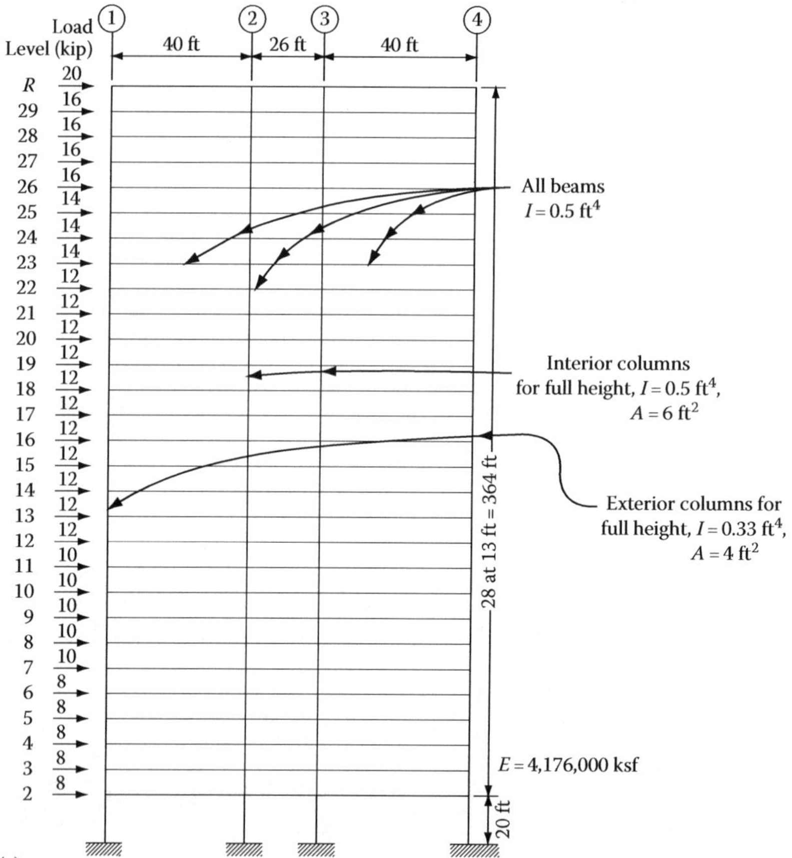

(a)

Level	Story Shear (kip)	Accumulated Shear (kip)	Level	Story Shear (kip)	Accumulated Shear (kip)
R	20	20	15	12	222
29	16	36	14	12	234
28	16	52	13	12	246
27	16	68	12	12	258
26	16	84	11	10	268
25	14	98	10	10	278
24	14	112	9	10	288
23	14	126	8	10	298
22	12	138	7	10	308
21	12	150	6	8	316
20	12	162	5	8	324
19	12	174	4	8	332
18	12	186	3	8	340
17	12	198	2	8	348
16	12	210			

(b)

FIGURE 7.7 Example frame: (a) dimensions and properties and (b) lateral loads.

TABLE 7.1
Lateral Loads for 30-Story Building Shown in Figure 7.11

Level	Story Shear (kip)	Accumulated Shear (kip)
R	20	20
29	16	36
28	16	52
27	16	68
26	16	84
25	14	98
24	14	112
23	14	126
22	12	138
21	12	150
20	12	162
19	12	174
18	12	186
17	12	198
16	12	210
15	12	222
14	12	234
13	12	246
12	12	258
11	10	268
10	10	278
9	10	288
8	10	298
7	10	308
6	8	316
5	8	324
4	8	332
3	8	340
2	8	348

As an example, assume that the columns of a three-bay framework are spaced 20 ft on centers and that the story height, measured between centerlines of floor beams, is 14 ft. Isolating the portion that lies above the points of inflection of the columns in the top story, we get Figure 7.8. The wind force at the roof is 8 kips. Assume that the cross-sectional area of each interior column is 1.5 times the cross-sectional area A_1 of the exterior column. Let the axial stress on the exterior column be f. Then on each interior column, the axial stress is $f/3$. The corresponding forces P_1 and P_2 are shown on the figure. From the equation of moments about the neutral axis O, we get

$$60P_1 + 20P_2 = 60fA_1 + 20 \times 0.5fA_1 = 8 \times 7$$

$$FA_1 = 0.8$$

Therefore, $P_1 - fA_1 = 0.8$ kip and $P_2 = 0.5fA_1 = 0.4$ kip.

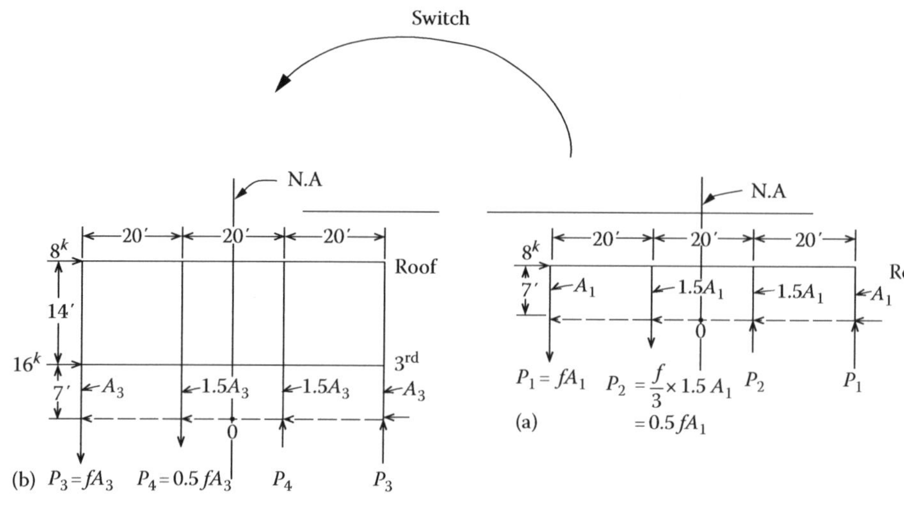

FIGURE 7.8 Cantilever method: (a) column axial forces below roof and (b) column axial forces below level 3.

If Figure 7.8a represents the top story of a multistory frame, we can find by proportion the axial forces in the columns of any other story for which the cross-sectional areas of the columns are in the same ratio as those in the top story. Thus, if we write the equation for moments at point O of Figure 7.8b we get the same equation as before except that A_3 replaces A_1 and the right member, $8 \times 7 = 56$ ft kips, becomes $8 \times 21 + 16 \times 7 = 280$ ft kips. Then, since $280/56 = 5$, we have $P_3 = 5P_1 = 4$ kips and $P_4 = 5P_2 = 2$ kips.

Once the axial forces in the columns are known, the shears in the girders are easy to determine. Thus, it is evident that the shear in the outside girder at the roof is equal to the axial force P_1 in the outside column while that in the interior girder is equal to the sum of the forces P_1 and P_2. Similarly, the shear in the outside girder at the top floor is equal to the difference between the axial forces P_3 and P_1 in the adjacent outside columns. With the shears in the girders known, end moments in the girders are determined by multiplying each shearing force by the half-span of the corresponding girder. This follows from the assumption that the points of inflection are at midspan. Moments in the columns may be found next, starting at the roof and taking advantage of the assumed equality of the end moments in each column of each story. Following this, determination of the shears in the columns and the axial forces in the girders completes the analysis.

The cantilever method is used less often than the portal method. It is suitable for buildings of moderate height–width ratio and not more than 25–35 stories high. It is considered to be superior to the portal method for high, narrow buildings. Story heights and girder spans should be approximately equal and the configuration reasonably symmetrical.

The deflections of a frame are a combination of flexural and shear distortions. Since the portal method is based on a distribution of lateral shear to the columns of a bent, it tends to produce a result that is primarily the effect of shear distortion. The distortion of frames with low height-to-width ratio is primarily that of shear. Therefore, the portal analysis is appropriate for such frames. On the other hand, the cantilever method is based on a distribution of column axial forces similar to the distribution of bending stress in a flexural member so that it tends to produce a result that is primarily a result of flexural distortion. The primary distortion of building frames with a high height-to-width ratio is due to bending as a cantilever. Therefore, the cantilever method is appropriate for such frames. A judicial combination of both methods would be ideal for preliminary sizing of frame members but rarely used in practice.

The frame analysis for horizontal loads by the so-called cantilever method is obtained by assuming that (1) inflection points, that is, hinges, form at midspan of each beam and at midheight of each column and (2) the unit direct stresses in the columns vary as the distance from the frame centrodial axis. It is forces will vary as the distance from the center of gravity of the bent. Using these assumptions, the frame is rendered statically determinate, and the direct forces, shears, and moments are determined by equilibrium considerations. The application of the method to an example frame will now be considered. To get a comparison with the results of the portal method, we shall apply the cantilever method to a three-bay portal frame.

The first assumption locates the points of contraflexure. Shown in Figure 7.9a, is a free-body diagram of the top story above the points of contraflexure in the columns. The frame axis of rotation is located at the center of gravity of the columns, which, for the example problem, coincides with the line of symmetry of the frame. The column axial forces for the top story are obtained by equating the moment of the

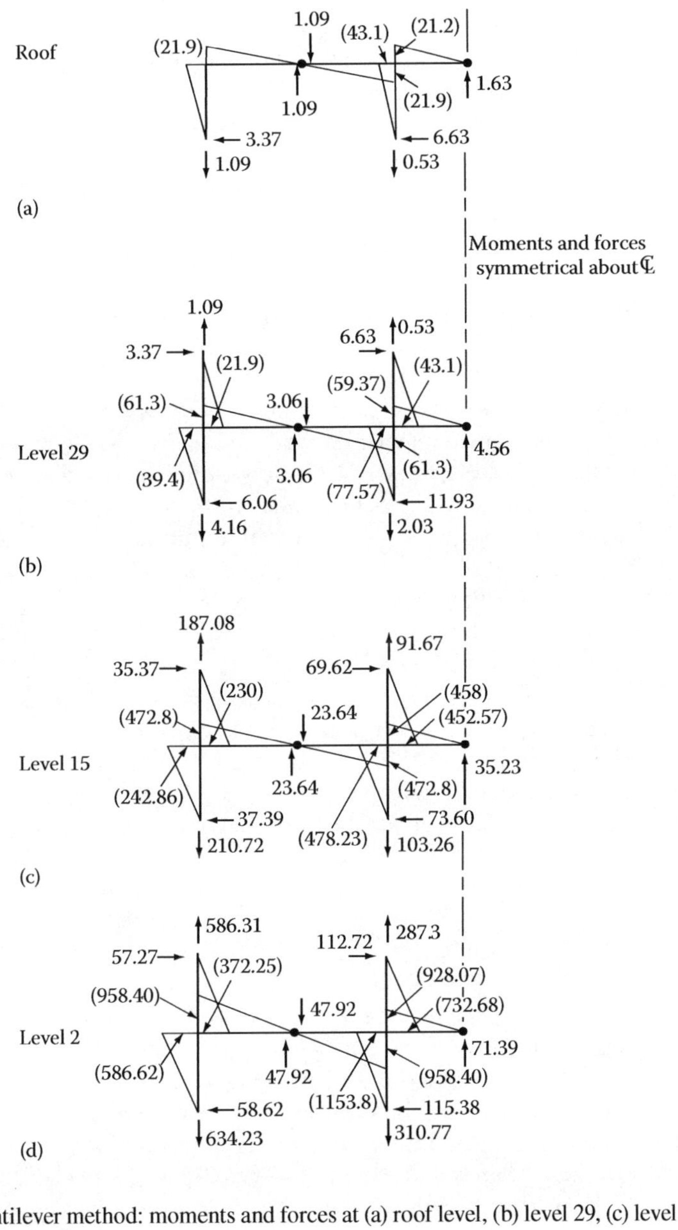

FIGURE 7.9 Cantilever method: moments and forces at (a) roof level, (b) level 29, (c) level 15, and (d) level 2. *Note*: Loads are in kips and moments in kip-ft.

column reactions about the frame axis to the moment of the wind forces taken about a horizontal plane through the assumed hinges of the top floor.

In a similar manner, the axial forces in the columns of other stories are computed by passing a section through the points of contraflexure of columns of each story and considering the moment equilibrium of the frame above the section.

After the column axial forces are found, the girder shears are determined at once. For example, in Figure 7.9c, the tension in the exterior windward column at the 15th level is 210.72 kips. Tension in the same column at the 14th level is 187.08 kips. Therefore, by the relation that the summation of the axial forces in the columns and the girder shear is equal to 0 at the joint where the 15th story girder joins the exterior windward column, the girder shear is 210.72 – 187.08 = 23.64. Figure 7.9c, shows the method of obtaining this and the remaining shears for the 15th level girder.

With the girder shears known, the girder moments follow directly. These equal the shear in the girder times one-half the span length. The study of the various joints will show that from the relation that $\Sigma M = 0$ at any joint, the sum of column moments must equal the sum of girder moments. Using this principle, the moments in the columns at the roof are obtained from roof girder moments (Figure 7.9a); since the points of contraflexure in the columns are at midheight, the column moments above the 29th level have the same value as at the roof level (Figure 7.9b). Moments in the columns below the 29th level are obtained from the relation EM = 0, and in a similar manner, column moments in other floors are found. The column shears are obtained by dividing column moments by half the height. Since the points of contraflexure in the columns are at midheight, the column moments above the 29th of columns. As a check, observe that the shear in the column of any level equals the sum of the horizontal external loads above the level. The moments are forces obtained by using the procedure discussed before for the example problem as shown in Figure 7.7.

To get a feel for the accuracy of the foregoing approximate procedures, the bent in Figure 7.9 has been analyzed by a plane-frame computer analysis and the results shown in Figure 7.10. As may be expected, the computer results vary considerably from either of the two methods. Chief among the reasons for the discrepancy are that (1) points of contraflexure in the lower stories are not at the midpoints and (2) the shears are greater in exterior girders than in the interior girders of that floor.

Before the advent of computers, it was common practice to use the portal or cantilever method with some modifications for the final design of structures. The modifications consisted of a number of assumptions. Chief among them are the following:

1. Locate point of contraflexure in exterior girders at 0.55 of their length.
2. Locate the points of contraflexure in the bottom-story columns at 0.6 heights from the base, in top-story columns at 0.65 heights from the top.
3. Use of rather complicated rules to divide the story shears among columns.

These approximations are no longer popular since the approximate methods are very rarely used in the final analysis of the structures.

7.2.6.2 Drift Assessment—Frame Structures

The lateral displacement of one floor relative to the floor below results from a combination of bending and shear deformation of the bent. The bending deformation of the chord drift, as it is sometimes called, is a consequence of axial deformation of the columns alone and is independent of the size, type, location, and arrangement of the web system. The shear deformation is due to the rotation of the joints in the frame, which causes bending of columns and girders of the frame. For relatively short frames with height-to-width ratios less than 3, the deflection due to axial shortening of columns can be neglected, and the deflection of the frame can be assumed

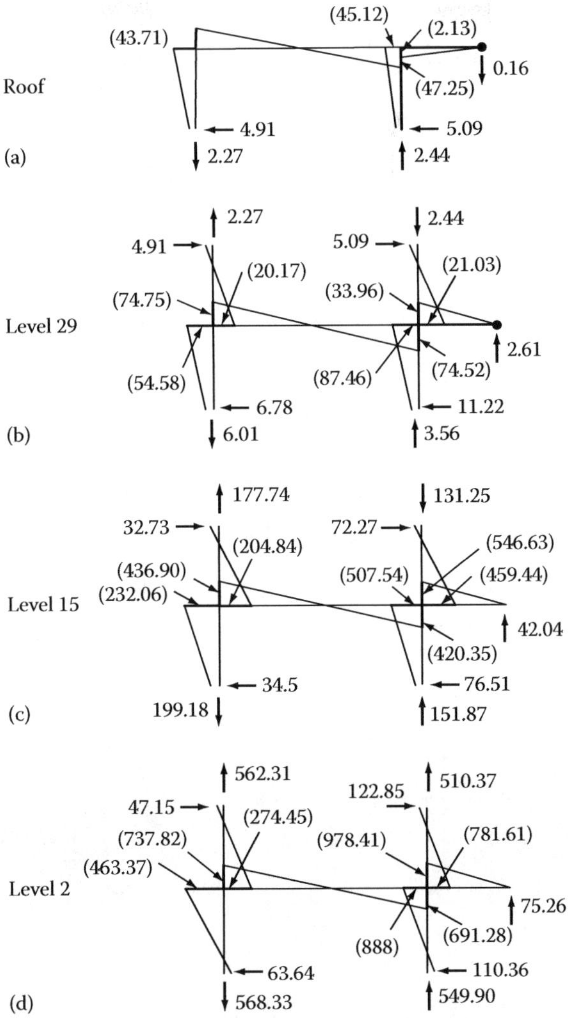

FIGURE 7.10 2D frame analysis: moments and forces at (a) roof level, (b) level 29, (c) level 15, and (d) level 2. *Note*: Loads are in kips and moments in kip-ft.

to be entirely due to joint rotations. Its contribution to deflection can, however, be obtained by considering the frame as a cantilever with an equivalent moment of inertia $I = 2ad^2$ where a is the area of exterior column and d is half the base of the portal frame. For taller frames, it is prudent to consider the axial deformation of the interior columns; the equivalent moment of inertia is determined by the relation $I = \Sigma n_1 a_1 d^2{}_1$, where $a_1, a_2, \ldots, a_n$ represent the areas of the columns and $d_1, d_2, \ldots, d_a$ represent the corresponding distances from the natural axis of the frame. To derive the equations for the shear deformations, let us consider a portal frame subjected to lateral shear forces as shown in Figure 7.11a. We isolate a representative portion of the frame consisting of a typical floor and column segments between the points of contraflexure above and below the floor as shown in Figure 7.11b. We will now consider the shear deformation, which is due to bending of columns and girders of the representative segment. First, we consider the contribution of columns by assuming the girders to be infinitely rigid; then we consider the girder contribution by assuming the columns to be infinitely rigid.

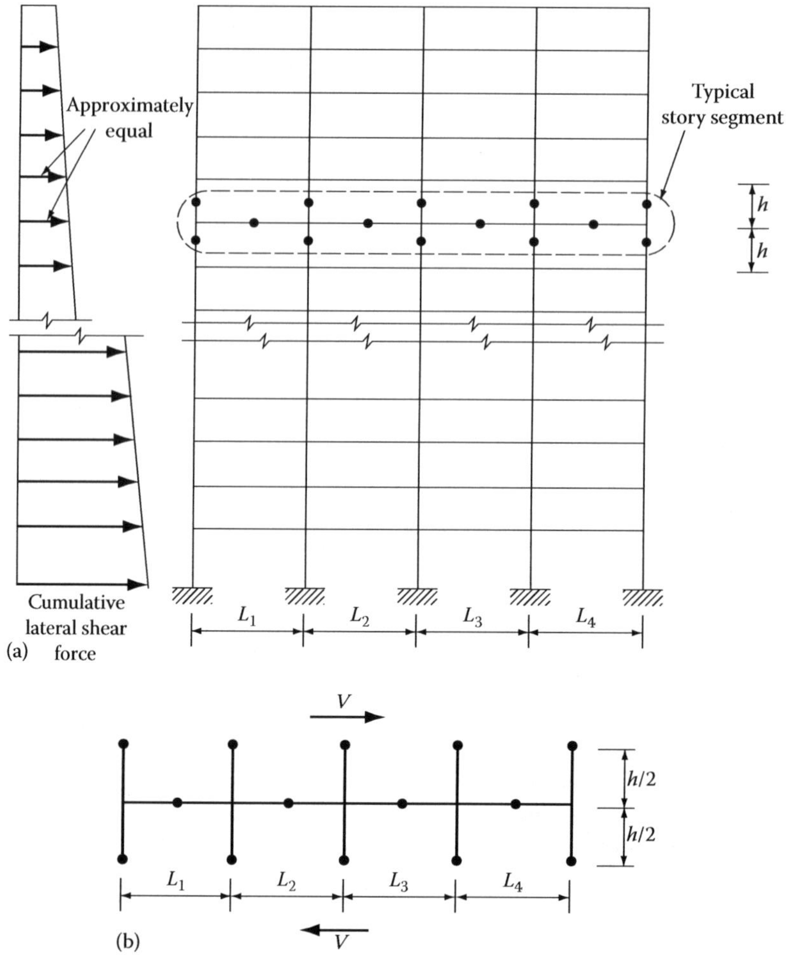

FIGURE 7.11 Portal frame shear deflections: (a) Frame subjected to lateral loads and (b) typical story segment.

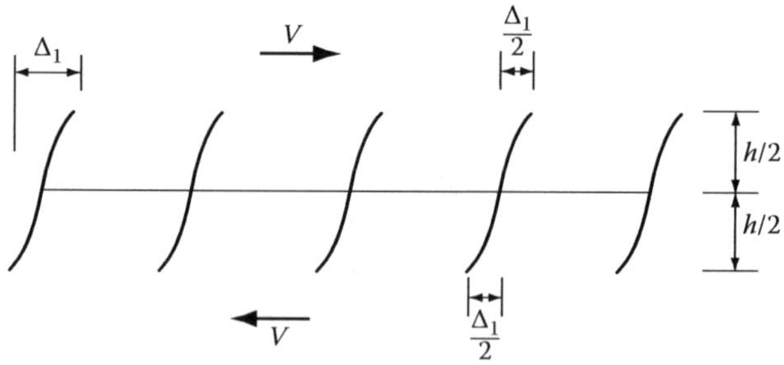

FIGURE 7.12 Lateral deflection due to bending of columns.

Deflection due to column rotations. Consider the free-body diagram of a typical story bounded between the points of contraflexure in the columns above and below the *i*th level as shown in Figure 7.12. When the number of stories is large, it is reasonable to assume that the shears in the columns above and below the floor do not differ appreciably. If the floor girders are rigid, the lateral deflection $\Delta_1/2$ of each column would be equal to the sum of the deflections of the two cantilevers of length $h/2$ under the action of wind shears V, giving for all columns $\Delta_1 = Vh^3/12E\Sigma I_c$

$$\frac{\Delta_1}{2} = \frac{V\left(\frac{h}{2}\right)^3}{3EI_c} \quad \text{or} \quad \Delta_1 = \frac{Vh^3}{12EI_c}$$

Deflection due to girder rotations. Next, consider the columns as rigid, giving rise to rotations of the girders as shown in Figure 7.13. Each girder undergoes a rotation equal to θ at each end, giving rise to an internal moment of $12EI\theta/L$ for each girder. The total internal moment is given by the summation of such terms for each girder. Thus, the total internal moment due to girder rotation

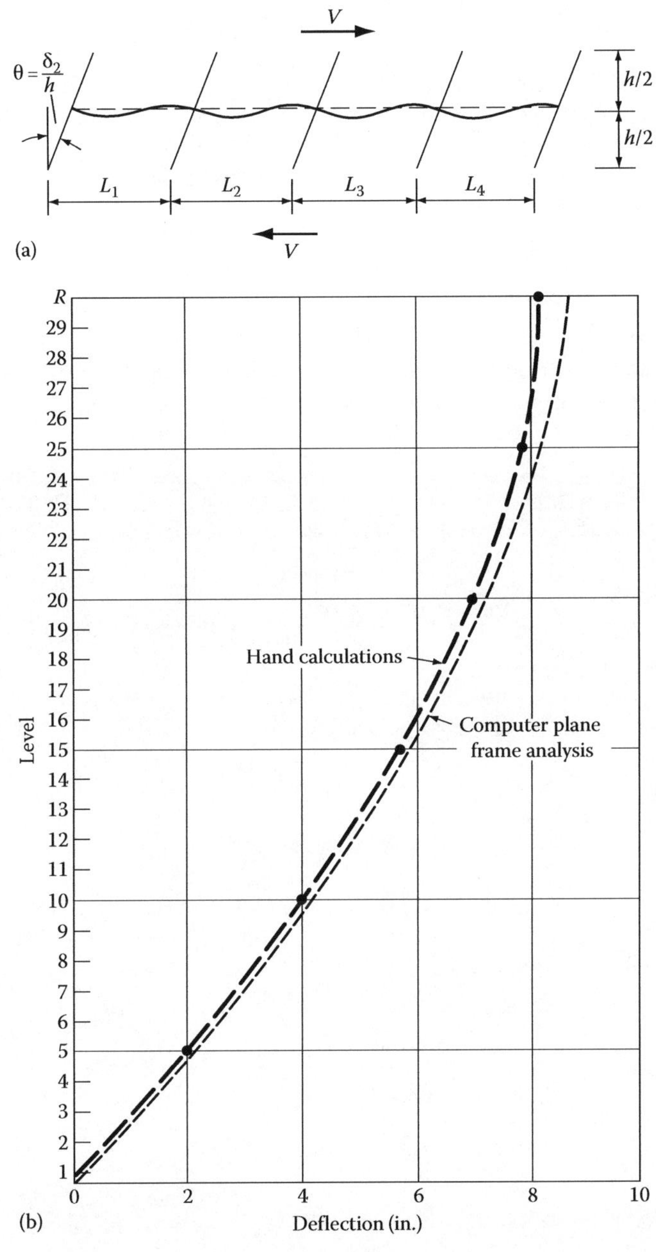

FIGURE 7.13 (a) Lateral deflection due to girder rotations and (b) deflection comparison (30-story frame).

is $12E\theta\Sigma(I_{bi}/L_i)$. The external moment due to wind shears V is given by $V \times h$. Equating external moment to internal moment and noting that e produces a displacement $\Delta_2 = \theta h$, we get

$$\Delta^2 = \frac{Vh^2}{12E\Sigma(I_{bi}/L_i)}$$

The total frame shear deflection Δs is given by

$$\Delta s = \Delta_1 + \Delta_2 = \frac{Vh^2}{12}\left\{\frac{h}{(\Sigma EI)_{col}} + \frac{1}{[\Sigma(EI/l)]_{bean}}\right\}$$

The deflection for the total number of stories is obtained by the summation of the deflections for each story.

An example of deflection calculations using the earlier procedure follows. To keep the presentation simple, we will consider the same example frame that was used for calculating moment and forces by the portal and cantilever methods (refer back to Figure 7.7).

7.2.6.3 Deflection Calculations

Cantilever deflection. The neutral axis for the frame lies on the line of symmetry. The moment of inertia of the frame about the neutral axis is given by $I = 2(a_1d_1^2 + a_2d_2^2)$ where a_1 and a_2 are the areas of the exterior and interior columns and d_1 and d_2 their distance from the neutral axis. Substituting $a_1 = 4$ ft^2 and $a_2 = 6$ ft^2, $d_1 = 53$ ft and $d_2 = 13$ ft, we get $I = 2(4 \times 532 + 6 \times 132) =$ 24,500 ft^4.

For the purpose of deflection calculation, we can assume that the frame is subjected to a uniformly distributed horizontal load = 12/13 = 0.9231 k/ft. The cantilever deflection at the top is given by

$$\Delta_{cant} = \frac{\omega l^4}{8EI} = \frac{0.9231 \times 384^4}{8 \times 4{,}176{,}000 \times 24{,}500} = 0.0245 \text{ ft (7.47 mm)}$$

Shear deflection due to column rotations. This is given by

$$\Delta_1 = \frac{Vh^3}{12E\sum I_c}$$

For the example problem, the moments of inertia for the exterior and interior columns are, respectively, equal to to 0.33 and 0.5 ft^4, giving $\sum I_c = 2 \times 0.33 + 2 \times 0.5 = 1.66$ ft^4. Using an average cumulative shear value of $V = 210$ kips and $h = 13$ ft,

$$\Delta_1 = \frac{210 \times 13^3}{12 \times 4{,}176{,}000 \times 1.66} = 0.0056 \text{ ft (1.70 mm)}$$

Shear deflection due to girder rotations. This is given by

$$\Delta_2 = \frac{Vh^2}{12E\,\Sigma(I/L)}$$

For the example problem, $\Sigma I/L$ of girders = 0.5/40 + 0.5/26 + 0.5/40 = 0.0442 ft, giving

$$\Delta_2 = \frac{210 \times 13^3}{12 \times 4{,}176{,}000 \times 0.0442} = 0.016 \text{ ft/floor } (4.87 \text{ mm/floor})$$

The total shear deflection $\Delta s = \Delta_1 + \Delta_2 = 0.0056 + 0.016 = 0.0216$ ft/floor. The shear deflection at the top of 30 stories is given by 30 × 0.0216 = 0.648 ft. Therefore, total deflection at top due to chord drift and shear deformation is 0.0245 + 0.648 = 0.6725 ft. A comparison of floor-by-floor deflections obtained by using the previous approach with those of a computer plane-frame analysis is given in Figure 7.13. The appropriateness of the method for preliminary design is obvious.

Another method with the objective of simplifying the numerical work involved in the calculation of frame deflection consists of representing the columns and beams as a single cantilever column bestowed with an equivalent flexural stiffness of I_e and shear stiffness of A_e to simulate the cantilever and shear modes of bending of the frame. The method is best explained with reference to Figure 7.14, which shows a 19-story, 3-bay unsymmetrical portal frame with columns of varying moments of inertia. We first locate x, the distance of frame axis of bending from the windward column by equating the moments of individual column areas to the moment of total area about the windward column. Using the values given in Figure 7.14, we get

$$4 \times 30 + 6 \times 50 + 6 \times 90 = (4 + 6 + 6 + 4) \times$$

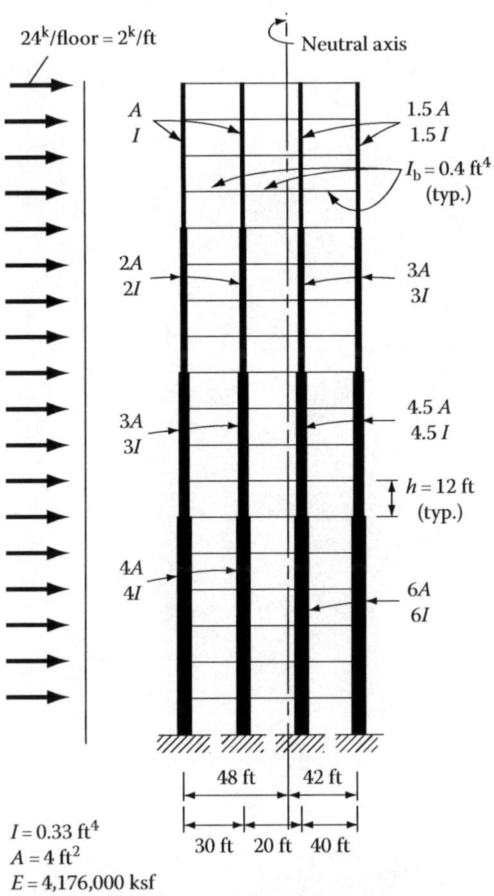

FIGURE 7.14 Example portal frame for deflection calculations.

giving

$$x = 48 \text{ ft}$$

from the windward column.

Calculate the moment of inertia of the frame about its axis of bending by the relation $I = \Sigma Ax^2$. Since the areas of the columns change at four locations, the corresponding four values of frame moment of inertia from the top work out equal to 21,120, 42,240, 63,360, and 84,480 ft^4, respectively.

Figure 7.15 shows the equivalent cantilever with varying moments of inertia. If the beams were infinitely rigid, the deflection calculated for the cantilever would have represented the total lateral deflection of the frame. Since, in reality, the beams are flexible, the deflection of the cantilever is increased by the racking component, which is equivalent to the shear deformation of the cantilever. This was shown equal to

$$\Delta s = \frac{Vh^2}{12}\left[\frac{h}{(EI/h)_{\text{col}}} + \frac{h}{(EI/h)_{\text{beam}}}\right]$$

Defining story stiffness as the deflection per unit of horizontal equivalent shear, the equivalent story stiffness is given by the relation

$$\frac{V}{\Delta s} = \frac{12}{h^2\{1/(\sum EI)_{\text{col}} + 1/[\sum(EI/L)]_{\text{beam}}\}}$$

An equivalent shear area for the cantilever is worked out as follows: Consider the shear deformation of the cantilever for unit height h subjected to horizontal forces V as shown in Figure 7.16. The shear deflections is given by

$$\Delta_s = \frac{Vh}{GA_\upsilon}$$

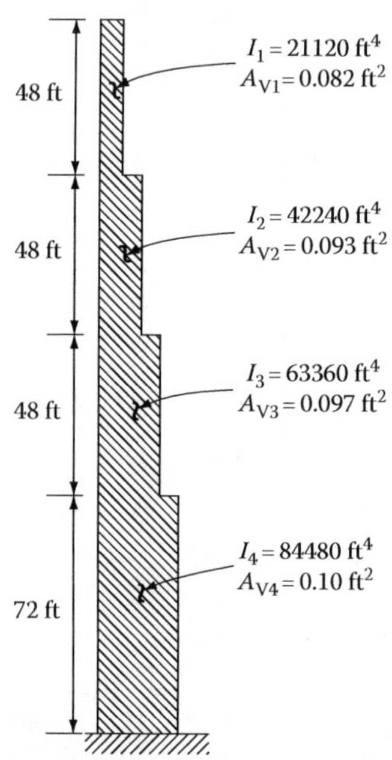

FIGURE 7.15 Equivalent cantilever.

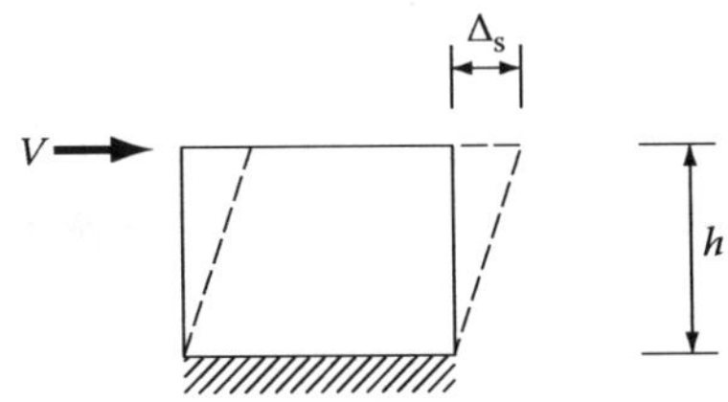

FIGURE 7.16 Shear deformation of a cantilever of unit height h.

The story stiffness $\Delta g/h$ works out equal to $0.4\,EA_\upsilon/h$ in which it is assumed that $G = 0.4E$. Equating story stiffness relations of Equations () and (), we get

$$\frac{0.4\,EA_\upsilon}{h} = \frac{12}{h^2\{1/(\sum E_c I)_{col} + 1/[\sum(E_b I/L)]_{beam}\}}$$

Assuming E is constant for beams and columns, that is, $E_c = E_b = E$, we get

$$A_\upsilon = \frac{12}{h\{1/(\sum I)_{col} + 1/[\sum(E_b(I/L)]_{beam}\}}$$

Using the numerical values shown in Figure 7.14, the equivalent shear areas at four vertical locations work out, respectively, equal to 0.082, 0.093, 0.097, and 0.1 ft^2 from the top. These values are shown schematically in Figure 7.15.

The deflection of the equivalent cantilever of varying moment of inertia can be obtained either by long-hand methods such as virtual work or by using a relatively simple stick computer model. In keeping with the approximate nature of analysis, reasonable results can be obtained by assuming average properties for the equivalent cantilever. The average values for I and A for the example problem work out equal to 56,320 ft^4 and 0.093 ft^2, respectively. Using a value of 216 kips for the average cumulative shear V, we get a total top deflection of 0.319 ft as compared with a value of 0.28 ft obtained from a stick computer model and a value of 0.24 ft as obtained from a plan-frame analysis. Comparison of deflections is shown in Figure 7.17.

The analysis presented thus far is based on the centerline dimensions, which in general overestimate the deflection. Although all structural members have finite widths, it is unnecessary, especially in view of the approximate nature of the analysis, to be overly concerned about the effect of joint widths on the stiffness of the structure. However, in those cases in which the dimensions of the members are large in comparison with story height and girder spans, it is possible to incorporate the effect of joints by assuming that no member deformation occurs within the joint. An approximate expression for the equivalent shear area for the equivalent column can be shown to be

$$A_\upsilon = \frac{30}{h^2\{h\alpha_1^{\,3}/(\sum I)_{col} + \alpha_2^{\,3}/(\sum(I/L))_{beam}\}}$$

where

α_1 is the average ratio of clear height to center to center heights of columns
α_2 is the average of the ratio of the clear span to the centerline spans of girders

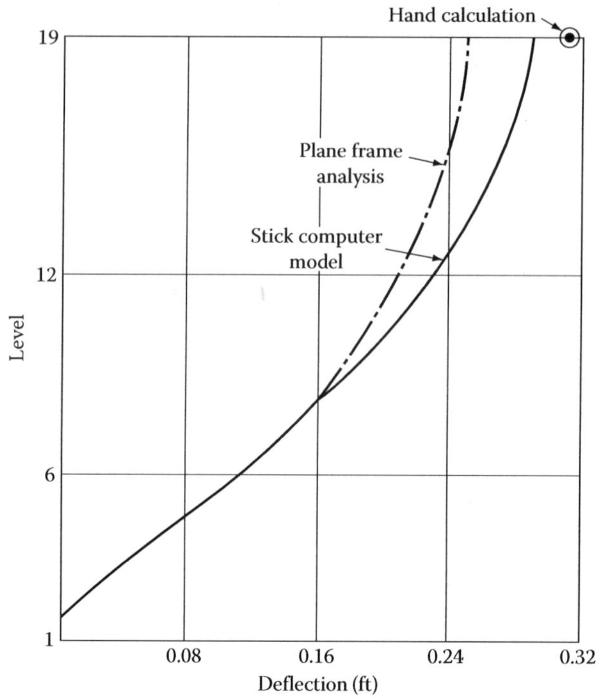

FIGURE 7.17 Deflection comparison.

Analytical and experimental investigations have shown that an analysis based on rigid offset lengths to the outer face of supports overestimates the stiffness of the structure. The analysis should, therefore, include some method of compensating the deformations that do exist in the panel zones. A rigid zone reduction factor can be used to reduce the lengths of rigid offsets—a method similar to that employed in many commercial computer programs. Arbitrary reductions are assigned to joint sizes in an effort to compensate for the joint deformation.

The underlying principle in both the portal and cantilever methods is the assumption that the point of contraflexure is located at midheight and midspan of columns and girders. Rigorous computer analyses almost invariably show that the fundamental assumption is violated in various degrees, especially at the top and bottom floors of a tall building. It is possible, however, to improve the results of the approximate analyses by assuming locations for points of contraflexture at locations representative of what is commonly found from computer analysis.

For example, it is fairly well known that the actual points of contraflexure in portal frames at the lower floors, especially at the first story, occur at a location closer to about $h/3$ below the second floor. Expressions for equivalent shear stiffness for the first story can be shown to work out as follows:

$$A_\upsilon = \frac{30}{h\{1/(\sum I)_{\text{col}} + 1/5[\sum (EI/L)_{\text{beam}}]\}}$$

Further refinement of the analysis is generally considered unnecessary in view of the approximate nature of the analysis and the availability of computer techniques.

7.2.6.4 Design Examples, Portal and Cantilever Methods

Consider a simply supported Vierendeel truss shown in Figure 7.18 carrying vertical loads at the top chord joints only. For purposes of preliminary analysis, let us assume that the loads are equal at the panel points. Determination of the preliminary analysis and moment diagrams for the truss is the purpose of this section.

Because the loads are equal at panel points, the truss behaves much like a beam carrying a uniformly distributed vertical load (Figure 7.18a). Further, by fixing the beam at the point of zero slope (i.e., at midspan of the beam), it can be shown that the Vierendeel truss symmetrically loaded by vertical loads behaves analogously to a cantilever rigid frame carrying lateral loads. Thus, the portal method can be used to determine the shear, axial and the moment diagrams by assuming hinges to form at midheight of truss columns and midspan of the top and bottom chords in each bay of the Vierendeel truss. It should be noted, however, that the equivalent cantilever is subjected simultaneously to vertical downward loads at the panel points, and a reaction at the support point acting vertically upward as shown in Figure 7.18c and d (see deflection calculations in Figure 7.18e).

To firm up our understanding of the behavior of the Vierendeel truss, consider an 80 ft span, 15 ft deep truss consisting of eight equal panels shown in Figure 7.19A. Assume the load at the panel point, P, is equal to 300 kips.

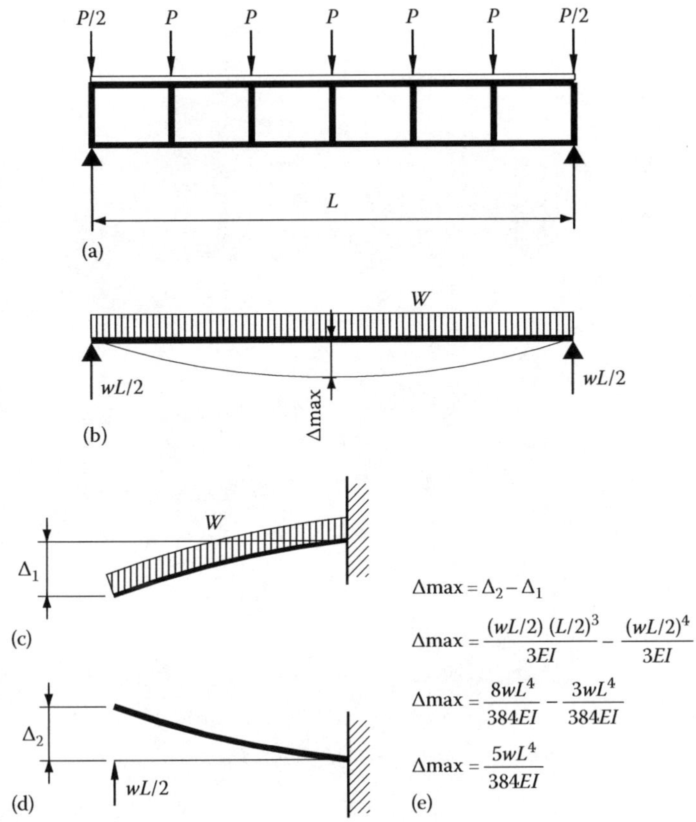

FIGURE 7.18 Vierendeel truss preliminary analysis: (a) loads at panel points, (b) equivalent simple beam, (c) tip deflection due to uniformly distributed load, (d) tip deflection due to end reaction, and (e) deflection calculations.

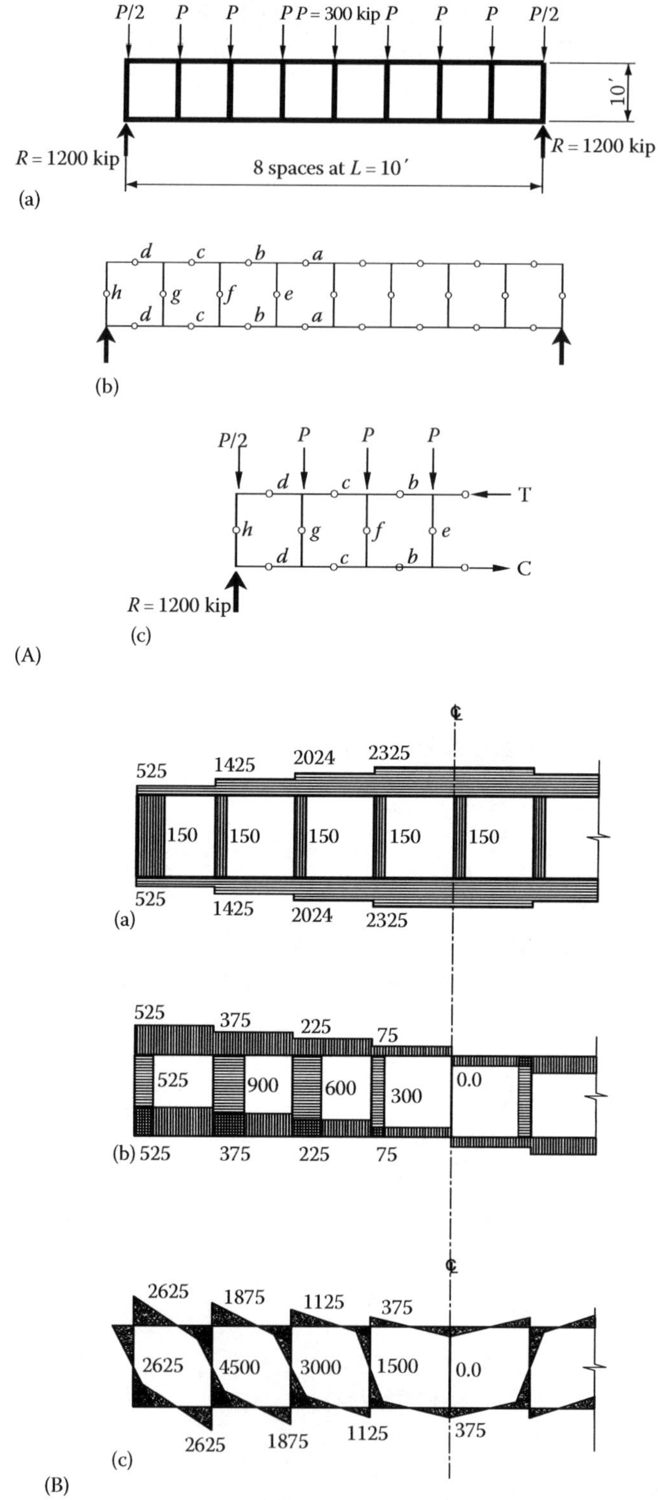

FIGURE 7.19 (A) (a) Truss geometry and loads, (b) points of contraflexure, (c) half-truss analytical model. (B) Schematics of flow of forces: (a) axial forces, (b) shear forces, and (c) bending moments. *Note*: Forces are in kips and bending moments in kip-ft.

Because of the symmetry in geometry and loading, only half the truss is considered. By cutting the free body of the truss at midspan of the chords, that is, at zero moment points, we ensure that only shear and axial forces will occur at that section:

$$EM_a = 1200 \times 35 - 150 \times 35 - 300 \times 25 - 300 \times 15 - 300 \times 5$$
$$= 23{,}250 \text{ kips-ft}$$

This moment is resisted by the equal and opposite axial tension and compression in the bottom and top chords, separated by the truss depth of 10 ft. Thus,

$$T = C = \Sigma M_a/10$$
$$= 23{,}250/10$$
$$= 2{,}325 \text{ kips (tension and compression)}$$

Vertical equilibrium yields the chord shear forces. At location "a,"

$$\Sigma V = 0 = 1\ 200\text{-}150\text{-}3(300) - 2V_4$$
$$V_4 = 75.0 \text{ kips}$$

A similar process is used to determine the shear and axial forces in other chord members. The shear and axial forces in the columns and bending moments in the top and bottom chords and columns are determined by equilibrium considerations. For this purpose, free-body diagrams cut at the assumed hinge locations are used. The results of the analysis with schematic flow of forces are shown in Figure 7.19B.

7.2.6.5 Framed Tubes

As mentioned earlier, the framed tube system in its simplest form consists of closely spaced exterior columns tied at each floor level by relatively deep spandrels. The behavior of the tube is in essence similar to that of a hollow perforated tube. The overturning moment under lateral load is resisted by compression and tension in the columns while the shear is resisted by the bending of columns and beams primarily in the two sides of the building parallel to the direction of the lateral load. The bending moments in the beams and columns of these frames, which are called web frames, can be evaluated using either of the two approximate procedures, namely, the portal or the cantilever analysis. It is perhaps more accurate to use the cantilever method because tube systems are predominately used for very tall buildings in the 40- to 80-story range in which the axial forces in the columns play a dominant role. The moments in spandrels and columns as well as the racking components of the tube deflection can be evaluated by using the cantilever method.

As mentioned earlier, because of the continuity of closely spaced columns and spandrels around the corners of the building, the flange frames are coaxed into resisting the overturning moment. Whether or not all the flange columns, or only a portion thereof, contribute to the bending resistance is a function of shear rigidity of the tube. A device normally used in approximate analysis is to reduce the tube configuration into two equivalent channels as shown in Figure 7.20. The determination of width of channel flange is subjected to engineering judgment and is usually limited to 15%–20% of the width of the building. It is a function of the shear lag across the windward and leeward sides of the tube, and the aforementioned rules of thumb give results sufficiently accurate for preliminary sizing of the tube system.

Shown in Figure 7.21 is the plan of a framed tube system delineating portions of the columns in the leeward and windward sides that were assumed to be part of the equivalent channel flanges. The axial forces were obtained on the basis of equivalent channel flanges. The axial forces were

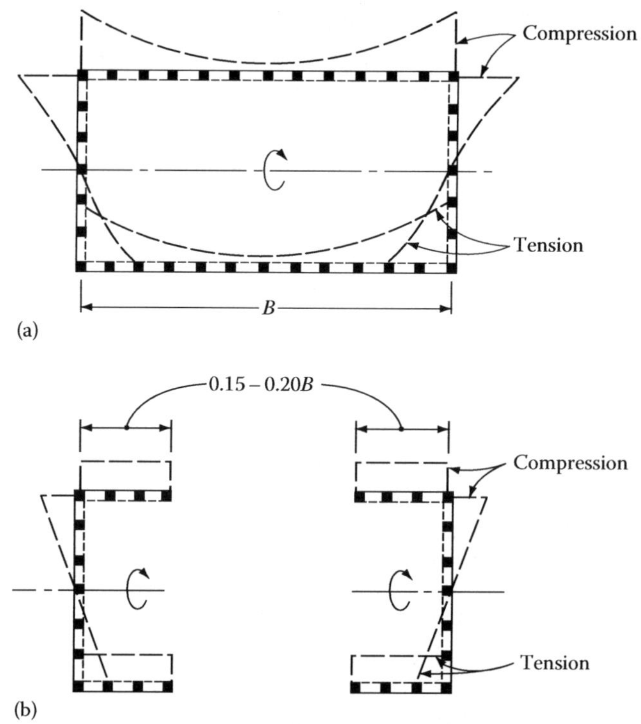

FIGURE 7.20 Framed tube: (a) axial stress distribution with shear lag and (b) axial stresses distribution in equivalent channels without shear lag.

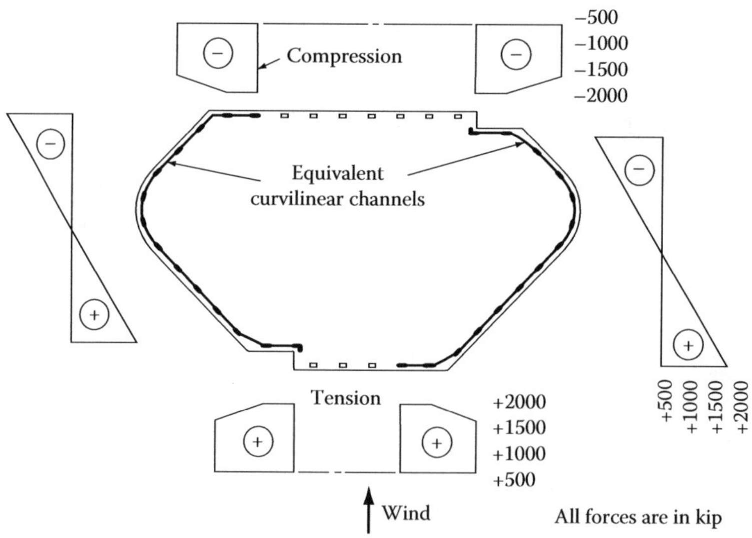

FIGURE 7.21 Axial forces in tube columns assuming two equivalent curvilinear channels.

obtained on the basis of equivalent structure, as shown. Shown in Figure 7.22 are the axial forces obtained from a three-dimensional computer analysis.

An equivalent column approach, as shown in the previous section, can be used to obtain approximate deflection values. In calculating the moment of inertia of the frame, it is only necessary to include the contribution of equivalent flange columns on the windward and leeward sides of the tube.

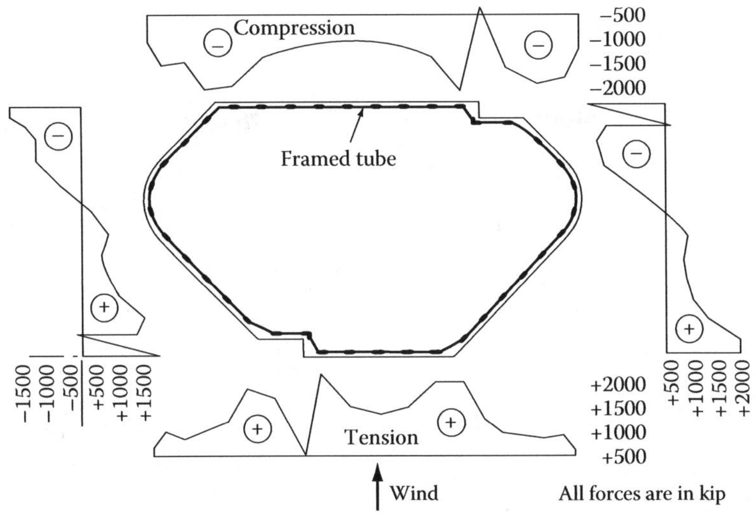

FIGURE 7.22 Axial forces in tube columns from three dimensional analysis of framed tube.

7.3 PRELIMINARY DESIGN: CONCRETE

Reinforced concrete structural components such as slabs, beams, columns, walls, and foundation are combined in various ways to devise structural systems for buildings. An important part of design is to select from many alternatives the best structural system for a given building.

The judicious choice of structural system is far more significant, in its effect on overall economy and serviceability, than meticulous refinements in proportioning individual members. Therefore, preliminary designs play a critical role in guiding us toward the selection of optimum structural systems.

7.3.1 Preliminary Design: Concrete Columns

The main reinforcement in columns is longitudinal, parallel to the direction of axial load, and consists of rebars arranged in a square, rectangular, or circular pattern, as shown in Figure 7.23. Oftentimes, the main reinforcement is bundled together to form a unit. Observe that ACI 318-11, like it predecessors, allows up to four bars to be placed in a bundle.

The ratio of longitudinal rebar A_{st} to gross concrete cross section is typically in the range of 0.01 (ACI minimum) to 0.08. However, for preliminary designs, and for reasons of economy, it is advisable to limit the ratio to about 3%–4% of the gross area A_g.

Columns may be divided into two categories, *short columns*, for which the strength is governed by the strength of the materials and the geometry of the cross section, and *slender columns*, for which strength may be reduced due to slenderness effects. Although reduced due to slenderness effects. Although slender columns may be more common now because of the wider use of high-strength materials, for example, $f_y = 100$ ksi and $f_c' = 10$ ksi or higher, only short columns are discussed in this section.

The nominal ultimate strength of an axially loaded column is obtained by summing the strength contributions of the two components of the column—the rebar and concrete. The useful design strength, $\Phi P_n(\max)$, is given by

$$\Phi P_n(\max) = 0.85\Phi[0.85 f_c'(A_g - A_{st}) + f_y A_{st}]$$

For tied columns with $\Phi = 0.70$,

$$\Phi P_n(\max) = 0.85\ \Phi[0.85 f_c'(A_g - A_{st}) + f_y A_{st}]$$

For spirally reinforced columns with $\Phi = 0.75$,

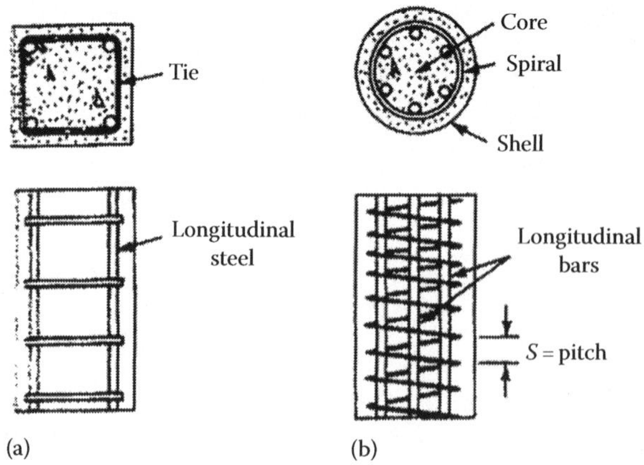

FIGURE 7.23 Ties of columns: (a) Tied column, minimum of four longitudinal bars required in a rectangular column. (b) Spiral column, minimum of six longitudinal bars required.

Observe that the values of Φ are lower for columns than beams, reflecting their greater importance in a structure. A beam failure would typically affect only a local region, where a column failure could result in partial collapse or tumble down of the entire building.

7.3.2 Preliminary Design of PT Floor Systems

The aim of post-tension design is to determine the required prestressing force and hence the number, size, and profile of tendons for satisfactory behavior at service loads. The ultimate capacity must then be checked at critical sections to ensure that prestressed members have an adequate factor of safety against failure.

The design method presented in this section uses the technique of load balancing in which the effect of prestressing is considered as an equivalent load. Take, for example, a prismatic simply supported beam with a tendon of parabolic profile, shown in Figure 7.24. The tendon exerts a horizontal force equal to $P \cos \varphi = P$ (for small values of φ) at the ends along with vertical components equal to $P \sin \varphi$. The vertical component is neglected in design because it occurs directly over the supports. In addition to these loads, the parabolic tendon exerts a continuous upward force on the beam along its entire length. By neglecting friction between the tendon and concrete, we can assume that (1) the upward pressure exerted is normal to the plane of contact and (2) tension in the tendon is constant. The upward pressure is exerted by the tendon profile. Due to the shallow nature of post-tensioned structures, the vertical component of the tendon force may be assumed constant. Considering one-half of the beam as a free body (Figure 7.24b), the vertical load exerted by the tendon may be derived by summing moments about the left support. Thus, the equivalent load $W_P = 8Pe/L^2$. Equivalent loads and moments produced by other types of tendon profile are shown in Figure 7.25.

The step-by-step procedure is as follows:

1. Determine the preliminary size of prestressed concrete members using the values given in Table 7.2 as a guide.
2. Determine the section properties of the member: area A, moment of inertia I, and section moduli S_t and S_b.
3. Determine tendon profile with due regard to cover and location of mild steel reinforcement.
4. Determine effective span L_e, by assuming $L_1 = {}^1/_{16}$ to ${}^1/_{19}$ of the span length for slabs and $L = {}^1/_{10}$ to ${}^1/_{12}$ of the span n length for beams. L_1 is the distance between the center line of support and the inflection point. The concept of effective length will be explained shortly.

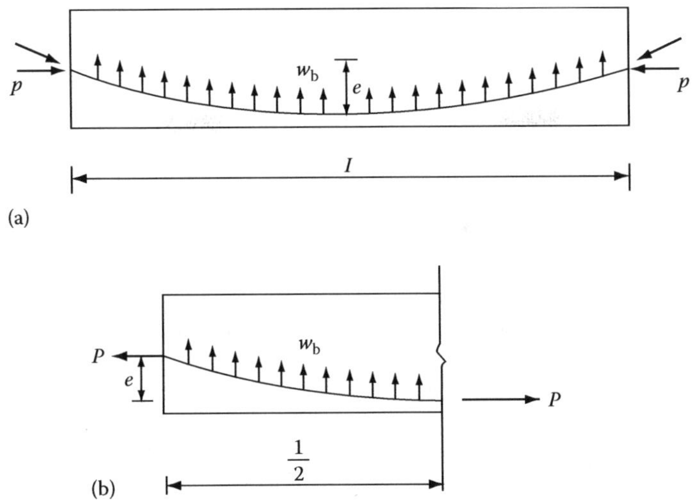

FIGURE 7.24 Load-balancing concept: (a) beam with parabolic tendon and (b) free-body diagram.

5. Start with an assumed value for balanced load W_P equal to, say, 0.7–0.9 times the total dead load.
6. Determine the elastic moments for the total dead plus live loads (working loads). For continuous beams and slabs, use a computer plane-frame analysis program, moment distribution method, or ACI coefficients, if applicable, in decreasing order of preference.
7. Reduce negative moments to the face of supports.

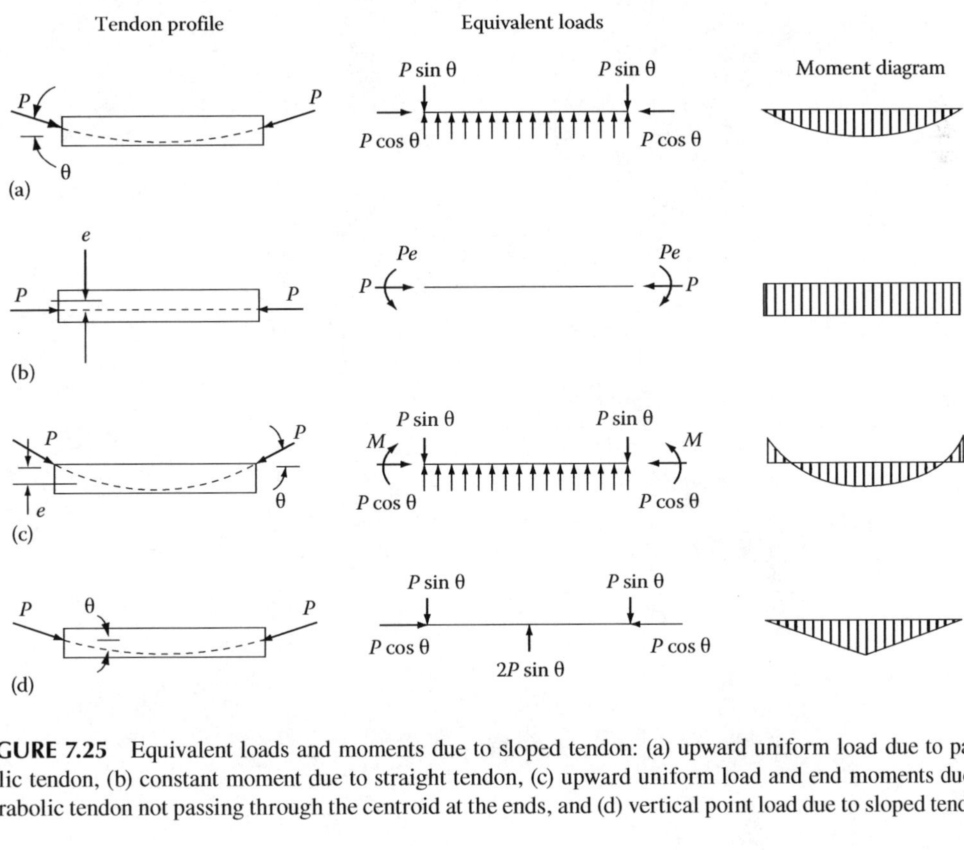

FIGURE 7.25 Equivalent loads and moments due to sloped tendon: (a) upward uniform load due to parabolic tendon, (b) constant moment due to straight tendon, (c) upward uniform load and end moments due to parabolic tendon not passing through the centroid at the ends, and (d) vertical point load due to sloped tendon.

TABLE 7.2
Sample Unit Quantities for Hotels

				Concrete Quantity Equiv. inches/S.F. of Superstructure				Reinforcing Quantity PSF of Tower Area								
								Slab								
Hold Location	No. of Floors	Wind Controls	Seismic Controls	Slab	Cols.	Shear Walls	Total	P.T. + 4	Rebar	Total	Cols.	Shear Walls	Total PSF	Lineal Ft. of Shear Wall per 15,000 sq ft of Typ. Floor Plate	Equivalent Rebar per cubic Yd.	w/o P.T.
Miami, FL	9	Yes		7.00	0.75	0.60	8.35	—	2.75	2.75	0.70	0.50	3.95	144	153lb	No PT
Miami, FL	14	Yes		7.00	0.68	0.53	8.21	—	4.08	4.08	1.01	0.84	5.93	161	234lb	No PT
Miami, FL	30	Yes		7.50	0.81	4.67	12.98	—	3.40	3.40	1.10	6.82	11.32	516	282lb	No PT
Nashville, TN	18	Yes		6.50	0.60	2.50	9.60	2.60	1.45	4.05	0.90	1.35	6.30	398	212lb	125
Houston, TX	25	Yes		7.00	0.80	3.40	11.20	2.80	2.37	5.17	1.25	2.93	9.35	314	270lb	189
El Paso, TX	6		Yes	6.50	0.40	1.00	7.90	2.60	1.40	4.00	1.40	0.70	6.10	184	250lb	144
Irvine, CA	16		Yes	6.50	1.08	4.25	11.83	2.60	2.10	4.70	0.58	4.30	9.58	423	262lb	191
Albuquerque, NM	16		Yes	7.80	1.00	3.50	12.30	2.84	2.80	5.64	1.00	2.40	9.04	310	238lb	163
Atlantic City, NJ	20	Yes		NA	NA	NA	NA	NA	NA	NA	NA	NA	NA	229	NA	NA
SAN Jose, CA	10		Yes	7.50	0.34	2.37	10.21	2.72	1.86	4.58	0.31	1.44	6.33	367	201lb	115
Santa Clare, CA	13		Yes	6.50	0.70	2.09	9.29	2.60	2.35	4.95	1.37	2.91	9.23	293	321lb	231

Note: NA, Not available.

8. By proportioning the unbalanced load to the total load, determine the unbalanced moments at M_{ub} at critical sections such as at the supports and the center of spans.
9. Calculate the bending stresses f_b and f_t at the bottom and top of the cross section due to M_{ub} at critical sections. Typically, at supports, the stresses f_t and f_b are in tension and compression, respectively. At the center of spans, the stresses are typically compression and tension at top and bottom, respectively.
10. Calculate the minimum required post-tension stress f_p by using the following equations. For negative zones of one-way slabs and beams,

$$F_p = f_t - 6\sqrt{f_c'}$$

For positive moments in two-way slabs,

$$F_p = f_t - 2\sqrt{f_c'}$$

11. Find the post-tension force P by the relation $P = f_p \times A$ where A is the area of the cross section of the beam.
12. Calculate the balanced load W_P due to P by the relation

$$W_P = 8 \times Pe/L_e^2$$

where
e is the drape of the tendon
L_e is the effective length of tendon between inflection points

13. Compare the calculated value of W_P from step 12 with the value assumed in step 5. If they are about the same, the selection of post-tension force for the given loads and tendon profile is complete. If not, repeat steps 9–13 with a revised value of $W_p = 0.75\ W_{p1} + 0.25\ W_{p2}$. W_{p1} is the value of W_P assumed at the beginning of step 5, and W_{p2} is the derived value of W_P at the end of step 12. Convergence is fast, requiring no more than three cycles in most cases.

7.3.2.1 Simple Span Beam

The concept of preliminary design discussed in this section is illustrated in Figure 7.26 where a parabolic profile with an eccentricity of 12 in. is selected to counteract part of the applied load consisting of a uniformly distributed dead load of 1.5 kip-ft and a live load of 0.5 kip-ft.

In practice, it is rarely necessary to provide a prestress force to fully balance the imposed loads. A value of prestress, often used for building system, is 75%–95% of the dead load. For the illustrative problem, we begin with an assumed 80% of the dead load as the unbalanced load.

First cycle. The load being balanced is equal to 0.80 × 1.5 = 1.20 kip-ft. The total service dead plus live load = 1.5 + 0.5 = 2.0 kips-ft, of which 1.20 kip-ft is assumed in the first cycle to be balanced by the prestressing force in the tendon. The remainder of the load equal to 2.0 – 1.20 = 0.80 kip-ft acts vertically downward, producing a maximum unbalanced moment M_{ub} at center span given by

$$M_{ub} = 0.80 \times 54^2/8$$

$$= 291.6 \text{ kips-ft}$$

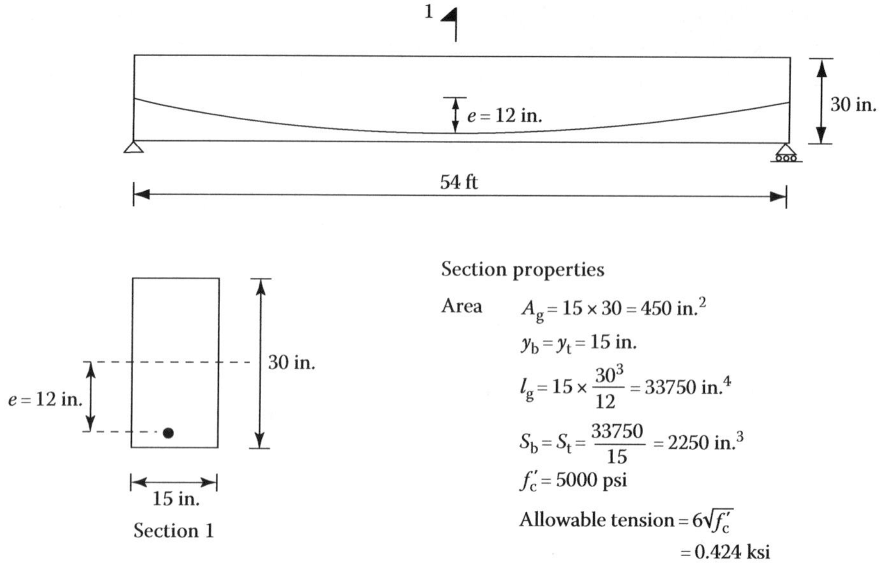

FIGURE 7.26 Preliminary design: simple span beam.

The tension and compression in the section due to M_{ub} is given by

$$f_c = f_b = 291.6 \times 12/2250$$

The minimum prestress required to limit the tensile stress to $6\sqrt{f_c'} = 0.424$ is given by

$$f_p = 1.55 - 0.424 = 1.13 \text{ ksi}$$

Therefore, the required minimum prestressing force P = area of beam × 1.13 = 450 × 1.13 = 509 kips. The load balanced by this force is given by

$$W_P \times 54^2/8 = Pe = 509 \times 1$$

and so $W_P = 1.396$ kip-ft compared to the value of 1.20 used in the first cycle. Since these two values are not close to each other, we repeat the previous calculations starting with a more precise value for W_p in the second cycle.

Second cycle. We start with a new value of W_p by assuming a new value equal to 75% of the initial value + 25% of the derived value. The new value of

$$W_p = 0.75 \times 1.20 + 0.25 \times 1.396 = 1.25 \text{ kip-ft}$$

$$M_{ub} = (2 - 1.25) \times 54^2/8 = 273.3 \text{ kips-ft}$$

$$f_b = f_t = 273.3 \times 12/2250 = 1.458 \text{ ksi}$$

The minimum stress required to limit the tensile stress to $6\sqrt{f_c'} = \sqrt{5000} = 0.424$ ksi is given by

$$F_p = 1.458 - 0.424 = 1.03 \text{ ksi}$$

Minimum prestressing force $P = 1.03 \times 450 = 465$ kips. The balanced load corresponding to the prestress value of 465 is given by

$$W_P = 8Pe/L^2 = 8 \times 465 \times 1/54^2 = 1.27 \text{ kip-ft}$$

Therefore, $W_p = 1.27$ kip-ft, nearly equal to the value assumed in the second cycle. Thus, the minimum prestress required to limit the tensile stress in concrete to $6\sqrt{f_c'}$ is 465 kips.

To demonstrate how rapidly the method converges to the desired answer, we will rework the problem by assuming an initial value of $W_P = 1.0$ kip-ft in the first cycle.

First cycle

$$W_p = 1.0 \text{ kip-ft}$$

$$M_{ub} = (2 - 1) \times 54^2/8 = 364.5 \text{ kips-ft}$$

$$f_b = f_t = 364.5 \times 12/2250 = 1.944 \text{ ksi}$$

$$f_p = 1.944 - 0.454 = 1.49 \text{ ksi}$$

$$P = 1.49 \times 450 = 670.5 \text{ kips}$$

$$W_p \times 54^2/8 = 670.5 \times l$$

$$W_p = 1.84 \text{ kip-ft}$$

Compared to 1.0 kip-ft used at the beginning of the first cycle.

Second cycle

$$W_P = 0.75 \times 1 + 0.25 \times 1.84 = 1.21 \text{ kip-ft}$$

$$M_{ub} = (2 - 1.21) \times 54^2/8 = 288 \text{ kips-ft}$$

$$f_b = f_c = 388 \times 12/2250 = 1.536 \text{ ksi}$$

$$f_p = 1.536 - 0.454 = 1.082 \text{ ksi}$$

$$P = 1.082 \times 450 = 486.8 \text{ kips}$$

$$W_p = 486.8 \times 1 \times 8/54^2 = 1.336 \text{ kip-ft}$$

Compared to the value of 1.21 used at the beginning of the second cycle.

Third cycle

$$W_p = 0.75 \times 1.21 - 1.21 \times 0.25 \times 1.336 = 1.24 \text{ kip-ft}$$

$$M_{ub} = (2 - 1.24) \times 54^2/8 = 276.67 \text{ kips-ft}$$

$$f_b = f_c = 276.47 \times 12/2250 = 1.475 \text{ ksi}$$

$$f_p = 1.475 - 0.454 = 1.021 \text{ ksi}$$

$$P = 1.021 \times 450 = 459.3 \text{ kips}$$

$$W_p = 459.3 \times 1 \times 8/54^2 = 1.26 \text{ kip-ft}$$

Compared to 1.24 assumed at the beginning of the third cycle. The value of 1.26 kip-ft is considered close enough for design purposes.

7.3.2.2 Continuous Beams

The earlier example illustrates the salient features of load balancing. Generally, the prestressing force is sleeted to counteract or balance a portion of dead load, and under this loading condition, the net stress in the tension fibers is limited to a value of $6\sqrt{f_c'}$. If it is desired to design the member for zero stress at the bottom fiber at center span (or any other value less than the code allowed maximum value of $6\sqrt{f_c'}$), it is only necessary to adjust the amount of post-tensioning provided in the member.

There are some qualifications to the foregoing procedure that should be kept in mind when applying the technique to continuous beams. Chief among them is the fact that it is not usually practical to install tendons with sharp break in curvature over supports, as shown in Figure 7.27. The stiffness of tendons requires a reverse curvature (Figure 7.27) in the tendon profile with a point of contraflexure some distance from the supports. Although this reverse curvature modifies the equivalent loads imposed by post-tensioning from those assumed for a pure parabolic profile between the supports, a simple revision to the effective length of tendon, as will be seen shortly, yields results sufficiently accurate for preliminary designs.

Consider the tendon profiles shown in Figure 7.28 for a typical exterior and an interior span. Observe three important features:

1. The effective span L_e, the distance between the inflection points that is considerably shorter than the actual span.
2. The sag or drape of the tendon is numerically equal to the average height of inflection points, less the height on the tendon midway between the inflection points.
3. The point midway between the inflection points is not necessarily the lowest point on the profile.

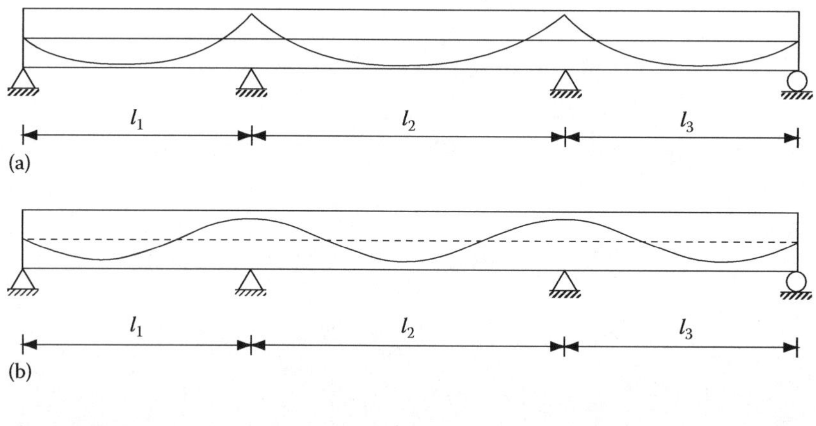

FIGURE 7.27 Tendon profile in continuous beam: (a) simple parabolic profile and (b) reverse curvature in tendon profile.

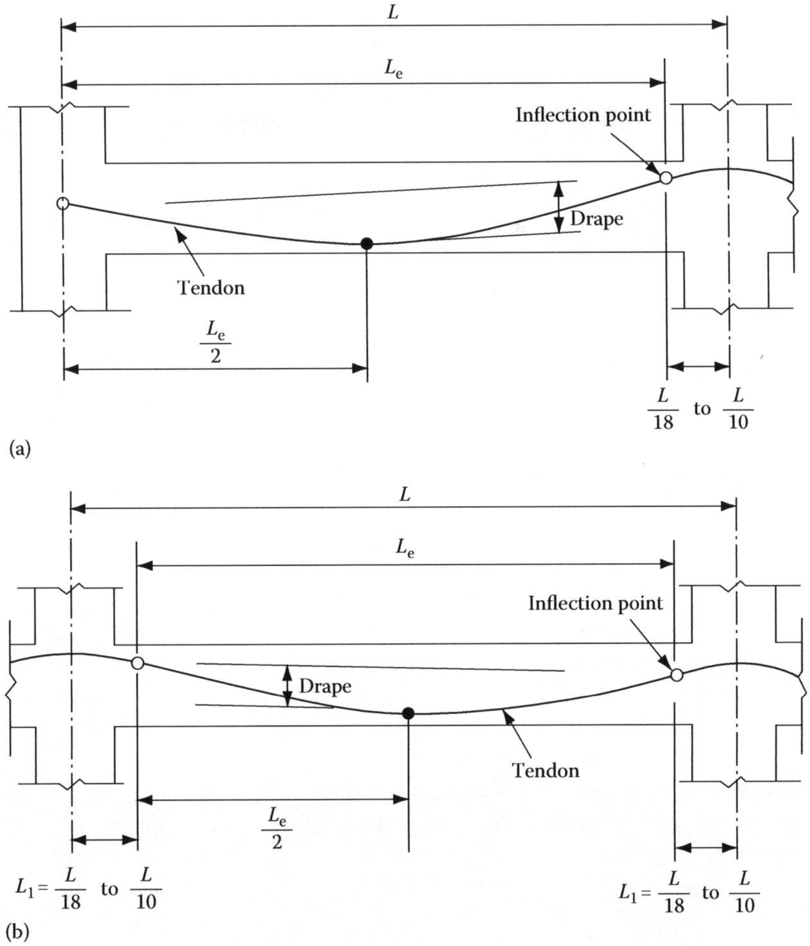

FIGURE 7.28 Tendon profile: (a) typical exterior span and (b) typical interior span.

The upward equivalent uniform load produced by the tendon is given by

$$W_p = 8Pe/L_e^2$$

where

W_p is the equivalent upward uniform load due to prestress
P is the prestress force
e is the cable drape between inflection points
L_e is the effective length between inflection points

Note that relatively high loads acting downward over the supports result from the sharply curved tendon profiles located within these regions (Figure 7.29).

Since the large downward loads are confined to a small region, typically $^1/_{10}$–$^1/_8$ of the span, their effect is secondary as compared with the upward loads. Slight differences occur in the negative moment regions between the applied load moments and the moment due to prestressing force. The differences are of minor significance and can be neglected in the design without losing meaningful accuracy.

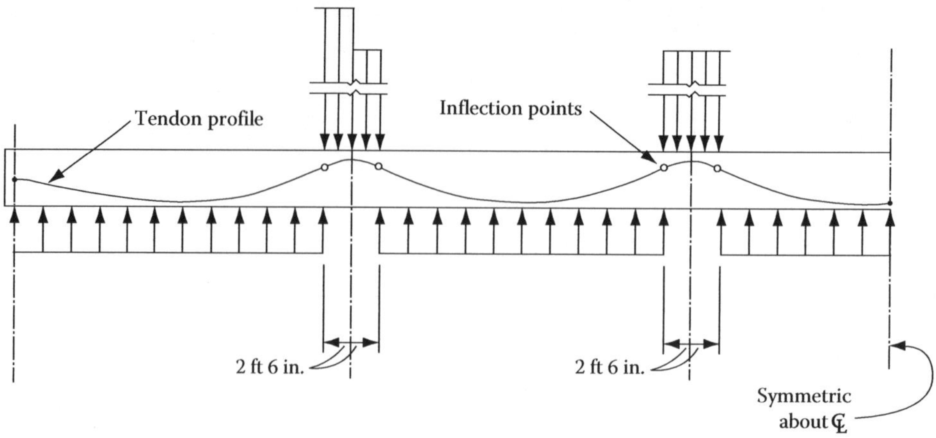

FIGURE 7.29 Equivalent loads due to prestress.

As in simple spans, the moments caused by the equivalent loads are subtracted from those due to applied loads, to obtain the net unbalanced moment that produces the flexural stresses. To the flexural stresses, the axial compressive stresses from the prestress are added to obtain the final stress distribution in the members. The maximum compressive and tensile stresses are compared with the allowable values. If the comparisons are favorable, an acceptable design has been found. If not, either the tendon profile or the force (and very rarely the cross-sectional shape of the structure) is revised to arrive at an acceptable solution.

In this method, since the moments due to equivalent loads are linearly related to the moments due to applied loads, the designer can bypass the usual requirement of determining the primary and secondary moments.

Example 1: One-Way PT Slab

Given a 30′–0″ column grid layout, design a one-way slab spanning between the beams shown in Figure 7.30.

Slab and beam depths:
Clear span of slab = 30 − 5 = 25 ft
Recommended slab depth = span/40 = 25 × 12,140 = 7.5 in.
Clear span for beams = 30 ft center-to-center span, less 2′–0″ for column width = 30 − 2 = 28 ft.
Recommended beam depth = span/25 = 28 × 12/25 = 13.44 in. Use 14 in.

Loading:

Dead load: 7.5″ slab	94 psf
Mech. and lights	6 psf
Ceiling	6 psf
Partitions	20 psf
Total dead load	126 psf
Live load: office load	100 psf
Code minimum	50 psf

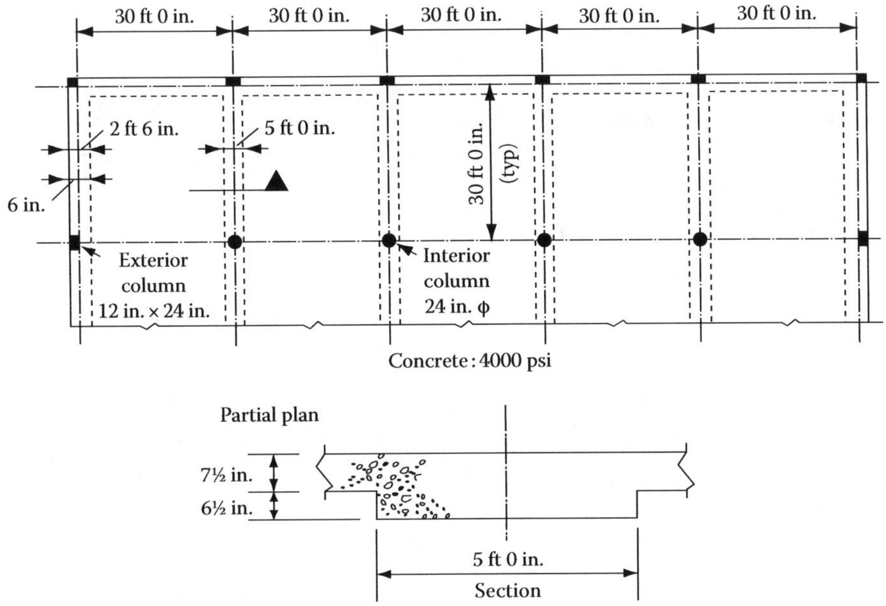

FIGURE 7.30 Example 1: one-way post-tensioned slab.

Use 100 psf per owner's request

Total $D + L = 226$ psf

Slab design: Slab properties for 1′–0″ wide strip:

$$I = bd^3/12 = 12 \times 7.53/12 = 422 \text{ in.}^4$$

$$S_{top} = S_{bat} = 422/3.75 = 112.5 \text{ in.}^3$$

$$\text{Area} = 12 \times 7.5 = 90 \text{ in.}^2$$

A 1 ft width of slab is analyzed as a continuous beam. The effect of column stiffness is ignored.

The moment diagram for a service load of 226 plf is shown in Figure 7.31

Moments at the face of supports have been used in the design instead of center line moments. Negative center line moments are reduced by a $Va/3$ factor (V = shear at that support, a = total support width), and positive moments are reduced by $Va/6$ using average adjacent values for shear and support widths. A frame analysis may be used to obtain more accurate results.

The design of continuous strands will be based on the negative moment of 10.6 kips-ft.

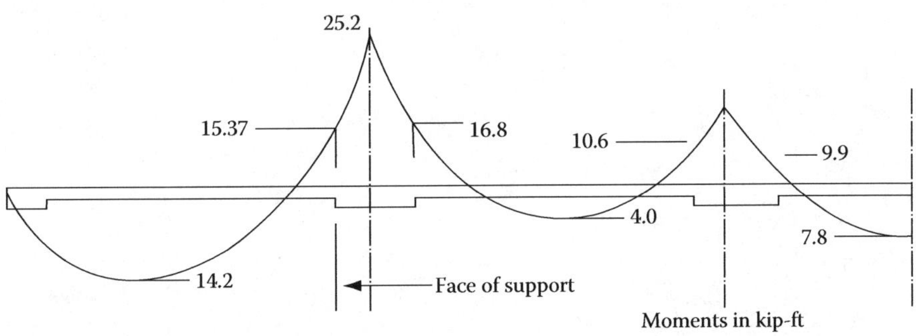

FIGURE 7.31 Example 1: one-way moment diagram.

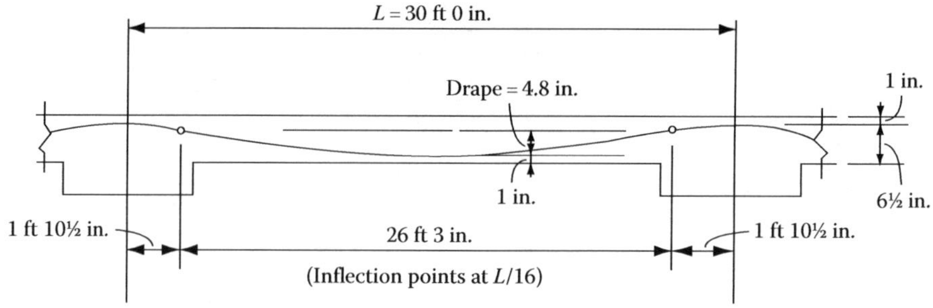

FIGURE 7.32 Example 1: one-way tendon profile—interior bay.

The additional prestressing required for the negative moment of 16.8 kips-ft will be provided by additional tendons in the end bays only.

Determination of tendon profile. Maximum tendon efficiency is obtained when the cable drape is as large as the structure will allow. Typically, the high points of the tendon over the supports and the low point within the span are dictated by concrete cover requirements and the placement of mild steel.

The high and low points of tendon in the interior bay of the example problem are shown in Figure 7.32. Next, the location of inflection points is determined. For slabs, the inflection points usually range within $^{1}/_{16}$ to 1119 of the span. The fraction of span length used is a matter of judgment and is based on the type of structure. For this example, we choose $^{1}/_{16}$ of span, which works out to 1′–10½″.

An interesting property useful in determining the tendon profile shown in Figure 7.33 is that if a straight line (chord) is drawn connecting the tendon high point over the support and the low point midway between, it intersects the tendon at the inflection point. Thus, the height of the tendon can be found by proportion. From the height, the bottom cover is subtracted to find the drape.

Slope of the chord line = $h_1 - h_2/(L_1 + L_2)$

$$h_s = h_2 + L_2 \times (\text{slope})$$

$$= h_2 + L_2(h_1 - h_2)/(L_1 + L_2)$$

This simplifies to $h_3 = (h_1L_2 + h_2L_1)/L_1 + L_2)$

The drape h_d is obtained by subtracting h_2 from the foregoing equation. Note that notion e is also used in these examples to denote drape h_d.

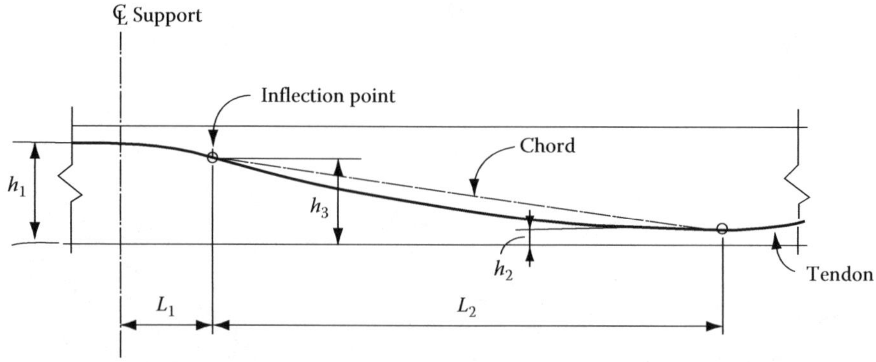

FIGURE 7.33 Dimensions for determining tendon drape.

In this case, the height of the inflection point is exact for symmetrical layout of the tendon about the center span. If the tendon is not symmetrical, the value is approximate but sufficiently accurate for preliminary design.

Returning to our example problem, we have $h_1 = 6.5''$, $h_2 = 1''$, $L_1 = 1.875'$, and $L_2 = 13.125'$.

Height of tendon at the inflection point:

$$h_3 = (h_1L_2 + h_2L_1)/L_1 + L_2$$

$$= 6.5 \times 13.125 + 1 \times 1.875/(1.875 + 13.125) = 5.812 \text{ in. Drape } h_d$$

$$= e = 5.813 - 1 = 4.813'' \text{ in. Use } 4.8''$$

Allowable stresses from ACI 318-11 are as follows:

$$f_t = \text{tensile stress } 6\sqrt{f'_c}$$

$$f_c = \text{compressive stress} = 0.45f'_e$$

For $f'_c = 4000$ psi concrete:

$$f_t = 66\sqrt{f'_c}\ 380 \text{ psi}$$

$$f_c = 0.45 \times 4000 = 1800 \text{ psi}$$

Design of through strands. The design procedure is started by making an initial assumption of the equivalent load produced by the prestress. A first value of 65% of the total dead load is used.

First cycle. Assume

$$W_P = 0.65\ W_d$$

where

W_P is the equivalent upward load due to post-tensioning, also denoted as W_{pt}

W_d is the total dead load

Therefore, $W_P = 0.65 \times 126 = 82$ plf.

The balancing moment caused by the equivalent load is calculated from

$$M_{pt} = M_s\ W_{pt}/W_s$$

where

M_{pt} is the balancing moment due to equivalent load (also indicated by notation M_b)

M_s is the moment due to service load, $D + L$

W_s is the total applied load, $D + L$

In our example, $M = 10.6$ kips-ft for the interior span

$$M_{pt} = 10.6 \times 821{,}226 = 3.85 \text{ kips-ft}$$

Next, M_{pt} is subtracted from M_s to give the unbalanced moment M_{ub}. The flexural stresses are then obtained by dividing M_{ub} by the section moduli of the structure's cross section at the point where M_s is determined. Thus,

$$f_t = M_{ub}/S_t$$

$$f_b = M_{ub}/S_b$$

In our case, $M_{ub} = 10.6 - 3.85 = 6.75$ kips-ft. The flexural stress at the top of the section is found by $f_t = M_{ub}/S_t = 6.75 \times 12/112.5 = 0.72$ ksi.

The minimum required compressive prestress is found by subtracting the maximum allowable tensile stress fa given in the following, from the tensile stresses calculated earlier. The smallest required compressive stress is

$$f_P = f_{ts} - f_a$$

where
- f_{ts} is the computed tensile stress
- f_a is $6\sqrt{f_c'}$ for one-way slabs or beams from the negative zones
- f_a is $2\sqrt{f_c'}$ for positive moments in two-way slabs

In our case,

$$f_P = 0.720 - 0.380 = 0.34 \text{ ksi}$$

and

$$P = 0.34 \times 7.5 \times 12 = 30.60 \text{ kips-ft}$$

Use the following equation to find the equivalent load due to prestress:

$$W_p = 8Pe/L_e^2$$

$$= 8 \times 30.6 \times 4.81/12 \times (26.25)^2 = 0.142 \text{ kif} = 142 \text{ plf}$$

This is more than 82 plf. N.G.

Since the derived value of W_P is not equal to the initial assumed value, the procedure is repeated until convergence is achieved. Convergence is rapid by using a new initial value for the subsequent cycle, equal to 75% of the previous initial value of W_{p1} plus 25% of the derived value W_{p2}, for that cycle.

Second cycle. Use the earlier criteria to find the new value of W_P for the second cycle.

$$W_p = 0.75W_{p1} + 0.25W_{p2} = 0.75 \times 82 + 0.25 \times 142 = 97 \text{ plf}$$

$$M_b = 97/226 \times 10.6 = 4.55 \text{ kips-ft}$$

$$M_{ub} = 10.6 - 4.55 = 6.05 \text{ kips-ft}$$

$$f_t = f_b = 6.05 \times 12/112.5 = 0.645 \text{ ksi}$$

$$f_p = 0.645 - 0.380 = 0.265 \text{ ksi}$$

$$P = 0.265 \times 90 = 23.89 \text{ kips}$$

$$W_p = 8 \times 23.89 \times 4.81/12 \times (26.25)^2 = 0.111 \text{ kif} = 111 \text{ plf}$$

This is more than 97 psf. N.G.

Third cycle

$$W_p = 0.75 \times 97 + 0.25 \times 111 = 100.5 \text{ plf}$$

$$M_b = 100.5/226 \times 10.6 = 4.71 \text{ kips-ft}$$

$$M_{ub} = 10.6 - 4.71 = 5.89 \text{ kips-ft}$$

$$f_t = f_b = 5.89 \times 12/112.5 = 0.629 \text{ ksi}$$

$$f_p = 0.629 - 0.380 = 0.248 \text{ ksi}$$

$$P = 0.248 \times 90 = 22.3 \text{ kips}$$

$$W_p = 8 \times 22.3 \times 4.81/12 \times (26.25)^2 = 0.104 \text{ klf} = 104 \text{ plf}$$

This is nearly equal to 100.5 plf. Therefore, satisfactory.

Check compressive stress at the section:

Bottom flexural stress = 0.629 ksi
Direct axial stress due to prestress = 22.3/90 = 0.46 ksi
Total compressive stress = 0.629 + 0.246 = 0.876 ksi is less than $0.45f'_c = 1.8$ ksi

Therefore, satisfactory.

End bay design. Design end bay prestressing using the same procedure for a negative moment of 15.37 kips-ft.

Assume that at the left support, the tendon is anchored at the center of gravity of the slab with a reversed curvature. Assume further that the center of gravity of the tendon is at a distance 1.75 in. from the bottom of the slab. With these assumptions, we have $h_1 = 3.75''$, $h_2 = 1.75''$, $L_1 = 1.875'$, and $L_2 = 13.125'$.

The height of the tendon inflection point at left end:

$$h_3 = 3.75 \times 13.125 + 1.75 \times 1.875/15 = 3.25 \text{ in.}$$

The height of the right end:

$$h_3 = 6.5 \times 13.125 + 1.75 \times 1.875/15 = 5.906 \text{ in.}$$

Average height of tendon = 3.25 + 5.906/2 = 4.578″. Use 4.6 in.
Drape $h_d = e = 4.6 - 1.75 = 2.85$ in.

First cycle. We start with the first cycle, as for the interior span, by assuming $W_{pt} = 82$ plf.

$$M_{pt} = 15.37 \times 82/226 = 5.58 \text{ kips-ft}$$

$$M_{ub} = 15.37 - 5.58 = 9.79 \text{ kips-ft}$$

$$f_t = f_b = 9.79 \times 12/112.5 = 1.04 \text{ ksi}$$

$$f_p = 1.04 - 0.380 = 0.664 \text{ ksi}$$

$$P = 0.664 \times 90 = 59.7 \text{ kips}$$

$$W_p = 8 \times 59.7 \times 2.85/12 \times (26.25)^2 = 0.165 \text{ kif} = 165 \text{ plf}$$

This is more than 82 plf. N.G.

Second cycle

$$W_p = 0.75 \times 82 + 0.25 \times 165 = 103 \text{ plf}$$

$$M_{pt} = 15.37 \times 103/226 = 7.0 \text{ kips-ft}$$

$$M_{ub} = 15.37 - 7.0 = 8.37 \text{ kips-ft}$$

$$f_t = f_b = 8.37 \times 12/112.5 = 0.893 \text{ ksi}$$

$$f_p = 0.893 - 0.380 = 0.513 \text{ ksi}$$

$$P = 0.513 \times 90 = 46.1 \text{ kips}$$

$$W_p = 8 \times 46.1 \times 2.85/12 \times (26.25)^2 = 0.127 \text{ klf} = 127 \text{ plf}$$

This is more than 103 plf. N.G.

Third cycle

$$W_P = 0.75 \times 103 + 0.25 \times 127 = 109 \text{ plf}$$

$$M_{pt} = 15.37 \times 1{,}091{,}226 = 7.41 \text{ kips-ft}$$

$$M_{ub} = 15.37 - 7.41 = 7.96 \text{ kips-ft}$$

$$f_t = f_b = 7.96 \times 12/112.5 = 0.849 \text{ ksi}$$

$$f_P = 0.849 - 0.380 = 0.469 \text{ ksi}$$

$$P = 0.469 \times 90 = 42.21 \text{ kips}$$

$$W_p = 8 \times 42.21 \times 2.85/12 \times (26.25)^2 = 0.116 \text{ klf} = 116 \text{ plf}$$

This is nearly equal to 109 plf used at the start of the third cycle. Therefore, satisfactory.
Check compressive stress at the section:

$$f_b = 0.849 \text{ ksi}$$

Axial stress due to prestress = 42.21/90 = 0.469 ksi
Total compressive stress = 0.849 + 0.469 = 1.38 ksi
This is less than 1.8 ksi. Therefore, design is OK.
Check the design against positive moment of 14.33 kips-ft:

$$W_p = 116 \text{ plf}$$

$$M_b = 14.33 \times 116/226 = 7.36 \text{ kips-ft}$$

$$M_{ub} = 14.33 - 7.36 = 6.97 \text{ kips-ft}$$

Bottom flexural stress = 6.97 × 12/122.5 = 0.744 ksi (tension)
Axial compression due to prestress = 42.21/12 × 7.5 = 0.469 ksi
Tensile stress at the bottom = 0.744 − 0.469 = 0.275 ksi
This is less than 0.380 ksi. Therefore, end bay design is OK.

Example 2: Continuous PT Beam

Refer to Figure 7.34 for dimensions and loading. Determine the flange width of beam using the criteria given in ACI.

The flange width bf is the least of

1. Span/4
2. Web width + 16 × (flange thickness)
3. Web width + ½ clear distance to next web

Therefore,

$$b_f = 30/40 = 7.5 \text{ ft (controls)}$$

$$= 5 + 16 \times 7.5/12 = 15 \text{ ft}$$

$$= 5 + 25/2 = 17.5 \text{ ft}$$

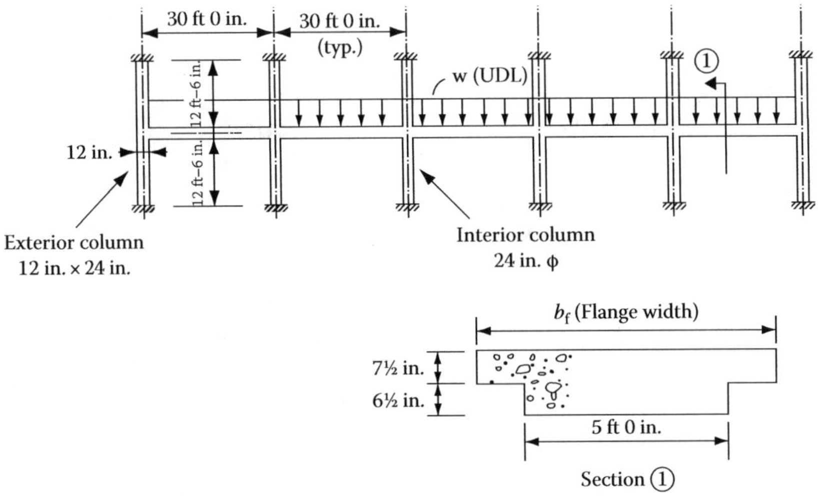

FIGURE 7.34 Example 2: post-tensioned continuous beam dimensions and loading.

Section properties:

$$I = 16{,}650 \text{ in.}^4$$

$$Y = 7.69 \text{ in.}$$

$$S_t = 2637 \text{ in.}^8$$

$$S_b = 2166 \text{ in.}^3$$

$$A = 1065 \text{ in.}^2$$

Loading:

Dead load of 7½ in. slab = 94 psf
Mech. and elec. = 6 psf
Ceiling = 6 psf
Partitions = 20 psf
Additional dead load due to beam self-weight = 615 × 60 × 150/144 × 30 = 13.5 = 14 psf
Total dead load = 140 psf
Live load at owner's request = 80 psf

$$D + L = 220 \text{ psf}$$

Uniform load per ft of beam = 0.220 × 30 = 6.6 klf. The resulting service load moments are shown in Figure 7.35. As before, we design for the moments at the face of supports.

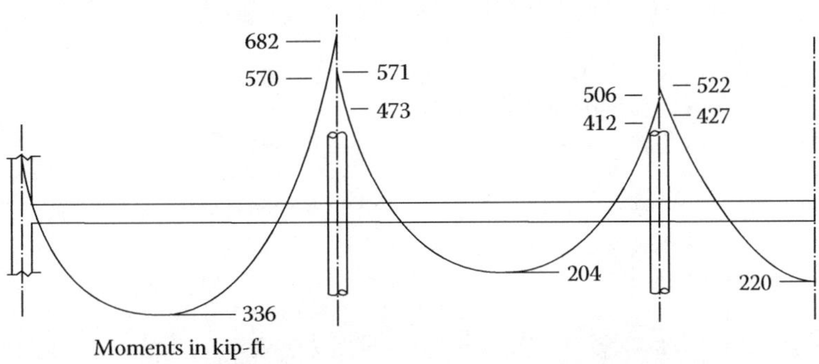

FIGURE 7.35 Example 2: post-tensioned continuous beam—service load moments.

*Inter*ior span. Calculate through tendons by using interior span moment of 427 kips-ft at the inside face of third column.

Assume $h_1 = 11.5$ in., $h_2 = 2.5$ in., $L_1 = 2.5$ ft, and $L_2 = 12.5$ ft.

The height of inflection point

$$h_3 = 11.5 \times 12.5 + 2.5 \times 2.5/15 = 10 \text{ in.}$$

$$h_d = e = 10 - 2.5 = 7.5 \text{ in.}$$

First cycle. Assume

$$W_p = 3.5 \text{ klf}$$

$$M_p = 3.5/6.6 \times 427 = 226 \text{ kips-ft}$$

$$M_{ub} = 427 - 226 = 20/\text{kips-ft}$$

$$f_t = 201 \times 12/2637 = 0.915 \text{ ksi}$$

$$f_p = 0.915 - 0.380 = 0.535 \text{ ksi}$$

$$P = 0.535 \times 106.5 = 570 \text{ kips}$$

$$W_p = 8 \times 570 \times 7.5/12 \times (26.25)^2 = 3.77 \text{ klf}$$

which is greater than 3.5 klf. N.G.

Second cycle. New value of

$$W_p = 0.75 \times 3.5 + 0.25 \times 377 = 3.57 \text{ klf}$$

$$M_p = 3.57/6.6 \times 427 = 231 \text{ kips-ft}$$

$$M_{ub} = 427 - 231 = 196 \text{ kips-ft}$$

$$f_t = 196 \times 12/2637 = 0.892 \text{ ksi}$$

$$f_p = 0.892 - 0.380 = 0.512 \text{ ksi}$$

$$P = 0.512 \times 1065 = 545 \text{ kips}$$

$$W_p = 8 \times 545 \times 7.5/12 \times (26.25)^2 = 3.60 \text{ klf}$$

which is nearly equal to 3.57 kif. Therefore, the design is satisfactory.

Check design against positive moment of 220 kips-ft:

$$M_p = 3.6 \times 220/6.6 = 120 \text{ kips-ft}$$

$$M_{ub} = 220 - 120 = 100 \text{ kips-ft}$$

$$F_{bot} = 100 \times 12/2166 = 0.554 \text{ ksi (tension)}$$

$$\text{Axial comp. stress} = 545/1065 = 0.512 \text{ ksi (comp)}$$

$$F_{total} = 0.554 - 0.512 = 0.045 \text{ ksi (tension)}$$

This is less than the allowable tensile stress of 0.380 ksi. Therefore, the design is satisfactory.

End span. Determine end bay prestressing for a negative moment of 570 kips-ft at the face of first interior column (Figure 7.35).

First cycle. As before, assume

$$W_p = 3.5 \text{ klf}$$

$$M_p = 3.5/6.6 \times 570 = 302 \text{ kips-ft}$$

$$M_{ub} = 570 - 302 = 2686 \text{ kips-ft}$$

$$f_t = 268 \times 12/2637 = 1.22 \text{ ksi}$$

$$f_p = 1.22 - 0.380 = 0.84 \text{ ksi}$$

$$P = 0.84 \times 1065 = 894 \text{ kips}$$

$$W_p = 8 \times 894 \times 7.5/12 \times (26.25)^2 = 5.912 \text{ kif}$$

which is greater than 3.5 klf. N.G.

Second cycle. New value of

$$W_p = 0.74 \times 3.5 + 0.25 \times 5.912 = 4.1 \text{ klf}$$

$$M_p = 4.1/6.6 \times 570 = 354 \text{ kips-ft}$$

$$M_{ub} = 570 - 354 = 216 \text{ kips-ft}$$

$$f_t = 216 \times 1{,}212{,}637 = 0.983 \text{ ksi}$$

$$f_p = 0.983 - 0.380 = 0.603 \text{ ksi}$$

$$P = 0.603 \times 1065 = 642 \text{ kips}$$

$$W_p = 8 \times 642 \times 7.5\ 1\ 12 \times (26.25)^2 = 4.24 \text{ klf}$$

This is nearly equal to 4.1 klf. However, a more accurate value is calculated as follows:

$$W_p = 0.75 \times 4.1 + 0.25 \times 4.24 = 4.13 \text{ klf}$$

Check the design against positive moment of 336 kips-ft:

$$M_p = 4.13/6.6 \times 336 = 210 \text{ kips-ft}$$

$$M_{ub} = 336 - 210 = 126 \text{ kips-ft}$$

Bottom flexural stress = 126 × 12/2166 = 0.698 ksi (tension)
Axial compressive stress due to post-tension = 642/1065 = 0.603 ksi (comp)

$$F_{total} = 0.698 - 0.603 = 0.095 \text{ ksi}$$

This is less than the allowable tensile stress of 0.380 ksi. Therefore, the design is OK.

Example 3: PT Flat Plate

Figure 7.36 shows a schematic section of a two-way flat plate system. Design of post-tension slab for an office-type loading is required.

Given:

Specified compressive strength of concrete $f_c' = 4000$ psi
Modulus of elasticity of concrete: $E_c = 3834$ ksi
Allowable tensile stress is precompressed tensile zone $= 6\sqrt{f_c'} = 380$ psi
Allowable fiber stress in compression $= 0.45f_c' = 0.45 \times 4000 = 1800$ psi

Tendon cover: Interior spans top 0.75 in.

Bottom 0.75 in.

Exterior spans top 0.75 in.

Bottom 1.50 in.

Tendon diameter = ½ in.

Minimum area of bonded reinforcement:

In negative moment areas at column supports:

$$A_s = 0.00075A_{cf}$$

where A_{cf} is the larger gross cross-sectional area of the slab beam strips of two orthogonal equivalent frames intersecting at a column of a two-way slab, in.^{2}

In positive moment areas where computed concrete stress in tension exceeds $2\sqrt{f_c'}$:

$$A_x = N_c/0.5f_y$$

where N_c is the tensile force in concrete due to unfactored dead load plus live load ($D + L$), in lb.

Rebar yield stress = 60 ksi. Max bar size = #4

Rebar cover 1.63 in. at top and bottom

Post-tension requirements:

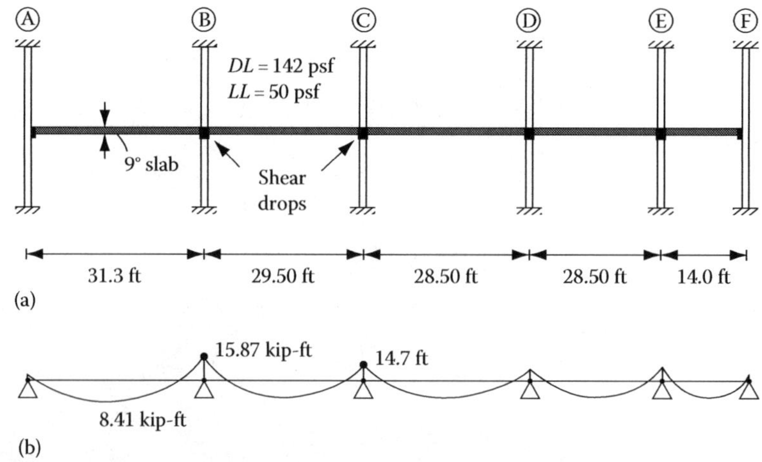

FIGURE 7.36 Example 3: flat plate design—(a) span and loading and (b) elastic moments due to dead load.

TABLE 7.3
Approximate Span Depth Ratios for Post-tensioned Systems

Floor System	Simple Spans	Continuous Spans	Cantilever Spans
One-way solid slabs	48–48	42–50	14–16
Two-way fl at slabs	36–45	40–48	13–15
Wide band beams	26–30	30–35	10–12
One-way joists	20–28	24–30	8–10
Beams	18–22	20–25	7–8
Girders	14–20	16–24	5–8

Note: The values are intended as a preliminary guide for the design of building floors subjected to a uniformly distributed superimposed live load of 50–100 psf (2394–4788 Pa). For the final design, it is necessary to investigate for possible effects of camber, deflections, vibrations, and damping. The designer should verify that adequate clearance exists for proper placement of post-tensioning anchors.

Minimum post-tensioned stress = 125 psi (see ACI 318-11, Sect. 18.12.4)

Minimum balanced load = 65% of total dead load

Design. The flat plate is sized using the span: depth ratios given in Table 7.3. The maximum span is 31′–4″ between grids A and B. Using a span: depth ratio of 40, the slab thickness is 31.33 × 12/40 = 9.4 in., rounded to 9 in.

The flat plate has *shear drops* intended to increase only the shear strength and flexural support width. The shear heads are smaller than a regular drop panel as defined in the ACI code. Therefore, shear heads cannot be included in calculating the bending resistance.

Loading	Dead load of 9″ slab	112 psf
	Partitions	20 psf
	Ceiling and mechanical	10 psf
	Reduced live load	50 psf

Total service load = 112 + 20 + 10 +50 = 192 psf

Ultimate load = 1.4 × 142 + 1.7 × 50 = 285 psf

Slab properties (for a 1 ft wide strip):

$$I = bh^3/12 = 12 \times 9^3/12 = 729 \text{ in.}^4$$

$$S_{\text{top}} = S_{\text{bot}} = 729/4.5 = 162 \text{ in.}^3$$

$$\text{Area} = 12 \times 9 = 108 \text{ in.}^2$$

The moment diagram for a 1-ft-wide strip of slab subjected to a service load of 192 psf is shown in Figure 7.37.

The design of continuous strands will be based on a negative moment of 14.7 kips-ft at the second interior span. The end bay prestressing will be based on a negative moment of 15.87 kips-ft.

Interior span. Calculate the drape of tendon using the procedure given for the previous problem. See Figure 7.33:

$$h_3 = h_1L_2 + h_2L_1/L_1 + L_2$$

$$L_1 = 1.84 \text{ ft}, h_1 = 8 \text{ in.}$$

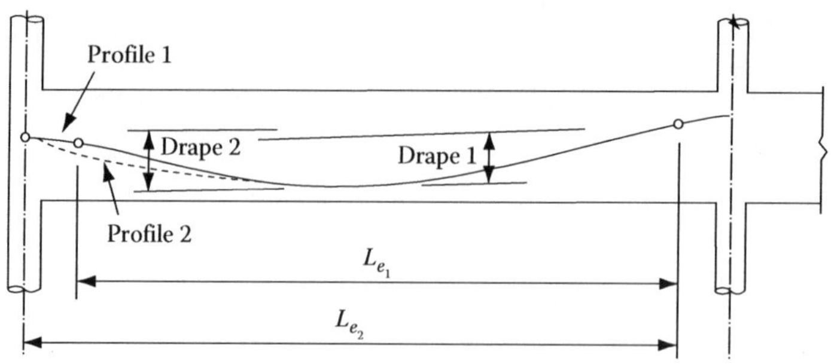

FIGURE 7.37 End bay tendon profiles.

$$L_2 = 12.90 \text{ ft}, h_2 = 1.25 \text{ in.}$$

$$L_e = 12.9 \times 2 = 25.8 \text{ ft}$$

$$h_3 = 8 \times 12.90 + 1.25 \times 1.84/14.75 = 7.153 \text{ in.}$$

$$\text{Tendon drape} = 7.153 - 1.25 = 5.90 \text{ in.}$$

First cycle

$$\text{Minimum balanced load} = 0.65 \times (\text{total DL})$$

$$= 0.65(112 + 10 + 20) = 92 \text{ psf}$$

$$\text{Moment due to balanced load} = \frac{92}{192} \times 14.7$$

$$= 7.04 \text{ kip-ft}$$

This is subtracted from the total service load moment of 14.7 kips-ft to obtain the unbalanced moment M_{ub}.

$$M_{ub} = 14.7 - 7.04 = 7.66 \text{ kips-ft}$$

The flexural stresses at top and bottom are obtained by dividing M_{ub} by the section moduli of the structure's cross section:

$$f_t = \frac{7.66 \times 12}{162} = 0.567 \text{ ksi}$$

$$f_b = \frac{7.66 \times 12}{162} = 0.567 \text{ ksi}$$

The minimum required compressive prestress f_p is found by subtracting the maximum allowable tensile stress $f_a = 6\sqrt{f_c}$ from the calculated tensile stress. Thus, the smallest required compressive stress is

$$f_p = f_t - f_a$$

$$= 0.567 - 380 = 0.187 \text{ ksi}$$

The prestress force is calculated by multiplying f_p by the cross-sectional area:

$$p = 0.187 \times 9 \times 12 = 20.20 \text{ kips-ft}$$

Determine the equivalent load due to prestress force P by the relation

$$W_p = \frac{8Pe}{L_e^2}$$

For the example problem,

$$P = 20.20 \text{ kips-ft}, e = 5.90''$$

$$L_e = 2 \times 12.90 = 25.8 \text{ ft}$$

Therefore,

$$W_p = \frac{8 \times 20.20 \times 5.90}{25.8^2 \times 12} = 0.120 \text{ klf} = 120 \text{ plf}$$

Comparing this with the value of 93 plf assumed at the beginning of the first cycle, we find the two values are not equal. Therefore, we assume a new value and repeat the procedure until convergence is obtained.

Second cycle

$$W_p = 0.75 \times 92 + 0.25\ (120) = 99 \text{ plf}$$

$$M_b = \frac{99}{192} \times 14.7 = 7.58 \text{ kip-ft}$$

$$M_{ub} = 14.7 - 7.58 = 7.12 \text{ kips-ft}$$

$$f_t = f_b = \frac{7.12 \times 12}{162} = 0.527 \text{ksi}$$

$$f_p = 0.527 - 0.380 = 0.147 \text{ ksi}$$

$$P = 0.147 \times 9 \times 12 = 15.92 \text{ kips-ft}$$

$$W_p = \frac{8 \times 15.92 \times 5.90}{25.8^2 \times 12} = 0.094 \text{ klf} = 94 \text{ plf}$$

This is less than 99 plf assumed at the beginning of the second cycle. Therefore, we assume a new value and repeat the procedure.

Third cycle

$$W_p = 0.75 \times 99 + 0.25\ (94) = 97.7 \text{ plf}$$

$$M_b = \frac{99.7 \times 14.7}{192} = 7.48 \text{ kips-ft}$$

$$M_{ub} = 14.7 - 7.48 = 7.22 \text{ kips-ft}$$

$$f_t = f_b = \frac{7.22 \times 12}{162} = 0.535 \text{ ksi}$$

$$f_p = 0.535 - 0.380 = 0.155 \text{ ksi}$$

$$P = 0.155 \times 9 \times 12 = 16.74 \text{ kips-ft}$$

$$W_p = \frac{8 \times 16.74 \times 5.90}{25.8^2 \times 2} = 0.99 \text{ klf} = 99 \text{ plf}$$

This is nearly equal to 97.7 plf assumed at the beginning of the third cycle. Therefore, OK.

Check compressive stress at the support:

$$M_p = \frac{99 \times 14.7}{192} = 7.58 \text{ kips}$$

$$M_{ub} = 14.7 - 7.58 = 7.12 \text{ kips-ft}$$

$$f_b = \frac{7.12 \times 12}{162} = 0.527 \text{ ksi} = 527 \text{ ksi}$$

$$\text{Axial compressive stress due to post-tension} = \frac{16.74 \times 1000}{9 \times 12} = 155 \text{ psi}$$

Total compressive stress = 527 + 155 = 682 psi

This is less than the allowable compressive stress of 1800 psi. Therefore, the design is satisfactory.

End bay design. The placement of tendons within the end bay presents a few problems. The first problem is in determining the location of the tendon over the exterior support. Placing the tendon above the neutral axis of the member results in an increase in the total tendon drape, allowing the designer to use less prestress than would otherwise be required. Raising the tendon, however, introduces an extra moment that effectively cancels out some of the benefits from the increased drape. For this reason, the tendon is usually placed at a neutral axis at exterior supports.

The second problem is in making a choice in the tendon profile: whether to use a profile with a reverse curvature over each support (see Figure 7.37, profile 1) or over the first interior support only (see Figure 7.37, profile 2). A profile with the reversed curvature over the first interior support only gives a greater cable drape than the first profile, suggesting a larger equivalent load with the same amount of prestress. On the other hand, the effective length L_c between inflection points of profile 1 is less than that of profile 2, which suggests the opposite. To determine which profile is in fact more efficient, it is necessary to evaluate the amount of prestress for both profiles. More usually, a tendon profile with reverse curvature over both supports is 5%–10% more efficient since the equivalent load produced is a function of the square of the effective length.

The last item addresses the extra end bay prestressing required in most situations. The exterior span in an equal span structure has the greatest moments due to support rotations. Because of this, extra prestressing is commonly added to end bays to allow efficient design to end spans. For design purposes, the extra end bay prestressing is considered to act within the end bay only. These tendons actually extend well into the adjacent span for anchorage, as shown in Figure 7.38. Advantage can be taken of this condition by designing the through tendons using the largest moment found within the interior spans, including the moment at the interior face of the first support. The end bay interior spans, including the moment at the interior face of the first support. The end bay prestress force is determined using the largest moment within the exterior span. The stress at the inside face of the first support is checked using the equivalent loads produced by the through tendons and the axial compression provided by both the through and added tendons. If the calculated stresses

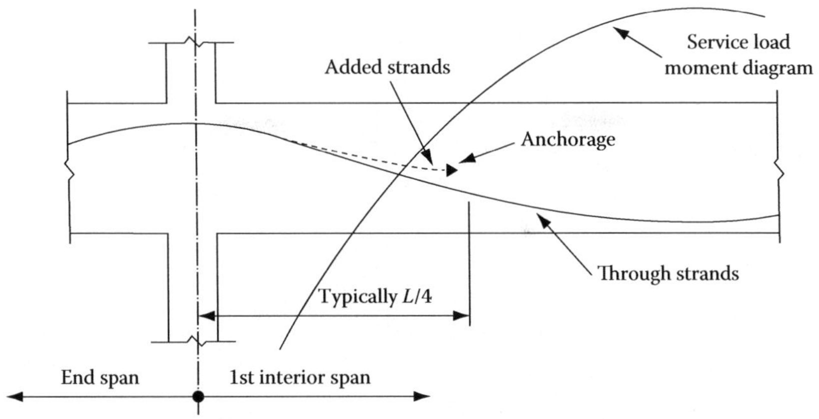

FIGURE 7.38 Anchorage of added tendons.

are less than the allowable values, the design is complete. If not, more stress is provided either by through tendons or added tendons or both.

The design of end bay using profiles 1 and 2 follows.

Profile 1: Reverse curvature at interior support only (Figure 7.39). Observe that the height of inflection point is exact if the tendon profile is symmetrical about the center span. If it is not, as in

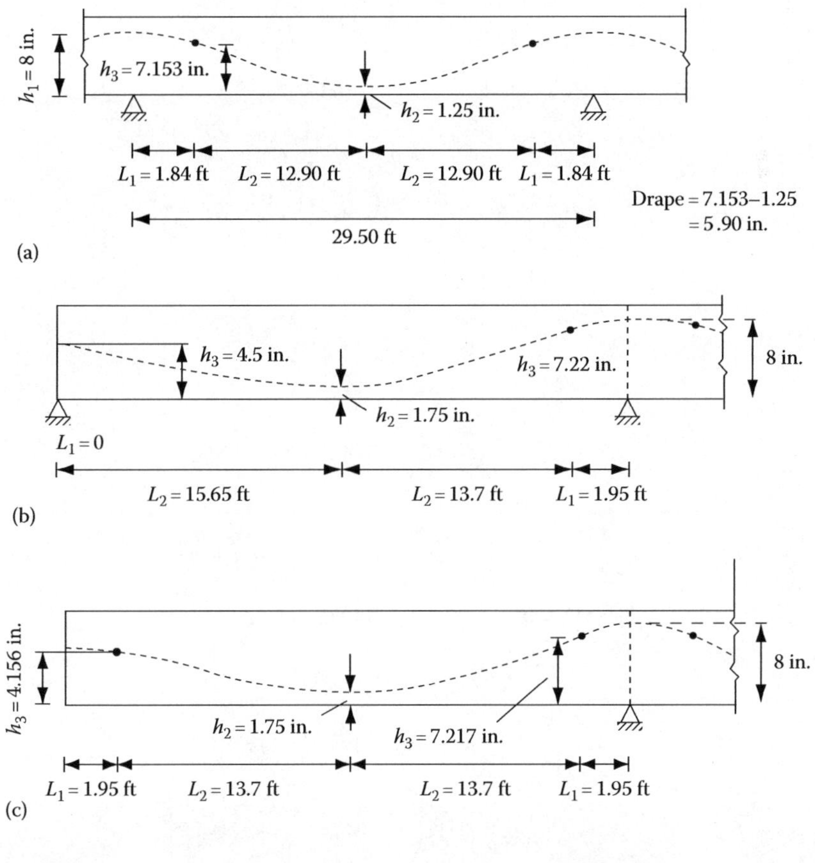

FIGURE 7.39 Example problem 3: flat plate, tendon profiles—(a) interior span, (b) exterior span, reverse curvature at right support, and (c) exterior span, reverse curvature at both supports.

span 1 of the example problem, sufficiently accurate value can be obtained by taking the average of the tendon inflection point at each end as follows:

Left end:

$$h_3 = \frac{4.5 \times 15.6 + 1.75 \times 0}{0 = 15.67} = 4.5 \text{ in.}$$

Right end:

$$h_3 = \frac{8 \times 13.7 + 1.75 \times 1.95}{(1.95 + 13.70)} = 7.22 \text{ in.}$$

$$\text{Average } h_3 = \frac{4.5 + 7.22}{2} = 5.86 \text{ in.}$$

Drape = 5.86 − 1.75 = 4.11 in.

First cycle. To show the quick convergence of the procedure, we start with a rather high value of

$$W_p = 0.75 \text{ DL} = 0.75 \times 142 = 106 \text{ plf}$$

$$M_b = \frac{106}{192} \times 15.87 = 8.76 \text{ kips-ft}$$

$$M_{ub} - 15.87 - 8.76 = 7.11 \text{ kips-ft}$$

$$f_t = f_b = \frac{7.11 \times 12}{162} = 0.527 \text{ ksi}$$

$$f_p = 0.527 - 0.380 = 0.147 \text{ ksi}$$

$$p = 0.147 \times 9 \times 12 = 15.87 \text{ kips-ft}$$

$$W_p = \frac{8Pe}{L_e^2}$$

$$= \frac{8 \times 15.87 \times 4.11}{29.35^2 \times 12}$$

$$= 0.050 \text{ klf} = 50.0 \text{ plf}$$

This is less than 106. N.G.

Second cycle

$$W_p = 0.75(106) + 0.25(50.0) = 92 \text{ plf}$$

$$M_b = \frac{92}{192} \times 15.87 = 7.60 \text{ kips-ft}$$

$$M_{ub} = 15.87 - 7.60 = 8.27 \text{ kips-ft}$$

$$f_t = f_b = \frac{8.27 \times 12}{162} = 0.612 \text{ ksi}$$

$$f_p = 0.612 - 0.380 = 0.233 \text{ ksi}$$

$$p = 0.233 \times 12 \times 9 = 25.16 \text{ kips-ft}$$

$$W_p = \frac{8 \times 25.16 \times 4.11}{29.35^2 \times 12}$$

$$= 0.080 \text{ klf} = 80.0 \text{ plf}$$

This is less than 91.5 psi used at the beginning of the second cycle. N.G.

Third cycle

$$W_p = 0.75 \times 92 + 025 \times 80 = 89 \text{ plf}$$

$$M_b = \frac{89}{192} \times 15.87 = 7.356 \text{ kips-ft}$$

$$M_{ub} = 15.87 - 0.356 = 8.5 \text{ kips-ft}$$

$$f_t = f_b = \frac{8.5 \times 12}{162} = 0.631 \text{ ksi}$$

$$f_p = 0.631 - 0.380 = 0.251 \text{ ksi}$$

$$P = 0.251 \times 12 \times 9 = 27.10 \text{ kips-ft}$$

$$W_p = \frac{8 \times 27.10 \times 4.11}{29.35^2 \times 12} = 0.086 \text{ klf} = 86 \text{ plf}$$

This is nearly equal to 89 plf used at the beginning of the third cycle. Therefore, OK.

Profile 2. Reverse curvature over each support (Figure 7.39)

Left end:

$$h_3 = \frac{4.5 \times 13.70 + 1.75 \times 1.95}{(13.70 + 1.95)} = 4.156 \text{ in.}$$

Right end:

$$h_3 = \frac{8 \times 13.70 + 1.75 \times 1.95}{(13.70 + 1.95)} = 7.221 \text{ in.}$$

$$\text{Average } h_3 = \frac{4.156 + 7.221}{2} = 5.689 \text{ in.}$$

$$e = h_d = 5.689 - 1.75 = 3.939 \text{ in.}$$

First cycle. We start with an assumed balanced load of 0.65 DL = 7.60 klp-ft

Balanced moment

$$M_b = 15.87 \times \frac{92}{192} = 7.60 \text{ klp-ft}$$

$$M_{ub} = 15.87 - 7.6 = 8.27 \text{ kips-ft}$$

$$f_t = f_b = \frac{8.27 \times 12}{162} = 0.613 \text{ ksi}$$

$$f_p = 0.613 - 0.380 = 0.233 \text{ ksi}$$

$$p = 0.233 \times 12 \times 9 = 25.16 \text{ kips}$$

$$W_p = \frac{8 \times 25.12 \times 3.937}{27.38^2 \times 12} = 0.088 \text{ klf} = 88 \text{ plf}$$

This is less than 92 plf. N.G.

Second cycle

$$W_p = 0.75 \times 92 + 0.25 \times 88 = 91 \text{ plf}$$

$$M_p = 15.87 \times \frac{91}{192} = 7.52 \text{ kips-ft}$$

$$M_{ub} = 15.87 - 7.52 = 8.348 \text{ kips-ft}$$

$$f_t = f_b = \frac{8.348}{162} \times 12 = 0.618 \text{ ksi}$$

$$f_p = 0.618 - 0.380 = 0.238 \text{ ksi}$$

$$P = 0.238 \times 9 \times 12 = 25.75 \text{ kips}$$

$$W_p = \frac{8 \times 25.75 \times 3.937}{(27.38)^2 \times 12} = 0.090 \text{ klf} = 90 \text{ klf}$$

This is nearly equal to the value at the beginning of the second cycle. Therefore, OK.

Check the design against the positive moment of 8.41 kips-ft:

$$W_p = 0.090 \text{ klf}$$

$$M_b = 8.41 \times \frac{0.090}{0.142} = 5.33 \text{ kips-ft}$$

$$M_{ub} = 8.41 - 5.33 = 3.08 \text{ kips-ft}$$

$$\text{Bottom flexural stress} = \frac{3.08}{162} = 0.288 \text{ ksi (tension)}$$

$$\text{Axial compression due to post-tension} = \frac{25.75}{12 \times 9} = 0.238 \text{ ksi}$$

Total stress at bottom = 0.228 − 0.238 = −0.10 ksi (compression).
This is less than the allowable tension of 0.380 ksi. Therefore, design OK.

7.3.3 Concept of Secondary Moments

In prestressed statically determinate beam, such as a single-span simply supported beam, the moment M_p due to prestress is given by the eccentricty e of prestress multiplied by the prestress P.

In prestressed design, the moment $M_p = Pe$ is commonly referred to as the primary moment. In a simple beam or any other statically determinate beam, no support reactions can be induced by prestressing. No matter how much the beam is prestressed, only the internal stresses will be affected by the prestressing. The external reactions, being determined by statics, will depend on the dead and live loads but are not affected by the prestress. Thus, there are no secondary moments in a statically determinate beam. The total moment in the beam due to prestress is simply equal to the primary moment $M_0 = Pe$.

The magnitude and nature of secondary moments may be illustrated by considering a two-span, continuous, prismatic beam that is not restrained by its supports but remains in contact with them. The beam is prestressed with a straight tendon with force P and eccentricity e. See Figure 7.40.

When the beam is prestressed, it bends and deflects. The bending on the beam can be such that the beam will tend to deflect itself away from B. Because the beam is restrained from deflection at B, a vertical reaction must be exerted to the beam to hold it there. The induced reaction produces secondary moments in the beam. These are called secondary because they are the by-products of prestressing and do not exist in a statically determinate beam. The term *secondary* is misleading because the moments are secondary in nature, but not necessarily in magnitude.

One of the principal reasons for determining the magnitude of secondary moments is because they are required in the computations of ultimate flexural strength. An elastic analysis of a prestressed beam offers no control over the failure mode or the factor of safety. To ensure that prestressed members will be designed with an adequate factor of safety against failure, ACI 318-11, like its predecessors, requires that M_u, the moment due to factored service loads including secondary moments, not exceed ΦM_n, the flexural design strength of the member. The ultimate factored moment M_u is calculated by the following load combinations:

$$M_u = 1.2M_D + 1.6M_L = 1.0\ M_{sec}$$

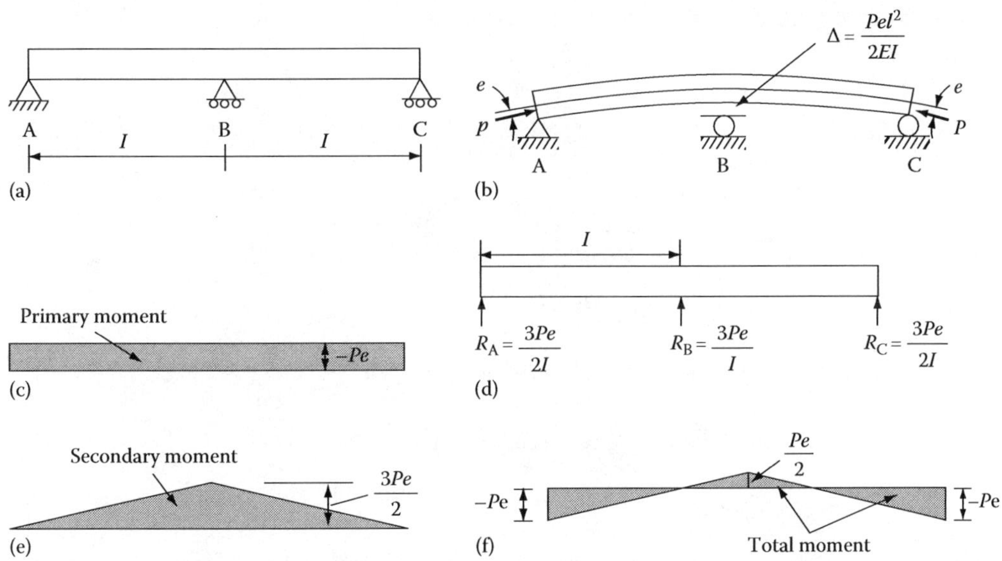

FIGURE 7.40 Concept of secondary moments: (a) two-span continuous beam, (b) vertical upward displacement due to PT, (c) primary moment, (d) reactions due to PT, (e) secondary moment, and (f) final moments.

Since the factored load combination must include the effects due to secondary examples are given here:

1. A two-span continuous beam with a prestressed tendon at a constant eccentricity *e*.
2. The same beam as in the preceding example except the tendon is parabolic between the supports. There is no eccentricity of the tendon at the supports.
3. The same as in Example 2, but the tendon has an eccentricity at the center support.

7.3.3.1 Secondary Moment

Design examples

Example 1

Given: A two-span prestressed beam with a tendon placed at a constant eccentricity e from the C.G. of the beam. The prestress in the tendon is equal to *P* (See Figure 7.40).

Required: Secondary moments in the beam due to prestress *P*

Solution: The beam is statically indeterminate to the first degree because it is continuous at the center support B. It is rendered determinate by removing the support at B. Due to the moments $M_0 = Pe$ at the ends, the beam bends and deflects upward. The magnitude of vertical deflection δ_B due to moment M_0 is calculated using standard beam formulas such as the one that follows.

Beam Deflection Formula:

Type of Load	Slope or Shown	Maximum Deflection	Deflection Equation
Simply supported beam		Bending moment applied at one end	
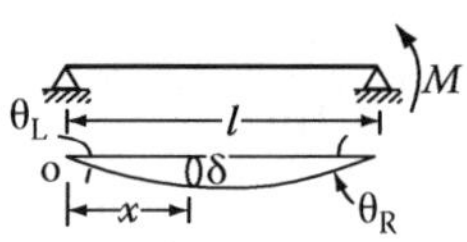	$\theta_L = \frac{Ml}{6EI}$ $\theta_R = \frac{Ml}{3EI}$	$\delta = \frac{Ml^2}{9\sqrt{3}EI}$ at $x = 1/\sqrt{3}$	$\delta = \frac{Mlx}{6EI}\left(1 - \frac{x^2}{l^2}\right)$

In our case, moment *M* is applied at both ends. Therefore,

$$\delta = Mlx/3EI(1 - x^2/1^2)$$

$$\delta_{1/2} = Ml \times 1/3EI \times 2\ (1 - 1^2/4^2)$$

$$= Ml^2/8EI$$

Also, for the example problem, $l = 2L$

Therefore, deflection δ_B at support $B = \delta_L = M \times (2L)^2/8EI$

$$= ML^2/12EI$$

Since the beam is restrained from deflecting upward at B, a downward reaction $R_{B,sec}$ must be exerted to the beam to hold it there. The reaction $R_{B,sec}$ is given by

$$\delta_B = R_{B,sec}(2L)^3/48EI$$

$$= 48EI/(2L)^3 \times M_0L^2/2EI$$

$$= 2M_0/L$$

The secondary moment induced due to the reaction $R_{B,sec}$ at the support B is given by

$$M_{B,sec} = R_{B,sec} \times 2L/4$$

$$= 3M_0/L \times 2L/4$$

$$= 3/2M_0$$

Observe that in this example, the secondary moment at B = 150% of the primary moment due to prestress. The secondary moment is thus secondary in nature, but not in magnitude.

Example 2A

Given: The two-span prestressed concrete beam shown in Figure 7.41 has a parabolic tendon in each span with zero eccentricity at the A and C ends, and at the center support B. Eccentricity of the tendon at the center of each span = 1.7 ft. The prestress force P = 263.24 kips.

Required: Secondary reactions and moments

Solution: The approach here is similar to that typically used in commercially available computer programs. However, in the computer programs, statically indeterminate structures such as the example problem are typically analyzed using a stiffness matrix approach. Here, we take the easy street: We use beam formulas to analyze the two-span continuous beam. It should be noted that the analysis could be performed using other classical methods such as the moment distribution method or slop-deflection method.

First, we determine the equivalent load due to prestress P = 263.24 kips acting at eccentricity e = 1.7 ft at the center of the two spans. The equivalent load consists of (1) an upward uniformly distributed load W_p due to drape in the tendon; (2) a horizontal compression P equal to 263.23 kips at the ends; (3) downward loads at A, B, and C to equilibrate the upward load W_p; and (4) additional reactions at A, B, and C due to the restraining effect of support at B. The last set of loads need not be considered for this example, because the loads are implicitly included in the formulas for the statically indeterminate beam.

Of the equivalent loads shown in Figure 7.41, only the uniformly distributed load W_p corresponding to P acting at eccentricity e induces bending action in the beam. W_P is determined by the relation

$$Pe = W_pL^2/8$$

$$W_p = Pe \times 8/L^2$$

$$= 263.24 \times 1.7 \times 8/54^2$$

$$= 1.227 \text{ kip-ft}$$

Having determined the equivalent loads, we can proceed to determine the bending moments in our statically indeterminate beam, as for any continuous beam. As mentioned earlier, we use the formulas for continuous beams given in standard textbooks. One such formula follows.

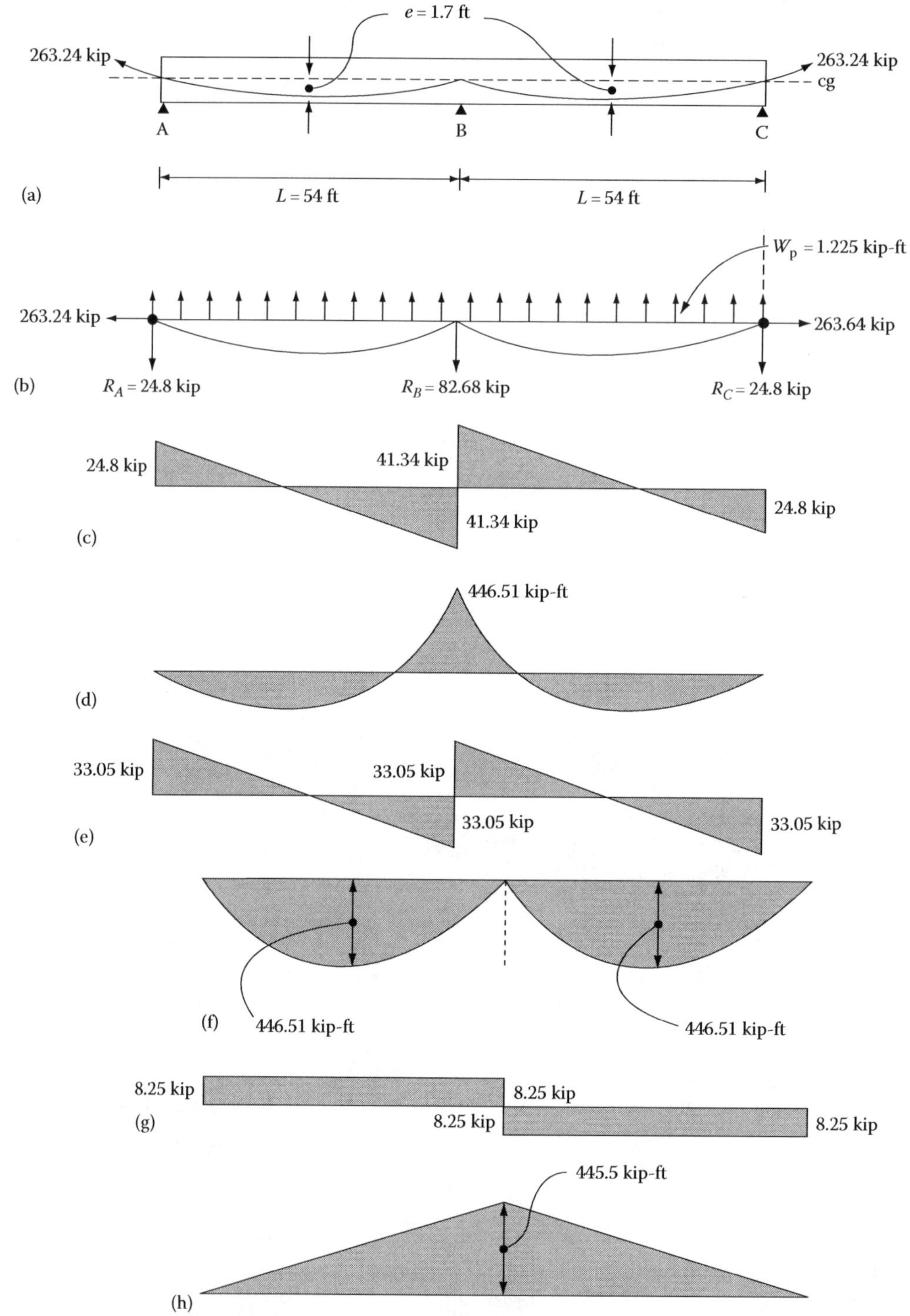

FIGURE 7.41 Secondary moment (Example 2A): (a) two-span continuous prestressed beam, (b) equivalent loads due to prestress, consisting of upward load, horizontal compression due to prestress W_p and downward loads at A, B, and C, (c) shear force diagram, statically indeterminate beam, (d) moment diagram, statically indeterminate beam, (e) primary shear force diagram, (f) primary moment diagram, (g) secondary shear force diagram, and (h) secondary moment diagram.

Continuous beam with two equal spans and uniform load on both spans

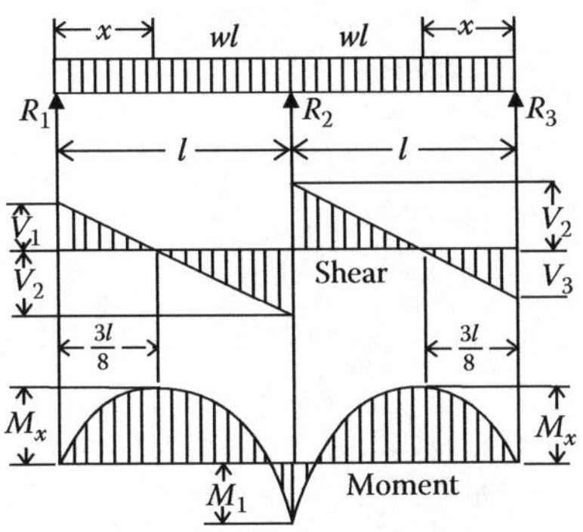

$$R_1 = V_1 = R_3 = V_3 = \frac{3}{8}wl$$

$$R_2 = 2V_2 = \frac{10}{8}wl$$

$$V_2 = \frac{5}{8}wl$$

$$M_X = R_1X - \frac{wX^2}{2}$$

$$M_X\left(\text{at } X = \frac{3l}{8}\right) = \frac{9}{128}wl^2$$

$$M_1 \text{ (at support } R_2) = -\frac{wl^2}{8}$$

$$\Delta\text{Max. } (0.4215l \text{ from } R_1 \text{ or } R_3) = wl^2/185EI$$

$$\Delta_X = \frac{wX}{48EI}(l^3 - 3l^2 + 2X^3)$$

In our case, $w = W_p = 1.225$ kip-ft, $l = L = 54$ ft. Therefore,

$$V_1 = 3/8W_pL$$

$$= 318 \times 1.225 \times 54$$

$$= 24.8 \text{ kips}$$

$$V_2 = 3/8W_pL$$

$$= 5/8 \times 1.225 \times 54$$

$$= 41.34 \text{ kips}$$

$$\text{MI} = \text{MB} = W_pL^2/8 = 1.225 \times 542/8 = 446.50 \text{ kips-ft}$$

The shear-force and bending-moment diagrams are shown in Figure 7.41 parts (c) and (d).

Since the formulas account for the beam continuity, the resulting shear force and bending moments shown in Figure 7.41c and d include the effect of secondary moments. The resulting moment due to prestress, then, is the algebraic sum of the primary and secondary moments. Once the resulting moments are determined, the secondary moments can be calculated by the relation

$$M_{bal} = M_p + M_{sec}$$

where

M_{bal} is the resulting moment, also referred to as the total moment in the redundant beam due to equivalent loads

M_p is the primary moment that would exist if the beam were a statically determinate beam (M_p is given by the eccentricity of the prestress multiplied by the prestress)

M_{sec} is the secondary moment due to redundant secondary reactions

With the known primary moment acting on the continuous beam, the secondary moment caused by induced reactions can be computed from the relation

$$M_{sec} = M_{bal} - M_p$$

A similar equation is used to calculate the shear forces.

The resulting secondary shear forces and bending moments are shown in Figure 7.41g and h, while the primary shear forces and bending moments are shown in Figure 7.41e and f.

Example 2B: Compatibility Method

To firm up our concept of secondary reactions and moments, perhaps it is instructive to redo the previous example using a compatibility approach. In this method, the beam is rendered statically determinate by removing the redundant reaction at B. The net vertical deflection (which happens to be upward in our case) is calculated at B due to W_p = 1.225 kip-ft acting upward and a vertical downward load = 1.225 × 54 = 66.15 kips acting downward at B. Observe that the reaction at B along with those at A and C equilibrates the vertical load of 1.225 kip-ft action on the tendon in its precise profile but does not necessarily guarantee compatibility at B.

Given: A two-span continuous beam analyzed previously, shown again for convenience in Figure 7.42.

Required: Secondary moments and shear forces using compatibility approach.

Solution: The equivalent loads required to balance the effect of prestressed, draped tendons are shown in Figure 7.42b. As before, W_p = 1.225 kip-ft. However, the reactions at A, B, and C do not include those due to secondary effects. The reactions are in equilibrium with load W_p and do not necessarily ensure the continuity of the beam at support B. (If continuity were established, their magnitudes would be the same as calculated in the previous example.)

In determining the equivalent loads, we have not considered the effect of continuity at support B. Therefore, the beam has a tendency to move away from the support due to the upward-acting equivalent loads. Because the beam, by compatibility requirements, stays attached to support B, another set of reactions is needed to keep the beam in contact with support B. These are the secondary reactions, and the resulting moments are the secondary moments. Of the loads shown in Figure 7.42, only the upward load W_p = 1.225 kip-ft and the downward reaction R_B = 66.15 kips influence the vertical deflection at B. The upward deflection of the beam at B due to W_p is given by the standard formula:

$$\Delta_{up} = 5wl^4/384WI \text{ (see Figure 7.42c)}$$

In our case, $w = W_p$ = 1.225 kip-ft, l = 3 × 54 = 108 ft. Therefore,

$$\Delta_{up} = 5 \times 1.225 \times 108^4/384EI \uparrow \text{ upward}$$

The downward deflection at B due to reaction R_B is given by

$$\Delta_{down} = R_B L^3/48EI$$

In our case, R_B = 66.15 kips, L = 108 ft.

$$\Delta_{down} = 66.15 \times 108^3/48EI$$

$$= 1{,}736{,}040/EI \downarrow \text{ downward (see Figure 7.42d)}$$

The net deflection at B

$$\Delta_B = 217{,}050 - 1{,}736{,}040/EI = 434{,}010/EI \uparrow \text{ upward (see Figure 7.42e)}$$

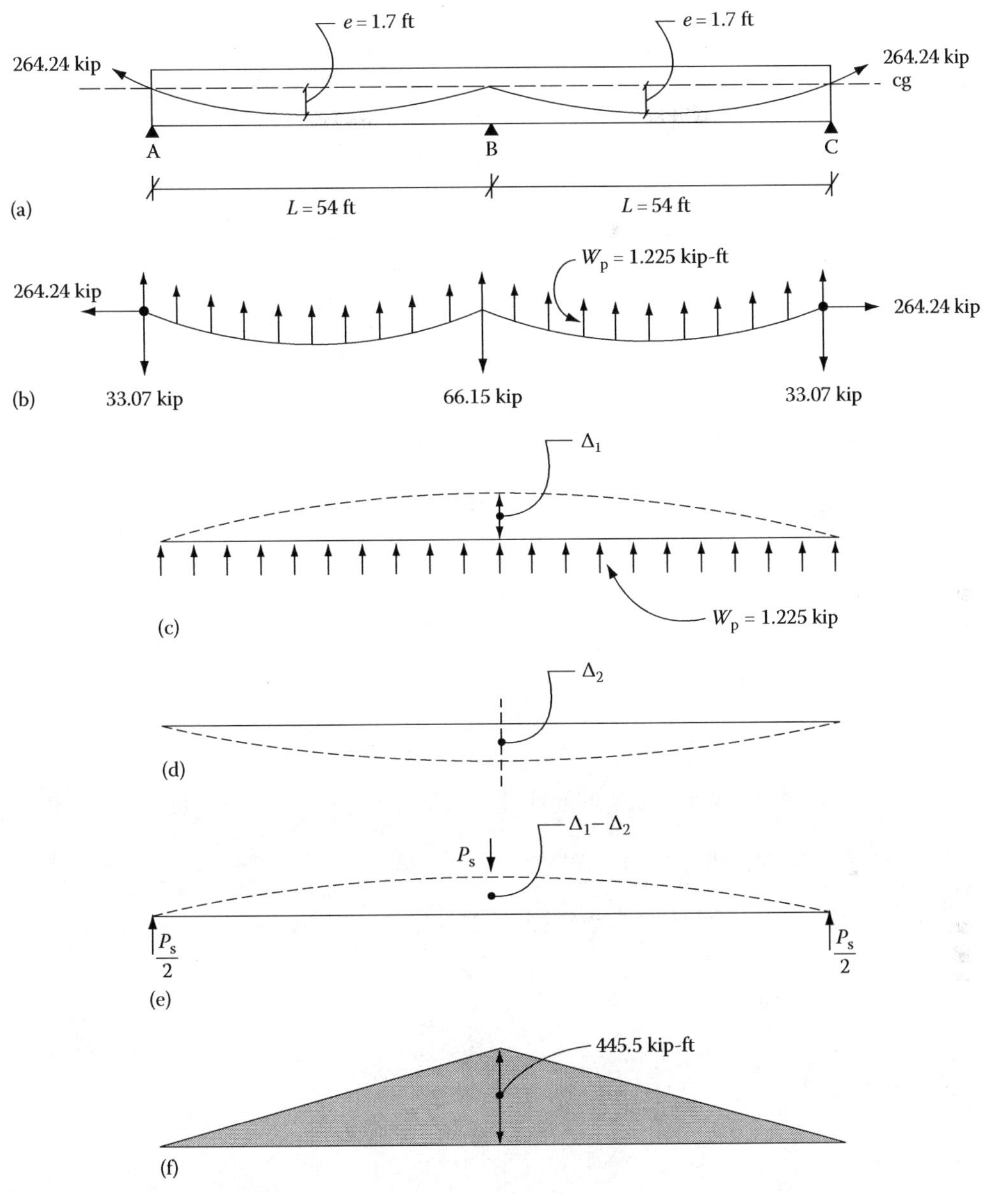

FIGURE 7.42 Secondary moment: compatibility method (Example 2B). (a) two-span continuous beam, (b) equivalent loads, (c) upward deflection Δ_1 due to W_p, (d) downward deflection Δ_2 due to a load of 66.15 kip at center span, (e) load Ps corresponding to Δ_1–Δ_2, and (f) secondary moments.

Because the beam is attached to support B, for compatibility, the vertical deflection at B should be zero. This condition is satisfied by imposing a vertically downward secondary reaction $R_{B,sec}$ at B given by the relation

$$R_{B,sec} \times 108^3/48EI = 4340/0/EI$$

$$R_{B,sec} = 16.54 \text{ kips}$$

The resulting secondary reactions and moments shown in Figure 7.81 are exactly the same as calculated previously.

Example 3

Given: Same data as in Example 2B. The only difference is that the tendon at the center support B has an eccentricity of 0.638′

Required: Secondary moments using a compatibility-type analysis

Solution: The equivalent loads balancing the effects of prestressed, draped tendons with eccentricities at the centers of spans and at the interior support B are shown in Figure 7.43

Notice the two equal and opposite moments equal to the prestress of 263.24 kips times the eccentricity of 0.68 ft at the center support (see Figure 7.43b and c). The solution follows the same procedure as used in the previous example, except that we include the effect of moments at B in deflection calculations.

As before, $W_p = 1.23$ kip-ft. Upward deflection at B due to W_p is given by

$$\Delta_{up} = 5/384 W_p (2L)^4\ EI$$

$$= 5 \times 1.23 \times (2 \times 54)^4/384 \times EI$$

$$= 2{,}170{,}050/EI \uparrow \text{ upward (see Figure 7.43d)}$$

The vertical reaction R_B at B to maintain vertical equilibrium is equal to 1.23 × 54 = 66.42 kips. The downward deflection at B due to this load is

$$\Delta_{down,RB} = R_B \times (54 \times 2)^3/48EI$$

$$= 66.42 \times (108)^3/48EI$$

$$= 1{,}743{,}126/EI \downarrow \text{ downward (see Figure 7.43e)}$$

In addition to the upward and downward deflections at B, there is a third component to the vertical deflection due to the moment at B = 263.24 × 0.638 = 167.95 kips-ft.

For purposes of deflection calculations, moment M_B at B may be replaced by an equivalent point load equal to $2M_B/L$

The downward deflection at B due to M_B, then, is

$$\Delta_{down,MB} = 2M_B \times (2L)^3/L\ 48EI$$

$$= M_B L2/3EI \text{ (see Figure 7.43f)}$$

For example, $M_B = 167.95$ kips-ft, $L = 54$ ft

$$\Delta_{down,MB} = 167.95 \times 54^2/3EI$$

$$= 163{,}150/EI \downarrow \text{downward}$$

The net upward deflection due to W_p, R_B, and M_B is

$$l/EI(2{,}170{,}050 - 143{,}126 - 163{,}150) = 263{,}774/EI \uparrow \text{ upward (see Figure 7.43)}$$

The secondary reaction to establish vertical compatibility at B is given by

$$R_{B,sec} \times (2L)^3/48EI = 263{,}774/EI$$

$$R_{B,sec} = 48 \times 263{,}774/(2 \times 54)^3 = 10.05 \text{ kips}$$

The secondary moments due to this redundant reaction are shown in Figure 7.43

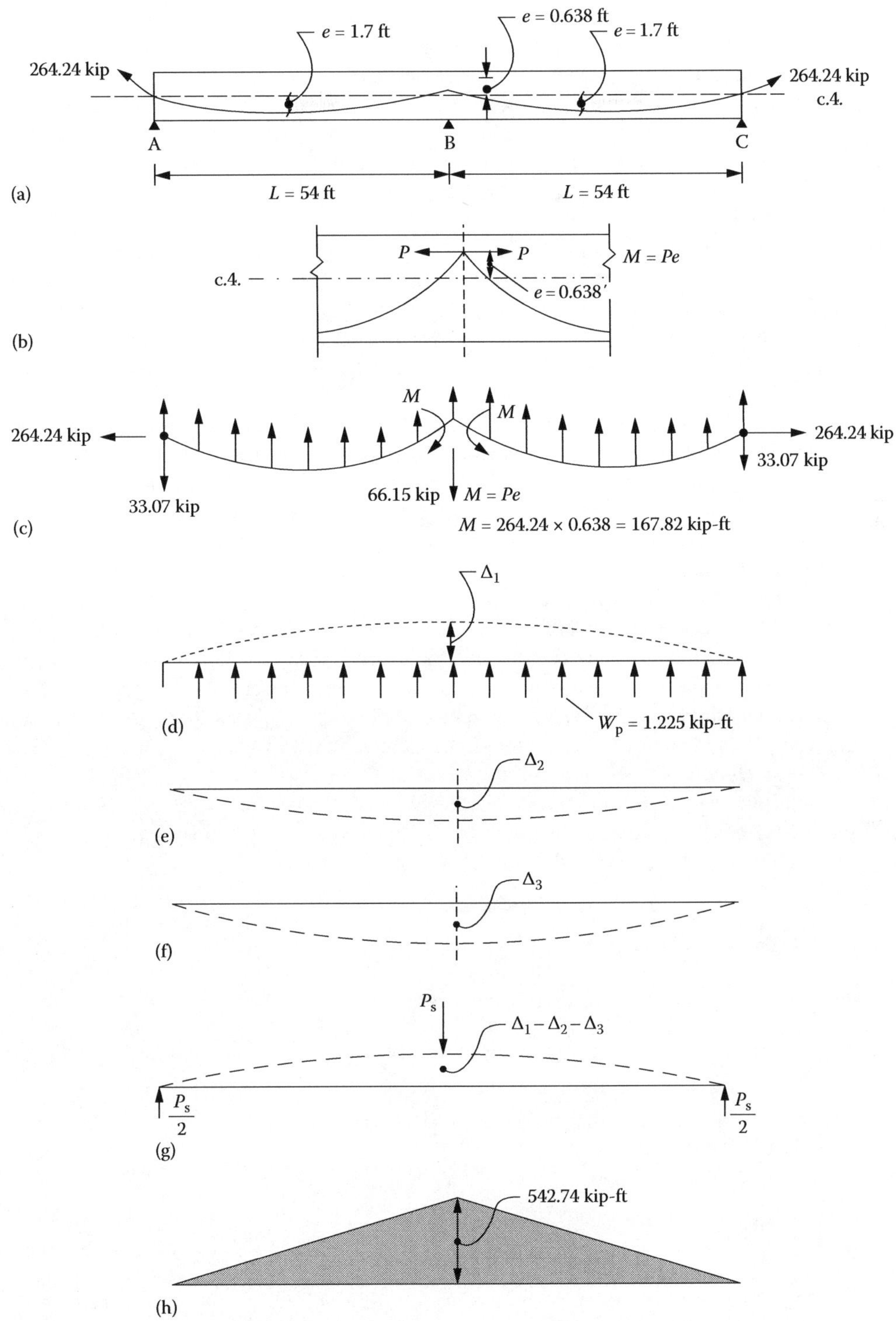

FIGURE 7.43 Secondary moment (Example 3): (a) two-span continuous beam, (b) equivalent moment $M = Pe$ at the center of span, (c) equivalent loads and moments, (d) upward deflection Δ_1 due to W_p, (e) downward deflection Δ_2 due to a load of 66.15 at the center of span, (f) downward deflection Δ_3 due to moments $M = Pe$ at the center of span, (g) load Ps corresponding to $\Delta_1 - \Delta_2 - \Delta_3$, and (h) secondary moments.

7.3.4 Strength Design for Flexure

In the design of prestress members, it is not enough to limit the maximum values of tensile and compressive stresses within the permitted values at various loading stages. This is because although such a design may limit deflections, control cracking, and prevent crushing of concrete, an elastic analysis offers no control over the ultimate behavior or the factor-of-safety of a prestressed member. To ensure that prestressed members will be designed with an adequate factor-of-safety against failure, ACI 318 requires that M_u, the moment due to factored service loads, not exceed ΦM_n, the flexural design strength of the member.

The nominal bending strength of a prestressed beam with bonded tendons is computed in nearly the same manner as that of a reinforced concrete beam. The only difference is in the method of stress calculation in the tendon at failure. This is because the stress–strain curves of high-yield-point steels used as tendons do not develop a horizontal yield range once the yield strength is reached. It continues upward at a reduced slope. Therefore, the final stress in the tendon at failure f_{ps} must be predicted by an empirical relationship.

The method of computing the bending strength of a prestressed beam given in the following section applies only to beams with bonded tendons. The analysis is performed using strain compatibility. Because, by definition, there is no strain compatibility between the tendon and concrete in an unbounded prestressed beam, this method cannot be used for prestressed beams with unbounded tendons; the empirical approach given in ACI 318-11, Section 18.7, is the recommended method.

The procedure for bonded tendons consists of assuming the location of the neutral axis, computing the strains in the prestressed and nonprestressed reinforcement, and establishing the compression stress block. Knowing the stress–strain relationship for the reinforcement, and assuming that the maximum strain in concrete is 0.003, the forces in the prestressed forces are computed. If necessary, the neutral axis location is adjusted on a trail-and-error basis until the sum of the forces is zero. The moment of these forces is then computed to obtain the nominal strength of the section. To compute the stress in the prestressing strand, the idealized curve shown in Figure 7.44 (adapted from *PCI Design Handbook*, 5th edition) is used.

The analysis presented here follows a slightly different procedure. Instead of assuming the location of the neutral axis, we assume a force in the prestressing strand and compare it with the derived value. The analysis is continued until the desired convergence is reached.

Example 1: Strength Design

Given: A rectangular prestressed concrete beam, as shown in Figure 7.45.

$$f'_c = 5000 \text{ psi}$$

Mild steel reinforcement = 4 # 5 bars at bottom, f_y = 60 ski
Prestressed strands = 4 – ½Φ, f_{ps} = 270 ksi

Required: Ultimate flexural moment capacity of the beam

Solution: A trial-and-error procedure is used.

First trial. For the first trial, assume the stress in the prestressed strands = 250 ksi, and the yield stress in the mild steel is 60 ksi.

The total tension T at the tension zone of the beam consists of T_1, the tension due to prestressed stands, plus T_2, the tension due to mild steel reinforcement.

Thus, $T = T_1 + T_2$

$$T_1 = \text{area of strands} \times \text{assumed stress in prestressing steel}$$

$$= 4 \times 0.153 \times 250 = 153 \text{ kips}$$

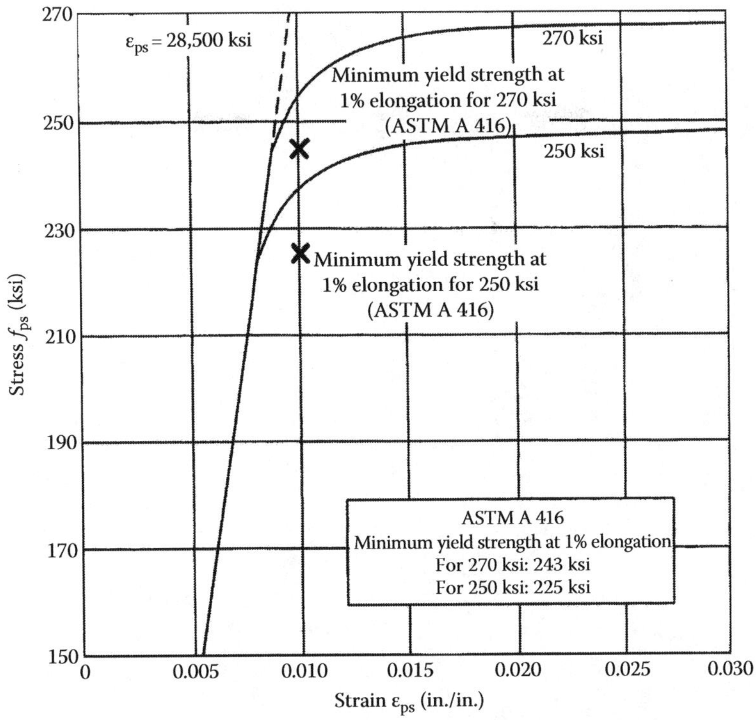

FIGURE 7.44 Idealized stress–strain curve. (Adapted from *PCI Design Handbook*, 5th edn., PCI, Chicago, 1999.) Typical stress–strain curve with seven-wine low-relaxation prestressing strand. These curves can be approximated by the following equations:

250 ksi

$$\varepsilon_{ps} \le 0.0076: f_{ps} = 28{,}500\ \varepsilon_{ps}(\text{ksi})$$

$$\varepsilon_{ps} > 0.0076: f_{ps} = 250 - \frac{0.04}{\varepsilon_{ps} - 0.0064}(\text{ksi})$$

270 ksi

$$\varepsilon_{ps} \le 0.0086: f_{ps} = 28{,}500\ \varepsilon_{ps}(\text{ksi})$$

$$\varepsilon_{ps} > 0.0086: f_{ps} = 270 - \frac{0.04}{\varepsilon_{ps} - 0.0007}(\text{ksi}).$$

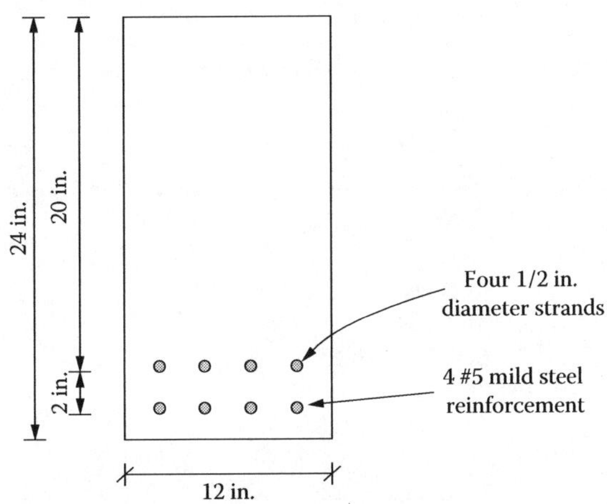

FIGURE 7.45 Strength design example 1: beam section.

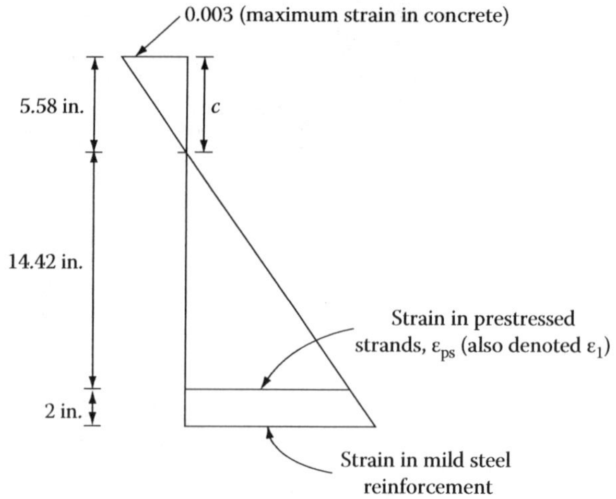

FIGURE 7.46 Example 1: strain diagram, first trial.

$$T_2 = \text{area of mild steel reinforcement} \times \text{yield stress}$$
$$= 4 \times 0.31 \times 60 = 74.40 \text{ kips}$$
$$T = 153 + 74.40 = 227.4 \text{ kips}$$

Draw a strain diagram for the beam at the nominal moment strength defined by a compressive strain of 0.003 at the extreme compression fiber. Using the strain diagram, find the compressive force $C = 0.85f_c'\,ab$. See Figure 7.46:

$$C = T = 227.4 \text{ kips}$$
$$a = 227.4/0.85f_c'b$$
$$= 227.4/0.85 \times f_c' \times 12 = 4.46 \text{ in.}$$
$$c = a/\beta_t \beta_t = 0.8 \text{ for } f_c' = 5 \text{ ksi}$$
$$= 4.46/0.8 = 5.58 \text{ in.}$$

Compute the strain in the prestressing steel and the corresponding stress:

$$0.003/5.58 = \varepsilon_1/14.42$$
$$\varepsilon_1 0.00775$$

Since the strain = 0.00775, the corresponding stress is in the elastic region of the stress–strain curve. The stress in the prestressed strand is given by

$$F_{ps} = 28{,}500 \times \varepsilon_1$$
$$= 28{,}500 \times 0.00775 = 221 \text{ ksi}$$
$$T_1 = 4 \times 0.153 \times 221 = 135.4 \text{ kips}$$
$$T_2 = 74.40 \text{ kips as before}$$
$$T = T_1 + T_2 = 135.4 + 74.40 = 209.8 \text{ kips}$$

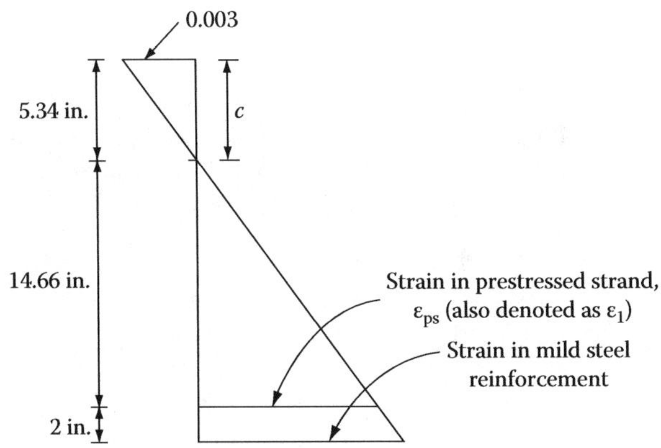

FIGURE 7.47 Example 1: strain diagram, second trial.

Comparing this with $T = 227.4$ kips, by inspection, we estimate that an improved value of $T = C =$ average of the two values

$$= (227.4 + 209.8)12 = 218.6 \text{ kips, say, } 218 \text{ kips}$$

Use this value for the second trial (see Figure 7.47).
Second trial

$$C = T = 218 \text{ kips}$$

$$a = 218/0.85f_c'b = 218/0.85 \times 5 \times 12 = 4.27 \text{ in.}$$

$$c = a/\beta_1 = 4.27/0.8 = 5.34 \text{ in.}$$

$$0.003/5.34 = \varepsilon_1/14.66$$

$$\varepsilon_1 = 0.0082 </0.0086$$

Therefore,

$$F_{ps} = 28{,}500 \times 0.00824$$

$$= 234.7 \text{ kips}$$

$$T_1 = 4 \times 0.153 \times 234.7 = 143.7$$

This is practically the same as the value we assumed in the second trial. Therefore, $T = 218$ kips may be used to compute the flexural strength of the beam.

Flexural strength. The nominal moment strength is obtained by summing the moments of T_1 and T_2 about the C.G. of compressive force C (see Figure 7.48):

$$M_n = 74.4 \times (20 - 2.14)\ 143.7\ (22 - 2.14)$$

$$= 4183 \text{ kips-in.} = 348.6 \text{ kips-ft}$$

Usable capacity of the beam $= \Phi M_n$

$$= 0.9 \times 348.6 \text{ kips-ft}$$

$$= 313.7 \text{ kips-ft}$$

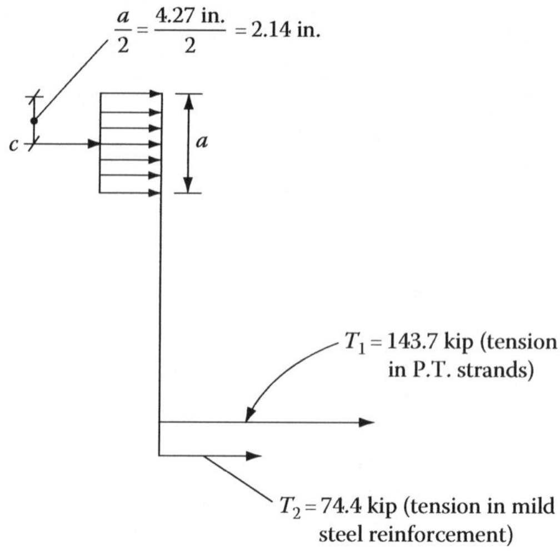

FIGURE 7.48 Example 1: force diagram.

Example 2: Strength Design

Given: Same data as for Example 1, except 3½″ Φ strands are used instead of 4½″ Φ strands. This example illustrates the calculation of stress in the strand in the nonelastic range of the stress–strain curve shown in Figure 7.44.

Required: Ultimate flexural capacity of the beam

Solution: As before, we use a trial-and-error procedure.
First trial. Assume stress in the strands = 240 ksi

$$\text{Total tension } T = T_1 + T_2$$

$$= 3 \times 0.0153 \times 240 + 0.31 \times 60$$

$$= 110.16 + 74.4 = 184.56 \text{ kips}$$

$$T = C = 0.85 \times 12 \times 5 \times a = 184.56 \text{ kips}$$

$$a = 184.56/0.85 \times 12 \times 5 = 3.62 \text{ in.}$$

$$c = 3.62/0.80 = 4.52 \text{ in.}$$

$$0.003/4.52 = \varepsilon_{ps}/15.48$$

$$\varepsilon_{ps} = 0.0102$$

$$f_{ps} = 270 - 0.04/0.0102 - 0.007 \text{ (see Figure 7.49)}$$

$$= 270 - 12.22 = 257.8 \text{ ksi}$$

$$T = T_1 + T_2$$

$$= 3 \times 0.0153 \times 257.8 + 74.4$$

$$= 192.7 \text{ kips compared with } 184.56 \text{ kips}$$

Use an average value $T = 192.7 + 184.56/2 = 188.6$, say, 189 kips for the second trial.

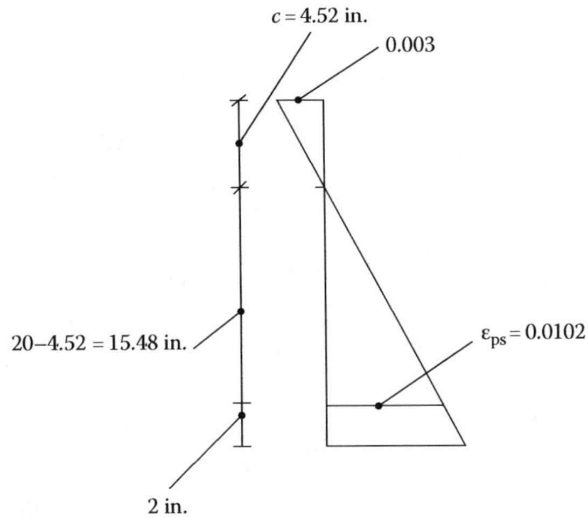

FIGURE 7.49 Example 2: strain diagram, first trial.

Second trial

$$C = T = 189 \text{ kips}$$

$$a = 189\ 10.85 \times 12 \times 5 = 3.71 \text{ in.}$$

$$c = 3.71/0.8 = 4.63 \text{ in. (see Figure 7.50)}$$

$$0.003/4.63 = \varepsilon_{ps}/15.37$$

$$\varepsilon_{ps} = 0.00996$$

$$f_{ps} = 270 - 0.04/0.00996 - 0.007$$

$$= 270 - 13.5 = 256.5 \text{ ksi}$$

$$T_1 = 0.459 \times 256.5 + 74.4$$

$$= 117.72 + 74.4 = 192 \text{ kips}$$

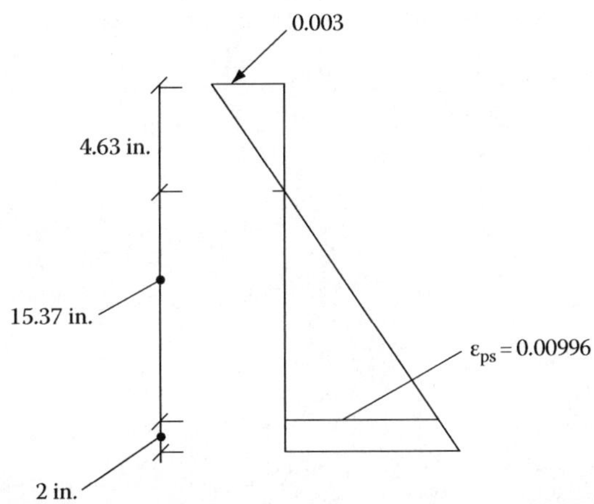

FIGURE 7.50 Example 2: strain diagram, second trial.

Compared with 189 kips used at the beginning of the second trial, this is considered sufficiently accurate for all practical purposes.

Calculate the nominal moment M_n by taking moments of T_1 and T_2 about the C.G. of compression block:

$$M_n = 117.72 \times 18.14 + 74.4 \times 20.14$$

$$= 2135.6 + 1498.42 = 3634 \text{ kips-in.}$$

$$= 302.84 \text{ kips-ft}$$

$$\Phi M_n = 0.9 \times 302.84 = 272.6 \text{ kips-ft}$$

Example 3: Prestressed T-Beam

Given: See Figure 7.51 for the beam geometry. The area of prestressed strands = 2.4 in. $2f_c' = 5$ ksi

Required: Ultimate flexural capacity of the T-beam

First trial. Assume $f_{ps} = 250$ ksi

$$T = T_1\ T_2 = 0, \text{ since there is no mild steel reinforcement}$$

$$= 2.4 \times 250 = 600 \text{ kips}$$

$$C = T = 600 \text{ kips}$$

$$a = 600/0.85 \times 48 \times 5 = 2094 \text{ in.}$$

$$c = a/\beta_1 = 2.94/0.8 = 3.68 \text{ in.}$$

$$0.003/3.68 = \varepsilon_1/20.32 \text{ (see Figure 7.52)}$$

$$\varepsilon_1 = 0.01657$$

$$f_{ps} = 270 - 0.04/0001657 - 0.007$$

$$= 270 - 4.18 \approx 265.8 \text{ ksi}$$

$$T = 265.8 \times 2.4 = 638 \text{ kips (compared to 600 kips)}$$

Use an average of two, 638 + 600/2 = 619 kips, for the second trial.

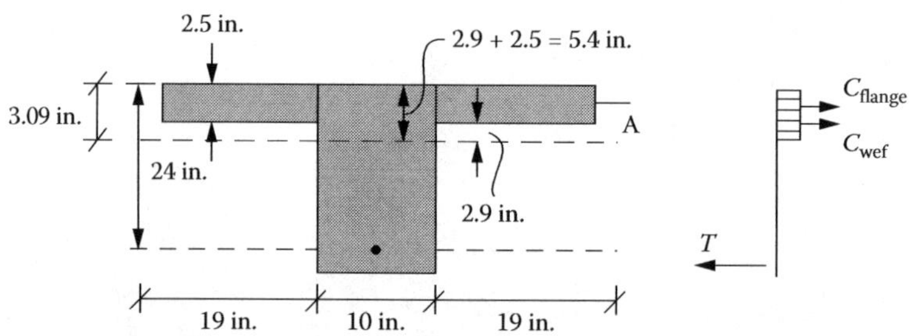

FIGURE 7.51 Example 3: prestressed T-beam.

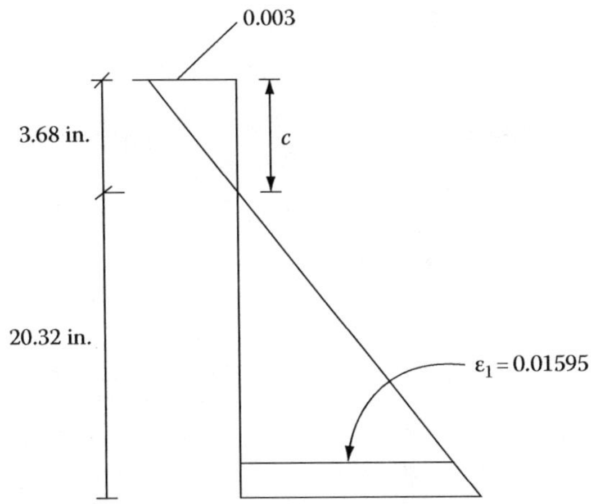

FIGURE 7.52 Example 3: strain diagram, first trial.

Second trial

$$T = 619 \text{ kips}$$

$$c = 3.68/600 \times 619 = 3.80 \text{ in.}$$

$$0.003/3.80 = \varepsilon_1/20.20 \text{ (see Figure 7.53)}$$

$$\varepsilon_1 = 0.01595$$

$$f_{ps} = 270 - 0.04/0.01595 - 0.007 = 265.5 \text{ ksi}$$

$$T = 265.5 \times 2.4 = 636 \text{ kips}$$

Use $T = 619 + 636/2 = 627.5$, say, 630 kips for the third trial.

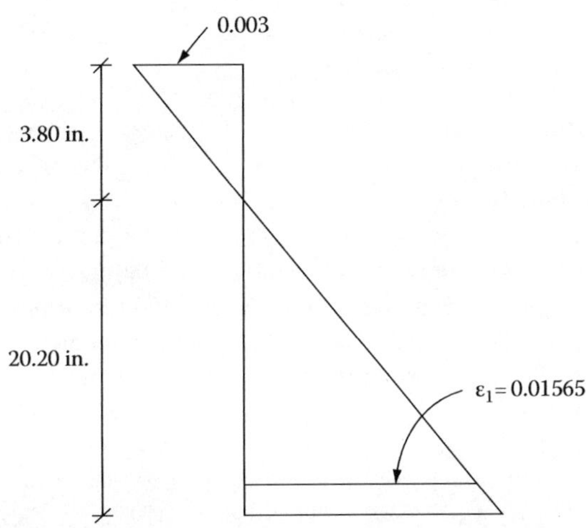

FIGURE 7.53 Example 3: strain diagram, second trial.

Third trial

$$T = 630 \text{ kips}$$

$$c = 3.68/600 \times 630 = 3.86 \text{ in.}$$

$$a = 0.8 \times 3.86$$

$$= 3.09 \text{ in.}$$

$$0.003/3.86 = \varepsilon_1/20.14$$

$$\varepsilon_1 = 0.01565$$

$$f_{ps} = 270 - 0.04/0.01565 - 0.007 = 265.37 \text{ ksi } T = 265.37 \times 2.4 = 637 \text{ kips}$$

This value is nearly the same as the value of 630 kips used at the beginning of the third iteration. However, use an average value equal to (630 + 637)/2 = 633.5 kips for calculating M_n.

Flexural strength

$$633.5 = A_c(0.85f_c')$$

$$A_c = 633.5/0.85 \times 5 = 149 \text{ in.}^2$$

Since the area of flange = 2.5 × 48 = 120 in.^{2} is less than 149 in.^{2}, the stress block extends into the web: 149 – 120 = 29/10 = 2.9 in. into the web (see Figure 7.51). Compute M_n by separating the compression zone into two areas and summing the moments of forces about the tendon force T:

$$M_n = C_{flg}(24 - 2.5/2) + C_{wlb}(24 - 5.4/2)$$

$$= 38 \times 2.5 \times 0.85 \times 5 \times 22.75 + 5.4 \times 10 \times 085 \times 5 \times 21.30$$

$$= 9185.3 + 4888 = 14{,}073 \text{ kips-in.}$$

$$= 1172.8 \text{ kips-ft}$$

Usable flexural capacity of beam = ΦM_n

$$= 0.9 \times 1172.8$$

$$= 1055.5 \text{ kips-ft}$$

7.3.5 Guidelines for Thinking on Your Feet

Throughout this book, we have discussed thumb rules for determining approximate member sizes for various structural components. In this chapter, we consolidate the same information using schematic illustrations (see Figure 7.54 through Figure 7.64). The design information given in these illustrations should prove useful in preliminary estimating, for establishing sizes and clearances, and for comparing different types of construction. It should be noted that the guidelines for the selection of member sizes given in here are typical as used in the building industry. They are not mandatory: For example, where vibration is not critical and live loads are light such as in parking structures, a slab thickness of 4.5 in. (120 mm) for a 17 ft (7 m) span has been successfully used.

7.3.6 Unit Quantities: Reinforced Concrete Buildings

Unit structural quantities such as cubic feet of concrete, pounds of mild steel reinforcement, and prestressed strands (if applicable) per square foot of building frames area are required for cost

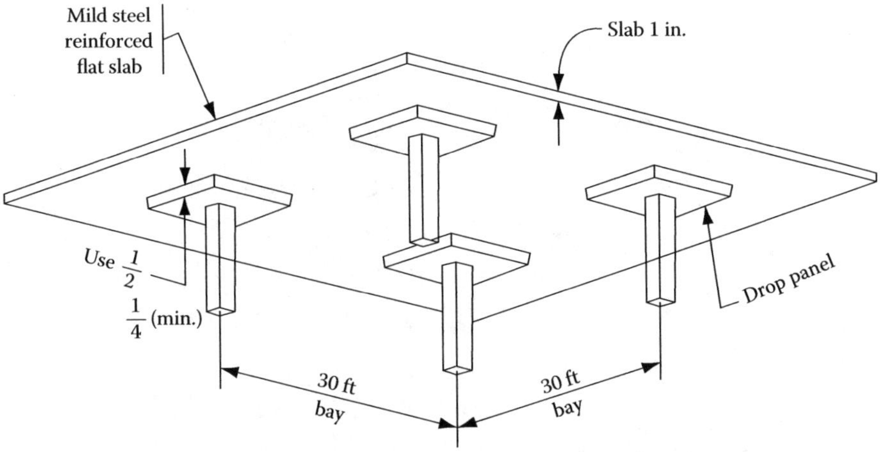

Preliminary depths for 30 ft × 30 ft bays
Slab 1 in. $= \frac{L\ (ft)}{3} = \frac{30\ ft}{3} = 10$ in.
Drop panel plan dimensions $\frac{30\ ft}{3} = 10\ ft \times 10\ ft$
Drop panel depth use $\frac{1}{2} = \frac{10}{2} = 5$ in.

FIGURE 7.54 Flat slab with drop panels.

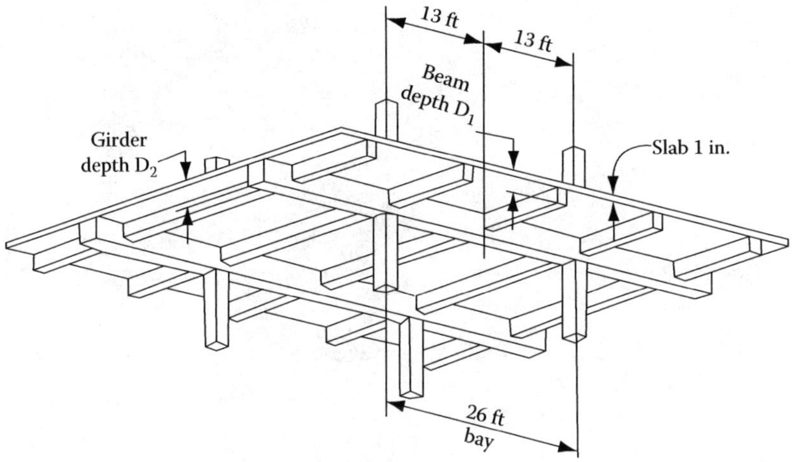

Preliminary depths for 26 ft × 26 ft bays
Slab t in. $= \frac{Span}{3} + 1 = \frac{13}{3} + 1 = 5.33$ soy 5½ in.
Beam $D_1 = \frac{Span}{1.5} = \frac{26}{1.5} = 17.3$ in. soy 18 in.
Girder $D_2 = \frac{Span}{1.5} + 2 = 19.3$ in. soy 20 in.

FIGURE 7.55 Beam and slab system 26 ft × 26 ft bays.

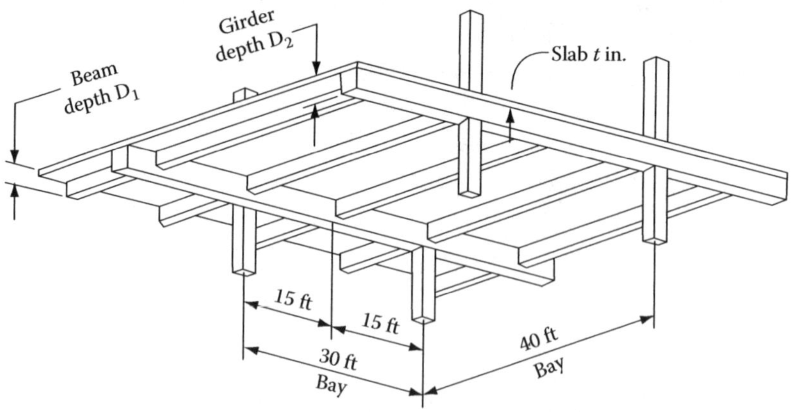

Preliminary depths for 30 ft × 40 ft bays
Slab t in. $= \frac{\text{Span}}{3} + 1$ $= \frac{15}{3} + 1 - 6\,\text{ft}$
Beam $D_1 = \frac{\text{Span}}{1.5} = \frac{40}{15} = 26.6$ in. use 26 in.
Girder $D_2 = \frac{\text{Span}}{15} = \frac{30}{15} = 20$ ft use 26 ft to match beam depth

FIGURE 7.56 Beam and slab system 30 ft × 40 ft bays.

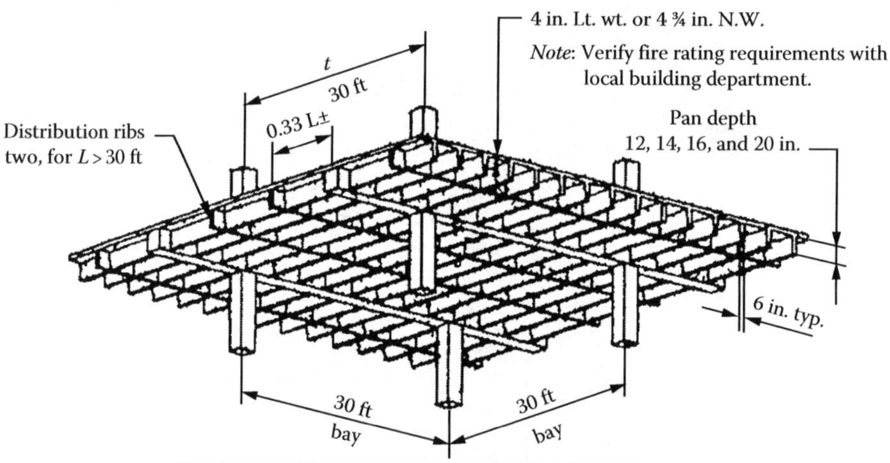

Preliminary depths for 30 ft × 30 ft bays
Joist depth $= \frac{L}{1.5} = \frac{30}{1.5} = 20$ in.
Girder depth $= \frac{L}{1.5} + 2 = 22$ in.

Note: Depth includes 4 in. for the thickness of slab.

FIGURE 7.57 One-way joist (par joist) system 30 ft × 30 ft bays.

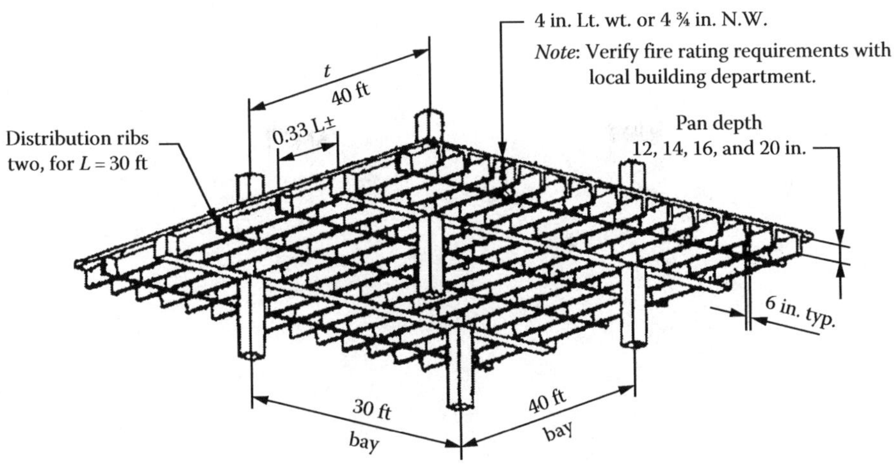

Preliminary depths for 30 ft × 40 ft bays	
Joist depth = $\frac{L}{1.5} = \frac{40}{1.5}$ = 26.6 in. bay 26 in.	
Girder depth = $\frac{L}{1.5} + 2 = \frac{30}{1.5} + 2 = 22$ in.	Use 26 in. to match joist depth

Note: Depth includes 4 in. for the thickness of slab.

FIGURE 7.58 One-way joist system 30 ft × 40 ft bays.

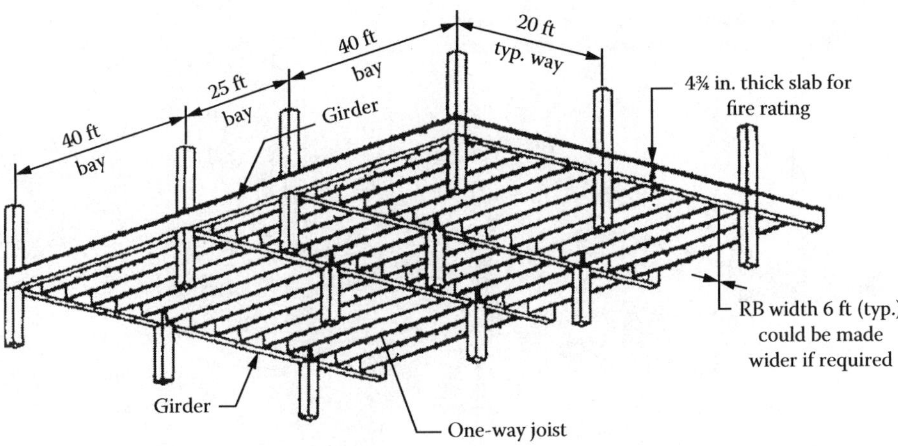

Preliminary depths for 30 ft × 40 ft bays
Girder depth = $\frac{t\ ft}{1.5} = \frac{30}{1.5}$ = 20.0 joy 24 ft
Joist depth = $\frac{t\ ft}{2} = \frac{40}{2}$ = 20 joy 22 ft

Note: Depth includes 4 ft for the thickness of slab.

FIGURE 7.59 One-way joist with constant depth girders 30 ft × 40 ft bays.

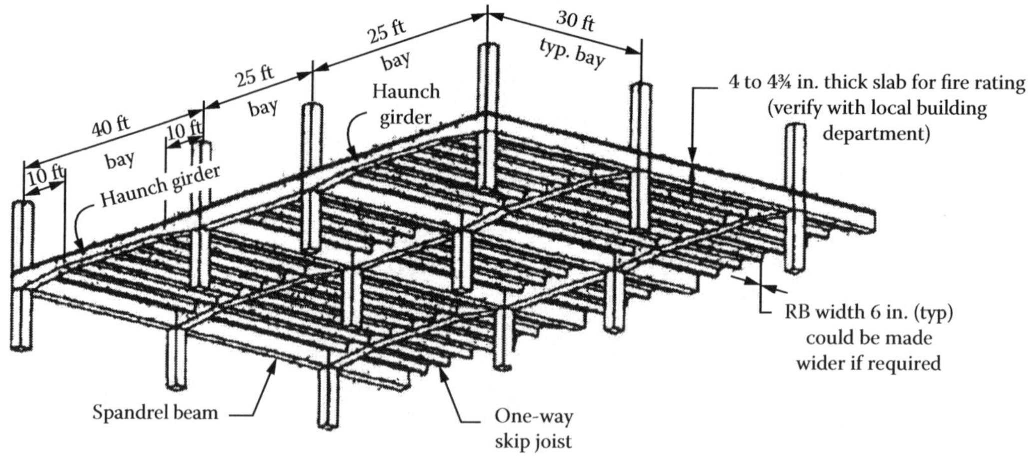

Preliminary depths for 30 ft × 40 ft bays
Prismatic girder depth $= \frac{40}{15} = 26.6$ in. joy 26 in.
Haunch girder depth at center = 26 in. − 6 in. −20 in.
Haunch girder depth at haunch = 26 in. + 6 in. −32 in.
Skip joist depth for 30 ft span $= \frac{L}{1.5} = \frac{30}{1.5} = 20$ in.

Note: Depth includes 4 in. for the thickness of slab.

FIGURE 7.60 One-way skip joist with haunch girders 30 ft × 40 ft bays.

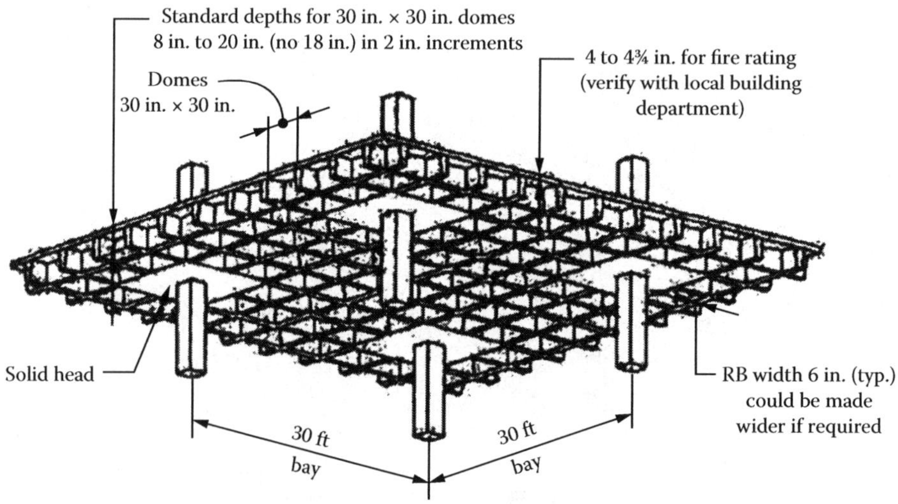

Preliminary depths for 30 ft × 30 ft bays
Waffle depth (in.) $= \frac{L}{2} = \frac{30}{2} = 15$ in. joy 16 in. includes slab depth

FIGURE 7.61 Waffle slab (two-way joist) system 30 ft × 30 ft bays.

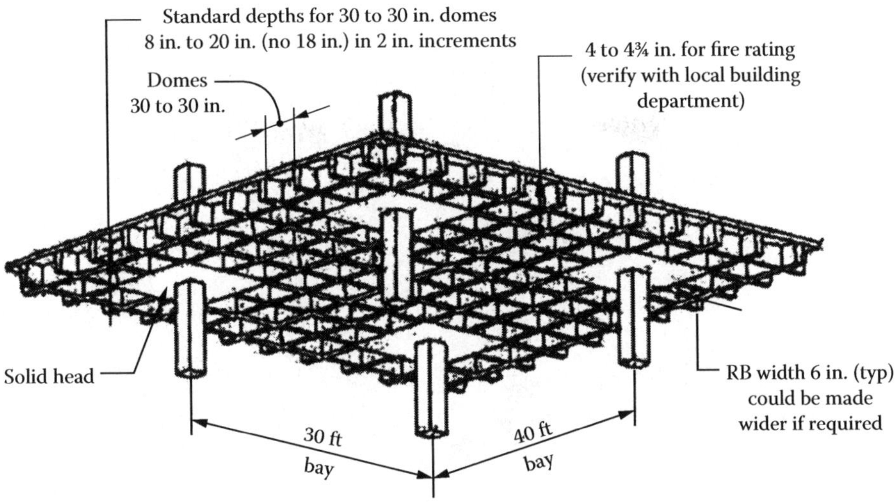

Preliminary depths for 30 ft × 40 ft bays
Waffle depth (in.) includes slab depth $= \frac{L}{2} = \frac{40}{2} = 20$ in.

FIGURE 7.62 Waffle slab (two-way joist) system 30 ft × 40 ft bays.

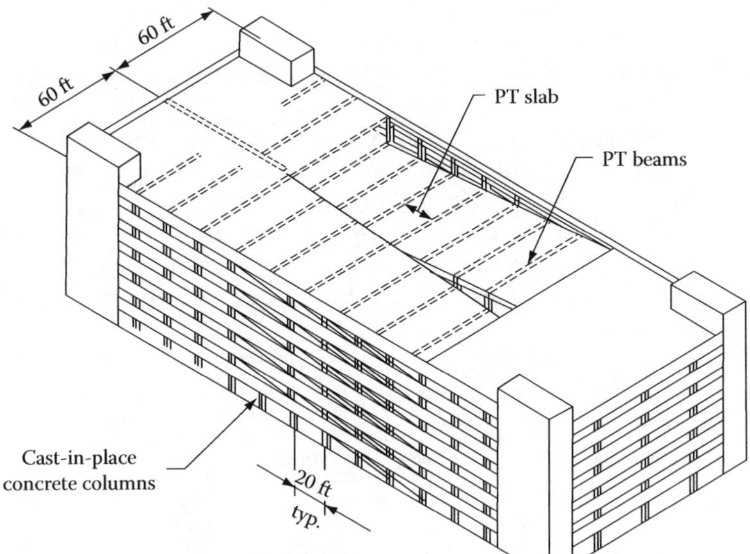

Preliminary depths 20 ft × 60 ft bays			
PT slab	$L = 20$ ft	$t = \frac{L}{3} = -J = 5.67$ ft	Use 6 in.
Regular slab	$L = 20$ ft	$t = \frac{L}{3} = +J = 7.67$ ft	Use 7.5 in.
PT beam	$L = 20$ ft	$D = \frac{L}{20} = 3$ ft	Use 36 in.

Note: Includes slab depth.

FIGURE 7.63 Post-tensioned system for parking structure 20 ft × 60 ft bays.

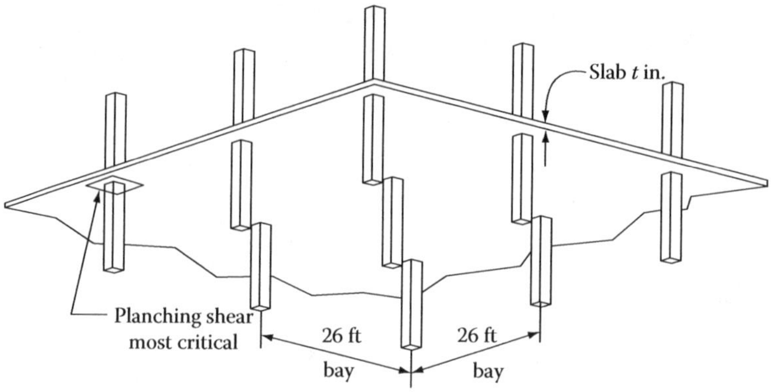

Preliminary depth for 26 ft × 26 ft bays		
PT Slab $L = 26$ ft	$t = \frac{L}{3} - 1$ $= \frac{26}{3} - 1 = 7.6$ ft	Use 7 or 8 in.

FIGURE 7.64 Post-tensioned flat plato 26 ft × 26 ft bays.

estimation of a given project. These quantities are items of construction to which unit costs are assigned to arrive at total construction cost. These are relatively easy to calculate once the working drawings and specifications have been prepared. Prior to this stage, however, the estimator must make a *conceptual estimate* to determine the approximate cost of the project. Conceptual estimates require considerable judgment to modify the so-called average unit cost to reflect the complexity of construction operations, expected time required for construction, etc., of the project under consideration.

Typically, in the United States, units of structural quantities for various concrete floor-framing systems are shown in Figures 7.65 through 7.70. Live loads shown are working loads and range from a typical office live load of 50 psf to a maximum of 200 psf appropriate for heavily loaded warehouse floors. The rebar quantities shown are for reinforcement required by design and do not include the reinforcement required for crack control, support bars, and additional lengths required for laps. The estimator should make allowances for these in the preliminary estimates, by discussing these items with the design engineer.

Table 7.3 gives estimated unit quantities of concrete, rebars, and post-tensioning steel for hotels built in various regions of the United States. Also included in the estimated length of shear walls for a given number of stories. Given in Table 7.4 are the unit quantities of reinforcement for mat foundations of several tall buildings constructed in Houston, Texas.

7.3.7 Unit Quantity of Reinforcement in Columns

During the preparation of preliminary schemes for purposes of conceptual estimates, engineers are often directed to provide unit quantities of materials for the structural elements proposed for the scheme. One such measure commonly used for vertical elements such as columns is the weight of mild steel reinforcement, typically expressed as *so many pounds of reinforcement per cubic yard of column concrete*. These quantities are relatively easy to calculate once the working drawings are complete. However, prior to this point, when we are still in the conceptual design stage, it is useful to have this information tied to the expected percentage of vertical reinforcement in columns. Table 7.5, and Figure 7.71 provided here serve that purpose. It should be noted that allowance for column ties has been made in the unit quantities shown in the table and in the graph.

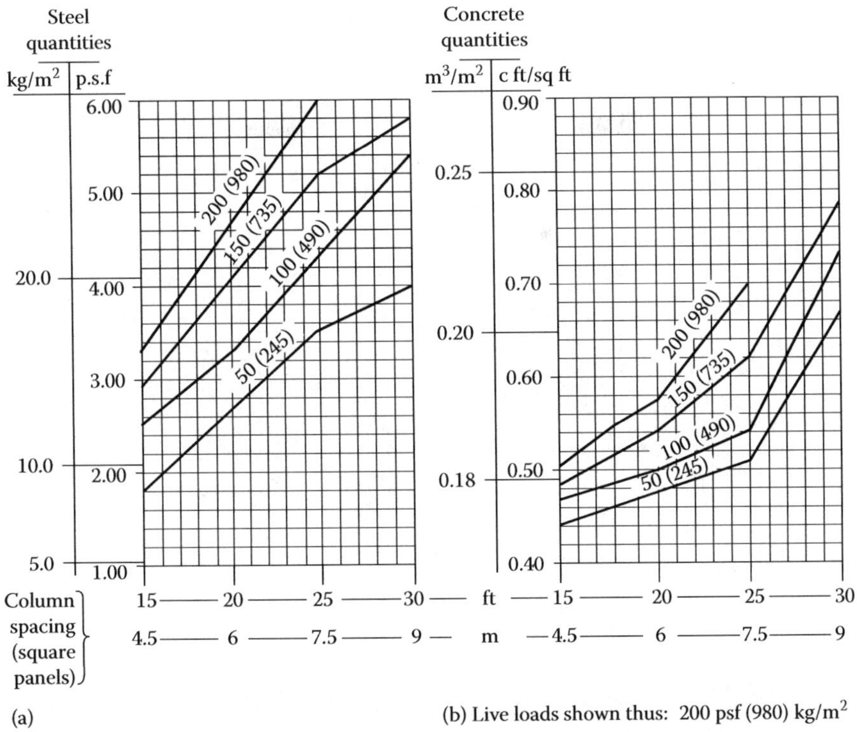

FIGURE 7.65 One-way solid slabs, unit quantities: (a) reinforcement and (b) concrete.

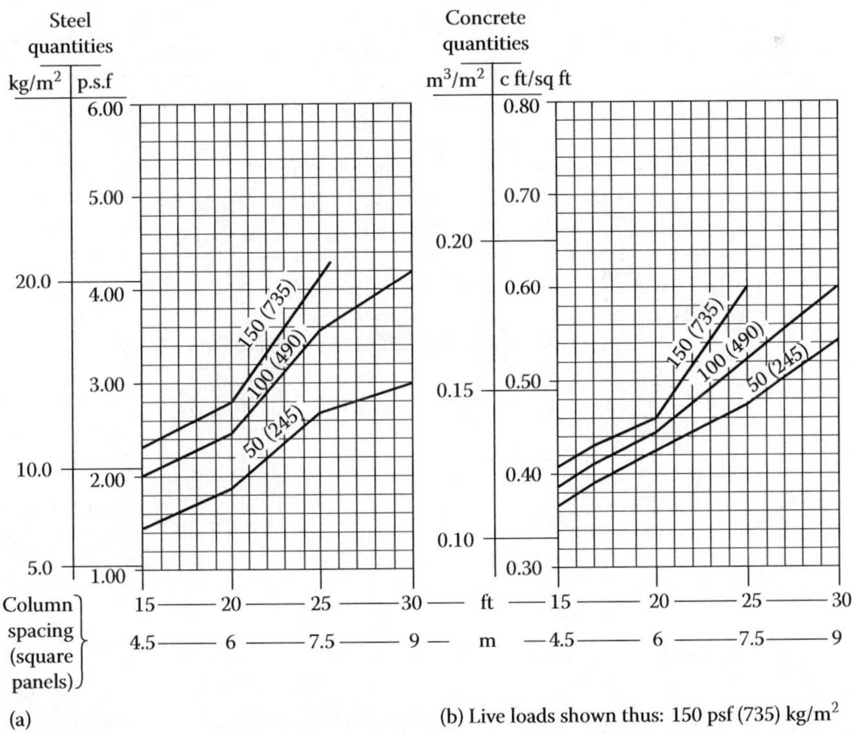

FIGURE 7.66 One-way pan joists, unit quantities: (a) reinforcement and (b) concrete.

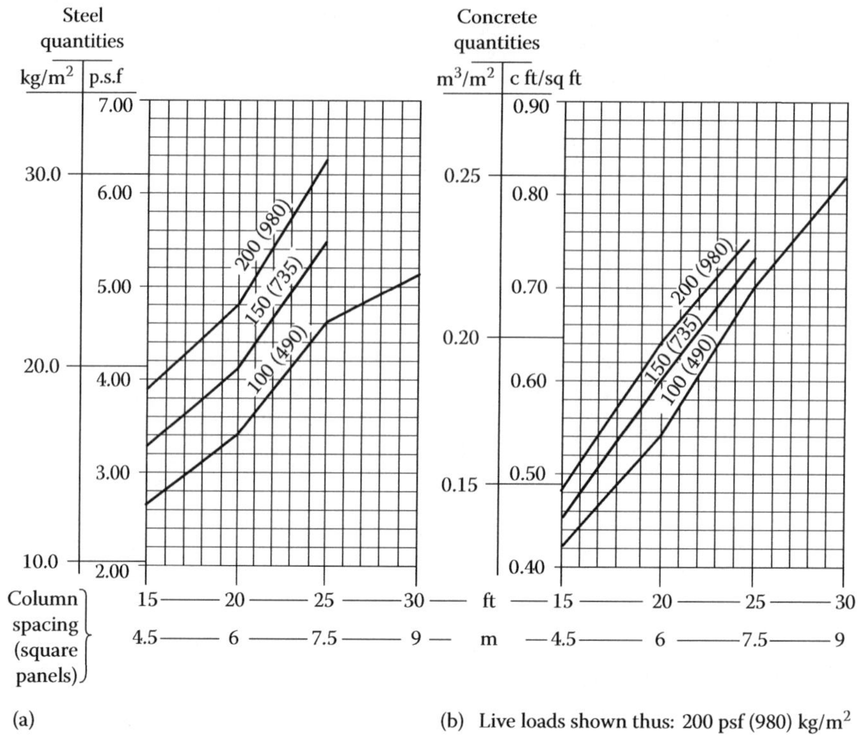

FIGURE 7.67 Two-way slabs, unit quantities: (a) reinforcement and (b) concrete.

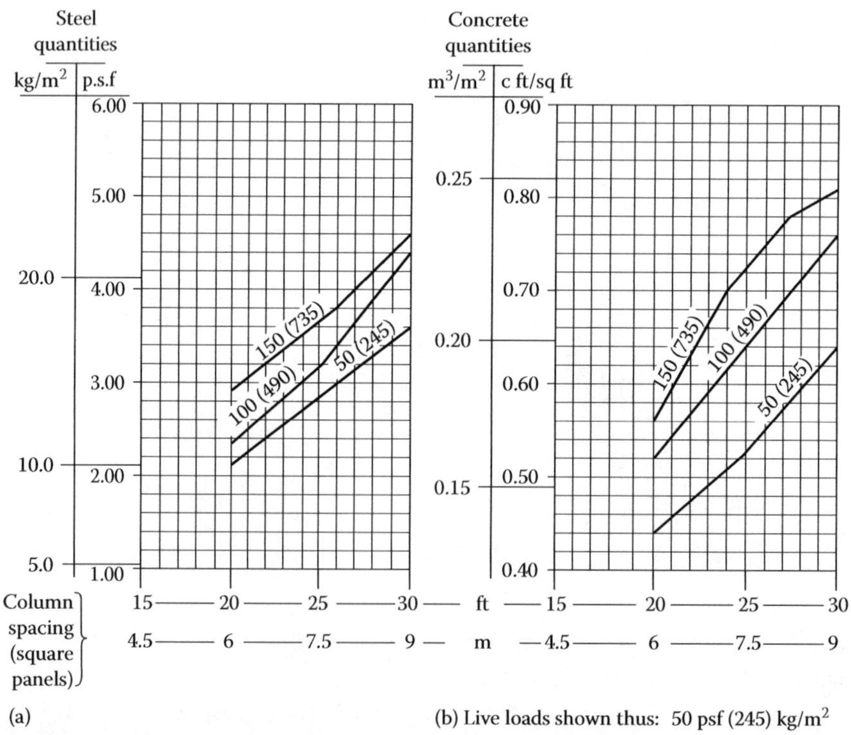

FIGURE 7.68 Waffle slabs, unit quantities: (a) reinforcement and (b) concrete.

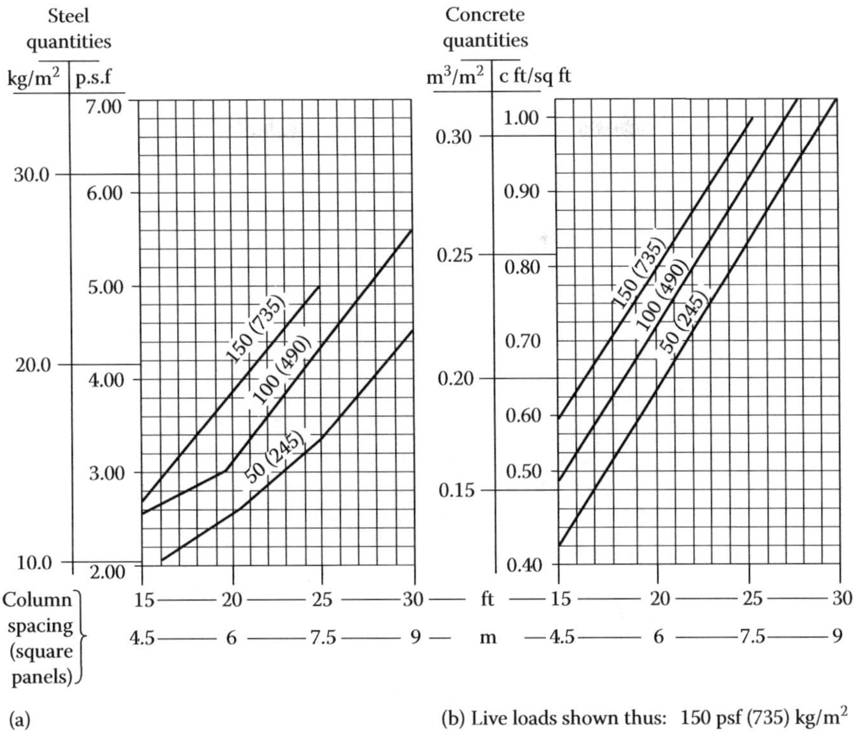

FIGURE 7.69 Flat plates, unit quantities: (a) reinforcement and (b) concrete.

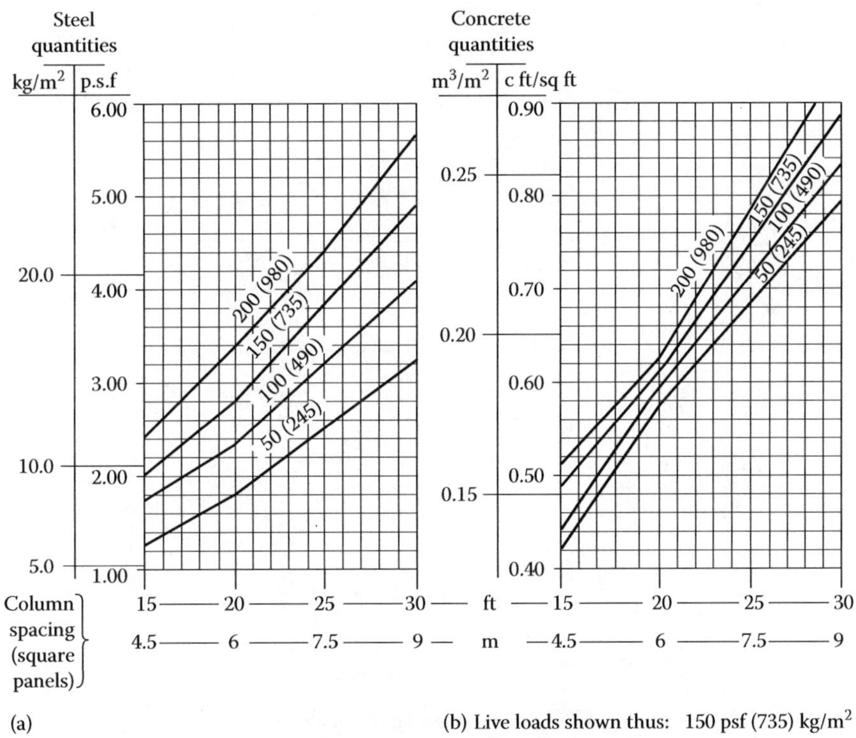

FIGURE 7.70 Flat slabs, unit quantities: (a) reinforcement and (b) concrete.

TABLE 7.4
Unit Quantity of Reinforcement in Mat Foundations

Building #	No. of Stories	Excavation Depth, ft	Mat Thickness, ft	Mat Area Square Foot	Unit Quantity of Rebar, lbs/CY
1	85 (NB)[a]	67	13.5 and 8	49720	260
2	75	63	9.75	43800	224
3	71	63	9.5	NA[b]	165
4	62	34	8	44370	179
5	56	53	8	33800	170
6	52	33	6.66	37560	148
7	50	30	8 and 6	41000	219
8	49	28	7	43490	153
9	44	35	6	NA[b]	165
10	25	25	6 and 4.5	25890	207

Note: Based on buildings constructed in Houston, Texas.
[a] NB, Not built.
[b] NA, Not available.

TABLE 7.5
Unit Quantity of Reinforcement in Columns

Percent reinforcement (includes allowance for ties)	1	1.25	1.5	1.75	2	2.5	3	3.5	4	4.5	5	5.5	6
Average steel reinforcement (lbs/Cu. Yd)	183.5	217.5	251.6	285.9	320.4	389.9	460.1	531.0	602.7	675.1	748.3	822.2	897.0

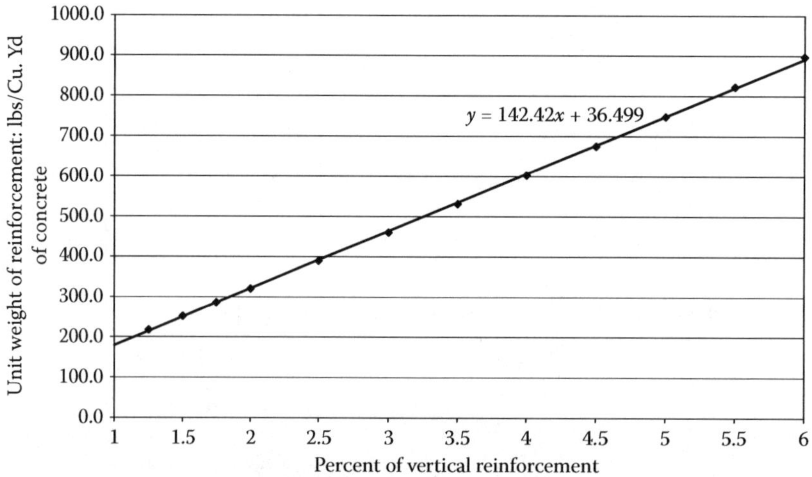

FIGURE 7.71 Unit quantity of reinforcement in columns.

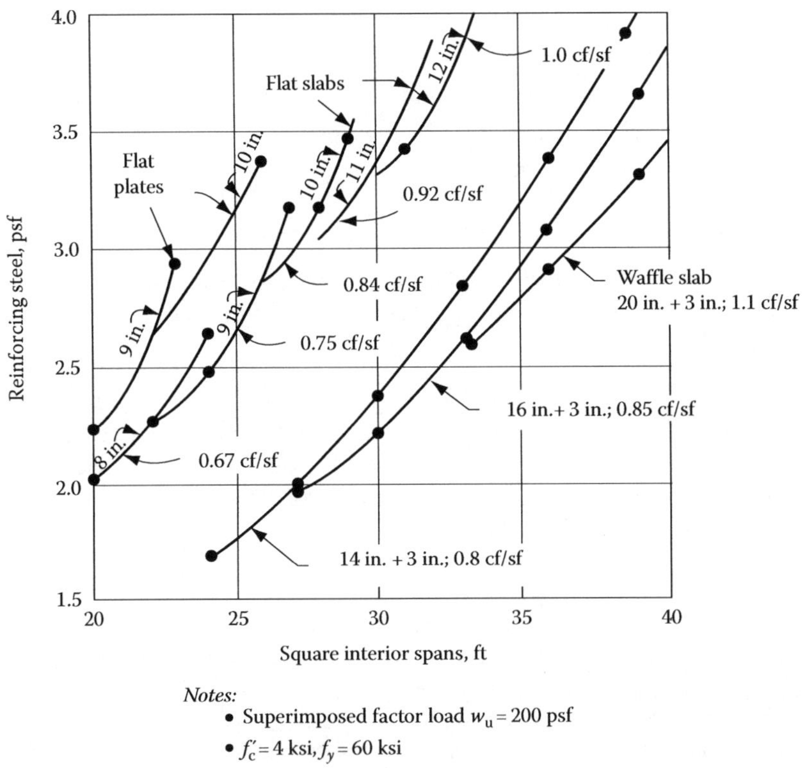

FIGURE 7.72 Preliminary design and material quantities for floor systems.

7.3.8 Unit Quantity of Reinforcement and Concrete in Floor-Framing Systems

For purposes of cost estimation, preliminary designs of floor systems are often compared using tables published by the concrete industry. Such a table published by the Concrete Reinforcing Steel Institute, CRSI, is shown in Figure 7.72 with slight modifications. It should be noted that integrity reinforcement, stirrups, shear reinforcement at columns, and welded wire reinforcement in the top slab of joist and waffle systems are not included. Similarly, the concrete quantities for distribution joists are also not included. The factored superimposed load consists of live loads, floor finishes, partition allowance, and everything except the weight of structural concrete multiplied by the approximate load factors.

7.4 ESTIMATION OF PRELIMINARY WIND LOADS, ASCE 7-10

Perhaps it is appropriate to recap the ASCE 7-10 Directional Procedure for determining wind loads on MWFRS, before discussing preliminary wind load estimation. Observe that there are seven distinct steps in ASCE 7-10:

Step 1: Determine risk category of building.

Step 2: Determine the basic wind speed, V, for the applicable risk category.

Step 3: Determine wind load parameters:

- Wind directionality factor, K_d
- Exposure Category, B, C, or D
- Topographic factor, K_{zt}
- Gust effect factor, G or G_f

Step 4: Determine velocity pressure exposure coefficient.

Step 5: Determine velocity pressure q_z on the windward surface and suction q_h on the leeward surface.

Step 6: Determine external pressure coefficient, C_p.

Step 7: Calculate wind pressure, P, on each surface as follows:

$$P = qGC_p - Qi(GC_p)\text{—rigid buildings}$$

$$P = qG_fC_p - q(GC_p)\text{—flexible buildings}$$

To simplify the procedure for assessing preliminary wind loads, we made the following assumptions, all reasonable, given the fact that we are indulging only in the determination of approximate, ball park values of wind loads:

1. Building Occupancy Category is II, a category typical of nonessential buildings.
2. Basic wind speed, V, is equal to 115 mpg. Note this wind speed is chosen here because it is applicable to the entire United States, except for coastal, hurricane-prone regions on the east cost of the United States. It is slightly conservative for some regions on the west coast, not by much, 110 mph vs. 115 mph.
3. A. Directionality factor $K_d = 0.85$. Observe this is not really an assumption. It is the suggested ASCE 7-10 value for all buildings.
   B. Exposure Category B, C, or D is accounted for in the graphs and tables presented shortly.
   C. We proceed making the daring assumption that our building is sited on a level terrain. Thus $K_{zt} = 1$.
   D. Determination of an appropriate value for gust factor G or G_f is rather involved. We take the easy road here, by assuming $G = 0.85$ for rigid buildings and $G_f = 1.XX$ for flexible buildings. Observe $G = 0.85$ is permitted in ASCE 7-10 for rigid buildings; however, G_f for flexible buildings depends on such parameters as exposure category, basic wind speed, V, building period, T, and damping coefficient assumed for the building. (The reader may want to look up the publication *Steel, Concrete & Composite Design of Tall Buildings* by the author for parametric studies of G_f.)
4. Velocity pressure coefficient K_z and K_h are given in Table 27.3-1 of ASCE 7-10.
5. External pressure coefficient, C_p, for determining velocity pressure, q_z, on the wind ward surface of a building, irrespective of the plan dimensions, is always = 0.8. On the other hand, C_p for determining the suction on the leeward surface, $q_n = q_{zrooflevel}$, varies from a high of 0.5 to as low as 0.2. We use this figure for calculating the values of C_p, since the approximate geometry and hence the plan dimension ratio L/B are known prior to estimating the preliminary wind loads.
6. External pressure coefficient, C_p: Wind loads, see item 5, given earlier.
7. Internal pressures are of no consequence in the determination of overall wind loads on MWFRS. Typical buildings are considered for analytical purposes, hermetically sealed, that is, completely enclosed. Therefore, wind pressure P may be calculated by using the following simplified equations:

$$P = G(C_p, q_{zwindward} + C_p q_{zleeward})\text{—rigid building}$$

$$P = G_f(C_p, q_{zwindward} + C_p q_{zleeward})\text{—flexible building}$$

Example 1

Given: A 40-story building with the following characteristics:

- Building height = 450 ft (137.15 m)
- Building plan dimensions = 185 × 125 ft (56.396 × 38.10 m)

- Exposure Category = C
- Basic wind speed = 110 mph (49 m/s)
- Topographic factor K_{zt} = 1.0
- The building is for typical office occupancy. Therefore, occupancy category is 11.
- Height to least-width ratio = 450/125 = 3.6. For purposes of wind load calculations, assume that the building is rigid. Use a gust factor, *G*, equal to 0.85.

Required: Using the graph given in Figure 7.73 or numerical values given in Table 7.6, determine design wind pressure for the MWFRS of the building for wind parallel to the short side.

Solution: From the building's plan dimensions, L/B = 125/180 = 0.694 < 1.0. Therefore, C_p for the windward surface = 0.8, and C_p for the leeward surface = −0.5. From Figure 7.73, select the curve for Exposure Category C. Use the graph (Figure 7.73) or the numerical table (Table 7.7) to determine the values of q_z and q_h at various heights. For example, at h = 150 ft, q_z = 37 psf.

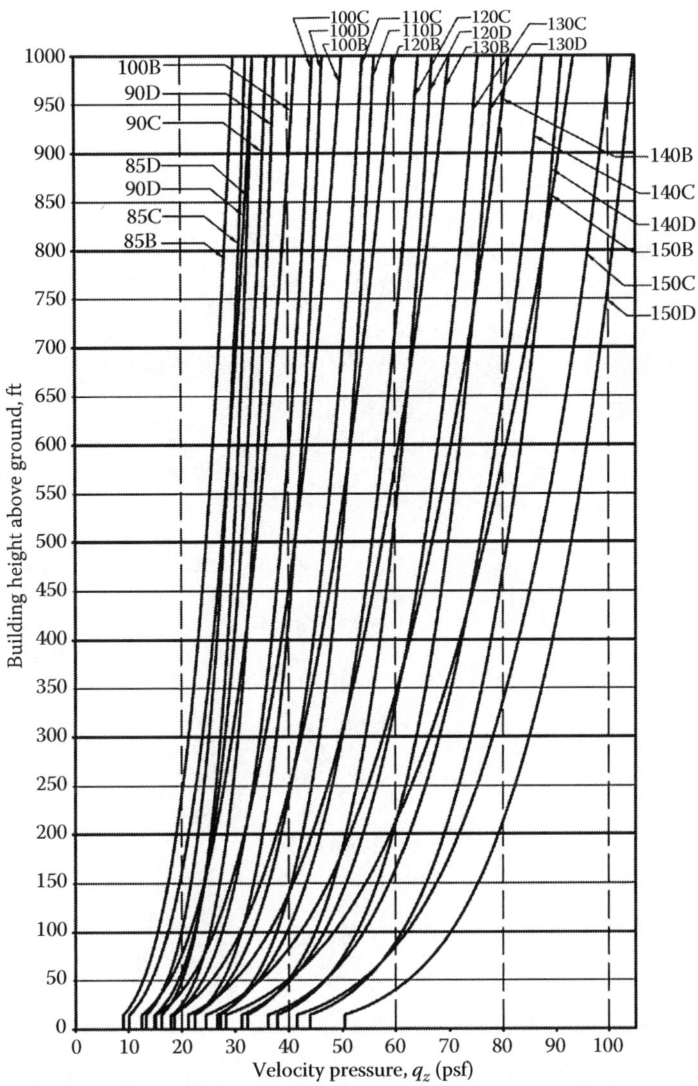

FIGURE 7.73 Variation of velocity pressure, q_z, versus wind speed and exposure categories. *Note*: $q_z = 0.00256\ K_z K_d K_{zt} V_2$, $K_d = 0.85$, $K_{zt} = 1.0$.

TABLE 7.6

Variation of Velocity Pressure, q_z, versus Wind Speed and Exposure Categories (Numerical Values) Velocity Pressure q_z (psf)

	85B	85C	85D	90B	90C	90D	100B	100C	100D	110B	110C	110D
0–15	9.04	13.35	16.20	10.13	14.96	18.16	12.51	18.47	22.42	15.13	22.35	27.13
25.0	10.46	14.86	17.70	11.72	16.66	19.85	14.47	20.57	24.50	17.51	24.89	29.65
50.0	12.75	17.20	19.97	14.29	19.28	22.38	17.64	23.80	27.64	21.34	28.80	33.44
100.0	15.54	19.90	22.53	17.42	22.31	25.26	21.50	27.54	31.18	26.02	33.32	37.73
150.0	17.44	21.67	24.17	19.56	24.30	27.10	24.15	20.00	33.46	29.22	36.29	40.48
200.0	18.94	23.02	25.41	21.23	25.81	28.49	26.21	31.87	35.18	31.72	38.58	42.56
250.0	20.19	24.13	26.42	22.63	27.05	29.62	27.94	33.40	36.57	33.81	40.41	44.25
300.0	21.27	25.08	27.27	23.84	28.11	30.57	29.43	34.71	37.74	35.61	41.99	45.67
350.0	22.22	25.90	28.01	24.91	29.04	31.40	30.76	35.85	38.77	37.22	43.38	46.91
400.0	23.09	26.64	28.67	25.88	29.87	32.14	31.95	36.87	39.68	38.67	44.62	48.01
450.0	23.88	27.31	29.26	26.77	30.62	32.81	33.05	37.80	40.50	39.99	45.74	49.01
500.0	24.61	27.92	29.80	27.59	31.30	33.41	34.06	38.65	41.25	41.21	46.76	49.91
550.0	25.29	28.49	30.30	28.35	31.94	33.97	35.00	39.43	41.94	42.35	47.71	50.75
600.0	25.92	29.01	30.76	29.06	32.53	34.49	35.88	40.16	42.58	43.41	48.59	51.62
650.0	26.52	29.51	31.20	29.73	33.08	34.97	36.71	40.84	43.18	44.42	49.42	52.24
700.0	27.09	29.97	31.60	30.37	33.60	35.43	37.50	41.48	43.74	45.37	50.20	52.92
750.0	27.63	30.41	31.98	30.98	34.09	35.86	38.24	42.09	44.27	46.27	50.93	53.56
800.0	28.14	30.83	32.34	31.55	34.56	36.26	38.95	42.67	44.77	47.13	51.63	54.17
850.0	28.64	31.22	32.69	32.10	35.00	36.64	39.63	43.21	45.24	47.96	52.29	54.74
900.0	29.11	31.60	33.01	32.63	35.43	37.01	40.29	43.74	45.69	48.75	52.92	55.29
950.0	29.56	31.96	33.32	33.14	35.83	37.36	40.91	44.24	46.12	49.51	53.53	55.81
1000.0	30.00	32.31	33.62	33.63	36.22	37.69	41.52	44.72	46.54	50.24	54.11	56.31
	120B	**120C**	**120D**	**130B**	**130C**	**130D**	**140B**	**140C**	**140D**	**150B**	**150C**	**150D**
0–15	18.01	26.60	32.28	21.13	31.22	37.89	24.51	36.20	43.94	28.14	41.56	50.44
25.0	20.84	29.62	35.28	24.46	34.76	41.41	28.36	40.32	48.02	32.56	46.28	55.13
50.0	25.40	34.27	39.80	29.81	40.22	46.71	34.58	46.65	54.17	39.69	53.55	62.19
100.0	30.97	39.66	44.90	36.34	46.54	52.69	42.15	53.98	61.11	48.38	61.96	70.16
150.0	34.77	43.19	48.18	40.81	50.69	56.54	47.32	58.79	65.58	54.33	67.49	75.28
200.0	37.75	45.89	50.65	44.30	53.85	59.45	51.38	62.46	68.94	58.98	71.70	79.14
250.0	40.23	48.10	52.66	47.22	56.44	61.80	54.76	65.46	71.67	62.86	75.15	82.28
300.0	42.38	49.98	54.35	49.74	58.65	63.79	57.69	68.02	73.98	66.22	78.09	84.93
350.0	44.29	51.63	55.83	51.98	60.59	65.52	60.29	70.27	75.99	69.21	80.66	87.23
400.0	46.01	53.10	57.14	54.00	62.32	67.06	62.63	72.27	77.78	71.90	82.96	89.28
450.0	47.59	54.43	58.32	55.85	63.88	68.45	64.77	74.09	79.39	74.36	85.05	91.13
500.0	49.04	55.65	59.40	57.56	65.31	69.72	66.75	75.75	80.85	76.63	86.96	92.82
550.0	50.40	56.78	60.40	59.15	66.64	70.88	68.60	77.28	82.20	78.75	88.72	94.37
600.0	51.67	57.83	61.32	60.64	67.87	71.96	70.32	78.71	83.46	80.73	90.36	95.81
650.0	52.86	58.81	62.18	62.04	69.02	72.97	71.95	80.05	84.63	82.60	91.89	97.15
700.0	53.99	59.74	62.98	63.37	70.11	73.92	73.49	81.31	85.73	84.36	93.34	98.41
750.0	55.07	60.61	63.74	64.63	71.13	74.81	74.95	82.50	86.76	86.04	94.70	99.60
800.0	56.09	61.44	64.46	65.83	72.11	75.65	76.35	83.63	87.74	87.64	96.00	100.72
850.0	57.07	62.23	65.15	66.98	73.03	76.46	77.68	84.70	88.67	89.18	97.23	101.79
900.0	58.01	62.98	65.80	68.08	73.92	77.22	78.96	85.73	89.56	90.64	98.41	102.81
950.0	58.92	63.70	66.42	69.14	74.76	77.95	80.19	86.71	90.40	92.06	99.54	103.78
1000.0	59.79	64.39	67.01	70.16	75.57	78.65	81.37	87.65	91.21	93.41	100.62	104.71

TABLE 7.7
Design Wind Pressures (psf)—Example 1

1	2	3	4	5	6
Height (ft)	K_z	K_h	q_z	q_h	Design Pressure $P = 0.85\ (0.8q_z + 0.5q_h)$
450	1.74	1.74	45.74	45.74	50.54
400	1.69	1.74	44.62	45.74	49.78
350	1.65	1.74	43.38	45.74	48.93
300	1.59	1.74	42.00	45.74	48.00
250	1.53	1.74	40.41	45.74	46.92
200	1.46	1.74	38.58	45.74	45.67
150	1.38	1.74	36.29	45.74	44.11
100	1.26	1.74	33.32	45.74	42.10
50	1.09	1.74	28.80	45.74	39.02
30	0.98	1.74	24.90	45.74	36.38
15	0.85	1.74	22.35	45.74	34.63

Observe the suction q_h referenced at roof height of 450 ft remains constant the entire height of leeward surface.

Column 6 in Table 7.7 gives the total design wind pressure. It is the summation of $0.8q_z$, the positive pressure on the windward surface, plus $0.5q_h$, the suction on the leeward surface, multiplied by the gust factor, G, equal to 0.85.

Example 2

Given:

- Building height = 1000 ft
- Building plan dimensions = 270 × 270 ft
- Exposure Category = C
- Basic 3 s gust speed = 110 mph
- Topographic factor $K_{zt} = 1.0$
- The building is for typical occupancy; therefore, occupancy category is II.
- Gust effect factor, G_f, based on exposure category, wind speed, building period, and damping = 1.15 (the reader may want to look up the publication *Steel, Concrete & Composite Design of Tall Buildings* by the author for parametric studies of G_f).

Required: Using Figure 7.73 or Table 7.7, determine wind pressures for the design of the MWFRS of the building.

Solution: From the building's plan dimensions, $L/B = 270/270 = l$. Therefore, C_p for the windward surface = 0.8 and 0.5 for the leeward surface. From Figure 7.73, we select the column for Exposure Category G. Observe that 115 mph is the 3 s gust speed that applies to most of the United States and C is the default exposure category that may be used in most designs, at least for preliminary estimate of wind loads.

The design pressures are obtained from Table 7.7 in exactly the same manner as for the previous problem. The only difference is we multiply the design pressures by the gust factor equal to 1.15. The results are shown in Table 7.8.

TABLE 7.8
Design Wind Pressures, Example 2

1	2	3	4	5	Design Pressure
Height (ft)	K_z	K_h	q_z	q_h	$P = G_f(0.8q_h + 0.5q_z)$
1000	2.06	2.06	36.22	36.22	54.15
900	2.01	2.06	35.43	36.22	53.42
800	1.96	2.06	34.56	36.22	52.62
700	1.91	2.06	33.60	36.22	51.74
600	1.85	2.06	32.53	36.22	50.75
500	1.77	2.06	31.30	36.22	49.62
400	1.69	2.06	29.87	36.22	48.31
300	1.59	2.06	28.11	36.22	46.69
200	1.46	2.06	25.81	36.22	44.57
100	1.26	2.06	22.31	36.22	41.35
50	1.09	2.06	19.28	36.22	38.56
25	0.94	2.06	16.66	36.22	36.15
15	0.85	2.06	14.96	36.22	34.59

Note: $G_f = 1.15$ as given in the statement of the problem.

Example 3

Given:

- Building height = 1000 ft. The building has the same dynamic characteristics as the one in Example 2.
- Building plan dimensions = 270 × 270 ft
- Exposure Category = D

TABLE 7.9
Design Wind Pressures (psf)—Example 3

1	2	3	4	5	Design Pressure
Height (ft)	K_z	K_h	q_z	q_h	$P = G_f\ (0.8\ q_z + 0.5\ q_h)$
1000	2.14	2.14	57	57	85
900	2.10	2.14	56	57	84
800	2.06	2.14	55	57	83
700	2.01	2.14	54	57	82
600	1.96	2.14	52	57	81
500	1.89	2.14	50.5	57	79
400	1.82	2.14	48	57	77
300	1.73	2.14	46	57	75
200	1.61	2.14	43	57	72
100	1.43	2.14	38	57	68
50	1.27	2.14	34	57	64
25	1.12	2.14	30	57	60
15	1.03	2.14	27	57	58

Note: $G_f = 1.15$ as given in the statement of the problem.

- Basic 3 s gust speed = 115 mph
- Topographic factor $K_{zt} = 1.0$
- The building is for typical office occupancy; therefore, occupancy category is II.
- Gust effect factor, 1.15. (Determination of gust factors is not discussed here. The reader is referred to the reference cited earlier.)

Required: Wind pressures for the design of the MWRFS of the building.

Solution: As for Example 2, $C_p = 0.8$ for the windward face and is equal to 0.5 for the leeward surface. We select he column with the heading Exposure Category D to calculate the wind loads. The results are shown in Table 7.9.

7.5 PRELIMINARY SEISMIC BASE SHEAR, *V*, AS A PERCENT OF BUILDING'S SEISMIC WEIGHT, *W*

We start this section by asking ourselves a question "what would be the approximate base shear for Seismic Design Category (SOC) D, building with a fundamental period of, say, 1.25 s, located in downtown San Diego, in terms of its seismic weight *W*?" What would be the base shear if the building were located in any of the other 48 cities in the United States, as shown in Figure 7.74.

Answers to general questions like this would be of great assistance in preliminary seismic designs and in ensuring that our computer-generated results are in the right ball park. Additionally, these answers provide an effective way of catching flawed analysis that may have resulted from computer modeling procedures and would be particularly beneficial to graduates entering the structural engineering profession. Of use they would be to peer reviewers who would want to eyeball designs using the proverbial back-of-the-envelope procedures before commencing a full-blown review.

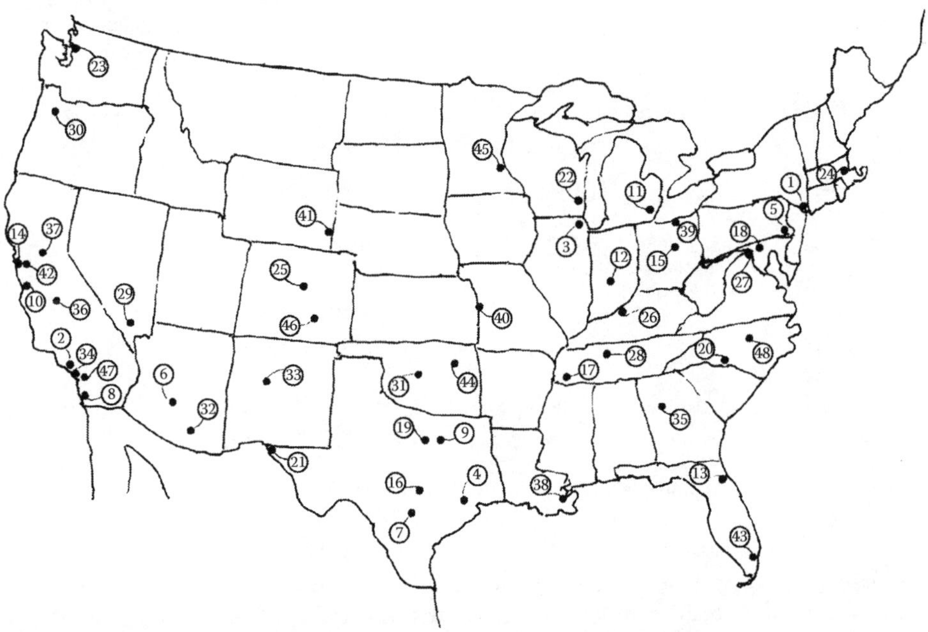

FIGURE 7.74 Locations of selected 48 U.S. cities cited in Tables 7.10, 7.11 and Figure 7.76.

The question we posed for ourselves, however, is deceptively simple because meaningful answers would require consideration of additional factors such as the following:

1. What is the seismic-force-resisting system of the building?
2. Is the building height within the limits given in ASCE Table 12.2-1 for the assigned SDC of the building?
3. What is the site class of the soil on which the building rests upon?
4. What is the value of the seismic response modification factor, *R*?

Because the building's dynamic characteristics have an impact on the base shear, it is unwieldy at best, if not impossible, to respond to the questions in general terms without addressing the dynamic properties of the building. The only response, it seems, would be to estimate the base shear using certain simplifying assumptions for the dynamic characteristics of the building and then to adjust the result as required for the given building.

The base shear values calculated for selected cities in the United States are given in Table 7.10 for site class D and Table 7.11 for site class C.

The assumptions used in preparing the tables are as follows:

1. Building height is 160 ft, corresponding to a 12-story office building approximately 13 ft floor-to-floor height or a 16-story residential building with a 10 ft floor-to-floor height. We have chosen 160 ft as the height because, with the exception of only a few structural systems (see ASCE 7-10, Table 12.2-1 for exceptions), all other systems are permitted up to this height for the buildings assigned to SDC D or lower.
2. The period T_a also denoted as TA in Tables 7.11 and 7.12 is calculated using ASCE 7-10 Section 12.8.2.1, Table 12.8-2 as follows:
   a. For steel moment frame,

$$T_a = C_t(h_n)^x$$

$$= 0.28 \times (160)^{0.8} = 1.62 \text{ s}$$

   b. For eccentrically braced steel frame,

$$T_a = 0.03 \times (160)^{0.75} = 1.34 \text{ s}$$

   c. For all other structural systems,

$$T_a = 0.02 \times (160)^{0.75} = 0.90 \text{ s}$$

   Since the lateral-load-resisting system for the building is presumably not known at this preliminary stage, we use an average value for $T_a = (1.62 + 1.34 + 0.9)/3 = 1.286$ s.

   However, we use a slightly conservative value of 1.22 s.
3. Site class C or D as noted in Tables 7.11 and 7.12.
4. Response modification coefficient, $R = 3$. Observe that AISC 341-10 provisions are not intended to apply to buildings with *R* equal to 3 or less. In the design of steel buildings, these not-so-ductile systems are permitted in SDC A, B, or C but not in SDC D, E, or F.
5. Importance factor $I_E = 1.0$.

TABLE 7.10
Approximate Base Shear as a Percent of Building Weight: Site Class D

	City	Map Location (Latitude, Longitude)	S_s	S_1	F_a	F_V	S_{MS} (g)	S_{MT} (g)	S_{DS} (g)	S_{D1} (g)	SDC	T_L, Long-Period Transition (Informational Purposes)	C_uT_a	C_s, Base Shear as a Percentage of Building Weight
1	New York, NY	40° 47′ N 73° 58′ W	0.364	0.070	1.509	2.400	0.549	0.169	0.366	0.113	C	6	1.51	3.06%
2	Los Angeles, CA	34° 3′ N 118° 14′ W	2.176	0.729	1.000	1.500	2.176	1.093	1.451	0.729	D	12	1.26	19.92%
3	Chicago, IL	41° 53′ N 87° 38′ W	0.162	0.059	1.600	2.400	0.260	0.142	0.173	0.095	B	12	1.53	2.57%
4	Houston, TX	29° 59′ N 95° 22′ W	0.090	0.038	1.600	2.400	0.144	0.091	0.096	0.060	A	12	1.53	1.33%
5	Philadelphia, PA	39° 53′ N 75° 15′ W	0.266	0.059	1.587	2.400	0.422	0.142	0.282	0.094	B	6	1.53	2.57%
6	Phoenix, AZ	33° 26′ N 112° 1′ W	0.180	0.061	1.600	2.400	0.288	0.147	0.192	0.098	B	6	1.53	2.66%
7	San Antonio, TX	29° 32′ N 98° 28′ W	0.103	0.030	1.600	2.400	0.165	0.073	0.110	0.049	A	6	1.53	1.33%
8	San Diego, CA	32° 44′ N 117° 10′ W	1.573	0.613	1.000	1.500	1.573	0.919	1.049	0.613	D	8	1.26	16.75%
9	Dallas, TX	32° 51′ N 96° 51′ W	0.118	0.050	1.600	2.400	0.188	0.121	0.125	0.081	B	12	1.53	2.19%
10	San Jose, CA	37° 22′ N 121° 56′ W	1.500	0.600	1.000	1.500	1.500	0.900	1.000	0.600	D	8	1.26	16.4%
11	Detroit, MI	42° 25′ N 83° 1′ W	0.118	0.044	1.600	2.400	0.190	0.106	0.126	0.070	B	12	1.53	1.92%
12	Indianapolis, IN	39° 44′ N 86° 17′ W	0.203	0.086	1.600	2.400	0.325	0.207	0.217	0.138	C	12	1.46	3.76%
13	Jacksonville, FL	30° 30′ N 81° 42′ W	0.151	0.064	1.600	2.400	0.242	0.152	0.161	0.102	B	8	1.53	2.80%s
14	San Francisco, CA	37° 46′ N 122° 26′ W	1.502	0.731	1.000	1.500	1.502	1.097	1.001	0.731	D	12	1.26	19.8%
15	Columbus, OH	40° 0′ N 82° 53′ W	0.145	0.058	1.600	2.400	0.232	0.140	0.155	0.093	B	12	1.53	2.53%
16	Austin, TX	30° 18′ N 97° 42′ W	0.082	0.033	1.600	2.400	0.131	0.079	0.087	0.053	B	12	1.53	1.33%
17	Memphis, TN	35° 3′ N 90° 0′ W	1.130	0.316	1.048	1.768	1.184	0.559	0.790	0.372	D	12	1.26	10.17%
18	Baltimore, MD	39° 11′ N 76° 40′ W	0.163	0.050	1.600	2.400	0.261	0.121	0.174	0.081	B	6	1.53	2.19%
19	Fort Worth, TX	32° 50′ N 97° 3′ W	0.14	0.049	1.600	2.400	0.183	0.118	0.122	0.079	B	12	1.53	2.19%
20	Charlotte, NC	35° 13′ N 80° 56′ W	0.318	0.106	1.546	2.377	0.491	0.251	0.328	0.168	C	8	1.41	4.59%
21	El Paso, TX	31° 48′ N 106° 24′ W	0.336	0.108	1.531	2.367	0.514	0.256	0.343	0.171	C	6	1.40	4.65%
22	Milwaukee, WI	42° 57′ N 87° 54′ W	0.110	0.045	1.600	2.400	0.176	0.109	0.117	0.073	B	12	1.53	1.97%
23	Seattle, WA	47° 39′ N 122° 18′ W	1.306	0.444	1.000	1.556	1.306	0.691	0.870	0.460	D	6	1.26	12.59%
24	Boston, MA	42° 22′ N 71° 2′ W	0.282	0.068	1.574	2.400	0.445	0.163	0.296	0.109	B	6	1.51	3.11%
25	Denver, CO	39° 45′ N 104° 52′ W	0.210	0.055	1.600	2.400	0.337	0.133	0.225	0.089	B	4	1.53	2.4%

(Continued)

TABLE 7.10 (*Continued*)
Approximate Base Shear as a Percent of Building Weight: Site Class D

	City	Map Location (Latitude, Longitude)	S_s	S_1	F_a	F_V	S_{MS} (g)	S_{MT} (g)	S_{DS} (g)	S_{D1} (g)	SDC	T_L, Long-Period Transition (Informational Purposes)	C_uT_a	C_s, Base Shear as a Percentage of Building Weight
26	Louisville, KY	38° 11′ N 85° 44′ W	0.247	0.103	1.600	2.389	0.395	0.245	0.263	0.164	C	12	1.41	4.48%
27	Washington, DC	38° 51′ N 77° 2′ W	0.153	0.050	1.600	2.400	0.244	0.121	0.163	0.081	B	8	1.53	2.19%
28	Nashville,TN	36° 7′ N 86° 41′ W	0.331	0.129	1.535	2.283	0.508	0.295	0.339	0.197	C	12	1.36	5.37%
29	Las Vegas, NV	36° 5′ N 115° 10′ W	0.554	0.171	1.357	2.115	0.752	0.362	0.501	0.241	D	6	1.31	6.59%
30	Portland, OR	45° 36′ N 122° 36′ W	0.907	0.317	1.137	1.767	1.031	0.559	0.687	0.373	D	16	1.26	10.2%
31	Oklahoma City, OK	35° 24′ N 97° 36′ W	0.354	0.076	1.516	2.400	0.537	0.182	0.358	0.121	C	12	1.49	3.32%
32	Tucson, AZ	32° 7′ N 110° 56′ W	0.282	0.080	1.575	2.400	0.444	0.192	0.296	0.128	B	6	1.48	3.49%
33	Albuquerque, NM	35° 3′ N 106° 37′ W	0.571	0.172	1.343	2.112	0.767	0.363	0.511	0.242	D	6	1.31	6.61%
34	Long Beach, CA	33° 49′ N 118° 9′ W	1.693	0.644	1.000	1.500	1.693	0.966	1.129	0.644	D	8	1.26	17.60%
35	Atlanta, GA	33° 39′ N 84° 26′ W	0.218	0.084	1.600	2.400	0.348	0.202	0.232	0.134	C	12	1.47	3.67%
36	Fresno, CA	36° 46′ N 119° 43′ W	0.488	0.219	1.410	1.962	0.688	0.430	0.459	0.287	D	12	1.27	7.83%
37	Sacramento, CA	38° 31′ N 121° 30′ W	0.629	0.251	1.297	1.898	0.816	0.477	0.544	0.318	D	12	1.26	8.68%
38	New Orleans, LA	29° 59′ N 90° 15′ W	0.111	0.048	1.600	2.400	0.177	0.115	0.118	0.077	B	12	1.53	2.09%
39	Cleveland, OH	41° 24′ N 81° 51′ W	0.179	0.051	1.600	2.400	0.289	0.123	0.191	0.082	B	12	1.53	2.23%
40	Kansas City, MO	39° 7′ N 94° 35′ W	0.127	0.058	1.600	2.400	0.204	0.140	0.136	0.094	B	12	1.53	2.53%
41	Omaha, NE	41° 18′ N 95° 54′ W	0.118	0.042	1.600	2.400	0.188	0.100	0.125	0.066	B	12	1.53	1.84%
42	Oakland, CA	37° 49′ N 122° 19′ W	1.500	0.600	1.000	1.500	1.500	0.900	1.000	0.600	D	8	1.26	16.4%
43	Miami, FL	25° 48′ N 80° 16′ W	0.051	0.020	1.600	2.400	0.082	0.047	0.055	0.031	A	8	1.53	1.33%
44	Tulsa, OK	36° 12′ N 95° 54′ W	0.159	0.067	1.600	2.400	0.255	0.160	0.170	0.107	B	12	1.52	2.93%
45	Minneapolis, MN	44° 53′ N 93° 13′ W	0.060	0.027	1.600	2.400	0.096	0.065	0.064	0.043	A	12	1.53	1.33%
46	Colorado Springs, CO	38° 49′ N 104° 43′ W	0.203	0.058	1.600	2.400	0.324	0.139	0.216	0.093	B	4	1.53	2.53%
47	Irvine, CA	33° 40′ N 117° 47′ W	1.476	0.522	1.000	1.500	1.476	0.783	0.984	0.522	D	8	1.26	14.39%
48	Raleigh, NC	35° 52′ N 78° 47′ W	0.202	0.079	1.600	2.400	0.323	0.189	0.215	0.126	B	8	1.48	3.45%

Note: h_o = 160ft. I_E = 1.0. T_a = 1.22. s (see text). Site Class = D (Stiff soil, see ASCE 7-05, Chapter 20), R = 3.0. For systems with differing R values, multiply base shear by the ratio 3.0/R.

TABLE 7.11
Approximate Base Shear as a Percent of Building Weight: Site Class C

	City	Map Location (Latitude, Longitude)	S_s	S_1	F_a	F_V	S_{MS} (g)	S_{MT} (g)	S_{DS} (g)	S_{D1} (g)	SDC	T_L, Long-Period Transition (Informational Purposes)	C_uT_a	C_s, Base Shear as a Percentage of Building Weight
1	New York, NY	40° 47′N 73° 58′W	0.364	0.070	1.200	1.700	0.437	0.120	0.291	0.080	B	6	1.53	2.17%
2	Los Angeles, CA	34° 3′N 118° 14′W	2.176	0.729	1.000	1.300	2.176	0.948	1.451	0.632	D	12	1.26	17.27%
3	Chicago, IL	41° 53′N 87° 38′W	0.162	0.059	1.200	1.700	0.195	0.101	0.130	0.067	A	12	1.53	1.33%
4	Houston, TX	29° 59′N 95° 22′W	0.090	0.038	1.200	1.700	0.108	0.064	0.072	0.043	A	12	1.53	1.33%
5	Philadelphia, PA	39° 53′N 75° 15′W	0.266	0.059	1.200	1.700	0.319	0.100	0.213	0.067	B	6	1.53	1.82%
6	Phoenix, AZ	33° 26′N 112° 1′W	0.180	0.061	1.200	1.700	0.216	0.104	0.144	0.069	B	6	1.53	1.88%
7	San Antonio, TX	29° 32′N 98° 28′W	0.103	0.030	1.200	1.700	0.124	0.052	0.083	0.034	A	6	1.53	1.33%
8	San Diego, CA	32° 44′N 117° 10′W	1.573	0.613	1.000	1.300	1.573	0.796	1.049	0.531	D	8	1.26	14.52%
9	Dallas, TX	32° 51′N 96° 51′W	0.118	0.050	1.200	1.700	0.141	0.086	0.094	0.057	A	12	1.53	1.33%
10	San Jose, CA	37° 22′N 121° 56′W	1.500	0.600	1.000	1.300	1.500	0.780	1.000	0.520	D	8	1.26	14.21%
11	Detroit, MI	42° 25′N 83° 1′W	0.118	0.044	1.200	1.700	0.142	0.075	0.095	0.050	A	12	1.53	1.33%
12	Indianapolis, IN	39° 44′N 86° 17′W	0.203	0.086	1.200	1.700	0.244	0.146	0.163	0.098	B	12	1.53	2.66%
13	Jacksonville, FL	30° 30′N 81° 42′W	0.151	0.064	1.200	1.700	0.181	0.108	0.121	0.072	B	8	1.53	1.98%
14	San Francisco, CA	37° 46′N 122° 26′W	1.502	0.731	1.000	1.300	1.502	0.951	1.001	0.634	D	12	1.26	17.26%
15	Columbus, OH	40° 0′N 82° 53′W	0.145	0.058	1.200	1.700	0.174	0.099	0.116	0.066	A	12	1.53	1.33%
16	Austin, TX	30° 18′N 97° 42′W	0.082	0.033	1.200	1.700	0.098	0.056	0.065	0.037	A	12	1.53	1.33%
17	Memphis, TN	35° 3′N 90° 0′W	1.130	0.316	1.000	1.484	1.130	0.469	0.753	0.313	D	12	1.26	8.55%
18	Baltimore, MD	39° 11′N 76° 40′W	0.163	0.050	1.200	1.700	0.196	0.086	0.130	0.057	A	6	1.53	8.27%
19	Fort Worth, TX	32° 50′N 97° 3′W	0.114	0.049	1.200	1.700	0.137	0.083	0.091	0.056	A	12	1.53	3.33%
20	Charlotte, NC	35° 13′N 80° 56′W	0.318	0.106	1.200	1.694	0.381	0.179	0.254	0.119	B	8	1.49	1.33%
21	El Paso, TX	31° 48′N 106° 24′W	0.336	0.108	1.200	1.692	0.403	0.183	0.269	0.122	B	6	1.49	3.3%
22	Milwaukee, WI	42° 57′N 87° 54′W	0.110	0.045	1.200	1.700	0.132	0.077	0.088	0.051	A	12	1.53	1.33%
23	Seattle, WA	47° 39′N 122° 18′W	1.306	0.444	1.000	1.356	1.306	0.602	0.870	0.401	D	6	1.26	10.96%
24	Boston, MA	42° 22′N 71° 2′W	0.282	0.068	1.200	1.700	0.339	0.116	0.226	0.077	B	6	1.53	2.11%
25	Denver, CO	39° 45′N 104° 52′W	0.210	0.055	1.200	1.700	0.253	0.094	0.168	0.063	B	4	1.53	1.71%

(Continued)

TABLE 7.11 (*Continued*)
Approximate Base Shear as a Percent of Building Weight: Site Class C

	City	Map Location (Latitude, Longitude)	S_s	S_1	F_a	F_v	S_{MS} (g)	S_{MT} (g)	S_{DS} (g)	S_{D1} (g)	SDC	T_L, Long-Period Transition (Informational Purposes)	C_uT_a	C_s, Base Shear as a Percentage of Building Weight
26	Louisville, KY	38° 11′N 85° 44′W	0.247	0.103	1.200	1.697	0.296	0.174	0.197	0.116	B	12	1.50	3.19%
27	Washington, DC	38° 51′N 77° 2′W	0.153	0.050	1.200	1.700	0.183	0.086	0.122	0.057	A	8	1.53	1.33%
28	Nashville, TN	36° 7′N 86° 41′W	0.331	0.129	1.200	1.671	0.397	0.216	0.265	0.144	C	12	1.45	3.92%
29	Las Vegas, NV	36° 5′N 115° 10′W	0.554	0.171	1.178	1.629	0.653	0.279	0.435	0.186	C	6	1.38	5.08%
30	Portland, OR	45° 36′N 122° 36′W	0.907	0.317	1.037	1.483	0.940	0.469	0.627	0.313	D	16	1.26	8.56%
31	Oklahoma City, OK	35° 24′N 97° 36′W	0.354	0.076	1.200	1.700	0.425	0.129	0.284	0.086	B	12	1.53	2.36%
32	Tucson, AZ	32° 7′N 110° 56′W	0.282	0.080	1.200	1.700	0.338	0.136	0.225	0.091	B	6	1.53	2.48%
33	Albuquerque, NM	35° 3′N 106° 37′W	0.571	0.172	1.172	1.628	0.669	0.280	0.446	0.187	C	6	1.37	5.11%
34	Long Beach, CA	33° 49′N 118° 9′W	1.693	0.644	1.000	1.300	1.693	0.837	1.129	0.558	D	8	1.26	15.25%
35	Atlanta, GA	33° 39′N 84° 26′W	0.218	0.084	1.200	1.700	0.261	0.143	0.174	0.095	B	12	1.53	2.6%
36	Fresno, CA	36° 46′N 119° 43′W	0.488	0.219	1.200	1.581	0.585	0.346	0.390	0.231	D	12	1.32	6.31%
37	Sacramento, CA	38° 31′N 121° 30′W	0.629	0.251	1.148	1.549	0.722	0.389	0.481	0.259	D	12	1.30	7.08%
38	New Orleans, LA	29° 59′N 90° 15′W	0.111	0.048	1.200	1.700	0.133	0.081	0.089	0.054	A	12	1.53	1.33%
39	Cleveland, OH	41° 24′N 81° 51′W	0.179	0.051	1.200	1.700	0.214	0.087	0.143	0.058	A	12	1.53	1.33%
40	Kansas City, MO	39° 7′N 94°35′W	0.127	0.058	1.200	1.700	0.153	0.099	0.102	0.066	A	12	1.53	1.33%
41	Omaha, NE	41° 18′N 95° 54′W	0.118	0.042	1.200	1.700	0.141	0.071	0.094	0.047	A	12	1.53	1.33%
42	Oakland, CA	37° 49′N 122° 19′W	1.500	0.600	1.000	1.300	1.500	0.780	1.000	0.520	D	8	1.26	14.21%
43	Miami, FL	25° 48′N 80° 16′W	0.051	0.020	1.200	1.700	0.062	0.033	0.041	0.022	A	8	1.53	1.33%
44	Tulsa, OK	36° 12′N 95° 54′W	0.159	0.067	1.200	1.700	0.191	0.113	0.127	0.076	B	12	1.53	2.08%
45	Minneapolis, MN	44° 53′N 93° 13′W	0.060	0.027	1.200	1.700	0.072	0.046	0.048	0.031	A	12	1.53	1.33%
46	Colorado Springs, CO	38° 49′N 104° 43′W	0.203	0.058	1.200	1.700	0.243	0.099	0.162	0.056	A	4	1.53	1.33%
47	Irvine, CA	33° 40′N 117° 47′W	1.476	0.522	1.000	1.300	1.476	0.679	0.984	0.452	D	8	1.26	12.36%
48	Raleigh, NC	35° 52′N 78° 47′W	0.202	0.079	1.200	1.700	0.242	0.134	0.161	0.089	B	8	1.53	2.44%

Note: $h_o = 160$ ft. $I_E = 1.0$. $T_a = 1.22$. Site Class = C (Very dense soil and soft rock, see ASCE 7-05, Chapter 20). $R = 3.0$. For systems with differing R values, multiply base shear by the ratio $3.0/R$.

TABLE 7.12
Example 2: Column Schedule

Level	Interior Column	Area (in.^{2})	Exterior Column	Area (in.^{2})
R	W14 × 68	20	W30 × 211	62
28	W14 × 90	26.5	W30 × 235	69
26	W14 × 120	35.3	W30 × 261	76.7
24	W14 × 132	35.8	W30 × 292	85.7
22	W14 × 145	42.7	W30 × 326	95.7
20	W14 × 193	56.7	W30 × 326	95.7
18	W14 × 233	68.5	W30 × 357	104
16	W14 × 257	75.6	W30 × 357	104
14	W14 × 283	83.3	W30 × 351	114
12	W14 × 311	91.4	W30 × 391	114
10	W14 × 342	101	W30 × 433	127
8	W14 × 370	109	W30 × 433	127
6	W14 × 398	117	W30 × 477	140
4	W14 × 426	125	W30 × 477	140
2	W14 × 455	134	W30 × 526	154
G	W14 × 500	147	W30 × 581	170

7.5.1 Building Height, h_n = 160 ft

Example 1

Given: A 160 ft tall building in Boston sited at the map location given in item #24, in Table 7.12. The building has an R equal to 3 and is located on site class C.

Required: Base shear as a percent of building weight W.

Solution: Because the site class is C as stated for the example building, we select Table 7.12. In item #24 of the table, the base shear is given as a percent of the building weight W and is equal to 2.11%. Therefore, $V = 0.211$ W. Observe the building period used in the calculation of base shear is equal to $C_u T_a$. See ASCE 7-10 Table 12.8-1 for values of coefficient C_u.

Example 2

Given: A 160 ft tall steel building with special steel concentrically brace frames. (See ASCE 7-10, Table 12.2-1 and item B3). The building is located on site class D, in San Diego, California.

Required: Base shear as a percent of building weight W.

Solution: Immediately, we notice our building is in a seismic area, California, with an assigned SDC D. A special steel concentrically braced frame is permitted up to a height of 160 ft. And because AISC Seismic Provisions mandate stringent detailing requirements for this system, the seismic response modification coefficient, R, is equal to 6 instead of 3 that was used in developing Tables 7.11 and 7.12.

From Table 7.11 item #8, the required base shear $V = 16.75\%$ W. However, as stated previously, the base shears in the tables are normalized using $R = 3$. Therefore, the base shear for our building is equal to $V = (16.75 \times 3/6)\%$ W = 8.38% W.

7.5.2 Buildings Taller than 160 ft

It is clear from a previous discussion that the base shear tables (Tables 7.11 and 7.12) are good for buildings of height h_n = 160 ft. If h_n is considerably different from 160 ft, we are out of luck; we cannot use the tables. The only recourse we have, it seems, is to revert back to using generalized response spectrum.

Example 3

Given: A steel building of height h_n equal to 300 ft with a dual system as seismic-force-resisting system. It consists of special moment frames capable of resisting at least 25% of seismic forces, and buckling-restrained braced frame.

The corresponding $R = 8$. The building is situated in Los Angeles, California, on a site classified as D, located at the intersection of latitude 34°3′N and 110°14′W.

Required: Base shear as a percent of building's seismic weight, W, using the response given in Figure 7.75.

Solution: $h_n = 300$ ft (given in the statement example)

Calculate T_a as follows:

$$T_a = 0.03(h_n)^{0.75}$$

$$= 0.03(300)^{0.75}$$

$$= 2.16 \text{ s}$$

From ASIC 7-10 Table 12.8-1

$$\text{For } S_{D1} = 0.729 > 0.4, C_u = 1.4$$

Therefore, the adjusted period for the determination of base shear $= C_uT_a = 1.4 \times 2.16 = 3$ s. Using the response spectrum for the given site in Los Angeles, site class D, shown in Figure 7.75.

$$\text{For } C_uT_a = 3\times, S_a = 0.25 \text{ g}$$

Using $R = 8$, $V = 0.25/8 = 0.0313$ g

In other words, $V = 3.13\%$ W

Granted, the procedure is somewhat trivial once the response spectrum is known. However, the response spectrum curves shown for site class D and C, in Figure 7.76, are expected to assist engineers in their preliminary designs.

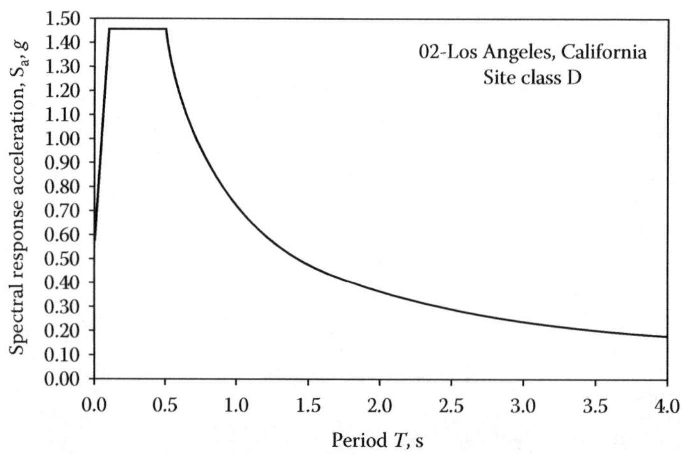

FIGURE 7.75 Design response spectrum for a selected site in downtown Los Angeles.

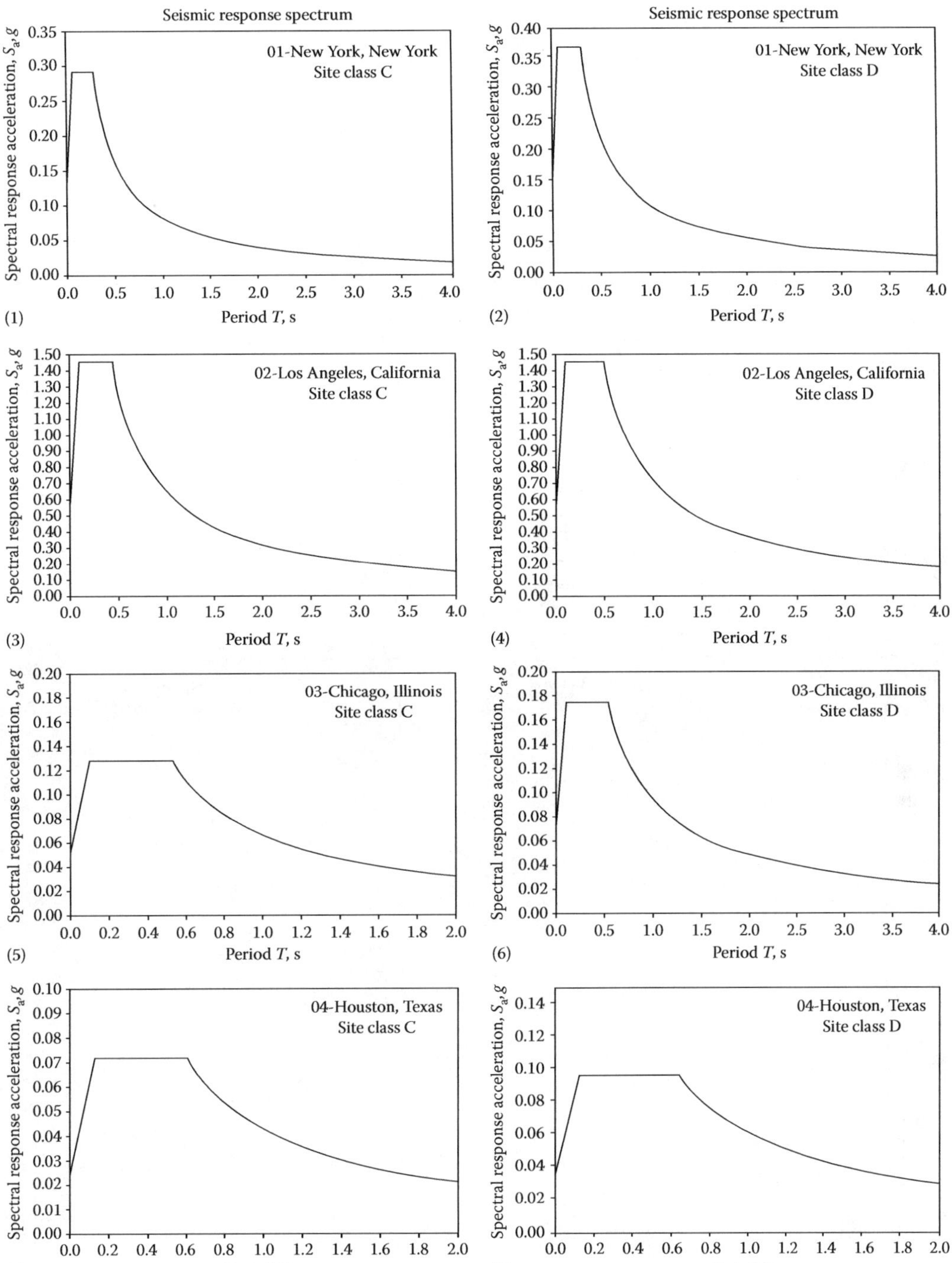

FIGURE 7.76 Design response spectrum for selected 48 U.S. cities. *(Continued)*

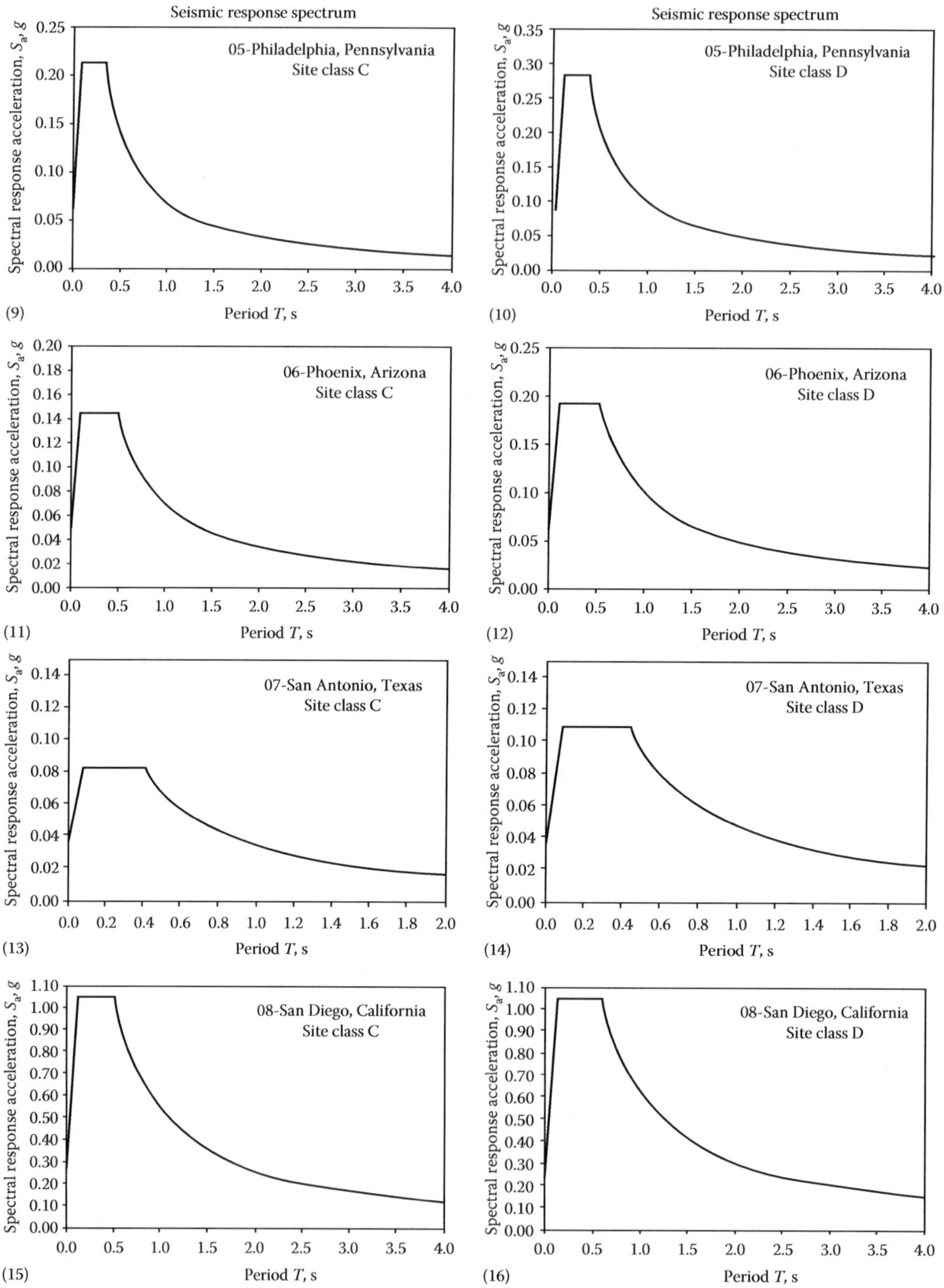

FIGURE 7.76 (*Continued*) Design response spectrum for selected 48 U.S. cities. (*Continued*)

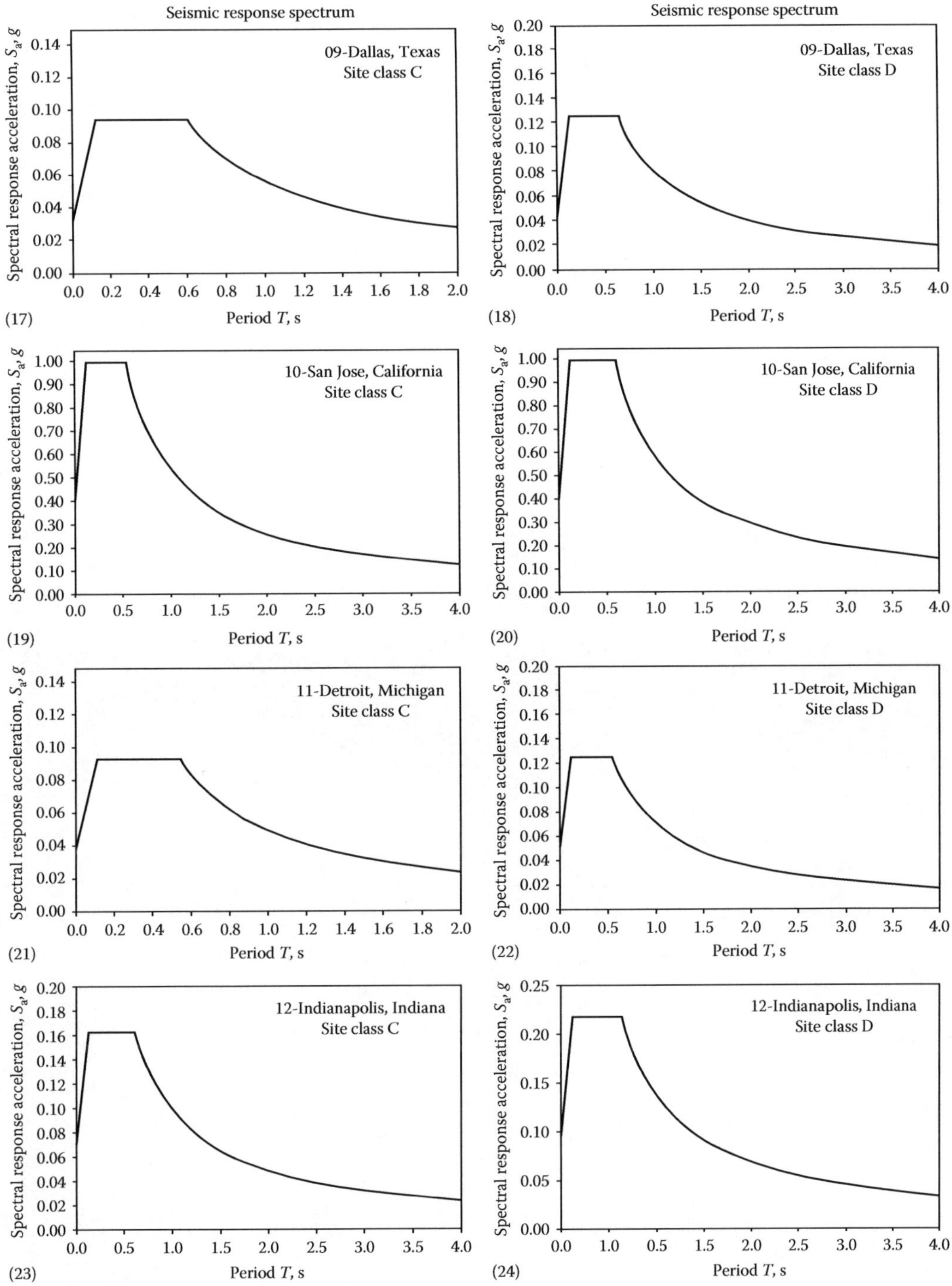

FIGURE 7.76 (*Continued*) Design response spectrum for selected 48 U.S. cities. (*Continued*)

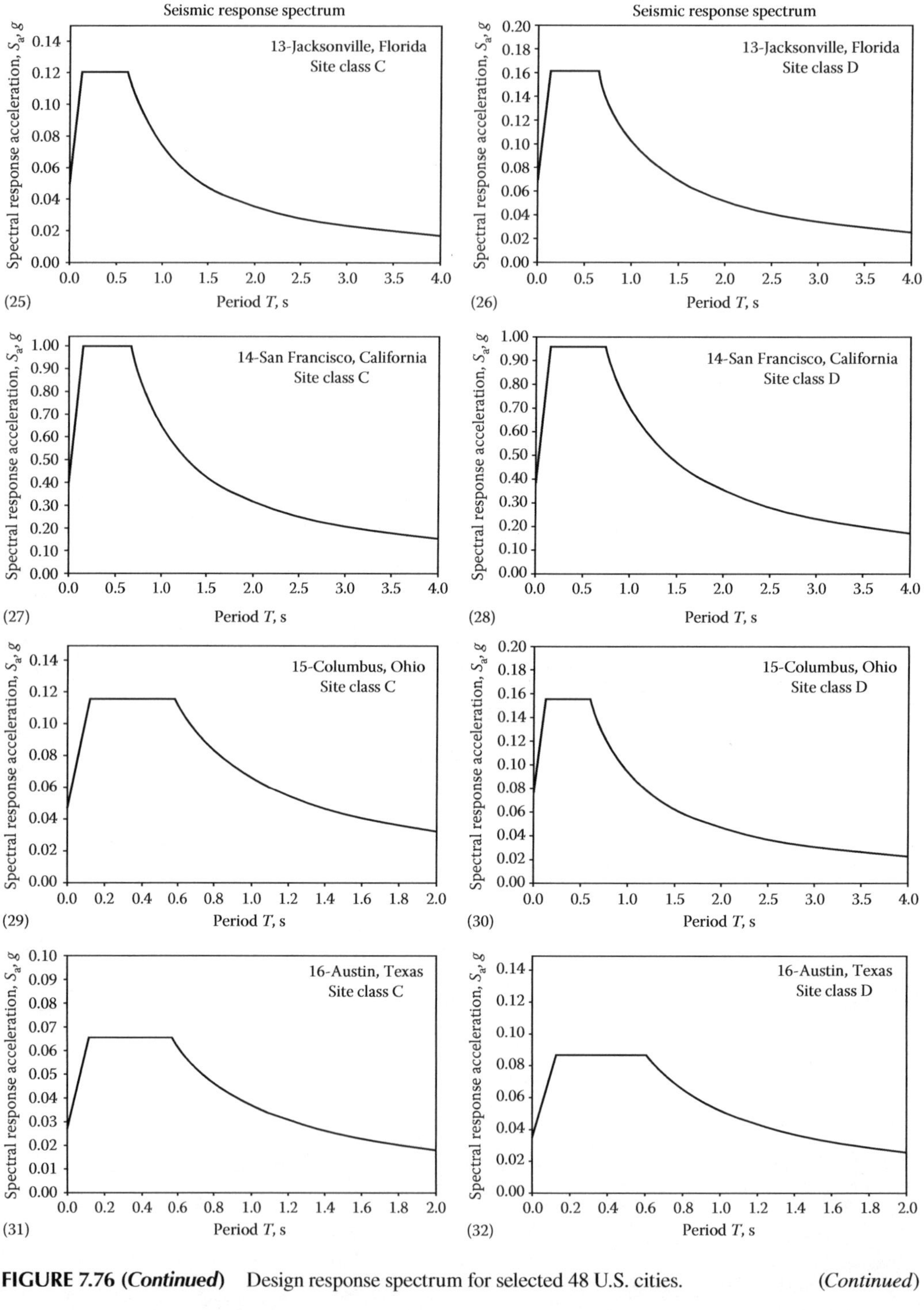

FIGURE 7.76 (*Continued*) Design response spectrum for selected 48 U.S. cities. (*Continued*)

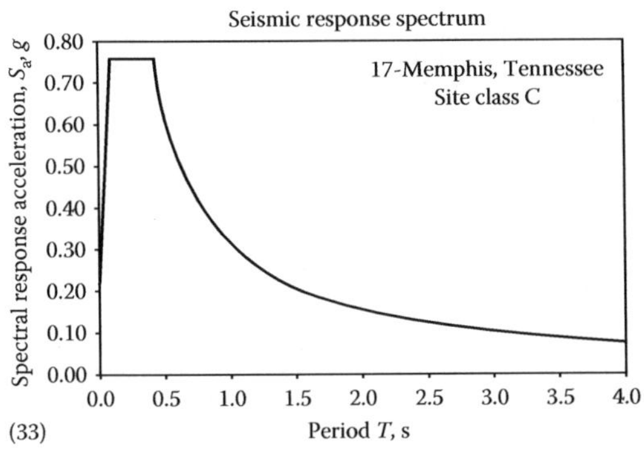

(33)

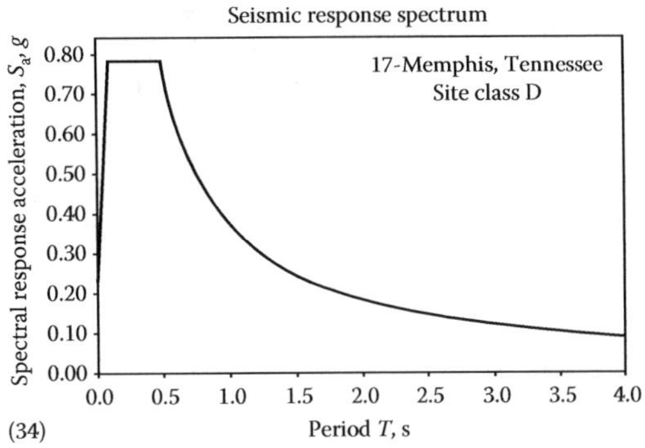

(34)

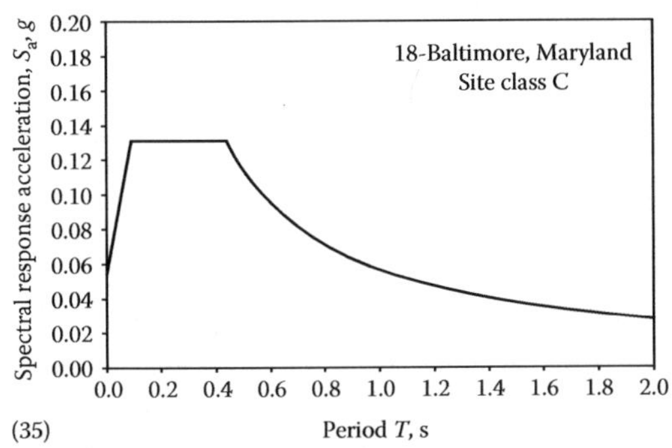

(35)

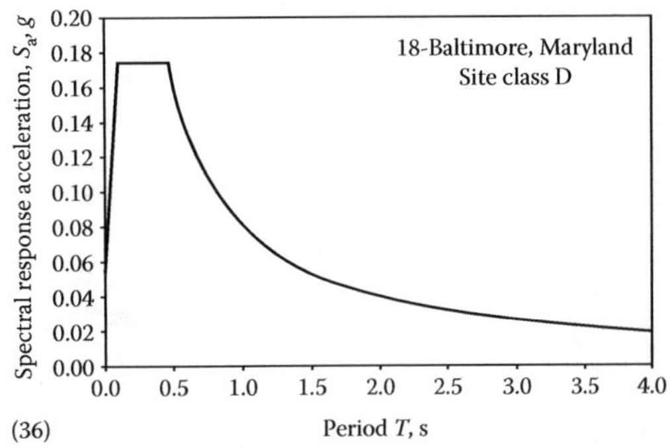

(36)

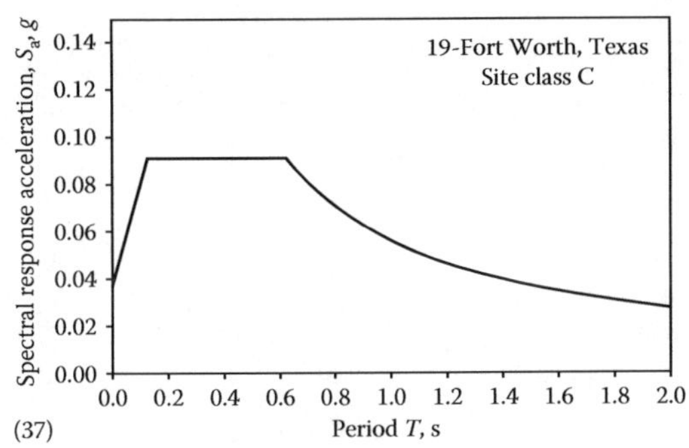

(37)

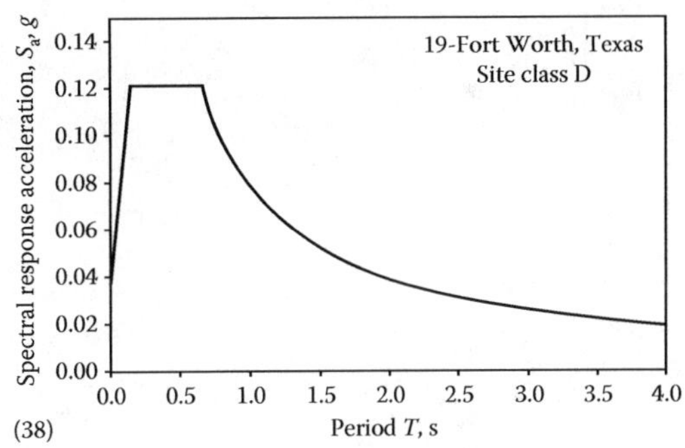

(38)

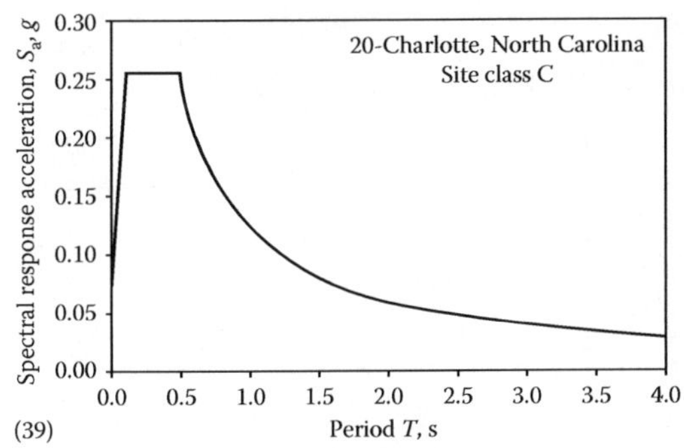

(39)

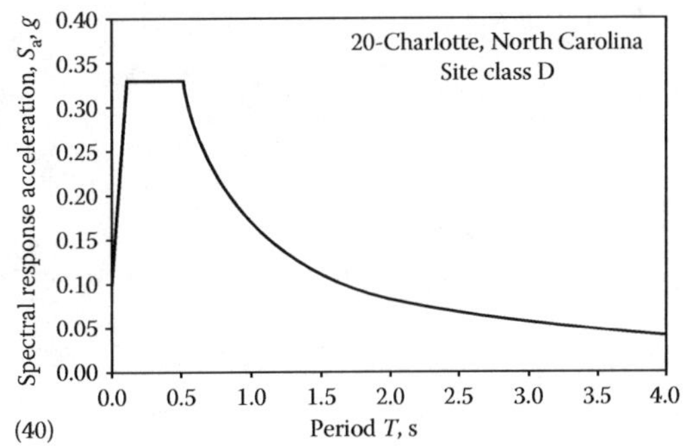

(40)

FIGURE 7.76 (*Continued*) Design response spectrum for selected 48 U.S. cities. (*Continued*)

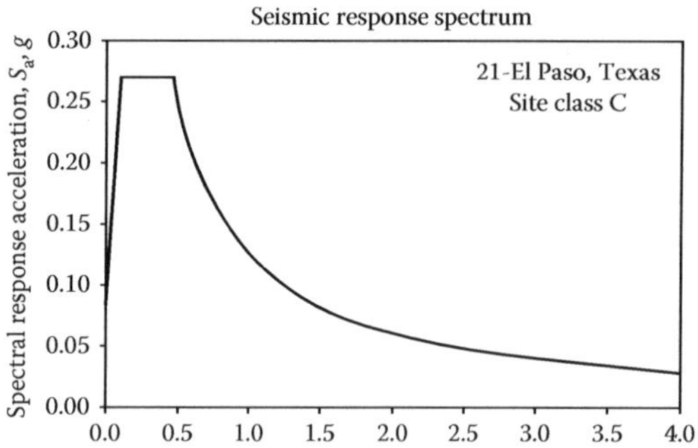

(41)

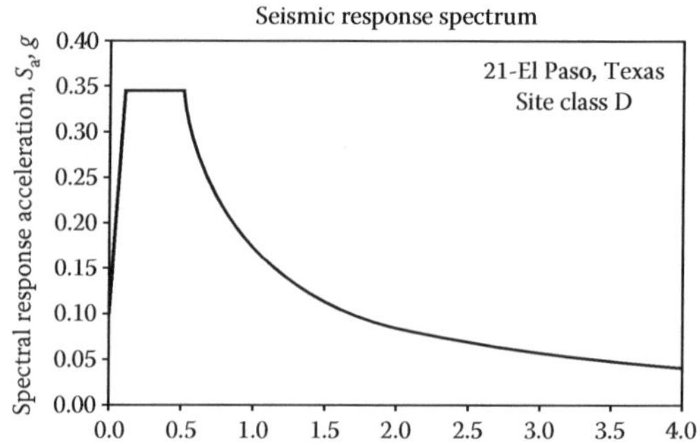

(42)

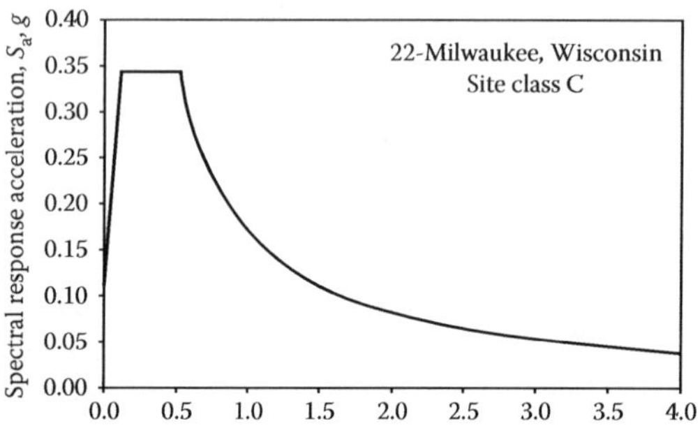

(43)

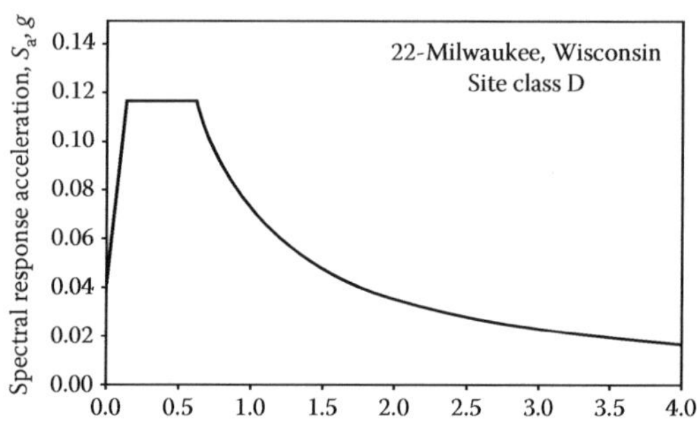

(44)

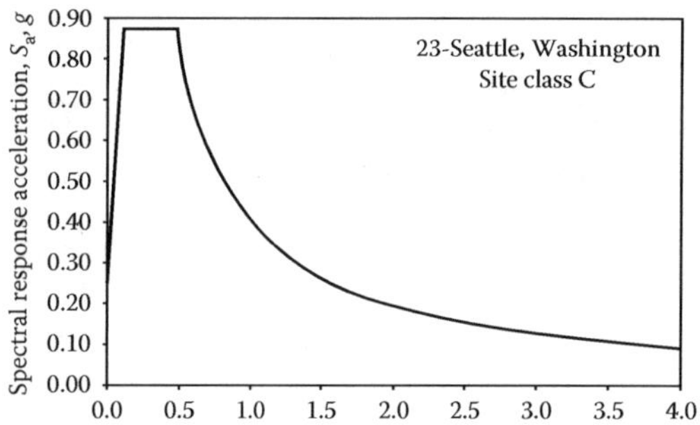

(45)

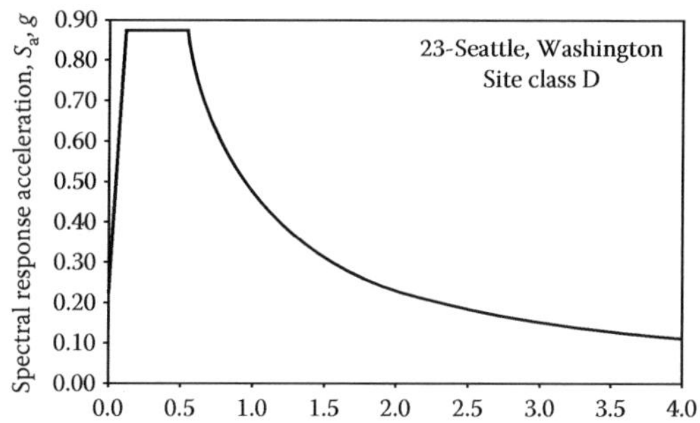

(46)

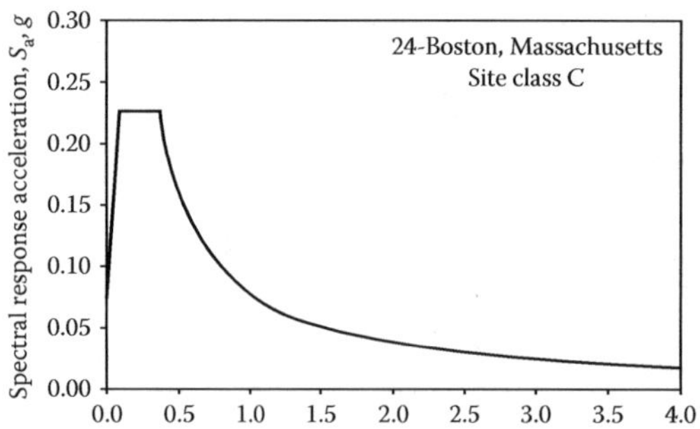

(47)

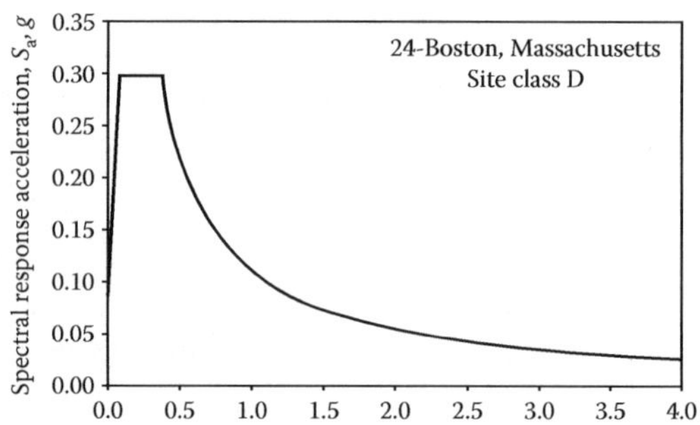

(48)

FIGURE 7.76 (*Continued*) Design response spectrum for selected 48 U.S. cities. (*Continued*)

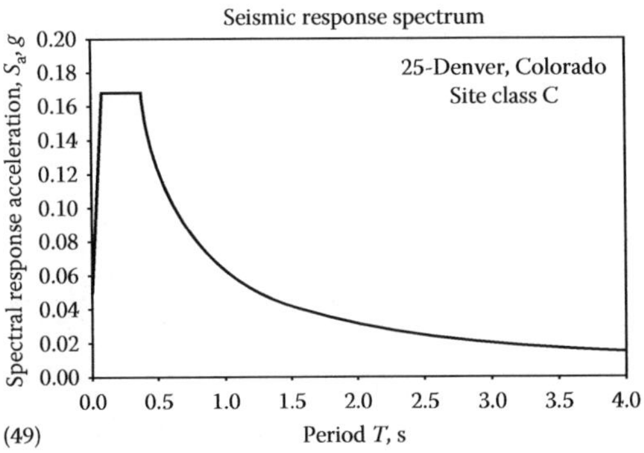

(49)

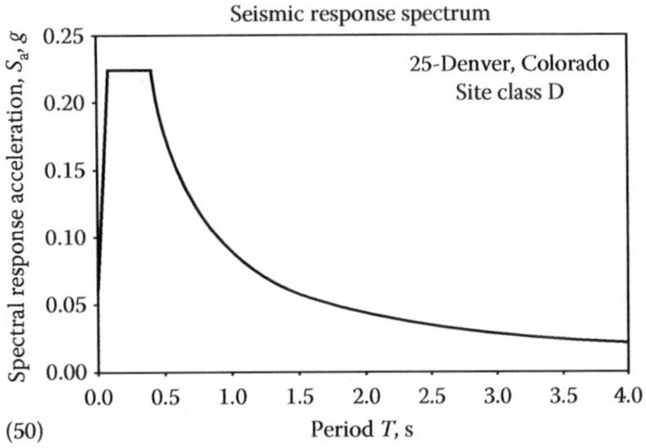

(50)

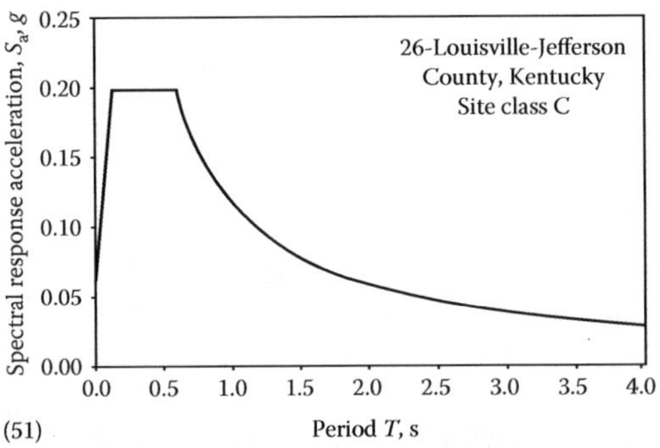

(51)

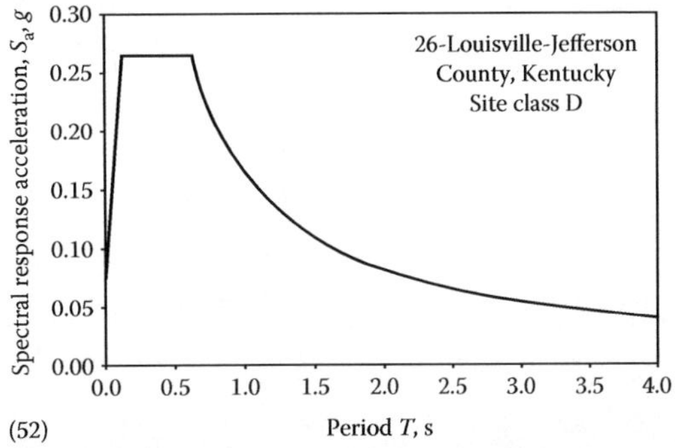

(52)

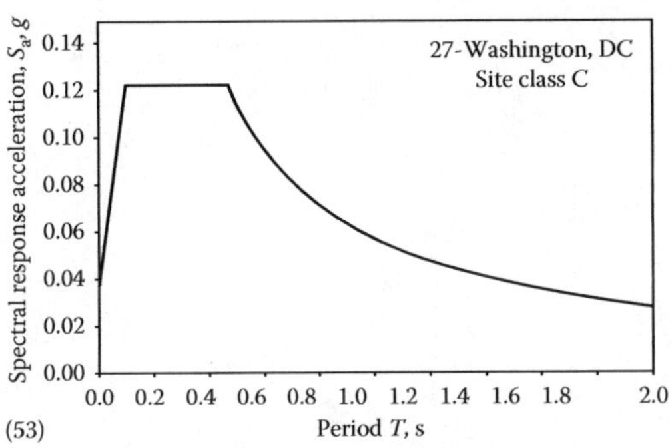

(53)

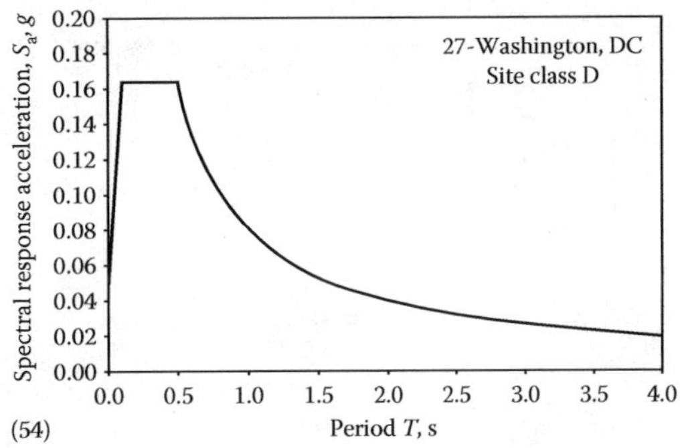

(54)

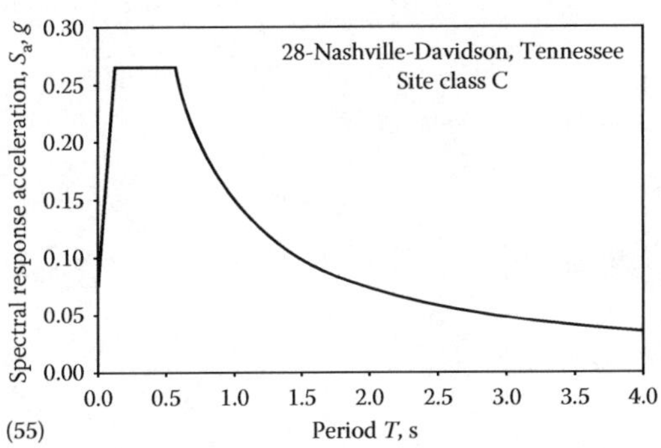

(55)

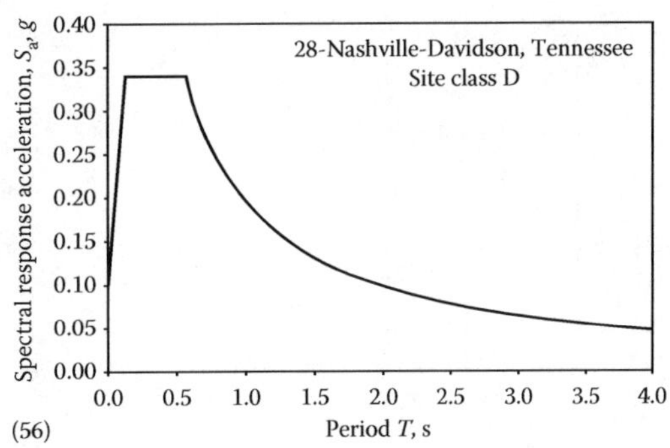

(56)

FIGURE 7.76 (*Continued*) Design response spectrum for selected 48 U.S. cities. (*Continued*)

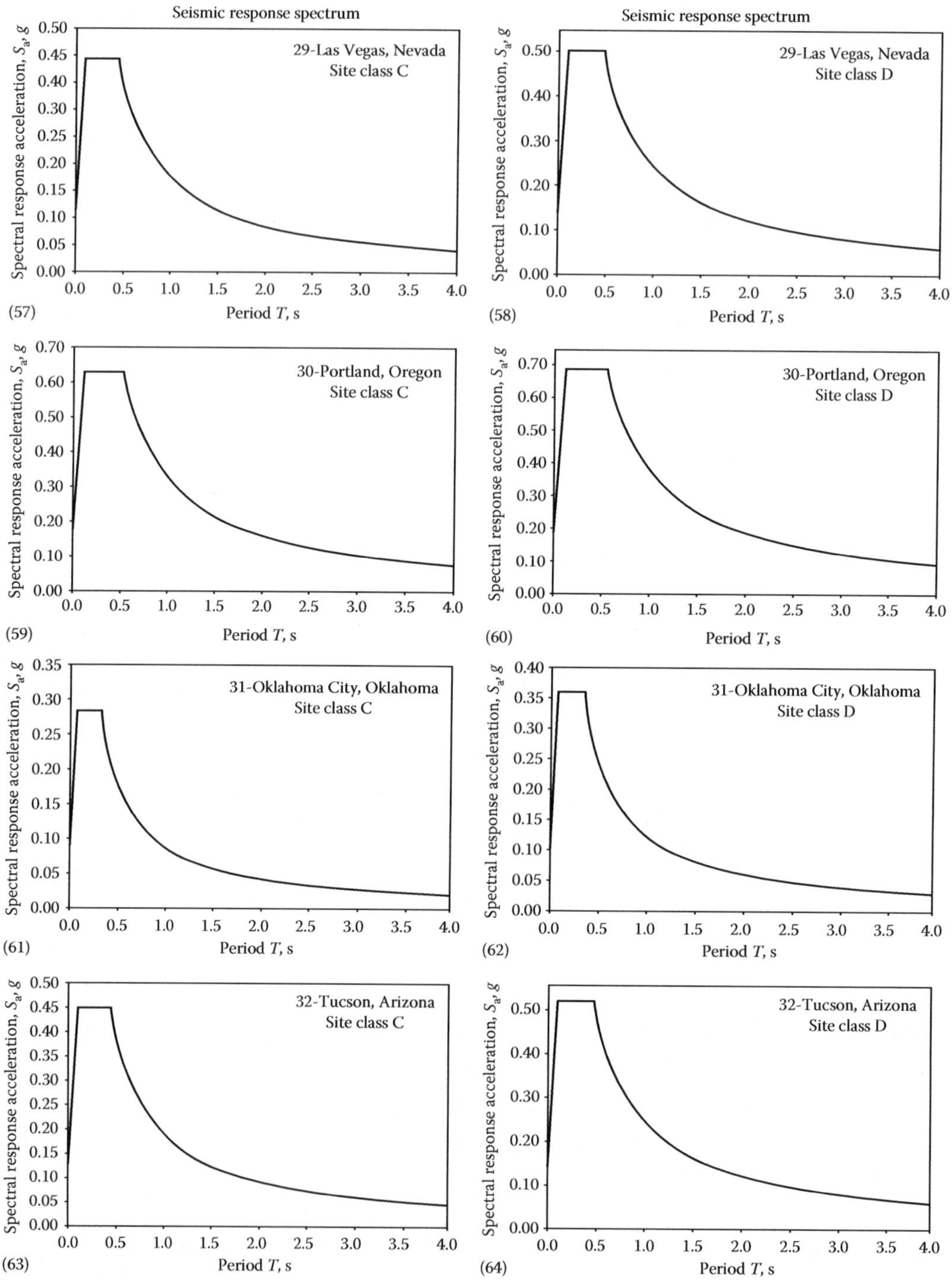

FIGURE 7.76 (*Continued*) Design response spectrum for selected 48 U.S. cities. *(Continued)*

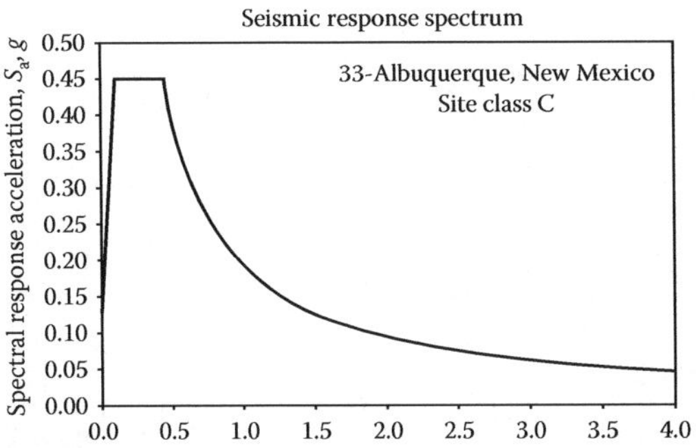

(65)

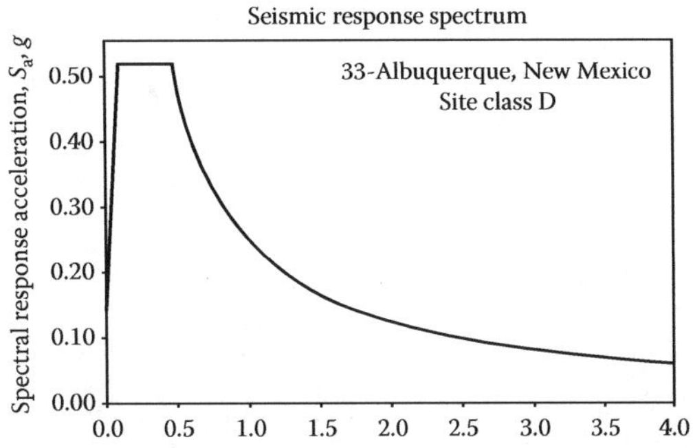

(66)

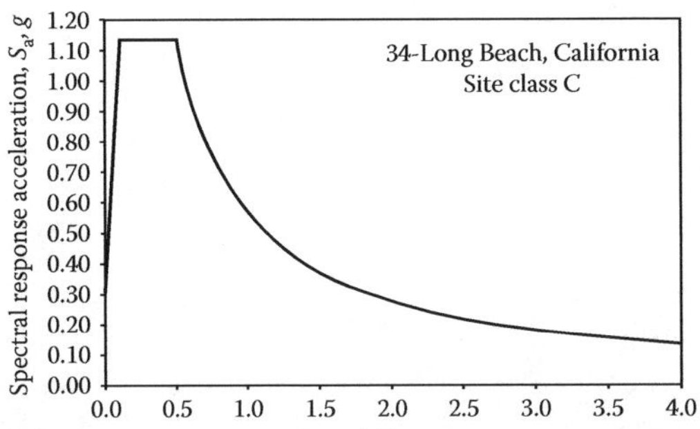

(67)

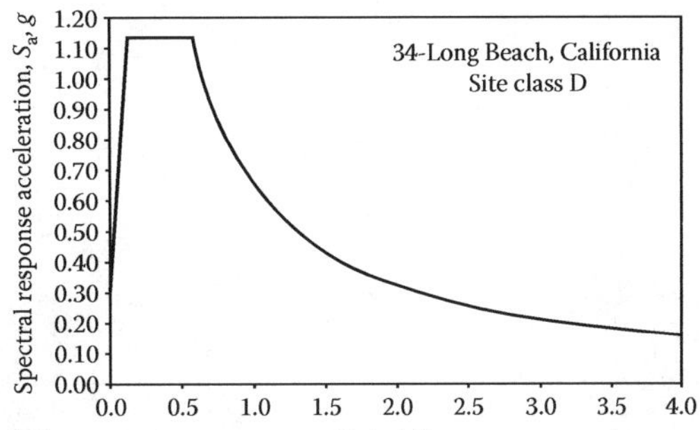

(68)

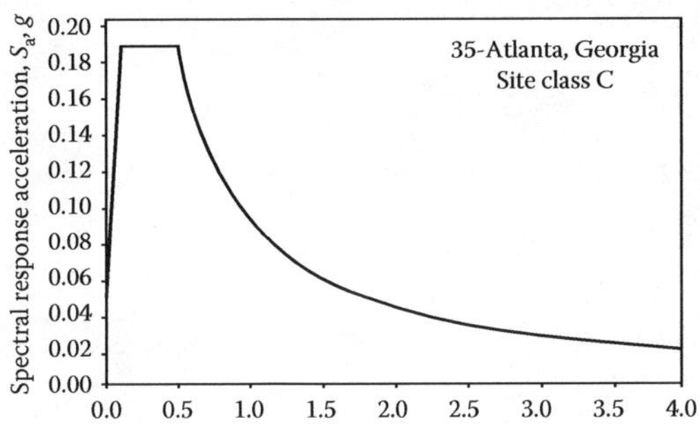

(69)

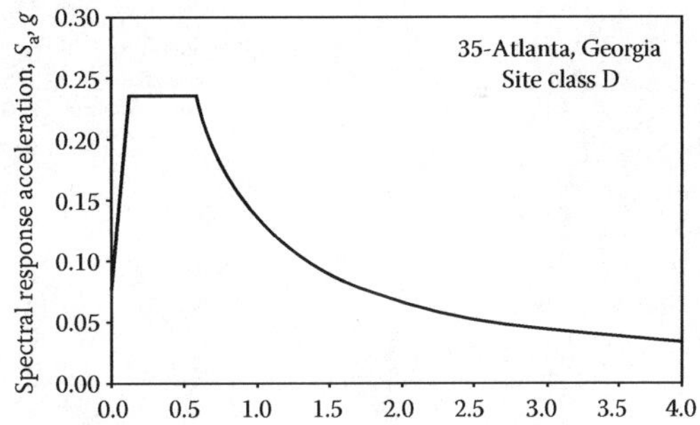

(70)

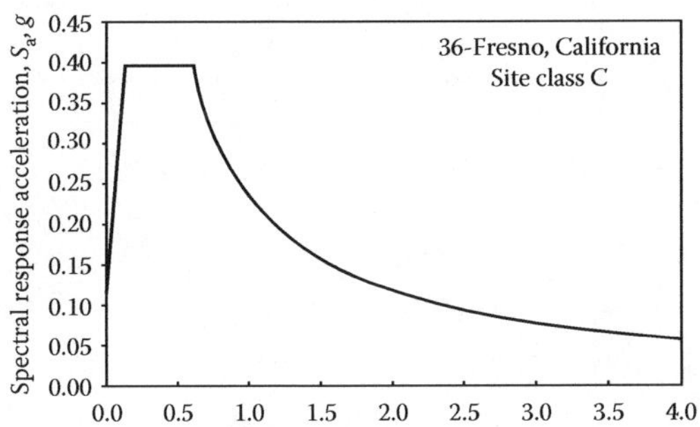

(71)

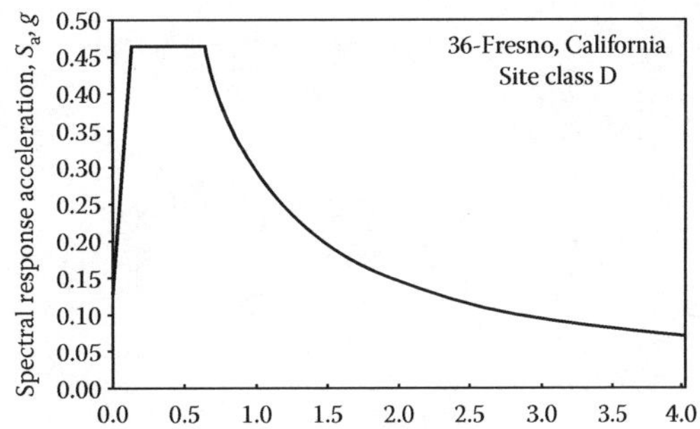

(72)

FIGURE 7.76 (*Continued*) Design response spectrum for selected 48 U.S. cities. *(Continued)*

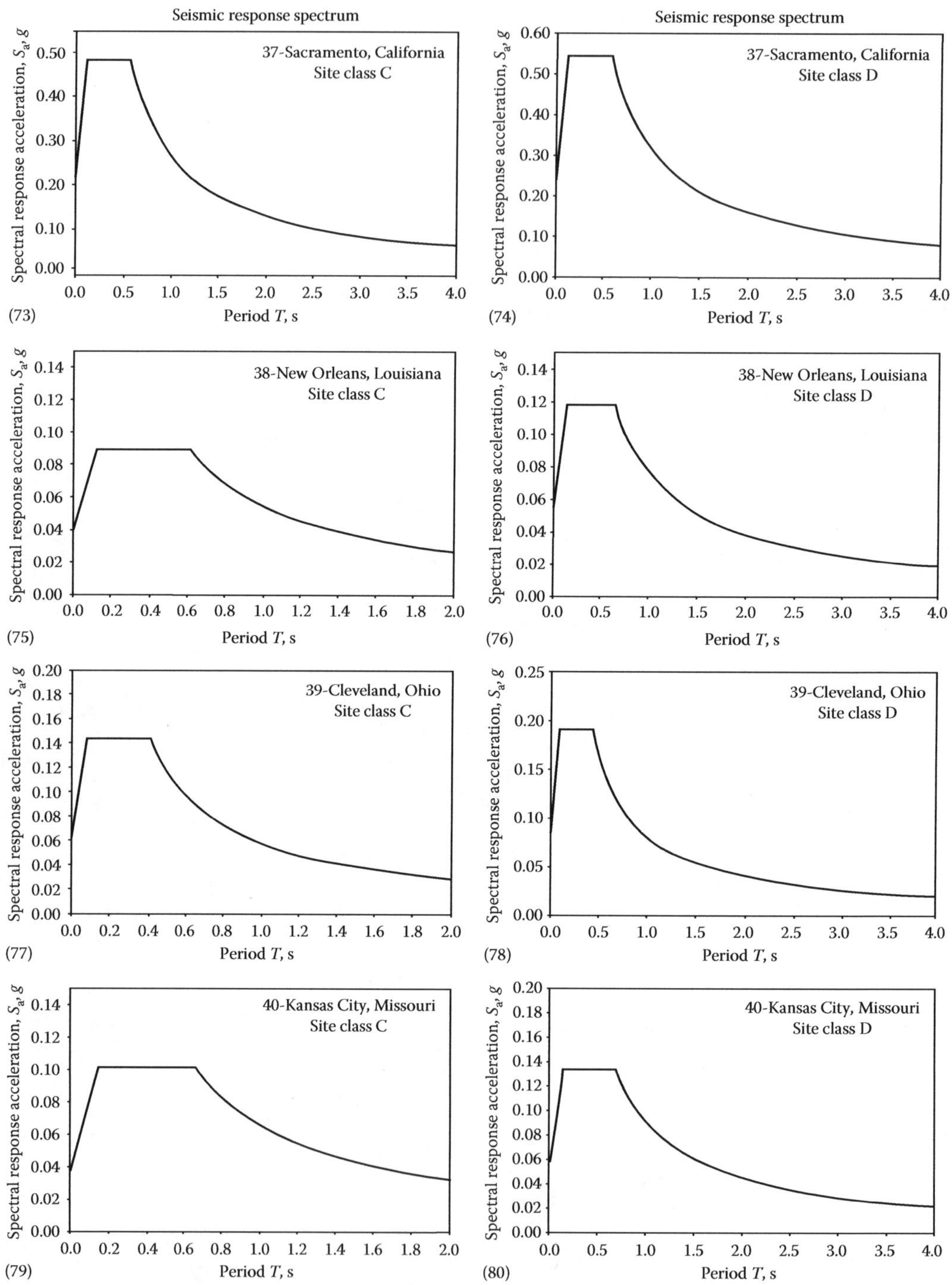

FIGURE 7.76 (*Continued*) Design response spectrum for selected 48 U.S. cities. (*Continued*)

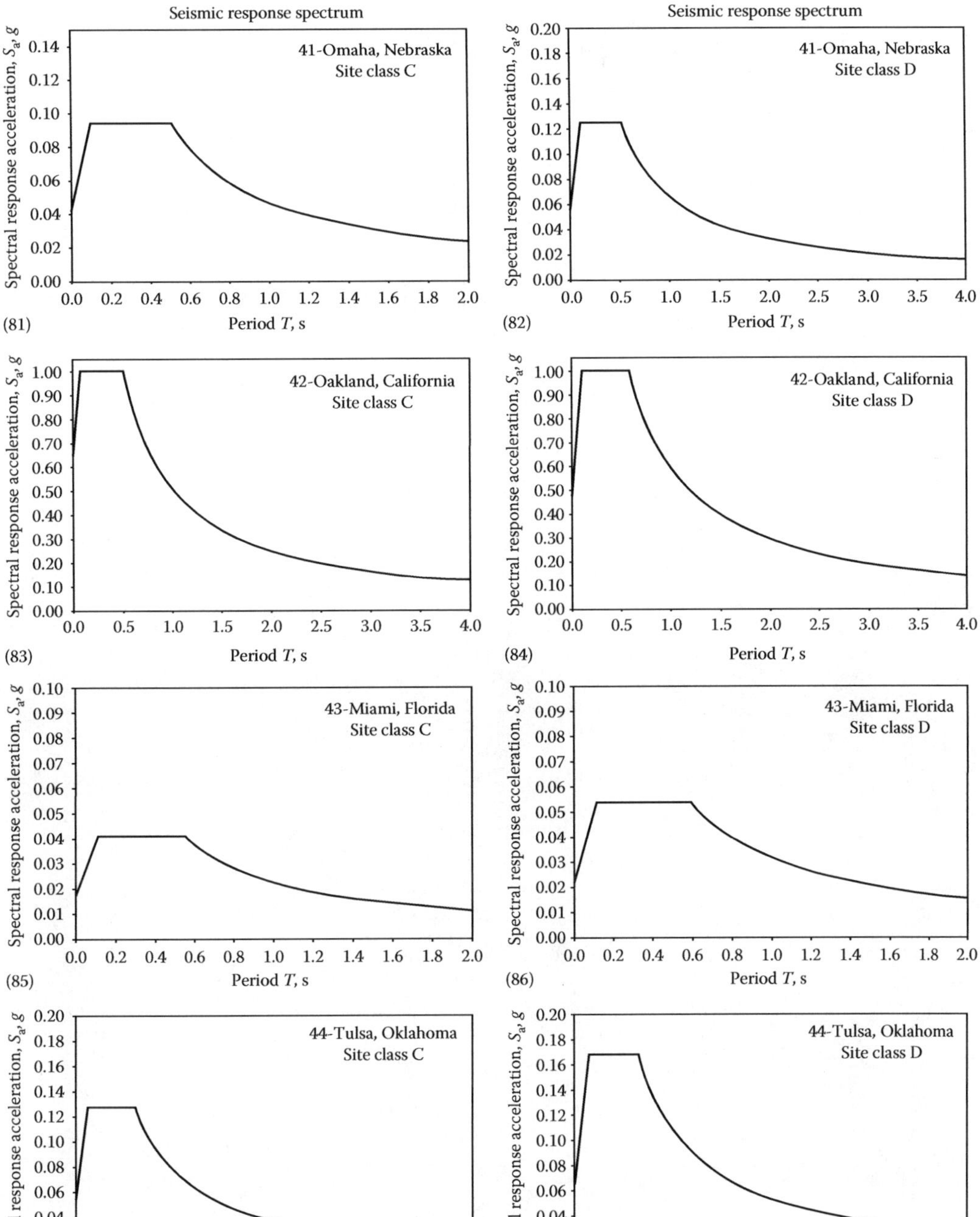

FIGURE 7.76 (*Continued*) Design response spectrum for selected 48 U.S. cities. (*Continued*)

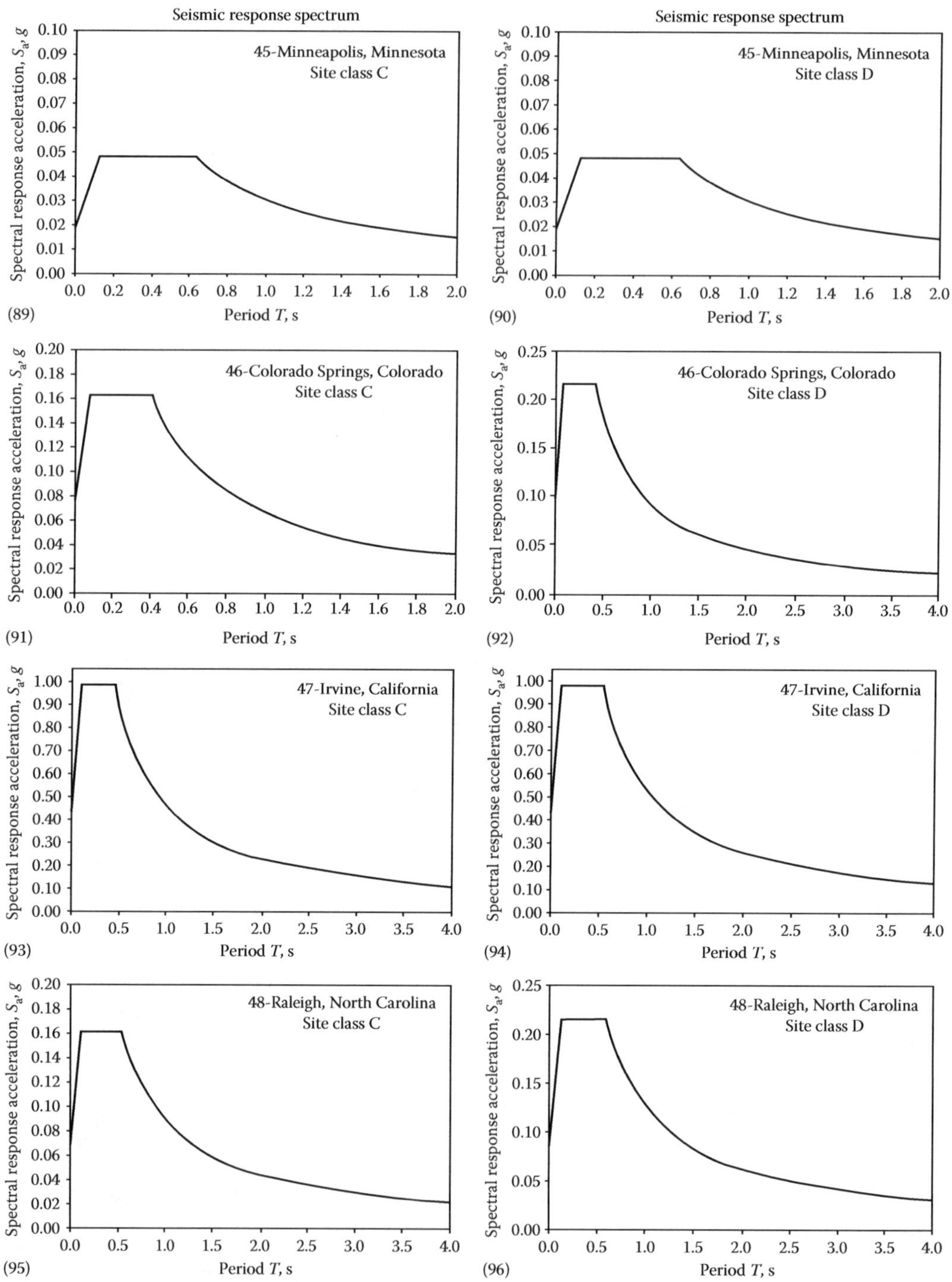

FIGURE 7.76 (*Continued*) Design response spectrum for selected 48 U.S. cities.

7.6 DIFFERENTIAL SHORTENING OF STEEL COLUMNS

Columns in tall buildings experience large axial displacements because they are relatively long and accumulate gravity loads from a large number of floors. A 60-story column of a steel building, for example, may shorten as much as 2–3 in. (50–76 mm) at the top. If such a shortening is not given due consideration, problems may develop in providing level floors and ensuring trouble-free performance of building cladding systems. Proper awareness of this problem is necessary on the part of structural engineer, architect, and curtain wall supplier to avoid unwelcome arguments and loss of time and money.

The maximum shortening of a column occurs at the roof level, reducing to zero at the base. Very little can be done to minimize the physical phenomenon of shortening, but the design team should be aware of the problem particularly at the building exterior so that soft joints are provided if need be, between the building frame and cladding, to prevent axial load being unwittingly transferred into the building's façade. Before the fabrication of cladding, the in-place elevations of structural frame should be verified, and if required, the cladding should be manufactured to fit the existing field condition of the steel frame. The design should provide for sufficient space between the cladding panels to allow for the movement of the structure. Insufficient space may result in bowed cladding components, or, in extreme cases, the cladding panels may even out of the building.

A similar problem occurs when mechanical and plumbing lines are attached rigidly to the structure. Frame shortening may force the pipes to act as structural columns, resulting in their distress. A general remedy is to make sure that nonstructural elements are not brought in to bear vertical loads by separating them from the structural elements.

The axial loads in all columns of a building are seldom the same, giving rise to the problem of so-called differential shortening. The problem is more acute in a composite structure because steel erection columns that are later encased in concrete are typically slender and are therefore subject to large axial loads during construction. Determining the magnitude of axial shortening in a composite system is complicated because many of the variables that contribute to the shortening cannot be predicted with sufficient accuracy. Consider, for example, the lower part of the composite column that is continually undergoing creep. The steel erection column during construction is partly enclosed in concrete at the lower floors, with the bar steel column projecting several floors above the concreted level. Other factors difficult to predict are the gravity load redistribution due to frame action of columns and, if the building is founded on compressible material, the relative settlement of the foundation.

The magnitude of load imbalance between any two columns is continually changing, making an accurate assessment of column shortening rather a challenging task.

Differential rather than the absolute shortening of column is more significant. Relative displacement between columns occurs because of the difference between the axial stresses, P/A ratios of columns. P is the axial load, and A is the area of the column under consideration. If all columns in a building were to have the same P/A ratio under gravity loads, there would be no relative vertical moment between the columns. In typical buildings, however, this condition is seldom present. This is because, in buildings, not all columns are designed of the same axial loads. For example, the design of frame column is governed by the combined gravity and lateral loads while nonframe columns are designed essentially for gravity loads only. This results in a rather large difference in the P/A ratios between the exterior and interior column. The differential column shortening between perimeter and interior columns may result in sloping floors leading to unwelcome problems in setting partitions, doors, and ceiling, in floors that slope excessively.

Consider, for example, a tubular system with closely spaced exterior columns and widely spaced interior columns. Because of their large tributary areas, the interior columns are more than likely to have large P/A ratios. The exterior columns, on the other hand, usually have a small P/A ratio for two reasons. First, their tributary areas are small because of their close spacing. Second, they are sized for stiffness to limit lateral displacements, resulting in areas much in excess of these required from strength consideration alone. Because of this imbalance in the gravity stress level, these two groups of columns undergo different axial shortenings; the interior columns shorten much more than the exterior columns.

A reversed condition occurs in buildings with interior-braced core columns and widely spaced exterior columns; the exterior columns experience more axial shortening than the interior columns. The behavior of columns in buildings with other types of structural systems, such as interacting core and exterior frames, tends to be somewhat in between these two limiting cases.

7.6.1 Simplified Method of Calculating Δ_z, Axial Shortening of Columns

In a steel building, typically, the cross-sectional area of a column increases in two-story increments from a minimum at the top to a maximum at the base. The incremental steps in column areas are due to the finite choice of available column shapes. Similarly, the axial load on a column increases at each floor in a stepwise manner up the building height. In tall buildings, the significance of these incremental steps diminishes rapidly, allowing us to make the following assumptions that can be used to derive a simplified formulation for axial shortening of columns. The first assumption relates to the variation of gravity loads that may be assumed to increase linearly from top to the bottom. The second is similar to the first but applies to the variation of column areas. However, a linear variation using the actual column area at the bottom appears to underestimate the actual shortening of columns. A slight modification in which the column area at the bottom is taken equal to 0.9 times the actual area ($0.9 \times A_b$) appears to work well in predicting axial shortening.

7.6.2 Derivation of Simplified Expression for A_z

Integral calculus in conjunction with the simplified assumptions stated earlier is used to derive the following equation for Δ_z:

$$\Delta_z = P_b/E[-1/a\ \text{In}(l = az/A_b)] - \beta/E[-1/a^2\{az + A_b\ \text{In}(1 - az/A_b)\}]$$

where

l is the height of the building
Δ_z is the axial shortening at height x (also denoted as z), above the foundation level
A_t is the column area at top
A_b is the modified column area at bottom equal to 0.9 × actual area of column at bottom $= 0.9 \times A_b$
A_x is the area of column at height x (also denoted as z) above foundation level
a is the rate of change of area of column
P_t is the axial load at top
P_b is the axial load at bottom
P_x is the axial load at height x above foundation
β is the rate of change of axial load
E is the modulus of elasticity of steel

The reader is referred to the publication *Wind and Earthquake Resistant Buildings, Structural Analysis and Design* for the derivation of Δ_z.

Definitions of the variable used in the equation are shown schematically in Figure 7.77A.

Example 1

Given (see Figure 7.77B):

Height of building: $L = 682$ ft = 8184 in. (207.8 m)
Modulus of elasticity: $E = 29{,}000$ ksi (200 × 103 m)
Axial of load at top: $P_t = 53$ kips (237.5 kN)
Area of column at top: $A_t = 12.48$ in.^{2} (8052 mm^2)
Axial load at base: $P_b = 2770$ kips (12.32 × 103 kN)
Actual column area at base: $A_b = 147$ in.^{2} (94.84×10^3 mm^2)
Reduced column area at base: $A_b = 0.9 \times 147 = 133.3$ in.^{2} (86.0×10^3 mm^2)

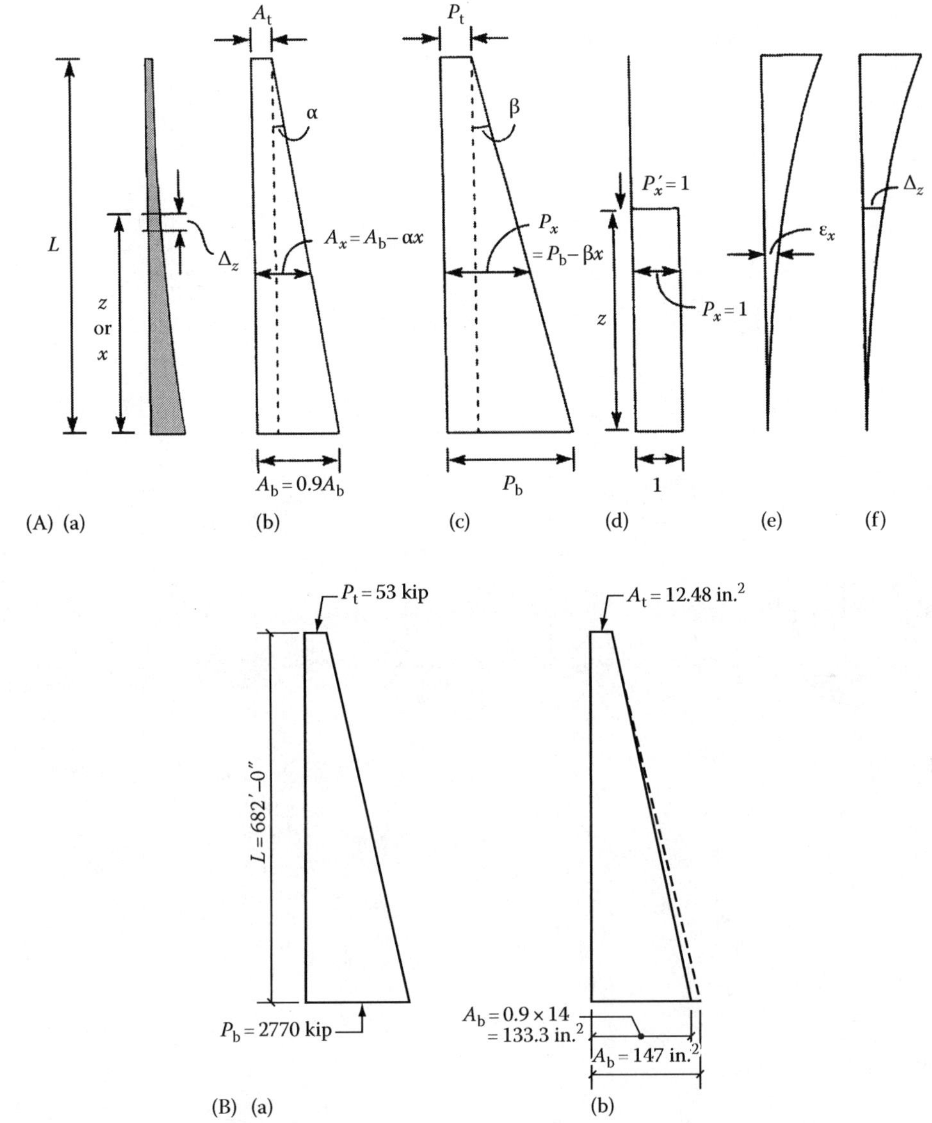

FIGURE 7.77 (A) Axial shortening of columns, closed-form solution: (a) axial shortening Δ_z; (b) variation of column areas; (c) axial load variation; (d) unit load at height z; (e) axial strain variation; (f) axial shortening. (B) Example 1: (a) axial load variation; (b) actual and assumed columns cross-sectional areas. *(Continued)*

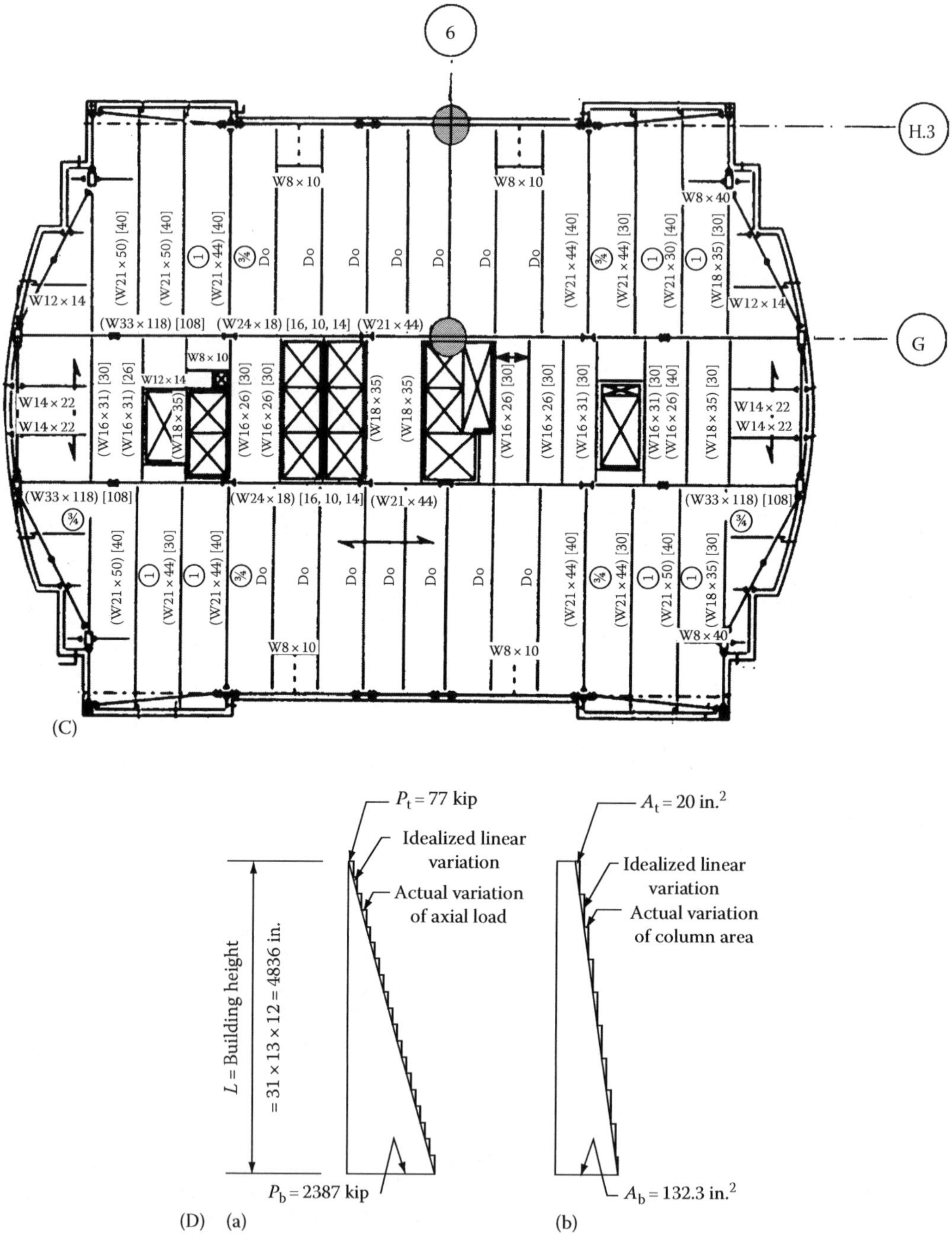

FIGURE 7.77 (*Continued*) (C) Schematic framing plan. (D) Example 2, interior column G-6: (a) axial load variation; (b) variation of column areas. (*Continued*)

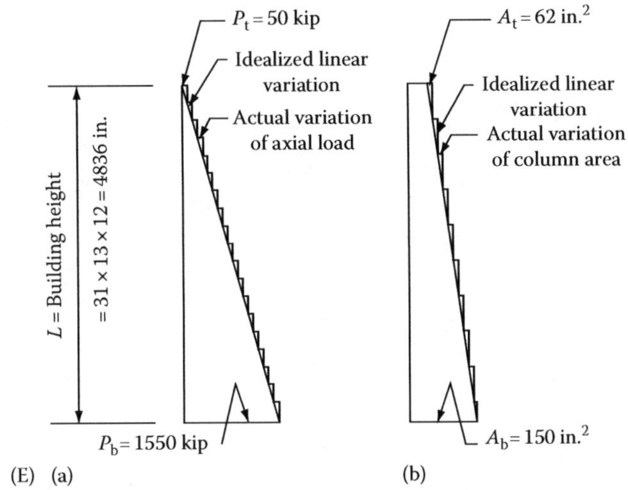

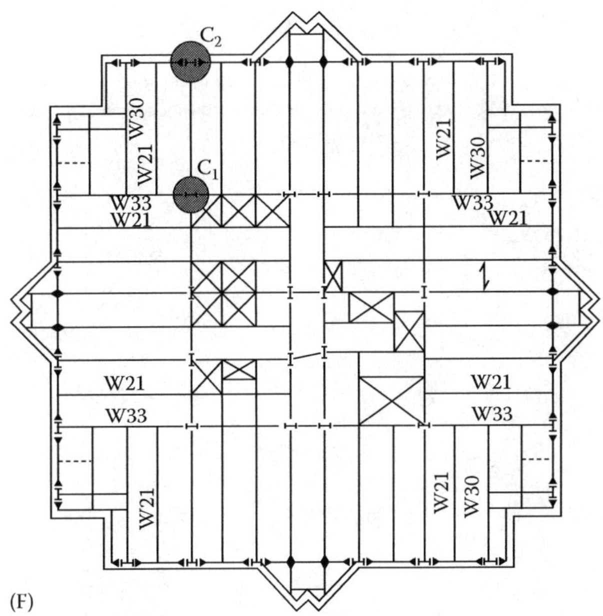

FIGURE 7.77 (*Continued*) (E) Example 2, exterior column H.3-6: (a) axial load variation; (b) variation of column areas. (F) Framing plan. Column C, designed for gravity loads only, shortens more than C_2, designed for both gravity and lateral loads. Compensating for relative elevation difference between columns is of importance in tall buildings. (*Continued*)

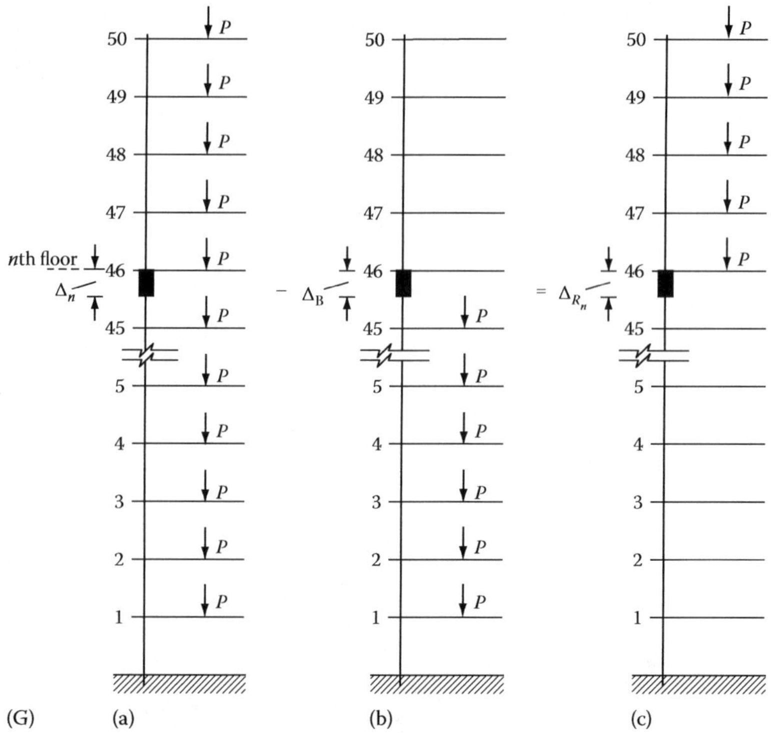

FIGURE 7.77 (*Continued*) (G) Interpretation of column overlength: (a) Δ_n = column shortening at *n*th level due to loads on the entire height of column; (b) Δ_B = column shortening at *n*th level due to loads imposed at and below nth level; (c) column shortening yet to occur due to loads above *n*th level.

Required: Axial shortening of column at top

Solution: Since column shortening is calculated at top, $z = L$

$$A = (A_b - A_t)/L = (133 - 12.48)/8184 = 0.01476 \text{ in.}^2/\text{in.}$$

$$B = (P_b - P_t)/L = (2770 - 53)/8184 = 0.322 \text{ kip/in.}$$

$$\ln(1 - aL/A_b) = \ln(1 - (0.0147 \times 8184)/133.3)$$

$$= \ln(0.09362)$$

$$= -2.36847$$

$$\Delta L \text{ at top} = 2770/29{,}000\{-1/0.01476 \times (-2.36847)\} - 0.332/29{,}000$$

$$X\{-4590.15(0.01476 \times 8184 + 133.3 \times -2.36847)\}$$

$$= 15.327 - 10.2$$

$$= 5.127 \text{ in.}$$

Similarly, the axial shortening is calculated at various heights by substituting appropriate values for *z*.

Example 2

Given: A steel building 403 ft tall with 31 framed levels, including the roof. Tributary area of gravity load calculations for the exterior H.3 = 472 ft^2 per floor, and the interior column G.6 is 810 ft^2. See Figure 7.77C for a schematic framing plan and Table 7.13 for an abbreviated column schedule. Typical loads for estimating axial shortening columns are as follows:

Interior Col G.6 3¼ lt. wt. on 3 in. deck = 50 psf
Partitions = 10 psf
Allowance for floor finishes
Ceiling, mech., etc. = 10 psf
Structural frame = 10 psf
Live load (reduced) = 15 psf

$$95 \text{ psf} \times 810/1000 = 76.9 \text{ kips/floor}$$

Use 77 kips/floor
See Figure 7.77D for load distribution up the building height
Exterior Col. H.3-6 3¼ lt. wt. on 3 in. deck = 50 psf
Partitions = 10 psf
Allowance for floor finishes
Ceiling, mech., etc. = 10 psf
Structural frame = 15 psf
Live load (reduced) = 15 psf
Exterior cladding = 5 psf

$$105 \text{ psf} \times 472/1000 = 49.56 \text{ kips/floor}$$

Use 50 kips/floor
See Figure 7.77E for load distribution

Required: Compute axial shortening of column H3-6 and G-6 at the roof due to gravity loads using the closed equation from the equation given earlier. Provide column length corrections at levels 8, 16, 24, and the roof.

Solution:
The variation of axial loads and cross-sectional areas for the two columns are shown in Figure 7.77D and E.

Observe that A_b is the actual area of the column at the foundation level multiplied by a factor of 0.9.

Thus, A_b for the interior column = 0.9 × 147 = 132.3 in.^{2}

A_b for the exterior column = 0.8 × 170 = 153 in.^{2}

The loads P_b for the exterior and interior columns at foundation level are as follows:

$$P_b = 31 \times 50 = 1550 \text{ kips (exterior column)}$$

$$P_b = 31 \times 77 = 2387 \text{ kips (interior column)}$$

Load P_t = 50 kips for the exterior and 77 kips for the interior column
Column-length-shortening computations for column G.6 (interior column)

$$L = 4836 \text{ in.}$$

$$E = 29{,}000 \text{ ksi}$$

$$P_t = 77 \text{ kips}$$

TABLE 7.13
Axial Shortening Computations

Level	$\sum P_i$ Accumulated Load, kip	Column Section	L_k Story Height, in.	Δ*n* Column Shortening, in.	Column Length Correction Each Level, in.	Lumped Column Length Correction, in.	Column Shortening, in.
50	53	W14x43	156	5.14	0.023	0.73	5.11
49	106	43	210	5.12	0.061		5.08
48	159	53	168	5.05	0.051		5.02
47	212	53	156	5.00	0.073		4.95
46	265	68	156	4.93	0.071		4.89
45	318	68	156	4.86	0.086		4.82
44	371	84	156	4.77	0.081		4.75
43	424	84	156	4.69	0.092		4.67
42	477	95	156	4.60	0.092		4.59
41	530	95	156	4.51	0.102		4.50
40	583	111	156	4.41	0.09	1.02	4.42
39	636	111	156	4.32	0.105		4.33
38	689	127	156	4.21	0.09		4.24
37	742	127	156	4.12	0.107		4.15
36	795	142	156	4.01	0.103		4.04
35	848	142	156	3.91	0.109		3.96
34	901	167	156	3.80	0.09		3.86
33	954	167	156	3.71	0.105		3.76
32	1007	176	156	3.62	0.105		3.66
31	1060	176	156	3.50	0.110		3.56
30	1113	202	156	3.39	0.101	1.07	3.46
29	1166	202	156	3.29	0.106		3.36
28	1219	211	156	3.19	0.106		3.26
27	1272	211	156	3.08	0.110		3.16
26	1325	228	156	2.97	0.106		3.12
25	1378	228	156	2.86	0.111		2.95
24	1431	246	156	2.75	0.107		2.85
23	1484	246	156	2.65	0.111		2.74
22	1537	264	156	2.53	0.107		2.64
21	1590	264	156	2.43	0.110		2.53
20	1643	287	156	2.32	0.104	1.06	2.42
19	1696	287	156	2.20	0.108		2.32
18	1749	314	156	2.10	0.101		2.21
17	1802	314	156	2.00	0.105		2.10
16	1855	314	156	1.90	0.108		1.99
15	1908	314	156	1.79	0.111		1.88
14	1961	342	156	1.68	0.104		1.78
13	2014	342	156	1.58	0.107		1.67
12	2067	370	156	1.47	0.101		1.56
11	2120	370	156	1.37	0.104		1.45
10	2173	370	156	1.26	0.107	0.53	1.34
9	2226	370	156	1.16	0.109		1.23
8	2279	398	156	1.05	0.104		1.12

(Continued)

TABLE 7.13 (*Continued*)
Axial Shortening Computations

Level	$\sum P_i$ Accumulated Load, kip	Column Section	L_k Story Height, in.	Δn Column Shortening, in.	Column Length Correction Each Level, in.	Lumped Column Length Correction, in.	Column Shortening, in.
7	2332	398	156	0.94	0.107		1.01
6	2385	398	156	0.84	0.11		0.89
5	2488	398	210	0.73	0.11		0.78
4	2491	426	168	0.62	0.11		0.67
3	2544	426	156	0.51	0.11		0.56
2	2597	500	156	0.40	0.09		0.45
Mezzanine	2650	500	240	0.31	0.15		0.17
1	2770	W14×500	240	0.16	0.16		0.17

NS = Number of stories = 50.

$$P_b = 2387 \text{ kips}$$

$$A_t = 20 \text{ in.}^2$$

$$A_b = 0.9 \times 147 = 132.3 \text{ in.}^2$$

$$a = A_b - A_t/L = 132.3 - 20/4836 = 0.02322 \text{ in.}^2/\text{in.}$$

$$\beta = P_b - P_t/L = 2387 - 77/4836 = 0.4777 \text{ kip/in.}$$

$$L_n(1 - aL/A_b) = \ln(1 - 0.02322/132.2 \times 4836)$$

$$= \ln(0.151233)$$

$$= -1.8889$$

$$\Delta L = 2387/29{,}000\{-1/0.2333 \times -1.8889\} - 0.4777/29{,}000\{1 - 4854.7(0.02322 \times 4836) + 132.2(-1.8889)\}$$

$$= 6.6958 - 4.1869$$

$$= 2.50 \text{ in.}$$

Column-length-shortening calculations for column H.3-6 (exterior column)

$$L = 31 \times 13 \times 12 = 4836 \text{ in.}$$

$$E = 29{,}000 \text{ ksi}$$

$$P_t = 50 \text{ kips}$$

$$P_b = 1550 \text{ kips}$$

$$A_t = 62 \text{ in.}^2$$

$$A_b = 0.9 \times 170 = 153 \text{ in.}^2$$

$$a = A_b - A_t/L = 153 - 62/4836 = 0.018817 \text{ in.}^2/\text{in.}$$

$$\beta = P_b - P_t/L = 1550 - 50/4836 = 0.3102 \text{ kip/in.}$$

$$L_n(1 - aL/A_b) = \ln(1 - 0.018817 \times 4836/153)$$

$$= \ln(0.4052)$$

$$= -0.90328$$

$$\Delta L = 1550/29{,}000\{-1/0.018817 \times -0.90328\}$$

$$= 0.3102/29{,}000\{-2824.2(0.018817 \times 4836) + 153(-0.90328)\}$$

$$= 2.56 - 1.42 = 1.134 \text{ in.}$$

7.6.3 Column Length Corrections, Δ*c*

After determining the axial shortening of building column, the next step is to assign a column length correction Δ*c* for each column. The objective is to attain as level a floor as practical. Δ*c* is thus the difference between the theoretical height of a given column and its actual height after is has shortened. The magnitude of correction Δ*c* in a tall building of, say, 60 stories is rather small, perhaps $^1/_8$ in. (3.17 mm) per floor, at the most. Therefore, instead of specifying these small corrections at each level, in practice, it is usual to specify lumped corrections at a few selected floors.

For example, in lieu of $^1/_8$ in. correction at each level, one would lump the correction, say, at every eighth floor. Thus, Δ*c* would be equal to $^1/_8 \times 8 = 1$ in. (25.4 mm); to emphasize this concept, let us consider Table 7.14 that shows axial shortening computations for an assumed column of a 60-story building.

The assumed column sections, lengths, and variation of axial loads are given in the table. The last column shows the lumped corrections at levels 2, 10, 20, 30, 40, and the roof. Basically, these corrections represent the lengths to be added to the theoretical lengths of the column to achieve the specified height after the column has shortened due to gravity loads. For example, Δ*c* = 1¼ in. (31.75 mm) at the 10th level means that the actual fabricated length of column from its base on the 10th level should be made 1¼ in. longer than the theoretical length. This overlength could be achieved by increasing the length of column in each tier by ¼ in. (6.35 mm) (10 stories equal 5 tiers; therefore, ¼ in. times 5 gives 1¼ in.). However, the fabricator may elect to increase the column length in each story by 1/8 in. (3.2 mm) instead of ¼ in. per tier. This and other similar options are, of course, permissible because the end result of achieving a desired Δ*c* will be the same.

The value of Δ*c* = 2 in. (50.8 mm) at the 20th floor means the overlength of columns between levels 1 and 20 should be 2 in. However, the overlength of 1¼ in. (31.75 mm) up to the 10th level has already been achieved by specifying Δ*c* = 1/14 in. at the 10th level. Therefore, the increment between the 10th and 20th levels should be 2 in. less 1¼ in. = ¾ in (19.0 mm).

Let us apply this procedure to the two columns G.6 and H.3-6 of Example 2. The calculated axial shortening for these two columns at top are 2.50 and 1.134 in. A practical method of specifying corrections Δ*c* for these columns is shown in Table 7.14.

7.6.4 Column Shortening Verification during Construction

This concept is best explained with reference to Figure 7.77F that shows a framing plan for a hypothetical building, say, some 48 stories tall. Identified therein are two columns: C_1, an interior column with a large tributary area; and C_2, an exterior column of framed tube with a relatively small tributary area. Under gravity loads, C_1 would shorten more than C_2 because (1) C_1,

TABLE 7.14
Column Length Correction

	Interior Column		Exterior Column	
Level	Column Shortening	Correction to Scheduled Length	Column Shortening	Correction to Scheduled Length
Roof	2.50 in.	4@ 3/16 in. = 0.75 in.	1.13 in.	2@ 1/8 in. = 0.25 in.
24	1.80 in.	4@ 1/8 in. = 0.50 in.	0.75 in.	2@ 1/8 in. = 0.25 in.
16	1.25 in.	4@ 3/16 in. = 0.75 in.	0.50 in.	2@ 3/16 in. = 0.375 in.
8	0.625 in.	4@ 1/8 in. = 0.50 in.	0.25 in.	2@ 1/8 in. = 0.25 in.

designed only for gravity loads, has a *P*/*A* ratio that is relatively high and (2) C_2, designed as a frame column, has its *P*/*A* ratio significantly less than that for C_1 because it is lightly loaded under gravity loads.

Assume that you, as the engineer for the project, have specified column length corrections to C_1 at levels 8, 16, 24, 32, 40, and 48 with correction of 2 in. specified at level 24. Let us say that when steel erection is at the level 24, the steel erector surveys the top elevations of columns, reports the top of column C_1 is 1 in. higher than the top of C_2, and requests you, the engineer, to confirm if this is acceptable in view of the fact that additional shortening of the column is yet to occur.

Further calculations are needed to verify that this 1 in. overlength of C_1 will indeed come down after the application of dead loads at levels 24 through roof. This concept of verifying the overlength of columns during construction is shown in Figure 7.77G. Note that Δ_{Rn} shown therein corresponds to the 1 in. discussed here for the hypothetical building.

7.7 GUIDANCE FOR PREPARING CONCEPTUAL ESTIMATES

The total quantity of structural steel required for a building divided by its gross area has always been, and will always be, an item of great interest to building developers and designers alike.

In the US practice, the unit quantity of steel is expressed in terms of pounds per square foot as shown in Table 7.15 for selected buildings built between 1930 and 1971. In selecting a

TABLE 7.15
Unit Quantity of Steel for Selected High and Medium-Rise Buildings

Height/Year	Stories	Width	psf	Building
1930	102	9.3	42.2	Empire State Building, New York
1968	100	7.9	29.7	John Hancock Center, Chicago
1972	110	6.9	37.0	World Trade Center, New York
1974	109	6.4	33.0	Sears & Roebuck, Chicago
1963	60	7.3	55.2	Chase Manhattan, New York
1969	60	5.7	38.0	First National, Bank, Chicago
1971	64	6.3	30.0	U.S. Steel Building, Pittsburgh
1971	57	6.1	17.9	I.D.S. Center, Minneapolis
1957	42	5.1	28.0	Seagram Building, New York
1970	41	4.1	21.0	Boston Co. Building, Boston
1965	30	5.7	38.0	Civic Center, Chicago
1969	26	4.0	26.0	Alcoa Building, San Francisco
1971	10	5.1	6.3	Low Income Housing, Brockton, MA

structural system for a given building, as stated previously, it is common practice to evaluate several possible structural schemes that meet the owner's requirements along with architectural aspirations. The deciding factor most often, then, is the unit quantity of material for the systems. Although the unit weight of steel required for a given building may not address all factors relevant to construction cost, it is a component that most often pushes a particular scheme to the forefront.

Unit quantities are relatively easy to evaluate once complete working drawings and specifications have been prepared. Prior to this point, however, one must use *conceptual estimating* to determine approximate costs. Conceptual estimate for a given building demands considerable engineering judgment in addition to determining unit quantities. This is because we need to adjust the so-called average unit costs to reflect the complexity of construction operations, expected time required for construction, cost of borrowing money for the length of construction, etc.

Prior to the advent of high-efficiency systems such as tubular and mega frame frames, most high-rises were designed using braced and/or moment frames as lateral load–resisting systems. Consequently, their poundage is relatively high compared with present-day buildings.

It is of interest for conceptual estimating purposes to assemble unit structural steel quantities for buildings that have been built in the past. Figures 7.78 through 7.80 and Table 7.15 show the general trend in the increase of unit weight of steel as the building height is increased.

For distinct regions, A, B, C, and D are shown in Figure 7.80. Region A is for buildings up to 30 stories, B for buildings between 30 and 50 stories, C for buildings 50 and 70 stories, and D for high-rises in excess of 70 stories. The use of the figure is best explained using the following examples.

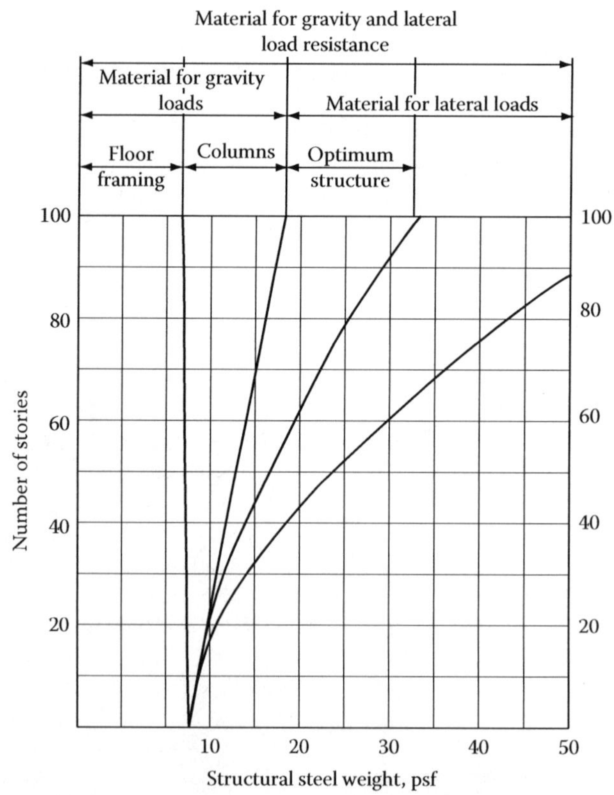

FIGURE 7.78 Steel quantities for gravity and lateral load resistance.

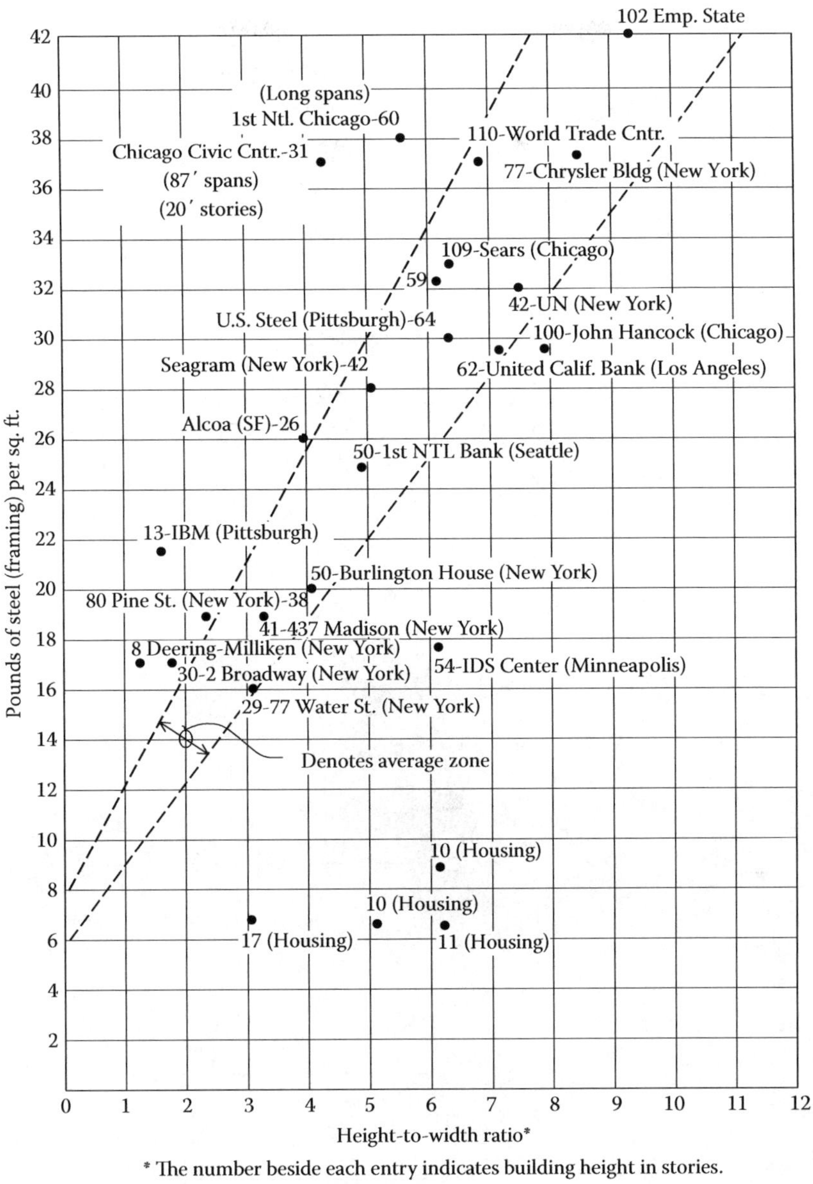

FIGURE 7.79 Unit weight of steel as it relates to building's height-to-width ratio.

Example 1

A proposed 10-story building in a wind-controlled low-seismic-risk area. The engineer is asked to come up with a unit quantity of structural steel for purposes of a conceptual cost estimate.

For a 10-story building, it is seen from Figure 7.80 that the lower and upper bounds for the unit weight are 10.5 and 15 psf, respectively, which is a rather wide range. The engineer now has to make some judgment calls, depending on what is known about the building at this stage of the game. Some relevant questions are as follows: (1) What are the typical spans for the floor framing? (2) Are there any undue restrictions for the beam and girder depths? (3) What is the likely lateral framing system? If none of these questions can be answered with any great certainty, it is perhaps prudent to use an average of two unit weights. Thus, the unit weight for our building would be equal to (10.5 + 15)/2 = 12.75, rounded to 13 psf.

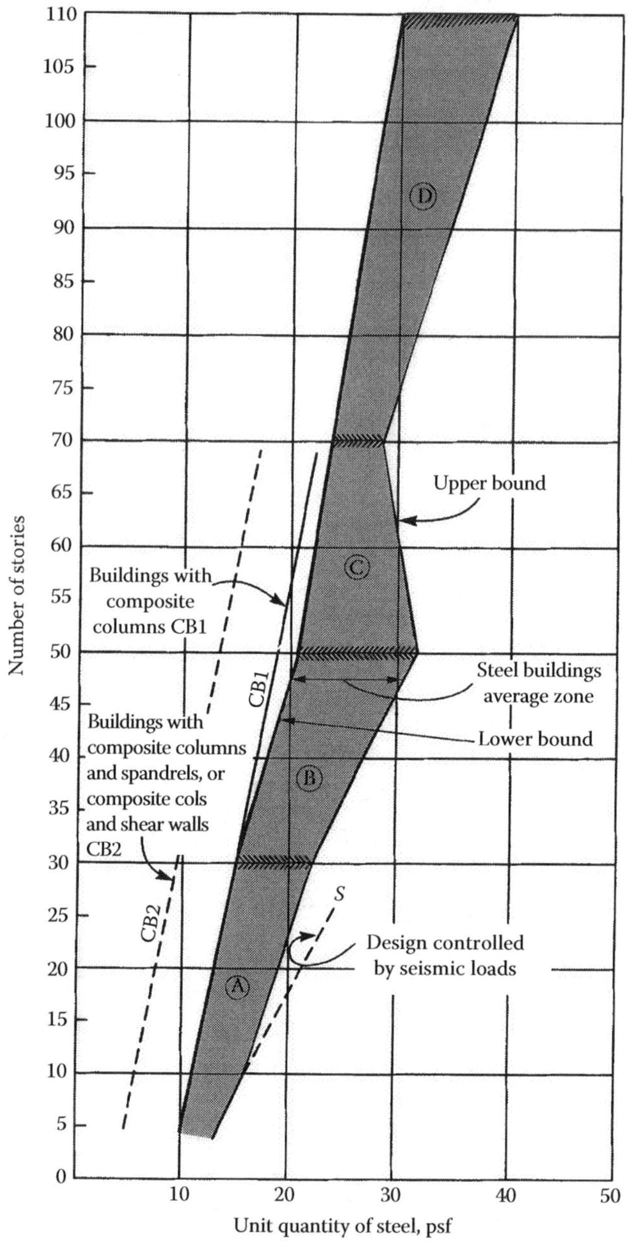

FIGURE 7.80 Structural steel unit quantities. A design aid for conceptual estimate.

Example 2

Fifty-story building. Using a similar procedure, the engineer determines that the lower- and upper-bound values are 21 and 32 psf, giving an average unit quantity of 26.5 psf.

Example 3

Twenty-two-story building, assigned to SDC D. Design controlled by seismic loads. The straight line designated S in Figure 7.80 represents an approximate unit weight of steel for buildings located in high-risk seismic zones. Observe that this line stops at 25 stories, implying that the design of tall buildings, in general, is controlled by wind loads. For the example building, the unit weight of steel as given by the straight line S is 22 psf.

Example 4

Now consider the same building in a seismically low-risk area such as Houston, Texas. We go to the region designated as an average zone on the graph. From the graph, the unit weight is seen to vary from a low 13.5 psf to a maximum of 19.5 psf. As in previous cases, we use an average unit quantity of 16.5 for the preliminary conceptual estimate.

Example 5

Seventy-story composite building. The line designated as CB 2 shows unit weight for steel buildings with composite frame columns. As a preliminary estimate, a unit quantity of 23 psf is appropriate for the building.

Example 6

Fifty-story composite building with composite columns and composite girders (buildings with composite columns and composite shear walls may be considered similar). From the graph noted as CB2, we get a unit quantity of 13.5 psf for this building.

7.8 CONCEPT OF PREMIUM FOR HEIGHT

If there were no lateral loads such as wind or earthquake, a given high-rise building could be designed primarily for gravity loads. Such a design would not impose a premium for height. Since there is no way to circumvent gravity loads, the absolute minimum quantity of material required for a building, irrespective of its height, would not be less than that required for gravity loads alone. From a structural point of view, this would correspond to the most efficient or optimum system.

Assuming equal bays, the material quantities required for gravity floor framing in low- and high-rise structures are essentially identical; it makes no difference in steel quantities, whether the floor being framed is at the 2nd level of a low-rise or at the 70th level or a high-rise building. The steel required for floor framing is a function of the column-to-column span and not the building height. However, the material required for columns in a high-rise is substantially more than that for a low-rise. In fact, the steel increases in the ratio $(n + 1)/2$, where n is the number of floors. This is because columns in taller buildings are designed to carry cumulative gravity loads, thus requiring more steel than a one-story structure of the same floor area.

The quantity of materials required for resisting lateral loads is even more pronounced and would soon far outstrip all other structural costs if rigid frame action were employed in very tall buildings. The graph shown in Figure 7.78 illustrates how the unit weight of structural steel increases as the number of floor increases. Wind begins to show its dominance at about 50 stories and becomes increasingly important with greater height. For example, in a steel building using rigid frame action, the total weight of, say, 24 psf (1149 Pa) of structural steel is split evenly at about 8 psf (393 Pa) for each of the three subsystems, namely, (1) floor framing, (2) gravity column, and (3) wind bracing system. Above 50 stories, lateral bracing ingenuity often makes the difference between an economical solution and an expensive one. The objective is to arrive at a lateral bracing system that keeps the additional material required for lateral loads to a reasonable amount.

Low- to medium-rise buildings, particularly those controlled by wind, are normally designed for gravity loads, then checked for their ability to resist lateral loads. For high-rise buildings, however, there is a drastic increase in the amount of material needed for resistance of lateral forces. With respect to gravity loads, the weight of the structures increases almost linearly with the number of stories. However, the amount of material needed for the resistance of lateral forces increases at an accelerating rate. For example, using the rigid frame principle, one would estimate about 55 psf of steel for a 90-story building instead of the tubular system with only 34 psf (e.g., Standard Oil Building, Chicago).

Weight-to-area ratios (which we referred to as unit weight, in the previous sections) for some typical high-rise buildings are given in Figure 7.79. The Empire State Building has a ratio of 42.2 psf in contrast to the 29.7 of the diagonally braced tubular John Hancock Center. An even larger contrast is the 60-story Chase Manhattan Building with 55 psf and the slight lower 54-story IDS Building with only 17.9 psf. The Chase Manhattan Building is a long-span rigid frame that needs huge girders to resist wind forces. The IDS Building owes its efficiency to the belt truss system. Disproportionate amounts of steel does not necessarily indicate that the structural design of the building is poor. For instance, the Civic Center Building in Chicago uses about twice as much steel as other buildings in its height range. However, it has to satisfy the functional requirements of size and location of courtrooms. Thus, the girder spans are 87 ft, and the floor heights are much larger than usual (30 stories for 640 ft).

8 Seismic Evaluation and Rehabilitation of Existing Buildings

PREVIEW

Not all older buildings are seismically at risk. If they were, the damage from several earthquakes in this country, including the 1971 San Fernando and the 1984 Northridge events, would have been devastating, because much of the inventory affected was 25 or more years old. Often, strong ground shaking from earthquakes significantly damages building types and configurations well known to be vulnerable and occasionally highlights vulnerabilities previously unrealized. For example, the Northridge earthquake caused damage to many wood-frame buildings—mostly apartments—and relatively modern steel moment-frame buildings, both previously considered to be of low vulnerability.

However, only in the most vulnerable building types does damage occur relatively consistently. For example, at higher levels of shaking, the exterior walls of unreinforced masonry bearing-wall buildings have consistently fallen away from their buildings in many earthquakes, ever since this building type was built in large numbers in the late nineteenth century.

Why seismic rehabilitation? Because it is a classic mitigation strategy, unlike preventive medicine.

The core argument for the seismic rehabilitation of buildings is that rehabilitated buildings will provide increased protection of life and property in future earthquakes, thereby resulting in fewer casualties and less damage than would otherwise be the case. More earthquake-resistant buildings will mean fewer deaths and injuries in an event and therefore lower demand on emergency medical services, urban search and rescue teams, fire and law enforcement personnel, utilities, and the providers of emergency shelter. In the commercial sector, less damage to structures will mean enhanced business survival and continued ability to serve customers and maintain market shares. More specifically, for commercial enterprises, seismic rehabilitation will better protect physical and financial assets, reduced inventory loss, shorten the business interruption period, avoid the need for relocation, and minimize secondary effects on suppliers, shippers, and other businesses involved in support services or product cycles. For governments, if their structures come through an earthquake with little or no damage, public officials can better respond to the immediate and long-term demands placed on them by the event. In short, seismic rehabilitation as a pre-event mitigation strategy actually will improve post-event response by lessening life loss, injury, damage and disruption.

Seismic rehabilitation also will help achieve other important goals that contribute to business and community well-being. For example, seismic rehabilitation will

- Reduce community economic and social impacts
- Minimize the need for getting disaster assistance, as in seeking loans or grants
- Help to protect historic buildings, structures, or areas that represent unique community values and that provide the residents with a sense of their unique histories
- Minimize impacts on such critical community services as hospitals and medical care facilities

- Support the community's post-earthquake need to return to a pattern of normal activities by helping to ensure the early reopening of business and civic facilities, which restores normal activities as soon as possible thus contributing greatly to the psychological well-being of a community
- Minimize the many, and often subtle, direct and indirect socioeconomic impacts of earthquakes, some of which emerge slowly but often last a long time and help marginal businesses to reopen, thus strengthening a community's economic and social fabric
- Reduce the environmental impacts of earthquakes (e.g., the need to dispose of large quantities of debris, the release of asbestos in damaged buildings, and the contamination of the air and water with spilled hazardous materials)

The rehabilitation of existing buildings significantly reduces future losses and, in economic terms, can be considered an investment to protect assets currently at risk.

Earthquake-vulnerable buildings exist nationwide, but the earthquake hazard is not uniform across the country. Moreover, awareness of the earthquake hazard, the precursor to any action, varies even more than the hazard itself. Therefore, tackling the earthquake-vulnerable building problem takes place in an incredibly diverse set of geographic, social, economic, and political environments. Further complicating the situation is the fact that no two buildings ever seem to present exactly the same problems. Each building has its own earthquake-vulnerability profile—location, architecture, structural system, occupancy, economic role, and financing. In other words, each building has its own story.

In sum, the intent of this chapter is to explain seismic rehabilitation and to offer a set of approaches or *models* as given in ASCE/SEI41-06 document.

Seismic rehabilitation of a building entails costs as well as disruption of its usage. In fact, the effects of a rehabilitation program are similar to those of an earthquake because strengthening, in terms of cost and the need to vacate the structure while strengthening is underway, is analogous to building repair after an earthquake. The crucial difference is that strengthening occurs at a specified time and no deaths or injuries will occur during the process.

In a seismic rehabilitation study, it is convenient to classify the damage within a building in two categories, structural and nonstructural. Structural damage refers to degradation of the building's support system, such as frames and walls, whereas nonstructural damage is any damage that does not affect the integrity of the building's physical support system. Examples of nonstructural damage are chimneys that collapse, broken windows or ornamental features, and collapsed ceilings. The type of damage a building experiences depends on its structural characteristics, age, configuration, construction, materials, site conditions, proximity to neighboring buildings, and the type of nonstructural elements.

An earthquake can cause a building to experience four types of damage:

1. The entire building collapses.
2. Portions of the building collapse.
3. Components of the building fail and fall.
4. Entry–exit routes are blocked, preventing evacuation and rescue.

Any of the aforementioned may result in unacceptable risk to human lives. It can also mean loss of property and interruptions of use or normal function.

Another type of damage that should be included in the rehabilitation study is the structural damage from the pounding action that results when two insufficiently separated buildings collide. This condition is particularly severe when the floor levels of the two buildings do not match, because the stiff floor framing of one building can badly damage the more fragile walls or columns of its neighbor,

A rehabilitation objective may be achieved by implementing a variety of measures, including

1. Local modification of deficient components
2. Removal or partial mitigation of existing irregularities
3. Global stiffening
4. Global strengthening
5. Reduction of mass
6. Seismic isolation
7. Installation of supplemental energy dissipation devices

Failure of nonstructural architectural elements can also create life-threatening hazards. For example, windows may break or architectural cladding such as granite veneer or precast with insufficient anchorage may separate from the building, causing injury to pedestrians. Consequently, a seismic retrofit program should explore techniques for dealing with nonstructural components such as veneers, light fixtures, glass doors and windows, raised computer access floors, and ceilings. Similarly, because damage to mechanical and electrical components can impair building functions that may be essential to life safety (LS), seismic strengthening should be considered for components such as mechanical and electrical equipment, ductwork and piping, elevators, emergency power systems, communication systems, and computer equipment.

We know by now that the forces experienced in a building during a major earthquake are much larger than the forces they are typically designed for. It is neither practical nor economically feasible to design buildings to remain elastic during major earthquakes. Instead, we design buildings to remain elastic under considerably lower forces, and by using certain stringent details, we expect them to ride out larger earthquakes without collapse. The rationale for this procedure is based on our observation of buildings that have lived through large earthquakes in the past.

In earthquake design, the ability of a building to withstand deformations well beyond the elastic range without losing its gravity carrying capacity is termed ductility. Ductile structures may deform excessively under seismic lateral loads, but by and large, they do not fall down. Therefore, in seismic design, providing capacity for lateral displacement well into inelastic range without causing collapse is the primary goal. What this does, in effect, is to offer engineers a range of design options to swap strength requirement for ductility.

The core argument for seismic rehabilitation is above all protection of life and property in future earthquakes. Other worthy goals include protecting investments, lengthening a buildings usable life, reducing demands on post-earthquake search and rescue resources, and shortening business interruption time. Because rehabilitation deals with existing and usually occupied buildings, the range of socioeconomic issues likely to be encountered can be formidable.

Provisions of ASCE/SEI 41-06 are based on PBD methodology. The performance levels are defined in terms of specifically limiting damage status. This tiered specification of building performance at predetermined earthquake intensities forms the basis for design.

8.1 CODE-SPONSORED DESIGN

In seismic design, it's well known over the past several decades that the forces experienced by buildings during a major earthquake are much greater than the forces they are designed for. This is because it is neither practical nor economically feasible to design a building to remain elastic during a major seismic event. Instead, we design the structure to remain elastic at a reduced force level, and by prescribing detailing requirements, we rely upon ductility of the structure to sustain post-yield displacements without collapse when subjected to higher levels of ground motion. The rationale for designing with lower forces is based on the premise that the special ductile detailing of the components is adequate to allow for additional deformation without collapse. Historically, this approach has produced buildings with a strength capacity adequate for the scaled-down seismic forces and, more importantly, with adequate performance characteristics beyond the elastic range. It is the consensus of the structural engineering profession that a building properly designed to both code-specified forces *and* detailing requirements will have an acceptable level of LS during a major seismic event.

The ability of a member to undergo large deformations beyond the elastic range is termed ductility. The same property in a building that allows it to absorb earthquake-induced damage and yet remain stable may be considered, in a conceptual sense, similar to ductility. Ductile structures may deform excessively under load, but they remain by and large intact. This characteristic prevents total structural collapse and provides protection to occupants of buildings. Therefore, providing capacity for displacement beyond the elastic range without collapse is a primary goal.

Aside from this implicit philosophy, no explicit earthquake performance objectives are stated in most building codes. However, building structures designed in conformance with modern codes such as the IBC-06 are expected to

1. Resist low-level earthquakes without damage
2. Resist moderate-level of earthquakes without structural damage while possibly experiencing some nonstructural damage
3. Resist high-level earthquakes of intensity equal to the strongest experienced or forecast for the building site without collapse while possibly experiencing some structural or nonstructural damage

It is expected that structural damage, even in a major earthquake, will be limited to a repairable level for structures that meet these requirements. However, conformance to these provisions does not ensure that significant structural damage will not occur in the event of a large earthquake. Therefore, additional requirements are given in the code to provide for structural stability in the event of extreme structural deformations.

The protection of life rather than prevention and repairability of damage is the primary purpose of the code; the protection of life is thus reasonably provided for but not with complete assurance.

Building codes deal primarily with design of new buildings. For seismic upgrade, the primary use of these documents is for determining existing building capacity. They do not, in general, provide guidance for evaluating and upgrading the seismic resistance of existing buildings.

Most codes allow existing buildings to use their current lateral-load-resisting systems if only trivial changes to the structure are proposed and the building's use remains unchanged. Codes require upgrading of buildings when major changes or tied-in additions are planned and when the proposed alterations reduce the existing lateral-load-resisting capacity. A lateral-load upgrade may also be required if the proposed changes move the building into the categories of *essential* or *hazardous* facilities.

The seismic provisions of the IBC-06 attempt to be more specific by quantifying the meaning of *significant change*. It requires that the addition itself be compliant with the code for new construction and requires a seismic upgrade of the existing building if the addition increases the seismic forces in any existing structural member by more than 5% unless that member is already strong enough to comply with the code. Similarly, the addition is not allowed to weaken the seismic capacity of any existing structural member to a level below that specified for new construction. However, there remain some questions as to how to interpret these provisions.

When building codes prescribe full compliance with their current seismic provisions, they are rarely explicit in telling users what measures to take to upgrade the building. There are exceptions, of course. On the US west coast, San Francisco's building code requires upgrading of existing structures to 75% of the strength required by the code for new construction. On the east coast, the *Commonwealth of Massachusetts Building Code* offers an elaborate path for determination of required remedial measures. In some cases, it allows lower seismic forces than those used for new construction. In some regions of high seismic activity, state and local codes and ordinances may require a seismic upgrade even for buildings that are not undergoing renovation. Perhaps the best known of these is California's Senate Bill 1953, a seismic retrofit ordinance adopted on February 24, 1994, in the wake of the Northridge earthquake. It required more than 450 acute care facilities to submit seismic evaluation and compliance plans showing how the facilities will withstand a code-level earthquake, defined as a seismic event with a 10% probability of being exceeded in 100 years.

In general, the process for seismic upgrade is somewhat disorderly. It is not uncommon to have one engineer declare that a building needs a complete seismic upgrade, while another states that none is needed. Sometimes the owner will *shop* for an engineer in whose opinion an upgrade is not needed and is willing to justify this interpretation of the code to building officials.

These real-life observations lead to the conclusion that guidance on this issue from an authoritative source is sorely needed. One source—the ASCE/SEI 41-06 publication—discussed in this chapter attempts to fill the void.

As compared to seismic upgrade of existing structures, design of a new structure for proper seismic performance is a *cinch*. This is because most structural characteristics important to seismic performance including ductility, strength, deformability, continuity, configuration, and construction quality can be designed and, to a certain extent, controlled.

Seismic rehabilitation of existing structures poses a completely different problem. First, until recently, there was no clear professional consensus on appropriate design criteria. That changed substantially with the publication of FEMA 356, *Prestandard and Commentary for the Seismic Rehabilitation of Buildings*—a predecessor to ASCE/SEI 41-06. Second, the building codes for new construction are not directly applicable because they incorporate levels of conservatism and performance objectives that may not be appropriate for use on existing structures due to economic limitations. Third, the material strengths and ductility characteristics of an existing structure will, in general, not be well defined. And finally, the details and quality of construction are frequently unknown and, because the structure has been in service for some time, deterioration and damage are often a concern.

The successful seismic upgrade of an existing structure therefore requires a thorough understanding of the existing construction, its limiting strength and deformation characteristics, qualification of the owner's economic and performance objectives, and selection of an appropriate design criterion to meet these objectives and also be acceptable to the building official. Most of the time, it includes the selection of retrofit systems and detailing that can be installed within the existing structure.

8.1.1 Building Deformations

The basic design procedure for new structures consists of the selection of lateral forces appropriate for design purposes and then providing a complete, appropriately detailed, lateral-force-resisting system to carry these forces from the mass levels to the foundations. Although deformations are checked, experience has shown that new structures with modern materials and ductile detailing can sustain large deformations while experiencing limited damage. Older structures, however, may not have the advantage of this inherent ductility. Therefore, control of deformations becomes an extremely important issue in the design of seismic retrofits.

Determination of the deformations expected in a structure, when subjected to the design earthquake, is the most important task in seismic rehabilitation design. There are three types of deformations that must be considered and controlled in a seismic retrofit design. These are

1. Global deformations
2. Elemental deformations
3. Interstructural deformations

Although they are all interrelated, for purposes of seismic upgrade, it is convenient to consider each of these separately.

Global deformations are the only type explicitly controlled by the building codes and are typically considered by reviewing interstory drift. The basic concern is that large interstory drifts can result in $P\Delta$ instabilities. Control of interstory drift can also be used as a means of limiting damage to nonstructural elements of a structure. However, it is less effective than elemental or interstructural deformations in limiting damage to individual structural elements.

Elemental deformation is the amount of seismic distortion experienced by an individual element of a structure such as a beam, column, shear wall, or diaphragm. Building codes have very few provisions that directly control these deformations. They rely on ductility to ensure that individual elements will not fail at the global deformation levels predicted for the structure. In existing structures with questionable ductility, it is therefore critical to evaluate the deformation of each element and to ensure that expected damage to the element is acceptable. This requirement extends to elements not normally considered as participating in the lateral-force-resisting system. A glaring example that is attracting much attention after the Northridge earthquake is the punching shear failure of flat slabs at interior columns, resulting from excessive rotation at the slab–column joint. Often, the slab system is not considered to participate in the lateral-force-resisting system. In fact, building codes indirectly prohibit the use of flat slab–frames in the lateral system of buildings in high seismic zones. However, in relatively flexible buildings such as those without shear walls, when flat slabs *go for a ride*, they bend and twist. In doing so, they fail if they do not have adequate ductility. Therefore, it is very important to limit the rotational deformation of these joints to prevent a punching shear failure.

Interstructural deformations are those that relate to the differential movement between elements of the structure. Failures that result from lack of such control include failures of masonry walls that have not been anchored to diaphragms and failures resulting from bearing connections slipping off beam seats. Building codes control these deformations, which may cause separation of one element from another, by requiring interconnection of all portions of structures. A similar technique should be considered in the retrofit of an existing structure.

Code methodologies rely on elastic dynamic analysis using base shears that are reduced by response modification coefficient R and then they are scaled up or down to 85% of base shear values computed on the basis of an equivalent lateral-load procedure. Therefore, design forces are significantly smaller than those likely to be experienced by the building. However, when it comes to deformations, it is explicitly recognized that the predicted elastic levels of deformation, termed δ_{xe}, are quite small compared to the actual deformations that may be experienced by the building. Hence, the amplified deformations $\delta = C_d\delta_{xe}/I_s$ are specified in the codes to evaluate the effects of seismic deformation. It is even more important to use a similar approach in evaluating existing structural elements in a retrofitted structure, because pre-1971 buildings rarely have the required ductility.

8.2 ALTERNATE DESIGN PHILOSOPHY

Although earthquake performance objectives are implicit in building codes, significant questions linger. Is the philosophy of inferring the behavior adequate to define the expected earthquake performance? Can the performance be actually delivered? Should the earthquake response objectives be explicitly stated in building codes? Is it feasible to make an existing nonductile building conform to current detailing and ductility provisions? If not, what level of upgrade will provide for minimum LS? How much more strengthening is required to achieve an *IO rating*?

Explicit answers to these and similar questions cannot be found in current building codes. Although a set of minimum design loads are prescribed, the loads may not be appropriate for seismic performance verification and upgrade design because of the following:

1. The code provisions do not provide a dependable or established method to evaluate the performance of non code compliant structures.
2. They are not readily adaptable to a modified criterion, such as one that attempts to limit damage.
3. Since the primary purpose is protection of LS, the code does not address some building owners' business concerns such as protection of property, the environment, or business operations.

To overcome these shortcomings, a procedure that uses a two-phase design and analysis approach has been in use for some time. The technique explicitly requires verification of serviceability and survival limit states by using two distinct design earthquakes: one that defines the threshold of damage and the other that defines collapse. The serviceability level earthquake is normally characterized as an earthquake that has a maximum likelihood of occurring once during the life of the structure. The collapse threshold is typically associated with the maximum earthquake that can occur at the building site in the presently known tectonic framework. This characterization can vary, however, to suit the specifics of the project, such as the nature of the facility, associated risk levels, and the threshold of damageability.

The principle behind the two-phase approach may be explained by recalling the primary goal in seismic design, which is to provide capacity for displacement beyond the elastic range. Any combination of elastic and inelastic deformations is possible to attain this goal. For example, we could design a structural system that would remain elastic throughout the displacement range. This system would have a high elastic strength but low ductility. Conversely, it is entirely possible to have a system with relatively low elastic strength but high ductility, meeting the same design objective of remaining stable. It may be easier to understand the methodology if it is recognized that a specific earthquake excitation causes about the same displacement in a structure whether it responds elastically or with any degree of inelasticity.

Figure 8.1 shows the behavior of an idealized structure subjected to three levels of earthquake forces F_L, F_U, and F_C corresponding to lower-level, upper-level, and collapse-level earthquakes. Also shown is an earthquake force F_E experienced by the structure if it were to remain completely elastic. The structure designed using the lower-level earthquake force F_L deforms elastically from 0 to E and inelastically from E to U. The same structure designed using the force F_U needs to deform 0 to U, responding elastically all through the displacement range. Both systems are capable of attaining the anticipated deformation of Δ_U. However, a building designed using the force F_L will require a

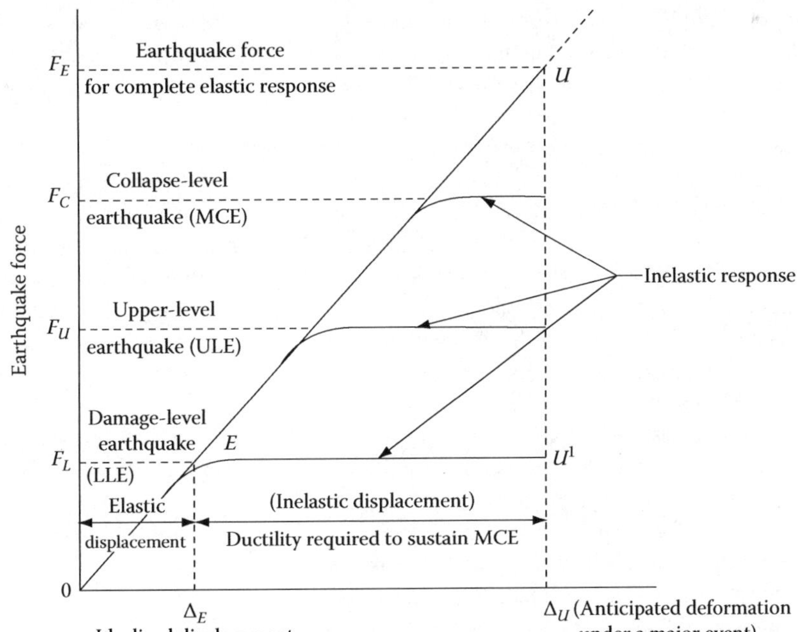

FIGURE 8.1 Idealized earthquake force–displacement relationships. Δ_U = anticipated deformation irrespective of ductility that can be achieved as a combination of elastic deflection OE + inelastic deformation EU^1 or by a totally elastic response of $OE + EU$.

more ductile system than a building designed for the fully elastic force F_E. More importantly, it will suffer heavier damage should the postulated event occur. Nevertheless, both systems achieve the primary goal: both remain stable without collapse under the expected deformation Δ_U. Therefore, it is possible to design the structure using any level of force between F_L and F_E with the understanding that a corresponding ductility is developed by the detailing of the system. For example, a structure designed for the force level F_U requires a higher strength but less ductility than if it were designed for force level F_L. Hence, it is a matter of choice as to how much strength can be traded off for ductility and, conversely, ductility traded for strength. Expressed another way, structural systems of limited ductility may be considered valid, provided they are capable of resisting correspondingly higher seismic forces.

This is the approach used in the seismic retrofit design of existing buildings. Since buildings of pre-1970 vintage do not have the required ductile detailing, the purpose is to establish the strength levels that can be traded off in part, for lack of required ductility.

Because rehabilitation deals with existing and usually occupied buildings, the range of socioeconomic issues likely to be encountered and need to be solved can be formidable. Moreover, the intensity, nature, and complexity of such problems will vary somewhat from building to building even though sections or neighborhoods of cities and towns slated for seismic rehabilitation. Accordingly, the rehabilitation process varies depending on the demographic and socioeconomic characteristics of the designated areas.

An overriding concern in seismic rehabilitation has to do with accommodating the building's intended use. Obviously, one has to accommodate the owner's intended uses of the candidate building. However, seismic rehabilitation projects often are technically tricky and part of their success depends on achieving an effective balance between improved earthquake safety and functionality.

The amount of rehabilitation needed is a function of the building type and existing conditions, the seismicity of the building site, and the rehabilitation objective desired by the owner.

Due to the advancement of earthquake engineering in recent decades, new codes and design standards are continually being developed and adopted. With each advance, however, the number of existing buildings that have not been designed consistent with the current technology increases. Although many of these buildings will perform well in future earthquakes, many others are likely to suffer damage, partial collapse, or complete collapse, causing injuries and deaths.

Due to the cost and impracticality of implementing design codes for new buildings in the rehabilitation of existing buildings, procedures for rehabilitation will be different from those for the design of new buildings. The length of remaining building life, the consequence of economic loss due to major earthquake damage, and the cost-to-benefit ratio for each structure must be considered when designing seismic rehabilitation of existing buildings and proceeding with the rehabilitation. Therefore, procedures and guidelines specifically for the seismic rehabilitation of existing buildings, different from codes for new buildings, are appropriate in order to permit the most effective use of existing elements and components that do not comply with modern seismic detailing requirements.

Evaluation methods are useful generally as criteria for the rehabilitation decision. Merely correcting deficiencies is a viable alternative for certain simple buildings, although it is often conservative and may be an expensive approach.

The seismic design for rehabilitation of an existing building must consider each seismic element and component of the building as well as the building system as a whole. Many of the desirable features of new buildings, such as a complete load path, redundancy, regular building configuration, and separation from adjacent buildings, which are all fundamental to satisfactory performance, may not be present in existing buildings and will have to be compensated for by the rehabilitation process.

A comprehensive rehabilitation design must include consideration of the nonstructural systems and components (such as architectural elements and mechanical and electrical systems) and the cost of their rehabilitation, as well as the structural components of the building.

The general steps in the seismic rehabilitation process are the following:

1. *Review initial status*
   Review characteristics that define the rehabilitation work for the building under investigation. This includes information such as global structural characteristics, seismic site hazards, occupancy, historic status, and estimated range of costs for the rehabilitation work. Much of this information may be available from past evaluations. If not, it must be determined before the rehabilitation procedures can be implemented efficiently.
2. *Select a rehabilitation objective*
   Together with the building owner, select the desired performance level and the level of ground sharing that define the rehabilitation objective to be used as the basis for rehabilitation design.
3. *Select a rehabilitation method*
   Determine the appropriate rehabilitation method preliminarily by using simplified rehabilitation method. Then proceed using the more comprehensive systematic rehabilitation method.
4. *Develop a rehabilitation strategy*
   Develop a plan to mitigate structural deficiencies that considers the global characteristics of the building system.
5. *Select an analysis procedure*
   Select an appropriate analysis procedure according to the provisions of ASCE/SEI 41-06.
6. *Perform preliminary rehabilitation design and analysis*
7. *Verify rehabilitation design*:
   Check system and component acceptability and review for economic acceptability.
8. *Iterate*
   Refine the rehabilitation design or reconsider the rehabilitation objective until the system and components meet the acceptability criteria.
9. *Construction Documents and Quality Assurance*
   Prepare construction documents, including plans, details, and specifications, with an appropriate quality assurance program.

8.2.1 Initial Considerations

Prior to embarking on the seismic rehabilitation process, the engineer must review and clearly understand the building characteristics that define the required rehabilitation work. This includes information such as global structural characteristics, seismic site hazards, occupancy, historical status, and range of costs for the rehabilitation work:

Global structural characteristics: The global structural characteristics include the primary lateral-force-resisting system, other structural and nonstructural systems, the building size, and the deficiencies that are to be mitigated. Once the lateral-force-resisting system is identified, the extent of the rehabilitation measures necessary to mitigate the deficiencies will provide insight into the cost needed to rehabilitate the structure.

In addition to the structural concerns, the scope of the nonstructural rehabilitation, if any, must be understood. As in structural rehabilitation costs, the cost associated with nonstructural rehabilitation is driven by existing conditions and both seismicity and performance level.

Seismic site hazards: Seismicity is defined in terms of the level of ground shaking expected at the site. Three seismicity zones are defined: low, moderate, and high. In general, the higher the seismicity, the greater the cost of the rehabilitation, as a higher level of strengthening will be needed to achieve the same performance. The potential for other seismic site hazards such as ground failure should be determined also.

Occupancy: The engineer needs to determine whether or not the building will be occupied during the rehabilitation, and if so, what level of disruption will be tolerated by the occupants. If the building is to remain occupied, measures may have to be taken during construction to minimize disruption and, as a result, the cost of the rehabilitation may significantly increase.

Historic status: The rehabilitation of historic buildings may involve additional concerns beyond those normally considered in the rehabilitation of a typical building. Limitations may be placed on the rehabilitation strategies used, and special care may have to be taken to preserve the historic fabric of the structure. All of these factors may contribute to increasing the cost of the rehabilitation. The advice of a specialist in historic preservation may be required.

8.2.2 Rehabilitation Objective

The building owner and the design team should meet in the early stages of a rehabilitation project to set forth the rehabilitation objective. The rehabilitation objective consists of expected performance for a defined level of ground sharing. In selecting the rehabilitation objective, the engineer should carefully consider the rehabilitation costs, disruption to the occupants, and the extent of architectural remodeling in order to incorporate all aspects of the project early in the planning stages.

8.2.2.1 Performance Levels

Four building performance levels are defined in ASCE 41-06. These are operational, immediate occupancy, life safety, and collapse prevention. A generalized description of the damage states associated with these performance levels is given later in this chapter.

8.2.2.2 Seismic Hazard

The most common and significant cause of earthquake damage to buildings is ground shaking. The structural response to ground shaking is dependent on many factors, such as the proximity to the fault, the soil type, and the period of vibration of the building. Ground shaking hazards for use in seismic rehabilitation can be determined in accordance with either the general procedure or the site-specific procedure. In addition to ground shaking, other seismic site hazards such as ground failure should also be considered.

8.2.2.3 Selecting a Rehabilitation Objective

In addition to the basic safety objective (BSO), there are descriptions of enhanced objectives and limited objectives. The BSO is a benchmark that some owners may choose to exceed by setting an enhanced objective, and others may choose to limit by specifying a limited objective. The rehabilitation objective is distinct to each project, and technical guidance from a structural engineer, experienced in seismic design, to assist the owners in their decision, may be necessary.

The selection of the rehabilitation objective is entirely the choice of the owner, unless the appropriate jurisdiction has the authority to influence the selection. A few factors that may persuade the owner to select an enhanced objective are the use of the building, the value of the building contents, the operation of the building, or the code mandated or legislation requirements affecting the building. An enhanced objective may be beneficial in structures that house hospitals, fire stations, and any post-earthquake response and recovery operations. It also may be desirable for museum or buildings with historic fabric, where the cost to replace these items, if replacement is even possible, far exceeds the cost of upgrading the building to this rehabilitation objective. An enhanced objective also may be beneficial for structures that house function essential to the business, where the cost due to downtime of the operation is critical. These structures may include data processing centers, manufacturing plants, or oil refineries,

Any rehabilitation objective intended to provide performance inferior to that of the BSO is termed a limited objective. A limited objective may consist of either partial rehabilitation or reduced rehabilitation. Partial rehabilitation may be chosen as a part of a program implemented as a series of incremental rehabilitations intended to result in finally achieving the BSO. Reduced rehabilitation may be selected for buildings with limited remaining life span or low or infrequent occupancy levels. In reduced rehabilitation, designers will typically prepare a rehabilitation scheme that will reduce the damage to the structure for moderate events that are likely to occur in the remaining life of the building.

8.2.2.4 Rehabilitation Method

Two basic methods for proceeding with the rehabilitation process are given in ASCE/SEI 41-06: the simplified rehabilitation method and the systematic rehabilitation method.

The systematic rehabilitation method involves a complete analysis of the building to the specified rehabilitation objective. This analysis could be linear or nonlinear, static or dynamic. Choice of the analysis method is generally left to the engineer. It is appropriate to use a nonlinear procedure for irregular or complex buildings, or both. The systematic rehabilitation method may be used for any structure and requires checking the acceptability of individual components in the building, including those that are part of the lateral-force-resisting system and others such as gravity framing and nonstructural components.

8.2.2.5 Rehabilitation Strategy

Once the building deficiencies have been identified, a mitigation program, or strategy, that considers the global characteristics of the structural system must be developed. General rehabilitation strategies include

- Local modification of components
- Correcting existing irregularities and discontinuities
- Global structural stiffening
- Global structural strengthening
- Mass reduction
- Seismic isolation
- Supplemental energy dissipation

These strategies may each be more effective for certain buildings and rehabilitation objectives and may result in substantially different rehabilitation costs. If possible, in every rehabilitation strategy, adding redundancy to the lateral-force-resisting system in the structure should be considered.

Historic buildings may require special consideration when developing a rehabilitation strategy. For example, the removal of an irregularity by adding a braced frame to balance the strength and/or stiffness of a structure may not be acceptable for a building that has historic features. Seismic isolation may be appropriate alternative to added bracing for limiting the interstory drifts or floor accelerations.

8.2.3 Analysis Procedures

An analysis procedure that is suitable for the building and selected rehabilitation objective must be chosen. The analysis should be used to verify the rehabilitation strategy selected for implementation and may need to be repeated several times with perhaps new strategies to refine the rehabilitation design.

There are four analysis procedures available for use with the systematic rehabilitation method: the linear static, linear dynamic, nonlinear static, and nonlinear dynamic procedures (NDPs).

Linear procedures are appropriate for buildings with regular configurations and those buildings that will behave in a predominantly elastic manner. If the demand–capacity ratio (DCR) of all components is less than 2.0, the linear procedure is an appropriate analysis method. Linear procedures are not recommended for those buildings for which energy dissipation or seismic isolation is a possible rehabilitation strategy or where extensive nonlinear behavior in the structural system is expected. The linear static procedure (LSP) should not be used when the height of the building is greater than 100 ft. or where there are significant vertical or torsional irregularities.

The nonlinear procedures may be used for any structure. When the NDP is used, however, the project should be subject to an independent third-party review by a registered structural engineer with substantial experience in seismic design.

8.2.4 Verification of Rehabilitation Design

The philosophy behind the analysis procedure for the systematic rehabilitation method is (1) to check the capacity of the rehabilitated lateral-force-resisting system, (2) to check the capacity of the gravity system, and (3) to check other potential areas of damage, such as the nonstructural components. The general analysis approach is to calculate the demand (force, moment, deformation or any combination) on an individual component, resulting from the design level of ground motion and to compare that demand with the calculated capacity of the component at the specified rehabilitation objective. If the demand is less than the capacity, the component is acceptable. If not, the rehabilitation design must be modified until the component demonstrates sufficient capacity.

Components are categorized as deformation-controlled or force-controlled, depending on which is most likely to be critical to the component's acceptability. If ductility is most likely to be critical, the component is deformation-controlled. Similarly, if a component is most likely to perform in a brittle manner, it is characterized as force-controlled.

8.2.5 Nonstructural Risk Mitigation

A significant portion of building performance is dependent upon the amount of nonstructural damage that may endanger lives or delay the reoccupancy of a building. The amount of nonstructural risk mitigation is more dependent on the occupancy type of a structure, than its structural building type. For instance, institutional buildings such as hospitals may have significantly more lighting and equipment bracing than a commercial building.

Cost may be incurred due to removing and replacing the various finishes that are affected by the installation of new structural members during a seismic rehabilitation effort. The architectural *cover-up* that may be needed is very difficult to determine, and a quantitative calculation method is not available.

8.2.5.1 Disabled Access Improvements

The requirements of the federally mandated Americans with Disabilities Act (ADA) may be triggered by the seismic rehabilitation work, or the owner may voluntarily choose to improve accessibility for the disabled, in and around the building.

8.2.5.2 Hazardous Material Removal

Hazardous materials such as asbestos, lead-based paint, or contaminated soil may be encountered when performing seismic rehabilitation. The removal of these materials is usually performed under contracts that are separate from those for the seismic rehabilitation, but the user may want to consider the impact of the added costs.

8.2.5.3 Design, Testing and Inspection, and Management Fees

Design fees cover the costs of design professionals such as architects, structural engineers, geotechnical engineers, civil engineers, surveyors, and cost estimators who may be required for studies and design efforts. Testing is often required to establish the strength of existing materials during the design phase. Testing and inspection are necessary during the construction phase to verify compliance with the building codes and project specifications.

8.2.5.4 Historic Preservation Costs

Historic buildings are typically placed on a city, state, or national register of historically significant buildings, and all construction and rehabilitation work on them is subject to review by a regulatory agency. Restrictions are usually placed on alteration to the historic facades, architectural columns, or special ceilings that may need rehabilitation. The historic status of a structure may influence the rehabilitation strategies employed in the seismic rehabilitation work.

8.3 SEISMIC REHABILITATION OF EXISTING BUILDINGS: ASCE/SEI STANDARD 41-06

This standard is the latest generation of performance-based seismic rehabilitation methodology. Developed from FEMA 356, Prestandard and Commentary for the Seismic Rehabilitation of Buildings, it represents the state-of-the-art knowledge in earthquake engineering. It serves as a *technical platform* of consensus criteria on how to deal with some of the major engineering aspects of the seismic rehabilitation of buildings.

The systematic rehabilitation method outlined in the standard allows practitioners to choose design approaches consistent with different levels of seismic safety as required by geographic locations, performance objective, type of building, use of occupancy, or other relevant considerations.

The goals of seismic rehabilitation are important. They include, above all, protecting life and property in future earthquakes as well as protecting investments, lengthening a building's usable life, reducing demands on post-earthquake search and rescue resources, protecting historic structures, shortening business interruption time, maintaining inventories and customers, and reducing relocation needs/demand. Other worthy goals include limiting the need for post-earthquake emergency shelter and temporary housing, minimizing the release of hazardous substances, conserving natural resources, avoiding the costly processes of settling insurance claims and applying for post-disaster aid, protecting savings and contingency funds, reducing the amount of debris to be removed, and facilitating earthquake-stricken community's return to normal patterns of activity.

A four-step iterative process may be used to outline a set of decision points to determine whether seismic rehabilitation efforts are needed and, if so, their potential scope.

The four-step decision process includes

- Defining the problem by conducting preliminary and, if needed, detailed analyses of the risk
- Developing and refining the alternatives for addressing seismic rehabilitation
- Adopting an approach and an implantation strategy
- Securing the needed resources and implementing the seismic rehabilitation measures

The economic, social, and political complexities and the varying seismic environments of the United States are such that seismic rehabilitation programs will have to be tailored to thousands of individual situations.

The core argument for the seismic rehabilitation of buildings is that rehabilitated buildings will provide increased protection of life and property in future earthquakes, thereby resulting in fewer casualties and less damage than would otherwise be the case. It is a classic mitigation strategy not unlike preventive medicine. On the human level, more earthquake-resistant buildings will mean fewer deaths and injuries in an event and therefore lower demand on emergency medical services, urban search and rescue teams, fire and law enforcement personnel, utilities, and the providers of emergency shelter. In the commercial sector, less damage to structures will mean enhanced business survival and continued ability to serve customers and maintain markets or market shares. More specifically, for commercial enterprises, seismic rehabilitation will better protect physical and financial assets, reduce inventory loss, shorten the business interruption period, avoid the need for relocation, and minimize secondary effects on suppliers, shippers, and other businesses involved in support serves or product cycles. In short, seismic rehabilitation as a pre-event mitigation strategy actually will improve post-event response by lessening life loss, injury, damage, and disruption.

As stated earlier, rehabilitation of existing buildings to better resist future damaging earthquakes truly is *preventative medicine*. While seismic rehabilitation costs money, it can significantly reduce future losses and, in economic terms, can be considered an investment to protect assets currently at risk.

Despite the fact that each building has *its own story* when it comes to seismic rehabilitation, the generic, four-step process outlined earlier is generally useful. As in many processes of this type, this generic four-step model emphasizes the feedback function at every step because no existing building seismic rehabilitation effort can possibly succeed in isolation, no matter how splendid the technical components. Seismic rehabilitation takes place in a wide variety of socioeconomic and political contexts, and continuous feedback and adjustments are necessary for success.

Step 1: *Defining the problem* actually comprises two substeps: *preliminary analysis* and *detailed analysis*. Preliminary analysis entails an initial and perhaps even cursory survey of the general issues raised by an identified earthquake threat. Because earthquake-induced life and property losses tend to be concentrated in building types already known to be vulnerable, once a relatively specific degree of seismic risk and likely consequences have been identified, the issue of seismic rehabilitation arises almost immediately.

Step 2: *Develop and refine alternatives* involves using the data assembled under Step 1 to develop and refine alternative approaches. It provides a kind of *menu* delineating seismic rehabilitation options and is usually a very long and involved process because it addresses key variables such as the desired performance levels and the scope of the approach.

Step 3: *Adopt an approach and implementation strategy* is the decision point at which the building owner receives input from other sources and weights the alternatives (not to be ignored is the alternative of doing nothing). It is a decision that allocates scarce resources, costs, and benefits. Finally, the decision to act sets in motion the necessary organizational routines to actually yield activity, in this case seismic rehabilitation.

Step 4: *Secure resources and implement* is the critical process that turns a decision to rehabilitate into its physical result—safer, more seismically resistant buildings. Without resources to carry out seismic rehabilitation, the adoption of an approach is simply *a piece of paper*.

Earthquake-vulnerable buildings exist nationwide, but the earthquake hazard is not uniform across the country. Moreover, awareness of the earthquake hazard, the precursor to any action, varies even more than the hazard itself. Therefore, tackling the earthquake-vulnerable building problem takes place in an incredibly diverse set of geographic, social, economic, and political environments. Further complicating the situation is the fact that no two buildings ever seem to present exactly the same problems. Each building has its own earthquake-vulnerability profile—location, architecture, structural system, occupancy, economic role, and financing. In other words, each seismically deficient building has its own story asking for one of a kind retrofit.

Abating the risk posed by earthquake-hazardous buildings often brings into play social, economic, psychological, and various other considerations that make seismic rehabilitation very complex. The most recent examples of unpleasant earthquake lessons came from the 1994 Northridge earthquake and the 1995 Kobe earthquake, which revealed as vulnerable steel frame buildings, long believed to be the most earthquake-resistant type of construction.

The damage to as many as 400 steel buildings raised many serious questions for the design profession. Because many damaged structures were designed using the latest building codes and building according to modern construction practices, seismic building codes for steel construction have been essentially invalidated.

In sum, earthquakes usually teach painful, if not tragic lessons, but the learning generates state-of-the-art advances in earthquake engineering. The reader is referred to the ASCE standard, *Seismic Rehabilitation of Existing Buildings* (ASCE/SEI 41–06).

ASCE/SEI 41-06 endorses the use of PBD solutions for seismic rehabilitation of buildings. The chosen performance of the building may vary from preventing collapse to a near-perfect building that would survive an expected earthquake without a scratch. The standard allows owners to select their desired performance objective and permits designers to choose their own approaches to achieve the desired results rather than strictly adhering to the prescriptive requirements of codes. Instead of dictating how to achieve a given design goal, PBD emphasizes the goals that must be met and sets the criteria for acceptance. This way, engineers are free to innovate without running afoul of specific code provisions, within certain limits.

ASCE/SEI 41-06 outlines criteria and methods for ensuring the desired performance of buildings at various performance levels selected by the owners with input from their design professionals. The guidelines allow owners to select a level of seismic upgrade that not only protects lives, a goal of all building codes, but also protects their investment.

ASCE/SEI 41-06 is a radical departure from current practice in that it seeks to provide the structural engineering profession with tools to explicitly, rather than implicitly, design for multiple, specifically defined, levels of performance. These performance levels are defined in terms of specifically limiting damage states, against which a structure's performance can be objectively measured. Recommendations are developed as to which performance levels should be attained by buildings of different occupancies and use. This tiered specification of performance levels at predetermined earthquake hazard levels becomes the design performance objective and a basis for design. It recognizes the importance of the performance of all the various component systems to the overall building performance and defines a uniform methodology of design to obtain the desired performance.

8.3.1 Overview of Performance Levels

ASCE/SEI 41-06 sets forth a menu of four rehabilitation objectives associated with four earthquake hazard levels. The rehabilitation objectives are

1. Operational performance
2. Immediate occupancy performance
3. Life safety performance
4. Collapse prevention performance

Each of these performance levels is associated with defined levels of damage to structural, architectural, mechanical, and electrical building components as well as tenant furnishings. The designer is referred to ASCE/SEI-41 Tables CI.3 through CI.7 for an overview of where each performance level falls within the overall spectrum of possible damage states. From these tables, the designer may infer, for example, a building designed for top-of-the-line performance using higher earthquake hazard levels is likely to come out scratch-free, delivering performance well above the code minimum for LS level. On the other hand, much less is expected of a building rehabilitated to a CP performance level. It is deemed to have fulfilled its obligations if it remains standing during and after a large earthquake: any other damage or loss is acceptable.

The four levels of earthquake hazard recognized in the development of design performance objectives are

1. Frequent earthquakes, having a 50% chance of exceedence in 30 years (43-year mean return period)
2. Occasional earthquakes, having a 50% chance of exceedence in 50 years (72-year mean return period)
3. Rare earthquakes, having a 10% change of exceedence in 50 years (475-year mean return period); also called basic safety earthquake (BSE-1) and design basis earthquake (DBE)
4. Very rare earthquakes, having a 10% chance of exceedence in 100 years (950-year return period); also called BSE-2 and MCE

In order to execute a PBD, a series of design parameters and acceptance criteria are given for each performance level for the various structural and nonstructural components. Design response parameters are defined at an element level in terms of element forces, interstory drifts, and plastic rotations. These can be derived from a structural analysis of building response to a particular design earthquake. Acceptance criteria are the limiting values for design parameters in order to attain a given performance level. For example, if interstory drift ratio is a design parameter used for a certain class of building, acceptance criteria would be the drift ratios defined for each performance level. Typical drift ratios normally considered in design are 0.020 for the near collapse level, 0.015 for the LS level, 0.01 for the operational level, and 0.005 for the fully operational level. A wide variety of potential design parameters may need to be defined including deformation, strength, and energy-based parameters. The purpose of ASCE/SEI 41-06 is to provide a consensus-backed, professionally accepted, nationally applicable, seismic rehabilitation standard. It can be used as a tool by design professionals, a reference document by building regulatory officials, and a foundation for the future development and implementation of building code provisions and standards related specifically to existing buildings. The absence of such a standard has been the primary barrier to widespread seismic upgrading of buildings in the United States.

In new buildings, the structural system can be controlled to fit a set of preconditions or a configuration to satisfy the design objectives prescribed by building codes. The degree of nonlinear behavior can be designed to be consistent throughout the structural system, allowing a single seismic reduction factor, R, to be used for the entire building.

Experience in seismic design over the past 100 years has shown that buildings designed to resist ground shaking from an earthquake with a 10% chance of exceedence in 50 years, at a LS level of performance, have been able to resist the strongest earthquake without collapse. This experience has given structural engineers enough confidence to design new structures in which ductile details are specified, properties of materials used in construction are controlled, and stringent requirements of testing and inspection are specified.

Assessing the seismic vulnerability of existing buildings is an entirely different problem. This is because, for existing buildings, structural details and the properties of materials must be confirmed or assumed from available information augmented by testing and inspection. Conservative assumptions consistent with the quality of the information available must be made prior to seismic evaluation. The engineer has no control over the structural system or its configuration. The existing building may not fit prescriptive details to permit code-type analysis. Nonlinear behavior of the components of the structural system will probably not be consistent. Thus, the properties of each component must be separately studied. Because of the inconsistent levels of reserve capacity in existing buildings and the differences between the 10% in a 50-year earthquake and the MCE in various regions of the country, it is inappropriate that rehabilitated buildings be designed to resist a single level of earthquake shaking. Therefore, using an entirely different approach, ASCE/SEI-41-06 provides a basis of rehabilitation designs for a variety of structural performance levels, ranging from enhanced performance to CP. It emphasizes the idea that seismic rehabilitation should be directed to controlling deformation in order to minimize damage. Use of all existing seismic resistance is permitted in the evaluation. Acceptance criteria tailored to recognize the deformation capacity of all existing as well as enhanced or new components are provided.

The seismic loads used in the evaluation are based on a suite of USGS-developed acceleration maps including four key maps. Two of these are BSE-1 maps of acceleration response spectra having a 10% probability of exceedence in 50 years. The other two are BSE-2 maps of acceleration response spectra for the MCE—modified 2% probability of exceedence in 50-year maps: both BSE-1 and BSE-2 maps are given for 0.2 s period (short period) and 1 s period buildings.

8.3.2 Permitted Design Methods

Two methods are permitted by ASCE/SEI 41-06, a simplified method and a systematic method. The simplified approach is for the rehabilitation design of small buildings of regular configuration and is intended to fulfill limited objectives. Partial rehabilitation measures that seek to eliminate high-risk building deficiencies such as exterior falling hazards are included in the technique.

The systematic rehabilitation method discussed at length in this section is applicable to any building. It is a component- and element-based design. In this method, global seismic response of the building is sought with unreduced seismic loads (i.e., with a global R-factor of unity). In the seismic evaluation, all components and seismic elements are considered with their individual deformation and force-resisting characteristics. It is a deformation-based design with the explicit rather than tacit acknowledgment that seismic elements and components behave in a nonlinear manner.

Any of the following analysis procedures may be used in the rehabilitation study and upgrade design:

- *Linear static procedure (LSP)*: This procedure replaces the equivalent lateral-force procedure included in most seismic design codes. It incorporates techniques for considering the nonlinear response of individual seismic elements. The distribution of forces is similar to equivalent lateral-force procedures for new buildings.
- *Linear dynamic procedure (LDP)*: In this method, the modeling and acceptance criteria are similar to those of LSP. However, calculations are carried out using modal spectra analysis or time-history analysis using response spectra or time-history records that are not modified to account for inelastic response for distribution of forces.
- *Nonlinear static procedure (NSP)*: This method is frequently referred to as a pushover analysis. It has been in use for some time without specific guidance from building codes and standards regarding modeling assumptions and acceptance criteria. This is now alleviated to some extent because previously FEMA 356 and currently ACSE/SEI 41-06 have set forth specific procedures.
- *Nonlinear dynamic procedure (NDP)*: The modeling approaches and acceptance criteria for this method are similar to those of NSP. It differs from NSP in that response calculations are made using inelastic time-history dynamic analysis to determine distribution of forces and corresponding internal forces and system displacements. Peer review by an independent engineer with experience in seismic design and nonlinear procedures is recommended and often times mandated because this method requires assumptions that are not included in ASCE/SEI 41-06.

8.3.3 Systematic Rehabilitation

The process of arriving at a systematic rehabilitation design includes the following steps:

1. Determination of seismic ground motions
2. Determination of as-built conditions
3. Classification of structural components into primary and secondary components
4. Setting up of analytical models and determination of design forces
5. Ultimate load combinations: combined gravity and seismic demand
6. Component capacity calculations, QCE, and QCI
7. Capacity versus demand comparisons
8. Development of seismic strengthening strategies

First, the seismic hazard for the site is established by determining the probable ground shaking (spectral acceleration) from either seismic hazard maps or a site-specific investigation. Other site hazards such as liquefaction, lateral spreading, and land sliding are determined from site reconnaissance, existing documentation, or a subsurface investigation.

The desired performance level is then established. This requires close communication with the client, using damage descriptions for each performance level as a tool to get ideas across. The damage descriptions associated with each performance level can be used to inform and assist the client to make a decision of the preferred performance level. This distinction is required because the acceptance criteria are different for each type of component. The primary components are parts of the building's lateral-force-resisting system, whereas the secondary components are those not required for lateral-force resistance, although they may actually resist some lateral forces. The analysis is performed by considering general requirements such as $P\Delta$ effects, torsion, overturning, continuity, integrity of elements, and building separations.

New or modified components are evaluated using the same standards as existing components, and the designs are completed by comparing capacities with demands for each component. The components and connections are redesigned where demand exceeds capacity and analysis is iterated to confirm the design. Nonstructural components are verified for the performance level and rehabilitation objective selected.

It should be noted that selection of a rehabilitation strategy follows confirmation of seismic deficiencies. From among many possible strategies, the strategy most likely to meet requirements is selected. Some possible strategies are modification of components, removal of irregularities and discontinuities, global strengthening and stiffening, mass reduction, seismic isolation, and energy dissipation.

8.3.3.1 Determination of Seismic Ground Motions

Two characteristic earthquakes, referred to as BSE-1 and BSE-2, are of particular importance. These generally correspond to return periods of 474 and 2475 years, respectively, and are commonly referred to as earthquakes with a 10% chance of exceedence in 50 years and a 2% chance of exceedence in 50 years, respectively. At sites close to major faults, the probabilistic estimates of ground motion are capped by deterministic ones. The engineer has three choices for determining the acceleration response spectra corresponding to these earthquakes: (1) use spectral response acceleration contour maps developed by the USGS, (2) use CD-ROM available from the USGS, or (3) engage a geotechnical engineer to develop site-specific response spectra based on the geologic, seismologic, and soil characteristics associated with the specific site. For some sites, option three may be the only permitted method. However, to define as precise a seismic demand as possible, it is common practice to engage a geotechnical engineer to perform a site-specific study for developing response spectra corresponding to specific return periods. The geotechnical report also typically addresses other seismic hazards such as liquefaction, lateral spreading, or potential for land sliding at the site.

It should be noted that acceleration response spectra for earthquake hazard levels corresponding to probabilities of exceedence other than the BSE-1 and BSE-2 earthquakes can be determined by following procedures specified in ASCE/SEI 41-06.

8.3.3.2 Determination of As-Built Conditions

In this step, the following tasks are performed:

- Field observation
- Review of available documents, including plans, specifications geotechnical reports, shop drawings, test records, and maintenance histories
- Review of information regarding material standards and construction practices for location and date of construction

- Destructive and nondestructive testing of selected building components for determination of material properties and configuration of details
- Interviews with people knowledgeable about the building (i.e., owners, tenants, maintenance personnel, architects, engineers, and builders)

As a measure of the knowledge gained from this investigation, engineers assign a numerical value to the knowledge coefficient k. (k = 1.0 if the available information is reliable; if not, k = 0.75.)

8.3.3.3 Primary and Secondary Components

Before setting up the analytical model, structural components are classified as either primary or secondary. Primary components are those that provide the structure's basic lateral resistance. Secondary components are those that do not and, as such, are permitted to experience more damage and displace more than the primary components. Additionally, components are further classified as either deformation-controlled, if they are capable of sustaining the loads when strained inelastically, or force-controlled, if they are not capable of sustaining load when strained inelastically.

8.3.3.4 Setting Up Analytical Model and Determination of Design Forces

An analytical model of the building is set up to represent the structure's dynamic behavior. Although 2D models may be adequate, current practice is to use 3D models to account for torsion, plan and vertical irregularities, and nonuniform distribution of building mass. Only the primary components are modeled, with the stipulation that the secondary elements, if used in the model, cannot exceed 25% of the total structural stiffness. If they do, then some of the secondary components must be reclassified as primary components.

8.3.3.4.1 Calculation of Building Period

The building period, T, is calculated by using (1) modal analysis procedure, Method 1, or (2) empirical equations, Method 2, or (3) Rayleigh's method, Method 3. Method 1 is the preferred method. The fundamental period T is obtained by an eigenvalue analysis using the analytical model. This is the more commonly used method, particularly in seismic vulnerability studies.

In Method 2, the period T is determined using the following equation:

$$T = C_t h_n^{\beta}$$

where

T is the fundamental period, in seconds, in the direction under consideration
C_t = 0.035 for steel moment frames
= 0.030 for steel eccentrically braced frames
= 0.020 for all other framing systems
h_n is the building height, in feet, above shear base
β = 0.80 for steel moment frames
= 0.75 for all other framing systems

However, there is a major difference worthy of note between building code procedures for new buildings and the ASCE/SEI 41-06 approach. Unlike the codes, there is no maximum limit on period calculated using Method 1. The intent of this omission is to encourage the use of more advanced analysis such as computer dynamic analysis. It is believed that sufficient controls on analysis and acceptance criteria are present within the ASCE/SEI 41-06 to provide reasonably conservative results even though there is no upper limit for the period obtained by Method 1.

8.3.3.4.2 *Determination of Pseudolateral Load (Base Shear)*

The base shear, also referred to as pseudolateral force, for use in the design of new components and in the verification of existing components of the lateral-force-resisting system is given by

$$V = C_1 C_2 C_m S_a W \tag{8.1}$$

where

- V is the pseudolateral load (the base shear)
- C_1 is the modification factor that accounts for the difference in the structure's elastic and inelastic displacement amplitude (its value is equal to 1.0 for $T > 1.0$ s [see ASCE/SEI 4-06 Section 3.3.1.3 for additional information])
- C_2 is the modification factor that represents the effect of strength and stiffness degradation of the components on maximum displacement response (for periods greater than 0.7 s, $C_2 = 1.0$)
- C_m is the effective mass factor to account for higher modal mass participation equal to 1.0 if building fundamental period T is greater than 1.0 s (see ASCE/SEI 4-06 Table 3.1 for additional information)
  = 0.9 for three or more story concrete frame buildings
  = 0.8 for three or more story concrete shear wall or pier-spandrel buildings
- S_a is the response spectrum acceleration at the fundamental period and damping ratio of the building
- W is the effective seismic weight of the building, including the total dead load and applicable portions of other gravity loads listed as follows:

- In storage and warehouse occupancies, a minimum of 25% of floor live load.
- An allowance for partition load equal to the actual partition weight or a minimum of 10 psf of floor area, whichever is greater.
- The total operating weight of permanent equipment.
- The effective snow load equal to 20% of the design snow load if design snow load exceeds 30 psf. If not, the effective snow load may be taken to be zero.

8.3.3.4.3 *Vertical Distribution of Base Shear*

The lateral force F_x, applied at any level x, is determined in accordance with the following equations:

$$F_x = C_{vx} V \tag{8.2}$$

$$C_{vx} = \frac{w_x h_x^k}{\sum_{i=1}^{n} w_i h_i^k} \tag{8.3}$$

where

- C_{vx} is the vertical distribution factor
- V is the pseudolateral force (base shear)
- w_i and w_x are the portion of the total gravity load of the building W located or assigned to level i or x
- h_i and h_x are the height in feet from the base to level i or x
- k is an exponent related to the building period as follows:

If the building period is 0.5 s or less, $k = 1$.
If the buildings period is 2.5 s or more, $k = 2$.

Linear interpretation is used for intermediate values of the period T.

8.3.3.4.4 Diaphragm Design Force F_{px}

Floor and roof diaphragms shall be designed to resist the combined effects of the inertial force F_{px} calculated at the diaphragm level with the horizontal combined forces resulting from offsets in the vertical seismic elements above and below the diaphragm. The design force shall not be less than

$$F_{px} = \frac{\sum_{i=x}^{n} w_{px}}{\sum_{i=x}^{n} w_i} + \rho V_{px} \tag{8.4}$$

The first term of the earlier equation need not exceed 0.4 S_{DS} I W_{px} but shall not be less than 0.20 S_{DS} I W_{pn}

where
- V_{px} is the forces due to transfer of vertical resisting elements above the diaphragm or changes in relative lateral stiffness in the vertical elements
- F_{px} is the total diaphragm inertial force at level x
- F_i is the lateral load at level i
- w_i is the portion of the effective seismic weight w
- w_{px} is the portion of the effective seismic weight w located at or assigned to floor level x
- ρ is the redundancy factor applicable to diaphragms in SDC D, E, or F buildings

8.3.3.5 Combined Gravity and Seismic Demand

In this step, the earthquake actions Q_E obtained from the analysis for the unreduced response spectra are combined with the gravity actions, Q_G, to determine the demand imposed on the component. When the effects of gravity and seismic loads are additive, an upper-bound value for gravity loads is estimated by using the following load combinations:

$$Q_G = 1.1(Q_D + Q_L + Q_s) \tag{8.5}$$

And when the effects of gravity and seismic loads are counteracting, a lower- bound value of the gravity load is estimated by using 90% of the dead load:

$$Q_G = 0.9Q_D \tag{8.6}$$

where
- Q_D is the dead-load action
- Q_L is the effective live-load action equal to 25% of the unreduced design live load but not less than the actual live load
- Q_s is the effective live-load action equal to 20% of the design snow load where the design snow load exceeds 30 psf (no part of the snow load need be included if the design snow load is less than 30 psf)

Next, the gravity and seismic loads are combined together to determine the demand using the following equations:

For deformation-controlled actions:

$$Q_{UD} = Q_G + Q_E \tag{8.7}$$

For force-controlled actions:

$$Q_{UF} = Q_G + \frac{Q_E}{C_1 C_2 J} \quad (8.8)$$

where

Q_{UD} is the deformation-controlled demand due to gravity loads and earthquake loads
Q_{UF} is the force-controlled demand due to gravity loads in combination with earthquake loads
J is the force-delivery reduction factor, greater than or equal to 1.0
This factor is used to estimate the actual forces delivered to force-controlled components by other yielding components. The values of J are
$J = 2.0$ in zones of high seismicity
$= 1.5$ in zones of moderate seismicity
$= 1.0$ in zones of low seismicity

Alternatively, J may be taken as the smallest DCR for the components in the load path delivering force to the component being designed. The minimum value of J, the force-delivery reduction factor, is 1.0. The reader is referred to ASCE/SEI 41-06 Section 3.4 for further definition of terms used in this section.

8.3.3.6 Component Capacity Calculations Q_{CE} and Q_{CL}

ASCE/SEI 41-06 specifies two different equations for evaluating component capacities depending upon whether the action of the component is deformation-controlled (Q_{CE}) or force-controlled (Q_{CL}). The subscript E in Q_{CE} stands for expected capacity, whereas L in Q_{CL} stands for lower-bound capacity. The subscript C in both Q_{CE} and Q_{CL} stands for capacity. The term design action is used to define forces and moments in the components due to seismic and gravity effects.

The two types of actions—deformation-controlled actions and force-controlled actions—are used to distinguish a ductile behavior from a brittle behavior.

8.3.3.6.1 Deformation-Controlled Actions

Deformation-controlled actions in simple terms refer to forces and moments in a component that has recognizable nonlinear deformation characteristics. Because of possible anticipated nonlinear response, the design forces and moments in the component are permitted to exceed their capacity. The acceptance criteria, discussed presently, take this overload into account through the use of an m-factor, which in a conceptual sense is an indirect measure of the nonlinear deformation capacity of the component.

8.3.3.6.2 Force-Controlled Actions

Force-controlled actions differ from deformation-controlled actions in that they do not have a recognizable inelastic response. Therefore, demands for force-controlled actions must not exceed the calculated capacity (i.e., there are no m-factors in the acceptance criteria). It should be noted, however, that the calculated design force (demand) itself is reduced by the C_1 C_2, C_3, and J factors before demand is compared to capacity.

An ideal procedure for determining the magnitude of force-controlled actions is by identifying an inelastic limit state for the component and then, by statics, evaluation of the corresponding force-controlled action. For example, seismic shear in a frame–beam is determined from equilibrium

TABLE 8.1
Default Lower-Bound Material Strengths for Archaic Materials[a,b]

Year	Material	Lower-Bound Yield Strength (ksi)	Lower-Bound Tensile Strength (ksi)
Pre-1900	Cast iron	18	—
Pre-1900	Steel	24	36

[a] Modified from unit stress values in AISC *Iron and Steel Beams from 1873 to 1952* (AISC 1983).

[b] Properties based on tables of allowable loads as published in mill catalogs.

considerations of a free-body diagram of the beam with a moment equal to the expected moment strength plus gravity moments. However, it is acceptable to determine force-controlled actions from the equation given earlier, where it is not possible to identify a well-defined limit state (Table 8.1; the reader is also referred to ASCE/SEI 41-06, Table 5-2 and Table 5-3.).

8.3.3.6.3 Capacity versus Demand Comparisons

In this step, the component capacities are compared with the demand due to earthquake and gravity loads. If the capacity of a component exceeds the demand imposed on it by the seismic and gravity load combinations, the component is judged to satisfy the performance criteria. If not, a more refined technique such as a pushover analysis is performed before declaring the component deficient.

Two equations are given in ASCE/SEI 41-06 for verifying the acceptance criteria.

$$\text{For deformation-controlled actions: } m_k\, Q_{CE} > Q_{UD}$$

$$\text{For force-controlled actions: } \kappa\, Q_{CL} > Q_{UF}$$

where

m is the modifier given in Tables 5-5 of ASCE 41-06 for linear procedure (the table takes into account the expected ductility of the component associated with the action being verified at the selected structural performance level)

Q_{CE} is the expected strength of the component at the deformation level under consideration for deformation-controlled actions

κ is the knowledge factor defined earlier

Q_{CL} is the lower-bound strength of a component for force-controlled actions

Q_{UD} is the deformation-controlled demand due to gravity and earthquake loads

Q_{UF} is the force-controlled demand due to gravity and earthquake loads

Numerical values of m for structural steel components verified using linear procedures are given in Table 5.5 of ASCE/SEI 41-06. Abbreviated version of the table for frame–beams and columns is given in Table 8.8 of this work.

8.3.4 Development of Concepts for Seismic Upgrading

The results of the detailed structural analysis will identify, for a given building, the deficiencies with respect to the acceptance criteria of the various structural elements or systems. These results will be carefully reviewed in the development of alternative design concepts to upgrade the structure to meet the acceptance criteria. Several alternative concepts will be developed for each building unless justification can be shown for fewer alternatives (e.g., it may be shown that the obvious cost-effective solutions for a deficient steel braced building are either to replace the existing bracing with stronger bracing members or to add new bracing so that only two alternatives need to be developed). Each concept will be developed to the extent that will permit a reasonable cost estimate to be made. The extent of removal of existing construction will be indicated; the sizes and locations of new, replaced, or strengthened structural members will be indicated; typical structural connections will be shown and the extent and schematic details for upgrading nonstructural elements will be indicated. The following considerations will be addressed in the development of the design concepts:

- Structural systems
- Configuration
- Horizontal diaphragms and foundation ties
- Eccentricity
- Deformation compatibility of new and existing materials
- Base isolation and energy dissipation

8.3.4.1 Structural Systems

Structural systems: The development of the structural upgrading concept requires a complete understanding of the existing vertical- and lateral-load-resisting systems of the existing building. The designer must be able to determine the consequences that the removal, addition, or modification of any structural or nonstructural element will have on the performance of the strengthened building.

Gravity load-resisting system: An evaluation of the existing vertical-load-carrying structural system will be made to determine the effects that the seismic upgrading may have on future performance of the building to resist dead and live gravity loads. The evaluation will include a description of the components of the vertical-load-carrying system and the load path from the source of the dead and live loads to the foundations.

a. *Floor and roof framing*: In most buildings, the horizontal framing systems (i.e., floors and roofs) will participate in the lateral-force-resisting system as a diaphragm in addition to supporting gravity loads. As part of the seismic upgrading, the floor and roof systems may require modifications (e.g., horizontal bracing) that will add to the dead load; thus, the capacity of the modified system must be evaluated for the new loading conditions.
b. *Vertical structural elements*: Vertical-load-resisting elements such as columns, bearing walls, and framing systems may also be affected by the seismic upgrading. In addition to the added weight that may be imposed due to the seismic strengthening, these elements may participate in the lateral-force-resisting system. For example, bearing walls may also be used as shear walls and frames may be strengthened or braced to resist seismic forces. If these framing elements are not used for the lateral-force-resisting system, they will have to be analyzed for deformation compatibility. This analysis will include the effects of the lateral displacements due to extreme seismic motion on the vertical-load-carrying capacity of the vertical structural elements.
c. *Foundations*: If the seismic upgrading adds weight or redistributes the existing gravity loads, the foundations must be analyzed for the additional gravity loads combined with the horizontal and overturning forces associated with the seismic lateral-force design. The structural system that is designed to resist the seismic forces basically relies on a complete 3D space frame, a coordinated system of shear walls or braced fames with horizontal diaphragms, or a combination of both. In the evaluation and upgrading of an existing

structure, it is sometimes difficult to identify an existing lateral-force-resisting system. Innovative analytical procedures and reliance on existing materials and systems that are not generally considered for new construction are required to determine the load paths and capacities of the existing structures. When an existing structure is not adequate to resist the prescribed lateral forces, as determined by the detailed structural analysis, strengthening of the existing lateral-force-resisting system will be required.

8.3.4.2 Configuration

If the existing building is highly irregular in plan configuration or is comprised of units with incompatible seismic response characteristics, severe problems that cannot be resolved by strengthening or upgrading could develop at the connection between two units. In such cases, consideration should be given to separating the two units with a structural expansion joint. Each unit should have a complete system for resisting vertical as well as lateral loads. Structural members bridging the joint with sliding supports on the adjacent unit should be avoided. Expansion joints should provide for 3D uncoupled response of each of the separate units of a building but need not extend through the foundations.

8.3.4.3 Horizontal Diaphragms and Foundation Ties

Every upgraded existing building will have either a rigid or semirigid horizontal floor diaphragm. Roof diaphragms may be flexible or semiflexible. Foundation ties between pile caps and caissons will be provided and diaphragms and foundation ties that do not comply with these requirements will be strengthened or replaced unless proper justification can be provided for waiving the deficiency.

8.3.4.4 Eccentricity

Provisions shall be made for the increase in shear resulting from the horizontal torsional moment due to an eccentricity between the center of mass and the cr. In the development of upgrading concepts for existing buildings, when the vertical shear-resisting elements must be strengthened, supplemented, or replaced with new elements, consideration will be given to location of new or strengthened elements so as to reduce any eccentricity between the cr and the center of mass.

8.3.4.5 Deformation Compatibility of New and Existing Materials

The compatibility of the deformation characteristics of the existing elements and the new strengthening elements will be considered in the strengthening design of the building:

a. *Relative rigidities*: When lateral forces are applied to a building, they will be resisted by the various elements in proportion to their relative rigidities. The lateral stiffness of a structure is calculated on the basis of the deformation characteristics of the lateral-force-resisting elements. If the structure is to be strengthened to resist seismic forces, the new structural elements must be more rigid than the existing elements if they are to take a major portion of the lateral forces and reduce the amount of force that is taken by the existing elements. Both the relative rigidities and strengths of all lateral-force-resisting elements must be considered. To illustrate, the following two examples are given:

   Existing steel moment frame strengthened by diagonal steel bracing. Assume an existing steel moment-frame building that has a one inch story displacement due to seismic forces. Diagonal bracing is added to the moment frames to reduce the lateral displacement to 0.1 in. for the same force level. Thus, it can be estimated that the bracing resists about 90% of the lateral force and the frame resists about 10%. If the original moment frame can safely resist 10% of the seismic forces, the bracing system is effective. (Note this example neglects the possible increase in magnitude of the seismic forces due to a decrease in the period of vibration.)

b. *Deformation characteristics of structural elements*: The accuracy of the relative rigidity calculations is dependent on the accuracy of the assumptions used for determining the stiffness characteristic of each element or system. When the entire lateral-force-resisting elements are of the same material and have similar deformation characteristics (e.g., flexural and/or shearing deformations), the relative rigidities can be calculated with reasonably good accuracy. However, when there is a variety of materials and cross-sectional shapes, the confidence level on the accuracy of the relative rigidities is greatly reduced. When the confidence level is low, the range of stiffness values should be enveloped and the distribution should be overlapped to account for the inaccuracies. Structural elements that require special consideration in determination of relative rigidities include

   Concrete: cracked versus uncracked

   Shear walls: participation of intersecting walls (e.g., effective flange widths) and effects of openings

   Steel frames: participation of concrete fireproofing, floor slab and framing, and infill walls (structural and nonstructural)

c. *Evaluation of structural elements not part of the lateral-force-resisting system*: Structural elements that are not part of the lateral-force-resisting system will be evaluated for the effects of the deformation that occur in the lateral-force-resisting system.

d. *Protection of existing brittle elements*: Brittle elements that are not part of the lateral-force-resisting elements are susceptible to damage if they are forced to conform to the deformation of the lateral-force-resisting system. In order to protect these elements from the possibility of being subjected to large distortions, provisions can be made to allow the structural system to distort without forcing distortion on the brittle elements. When rigid walls are locked in between columns, a method of isolation may be required at each end of the wall.

8.3.4.6 Base Isolation and Energy Dissipation

a. *Base isolation*: Design strategies that significantly modify the dynamic response of a structure at or near the ground level are generically termed base isolation. This is usually achieved by introduction of additional flexibility at the base of the structure. The objective is to force the entire superstructure to respond to vibratory ground motion as a rigid body with a new fundamental mode based on the stiffness of the isolation devices. This strategy is particularly effective for short stiff buildings (i.e., with a fundamental mode less than 1 s). For these buildings, it is feasible with the isolation devices to develop a new fundamental mode with a period of about 2 s. For most sites (e.g., those with a predominant site period less than 1 s), the new fundamental mode period will be beyond the portion of the response spectrum that is subject to dynamic amplification and the response of the structures will be greatly reduced. The concept of base isolation is not new; for many years, it has been used to reduce the vibration of equipment and machinery with springs, resilient mounting, and shock absorbers. Similarly, bridges and other simple structures have been isolated to reduce vibration and noise, and in some instances, to reduce the seismic such as buildings has been made possible in recent years due to greatly improved computational capability and development and marketing of the isolation assemblies. A typical installation consists in large pads of natural or synthetic rubber layers bounded to steel plates in a sandwich assembly. The isolator assembly, as well as all connecting elements and building services, must be capable of resisting the design spectral displacement corresponding to the new fundamental mode (a recent California installation has base-isolation assemblies that can be deflected elastically up to 18 in.). Some base-isolation assemblies may have a lead core or other devices to increase damping and thus decrease the response at the isolator. Because of the uncertainties associated with ground motion predictions, most seismic base isolators are designed with fail–safe provisions to arrest the motion of the building prior to development of instability due to excessive displacement of the isolator. Base isolation can be an effective strategy to reduce the seismic response of buildings provided careful consideration is given to the amplitude and frequency content of the expected ground motion, the design of the connecting services to accommodate the expected displacements, and provision of fail–safe mechanisms as described earlier. The ability of base isolation to reduce seismic response is even more attractive in application to existing buildings with inadequate seismic resistance. However, in addition to the considerations described earlier, installation of base isolation in an existing building entails accurate determination of the magnitude and location of the vertical loads, a rigid diaphragm above the isolators to collect and distribute the lateral loads, and careful underpinning and jacking of the existing structure in order to effect a systematic transfer of the existing foundation loads to the base-isolation device.

b. *Energy dissipation*: An effective means of providing a substantial level of damping is through hysteretic energy dissipation. Some structures (e.g., properly designed ductile steel and concrete frames) exhibit additional damping and reduced dynamic response as a result of the limited yielding of structural steel or concrete reinforcement. Mechanical devices, designed to increase structural damping, have been developed using mild steel in flexure or torsion and the deformation of lead by extrusion or shear. Viscoelastic materials in shear have been used successfully to control wind vibration in tall buildings and similar installations have been proposed for reducing the seismic response of buildings. Friction is another source of energy dissipation that can be used to reduce dynamic response and limit deflections. However, frictional resistance is difficult to quantify and its reliability may be questionable. Hydraulic damping has been successfully used on machinery and bridge structures, but there are no known applications used to modify building response. The use of structural dampers to reduce the seismic response of existing buildings may be feasible and cost-effective.

8.3.4.7 Selection of Strengthening Technique

The selection of an appropriate strengthening technique for the upgrading of an existing building that does not comply with the acceptance criteria will depend upon the type of structural systems in the existing building and the nature of the deficiency. In some cases, the selection may be influenced by other than structural considerations. For example, a requirement that the building be kept operational, with minimal interference to the functions that it provides during the structural modifications, may dictate that the modifications be restricted to the periphery of the building with as much work as possible accomplished on the exterior of the buildings. On the other hand, it may be possible temporarily to relocate the function and occupants of an existing building that is to be upgraded. This, of course, provides more latitude in the selection of appropriate and cost-effective strengthening techniques. In many cases, seismic upgrading is accomplished concurrently with functional alterations, renovations, and/or energy retrofits. In these cases, the selected structural modification scheme should be the one that best suits the requirements of all the proposed alterations.

8.3.5 Seismic Risk Reduction Strategies

The likelihood of earthquake damage is a function of the seismic hazard at the site and the seismic vulnerability of the building. Seismic hazard is addressed in the context of site selection and evaluation of site-specific earthquake-related hazards. Seismic vulnerability is evaluated in the context of PBD, a relatively new tool that engineers can use to adjust up or down the earthquake-resisting capacity of a building, depending on the desired performance in future earthquakes. Seismic risk reduction strategies are techniques that reduce the likelihood of damage to a structure. The damage potential is minimized by either reducing the site hazards associated with a building or by increasing the expected performance of the building.

8.3.5.1 Reduce Site Hazards

An owner can reduce site hazards by reducing the intensity of earthquake shaking expected at the building site over the life of the structure and by eliminating or reducing the potential for other seismic hazards, such as fault rupture, liquefaction, landslide, and inundation. Several techniques for accomplishing this are described in the following:

Locate the building in a region of lower seismicity, where earthquakes occur less frequently or with typically smaller intensities. This option is generally the most effective strategy solely in terms of reducing the potential for earthquake damage to a facility, whether it is caused by ground shaking, fault rupture, liquefaction, landslide, or inundation. Locating a building in Dallas, Texas, for example, will almost guarantee that it will never be damaged by an earthquake. Of course, this option isn't possible for many building owners.

It is however, fairly common for high-technology manufacturing plants to be located far from their headquarter locations, at sites with low seismicity such as Texas, Massachusetts, or Idaho. While it would be very rare for a retail establishment to make a siting decision based on seismic risk over the demographics of the market in a particular note that moving a facility even a few miles in some cases can make a measurable difference in seismic hazard, for example, moving a proposed building location from within a mile of a major fault to 5 miles away.

Locate the building on a soil profile that reduces the hazard. Local soil profiles can be highly variable, especially near water, on sloped surfaces, or close to faults. In an extreme case, siting on poor soils can lead to liquefaction, land sliding, or lateral spreading. Often, as was the case in 1989 Loma Prieta earthquake near San Francisco, similar structures located less than a mile apart each performed in dramatically different ways because of differing soil conditions. Even when soil-related hazards are not present, the amplitude, duration, and frequency content of earthquake motions that have to travel through softer soils can be significantly different than those traveling through firm soils or rock.

Increase the capacity of the building to resist earthquake forces. The most traditional method for decreasing vulnerability of buildings is to make them *stronger.* By increasing the forces that a building can resist, such as by providing larger structural systems, less damage would be expected. This strategy can be costly and, in some cases, may not be the most efficient means of increasing performance. Another option is to increase the ductility of the structural or nonstructural systems, improving their ability to absorb the energy of the earthquake without permanent damage.

Engineer the soil profile to increase building performance and reduce vulnerability. If relocating to a region of lower seismicity or to an area with a better natural soil profile is not a cost-effective option, the soil at the designated site can often be reengineered to reduce the hazard. On a liquefiable site, for instance, the soil can be grouted or otherwise treated to reduce the likelihood of liquefaction occurring. Soft soil can be excavated and replace or combined with foreign materials to make them stiffer. The building foundation itself can be modified to account for the potential effects of the soil, reducing the building's susceptibility to damage even if liquefaction or limited land sliding does occur.

8.3.5.2 Improve Building Performance

Seismic vulnerability may be reduced by increasing the performance of the building, thereby reducing the damage expected in earthquakes. There are two methods by which this is typically accomplished:

Reduce the response of the building to earthquake shaking: An earthquake generates inertial forces in a building that are a function of the structure's mass, stiffness, and damping and of the acceleration and frequency of the earthquake motion. While the actual mass of the building (by weight of the structure, contents, and people) typically cannot be significant altered, the effective mass can be changed by providing special devices, such as passive or active mass dampers, that can effectively reduce the overall mass that is accelerated by the earthquake. Stiffness can be altered by modifying the structural system (e.g., concrete shear wall, steel moment frames) or by using braces and seismic dampers. The building's fundamental period can be significantly increased (and resulting forces reduced) by providing seismic isolating devices at the building foundation.

Capacity of the structure: The value for capacity is a simplified representation of the capacity of the overall building for a specified level of stressor distortion such as when yielding of major structural members occur or when lateral displacements reach a prescribed limit. The capacity of the structure to resist lateral forces estimated by means of the calculations, usually using highly sophistication analysis techniques to evaluate the best possible scenario.

Determination of the lateral-force capacity of a structure will include consideration of all elements, structural and nonstructural, that contribute to the resistance of lateral forces. Physical properties for existing buildings are generally obtained from existing available data; otherwise, assumptions and/or tests must be made.

The analysis must include the evaluation of the most rigid elements resisting the lateral loads, as well as the more flexible elements that resist the lateral distortions after the rigid elements yield or fail. Considerations must also be given to the interaction of various combinations of the structural framing systems and elements which will contribute to the resistance of the lateral loads.

8.3.6 ASCE/SEI 41-06: Seismic Evaluation Example: Steel Building

Steel building with pre-Northridge moment frames and concentric bracing

Given: An existing 5-story office building located in downtown Los Angeles, California. The lateral system consists of a combination of moment frames and concentric braces. Built in 1990, the buildings moment connections suffered damage during the 1994 Northridge earthquake. The damaged connections were repaired by using notch-tough welds for the beam bottom flanges. No seismic evaluations were made at the time of moment connection repairs.

In 2011, the building is being acquired by new owners who desire an assessment of the building's expected seismic performance before making the final deal. A structural engineer has been hired to evaluate the seismic vulnerability of the building, and, if required, to come up with a seismic upgrade scheme. The selected design basis is that the building should be operational after a major earthquake, that is, the performance criterion is IO.

The engineer has selected the following components for a preliminary seismic evaluation:

1. Frame beams
2. Frame columns
3. Beam–column connection, including column panel zones
4. Braces in concentric braced frames
5. Diaphragm components
6. Frame column-to-foundation connections

However, for purposes of this example, compare the bending capacity of one sample frame–beam to the demand imposed by seismic and gravity loads. Assume the following:

The sample frame–beam is a W30 × 116, $F_y = 50$ ksi, subjected to earthquake and gravity loads as noted in the following:

Q_E is the action due to unreduced earthquake loads (i.e., $R = 1$), determined using a LDP. In our case, the beam action is bending. Assume $M_E = Q_E = 3280$ k-ft.
Q_D is the dead-load action (i.e., dead-load moment) Assume $Q_D = M_D = 580$ k-ft
Q_L is the effective live-load action (equal of 25% of unreduced design live load but not less than actual live load) Assume $Q_L = M_L = 145$ k-ft.
Q_S is the effective snow load contribution = 0.

Required: Building seismic evaluation and possibly an upgrade scheme that meets the requirements of SEI 41-06 for enhanced performance objectives. To keep the numerical work simple, verify the acceptability of our W30 × 116 beam. Consider only bending action.

Solution: *Step 1* is the determination of the characteristics of the ground motion likely to be experienced at the building site. Because ground motions are the most common and significant cause of earthquake damage, rehabilitation objectives are commonly established using earthquake ground shaking hazards, usually defined on a probabilistic basis. Observe that the performance characteristics are functions of the severity of specified earthquakes and are directly related to the extent of damage sustained by the building.

The owners of the example building have commissioned the engineer to determine the seismic vulnerability of the building and design a rehabilitation, if required, for an enhanced performance objective. Their intent is to have the building operational during and after the seismic events specified in ASCE 41-06 for the given rehabilitation objective. The engineer has, in nontechnical terms, described to the owners the broad range of expected building performance characteristics in terms of possible damage to both structural and nonstructural building components. Communication with the owner in lay terms is perhaps the most important step in a seismic rehabilitation study. The owners of our subject building are now well informed about probable post-earthquake scenarios. Because the selected design is based on an operational level, they expect

- Overall damage to the building to be light
- Structure to have no permanent lateral displacement
- Structure's original strength and stiffness to remain substantially unchanged
- Minor cracking in façade partitions and ceilings
- Minor local yielding of structural elements at a few places, without fracture
- Elevator and fire protection system remaining operable
- In terms of nonstructural components, equipment and contents to be generally secure but perhaps not operable due to mechanical failure or lack of utilities

As stated previously, prediction of building performance in a future earthquake is a dangerous and risky business. Consider, for example, the top-of-the-line performance as set forth in ASCE/SEI 41-06. It implies that a building designed for this performance level is likely to come out scratch free after a high seismic event, giving the impression that the building is invincible. This may not be so even when the building is designed or retrofitted to IO level. This important difference should be brought to the attention of building owners before they embark on a seismic upgrade.

ASCE/SEI 41-06 Table Cl-1 displays in matrix format the characteristics of ground motion for three distinct earthquakes, represented by notations $k + p + e$. These are

1. BSE-1 earthquake with a 10% probability in 50 years (mean return period of 474 years, rounded to 500 years, used in most building codes)
2. BSE-2 earthquake with a 2% probability in 50 years (mean recurrence interval of 2475 years, rounded to 2500 years)
3. An earthquake with a 20% probability in 50 years (mean recurrence interval of 225 years)

For the example building, we assume that the project geotechnical engineer has, after conducting site-specific studies, developed acceleration response spectra for the earthquakes listed in the foregoing. Since the methodology of seismic vulnerability study is the same, we will, for purposes of illustration, verify the performance of the frame–beam for only the BSE-2 earthquake.

Step 2 is the determination of as-built conditions in order to arrive at a value for the reliability coefficient κ. The building is fairly new, built in 1990. As-built information pertinent to its seismic performance, including construction documents and material test reports, is available, and a visual survey has indicated that there are no site-related concerns such as pounding from neighboring structures. Because of the abundance of as-built information, the engineer is able to gain a comprehensive knowledge and understanding of the behavior of structural components allowing a value of $\kappa = 1.0$ in the analyses, where κ is a reliability coefficient used to reduce the component strength value for existing components. Because $\kappa = 1.0$, there is no need to reduce the computed strength values of the component when making demand capacity comparisons.

Step 3, the classification into primary and secondary components, is a straightforward task for the example building. Both frame–beams and columns are classified as primary because they are essential for providing the structure's basic lateral resistance. This step also includes the classification of the response of lateral-resisting components into either deformation-controlled or force-controlled actions. In our case, it is evident that the flexural action of the frame–beams is deformation-controlled. However, the classification of frame columns subject to combined compression and bending is not so obvious. It could be either deformation-controlled or force-controlled, depending on the ratio of axial load in the column and its axial strength.

Step 4 entails setting up the analytical model, calculating the building period, and determining the base shear, its vertical distribution up the building height, and the forces in the floor and roof diaphragms. This task, an everyday occurrence in a design office, does not here require further explanation except to point out that

1. If an LSP is used for seismic analysis, then the base shear V, also referred to as the pseudolateral load, is calculated using the unreduced spectral acceleration S_a without an upper limit on the building period, T
2. If an LDP such as modal superposition is used, then the analysis is carried out using a response spectrum that is not reduced to account for the anticipated inelastic response of the building (i.e., $R = 1.0$)

The purpose of an LSP or LDP is to determine the distribution of forces and deformations induced in a structure by the design ground motion. Although an LSP is permitted for simple buildings less than 100 ft. in height, prevailing practice in most design offices is to use an LDP and the modal superposition method. Hence, we will assume that the forces and moments in the frame–beam have been evaluated by performing linear dynamic analyses for each of the three response spectra selected for the study.

Step 5 is where the seismic and gravity loads are combined to determine Q_{UD}, the demand imposed on elements due to seismic and gravity loads. Because the actions of both frame–beams and frame columns of the example building are considered deformation-controlled, we use the following equation to calculate demand:

$$Q_{UD} = Q_G + Q_E \tag{8.9}$$

If, on the other hand, the action of an element under consideration is force-controlled, the corresponding equation for verifying the acceptance criteria would have been

$$Q_{UF} = Q_G + \frac{Q_E}{C_1 C_2 C_3 J} \tag{8.10}$$

where

Q_{UF} is the design action due to combinations of gravity and seismic loads

J is the coefficient that estimates the maximum earthquake force that a component can sustain and deliver to other components

The parameter J is in the denominator of the second term of the equation given earlier. Therefore, it is possible to recognize that in a nonlinear response, the actual force sustained by the components is likely to be less than earthquake force Q_E, determined by elastic analysis. The maximum value of J is 2. It is related to the seismicity of the site as follows:

$J = 1$ in zones of low seismicity
$J = 1.5$ in zones of moderate seismicity
$J = 2$ in zones of high seismicity
C_1, C_2, and C_3 are the modification coefficients explained earlier in Section 8.3.3.5.

Step 6 is where the component capacities Q_{CE} or Q_{CL} are calculated, depending upon whether the action considered is deformation—or force-controlled.

Returning to the example frame–beam, since bending is categorized as a deformation-controlled action, the demand Q_{UD} due to gravity and seismic is given by

$$Q_{UD} = Q_{UD} + Q_G + Q_E$$

$$Q_G = 1.1\ (Q_D + Q_L + Q_S)$$

$$= 1.1\ (580 + 145 + 0) = 797.5 \text{ k-ft (say, 800 k-ft)} \qquad (8.11)$$

$$Q_E = 3820 \text{ k-ft}$$

$$Q_{UD} = 3820 + 800 = 4620 \text{ k-ft}$$

Step 7 is the final step in which the acceptance criterion is verified for each component. Although we earmarked six distinct components for seismic assessment, to keep the presentation simple, we will check the acceptance criterion for the frame–beam only.

In step 3, it was determined that the limit state for the example beam was flexure. Beam flexure (and, for that matter, beam shear) with negligible axial loads is considered a deformation-controlled action.

The design properties for the example frame–beam are as follows:

W30 × 116 $F_y = 50$ ksi, group 3, manufactured after the year 1990
Default lower-bound yield strength = 52 ksi (ASCE/SEI 41-06, Table 5-2)
F_{ye} = expected steel strength = 1.05 × 52 = 54.6 ksi (ASCE/SEI 41-06, Table 5-3)

Expected capacity $Q_{CE} = ZF_{ye}$
= 378 × 54.6 (note z for W30 × 116 = 378 in.^{3})
= 20,639 k-in.
= 1720 k-ft

$A_s = 34.2$ in.^{2}, $d = 30.01$ in., $t_w = 0.565$ in., $b_f = 10.495$ in., $t_f = 0.85$ in.

As previously discussed, the action of the beam is deformation-controlled, with $Q_{CE} > M_{PCE}$ due to lateral torsional buckling. Therefore, the m-factors in the ASCE/SEI 41-06 Table 5-5 may be used directly, without calculating and equivalent value of m. Since we are verifying the acceptance criterion for IO performance, from Table 5-5 m is either 2 or 1.25, depending upon the ratio $b_f/2t_f$ and h/t_w:

$$b_f/2t_f = 10.495/2 \times 0.85 = 6.17$$

$$52/\sqrt{F_{ye}} = 52/\sqrt{53.55} = 7.10 > 6.17\ b_f/2t_f \qquad (8.12)$$

$$h/t_w = 28.31/0.565 = 50.1, \quad 418/\sqrt{F_{ye}} = 418/\sqrt{53.55} = 57.12 > 50.1$$

Therefore, the *m*-factors are

IO	LS	CP
2	6	8

The knowledge factor κ = 1.0 because the quality and extent of available information, as stated at the beginning of the design example, is comprehensive. Since we are verifying the component's acceptability for IO performance, $m = 2$, and

$$mk\, Q_{CE} = 2 \times 1 \times 1720 = 3440 \text{ k-ft}$$

$$= 6 \times 1 \times 1416.7 = 8500 \text{ k-ft} \tag{8.13}$$

The demand Q_{UD} from step 6 is 4620 k-ft, which is greater than the expected capacity of 3440 k-ft, indicating noncompliance. A similar evaluation is made for other actions of the beam. The analysis is repeated for the other two earth quakes. The results are reviewed, keeping in mind that values of *m* are only an approximate indicator of seismic performance. Reevaluation of the building using nonlinear analysis procedures with reevaluated gravity and lateral loads using a sharper pencil is a prudent course of action before deciding on a seismic rehabilitation.

The procedure for evaluating acceptance criteria is conceptually the same for the other components of the moment frame, listed earlier, such as columns, panel zones, beam–column connections, and column-to-foundation connections. Hence, they are not discussed here.

Suppose the objective of the seismic evaluation of our example building is basic LS, instead of IO. Does the method of seismic study differ from the preceding procedure for IO performance? What if the target performance is CP instead of IO or LS?

The procedure is generally the same, irrespective of the selected target performance. The differences are in ground motion and in the values of the m-factors used in the acceptance criteria.

Consider, for example, the basic LS objective as the target performance. A building that satisfies this objective will pose approximately the same earthquake risk for life and safety traditionally considered acceptable in the United States. Therefore, an LS study includes verification of the building's behavior for two distinctly different earthquakes:

1. An earthquake with a 10% probability of occurrence in 50 years to satisfy LS
2. An earthquake with a 2% probability of occurrence in 50 years to satisfy CP

These two earthquakes, as stated earlier, are defined as building safety earthquake BSE-1 and BSE-2, in ASCE/SEI 41-06.

CP refers to the building in the post-earthquake damage state that is on the verge of partial or total collapse but has not yet collapsed. Substantial damage to the structure has occurred, with considerable loss of stiffness and strength in the lateral- force-resisting system. However, all significant components of the gravity-load-resisting system continue to function.

If the building does not collapse, some engineers may wrongly consider that the LS objective has been met. The LS performance level includes a margin of safety against collapse for the lower-level earthquake. Significant risk of injury from falling structural debris may exist. It may not be practical to repair the structure and it may not be safe for reoccupancy, because aftershock activity may induce collapse.

To satisfy the limited safety objective of CP, the building may be evaluated for a single earthquake chosen from a range of specified earthquake hazard levels. Building safety earthquake BSE-2, with a 2% probability of occurrence in 50 years, is one example. See ASCE/SEI 41-06 Table C1-1 for other specified earthquakes.

8.3.7 ASCE/SEI 41-06: Seismic Evaluation: Concrete Building

Dual system: Moment frames and shear walls

Given: An existing 15-story office building located in downtown Los Angeles, CA. The lateral system consists of moment frames and shear walls. The building was built in 1972 and suffered cosmetic damage during the 1994 Northridge earthquake. The damage was repaired by patching spalled concrete. No seismic evaluations were made at the time of repairs.

In 2008, the building is being acquired by new owners who desire an assessment of the building's expected seismic performance before making a final deal. A structural engineer has been hired to evaluate the seismic vulnerability of the building, and, if required, to come up with a seismic upgrade scheme. The selected design objective, as desired by the owners, is that the building should be operational after major earthquakes, that is, the performance criterion is IO.

The engineer has selected the following components for a preliminary seismic evaluation:

1. Frame beams
2. Frame columns
3. Shear walls
4. Diaphragm components
5. Frame column-to-foundation connections

However, to avoid explanation in excruciating details, seismic evaluation of only one frame–beam will be performed.

Assume the following:

Typical frame–beam 24 in. wide × 36 in. deep.

Q_E is the action due to unreduced earthquake loads (i.e., R = 1), determined using a LDP. In our case, the beam action is bending; therefore, $M_E = Q_E = 3820$ k-ft.

Q_D is the dead-load action (i.e., dead-load moment) $Q_D = M_D = 580$ k-ft.

Q_L is the effective live load action (equal of 25% of unreduced design live load but not less than actual live load) $Q_L = M_L = 145$ k-ft.

Q_s is the effective snow load contribution = 0.

Required: Building seismic evaluation and possibly an upgrade scheme that meets the requirements of ASCE/SEI 41-06 for enhanced performance objectives. To keep the numerical work simple, verify the acceptability of typical frame–beam, 24 in. wide × 36 in. deep procedures. Consider only the bending action.

Solution:

Step 1 is the determination of the characteristics of the ground motion likely to be experienced at the building site, because ground motions are the most common cause of earthquake damage. Consequently, rehabilitation objectives are commonly established using earthquake ground shaking hazards, typically defined on a probabilistic basis. Performance characteristics that are functions of the severity of specified earthquakes are directly related to the extent of damage sustained by the building.

As stated earlier, owners of the building desire to have an evaluation of seismic vulnerability of the building and, if required, a seismic rehabilitation design for an enhanced performance objective. Their intent is to have the building operational during and after seismic events postulated for IO. The engineer has, in nontechnical terms, described to the owners the broad range of expected building performances in terms of possible damage to both structural and nonstructural building components. Communication with the owner in lay terms is perhaps the most important step in a seismic rehabilitation study. The owners of our subject building are now well informed about probable post-earthquake scenarios. Because retrofit objective will be for IO, they expect

- Overall damage to the building to be light
- Structure to have no permanent lateral displacement
- Structure's original strength and stiffness to remain substantially unchanged
- Minor cracking in facade partitions and ceilings
- Minor local yielding of structural elements at a few places, without fracture
- Elevators and fire protection system remaining operable
- In terms of nonstructural components, equipment and contents to be generally secure, but perhaps not operable due to mechanical failure or lack of utilities

Prediction of building performance in a future earthquake is a dangerous and risky business. Consider, for example, the top-of-the-line performance as set forth in ASCE/SEI 41-06. Thence, it implies that a building designed for IO performance level is likely to come out search-free after a high seismic event, giving the impression that the building is earthquake-proof. Structural engineers understand that buildings designed using the principle of ductile design is earthquake-resistant and not earthquake-proof. This important difference should be brought to the attention of building owners before they embark on a seismic upgrade.

ASCE/SEI 41-06 Table Cl-1 displays in matrix format the characteristics of ground motion for three distinct earthquakes, represented by notations $k + p + e$. These are

- BSE-1 earthquake with a 10% probability in 50 years (mean return period of 474 years, rounded to 500 years, used in most building codes)
- BSE-2 earthquake with a 2% probability in 50 years (mean recurrence interval of 2475 years, rounded to 2500 years)
- An earthquake with a 20% probability in 50 years (mean recurrence interval of 225 years)

For the example building, we assume that the project geotechnical engineer has, after conducting site-specific studies, developed an acceleration response spectra for the earthquakes listed in the foregoing. Since the methodology of seismic evaluation is the same for all postulated earthquakes (except for numerical values of demand and in factors), we will verify the performance of the frame–beam for only BSE-2 earthquake. This shortcut is to keep the explanation simple.

Step 2 is the determination of as-built conditions in order to arrive at a value for the reliability coefficient κ the building is fairly old, built in 1972. For simplicity we will assume that as-built information pertinent to its seismic performance, including construction documents and material test reports is not available. Further, a visual survey has indicated that there are site-related concerns such as pounding from neighboring structures. Because of the absence of as-built information, the engineer is not able to gain a comprehensive knowledge and understanding of the behavior of structural components. Therefore, a value of $\kappa = 1.0$ cannot be used in the analyses, where κ is a reliability coefficient used to reduce the component strength value for existing components. Because κ is less than 1.0, there is a need to reduce computed strength values of the component when making demand capacity comparisons. For our example, we will assume $\kappa = 0.9$.

Step 3, the classification into primary and secondary components, is a straightforward task for the example building. Both frame–beams and columns are classified as primary because they are essential for providing the structure's basic lateral resistance. This step also includes the classification of the response of lateral-resisting components into either deformation-controlled or force-controlled actions. In our case, it is obvious that the flexural action of the frame–beams is deformation-controlled. However, the classification of frame columns subject to combined compression and bending is not so obvious. It could be either deformation-controlled or force-controlled, depending on the ratio of axial load in the column and its axial strength.

Step 4 entails setting up the analytical model, calculating the building period, and determining the base shear, its vertical distribution up the building height, and the forces in the floor and roof diaphragms. This task—an everyday occurrence in a design office—does not here require explanation except to point out that

1. If an LSP is used for seismic analysis, then the base shear V, also referred to as the pseudolateral load, is calculated using the unreduced spectral acceleration S_a without an upper limit on the building period, T
2. If an LDP such as a modal superposition is used, then the analysis is carried out using a response spectrum that is not reduced to account for the anticipated inelastic response of the building (i.e., $R = 1.0$)

The purpose of an LSP or LDP is to determine the distribution of forces and deformations induced in a structure by the design ground motion. Although an LSP is permitted for simple buildings less than 100 ft. in height, prevailing practice in most design offices is to use an LDP and the modal superposition method. Hence, we will assume that the forces and moments in the frame–beam have been evaluated by performing linear dynamic analyses for each of the three response spectra selected for the study.

Step 5 is where the seismic and gravity loads are combined to determine Q_{UD}, the demand imposed on elements due to seismic and gravity loads. Because the actions of both the frame–beams and frame columns of the example building are considered deformation-controlled, we use the following equation to calculate demand:

$$Q_{UD} = Q_G + Q_E \tag{8.14}$$

If, on the other hand, the action of an element under consideration is force-controlled, the corresponding equation for verifying the acceptance criteria would have been

$$Q_{UF} = Q_G + \frac{Q_E}{C_1 C_2 J} \tag{8.15}$$

where

Q_{UF} is the design action due to combinations of gravity and seismic loads

J is the coefficient that estimates the maximum earthquake force that a component can sustain and deliver to other components

It is in the denominator of Equation 8.15, related to earthquake force Q_E, that it is possible to recognize that in a nonlinear response, the actual force sustained by the component is likely to be less than earthquake force Q_E, determined by elastic analysis. The maximum value of J is 2. It is calculated by the relation

$J = 1$ in zones of low seismicity

$J = 1.5$ in zones of moderate seismicity

$J = 2$ in zones of high seismicity

C_1 and C_2 are the modification coefficients explained earlier in Section 8.3.3.4.2.

Step 6 is where the component capacities Q_{CE} or Q_{CL} are calculated, depending upon whether the action considered is deformation- or force-controlled. We will not dwell on this here, because it was explained in detail in the previous section.

Returning to the example, since the bending of the frame–beam is categorized as a deformation-controlled action, the demand Q_{UD} due to gravity and seismic is given by

$$
\begin{aligned}
Q_{UD} &= Q_{UD} + Q_G + Q_E Q_G \\
&= 1.1\ (Q_D + Q_L + Q_S) \\
&= 1.1\ (580 + 145 + 0) \\
&= 797.5 \text{ k-ft (say, 800 k-ft) } Q_E \\
&= 3820 \text{ k-ft } Q_{UD} \\
&= 3820 + 800 = 4620 \text{ k-ft}
\end{aligned}
\tag{8.16}
$$

Step 7 is the final step, in which the acceptance criterion is verified for each component. Although we earmarked six distinct components for seismic assessment, to keep the presentation simple, we will check the acceptance criterion for the frame–beam only.

In step 3, it was determined that the limit state for the example beam was flexure. Beam flexure (and, for that matter, beam shear) with negligible axial loads is considered deformation-controlled action.

The design properties for the example frame–beam are as follows:

Size 24 in. wide × 36 in. deep.

y

Default lower-bound yield strength of mild steel reinforcement = 40 ksi (ASCE/SEI 41-06, Table 5-2)

F = expected steel strength = 1.25 × 40 = 50,000 psi (ASCE/SEI 41-06, Table 5-3)

As previously discussed, the action of the beam is deformation-controlled, with $Q_{CE} > M_{PCE}$ due to lateral torsional buckling. Therefore, the m-factors given in ASCE/SEI 41-06 Table 5-3 may be used directly, without calculating an equivalent value of m_e. Since we are verifying the acceptance criterion for IO performance, from ASCE/SEI 41-06, Table 5-5, m is either 2 or 1.25, depending upon the ratio $b_f/2t_f$ and h/t_w.

In our case, the knowledge factor $\kappa = 0.9$ because the quality and extent of available information, as stated at the beginning of the design example, is less than comprehensive. Since we are verifying the component's acceptability for IO performance, $m = 2$, and

$$
\begin{aligned}
mkQ_{CE} &= 2 \times 1 \times 1720 = 3440 \text{ k-ft} \\
&= 6 \times 1 \times 14.16.7 = 8500 \text{ k-ft}
\end{aligned}
\tag{8.17}
$$

The demand Q_{UD} from step 6 is 4620 k-ft, which is greater than the expected capacity of 3440 k-ft, indicating noncompliance. A similar evaluation is made for other actions of the beam. The analysis is repeated for the other two earthquakes. The results are reviewed, keeping in mind that values of m are only an approximate indicator of seismic performance. Reevaluation of the building using nonlinear analysis procedures with reevaluated gravity and lateral loads is a prudent course of action before deciding on seismic rehabilitation.

The procedure for evaluating acceptance criteria is conceptually the same for other components of the moment frame, such as columns, panel zones, beam–column connections, and column-to-foundation connections.

Suppose the objective of the seismic evaluation of our example building is basic LS, instead of IO. Does the procedure for seismic study differ from the preceding procedure for IO performance? What if the target performance is CP instead of IO?

The procedure is generally the same, irrespective of the selected target performance. The differences are in ground motion and in the values of the m-factors used in the acceptance criteria.

Consider, for example, the BSO as the target performance. A building that satisfies BSO will pose approximately the same earthquake risk for LS traditionally considered acceptable in the United States. Therefore, a BSO study includes verification of the building's behavior for two distinctly different earthquakes:

1. An earthquake with a 10% probability of occurrence in 50 years to satisfy LS
2. An earthquake with a 2% probability of occurrence in 50 years to satisfy CP

These two earthquakes, defined as building safety earthquake (BSE)-1 and -2, are shown in ASCE/SEI 41-06, Table C1-1.

The *m*-factors corresponding to BSO are typically larger than those for IO. For instance, *m* would equal 6, instead of 2, for the deformation-controlled steel beam studied in the illustrative example.

CP refers to the building in the post-earthquake damage state that is on the verge of partial or total collapse but has not yet collapsed. Substantial damage to the structure has occurred, with considerable loss of stiffness and strength in the lateral- force-resisting system. However, all significant components of the gravity-load-resisting system continue to function.

If the building does not collapse, some engineers may wrongly consider that the LS objective has been met. The LS performance level includes a margin of safety against collapse for the lower-level earthquake. Significant risk of injury from falling structural debris may exist. It may not be practical to repair the structure and it may not be safe for reoccupancy, because aftershock activity may induce collapse.

To satisfy the limited safety objective of CP, the building must be evaluated for a single earthquake chosen from a range of specified earthquake hazard levels. BSE-2 with a 2% probability of occurrence in 50 years is one example. See ASCE/SEI 41-06, Table C1-1 for other specified earthquakes.

The *m*-factors corresponding to CP are larger than those for LS. For instance, *m* would equal 8 instead of 6 for the beam investigated in the design example.

8.4 COMMON DEFICIENCIES AND UPGRADE METHODS: CONCRETE BUILDING

Seismic upgrade of buildings typically involves strengthening of their horizontal and vertical lateral-load-resisting elements. This can be done by reinforcing the existing elements or by adding new elements. If the existing lateral-load-resisting structure is grossly deficient, it can be replaced. Whenever buildings are upgraded to resist a larger seismic load, their foundations must be checked for the new loading and be reinforced if necessary.

Prime candidates for renovation and strengthening are

- Buildings with irregular configurations, such as those with abrupt changes in stiffness, large floor openings, very large floor heights, reentrant corners in plan, and soft stories
- Buildings with walls of unreinforced masonry, which tend to crack and crumble under severe ground motions
- Buildings with inadequate diaphragms lacking ties between walls and floors or roofs
- Buildings with nonductile concrete frames, in which shear failures at beam–column joints and column failures are common
- Concrete buildings with insufficient lengths of bar anchorage and splices
- Concrete buildings with flat-slab framing, which can be severely affected by large story drifts
- Buildings with open storefronts; buildings with clear-story conditions

- Buildings with elements that tend to fail during ground shaking (examples are unreinforced masonry parapets and chimneys and nonstructural building elements, which may fall, blocking exits and injuring people)

8.4.1 Diaphragms

The floors and roofs of buildings must act as a diaphragm—a deep horizontal beam capable of transferring lateral-load generated by the floor/roof mass to the vertical elements resisting lateral loads. To do so, it must have the following:

1. The ability to resist horizontal shear forces, meaning that it must possess a certain degree of strength and rigidity in its plane. In other words, it must be able to function as a web of the horizontal beam that does not break or deflect excessively under horizontal load.
2. Flanges at opposite ends of the diaphragm perpendicular to the applied forces. These flanges, called chords, must be attached to the diaphragm's web with connections capable of transmitting the seismic forces.
3. Drag struts, also called collector elements, to deliver the seismic load from the diaphragm to the vertical lateral-load-resisting elements. However, drags are required only when the horizontal distribution of load among the walls or frames depends on the types of floor and roof diaphragms in the building. Flexible systems are assumed to distribute lateral loads to the walls or frames in proportion to their tributary areas. In contrast, rigid diaphragms are assumed to distribute lateral loads to the walls or frames in proportion to their relative rigidities. Rigid diaphragms can distribute horizontal forces by developing torsional resistance. This is helpful in buildings with irregular wall layout. Flexible diaphragms are considered too supple to work in torsion. The majority of real-life floor structures fall between the two categories; engineering judgment is required to predict the behavior of these semirigid or semiflexible diaphragms. However, prevailing practice allows the assumption of rigid diaphragms for concrete slabs, unless diaphragm span to depth are very large, typically in excess of 3.

The type and function of existing diaphragms must be evaluated prior to making a decision on how to strengthen the vertical lateral-load-resisting elements of the building. For example, it is unwise to add shear walls in an asymmetric manner if this introduces additional torsion into the existing diaphragm and may lead to its possible distress. If shear walls are placed in the interior of the building, collector elements must be present in the diaphragm to carry the inertial forces to them.

Methods of strengthening diaphragms depend on their composition and the nature of their weaknesses. Deficiencies of existing diaphragms typically fall into two categories: insufficient strength or stiffness and the absence of chords and collectors. Replacing a diaphragm, which involves taking out the building floor, is reserved for the most critical conditions.

8.4.1.1 Cast-in-Place Concrete Diaphragms

Cast-in-place diaphragms are sturdy elements that rarely require major upgrade except at their connections to the chord. However, common deficiencies at diaphragm openings or plan irregularities include inadequate shear capacity, inadequate chord capacity, and excessive shear stresses.

Two alternatives may be effective in correcting the deficiencies: either improve strength and ductility or reduce demand. Providing additional reinforcement and encasement may be an effective measure to strengthen or improve individual components. Increasing the diaphragm thickness may also be effective, but the added weight may overload the footings and increase the seismic loads. Lowering seismic demand by providing additional lateral-force-resisting elements, introducing additional damping, or base isolating the structure may also be effective rehabilitation measures.

Inadequate shear capacity of concrete diaphragms may be mitigated by reducing the shear demand on the diaphragm by providing additional vertical lateral-force-resisting elements or by increasing the diaphragm capacity by adding a concrete overlay. The addition of a concrete overlay is usually quite expensive, since this requires the removal of existing partitions and floor finishes and may require the strengthening of existing beams and columns to carry the added dead load. Adding supplemental vertical lateral-force- resisting elements will provide additional benefits by reducing demand on other elements that have deficiencies.

Increasing the chord capacity of existing concrete diaphragms can be realized by adding new concrete or steel members or by improving the continuity of existing members. A common method for increasing the chord capacity of a concrete diaphragm with the addition of a new concrete member is shown in Figures 8.2 and 8.3. This member can be placed above or below the diaphragm. Locating the chord below the diaphragm will typically have less impact on floor space (Figure 8.4). Figure 8.5 shows addition of collectors at reentrant comers of a diaphragm.

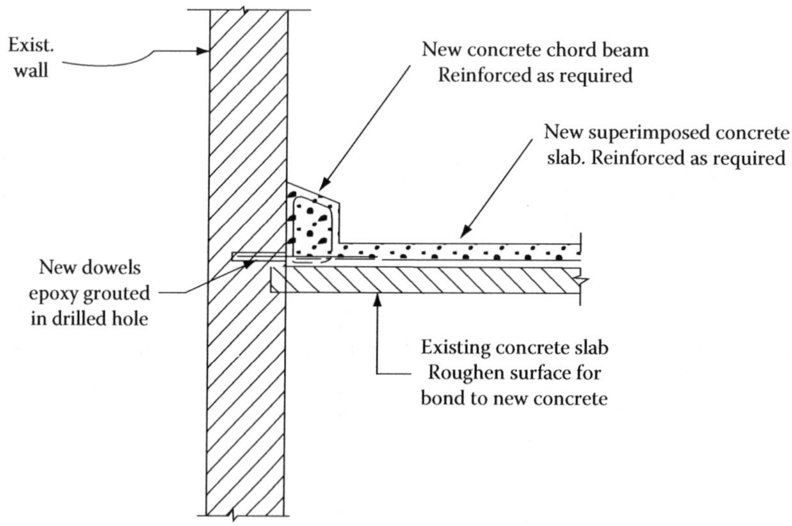

FIGURE 8.2 Superimposed diaphragm slab at an existing concrete wall.

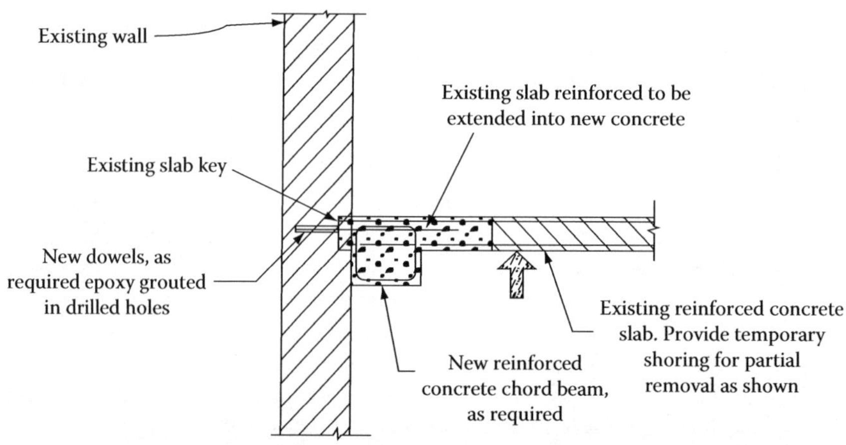

FIGURE 8.3 Diaphragm chord for existing concrete slab.

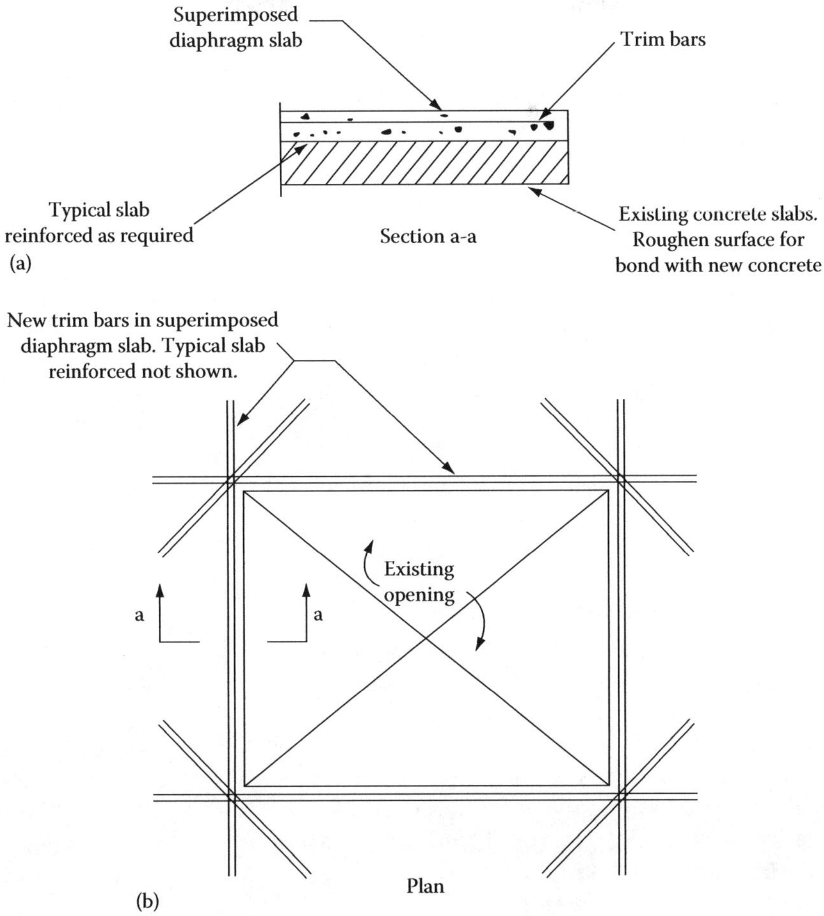

FIGURE 8.4 Strengthening of openings in a superimposed diaphragm: (a) section; and (b) plan.

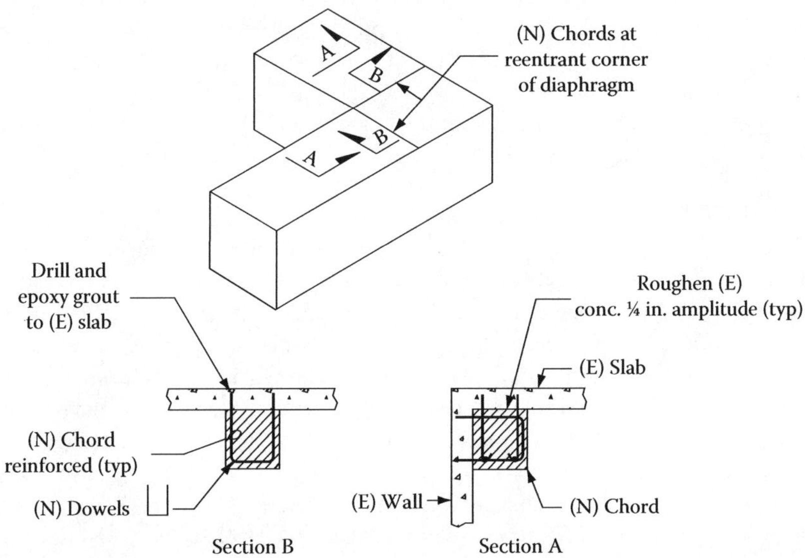

FIGURE 8.5 New chords at reentrant corners.

The following measures may be effective in rehabilitating chord and collector elements:

1. Strengthening the connection between diaphragms and chords and collectors
2. Strengthening steel chords or collectors with steel plates attached directly to the slab with embedded bolts or epoxy and strengthening slab chord or collectors with added reinforcing bars
3. Adding chord members

8.4.1.2 Precast Concrete Diaphragms

Common deficiencies of precast concrete diaphragms include inadequate shear capacity, inadequate chord capacity, and excessive shear stresses at diaphragm openings or plan irregularities. Existing precast concrete slabs constructed using precast tees or cored planks commonly have inadequate shear capacity. Frequently, limited shear connectors are provided between adjacent units, and a minimal topping slab with steel mesh reinforcement is placed over the planks to provide an even surface to compensate for irregularities in the precast elements. The composite diaphragm may have limited shear capacity.

Strengthening the existing diaphragm is generally not cost-effective. Adding a reinforced topping slab is generally not feasible because of the added weight. Adding mechanical connectors between units is generally not practical, because the added connectors are unlikely to have sufficient stiffness, compared to the topping slab, to resist an appreciable load. The connectors would therefore need to be designed for the entire shear load assuming the topping slab fails. The number of fasteners, combined with edge distance concerns, typically makes this impractical. The most cost-effective approach is generally to reduce the diaphragm shear forces through the addition of supplemental shear walls or braced frames.

Inadequate chord capacity in a precast concrete deck can be mitigated by adding new concrete or steel members, as discussed earlier for a cast-in-place concrete diaphragm. A new chord member can be added above or below the precast concrete deck. Excessive stresses at diaphragm openings or plan irregularities in precast concrete diaphragms can also be mitigated by introducing drag struts, as described earlier for cast-in-place concrete diaphragms.

8.4.2 Shear Walls

The problems that are most difficult to fix are those caused by the irregular configuration of a building (e.g., abrupt changes in stiffness, soft stories, large floor openings, and reentrant floor corners). These cases may require the addition of vertical or horizontal rigid structural elements, as well as strengthening of existing foundations or addition of new ones.

There are several approaches to increasing the capacity of existing concrete shear walls. These are discussed in the following sections.

8.4.2.1 Increasing Wall Thickness

Wall thickness is increased by applying reinforced concrete to the wall surface. Shotcrete, a mixture of aggregate, cement, and water sprayed by a pneumatic gun at high velocity, is widely used for strengthening walls because it bonds well with concrete. Some prefer application by the dry-mix method (sometimes called gunite) because the slump and stiffness can be better controlled by the nozzle operator and because gunite is applied at higher nozzle velocities, promoting superior bonding.

Concrete shear walls that lack ductility may fail by crushing of their boundary elements, horizontal sliding along construction joints due to shear, or diagonal cracking caused by combined flexure and shear. Among the most common areas of damage are the coupling beams. These can be repaired by through-bolted side plates extending onto the faces of the walls. Short and rigid piers between wall openings also tend to attract an inordinate amount of seismic loading and are therefore prone to damage.

The key to shotcreting walls lies in the surface preparation of the wall because existing concrete may be counted as part of the strengthened wall. All loose and cracked concrete must be removed from the existing wall, and its surface cleaned and roughened by sandblasting or other means. To assure composite action, the overlay is mechanically connected to the wall by closely spaced shear dowels. In addition, steel reinforcement placed in shotcrete is developed at the ends by grouted-in dowels or by continuation into an adjacent overlay space. This involves drilling through the perimeter beams or columns, filling the drilled openings with epoxy, and splicing the bars with those in the adjoining overlay areas. If the existing wall openings must be filled, the infill should be connected to the roughened edges of the opening with perimeter dowels set in epoxy.

When interior shotcreting is used, attention must be directed toward stabilizing the exterior walls and any exterior ornamental elements of the structure. These may have to be tied back into the new shotcrete by drilled-in dowels set at regular intervals. Dowels placed in exterior elements that are exposed to moisture should be given a measure of corrosion protection, such as galvanizing.

In cases where it is desirable not to increase the wall size, the outer course of bricks can be removed and replaced with shotcrete. The same can be done with interior shotcreting, except that any members framing into the wall may have to be shored during this operation. The added bonus of this approach is that the vertical load on the existing wall foundations changes very little, and they may not require the otherwise necessary enlargement.

8.4.2.2 Increasing Shear Strength of Wall

Increasing the shear strength of the web of a shear wall by casting additional reinforced concrete adjacent to the wall web may be an effective rehabilitation measure. The new concrete should be at least 4 in. thick, contain horizontal and vertical reinforcement, and be properly bonded to the existing web of the shear wall. The use of composite fiber sheets, epoxied to the concrete surface, is another method of increasing the shear capacity of a shear wall. The use of confinement jackets as a rehabilitation measure for wall boundaries may also be effective in increasing both the shear capacity and deformation capacity of coupling beams and columns supporting discontinuous shear walls.

8.4.2.3 Infilling between Columns

Where a discontinuous shear wall is supported on columns that lack either sufficient strength or deformation capacity, making the wall continuous by infilling the opening between these columns may be an effective rehabilitation measure. The infill and existing columns should be designed to satisfy all the requirements for new wall construction, including any strengthening of the existing columns required by adding a composite fiber jacket or a concrete or steel jacket for strength and increased confinement. The opening below a discontinuous shear wall may also be infilled with steel bracing. The bracing members should be sized to satisfy all design requirements for new construction and the columns should be strengthened with steel or a concrete jacket. All of these rehabilitation measures require an evaluation of the wall foundation, diaphragms, and connections between existing structural elements and any elements added for rehabilitation purposes.

Adding new shear walls or braced frames conforming to current code detailing provisions is among the most common steps taken to strengthen the lateral-load-resisting systems of buildings. The new walls and frames can either (1) complement the existing elements or (2) be designed as the sole means of providing vertical rigidity in the building. In the first case, analysis of comparable rigidities must be done to determine what percentage of the total lateral loading the new construction will carry. In the second case, the existing rigid elements that are now considered to be nonstructural must be checked for inelastic deformation compatibility. In any case, new foundations must be provided under the new elements and dowels placed around them for proper transfer of loads.

A common complication of adding shear walls and braced frames is that they tend to interfere with the building layout, circulation, or fenestration. Quite often, shear walls with openings or braced frames of unusual configurations may be needed to accommodate window or door openings. In some rare cases, exterior buttresses or counterforts may be considered.

8.4.2.4 Addition of Boundary Elements

Addition of boundary members may be an effective measure in strengthening shear walls or wall segments that have insufficient flexural strength. These members may be either cast-in-place reinforced concrete elements or steel sections. In both cases, proper connections should be made between the existing wall and the added members. The shear capacity of the rehabilitated wall should be reevaluated.

8.4.2.5 Addition of Confinement Jackets

Increasing the confinement at wall boundaries with the addition of a steel or reinforced concrete jacket may be effective in improving the flexural deformation capacity of a shear wall. The minimum thickness for a concrete jacket should be 3 in. A composite fiber jacket may be used to improve the confinement of concrete in compression.

8.4.2.6 Repair of Cracked Coupling Beams

These can be repaired by adding side plates extending on the faces of the walls. In this procedure, the plates are attached with both epoxy adhesive and anchor bolts. The plates may be attached to only one face of the wall or can be placed at both faces for extra strength, with the opposite plates through-bolted together. Another possibility for improving coupling beams is by using composite fiber wrap. This method is least intrusive because the wrapping and the epoxy combined are only 0.25 in. thick.

8.4.2.7 Adding New Walls

Adding new shear walls at a few strategic locations can be a very cost-effective approach to a seismic retrofit. The new wall is connected to the adjoining frame by drilled-in dowels. Its foundations are similarly doweled into the existing column footings. To accommodate wall shrinkage, the wall can stop short some distance—2 in., for example—from the existing concrete at the top. The space can be filled later with nonshrink grout.

8.4.2.8 Precast Concrete Shear Walls

Precast concrete shear wall systems may suffer from some of the same deficiencies as cast-in-place walls. These may include inadequate flexural capacity, inadequate shear capacity with respect to flexural capacity, lack of confinement at wall boundaries, and inadequate splice lengths for longitudinal reinforcement in wall boundaries. Deficiencies unique to precast wall construction are inadequate connections between panels, to the foundation, and to floor or roof diaphragms.

The rehabilitation measures previously described for concrete buildings may also be effective in rehabilitating precast concrete shear walls. In addition, the following rehabilitation measures may be effective:

- *Enhancement of connections between adjacent or intersecting precast wall panels*: Mechanical connectors such as steel shapes and various types of drilled-in anchors, cast-in-plane strengthening methods, or a combination of the two may be effective in strengthening connections between precast panels. Cast-in-place strengthening methods include exposing the reinforcing steel at the edges of adjacent panels, adding vertical and transverse reinforcement, and placing new concrete.

- *Enhancement of connections between precast wall panels and foundations*: Increasing the shear capacity of the wall panel-to-foundation connection by using supplemental mechanical connectors or a cast-in place overlay with new dowels into the foundation may be effective rehabilitation measures. Increasing the overturning moment capacity of the panel-to-foundation connection by using drilled-in dowels within a new cast-in-place connection at the edges of the panel is another effective rehabilitation measure. Adding connections to adjacent panels is also an effective rehabilitation measure, eliminating some of the forces transmitted through the panel-to-foundation connection.

8.4.3 Infilling of Moment Frames

In many cases, the existing concrete skeleton is stiffened by filling in the space between the beams and columns with masonry or cast-in-place concrete. These infill walls can be a cost-effective method of increasing the lateral strength and rigidity of the building.

Designers should avoid counting on some of the infill walls in structural analysis but not on others, because the stiffness of the frames filled with this nonstructural masonry will increase, whether the designers realize this fact or not. In an earthquake, these panels attract large lateral forces and are damaged, or the perimeter columns, beams, and their connections fail. When a frame, however well designed, is filled with rigid material, however brittle and weak, the fundamental behavior of this structural element is changed from that of a frame to that of a shear wall.

Rehabilitation measures commonly used for concrete frames with masonry infills may also be effective in rehabilitating concrete frames with concrete infills. Additionally, application of shotcrete to the face of an existing wall to increase the thickness and shear strength may be effective. For this purpose, the face of the existing wall should be roughened, a mat of reinforcing steel doweled into the existing structure, and shotcrete applied to the desired thickness.

8.4.4 Reinforced Concrete Moment Frames

Earthquake damage sometimes results in sheared-off columns that formerly were parts of a frame. Typically, the concrete cover is spalled, column bars buckled, and concrete inside broken up. Most problems in concrete frames involve bar splices and failures of beam–column joints that lack confinement and in which reinforcement is stopped prematurely.

Many old buildings with flat-slab and flat-plate floor systems, even those constructed after 1973 (and presumably reflecting the post-San Fernando earthquake code changes), are vulnerable to earthquakes.

Methods available for strengthening traditional concrete frames include encasing the beam–column joints in steel or high-strength fiber jackets. One such design uses jackets consisting of four U-shaped corrugated-metal parts, two around the beam and two around the column. The column jackets are bolted to the end of the beam, the pieces are welded together, and the space between the jackets and the frame is filled with grout.

Frame joints damaged during earthquakes can be repaired with epoxy injection, and badly fractured concrete can be removed and replaced. To minimize shrinkage, the replacement concrete should be made with shrinkage-compensating (type K) cement or should utilize a shrinkage-reducing admixture. Frame members that have been pushed out of alignment during an earthquake should be jacked back into the proper position before repair. Damaged columns can also be strengthened with fiber-reinforced wraps or other methods of exterior concrete confinement. This is common practice for seismic strengthening of building and bridge columns in California. Another structural issue that requires consideration is the transfer of load from the floor diaphragms to the frames and walls. This may require new drag struts. These elements can be added by attaching new concrete or structural steel sections to the underside of existing floors. They are typically placed against cleaned and roughened concrete surfaces and anchored to the floors and to frames by drilled-in dowels or through bolts.

Connections between new and existing materials should be designed to transfer the forces anticipated for the design load combinations. Where the existing concrete frame columns and beams act as boundary elements and collectors for the new shear wall or braced frame, these should be checked for adequacy, considering strength, reinforcement development, and deformability. Diaphragms, including drag struts and collectors, should be evaluated and rehabilitated, if necessary to ensure a complete load path to the new shear wall or braced frame element.

Another method of seismic rehabilitation is to jacket existing beams, columns, or joints with new reinforced concrete, steel, or fiber-wrap overlays. The new materials should be designed and constructed to act compositely with the existing concrete. Where reinforced concrete jackets are used, the design should provide detailing to enhance ductility and the jackets should be designed to provide increased connection strength and improved continuity between adjacent components.

Post-tensioning existing beams, columns, or joints using external post-tensioned reinforcement is an effective strategy of seismic rehabilitation. Post-tensioned reinforcement should be unbounded within a distance equal twice the effective depth from sections where inelastic action is expected. Anchors should be located away from regions where inelastic action is anticipated and be designed considering possible force variations due to earthquake loading.

8.4.5 Open Storefront

The deficiency in a building with an open storefront is the lack of a vertical line of resistance along one or two sides of a building. This results in a lateral system that is excessively soft at one end of the building, causing significant torsional response and potential instability.

The most effective method of correcting this deficiency is to install a new stiff vertical element in the line of the open-front side or sides. If the open-front appearance is desired, the steel frames may be located directly behind the storefront windows. Shear walls may also be used to provide adequate strength. In both cases, collectors are required to adequately distribute the loads from the diaphragm into the vertical lateral-load-resisting element. Adequate anchorage of vertical elements into the foundation is also required to resist overturning forces. Steel moment frames instead of brace frames can also be utilized to provide adequate strength, provided that inelastic deformations of the frame under severe seismic loads are carefully considered to ensure that displacements are controlled. Common methods for upgrading buildings with open storefronts are shown in Figure 8.6.

8.4.6 Clerestory

A clerestory, typically designed to produce an open airy feeling, can result in significant discontinuity in a horizontal diaphragm. A common method of correcting the diaphragm discontinuity is to add a horizontal steel truss. Steel members can be designed to transfer diaphragm shears while minimizing the visual obstruction of the clerestory.

An alternate approach is to reduce the demands on the diaphragm through the addition of new vertical lateral-force-resisting elements such as shear walls or braced frames.

8.4.7 Shallow Foundations

The following rehabilitation measures may be considered for shallow foundations:

1. Enlarging the existing footing to resist the design loads. Care must be taken to provide adequate shear and moment transfer capacity across the joint between the existing footing and the additions.

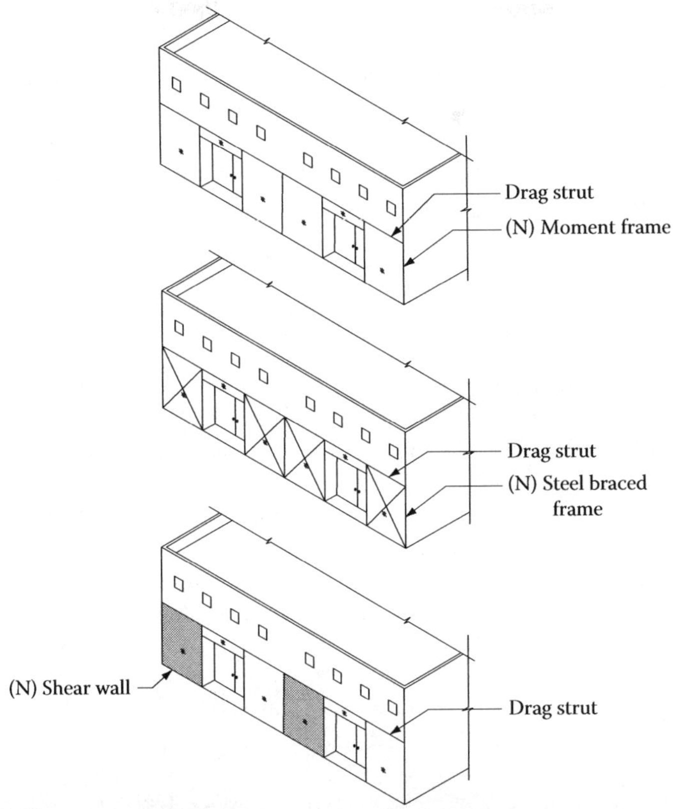

FIGURE 8.6 Common methods for upgrading buildings with open storefronts.

2. Underpinning the existing footing, removing of unsuitable soil underneath, and replacing it with concrete, soil cement, or another suitable material. Underpinning should be staged in small increments to prevent endangering the stability of the structure. This technique may be used to enlarge an existing footing or to extend it to a more competent soil stratum.
3. Providing tension hold-downs to resist uplift. Tension ties consisting of soil and rock anchors with or without prestress may be drilled and grouted into competent soils and anchored in the existing footing. Piles or drilled piers may also be effective in providing tension hold-downs for existing footings.
4. Increasing the effective depth of the existing footing by placing new concrete to increase shear and moment capacity. The new concrete must be adequately doweled or otherwise connected so that it is integral with the existing footing. New horizontal reinforcement should be provided, if required, to resist increased moments.
5. Increasing the effective depth of a concrete mat foundation with a reinforced concrete overlay. This method involves placing an integral topping slab over the existing mat to increase shear and moment capacity.
6. Providing pile supports for concrete footings or mat foundations. Adding new piles may be effective in providing support for existing concrete footing or mat foundations, provided the pile locations and spacing are designed to avoid overstressing the existing foundations.

7. Changing the building structural characteristics to reduce the demand on the existing elements. This may be accomplished by removing mass or height from the building or adding other elements such as energy dissipation devices to reduce the load transfer at the base. New shear walls or braces may be provided to reduce the demand on foundations.
8. Adding new grade beams to tie existing footings together when soil conditions are poor. This method is useful for providing fixity to column bases and to distribute lateral loads between individual footings, pile caps, or foundation walls.
9. Grouting techniques to improve existing soil.

8.4.8 Rehabilitation Measures for Deep Foundations

The following rehabilitation measures may be considered for deep foundations:

1. Providing additional piles or piers to increase the load-bearing capacity of the existing foundations.
2. Increasing the effective depth of a pile cap by adding concrete and reinforcement to its top. This method is effective in increasing its shear and moment capacity, provided the interface is designed to transfer loads between the existing and new materials.
3. Improving the soil adjacent to an existing pile cap by injection grouting.
4. Increasing the passive pressure-bearing area of a pile cap by addition of new reinforced concrete extensions.
5. Changing the building system to reduce the demands on the existing elements by adding new lateral-load-resisting elements.
6. Adding batter piles or piers to the existing pile or pier foundation to increase resistance to lateral loads. It should be noted that batter piles have performed poorly in recent earthquakes when liquefiable soils were present. This is especially important to consider near wharf structures and in areas with a high water table.
7. Increasing tension tie capacity from a pile or pier to the superstructure.

8.4.9 Nonstructural Elements

Two terms frequently used in earthquake engineering are structural and nonstructural items. The structural items of a building are those that resist gravity, earthquake, wind, and other types of loads. These are called structural components and include columns, braces, floor and roof slabs, load-bearing walls (i.e., walls designed to support the building weight and/or provide lateral resistance), and foundations (mat, spread footings, piles). The structure is typically designed and analyzed in detail by a structural engineer.

The nonstructural items of a building include every part of the building and all its contents with the exception of the structure. In other words, everything except the columns, floors, beams, etc. Common nonstructural components include ceilings, windows, office equipment, computers, inventory stored on shelves; file cabinets, heating, ventilating, and air conditioning (HVAC) equipment, electrical equipment, furnishing; and lights. Typically, nonstructural items are specified by architects, mechanical engineers, electrical engineers, or interior designers. Note that most of the structural components of a typical building are concealed from view by nonstructural materials.

Why is nonstructural earthquake damage of concern? What are the direct effects of damage to nonstructural items? What are the secondary effects or potential consequences of damage?

The following discussion covers three types of risk associated with earthquake damage to nonstructural components: LS, property loss, and interruption or loss of essential functions. Damage to a particular nonstructural item may have differing degrees of risk in each of these three categories.

In addition, damage to the item may result in direct injury or loss, or the injury or loss may be a secondary effect or consequence of the failure of the item.

8.4.9.1 Life Safety

The first type of risk is that people could be injured or killed by damaged or falling nonstructural components. Even seemingly innocuous items can be lethal if they fall on an unsuspecting victim. If a 25 lb fluorescent light fixture not properly fastened to the ceiling breaks loose during an earthquake and falls on someone's head, the potential for injury is great. Examples of potentially hazardous nonstructural damage that has occurred during past earthquakes include broken glass, overturned tall and heavy cabinets or shelves, falling ceilings or overhead light fixtures, ruptured gas lines or other piping containing hazardous materials, damaged friable asbestos materials, falling pieces of decorative brickwork or precast concrete panels, and collapsed masonry walls.

8.4.9.2 Property Loss

For most commercial buildings, the foundation and superstructure account for approximately 20%–25% of the original construction cost, while the mechanical, electrical, and architectural elements account for the remaining 75%–80%. Contents belonging to the building occupants, such as movable partitions, furniture, files, and office or medical equipment, represent a significant additional expense. Damage to the nonstructural elements and contents of a building can be costly, since these components account for the vast majority of building costs. Immediate property losses attributable to contents alone are often estimated to be one-third of the total earthquake losses.

8.4.9.3 Loss of Function

In addition to the LS and property loss considerations, there is the additional possibility that nonstructural damage will make it difficult or impossible to carry out the functions normally accomplished in a facility. After the serious LS threats have been dealt with, the potential for post-earthquake downtime or reduced productivity is often the most important risk.

The following are two examples of nonstructural damage that resulted in interruptions to post-earthquake emergency operations or to business.

During the 1994 Northridge earthquake, nonstructural damage caused temporary closure, evacuation, or patient transfer at 10 essential hospital facilities. These hospitals generally had little or no structural damage but were rendered temporarily inoperable, primarily because of water damage. At over a dozen of these facilities, water leaks occurred when fire sprinkler, chilled water, or other pipelines broke. Hospital personnel were apparently unavailable or unable to shut off the water, and in some cases water was flowing for many hours. At one facility, water up to 2 ft deep was reported at some locations in the building as a result of damage to the domestic water supply tank on the roof. At another, the emergency generator was disabled when its cooling water line broke where it crossed a separation joint. Other damage at these facilities included broken glass, dangling light fixtures, elevator counterweight damage, and lack of emergency power due to failures in the distribution or control systems. Two of these facilities, Los Angeles County Olive View Medical Center and Holy Cross Medical Center, both in Sylmar, California, had suffered severe structural damage or collapse during the 1971 San Fernando earthquake and had been demolished and entirely rebuilt.

The 1971 San Fernando earthquake caused extensive damage to elevators in the Los Angeles area, even in some structures where no other damage was reported. An elevator survey indicated 674 instances where counterweights came out of the guide rails, in addition to reports of other types of elevator damage.

These elevators were inoperable until they could be inspected and repaired. Many thousands of businesses were temporarily affected by these elevator outages. The State of California instituted seismic elevator code provisions in 1975, and while these provisions appear to have helped reduce the damage, there were still many instances of counterweight damage in the San Francisco area following the 1989 Loma Prieta earthquake and 688 cases in the Northridge earthquake.

In some cases, cleanup costs or the value of lost employee labor are not the key measures of the post-earthquake impact of an earthquake. For example, data processing facilities or financial institutions must remain operational on a minute-by-minute basis to maintain essential services and monitor transactions at distant locations. In such cases, spilled files or damage to communications and computer equipment may represent less tangible but more significant outage costs. Hospitals and fire and police stations are all facilities with essential functions that must remain operational after an earthquake; damage to their nonstructural elements can be a major cause of loss of functionality.

8.4.9.4 Causes of Nonstructural Damage

Earthquake ground shaking has three primary effects on nonstructural elements in buildings. These are inertial or shaking effects on the nonstructural components when the building structure sways back and forth and separation or pounding at the interface between adjacent structures:

a. *Inertial forces*: When a building is shaking during an earthquake, the base of the building moves in unison with the ground, but the entire building and building contents above the base will experience inertial forces. These inertial forces can be explained by using the analogy of a passenger in a moving vehicle. As a passenger, you experience inertial forces whenever the vehicle is rapidly accelerating or decelerating. If the vehicle is accelerating, you may feel yourself pushed backward against the seat, since the inertial force on your body acts in the direction opposite that of the acceleration. If the vehicle is decelerating or braking, you may be thrown forward in your seat. Although the engineering aspects of earthquake inertial forces are more complex than a single principle of physics, the law first formulated by Sir Isaac Newton, $F = ma$ or force is equal to the mass times the acceleration, is the basic principle involved. In general, the earthquake inertial forces are greater if the mass is greater (if the building or object within the building weights more) or if the acceleration or severity of the shaking is greater.

   File cabinets, emergency power-generating equipment, freestanding bookshelves, office equipment, and items stored on shelves or racks can all be damaged because of inertial forces. When unrestrained items are shaken by an earthquake, inertial forces may cause them to slide, swing, strike other objects, or overturn. Items may slide off shelves and fall to the floor. One common misconception is that large, heavy objects are stable and not as vulnerable to earthquake damage as lighter objects, perhaps because we may have difficult moving them. In fact, many types of objects may be vulnerable to earthquake damage caused by inertial forces: since inertial forces during an earthquake are proportional to the mass or weight of an object, a heavily loaded file is more subject to sliding or overturning than a light one with the same dimensions.

b. *Building distortion*: During an earthquake, building structures distort, or bend, from side to side in response to the earthquake forces. For example, the top of a tall office tower may lean over a few feet in each direction during an earthquake. The distortion over the height of each story, known as the story drift, might range from a percentage of an inch to several inches, depending on the size of the earthquake and the characteristics of the particular building structure. Windows, partitions, and other items that are tightly locked into the structure are forced to go along for the ride. As the columns or walls distort and become slightly out of square, if only for an instant, any tightly confined windows or partitions must also distort the same amount. The more space there is around a pane of glass where it is mounted between stops or molding strips, the more distortion the glazing assembly can accommodate before the glass itself is subjected to earthquake forces. Brittle materials like glass, plaster or drywall partitions, and masonry infill or veneer cannot tolerate any significant distortion and will crack when the perimeter gaps close and the building structure pushes directly on the brittle elements. Most architectural components such as glass panes, partitions, and veneer are damaged because of this type of building distortion, not because they themselves are shaken or damaged by inertial forces.

There have also been notable cases of structural–nonstructural interaction in past earthquakes, where rigid nonstructural components have been the cause of structural damage or collapse. These cases have generally involved rigid, strong architectural components, such as masonry infill or concrete spandrels that inhibit the movement or distortion of the structural framing and cause premature failure of column or beam elements. While this is a serious concern for structural designers, the focus of this guide is on earthquake damage to nonstructural components.

c. *Building separations*: Another source of nonstructural damage involves pounding or movement across separation joints between adjacent structures. A separation joint is the distance between two different building structures, often two wings of the same facility that allows the structures to move independently of one another. A seismic gap is a separation joint provided to accommodate relative lateral movement during an earthquake. In order to provide functional continuity between separate wings, building utilities must often extend across these building separations, and architectural finishes must be detailed to terminate on either side. For base-isolated buildings, a seismic isolation gap occurs at the ground level, between the foundation and the base of the superstructure. The separation joint may be only an inch or two in older construction or as much as a foot in some newer buildings, depending on the expected horizontal movement or seismic drift. Flashing, piping, fire sprinkler lines, HVAC ducts, partitions, and flooring all have to be detailed to accommodate the seismic movement expected at these locations when the two structures move closer together or drift apart. Damage to items crossing seismic gaps is a common type of earthquake damage. If the size of the gap is insufficient, pounding between adjacent structures may result in damage to structural components but often causes damage to nonstructural components, such as parapets, veneer, or cornices on the facades of older buildings.

8.4.9.5 Design Procedure for Nonstructural Components

By and large, advances in earthquake engineering made in recent decades have been successfully applied to the task of making building structures safer. In comparison, application of this knowledge to the design of nonstructural components has been slow and gradual. Design professionals and building owners have now recognized that the seismic resistance of critical nonstructural components must be addressed as part of the design process, since failures of nonstructural components may threaten the safety of building occupants and result in significant financial loss.

The following is a brief description of the engineering design procedure for nonstructural components. Minimum design levels for architectural, mechanical, and electrical systems and components are described in the ASCE 7-10, Chapter 13 provisions. The provisions specify horizontal seismic force factors to be used for the design of specific items, such as partitions, parapets, chimneys, ornaments, tank supports, storage racks over 8 ft tall, equipment or machinery, piping, and suspended ceilings. According to the procedure, the design force depends on a variety of factors such as, the SDC of the building, type of component, the location of the item within a building, and the type of occupancy. Design forces are generally greater for emergency generators than for HVAC equipment, greater for police and fire stations than for ordinary office buildings, and greater at the roof than at the ground.

8.4.9.6 Seismic Hazard

The seismic risk for a particular nonstructural component at a particular facility is governed by a variety of factors, including the SDC assigned to the building, the regional seismicity, the proximity to an active fault, the local soil conditions, the dynamic characteristics of the building structure, the dynamic characteristics of the nonstructural components and any connections to the structure, the location of the nonstructural component within the building, the function of the facility, and the importance of the particular component to the operation of the facility.

The seismic hazard in a given region or geographic location is related both to the severity of ground shaking expected in the area and to the likelihood, or probability, that a given level of shaking will occur. Seismologists review historical earthquake activity, locations and characteristics of mapped faults, and regional geology to estimate the seismic hazard. This information is often depicted on a seismic hazard map.

For the purposes of this discussion, seismic hazard may be broadly characterized in terms of three levels of shaking intensity: namely, light, moderate, and severe. USGS seismic hazard maps show the geographic areas in the United States where light, moderate, and severe shaking are likely to occur in future earthquakes.

For engineering purposes, earthquake shaking is often characterized by spectral acceleration parameters S_s and S_1 accelerations, measured as a percentage of the acceleration of gravity. The EPA is often less than the maximum acceleration recorded during an earthquake.

Several examples of earthquake motion recorded during the 1994 Northridge earthquake may help to put these accelerations ranges in perspective. The Northridge earthquake had a magnitude of 6.7. The magnitude of an earthquake is a measure of the amount of energy released by the fault rupture or ground shaking; it does not provide any information about the intensity of shaking at any particular location. During this earthquake, the intensity of ground shaking was severe in Northridge, near the earthquake epicenter. Several stations recorded ground motion with maximum accelerations in excess of 0.9*g*. Nevertheless, the majority of stations in downtown Los Angeles, at distances of approximately 30–40 km from the epicenter, recorded moderate ground shaking (between 0.15*g* and 0.30*g*), while stations in Long Beach, approximately 60 km from the epicenter, recorded light ground shaking (less than 0.15*g*). The EPAs from the Northridge earthquake records are lower than the maximum recorded accelerations indicated above, but the EPAs also shown the same trend, that is, the shaking intensity became less severe as the distance from the epicenter increased. While few people outside California may need to worry about the intensity of shaking experienced in Northridge, many areas of the country may experience the moderate or light shaking that was felt in downtown Los Angeles and Long Beach during the Northridge earthquake.

One further note regarding shaking intensity will serve to illustrate what appears to be one of the most extreme cases to date of recorded earthquake shaking intensity. A peak horizontal acceleration of 1.7*g* was recorded at the roof level of the Los Angeles County Olive View Medical Center in Sylmar during the Northridge earthquake. The roof acceleration was 2.6 times higher than the ground acceleration (0.65*g*) measured near the building. The horizontal forces on items at the roof level were 170% of their weight, if only for an instant. Although some roof-mounted items at the hospital were severely damaged, most of the anchored items performed well because they were designed using the special seismic requirements of the California Hospital Seismic Safety Act.

8.4.9.7 Non-Load-Bearing Walls

The performance of buildings with nonstructural walls that adversely affect the seismic response of a building may be improved by removing and replacing them with walls constructed of relatively flexible materials such as gypsum board sheathing or modifying the wall connections so that they will not resist lateral loads. Removal and replacement of existing hollow clay tile, concrete, or brick masonry partitions is the preferred method of addressing the inadequate out-of-plane capacity of nonstructural partitions. Alternatively, steel strongbacks can provide the out-of-plane support. Steel members are installed at regular intervals and secured to the masonry with drilled and grouted anchors. The masonry spans between the steel members, which span either vertically between floor diaphragms or horizontally between columns. A third method for mitigating masonry walls with inadequate out-of-plane capacity is to provide a structural overlay. The overlay may be constructed of plaster with welded wire mesh reinforcement or concrete with reinforcing steel or welded wire mesh. This approach is used at times merely to provide containment of the masonry. Nonstructural masonry walls are frequently used as firewalls around means of egress. Egress walls with deficient out-of-plane capacity can fail, resulting in rubble blocking the egress. Containment of the masonry with a plaster or concrete overlay can maintain egress, although the walls may need to be replaced following a major seismic event.

8.4.9.8 Precast Concrete Cladding

Precast concrete cladding panels with rigid connections may not have the flexibility or ductility to accommodate large building deformations. Failure of the connection may result in heavy panels falling away from the building. Complete correction of this deficiency is likely to be costly, since numerous panel connections would need to be modified to accommodate anticipated building drifts. This may require removal and reinstallation or replacement of the panels. A more economical solution is to install redundant flexible/ductile connections that will keep the panels from falling, should the existing connections fail.

Improper design or installation of precast concrete cladding may also be more than just a connection problem. The cladding may act as an unintended lateral-load-resisting element, should the connections be rigid or if insufficient gaps are present between panels. Correcting this deficiency can be accomplished by installing occasional seismic joints in the panels to minimize their stiffness or by stiffening the existing lateral-force-resisting system.

If an entirely new precast cladding system is installed, the connections should be designed to

- Carry gravity loads of precast panels
- Transfer the in-plane and out-of-plane inertia forces of the panels into the building
- Isolate the panels from the inelastic drift likely to be experienced by the building in a large earthquake

8.4.9.9 Stone or Masonry Veneers

Stone or masonry veneers may become falling hazards unless their anchorage can accommodate the inelastic deformation of the building. Removal and replacement by veneer with adequate anchorage is one option. A second option is to decrease the deformation of the supporting wall by adding stiffness to the structure.

8.4.9.10 Building Ornamentation

Building ornamentation such as parapets, cornices, signs, and other appendages are another potential falling hazard during strong ground shaking. Unreinforced masonry parapets with heights greater than $1^1/_2$ times their width are particularly vulnerable to damage. Parapets are commonly retrofitted by providing bracing back to the roof framing.

Cornices and other stone or masonry appendages may be retrofitted by installing drilled and grouted anchors at regular intervals. Sometimes they may be replaced with a lightweight substitute material such as plastic, fiberglass, or metal.

8.4.9.11 Acoustical Ceiling

Unbraced suspended acoustical tile ceilings are significantly more flexible than the floors or roofs to which they are attached. The ceilings sway independently from the floor or roof, typically resulting in their connections being broken. This deficiency can be reduced by stiffening the suspended ceiling system with diagonal wires between the ceiling grid and the structural floor or roof members. Vertical compression struts are also required at the location of the diagonal wires to resist the upward component of force caused by the lateral loads. Current code standards can be used for the upgrade of existing ceiling systems.

8.4.10 Fiber-Reinforced Polymer Systems for Strengthening of Concrete Buildings

Composite materials made of fibers in a polymeric resin—also known as fiber-reinforced polymers (FRP)—have come into use as an alternative to traditional strengthening techniques such as steel plate bonding, section enlargement, and external post-tensioning. This technique has been used to strengthen many bridges and buildings around the world and was first applied to concrete columns in Japan for providing additional confinement. The development of codes and standards for externally bonded FRP systems is ongoing in Europe, Japan, Canada, and the United States. Within the last 10 years, several documents related to the use of FRP materials in concrete structures have been published.

The FRP systems come in a variety forms including wet lay-up and precured systems. Wet lay-up systems consist of dry unidirectional or multidirectional fiber sheets impregnated with a saturating resin on-site. The saturating resin along with the compatible primer and putty is used to bond the FRP fabric to the concrete surface.

Prepregnation systems consist of uncured unidirectional or multidirectional fiber sheets or fabrics that are preimpregnated with a saturating resin in the manufacturing facility. They are bonded to the concrete surface with or without an additional resin application, depending upon specific system requirements.

Precured systems consist of a wide variety of manufactured composite shapes. The precured shapes are typically bonded to the concrete surface by an adhesive along with a primer and putty. There are three common types of precured systems:

- Unidirectional laminate sheets typically delivered to the site as thin ribbon strips coiled on a roll
- Multidirectional grids, also typically delivered to the site coiled on a roll
- Shell segments cut longitudinally so they can be opened and fitted around columns, beams, or other components of buildings

8.4.10.1 Mechanical Properties and Behavior

Unlike steel reinforcement, FRP materials do not exhibit plastic behavior when loaded in tension. The stress–strain relationship is linearly elastic until failure, which is sudden and can be catastrophic.

The tensile property of the FRP material is governed by the type of fiber and its orientation and quantity. The tensile property of an FRP system should be characterized as a composite, based on not just the material properties of the individual fibers but also on the efficiency of the fiber–resin system, the fabric design, and the method used to create the composite. The mechanical properties should be based on the testing of laminate samples with known fiber content.

Externally bonded FRP systems should not be used as compression reinforcement. There has been very little testing to validate their use in resisting compressive forces. The failure mode for FRP laminates subjected to longitudinal compression can include transverse tensile, fiber microbuckling, or shear failure.

The FRP materials subject to a constant load over time suddenly fail after a period referred to as endurance time, also referred to as creep–rupture. In general, carbon fibers are the least susceptible to creep–rupture, aramid fibers are moderately susceptible, and glass fibers are most susceptible.

Many FRP systems exhibit reduced mechanical properties after exposure to certain environmental factors, including temperature, humidity, and chemicals. The tensile properties reported by the manufacturers are based on tests conducted in a laboratory and do no reflect the effects of environmental exposure. Therefore, the properties should be adjusted to account for the anticipated service environment.

8.4.10.2 Design Philosophy

The design of FRP systems is based on traditional reinforced concrete design principles. The FRP strengthening systems are designed to resist tensile forces while maintaining strain compatibility with the concrete substrate. Unlike mild steel reinforcement, FRP systems should not be relied on to resist compressive forces. It is permissible, however, for the FRP tension reinforcement to experience compression due to changes in moment patterns or moment reversals, as in members subjected to seismic forces, with the proviso that the compressive strength of the FRP system is neglected in calculating the member capacities.

In FRP design, certain limits are imposed to guard against collapse of the structure, should bond or some other type of failure occur due to vandalism, fire, or other causes. The designer is referred to ACI committee 440 recommendations for further details.

8.4.10.3 Flexural Design

An increase in the flexural strength of a concrete member can be achieved by bonding FRP reinforcement to its tension face with fibers oriented along the member's length. Although higher-strength increases are reported in test results, an increase of up to 40% of the original strength is considered reasonable in view of ductility and serviceability limits.

Flexural strengthening using FRP systems is not recommended for enhancing flexural capacity of members in the expected plastic regions. Cases in point are the plastic hinge regions of ductile moment frames resisting seismic loads. For such cases, the effect of cyclic load reversal on the FRP system should be investigated.

See Figure 8.9 for typical application of FRP for a transfer girder.

8.5 CONCLUDING REMARKS

Before concluding this chapter, perhaps it is beneficial to reflect on some of the performance scenarios presented in ASCE/SEI 41-06, particularly for those buildings deemed to be top-of-the-line performers. It is the opinion of many engineers including the authors that high-end building performance cannot be promised or achieved with 100% certainty. Top-of-the-line performance implies a near-perfect earthquake-proof building. Therefore, building owners and the public are likely to ask for it more frequently. They should, but with the understanding that there is no iron-clad guarantee that the building will be earthquake-proof in the future. It is therefore the structural engineers' responsibility to make this fact clear to the owners and to the public so that their expectations for building performance do not exceed those implied in the ASCE/SEI 41-06 performance definition. Although major advances have been made in predicting analytical capability and in the synthesizing of experimental and earthquake performance data, prediction of building performance, relative to future earthquakes, is still a risky and dangerous business. Thus, seismic rehabilitation continues to challenge the very core of conventional thinking.

Generally, seismic retrofit should be considered only if its entire cost is less than 70%–80% of replacement cost. It should be noted that it is impossible to bring an existing structure into conformance with current code requirements. Therefore, the cost assigned to the retrofit should not out weight the seismic performance expected of the building.

It behooves the designer to consider more than one seismic retrofit strategy. One of these should be a conventional one. This ensures that designers are not carried away with a high-tech new approach when a more conventional retrofit strategy is more cost-effective.

Seismic retrofit design is invariably more expensive than new construction design. The extra design effort required for retrofit design should be communicated to the owner at the onset of the project. The cost of the retrofit design should be pegged to the complexity of the analysis required. Many designers do not assign sufficient design hours to projects that require NSP or NDP. The cost of developing and implementing material test recommendations should be considered at the start of the project. It is recommended that material test results be available to the designer before the design development phase is started. The design basis should already be stated and discussed with the owner and peer reviewers at the onset of the project.

It is emphasized that a number of parameters used in ASCE/SEI 41-06 remain a matter of discussion and research. It must remain at the discretion of engineers to modify the parameters as they deem appropriate. Agreement on key issues must be reached with peer reviewers as the analysis progresses.

ASCE/SEI 41-06 permits both linear and nonlinear analysis procedures for evaluation of existing buildings and evaluation of rehabilitated construction. Whether or not ASCE/SEI 41-06 is applicable to design of new construction is a matter of discussion between the engineers of record (EOR), peer reviewers, and the building official. Observe that the emergence of PBD does strongly promote such a procedure.

Rehabilitation strategies described in ASCE/SEI 41-06 include the following:

1. Global modifications such as
   a. Increasing stiffness and strength by adding new elements
   b. Increasing damping using supplemental damping devices
   c. Isolating the structure from seismic ground motions by using seismic isolation
   d. Decreasing mass
2. Local modification of components consisting of
   a. Local strengthening
   b. Jacketing

For a list of typical deficiencies and rehabilitation measures, the reader is referred to Table 8.2.

The LSP is similar to the equivalent lateral procedure included in most seismic standards. However, the pseudolateral loads $V = C_1C_2C_mS_aW$ incorporates techniques for considering the nonlinear response of individual elements and components and is based on the unreduced spectral acceleration S_a. In a manner of speaking, we are using a value of unity for the seismic response modification, R.

The LDP may be used for linear modal spectral analysis or linear time-history analysis. In both cases, the results are modified with coefficients similar to those in the LSP. The acceptability criteria are the same as for LSP, including separating force- and displacement-controlled actions. The acceptance of performance is judged for each component by defining component action as either deformation-controlled or force-controlled.

In an NSP, the analytical model consists of all elements having significant strength or stiffness. An analysis, commonly referred to as a pushover analysis, is performed to develop the relationship between lateral forces and displacement at the roof.

The elements that do not have significant lateral resistance may be designated secondary and removed from the model. Generally, a computer program with nonlinear analysis capability is used

TABLE 8.2
Typical Deficiencies and Rehabilitation Measures

Deficiency in Existing Buildings	Typical Rehabilitation Measures
General structural attributes	• Complete discontinuous frames
1. Irregularities: vertical and horizontal	• Add additional elements to eliminate irregularities and add redundancy
2. Lack of redundancy	
3. Pounding potential	• Stiffen elements (e.g., add cover plates or concrete, encasement)
Steel moment frames	• Add supplemental lateral-force-resisting elements to stiffen building
4. Insufficient framing strength	• Add damping to reduce drift
5. Noncompact sections	• Increase capacity of elements (e.g., add cover plates or lateral bracing)
6. Inadequate girder-column connections	• Add bracing to frames or additional frames to reduce loads on flexural/shear members
7. Strong beam–weak column	
8. Inadequate columns splices	• Increase capacity of elements (e.g., add cover plates or lateral bracing)
9. Nonductile joints	• Strengthen connections and panel zones (e.g., add plates or stiffeners)
10. Large drifts	• Increase capacity of columns (e.g., add cover plates or lateral bracing)
	• Add full penetration welds to column splices or splice plates
Braced frames	• Strengthen connections and panel zones (e.g., add plates or stiffeners)
11. Insufficient frame strength	• Stiffen elements (e.g., add cover plates or concrete)
12. Improper brace configuration slender braces	• Add supplemental lateral-resisting elements to stiffen building
	• Add damping to reduce drift
13. Insufficient connection strength. Inadequate column splices	• Add lateral-force-resisting elements to reduce loads on existing members
	• Replace or add braces in proper configuration
	• Reduce unbraced lengths of bracing members
	• Strengthen connection (e.g., add plates, stiffeners, bolts, welds)
	• Add full penetration weld column splices or splice plates
Diaphragms and diaphragm connections	• Overlay existing slab with new reinforced topping slab with trim bars and collectors
14. Inadequate diaphragm strength/stiffness	
15. Inadequate collectors	• Add drag struts or collectors at reentrant corners and openings
16. Lack of diaphragm tensile capacity at reentrant corners	• Strengthen connections to vertical elements
17. Inadequate shear transfer to vertical elements	
Foundation and foundation connections	
18. Inadequate foundations	• Strengthen foundations
19. Inadequate column-to-foundation connections	• Strengthen base plate or column connections to foundations

or a linear analysis with incremental loading is performed. Static lateral loads are applied incrementally and the element properties are adjusted for yielding or failure. The seismic global displacement demand is determined and deformation-controlled components are judged acceptable if their gravity plus earthquake deformation demand is less than or equal to the expected to permissible deformation capacity given in ASCE/SEI 41-06.

If all of the components in the structure meet the basic acceptance criteria associated with their actions, no further analysis is necessary, and the building may be judged to meet the evaluation criteria. If not, typically a more refined study (including a pushover analysis) would be considered before deciding on a seismic rehabilitation program. The final evaluation should be based on a review of the qualitative and quantitative results. The evaluating engineer is urged to consider the issues carefully, to refrain from penalizing the building due to fine technical points beyond those contained in the ACSE 41-06 evaluation methodology, and to visualize the building in its ultimate condition in an earthquake, being aware of the risks of brittle failure and buckling. Due consideration should be given to the mitigating influences of good workmanship, structural integrity, and the strengths and redundancies that are not explicitly considered to be part of the lateral-force-resisting system. Most importantly, engineering judgment based on sound seismic design principles should be exercised before pronouncing a building unsafe. The questions that review engineers should ask themselves before declaring a building noncompliant are many. Some are as follows:

1. What if the material properties are higher than assumed in the analysis?
2. What if we allow for a small amount of rocking and sliding at the base to absorb excess earthquake energy at little harm to structure?
3. What if we use composite properties for frame–beams?
4. For a moment frame building, what if we reanalyze the frame using different size rigid joints in the frame model? Does inclusion of an elastic spring to represent the stiffness of the joint result in a more favorable DCR?
5. What if we use slightly higher values for the ductility factor m in verifying the acceptance criteria?

Although ASCE/SEI 41-06 has procedures to answer some of these questions, the author however recommends a parametric study of the acceptance criteria before declaring the building noncompliant. This recommendation should not be constructed as sanctioning indiscriminate manipulation of the ASCE/SEI 41-06 procedure but as a reminder for engineers to use that nonquantifiable, mysterious branch of engineering often called the art of design. It should be kept in mind that no matter how sophisticated an analysis is, it is hard to justify that its seismic behavior will be satisfactory if it has large vertical and horizontal discontinuities. Experience has taught us time and again that unfavorable seismic characteristics mostly arise in a poorly balanced structural system. The seismic retrofit should, then, focus on removing irregularities and discontinuities.

In the evaluation and upgrading of an existing structure, it is sometimes difficult to identify an existing lateral-force-resisting system. Innovative analytical procedures and reliance on existing materials and systems that are not generally considered for new construction are required to determine the load paths and capacities of the existing structures. When an existing structure is not adequate to resist the prescribed lateral forces, strengthening of the existing lateral-force-resisting system will be required.

The selection of an appropriate strengthening technique for the upgrading of an existing building that does not comply with the acceptance criteria will depend upon the type of structural systems in the existing building and the nature of the deficiency. In some cases, the selection may be influenced by other than structural considerations. For example, a requirement that the building be kept operational during the structural modifications may dictate that the modification be restricted to the periphery of the building. On the other hand, it may be possible to temporarily relocate the occupants of a building that is to be upgraded. This, of course, provides more latitude in the selection of appropriate and cost-effective strengthening techniques. In many cases, seismic upgrading is accomplished concurrently with functional alterations, renovation, and/or energy retrofits. In these cases, the selected structural modification scheme should be the one that best suits the requirements of all the proposed alterations.

Determination of the seismic capacity of a structure includes consideration of all elements, structural and nonstructural, that contribute to the resistance of lateral forces.

Physical properties are generally obtained from available data; otherwise, assumptions and/or tests must be made. The analysis must include the evaluation of the most rigid elements resisting the initial lateral forces, as well as the more flexible elements that resist the lateral distortions after the rigid elements yield. Consideration must also be given to the interaction of various combinations of the structural framing systems and elements, which will contribute to the resistance of the lateral loads.

The results of the detailed structural analysis will identify the deficiencies with respect to the acceptance criteria of the various structural components and systems. These results should be carefully reviewed in the development of alternative upgrade concepts unless justification can be shown for a single solution.

Each concept should be developed to the extent that will permit a reasonable cost estimate to be made. The extent of removal of existing construction should be considered, including the sizes and locations of new, replaced, or strengthened structural members. Typical structural connections with schematic details for upgrading nonstructural elements should be included in the study.

8.6 SEISMIC STRENGTHENING DETAILS

A thorough understanding of existing construction and seismic retrofit objectives acceptable to owners and to the building official is an important consideration before a seismic retrofit is undertaken. The importance of considering global and elemental deformations at expected levels of seismic forces, not at code or design levels, cannot be overstressed. This is because even with the use of amplification factors, the deformations are at best an approximation, particularly when applied to complex multistory and multidegree-of-freedom systems. It should be kept in mind that detailing in existing buildings often does not meet the requirements of new construction and that the strength and stiffness of existing elements may not be comparable with new upgraded systems and elements. Thus, verification of elements for deformation compatibility becomes even more important. This criterion is secondary only to the requirement of providing a continuous load path that is sufficiently stiff and strong to resist realistic earthquake forces. Suggested rehabilitation measures listed by deficiencies are given in subsequent paragraphs:

1. *Load path*: Add elements to complete the load path. This may require adding new shear walls or frames to fill gaps in existing shear walls or frames that are not continued to the foundation. It also may require the addition of elements throughout the building to pick up loads from diaphragms that have no path into existing vertical elements.
2. *Redundancy*: Add new lateral-force-resisting elements in locations where the failure of a single element will cause instability in the building. The added lateral-force-resisting elements should be of comparable stiffness as the elements they are supplementing.
3. *Vertical irregularities*: Provide new vertical lateral-force-resisting elements to eliminate vertical irregularity. For weak stories, soft stories, and vertical discontinuities, add new elements of the existing type.
4. *Plan irregularities*: Add lateral-force-resisting bracing elements that will support major diaphragm segments in a balanced manner. Verify whether it is possible to allow the irregularity to remain and instead strengthen those structural elements that are overstressed.
5. *Adjacent buildings*: Add braced frames or shear walls to one or both buildings to reduce the expected drifts to acceptable levels. With separate structures in a single building complex, it may be possible to tie them together structurally to force them to respond as a single unit. The relative stiffness of each and the resulting force interactions must be determined to ensure that additional deficiencies are not created. Pounding can also be eliminated by demolishing a portion of one building to increase the separation.

6. *Lateral-load path at pile caps*: Typically, deficiencies in the load path at the pile caps are not a LS concern. However, if it is determined that there is a strong possibility of a LS hazard, piles and pile caps may be modified, supplemented, repaired, or, in the most severe condition, replaced in their entirety.
7. *Deflection compatibility*: Add vertical lateral-force-resisting elements to decrease the drift demand on the columns or increase ductility of the columns. Jacketing the columns with steel or concrete is one way to increase their ductility.
8. *Drift*: The most direct mitigation approach is to add properly placed and distributed stiffening elements—new moment frames, braced frames, or shear walls—that can reduce the interstory drifts to acceptable levels. Alternatively, the addition of energy dissipation devices to the system may reduce the drift.
9. *Frame and nonductile concerns*: Add properly placed and distributed stiffening elements, such as shear walls, to supplement the moment frame system with a new lateral-force-resisting system. For eccentric joints, columns and beams may be jacketed to reduce the effective eccentricity. Jackets may be also provided for shear critical columns.
   - Short captive columns: Columns may be jacketed with steel or concrete such that they can resist the expected forces and drifts. Alternatively, the expected story drifts can be reduced throughout the building by infilling openings or adding shear walls.
10. *Cast-in-place concrete shear walls*: Add new shear walls and/or strengthen the existing walls to satisfy seismic shear stress demand. New and strengthened walls must form a complete, balanced, and properly detailed lateral-force-resisting system for the building. Special care is needed to ensure that the connection of the new walls to the existing diaphragm is appropriate and of sufficient strength such that yielding will occur in the wall first. All shear walls must have sufficient shear and overturning resistance.
    - Overturning, lengthening, or adding shear walls can reduce overturning demand.
    - Coupling beams. Strengthen the walls to eliminate the need to rely on the coupling beam. The beam should be jacketed only as a means of controlling debris. If possible, the existing opening should be infilled.
    - Boundary component detailing. Splices may be improved by welding bars together after exposing them. The shear transfer mechanism can be improved by adding steel studs and jacketing the boundary components.

Techniques for strengthening or upgrading existing buildings will vary according to the nature and extent of the deficiencies the expected future performance of the building, the configuration of the structural systems, and the structural materials used in construction. Typical details commonly used for seismic upgrading of structural members and systems are given in this section to provide guidelines to engineers (Figures 8.7 through 8.88). In using these details, keep in mind that your judgment and ingenuity in addressing specific situations is an important prerequisite for appropriate use of these details. Good detail is at best a compromise between what's currently available and how it's being applied. Trying to achieve that balance between what's available, affordable, reliable, and functional is the exciting part. Ultimately, if the details manage to most appropriately meet the intended seismic performance, then that's the best design.

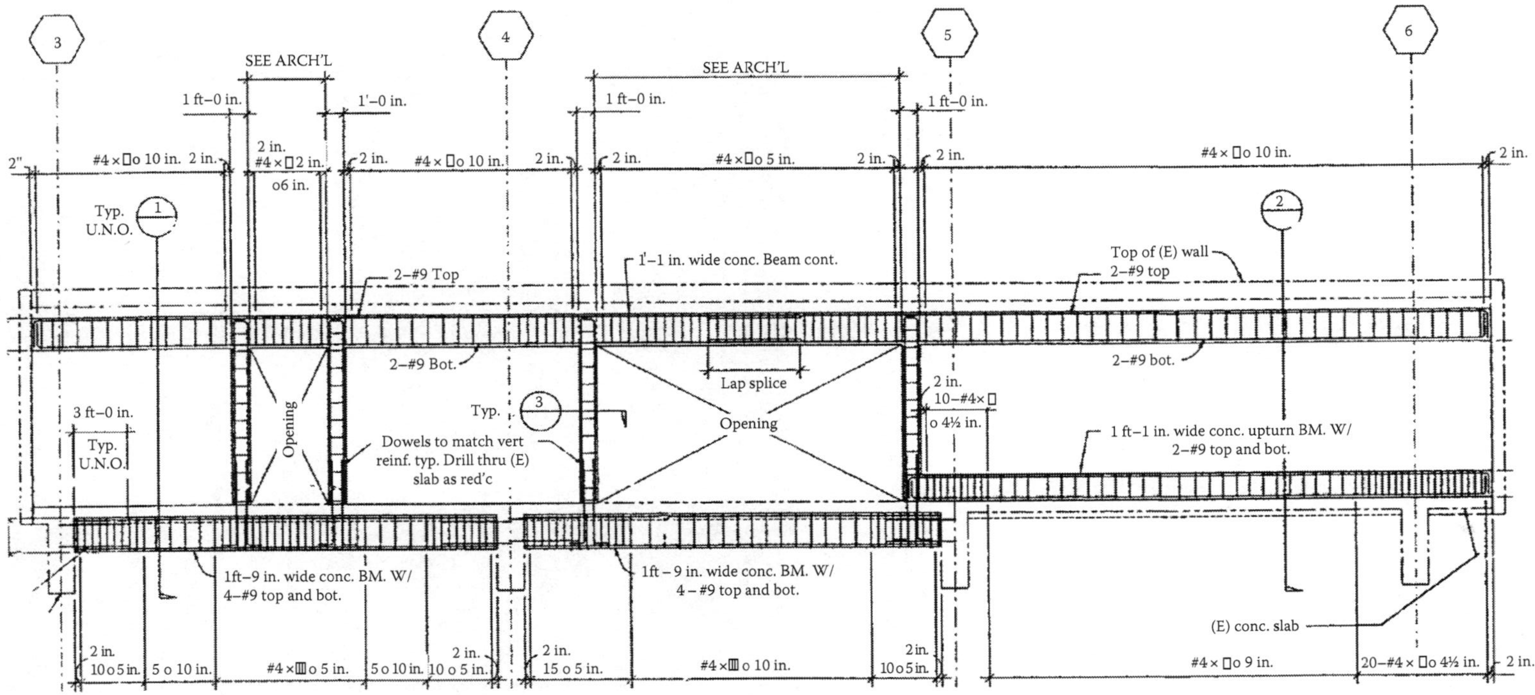

(a)

FIGURE 8.7 (N) openings in an (E) 3-story concrete shear wall building. The seismic upgrade consisted of providing concrete overlay to restore shear capacity of walls and adding boundary elements around (N) openings: (a) wall elevation. *(Continued)*

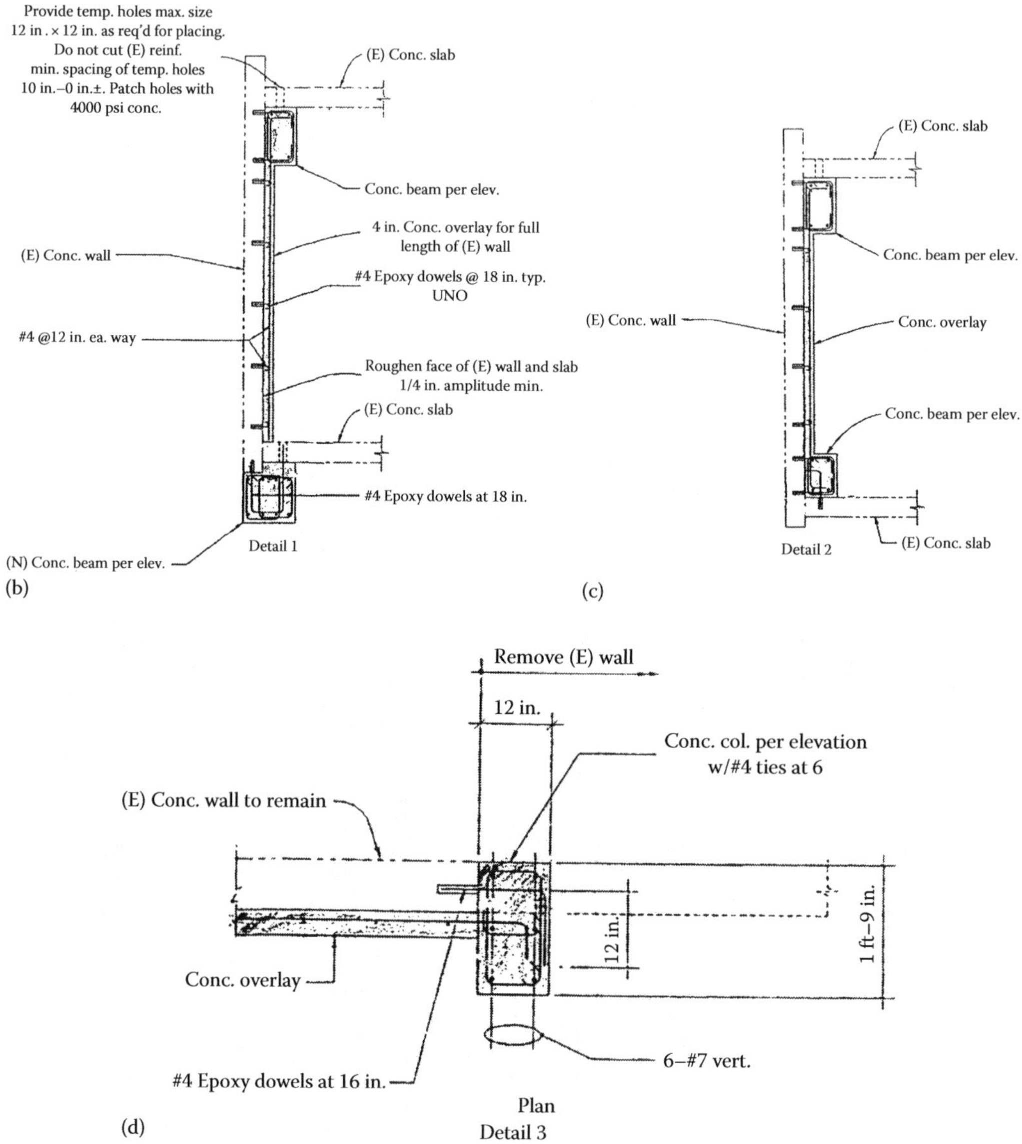

FIGURE 8.7 (*Continued*) (N) openings in an (E) 3-story concrete shear wall building. The seismic upgrade consisted of providing concrete overlay to restore shear capacity of walls and adding boundary elements around (N) openings: (b) concrete overlay with (N) beam below (E) slab, (c) concrete overlay with (N) beam above (E) slab, and (d) plan detail at (N) boundary element. *Note*: (E), existing; (N) new.

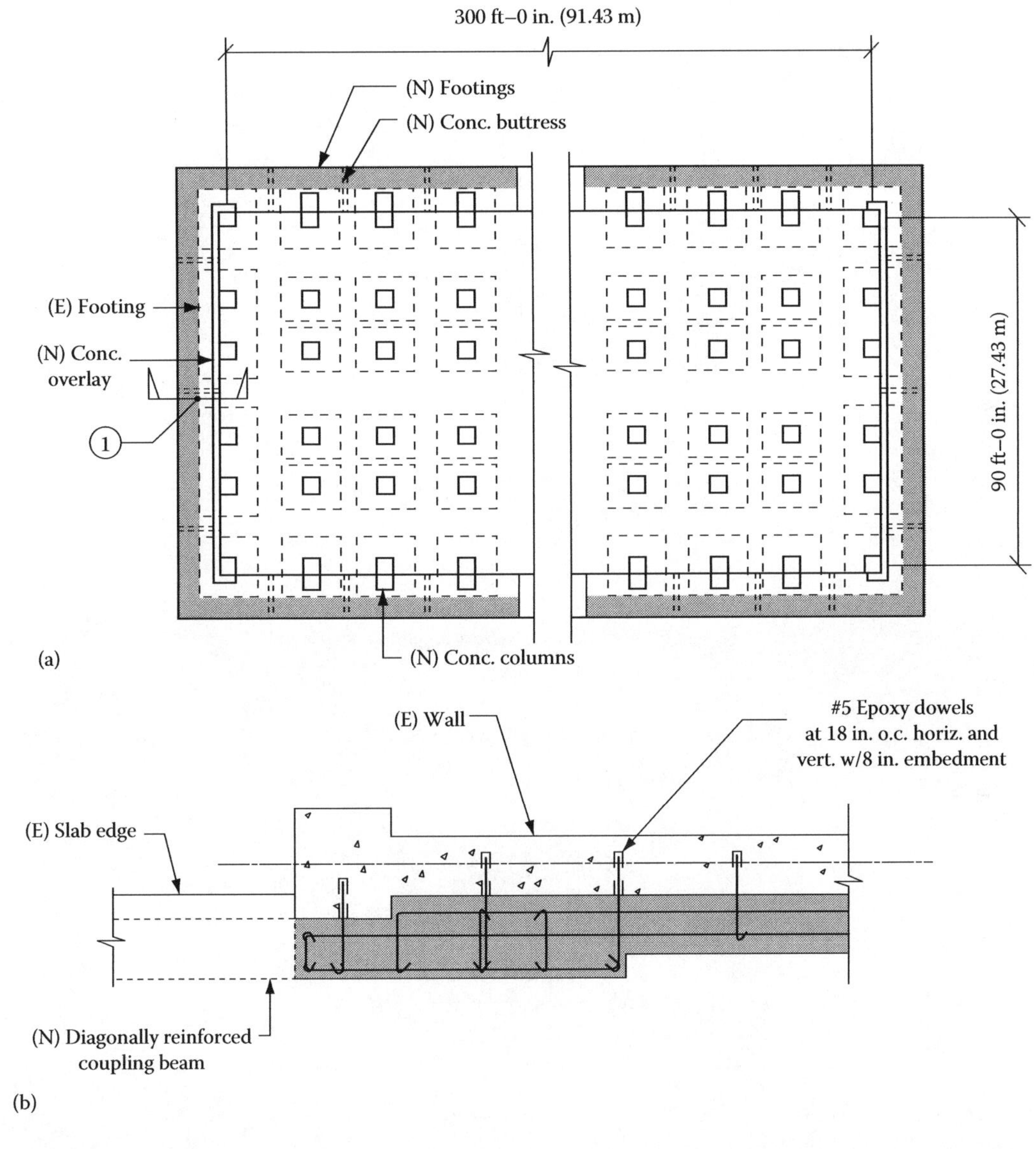

FIGURE 8.8 Seismic upgrade of a concrete hospital building with an external concrete moment frame. Modifications are restricted to the periphery of the building to keep the building operational with minimal interference to its functionality: (a) plan showing (N) foundation, (N) concrete overlay in the transverse direction, and (N) moment frames in the longitudinal direction; (b) enlarged plan at (N) coupling beam and shear wall overlay. *(Continued)*

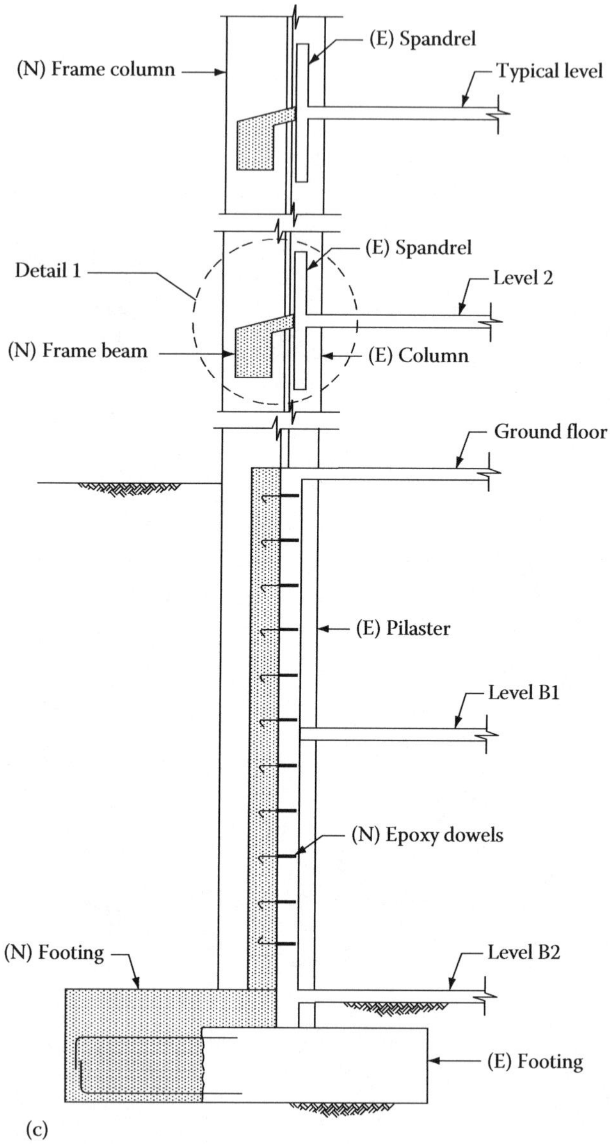

FIGURE 8.8 (*Continued*) Seismic upgrade of a concrete hospital building with an external concrete moment frame. Modifications are restricted to the periphery of the building to keep the building operational with minimal interference to its functionality: (c) section through longitudinal frame. (*Continued*)

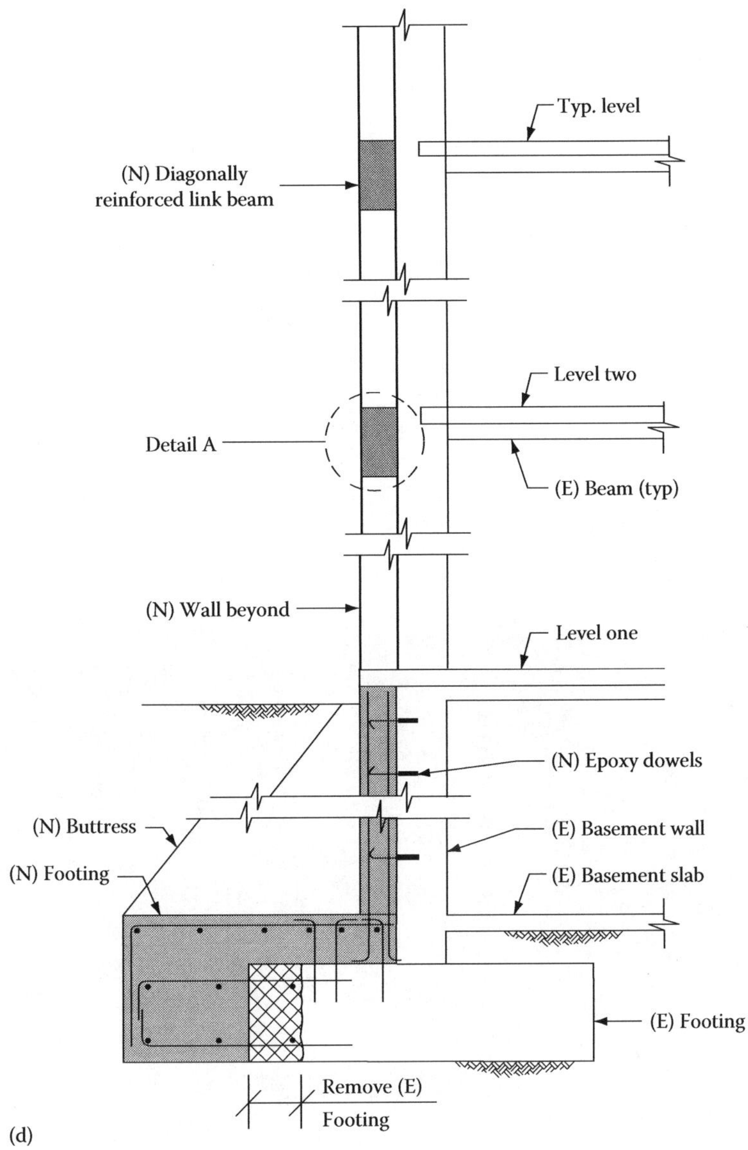

FIGURE 8.8 (*Continued*) Seismic upgrade of a concrete hospital building with an external concrete moment frame. Modifications are restricted to the periphery of the building to keep the building operational with minimal interference to its functionality: (d) section through transverse wall; (e) connection between (N) and (E) frame.

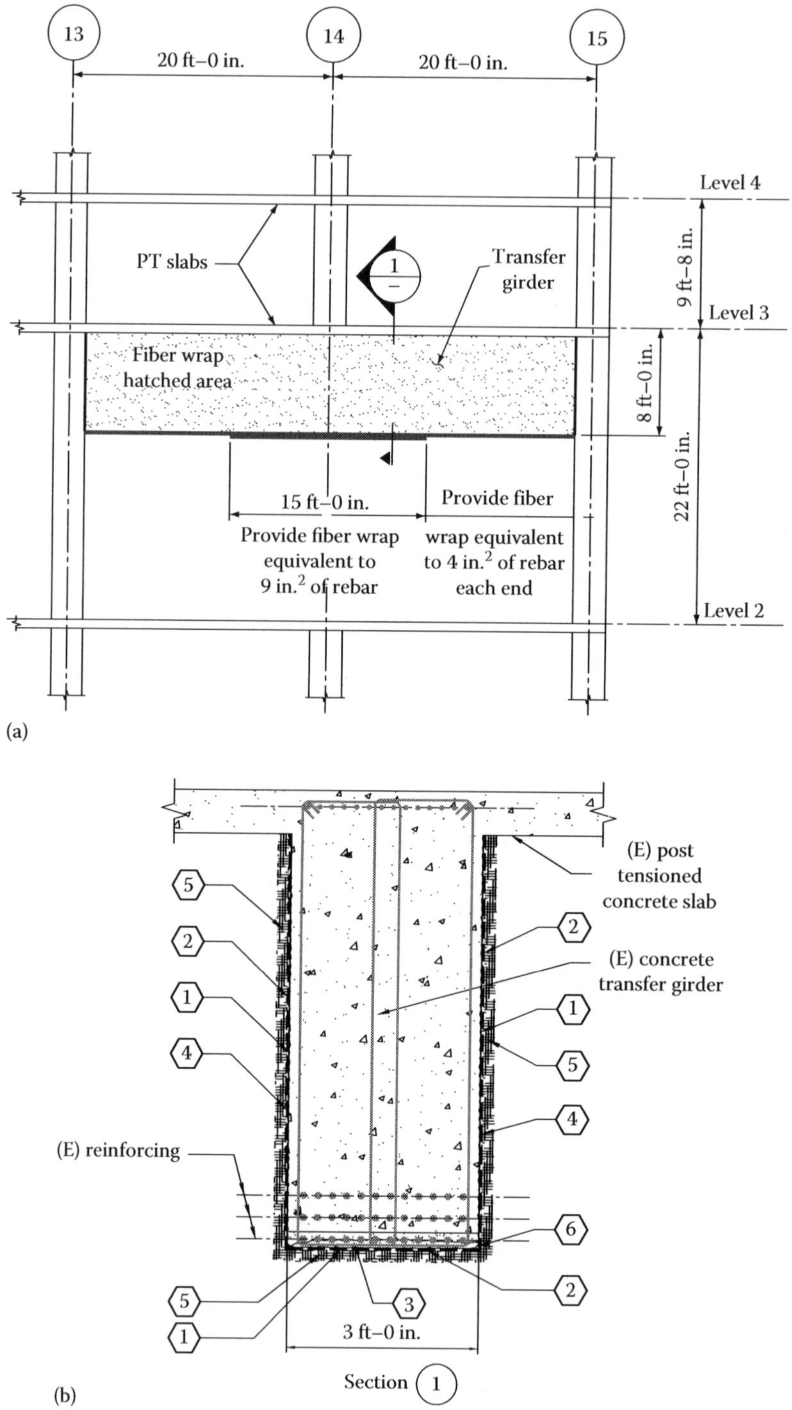

FIGURE 8.9 Fiber wrap of a transfer girder: (a) elevation and (b) section. Suggested repair procedure: (1) sand blast girder soffit and sides; (2) install fiber-wrap material at the bottom and sides of the girder for the entire length; (3) design fiber wrap at the bottom of the girder to compensate for the rebars as specified by the EOR (see elevation [a]); (4) the fiber wrap at the sides of the girder shall provide a tensile strength equal to 3 kip/in. width; (5) fireproof fiber-wrap material as required; (6) ensure that a minimum chamfer of 3/4 in. exists at corners. If not, provide a radius as required by the fiber-wrap design.

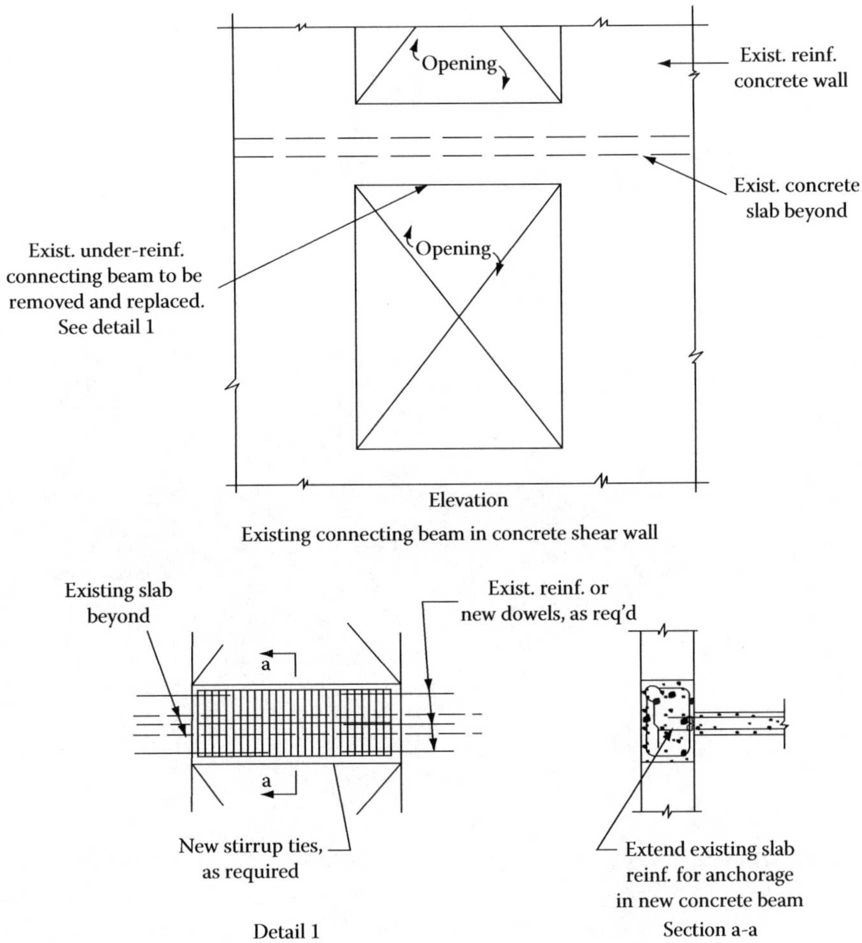

FIGURE 8.10 Strengthening of existing connection beams in reinforced concrete walls.

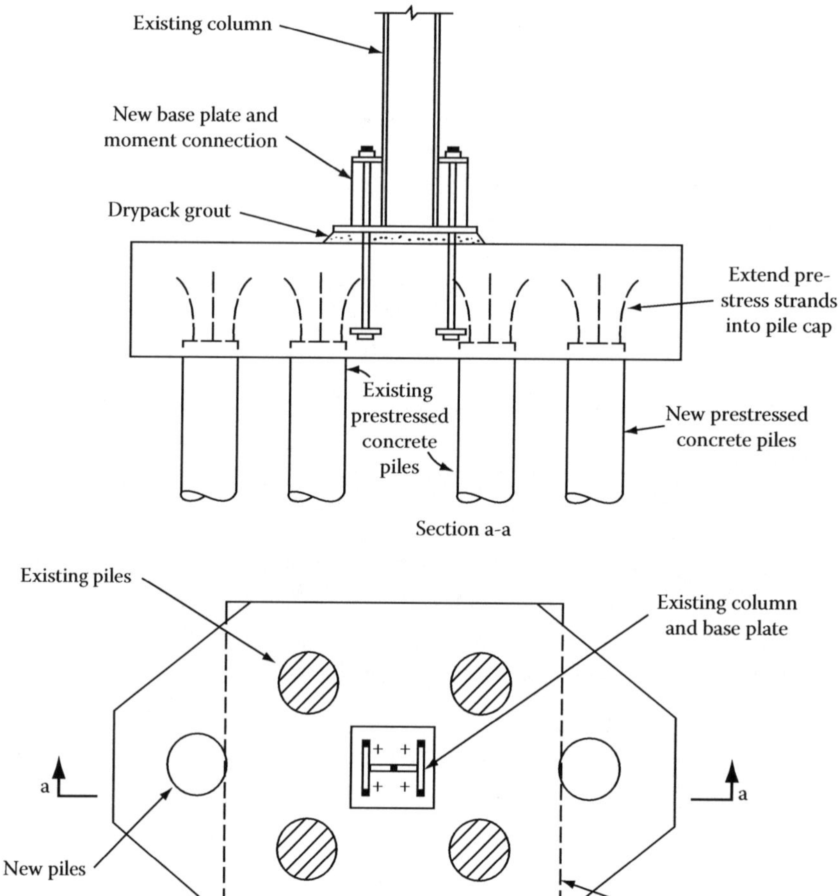

FIGURE 8.11 Upgrading of an existing pile foundation. Add additional piles or piers, remove, replace, or enlarge existing pile caps. *Note*: Existing framing to be temporarily shored to permit removal of existing pile cap and column base plate. Drive new piles; weld new base plate and moment connection to column; pour new pile cap; and dry pack under base plate.

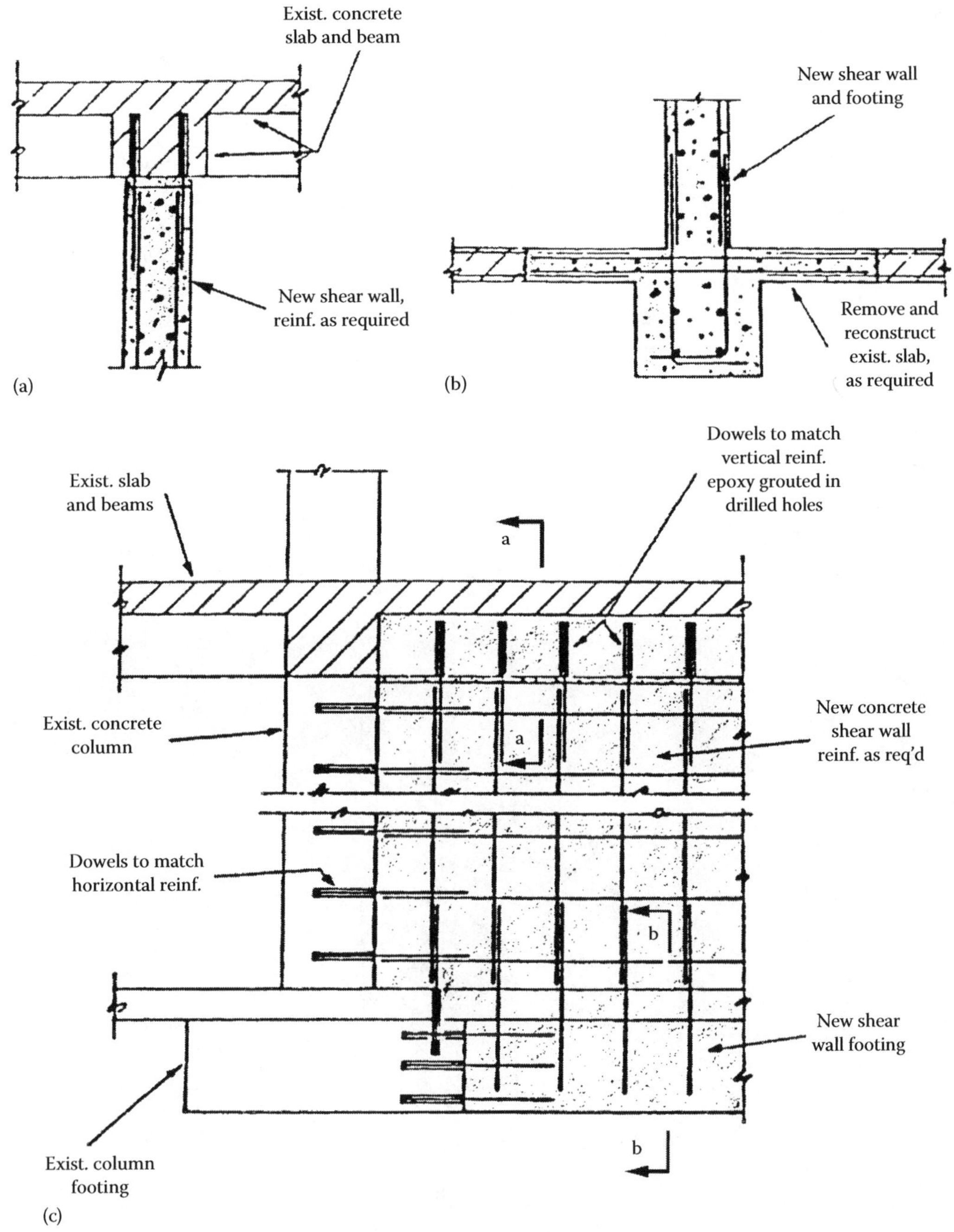

FIGURE 8.12 Strengthening of an existing concrete frame building by adding (N) a reinforced concrete shear wall: (a) section a-a, (b) section b-b, and (c) elevation.

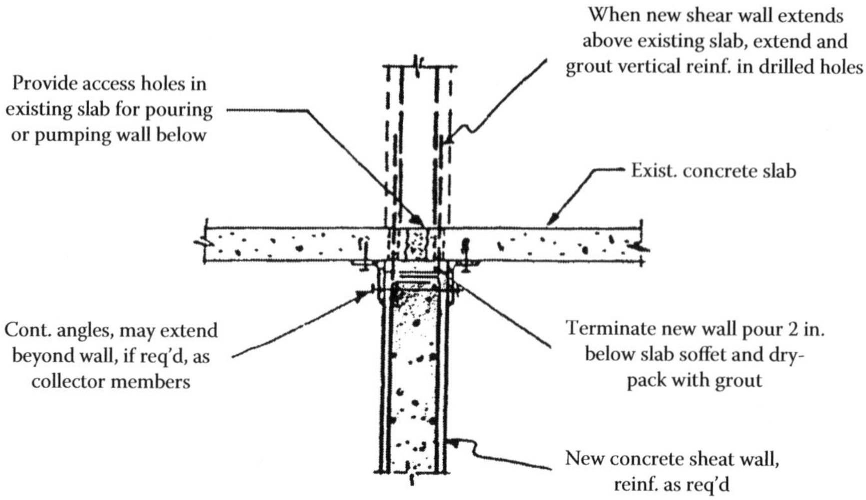

FIGURE 8.13 New concrete shear wall at existing slab.

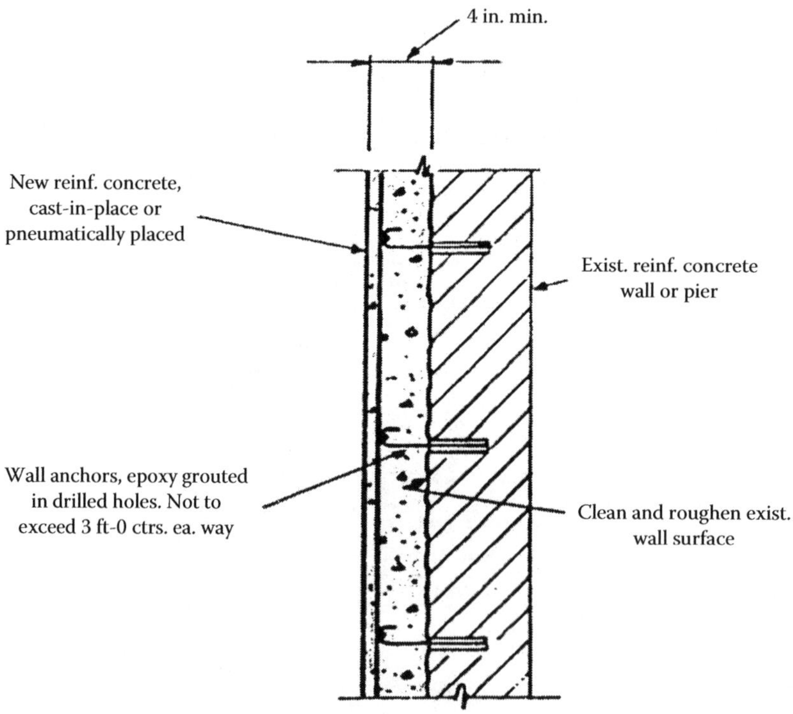

FIGURE 8.14 Strengthening of existing reinforced concrete wall or piers.

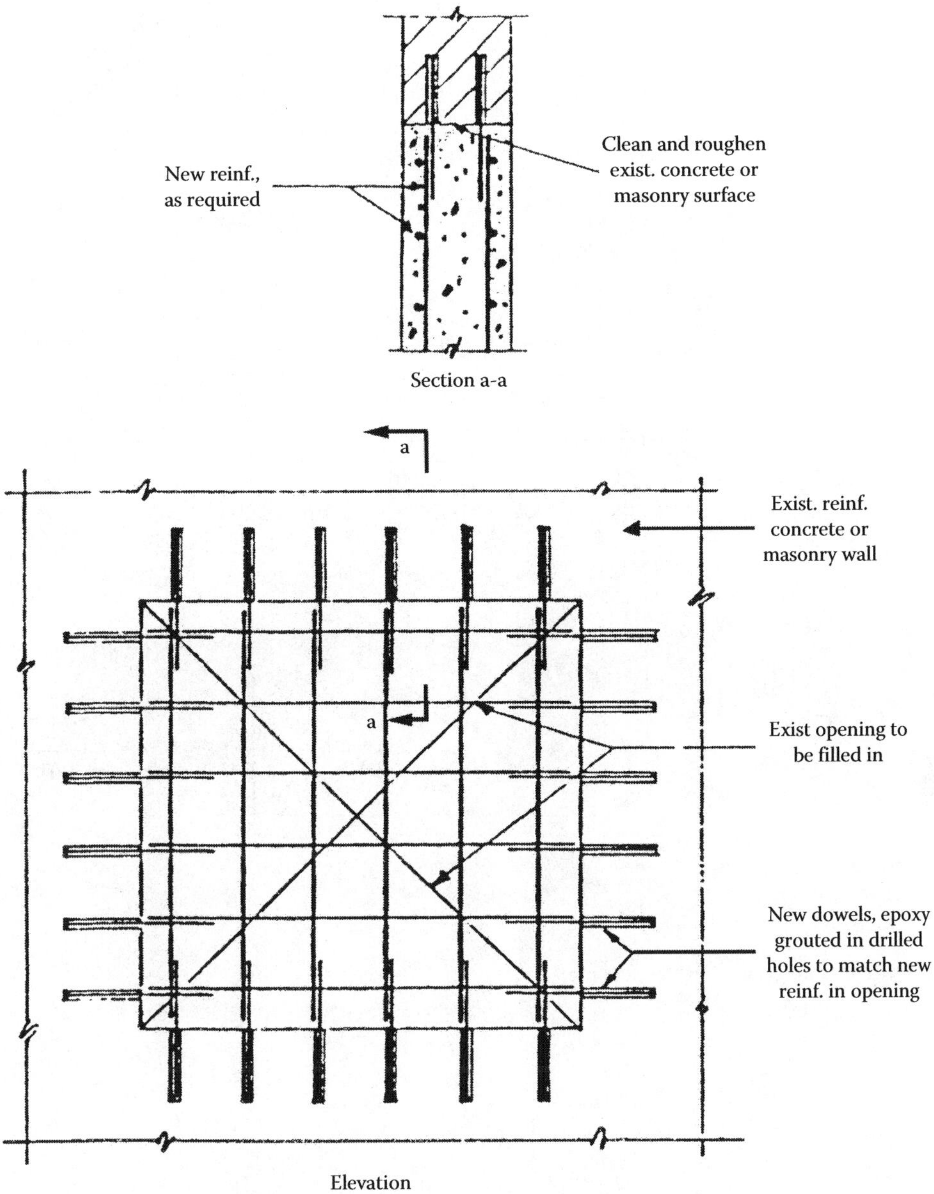

Existing reinforced concrete shear wall with opening to be filled in

FIGURE 8.15 Strengthening of existing reinforced concrete walls by filling in openings.

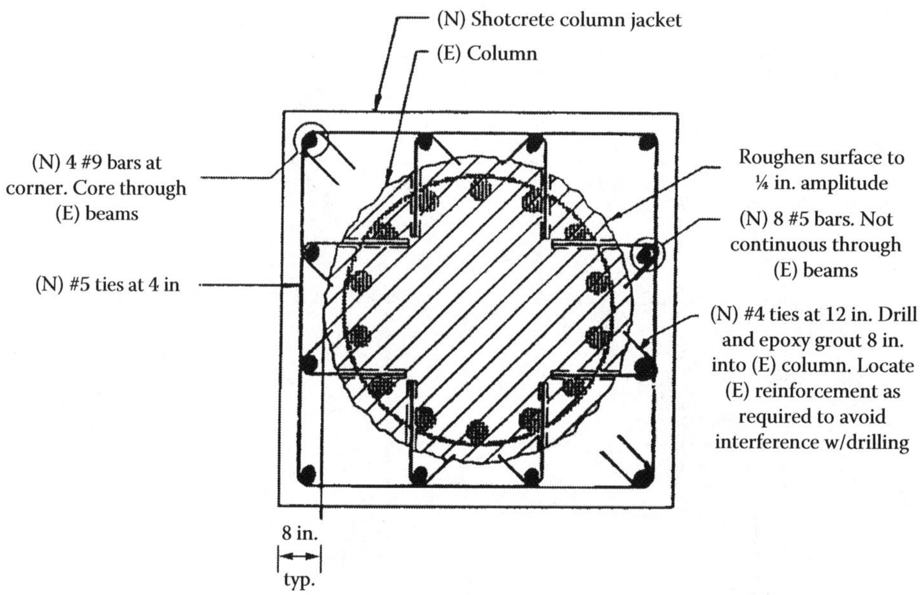

FIGURE 8.16 Jacketing of circular column.

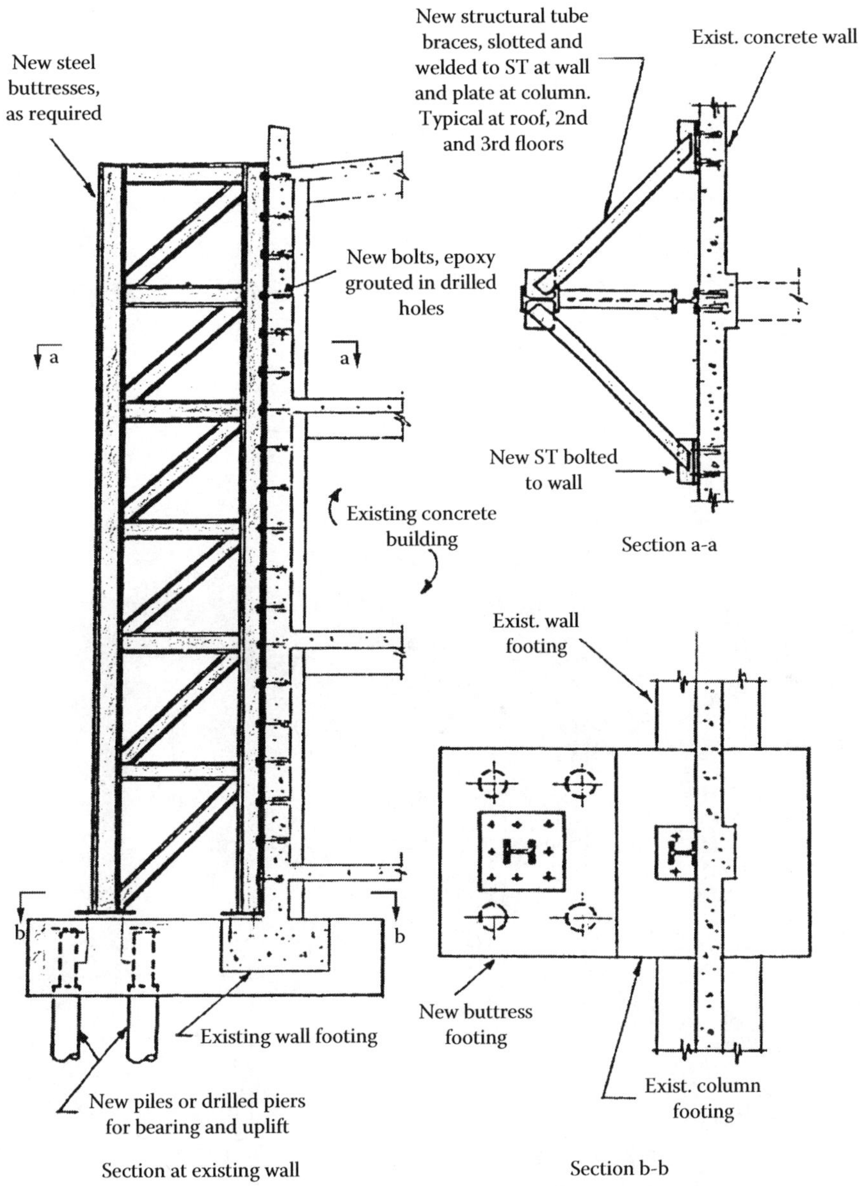

FIGURE 8.17 Braced structural steel buttresses to strengthen an existing reinforced concrete building.

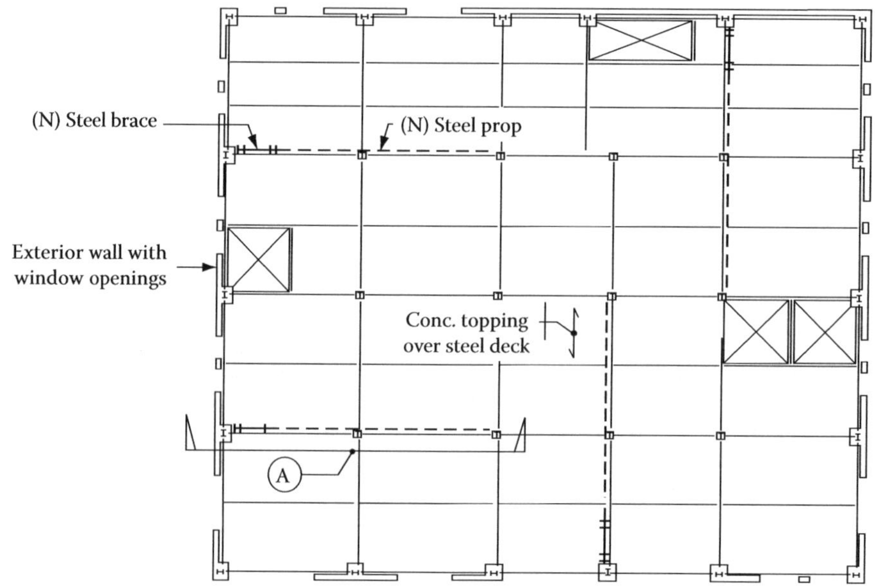

(a)

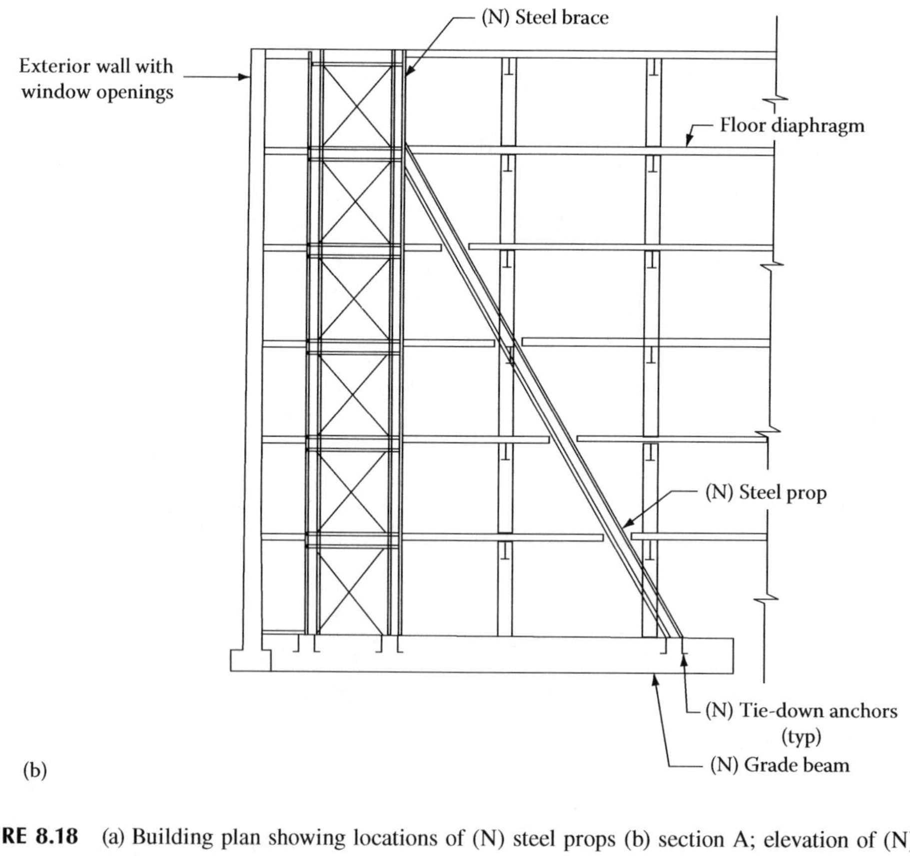

(b)

FIGURE 8.18 (a) Building plan showing locations of (N) steel props (b) section A; elevation of (N) steel prop.

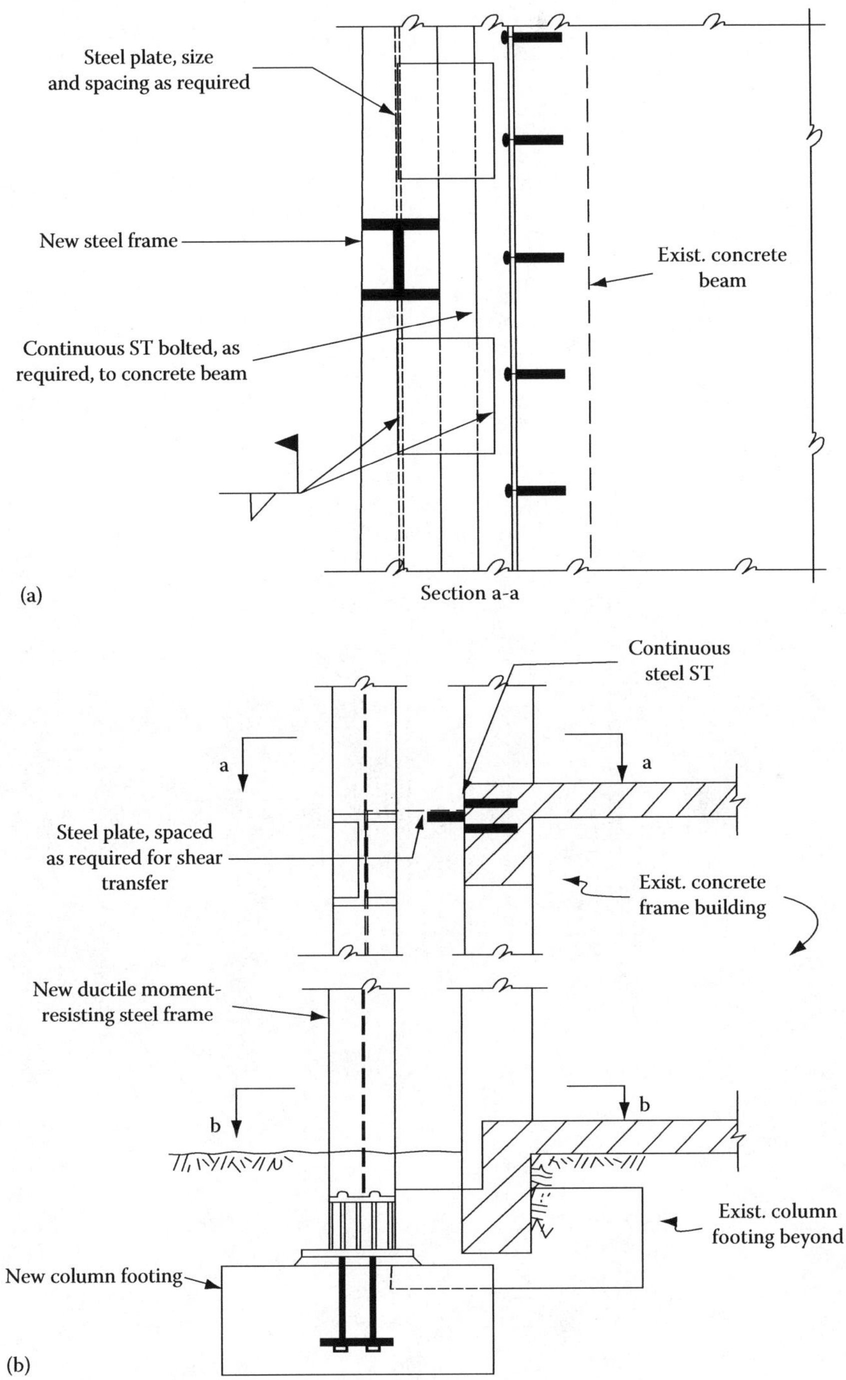

FIGURE 8.19 Upgrading an existing building with external frames. (a) Plan and (b) section at existing wall.

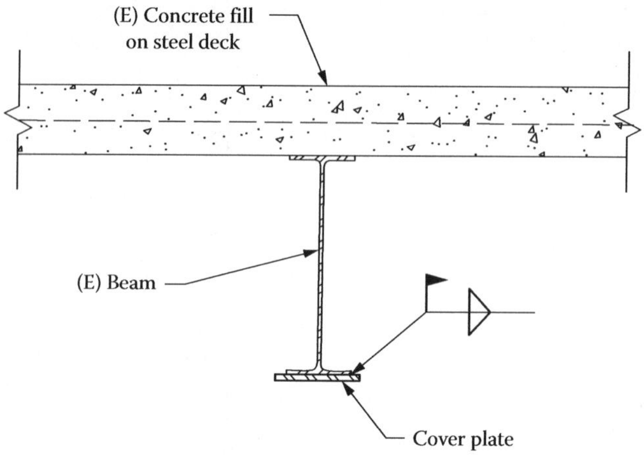

FIGURE 8.20 Cover plate at existing beam.

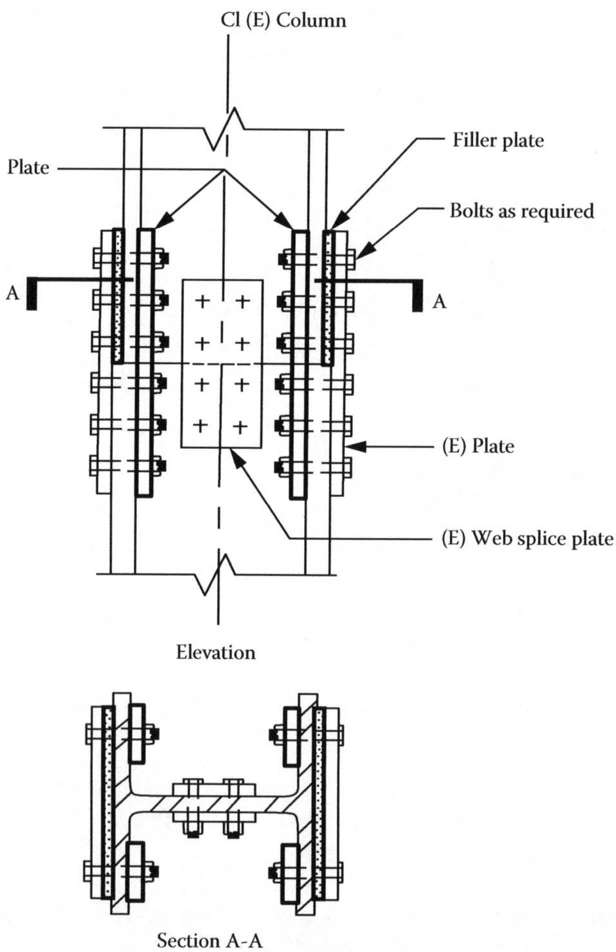

FIGURE 8.21 Bolted splice upgrade at existing column.

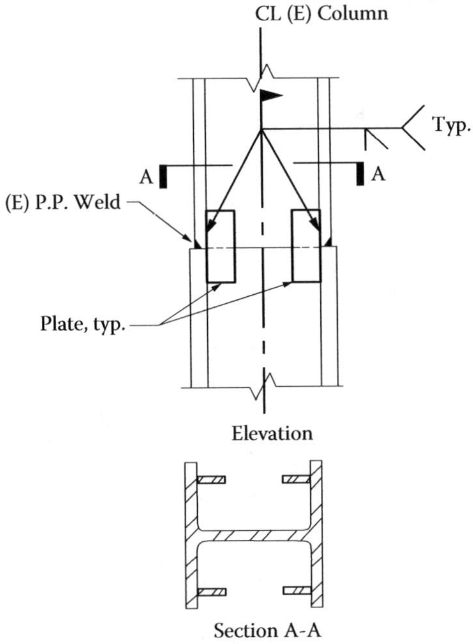

FIGURE 8.22 Welded splice upgrade at existing column.

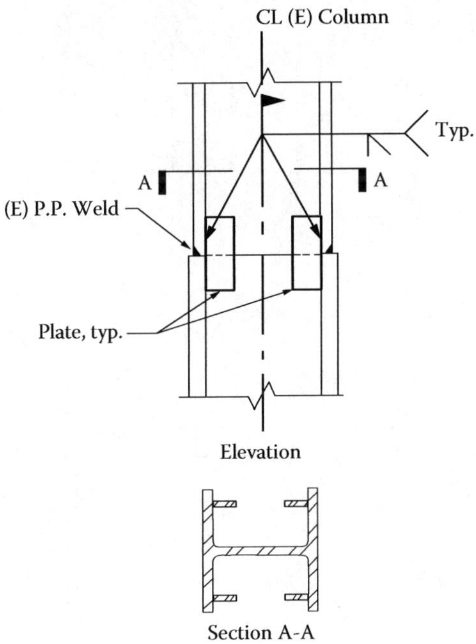

FIGURE 8.23 Welded splice upgrade at existing column.

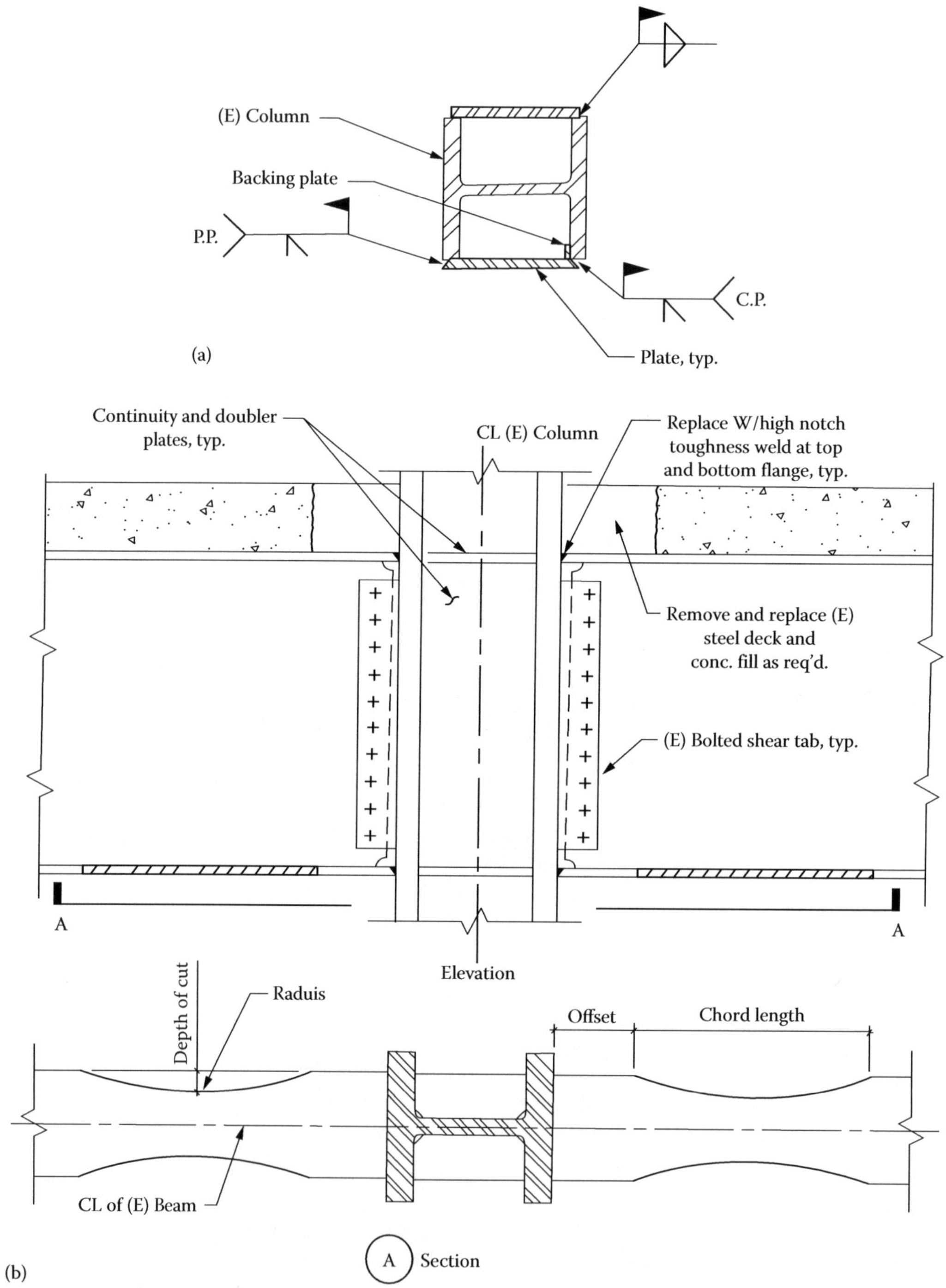

FIGURE 8.24 (a) Box section at existing column. (b) Reduced beam section at bottom flange of existing beam. *Note*: Welds shown indicate alternate possibilities of plate attachment.

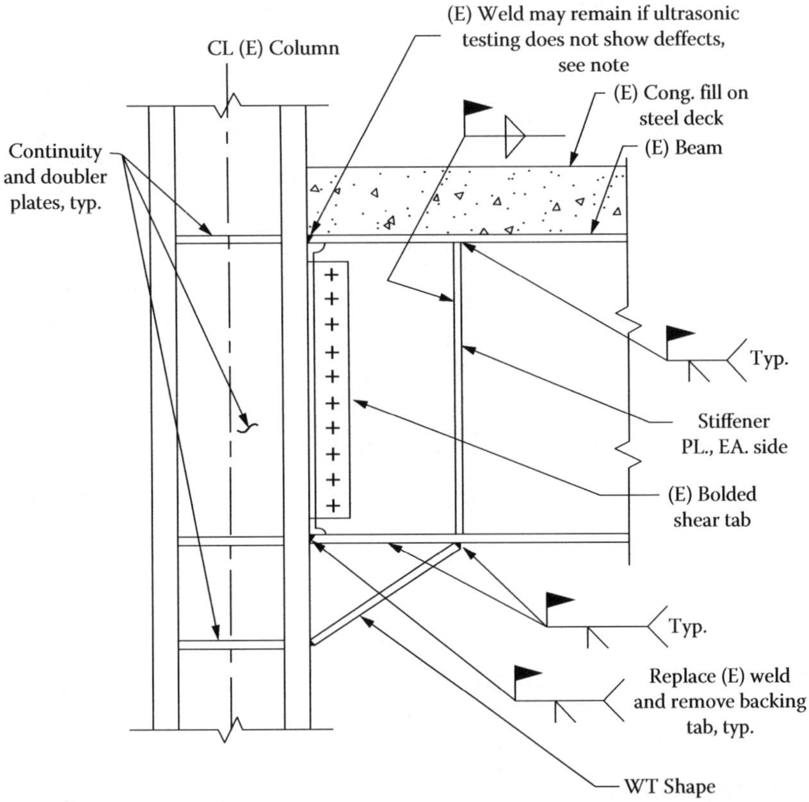

FIGURE 8.25 Welded haunch at bottom flange of existing beam. *Note*: Ultrasonic testing results are highly dependent on the skill of the technician. If a qualified technician is available, testing of the existing weld at the top flange can be performed from below. Otherwise, chipping of the existing slab may be required to perform the test.

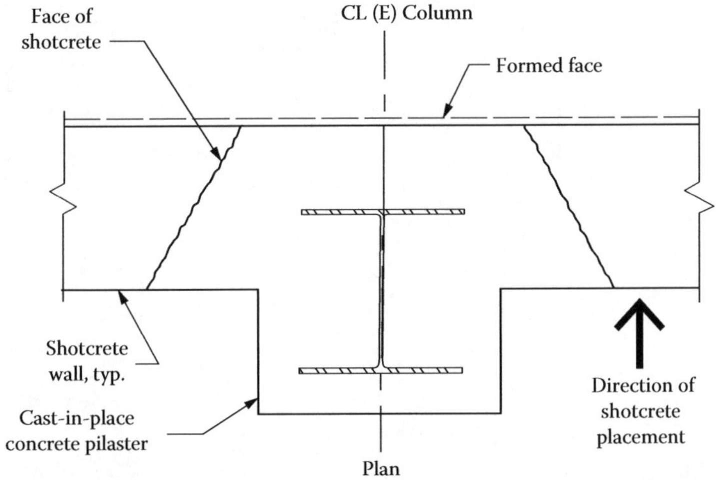

FIGURE 8.26 Combined shotcrete and cast-in-place construction. *Notes*: 1. Wall reinforcing steel not shown for clarity. 2. Shotcrete shadowing restrictions at steel column prevents use of shotcrete for pilaster.

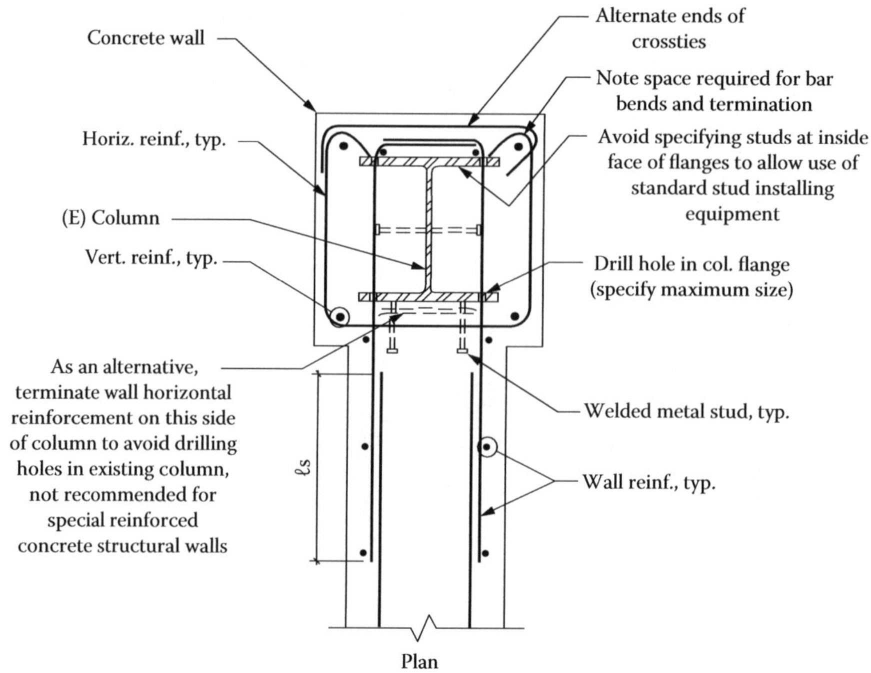

FIGURE 8.27 Cast-in-place concrete wall encasing existing column.

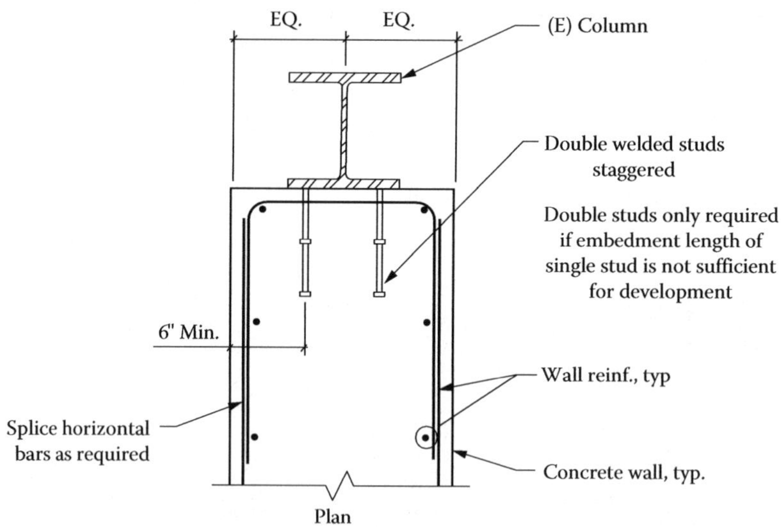

FIGURE 8.28 Wall at existing column. *Note*: Detail not appropriate for special reinforced concrete structural walls due to lack of confinement for boundary element.

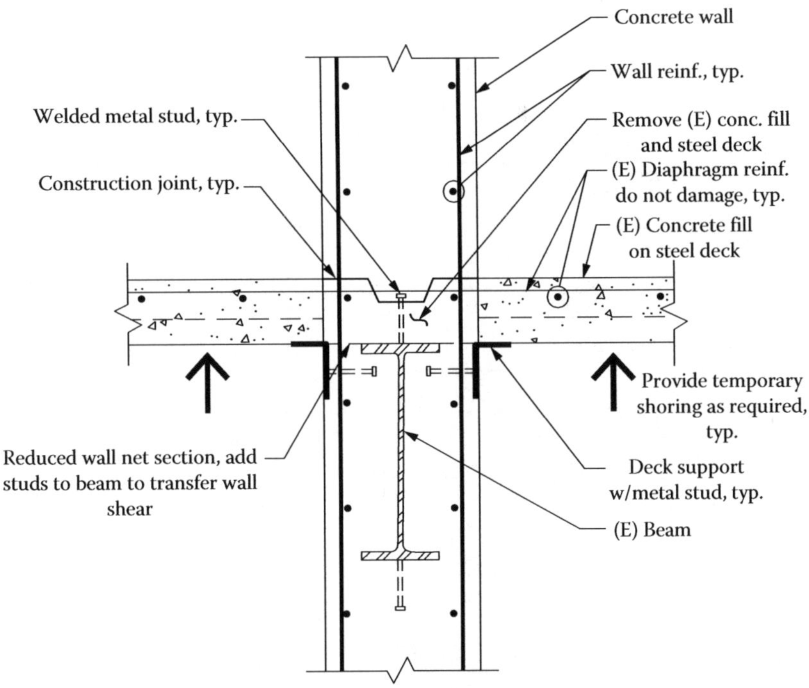

FIGURE 8.29 Cast-in-place concrete wall at existing beam. *Note*: Offset wall to one side of beam if additional room required for vibrating equipment of for shotcrete construction.

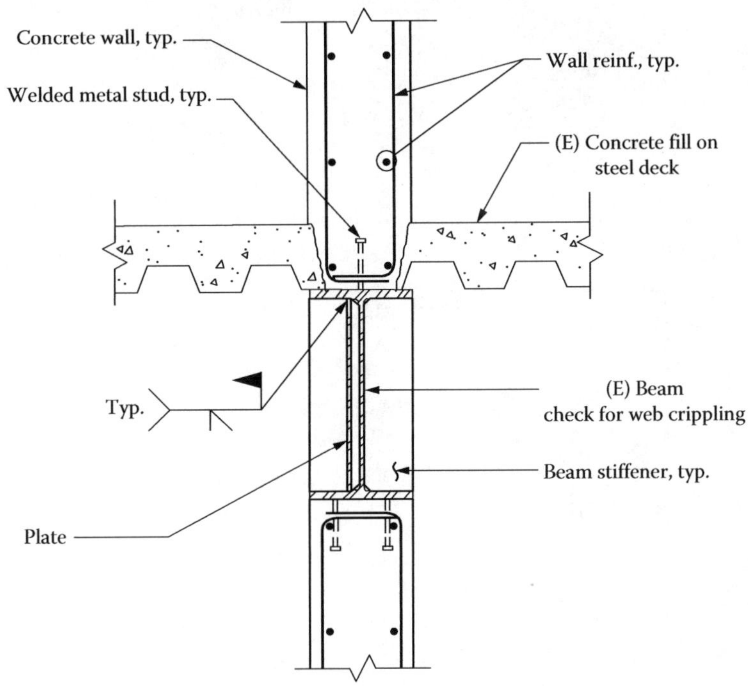

FIGURE 8.30 Discontinuous wall at existing beam. *Notes*: 1. Detail is appropriate for shotcrete construction and more problematic for cast-in-place concrete construction due to concrete placement and vibrating challenges. 2. Shear in wall is transferred entirely through studs. 3. For deck perpendicular to beam condition, number of metal studs limited by deck flute spacing.

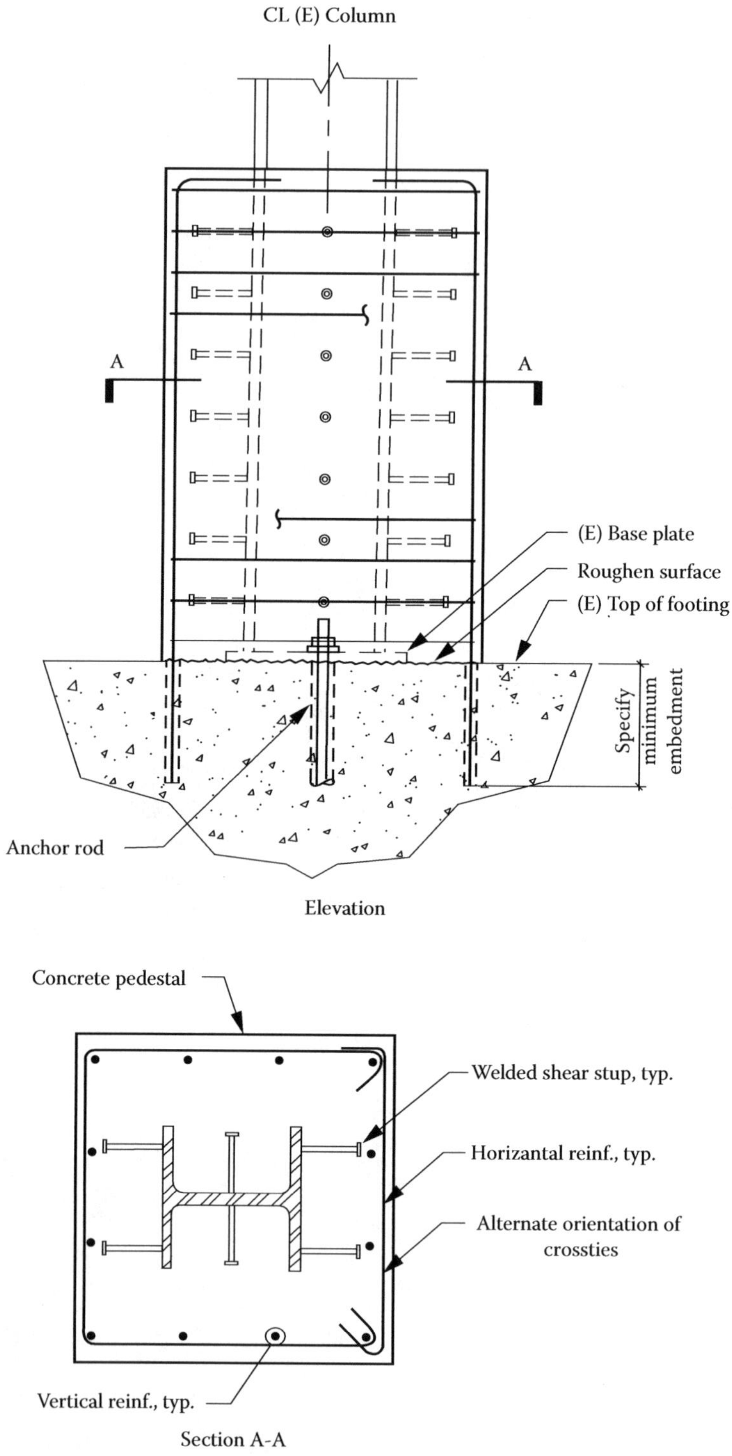

FIGURE 8.31 Concrete pedestal at existing column.

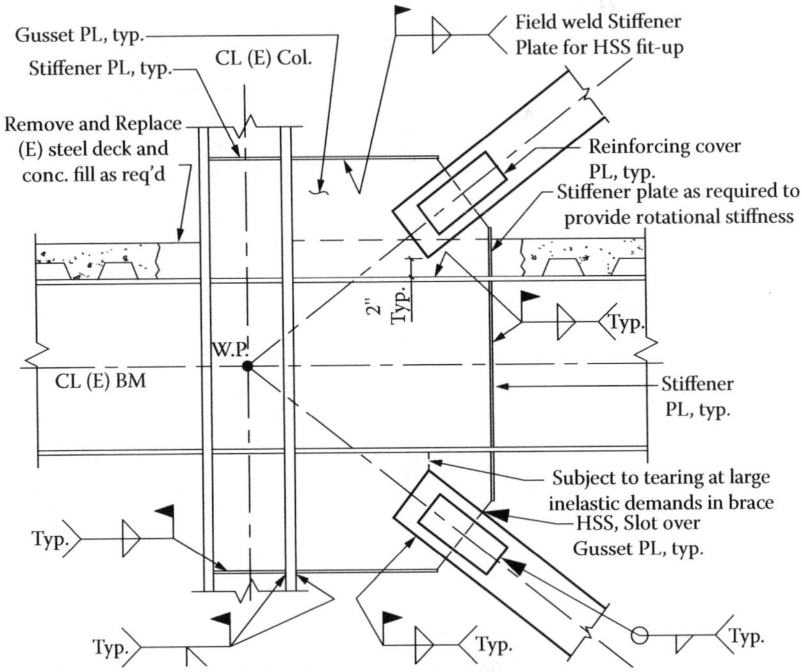

FIGURE 8.32 HSS brace at existing beam–column connection in ordinary CBF. *Note*: This detail, though compact, provides limited ductility and is only intended for use in low to moderate seismic applications.

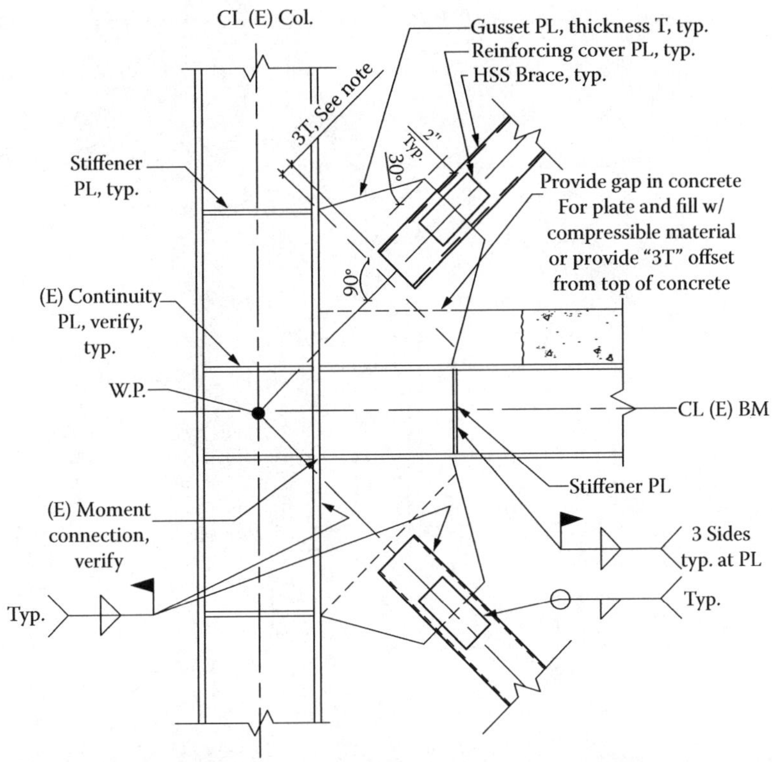

FIGURE 8.33 HSS brace at existing beam–column connection in SCBF. *Note*: AISC recommends 2T to allow for restraint-free plastic rotations. 3T is shown here to accomodate overcutting of HSS slots.

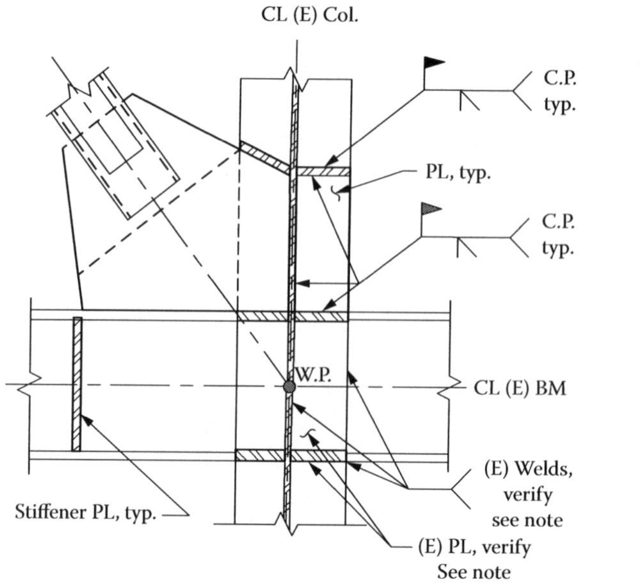

FIGURE 8.34 HSS braced at existing beam–column connection. *Notes*: Capacity of existing moment connection should be checked for new forces and strengthened if required.

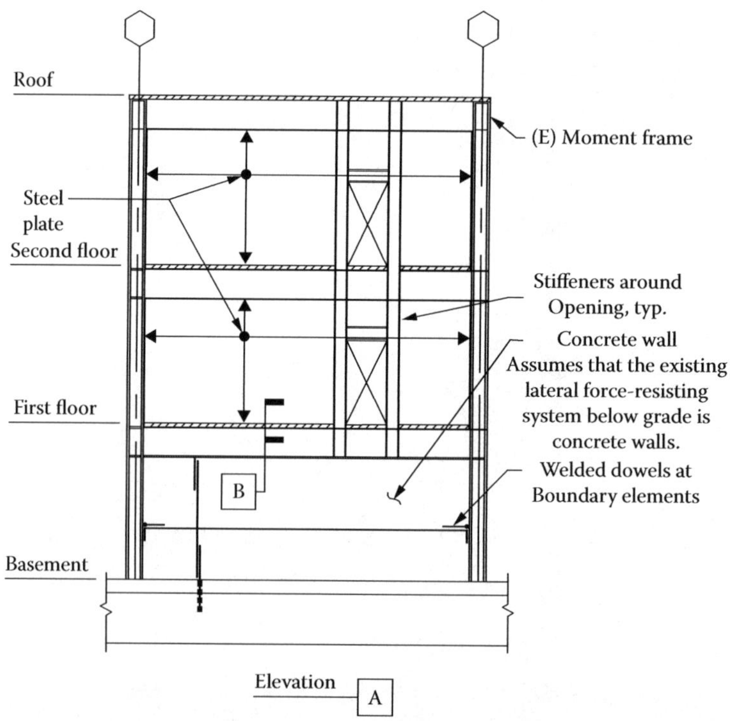

FIGURE 8.35 Unstiffened steel plate shear wall.

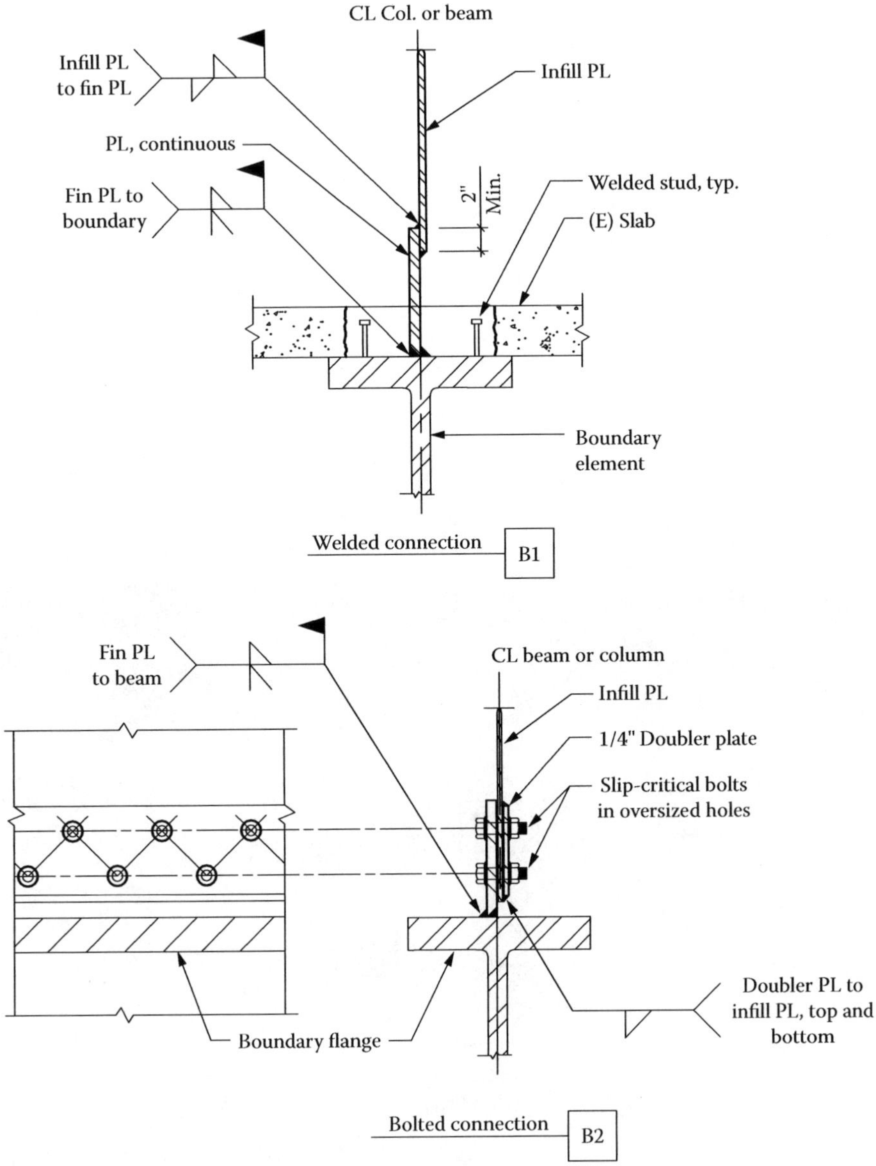

FIGURE 8.36 Fin plate connection options.

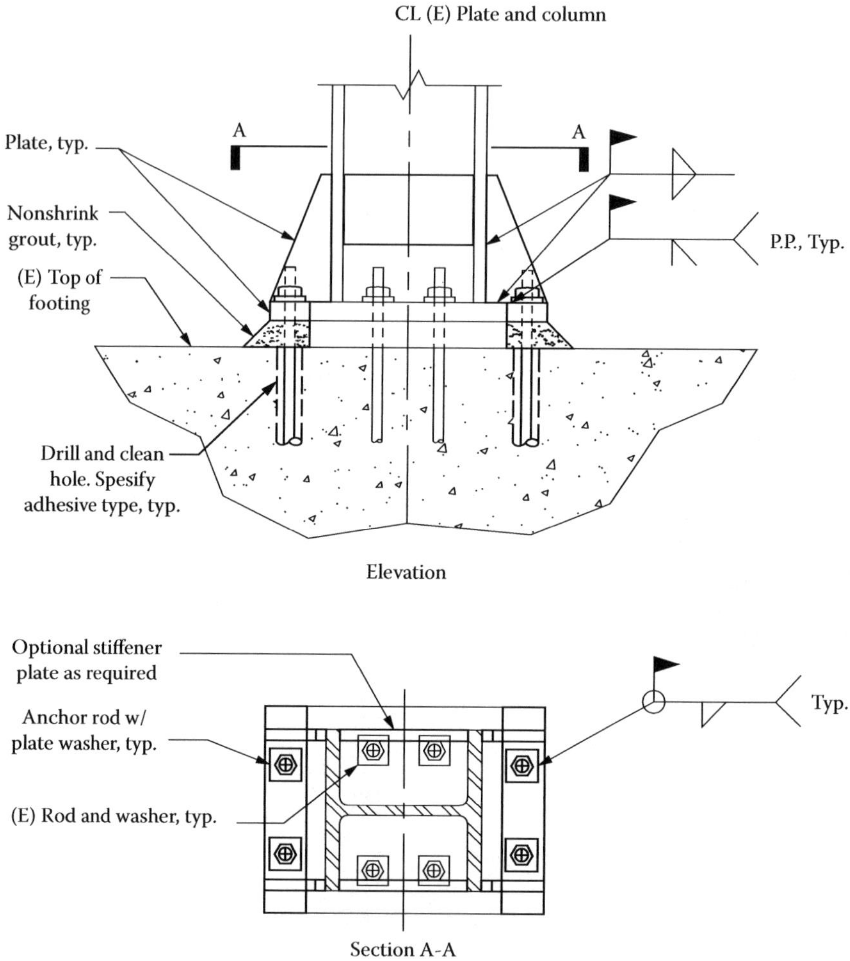

FIGURE 8.37 Modified base plate to increase uplift capacity.

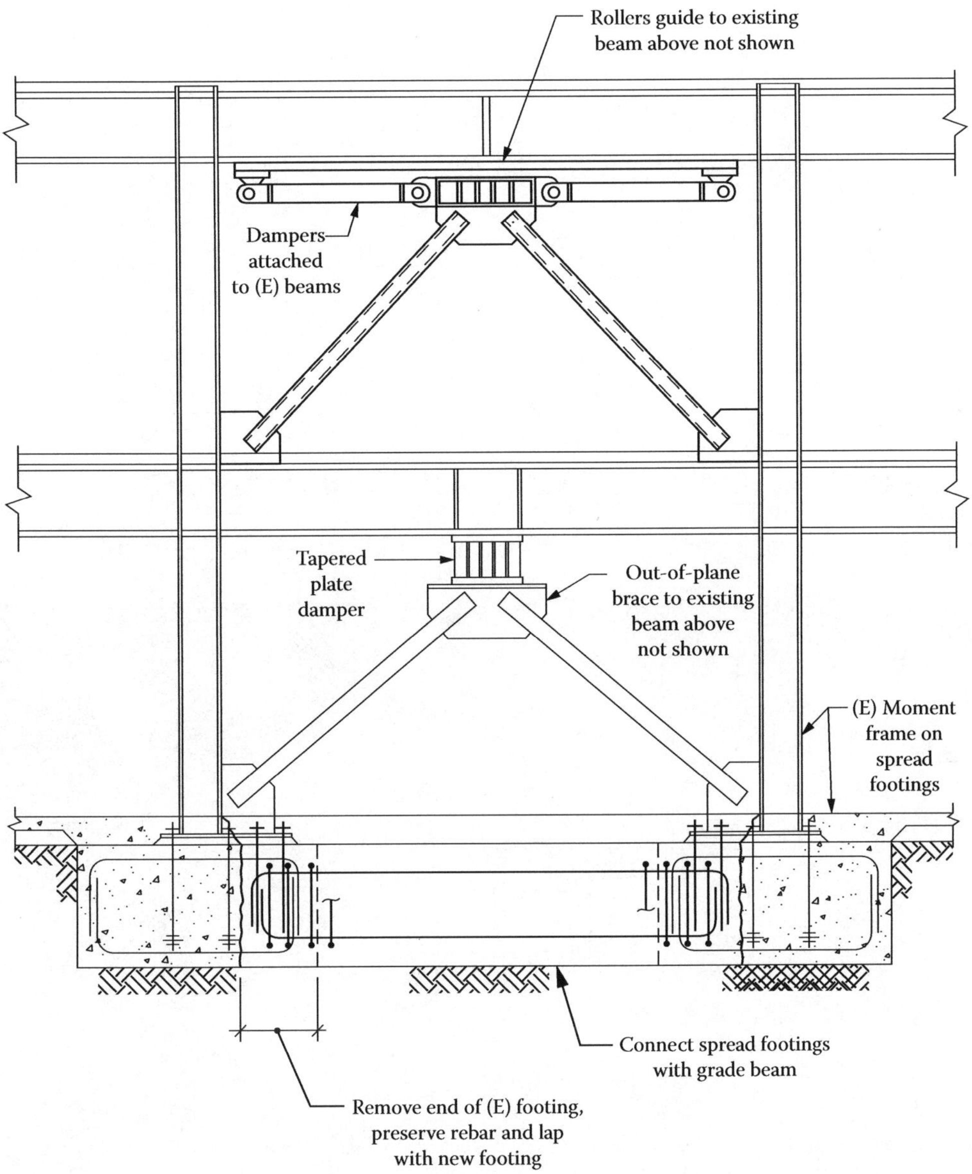

FIGURE 8.38 Additional damper alternatives for rehabilitating an existing moment frame.

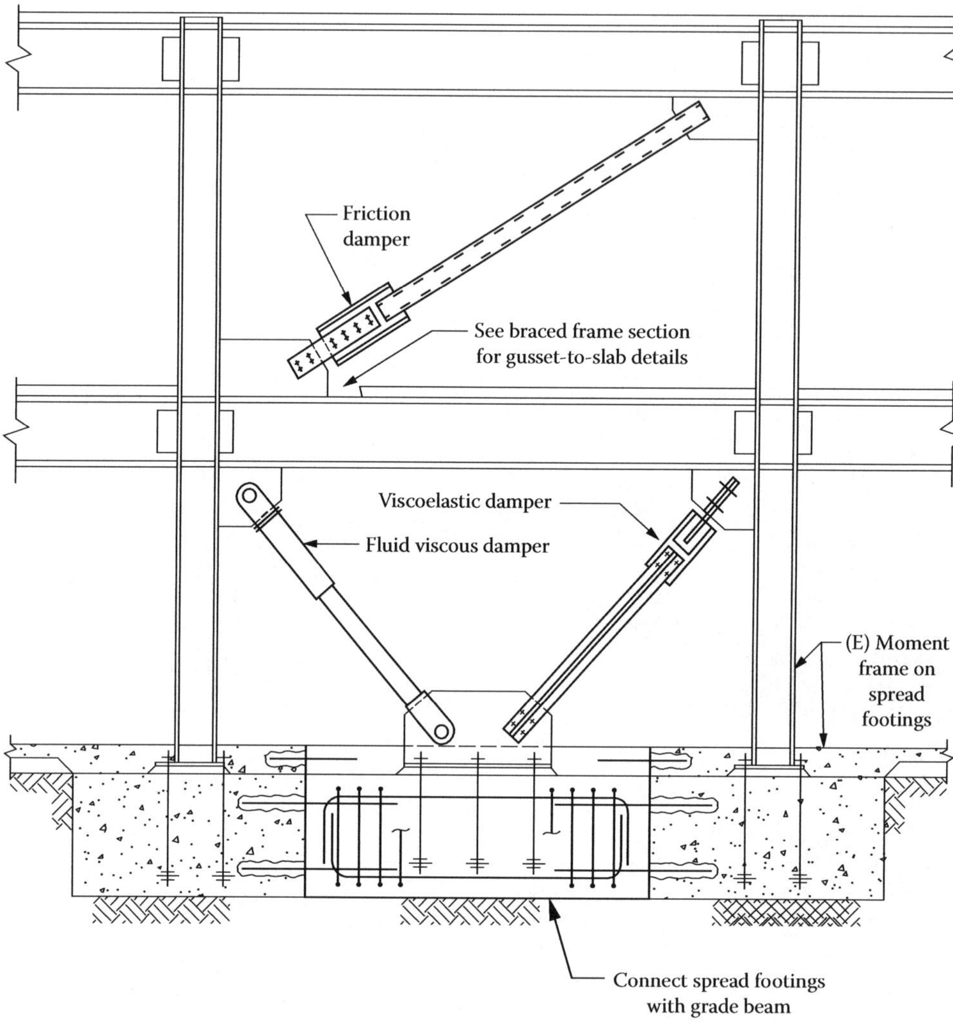

FIGURE 8.39 Damper alternatives for rehabilitating an existing moment frame.

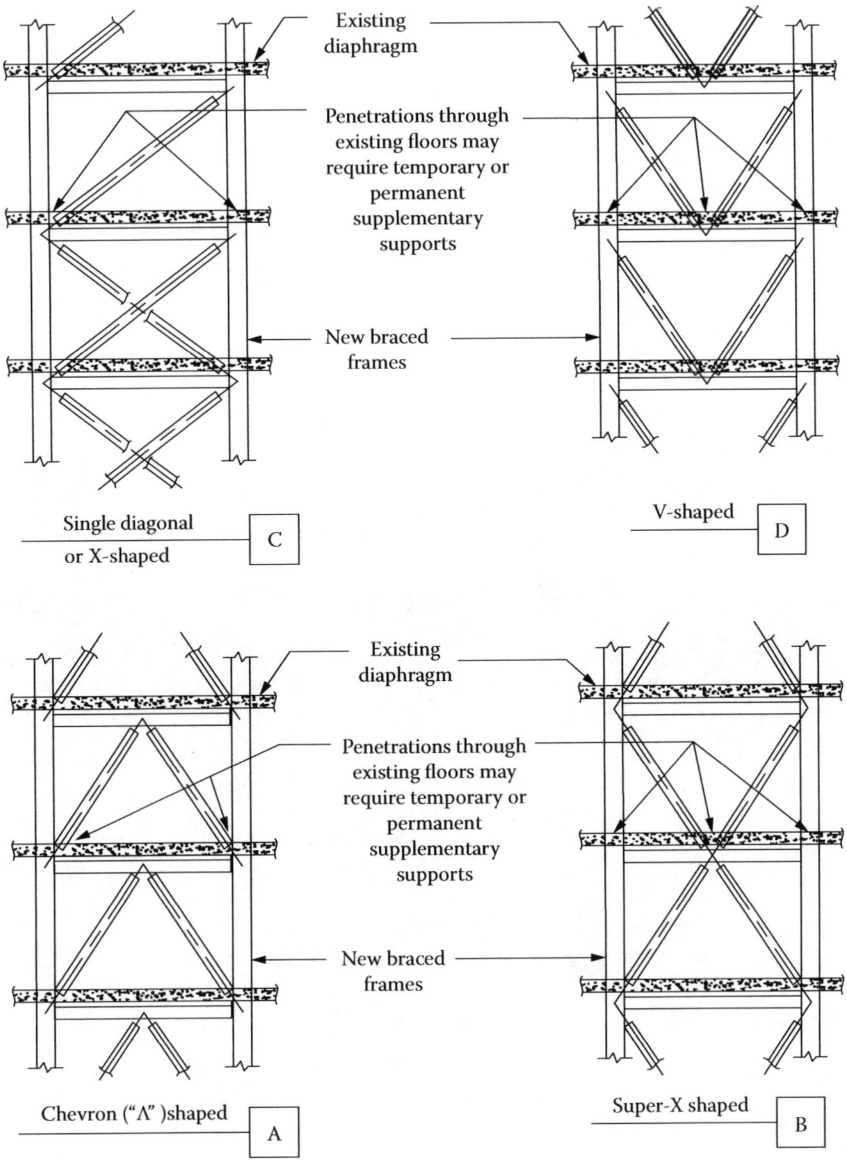

FIGURE 8.40 Typical braced frame configurations.

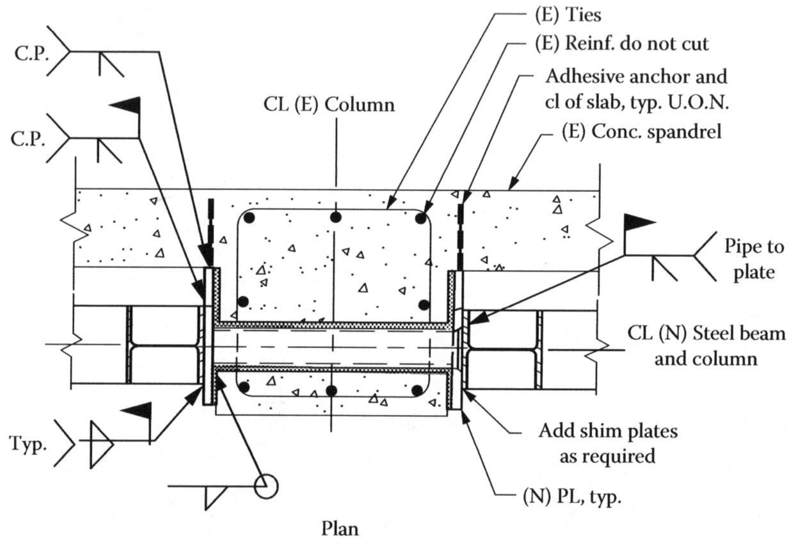

Installation procedure:

1. Install adhesive anchor and plate on each side of column.
2. Core hole for pipe through column.
3. Install pipe with shop welded plate on one side.
4. Install plate with hold and weld to pipe.
5. Weld cap plate to pipe.
6. Weld plate attached to pipe to plate attched to adhesive anchor.
7. Fill pipe to annular space solid with nonshrink grout.

FIGURE 8.41 Braced frame to concrete column connection.

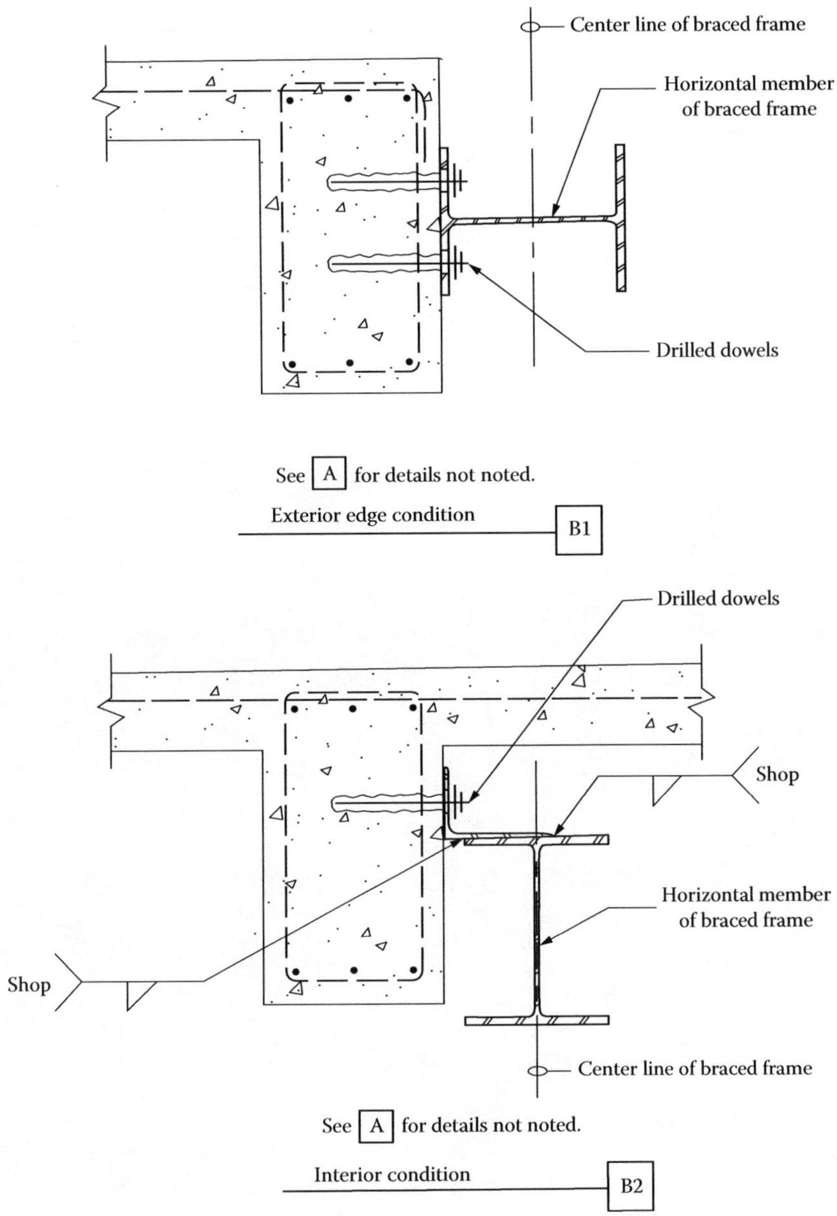

FIGURE 8.42 Typical connection to existing concrete beam.

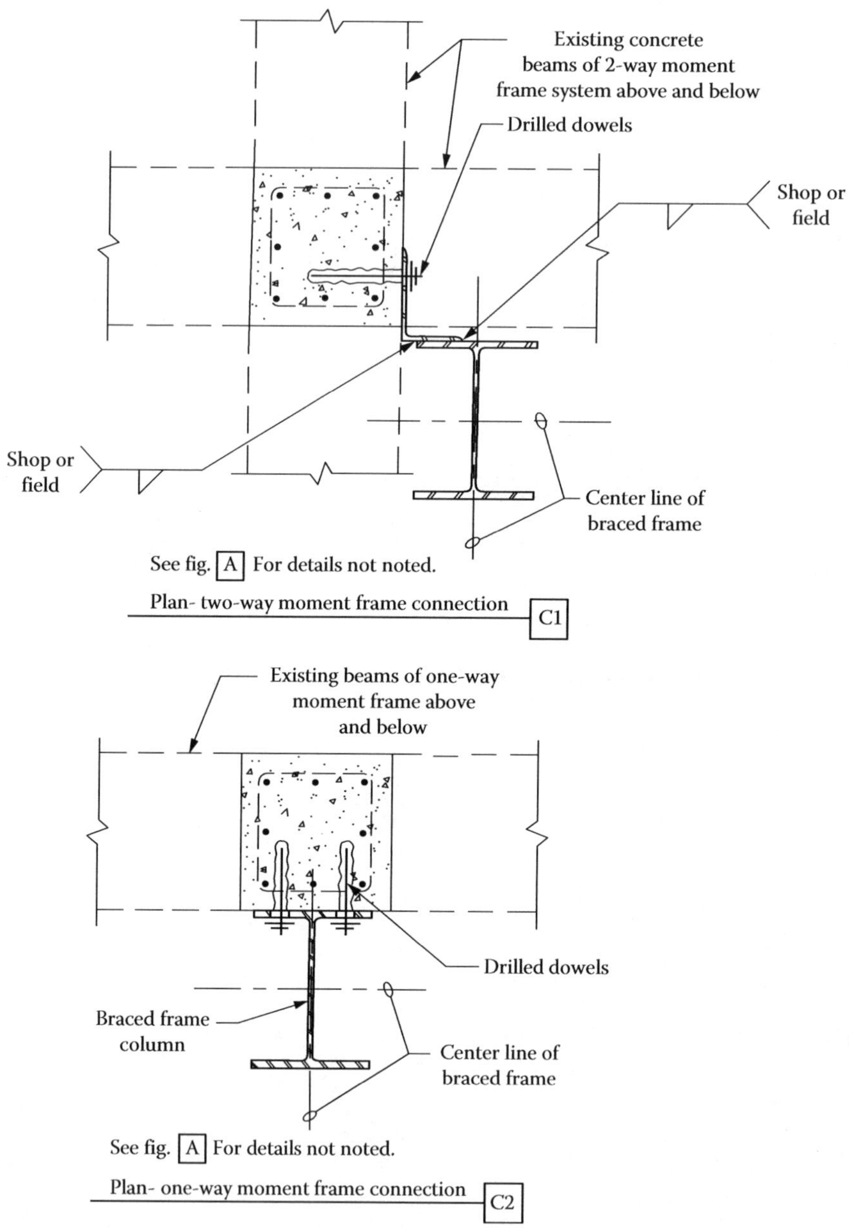

FIGURE 8.43 Typical connection to existing concrete column.

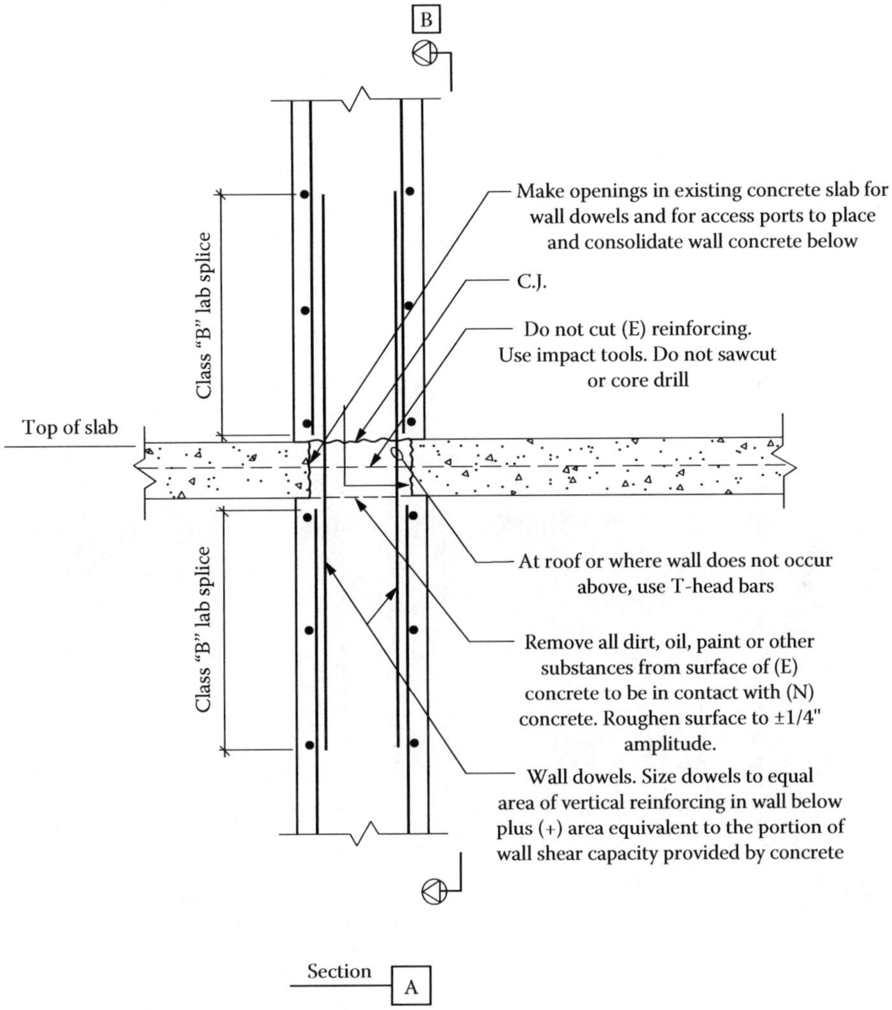

FIGURE 8.44 Concrete wall connection to concrete slab.

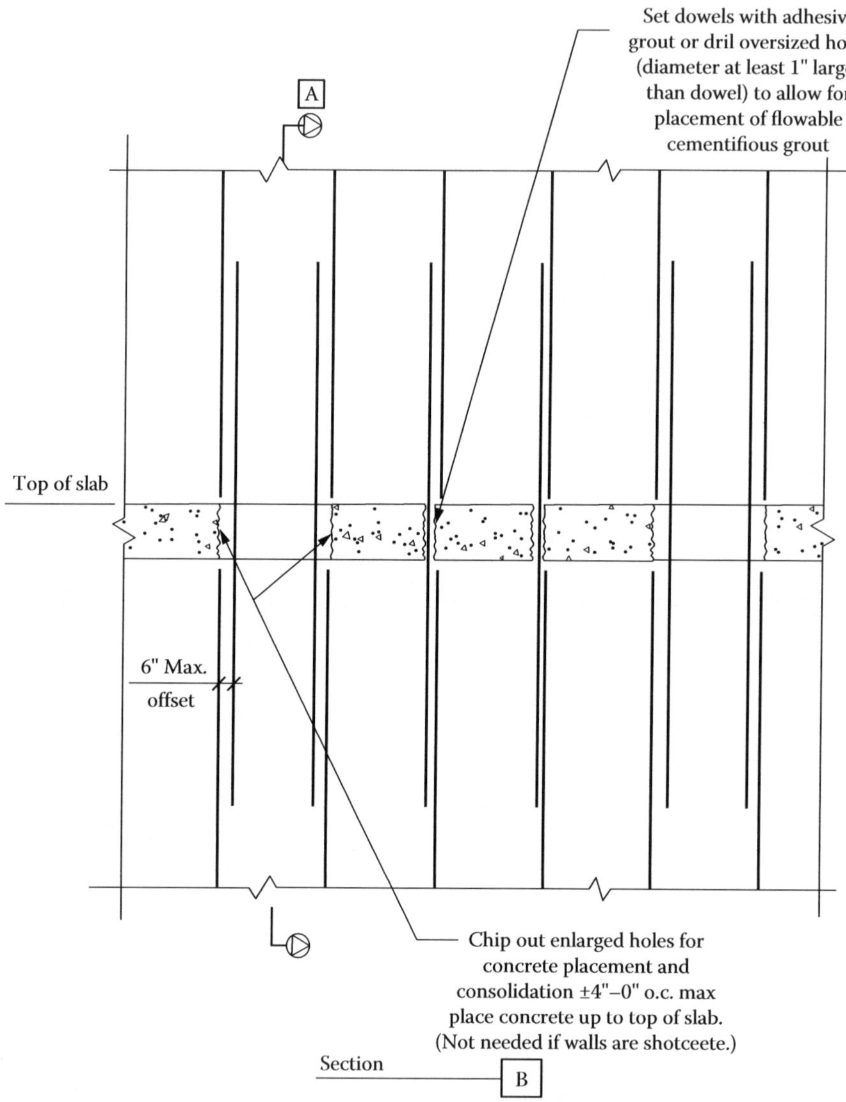

FIGURE 8.45 Concrete wall connection to concrete slab—partial elevation view.

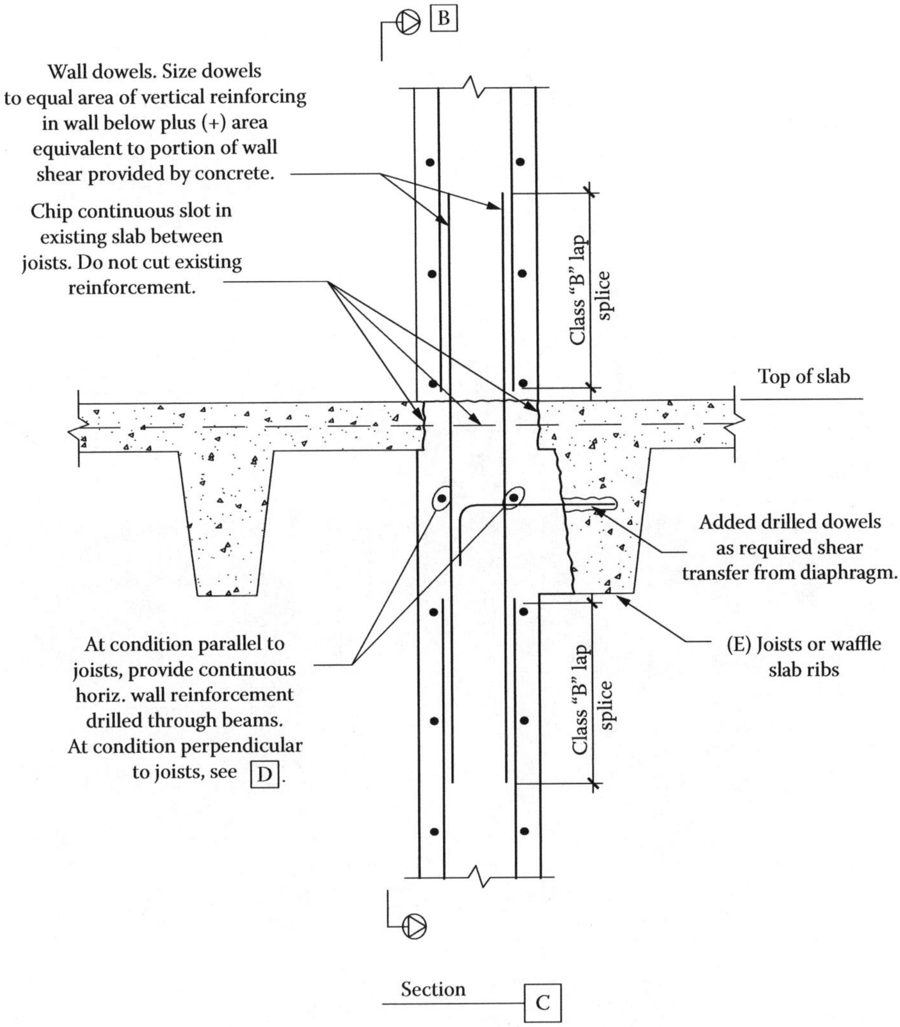

FIGURE 8.46 Connection to concrete wall to concrete joists or waffle slab.

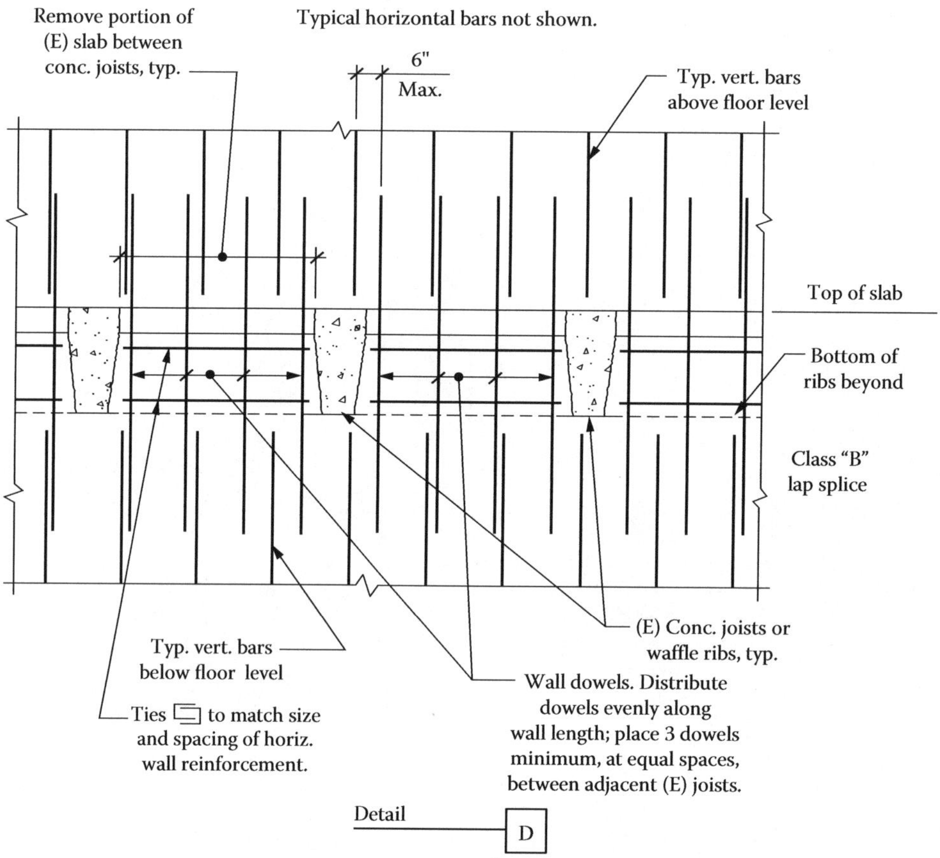

FIGURE 8.47 Concrete wall connection to waffle slab—partial elevation view.

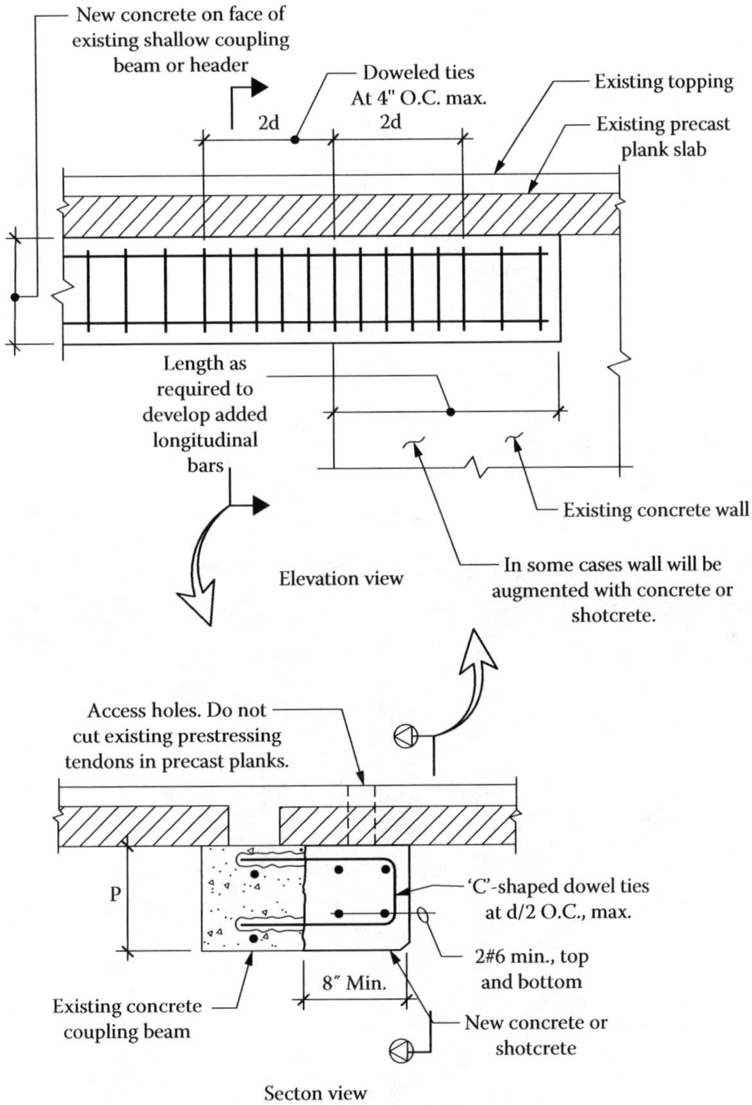

FIGURE 8.48 Typical strengthening of shallow coupling beam.

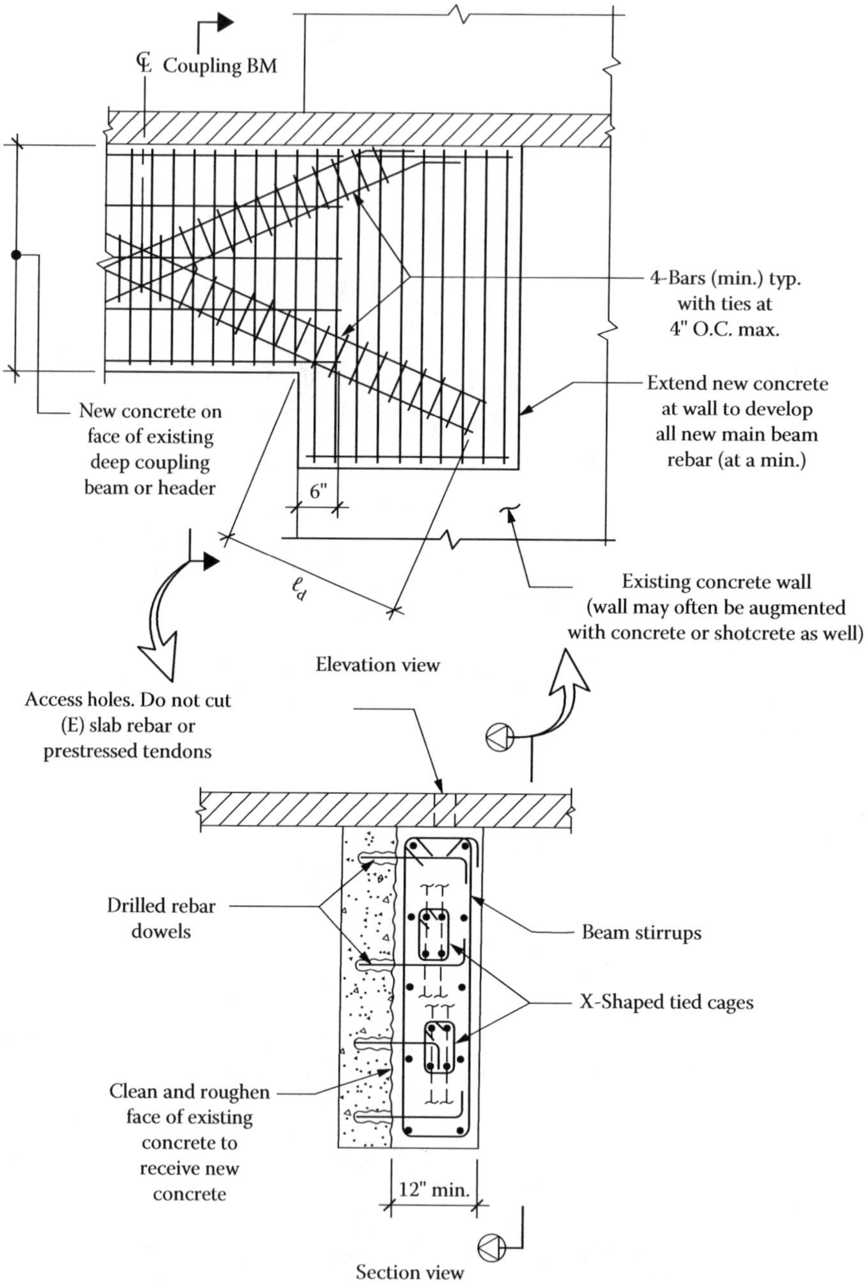

FIGURE 8.49 Strengthening of deep coupling beam.

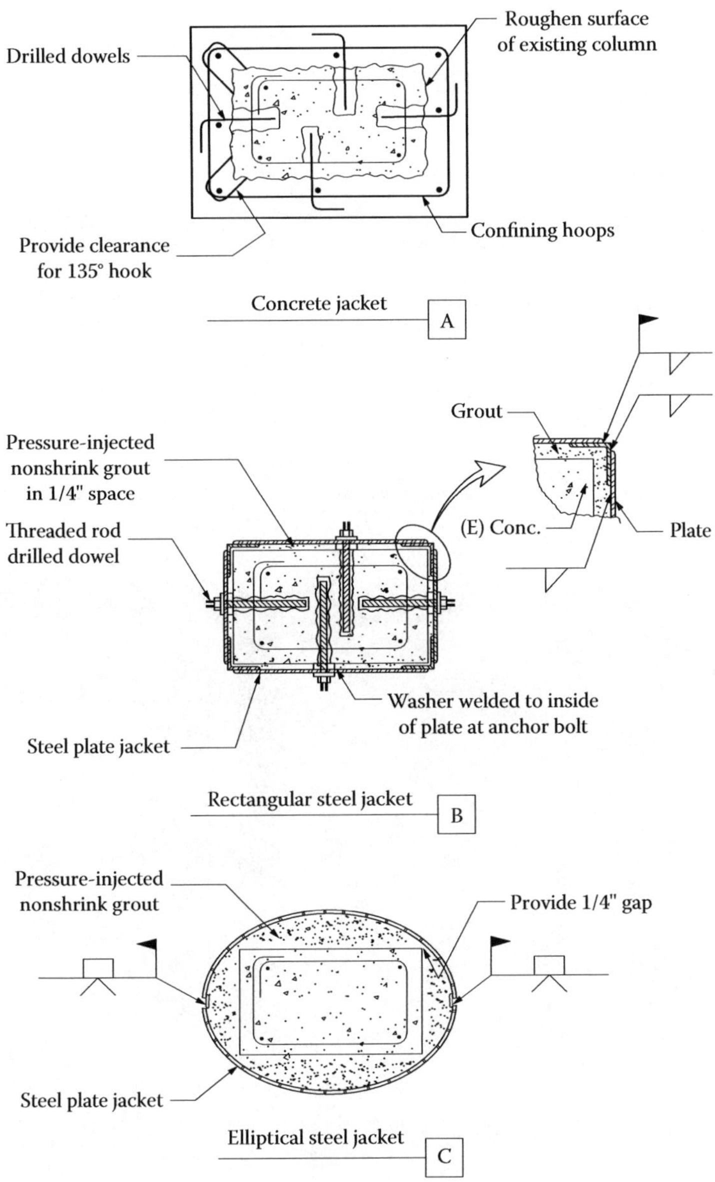

FIGURE 8.50 Concrete and steel overlays for concrete columns.

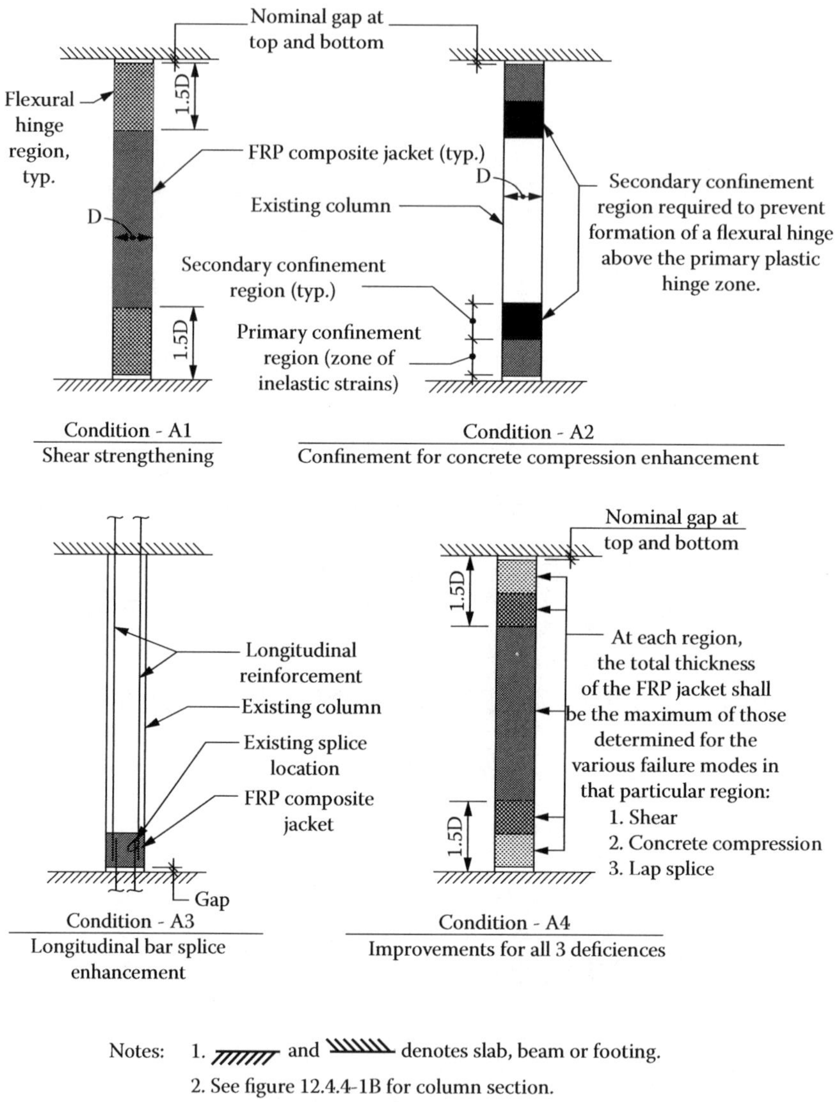

FIGURE 8.51 Seismic retrofit of columns using FRP composites.

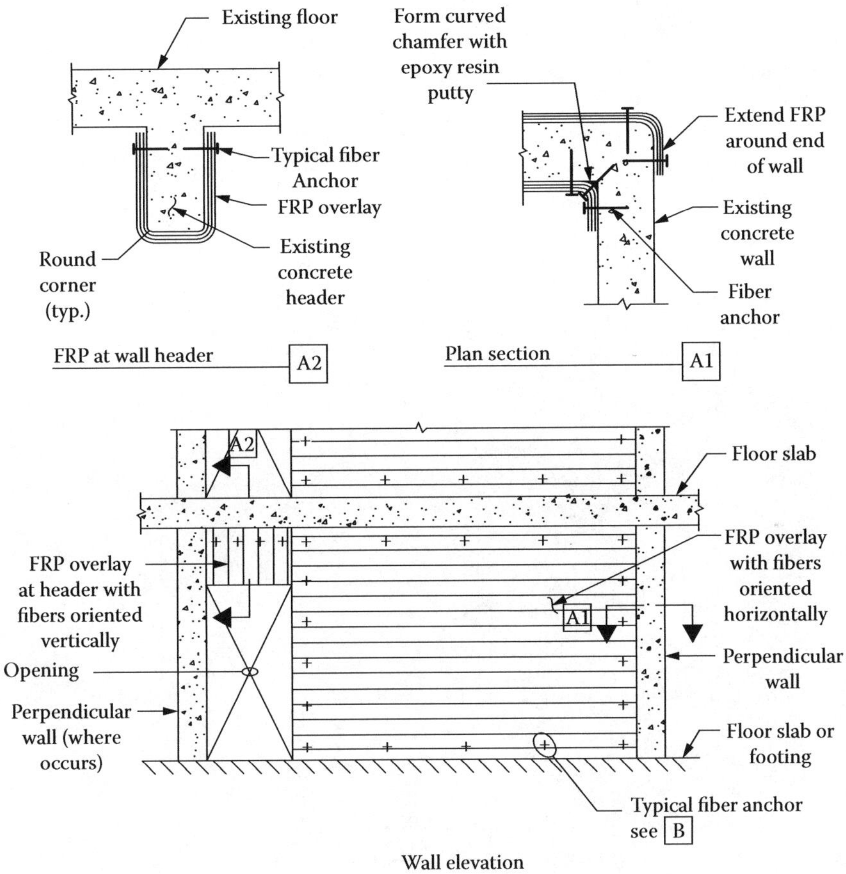

FIGURE 8.52 Shear strengthening of concrete shear walls using FRP composite.

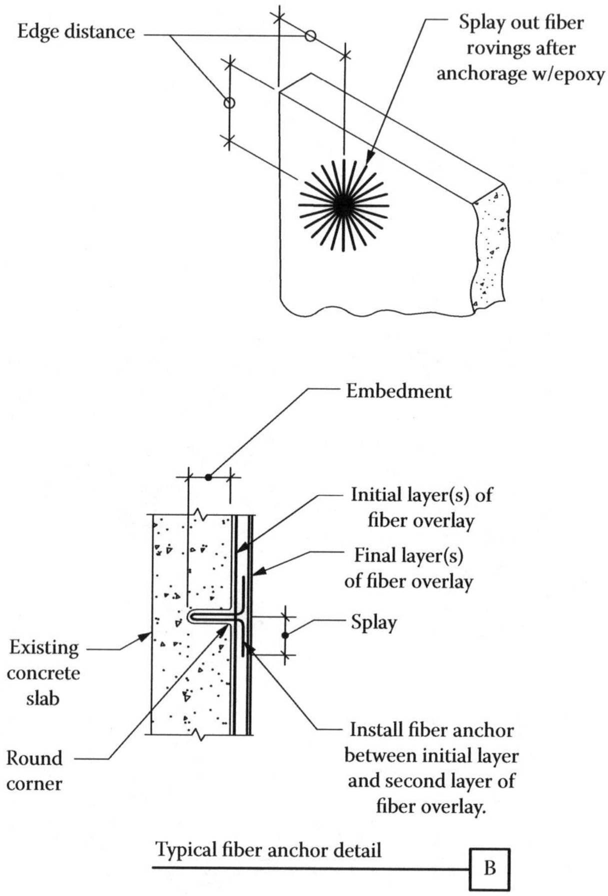

FIGURE 8.53 Fiber anchor details.

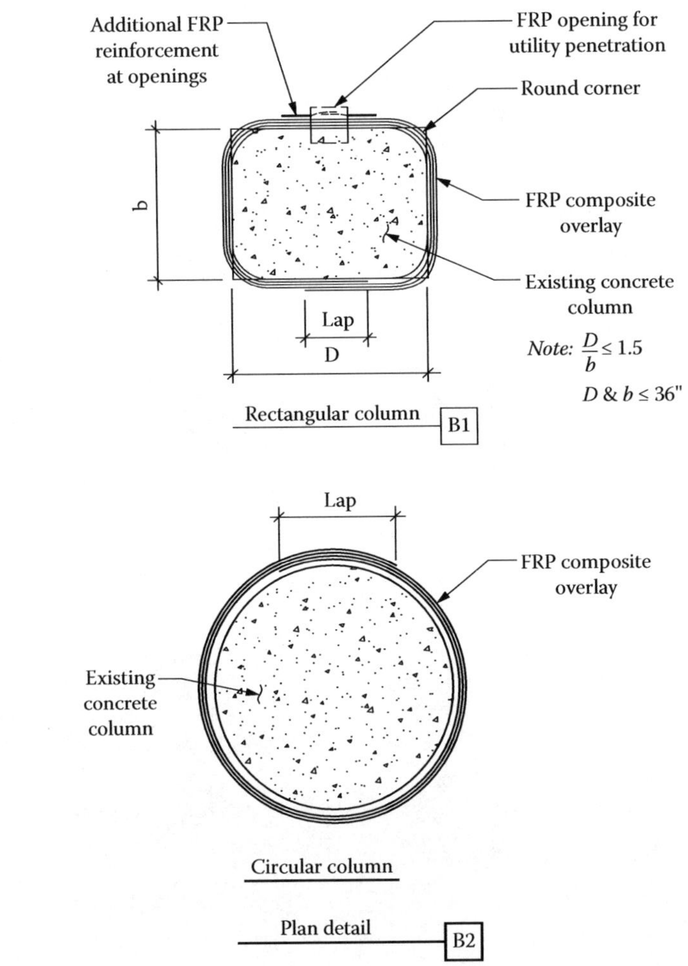

FIGURE 8.54 Seismic retrofit of columns using FRP composites.

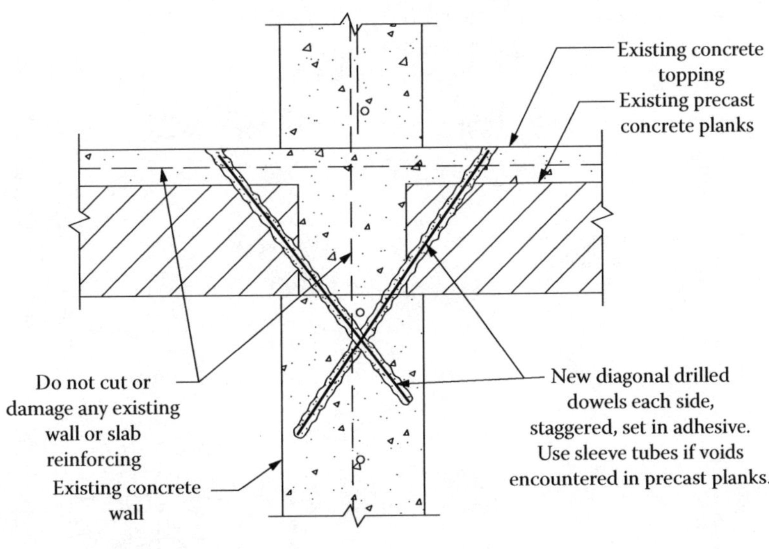

FIGURE 8.55 Added shear capacity at slab–wall joint.

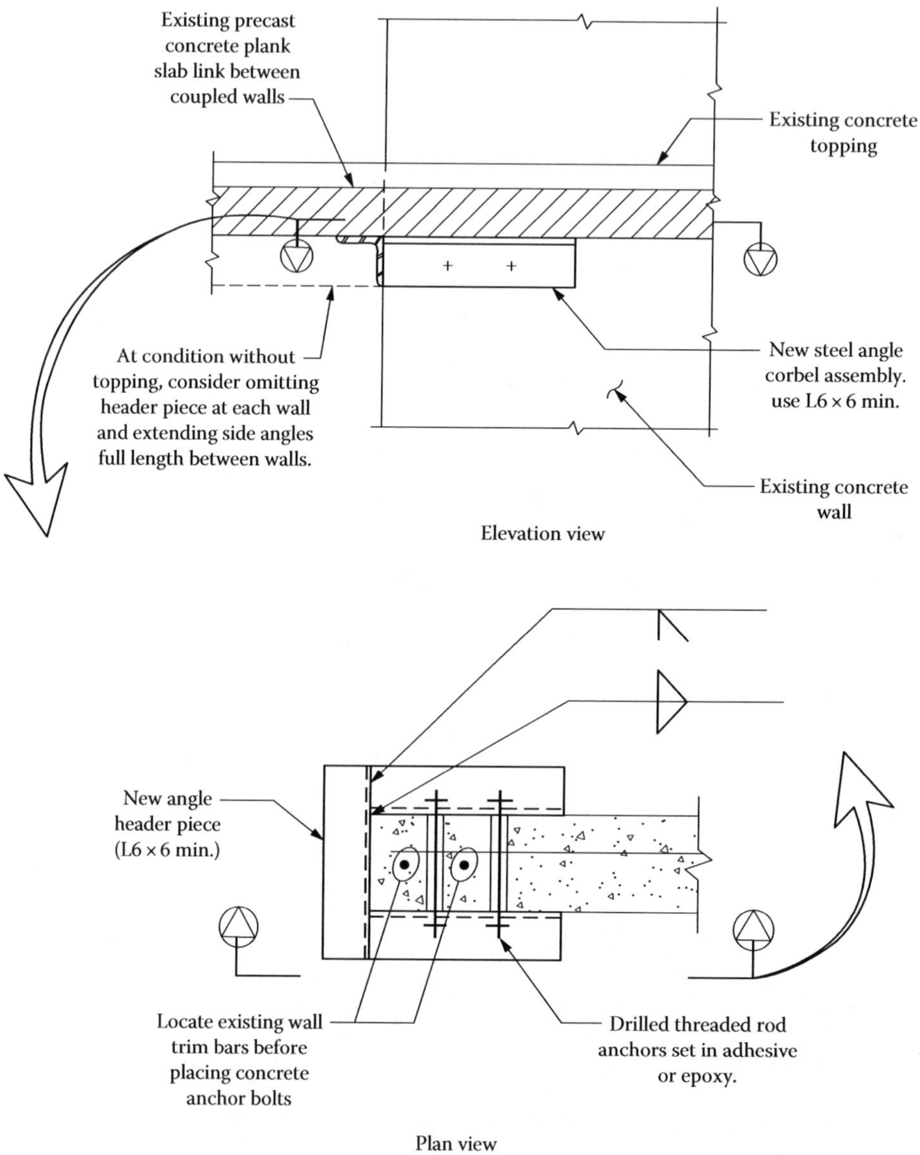

FIGURE 8.56 Typical corbel at linking slab.

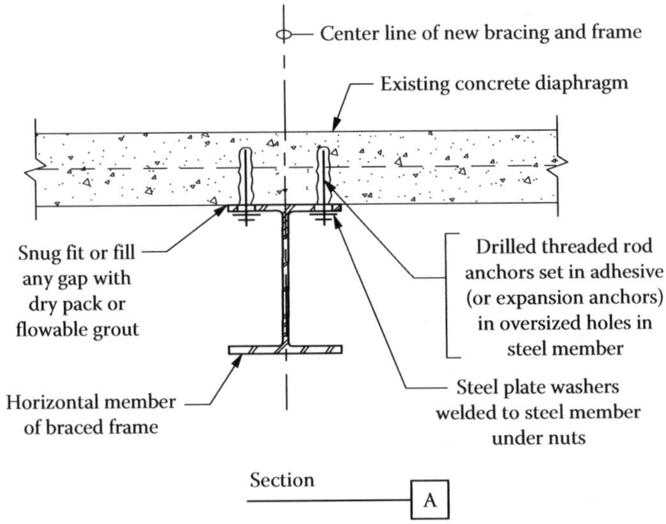

FIGURE 8.57 Typical connection to concrete diaphragm.

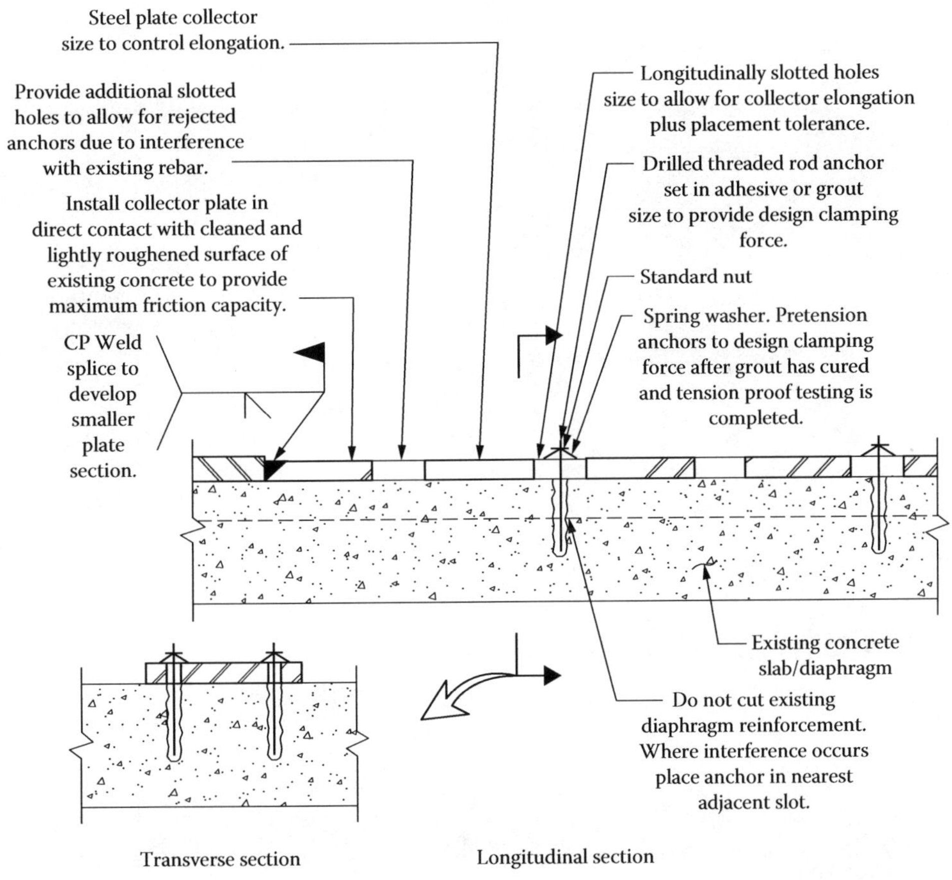

FIGURE 8.58 Steel plate collector.

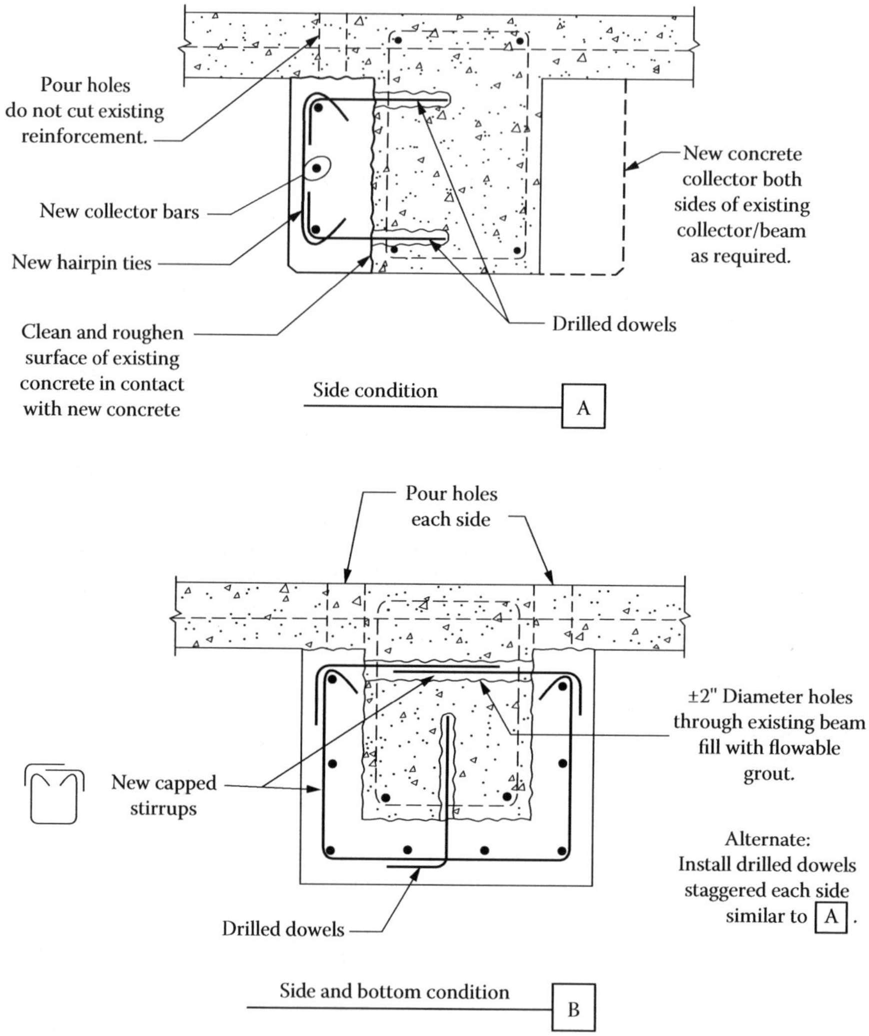

FIGURE 8.59 Concrete collector at existing beam.

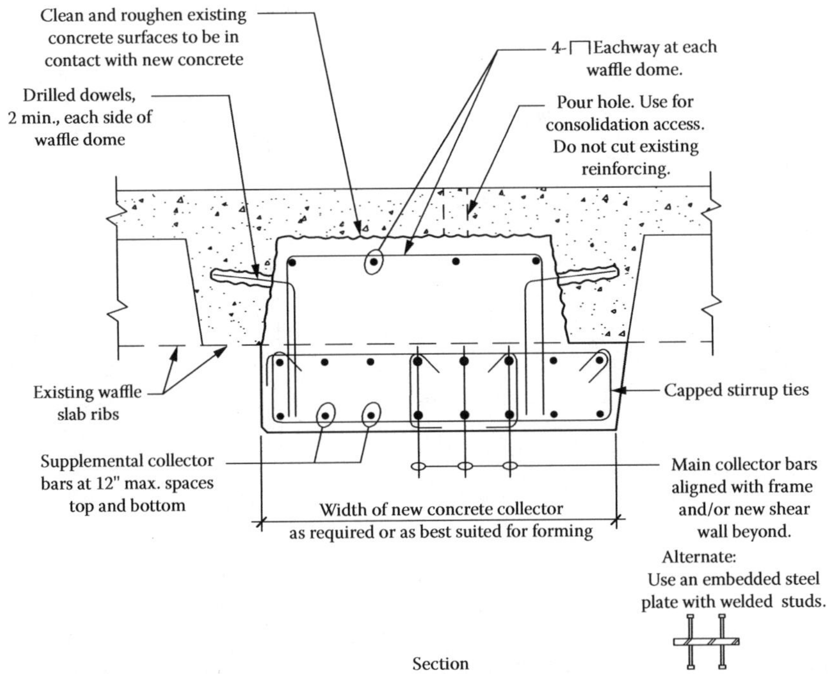

FIGURE 8.60 Concrete collector at waffle slab.

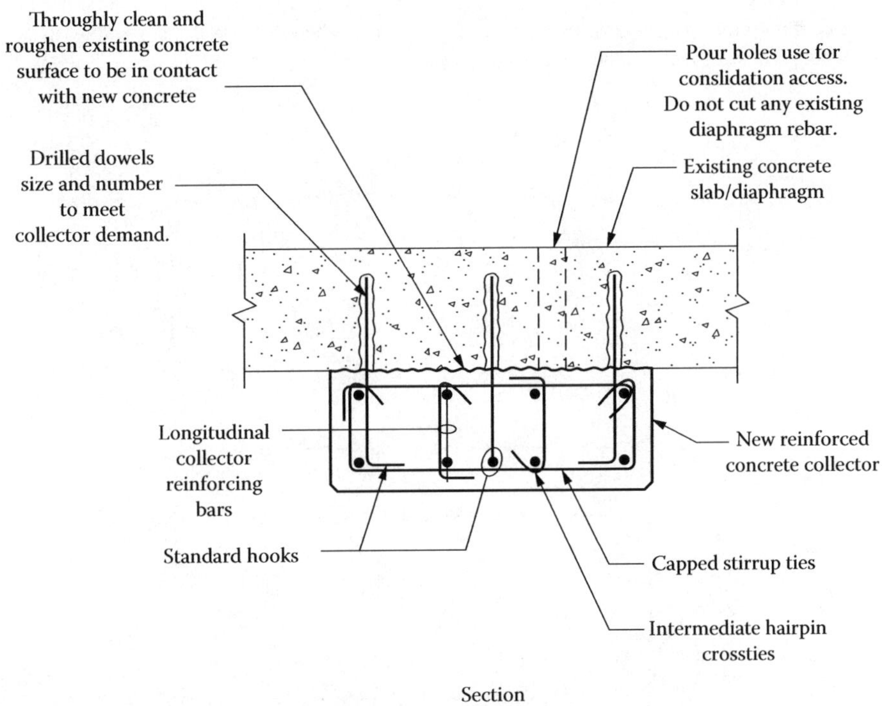

FIGURE 8.61 Concrete collector at concrete slab.

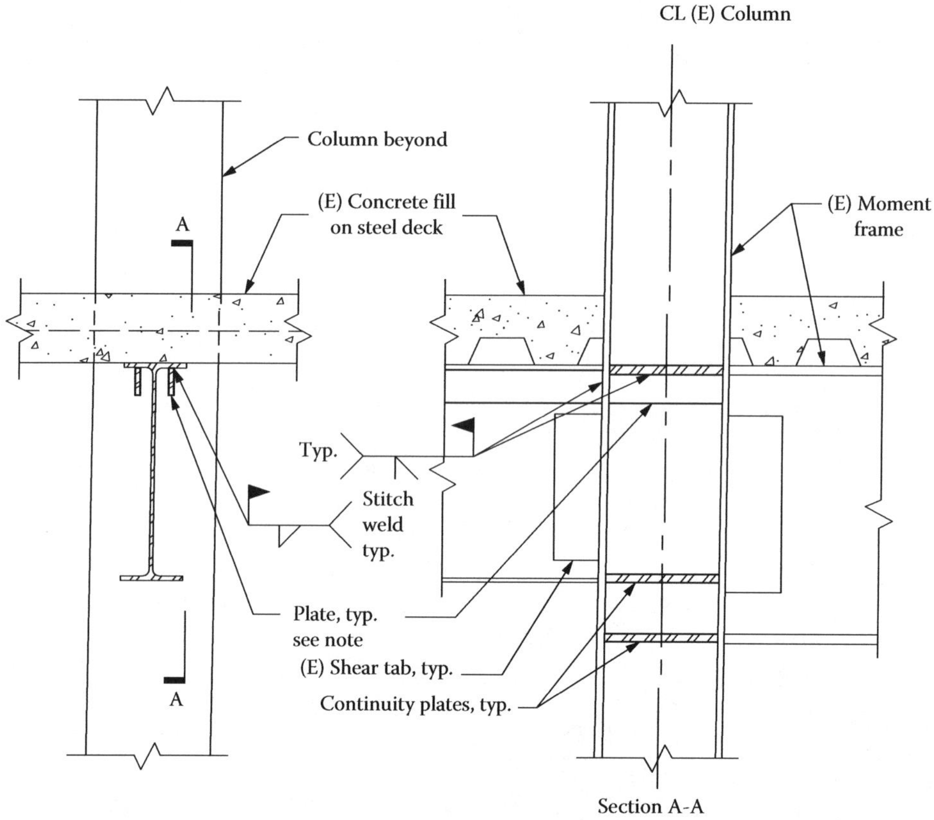

FIGURE 8.62 Plate collectors at existing beam. *Note*: Plates may be interrupted at column by shear tab from transverse beam. Provide complete joint penetration welds from plates to shear tab.

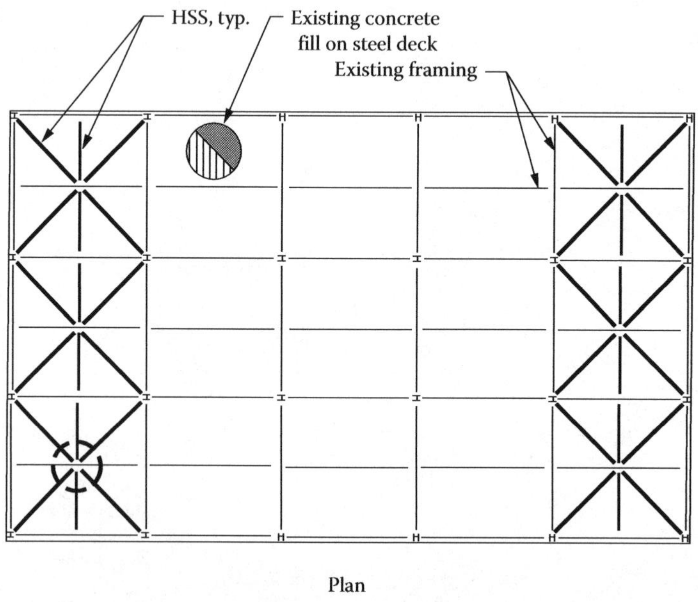

FIGURE 8.63 Diaphragm strengthening using horizontal braced frame.

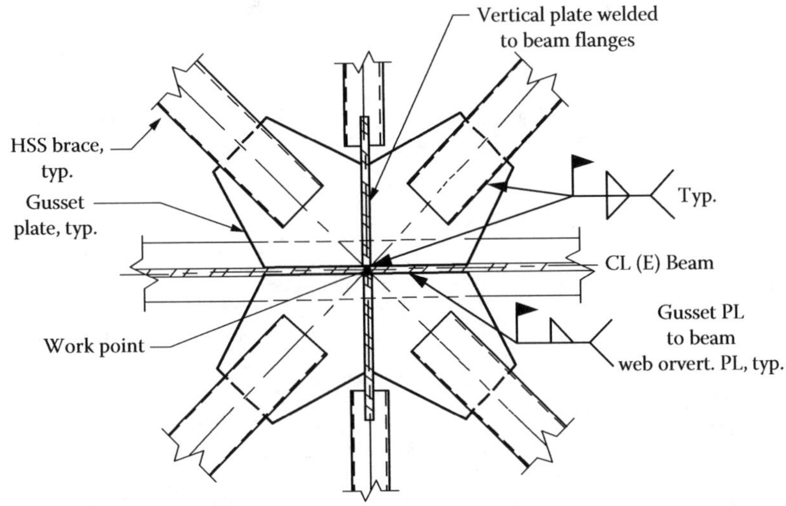

FIGURE 8.64 Horizontal braced frame connection. *Note*: Elevation of horizontal braced frame with respect to diaphragm is a balance between minimizing eccentric forces and allowing for construction access from below.

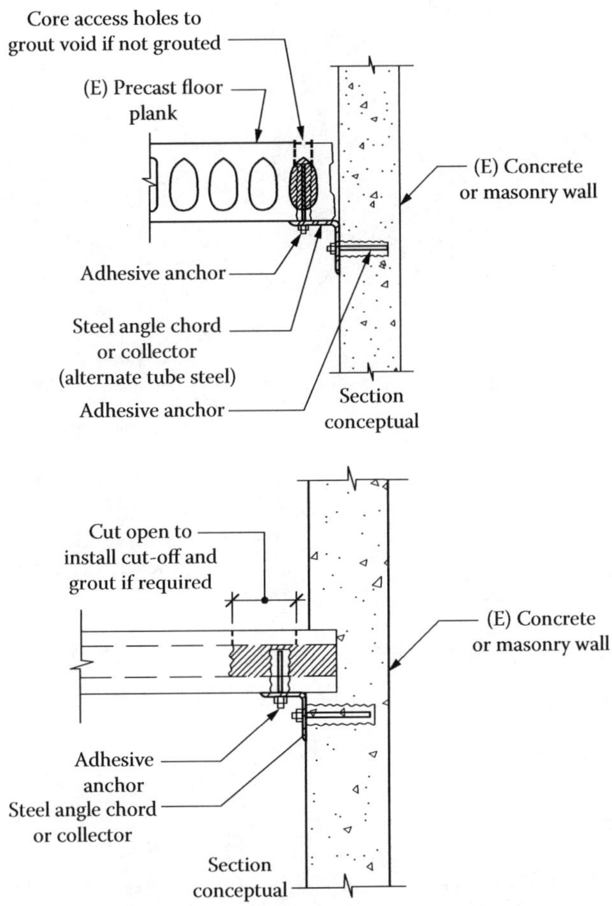

FIGURE 8.65 Steel chord or collector at floor perimeter without cast-in-place topping.

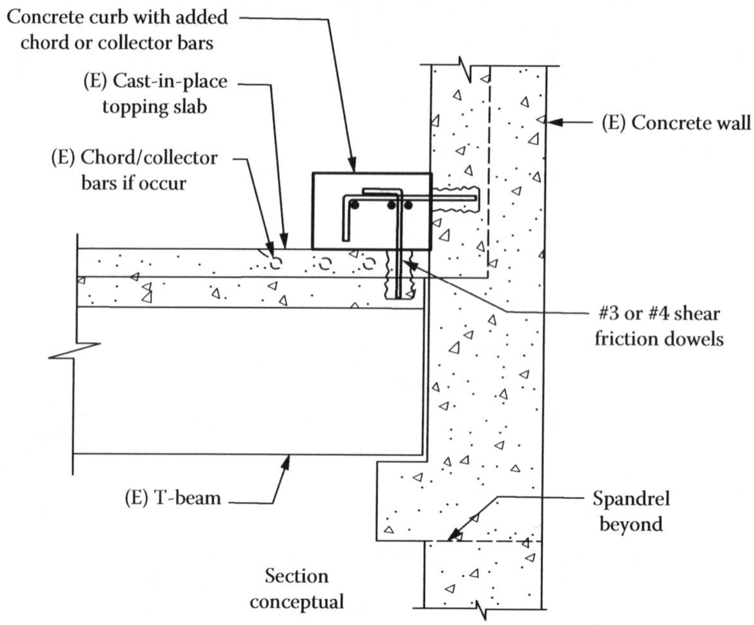

FIGURE 8.66 Added collector anchorage at shear wall with cast-in-place topping slab.

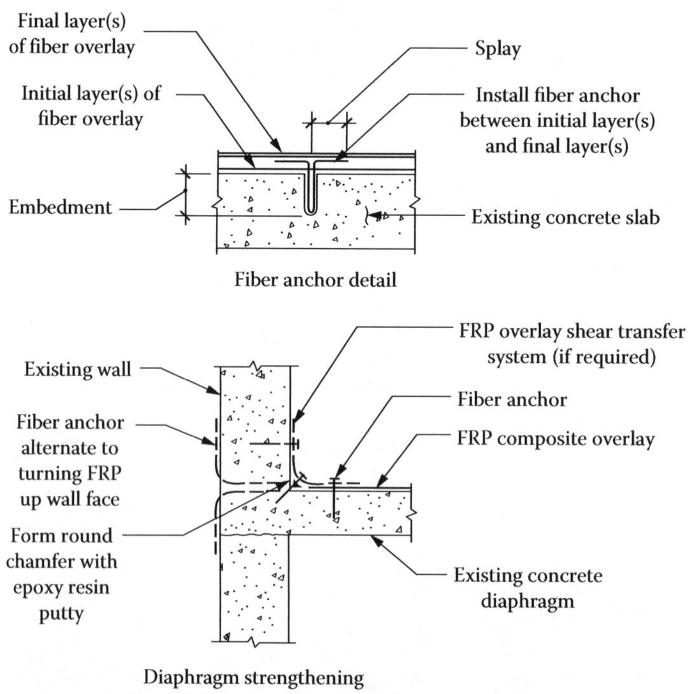

FIGURE 8.67 Shear strengthening of concrete diaphragm using FRP composite.

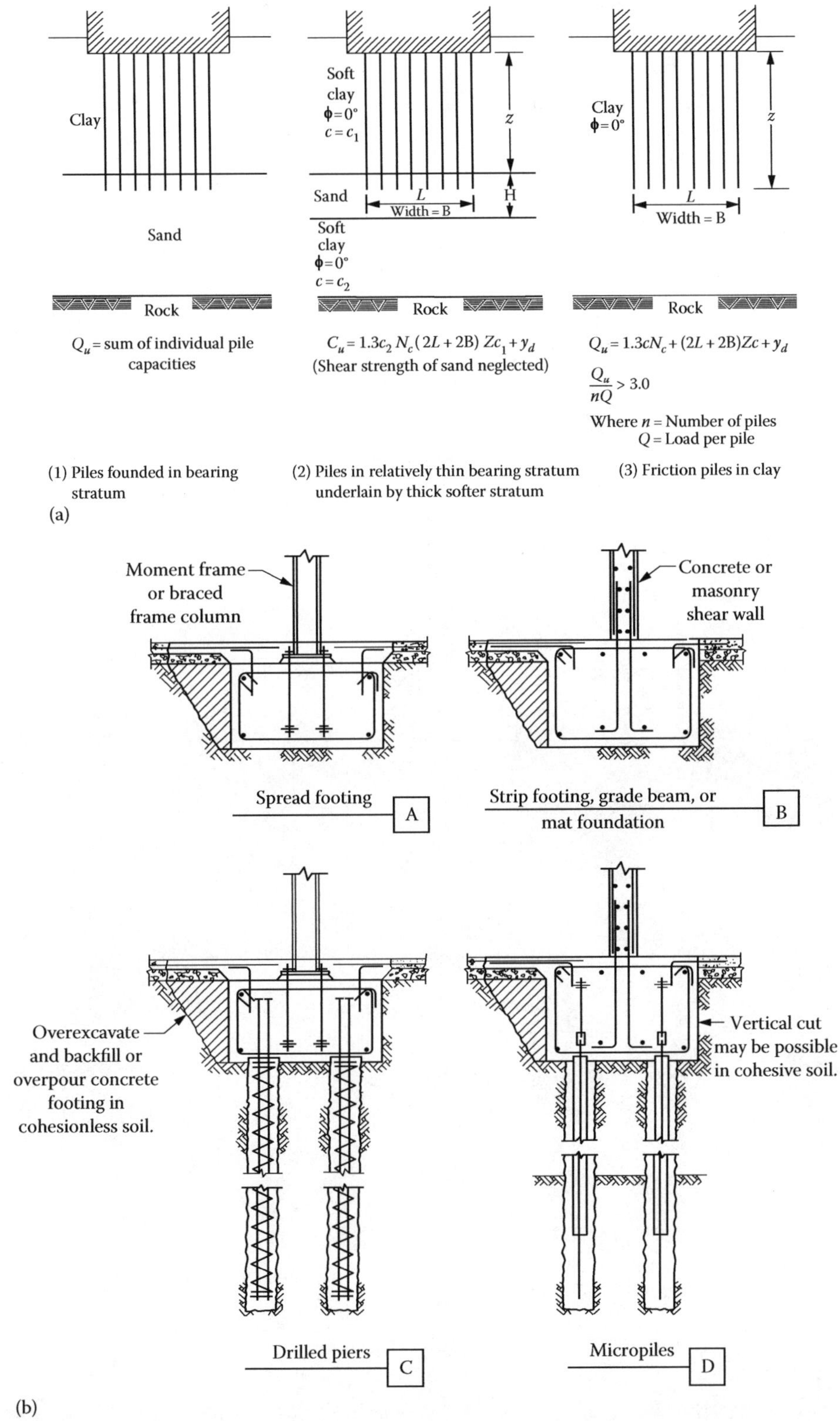

FIGURE 8.68 (a) Bearing capacity of pile groups. (b) Types of new foundations commonly used in seismic rehabilitation.

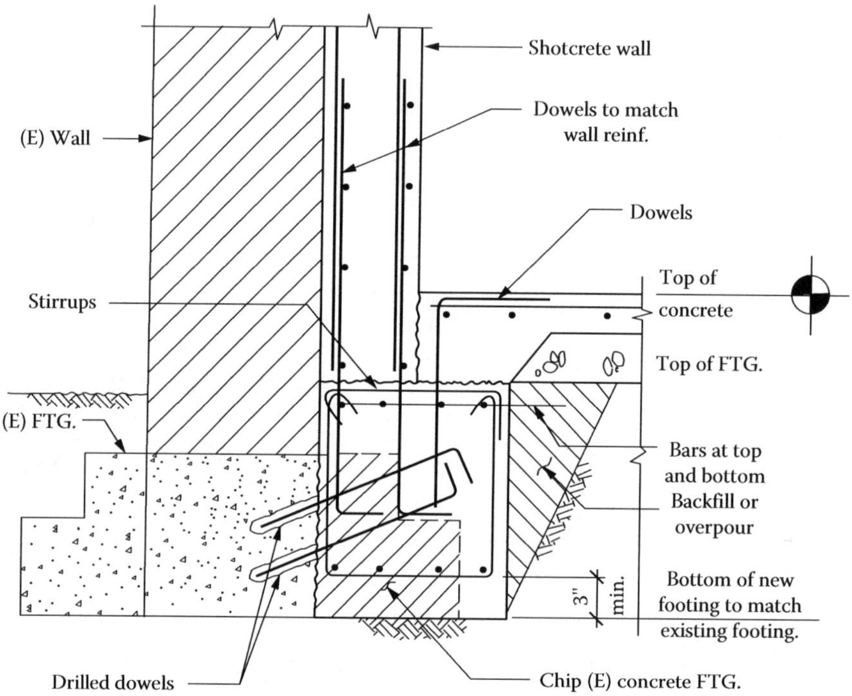

FIGURE 8.69 New concrete strip footing next to existing strip footing.

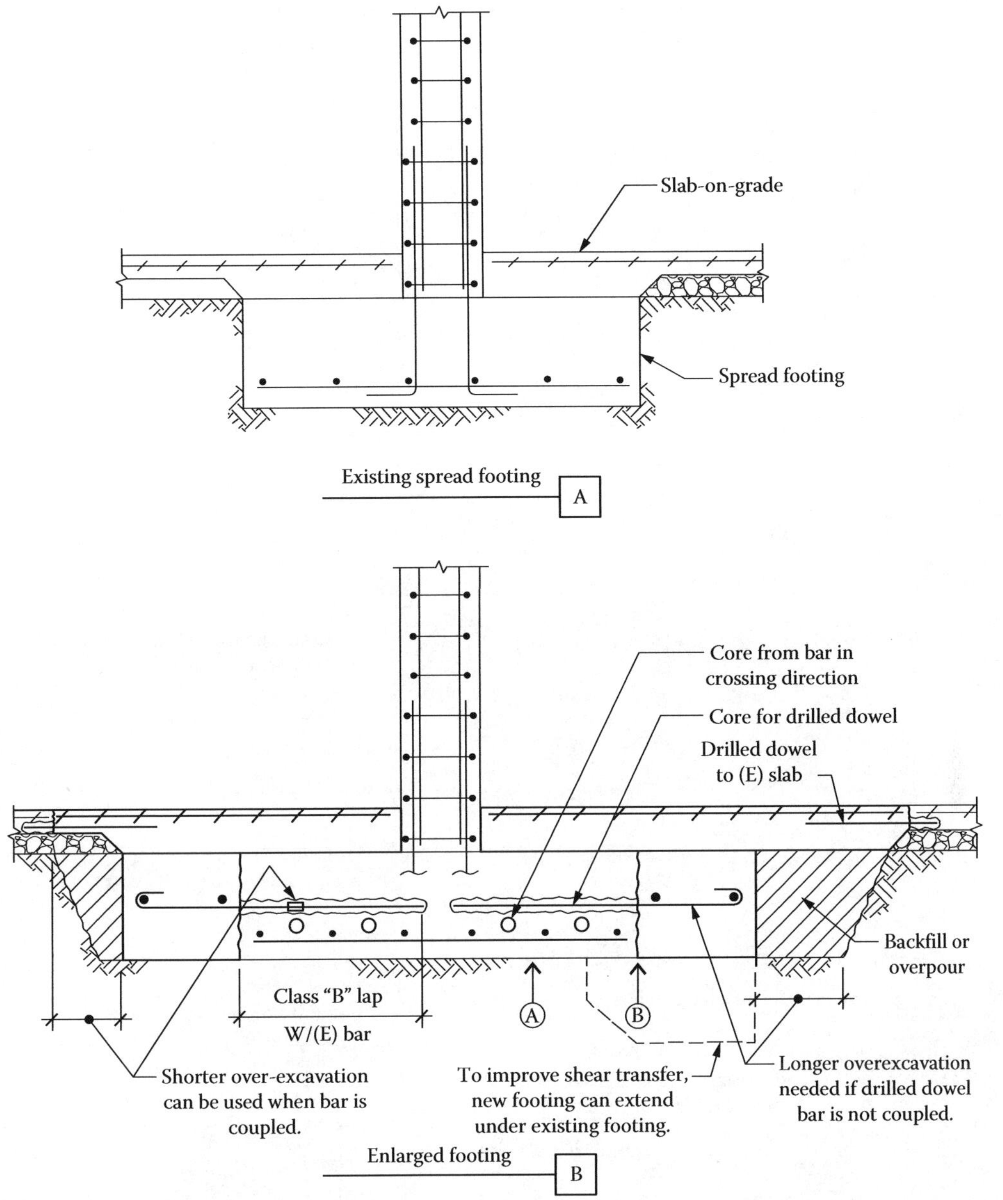

FIGURE 8.70 Enlarge existing spread footing. *Note*: 1. Establish existing layers and organize with new bars so existing bars are not cut. 2. Hole for bar without coupler can be smaller. 3. Check new bar capacity at location B. Check existing bar capacity at location A.

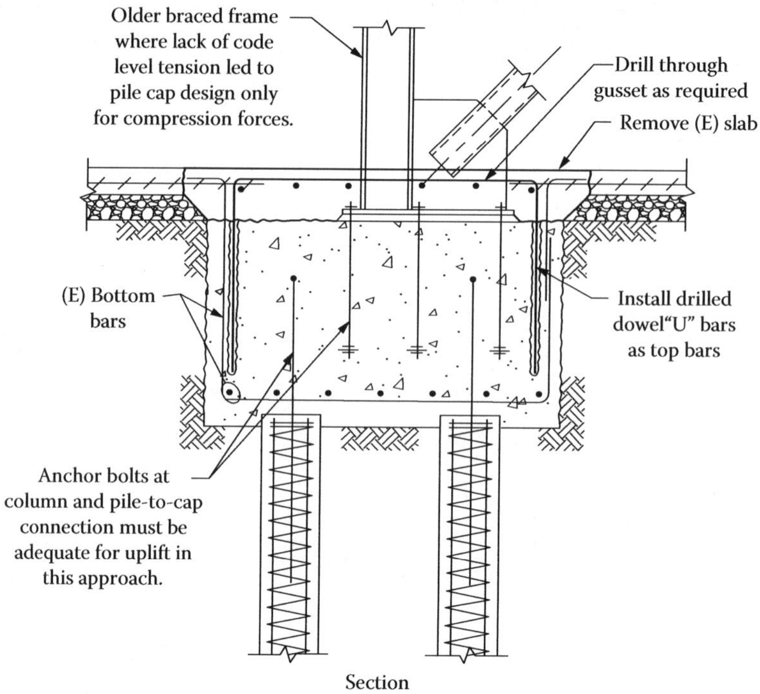

FIGURE 8.71 Adding top bars to an existing pile cap.

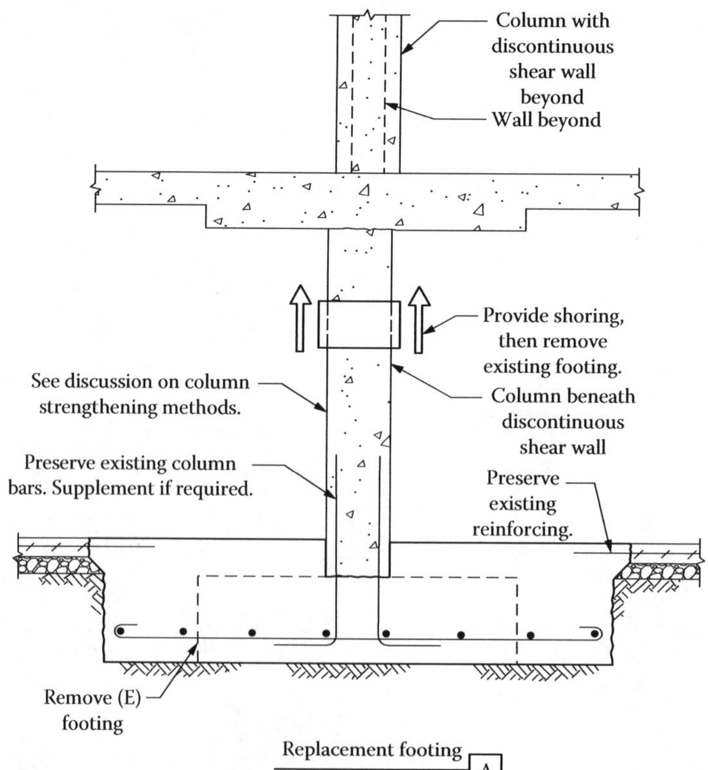

FIGURE 8.72 Replace existing spread footing.

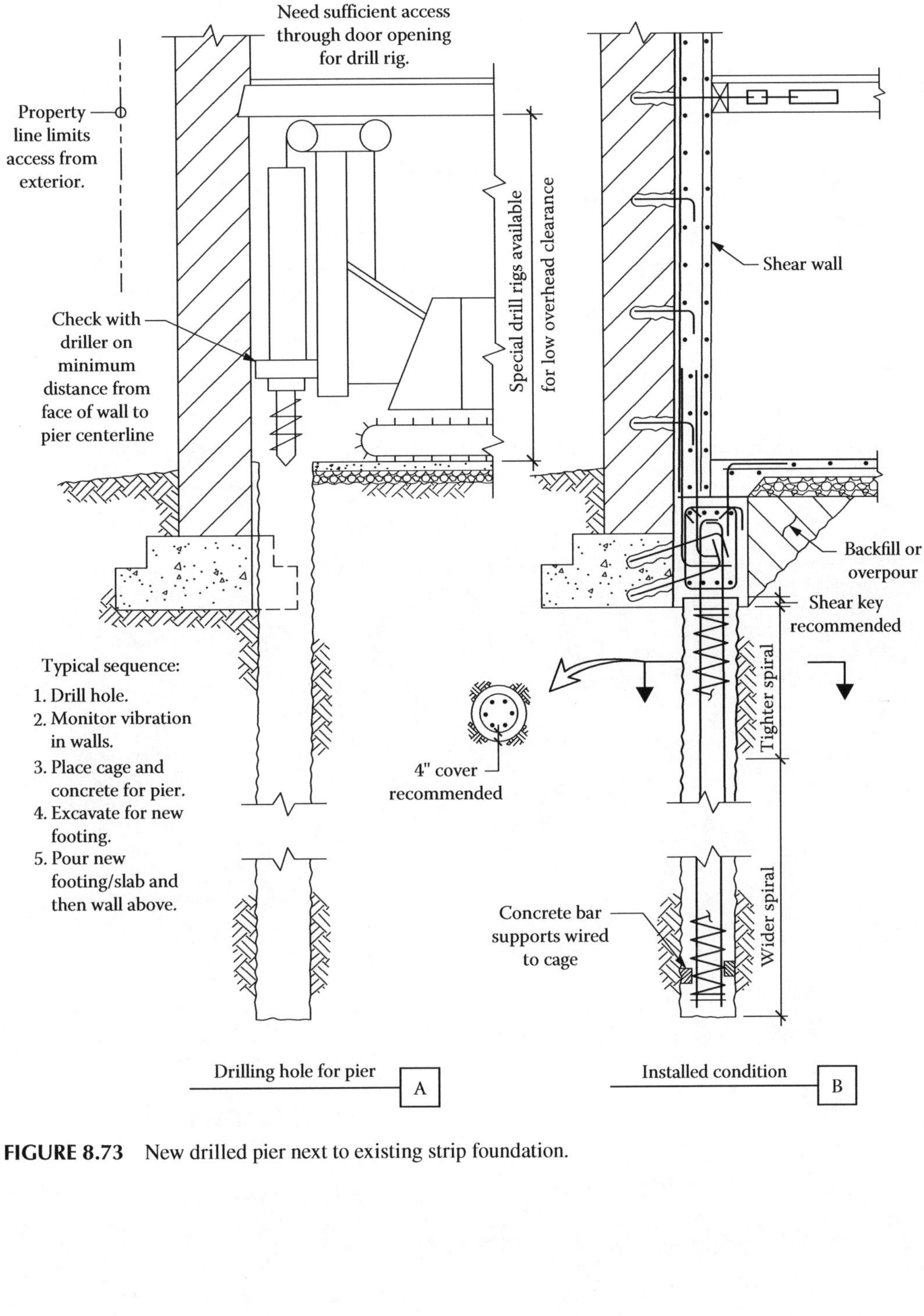

FIGURE 8.73 New drilled pier next to existing strip foundation.

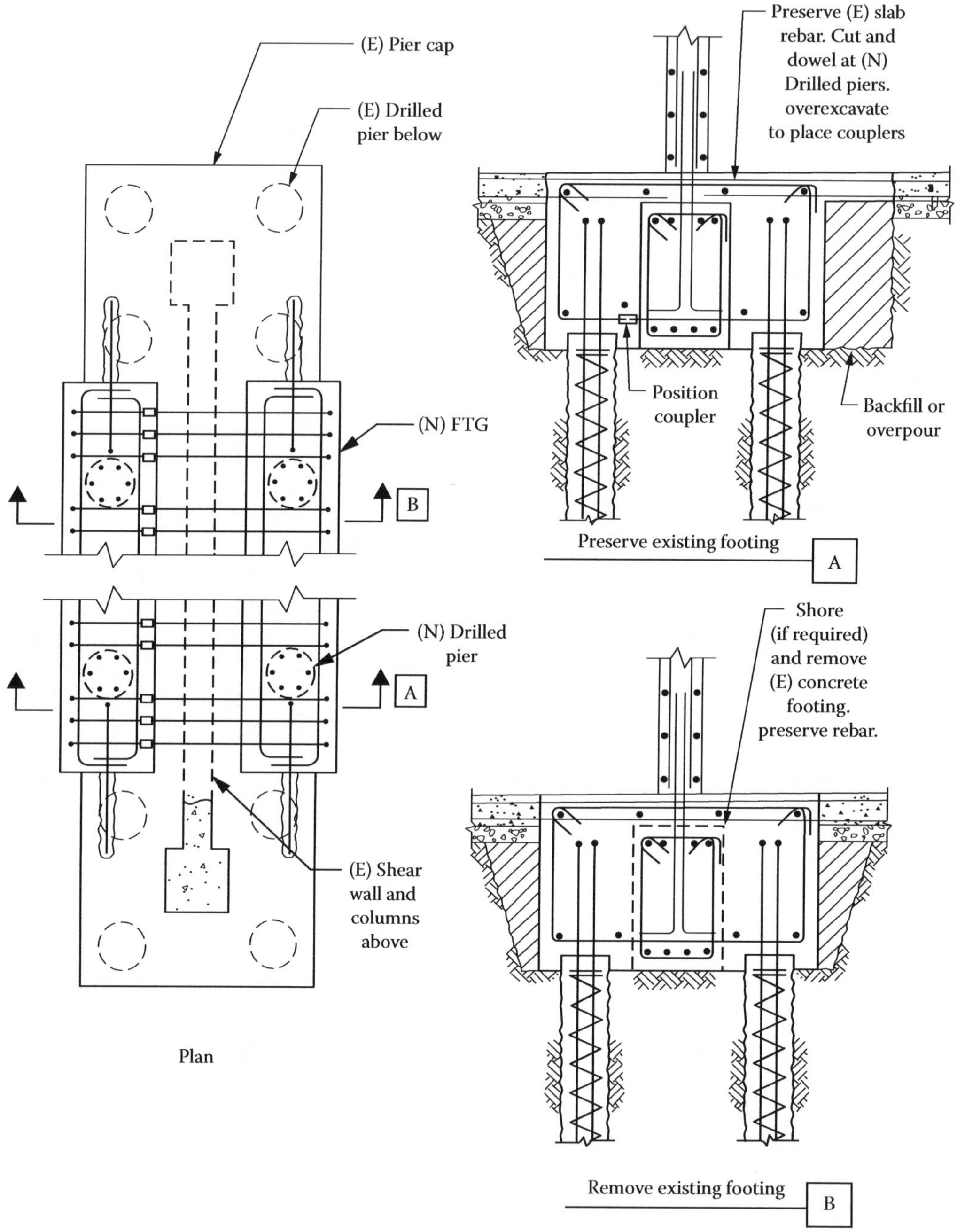

FIGURE 8.74 Adding drilled piers to an existing drilled pier foundation.

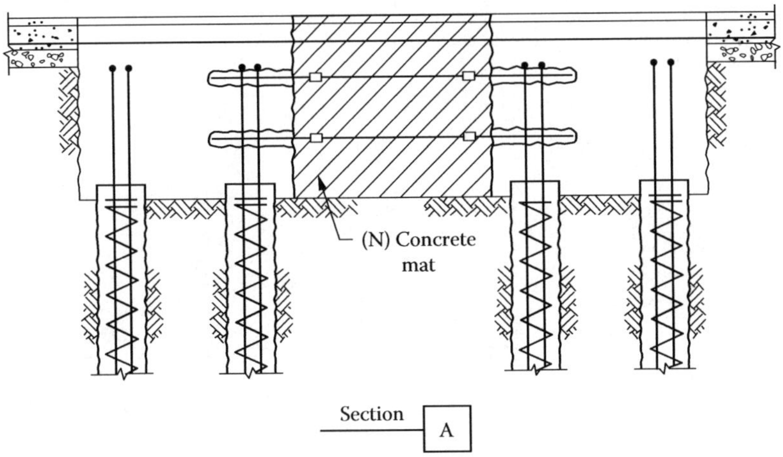

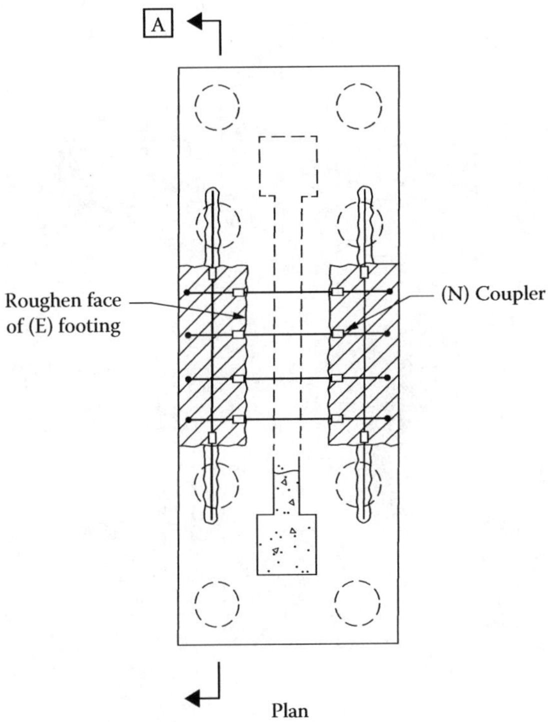

FIGURE 8.75 New mat foundation between existing drilled piers.

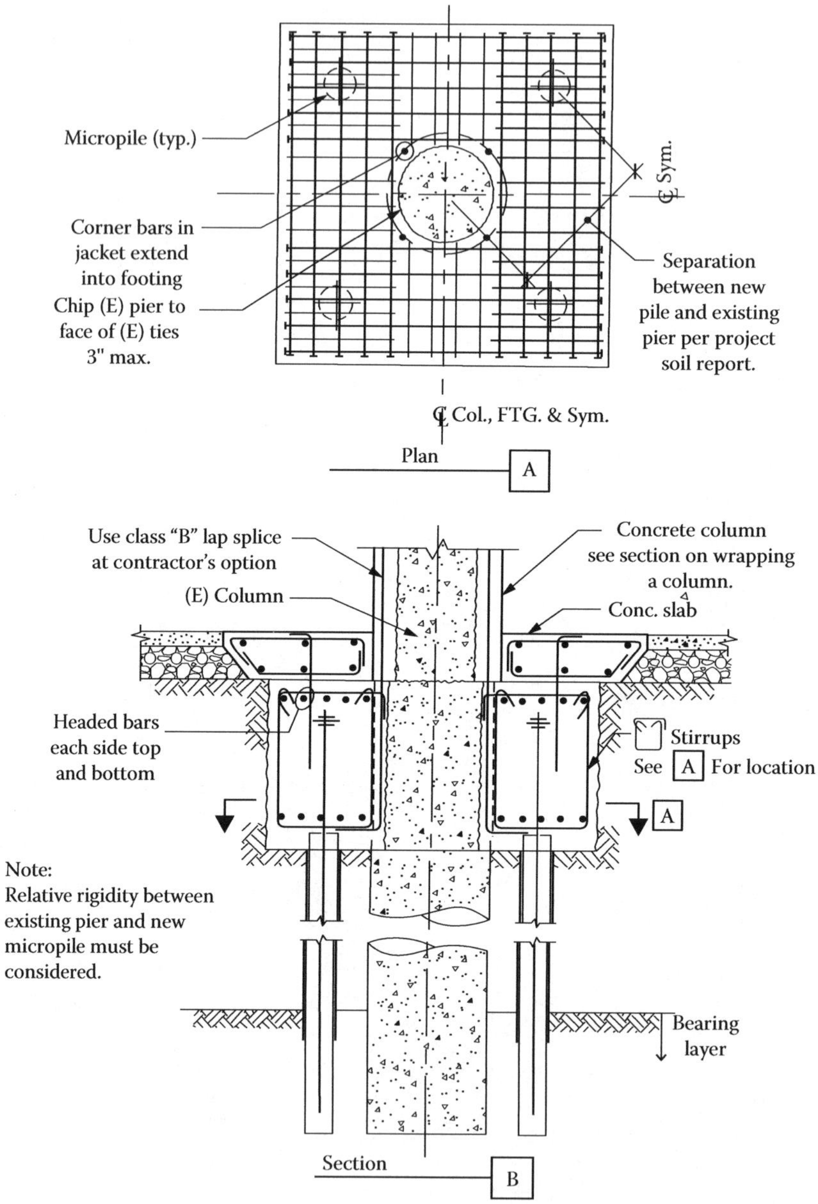

FIGURE 8.76 Micropile enhancement of an existing drilled pier footing.

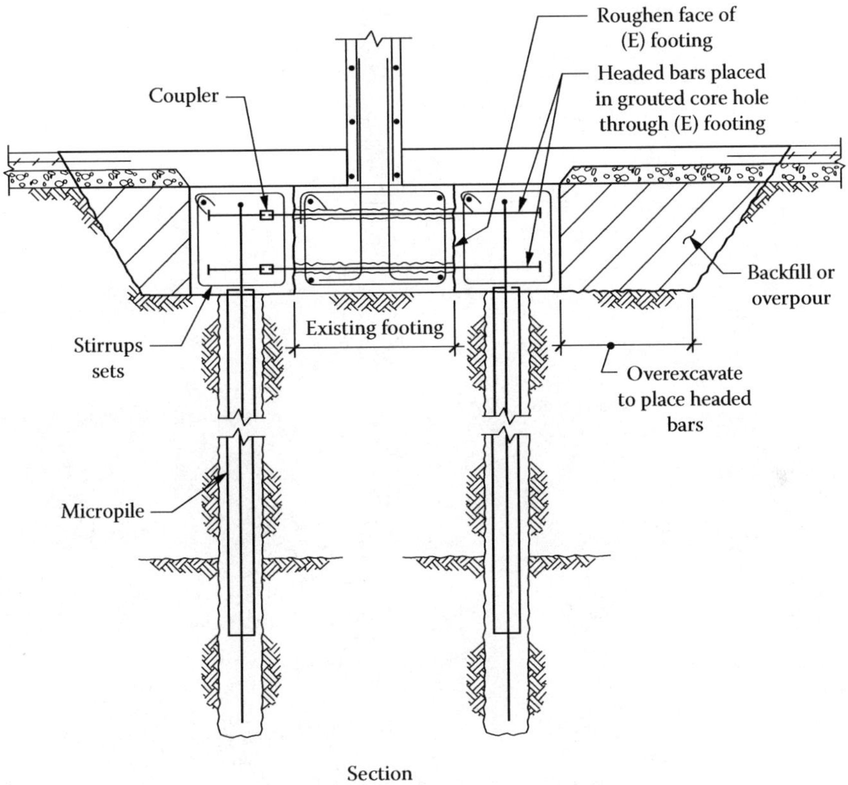

FIGURE 8.77 Micropile enhancement to existing strip footing.

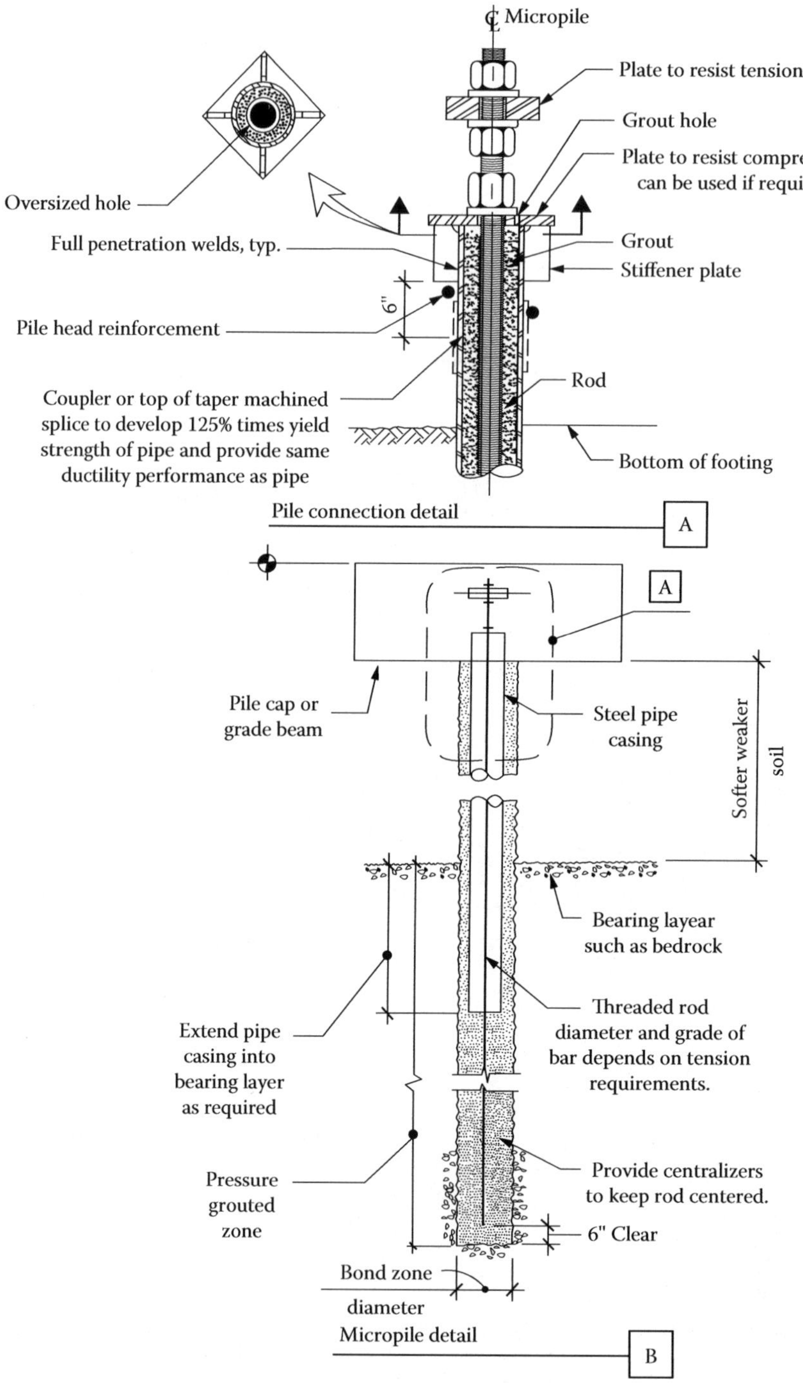

FIGURE 8.78 Micropile details. *Notes*: 1. Micropiles have performance design criteria, substantiated by proof load testing, for tension and compression loads and for elongation. 2. The pile hole should not be left open without casing or grout. Avoid splices at soil transition and bottom of pile.

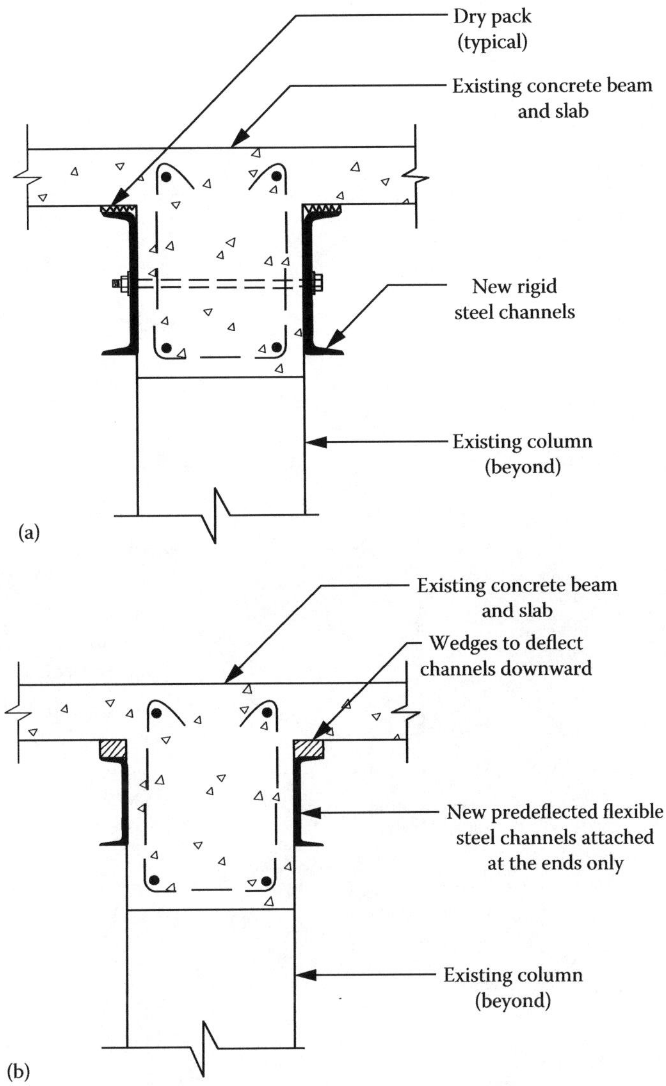

FIGURE 8.79 Adding steel beams on each side of an existing concrete beam. (a) Rigid channels designed for strain compatibility; and (b) flexible channels deflected downward to remove some load from the existing beam.

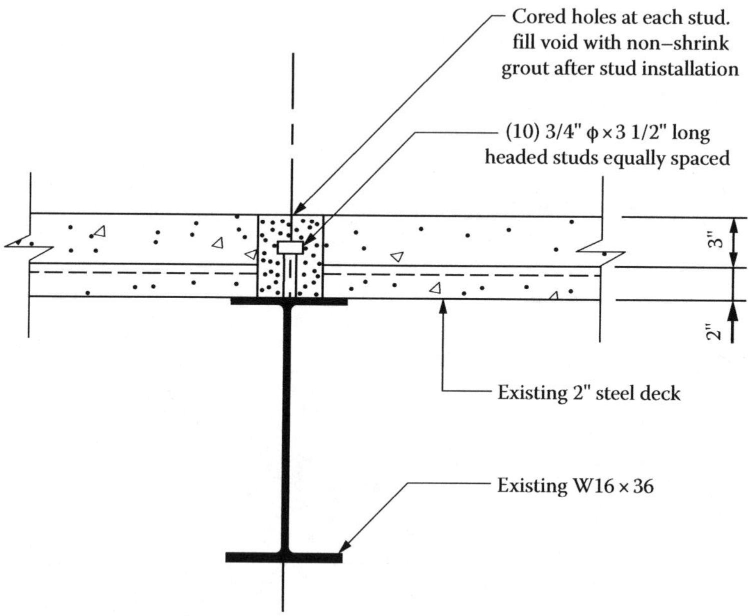

FIGURE 8.80 Beam reinforced by welding shear connectors for Example 3.3.

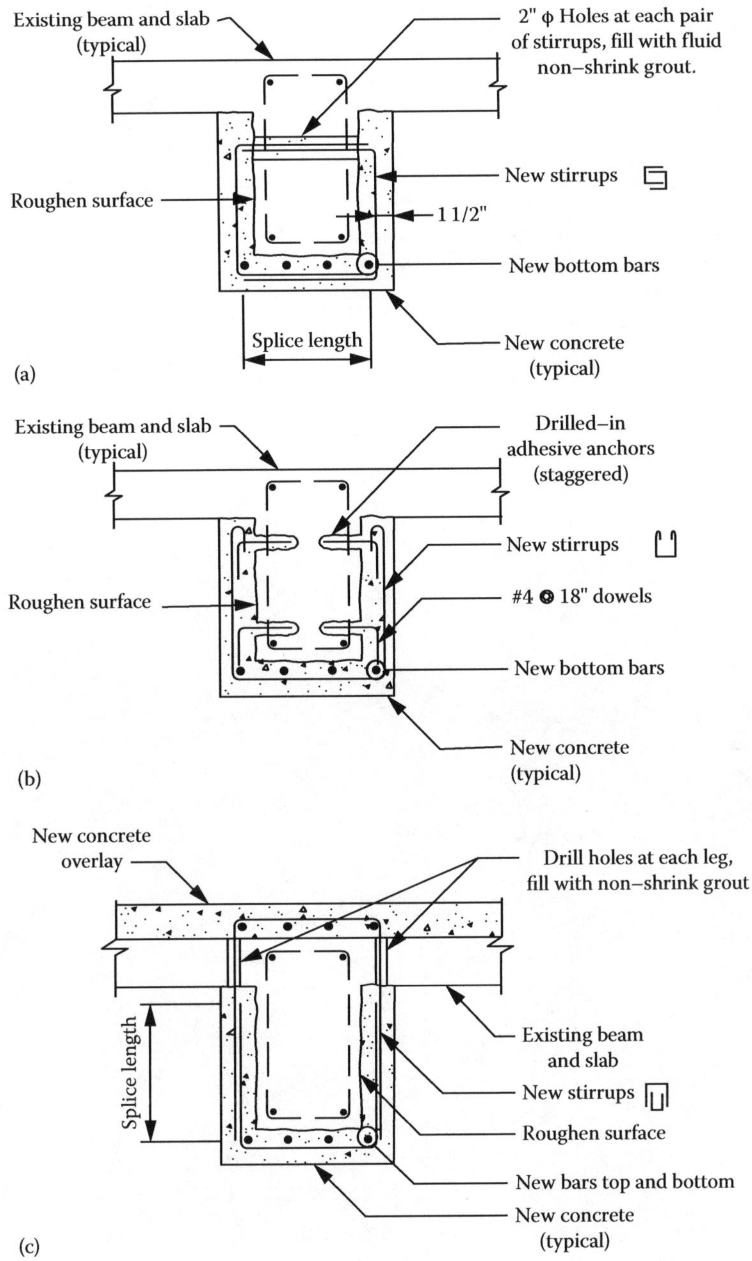

FIGURE 8.81 Enlarging a section of an existing concrete beam. The new and existing concrete can be tied together by (a) stirrups placed in horizontally drilled holes, (b) short dowels placed in drilled-in adhesive anchors, or (c) enveloping the existing beam.

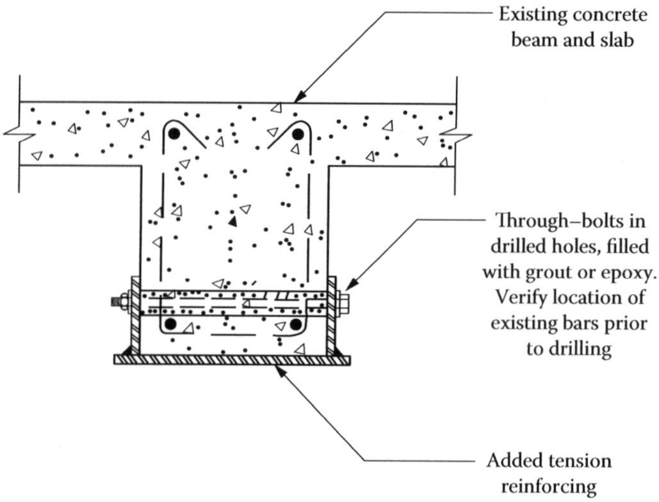

FIGURE 8.82 Bolting a built-up steel member to improve the positive moment capacity of an existing concrete beam.

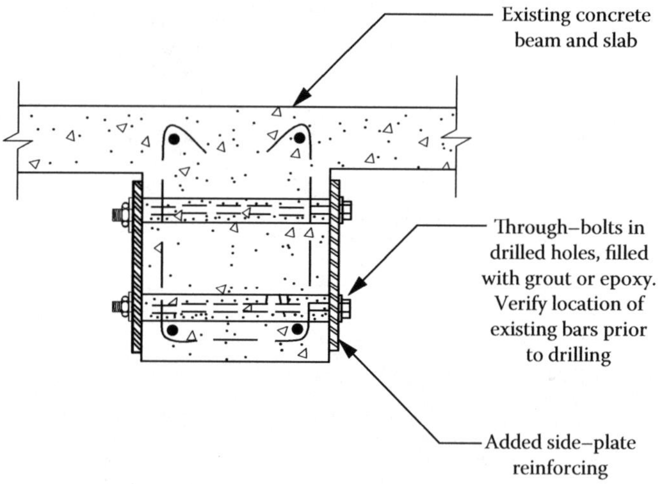

FIGURE 8.83 Adding steel plates to improve the shear resistance of an existing concrete beam.

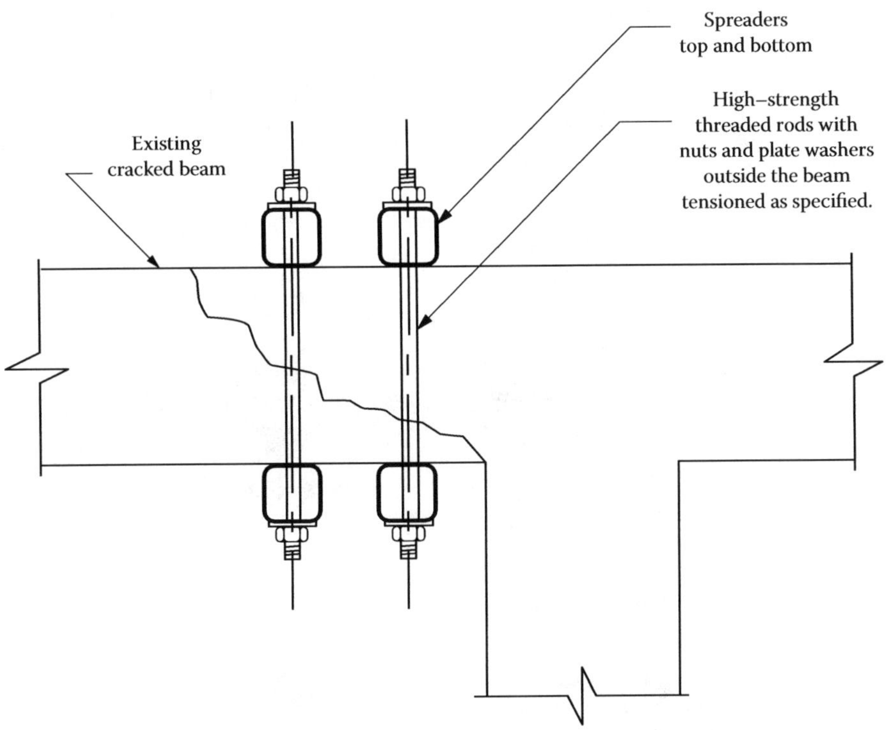

FIGURE 8.84 Upgrading the shear capacity of existing concrete beams by through bolting.

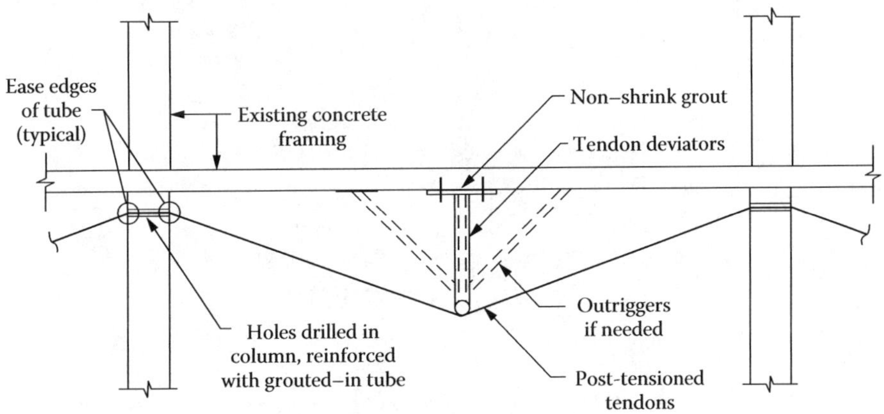

FIGURE 8.85 Strengthening concrete slabs by post-tensioning.

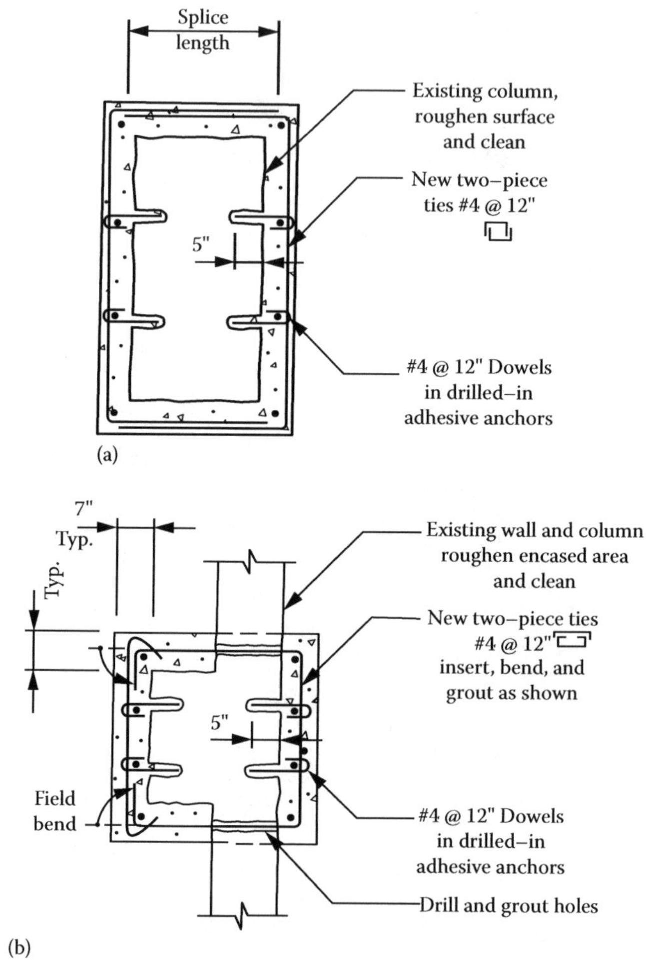

FIGURE 8.86 Strengthening concrete columns by section enlargement. (a) Interior isolated column; (b) exterior column forming part of a wall.

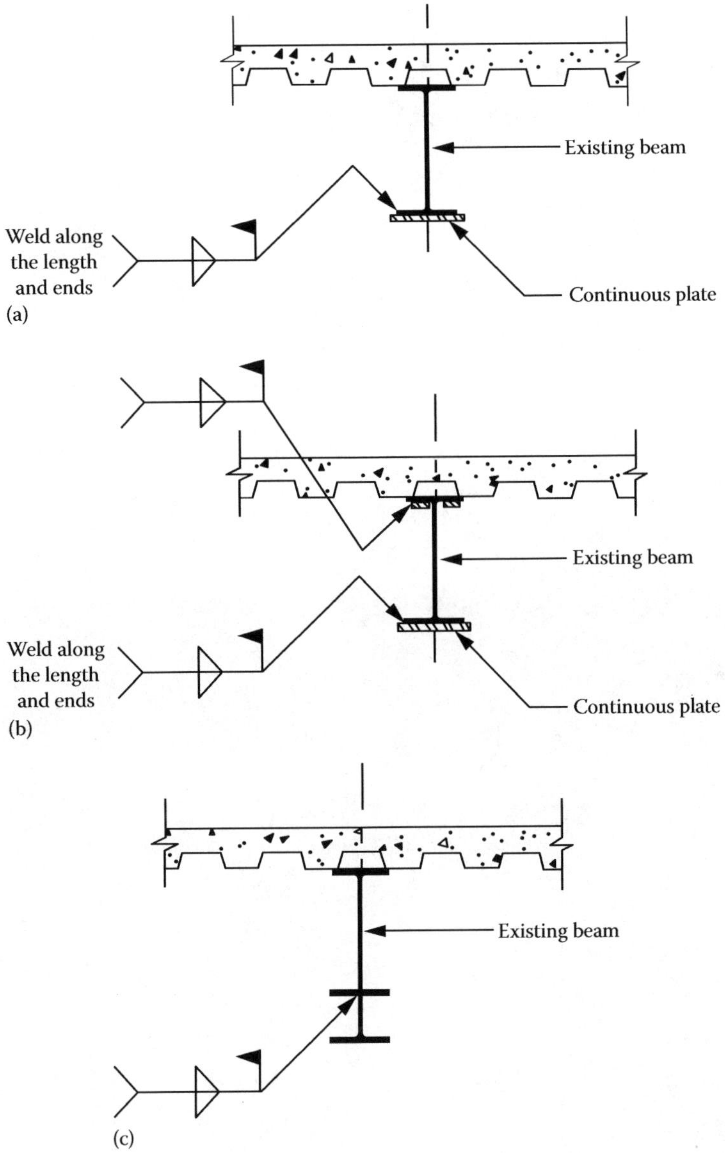

FIGURE 8.87 Reinforcing existing beams by welding: (a) cover plate welded to the bottom flange, (b) cover plates welded to both flanges, and (c) WT section welded to the bottom flange.

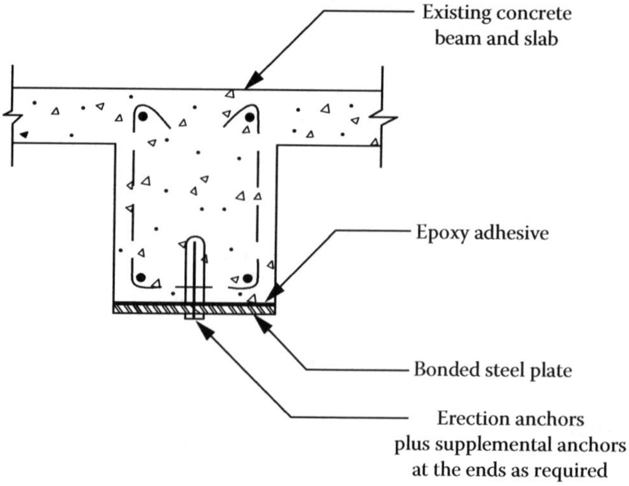

FIGURE 8.88 Adding a bonded steel plate to improve the positive moment capacity of an existing member.

9 Special Topics

9.1 SERVICEABILITY CONSIDERATIONS

Buildings must satisfy *strength* limit states in which members are proportioned to carry the design loads safely to resist buckling, yielding, fracture, and so forth. In addition to the strength limit states, buildings must also satisfy *serviceability* limit states that define functional performance and behavior under load and include such items as deflection and vibration. Strength limit states have traditionally been specified in building codes because they control the safety of the structure. Serviceability limit states, on the other hand, are usually noncatastrophic and involve the perceptions and expectations of the owner or user and are a contractual matter between the owner or user and the designer and builder. It is for these reasons, and because the benefits are often subjective and difficult to define or quantify, that serviceability limit states for the most part are not included within the model US building codes.

Serviceability limit states are conditions in which the functions of a building are impaired because of local damage, deterioration, or deformation of building components or because of occupant discomfort. Although safety generally is not an issue with serviceability limit states (one exception would be for cladding that falls off a building due to excessive story drift under wind load), they nonetheless may have severe economic consequences. The increasing use of stronger (but not stiffer) construction materials, the use of lighter architectural elements, and the uncoupling of the nonstructural elements from the structural frame have resulted in building systems that are relatively flexible and lightly damped. Therefore, verifying serviceability criteria is essential to ensure functional performance of a given design for such building structural systems.

In general, serviceability is diminished by

1. Excessive deflections or rotation that may affect the appearance, functional use, or drainage of the structure or may cause damaging transfer of load to non-load-supporting elements and attachments
2. Excessive vibrations produced by the activities of building occupants, mechanical equipment, or the wind, which may cause occupant discomfort or malfunction of building service equipment
3. Deterioration, including weathering, corrosion, rotting, and discoloration

In checking serviceability, appropriate service loads are used to evaluate the response of the structure and the reaction of the building occupants.

Service loads typically include static loads from the occupants and their possessions, snow or rain on roofs, temperature fluctuations, and dynamic loads from human activities, wind-induced effects, or the operation of building service equipment. The service loads are those loads that act on the structure at an arbitrary point in time and are therefore, in contrast, the nominal loads that have a small probability of being exceeded in any year; factored loads are only a fraction of the nominal loads.

The response of the structure to service loads normally can be analyzed assuming linear elastic behavior. However, members that accumulate residual deformations under service loads may require examination with respect to this long-term behavior. Service loads used in analyzing creep or other long-term effects may not be the same as those used to analyze elastic deflections or other short-term or reversible structural behavior.

Serviceability limits depend on the function of the building and on the perceptions of its occupants. In contrast to the ultimate limit states, it is difficult to specify general serviceability limits that are applicable to all building structures. The serviceability limits provide general guidance and have usually led to acceptable performance in the past. However, serviceability limits for a specific building should be determined only after a careful analysis of all functional and economic requirements and constraints in conjunction with the architect and the building owner. It should be recognized that building occupants are able to perceive structural deflections, motion, cracking, and other signs of possible distress at levels that are much lower than those that would indicate that structural failure was impending. Such signs of distress may be taken incorrectly as an indication that the building is unsafe and diminish its commercial value.

9.1.1 Deflections

Deformations of floor and roof members and systems due to service loads shall not impair the serviceability of the structure.

Excessive vertical deflections and misalignment arise primarily from three sources: (1) gravity loads, such as dead, live, and snow loads; (2) effects of temperature, creep, and differential settlement; and (3) construction tolerances and errors. Such deformations may be visually objectionable; may cause separation, cracking, or leakage of exterior cladding, doors, windows, and seals; and may cause damage to interior components and finishes. Appropriate limiting values of deformations depend on the type of structure, detailing, and intended use. Historically, common deflection limits for horizontal members have been 1/360 of the span for floors subjected to full nominal live load and 1/240 of the span for roof members. Deflections of about 1/300 of the span (for cantilevers, 1/150 of the length) are visible and may lead to general architectural damage or cladding leakage. Deflections greater than 1/200 of the span may impair operation movable components such as doors, windows, and sliding partitions.

In certain long-span floor systems, it may be necessary to place a limit independent of span on the maximum deflection to minimize the possibility of damage of adjacent nonstructural elements. For example, damage to non-load-bearing partitions may occur if vertical deflections exceed more than about 10 mm (3/8 in.) unless special provision is made for differential movement; however, many components can and do accept larger deformations.

Load combinations for checking static deflections can be developed using first-order reliability analysis. Current static deflection guidelines for floor and roof systems are adequate for limiting superficial damage in most buildings. A combined load with an annual probability of 0.05 of being exceeded would be appropriate in most instances. For serviceability limit states involving visually objectionable deformations, repairable cracking or other damage to interior finishes, and other short-term effects, the suggested load combinations are

$$D + L$$

$$D + 0.5S$$

For serviceability limit states involving creep, settlement, or similar long-term or permanent effects, the suggested load combination is

$$D + 0.5L$$

The dead load effect, *D*, used in applying the equations earlier may be that portion of dead load that occurs after attachment of nonstructural elements. For example, in composite construction, the dead load effects frequently are taken as those imposed after the concrete has cured; in ceilings, the dead load effects may include only those loads placed after the ceiling structure is in place.

9.1.2 Building Drift

Lateral deflection of drift of structures and deformation of horizontal diaphragms and bracing systems effects shall not impair the serviceability of the structure.

Drifts (lateral deflections) of concern in serviceability checking arise primarily from the effects of wind. Drift limits in common usage for building design are on the order of 1/600–1/400 of the building or story height.

These limits generally are sufficient to minimize damage to cladding and nonstructural walls and partitions. Smaller drift limits may be appropriate if the cladding is brittle.

An absolute limit on story drift may also need to be imposed in light of evidence that damage to nonstructural partitions, cladding, and glazing may occur if the story drift exceeds bout 10 mm (3/8 in.) unless special detailing practices are made to tolerate movement. It should be noted, however, that many components can accept deformations that are significantly larger.

The use of the nominal (700-year MRI) wind load in checking serviceability is excessively conservative. The following load combination can be used to check short-term effects:

$$D + 0.5L + W_a$$

in which W_a is wind load based on serviceability wind speeds. Some designers have used a 10-year MRI (annual probability of 0.1) for checking drift under wind loads for typical buildings, whereas others have used a 50-year MRI (annual probability of 0.02) or a 100-year MRI (annual probability of 0.01) for more drift-sensitive buildings. The selection of the MRI for serviceability evaluation is a matter of engineering judgment that should be exercised in consultation with the building client.

The serviceability maps are appropriate for use with serviceability limit states and should not be used for strength limit states. Because of its transient nature, wind load need not be considered in analyzing the effects of creep or other long-term actions.

Deformation limits should apply to the structural assembly as a whole. The stiffening effect of nonstructural walls and partitions may be taken into account in the analysis of drift if substantiating information regarding their effect is available.

9.1.3 Vibrations

Floor systems supporting large open areas free of partitions or other sources of damping, where vibration due to pedestrian traffic might be objectionable, shall be designed with due regard for such vibration.

Mechanical equipment that can produce objectionable vibrations in any portion of an inhabited structure shall be isolated to minimize the transmission of such vibrations to the structure.

Building structural systems shall be designed so that wind-induced vibrations do not cause occupant discomfort or damage to the building, its appurtenances, or contents.

Structural motions of floors or of the building as a whole can cause discomfort to the building occupants.

Traditional static deflection checks are not sufficient to ensure that annoying vibrations of building floor systems or buildings as a whole will not occur. Whereas control of stiffness is one aspect of serviceability, mass distribution and damping are also important in controlling vibrations. The use of new materials and building systems often requires that the dynamic response of the system be considered explicitly.

Excessive structural motion is mitigated by measures that limit building or floor accelerations to levels that are not disturbing to the occupants. Perception and tolerance of individuals to vibration is dependent on their expectation of building performance (related to building occupancy) and to their level of activity at the time the vibration occurs. Individuals find continuous vibrations more objectionable than transient vibrations. Continuous vibrations (over a period of minutes) with acceleration on the order of $0.005g$–$0.001g$ are annoying to most people engaged in quiet activities, whereas those engaged in physical activities or spectator events may tolerate steady-state accelerations on the order of $0.02g$–$0.05g$. Thresholds of annoyance for transient vibrations (lasting only a few seconds) are considerably higher and depend on the amount of structural damping present. For a finished floor with (typically) 5% damping or more, peak transient accelerations of $0.05g$–$0.1g$ may be tolerated.

Many common human activities impart dynamic forces to a floor at frequencies in the range of 2–6 Hz. If the fundamental frequency of vibration of the floor system is in this range and if the activity is rhythmic in nature (e.g., dancing, aerobic exercise, or cheering at spectator events), resonant amplification may occur. To prevent resonance from rhythmic activities, the floor system should be tuned so that its natural frequency is well removed from harmonics of the excitation frequency. As a general rule, the natural frequency of structural elements and assemblies should be greater than 2.0 times the frequency of any steady-state excitation to which they are exposed unless vibration isolation is provided. Damping is also an effective way of controlling annoying vibration from transient events because studies have shown that individuals are more tolerant of vibrations that damp out quickly than those that persist.

Several studies have shown that a simple and relatively effective way to minimize objectionable vibrations to walking and other common human activities is to control the floor stiffness, as measured by the maximum deflection independent of span. Justification for limiting the deflection to an absolute value rather than to some fraction of span can be obtained by considering the dynamic characteristics of floor system modeled as a uniformly loaded simple span. The fundamental frequency of vibration, f_o, of this system is given by

$$\frac{\pi}{2L^2}\sqrt{\frac{EI}{\rho}}$$

in which

EI is the flexural rigidity of the floor
L is the span
$\rho = w/g$ = mass per unit length (g is the acceleration due to gravity (9.81 m/s^2), and w is the dead load plus participating live load)

The maximum deflection due to w is

$$\delta = (5/384)\,(wl^4/EI)$$

Substituting EI from this equation into the first equation, we obtain

$$f_o \approx 18/\sqrt{\delta}\ (\delta \text{ in mm})$$

This frequency can be compared to minimum natural frequencies for mitigating walking vibrations in various occupancies. For example, the static deflection due to uniform load, w, must be limited to about 5 mm, independent of span, if the fundamental frequency of vibration of the floor system is to be kept above about 8 Hz. It should be noted, however, that many floors not meeting this guideline are perfectly serviceable.

9.1.4 Design for Long-Term Deflection

This is because, under sustained loading, structural members may exhibit additional time-dependent deformations due to creep, which usually occur at a slow but persistent rate over long periods of time. In certain applications, it may be necessary to limit deflection under long-term loading to specified levels. This limitation can be done by multiplying the immediate deflection by a creep factor, which ranges from about 1.5–2.0. This limit state should be checked using $D + 0.5L$ load combination.

9.1.5 Camber

Where required, camber should be built into horizontal structural members to give proper appearance and drainage and to counteract anticipated deflection from loading and potential ponding.

9.1.5.1 Recommended Camber Criteria

1. Camber for 75% of construction dead load because even in a simple connection, there is always some end restraint.
2. Minimum camber of ¾ in. is reasonable in building construction.
3. Provide camber in ¼ in. increments.
4. Always round down the calculated camber.
5. As a rule of thumb, $L/360$ is a reasonable camber.
6. Cambers in excess of $L/180$ or 2½ in. should be investigated further. Live load deflection and vibration criteria are likely to control the design.

9.1.6 Expansion and Contraction

Dimensional changes in a structure due to variation temperature, relative humidity, or other effects shall not impair the serviceability of the structure.

Provision shall be made either to control crack widths or to limit cracking by providing relief joints.

Structural distress in the form of wide cracks is caused by restraint of thermal, shrinkage, and prestressing deformations. Such effects should be minimized by using expansion joints or by controlling crack widths.

9.1.7 Durability

Buildings may deteriorate in certain service environments causing concerns that may be visible upon inspection (e.g., weathering, corrosion, and staining). The designer should specify adequate protection systems and/or planned maintenance to minimize the likelihood that such problems will occur. For portions of buildings exposed to weather, the design should eliminate pockets in which moisture can accumulate.

A final note on serviceability considerations given in the ASCE 7-10, Appendix C and its commentary, they are not a mandatory part of the standard but provide guidance for design for serviceability in order to maintain the function of a building and the comfort of its occupants during normal usage. Serviceability limits (e.g., maximum static deformations, accelerations) shall be chosen with due regard to the intended function of the structure.

Although not mandatory, if you stick to the recommendations given therein, your design is more than likely to be fine.

9.1.8 Serviceability Considerations: Concrete Systems

Strength design permits the utilization of a larger percentage of tension reinforcement without compression steel and allows advantageous use of higher-strength steels and concrete. Shallower construction than that desirable can result. It is necessary that the designer check that both deflection and cracking are within tolerable limits at service load levels, once the safety of the member or floor system is determined by strength design. It will generally be found that a member designed for maximum overall economy will have about the same span–depth ratio as obtained with working stress design but will contain less reinforcement. Thus, deflections will be in the same order of magnitude. If the reinforcing is supplied as small bars spaced close together, crack control will also not be a particular problem.

Deflection control: Deflections must be computed and compared to limits allowable for particular loading combinations when the thickness used is less than the minimums tabulated in the following.

For *two-way flat plate construction without beams*, a rule of thumb is to limit clear span, $l_n \leq 32.7h$, where h is the thickness of slab. For the effects of edge beams, drop panels, rectangular panels, edge panels, and corner panels, see ACI 318-11.

Computed deflections are limited to values that are acceptable for various conditions of loading and design requirements. These acceptable limits are given as fractions of the span lengths and vary with the use of a structure. For example, a floor carrying partitions or supporting a ceiling below can tolerate much less deflection, without cracking of the supported elements, than the same floor without a ceiling or partitions. If the problem is that of establishing clearance under a flexural member and over, say, a plate glass store front, many factors, such as age, moisture content, actual E_c, settlement of supports, and realistic end restraints, have an influence on the actual deflection.

When the computed long time deflection exceeds the allowable, one of the most satisfactory means of decreasing these creep-caused deflections is compressive reinforcement, provided by extending bottom bars into the support.

Deflection computations: The calculation of deflections requires the computation of E_c, the modulus of elasticity of the concrete, and I_e, the effective moment of inertia of the cross section.

The modulus of elasticity for concrete $E_c = w^{1.5}\, 33\sqrt{f_c'}$. Taking a density of 145 pcf for normal-weight concrete and f_c' as 3000 psi, $E_c = 3{,}160{,}000$ psi; and for $f_c' = 4000$ psi, $E_c = 3{,}640{,}000$ psi.

The effective moment of inertia, I_e, is a function of the gross moment of inertia, I_g, the cracked moment of inertia, I_{cr}, and the ratio of the cracking moment, M_{cr}, to the actual (service load) moment, M_a.

The value of M_{cr} for solid slabs and beams is

$$M_{cr} = f_r I_g / y_t$$

where

$f_r = 7.5\sqrt{f_c'}$ for normal-weight concrete

y_t is the distance from the neutral axis of the solid section to the tension fiber

The effective moment of inertia is

$$I_e = \left(\frac{M_{cr}}{M_a}\right)^3 I_g + \left[1 - \left(\frac{M_{cr}}{M_a}\right)^3\right] I_{cr}$$

It should be emphasized that these calculations illustrate results of a computed deflection only. The results are useful to determine compliance with minimum code limits. These results may be used as a starting point to estimate actual deflections but must be modified to reflect job conditions by the designer. Only the designer can properly assess effects of restraint, shrinking, probable actual load patterns and duration, probable curing and construction sequences, etc., all of which determine actual deflection. The computed deflections require judgment of the structural engineer to determine adjustments necessary for use in preparing estimates of actual probably deflection.

9.1.9 Tall Building Motions

The design of tall, slender buildings can be strongly influenced by the need to keep the wind-induced motions within levels that are acceptable to the occupants. Perception of building motions under the action of wind can be described by various physical quantities including maximum values of velocity, acceleration, and the rate of change of acceleration often referred to as *jerk*. Human response to motion in buildings is a complex phenomenon involving several psychological and physiological factors. It is unlikely that human beings are directly sensitive to velocity if isolated from visual effects because once in motion with a constant velocity, no forces act on the human body to keep it in such motion. Acceleration, on the other hand, requires a force to act that stimulates various body organs and senses. This changing acceleration is an important component of motion perception in tall, slender buildings and has become the widely used parameter for the evolution of motion perception in buildings. It has become the standard for comparison and establishment of motion perception guidelines of various countries and international organizations.

From motion simulator studies and full-scale experience with wind-induced motions in tall buildings, criteria are defined as limits that should not be exceeded more than once in a particular return period. The Council on Tall Buildings and Urban Habitat (CTBUH) recommends 10-year peak resultant accelerations of 0.01–0.015 g for residential buildings, 0.015–0.02 g for hotels, and 0.02–0.025 g for office buildings. Generally, more stringent requirements are suggested for residential buildings, which would have continuous occupancy in comparison to office buildings usually occupied only part of the time and whose occupants have the option of leaving the building in advance of a storm.

In the early 1980s, the target criterion used by many engineers for residential buildings was for the 10-year return period peak acceleration not to exceed 0.015 g. However, on some of the extremely slender towers, this may prove difficult to achieve structurally even after the structural designer had done all that was practically possible in terms of adding stiffness and mass. The remaining measure that could be taken is to install a supplementary damping system.

However, experience since then has shown that 0.015 g criteria can be relaxed to about 0.018–0.020 g ranges in recognition of the subjective nature and uncertainty in any such criterion involving human response.

Obtaining reliable quantitative measurements through instrumentation after construction of a building is perhaps the only way to gather evidence of the buildings' true dynamic behavior. Unfortunately, this task is quite challenging, mainly because there is resistance from many of the owners to funding such monitoring programs. Also, there is the perception that occupants may mistakenly take the presence of instrumentation and technicians in the building as a sign that that all is not quite right with the structure, and owners are very conscious of the image of their projects.

It should be noted that the prediction of peak accelerations for a given return period depends on not only wind tunnel data but also on the meteorological environment and the way the two are combined. The process by which wind tunnel and meteorological data are combined is critical in at arriving at meaningful results.

There is a trend toward setting motion criteria in terms of the 1-year rather than the 10-year return period. For example, the recently published ISO criteria are in terms of the 1-year return period, because it is felt that human comfort issues need to be assessed based on more common events than once in 10 years. However, past experience from the 10-year return period can still serve as a useful benchmark when using criteria based on shorter return periods appropriate for rescaling of the results.

As wind separates and flows around a building, vortices are formed at the windward edge of the building. These vortices then roll in an alternating pattern along the sides of the building. When the frequency of the vortex shedding approaches the natural frequency of the building, the building will move in tune with the vortices. Vortex shedding is considered a problem when the accelerations caused by the motions disturb or startle the building occupants. A second concern is long-term fatigue if the deflections and their frequency of occurrence are excessive.

Because of the catastrophic consequences, the flutter characteristics of a long-span bridge are without questions, the most critical wind engineering issue. Flutter occurs when the wind forces and the dynamic characteristics of a structure combine in such a way that the wind feeds energy into the oscillations. The bigger the oscillations, the more energy is fed into them. As there are no limits to these motions, the cycle may continue until the structure ultimately fails.

The failure of the Tacoma Narrows Bridge in 1940 is a classic example. It brought to the foreground the importance of wind in the design of long-span bridges. The bridge collapsed in 40 mph winds due to an aerodynamic instability known as flutter. Wind tunnel testing formed an important part of the investigation into that failure and has since become an integral component of long-span bridge design.

Oscillation of a structure in wind can rapidly lead to structural failure if the oscillations grow to large amplitude. Certain types of oscillation, while they are not large enough to cause structural problems, may cause problems of human discomfort in tall buildings and on long-span bridges.

Building motions under action of commonly occurring winds need to be kept within comfortable limits for the building occupants. For lightweight modern structures, keeping the building motions within acceptable limits can be more of a challenge than ensuring that they have sufficient structural strength.

Acceleration has emerged as the most common index of motion effects. The horizontal force felt on the human body is directly proportional to the horizontal acceleration. People are sensitive to accelerations as small as a few milli-g.

It is unrealistic to demand that no perceptible motions ever occur. Therefore, how much horizontal acceleration is acceptable and how often can it take place? ISO has published guidelines in terms of the RMS value of the acceleration on the top floor that should not be exceeded more than once in 5 years on average. For example, for a building with a 5 s period, the 5-year acceleration should not exceed 0.018 g. For a building with an 8 s period, a higher 5-year acceleration not exceeding 0.022 g is acceptable. These accelerations will be perceived, but if they only occur once every 5 to 10 years, the functioning and commercial viability of the building will not be adversely affected.

As stated earlier, more recently there has been a trend toward setting at the acceleration set about 30% lower. This has been prompted by the increasing number of buildings being constructed in hurricane and typhoon areas. Typically, in these areas, there is ample warning of the 5- or 10-year winds caused by these storms. Buildings are usually evacuated before the storm hits, and occupants who stay are not expecting normal comfortable conditions. Therefore, in such areas, it is more meaningful to consider the 1-year wind event.

The criteria discussed earlier apply primarily to office buildings since the most data are available for that situation. Target accelerations for residential buildings are often set 20% to 30% lower.

Long-span bridges can also move enough in wind to disturb people. Some bridges have experienced severe oscillations due to vortex shedding. A guideline frequently used in North America is that the peak acceleration not exceeds 5% of gravity, that is, 0.005 g, for wind speeds below 30 mph (13 m/s) and not above 10% of gravity, that is, 0.1 g, for winds higher than 30 mph. The British use a criterion of 0.04 g for wind below 20 m/s (45 mph) with no limit for higher wind.

These criteria are clearly much higher than those for horizontal motions in buildings. This difference is because of several factors, including

- The direction of motion.
- The outdoor environment.
- People on bridges are typically moving in vehicles and notice bridge motions less: at the higher wind speeds, the buffeting action of wind on the vehicle will be more noticeable than the bridge motions.
- At very high wind speeds, the bridge will probably be closed to traffic.

9.1.10 Building Motion Perception

A constantly buffeting wind will exert a fluctuating pressure on a building that can be approximated by a mean pressure. Such is not the case for wind gusts that buffet the building surface with fluctuating pressures. Wind gustiness is dynamic by nature and its effects, in a broad sense, may be considered similar to those resulting from earthquakes. Just as a building experiences accelerations due to sudden ground motions, so it does under gusty winds, particularly at top floors. Wind-induced oscillations of the Tacoma Narrows Bridge are a vivid example of dynamic nature of wind.

Human tolerance to wind-induced motions has become increasingly important as building structures are built to ever-increasing heights. It is known for quite some time that the prediction of dynamic properties of buildings such as building fundamental period, mode shapes, and damping, particularly for those with an aspect ratio greater than 8:1, has a great effect on the predicted accelerations. In spite of the great strides made in wind engineering over the past two decades, the subject of occupant tolerance to building motion is still not an exact science. It is rather subjective being a highly complex mix of engineering, psychology, and physiology and not well understood, even by wind engineering specialists.

Because acceleration limits have been developed from subjective criteria, it is perhaps prudent to have building owners and stakeholders participate in the selection of the acceleration design criteria. Having the owners experience building accelerations in a motion simulator such as the one at the Hong Kong University of Science may be something worth considering at the very beginning of a tall building project.

9.1.11 Structural Damping

The estimation of structural damping has been of great interest to structural engineers. This is especially the case in more recent years, since the number of tall buildings and structures throughout the world has increased dramatically. Many model and full-scale studies of these structures have been performed and are still being done today. The motivation behind this interest is usually the comfort of the occupants of the building, as well as peak loads and deflections. As buildings become more slender, the damping characteristics often become the most important variable. For these structures, the dynamic response tends to be the dominant structural action, and excessive vibrations could lead to unhappy owners and occupants alike. The majority of the structures, however, have sufficient inherent damping to control excessive vibrations with respect to the ultimate strength limit state. The additional criteria that are of more concern are the serviceability limit state. The serviceability limit state relates to the occupant comfort or the usability of the building when subjected to design loading rather than its behavior in extreme events.

The loads and deflections in the strength limit state are defined by the physical properties of the construction materials and the structural system that have been in existence for quite some time. They are well understood and used without much disagreement in practice. The serviceability limit state with respect to accelerations, however, is a more recent development. As the development of the skyscraper progressed, engineers recognized the difficult nature of describing the effect of motion on the occupants of buildings. In the early 1930s, engineers began to report on the dynamic performance of tall buildings. Heavy and stiff buildings were noted to inspire the occupants with confidence. The success of commerce throughout the world and the progression in construction methods and materials led to higher and lighter structures. The importance of defining the dynamic characteristics was noted for these taller buildings. The researchers realized that the perception of motion was subjective and depended on the individual experiencing building motions. This led to moving room experiments with people, to determine thresholds for tolerance of building accelerations without undue discomfort. These experiments consisted of a moving *room*, with occupants either sitting, standing, or lying down. Various recommendations resulted from the experiments. These experiments, however, do not fully define the experience of the occupants of a building. There are various other factors that have not been included, such as audio cues (creaking of building) and visual cues (swinging of the horizon). Further research on the combined effects of the physiological response to motion and additional audio/visual cues is continuing.

Guidelines for maximum accelerations have been suggested by many researchers and code committees. In most early skyscrapers, the dynamic effects due to the effect of wind were frequently not investigated. This did not lead to any severe problems, as the construction materials and methods of the day, made for very stiff and massive structures. The Empire State Building is a classic example. The amount of steel and masonry that it was built with is several times that of current buildings. For example, just the unit gravity structural steel in the Empire State Building is 42 psf as compared to the 33 psf for Sears Tower (now called Willis Tower) and 29 psf for the John Hancock Tower, both in Chicago, Illinois. This large quantity of material used in the earlier skyscrapers led to a higher stiffness, as well as contribution to the damping from masonry infills and non-load-bearing partitions. This reduced the dynamic amplification of the wind-induced response and as a result, occupant comport was achieved without much effort from engineers. As buildings became more flexible and lighter, however, more attention to the dynamic amplification was required.

If a structure suffers from excessive vibrations, that is, peak accelerations that are higher than the recommended threshold, the most effective way of reducing the vibration is by adding damping. There are two other ways of modifying the dynamic characteristics, by changing the mass and stiffens, but increasing the damping is generally the only economically viable alternative. In present-day wind-sensitive tall buildings, due to the method of construction and/or design, there is not enough inherent damping in the structural system itself. This led to the notion of increasing the total damping by additional damping.

There are many different systems of adding damping to a structure. They can be grouped in two classes: active systems and passive systems. Active systems rely on the input of energy from a mechanical and electrical system to effectively add damping. A TMD, such as was installed in the First National City Corporation headquarters in New York City, is an example of such an active system. This system requires several thousand kilowatts of electricity to operate the damper. Passive systems, on the other hand, depend only on the dynamic characteristics of the structure and the damper–structure interaction to provide additional damping. The grouping of passive systems can be further broken into two further classes. The first subclass is those systems that change the nature of the response of the structures. One example of this type would be sloshing water damper, which modifies the mechanical characteristics of the structural system. The second subclass would be those systems that remove energy from the system by the adding friction. A viscoelastic damper is an example of a damper that removes energy by the shearing action of a layer(s) of rubberlike material sandwiched in a mechanism attached to the structure at various locations.

Various buildings and structures throughout the world have added damping system installed in them. The First National City Corporation building in New York City is a 910 ft tall building that is elevated above a plaza with four 112 ft columns. Early in the design stage, it was realized that additional damping would be required. A TMD was used to reduce vibrations to an acceptable level. The now nonexistent WTC Towers in New York City had added dampers as well. Nearly 10,000 viscoelastic dampers had been installed in each tower of the WTC.

The effect of damping is less dominant for seismic response than it is for dynamic wind response. As mentioned previously, the choice of damping parameter is not easy, if only because there have been very few measurements on tall buildings. The current recommendations are based on measurements made over many years on relatively less tall buildings.

Based on a study of the Japanese database of published and unpublished measurements, it appears that the inherent damping of buildings has high variability with a trend to reduce with building height.

Typical structures posses 1%–5% inherent damping contributed by nonstructural elements. Engineered damping system can be built into new construction or used for retrofit of existing structures to significantly increase the structural damping. Many damped structures have critical damping of 15%–35%. The four major types of dampers used in the United States are fluid viscous dampers, viscoelastic dampers, friction dampers, and metallic-yielding dampers. Each type has specific characteristics, advantages, and disadvantages for structural applications, and each type is made by several different manufacturers. To design with dampers, it is critical to capture the dynamic characteristics and the nonlinear behavior of the structure. This permits the designer to determine the amount of additional damping and stiffness that is required to achieve the desired performance. Engineered damping systems are, therefore, designed on the basis of performance-based design.

9.2 DAMPING DEVICES FOR REDUCING MOTION PERCEPTION

Engineers have learned from building occupants and owners, and from wind tunnel studies, that designing a tall building to meet a given drift limit under code-specified equivalent static loads is not enough to make occupants comfortable during windstorms. However, they have only limited control over three intrinsic factors, namely, the height, the shape, and the mass, that influence the dynamic response of buildings. Additionally, the behavior of a tall building subjected to dynamic loads such as wind or seismic activity is difficult to predict with any accuracy because of the uncertainty associated with the evaluation of a building's damping and stiffness, as well as the complicated nature of loading.

The present state of the art is such that an estimate of structural damping can be made with a plus or minus accuracy of only 30% until the building is constructed and the nonstructural elements are fully installed. It is well known that wind-induced building response is inversely proportional to the square root of total damping, consisting of aerodynamic plus structural damping. So, if damping is quadrupled (increased by four times), a 50% response reduction is achieved, and if damping is doubled, the dynamic response is reduced by 29%. Because of the inherent damping of a building responding elastically to wind loads in the range of 0.5%–1.5% of the critical response, it is impractical to increase the damping to, say, four times as much by use of modified structural materials.

Suppression of excessive vibrations can be dealt with limited success in a variety of ways. Additional stiffness can be provided to reduce the vibration period of a building to a less sensitive range. Changes in mass of a building can be effective in reducing excessive wind-induced excitation. Aerodynamic modifications to the building's shape, if agreeable to the building's owner and architect, can result in reduced vibrations caused by wind. However, these traditional methods can be implemented only up to a point beyond which the solutions may become unworkable because of other design constraints such as cost, space, or aesthetics. Therefore, to achieve reduction in response, a practical solution is to supplement the damping of the structure with a mechanical damping system external to the building's structure.

9.2.1 Passive Viscoelastic Dampers

Figure 9.1 shows schematics of a viscoelastic polymer damper. An early example of application of this type of damper is the now nonexistent WTC Towers, conceived in the 1960s, constructed in the early 1970s, and destroyed by terrorists on September 11, 2001. These buildings were designed with viscoelastic dampers distributed at approximately 10,000 locations in each building. The dampers extended between the lower chords of the floor joists and gusset plates mounted on the exterior columns beneath the stiffened seats (Figure 9.2).

Viscoelastic dampers (Figure 9.3) dissipate energy through deformation of polymers sandwiched between relatively stationary steel plates. Their energy dissipation depends on both relative shear deformation of the polymer and relative velocity within the device. The device is typically used to reduce occupants' perception of wind-induced motions. It does not require constant operational monitoring and is not dependent on electric power.

The Columbia Seafirst Center, a 76-story building in Seattle built in 1984, is another example of using this technology to reduce occupant perception of wind-induced building motion.

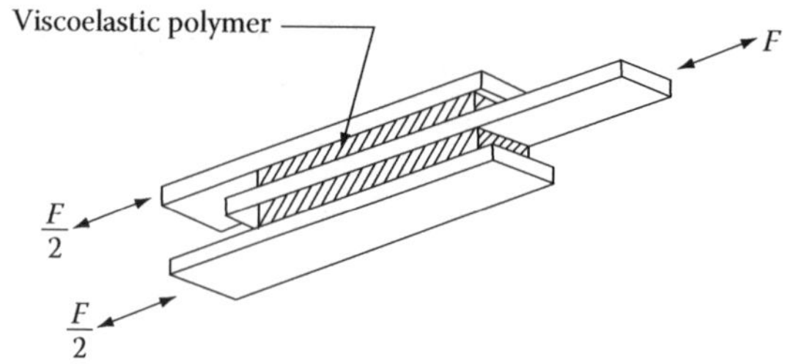

FIGURE 9.1 Schematics of viscoelastic damper.

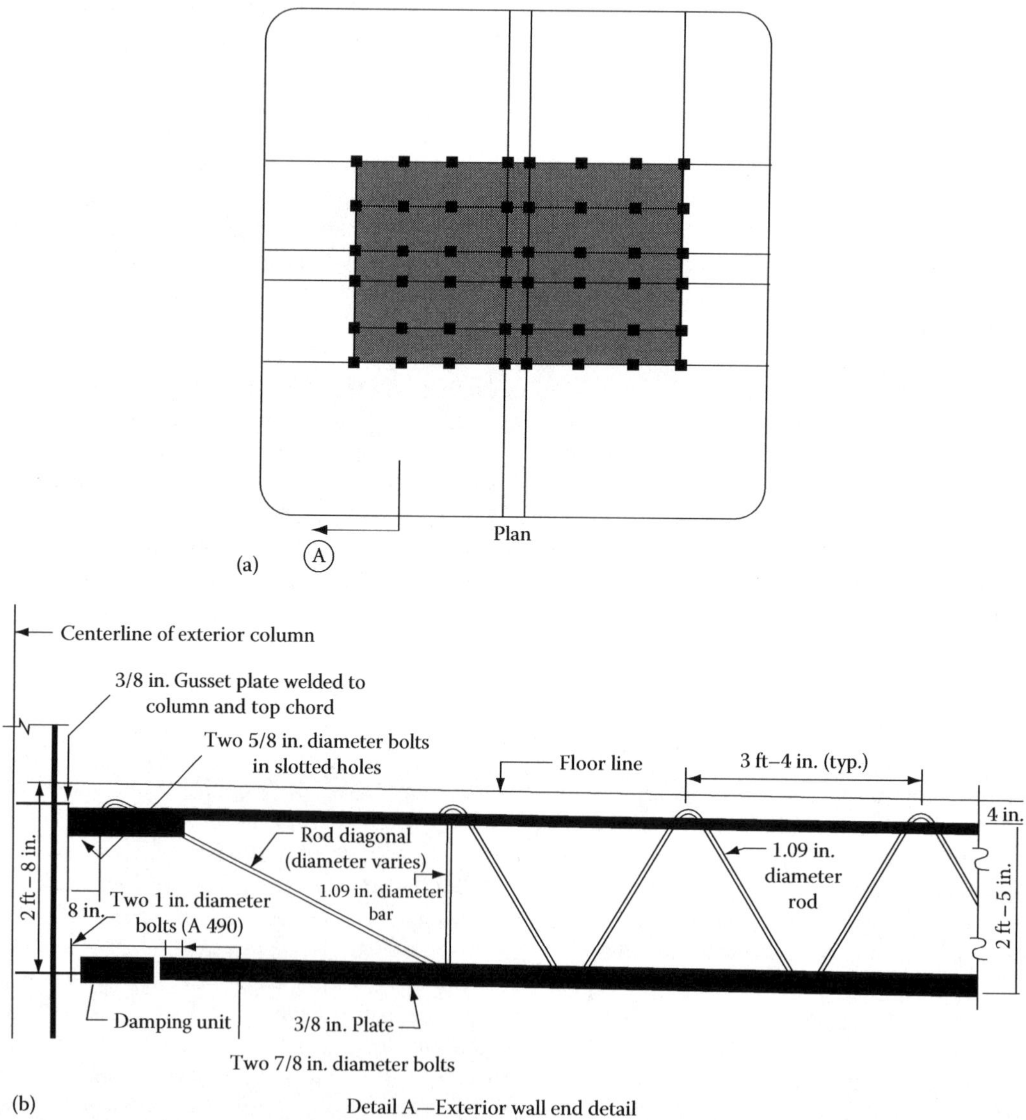

FIGURE 9.2 Viscoelastic dampers in World Trade Center towers.

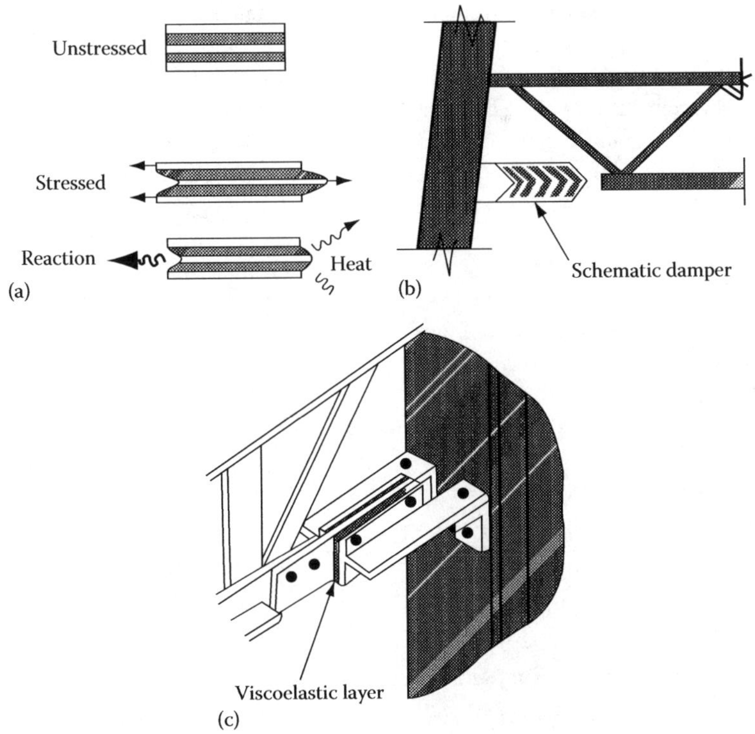

FIGURE 9.3 Viscoelastic dampers, schematics.

The dampers used in this building consist of steel plates coated with a polymer compound. The plates are sandwiched between a system of relatively stationary plates. As the building sways under the action of wind loads, the steel plates that are attached to structural members are subjected alternately to compression and tension. In turn, the viscoelastic polymer subjected to shearing deformations absorbs and dissipates much of the strain energy into heat, thus reducing wind-induced motions.

9.2.2 Tuned Mass Damper

A typical application of a TMD consists of a heavy mass installed near a building's top in such a way that it tends to remain still while the building moves beneath it. This strategy allows the mass at top to transmit its inertial force to the building in a direction opposite to the motions of the building itself, thereby reducing the building's oscillations.

The mass itself need weigh only a small fraction—0.25%–0.7%—of the building's total weight, which corresponds to about 1% to 2% of first modal mass. *Tuned* simply means the mass can be adjusted to move in a fundamental period equal to the building's natural period so that it will be more effective in counteracting the building oscillations. In addition to the initial tuning when it is first installed, the TMD may be fine-tuned as the building period changes with time. The period may increase as the building occupancy changes, as nonstructural partitions are added, or as elements contributing nonstructural stiffness *loosen up* after initial wind storms. Thus, a TMD may be considered as a small damped mass of SDOF system riding *piggyback* atop a building. Although its mass is a small fraction of the building's mass, its vibration characteristics are adjusted to mimic those of the building's. For example, if a tall building sways, say, 24 in. to the right at a fundamental frequency of 0.16 Hz, the TMD is designed to move to the left at the same frequency. The idea of using the inertia of a floating mass to tame the sway of a tall building is not entirely new. In fact, the invention of the TMD as an energy-dissipative vibration absorber is credited to Frahm, who developed the concept in 1909. The theory was later described by Den Hartog in his classic textbook in 1956 and since then has been applied in automotive and aircraft engines to reduce vibrations. Since the wind force–time relationship is not harmonic (sinusoidal), the basic ideas developed by Den Hartog have been modified in building applications to account for the random nature of wind. When activated during windstorms, the TMD becomes free floating by rising on a nearly frictionless film of oil. To dissipate energy, the TMD must be allowed to move with respect to the building. In the earlier TMDs installed in tall buildings, springlike devices connecting the mass to the building pull the building back to center, as the building sways away from its equilibrium position. The mass is also connected to the building with a damping device, in the form of a hydraulic actuator, which is controlled to provide a predetermined percentage of critical damping. This limits the lateral displacements of the mass relative to the building.

The TMD's advantages become academic in a power failure. It needs electricity to work and if that's lost in a heavy windstorm, when the TMD would most be needed, it wouldn't work. So it is advisable to have the TMD wired to an emergency power system.

In a very tall building, during a major wind storm, the mass will move in relation to the building some 2–5 ft. The system is controlled to activate when a predetermined building lateral acceleration occurs. This motion is registered on an accelerometer and, if the allowable limit is reached, the mass is activated automatically.

9.2.2.1 Tuned Mass Damper: Simple Pendulum Type

A passive TMD is a mass that is supported by a pendulum arrangement ample right, which is designed to reduce building motions by applying inertia and damping forces opposite to the direction of building motions.

The mass can be made of any material (typically concrete and/or steel), while damping is typically provided by viscous damping devices (such as large shock absorbers).

The space envelope for a TMD is a combination of the physical size of the TMD mass plus the additional space required to accommodate the necessary amplitude of the swinging mass, the supporting structure, and the viscous damping devices.

9.2.2.2 Tuned Mass Damper: Linked Pendulum Type

The main advantage of a linked pendulum TMD arrangement is that it requires a smaller vertical space envelope than a simple pendulum TMD. It can also be tuned to a wider range of structural frequencies.

9.2.2.3 Citicorp Tower, New York

The Citicorp Tower shown in Figure 9.4 consists of a unique structural system of perimeter-braced tubes elevated on four 112 ft high columns and a central core. It rises approximately 914 ft above grade. The tower is square in cross section with plan dimensions of approximately 157 by 157 ft The top 140 ft portion of the tower slopes downward from north to south.

The TMD designed for the building consists of a concrete block 29 × 29 × 9 ft that weighs 410 tons (820 kips). It is attached to the building with two nitrogen-charged pneumatic spring devices and two hydraulic actuators that are controlled to provide damping to the TMD and linearize the *springs*. One set counters north–south building dynamic motion and the other set counters east–west motion. The spring stiffness, and thereby the TMD frequency, is adjusted (tuned) by changing the pneumatic pressure. It also has an anti-yaw device to prevent twisting of the block and snubbers to prevent excessive motion of the block.

The TMD is capable of a 45 in. operating stroke in each orthogonal direction. The operating period is adjustable independently in each axis. The mass block is supported with twelve 22 in. diameter pressure-balanced bearings connected to a hydraulic pump.

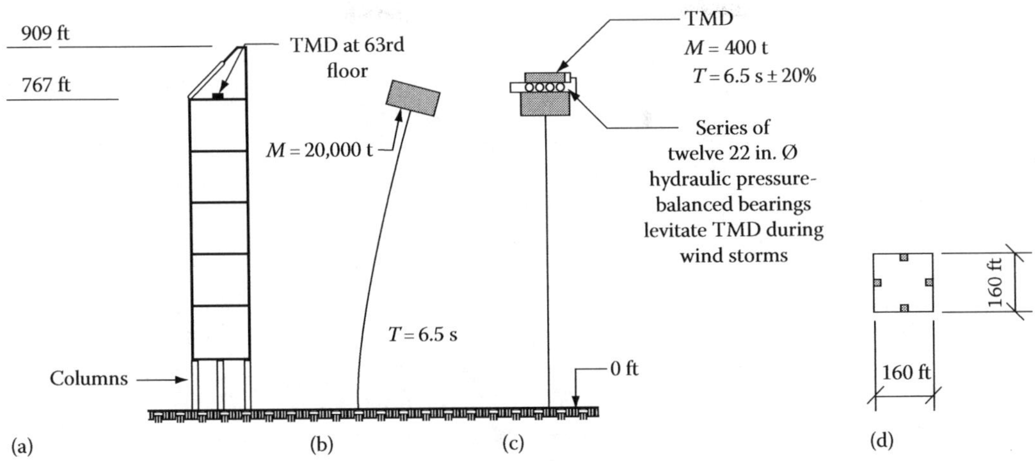

FIGURE 9.4 Tuned mass damper (TMD) for Citicorp, New York: (a) building elevation, (b) plan, (c) first-mode response, and (d) TMD atop the building.

The block positioned at the building's 63rd floor (780 ft high) represents approximately 2% of first-period modal mass of the building. The motions of the block are controlled by pneumatic devices and servo hydraulics resulting in a system that has the characteristics of a spring–mass damper system, as shown schematically in Figure 9.5.

To dissipate energy, the TMD is allowed to move with respect to the building. It is continuously on standby and is designed to start up automatically whenever the accelerations exceed a predetermined value. The TMD is activated whenever the accelerations for two successive cycles of

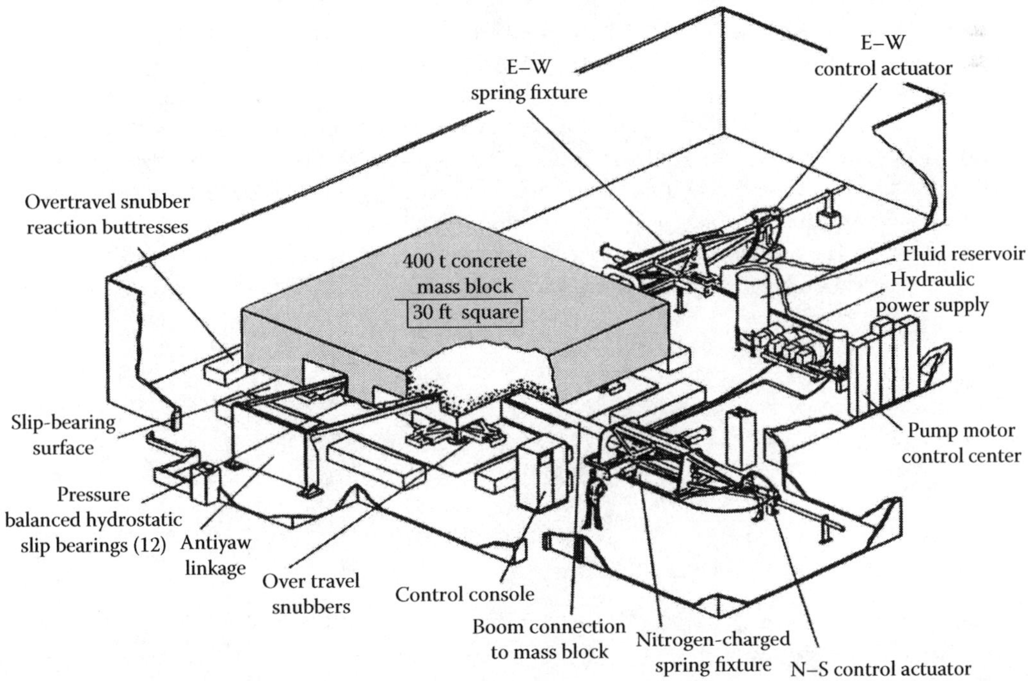

FIGURE 9.5 Schematics of TMD, Citicorp Center, New York.

building motion exceed 0.003 g (0.001 g = 1/1000 of acceleration due to gravity; therefore, 0.003 g corresponds to an acceleration of approximately 1.16 in./s^2).

The system continues to operate as long as building motions continue and stops only a half hour after the last pair of building cycles for which maximum acceleration is greater than 0.0075 g. The TMD provides the building with an effective structural damping of about 4% of critical. This is a significant increase above the inherent damping estimated to be just less than 1% of critical. Since wind-induced accelerations of a building are approximately proportional to the inverse of the square root of the damping, when in operation, the TMD reduces the building sway oscillations by over 40%.

The Citicorp TMD is installed on the 63rd floor. At this elevation, the building may be represented by an SDOF system with a modal mass of 40,000 kips resonating biaxial at a 6.8 s period with a critical damping factor of 1%. The TMD is designed with a moving mass of 820 kips, biaxial resonant with a period of 6.7 s plus or minus 20%, and an adjustable damping of 8%–14% of critical. Observe that the moving mass represents approximately 2% of the first-period modal mass, which typically corresponds to about 0.6%–0.7% of the total mass.

9.2.2.4 John Hancock Tower, Boston, MA

The TMD for the John Hancock Mutual Life Insurance Co.'s glass-clad landmark in Boston is somewhat different from that for Citicorp Tower. It was added as an afterthought to prevent occupant discomfort. Second, Hancock Tower is rectangular in plan and consists of moment frames unlike Citicorp's diagonally braced frame (Figure 9.6). Because of the building's shape, location, and vibration properties, its dynamic wind response is mainly in the east–west direction and in torsion about its vertical axis. There is a TMD near each end of an upper floor. They are tuned to a vibration period of approximately 7.5 s. The total east–west moving mass represents about 1.4% of the building's first-mode generalized mass, while in the twist direction, the moving masses represent about 2.1% of the building's generalized torsional inertia. The dampers, then, move only in an east–west direction and work together to resist sway motions in the short direction or in opposition to stabilize torsional rotations of the building. They are located 220 ft apart and when moving in opposition act in effect as a 220 ft lever arm to resist twisting. Hancock's dampers each have a 300-ton mass consisting of lead blocks contained in a steel coffer box. They also activate at 0.003 g of acceleration. In operation, the masses may move up to six feet with an operating cycle of about 7.5 s. Each mass block is supported on sixteen 22 in. diameter pressure-balanced bearings connected to a hydraulic pump. The TMDs in both the Citicorp and John Hancock towers are used only to assure occupants' comfort. Their beneficial effects in reducing wind-induced dynamic forces are not relied upon for structural integrity under extreme wind loads.

Both the John Hancock Tower and Citicorp Tower TMDs are called passive-powered because, although the reduction in the buildings sway response comes from the initially force of the dampers,

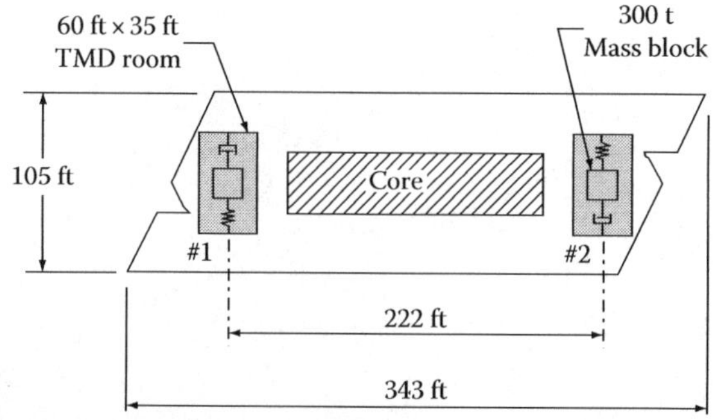

FIGURE 9.6 Dual TMD: John Hancock Tower, Boston, MA.

initially, power is required to activate the masses. The sliding masses installed in these towers cannot move until their oil bearings are pressurized to levitate the masses.

9.2.2.5 Design Considerations for Tuned Mass Damper

There are a number of practical considerations in the design of the TMD. One of these is the need to limit the motions of the TMD mass under very high wind loading such as will occur in the design storm or under ultimate load conditions. One way of doing this is to use a nonlinear hydraulic damper in the TMD. By employing such a damper, the motions of the TMD mass can be greatly reduced under very high wind loading conditions or under strong seismic excitation. A further safeguard against excessive TMD motion is to install hydraulic buffers around the mass. When the mass comes into contact with the buffers, high velocities are quickly reduced.

Both the Citicorp and John Hancock TMD systems have sensors and feedback and electronic control systems, but these were designed to make the TMD operate like a passive TMD. TMDs can in principle be readily converted to be an active system by incorporating sensors and feedback systems that can drive the TMD mass to produce more effective damping than is possible in a purely passive mode. As a result, a larger effective damping can be obtained from a given mass. This approach has been used in several commercially available ready-to-install systems. The TMD is thus made more efficient, a benefit to be weighed against the increased cost, complexity, and maintenance requirements that are entailed with an active system.

9.2.3 Tuned Sloshing Damper

A tuned sloshing damper (TSD) is a dynamic vibration absorber, which utilizes the motion of shallow liquid in a partially filled container for absorbing and dissipating vibration energy.

The liquid free surface wave is designed to have a frequency near the fundamental structural frequency of the building. When the building moves, the liquid resonates out of phase with the structure and energy is dissipated from the liquid by flow-damping devices such as screens and/or posts in the container. The frequency of the liquid is determined by the characteristic length and depth of the liquid.

Different shapes of container, such as rectangular or circular, can be used in TSD implantations. A rectangular-type tank can be tuned to two different frequencies in two orthogonal directions.

A simple sloshing type of damper consists of a tuned rectangular tank filled to a certain level with water. The tuning of the system consists of matching the tank's natural period of wave oscillation to the building's period by appropriate geometric design of the tank. If obstacles such as screens and baffles are placed in the tank, dissipation of the waves takes place when water sloshes across these obstructions resulting in a behavior similar to that of a TMD, and the result is again that the tank behaves as a TMD. However, analysis indicates that a sloshing water tank does not make as efficient use of the water mass as a tuned liquid column damper (TLCD).

9.2.4 Tuned Liquid Column Damper

For dynamically sensitive structures, structural dynamic performance can be highly dependent upon the initial assumption about the inherent (natural) damping in the structure. This energy absorption and dissipation characteristic is key to reducing dynamic effects as it increases the structure's ability to absorb the energy imparted to it from external excitation. Damping of structures is generally expressed as a ratio to *critical damping*, the energy absorption value that would cause the structure to simply return to center without oscillating if pulled to one side and released.

In structural design practice, the inherent damping value is chosen based on average values recommended in technical literature, codes, and standards. These values typically are assumed to be in the range of 1%–2% of critical damping for buildings. However, since each structure is unique, the actual inherent damping exhibited by the finished structure is also unique due to specific architectural and structural system layout, structural detailing, and cladding. Changes of only 0.5% of critical or less in the assumed level of damping can have a significant impact on wind-induced loads and motions of a structure.

One proven way to achieve a specific overall damping level is to incorporate an SDS into the structure.

An SDS is essentially a supplemental energy dissipation system that is designed to absorb vibration energy from a structure, thereby reducing energy dissipation demand on the structure. There are many ways to add energy absorption to a structural system. Technologies in common use today can be broadly classed as distributed, impact, active mass, semiactive mass, mechanical passive tuned mass, and liquid-based passive tuned mass. The trend has been to use passive damping systems, such as SDS, as these systems need no control electronics or powered drive mechanism and can be counted on to absorb energy when needed most, even during storms when power outages are most likely to occur.

Liquid TMDs typically have one of two forms, TLCDs or TSDs. The focus here is on TSDs due to their attractive qualities of simplicity, low cost, and dependability with little or no maintenance.

When the structure begins to move under wind forces, the liquid resonates out of phase with the structure, and energy is dissipated from the liquid by flow-damping devices such as screens, louvers, or posts in the tank that resist the wave action. See Figure 9.7 schematics. Different shapes of tanks, such as rectangular or circular, can be used in TSD to achieve certain goals. A rectangular tank can be tuned to two different frequencies in two orthogonal directions. The tank size for a 60-story building would be 28 ft long, 24 ft wide, and 15 ft deep, and nominal water depth would be about 3 ft with a frequency of 0.185 Hz and 0.20 Hz in the x and y directions. A TSD utilizes liquid waves to absorb energy from vibrating structures through wave travel and viscous action in a partially filled tank of liquid. The tank is designed so that the liquid surface wave has a frequency *tuned* to be near the fundamental frequency of the building. The frequency of the liquid is determined by the density, length, width, and depth of the liquid.

A TSD is an efficient SDS option for structures with a moderate need for additional damping. For typical midrise to high-rise structure, total effective damping levels of 3%–4% of critical can often be achieved. TSDs are attractive due to their qualities of simplicity, low cost, and dependability with little or no maintenance. A passive TLCD is a mass of liquid (typically water), enclosed in a custom U-shaped tank, which is designed to reduce building motions by applying inertial and damping forces opposite to the direction of building motions. The U-shaped tank is designed to allow the liquid to oscillate freely at a frequency that optimally matches one (or more) of the structure's natural frequencies. Damping is provided by adjusting the turbulence levels in the moving water.

Generally, a single TLCD is capable of providing damping along a single building axis.

A TLCD is in many ways similar to a TMD that uses a heavy concrete block or steel as the tuned mass. The difference is that the mass is now water or some other liquid. The damper is essentially a tank in the shape of a U. It has two vertical columns connected by a horizontal passage and filled up to a certain level with water or other liquid. Within the horizontal passage, screens or a partially closed sluice gate is installed to obstruct flow of water, thus dissipating energy due to motion of water. The TLCD is mounted near the top of a building, and when the building moves, the inertia of

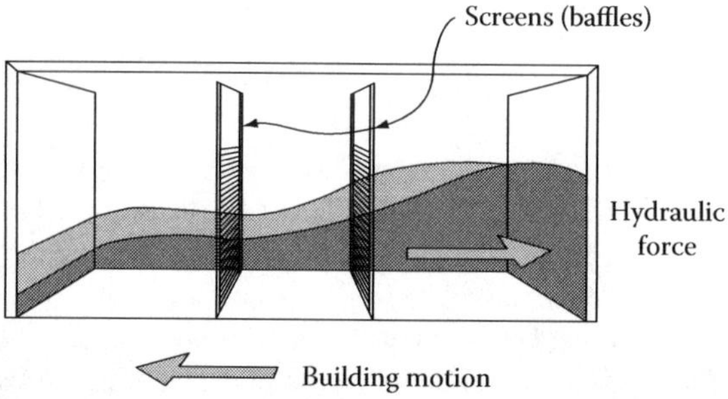

FIGURE 9.7 Sloshing water damper, schematics.

the water causes the water to oscillate into and out of the columns, traveling in the passage between them. The columns of water have their own natural period of oscillation, which is determined purely by the geometry of the tank. If this natural period is close to that of the building's period, then the water motions become substantial. Thus, the building's kinetic energy is transferred to the water. However, as the water moves past the screens or partially opens sluice gate in the horizontal portion of the tank, the drag of these obstacles to the flow dissipates the energy of the motion. The end result is added damping to reduce building oscillations.

9.2.4.1 Wall Center, Vancouver, BC

Shown in Figure 9.8 is the plan for the mechanical penthouse of a building called Wall Centre, a 48-story residential tower in Vancouver, British Columbia. From wind tunnel tests, predicted 10-year accelerations were in the range of 0.04 g, depending on the structural systems considered in the preliminary design. To minimize occupants' perception of motion due to wind excitations, a limit of 0.015 g was chosen as the design criterion for a 10-year acceleration. A damper using water serves a dual purpose by also providing a large supply of water high up in the tower for fire suppression. Initially, a sloshing water damper was considered but the TLCD was found preferable due to its greater efficiency in using the available water mass. The design turned out to be a remarkably economical solution considering the saved cost of having to install a high-capacity water pump and emergency generator in the base of the building as initially required by fire officials. The total mass required was on the order of 600 tons that corresponds to a large volume of water. However, sufficient space was available. Also a helpful factor was that the motions of the tower were primarily in one direction only. Therefore, only motions in one direction needed to be damped, which simplified the design. Figure 9.9 illustrates the TLCD design consisting of two identical U-shaped concrete tanks. Since the building was concrete, it was relatively easy to incorporate the tanks into the design and to construct them as a simple addition to the main structure. The structural design is by Glotman Simpson Engineers, Vancouver, British Columbia, Canada. The design of the TLCD is by Rowan, Williams, Davis, and Irwin, Inc., Guelph, Ontario, Canada.

9.2.4.2 Highcliff Apartment Building, Hong Kong

Another example of a tall building that uses TLMD to control accelerations and provide enhanced structural performance during typhoon conditions is the 73-story Highcliff apartment building in Hong Kong, one of the windiest places on earth (Figure 9.10). The building soars to a height of 705 ft (215 m) with an astonishing slenderness ratio of 20:1. A unique structural system that incorporates all vertical elements as part of the lateral system, in combination with a series of tuned liquid mass dampers, ensures the safety and comfort of the building occupants.

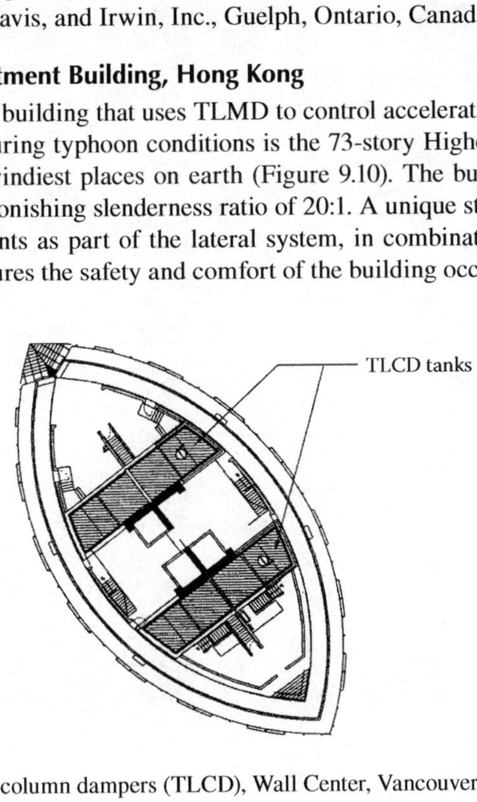

FIGURE 9.8 Tuned liquid column dampers (TLCD), Wall Center, Vancouver, BC.

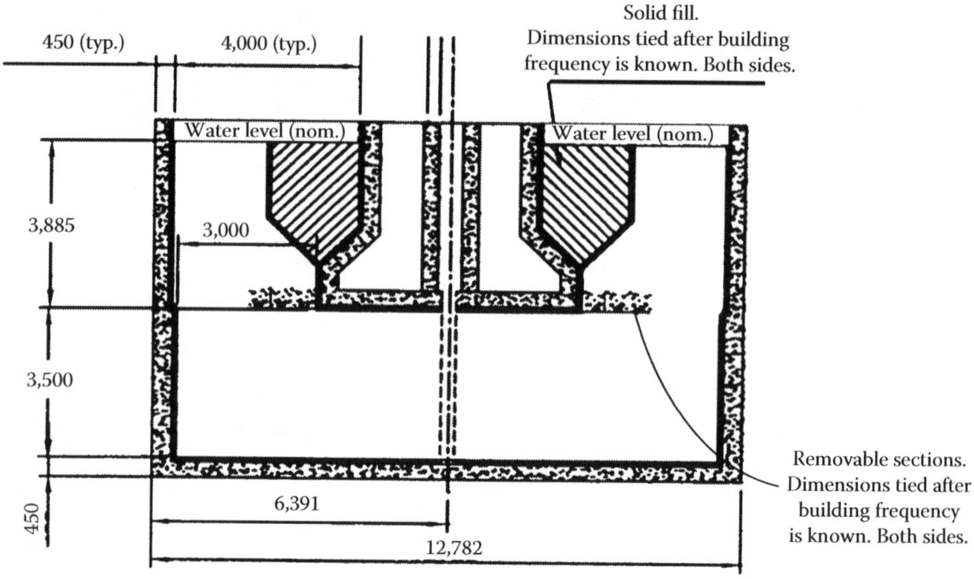

FIGURE 9.9 TLCD, Wall Center, Vancouver, BC.

FIGURE 9.10 Highcliff Apartments, Hong Kong.

(a) (b)

FIGURE 9.11 (a) Simple pendulum damper and (b) hydraulic dampers attached to mass block.

9.2.5 Simple Pendulum Damper

The principal feature of the system shown in Figure 9.11 is a mass block slung from cables with adjustable lengths. The mass typically represents approximately 1.5%–2% of the building's generalized mass in the first mode of vibration. The mass is connected to hydraulic dampers that dissipate energy while reducing the swinging motions of the pendulum.

The adjustable frame is used as a tuning device to tailor the natural period of vibration of the pendulum. The frame can be moved up and down and clamped on the cables to allow the natural period of the pendulum to be adjusted. The mass is connected to an anti-yaw device to prevent rotations about a vertical axis. Below the mass, there is a bumper ring connected to hydraulic buffers to prevent travel beyond the hydraulic cylinder's stroke length.

9.2.5.1 Taipei Financial Center

An example of a tuned mass pendulum damper (TMPD) architecturally expressed as a building feature is shown in Figure 9.12. The Taipei Financial Center has 101 stories and a height of

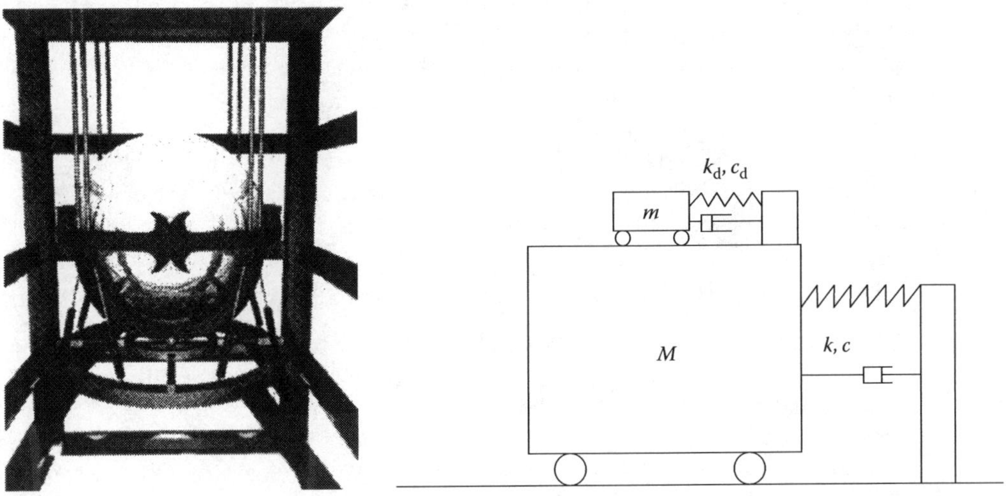

(a) (b)

FIGURE 9.12 (a) Spherical tuned mass pendulum damper (TMPD), Taipei Financial Center; (b) schematic.

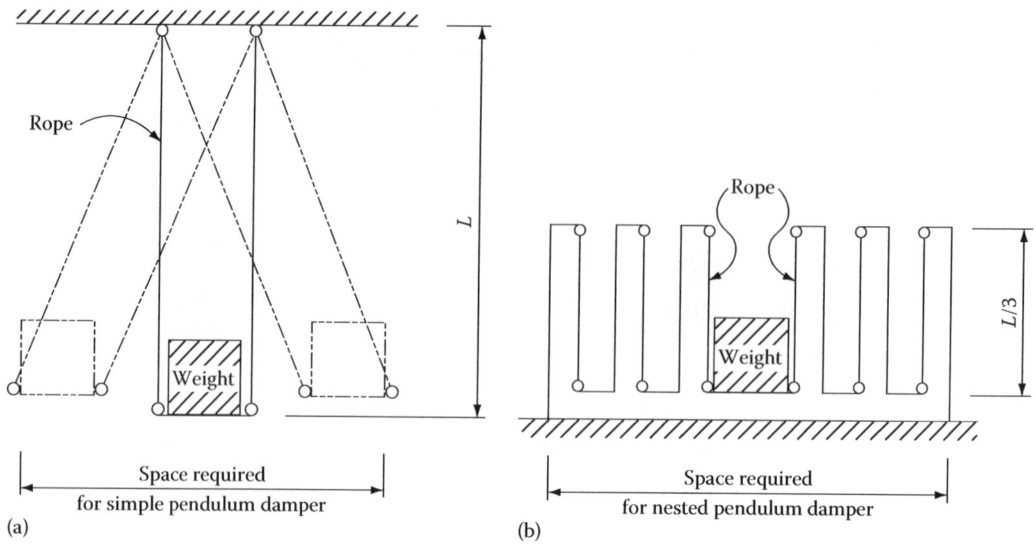

FIGURE 9.13 (a) Simple pendulum damper and (b) nested pendulum damper.

1,667 ft (508m). A special space has been allocated for the TMPD near the top of the building and people will be able to walk around it and view it from a variety of angles. The TMPD, consisting of a 730-ton steel ball, will be brightly colored, and special lighting effects are planned. The architecture of the building is by C.Y. Lee and Partners, Taiwan; structural engineering is by Evergreen Consulting Engineering, Inc., Taipei, Taiwan, and Thornton Tomasetti Engineers, New York; and the design of the TMPD is by Motioneering, Inc., a company in Ontario, Canada, that specializes in designing and supplying damping systems for dynamically sensitive structures.

9.2.5.2 Nested Pendulum Damper

In situations where the height available in a building is insufficient to allow installation of a simple pendulum system, a nested TMD may be designed as illustrated in Figure 9.13. The design shown is for a North American residential tower. The total vertical space occupied by the damper, which has a natural period of about 6 s and a mass of 600 tons, is less than 25 ft (7.62 m), as compared to 30 ft (9.14 m) required for a simple pendulum. The design of the damper is by Rowan, Williams, Davis, and Irwin, Inc., Guelph, Ontario, Canada.

A nested pendulum damper is installed at the top of the 70-story, 971 ft tall Landmark Tower, Yokohama, Japan. The damper requires only a one-story-high space and is semiactively controlled. Wind-induced lateral accelerations are expected to be reduced by at least 60%. The damper design is by Mitsubishi Heavy Industries, Ltd., Tokyo, Japan.

9.3 SEISMIC ISOLATION

Seismic isolation is a viable design strategy that has been used for seismic rehabilitation of existing buildings and in the design of a number of new buildings. In general, this system will be applicable to the rehabilitation and design of buildings whose owners desire superior earthquake performance and can afford the special costs associated with the design, fabrication, and installation of seismic isolators. The concepts are relatively new and sophisticated and require more extensive design and detailed analysis than do most conventional schemes. In California, peer review of these new concepts is required for all designs that use seismic isolation.

Conceptually, isolation reduces response of the superstructure by *decoupling* the building from seismic ground motions. Typical isolation systems reduce seismic forces transmitted to the superstructure by lengthening the period of the building and adding some amount of damping. Added damping is an inherent property of most isolators but may also be provided by supplemental energy dissipation devices installed across the isolation interface. Under favorable conditions, the isolation system reduces drift in the superstructure by a factor of at least two—and sometimes by as much as a factor of five—from that which would occur if the building were not isolated. Accelerations are also reduced in the structure, although the amount of reduction depends on the force–deflection characteristics of the isolators and may not be as significant as the reduction of drift.

Reduction of drift in the superstructure protects structural components and elements as well as nonstructural components sensitive to drift-induced damage. Reduction of acceleration protects nonstructural components that are sensitive to acceleration-induced damage.

To understand the design principles for base-isolated buildings, consider Figure 9.14, which shows four distinct response curves A, B, C, and D. Let us examine the design of a building, say, some five stories tall, with a fixed-base fundamental period of 0.6 s. Curve A, the lowest, shows lateral design demand resulting from loads prescribed in building codes such as IBC 2005 (ASCE 7-10). Curve B, the second lowest, represents the probable strength of the structure. This strength is generally greater than the design strength because of several factors. This is because of the following:

1. Actual material strengths are almost always higher than those assumed in design.
2. Use of load factors typically overestimates the actual loads imposed on the structure.
3. Some conservatism is used in sizing of structural members.
4. Designs are often based on drift limits.
5. Members are designed to have at least some ductility. It is estimated that the probable strength of a structure designed to code-level forces is about 1.5 to 2.0 times larger than the design strength.

Curve D at top shows the forces our fixed-base building would experience if it were to remain elastic for the entire duration of a design earthquake. However, in earthquake-resistant design, it is assumed that the lateral-force-resisting system will make excursions well into the nonlinear inelastic capacities of the structural materials. Therefore, typical buildings are designed to resist only a

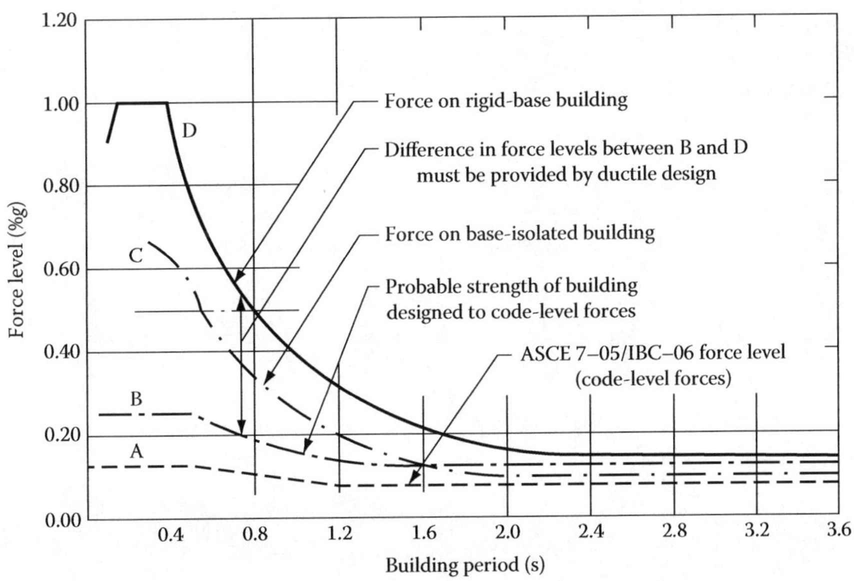

FIGURE 9.14 Design concept for base-isolated buildings.

fraction of the full linear elastic demands of major earthquakes. Heavy reliance is placed on special prescribed details to provide ductility for the extreme nonlinear inelastic demands. The difference between the linear elastic demand, curve D, and the probable capacity of the building, curve B, conceptually represents the magnitude of energy dissipation expected of the structure.

Let us compare this to the energy dissipation required of the building, if it is seismically isolated. The elastic forces experienced by a seismically isolated building are significantly reduced for two reasons. First, the flexibility of the base isolators shifts the period of the building toward the low end of the spectrum. For instance, our example building with a fixed-base period of 0.6 s would probably now have a period in the neighborhood of, say, 2–2.5 s. The drop in the elastic design force, as seen in the graph, is considerable.

The second factor contributing to the reduction in force level is the additional damping provided by the dampers. Depending on the type of base isolator and supplemental viscous damper (if any) chosen for the building, the damping may increase from a generally assumed value of 5% of critical to as much as 20% or more. Together, these two factors help to reduce the ductility demand expected of the structure during a large seismic event. In fact, it is quite likely that our base-isolated structure may never be pushed beyond its elastic limit. In other words, in the 2.0–2.5 s period range, the probable strength of the building is very nearly the same as the maximum unreduced elastic demand.

Therefore, the building need not take excursions into nonlinear inelastic range and can remain elastic for the entire duration of a design earthquake.

In simple terms, seismic isolation involves placing a building on isolators that have large flexibility in the horizontal plane (Figure 9.15). The system consists of

- A flexible mounting to increase the building period, which, in turn, reduces seismic forces in the structure mentioned earlier
- A damper or energy dissipater to reduce relative deflections between a building and the ground it rests upon
- A mounting that is sufficiently rigid to control the building lateral deflection during minor earthquakes and wind storms

Flexibility can be introduced at the base of a building by many devices. These include

- Elastomeric bearings rollers
- Sliding plates
- Cable suspension
- Sleeved piles
- Rocking foundations

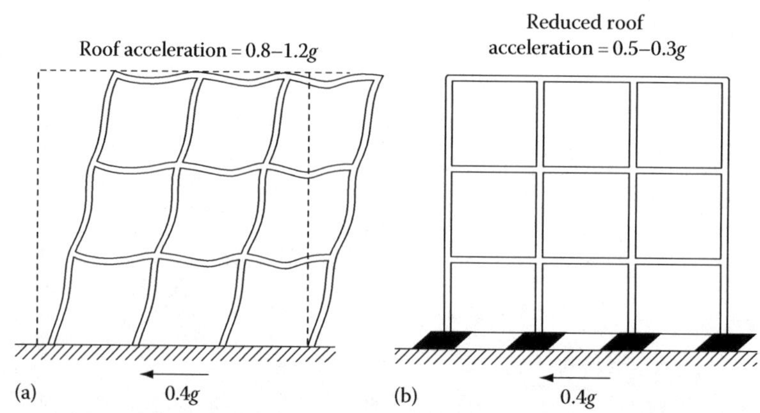

FIGURE 9.15 Comparison of response of (a) fixed-base and (b) base-isolated buildings.

However, decrease in base shear due to lengthening of a building's period comes at a price; the flexibility at the base gives rise to large relative displacements across the flexible mount, hence the necessity of providing additional damping at the base-isolation level.

While a flexible mounting is required to isolate a building from seismic loads, its flexibility under frequently occurring wind and minor earth tremors is undesirable. Therefore, the device at the base must be stiff enough at these loads, such that the building's response is as if it were on a fixed base.

Generally one isolator per column is used. However, more than one isolator may be required in certain type of buildings. For isolation of shear walls, one or more isolators are used at each end, and if the wall is long, isolators may be placed along its entire length, the spacing depending upon the spanning ability of the wall between the isolators.

9.3.1 Salient Features

It is important to consider the following features in the design of base-isolated buildings:

1. Access for inspection and replacement of bearings should be provided at bearing locations.
2. Stub walls or columns to function as backup systems should be provided to support the building in the event of isolator failure.
3. A diaphragm capable of delivering lateral loads uniformly to each bearing is preferable. If the shear distribution is unequal, the bearings should be arranged such that larger bearings are under stiffer elements.
4. A moat to allow free movement for the maximum predicted horizontal displacement must be provided around the building (Figures 9.16 and 9.17).
5. The isolator must be free to deform horizontally in shear and must be capable of transferring maximum seismic forces between the superstructure and the foundation.
6. The isolators should be tested to ensure that they have lateral stiffness properties that are both predictable and repeatable. The tests should show that over a wide range of shear strains, the effective horizontal stiffness and area of the hysteresis loop are in agreement with values used in the design.

When earthquakes occur, the elastomeric bearings used for base isolation are subjected to large horizontal displacements, as much as 15 in. or greater in a 10-story steel-framed building. They must therefore be designed to carry the vertical loads safely at these displacements.

Isolation systems should be considered if the desired performance objective is IO. Conversely, isolation will likely not be an appropriate design strategy if the desired performances objective is CP. In general, isolation systems provide significant protection to the building structure, nonstructural components, and contents but at a cost that precludes practical application when the budget and design objectives are modest.

9.3.2 Mechanical Properties of Seismic Isolation Systems

A seismic isolation system is the collection of all individual seismic isolators and may be composed entirely of one type of seismic isolator, a combination of different types of seismic isolators, or a combination of seismic isolators acting in parallel with energy dissipation devices (i.e., a hybrid system).

The most popular devices for seismic isolation in the United States may be classified as either elastomeric or sliding. Examples of elastomeric isolators include high-damping rubber bearings (HDR), low-damping rubber bearings (RB), or low-damping rubber bearings with a lead core (LRB). Sliding isolators include flat assemblies or those with a curved surface, such as the friction pendulum system (FPS).

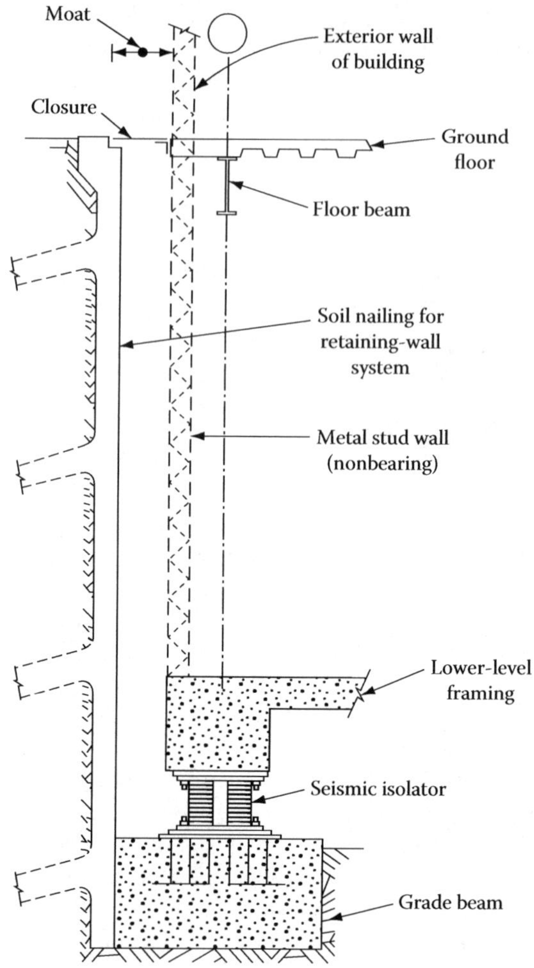

FIGURE 9.16 Moat around base-isolated building.

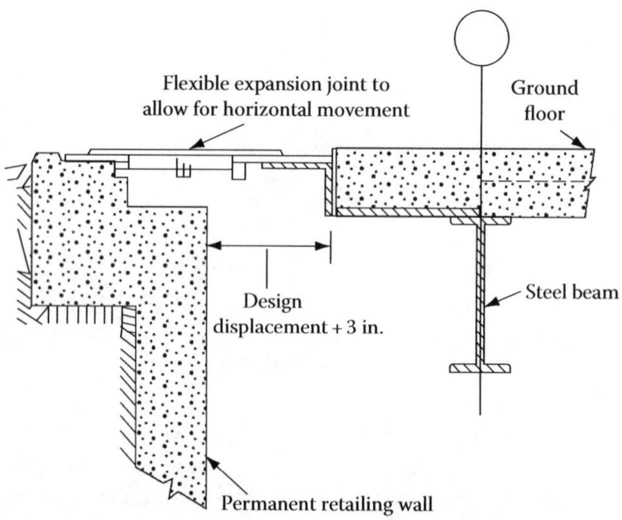

FIGURE 9.17 Moat detail at ground level.

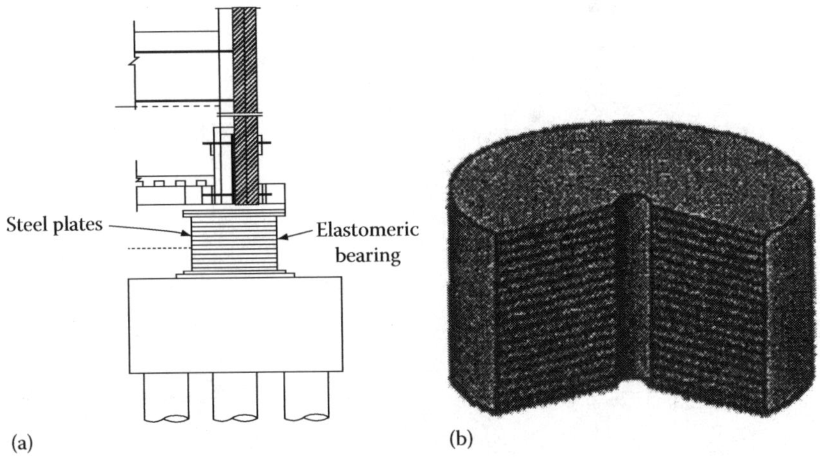

FIGURE 9.18 Elastomeric isolators. (a) High-damping rubber bearing made by bonding rubber sheets to steel plates and (b) high-damping rubber.

9.3.3 Elastomeric Isolators

Elastomeric bearings are a common means for introducing flexibility into structure. They consist of thin layers of natural rubber that are vulcanized and bonded to steel plates (see Figure 9.18). Natural rubber exhibits a complex mechanical behavior that can be described simply as a combination of viscoelastic and hysteretic behavior. Low- damping natural rubber bearings exhibit essentially linearly elastic and linearly viscous behavior at large shear strains. The effective damping is typically less than or equal to 0.07 for shear strains in the range of 0–2.0.

Lead-rubber bearings are generally constructed of low-damping natural rubber with a preformed central hole into which a lead core is press-fitted (see Figure 9.19). Under lateral deformation, the lead core deforms in almost pure shear, yields at low levels of stress of approximately 1160–1450 psi (8–10 MPa) in shear at normal temperature), and produces hysteretic behavior that is stable over many cycles. Unlike mild steel, lead recrystallizes at normal temperature (about 20°C), so that repeated yielding does not cause fatigue failure. Lead-rubber bearings generally exhibit characteristic strength that ensures rigidity under service loads.

HDRs are made of specially compounded rubber that exhibits effective damping between 0.10 and 0.20 of critical. The increase in effective damping of HDR is achieved by the addition of chemical compounds that may also affect other mechanical properties of rubber.

Scragging is the process of subjecting an elastomeric bearing to one or more cycles of large amplitude displacement. The scragging process modifies the molecular structure of the elastomer and results in more stable hysteresis at strain levels lower than that to which elastomer was scragged. Although it is usually assumed that the scragged properties of an elastomer remain unchanged with time, recent studies suggest that partial recovery of unscragged properties is likely. The extent of this recovery is dependent on the elastomer compound.

9.3.4 Sliding Isolators

Sliding isolators with either a flat or a single-curvature spherical sliding surface are typically made of PTFE or PTFE-based composites in contact with polished stainless steel. The shape of the sliding surface allows large contact areas that, depending on the materials used, are loaded to average bearing pressures in the range of 1015–10150 psi (7–70 MPa).

Sliding isolators tend to limit the transmission of force to an isolated structure to a predetermined level. While this is desirable, the lack of significant restoring force can result in significant variations in

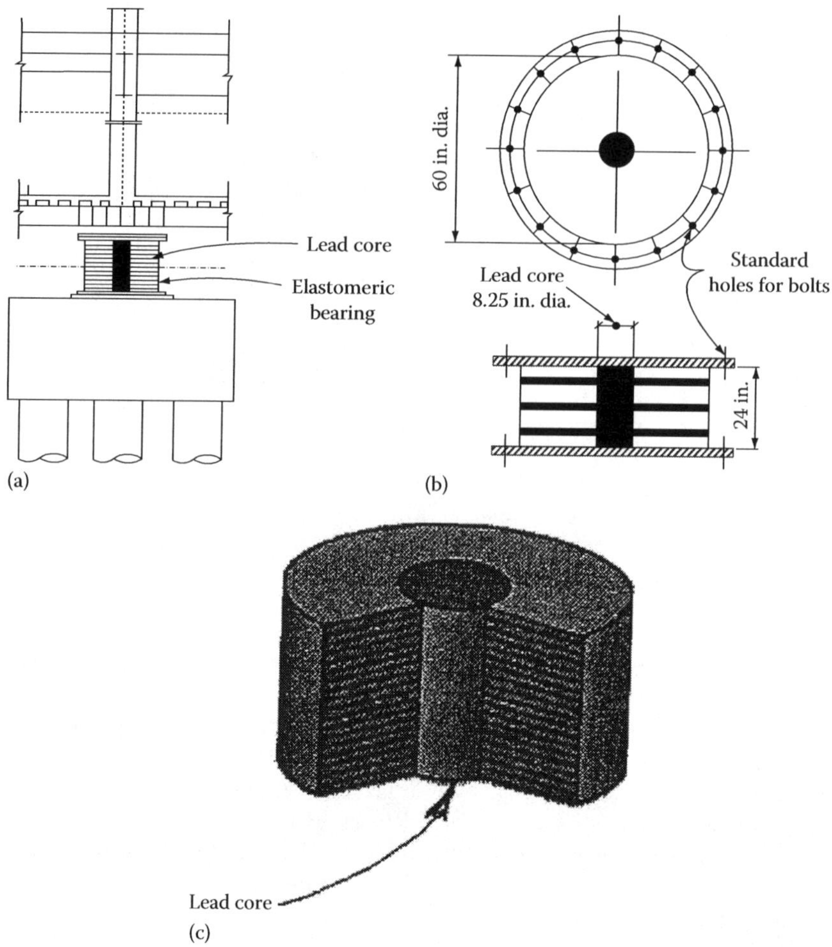

FIGURE 9.19 (a) Lead-rubber bearing under interior columns. (b) Lead-rubber bearing for an interior column for a five-story steel frame building: approximate dimensions. (c) Natural rubber bearing with press-fit lead core.

the peak displacement response and can result in permanent offset displacements. To avoid these undesirable features, sliding isolators are typically used in combination with a restoring force mechanism.

Combined elastomeric-sliding isolation systems have been used in buildings in the United States. Japanese engineers have also used elastomeric bearings in combination with mild steel elements designed to yield in strong earthquakes and enhance the energy dissipation capability of the isolation systems.

Details of a spherical sliding system commonly referred to as an FPS are shown in Figures 9.20 through 9.22. Figure 9.23 shows a schematic of a base-isolation devices acting in conjunction with viscoelastic dampers.

9.3.5 Seismically Isolated Structures: ASCE 7-10 Design Provisions

The procedures and limitations for the design of seismically isolated structures are determined considering zoning, site characteristics, vertical acceleration, cracked section properties of concrete and masonry members, seismic use group, configuration, structural system, and height. Both the lateral-force-resisting system and the isolation system must be designed to resist the deformations and stresses produced by the effects of ground motions. The stability of the vertical-load-carrying elements of the isolation system must be verified by analysis and tested for lateral seismic displacement

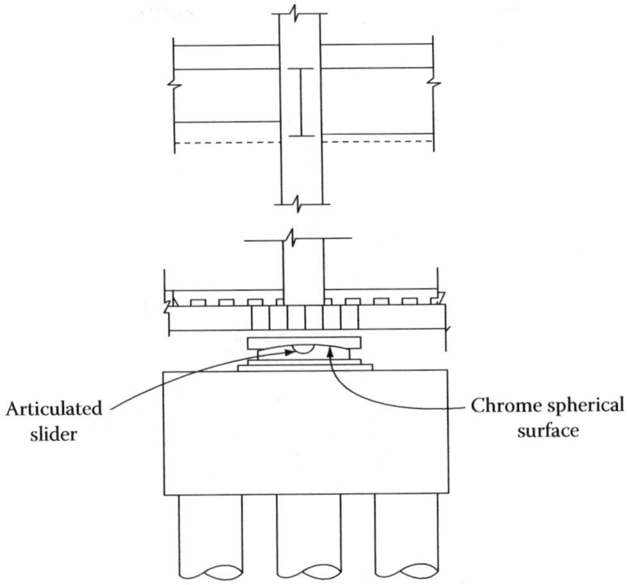

FIGURE 9.20 Sliding bearing: friction pendulum system base-isolation acting in conjunction with viscoelastic dampers.

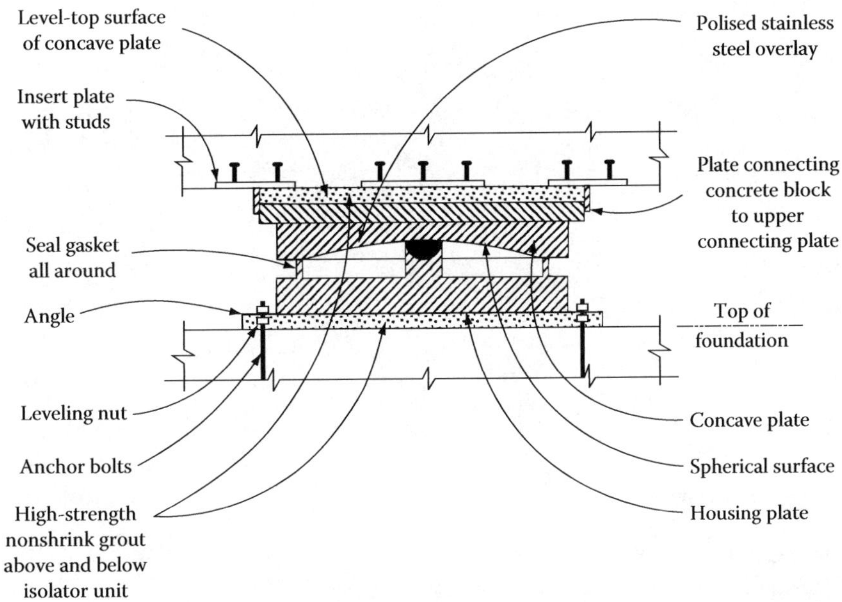

FIGURE 9.21 Friction pendulum system, details.

equal to the total maximum displacement. All portions of the structure, including the structure above the isolation system, must be assigned a seismic use group based on ASCE 7-10 provisions with an occupancy importance factor taken as 1.0 regardless of its seismic use group categorization. Each structure must be designated as being regular or irregular on the basis of the structural configuration above the isolation system.

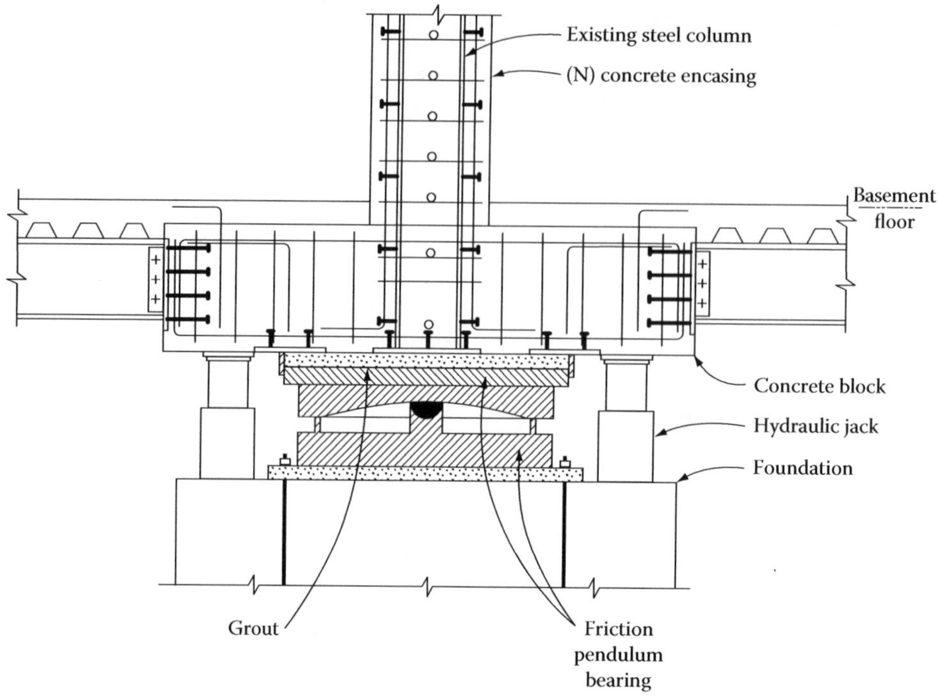

FIGURE 9.22 Installation details, FPS under existing interior columns.

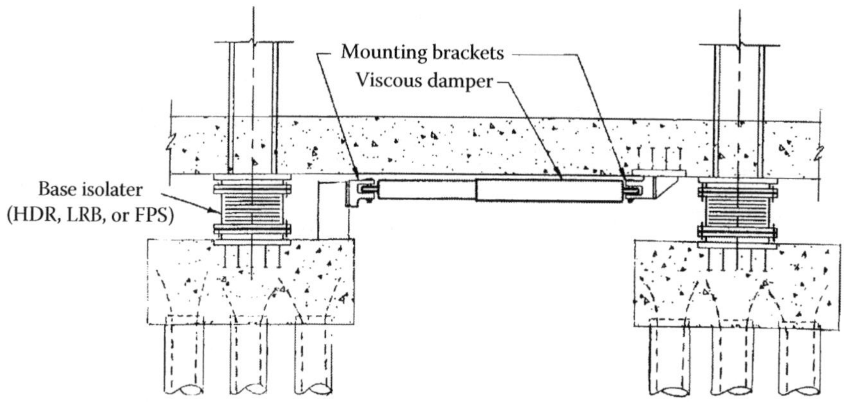

FIGURE 9.23 Base isolator operating in concert with viscous damper.

Three procedures are permissible: static analysis, response spectrum analysis, and time-history analysis. The static analysis procedure is generally used to start the design process and to calculate benchmark values for key design parameters (displacement and base shear) evaluated using either response spectrum or time-history analysis procedures.

The static analysis procedure is straightforward. However, the procedure cannot be used when the spectral demands cannot be adequately characterized using the assumed spectral shape. Typically, this occurs at

1. Isolated buildings located in the near field
2. Isolated buildings on soft soil sites
3. Long-period isolated buildings (beyond the constant velocity domain)

Further, the static procedure cannot be used for nonregular superstructures or for highly nonlinear isolation systems.

Response spectrum analysis is permitted for the design of all isolated buildings except for those buildings located on very soft soil sites (for which site-specific spectra should be established), buildings supported by highly nonlinear isolation systems for which the assumptions implicit in the definitions of effective stiffness and damping break down, or buildings located in the very near field of major active faults where response spectrum analysis may not capture pulse effects adequately.

Time-history analysis is the default analysis procedure: it must be used when the restrictions set forth on static and response spectrum analysis cannot be satisfied and may be used for the analysis of any isolated building. Arguably the most detailed of the analysis procedures, the results of time-history analysis must be carefully reviewed to avoid any gross design errors.

9.3.5.1 Illustrative Example: Static Procedure

The reader should become familiar with the basic principles of seismic isolation and the ELF procedure of ASCE 7-10. Although not discussed here, it should be noted that a dynamic analysis is mandatory for most buildings because buildings that meet the requirements for the use of ELF procedure such as regularity are indeed rare, even in high-seismic zones. However, the design principles are best understood by working through a static example as given in the following section. Ample interpretation of ASCE 7-10 provisions is repeated to present the solution in a stand-alone format.

Given: A new four-story hospital building to be located in the outskirts of Los Angeles, CA. The owners of the facility have desired a building of superior earthquake performance and are willing to

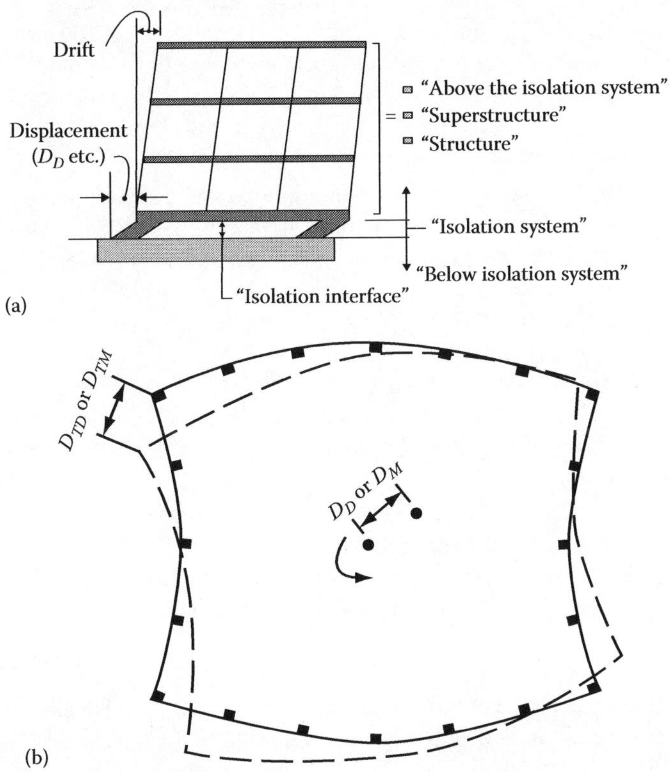

FIGURE 9.24 (a) Base isolation: ASCE 7-05 nomenclature and (b) Isolator displacements: ASCE 7-05 definitions.

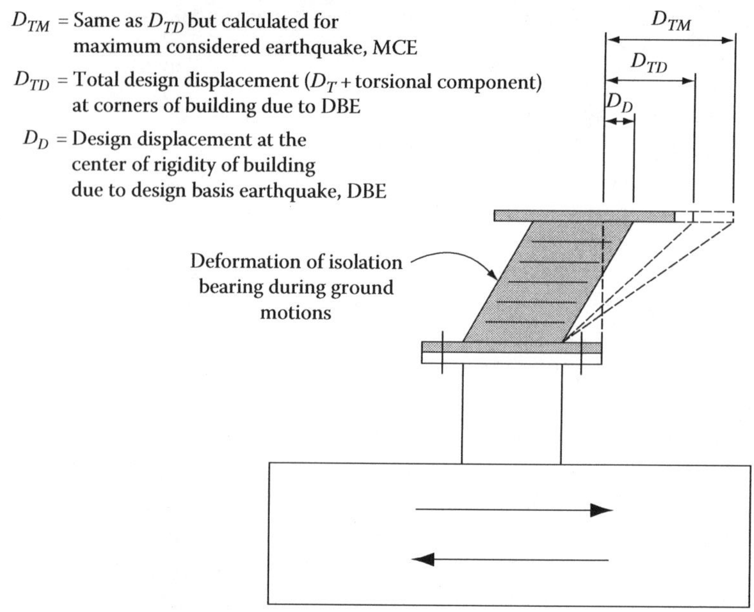

FIGURE 9.25 Isolator displacement terminology.

incur the special costs associated with the design, fabrication, and installation of seismic isolators. A target building performance level of IO or better is sought.

The structure is expected to outperform a comparable fixed-base building in moderate and large earthquakes. The intent is to limit damage to the structure and its contents by using seismic isolation that, in effect, permits an elastic response of the structure while limiting the floor accelerations to low levels even in a large earthquake.

9.3.5.2 Building Characteristics

- A single basement, four-story, regular configuration concrete building. The building has no vertical or plan irregularities. The lateral system is a dual system consisting of special reinforced concrete shear walls and special moment frames capable of resisting at least 25% of prescribed seismic force.
- Response modification coefficient $R = 7$ (ASCE7-10, Table 12.2-1).
- Building is located in the outskirts of Los Angeles, CA.
- From seismic hazard maps $S_s = 1.5$ g and $S_1 = 0.60$ g for the building site.
- Importance factor $I = 1.0$. Observe that the importance factor I for a seismic-isolated building is taken as 1.0, regardless of the occupancy category, since there is no design ductility demand on the structure.
- Building period calculated as a fixed-base building = 0.9 s.
- Building plan dimensions are 120 × 120 ft.
- Calculated distance between the center of mass and the center of rigidity is 5 ft at each floor and at the roof. The project geotechnical engineer has established the building site as site class D.
- Building weight for seismic design = 7200 kips.
- The project structural engineer has established that, to achieve IO performance goals, the isolation system should provide effective isolated periods of $T_D = 2.5$ and $T_M = 3.0$ s and a damping of 20% of the critical. A margin of ±15% variation in stiffness of isolators from the mean values is considered acceptable.

Required: A preliminary design using the provisions of ASCE 7-10 for base isolation of the building. For purposes of illustration, a FPS is selected as the base-isolation system. It should be noted that, in practice, building ownership, particularly if it is a public entity, requires that the design accommodate alternative systems to secure competitive bids. However, for illustration purposes, we will consider only the FPS, it being understood that other isolation systems such as HDR and lead-rubber isolators are equally viable alternatives.

As part of preliminary design, determine

- Minimum design displacements D_D and D_M under DBE and MCE and also total displacements D_{TD} and D_{TM}, which include effects of torsion
- Base shear V_b for designing the structure below the isolation surface
- Base shear V for designing the structure above the isolation surface
- Maximum dimension of the isolators

Solution: The restrictions placed on the use of the static lateral response procedures effectively require dynamic analysis for most isolated structures. Therefore, one might ask, "Why perform, in this day and age of computers, a static analysis of a building with a sophisticated system such as base isolation?" The answer is quite simple: to establish a minimum level of design forces and displacement. Lower-bound limits on design displacements and design forces are specified in ASCE 7-10 as a percentage of the values prescribed by the static procedure. These lower-bound limits on key design parameters ensure consistency in the design of isolated structures and serve as a safety net against gross undersign.

As mentioned previously, seismic isolation, also referred to as base isolation, is a design concept based on the premise that a structure can be substantially *decoupled* from potentially damaging earthquake ground motions. By decoupling the structure from ground shaking, isolation reduces the level of response in the structure from a level that would otherwise occur in a conventional fixed-base building. Typically, decoupling is accomplished using an isolation system that makes the effective period of the isolated structure several times greater than the period of the structure above the isolation system.

In our case, the four-story example building with a fixed-base period of 0.9 s and a standard damping of 5% would have experienced a first-mode acceleration of 0.48*g* (see Figure 9.14). By decoupling the building from the ground, the period of the building is expected to increase to 2.7 s. Additionally, the base isolation is counted upon to increase the damping from a standard 5% to about 20% of the critical. Together, these two factors reduce the first-mode acceleration to 0.12*g*, as shown in Figure 9.14.

The underlying philosophy behind isolated structures may be characterized as a combination of primary performance objective for fixed-base buildings, which is the provision of life safety in a major earthquake and the additional performance objective of damage protection, an attribute provided by isolated structures. The design criteria are then a combination of life safety and damage protection goals summarized as follows:

Two levels of earthquake, the DBE and the MCE, are typically considered in the design of isolated structures. The DBE is the same level of ground shaking as that recommended for design of fixed-base structures. The MCE is a higher level of earthquake ground motion defined as the maximum level of ground shaking that may be expected at the building site within the known geological framework.

The isolators must be capable of sustaining loads and displacements corresponding to the MCE without failure.

The structure above the isolation system must remain *essentially elastic* for the DBE.

From the criteria given earlier, it is seen that the performance objectives and design requirements for fixed- base and isolated buildings vary significantly. The performance objective for fixed-base construction is life safety in a DBE; the intent is to prevent substantial loss of life rather than control damage. For isolated buildings, the performance objectives are

1. Minimal to no damage in the design earthquake (thus providing life safety)
2. A stable isolation system in the maximum capable earthquake

The performance of an isolated building in a DBE will likely be much better (less interstory drift, smaller floor accelerations) than its fixed-base counterpart. Further, isolated buildings can be designed to provide continued function following a design earthquake: a level of performance that is very difficult to achieve with conventional fixed-base construction.

Fixed-base buildings are generally designed using large response modification factors to reduce elastic spectral demands to a design level, a strategy predicated on significant inelastic deformation of the framing system and damage to nonstructural building element. Such buildings are checked for response in the design earthquake only; there is no design check for the MCE. In contrast, isolated buildings are designed using a dual level approach, namely, the framing system is designed to remain essentially elastic (no damage) in the design earthquake, and the isolators are designed (and tested) to remain stable in the MCE.

The subject building is a concrete dual system building. Using the post-earthquake scenario given in ASCE/SE1 41-06 as a guide, our building is expected to have the following:

- No permanent drift. Structure substantially retains original strength and stiffness.
- Transient drift less than 0.5%.
- Negligible damage to nonstructural components.
- Minor hairline cracking in concrete walls, less than 1/16 in. wide. No crushing of concrete.
- Some evidence of sliding at construction joints.
- Coupling breams experience cracks less than 1/8 in. wide.
- Minor settlement and negligible tilting of foundations.
- Diaphragm experiences hairline cracking. Some minor cracks of larger size but less than 1/8 in.
- Cladding connections may yield. No failure.
- Some cracked panes in glazing. None broken.
- Negligible damage in stairs and fire escapes.
- Elevators operate.
- Fire alarm systems and electrical equipment functional.
- Computer units undamaged and operable.

Before proceeding with the illustrative example, certain design requirements touched upon briefly in the preceding sections will be explained in greater detail. The purpose is to delve into the design intent behind these provisions.

9.3.5.2.1 *Effective Stiffness of Isolators*

Typically, isolation systems are nonlinear, meaning that their effective stiffness is displacement and/or velocity dependent, as shown by an idealized force–deflection relationship in Figure 9.26.

The effective stiffness k_{eff} of a seismic isolator is calculated using the forces in the isolator at the maximum and minimum displacements as given in the following equation:

$$k_{eff} = \frac{|F^+| + |F^-|}{|\Delta^+| + |\Delta^-|}$$

where F^+ and F^- are the positive and negative forces at Δ^+ and Δ^-, respectively.

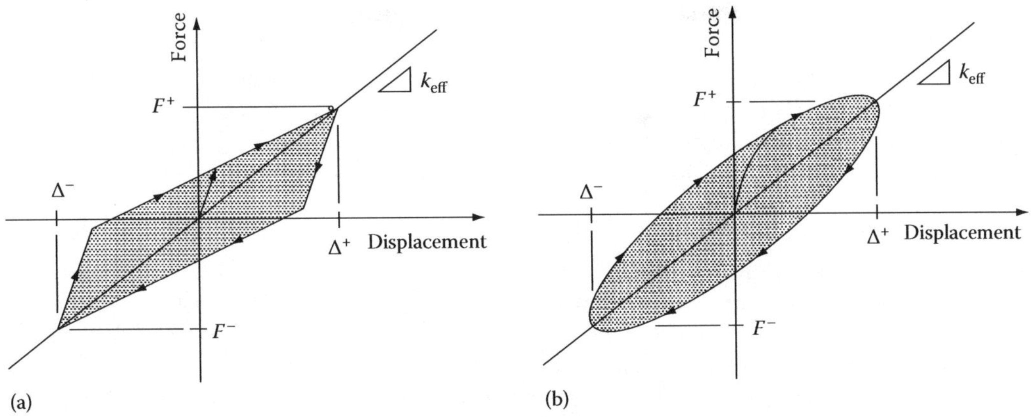

FIGURE 9.26 Idealized force-displacement relationship for base-isolation systems: (a) hysteretic system and (b) viscous system.

For isolators whose properties are independent of velocity, the forces in the isolator at the maximum and minimum displacements will generally be maximum and minimum forces, respectively. For isolators whose properties exhibit velocity dependence, the forces in the isolator at the maximum and minimum displacements will generally be less than the maximum and minimum forces, respectively. However, it is usually assumed that maximum and minimum forces in an isolator are attained at maximum and minimum displacements, respectively. For most types of isolator, this assumption is reasonable.

The deformational characteristics of an isolation system determine (1) the design displacements and (2) the maximum forces transmitted to the isolated structure. Deformational characteristics are represented by the effective (secant) stiffness of the isolation system. Recognizing that force–displacement hysteresis of an isolation system may change over the course of an earthquake, the maximum effective stiffness is used to calculate the maximum force transmitted by the isolators, and the minimum effective stiffness is used to calculate the fundamental period of the isolated building. The reason for using minimum is to arrive at a conservative estimate of the design displacement. The limiting values are generally established in the design phase and are required to be confirmed by testing.

Effective stiffness of the isolation system is determined from the force–displacement (hysteresis) loops based on the results of cyclic testing of a selected sample of isolator. The values of maximum effective stiffness and minimum effective stiffness can be calculated, as shown in Figure 9.26, for both design and maximum displacement levels.

9.3.5.2.2 Effective Damping

The effective damping β_{eff} is used to quantify the energy dissipation furnished by the isolation system. The maximum effective stiffness of the isolation system is used to provide a lower-bound, that is a conservative estimate of the effective damping.

For the purpose of design, energy dissipation is characterized as an equivalent viscous damping. The following equation defines the equivalent viscous damping β_{eff} for a single isolator:

$$\beta_{eff} = \frac{2}{\pi} \frac{E_{loop}}{k_{eff}\left(\left|\Delta^{+}\right| + \left|\Delta^{-}\right|^{2}\right)}$$

where

β_{eff} is the effective damping of the isolation system and isolator unit

E_{loop} is the area enclosed by the force–displacement loop of a single isolator in a complete cycle of loading to maximum positive and maximum negative displacement, Δ^+ and Δ^-

Δ^+ is the maximum positive displacement of isolator during prototype testing

Δ^- is the maximum negative displacement of isolator during prototype testing

9.3.5.2.3 *Total Design Displacement*

The design of isolated structures must consider additional displacements due to actual and accidental eccentricity, similar to those prescribed for fixed-base structures. The equations given in the following provide a simple means to combine translational and torsional displacement in terms of the gross plan dimensions of the building (i.e., dimensions *b* and *d*), the distance from the center of the building to the point of interest (i.e., dimension *y*), eccentricity, as follows:

$$D_{TD} = D_D\left[1 + y\frac{12e}{b^2 + d^2}\right]$$

$$D_{TM} = D_M\left[1 + y\frac{12e}{b^2 + d^2}\right]$$

where *e* is the sum of the actual and accidental eccentricities.

Notice that the design displacement D_D at the center of the building has been modified to account for additional displacement at the corners or edges of the building due to torsion. It is assumed that the stiffness of the isolation system is distributed in plan proportional to the distribution of the supported weight of the building.

Smaller values of D_{TD} can be used for design if the isolation system is configured to resist torsion (e.g., if stiffer isolator units are positioned near the edges and corners of the building). However, the minimum value of D_{TD} is set equal to $1.1D_D$ for all types of isolation systems. The total displacement D_{TM} is calculated in a manner similar to the calculation of D_{TD}. The eccentricity *e* used for calculating torsional displacements is the actual eccentricity of the isolation system plus an allowance of 5% of the width of building to account for accidental torsion. The parameter *y* is the distance between the center of rigidity of the isolation system and farthest corners of the building.

It should be noted that the stiffness values $K_{D\min}$ and $K_{m\min}$ are not known to the designer during the preliminary design stage but are derived from the known or expected values of periods of the building. Since the expected periods may not turn out to be equal to the final values, the derived stiffness values are also preliminary. After completing a satisfactory preliminary design, typically prototype isolators are tested to obtain values of $K_{D\min}$, $K_{D\max}$, $K_{M\min}$, and $K_{M\min}$.

9.3.5.2.4 *Minimum Design Lateral Forces*

9.3.5.2.4.1 *Isolation System and Structural Elements at or below Isolation Interface* The design actions for elements at or below the isolation interface are based on the maximum forces delivered by the isolation system during the DBE. The building's foundation, the isolation system, and all structural elements at or below the isolation interface are required to be designed and constructed to withstand a minimum lateral force:

$$V_b = K_{D\text{Max}} D_D$$

The maximum force V_b is the product of the maximum stiffness of the isolation system at the design displacement $K_{D\max}$ and the design displacement D_D. The design force V_b presents strength level forces.

The previous equation for V_b is for use in regions of high seismicity, such as UBC zone 4, wherein the difference between the total design displacement and total maximum displacement is relatively small; that is, if a supporting element was designed for DBE forces at the strength level, it is probable that such a supporting element could resist the forces associated with the MCE without failure.

There are significant differences in values of M_M between regions of high and low seismicity: values of M_M may be less than 1.25 in regions of high seismicity but may exceed 2.5 in regions of low seismicity. As such, in a region of low seismicity, a supporting element designed for DBE-induced forces may be unable to sustain forces associated with the MCE without significant distress or failure. Therefore, in these regions, it may be prudent to consider MCE-level forces to check the design of the isolation system and the structural elements at or below the isolation interface.

Isolation interface is the boundary between the upper portion of the building, which is isolated, and the lower portion, which is rigidly attached to the foundation or ground. The isolation interface can be assumed to pass through the midheight of elastomeric bearings or the sliding surface of sliding bearings. Observe that the isolation interface need not be a horizontal plane but could change elevation if the isolators are positioned at different elevations throughout the building.

The isolation system includes the isolator units, connections of isolator units to the structural system, and all structural elements required for isolator stability. Isolator units include bearings that support the building's weight and provide lateral flexibility. Typically, isolation system bearings provide damping and wind restraint as an integral part of the bearing. Isolator systems may also include supplemental damping devices. For example, an FPS of basic isolation may include viscous dampers.

Structural elements that are required for structural stability include all structural elements necessary to resist design forces at the connection of the structure to isolator units. For example, a column segment and a beam immediately above an isolator constitute elements of the isolation system because they are necessary to resist forces due to the lateral earthquake displacement of the isolators.

9.3.5.2.4.2 Structural Elements above Isolation System The design of the framing above the isolation system is based on the maximum force delivered by the isolation system divided by a response reduction factor, *R*. The values assigned to R_I reflect system overstrength only and no expected ductility demand. By using these values for R_I, a significant measure of damage control is afforded in the design earthquake, since the structure remains essentially elastic.

The minimum base shear for the design of the structure above the isolation is given by

$$V_s = K_{D\text{Max}} D_D / R_I$$

Three limits are imposed for the calculation of V_s:

- V_s shall not be less than the base shear required for a fixed-base structure of the same weight *w* and a period equal to the isolated period.
- V_s shall not be less than the total shear corresponding to the design wind load. (In wind design, engineers seldom use the term base shear to define the total shear due to wind. However, base shear and total shear are one and the same.)
- V_s shall not be less than 150% of the lateral seismic force required to fully activate the system.

Thus, there are three lower-bound limits set on the minimum seismic shear to be used for the design of the framing above the isolation system. The first limit requires design base shear to be at least that of a fixed-base building of comparable period. The second limit ensures that the elements above the isolation system remain elastic during a design windstorm. The third limit is designed to prevent the elements above the isolation system from deforming inelastically before the isolation system is activated.

9.3.5.2.4.3 Vertical Distribution of V_s The vertical distribution of the seismic base shear is similar to that used for fixed-base buildings, namely, a distribution that approximates the first-mode shape of the fixed-base building. This distribution conservatively approximates the inertia force distributions measured from time-history analyses.

Continuation of illustrative problem: The effective periods T_D and T_M of the isolated building are

$$T_D = 2\pi\sqrt{\frac{W}{K_{D\min}g}}$$

$$T_D^2 = \frac{4\pi^2 W}{K_{D\min}g}$$

where

$T_D = 2.5$ s (given)
$W = 720$ kips (given)
$g = 3.86.4$ in./s^2

$$K_{D\min} = \frac{4\pi^2 w}{T_D^2 g}$$

$$= \frac{4 \times \pi^2 \times 7200}{2.5^2 \times 386.4}$$

$$= 117.7\,\text{kips/in.,} \quad \text{say, } 118\,\text{kips/in.}$$

Similarly,

$$K_{M\min} = \frac{4 \times \pi^2 \times 7200}{3^2 \times 386.4} \quad T_M = 3\,\text{s (given)}$$

$$= 81.7\text{ kips/in.,} \quad \text{say, } 82\text{ kips/in.}$$

As stated in the problem, a plus or minus 15% variation in stiffness from the mean values is permitted. Therefore, use a factor of 0.85 to determine $K_{D\text{Max}}$ and $K_{M\text{Max}}$:

$$K_{D\max} = \frac{1.15 \times 118}{0.85} = 159.5\text{ kips/in.,} \quad \text{say, } 160\text{ kips/in.}$$

$$K_{M\max} = \frac{1.15 \times 82}{0.85} = 111\text{ kips/in.}$$

From ASCE-10, Table 17.5-1, for a 20% effective damping, that is, B_D or $B_M = 20\%$, the value of damping coefficient $B = 1.5$.

Observe that the same damping coefficient is applied to both DBE and MCE events. The value of F_v as a function of site class and mapped 1 s period MCE spectral acceleration is given in Table 11.4.2 of ASCE7-10. From this table, for site class D and $S_1 = 0.60 > 0.50$, we get $F_v = 1.5$. The spectral response acceleration S_{M1} at a period of 1 s, adjusted for site class *D*, is equal to

$$
\begin{aligned}
S_{MI} &= F_V S_1 \\
&= 1.5 \times 0.6 \\
&= 0.9g
\end{aligned}
$$

The design spectral response acceleration S_{D1} is given by

$$
\begin{aligned}
S_{D1} &= 2/3 S_{MI} \\
&= 2/3 \times 0.9 \\
&= 0.6g
\end{aligned}
$$

Similarly,

$$S_s = 1.5$$

$$
\begin{aligned}
S_{MS} &= F_a S_S \\
&= 1 \times 1.5 \\
&= 1.5
\end{aligned}
$$

$$
\begin{aligned}
S_{DS} &= 2/3 \times 1.5 \\
&= 1.0g
\end{aligned}
$$

The minimum design displacements are obtained as follows:

$$
\begin{aligned}
D_D &= \left(\frac{386.4}{4\pi^2}\right)\frac{0.60 \times 2.5}{1.5} \\
&= 9.79\,\text{in.}
\end{aligned}
$$

$$
\begin{aligned}
D_M &= \left(\frac{386.4}{4\pi^2}\right)\frac{0.9 \times 3}{1.5} \\
&= 17.62\,\text{in.}
\end{aligned}
$$

The eccentricity for calculating torsional effects is equal to the actual eccentricity plus 5% of the building width. Thus,

$$
\begin{aligned}
e &= 60 + 0.05 \times 120 \times 12 \\
&= 132\text{ in.}
\end{aligned}
$$

The displacements including the torsional effects are

$$
\begin{aligned}
D_{TD} &= D_D\left(1 + y\frac{12e}{b^2 + d^2}\right) \\
&= 9.79\left\{1 + \frac{120}{2} \times \frac{132}{\left(120^2 + 120^2\right)}\right\} \\
&= 12.48\,\text{in.}
\end{aligned}
$$

$$
\begin{aligned}
D_{TM} &= 17.62 \times 1.275 \\
&= 22.47\,\text{in.}
\end{aligned}
$$

$$
\begin{aligned}
V_b &= K_{D\max} D_D \\
&= 160 \times 9.79 \\
&= 1566.4\,\text{kips}
\end{aligned}
$$

Given a seismic weight of $W = 7200$ kips, the seismic base shear coefficient for the design of isolation system and structural elements below its corresponds to

$$1566.4/7200 = 0.218 \text{ or } 21.8\% \text{ of gravity}$$

Before calculating the base shear for the super structure, we need to calculate R_I. However, R_I need not be greater than 2.0.

Therefore,

$$R_I = 3/8R = 3/8 \times 7 = 2.657 > 2.0$$

$$R_I = 2.0$$

$$\begin{aligned} V_s &= K_{DMax}\, D_D/R_I \\ &= V_D/2 \\ &= 1566.4/2 \\ &= 783.2 \text{ kips or } 783.2/7200 \times 100 = 10.9\% \text{ of gravity.} \end{aligned}$$

Using the equivalent lateral procedure of ASCE 7-05, we now calculate the base shear required for a fixed-base structure of weight $W = 7200$ kips and a period $T = T_D = 2.5$ s, equal to the period of the isolated building:

1.

$$V = \frac{S_{DS}}{\left(\frac{R}{I}\right)} W$$

$$= \frac{1}{\left(\frac{7}{1}\right)} W = 0.143W = 14.3\%g$$

2.

$$V_{max} = \frac{S_{D1}}{T\left(\frac{R}{I}\right)} W$$

$$= \frac{0.6}{2.5 \times 7} W = 0.034W = 3.4\%g$$

3.

$$V_{max} = 0.01W$$

$$= 1.0\%g$$

4.

$$V_{min} = \frac{0.5S_1}{\left(\frac{R}{I}\right)}$$

$$= \frac{0.5 \times 0.6}{\left(\frac{7}{1}\right)} = 0.043W = 4.3\%g$$

The base shear $V_{min} = 4.3\%$ g, obtained from the fourth equation, yields the design base shear for the fixed-base building. However, a base shear equal to 10.9% of gravity, obtained from the calculations for a base-isolated building, controls the design of the subject building. Using this base shear, the structural elements above the isolation system are designed by applying the appropriate provisions of a nonisolated structure.

9.3.5.2.5 *Preliminary Design of Friction Pendulum System*

Recall that the period T of a pendulum is inversely proportional to the square root of its length and does not depend on the mass $m = w/g$. Similarly, the period of the FPS depends only on the square root of its radius R of the dish and not the supported mass of the building. To increase the period of a pendulum, we increase the length; to increase the apparent period of the building, we increase the radius of the disk.

If the weight of the building is W, and the radius of FPS dish is R, then the horizontal stiffness of the isolator is given by

$$K_h = W/R$$

The period of the isolated system is a function of its radius R only and is given by

$$T = 2\pi\sqrt{\frac{R}{g}}$$

For our building, the effective isolated period $T_D = 2.5$ s, as given in the statement of the problem.

Therefore,

$$2.5 = 2\pi\sqrt{\frac{R}{386.4}} \quad \text{giving}$$

$$R = 61.23 \text{ in.}$$

The effective stiffness of a FPS is given by

$$K_{eff} = \frac{W}{R} + \frac{\mu W}{D}$$

where the new term μ is the friction coefficient.

The friction coefficient μ for an FPS may be assumed to be independent of velocity for pressures of 20 ksi or more. The damping β provided by the system is given by

$$\beta = \frac{2}{\pi}\frac{\mu}{\mu + \frac{D}{R}}$$

Assuming $\mu = 0.06$ and a design displacement of 10 in., the effective damping is calculated from

$$\beta_{eff} = \frac{2 \times 0.06}{\pi\left(0.06 + \left[\frac{10}{61.23}\right]\right)}$$

$$= 0.10 \text{ or } 10\% \text{ of the critical}$$

The selected value of $D = 10$ in. satisfies the minimum code displacement of $D_D = 9.79$ in., calculated earlier for $T = 2.5$ s, $\beta = 20\%$, and $B = 1.5$.

The effective stiffness is calculated from

$$K_{eff} = W/R + \mu W/D$$

$$= 7200/61.23 + 0.06 \times 7200/10 = 161 \text{ kips/in.}$$

This is almost exactly the same as K_{DMax} of 160 kips/in. derived earlier. Therefore, no further iterations are necessary.

With regard to the example problem, the following observations are appropriate for preliminary design purposes:

- An FPS of approximately 5 ft radius is required underneath each column.
- The required stiffness of each FPS is approximately equal to 100 kips/in.
- A moat about 23 in. around the building is required to accommodate the calculated displacement $D_{TM} = 22.47$ in.
- The torsional contribution to the displacement is equal to $D_{TM} - D_M$. In our case, this is equal to (22.47 − 17.62) = 4.85 in. A possible solution to reducing the torsion contribution is to use a stiffer FPSD at the building perimeter.
- As mentioned previously, other competing isolation systems are generally evaluated to achieve competitive bids. Usually, a performance type of specifications for a base-isolation system accompanies structural drawings to encourage competitive bids.

9.3.5.3 Triple Pendulum Bearing

So far our discussion has been confined to a conventional, *single pendulum bearing* in which the bearing surfaces maintain constant friction, lateral stiffness, and dynamic period irrespective of the expected earthquake motion. This is in line with the current design practice of designing the isolation system to have sufficient displacement capacity to meet the demands of the maximum considered earthquake (MCE) (an event having a 2% probability of occurrence in 50 years, corresponding approximately to a 2500-year recurrence interval). While the design meets the objectives of MCE, its performance is less than optimum for the DBE, which is two-thirds of the MCE spectrum. Moreover, the performance of the isolation system in more frequent events of smaller magnitude is typically not considered in the design process of base isolation. Generally, low-level shaking is not a design issue in terms of strength or displacement capacity. However, it can be a performance issue. Isolation systems designed with sufficient damping and flexibility for large earthquakes may not activate in more frequent minor events. To overcome this issue, a new type of FPS referred to as an adaptive or smart seismic isolation system has recently (2006) been introduced. Developed by Earthquake Protection Systems (EPS), the new device permits the isolation system to be separately optimized for low intensity, design level, and maximum ground motions. To satisfy the dual requirement of controlling displacements in large earthquakes while still maintaining good performance in low to moderate earthquakes, one needs to design an isolation system that

1. Is very stiff with low damping at low-level shaking
2. Softens with increasing damping in the DBE
3. Softens even more and increases damping in the MCE
4. Stiffens beyond MCE

This methodology has been incorporated by EPS in a new device referred to as triple pendulum bearing system, which has three independently designed friction pendulum bearings incorporated in a single pendulum bearing. The properties of each of the bearing's three pendulums are chosen to become sequentially active as the earthquake motions become stronger.

The triple pendulum bearing's inner isolator consists of an inner slider along two inner concave spherical surfaces. Properties of the inner pendulum are chosen to reduce structure shear forces that occur during service-level earthquakes.

The two slider concaves, sliding along the two main concave surfaces, comprise of two more isolators designed to minimize the structure shear forces that occur during the DBE. And, finally, the properties of the pendulum are chosen to minimize bearing displacements that occur during the maximum credible earthquake.

According to the manufacturer, when designed for a severe maximum credible earthquake, the plan dimensions of the triple pendulum bearing are approximately 60% that of the single pendulum bearing.

9.3.5.4 Additional Notes on Friction Pendulum Systems

Friction pendulum bearings are seismic isolators that are installed between a structure and its foundation to protect the supported structure from earthquake ground shaking. Using friction pendulum technology, it is cost-effective to design and build buildings to elastically resist earthquake ground motions without structural damage.

Friction pendulum bearings use the characteristics of a pendulum to lengthen the natural period of the isolated structure so as to avoid the strongest earthquake forces. During an earthquake, the supported structures move with the motions of a pendulum. Since earthquake-induced displacements occur primarily in the bearings, lateral loads transmitted to the structure are greatly reduced.

The single pendulum bearing is the original friction pendulum bearing. The single slider maintains the vertical load support at the center of the structural member. The bearing also has a low height, which can be advantageous in some installations.

The period of friction pendulum bearing is selected simply by choosing the radius of curvature of the concave surface. It is independent of the mass of the supported structure. The damping is selected by choosing the friction coefficient. Torsion motions of the structure are minimized because the center of stiffness of the bearings coincides with the center of mass of the supported structure. The bearing's period, vertical load capacity, damping, displacement capacity, and tension capacity can all be selected independently. For the triple pendulum bearing, three effective radii and three friction coefficients are selected to optimize performance for different strengths and frequencies of earthquake shaking. This allows for maximum design flexibility to accommodate both moderate and extreme motions, including near-fault pulses.

The tension-capable bearing can accommodate structure vertical loads that vary from compression to tension during seismic movements. This bearing prevents uplift of a primary structural member and can eliminate concerns regarding potential structure overturning or large vertical earthquake motions.

The triple pendulum bearing offers better seismic performance as compared to seismic isolation technology.

It incorporates three pendulums in one bearing, each with properties selected to optimize the structure's response for different earthquake strengths and frequencies. The properties of each of the bearing's three pendulums are chosen to become sequentially active at different earthquake strengths. As the ground motions become stronger, the bearing displacements increase. At greater displacements, the effective pendulum length and the effective damping increase, resulting in lower seismic forces and bearing displacements.

The triple pendulum bearing's inner isolator consists of an inner slider that slides along two inner concave spherical surfaces. Properties of the inner pendulum are typically chosen to reduce the peak accelerations acting on the isolated structures and its contents, minimize the participation of higher structure modes, and reduce structure shear forces that occur during service-level earthquakes.

The two slider concaves, sliding along the two main concave surfaces, comprise two more independent pendulum isolators. Properties of the second pendulum are typically chosen to minimize the structure shear forces that occur during the DBE. This reduces construction costs of the structure. Properties of the third pendulum are typically chosen to minimize bearing displacements that occur during the maximum credible earthquake. This reduces the size and cost of the bearings and reduces the displacements required for the structure's seismic gaps.

The single pendulum bearing maintains constant friction, lateral stiffness, and dynamic period for all levels of earthquake motion and displacements. In the triple pendulum bearing, the three pendulum mechanisms are sequentially activated as the earthquake motions become strong. The small displacement, high-frequency ground motions are absorbed by the low-friction and short-period inner pendulum. For the stronger design-level earthquake, both the bearing friction and period increase, resulting in lower bearing displacements and lower structure base shears. For the strongest maximum credible earthquakes, both the bearing friction and lateral stiffness increase, reducing the bearing displacement. When designed for a server maximum credible earthquake, the plan dimensions of the triple pendulum bearing are approximately 60% that of the equivalent single pendulum bearing.

9.4 PASSIVE ENERGY DISSIPATION

Passive energy dissipation is an emerging technology that enhances the performance of a building by adding damping (and in some cases, stiffness) to the building. The primary use of energy dissipation devices is to reduce earthquake displacement of the structure. Energy dissipation devices will also reduce force in the structure, provided the structure is responding elastically, but would not be expected to reduce force in structure that is responding beyond yield.

For most applications, energy dissipation provides an alternative approach to conventional stiffening and strengthening schemes and would be expected to achieve comparable performance levels. In general, these devices are expected to be good candidates for projects that have a target building performance level of life safety or perhaps IO but would be expected to have only limited applicability to projects with a target building performance level of CP. Other objectives may also influence the decision to use energy dissipation devices, since these devices can also be useful for control of building response to small earthquakes and wind loads.

A wide variety of passive energy dissipation devices is available, including fluid viscous dampers, viscoelastic materials, and hysteretic devices. Ideally, energy dissipation devices dampen earthquake excitation of the structure that would otherwise cause higher levels of response and cause damage to components of the building. Under favorable conditions, energy dissipation devices reduce drift of the structure by a factor of about two to three (if no stiffness is added) and by larger factors if the devices also add stiffness to the structure.

Unlike base isolation, passive energy dissipation does not intercept earthquake energy entering the structure. It allows earthquake energy into the building. However, the energy is directed toward energy dissipation devices located within the lateral resisting elements. Earthquake energy is transformed into heat by these devices and dissipated into the structure.

A fluid viscous damper shown in Figure 9.27 is one such energy dissipation device. It dissipates energy by forcing a fluid through an orifice, similar to the shock absorbers of an automobile. The fluid used is usually of high viscosity, such as a silicone. The unique feature of these devices is that their damping characteristics, and hence the amount of energy dissipated, can be made proportional to the velocity. The response of a fluid viscous damper is considered to be out of phase with those due to seismic activity. This is because the damping force provided by the device varies inversely with the dynamic lateral displacements of a building. To understand the concept, consider a building shaking laterally back and forth during a seismic event. The stress in a lateral-load-resisting element such as a frame column is at its maximum when the building deflection is also at maximum. This is also the point at which the building reverses direction to move back in the opposite direction. The damping force of a fluid viscous damper will drop to zero at this point of maximum deflection. This is because the damper stroking velocity goes to zero as the building reverses direction. As the building moves back in the opposite direction, a maximum damper force occurs at the maximum velocity, which happens when the building goes through its normal upright position. This is also the point when the stresses in the lateral-load-resisting elements are at a minimum. Therefore, the damping provided by the device varies from a maximum to a minimum as the building moves from an at-rest position to its maximum lateral deflection position. This out-of-phase response is considered a desirable feature in seismic designs.

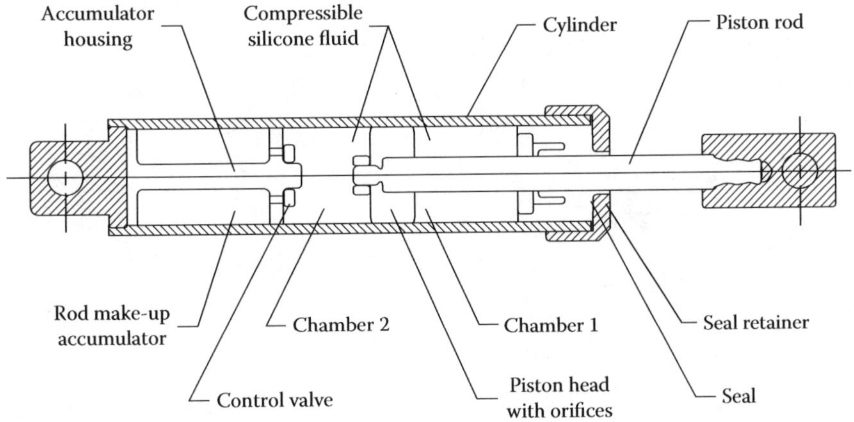

FIGURE 9.27 Fluid viscous damper consisting of a piston in a damping housing filled with a compound of silicone or similar type of fluid.

9.5 BLAST-RESISTANT DESIGN

It is worth noting that lateral loads caused by wind and earthquakes are not a result of human activity. In a manner of speaking, they are acts of God, nature-made. We, humans, have nothing in our resources to eliminate their effect. We have, however, learned from past experience how to engineer our buildings to safely withstand their effects, albeit uncertainties do exist in our complete understanding of their origin and potency.

Blast effects, on the other hand, are human-made. Speaking philosophically, they can be eliminated entirely but how this can be done in the messed up world we live in is a political subject, not appropriate for our technical book.

Unlike seismic and wind loads, blast loads have a short duration, typically milliseconds. Therefore, the building lateral framing system does not need to be strengthened. The fact that the blast wave applies positive pressure on all sides of the building also limits the effect on the lateral framing system, albeit the pressure is much higher on the sides facing the explosive source.

The pressure exerted on buildings by explosions may be many orders of magnitude higher, 5000 psi plus, than normal design pressures, but their duration is in milliseconds. The resulting pressures are inversely proportional to the cube of the distance from the center of the source. Damage to structures may be severe, but it is only a fraction of what a proportional static pressure would cause. When large surfaces are engaged by blast pressures, they will be moved as the shock wave passes, but the direction of the net force (initial uplift–overpressure) will be determined by the complexities of the wave path and time. Heavy columns tend to survive but may have problems if some of the floors that laterally brace them are removed. The wall and floor planes in buildings have large surfaces that will receive most of the blast pressure. They likely will be ripped away from their connections, leading to collapse of at least part of the structure (see Figures 9.28 through 9.30).

Events of the last fifteen years have greatly heighted our awareness of the threat of terrorist attacks using explosives. The private sector is increasingly considering protective design methods, especially for so-called icon buildings that are perceived to be prime targets, as well as nearby structures that are vulnerable to collateral damage.

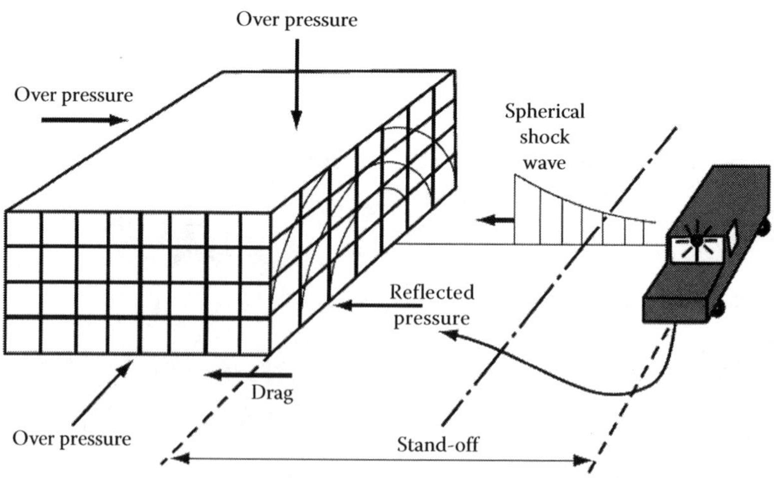

FIGURE 9.28 Exterior explosion.

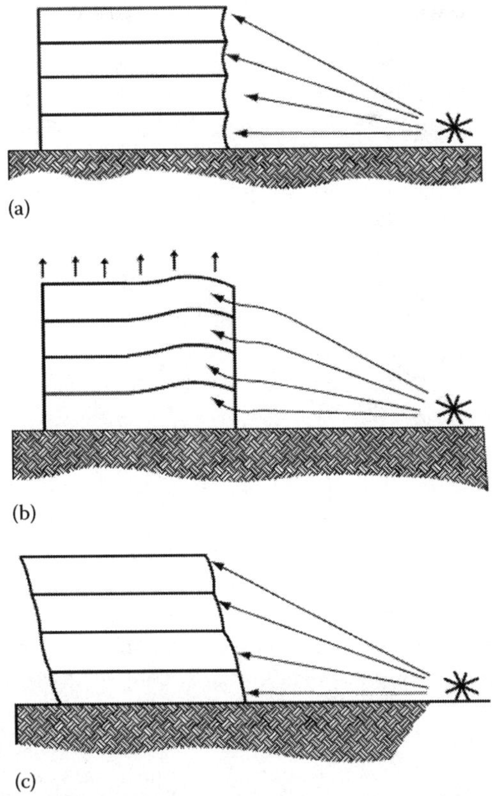

FIGURE 9.29 Damage due to exterior explosion. (a) Exterior windows, columns, and walls. (b) Roof and floor slabs. (c) Building sway due to ground shock.

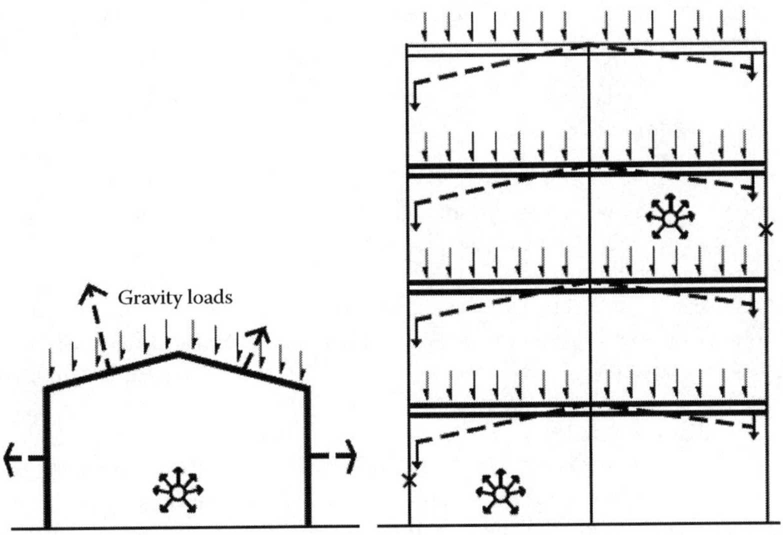

FIGURE 9.30 Interior explosion: when explosions occur within structures, pressures can build up within confined spaces, causing lightly attached wall, floor, and roof surfaces to be blown away.

9.5.1 Design Criteria

The first design criteria that must be established for a structure that is intended to survive one are various combinations of standoff distance (R) and explosive charge size (W). R measures how close to the building a bomb could explode and is therefore a function of the physical characteristics of the surrounding site. W is expressed in weight or mass of TNT in order to correlate with tests; the equivalent W of any other explosive material is based on experimentally determined factors or the ratio of its heat of detonation to that of TNT. The effects of any blast are then normalized by the scaled distance parameter $Z = R/W^{1/3}$.

The Department of Defense (DoE), Department of State (DoS), and General Services Administration (GSA) have developed specific requirements for military, embassy, and federal buildings, respectively. However, key portions of these criteria are only available to designers of specific projects to which they apply. The 1999 ASCE report by Conrath et al. provides some recommendations for private-sector facilities.

9.5.2 Load Criteria

The shock wave from an external explosion causes an almost instantaneous increase in pressure on nearby objects to a maximum value. This is followed by a brief positive phase during which the pressure decays back to its ambient value and a somewhat longer but much less intense negative phase during which the pressure reverses direction. For most structures, this phenomenon can be approximated decay. The parameters of this equivalent load are calibrated to match the maximum reflected pressure (p_r) and total reflected impulse (i_r) of the actual load's positive phase, so that the design duration $t_d = 2i_r/p_r$. Because it usually has little effect on the maximum response.

Structural elements that must withstand external blast pressure must be analyzed and designed accordingly. The same is true of internal elements; particularly elevate floor slabs, if windows or doors are not expected to remain intact during a blast event. Failure of these components will permit the blast pressures to propagate within the building. Although the actual blast load on an exposed element is typically taken as the product of this area and either the maximum pressure or a spatially averaged value. This is analogous to the manner in which design wind loads for components and cladding are routinely calculated. Blast loads need not be factored since they already represent an ultimate design condition.

9.5.3 Analysis Procedure

An element loaded by a blast can be modeled approximately as an elastic–plastic dynamic system with an SDOF corresponding to its maximum blast deflection. The element's effective mass, elastic and elastic–plastic stiffnesses, and available yield and ultimate strengths are derived from its actual physical configuration and properties.

Most materials used in actual construction have strengths that exceed their specified minimum values. In addition, the short duration of a blast load results in high strain rates that increase the design strength. Consequently, for dynamic design, the specified strength can be increased by a factor of 10%–25%.

The designer can also take advantage of the increase in concrete strength with age, which is on the order of 10% at 6 months and 15% at 1 year or more. Material-specific interaction equations account for the reduced moment capacities available to withstand a blast load because of the stresses already present due to the dead load and a realistic portion of the live load. For dynamic analysis and design, all strength reduction or resistance factors are set to unity.

Reinforced concrete, properly detailed, is generally preferred for blast-resistant structures. Concrete masonry may also be used for exterior walls but must always be reinforced and even then has a considerably higher potential for unacceptable brittle failure and subsequent fragmentation, especially if only cells containing reinforcing bars are grouted. Cavity walls are more effective than single wythes because the outer layer of brick will contribute additional mass and absorb many of the casing fragments produced by an external explosion.

Structural steel, especially when utilized in moment-resisting frames, can tolerate a considerable amount of deflection during a blast without collapse. However, exterior cold-formed steel wall panels or sheathed studs are often not practical for blast-resistant structures and can increase fragment hazards to building occupants. For strong-axis bending of open-section structural or cold-formed steel elements, lateral bracing of the compression flange or torsional bracing of the cross section is required at plastic hinge locations and at a spacing small enough to preclude lateral torsional buckling.

The designer can calculate the expected response of an element to a triangular blast impulse using published curves or a computer program capable of performing a nonlinear time-history analysis of the SDOF system. The relevant parameters for each element include the following ratios:

- Blast impulse duration to natural period of vibration (t_d/T)
- Maximum dynamic load to available ultimate strength (F_o/R_u)
- Maximum expected deflection to yield deflection (ductility ratio u)
- Span length to maximum expected deflection (deflection ration D)

μ and D correlate with the expected amount of damage to an element in a blast event, which is restricted by the level of protection that the structure must provide to its occupants and contents based on their nature, quantity, function, and importance. D_{min} is related to the maximum end rotation (θ_{max}), for example, for one-way elements other than cantilevers, assuming plastic hinge formation at midspan and fixed ends and a linear deflected shape between hinges, $D_{min} = 2/\tan \theta_{max}$. Several of the references given in this book describe the damage associated with qualitative levels of protection and suggest corresponding θ_{max} and D_{min} (or θ_{max}) values for various combinations of element type and material.

When $\mu > 1$, the element must actually be capable of undergoing the plastic deformation associated with its calculated μ and D values without suffering unacceptable damage. This requires careful detailing of members and especially connections. Although code requirements and industry guidelines for structures in high-seismic regions are helpful, they are not sufficient for blast design. Because of the localized nature of an explosion, such provisions must be followed even for elements that are not part of the lateral-force-resisting system, especially on the exterior. In case a primary supporting element does fail because of a blast, the structural system should include alternative load paths so that progressive collapse of additional bays will not follow. Multistory buildings are especially vulnerable in this respect and should have enough inherent redundancy to survive a local failure at the ground floor level.

For shear design, it is usually adequate to check an element's end connections for the equivalent static reactions produced by a uniformly distributed load with a total magnitude of R_u or $2F_o$, whichever is smaller. Supporting elements can then be conservatively designed to have ultimate strengths adequate to resist these loads. Since the shear failure mode of concrete and masonry elements is relatively brittle, it is essential to provide appropriate reinforcement at and near supports. The designer must also account for the elastic rebound of an element subsequent to its maximum deflection, which will induce stresses opposite to those caused by the blast pressure itself. Appropriate provisions for this effect will also improve the element's ability to withstand a load reversal, which may occur if an adjacent or supporting element fails during a blast event.

9.5.4 Difference between Seismic and Blast-Resistant Design

At first glance, designing for blast resistance may appear similar to designing for seismic resistance because both are nonstatic and rely on post-yield response. However, there is one important difference—blast loading is impulsive, not simply dynamic.

Conventional design for common time-varying loads, including wind and seismic, relies on techniques that permit conversion of these dynamic phenomena into equivalent static events. For wind loads, in particular, relatively simple adjustments are specified to the quasistatic design loads to account for dynamic response.

Seismic design involves a very elaborate conversion of the dynamic loading into a quasistatic problem. Building systems characterized for stiffness, ductility, and site conditions are evaluated for seismic exposures and characteristics of shaking. On the basis of extensive experience in evaluating the actual earthquake response of designed structures, seismic loads are idealized as a series of externally applied loads that are thought to mimic the loading effects of an earthquake. Complicated though the approach is, many buildings can be designed for earthquakes by engineers with little familiarity with dynamic behavior.

The first difference between seismic and blast-resistant design is in the way a given structure is loaded. In the case of an earthquake, the structure is subject to ground motions that shake the structure from the base up. In the case of an explosion produced by an air or a surface burst, the structure is loaded by means of a compression wave (shock wave) over some area. Since a portion of the blast energy is coupled into the ground, the structure is also subject to ground motions similar to an earthquake, though much less intense.

A second difference is the duration of loading. For earthquakes, the duration of induced motions (shaking) can range from seconds to minutes. Additional loadings are produced by *aftershocks*, which are generally less intense than the initial shaking. For conventional explosives, the duration of a pressure wave is on the order of milliseconds.

For example, in the Oklahoma City event, the yield of the weapon was approximately 4000 lb. TNT equivalent. The truck containing the explosive was positioned about 10′ from the building. The peak pressure at the face of the buildings was about 1900 psi, and the duration of the positive phase of the pulse was approximately 3 ms. Judging by the size of the crater, a fair portion of the energy coupled into the ground, producing ground shock. However, judging by the damage, clearly air blast was the primary damage mechanism. Further, earthquakes shake an entire building but produce mostly horizontal loads at floor-slab levels concentrating in the specifically designed, laterally stiffer structural systems. Blasts usually do not attack the entire structure uniformly but produces the most severe loads to the near structural elements, both vertical and horizontal with little regard to their stiffness. Uplift pressure load on floors is also a specific blast effect.

The ductility property of the material plays an important role both in seismic and blast resistance design. The term ductility refers to the ability of the material to absorb energy elastically without failure—the greater the ductility, the greatest the resistance to failure.

Ductile inelastic structural response can be expected during both severe blast and severe earthquake events. However, it is generally recognized that plastic hinge zones and ductility demands in the two events do not necessarily match because of the differences in the loading patterns and effects.

As in seismic, the mass of the building plays a significant role in blast-resistant design. For ordinary buildings such as apartments and offices, building codes do not require blast resistance. However, for buildings that house hazardous processes, some building codes require special safety considerations requiring that walls, floors, and roofs separating a use from an explosion exposure shall be designed to resist a minimum internal pressure of 100 psi in addition to other conventional loads.

9.5.5 Selection of Design Blast Load

Selection of the blast charge size W is based on the perceived risk to the design building and any buildings nearby. Various factors play a role here, such as the social and economic significance of the building, security measures that deter terrorists, and data from previous attacks on similar facilities. The minimum standoff distance R is determined from the layout of a building's surroundings and reflects the expectation of how close to the building the design charge could explode.

W and R are two necessary inputs for the scaled distance parameters $Z = R/W^{1/3}$ that is used to determine *equivalent* design pressure impulses using established curves.

Conventional structures, in particular those above grade, are susceptible to damage from explosions because the magnitudes of design loads are significantly lower than those produced by most explosions. For example, design snow loads in the Midwest range from about 5 psf to about 50 psf. The peak pressure in the blast pulse produced by 10 lb. of TNT at a range of about 50′ is approximately 2.4 psi with a duration of the positive phase of 7.7 ms. Conventional structures are not normally designed to resist blast loads.

Recent terrorist attacks demonstrate the types of damage that can be produced. The 1993 terrorist attack on the WTC in New York City removed several thousand square feet of concrete floor slabs in the general area of the explosion and severely damaged several buildings' communication, transportation, and utility systems. Due to the inherent redundancy of the street frames, the structures did not collapse.

The 1995 attack on the Alfred P. Murrah Federal Building in Oklahoma City revealed the vulnerability of conventional structural designs when subjected to blast loads. When a weapon is located at street level, the blast shock wave acts up against the underside of the floor slabs at upper stories. Floor slabs are not designed for this magnitude and direction of load—for this direction of load, the reinforcement is in the wrong place.

The first step in blast-resistant design is to establish the probable risk and the parameters of the threat to a facility. The risk of *collateral damage* to nearby buildings should also be considered. It is then possible to consider countermeasures (defensive strategies) to the threat. Common external blast threats are car, van, or truck bombs. Internal blast scenarios involve a smaller explosive charge packed in a letter or a brief case or a car bomb in a parking garage.

One way to and perhaps the best way to protect a building from a possible attack is to make weapon delivery difficult. A setback distance and a secure fence around the building can serve this purpose. However, this approach often is not viable in a city where buildings adjoin other buildings along busy streets. In these cases, measures such as surveillance, limits on traffic movement, and guards can enhance protection.

In the design of upgrades and retrofits of existing facilities, countermeasures that involve establishing a defensive perimeter (fences, bollards, etc.) and positioning the building at some distance from this secure perimeter often are not possible. Instead, threat countermeasures include the relocation of important functions to safer areas of the building. Other measures include hardening the mail area, moving people from external to inner offices, replacing or strengthening windows and window frames, hardening safety rooms, hardening portions of the building, or moving the entire operation to a more secure facility. In all circumstances, defensive strategies must incorporate some measures of facility access control contingency planning and emergency training for all occupants.

Blast loads are applied to external building cladding and supporting structural elements. Where windows, doors, and external walls are not expected to sustain blast loads, the loads also should be applied to internal structural elements. Floor slabs especially should be checked for uplift pressure.

Some level of blast resistance is required for new federal buildings. Existing federal buildings undergoing expansion also must include blast resistance. In each case, the GSA establishes design requirements. Specific actions can involve protecting windows, installing a secure perimeter fence and/or hardening a portion of the building, and determining the likelihood of progressive collapse and designing against it. There is no comparable universal guidance in the civilian sector. However, some of the guidance developed by the federal government is available to the general public.

An example of a steel-framed building subjected to an internal explosion was the WTC on Feb. 26, 1993. A van containing approximately 1800 lb of fertilizer-based explosive was parked on an exit ramp just south of column 324, one of the main street columns supporting the 110-story tower structure. The column measured about 4′ by 4′, across. It and six adjacent columns lost their fireproofing and lateral restraint (the bracings provided by the concrete floors there were blown out around them) but otherwise were not damaged by the explosion. The fact that the column did not buckle from the significant increase in its effective length speaks well for the redundancy in a building that was not designed for blast loading.

It is prudent for practical design purposes to adopt approximate methods that permit rapid analysis of complex structures with reasonable accuracy. These methods usually require that both the structure and the loading be idealized to some degree.

During the 1950s and 1960s, much work was done to develop simple methods for the design of structures subjected to blast loads produced by blast from nuclear weapons. (The book by J.M. Biggs contains an excellent introductory presentation of such methods.)

Although it is not practical to design buildings to withstand any conceivable terrorist attack, it is possible to improve performance of structures should one occur.

9.5.6 Design Summary

Blast-resistant design, in essence, is a design to resist shock waves without collapse.

An explosion is an instantaneous release of stored energy from an explosive material consisting of substances that are inert under ordinary conditions but undergoes a fast chemical change upon detonation. The rapid expansion of hot gases resulting from the detonation gives rise to a shock wave. Behind the shock wave front, the pressure decreases from its initial peak value, and at some distance from the charge, the pressure behind the shock front falls to a value below that of the atmosphere and raises again to a value equal to that of the atmosphere. The pressure over and above atmospheric pressure is called overpressure, while the pressure less than that of the atmosphere is called negative or suction pressure.

The blast wave duration, t_b, is typically in the range of 0.1 to 0.001 seconds. This is much shorter than the natural period, T_s, of typical structural elements. Hence, the DLF is much larger than unity close to, perhaps, 2.0 (for preliminary calculations).

A structure as a whole generally is not pushed over by an explosion. The overall mass of a structure is large enough to keep the kinetic energy imparted to the structure small enough so it can be absorbed by the multiple elements.

In many explosions that cause extensive destruction, the damage develops in two phases: (1) the energy released by the explosion degrades or destroys important structural elements, and (2) the damaged structure is unable to resist gravity and collapses beyond the area of initial damage. In some of the most devastating explosions, most of the structural damage has been caused by gravity.

Normally, individual elements fail, necessitating the activation of alternative load paths within the structure to carry the gravity loads. Studies that assess these alternate load paths need to consider the dynamic application of the redirected internal forces.

Hence, the evaluation of the full effect of a blast does not end with calculations of blast damage to individual elements or limited structural systems. We need to consider the ongoing effects in the damage structure, under the influence of gravity.

Distance is the single most important asset to a structural engineer designing for blast resistance. The farther the explosion is from the structure, the lower are the effects that the structure must resist. Further, there often is merit to the construction of blast walls to add protection to a facility.

An explosion is a violent thermochemical event. It involves supersonic detonation of the explosive material, violently expanding hot gases, and radiation of a shock front that has peak pressures that are several orders of magnitude higher than those that buildings normally experience under any other loadings.

Designers need to understand that the magnitudes of the pressures that an explosion imparts to a structure are highly dependent on several factors: (1) the nature of the explosive material, (2) the shape and casing of the device, (3) the size and range of the explosion, (4) the presence of nearby surfaces that restrict the expansion of hot gases, and (5) the robustness of the impacted surface itself. Designers also need to understand that the durations of the pressures induced by an external explosion are extremely short compared to the durations of other loads and compared to natural periods of structures.

9.5.7 Progressive Collapse

A prominent case of local collapse that progressed to a disproportionate part of the whole building was the Ronan Point disaster, which brought the attention of the profession to the matter of general structural integrity in buildings. This 22-story apartment building, discussed later in this section, was built using precast-concrete, load-bearing panels in Canning Town, England. In March 1968, a gas explosion in an 18-story apartment blew out a living room wall. The loss of the wall led to the collapse of the whole corner of the building. The apartments above the 18th story, suddenly losing support from below and being insufficiently tied and reinforced, collapsed one after the other. The falling debris ruptured successive floors and walls below the 18th story, and the failure progressed to the ground. Better continuity and ductility might have reduced the amount of damage.

Another example of progressive collapse is the Alfred P. Murrah Federal Building, Oklahoma. On April 19, 1995, a truck containing approximately 4000 lb of fertilizer-based explosive was parked near the sidewalk next to the nine-story reinforced concrete office building. The side facing the blast had corner columns and four other perimeter columns. The blast shock wave disintegrated one of the 20 × 36 in. perimeter columns and caused brittle failures of two others. The transfer girder at the third level above these columns failed, and the upper-story floors collapsed in a progressive fashion. Approximately 70% of the buildings experienced dramatic collapse. One hundred sixty-eight people died, many of them as a direct result of progressive collapse. Damage might have been less had this structure not relied on transfer girders for support of upper floors, if there had been better detailing for ductility and greater redundancy, and if there had been resistance for uplift loads on floor slabs.

9.5.7.1 Design Alternatives for Reducing Progressive Collapse

There are a number of ways to obtain resistance to progressive collapse; chief among them are the following:

1. During the design process, consider resistance to progressive collapse through the provision of minimum levels of strength, continuity, and ductility.
2. Provide alternate load paths so that the damage is absorbed and major collapse is averted.
3. Provide sufficient strength to resist failure from accidents or misuse.
4. Provide specific local resistance in regions of high risk, to have sufficient strength to resist abnormal loads in order for the structure as a whole to develop alternate paths.

9.5.7.2 Guidelines for Achieving Structural Integrity

1. Generally, connections between structural components should be ductile and have a capacity for relatively large deformations and energy absorption under the effect of abnormal conditions.
2. Good plan layout. An important factor in achieving integrity is the proper plan layout of walls and columns. In bearing-wall structures, there should be an arrangement of interior longitudinal walls to support and reduce the span of long sections of cross wall, thus enhancing the stability of individual walls and of the structures as a whole. In the case of local failure, this will also decrease the length of wall likely to be affected.
3. Provide an integrated system of ties among the principal elements of the structural system. These ties may be designed specifically as components of secondary load-carrying systems, which often must sustain very large deformations during catastrophic events.

4. Returns on walls. Returns on interior and exterior walls will make them more stable.
5. Changing directions of span of floor slab. Where a one-way floor slab is reinforced to span in the main direction, provide spanning capability in its secondary direction also, perhaps using a lower safety factor. With this approach, the collapse of the slab will be prevented and the debris loading of other parts of the structure will be minimized. Often, shrinkage and temperature steel may be enough to enable the slab to span in the secondary direction.
6. Load-bearing interior partitions. The interior walls must be capable of carrying enough loads to achieve the change of span direction in the floor slabs.
7. Catenary action of floor slab. Where the slab cannot change span direction, the span will increase if an intermediate supporting wall is removed. In this case, if there is enough reinforcement throughout the slab and enough continuity and restraint, the slab may be capable of carrying the loads by catenary action, though very large deflections will result.
8. Beam action of walls. Walls may be assumed to be capable of spanning an opening if sufficient tying steel at the top and bottom of the walls allows them to act as the web of a beam with the slabs above and below acting as flanges.
9. Redundant structural systems. Provide a secondary load path (e.g., an upper-level transfer girder system that allows the lower floors of a multistory building to hang from the upper floors in an emergency) that allows framing to survive removal of key support elements.
10. Ductile detailing. Avoid low-ductility detailing in elements that might be subject to dynamic loads or very large distortions during localized failures. Consider the implications of shear failures in beams or supported slabs under the influence of building weights falling from above.
11. Provide additional reinforcement to resist blast and load reversal when blast loads are considered in design.
12. Consider the use of compartmentalized construction in combination with special moment-resisting frames in the design of new buildings when considering blast protection.

While not directly adding to structural integrity for the prevention of progressive collapse, the use of special, nonfrangible glass for fenestration can greatly reduce risk to occupants during exterior blasts. To the extent that nonfrangible glass isolates a building's interior from blast shock waves, it can also reduce the likelihood of slab failure.

9.6 FAILURES AND DISTRESSES

When one considers the vast number of buildings in use today, the number of reported failures is few indeed. It is granted that such a situation is of satisfaction to those suffering loss of their loved ones or property damage. Failures that do occur can serve as significant source of knowledge. Investigation of failures almost always produces an insight into structural behavior that cannot be gained otherwise. Unfortunately, a great number of failure investigation reports are buried because of legal entanglements in the files of investigating engineers, insurance companies, and building owners. The purpose of this section is to describe in general terms several of the many failures that have been the subject of investigation.

It is rather amusing, but at the same time sad, to see a thick report of the investigation following a structural collapse containing a large volume of material identifying structural discrepancies that have nothing to do with the failure. The features contributing to the distress should be addressed in utmost detail, but dwelling in detail on unrelated deficiencies only results in muddying the water, causing undue embarrassment to the designers, contractor, or material supplier. This is an issue that certainly needs to be addressed by the engineering societies.

Common causes of failure: Aside from failures resulting from catastrophic wind forces or earthquakes forces, one of the most common causes of roof collapse is ponding of rainfall water. The circumstances leading to ponding failures are in general as follows:

1. Relative flat roof.
2. Improper or inadequate drainage.
3. Load-deflection characteristics of the roof structural system such that the deflection resulting from water accumulations means that more water can be accommodated, with resulting greater deflection and still more room for water. This relationship may be described mathematically in the form of a series that must be divergent for the mechanism so described result in failure.

Other common causes of failures are as follows:

1. Inadequate or improperly detailed or constructed bearing of precast beams or slabs on supporting members.
2. Shear failure in reinforced concrete flat slab or flat plate. It is of interest to note that in a typical flat slab or flat plate construction, you can't make the slab fail in flexure even if you want you—it is always by punching of the column through the slab.
3. Stability. In many cases, incipient failure is initiated in one way, but collapse actually takes place when large deflections produce instability.

9.6.1 Kemper Arena Roof Collapse

The Kemper Arena, Kansas City, Kansas, stood on a high isolated site at the outskirts of the city, its four-acre roof hanging from three, 3D trusses of steel tubing, enclosing the interior space, 360 ft (108 m) long, 324 ft (97 m) wide, and 60 ft (18 m) high. The arena built in 1973 collapsed after only 6 years.

On June 4, 1979, at about 6.45 pm, a downpour of 4.25 in. (108 mm) of water per hour began falling accompanied by a north wind gusting to 70 mph (112 km/h). Twenty-five minutes later, the central portion of the hanging roof, a one-acre area, 200 ft of 215 ft (60 × 65 m), collapsed and, acting like a giant piston, raised the interior pressure in the arena that blew out some of the exterior walls.

The structural system for the roof consisted of three giant, 300 ft (108 m) long space frame trusses spaced 153 ft (46 m) in the north–south direction. The space frame consisted of steel tubes forming an equilateral triangular truss of 54 ft (16 m) wide and 81 ft (24 m) high. Framing in between the giant trusses was a system of north–south trusses consisting of cantilever segments 54 ft (16 m) long with a 99 ft (30 m) long dropping trusses in between. Open-web joists with steel angle cords and bent rod diagonals 54 ft (16 m) long and space 9 ft (2.7 m) completed the roof framing with concrete topping on steel deck. The entire truss system in the north–south direction was hung from the bottom chords of three space frame trusses by 42 connectors or hanger assemblies, in an arrangement of 7 rows in the east–west direction by 6 columns in the north-south direction.

This schematic description together with the diagrammatic layout shown in Figures 9.31 through 9.33 should help in understanding the basic structural system of the roof.

The hanger assembly, because of its two essential functions, does not consist of a simple rod as the word *hanger* may imply (Figure 9.33). Since the weight of the roof itself is 26 psf (1.3 kN/m^2), or about 1500 tons, and since the roof was designed to carry 25 psf (1.25 kN/m^2) of additional load caused by rain, the load of the mechanical systems, and other hanging loads (or about another 1,500 tons), then each of the 42 hangers was supported in *tension* 140,000 lb (622 kN). Moreover, they had to resist the horizontals, variable wind forces (exerted on the roof and on the upper part of the walls hanging from it), which tended to move the roof horizontally as a gigantic pendulum.

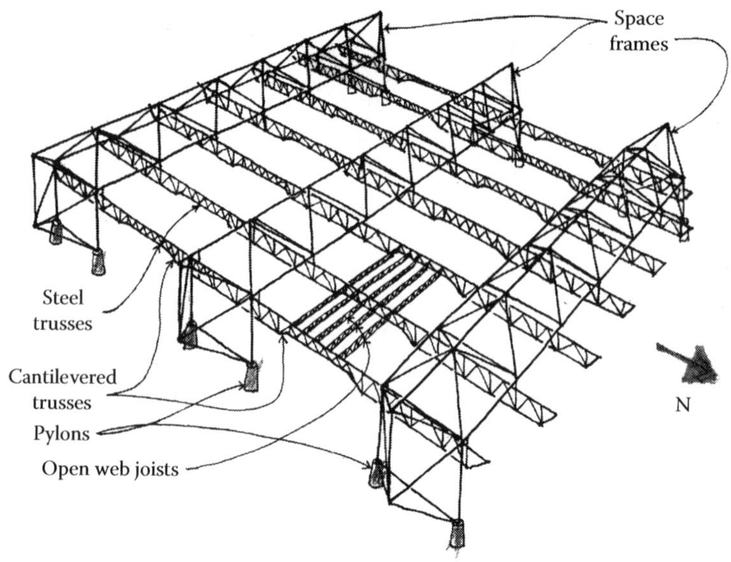

FIGURE 9.31 Kemper arena roof: schematics.

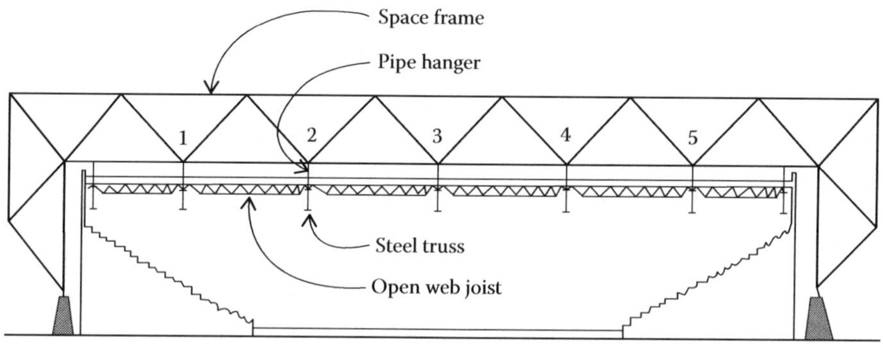

FIGURE 9.32 Schematic cross section.

For this purpose, six of the hangers were hinged both at their top and about halfway down their length, while the remaining 36 were hinged only at the top and connected rigidly to the top chord of the large trusses, while the two-hinge hangers limited the horizontal rod motions to the *bending* deflections of these hangers.

The connection between the bottom of the hangers and the top chords of the north–south trusses consisted of a steel base plate with four vertical stiffeners and four holes through which four high-strength steel bolts were tightened (see Figure 9.33). To guarantee good contact between the base plate and the trusses' top chord, a thin plate of a plastic composite was sandwiched between these two elements so that this connection was not entirely rigid. It has been estimated that during the 6 years preceding the failure, these connections were subjected to at least 24,000 oscillations, which in turn introduced oscillating variations in the initial tension of the bolts. Steel subjected to stress oscillations suffers from *fatigue* and fails at lower load values than under steady loads. Because of the type of steel used in the high-strength bolts (ASTM A490), steel codes warn against their use under variable loads, something that may not have been taken into account in the arena design.

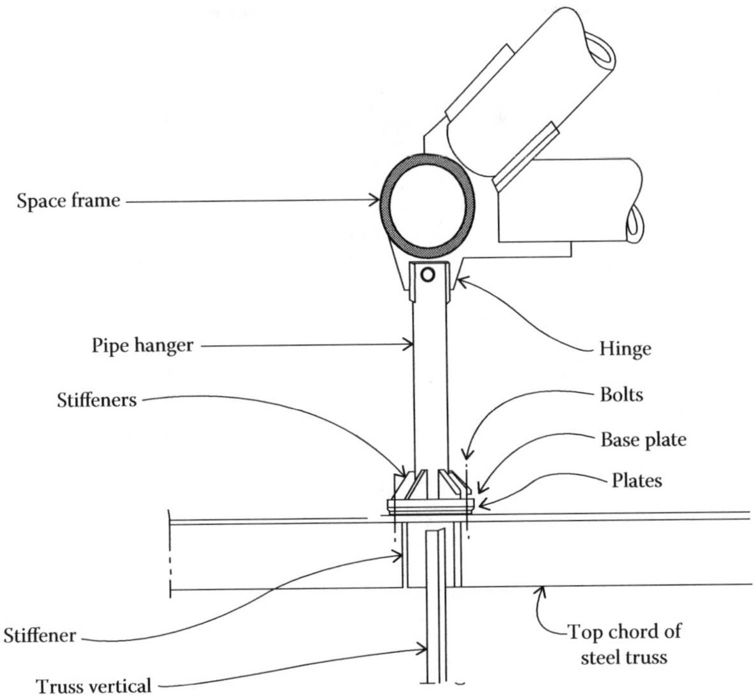

FIGURE 9.33 Kemper arena roof, roof hanger.

The 129,000 ft^2 (12,000 m^2) roof of the arena had been provided with only eight 5 in. (130 mm) diameter drains deliberately prevented from discharging more than at a modest rate (one-tenth of a cubic foot of water per second [0.0015 m^3/s]) when the water on the roof reaches 2 in. (50 mm) of depth.

The drains allowed substantial water accumulation on the roof, limited only by scuppers (openings) along the perimeter of the roof, through which the rainwater could fall directly to the ground in an emergency. These were set 2 in. (50 mm) above the roof level and enabled the roof to store at least that much water on the periphery and more in the interior of the roof, where the roof deflection would result in a greater depth of water.

The water accumulation on the roof was aggravated by two wind actions: the 70 mph (112 km/h) gusts that pushed the accumulated water from the north to the south portion of the roof and the upward suction, decreasing from north to south, created by the wind in turning from a vertical direction along the north wall to a horizontal direction on the roof. The suction also propelled the water from the north to the south portion of the roof.

The collapse of the Kemper Arena could not be explained by a single cause. It is attributed to intensity of rain downpour, drain deficiencies, wind effects, fatigue of bolts, and lack of redundancy and ponding. By September 1979, the reconstruction was in full swing, and by 1981, the *new* arena was inaugurated, just as another tragedy struck Kansas City, the Hyatt Regency walkways. When in 1983 the collapse case of the Kemper Arena reached the court and was settled in two days, the disaster had been practically forgotten.

9.6.2 Hartford Arena Roof Collapse

Throughout that winter of 1978, hundreds of roofs fell under the weight of unusually heavy snowfalls, but none was as dramatic as the Hartford collapse. Had the roof fallen six hours earlier, many the five thousand fans watching a basketball game might have been killed or injured.

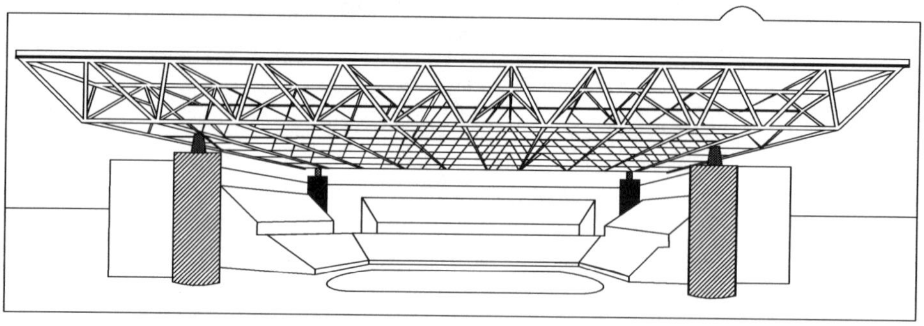

FIGURE 9.34 Hartford Center space frame, schematics.

Luckily, the fourteen hundred tons of twisted steel, gypsum roofing panels, and insulations fell on ten thousand empty seats.

The arena roof (Figure 9.34) measured 300 by 360 ft (91 × 110 m) and was constructed as a *space frame*, 21 ft (6.4 m) deep, a structure consisting of top and bottom square grids of horizontal steel bars with *nodes*, 30 ft (9 m) on center connected by diagonal bars between the horizontally staggered nodes on the upper and lower grids. The resulting space frame looked like a series of linked pyramidal trusses. The 330 ft (9 m) long top horizontals were braced by intermediate diagonals, and the main diagonals were braced at their midpoints by an intermediate layer of horizontal bars.

The top horizontal bars of most space frames perform a double function: They support the roofing panels, and they act as upper structural members of the space frame. In the Hartford roof, however, the roofing panels were supported on short vertical posts above the top nodes of the space frame (Figure 9.34). The designers claimed two advantages for this scheme: (1) If the height of the posts was varied, the roof could be sloped to provide positive drainage independently of the original level and the deflections of the top bars of the space frame, and (2) the top bars of the frame would not be subjected to bending stresses from roof loads.

Additionally, three unusual concepts characterized the design of the Hartford roof: (1) the frame's top horizontal bars were configured in the shape of a cross built up of four steel angles. (2) A truss *node* is usually the theoretical point where the center lines of all the bars connected at the node intersect, but in the Hartford frame, the top horizontal bars intersected at one point and the diagonal bars at another, somewhat below the first. Thus, the force transmitted between diagonal and horizontal bars caused bending stresses in these bars. (3) The overall space frame roof was supported on four pylon legs located 45 ft (13.7 m) *inboard* of the four edges of the space frame, rather than on boundary columns or walls.

In spite of the unusual aspects of its design, the frame appeared to be sturdy. For 5 years, it withstood the harsh Hartford weather before suddenly failing on that winter night of January 17, 1978.

A unique aspect of the construction procedure concerned the method of erection. Instead of the frame being assembled in place almost 100 ft (30 m) above the ground, a costly, time-consuming, and somewhat dangerous procedure, it was completely assembled on the ground. Not only was the structure bolted together, but the heating and ventilation ducts, the drain pipes, and the electrical conduits, as well as the service catwalks, were assembled while the structure sat on the ground. Assembly of the roof frame, which begun on the floor of the arena in February 1972, was completed by July of that year.

The lifting process was completed in two weeks by means of hydraulic jacks fixed to the top of the four pylons.

In January 1973, the roof, in its final position but not yet loaded by the weight of the roof deck, was measured to have a deflection at the center *twice* that predicted by the analysis. When notified of this condition, the *engineers* apparently *expressed no concern*.

By mid-1974, the roof was completed. Five years later, it collapsed. The first questions explored by the experts concerned the weight of snow and ice that has accumulated on the roof the night of January 17. Accurate measurements showed that the actual live load due to the accumulated snow from the two storms preceding the collapse was about half the live load specified by the code, but the weight of the roof (the dead load) turned out to be 25% greater than that assumed in the design. However, the sum of the dead and live loads was less than the total load assumed in the design.

Attention was directed to the configuration of the actual structure in comparison with the mathematical model postulated by the designers. The structural model had assumed that all the top chords were braced laterally by the inclined secondary diagonals, and this was the case in the interior of the space frame where diagonals form a pyramid. But along the frame ledges, the diagonals and top bars were in the same inclined plane; hence, buckling out of this plane was not prevented. The top bars were free to buckle, in a direction perpendicular to that plane. The prevention of buckling would have required the outer top horizontals to be four times stiffer than the typical interior top horizontals because the outer horizontals had twice the unbraced length. Since the top horizontals were the same size as the interior horizontals, they were destined to buckle.

Once the wreckage was cleared away, true to the promise of the city fathers, a bigger stadium seating 4000 more spectators than the old one was built. Its roof was simpler, with two ordinary parallel vertical trusses sitting on the same four pylons raised up 12 ft (3.6 m) to fit the grander facility. Secondary trusses were framed into these primary trusses at six locations, and tertiary trusses framed into the secondary ones, resulting in a grid of trusses bearing a family resemblance to the original roof. The revamped coliseum began to take shape 16 months after the collapse and by the spring of 1980 was ready.

9.6.3 Ronan Point: Progressive Collapse

Ronan Point (Figure 9.35) was a 22-story residential building in Newham, East London, which suffered a partial collapse when a gas explosion blew out a load-bearing wall, causing the collapse of one entire corner of the building. Four people were killed in the accident and seven

FIGURE 9.35 Ronan Point, a 22-story large precast panel building suffered a partial collapse when a gas explosion demolished a load bearing corner wall.

were injured. The tower destroyed by the explosion was the second of a planned nine identical building complex to be built on the site. The construction system used room-size panels of reinforced concrete for load-bearing walls and floors, stacked them like a house of cards to create housing units, and was one of many similar systems brought forth after the end of World War II. A severe housing shortage had been brought about throughout Europe by the extensive destruction caused by the war that led to the welcome introduction of many large concrete panel prefabrication systems. The increased productivity obtained by shifting a substantial part of the building process from the site to the factory meant savings in cost, reduction of manpower, and shortening of construction time, desirable goals at any time but particularly after the war. Although differing in details, all these systems involved floor panels sitting on wall panels. The joints between these two types of panels were usually filled with grout, a cement–sand and water mixture, and sometimes strengthened by reinforcing steel so placed as to lock the panels together, providing continuity and mutual interaction. The Larsen–Nielsen system had joints with grout between tooth-edged floor panels but no steel reinforcing to provide a sound connection between these panels and the wall panels above and below. Thus, a sufficient horizontal force, like that of an explosion, could easily push a wall panel off the floor. The explosion was deemed to have caused a pressure against the wall panels of less than 10 psi (0.07 N/mm^2). Yet tests showed that this pressure was high enough to cause the failure of a reinforced concrete wall either by bending it or by overcoming the friction resulting from the gravity loads and kicking it out. Further tests showed that the wall panel would slide out against the floor panel at a pressure of 2.8 psi (0.02 N/mm^2), less than one-third that of the explosion.

Once one wall panel blew out, the wall panels above it were left unsupported and fell. The floor panels that were consequently left virtually unsupported then crashed down on the floor below, overloading it and causing a progressive collapse of all the walls and floors below. This highly unusually mode of failure led to a reevaluation of building regulations around the world, in terms both of safety and of unusual loads. The importance of continuity in joints of buildings and of redundancy in structures was awakened in the profession. Redundancy implies that a structure can carry loads by more than one mechanism—that is, that the forces on it can follow alternate paths to the ground. It guarantees that if one mechanism fails, loads can still be carried by other mechanisms.

An explosion is such a rare event that codes do not make it the basis for design, except for certain military buildings, but as a consequence of the Ronan Point catastrophe, a design philosophy became accepted that considers the possibility of an explosion capable of destroying its surrounding area without causing substantial damage elsewhere in the structure. This approach guarantees that a building will not fail in a progressive manner even if some structural elements are severely strained.

But what was the real cause of the explosion at Ronan Point? It was uncovered by the tribunal only after interviewing dozens of witnesses. Some weeks before the disaster, a friend of Miss Hodge, Charley Pike, had offered to install the stove in her kitchen. Since he was not a professional plumber, he did not pay particular attention to the fittings required to make the connection between the pipe behind the stove and the gas riser that distributed gas throughout the building. A brass nut connecting the two pipes (later found to be below the standards set by the British Gas Board) could have been easily fractured when overtighted by a wrench and have caused a slow leak of gas. In fact, this is exactly what happened, although Miss Hodge did not smell the gas, possibly because of her half-awake state when she lit the match. The immediate consequence of the Ronan Point disaster was to cause the discontinuance of gas service to all similarly designed large concrete panel structures, of which there were more than 600 throughout Great Britain.

The need to strengthen Ronan Point was obvious, but discussion about what to do and who should pay for the remedial work moved the matter into the sociopolitical arena. Within a year of the disaster, the debris was cleared away and the wing rebuilt with a blast angle, a reinforced joint detail preventing the separation of the wall from the floor. The tower was reoccupied; but in 1984, cracks began appearing in other walls, and the entire tower was evacuated. Eventually, in May 1986, the building was swathed in a coat of reinforced polystyrene to contain the dust of demolition, and the tower was demolished, floor by floor, in a procedure that reversed that used in construction.

In a startling afterword to the disaster, evidence of incredibly shoddy workmanship was revealed as the tower was demolished. The tower had been dismantled rather than blown up because the joints between the walls and slab, supposedly packed with mortar, were discovered to be full of voids and rubbish. Upon this revelation, hundreds of similarly built apartment towers were deemed unsafe and also demolished. As late as 1991, six were blown up simultaneously in Salford, England, closing a sad chapter unjustly blamed on the lack of conscience of the 1960s.

9.6.4 Standard Oil of Indiana Building, Chicago: Curtain Wall Distress

The Standard Oil of Indiana office building in Chicago was started in 1971 and inaugurated in 1974. With 82 stories above ground on one-quarter of the site, the building, now renamed the Amoco Tower, reached 1123 ft (342 m) into the sky from a square base 186 ft (57 m) on a side and was entirely clad in white Carrara marble. The technology developed in the 1970s for cutting thin marble panels had allowed the delivery from Italy of forty-three thousand panels, 50 by 45 in. (1.27 × 1.14 m), $1^1/_4$ to $1^1/_2$ in. (32–38 mm) thick, to be connected to the steel structure through bolts. The shiny appearance of the tower was enhanced by four cutout corners, 14 ft (4.4 m) on a side that made it look like an even more slender than the tower's height-to-width ratio of 6.2.

The design engineers minimized the cost of the steel structure needed to resist gravity, wind, and earthquake loads by means of an ingenious structural design. They used a perimeter tube system by building the four walls of the tower out of thin steel plates. Slit windows make up less than 50% of the façade without unduly decreasing its stiffness. The tower oscillates horizontally with a period of 8.3 s, slow enough to prevent drift problems to the tenants.

The steel plate of the tube tower was too thin to carry to the ground, without buckling, the loads usually supported by the outside columns of a frame. In the Amoco Tower, thick-plate *chevrons* of steel, connected to the tube in the spaces between the windows, constitute hollow triangular columns capable of carrying vertical loads to the ground, besides incorporating the vertical water pipes and the ducks of the mechanical systems. Fireproofing and thermal insulation were also attached to the chevrons, and marble veneer was bolted over this structural sandwich.

Shortly after the building was completed, it was noticed that the marble panels were beginning to buck outward. By 1988, 30% of the panels had bowed out more than $^1/_2$ in. (13 mm) and some as much as $1^1/_2$ in. (38 mm). For safety reasons, all were immediately bolted through a steel clip attached to the structural frame. Tests then proved that many panels had lost a large part of their strength. Marble is a limestone, a sedimentary rock wholly or in large part composed of calcium carbonate and mostly formed by the deposition and consolidation of the skeleton of marine invertebrates, successively metamorphosed into solid rock by heat and pressure. Like all limestones, marble is corroded by the humidity and the acid fumes of a polluted atmosphere and loses strength, buckling out seven under its own weight.

In addition to the chemical action of corrosion, the panels of the Amoco Tower were subjected to the compressive stress resulting from the partial prevision of free thermal expansion by the semirigid bolt connections. These stresses were particularly high in the Windy City, where temperatures extremes go from −27°F (−35°C) in winter to +102°F (+39°C) in summer. The horizontal displacements of the tower under the high winds of Chicago may also have contributed in a minor way to the panel's failure.

On the other hand, there is no question about the reduced thickness of the panels being a major contributory cause of their failure. The buckling strength of a compressed, thin structural member is measured by the dimensionless geometrical parameter L/r, the slenderness ratio, where L is the length of the member and r, the radius of gyration. In the Amoco Tower, the L/r of the panels varied between 104 and 138, while most building codes, in order to guarantee against buckling failures, required values of L/r not greater, and often smaller, than 120.

In 1989, the Amoco Corporation decided to substitute all the marble panels with 2 in. (50 mm) thick granite panels from Mount Airy in North Carolina. Besides being much stiffer, these panels are less subject to chemical corrosion by polluted air because granite is a granular rock composed chiefly of hard crystals of quartz solidified from molten rock or magma, from the hot core of the earth.

There is a sad moral to this story. The unusual care of the architectural and engineering designers of the Amoco Tower, addressed to both the safety and the economy of construction of their splendid building, could be said to have been diminished by a minor saving in the choice of the cladding (less than 1% of the original cost).

9.6.5 Hancock Tower, Boston: Curtain Wall Distress

The Hancock Tower is a 790 ft (234 m) tall building with a conventional steel frame clad with 41/2 by $11^1/_2$ ft (1.35 × 3.45 m) double-glazed glass panels. The panels, available since the 1960s but never used on such a tall building before, had the inside face of the outer glass sheet (or light) coated with a thin layer of reflective material and a lead spacer around the edges to separate the inner from the outer glass lights. The panels' exceptional vertical dimension allowed for the first time continuous glass surfaces over the entire facades of the tower.

The tower satisfied all the requirements of the governing codes, but it must be noticed that one of the most essential requirements of high-rise construction—the limitation of wind sway at the top of the building—was not then, nor is it now, regulated by the code; it is left, instead, to the judgment of the structural designer. In the Hancock Tower, a building with an asymmetrical plan 300 ft (92.4 m) by 104 ft (30 m) in the shape of a thin rhomboid (Figure 9.36), the wind sway in the short direction was within the dictates of good engineering practice but larger than that usually chosen by conservative structuralisms.

The first of what were to be called four difficulties in the construction of the tower became apparent during the excavation of the site. It produced serious settlements and structural damage to the adjacent Trinity Church, built from 1872 to 1877 across the street from the tower. The sheet piling and the lateral braces of the excavation, designed to prevent the collapse of its vertical cuts, moved laterally as much as 3 ft (0.9 m) and careful monitoring of Trinity Church's movements showed them to be perfectly correlated with the progress of the excavation. The Hancock Company accepted full responsibility for this damage.

Then, on the night of January 20, 1973, a windstorm of unusual severity hit Boston, and a number of the glass panels being erected on the tower were blown out. The falling panels, caught in the air turbulence around the building, hit and cracked numerous other panels that were dismantled to avoid further breakage. Eventually, about one-third of the erected panels were replaced with temporary plywood panels. The facade failure started a series of investigations that eventually required radical changes in both the structure and the curtain wall of the tower.

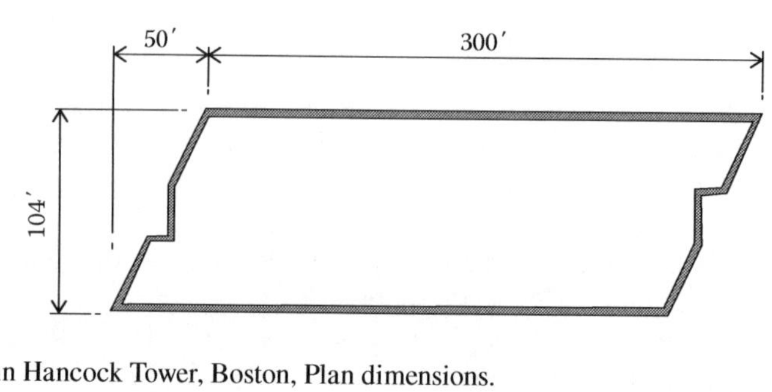

FIGURE 9.36 John Hancock Tower, Boston, Plan dimensions.

When glass panel damage occurs in a high-rise building, structuralists assume that it is due to the lateral deflection of the steel frame under the action of the wind, unless it is caused by unusual settlements of the soil. Experts from the faculty of the civil engineering department of the Massachusetts Institute of Technology (MIT) were called to give their opinion on the tower's facade failure, but their measurements of the wind displacements of the frame, coordinated with measurements of wind speeds at the top of the tower, did not explain the damage to the panels. Instead, they raised doubts about the magnitude of the displacements and the safety of the tower itself during a windstorm. As pointed out elsewhere in this text, the acceleration of the wind oscillations has an influence on the physiological reactions of the occupants: Resonant accelerations make people airsick and from this point of view, the accelerations measured on the Hancock Tower seemed to have unacceptable values. Since wind effects on buildings depend not only on their structure but on their shape, the shape of the surrounding buildings, and the lay of the land, wind tunnel tests were conducted, which confirmed that the wind sway was not responsible for the panel failures but rather was the cause of motions unacceptable for the comfort of the occupants. As expected, these motions were characterized not only by large displacement in the short direction of the tower but also by twisting motions caused by the narrow dimension of the building in the short direction and partially by the rhomboidal shape of its plan.

William LeMessurier, following the wind engineer's advice, designed two TMDs consisting of two masses of lead, each weighing 300 tons, to be set on a thin layer of oil near the opposite ends of the 58th floor of the tower. The dampers were connected to the structure by springs and shock absorbers, which allowed oscillations of their masses in the short direction of the building. When the tower oscillated mainly in the short direction, both dampers oscillated together also in the short direction but in opposition to the building, thus damping the bending motions of the tower (Figures 9.37 and 9.38); when the tower developed essentially twisting oscillations, the dampers moved in opposite directions to each other and counteracted the opposite oscillations of the two ends of the tower.

Through all this, the glass panel failure remained unexplained. Research proved that the panels had been correctly installed, while the wind tunnel tests had shown that the cracks in the glass were not due to the wind motions, and the panels had not cracked at the *hot spots* located on the facade by the wind tests. The true cause of the panels' failure was revealed from tests results obtained during this time by an independent laboratory that simulated the thousands of wind oscillations and thermal stress cycles caused by the expansion and contraction of the air between the two glass lights of the panels. It was noticed that while the two lights were identically supported and designed to share equally the wind pressures and suctions, the wind loads in most cases first cracked the outer light both in the lab tests on the tower facades. The lab researchers, looking at the edge of the lead solder binding the reflective material sheet to the spacer, found that it had chips of glass in it. The researchers were finally able to prove that the lead connection around the panels' edge had developed fatigue between the reflective coating and the outside light, as well as that between the reflective coating and the lead sealer.

Unfortunately too late, a last investigation showed that the material of the reflective panels had given the same kind of trouble in previous installations on smaller buildings. All the 10,344 identical panels of the Hancock Tower were replaced with single-thickness tempered glass.

9.6.6 Hyatt Regency Walkways Collapse

The Hyatt Regency Complex consists of three connected buildings: a reinforced concrete tower on the north end, housing the guests' bedrooms and suites; a 117 by 145 ft (34 × 44 m) atrium with a steel and glass roof 50 ft (15 m) above the floor; and at the south end, a four-story reinforced concrete *function block*, containing all the service areas—meeting rooms, dining rooms, kitchens, etc. The tower was connected to the function block by three pedestrian bridges, or walkways, hung from the steel trusses at the atrium roof: two, one above the other, at the second- and fourth-floor

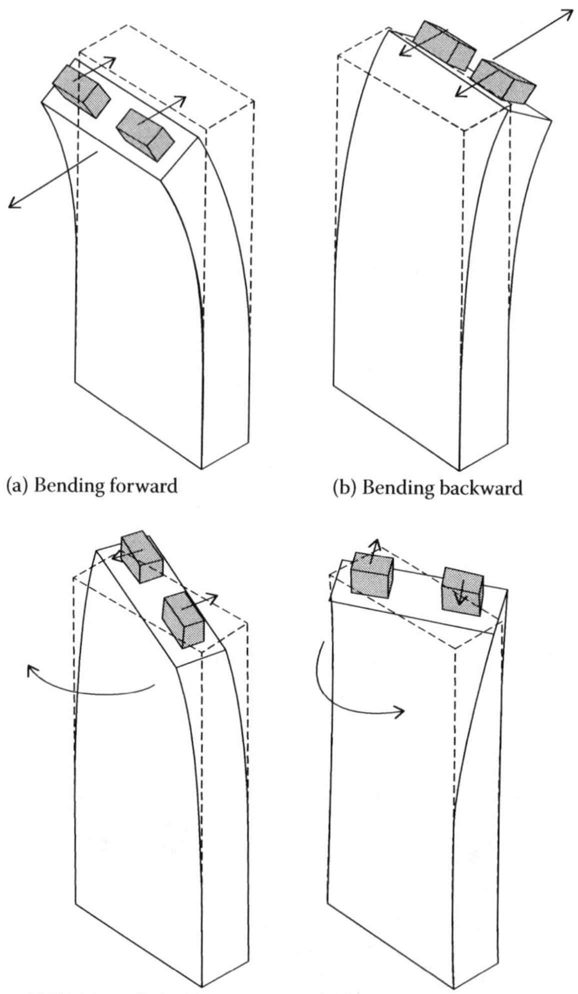

(a) Bending forward (b) Bending backward

(c) Twisting clockwise (d) Twisting counterclockwise

FIGURE 9.37 Tuned dynamic dampers, schematics.

levels near the west side of the atrium and one at the third-floor level near the east side of the atrium (Figure 9.39). The main purpose of the walkways was to permit people to pass between the tower and the function block without crossing the often crowed atrium

At 7:05 pm on Friday, July 17, 1981, when the atrium was filled with more than sixteen hundred people, most of them dancing to the music of a well-known band, the fourth-floor walkway suddenly dropped from the hangers holding it to the roof structure, leaving the hangers dangling like invalid stalactites. Since the second-floor walkway was hung from the fourth-floor walkway, the two began to fall together. There was a large roar as the concrete decks of the steel-framed walkway cracked and crashed down.

The final count due to one of the worst structural failures to occur in the United States resulted in 11 dead and over 200 injured, many maimed for life.

Within a few hours of the accident, rumors about the cause of the failure began to fly. Then technical opinions blossomed: Since the people on the two walkways were stomping in rhythm with the music, the up and down vibrations of the walkway must have had exactly the same rhythm; technically, they were in resonant with the impacts of the stomping people and therefore cause the failure.

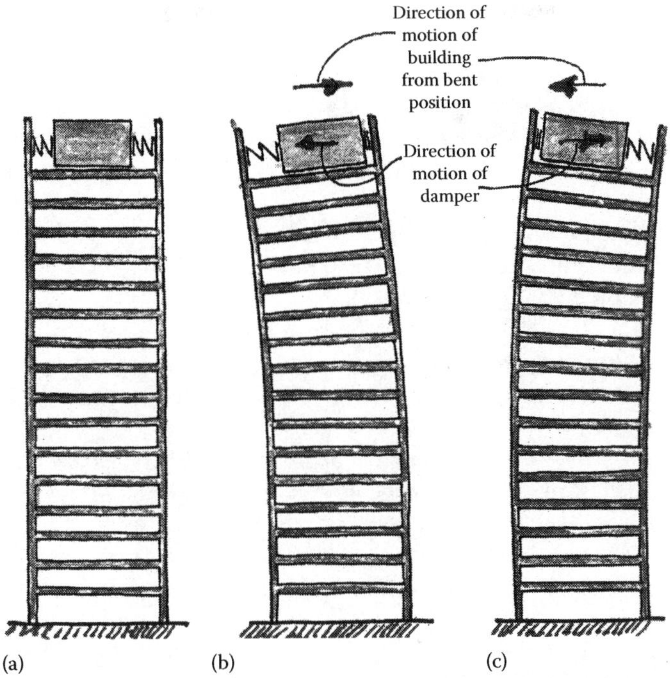

FIGURE 9.38 Tuned mass damper, (TMD), schematics. (a) At rest, (b) building bending to the left, and (c) building bending to the rig.

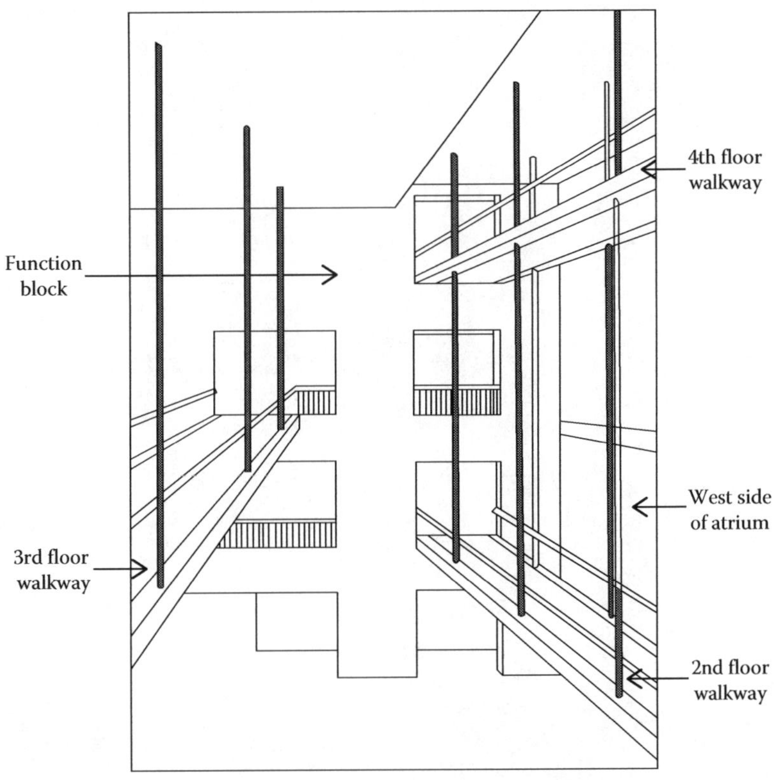

FIGURE 9.39 Atrium of Hyatt Regency Hotel, Kansas City, Missouri, schematics.

Continued resonance can quickly destroy even a sound structure. Then the quality of the materials used in the walkways or the skill of the workers who welded and bolted them together became suspects. For a relatively long time, the only unsuspected members of the construction team were the architects and the design engineers.

The management company of the hotel was the first to take action. It asked the design team of the hotel to prepare the drawings for the second-floor walkway supported by columns and authorized its immediate construction.

Shortly thereafter, at the request o the Kansas City mayor, the federal government authorized the National Bureau of Standards to perform an official investigation with *the objective of determining the most probable cause of the collapse.*

How could this tragedy have occurred in the year 1981 in the most advanced technical country in the world and after two years of design and two of construction? In order to clarify this mystery, we must understand how the walkways were originally designed and how they were eventually built.

The two walkways on the west side of the atrium involved in the collapse (the third-floor walkway that was separately hung remained in place) consisted of four 30 ft (9 m) long spans on each side, consisting of two longitudinal wide-flange steel beams each 16 in. (400 mm) deep. The four 30 ft (9 m) beams were connected by steel angle bolted to the upper flanges at the beams' ends, thus spanning the 120 ft (36 m) atrium width. The south ends of the walkways were welded to plates in the floors of the function block, and their north ends were supported on sliding bearings in the floors of the tower. The purpose of the sliding supports was to allow the beams to expand or constrict with temperature changes without giving rise to thermal stresses.

Intermediate supports of the walkways at each end of the 30 ft (9 m) beams consisted of transverse box beams, fabricated by butt welding along their entire length two 8 in. (200 mm) deep channels (Figure 9.40). In the original working drawings, each box beam had single holes

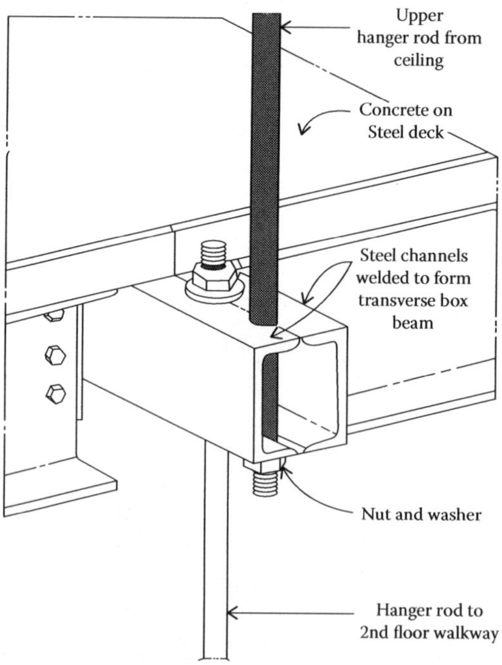

FIGURE 9.40 Box beam hanger detail: as built, Hyatt Regency Hotel.

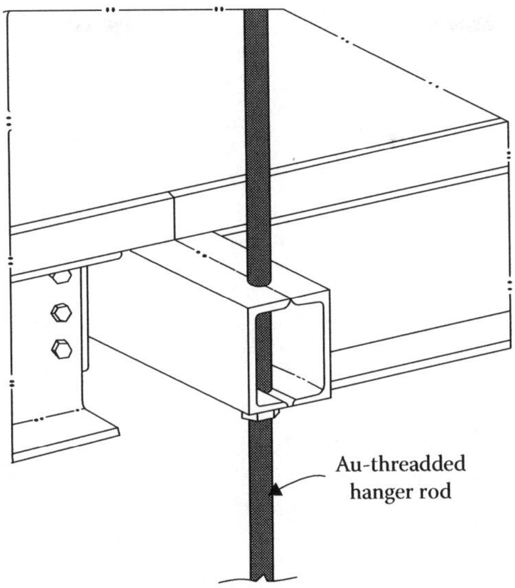

FIGURE 9.41 Box beam hanger detail: as designed, Hyatt Regency Hotel.

at both ends of the flanges (Figure 9.41), though each of which was threaded a single 1¼ in. (32 mm) steel rod that served as hanger for both the second- and fourth-floor walkways. In this design, the load of both walkways was supported every 30 ft by means of nuts screwed into a single rod on each side of the walkways at the level of the second-floor and the fourth-floor box beams. Thus, the single rods hung from the steel trusses of the atrium's roof supported the weights of both walkways, but the box beams of each walkway supported only the loads on that single walkway.

In the shop drawings (the final drawings submitted by the contractor to the design engineers and the architects), each end of the fourth-floor box beams had two holes through both flanges, one at 2 in. (50 mm) from the end and the other at 6 in. (150 mm) from the end (Figure 9.40). Two upper hangers, ending at the fourth-floor level and consisting of 1¼ in. (32 mm) rods, went through the outer hole in each box beam of the fourth floor and supported the fourth-floor walkway only by means of nuts and washers at their lower end, that is, below the box beams of the fourth-floor level, and went through the inner hole of each fourth-floor box beam, supported by a nut and washer at the upper ends, that is, above the fourth-floor box beam, and supported at their lower ends the second-floor walkway. This design was changed as suggested by the contractor in the shop drawings and stamped *approved* by the architects and *reviewed* by the structural engineers. In the contractor's design, the loads of both walkways were transmitted to the roof trusses by the shorter upper rods, which passed through only the fourth-floor box beams and supported the second-floor walkway by two additional shorter rods hanging from the fourth-floor box beams. Thus, in this design, the fourth-floor transverse box beams supported the loads of two walkways, rather than the one of the original design.

The investigation revealed that the weak elements in the chain of structural elements were the box beams of the fourth floor. But since the stress analysis of the complex beams could not be accurately obtained by theoretical calculations, the beams were tested in the laboratory using both brand-new duplicates of the box beams and some of the undamaged actual box beams.

The six upper hanger rods, carrying the load of the two walkways, pulled up on the thin lower flanges of the fourth-floor box beams through a single nut and bolt connection. Under this load (twice the design load), the bolt first bent the lower flange of the box beams, then broke through the lower hole in it, pulled out of the hole in the upper flange, and became disconnected from the

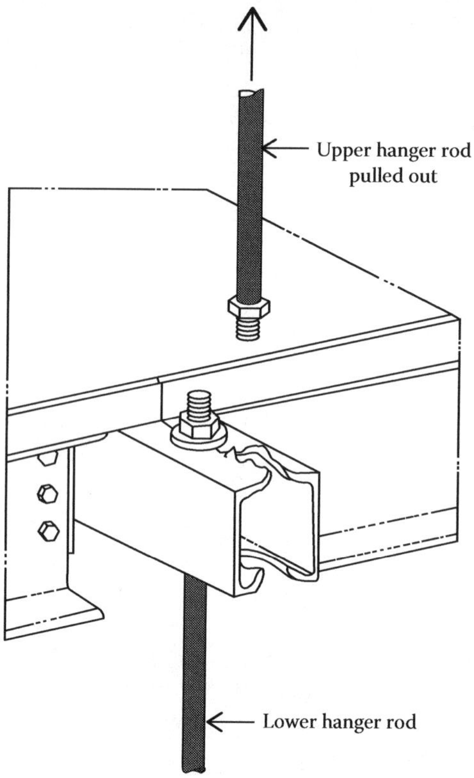

FIGURE 9.42 Pulled-out rod at fourth-floor box beam.

box beam (Figure 9.42). This first happened at the midspan upper hanger rod; the remaining upper rods, incapable of taking over the load unsupported by the failed rod, pulled out of their holes, and both walkways fell down. The walkway system not only was underdesigned but also lacked redundancy, a most prudent reserve of strength structures. The suggestion of the contractor, aimed at simplifying the construction of the walkways, was fatal because it went unnoticed by the design engineers.

An abbreviated form the conclusions of the National Bureau of Standards report is as follows:

1. The walkways collapsed under loads substantially less than those specified by the Kansas City Building Code.
2. All the fourth-floor box beam–hanger connections were candidates of initiation of walkway collapse.
3. The box beam–hanger rod connections, the fourth-floor floor-to-ceiling hanger rods, and the third-floor walkway hanger rods did not satisfy the design provisions of the Kansas City Building Code.
4. The box beam–hanger rod connections under the original hanger rod detail (continuous rod) would not have satisfied the Kansas City Building Code.
5. Neither the quality of workmanship nor the materials used in the walkway system played a significant role in initiating the collapse.
6. Under the original engineering design of the connections, which did not satisfy the code, the walkways might not have collapsed under the actual loads on them on July 17, 1981.

9.7 BUCKLING OF BUILDING UNDER ITS OWN WEIGHT

Let us consider the tantalizing possibility of a prismatic tall building buckling under its own weight. Can this phenomenon ever occur in a particle, super slim, super-tall building? Let us examine this probability by considering an equivalent cantilever with its lower end vertically built-in and the upper end free. We can, for analytical purposes, assume the building weight is uniformly distributed along the height of the equivalent vertical cantilever (Figure 9.43a). The cantilever shown in

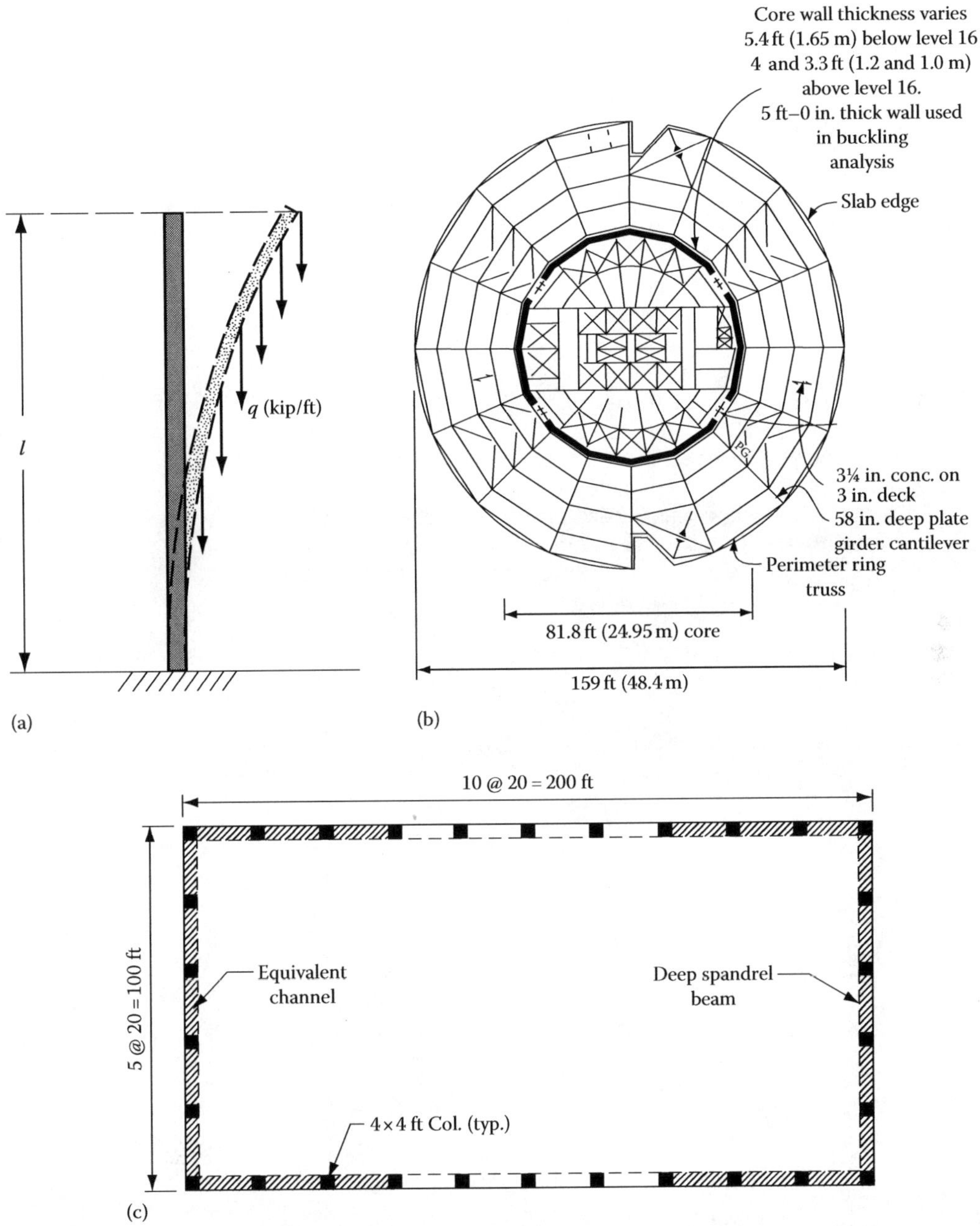

FIGURE 9.43

the figure has a tendency to buckle under its own weight as shown by the dotted line. This problem involving solution of a differential equation of the deflected curve was discussed first by Leonard Euler (1707–1783) and then eventually solved by mathematician A.G. Greenhill (1793–1841). He established that the lowest buckling load, q_{cr}, per unit height of the cantilever is given by

$$q_{cr} = 7.737\ EI/L^3$$

where

E is the modulus of elasticity of the construction material of the building
I is the moment of inertia of the building gin the direction of bending
L is the height of the building

9.7.1 Circular Building

To get a feel for what this equation means for a contemporary tall building, let us consider a 60-story building shown in Figure 9.43b. The lateral resistance is provided entirely by the interior core walls, assumed here for analytical purposes to be 5 ft thick. For purposes of analysis, we assume uncracked properties.

Example 1

1. Building plan area = $\pi\ 159^2/4$ = 19,855 ft^2.
2. Building height at 12 ft floor-to-floor—12 × 50 = 600 ft.
3. Unit dead load including self-weight of structural and nonstructural elements, curtain walls, interior partitions, finishes, and allowance for sustained live load = 225 psf or floor area. Therefore, the self-weight, $q = 225 \times 19{,}855/12 = 372.3$ kip/ft.
4. Moment of inertia (I) about the axis through the center = 0.049087 ($81.8^4 - 71.8^4$) = 893,196 ft^4
5. Modulus of elasticity € for concrete shear walls = 4415 KSI = 638,760 ksf.
6. The lowest buckling load q_{cr} is given by

$$q_{cr} = 7.837EI/1^3$$

$$= 7.837 \times 635{,}760 \times 893{,}196/600^3 = 20{,}603 \text{ kips/ft.}$$

Comparing this to the estimated self-weight of 372.3 kips/ft, we notice we have quite a comfortable margin of safety against buckling = 20,603/372.3 = 55.34. Even with an overly pessimistic assumption of fully cracked walls, that is, $I_{cracked} = I_{gross}/2$, our margin of safety is quite healthy = 55.34/2 = 27.67, which is pretty large indeed.

Example 2

Building height 40 stories at 13.5 ft floor-to-floor = 40 × 13.5 – 540 ft

Typical floor area = 100 × 200 feet = 20,000 ft^2

Floor weight at 225 psf = 20,000 × 225/1000 = 4500 kips

Unit weight, $q = 4500/13.5 = 333.3$ kips/ft

$E = 635{,}760$ ksf
I = 704,512 ft^4 (assuming two equivalent channels as shown in Figure 9.43c and uncracked properties)

Critical buckling load, $q_{cr} = 7.837EI/l^3$

$$= 7.837 \times 635{,}760 \times 704{,}512/540^3$$

$$= 22{,}292 \text{ kips/ft}$$

Comparing this to the estimated weight $q = 333.3$ kip/ft, it is seen that we have a lavish factor of safety against buckling of the building under its own weight.

Based on these two examples (Examples 1.2 and 1.3), it is reasonable to conclude that an adequate factor of safety exists even for super-tall buildings in the 2000 ft plus range.

9.8 FOUNDATIONS

9.8.1 Footings, Mats, and Piles

Structural requirements for footings, foundation mats, and piles are given in Sections 21.10.2.1 through 21.10.2.5 and in Section 22.10 of ACI 318-11. They are summarized as follows:

- Longitudinal reinforcement of columns and structural walls resisting earthquake-induced forces shall extend into the footing, mat, or pile cap and shall be fully developed for tension at the interface.
- Columns designed assuming fixed end conditions at the foundation shall comply with Section 21.10.2.1.
- If longitudinal reinforcement of a column requires hooks, the hooks shall have a 90° bend and shall be located near the bottom of the foundation with the free end of the bars oriented toward the center of the column.
- Transverse reinforcement in accordance with Section 21.4.4 shall be provided below the top of a footing when columns or boundary elements of special reinforced concrete structural walls have an edge located within one-half the footing depth from an edge of a footing. The transverse reinforcement shall extend into the footing a distance greater than or equal to the smaller of the depth of the footing, mat, or pile cap, or the development length in tension of the longitudinal reinforcement.
- Flexural reinforcement shall be provided in the top of a footing, mat, or pile cap supporting columns or boundary elements of special reinforced concrete structural walls subjected to uplift forces from earthquake effects. Flexural reinforcement shall not be less than that required by Section 10.5.
- The use of structural plain concrete in footings and basement walls is prohibited, except for specific cases cited in Section 22.10.

9.8.2 Grade Beams and Slab on Grade

Requirements for grade beams and slabs on grade given in Sections 21.10.3.1 through 21.10.3.4 are summarized as follows:

- Grade beams acting as horizontal ties between pile caps or footings shall have continuous longitudinal reinforcement that shall be developed within or beyond the supported column. At all discontinuities, the longitudinal reinforcement must be anchored within the pile cap or footing.
- Grade beams acting as horizontal ties between pile caps or footings shall be proportioned such that the smallest cross-section dimension is greater than or equal to the clear spacing between connected columns divided by 20, but need not be greater than 18 in.
- Closed ties shall be provided at a spacing not to exceed the lesser of one-half the smallest orthogonal cross-section dimension or 12 in.
- Grade beams and beams that are part of the lateral-force-resisting system shall conform to Section 21.3.
- Slabs on grade that resist seismic forces from columns or walls that are part of the lateral-force-resisting system shall be designed as structural diaphragms per Section 21.9.
- The design drawings shall clearly state that the slab on grade is a structural diaphragm and is part of the lateral-force-resisting system.

9.8.3 Piles, Piers, and Caissons

Requirements for piles, piers, and caissons are given in Sections 21.10.4.2 through 21.10.4.7. They must be summarized as follows:

- Piles, piers, or caissons resisting tension loads shall have continuous longitudinal reinforcement over the length resisting the design tension forces. The longitudinal reinforcement shall be detailed to transfer tensile forces between the pile cap and the supported structural members.

9.8.4 Effect of Seismic Forces on Foundation Design

Path of seismic forces: The base shear of the building, *W*, is a measure of the horizontal force transmitted from the ground to the building. The media used for this transmission of horizontal force may be friction between floor slab and ground, friction between bottom of footing and ground, and passive resistance of earth against vertical surfaces of footings, grade beams, or basement walls. The exact path that this transmission of forces will follow is not readily subject to a mathematical solution.

9.8.4.1 Footing and Raft Foundations

For ordinary nonseismic condition, a suitable spread footing or raft (mat) foundation must (1) be placed on a suitable depth, (2) be safe with respect to a bearing-capacity failure, and (3) not undergo settlements or heaving that will be detrimental to the structure. Primary damage to structures can be caused by differential settlements. An estimate of total and differential settlements that the structure can safely tolerate shall be made by the structural engineer. The foundation engineers generally provide criteria for proportioning footings to equalize settlements and size so as not to exceed allowable bearing pressures. In case of an earthquake, the overturning effect must be carried into the foundations and a careful analysis of permissible overloads for combined effect of vertical and lateral loads shall be made as part of the foundation design; the unbalanced tensile force from the supported structure, if any, must be resisted by anchorage into the foundation. The problem is to provide stability against overturning for the short-time loading during an earthquake (or wind) without imposing such restrictions as to create wide disparity in foundation settlements under normal loading. This disparity could create more damage to the structure than that which might occur in an earthquake under highly increased soil pressures. The soil pressure resisting combined static prescribed seismic loads can generally exceed the normal allowable pressure for static loads by 1/3. However, the various types of soils react differently to short-time seismic loading and any increase over normal allowable static loading will be confirmed by a soil analysis. In no case will the footing size be less than that required for static loads alone.

9.8.4.2 Pile Foundation

Earthquake vibrations may cause consolidation or liquefaction of loose soils, and the resulting settlement of building foundations is rarely uniform. In the use of rigid structures supported on individual spread footings, bearing on such material differential settlements can result in damage to the superstructure. Either the stabilization of the soil prior to construction or the use of piles caissons or deep piers bearing on a firm stratum is a solution to this problem. For pile-supported structures subjected to horizontal loads, it must be decided whether the lateral-load-carrying capacity of the vertical piles is adequate or whether batter piles should be used. The lateral-load-carrying capacity of vertical piles is dependent on the properties of the soil; the size, length, and material of the pile; the pile grouping and spacing; and the duration and frequency of loading. These factors should be taken into consideration in estimating the ability of vertical piles to withstand the horizontal loads. Some values that have been used for the allowable horizontal loads at the tops of piles vary from 1000 to 1500 lb per pile. In analyzing groups of battered piles, it may be assumed that the pile caps are rigid, that the piles are end bearing and pinned at the top, and that analysis of any arrangement is possible by simple statics. However, the moments resulting from this assumption due to eccentricity of application of pile loads should be resisted by the pile caps and members interconnecting the pile caps. For piles subject to both vertical and horizontal loads, it may be assumed that the applied horizontal thrust is equally shared between the horizontal components of the batter piles.

9.8.4.3 Load Capacity of Piles

Groups of piles that derive their bearing capacity mainly from point resistance will have capacity equal to the summation of the bearing capacity of the individual piles in the group. If the pile group is founded in a bearing stratum underlain by softer material, the bearing capacity of the entire pile group should be checked. Both the skin friction along the periphery of the pile group and pint resistance should be considered. For groups of piles that derive their resistance primarily from friction, the resistance of the group will be equal to the summation of the resistance of the individual piles or the resistance of the group acting as a pier. The skin friction along the periphery of the pile group should be considered in addition to the point resistance. The resistance of group of piles to uplift is limited to the weight of the mass of soil that clings to the piles regardless of the skin friction developed by a single pile.

9.9 EVOLUTION OF HIGH-RISE ARCHITECTURE

The high-rise architecture in the twentieth century shows such a diversification as to defy distinct classification. Nevertheless, its development can be traced, albeit imprecisely, in five phases. In the early 1940s before the advent of air conditioning and fluorescent lights, the building form was controlled by the need for natural daylight and ventilation and thus required a layout somewhat similar to those of contemporary apartments and hotels. The building width was limited to ensure that light and air reached all parts of the building, resulting in a width of 55 to 60 ft (16.7 to 18.3 m) that offers office space on either side of a double-loaded corridor. To achieve more leasable space on a given site, plan configurations with a central core and radiating wings were conceived.

The second phase of building configuration is the result of the interactions between the desire to create an increasing range of rentable area in a given space and the advent of air conditioning and fluorescent lighting. This period is also characteristic of the modern movement in architecture stressing the aesthetic value of simplicity in facade treatment and simple cubic shapes, such as rectangles, squares, and circles. At the exterior, the curtain wall was stretched tightly over the skin and the building shot up skywards in one regular prismatic shape. In keeping with the International Style, it was not offensive—in fact, it was highly desirable to display the structural muscle. Glass boxes with exposed structural members constituted the backbone of the International Style.

The third phase of the high-rise architectural development is a result of interaction between marketing experts and a mild boredom of the public toward the repetitive nature of the boxes on the cityscapes. The simplest prismatic shape has only four corners and therefore could offer at best only four corner offices. Even those corner offices more than likely displayed corner columns. Now, however, corner offices have become one of the most sought after lease space. To capture this market, the trend in planning is to have as many corner offices as possible. This is achieved by undulating the exterior, providing nicks, notches, and other convolutions at the perimeter. Additionally, to create visual identity and interest, setbacks are provided at selected levels. An otherwise simple shape is sliced and diced to create vertical lines to emphasize the verticality of the building while simultaneously providing for additional corner offices.

A fourth phase of office design known as postmodern architecture is bringing in daringly articulated buildings. These buildings not only have step backs, notches, and curves, but the resulting articulations are so severe as to preclude use of any single structural system for the entire building. This phase of design, which began around 1970, is considered primarily an aesthetic reaction to the cubism period. It has evolved gradually in three stages. First, the flat roof, which is all that is necessary from the functional point of view, started receiving architectural attention to gain identity in the city skyline. Today, many buildings' tops take on any number of articulations. The second stage is characterized by the elaborate entrances to the building in an effort to give it a street-level identity and allure. The third stage is a continuation in the battle for identity. Articulations at the extremities are no longer sufficient to create a building's identify. The entire architectural facade needs to proclaim the identity of the building. To this end, terracing of building plans, cutouts, slicing and dicing, and overhanging features are added to the building throughout its height.

The fifth phase can be looked upon as a modification to building shapes in the energy conservation context. We are witnessing buildings suitable for natural daytime lighting with courtyards, light wells, and skylights. Energy conservation efforts have brought about an understanding of spaces as a whole, especially in relation to how light influences the space. Instead of depending totally on mechanical heating and cooling and electric light, architects are considering the possible solar controls outside and inside as an integral part of architectural design.

It is however reassuring to know that the present-day flamboyance in high-rise architecture has not intimated structural engineers from offering support systems that are at once elegant and economical. In fact, it has stimulated the structural engineering profession to give almost total freedom to architectures in their quest of new and exciting avatars for tall and ultra-tall buildings. Today, with the use of computers, buildings are planned and designed that have little or no historic precedent. New structural systems are conceived and applied to extremely tall buildings in a practical demonstration of the engineer's confidence in the predictive ability and reliability of computer solutions.

Nine years after many soothsayers predicted the indelible images of September 11, 2001, would stifle humanities' enthusiasm for iconic skyscrapers, it is now certain they are wrong. Contrary to their prophecy, the landscape not only in the United States but in the entire world is moving up. Buildings that would either eclipse or stand spire to spire with the 1,250 ft tall Empire State Building of the 1930s are promising to reshape skylines. The collective introspective analysis of the topology that begged the question "Are tall buildings a viable part of our cities or they are not?" seems to have resulted in a resounding *yes*.

Compared to the skyscrapers that were erected in the 1970s and 1980s, the new crop of high rises is more likely to be expressions of civil or even national pride than symbols of corporate wealth. Their purpose is to project a certain status for a city on a world stage. This has become especially true as developing countries in Asia and the Middle East started erecting super-tall skyscrapers that made the iconic American Towers, the Empire State Building, the Sears Tower (now called Willis Tower), and the nonexistent WTC Towers look average in comparison.

In today's high-rise architectural vocabulary, remarkably, the repetition of structural and architectural elements is losing importance. Many nonorthogonal skyscrapers are designed to be iconic, to proclaim their presence in cities, regions, and even countries on the global map. Their facades are curved, or organically shaped for aesthetic reasons, and often the entire building sports a twisted elevation. Although complexity of fabricating and constructing superstructures of nonprismatic shapes increases, various major projects constructed within the past decade demonstrate the feasibility of large-scale application of nonstandard products. High-rise shaping is strongly influenced by the CAD modeling tools that architects have available. The conventional rectilinear box form of skyscrapers is gone for good and unlikely to make a comeback any time soon. Most developers and architects are very focused in creating exciting shapes and forms that would have once made these projects unbuildable. Today, these geometric complexities are better understood and can be dealt with efficiently. Probably the most significant factor in the cost of a tall building is the choice of the structural system chosen for resisting lateral loads. Tall buildings are flexible structures. Therefore, their dynamic response to wind excitation is critical in assessing their loading and performance particularly with respect to building deflections and accelerations at the top occupied floors. In this context, it is interesting to note that the across-wind response rather than the along-wind response dominates its behavior.

Although analytical methods exist for predicting wind response, the only method for predicting their dynamic behavior with any degree of certainty is through a wind tunnel study.

There are a number of methods designers can use to minimize building response to wind loads:

- Use openings at top floors integrated with architecture features of the building to break up eddies that may cause vortex shedding.
- Orient the building so that its weak direction does not line up with the strongest wind direction.
- Use rounded plan shapes instead of sharp-edged plans.
- Use step backs along the building's height to break up flow of wind that may cause vortex shedding.

In spite of the earlier methods known for some time to reduce adverse effects of winds, the current trend in tall building design has embraced shapes and forms that would have once made these projects unbuildable. Today, these complexities are better understood by using computer-generated geometries. Only a few years ago, many of the high-rise shaping would have seemed mere figments of architect's imagination—now many have been built and new ones planned. It seems CAD-based modeling tools and structural calculations have changed the shaping of high rises forever.

Innovative solution in a broad sense is one that achieves desired function at the lower cost, without sacrificing quality. Our problem with innovation is that it has been an elusive quality and one rarely captured in our educational process. Our education is mostly devoted to developing analytical skills, mainly through the use of computers and very little to developing their opposite, inductive skills. Advancement in commercial computer software has contributed heavily to the development of sophisticated methods of analysis and to optimization methods for well-defined structural problems. However, the conceptual selection of an engineering system is by and large still an elusive skill with no established learning procedure. The ability to conceive and generate design alternatives requires both a feeling for behavior and approximate design skills, best acquired by carefully examining structural behavior of buildings of the past through discussions and techniques such as *load path* concept.

The purpose of presenting case studies of certain selected tall buildings is to highlight the aspects of conceptualization that have been timeless constants in the design of tall and super tall buildings.

9.9.1 Architectural Review

Building architecture in the twentieth century can be traced to several nineteenth-century roots. One was the advent of new forms of structural and other materials so strikingly displayed in the building technology of skyscraper. This has allowed greater scope of aesthetic expression and innovation in architectural practice. Another was the development of expressive new idioms of form and space. The development of metal trusses made it possible to roof column-free interior spaces easily and economically. Such roofs were used for railroad stations, market halls, exhibition palaces, and domes.

The nineteenth century was one of the most technically inventive centuries. It witnessed the application of new techniques and of new mechanical means in virtually every human activity. It became clear in time that the innovation in architecture would come from those who grasped the possibilities of the new materials and techniques. Revolutionary methods of building with wood were developed in the 1830s to meet the demands for speedy construction and to overcome the shortage of skilled labor. Cast iron was then developed into a building material. Lighter and more adaptable than masonry and, combined with other inventions, notably the elevator, cast iron paved the way for tall buildings unprecedented not only for height but ease of construction.

In Chicago, during the later part of the nineteenth century, a school of architects, of whom Louis Sullivan and Frank Lloyd Wright were the most famous members, originated a new American style of domestic architectures. Their designs, primarily houses, were long, informal, and organic. Their ideas were ignored for more than a decade in America but were taken up aboard and developed into the so-called International Style. The International Style was also influenced by the German Bauhaus school, founded by Walter Adolf Gropius, and by the abstract artists interested in using pure forms for buildings. This style was the architectural response to the machine age. Simplicity meant elegance derived from *pure* forms. The display of the structural muscle beyond the tightly stretched curtain wall was widely accepted as the *in thing*. Structures designed in this era incorporated three distinct elements of the new style: (1) a new vocabulary of forms borrowed largely from abstract art, consisting of planes, lines, and rectangles without ornaments or moldings, (2) the representation of interior space and exterior facade as a cohesive unit, and (3) the use of new structural materials such as steel.

The new style, with its angular forms, plane surfaces, and lack of conventional ornament, met with some resistance from the public, which tended to regard it as bare and inhuman. But by the middle of the twentieth century, the style had become dominant across the country. Bold use of modern construction methods and structural materials became common. Noteworthy among the latter are glass tinted to reduce glare; glass brick designed to admit additional light while preventing glare and furnishing effective insulation against heat, cold, and noise; artificial stone; plastics; chromium, aluminum, and other metals; and above all steel.

The early stages of American architecture lacked truly monumental structures. The monumental idea was gradually added to American architectural forms, reaching its apex with the construction of Rockefeller Center in New York City. The center represented a new concept of building a city within a city, containing a towering 60-story structure surrounded by a number of smaller high-rise office buildings and recreational facilities. This complex of skyscrapers has exercised increased influence since 1931, the year work on the center was started. The building represents a departure in architectural thinking from a single-use, single-building concept to multiuse, multicomplex structures on a community scale. Because of that practical example, architects have responded more and more creatively to such demands of contemporary lifestyle as rapid intercommunication and integration of city and surrounding region. Another example of multi-building planning is the now nonexistent WTC in New York City, consisting of two 110-story towers and four smaller buildings grouped around the plaza.

During the period of 1950 to the mid-1960s, the International Style of architecture was embraced by prominent American architects and resulted in sleek boxlike glass and steel high rises that integrated the concept of purity of design into the architecture of the structure. Notable examples are the Seagram Building (1950) and the Whitney Museum (1966), both in New York City, and the John Hancock Center (1968) in Chicago.

During the mid-1960s, a reaction developed to the International Style that emphasized greater freedom of design. Figuratively speaking, the concept of the glass box was beginning to shatter. It was no longer a misdeed to hide a structure behind a more aesthetic exterior. The building and construction industry saw the advent of new forms of structural and other materials that allowed greater scope for aesthetic expression and innovation. Within the last decade, many cities, outside of the United States, have had imaginative new forms thrusting above their skylines using plan configurations that are other than prismatic. Large corporations have built a new generation of flamboyant headquarters buildings that are altering the urban skyline and brining new vigor to cities around the world. Many are giant architectural logos that draw enormous public attention with an expected increase in revenues to the companies that build them. These new buildings are emerging as good investments, serving as advertising symbols and marketing tools. The distinguishing architectural features for this new generation of buildings are sculptural shapes at the tops and elaborate articulation at the bases.

9.9.2 Prototype of Today (2013)

In truth, once the criteria are established, the engineer and, his right arm, the computer have clear sailing.

At the outset, we should establish just what is a tall building. The dividing line should perhaps be drawn where the design of the structure moves from the field of statics into the field of structural dynamics.

The first of our criteria, speed of construction, is our most serious problem. The designer's program for the Empire State Building was short and simple—a fixed budget, no space more than 28 feet from window to corridor, as many stories of such space as possible, an exterior of limestone, and, here is the astonishing part, completion by May 1, 1931, which meant a year and 6 months from the beginning of the architect's sketches to the finished and occupied building. The distribution of light, power and communication services, water, and particularly heat and air is infinitely more complex in today's modem structures than was even dreamed during the design stages of the Empire State Building. As such, the designers of today would have great difficulty completing their work within that 18 month period much less accomplishing the construction as well.

The structural steel work at the site began in April of 1930 and ended in September of that same year, a period of less than 6 months. 57,000 tons of structural steel was placed. During the month of July, 22 stories of steel were placed in 22 working days, involving regular hours and no night work. The steel work was finished 12 days ahead of schedule.

While we have seen much taller buildings, the design and construction record of the Empire State Building has not been broken.

Tall buildings, in order to be economically competitive with their lower brothers, must offer savings other than in the structural system. The cost of a structure represents far from an overwhelming portion of the total construction dollar. The cost of the exterior wall alone can equal or exceed that of the structural steel. The HVAC system—heating, ventilating, and air conditioning—dominates the cost picture.

Our cost criterion, then, is not necessarily to reduce the cost of the structure but rather to look to the structure as a system to allow a reduction in the overall cost of the building. It follows that if you are to develop new and efficient structural systems, you must have a working knowledge of the mechanical, electrical, architectural, and other systems that make up the total building.

Not too long, some 20 years ago, most owners used to desire an economical and unobtrusive building that will satisfy the local planning department and look nice but not unusual. However, that trend has changed. Now, the aspiration for the architect to provide a distinctive image for the building is very powerful and is the source of continued evolution in architectural style and art. This thrust comes from today's *marketing* demand for spectacular forms. The history of architecture shows that design innovation has its own life, fed by brilliant form givers who provide prototypes that keep architecture alive and exciting as an art form. Thus, like economics, architectural design has its *supply and demand sides* that each reinforce one another. To be sure, the International Style still exists as a vernacular and can range from everyday economical buildings to refine symbols of prestige. However, aesthetic tents of the International Style—particularly the metal/glass cubistic building—began to be seriously questioned by the mid-1970s. This questioning finally bore fruit in an architectural style known broadly as postmodern. Among other characteristics, postmodernism embraced

- The use of classical forms such as arches, decorative columns, pitched roofs in nonstructural ways, and generally in simplified variation of the original elements
- The revival of surface decoration on buildings
- A return to symmetry in configuration

In lateral-load-resisting terms, particularly for seismic, these changes in style were, if anything, beneficial. The return to classical forms and symmetry tended to result in regular structural/architectural configurations, and almost all of the decorative elements were nonstructural.

A conventionally engineered steel or concrete member that was supporting the building could be found inside every classical postmodern column. It is clear that an interest in seismic design or structure in general had no influence on the development of postmodernism; it was strictly an aesthetic and cultural movement.

At the same time that postmodernism was making historical architectural style legitimate again, another style began to flourish, to some extent in complete opposition. This style (originally christened *hi-tech*) returned to the celebration of engineering and new industrial techniques and materials as the stuff of architecture. This style originated primarily in Europe, notably in England and France.

Although wind and seismic concerns had not direct influence on the origin and development of this style, it became relevant because it revived an interest in exposing and celebrating structure as an aesthetic motif.

Postmodernism died a quick death as an avant-garde style, but it was important because it legitimized the use of exterior decoration and classically derived forms. These became common in commercial and institutional architecture. The notion of *decorating* the economical cube with inexpensive simplified historic or idiosyncratic nonstructural elements had become commonplace.

At the same time, in much everyday commercial architecture, evolved forms of the International Style still predominate, to some extent also representing simplified (and more economical) forms of the high-tech style. The use of new lightweight materials such as glass fiber-reinforced concrete and metal-faced insulated panels has a beneficial effect in reducing earthquake forces on the building, though provision must be made for the effects of increased drift on nonstructural components or energy-dissipating devices used to control it.

The importance of well-publicized designs by fashionable architects is that they create new prototypical forms. Architects are very responsive to form and design, and once a new idiom gains credence, practicing architects the world over begin to reproduce it. Today's architecture for a New York corporate headquarters high-rise building may become tomorrow's suburban savings and loan office.

Today, however, unlike the era of the International Style and the adoption of *modern* architecture, there is no consensus on a set of appropriate forms. At present, spectacular architectural design is in fashion and sought after by municipalities, major corporations, and institutions. So, it is useful to look at today's cutting-edge architecture, because among it will be found the prototypes of the vernacular forms of the future.

A story of the architectural form of the high-rise building from the 1920s to today shows that there is a steady evolution in which International Style dominates the scenes from about 1945 to 1985. For a brief interlude, postmodern architecture is fashionable, in company with *high-tech*. Toward the end of the century, architectural forms become more personal and idiosyncratic, and evolution is replaced by competition. The first five years of the millennium have seen the emergence of a number of very personal styles, from the jagged forms to the warped surfaces.

In general, today's high-rise buildings remain vertical and have direct load paths, and their exterior walls are reasonably planar. Some high-rise towers have achieved a modest nonverticality by the use of nonstructural components. A more recent development is that of the *torqued* tower, as in the Freedom Tower at the WTC and Santiago Calatrava's *Turning Torso* tower in Malmo, Sweden. For very tall buildings, it is claimed that these twisted forms play a role in reducing wind forces, besides their visual appeal, but their forms are not of significance seismically.

In relatively lower buildings where there is more freedom to invent forms than in the high rise, planning irregularities and corresponding 3D forms are now fashionable.

Highly fragmented facades now common, serve as metaphors for the isolated and disconnected elements of modern society. Often-repeated design motifs include segmental, undulating, or barrel-vaulted roofs and canopies and facades that change arbitrarily from metal and glass curtain wall to punched-in windows.

In all this ferment, there is much originality and imagination and often high seriousness. It remains to be seen whether any of these forms become attractive to the typical architectural practitioner and their more conservative clients; however, indications of the influence of some of these motifs can now be discerned in more commonplace buildings along the highways and in schools and universities. To be sure, the kinds of configuration irregularities that manifest themselves in the International Style era are on the increase.

This is because much new architecture is clearly conceived independently of structural concerns or in the spirit of theatrical set design, with the engineer in the role of an enabler rather than collaborator.

In the search for meaning in architecture that supersedes the era of International Style and the superficialities of fashion exemplified by much of postmodernism and after, perhaps architects and engineers in the seismic regions of the world might develop an *earthquake architecture*. One approach is an architecture that expresses the elements necessary to provide seismic resistance in ways that would be of aesthetic interest and have meaning beyond mere decoration. Another approach is to use the earthquake as a metaphor for design.

For the low- and midrise building, the only structural system that clearly expresses lateral resistance is the use of exposed bracing. There are historical precedents for this in the half-timbered wood structures of medieval Germany and England. This was a direct and simple way of bracing rather than an aesthetic expression, but none of these buildings are much prized for their decorative appearance. Indeed, the *half-timbered* style has become widely adopted as an applied decorative element on US architecture, though for the most part at a modest level of residential and commercial design.

Two powerful designs in the 1960s, both in the San Francisco Bay Area, used exposed seismic bracing as a strong aesthetic design motif. These were the Alcoa Office Building and the Oakland Coliseum, both designed in the San Francisco office of Skidmore, Owings, and Merrill.

In spite of these two influential designs and others that used exposed wind bracing, the subsequent general trend was to de-emphasize the presence of lateral-resistance systems. Architects felt that they conflicted with the desire for purity in geometric form, particularly in glass *box* architecture and also possibly because of a psychological desire to deny the prevalence of lateral forces. However, in the last two decades, it has become increasingly acceptable to expose lateral-bracing systems and enjoy their decorative but rational patterns.

This new acceptability is probably due to boredom with the glass cube and the desire to find a meaningful way of adding interest to the facade without resorting to the applied decoration of postmodernism. In addition, greater understanding of the lateral-load effects has led to realization that exposed bracing may add interest and reassurance rather than alarm.

Exposed bracing is also used as an economical seismic retrofit measure on buildings for which preservation of the façade appearance is not seen as important. A possible advantage of external bracing is that often the building occupants can continue to use the building during the retrofit work, which is a major economic benefit. External bracing retrofits have also sometimes had the merit of adding visual interest to a number of dull 1960s rectilinear-type facades.

The movement toward exposed lateral bracing has some parallels with the aesthetic movement of exposing the building's mechanical systems. Designers who had become bored with expanses of acoustical ceiling realized that mechanical systems, particularly when color-coded, were of great visual interest and also intrigued those who are fascinated by mechanical systems and devices. Another parallel with seismic design is that, when mechanical systems were exposed, their layout and detailing had to be much more carefully designed and executed, from an aesthetic viewpoint. In a similar way, exposed bracing has to be more sensitively designed, and this has seen the development of some elegant design and material usage.

New innovations, such as base isolation and energy-absorbing devices, are yet to be fully exploited for aesthetics and reassurance.

For the expression of seismic resistance and architectural statement to occur simultaneously, one has to look at building architectural design through a seismic *filter*. It certainly appears that the many flamboyant architectural forms are in conflict with seismic design needs.

The ultimate solution to these conflicts depends on the architect and engineer working together on building design from the outset of the project and engaging in knowledgeable negotiation in which the virtues of seismic design such as simplicity, symmetry, and regularity are discussed.

What is the motivation behind the tall building race? To be candid, the reasons are the same today as they were some 70 years ago; height now, as then, is an exhibition of technology and power.

Nothing is more expressive than an upright symbol, particularly the one with high-tech items such as pressurized double-decker elevators, external damping devices to reduce sway caused by windstorms, and fiber optics incorporated into curtain walls that transform buildings into giant billboards. Tall buildings become instant icons, putting their cities on the world map.

Given humanity's competitive nature, it is hard to believe that Burj Kalifa at 2717 ft (818 m) will wear the crown long. The quest for the title of the world's tallest building is alive and well. This begs the question, how tall can buildings go? Answer: No limits are in sight, at least from structural considerations. Humanity has an obsession with super-tall structures, particularly when humans can live and work in them. While there are indeed lessons to be learned from the WTC catastrophe, the skyscraper will remain viable well into the foreseeable future.

9.9.3 Structural Systems for Selected Tall Buildings

The purpose of presenting case studies of certain selected tall buildings is to highlight the aspects of conceptualization that have been timeless constants in the design of tall and super-tall buildings. It is as important to understand how engineers in the past have conceptualized the building skeleton using such ideas as mega frames, interior and exterior braced tubes, and spine-wall structures to name a few. Many of the examples given here are once-in-a-lifetime, high-profile, dream projects that you would give your right hand to design. Let us experience at once the architectural grace and the elegance of structures supporting tall and ultra-tall buildings on our planet.

9.9.3.1 Taipei 101

The 101-story Taipei financial center in Taipei, Taiwan, shown in Figure 9.44 rises to a height of 1668 ft (508 m). The lowest 25 floors of the building taper gradually inward, forming a truncated pyramid. Above is a stack of eight 8-story high modules with outward sloping wall, creating a *waist* at the 26th floor and setbacks at floors 34, 42, etc. (see Figure 9.45). The modules also have double-notched corners. A narrower tower segment and an architectural pinnacle top the eighth module.

FIGURE 9.44 Taipei 101 financial center. (Photograph courtesy of Mr. Hung Lee of John A. Martin & Associates, and Mr. David Lee.)

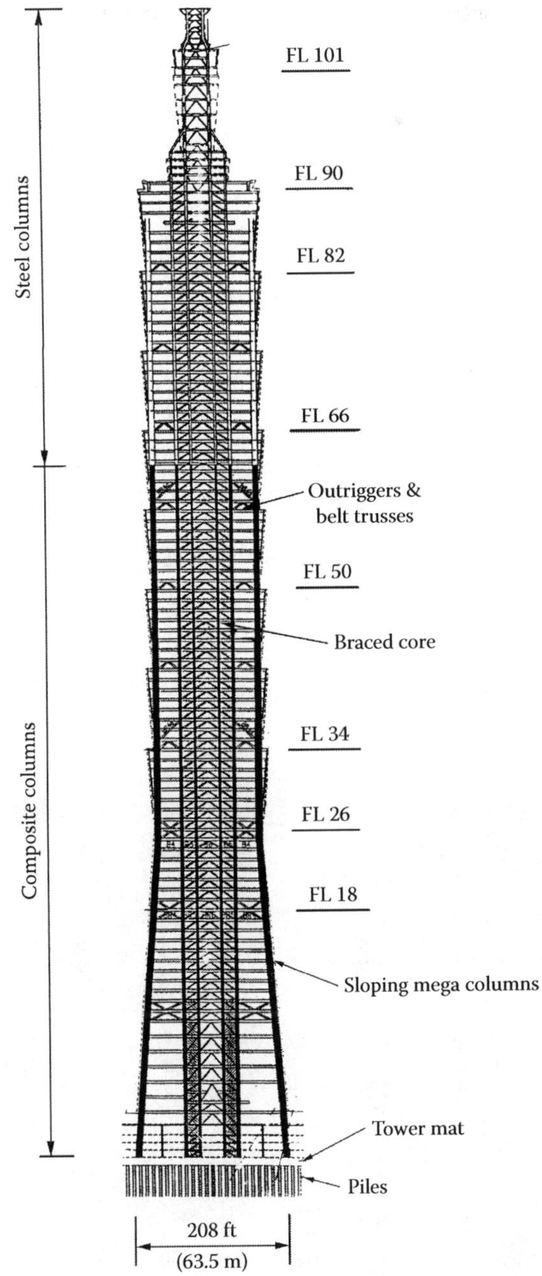

FIGURE 9.45 Taipei 101 schematic cross section.

Because façade slopes and setbacks interrupt vertical continuity of columns, and doubly notched floor plans reduce the efficiency of an exterior moment frame around corners, a perimeter tubular-frame system was not used for this project.

Instead, the lateral bracing consists of a dual system comprised of a braced core interconnected to a planar moment frame at each sloping face, through a system of outriggers and belt trusses. The braced core offers high-shear stiffness with chevron and diagonal braces of I-shaped sections in four planes in each direction. A mix of single-, double-, and triple-story outriggers is distributed every eight to ten floors along the building height. Typically on each building face, they engage two vertical super columns. Below the 26th floor, additional outriggers engage two more columns on each face with the belt trusses engaging corner columns as well. Steel box- core columns and perimeter super columns are filled with concrete to provide additional stiffness. The size of super columns' steel shell at the base is 8 × 10 ft (2.4 × 3.0 m) (See Figure 9.46).

The core is designed as a concentric braced frame (CBF). To clear architecturally required openings, some work points for adjacent braces are spread apart, creating eccentric links. But the system is not designed as an EBF. The design is as for a CBF with the braces, not links, controlling the system's strength. Reduced beam section (RBS), also referred to as a dogbone connection, is used at locations where the analyses showed plastic rotation demand in excess of 0.005 rad in a 950-year return period earthquake. For added strength, beam to column connections within the braced core system are detailed as moment connections.

Wind engineering studies indicated that accelerations of the building's upper floors would be 30%–40% higher than desired for this office building. Therefore, to improve the structure's ability to dissipate dynamic energy, a passive damping system consisting of a 730-ton TMD has been installed near the top of the tower. The TMD consists of a massive steel sphere suspended by steel cables.

The design of TMD is by Motioneering, Inc., Ontario, Canada. The building structural design is by Evergreen Consulting Engineering, Taipei, and Thornton Tomasetti Engineers, New York City.

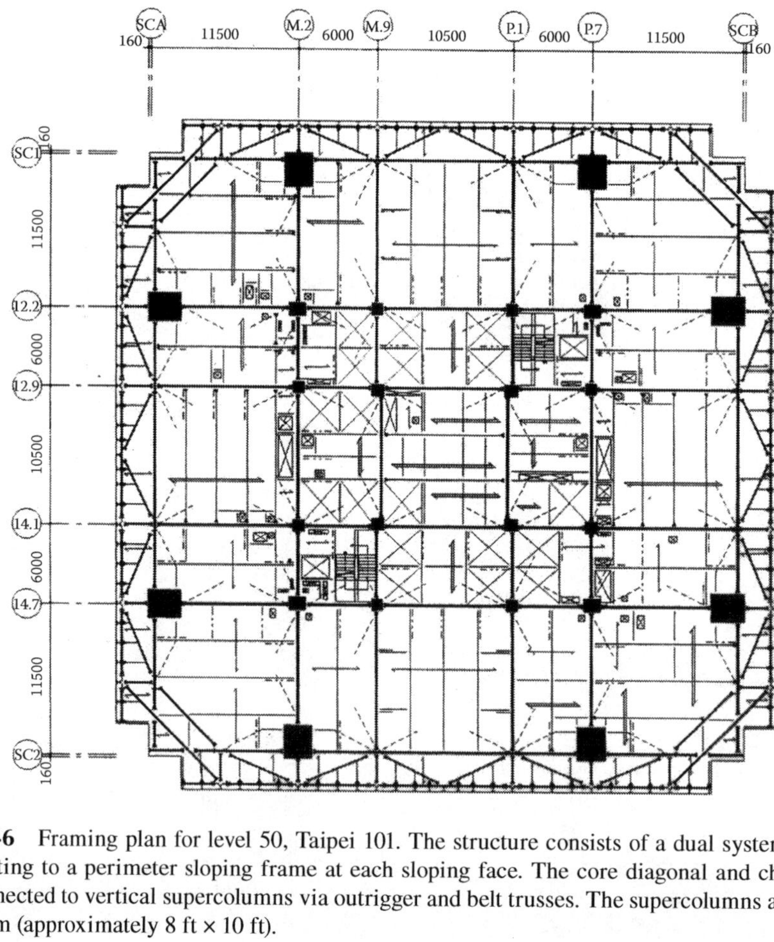

FIGURE 9.46 Framing plan for level 50, Taipei 101. The structure consists of a dual system of a braced core connecting to a perimeter sloping frame at each sloping face. The core diagonal and chevron braces are interconnected to vertical supercolumns via outrigger and belt trusses. The supercolumns at the base are 2.4 m × 3.0 m (approximately 8 ft × 10 ft).

9.9.3.2 Jin Mao Tower, Shanghai, China

This building consists of a 1381 ft (421 m) tower and attached low-rise podium for a total gross building area of approximately 3 million square feet (278,682 m^2). The building includes 50 stories of office space topped by 36 stories of hotel space with two additional floors for a restaurant and an observation deck. Parking for automobiles and bicycles is located below grade. The podium consists of a retail space as well as an auditorium and exposition spaces.

The superstructure is a mixed use of structural steel and reinforced concrete with many major structural members composed of both steel and concrete. The primary components of the lateral system include a central reinforced concrete core linked to exterior composite mega-columns by outrigger trusses (Figure 9.47). A central shear-wall core houses the primary buildings functions including elevators, mechanical fan rooms, and washrooms. The octagon-shaped core, nominally 90 ft (27.43 m) from centerline to centerline of perimeter flanges, is present from the foundation to level 87. Flanges of the core typically vary from 38 in. (84 m) thick at the foundation to 18 in. (46 m) at level 87 with concrete strengths varying from 7500 to 5000 psi (51.71 to 34.5 mPa). Four 18 in (46 cm) thick interconnecting core wall webs exist through the office floors. The central area of the core is open throughout the hotel floor, creating an atrium that leads into the spire with a total height of approximately 675 ft (206 m). The size of composite mega-columns varies from 5 × 16 ft (1.5 × 4.88 m) with a concrete strength of 7500 psi (51.71 mPa) at the foundation to 3 × 11 ft (0.91 × 3.53 m) with a concrete strength of 5000 psi (34.5 mPa) at level 87.

The shear-wall core is directly linked to the exterior composite mega-columns by structural steel outrigger trusses. The outrigger trusses resist lateral loads by maximizing the effective depth of the structure. Under bending, the building acts as a vertical cantilever with tension in the windward

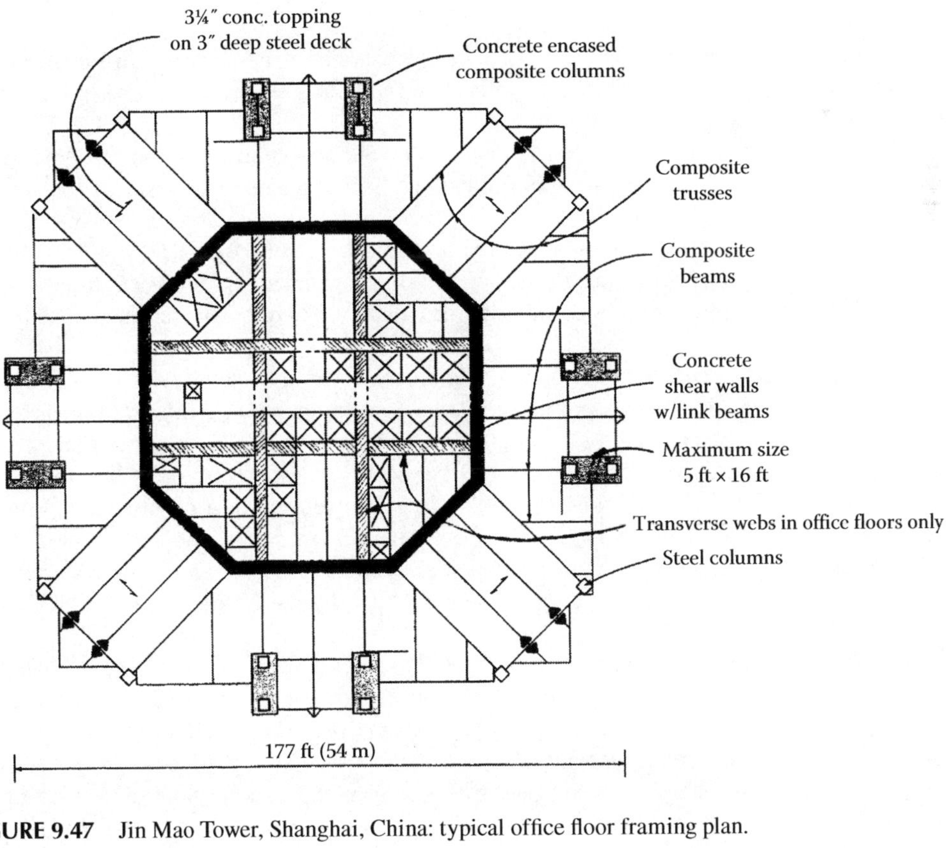

FIGURE 9.47 Jin Mao Tower, Shanghai, China: typical office floor framing plan.

columns and compression in the leeward columns. Gravity load framing minimizes uplift in the exterior composite mega-columns. The octagon-shaped core provides exceptional torsional resistance, eliminating the need for any exterior belt or frame systems to interconnect exterior columns.

The outrigger trusses are located between levels 24 and 26, 51 and 53, and 85 and 87. The outrigger truss system between levels 85 and 87 is capped with a 3D steel space frame that provides for the transfer of lateral loads between the core and the exterior composite columns. It also supports gravity loads of heavy mechanical spaces located in the penthouse floors.

The structural elements for resisting gravity loads include eight structural steel built-up columns. Composite wide-flange beams and trusses are used to frame the floors. The floor-framing elements are typically 14 ft 6 in. (4.4 m) on center with a composite 3 in. (7.6 cm) deep steel deck and a $3^1/_4$ in. (8.25 cm) thick normal-weight concrete topping slab spanning between the steel members.

The foundation system for the tower consists of high-capacity piles capped with a reinforced concrete mat. High-water conditions required the use of a 3 ft 3 in. (1 m) thick, 100 ft (30 m) deep, continuous reinforced concrete slurry wall diaphragm along the 0.5 mile (805 m) perimeter of the site.

The high-capacity pile system consists of a 3 ft (0.91 m) diameter structural steel open-pipe pile with a 7/8 in. (2.22 cm) thick wall typically spaced 9 ft (2.75 m) on center capped by a 13 ft (4 m) deep reinforced concrete mat. Since soil conditions at the upper strata are so poor, the piles were driven into a deep, stiff sand layer located approximately 275 ft (84 m) below grade. The individual design-pile capacity is 1650 kips (7340 kN).

Wind speeds can average 125 mph (56 m/s) at the top of the building over a 10 min time period during a typhoon event. The earthquake ground accelerations compare to 1994 UBC zone 2A. The overall building drift index for a 50-year return wind with a 2.5% structural damping is 1/1142. This increased to 1/887 for a future developed condition in which two tall structures are proposed adjacent to the Jin Mao Building. The drift index based on specific Chinese code-defined winds, which were equivalent to a 3000-year wind, is 1/575.

The structural design for the tower is governed by its dynamic behavior under wind and not by its strength or its overall or interstory drift. The calculated fundamental translational periods are 5.7 s for each principal axis. The torsional period is 2.5 s.

In a force-balance and aeroelastic wind tunnel study, the accelerations at the top floors were evaluated using a value of 1.5% for structural damping. The accelerations measured in the wind tunnel were between 0.009–0.013 g for a 10-year return period and between 0.003–0.005 g for a one-year return period—well within the generally accepted range of 0.020–0.025 g for a 10-year return. Only the passive characteristics of the structural system including its inherent mass, stiffness, and damping are required to control the dynamic behavior. Therefore, no mechanical damping systems are used.

Since the central core and composite mega-columns are interconnected by outrigger trusses at only three 2-story levels, the stresses in the trusses due to differential shortening of the core relative to the composite columns were of concern. Therefore, concrete stress levels in the core and mega-columns were controlled in an attempt to reduce relative movements. To further reduce the adverse effect of differential shortening, slotted connections were used in the trusses during the construction period of the building. Final bolting with hard connections was done after completion of construction to relieve the effect of differential shortening occurring during construction. The architecture and structural engineering of the building is by the Chicago office of Skidmore, Owings, and Merrill.

The Jin Mao Tower is a landmark skyscraper in the Lujiazui area of Pudong District of Shanghai. Its architecture draws the traditional architectures such as tiered pagoda, gently sloping back to create a rhythmic pattern as it raises. As stated previously, the tower is built around an octagon-shaped shear-wall core surrounded by eight (8) exterior composite super columns and eight (8) steel columns. Three sets of two-story-high outrigger trusses connect the columns to the core at six of the floors to provide additional resistance to overturning.

9.9.3.3 Petronas Towers, Kuala Lumpur, Malaysia

Two 1476 ft (450 m) towers and a sky bridge connecting the twin towers characterize the buildings in Kuala Lumpur, Malaysia (Figure 9.48).

The towers have 88 numbered levels but are in fact equal to 95 stories when mezzanines and extra-tall floors are considered. In addition to 6,027,800 ft^2 (560,000 m^2) of office space, the project includes 1,501,000 ft^2 (140,000 m^2) of retail and entertainment space in a six-story structure linking the base of the towers, plus parking for 7000 vehicles in five belowground levels.

The lateral system for the towers is of reinforced concrete consisting of a central core, perimeter columns, and ring beams using concrete strength up to 11,600 psi (80 mPa). The foundation system consists of pile and friction barrette foundations with a foundation mat.

The typical floor system consists of wide-flange beams spanning from the core to the ring beams. A 2 in. deep composite steel deck system with a $4^1/_4$ in (110 mm) concrete topping completes the floor system.

Architecturally, the towers are cylinders 152 ft (46.2 in) in diameter formed by 16 columns. The facade between columns has pointed projections alternating with arcs, giving unobstructed views through glass and metal curtain walls on all sides. The floor plate geometry is composed of two rotated and superimposed squares overlaid with a ring of small circles. The towers have setbacks at levels 60, 72, 82, and 88 and circular appendages at level 44. Concrete perimeter framing is used up to level 84. Above this level, steel columns and ring beams support the last few floors and a pointed pinnacle.

The towers are slender with an aspect ratio of 8.64 (calculated to level 88). The design wind speed in Kuala Lumpur area is based on 65 mph (35.1-m/s) peak, 3 s gusts at 33 ft (10 m).

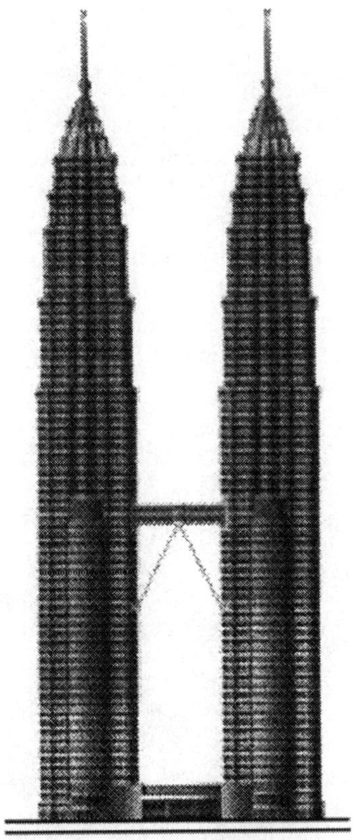

FIGURE 9.48 Petronas Twin Towers, Kuala Lumpur, Malaysia: elevation.

9.9.3.4 World Trade Center Towers, New York

Of the seven buildings of the WTC complex of New York City, the WTC towers, known as WTC 1 and WTC 2, were the most visible and recognized tall buildings throughout the world. Each of the towers encompassed 110 stories above plaza level and seven levels below. WTC 1, the north tower, had a roof height of 1368 ft, while WTC 2 stood nearly as tall at 1362 ft. Each building had a square floor plate 207 ft 2 in. (63.14 m) long on each side, with chamfered corners measuring 6 ft 11 in. (2.10 m). A rectangular service core of approximately 137 by 87 ft (41,75 × 26.52 m) was present at the center of each building.

The buildings' signature architectural design feature was the vertical fenestration that featured a series of closely spaced built-up box columns, as shown in Figure 9.49. At typical floors, a total of 59 of these columns were present on each of the flat faces of the building, placed at 3 ft 4 in. (1.0 m) on centers. Adjacent perimeter columns were connected at each floor level typically by 52 in. (1.32 m) deep spandrel plates (see Figure 9.50). In alternate stories, an additional column was present at the center of each of the chamfered building corners. The resulting configuration of closely spaced columns interconnected with deep spandrel plates created a perforated perimeter tube (see Figure 9.51).

Twelve grades of steel, having yield strengths varying from 42 to 100 ksi (191 to 455 kN), were used to fabricate the perimeter columns and spandrel plates. In the upper stories of the buildings, plate thickness in the exterior wall was generally $^1/_4$ in (6.35 m), and at the base, column plates as thick as 4 in. (101.6 mm) were used.

The structural system was considered to constitute a tubular system, acting essentially as a cantilevered hollow tube with perforated walls. The side walls acting as stiff webs transfer shear between the windward and leeward walls, thus creating an efficient 3D structure for resisting lateral loads. In the lower seven stories of the towers where there are fewer columns, vertical diagonal bracing in the building cores provided the lateral stiffness.

Floor construction typically consisted of 4 in. (101.6 mm) of concrete on $1^1/_2$ in. (38.1 mm) deep, 22-gage noncomposite steel deck. The slab thickness was 5 in. (127 mm) in the core area. Outside the central core, the floor deck was supported by a series of composite floor trusses, 29 in (0.74 m) deep, open-web joist-type trusses with ASTM A36 steel chord angles and steel rod diagonals. Composite behavior of the truss with the floor slab was achieved by extending the diagonal truss members above the top chord so that they would act much like shear stud (see Figure 9.52). Detailing of theses trusses was similar to that typically used in open-web joist fabrication, but the floor system design was not typical of open-web joist floor systems. It was considerably more redundant and was well braced with transverse members. Trusses placed in pairs, at 6 ft 8 in. (2.03 m) spacing, spanned approximately

FIGURE 9.49 WTC vertical fenestration: box columns.

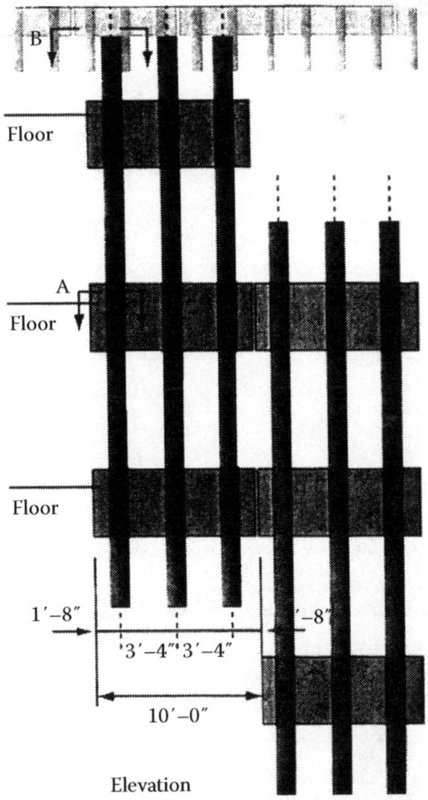

FIGURE 9.50 Columns connected by spandrel plates.

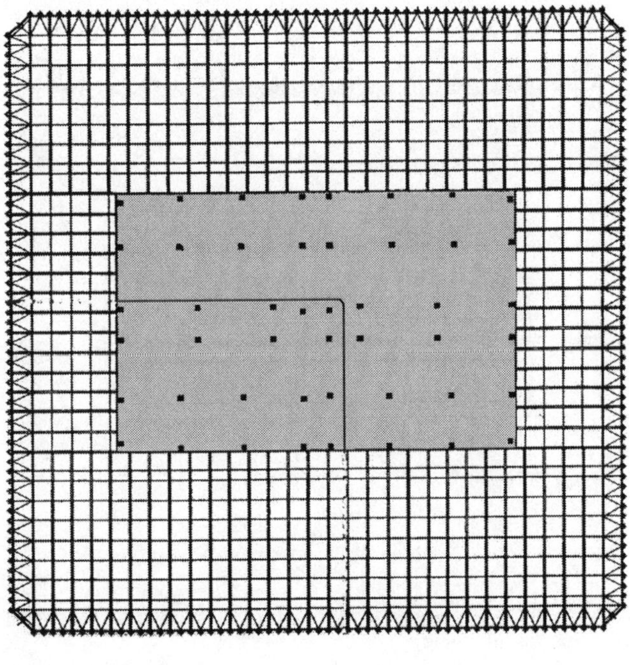

FIGURE 9.51 Perforated perimeter tube.

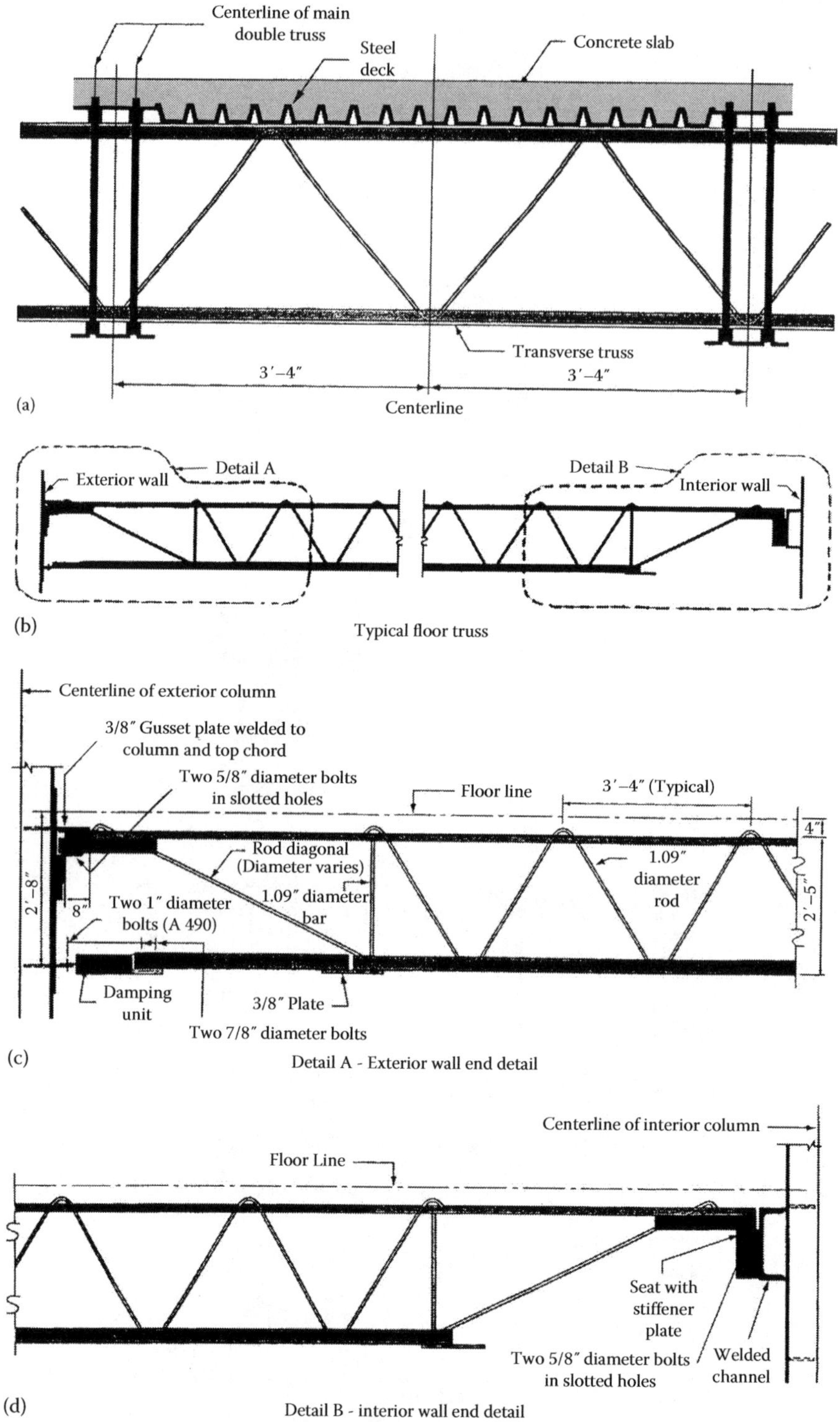

FIGURE 9.52 WTC floor construction. (a) Metal desk and truss detail; (b) truss attachment to exterior and interior walls; (c) exterior end wall detail; (d) interior end wall detail.

60 ft (18.29 m) to the sides and 35 ft (10.67 m) at the ends of the central core. Steel deck spanned parallel to the main trusses and was supported by continuous transverse bridging trusses spaced at 13 ft 4 in. (4.06 m) from the transverse trusses.

In approximately 10,000 locations in each building, viscoelastic dampers extended between the lower chords of the trusses and gusset plates attached to exterior columns (see Figure 9.52c). These dampers were provided to reduce occupant perception of wind-induced motion. Pairs of flat bars extended diagonally from the exterior wall to the top of chord of adjacent trusses. These diagonal flat bars, which were typically provided with shear studs, provided horizontal shear transfer between the floor slab and exterior wall, as well as out-of-plane bracing for perimeter columns not directly supporting floor trusses.

The core framing consisted of 5 in. (127 mm) concrete fill on steel deck supported by floor framing of rolled structural shapes, in turn supported by wide-flange shape and built-up box section columns. Some of these columns measured 14 by 36 in. (0.35 × 0.91 m). For the upper levels, these box columns transitioned into wide-flange shapes. At the top, a total of 10 outrigger trusses were present, six extending across the long direction of the core and four extending across the short direction (see Figure 9.53). In addition to providing support for a transmission tower (WTC 1 had a transmission tower; WTC 2 did not but was designed to support such a tower), this outrigger system provided stiffening of the frame for wind resistance.

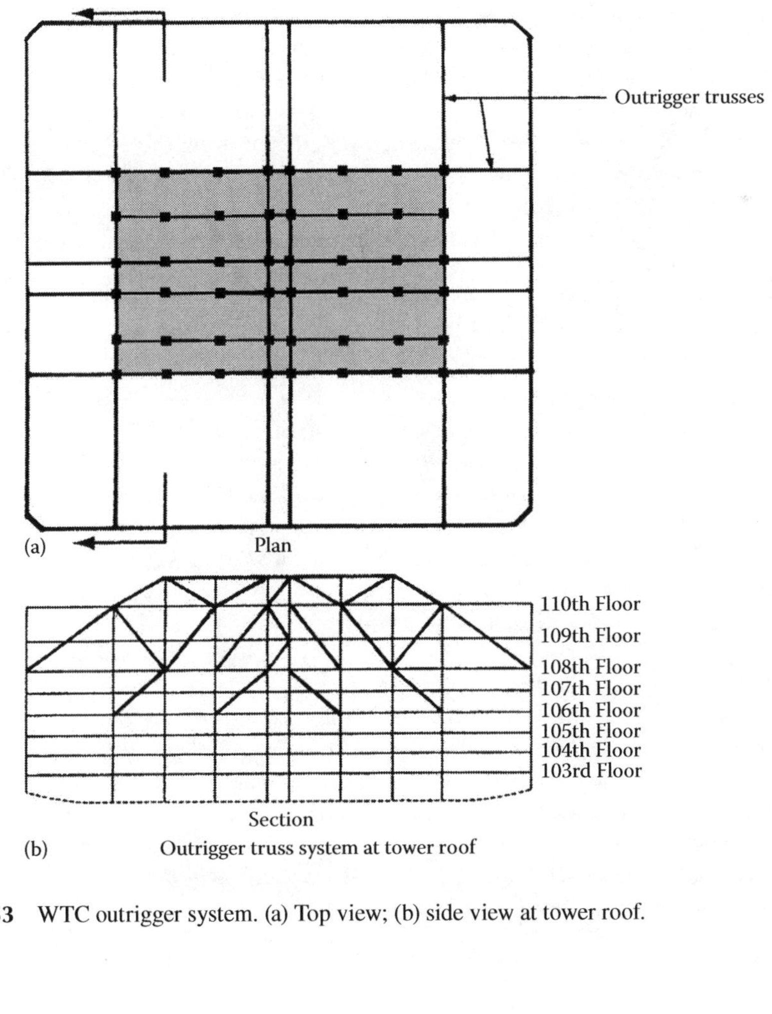

FIGURE 9.53 WTC outrigger system. (a) Top view; (b) side view at tower roof.

Prior to construction, the site was underlain by deep deposits of fill material, placed over a period of several hundred years to reclaim the shoreline. In order to construct the towers, perimeter walls for the subterranean structure were constructed using slurry wall and tieback techniques.

Floors within the substructure were of reinforced concrete flat-slab construction, supported by structural steel columns. These floors also provide lateral support for the perimeter walls, holding back the earth and water pressures from the unexcavated side of the excavation. The tiebacks, which had been installed as temporary stabilizers, were decommissioned by cutting off their end anchorage hardware and repairing the pockets in the slurry wall where these anchors had existed.

In slurry wall construction, a trench is dug in the eventual location of the perimeter walls. A bentonite slurry is pumped into the trench as it is excavated, to keep the trench open against caving of the surrounding earth. Prefabricated reinforcing steel is lowered into the trench, and concrete is placed through a tremie to create a reinforced concrete wall around the site perimeter. After the concrete is cured, excavation of the structure begins. As the excavation progresses below surrounding grade, tiebacks are drilled through the exposed concrete wall and through the surrounding soil into the rock below to provide stability for the excavation.

Tower foundations beneath the substructure consisted of massive spread footings, socketed into and bearing directly on the massive bedrock. Steel grillages, consisting of layers of orthogonally placed steel beams, were used to transfer the column loads in bearing to the reinforced concrete footings.

On September 11, 2001, two commercial airlines were hijacked, and one was flown into each of the towers. The structural damage sustained by each tower from the impact, combined with the ensuing fires, resulted in the total collapse of both buildings. The north tower was struck between floors 94 and 98, with the impact roughly centered on the north face. The south tower was hit between floors 78 and 84 toward the east side of the south face. Both planes banked steeply with estimated speeds of 470 mph and 590 mph at the time of impacting the north and south towers, respectively. The population on September 11, 2001, of the seven buildings of the WTC complex has been estimated at 58,000 people. Almost everyone in WTC 1 and 2 who was below the impact area was able to evacuate the buildings, due to the length of time between the impact and collapse of the individual towers.

In each case, the aircraft impacts resulted in severe structural damage, including some localized partial collapse, but did not result in the initiation of global collapse. In fact, WTC 1 remained standing for a period of 1 h 43 min following the initial impact and WTC 2 for approximately 56 min. The fires heated the structural systems and over a period of time resulted in additional stressing of the damaged structure, as well as additional damage and strength loss to initiate a progressive sequence of failures that culminated in total collapse of both structures.

Design experts from ASCE and FEMA who investigated the WTC destruction have agreed that it would be futile to create a *terror code* to try to outdesign terrorists. The WTC buildings were not required to protect their occupants during the disaster, but on 9/11 did so stunningly. Despite being subjected to stresses that never could have been anticipated, the structural design of the towers kept them standing long enough for more than 20,000 people to evacuate.

9.9.3.5 Empire State Building, New York

The Empire State Building was the tallest building in the world for more than 40 years, from the day of its completion in 1931 until 1972 when the Twin Towers of New York's WTC exceeded its 1280 ft (381 m) height by almost 120 ft (37 m). The structural steel frame consisting of moment and braced frames with riveted joints, although encased in concrete, was designed to carry 100% of the gravity and wind loads. The concrete encasement, although neglected in strength analysis, stiffened and frame considerably against wind loads. Measured frequencies of the building have estimated the actual stiffness at 4.8 times the stiffness of the bare frame.

The Empire State Building was completed in the year 1931 and with 102 stories stood no less than 1250 ft high. The height was increased by the addition of a 222 ft TV aerial in 1951. Among the many astonishing features of this building is the fact that it has seventy-three elevators although not all traverse the full height. Amazingly, it was built in a relatively short period of 410 days, this being made possible by the extensive use of prefabricated steel sections so that at one point it was raising by four or five stories a week.

9.9.3.6 Bank of China Tower, Hong Kong

This prism-shaped building (shown in Figure 9.54), designed by the architectural firm of I. M. Pei and Partners and structural engineer Leslie E. Robertson, is a 76-story building consisting of four quadrants. Each of the quadrants rises to a different height, and only one reaches the full 76 stories. The lateral bracing consists of a space truss spanning between the four corner columns. From the top quadrant down, the gravity loads are systematically transferred out to the building corner columns by truss action. Transverse trusses wrap around the building at selected levels. At the 25th floor, the center column is transferred to the corners by the space truss, providing for an uninterrupted 158 ft (48 m) clear span at the lobby. At the fourth floor, the horizontal shear forces

FIGURE 9.54 Bank of China Tower, Hong Kong.

are transferred from the space truss to the interior composite core walls through $^1/_2$ in. (12 mm) thick steel plate diaphragms acting compositely with the floor slab. The foundation for the building consists of caissons as large as 30 ft (9.1 m) in diameter hand-dug to bedrock.

The Bank of China Tower is one of the most recognized skyscrapers in central Hong Kong. The building is 1033ft (315m) tall with two masts reaching 1209 ft (369m) high. The structural expression adapted in the design of the building resembles growing bamboo shoots, symbolizing livelihood and prosperity. The entire structure is supported by four columns at the corner of the building, with the triangular frameworks transferring the weight of the structure to these columns.

9.9.3.7 Standard Oil of Indiana Building, Chicago

The building project consists of a base structure extending five levels below a plaza, covering the entire site measuring 350 ft by 385 and a tower structure extending through and 82 stories above the plaza.

There are about three million square feet of floor area in the tower and an additional half-a-million square feet below the plaza surrounding the tower. The tower height above the plaza is 1123 ft and the height above the lowest basement is 1179 ft.

The lateral-load-resisting system for the tower is a tube system achieved by an arrangement of steel plate walls with openings only large enough to accommodate architecturally acceptable vision panels. Thus, the steel for the tube system providing lateral resistance is distributed continuously around the perimeter walls, rather than in intermittent exterior columns and spandrels.

The plates in the exterior wall are arranged in a pattern as shown in Figures 9.55a–e. The vertical columns are V-shaped consisting of two plates and are welded to horizontal plates that serve as spandrel beams. The balance of the structural system includes the framing for the central core and the floor system. Figure 9.55b shows the floor plan for floors 3 through 19, with 16 interior columns and 64 exterior V-shaped columns.

The reentrant corners made of stiffened steel plates serve to transfer the column axial loads from one face of the building to the other, under the action of the lateral loads.

The main floor span is 45 ft from the core to the exterior and is framed by steel trusses 38 in. deep overall, spaced 10 ft apart. The variable span corner floor beams are wide-flange sections, as are the four diagonals from the corner columns of the core to the center of the reentrant corners. The floor's framing is a blended steel deck system consisting of 3 ft wide panels of $1^1/_2$ in. deep, 18-gage deck alternating with 2 ft wide panels of 18-/20-gage cellular deck. Topping consists of 5000 psi lightweight concrete, 4 in. thick for a total diaphragm thickness of $5^1/_2$ in.

The tower is founded on caissons extending to rock. There are 40 perimeter caissons supporting a transfer girder that carries 64 V-shaped columns. See Figure 9.55e. The caissons vary from 5 ft to 6 ft 9 in. in diameter. The 16 core columns are supported on caissons varying from 8 ft -9 in. to 10 ft-3 in. in diameter. All caissons are constructed of 6000 psi concrete.

One of the main objectives in developing tubular behavior of building structures is to minimize the steel tonnage used in the structural frame. The steel weight per square foot of floor area for the building is 33 psf. The building completed as the Standard Oil Building in 1973 was renamed Amoco Building in 1985, then as the Aon Center in 1999.

When completed in 1973, the entire building was sheathed with slabs of Italian marble. In 1974, just a year after completion, one of the marble slabs detached from the facade and penetrated the roof of a nearby building. Further inspection found numerous cracks and bowing in the marble cladding. To prevent the problem, stainless steel straps were added to hold the marble in place. Later, from 1990 to 1992, the entire building was resurfaced with white granite.

The project architects and structural engineers were Edward Durell Stone and Associates and the Perkins and Will Corporation.

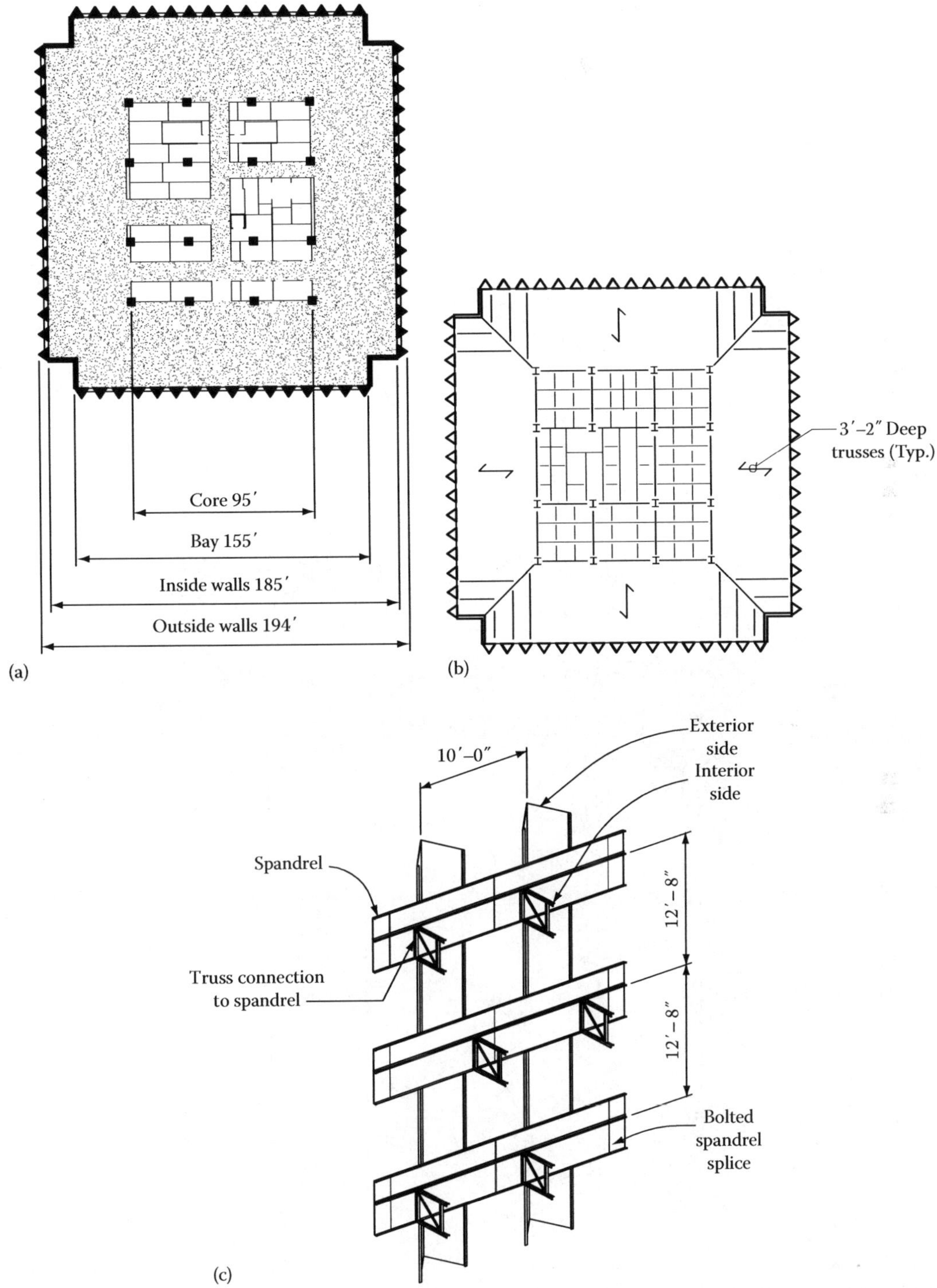

FIGURE 9.55 Standard Oil Building of Indiana, Chicago; (a) building floor plan; (b) floor framing plan, levels 3–19; (c) steel plate arrangement of exterior wall. *(Continued)*

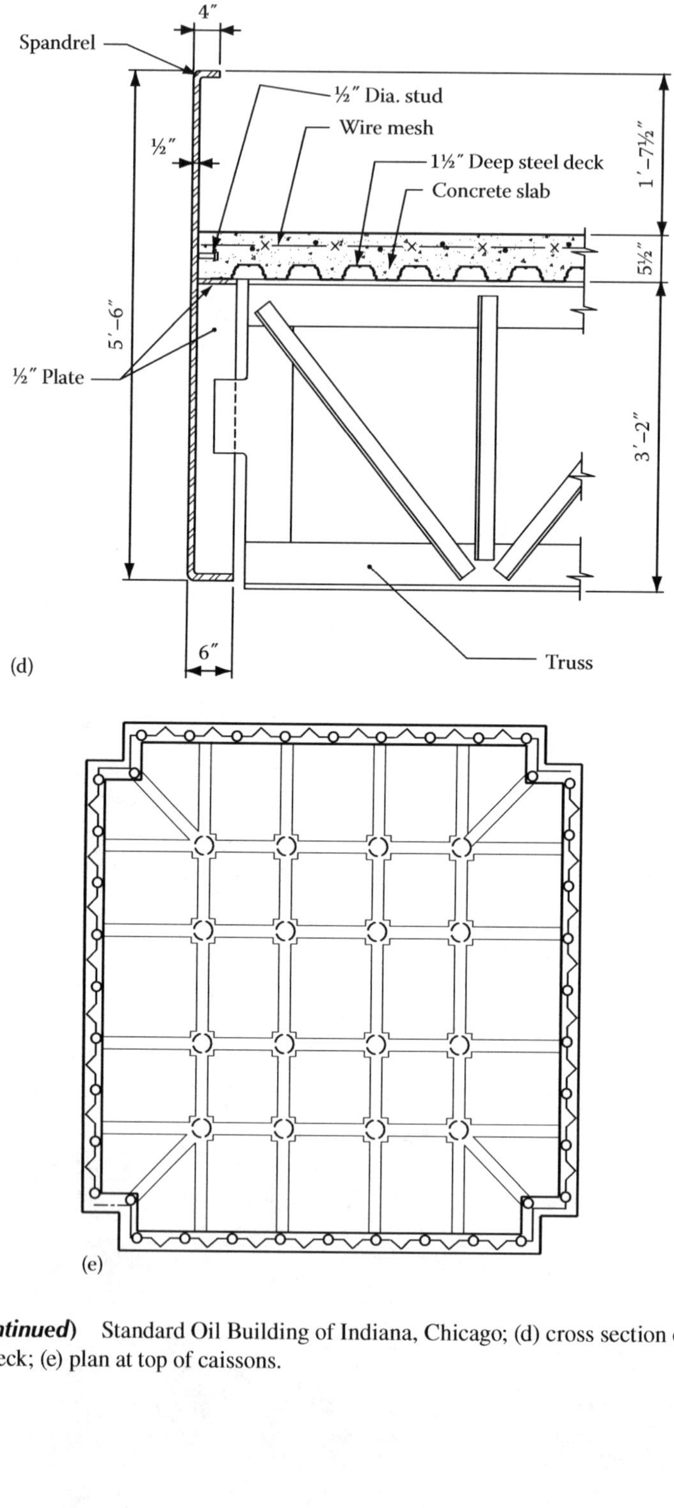

FIGURE 9.55 (*Continued*) Standard Oil Building of Indiana, Chicago; (d) cross section of spandrel beam, floor trusses, and deck; (e) plan at top of caissons.

9.10 POST-TENSION STRENGTHENING OF EXISTING STRUCTURES

Post-tensioning can be used for repairs, modification, and strengthening of existing structures, particularly so for reinforced and post-tensioned structures. It can be a cost-effective solution, because new, external high-strength steel tendons can be easily installed in the structure so that strengthening forces are applied where needed. The tendon geometry can easily be adjusted to work around the existing utility pipes and ducts and apply the forces at specific points of the structure. Applications of post-tensioning for strengthening of structures are diverse and only limited by the engineer's imagination.

Structures are strengthened for a variety of reasons such as change of loading, modification of usage, incorrect construction practices, damage to the structure, or even a design error. Post-tensioning using external tendons can easily be attached to structures, or the added tendons can be encased in concrete.

9.10.1 Tendon Profiles

Tendon profiles are typically made up of parabolic segments and straight lines. In new construction, a reverse parabola or partial parabola is generally utilized. For strengthening with external tendons, straight-line profiles are more appropriate.

Two concepts of external post-tensioning are as follows:

1. Changing the tendon force will alter the vertical force exerted on the structure.
2. Changing the tendon eccentricity will alter the vertical force exerted on the structure.

By utilizing these two factors, the engineer can fine-tune the force applied to the structure. A large amount of post-tensioning force in the horizontal direction would require a small amount of vertical offset, while increasing the depth of the tendons would require a smaller tension force in the tendons to achieve the same vertical force.

When strengthening tendons are encased in concrete to make them integral with the existing concrete, the design concepts used for the strengthening are an extension of the original design. The design concept is to have the existing concrete and the new concrete act together as if they are one. Parabolic or straight-line tendons can be used when encased in concrete. Many times, strengthening tendons are stressed prior to concrete placement to confirm the uplift force.

9.10.2 Supports

Supporting tendons used to strengthen structures are an important aspect of the strengthening design. They involve the mechanics of transferring the force into the structure. Sharp, angular changes in the tendon profile can exert significant concentrated forces on the structure. Supports transfer this force to the main structure. The number of support points depends on the method used in designing the transfer of force to the structure. Several methods are available, such as the following:

- *Single point*: Usually used to strengthen structures at point loads
- *Multiple points*: Used to strengthen structures at multiple point loads or to strengthen elements that can span between support points
- *Uniform points*: Used when the strengthening load needs to be uniformly applied; also used when strengthening tendons are encased in concrete and stressed after the new concrete is placed

9.10.3 Anchorages

The design of anchorages is critical in external post-tensioning applications. Their location must be carefully considered to not affect the structure adversely. The anchorage points apply additional

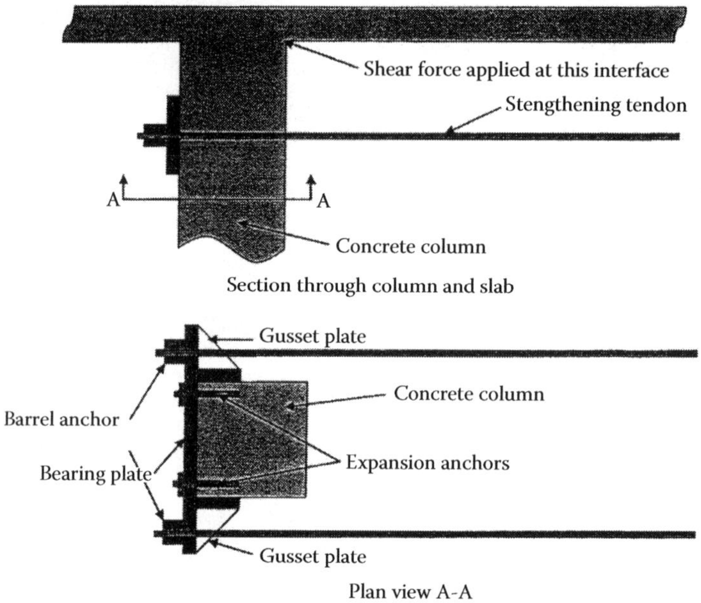

FIGURE 9.56 Anchorage bracket at column.

horizontal force and vertical forces to the existing structure. It is important that the structure be checked to ensure that it can support these additional forces.

Anchorages are attached to a soffit (slab or beam), passed through a beam or column and anchored on the vertical surface, or attached to a bracket that envelops a column. Some factors that need to be considered in the design of anchorages are as follows:

- *Attached to soffit*: When the forces are relatively small, the tendons can be attached to the slab soffit with drilled anchors.
- *Through a beam or column*: External tendons can be attached to existing beams. The bearing plate of the anchorage must be designed to safely distribute the forces from the anchor to the concrete. The designer should also check the existing reinforcement in the beam to ensure that it can safely transfer the forces to the existing structure. Also, when passing through a beam, the existing beam must be able to resist the sideway bending induced by the tendon force (Figure 9.56).
- *Bracket at column*: It may sometimes be possible to drill through the column or attach a bracket to support the anchorage at the columns. The bracket must be designed to have sufficient bearing to distribute the loads from the tendon to the concrete. The column must be checked for shear capacity at the anchorage location. Shear reinforcing may be used to reinforce the column.

External strengthening applies axial forces on the structure. The axial capacity of the structure should be evaluated to confirm that axial forces can safely be transferred from the tendon to the structure.

9.10.4 Tendon Protection

Internal strengthening tendons encased in concrete are protected in similar fashion as new construction. Concrete cover for fire and corrosion protection should be verified with applicable code requirements.

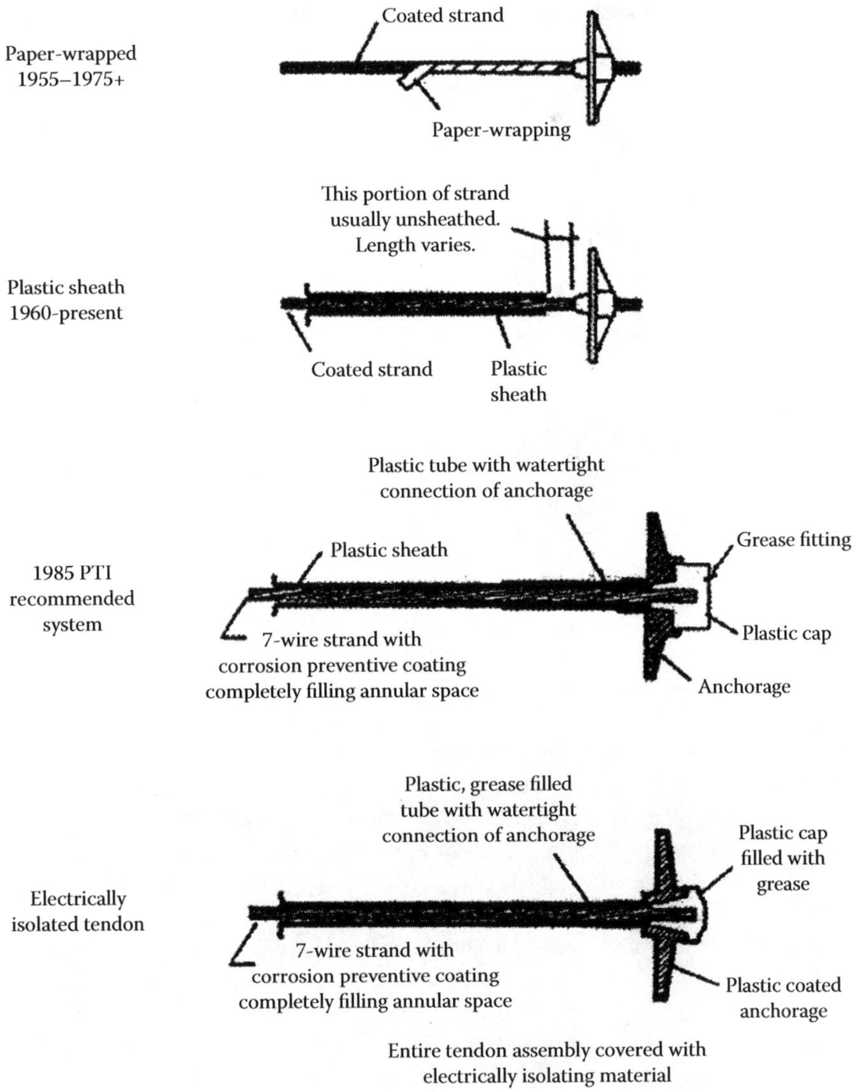

FIGURE 9.57 Evolution of corrosion protection for unbonded strand tendons for buildings.

Because of the nature of structure strengthening, encasing tendons integral with concrete is not always possible. External strengthening tendons can be encased in concrete or grout-filled pipes or encased by precast channels for fire protection when required. Corrosion protection of the tendons can be provided by encasement, greased and sheathed strand, galvanized strand, or epoxy-coated strands. Brackets, pipes, posts, and anchorages can be similarly protected (See Figure 9.57 and Figure 9.58).

9.10.5 Beams

Concrete, precast, and steel beams can also be strengthened with external or encased post-tensioning tendons.

Prior to determining the tendon force, the tendon profile (straight line or parabolic) and the required uplift loading should be established. The engineer should also decide whether the tendons

FIGURE 9.58 Plastic sheathing techniques for unbonded single strand tendons.

will be external or encased. Similarly, any fire or corrosion protection requirements should be recognized. Any height limitations affecting the tendon eccentricity should be identified.

Strengthening tendons encased in concrete will generally have a parabolic profile. The profile can be accomplished with dowels inserted into the existing beam. When new concrete and existing concrete are to act integrally, the details of attaching the new concrete to the existing for proper transfer of forces must be designed. Minimum amounts of non-prestressed reinforcement in new concrete, dowels to existing concrete, bonding agent, and roughness of the existing concrete are some of the items to be considered in the design of the interface between the new and existing concrete.

When using external tendons, the support points and anchorages must be analyzed. For beams, support points can be made with steel pipes or brackets. Steel pipes can be placed against the beam soffit extending out on either side to support the tendon, or the existing beam may be cored to insert the pipe. Pipes need to be designed to transfer the load to the structure. Brackets at the soffit of the beams offer more flexibility than pipes (Figure 9.59). Tendon geometry is easily adjusted with brackets, and they can be placed at multiple locations along the beam (Figure 9.60).

9.10.6 Floors

Floors are typically strengthened with external post-tensioning tendons.

Straight-line profiles for floor strengthening tendons are generally used with single support points at the center of the slab span (Figure 9.61). Structures with beams supporting one-way slabs are the easiest for slab strengthening because holes can be drilled through the beams at high points. Without a beam, some kind of a bracket needs to be devised that would hold the tendon close to the slab soffit. Support points at midspan of the slab are normally constructed out of a steel tube with a base sized to transfer the uplift force into the slab (Figure 9.62). The tendon typically rests in a saddle with side retainers. The tendon eccentricity is adjusted with the length of the support post.

Single strands are normally used for slab strengthening (Figure 9.63). Tendon spacing is determined by the required amount of uplift and the height of the support post. Tendon anchorages can either be placed against a beam or attached to the slab soffit along the perimeter (Figure 9.64).

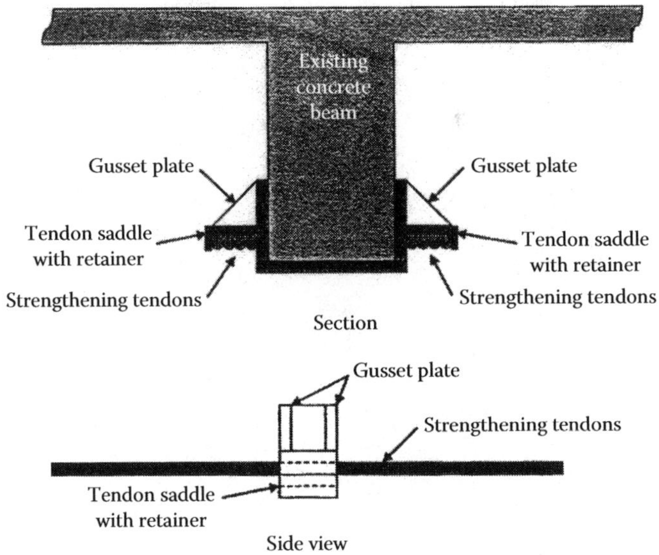

FIGURE 9.59 Beam bracket.

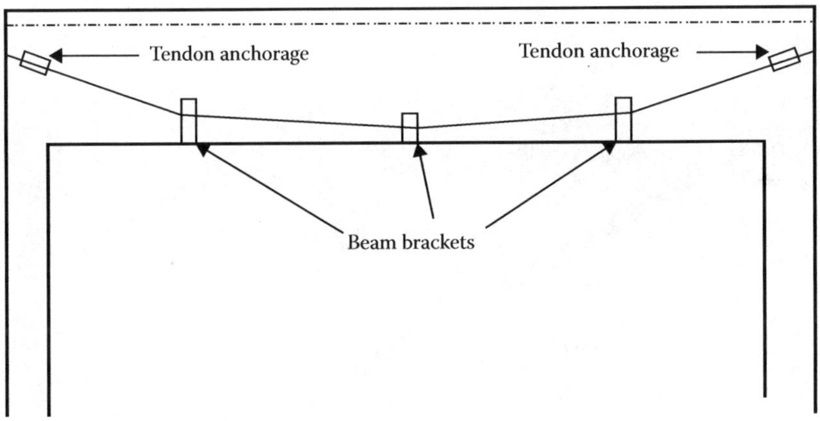

FIGURE 9.60 Tendon geometry using beam brackets.

9.10.7 Removing Columns

Removing columns in existing structures requires strengthening of the structure to carry the column load to the foundation. The column is typically removed to add open space to an area of the building. The column may be removed on one floor with columns remaining above/below or may be removed on each floor. Post-tensioning tendons can be used to transfer the column load to other supports that may exist or to new columns, beams, or walls (Figure 9.65). The strengthening tendon would most likely be a multiple-strand tendon depending on the required loading and allowable eccentricity.

When columns are removed on each floor, the strengthening tendons could run in one direction or they may be placed orthogonally depending on the structure's analysis.

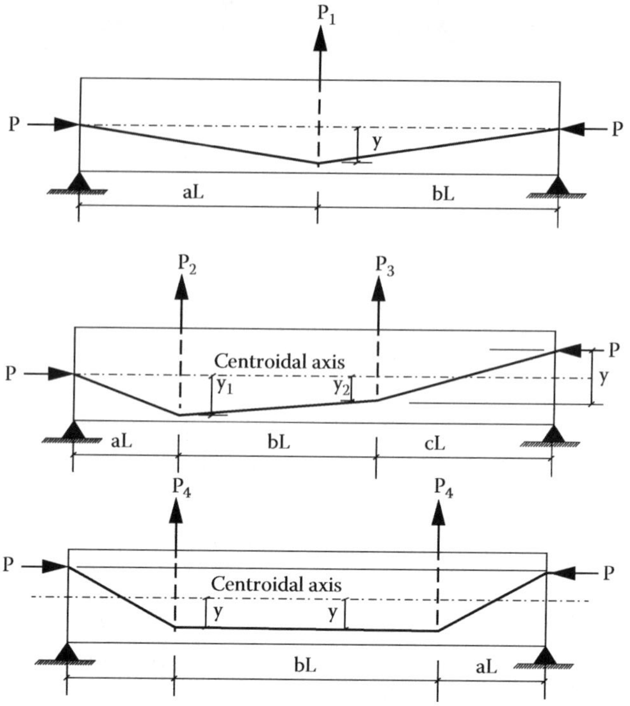

FIGURE 9.61 Straight line tendons.

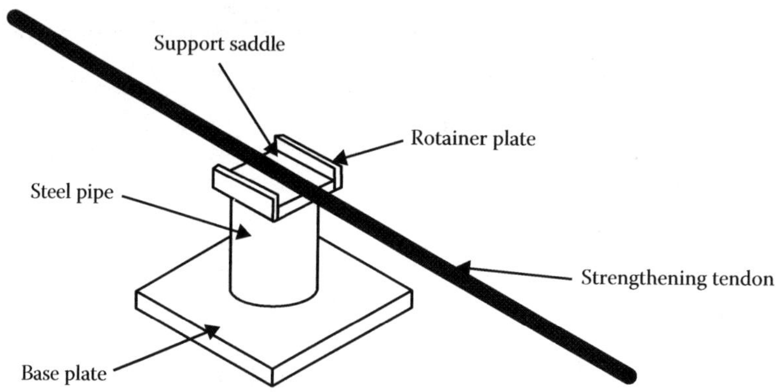

FIGURE 9.62 Example of slab support post.

9.10.8 Closing Comments

The use of externally applied post-tensioning tendons (EPT) to strengthen and repair existing buildings is extremely effective. This technique has been used in the past to strengthen all types of members and all types of materials, wood, structural steel, and concrete, both prestressed and nonprestressed. In some cases, the use of EPT has made the difference between saving and demolishing an existing building.

In most cases, the use of EPT in an existing building involves the construction of a *king-post* truss under or alongside an existing member. Holes are drilled through existing members, beams, columns, or walls, and the tendons are anchored against the outboard surfaces of these members.

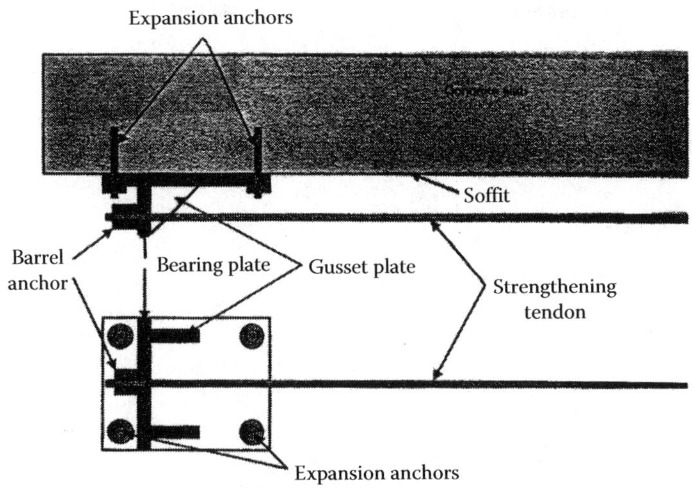

FIGURE 9.63 Anchorage device for single strand strengthening tendon.

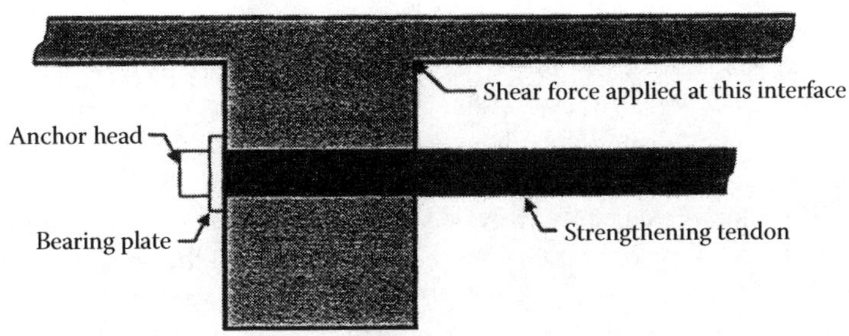

FIGURE 9.64 Cut through beam indicating critical location of shear force.

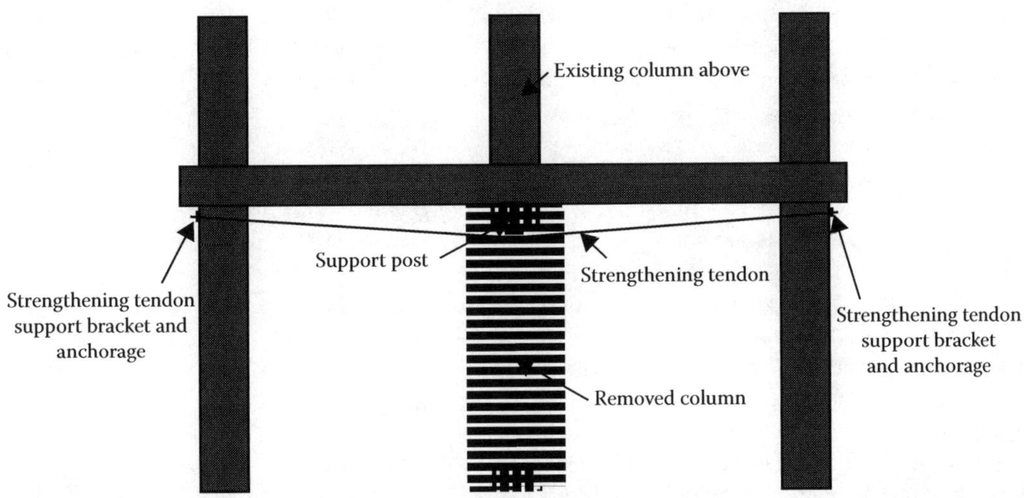

FIGURE 9.65 Example of post tensioning used when removing a column.

The king post, or *saddle*, is normally a vertical structural steel tube located in the most beneficial position on the member strengthened. The lower (tension) chord of the truss is the stressed tendon. The upper (compression) chord is the existing structure itself. The tendons pass under the tube, and when they are stressed, they exert a large, upward force on the member at the desired location. The use of high-strength prestressing steel for the tension chord allows the application of large forces on the structure with minimum structural depth. With increased awareness of the restoration of existing buildings, the use of EPT should increase substantially in future years.

9.10.9 Historical Recap of Post-Tensioned Concrete

The mid- to late 1950s were heady times in the history of post-tensioned concrete. It had been used in building construction for only a few years in the 1950s, primarily in lift-slab buildings and a few parking structures. Prestressed concrete had just appeared, for the first time, in the 1963 edition of the ACI Building Code.

The US post-tensioning industry owes its existence to lift-slab construction. The first lift-slab buildings were built in the United States in the mid-1950s using non-prestressed slabs. Problems were encountered during lifting in these early slabs because of their weight, and large deflections developed after construction due to flexural creep. Post-tensioning was being widely used in European bridges at that time, and the first post-tensioned bridges had been built in the United States and were functioning well. Post-tensioning offered a potential solution to the problem of weight and deflection in lifted slabs in buildings. The problem was that all of the existing post-tensioning systems available were in Europe and most of those systems were heavy bonded multi-strand systems not suitable for slab construction. One of the European systems, however, held some promise for use unbounded in thin slabs. That was the *button-headed* tendon system (Figure 9.66). This system consisted of parallel-lay ¼" diameter high-strength (240 ksi) wires that passed through a steel bearing plate and an externally threaded stressing washer, with *buttons* cold-formed by impact on the ends of the wires. The buttons were anchored against the outside face of the stressing washer, which attached to a hydraulic ram that elongated the wires and applied the stress. The prestress force was held by steel shims inserted between the stressing washing and the bearing plate.

Post-tensioning slabs in lift-slab buildings reduced their weight by about 30%, making lifting easier, and solved the deflection problems. For a short time, the lift-slab industry thrived and many quality lift-slab buildings were built. However, while solving some problems, the button-headed tendon system created others. First, since both dead-end and stressing-end anchorages were attached in the factory, button-headed tendons had to be fabricated to a precise length between slab edge forms, with very little tolerance. If the as-delivered tendon length was shorter or longer than the length between edge forms, either the tendon had to be replaced with another one of the correct *exact* lengths, or the edge forms had to be moved.

Next, button-headed tendons required some type of stressing pocket at their stressing end to cover the shims and stressing washer that protruded out from the bearing plate. Some contractors used a continuous edge strip to cover the anchorages; others preferred a *sawtooth* arrangement with a pocket at each anchorage. But in both cases, a second concrete pour was required to fill the pockets or the continuous edge strip.

Finally, button-headed tendons required bulky and expensive couplers when intermediate stressing was required. The coupler was usually provided in the form of a large high-strength steel stud, externally threaded, that screwed into an internally threaded hole in the stressing washer. Tendon friction in wire tendons at that time limited stressing lengths to about 80 ft from one end and twice that, or about 160 ft, from two ends. Any building longer than 160 ft in either direction therefore required an intermediate construction joint, intermediate tendon stressing, and expensive couplers. Most buildings required such a joint.

The first strand post-tensioning system was introduced to the construction market in the United States in the early 1960s. Although competition with the button-headed tendon firms was fierce, the

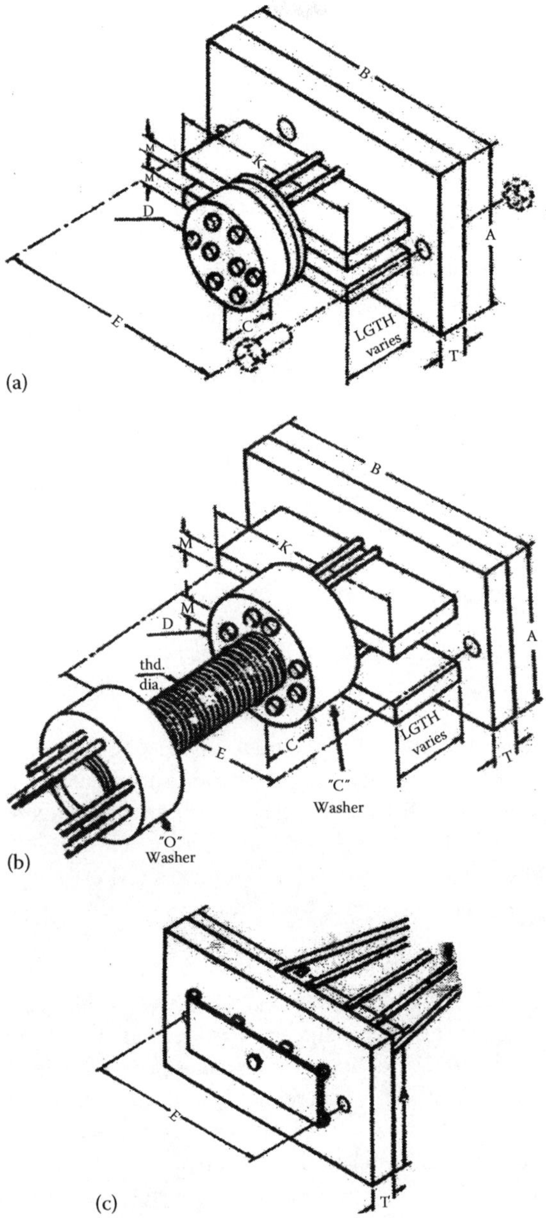

FIGURE 9.66 Fixed end anchorage assembly for unbonded button-headed wire tendons.

strand system met with much success because the strand system eliminated all of the construction problems inherent in the button-headed tendons. The strand system did not require exact length; the strand could be cut a few feet longer than the finished slab length; and the excess strand was simply trimmed off after stressing. The strand anchorages did not require formed stressing pockets or edge strips. A small two-piece round rubber *grommet*, positioned between the anchorage and the finished edge form, recessed the anchorage a few inches back inside the slab from the edge. When the grommet was removed after concrete placement, it formed a round hole into which the jack nosepiece was placed when the strand was stressed. A portion of the grommet also filled up the space inside the anchorage, preventing ingress of cement paste from the back of the anchorage during concrete placement. After stressing and cutting off the excess strand just inside the finished face of the concrete, the hole was simply filled with grout and finished flush with the slab edge. Stressing at intermediate construction joints was easy, the strand was cut to the full length of the slab, and an intermediate anchorage was simply slid onto the strand and stressed at the intermediate construction joint using open-throated stressing jacks. The remaining length of tendon was then rolled out into the next pour.

There was no bearing plate used with this anchorage. A small steel plate was used only to attach the anchorage to the forms with nails passing through the nail holes. The prestressing force was transferred to the concrete not by bearing but by the direct tensile resistance of the concrete to the lateral forces generated by the wedges on the inside surface of the coil. This required significant concrete tensile strength in the anchorage zone. Many concrete breakouts occurred when coil anchorages were stressed. These breakouts were particularly prevalent in lightweight concrete, which was widely used in California in the 1960s. It became obvious that the coil anchorage had to be replaced with a bearing-type anchorage. This resulted in the development of the first ductile iron casting. It went into service for the first time in 1965. The use of ductile iron, a casting material with ductile properties, permitted a bearing plate surface to be combined with the *barrel* ring containing the tapered hole housing the wedges in a single casting piece. The development of the ductile iron casting was a huge event in the history of post-tensioned building construction, and ductile iron castings similar to the original design continue to be used as the industry standard today.

9.10.10 Landmarks in Post-Tensioned Buildings

The following are the most significant developments affecting the growth and use of post-tensioned concrete in US building construction:

- The introduction of the strand/wedge system to replace the button-headed tendon system
- The development of the ductile iron casting for single-strand unbonded tendons
- The introduction of the *load-balancing* method for the design and analysis of post-tensioned concrete members
- The introduction of the *banded* tendon system for two-way post-tensioned slab systems

9.10.11 Load Balancing

Perhaps the most important single event in the history of post-tensioned concrete building construction was the introduction of a simplified method for the design and analysis of complex, indeterminate post-tensioned concrete members called *load balancing*. This was done in a paper written by T.Y. Lin himself, published in 1963 in an ACI Journals paper. It involved mentally removing the tendon from the concrete member and replacing it with all of the forces that tendon exerts on the concrete. The concept was brilliant, easy to understand, and greatly reduced the mathematical drudgery involved in other design and analysis methods. It made the design of post-tensioned concrete members as easy for the practicing engineer as the design of non-prestressed concrete members. This design simplicity encouraged structural engineers to select post-tensioned concrete as the preferred framing method.

9.10.12 Banded Tendons

Two-way post-tensioned slabs have been a popular type of framing in concrete building construction. When this type of framing started to be commonly used, tendons in two-way slabs were installed in each of two orthogonal directions with some located in the *column strip*, an imaginary area centered on the column lines and extending one-quarter of the bay width on either side of the column. The remaining tendons were installed in the *middle strip*, the area located between the column strips. Since the tendons were *draped* in a curved vertical profile (generally parabolic), high at the column lines and low at midspans, each individual tendon would typically have some perpendicular tendons above it and some below it. This tendon arrangement was generically known as a *basket-weave* system.

In order to install such a system of woven tendons, the tendon detailer had to locate and identify the single tendon that was below all other perpendicular tendons. That tendon, or group of tendons, was identified on the placing drawings as tendon sequence #1. Next, the detailer found the tendon in the other direction that was below all other perpendicular tendons, with the exception, of course, of tendon sequence #1. That tendon, or group of tendons, was identified as tendon sequence #2. All tendons in the slab were identified in this manner with a sequence number. Each tendon had to be installed with the precise sequence number, or a bird's nest of tendons would result and the tendons could not be chaired at the proper heights. Often, slabs would have 30–40 sequence numbers.

9.10.13 Irregular Column Layout

In 1968, the most famous post-tensioned building in history was built. Its primary fame was not because it was post-tensioned but because of what eventually happened in it. It was the Watergate Apartments in Washington, DC. This building, also famous for another reason, was the first building ever built using a two-way post-tensioned slab with a new and innovative tendon distribution that came to be known as the *banded* tendon distribution.

In the architectural design of the Watergate building, the floor plan was curved and columns were located randomly in areas that substantially hid them, including walls, duct spaces, and closets. The resulting column layout did not line up in either direction. No column was spaced any farther than about 22 ft from any other column; however, the concepts of gridlines, column strips, and middle strips were meaningless. The structural design was challenging because, using conventional two-way slab techniques, there was no obvious path for slab loads to columns. The solution was connecting columns in one axis of the building with imaginary straight lines between individual columns, and thinking of those lines as a series of beams, or hard points. A *band* of tendons could be run along that line connecting columns in one direction; then in the other direction, tendons could be spaced uniformly over bands. With this concept, the load path became obvious, and the forces and profiles for both the band tendons and the uniform tendons could be easily calculated.

This tendon layout, with all of the post-tensioning tendons in one direction located in a narrow band over columns and tendons distributed uniformly with no regard for imaginary column strips and middle strips, had never been done before. The performance of the slabs appeared to be good, with a significant savings in tendon placing costs when compared to the conventional *basket-weave* system. The primary labor savings resulted from elimination of tendon sequencing. In this new banded layout, all of the band tendons were placed first, then all of the uniform tendons. Ironworkers did not have to place individual series of tendons, alternating in each direction, according to a complex sequence.

Since the Watergate Apartment building, built almost 50 years ago, the banded tendon layout has become the standard method for placing tendons in two-way post-tensioned slabs. The adequacy of the banded tendon layout has been confirmed by the functional performance of hundreds of millions of square feet in service and numerous laboratory tests.

9.10.14 Cracking in Post-Tensioned Slabs

Restraint to concrete volume change (shortening) was the first big pervasive problem faced by the industry. The mechanics of volume change and the restraint to that volume change is different in post-tensioned concrete members than in non-prestressed members. In non-prestressed concrete beams and slabs, restraint to shortening provided by connected elements (walls and columns) results in many closely and uniformly spaced cracks throughout the length of the member. The ends of a non-prestressed member tend to stay in their original positions, and the total shortening is simply the sum of the crack widths along its length. Because of this, shears and moment induced into restraining connected elements are relatively small.

In post-tensioned concrete members, however, the effect of the axial prestress force tends to close most of the restraint to shortening cracks, which would otherwise form between the ends of the member. Unlike non-prestressed members, the total volume change along the length of the post-tensioned concrete member is reflected in significant movement inward at the ends. This induces large shears and moments into the connected walls and columns. These shears and moments can result in large and unsightly cracks in the post-tensioned member and in the walls and columns themselves.

Engineers had to learn how to design post-tensioned concrete floor systems with levels of cracking normally accepted in non-prestressed floor systems. This was accomplished over the years largely with the use of joinery details (slip joins and pour strips) and the use of properly located and sized non-prestressed reinforcement. Mitigation of restraint to shortening effects in the design of post-tensioned buildings has become as large a part of the design process as the selection of the forces and profiles themselves.

But without a doubt, the biggest problem ever faced by the industry was tendon corrosion. The early unbonded tendon sheathings and coatings (grease) were inadequate for aggressive environments, such as those where de-icing salts are applied to exposed concrete surfaces. Serious corrosion problems began to be apparent in such buildings within about 10 years of service. Most were repairable, and several companies thrived by specializing in corrosion-related repairs.

Improved material specifications developed and enforced by PTI through certification and informational programs have largely solved these corrosion problems. These include improvements in the quality and performance of sheathing material, coatings, and the development of a complete tendon encapsulation system for use in the most severe environments.

9.11 REINFORCED CONCRETE SPECIAL MOMENT FRAMES

Reinforced concrete special moment frames are used as part of seismic-force-resisting systems in buildings that are designed to resist earthquakes. In these frames, beams, columns, and beam–column joints are proportioned and detailed to resist flexural, axial, and shearing actions that result as a building sways through multiple displacement cycles during strong earthquakes. Special proportioning and detailing requirements result in a frame capable of resisting strong earthquake shaking without significant loss of stiffness or strength. These moment-resisting frames are called *special moment frames* because of these additional requirements, which improve the seismic resistance in comparison with less stringently detailed intermediate and ordinary moment frames.

The design requirements for special moment frames are presented in the ACI 318 *Building Code Requirements for Structural Concrete*. The special requirements relate to inspection, materials, and design of framing members (beams, columns, and beam–column joints) and construction procedures. The numerous interrelated requirements are not necessarily arranged in a logical sequence, making their application challenging for all but the most experienced designers. It should be noted that most special moment frames use cast-in-place, normal-weight concrete having rectilinear cross sections without prestressing.

Historically, reinforced concrete special moment frame concepts were introduced in the United States starting around 1960. Their use at that time was essentially at the discretion of the designer, as it was not until 1973 that the Uniform Building Code first required use of the special frame details in regions of high seismicity.

In most early applications, special moment frames were used in all framing lines of a building. A trend that was developed in the 1990s was to use special moment frames in fewer framing lines of the building, with the remainder comprising gravity-only framing that was not designated as part of the seismic-force-resisting system. Some of these gravity-only frames did not perform well in the 1994 Northridge earthquake, leading to more stringent requirements for proportioning and detailing these frames. The provisions for members not designated as part of the seismic-force-resisting system are contained in ACI 318 and apply wherever special moment frames are used in SDCs D, E, or F. Because the detailing requirements for the gravity-only elements in those cases are similar to the requirements for the special moment frames, some economy may be achieved if the gravity-only frames can be made to qualify as part of the seismic-force-resisting system.

Special moment frames also have found use in dual systems that combine special moment frames with shear walls. In current usage, the moment frame is required to be capable of resisting at least 25% of the design seismic forces, while the total seismic resistance is provided by the combination of the moment frame and the shear walls in proportion with their relative stiffnesses.

Moment frames are generally selected as the seismic-force-resisting system when architectural space planning flexibility is desired. When concrete moment frames are selected for buildings assigned to SDCs D, E, or F, they are required to be detailed as special reinforced concrete moment frames. Proportioning and detailing requirements for a special moment frame will enable the frame to safely undergo extensive inelastic deformations that are anticipated in these SDCs. Special moment frames may be used in SDCs A, B, and C, though this may not lead to the most economical design.

9.11.1 Frame Proportioning

Typical economical beam spans for special moment frames are in the range of 20–30 ft. In general, this range will result in beam depths that will support typical gravity loads and the requisite seismic forces without overloading the adjacent beam–column joints and columns. The clear span of beams must be at least four times its effective depth per ACI 318. They are allowed to be wider than the supporting columns. The provisions for special moment frames exclude the use of slab–column framing as part of the seismic-force-resisting system.

Special moment frames with story heights up to 20 ft are not uncommon. For buildings with relatively tall stories, it is important to make sure that soft and/or weak stories are not created.

The ratio of the cross-sectional dimensions for columns shall not be less than 0.4 per ACI 318. This limits the cross section to a more compact section rather than a long rectangle. ACI 318 sets the minimum column dimension to 12 in., which is often not practical to construct. A minimum dimension of 16 in. is perhaps more appropriate, except for unusual cases or for low-rise buildings.

9.11.2 Strength and Drift Limits

Both strength and stiffness need to be considered in the design of special moment frames. According to ASCE 7, special moment frames are allowed to be designed for a force reduction factor of $R = 8$. That is, they are allowed to be designed for a base shear equal to one-eighth of the value obtained from an elastic response analysis. Moment frames are generally flexible lateral systems; therefore, strength requirements may be controlled by the minimum base shear equations of the code. Base shear calculations for long-period structures, especially in SDCs D, E, and F, are frequently controlled by the approximate upper limit period as defined in ASCE 7. Wind loads must also be checked and may govern the strength requirements of special moment frames. Regardless of whether gravity, wind, or seismic forces are the largest, proportioning and detailing provisions for special moment frames apply wherever special moment frames are used.

The stiffness of the frame must be sufficient to control the drift of the building at each story within the limits specified by the building code. Drift limits in ASCE 7 are a function of both occupancy category and the redundancy factor ρ.

The drift of the structure is to be calculated using the factored seismic load, amplified by C_d, when comparing it with the values listed in the ACI table. Furthermore, effective stiffness of framing members must be reduced to account for effects of concrete cracking. The allowable wind drift limit is not specified by ASCE 7 or ACI 318; therefore, engineering judgment is required to determine the appropriate limit. Consideration should be given to the attachment of the cladding and other elements and to the comfort of the occupants.

P-delta effects, discussed elsewhere in this book, can be significant in a special moment frame and must be checked.

9.11.3 Design Principles

The design base shear equations of building codes incorporate a seismic force reduction factor R that reflects the degree of inelastic response expected for design-level ground motions, as well as the ductility capacity of the framing system. Because the R factor for special moment frames is 8, a special moment frame should be expected to sustain multiple cycles of inelastic response if it experiences design-level ground motion.

The proportioning and detailing requirements for special moment frames are intended to ensure that inelastic response is ductile. Three main goals are (1) to achieve a strong-column/weak-beam design that spreads inelastic response over several stories, (2) to avoid shear failure, and (3) to provide details that enable ductile flexural response in yielding regions.

9.11.3.1 Strong-Column/Weak-Beam Design

When a building sways during an earthquake, the damage over height depends on the distribution of lateral drift. If the building has weak columns, drift tends to concentrate in one or a few stories and may exceed the drift capacity of the columns. On the other hand, if columns provide a stiff and strong element over the building height, drift will be more uniformly distributed, and localized damage will be reduced. Additionally, it is important to recognize that the columns in a given story support the weight of the entire building above those columns, whereas the beams only support the gravity loads of the floor of which they form a part; therefore, failure of a column is of greater consequence than failure of a beam. Recognizing this behavior, building codes specify that columns be stronger than the beams that frame into them. This strong-column/weak-beam principle is fundamental to achieving safe behavior of frames during strong earthquake ground shaking.

ACI 318 adopts the strong-column/weak-beam principle by requiring that the sum of column strengths exceed the sum of beam strengths at each beam–column connection of a special moment frame. Studies have shown that the full structural mechanism can only be achieved if the column–beam strength ratio is relatively large (on the order of four). As this is impractical in most cases, a lower strength ratio of 1.2 is adopted by ACI 318. Thus, some column yielding associated with an intermediate mechanism is to be expected, and columns must be detailed accordingly.

9.11.3.2 Avoid Shear Failure

Shear failure is avoided through the use of a capacity-design approach. The general approach is to identify flexural yielding regions, design those regions for code-required moment strengths, and then calculate design shears based on equilibrium assuming the flexural yielding regions develop probable moment strengths. The probable moment strength is calculated using procedures that produce a high estimate of the moment strength of the as-designed cross section.

Ductile response requires that members yield in flexure and that shear failure be avoided. Shear failure, especially in columns, is relatively brittle and can lead to rapid loss of lateral strength and axial load-carrying capacity. Column shear failure is the most frequently cited cause of concrete building failure and collapse in earthquakes.

9.11.3.3 Detail for Ductile Behavior

Ductile behavior of reinforced concrete members is based on

1. Confinement of heavily loaded sections
2. Adequate shear reinforcement
3. Prevention of anchorage on splice failure

9.11.3.3.1 Confinement for Heavily Loaded Sections

Plain concrete has relatively small usable compressive strain capacity (around 0.003), and this might limit the deformability of beams and columns of special moment frames. Strain capacity, however, can be increased by as much as 10-fold by confining the concrete with reinforcing spirals or closed hoops. The hoops act to restrain puffing out of the core concrete as it is loaded in compression, and this confining action leads to increased strength and strain capacity. Hoops typically are provided at the ends of columns, as well as through beam–column joints.

To be effective, the hoops must enclose the entire cross section and must be closed by 135° hooks embedded in the core concrete; this prevents the hoops from opening if the concrete cover spalls off. Crossties should engage longitudinal reinforcement around the perimeter to improve confinement effectiveness. The hoops should be closely spaced along the longitudinal axis of the member, both to confine the concrete and restrain buckling of longitudinal reinforcement. Crossties, which typically have 90° and 135° hooks to facilitate construction, must have 90° and 135° hooks alternated along the length of the member to improve confinement effectiveness.

9.11.3.3.2 Adequate Shear Reinforcement

Shear strength degrades in members subjected to multiple inelastic deformation reversals, especially if axial loads are low. In such members, ACI 318 requires that the contribution of concrete to shear resistance be ignored, that is, $V_c = 0$. Therefore, shear reinforcement is required to resist the entire shear force.

9.11.3.3.3 Prevention of Anchorage or Splice Failure

Severe seismic loading can result in loss of concrete cover, which will reduce development and lap-splice strength of longitudinal reinforcement. Lap splices, if used, must be located away from sections of maximum moment (i.e., away from ends of beams and columns) and must have closed hoops to confine the splice in the event of cover spalling. Bars passing through a beam–column joint can create severe bond stress demands on the joint; for this reason, ACI 318 restricts beam bar sizes. Bars anchored in exterior joints must develop yield strength (f_y) using hooks located at the far side of the joint. Finally, mechanical splices located where yielding is likely must be type II splices capable of developing at least the specified tensile strength of the bar.

9.11.4 Analysis

As for other lateral systems, ASCE 7 allows the seismic forces within a special moment frame to be determined by three types of analysis: ELF analysis, modal response spectrum analysis, and seismic response history analysis. The ELF analysis is the simplest and can be used effectively for low-rise structures. This analysis procedure is not permitted for long-period structures (fundamental period T greater than 3.5 s) or structures with certain horizontal or vertical irregularities.

The base shear calculated according to ELF analysis is based on an approximate fundamental period, T_a, unless the period of the structure is determined by analysis. Generally, analysis will show that the building period is longer than the approximate period, and, therefore, the calculated base shear per ASCE 7 can be lowered. The upper limit on the period (C_uT_a) will likely limit the resulting base shear, unless the minimum base shear equations control.

A modal analysis is often preferred to account for the overall dynamic behavior of the structure and to take advantage of calculated, rather than approximated, building periods. Another advantage of this analysis is that the combined response for the modal base shear can be less than the base shear calculated using the ELF procedure. In such cases, however, the modal base shear must be scaled to a minimum of 85% of the ELF base shear.

A 3D model, typically used in design offices, is effective in identifying the effects of any inherent torsion in the lateral system as well as combined effects at corner conditions.

ASCE 7 specifies the requirements for the directions in which seismic loads are to be applied to the structure. The design forces for the beams and columns are independently based on the seismic loads in each orthogonal direction. It is common to apply the seismic loads using the orthogonal combination procedure of ASCE 7 in which 100% of the seismic force in one direction is combined with 30% of the seismic force in the perpendicular direction. Multiple load combinations are required to bound the orthogonal effects in both directions. The design of beam and column is then based on an axial and biaxial flexural interaction for each load combination.

ACI 318 requires that the interaction of all structural and nonstructural members that affect the linear and nonlinear response of the structure to earthquake motions be considered in the analysis. This can be especially important for special moment frames, which may be flexible in comparison with other parts of the building, including parts intended to be nonstructural. Important examples include interactions with masonry infills (partial height or full height), architectural concrete walls, stair wells, cast-in-place stairways, and inclined parking ramps.

While permitting the use of rigid members assumed not to be part of the seismic-force-resisting system, ACI 318 requires that effects of these members be considered and accommodated by the design. Furthermore, effects of localized failures of one or more of these elements must be considered. For example, the failure of a rigid architectural element in one story could lead to formation of a story mechanism. Generally, it is best to provide an ample seismic separation joint between the special moment frame and rigid elements assumed not to be part of the seismic-force-resisting system. If adequate separation is not provided, the interaction effects specified must be addressed.

9.11.4.1 Stiffness Recommendations

When analyzing a special moment frame, it is important to appropriately model the cracked stiffness of the beams, columns, and joints, as this stiffness determines the resulting building periods, base shear, story drifts, and internal force distributions. More detailed analysis may be used to calculate the reduced stiffness based on the applied loading conditions.

Note that for beams, this produces $I_e/I_g = 0.30$. *When considering serviceability under wind loading, it is common to assume gross section properties for the beams, columns, and joints.*

ACI 318 does not contain guidance on modeling the stiffness of the beam–column joint. In a special moment frame, the beam–column joint is stiffer than the adjoining beams and columns, but it is not perfectly rigid. As described in ASCE 41, the joint stiffness can be adequately modeled by extending the beam flexibility to the column centerline and defining the column as rigid within the joint.

9.11.4.2 Foundation Modeling

Modeling pinned restraints at the base of the columns is typical for frames that do not extend through floors below grade. This assumption results in the most flexible column base restraint. The high flexibility will lengthen the period of the building, resulting in a lower calculated base shear but larger calculated drifts. Pinned restraints at the column bases will also simplify the design of the footing. Where pinned restrains have been modeled, dowels connecting the column base to the foundation need to be capable of transferring the shear and axial forces to the foundation.

One drawback to the pinned base condition is that the drift of the frame, especially the interstory drift in the lowest story, is more difficult to control within code-allowable limits. This problem is exacerbated because the first story is usually taller than typical stories. In addition, a pinned base may lead to development of soft or weak stories, which are prohibited in certain cases as noted in ASCE 7.

If the drift of the structure exceeds acceptable limits, then rotational restraint can be increased at the foundation by a variety of methods. Regardless of which modeling technique is used, the base of the column and the supporting footing or grade beam must be designed and detailed to resist all the forces determined by the analysis. The foundation elements must also be capable of delivering the forces to the supporting soil.

ASCE 7 outlines requirements where special moment frames extend through below-grade floors. The restraint and stiffness of the below-grade diaphragms and basement walls needs to be considered. In this condition, the columns would be modeled as continuous elements down to the footing. The type of rotational restraint at the column base will not have significant effect on the behavior of the moment frame.

10 Torsion

PREVIEW

The stated objective of this book is to promote the ability of the engineers to sense when an answer obtained through a computer is not correct. To fulfill that desire in a simple manner, we will revisit some approximate analysis techniques and, yes, even dust off those classic methods that were so precious to engineers of yesteryear. One such method useful in understanding the torsional behavior of shear walls is the so-called bimoment theory also known as nonuniform torsion or warping theory. Developed by Vlasov in 1941, this elegant method paves the way for developing a feel for the structural behavior of shear walls, particularly those subjected to torsional loads.

The theory is quite simple, no more complicated than the engineer's theory of bending (ETB). At first glance, the related bimoment equations may look formidable, but in practice, they can be further simplified into a format well suited for preliminary back-of-the-envelope calculations.

In this section, we study in some detail the torsional response of open-section shear walls. After a brief introduction to torsion, we examine the warping behavior of shear walls and then discuss warping properties that are used in calculating the warping stresses. Then we introduce the general theory of warping torsion. The chapter closes with a discussion of worked examples that demonstrate the usefulness of the procedure for verifying computer results.

We begin this section with a discussion of the concept of *shear center*. Let us consider a cantilever shear wall of singly symmetric cross section supporting a load P at the free end (Figure 10.1a). The force P is perpendicular to the y–z plane, which is the plane of symmetry of the shear wall. Note that the y-axis is an axis of symmetry of the cross section and that the origin of coordinates is taken at the centroid C of the cross section. Therefore, the x- and y-axes are principal centroidal axes.

Under the action of the load P, the wall bends with the y-axis as the neutral axis. Two stress resultants exist on every cross section of the wall—the bending moment M_y acting about the y-axis and the shear force V_x (equal to P) acting in the x-direction.

The moment M_y is the resultant of the bending stresses that vary linearly with the distance from the neutral axis. We obtain the shear stresses from the bending stresses by means of static equilibrium. The ensuing shear stresses have a specific resultant force that has its line of action through a point S lying on the y-axis (Figure 10.1b). This point is known as the *shear center* (or the center of flexure) of the cross section. It does not coincide with the centroid C except in special cases.

From this discussion, we see that load P must act through the shear center S if the only resultant of the shear stress is to be the force P itself. If P does not act through S, it can be replaced by a force P through S plus a twisting couple (Figure 10.1c). The effect of the force is to produce bending about the y-axis as just described, and the effect of the couple is to produce torsion of the wall. Thus, we observe that a lateral load acting on a wall will produce bending without twisting only if it acts through the shear center. As a consequence, locating the shear center S is an important aspect of wall design whenever the plane of bending is not a plane of symmetry.

The shear center of the singly symmetric shear wall shown in Figure 10.1a be determined as follows. We consider the cross section to consist of three rectangular parts, namely, the two flanges and

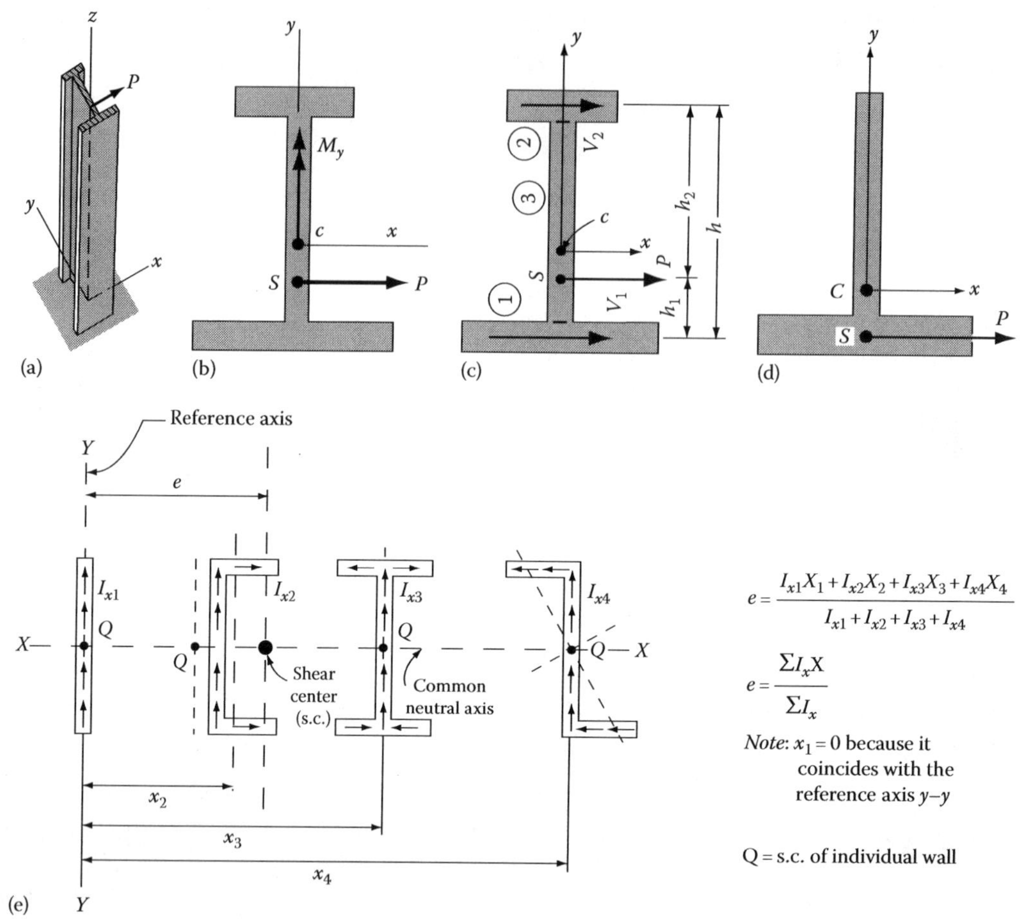

FIGURE 10.1 (a) Singly symmetrical shear wall. (b) Shear center S as it relates to centroid C. (c) Shear forces in flanges. (d) Shear center in a T-section shear wall. (e) Shear center concept; shear walls bending about a common neutral axis.

the web (Figure 10.1c). All three parts are subject to the same curvature when bending takes place, because they are integral parts of the same cross section. Therefore, the bending moment carried by each part is in proportion to its moment of inertia about the y-axis:

$$K = \frac{M_1}{EI_1} = \frac{M_2}{EI_2} = \frac{M_3}{EI_3} \tag{10.1}$$

in which

M_1, M_2, and M_3 are the moments resisted by parts 1, 2, and 3, respectively
I_1, I_2, and I_3 are their respective moments of inertia about the y-axis

If the web is thin, as is the case in practical shear walls, its moment of inertia I_3 will be very small compared to I_1 and I_2. Then we may disregard the effect of the web and assume that all of the moment is resisted by flanges:

$$M_y = M_1 + M_2$$

Also, from Equation 10.1, we get

$$\frac{M_y}{I_1 + I_2} = \frac{M_1}{I_1} = \frac{M_2}{I_2}$$

Combining the preceding two equations, we obtain

$$M_1 = \frac{M_y I_1}{I_1 + I_2} \quad M_2 = \frac{M_y I_2}{I_1 + I_2} \tag{10.2}$$

The shear forces V_1 and V_2 in the flanges are in the same proportions as the bending moments (because $V = dM/dx$); hence,

$$\frac{V_1}{V_2} = \frac{M_1}{M_2}$$

Also, the total shear force $V_x = V_1 + V_2$

By comparing this equation with Equation 10.2, we see that the shear forces are

$$V_1 = \frac{V_x I_1}{I_1 + I_2} \quad V_2 = \frac{V_x I_2}{I_1 + I_2} \tag{10.3}$$

The line of action of the resultant of these two shear forces determines the location of the shear center S.

To locate the shear center S, we determine the distances h_1 and h_2 from the center lines of the flanges to the shear center S. In as much as P is the resultant of V_1 and V_2, the forces V_1 and V_2 must produce no moment resultant about point S; therefore, the shear center distance is determined by

$$V_1 h_1 = V_2 h_2$$

Or, using Equation 10.3,

$$\frac{h_1}{h_2} = \frac{I_2}{I_1}$$

A special case occurs when the shear wall has only one flange (Figure 10.1d). For this shape, we obtain

$$h_1 = O h_2 = h$$

This result shows that the shear center is located at the intersection of the center lines of the flanges and web. We could have anticipated this result because in the derivation, we assumed that the web was very thin so that the shear force was carried entirely by the flanges.

A similar procedure may be used to find the shear center of a group of shear walls bending about a common neutral axis (Figure 10.1e).

The most common elements of concrete construction resisting torsion are the shear walls that enclose the elevator, stairs, and mechanical shafts. These typically are of *I* and *C* cross sections. We will develop torsion theory for these elements by starting with a discussion of circular cross-sectional shafts.

Recall that the shear stress in a circular cross section is directed tangentially. Its magnitude is

$$V_t = T_r/I_P = T_r/\pi R^4/2$$

where

R is the outside radius of the shaft
r is the radius at which the shear stress is measured
T is the twisting moment

Also, recall that the angle of twist θ is determined by

$$\theta = \frac{Tl}{GI_p}$$

In deriving these two equations, two assumptions are necessary: (1) plane cross sections in the untwisted state remain plane when torque is applied and (2) the cross sections remain undistorted in their own plane.

The first assumption does not remain true for noncircular sections and particularly so for *I* and *C* sections often referred to as open sections. Without giving a proof, suffice it to state here that nature resists a given action (here a torque) always with the simplest possible stresses, and thus makes it plausible that the plane cross section in an open section does not remain plane but becomes warped vertically. Using this argument, it can be shown that the torsional shear stress, V_t, in such a cross section is

$$V_t = \frac{3T}{bt^2}$$

The torsional stiffness, *C*, is

$$C = \frac{Gbt^3}{3}$$

These equations are reasonably true for $b/t > 10$ or so.

There are two remarkable facts that are important in understanding the torsional behavior of shear walls. First, the maximum shear stress occurs at that point of the periphery that is closest to the center of the section, whereas the peripheral point farthest away from the center, that is, the corner, has zero stresses. This is in complete opposition to what happens in the bending of rectangular beams and in the torsion of a circular shaft. The second point of importance is that the torsional stiffness, *C*, of the shear all is proportional to the first power of *b* only. Compare this to the torsional stiffness (polar moment of inertia) of a circular shaft, which is proportional to the cube of *b*.

Thus, if we should extrapolate the stiffness formula GI_p for the circular cross section on to the narrow rectangle section, we would be in complete error.

TABLE 10.1
Torsion Terminology

Uniform (St. Venant) Torsion	Warping Torsion
Torsional shear stress	Shear center
Twist	Open section
Polar moment of inertia	Warping deformation
Membrane analogy	Sectorial coordinate
Shear flow	Warping moment of inertia
Cellular sections	Bimoment
	Normal stress
	Tangential stress

TABLE 10.2
Analogy between Bending and Warping Torsion

Elementary Bending Theory	Warping Theory
Plane sections remain plane	Profile warps
$I_x = \int_A y^2 dA$	$I_\omega = \int_A \omega^2 dA$
$M_x = -EI_x \frac{d^2 x}{dz^2}$	$B = -EI_\omega \frac{d^2\theta}{dz^2}$
$\sigma_x = \frac{MC}{I_x}$	$\sigma_\omega = \frac{B_\omega}{I_\omega}$
$\tau_x = -\frac{VQ_x}{I_x t}$	$\tau_\omega = -\frac{HQ_\omega}{I_\omega t}$

This section gives an overview of analysis of structural systems for torsion with particular emphasis on the torsion of elements such as circular, noncircular, and cellular sections and later on discuss warping torsion of structural systems that consists of open cores.

The terminology used in torsion analysis may be conveniently grouped under two headings: uniform or St. Venant's torsion and warping torsion, oftentimes referred to as constrained torsion or torsion bending. The terms for uniform torsion are well established and given in most textbooks on structural mechanics. The purpose of recalling them here is to show how they relate to the warping theory.

The terms shown on the right-hand side of Tables 10.1 and 10.2 relating to warping torsion have, in the past, been given little attention. Consequently, designers are generally not at ease either with the concepts of warping behavior or with its methods of analysis. The aim here is to introduce the concept of importance of considering warping in practical cases.

Torsional effects on buildings as a whole are enhanced when the center of twist is eccentric from the center of gravity for inertial loading, or from the center of area for wind loading. Minimum eccentricities are prescribed by building codes to account for accidental seismic torsion. And to reflect the observed torsional behavior of a building in turbulent wind, ASCE 7-10 requires that buildings be designed for partial as well as full wind loading.

Consider the twisting of a circular shaft, as shown in Figure 10.2a. The twisting of the shaft does not produce any longitudinal stress, that is, axial compression nor tension, but only pure shear stresses. The shear stresses vary from zero at the center of the shaft to maximum value at the perimeter. In the absence of axial deformation, a cylindrical layer peeled off of the shaft changes its shape under the twisting action, from a rectangle to a parallelogram (Figure 10.2b). The absence of longitudinal stresses indicates that the surfaces at the ends of the shafts remain plane. In other words, no warping will take place. The work done by the twisting moment is expended in developing shear stresses, as shown in Figure 10.3.

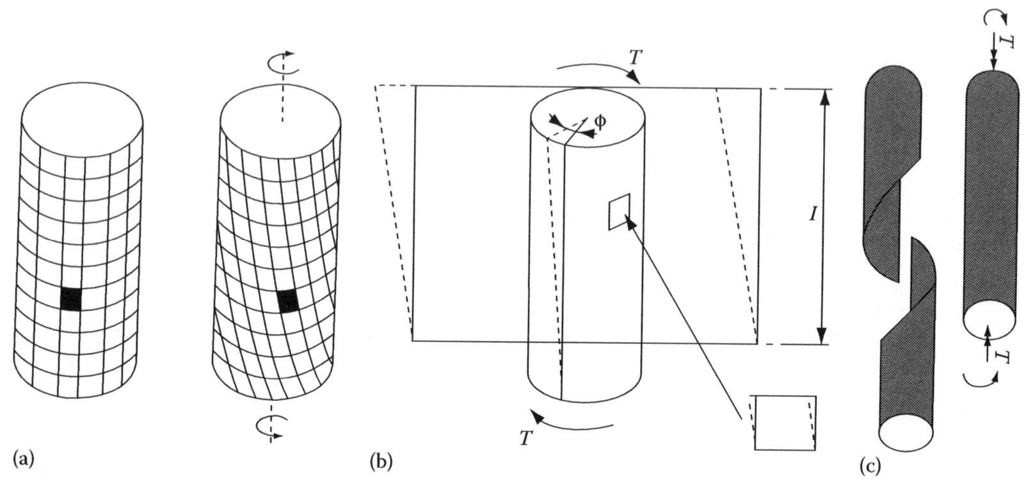

FIGURE 10.2 (a and b) Twisting of circular shaft. (c) Torsion failure of a brittle material by tension cracking along a 45° helical surface.

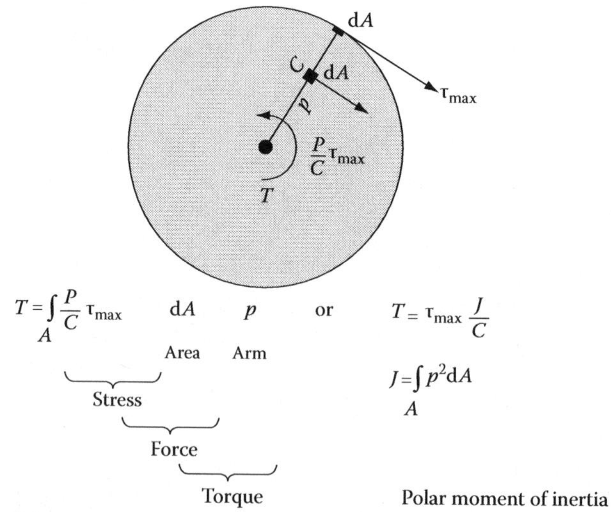

FIGURE 10.3 Variation of torsional shear stresses in circular shaft.

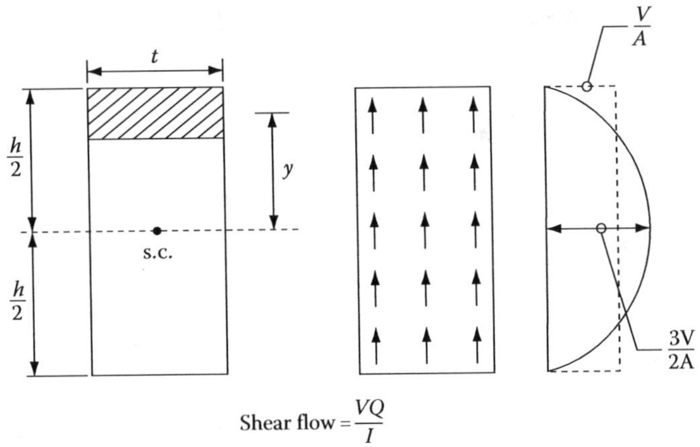

FIGURE 10.4 Shear flow in a rectangular section.

Consider a rectangular beam section subjected to the action of a vertical load at the center of gravity of the section (Figure 10.4). To find shear stress at any horizontal section, we introduced an imaginary horizontal cut at the section and obtained the shear stresses by the relation *VQ/It*. By inspection, the resultant of the vertical shear stresses is at the center of gravity of the beam.

Next, we take a look at the torsional behavior of a thin-walled section. The main reason why a thin-walled section must be given special consideration is that the shear stresses and strains in it are much larger than those in solid sections. An examination of distribution of shear stresses through the cross section shows that the shear stresses through the cross section are as if they were a fluid: hence the name shear flow (Figure 10.5).

Now consider a flanged section such as a C-shaped shear wall (Figure 10.6). To find the shear flow, we abandon the idea of the horizontal cut. Instead, we consider a cut perpendicular to the profile and find the shear along the profile. The shear R_2 in the web is in equilibrium with the vertical load *V*, and while the horizontal shears R_1 and R_3 in the webs result in no net horizontal load, the

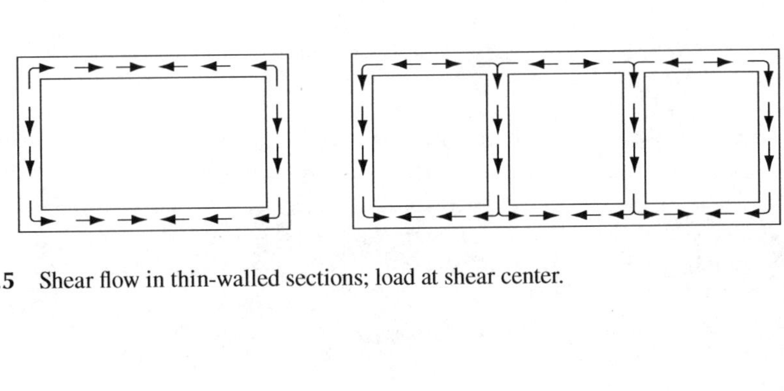

FIGURE 10.5 Shear flow in thin-walled sections; load at shear center.

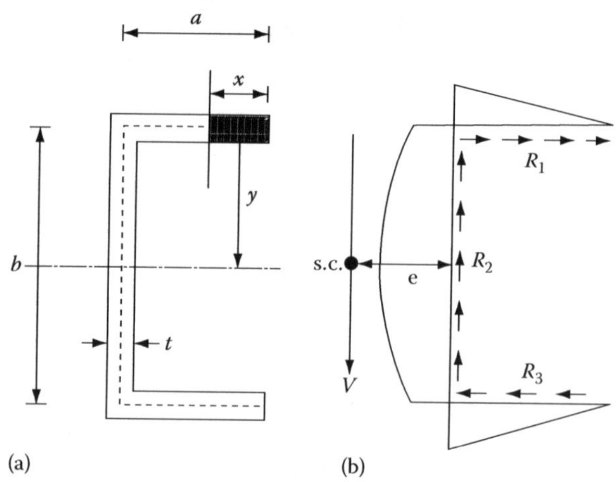

FIGURE 10.6 Shear center in C-section.

resulting moment requires offsetting of the vertical load to a location left of the web. The resultant forces from shear stresses are

$$R_1 = R_3 = \int_0^a \tau t dx_1$$

$$= \frac{Vbta^2}{4I_y}$$

$$= \frac{3Va^2}{b^2(1+(6a/b))}$$

For vertical equilibrium,

$$R_2 = V$$

For zero rotational effect,

$$R_1 b = R_2 e$$

Hence,

$$e = \frac{R_1 b}{R_2} = \frac{3a^2}{b(1+(6a/b))}$$

To find shear stresses in a cellular section, a two-step approach is required because the problem is statically indeterminate (Figure 10.7). First, the section is rendered statically determinate by inserting a horizontal cut along the length of the section, and the shear flow in the section is evaluated by the relation *VQ/I*. Next, the shear flow required to close the gap is evaluated. The final shear stress is evaluated by combining the two.

As an example, Figure 10.8 shows schematically the final shear stresses in a hollow rectangular section. The section consists of webs of unequal thickness and is subjected to a vertical load at its shear center.

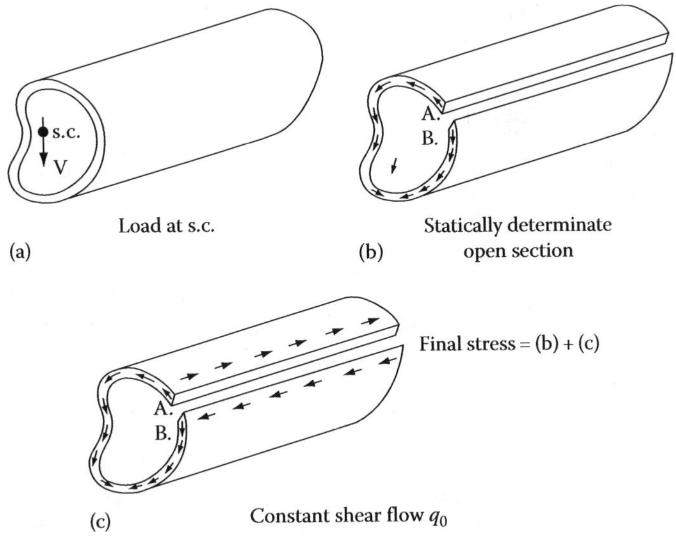

FIGURE 10.7 Shear stresses in hollow section: load at shear center.

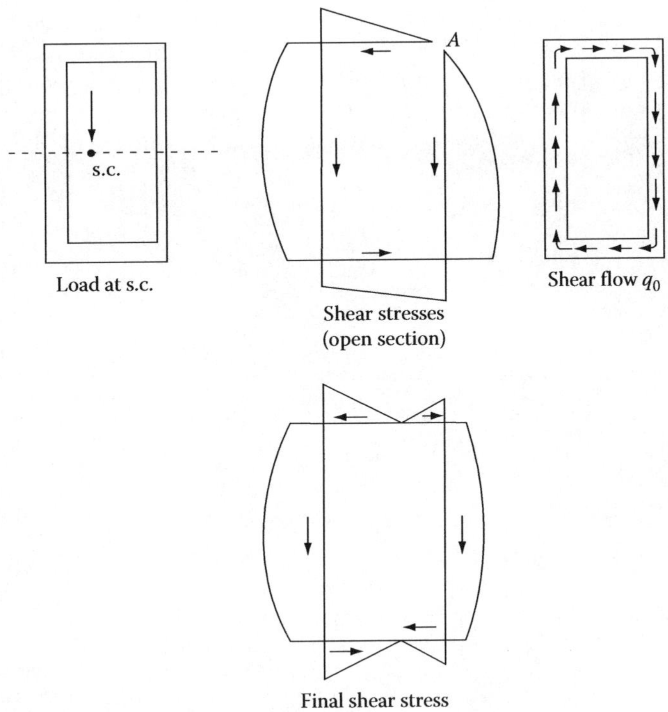

FIGURE 10.8 Shear stresses in hollow rectangular section.

If we have a multiple cellular section, the procedure is similar to that for a single-cell section. The only difference is that the problem is statically indeterminate to the nth degree where n represents the number of cells. The example in Figure 10.9 has two cells; hence, $n = 2$. Two cuts are made to render the section open. The shear flows q_1 and q_2 are evaluated by solving two simultaneous equations, and the final shear stress is obtained by superposition.

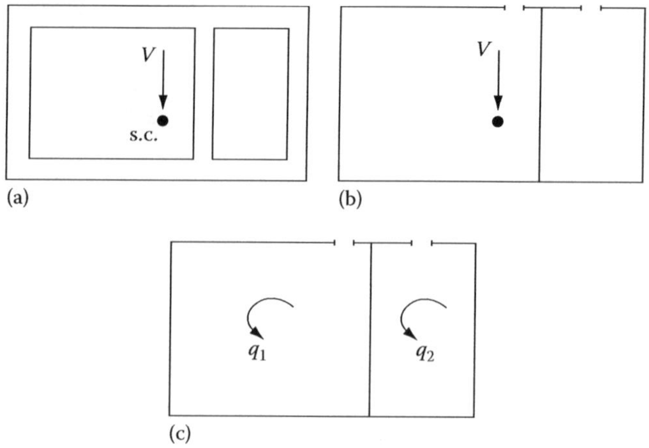

FIGURE 10.9 Shear flow in cellular sections. (a) Load at shear center, (b) section rendered open with two cuts, (c) shear flows required for compatibility, and (d) final shear flow = $b + c$.

The theory of torsion and related formulas discussed earlier are commonly referred to as St. Venant's torsion formulas and are valid for beams of circular cross sections. This formula can be accepted for noncircular sections only when the additional stress caused by warping deformation is ignored. Consider, for example, a rectangular section shown in Figure 10.10a. The vertical fibers of the section are moving up and down from their initial position in space due to torsion. The top and the bottom of the beam do not remain plane, but become warped. However, no additional stresses are induced because the warping deformations are not restrained either at the ends or at any section along its length.

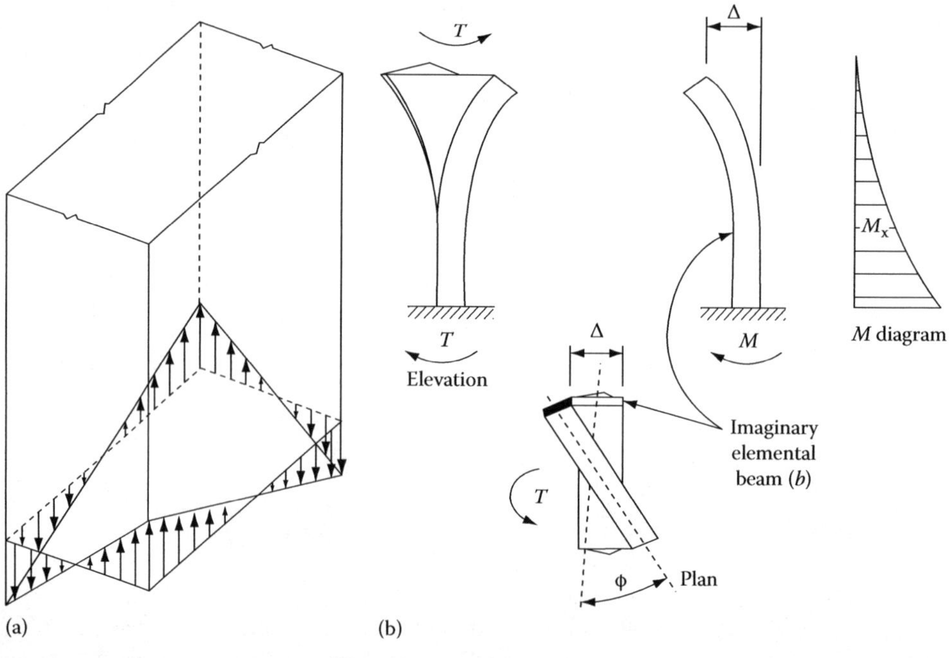

FIGURE 10.10 (a) Warping of solid beams. (b) Thin rectangular beam: bending moment due to warping restraint.

Let us examine the case when the bottom of the beam is fixed. The warping of the bottom surface of the beam is restrained resulting in longitudinal strains and stresses. If we separate an imaginary elemental beam, as shown in Figure 10.10b, it can be seen that the deflected shape is similar to that of a laterally loaded cantilever. It is obvious that bending stresses manifest at the fixed end of the beam.

The presence of bending stresses implies that part of the work done by the twisting moment is used up in bending the beam and only the remainder will develop shear stresses associated with the St. Venant's twist. Hence, the resistance to external twisting moment is offered as the sum of pure torsion plus some additional torsion, which causes bending of the section. This second part is called *warping torsion, nonuniform torsion, or flexural twist.*

For the thin rectangular beam shown in Figure 10.10b, very little energy is expended to cause elemental bending about the weak axis. For such beams, we can safely neglect the warping component of the twisting moment because the effect of constraining warping is usually restricted to the vicinity of the restraint. This phenomenon is valid, to a lesser extent for thin-walled closed sections. On the other hand, the effect of constraining warping of thin-walled open sections does not diminish rapidly and has a considerable influence on the stress distribution over a greater portion of the section.

Flexural twist causes a pair of moments. Such a pair of moments, called *bimoment*, is a mathematical function that can be visualized in most practical cases. For example, consider a two-span continuous beam supported by an interior column as shown in Figure 10.11. Since the two channels frame into opposite flanges of the column, a bimoment is introduced at the top of the column.

Restrained warping behavior involves a set of so-called sectorial parameters, each of which is counterpart in the theory of bending of beams. Since the sectorial parameters are generally unfamiliar to practicing engineers, it is perhaps appropriate to review them briefly here.

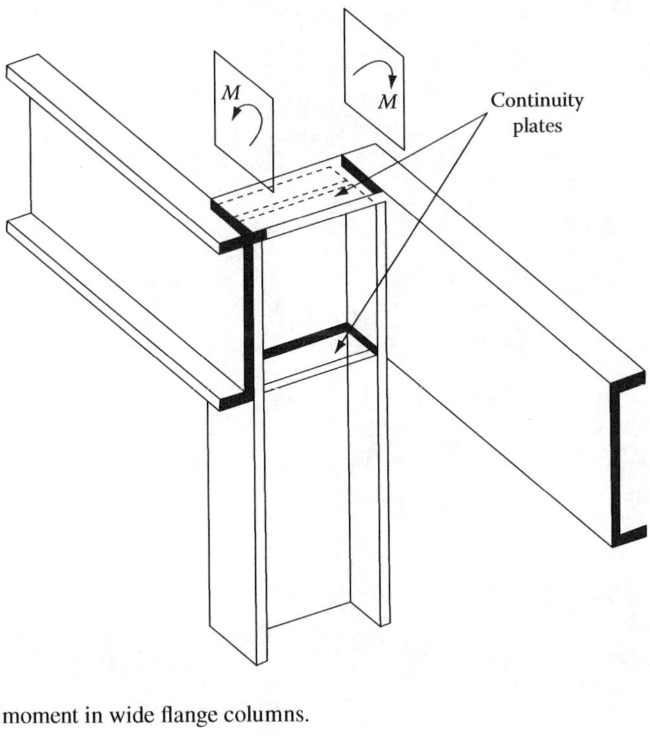

FIGURE 10.11 Bimoment in wide flange columns.

The sectorial coordinate, w, at a point on the profile of a warping core is the parameter that expresses the axial response such as axial stress and strain at that point relative to other points around the core. The w diagram can be constructed with the known location of the shear center and a point of zero warping deflection as an origin. The principal sectorial coordinate in warping theory is analogous to the distance c of a point from the neutral axis of a section in bending. Just as the parameter c is used in developing the well-known bending theory, the parameter w is used in developing the warping theory.

A great advantage of the theory of bimoment is that internal strains and stresses can be found from formulas as simple as those used in the ETB. The bimoment and flexural twist can be solved in a manner similar to bending moment and shearing forces. The procedure differs in that we use the sectorial coordinate w instead of the linear coordinate c to calculate the physical properties related to warping torsion.

To a beginner, the thin-walled beam theory with its differential equations presented later in this chapter may look too academic for use in a down-to-early practical design. In reality, once the idea of bimoment is assimilated, its use is not much more difficult than the use of bending moments or shear forces. It provides the engineer with a means for verifying the behavior of tall shear wall buildings subjected to torsion.

10.1 CONCEPT OF WARPING BEHAVIOR

Perhaps the easiest model to describe the warping theory is an I-shaped shear wall with unequal flanges, as shown in Figures 10.12 and 10.13. In most shear wall buildings, the core around elevators and stairs consists of a series of I- and C-shaped shear walls. Therefore, the model chosen has practical significance. Since torsion is the subject of discussion, the location of shear center of the cross section is of importance. Its location is determined in a manner similar to the location of the center of gravity of the section. The only difference is that instead of dealing with the areas of the segments, we use their moments of inertia.

If an axial force is applied to the center of area, only axial deformations and stresses will occur. If, however, the axial force is applied through a point other than the center of gravity, bending about

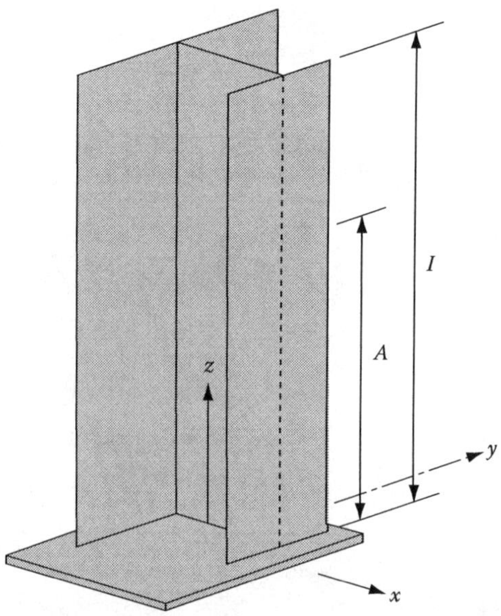

FIGURE 10.12 I-section core.

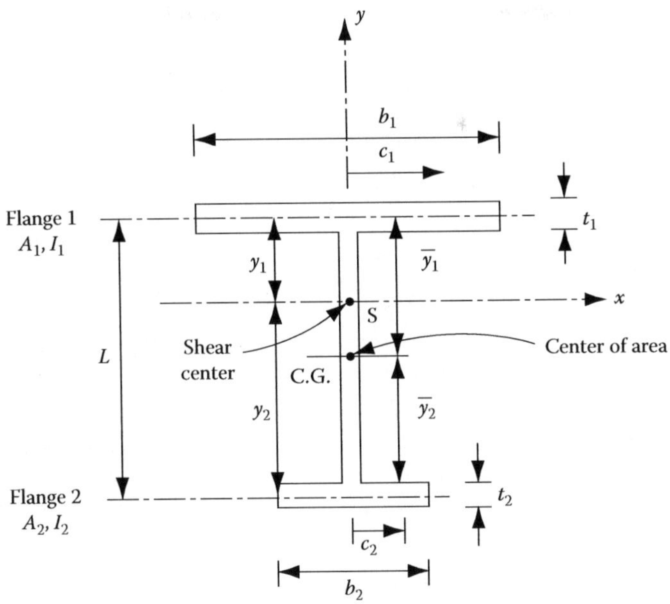

FIGURE 10.13 Core properties.

the transverse axes, and possibly warping, can also occur. Neglecting the web, the position of the center of gravity also called the center of area is given by

$$y_1 = \frac{A_2 L}{A_1 + A_2} \quad \text{and} \quad y_2 = \frac{A_1 L}{A_1 + A_2} \tag{10.4}$$

The location of the center of gravity is important in relation to vertical axial forces. The shear center s on the other hand is important in relation to transverse forces. If a transverse force acts through *S*, the member will only bend. If, however, a transverse force acts elsewhere than through *S*, the member will twist and warp as well as bend. The shear center in this case is located along the *y*-axis by

$$y_1 = \frac{I_2 L}{I_1 + I_2} \quad \text{and} \quad y_2 = \frac{I_1 L}{I_1 + I_2} \tag{10.5}$$

An inspection of Equations 10.4 and 10.5 indicates that the center of the area and the shear center generally will not coincide unless the section is doubly symmetric, in which case both points lie at the center of symmetry.

When a torque T is applied to the top of the member shown in Figure 10.14a, it twists about the shear center axis causing the flanges to (1) bend in opposite directions about the y-axis and (2) twist about their vertical axes. The effect of the flange bending is to cause the flange sections to rotate in opposite direction about their y-axes so that initially plane sections through the member become nonplanar or warped. Diagonally opposite corners 1 and 4, in Figure 10.14a, displace downward while 2 and 3 displace upward. At any level z up the height of the core, the torque $T = T_z$ is resisted internally by a couple $T_w(z)$ resulting from the shears in the flanges and associated with their in-plane bending, and a couple $T_v(z)$ resulting from shear stresses circulating within the section and associated with the twisting of the flanges and the web. Then

$$T_w(z) + T_v(z) = T_z \tag{10.6}$$

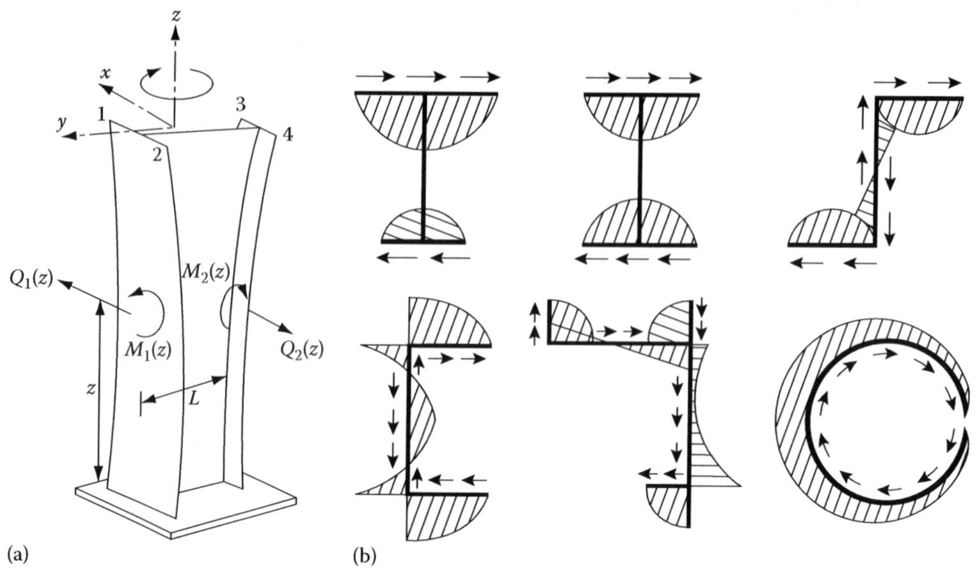

FIGURE 10.14 (a) Bending of flanges due to torque. (b) Shear forces due to warping torsion.

The rotation of the member about its shear center axis at a height z from the base is θ_z; hence, the horizontal displacement of flange #1 at that level is

$$x_1(z) = y_1\theta(z) \tag{10.7}$$

And its derivatives are

$$\begin{aligned} \frac{dx_1}{dz}(z) &= y_1\frac{d\theta}{dz}(z) \\ \frac{d^2x_1}{dz^2}(z) &= y_1\frac{d^2\theta}{dz^2}(z) \\ \frac{d^3x_1}{dz^2}(z) &= y_1\frac{d^3\theta}{dz^3}(z) \end{aligned} \tag{10.8}$$

Similar expressions may be written for flange #2.

The shear associated with the bending in flanges #1 and #2 can be expressed by

$$Q_1(z) = -EI_1\frac{d^3x_1}{dz^3}(z) = -EI_1y_1\frac{d^3\theta}{dz^3}(z) \tag{10.9}$$

$$Q_2(z) = -EI_2\frac{d^3x_2}{dz^3}(z) = -EI_2y_2\frac{d^3\theta}{dz^3}(z) \tag{10.10}$$

Multiplying the shear forces Q_1 and Q_2 by their respective distances from the shear center, we obtain the torque resisted by these factors. Therefore, the torque contributed by these shear forces is

$$\begin{aligned} T_\omega(z) &= Q_1y_1 + Q_2y_2 \\ &= -\left(EI_1y_1^2 + EI_2y_2^2\right)\frac{d^3\theta}{dz^3}(z) \end{aligned} \tag{10.11}$$

Or

$$T_{\omega}(z) = EI_{\omega}\frac{d^3\theta}{dz^3}(z) \tag{10.12}$$

where

$$I_{\omega} = I_1 y_1^2 + I_2 y_2^2 \tag{10.13}$$

I_{ω} is a geometric property of the section similar to the moments of inertia I_x and I_y and is called the warping moment of inertia or warping constant. It expresses the capacity of the section to resist warping torsion. Neglecting the web, the torque resisted by the twisting of the section is

$$T_v(z) = GJ_1\frac{d\theta}{dz}(z) \tag{10.14}$$

where J_1 is the torsion constant of the section given by

$$J_1 = \frac{b_1 t_1^3}{3} + \frac{b_2 t_2^3}{3} \tag{10.15}$$

in which b_1 and b_2 are the widths and t_1 and t_2 are the thickness of flanges #1 and #2, respectively.

Summing the two internal torques, Equations 10.12 and 10.14, and equating the sum to external torques as in Equation 10.6,

$$-EI_{\omega}\frac{d^3\theta}{dz^3}(z) + GJ_1\frac{d\theta}{dz}(z) = T \tag{10.16}$$

Equation 10.16 is the fundamental equation for restrained warping torsion. It simply states that an external torque applied to an open core is resisted by a combination of internal torque due to St. Venant's shear stresses and a couple due to equal and opposite shear forces in the flanges. The distribution of shear forces due to the torsion in typical shear profiles is shown in Figure 10.14b.

Considering the stresses in the flanges due to bending, the compressive stress in flange #1 at c_1 from the y-axis and z from the base is

$$\sigma_1(c_1, z) = \frac{M_1(z)c_1}{I_1} \tag{10.17}$$

The tensile stress in flange #2 at c_2 from the y-axis is

$$\sigma_2(c_2, z) = \frac{M_2(z)c_2}{I_2} \tag{10.18}$$

Multiplying the right-hand side of Equation 10.5 by the expressions

$$\frac{L}{y_1 + y_2} \quad \text{and} \quad \frac{y_1}{y_1} \tag{10.19}$$

which is equal to unity, and since $Q_1 = Q_2$ and the flange moments $M_1 = M_2 = M$, the equation 10.17 becomes

$$\sigma_1(c_1, z) = \frac{M_{(z)} L y_1 c_1}{I_1 y_1^2 + I_1 y_1 y_2} \tag{10.20}$$

And since from Equation 10.5

$$I_1 y_1 y_2 = I_2 y_2^2 \tag{10.21}$$

substituting Equation 10.21 in Equation 10.20,

$$\sigma_1(c_1, z) = \frac{M_{(z)} L y_1 c_1}{I_1 y_1^2 + I_2 y_2^2} \tag{10.22}$$

$$\sigma_1(c_1, z) = \frac{B_{(z)} \omega(c_1)}{I_\omega} \tag{10.23}$$

in which $B(z) = M(z)\,L$ is an action termed a bimoment and $\omega(c) = y_1 c_1$, a coordinate termed the sectorial area, or principal sectorial ordinate, for that point of the section. In its simplest form, as considered here, a bimoment consists of a pair of equal and opposite couples acting in parallel planes. Its magnitude is the product of the couple and the perpendicular distance between the planes.

The aforementioned simple treatment of torsion of an I section explains the concept of warping and how the equations of torsion bending, also called restrained warping, are related to the simple bending theory. The analogy is perhaps even more obvious by comparing the terms given in Table 10.2.

10.2 SECTORIAL COORDINATE ω'

The sectorial coordinate, also called the warping function at a point on the profile of a warping core, is the parameter that expresses the axial response (i.e., displacement, strain, and stress) at that point, relative to the response at other points around the section. Conceptually, this is similar to the distance c we use in the bending formula $f = Mc/I$ to find the bending stress f at a point in the cross section located at a distance c from the neutral axis.

The warping coordinate is defined in relation to two points: a pole O' at an arbitrary position in the plane of the section and an origin P_o at an arbitrary location on the profile of the section (Figure 10.15a and b). The value of the sectorial coordinate at any point P on the profile is then given by the area:

$$\omega'_{(s)} = \int_0^s h\,ds \tag{10.24}$$

where

h is the distance from the pole O' to the tangent to the profile at P
s is the distance to P along the profile P_o

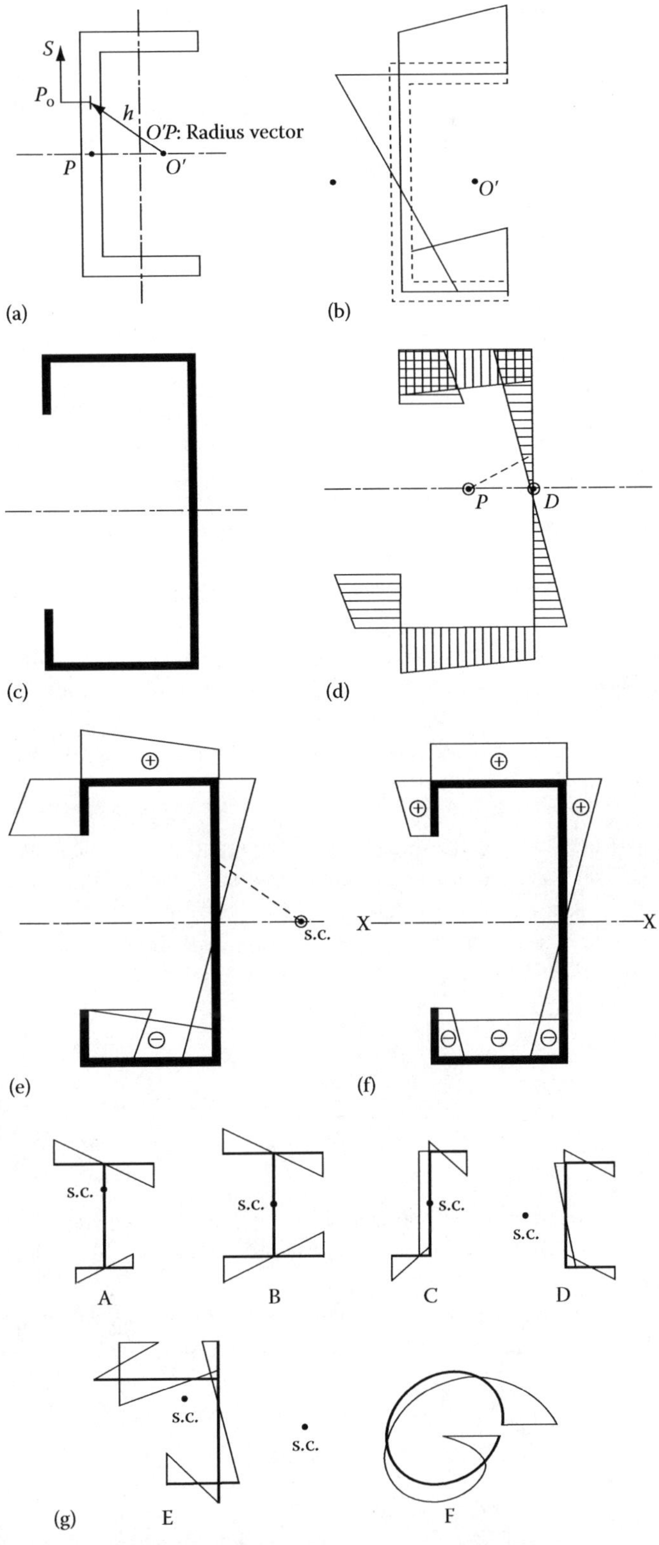

FIGURE 10.15 (a) Section profile, (b) sectorial coordinate ω_s, diagram, (c) singly symmetric curve, (d) ω_s, diagram, (e) y-coordinate diagram, (f) principal sectional coordinates, and (g) sectorial coordinates for common profiles.

It is evident that the warping function is an area and its magnitude depends on the location of the pole and the point in the profile from which the integration is started.

In effect, the sectorial coordinate ω' is equal to twice the area swept out by the radius vector $O'P$ in moving from P_o to P. The sectorial coordinate diagram (Figure 10.15b) indicates the values of ω' around the profile. When the sectorial coordinates are related to the shear center as a pole and to the origin of known zero warping displacement, Equation 10.21 gives the principal sectorial coordinate values, ω, and their plot is the principal sectorial coordinate diagram. The principal sectorial coordinate of a section in warping theory is analogous to the distance c of a point from the neutral axis of a section in bending. The parameters ω and c are used in developing the corresponding warping and bending stiffness properties of the sections and in determining the axial displacements and stresses. Sectorial coordinates for common profile are shown in Figure 10.15g.

10.3 SHEAR CENTER

The shear center of a section is a point in its plane through which a load transverse to the section must pass to avoid causing torque and twist. It is also the point in which warping properties of a section are related, in the way that bending properties of a section are related to the neutral axis.

Tall building cores are often singly or doubly symmetric in plan, which simplifies the location of the shear center. In doubly symmetric sections, the shear center lies at the center of symmetry, while in singly symmetric sections, it lies on the axis of symmetry.

The procedure for determining the location of shear center for a singly symmetric section (Figure 10.15c) is as follows:

1. Construct the ω_p diagram (Figure 10.15d) by taking an arbitrary pole P on the line of symmetry and origin D where the line of symmetry intersects the section and by sweeping the ray PD around the profile.
2. Using the ω_p and the y diagrams for the section (Figure 10.15d and e), respectively, calculate the product of inertia of the ω_p diagram about the x-axis $I_{\omega_p x}$ using

$$I_{\omega_p x} = \int^{A} \omega_p y dA \tag{10.25}$$

   in which $dA = tds$, the area of the segment of the profile of thickness t and lengths ds. The integral in Equation 10.25 may be evaluated simply by using the product integral table (Table 10.3).
3. Calculate I_{xx}, the second moment of area of the section about the axis of symmetry.
4. Finally, calculate the distance α_x, of the shear center O form O', along the axis of symmetry using

$$\alpha_x = \frac{L_{\omega_p x}}{I_{xx}} \tag{10.26}$$

TABLE 10.3
Product Integral Tables

M / m	Linear M Diagrams				Parabolic M Diagrams		
	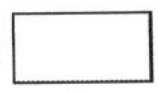 M, L	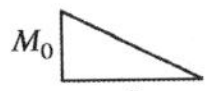 M_0, L	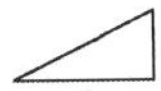 M_1, L	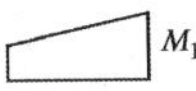 M_0, M_1, L	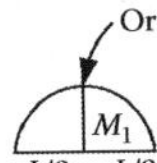 Origin, M_1, $L/2$ $L/2$	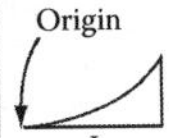 Origin, M_1, L	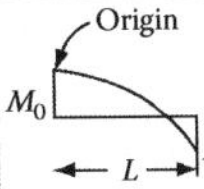 Origin, M_0, M_1, L
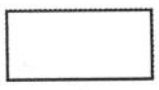 m, L	mML	$\frac{1}{2}m_0ML$	$\frac{1}{2}mM_1L$	$\frac{1}{2}mL(M_0+M_1)$	$\frac{2}{3}mM_1L$	$\frac{1}{3}mM_1L$	$\frac{1}{3}mL(2M_0-M_1)$
 m_0, L	$\frac{1}{2}m_0ML$	$\frac{1}{3}m_0M_1L$	$\frac{1}{6}m_0M_1L$	$\frac{1}{6}m_0L(2M_0+M_1)$	$\frac{1}{3}m_0M_1L$	$\frac{1}{12}m_0M_1L$	$\frac{1}{12}m_0L(5M_0-M_1)$
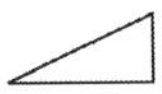 m_1, L	$\frac{1}{2}m_1ML$	$\frac{1}{6}m_1M_0L$	$\frac{1}{3}m_1M_1L$	$\frac{1}{6}m_1L(2M_1+M_0)$	$\frac{1}{3}m_1M_1L$	$\frac{1}{4}m_1M_1L$	$\frac{1}{4}m_1L(M_0-M_1)$
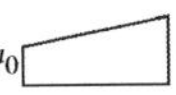 m_0, m_1, L	$\frac{1}{2}ML(m_0+m_1)$	$\frac{1}{6}M_0L(2m_0+m_1)$	$\frac{1}{6}M_1L(m_0+2m_1)$	$\frac{L}{6}m_0(2M_0+2M_1)$ $+m_1(2M_0+M_1)$	$\frac{1}{3}M_1L(m_0+m_1)$	$\frac{1}{12}M_1L(m_0+3m_1)$	$\frac{L}{12}m_0(5M_0-M_1)$ $+3m_1(M_0-M_1)$

10.4 EVALUATION OF PRODUCT INTEGRALS

When evaluating the integrals in Equation 10.25, we usually are dealing with members for which the material properties and cross-sectional dimension are constant from one end of the member to the other. The integrals are in the form of a product such as

$$\int_0^{\ell} Mmd_x \tag{10.27}$$

These product integrals must be evaluated over the length of each member and then added for all members. For any particular member, each quantity (such as m or M) is a function of the distance x measured along the axis of the member; specifically, the quantity may be constant along the length, may vary linearly along the length, or may be a function of higher order, such as quadratic or cubic. To save time when performing calculations, these produce integrals can be evaluated in advance and the results tabulated for ready use. A compilation of product integrals, covering the most commonly encountered functions, is given in Table 10.3. The table is presented in terms of the functions M and m, but it is apparent that these functions can be replaced by others, such as ω_p and y. Illustrations of the use of the table are given in some of the examples.

10.5 PRINCIPAL SECTORIAL COORDINATE ω_S DIAGRAM

The ω diagram is related to the shear center 0 as its pole and a point of zero warping deflection as its origin. In a symmetrical section, the intersection of the axis of symmetry with the profile at *D* defines a point of antisymmetrical behavior and hence of zero warping: therefore, it may be used as the origin.

Values of ω can be found by sweeping the ray *OD* around the profile and taking twice the values of the swept areas. For the section of Figure 10.15c, the principal sectorial coordinate diagram is shown in Figure 10.15f.

10.5.1 Sectorial Moment of Inertia I_ω

This geometric parameter expresses the warping torsional resistance of the core's sectional shape. It is analogous to the moment of inertia in bending.

The sectorial moment of inertia is derived from the principal sectorial coordinate distribution using the relation:

$$I_\omega = \int_0^A \omega^2 dA \qquad (10.28)$$

Note the similarity with the expression of the moment of inertia:

$$I_{yy} = \int_0 x^2 dA \qquad (10.29)$$

10.5.2 Torsion Constant *J*

When a beam is twisted, its fibers must undergo a shear strain to accommodate the twist. Associated with the strain are the shear stresses called St. Venant's shear stresses. When an open-section core is subjected to torque (Figure 10.15a), each wall twists, developing St. Venant's shear stresses within the thickness of the wall. The stresses are distributed linearly across the thickness of the wall, acting in opposite directions on opposite sides of the wall's middle line. As the effective lever arm of these stresses is equal to only two-thirds of the wall thickness, the torsional resistance of these stresses is low. The torsion constant for this plate-twisting action is

$$J = \frac{1}{3} k \sum^{n} bt^3 \qquad (10.30)$$

in which *b* is the width and *t* is the thickness of a wall. The summation includes the *n* walls that comprise the section. The plate-twisting rigidity of an open-section core is given by *GJ*.

k is a factor that makes allowance for small fillets within the cross section. If there are no fillets in, its value is equal to 1.00.

10.6 CALCULATION OF SECTORIAL PROPERTIES: WORKED EXAMPLE

Consider again the shear core with unequal flanges, as shown in Figure 10.16. To determine the core, the following is required:

1. The location of the shear center
2. The principal sectorial coordinate, ω_s, diagram
3. The sectorial moment of inertia I_ω
4. St. Venant's torsion constant *J*

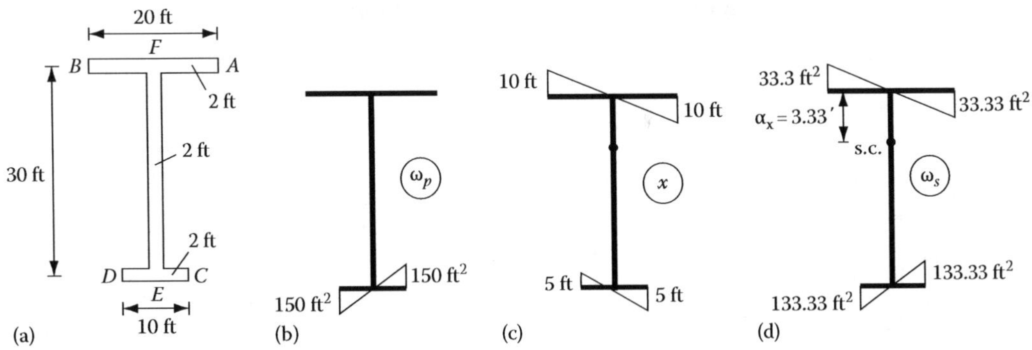

FIGURE 10.16 Calculation of sectorial properties: (a) cross section, (b) ω_p diagram, (c) x-coordinate diagram, and (d) principal sectorial coordinate ω_s diagram.

1. *Location of the shear center*
   The axis of symmetry of the section is OY, therefore the shear center, lies on the OY axis. We select an arbitrary pole P at the junction of the web and the upper flange of the core. The ω_p diagram is constructed, as shown in Figure 10.16, by taking an arbitrary point on the web as the sectorial origin. The sectorial areas for the section of the upper flange and the web are equal to zero while they are distributed skew symmetrically for the lower flange.

   Using the ω_p and the Y coordinate diagrams, Figure 10.16(b and c), we calculate the integral $\omega_p dA$ by using the product integrals given in Table 10.3.
   A summary of the calculations is given in Table 10.4. For the whole section, $I_{\omega_p} = 2500 \times 2 =$ 6500 ft⁵. The moment of inertia of the section about y-axis is

$$I_{yy} = \frac{1}{12}\left(2\times10^3 + 2\times20^3\right) = 1500\ \text{ft}^4 \tag{10.31}$$

TABLE 10.4
Calculations for Integral $\omega_p dA$

Segment	ω_p	x	$\int_0^s \omega_p x t\, ds$
DE	150, 5	5, 5	$\frac{1}{3}\times5\times150\times5\times2 = 2500\ \text{ft}^5$
EC	150, 5	5, 5	$\frac{1}{3}\times5\times150\times5\times2 = 2500\ \text{ft}^5$
AF	0	10, 10	0
BF	0	10, 10	0

From Equation 10.26, the distance of the shear center from the center of web is

$$\alpha_x = \frac{I_{\omega px}}{I_{yy}} = \frac{5000\,\text{ft}^5}{1500\,\text{ft}^4} = 3.33\,\text{ft} \tag{10.32}$$

2. *Principal sectorial coordinate diagram*
   This is constructed by using the shear center (s.c.) as the pole and sweeping the ray from the middle of the web, around the profile (Figure 10.16d).
3. *Sectorial moment of inertia I_ω*
   From Equation 10.28,

$$I_\omega = \int \omega^2 dA = \int^{s} \omega^2 t ds \tag{10.33}$$

Using the ω diagram (Figure 10.16) and the product integral table (Table 10.3), the calculations for evaluating I_ω are as shown in Table 10.5.

I_ω for the whole section 59,260 × 2 + 7,406 × 2 = 133,332 ft^6

4. *Torsion constant J*
   For the I-section core, using Equation 10.30,

$$J = \frac{1}{3}\sum^{n} bt^3 = \frac{1}{3}\times 2^3\left(20+10+30\right) = 160\,\text{ft}^4 \tag{10.34}$$

TABLE 10.5
Calculations for Sectorial Moment of Inertia

Segment	Variation of ω	$\int \omega^2 t\,ds$
DE	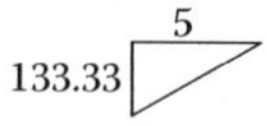	$\frac{1}{3}\times 5\times 133.33^2\times 2 = 59{,}260\,\text{ft}^6$
EC	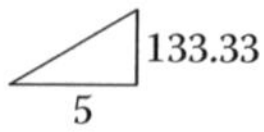	$\frac{1}{3}\times 5\times 133.33^2\times 2 = 59{,}260\,\text{ft}^6$
BF	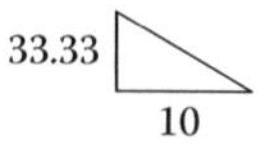	$\frac{1}{3}\times 10\times 133.33^2\times 2 = 7{,}406\,\text{ft}^6$
AF		$\frac{1}{3}\times 10\times 133.33^2\times 2 = 7{,}406\,\text{ft}^6$

Note: Therefore I_ω for the whole section 59,260 × 2 + 7,406 × 2 = 133,332 ft^6.

10.7 GENERAL THEORY OF WARPING TORSION

Before derivation of general warping torsion equations, it is instructive to consider qualitatively the difference between the behavior of thin-walled open sections and solid sections. A major difference lies in the matter in which the stresses attenuate along their length. Consider a square cantilever column loaded at the top corner by a vertical load P, as shown in Figure 10.17. The load can be replaced by four sets of loads acting at each corner, which together constitute a system of loads statically equivalent to the applied force P. The first set represents axial loading; the second and third sets represent bending about the x- and y-axes. The resulting axial and bending stresses can be computed by the usual ETB, which assumes that Bernoulli hypothesis is valid. In the last loading case, the cross sections do not remain plane because the two pairs of loads on opposite faces of the column tend to twist the cross section in opposing directions. This equal and opposite twisting results in warping of the cross section. The last set of loads is, however, statically equivalent to zero and can be ignored by invoking St. Venant's principle, which states that the perturbations imposed on a structure by a set of self-equilibrating system of forces affect the structure locally and will not appreciably affect parts of the structure, which simply means that the effect of the self-equilibrating system of forces can be neglected in the analysis. The stresses caused by these forces equal to the characteristic dimension of the cross section. The stresses due to the self-equilibrating system of forces can be ignored throughout the whole length of the cantilever except at the very top region.

Now consider an I-shaped shear wall, as shown in Figure 10.18, which has the same overall dimensions as the column with the exception that it is composed of thin plates of thickness t. The first three sets of loads result in stress distributions, which can be obtained as before by using the Bernoulli hypothesis. Although the fourth loading is self-equilibrating as before, its effect is far from local. The flanges, which are bending in opposite directions, do so as thought they were independent of each other. The web acts as a decoupler separating the self-equilibrating load into two subsets, one in each flange. Each subset is not self-equilibrating and causes bending in each flange. The bending action of the flanges can be thought of as being brought by equal and opposite horizontal forces parallel to the flanges. The compatibility condition between the web and flanges results in a twisting of the cross section, as shown in Figure 10.19. Although the cross section of each of the flanges remains plane, the wall as a whole is subjected to warping deformations. The restraint at the foundation prevents free warping at this end and sets up warping stresses.

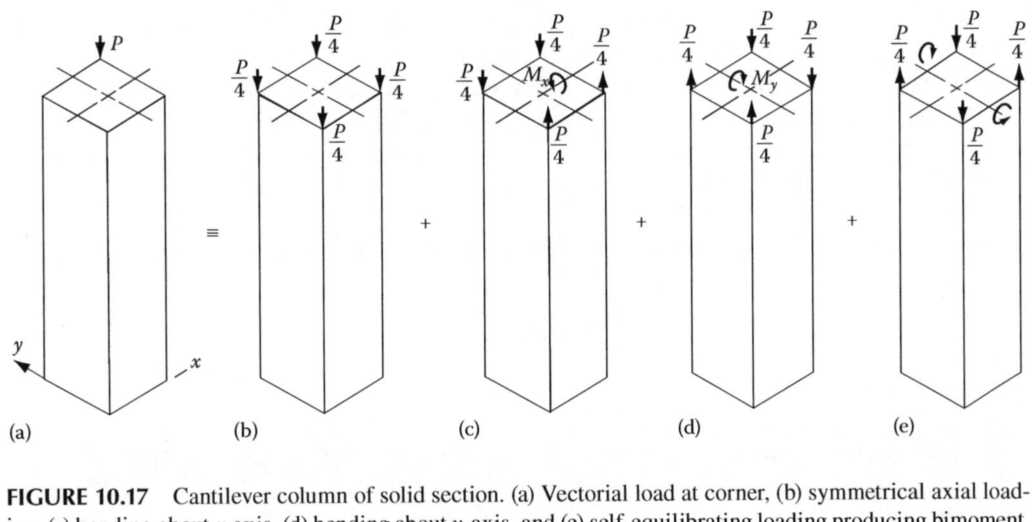

FIGURE 10.17 Cantilever column of solid section. (a) Vectorial load at corner, (b) symmetrical axial loading, (c) bending about x-axis, (d) bending about y-axis, and (e) self-equilibrating loading producing bimoment.

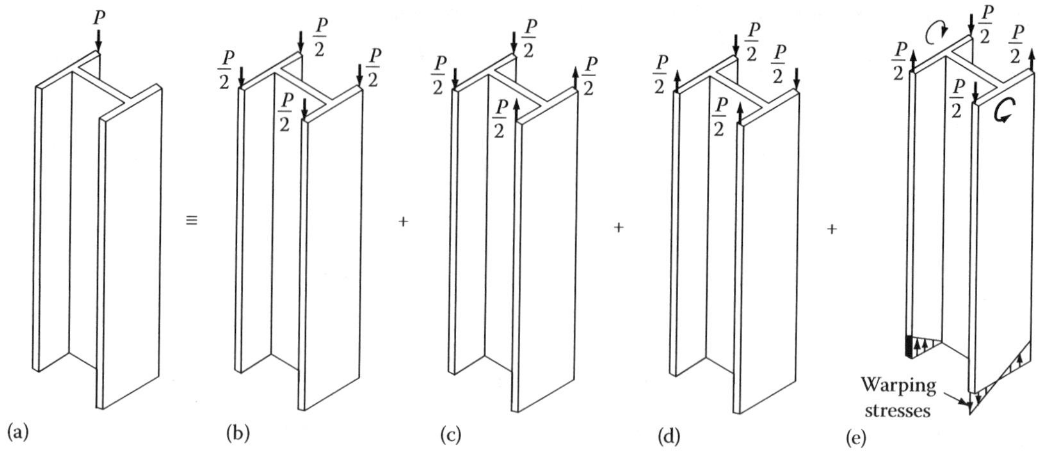

FIGURE 10.18 I-shaped cantilever beam: (a) vertical load at corner, (b) symmetrical axial loading, (c) bending about x-axis, (d) bending about y-axis, and (e) self-equilibrating loading producing bimoment.

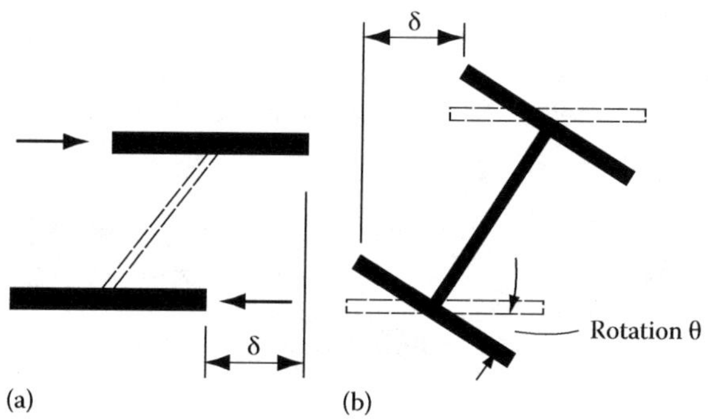

FIGURE 10.19 Plan section of I-shaped column: (a) displacement of flanges due to bimoment load and rotation due to geometric compatibility between flanges and web.

The system of skew-symmetric loads, which is equivalent to an internally balanced force system arising out of warping of cross section, is termed a bimoment in thin-walled beam theory. Mathematically, it can be construed as a generalized force corresponding to the warping displacement, just as moment and torsion are associated with rotation and twisting deformation, respectively. In the present example, bimoment can be visualized as a pair of equal and opposite moments acting at a distance e from each other. Its magnitude is equal to M times e and has units of force times the square of the distance (lb · in.^{2}, kip · ft^2, etc.).

Presently, it will be shown that the warping stresses can be calculated by the relation

$$\sigma_{\omega} = \frac{B_{\omega}\omega_s}{I_{\omega}} \tag{10.35}$$

where

B_{ω} is the bimoment, a term that represents the action of a set of self-equilibrating forces
ω_s is the warping function
I_{ω} is the warping moment of inertia

The three terms B_ω, ω_s, and I_ω are conceptually equivalent to moment M, linear coordinate or x or y, and moment of inertia I countered in bending problems. Note the similarity between the bending stress as calculated by the familiar relation $\sigma_b = My/I$ and the warping stress formula given in Equation 10.35.

10.8 TORSION ANALYSIS OF SHEAR WALL BUILDING: WORKED EXAMPLE

Example 9.1

Consider a 25-story, 300 ft (91.44 m) building consisting of two cores as shown in Figure 10.20a and b. To keep the analysis simple, assume that the resistance to lateral loads and torque is provided solely by the core. The building is subjected to a uniform wind load of 25 psf (1.197 kN/m^2) in the x-direction.

It is required to determine the maximum deflection and rotation at the top and the vertical stresses at the base due to bending and twisting. An elastic modulus E = 3,600 ksi (24,882 MPa) and

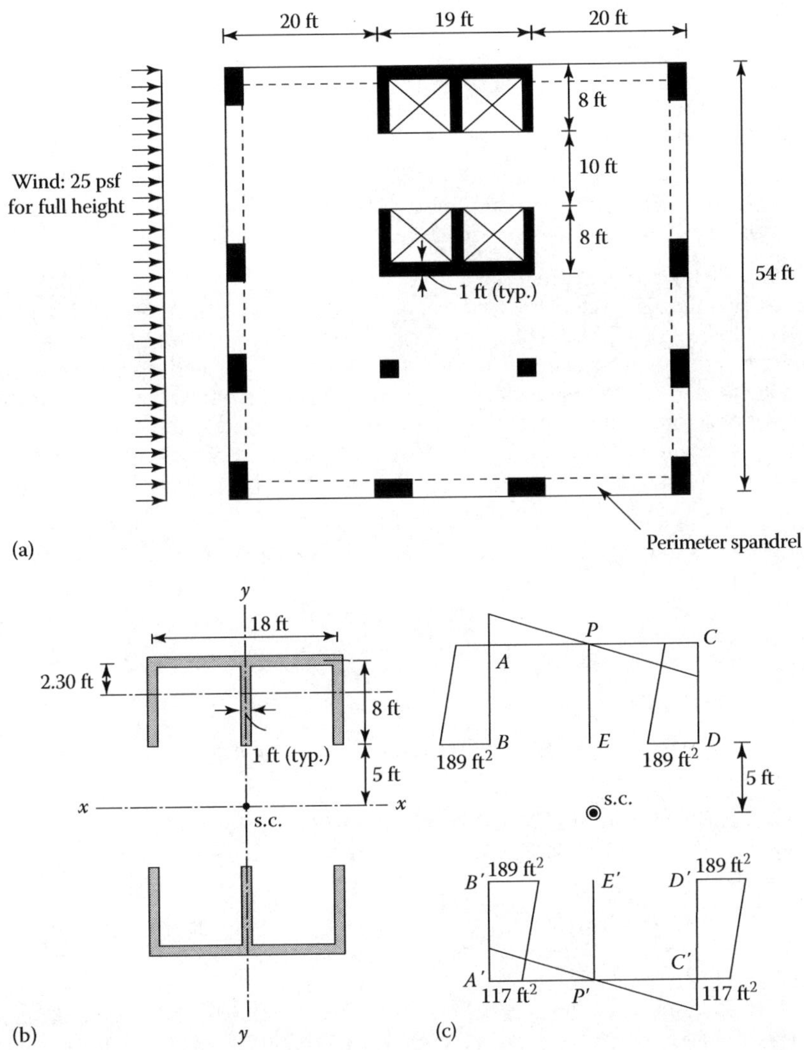

FIGURE 10.20 (a) Twin-core example. (b) Core properties. (c) ω_s diagram (sectorial coordinates). (*Continued*)

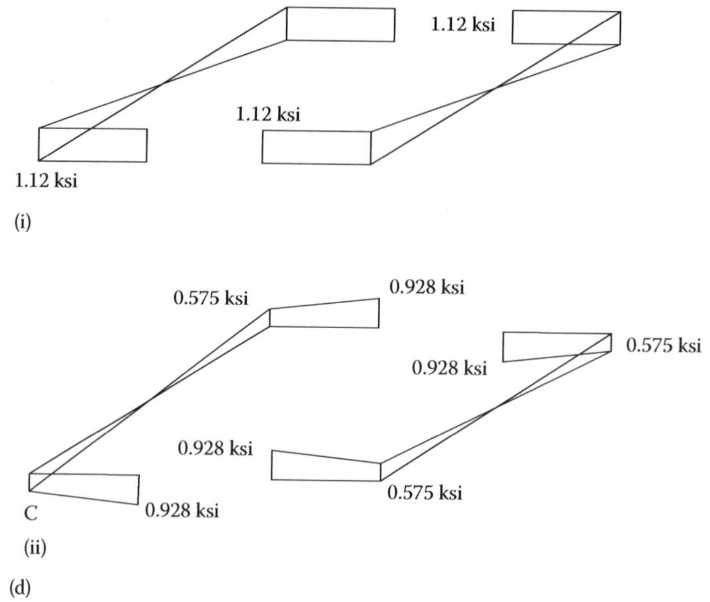

FIGURE 10.20 (*Continued*) (d) Comparison of stresses: (i) bending stress σ_b and (ii) warping stress σ_ω.

a shear modulus G = 1,565 ksi (10,791 MPa) are assumed for the concrete properties. The procedure is first described and then illustrated numerically.

Step 1: Determine the sectorial properties.
For the given structure, by inspection, the location of shear center O is determined at a point midway between the two cores. The ω diagram is related to the shear center O as its pole and a point of zero warping deflection as its origin. In a symmetrical section, as in the example probable, the intersection of the axis of symmetry with the profile at D defines a point of antisymmetrical behavior and hence of zero warping deflection: therefore, it may be used as the origin.

Values of ω are found from first principles, by sweeping the ray OD around the profile and taking twice the values of the swept areas. For the example problem, the principal sectorial coordinate diagram is shown in Figure 10.20c.

Step 2: Determine the sectorial moment of inertia I_ω from equation 10.33, again:

$$I_\omega = \int_0^A \omega^2 dA = \int_0^s \omega^2 t ds \qquad (10.36)$$

Using the ω_s diagram (Figure 10.20c) and the product integral table (Table 10.3), the value for I_ω is evaluated, as shown in the worksheets.

Step 3: Torsion constant J for the core is determined from Equation 10.30.
For one core, $J = l/3\Sigma bt^3 = 1/3 \times 1^3 (3 \times 7.5 + 1 \times 19) = 13.834$ ft^4, and for two cores, $J = 13.834 \times 2 = 27.671$ ft^4.

Step 4: Determine eccentricity e of the line of action of wind resultant for the shear center.
The resultant wind force per unit height of the building is equal to 25 × 54 = 1.35 kip/ft (1.83 kN/m), acting at 13.5 ft (4.12 m) to the south of shear center. Therefore, the eccentricity, e, from the shear center is 13.5 ft (4.12 m). Since the external torque is the product of the horizontal loading and its eccentricity, the torsion due to wind is 1.35 × 13.5 = 18.225 kft/ft (24.70 kN m) per unit height, anticlockwise.

Step 5: Determine bending deflection at the top and stresses at the base.
A bending analysis is now performed to determine the maximum lateral deflection at the top and the bending stresses at the base. Deflection at the top due to bending is calculated as follows:

$$\Delta_{y(\max)} = \frac{wl^4}{8EI_{yy}} = 0.737\,\text{ft}. \tag{10.37}$$

The deflection is 1/406 of the height, as compared to the generally accepted limit of 1/400 and therefore is acceptable. Observe that the building is very flexible in the *y*-direction because the moment of inertia of the core I_{xx} is about one-sixth of I_{yy}, indicating that supplemental bracing is required in the *y*-direction. One solution is to add perimeter rigid frames, as indicated in Figure 10.20c. However, we continue the problem with the assumption made earlier, namely, that the core resists all the lateral loads:

$$\begin{aligned}\text{Maximum bending stress} = \frac{MC}{I} &= \frac{60{,}750 \times 9.5}{1788 \times 2}\\ &= 161.39 \text{ ksf}\\ &= 1.12 \text{ ksi}\left(8.39 \text{ MPa}\right)\end{aligned} \tag{10.38}$$

The bending stress diagram is given in Figure 10.20d.

Step 6: Determine the parameter *k* using Equation 10.39.

$$\begin{aligned}k &= 1\sqrt{\frac{GJ}{EI_\omega}}\\ &= 300\sqrt{\frac{1565 \times 144 \times 28}{927{,}180 \times 2.3}}\\ &= 1.08\end{aligned} \tag{10.39}$$

Step 7: Determine the rotation and total deflection at the corner of the top floor.
The rotation at any level of the building for a uniformly distributed torque *m* may be obtained from Equation 10.37.

At the top, $z = 1$. Substituting $z = 1$ in Equation 10.42, we get

$$\theta_l = \frac{-ml^2}{GJ\cosh k}\left[-\frac{1}{k^2} - \frac{1}{k}\sinh k + \frac{1}{2}\cosh k + \frac{1}{k^2}\cosh k\right] \tag{10.40}$$

Substituting for the various parameters, we get $\sigma_H = 0.0196$ rad, anticlockwise. Therefore, the additional deflection at the southeast corner c of the top floor due to torsion is

$$\begin{aligned}\Delta_t &= \sigma_H \times \text{distance of } c \text{ from shear center}\\ &= 0.196 \times 50.10 = 0.9821 \text{ ft } (0.30 \text{ m})\end{aligned} \tag{10.41}$$

The total deflection at c due to bending and torsion = $\Delta_b + \Delta_t$, = 1.47 + 0.9821 = 2.45 ft (0.75 m). This value represents 1/122 of the building height, an unacceptably large value, confirming the earlier observation that the building is too flexible in the *y*-direction.

Step 8: Determine bimoments and warping stresses.

The warping stresses σ_w at the base are determined from the bimoment B at that level. The bimoment is obtained from a uniformly distributed torque from Equation 10.6. Then, at any point on the section where the principal sectorial coordinate is ω_s, the vertical warping stress is obtained from Equation 10.7. The total axial stresses due to horizontal loading is obtained by combining the warping stresses with the bending stresses.

The vertical stresses at the base due to warping are determine from the bimoment at the base, as given in Equation 10.42:

$$B_z = \frac{-ml^2}{k^2 \cosh k}\left[\cosh k - \cosh\frac{k_z}{l} - k\sinh\frac{k}{l}(l-z)\right] \tag{10.42}$$

At the base,

$$z = 0, \quad B_0 = \frac{-ml^2}{k^2}\cosh k\left[\cosh k - 1 - k\sinh\frac{k}{l}(l-z)\right] \tag{10.43}$$

$$k = 1.08 \qquad \sinh k = 1.3025 \qquad \cosh k = 1.642$$

$$\sinh 0 = 0 \qquad \cosh 0 = 1$$

Substituting the aforementioned values,

$$\begin{aligned} B_0 &= \frac{-ml^2}{1.08^2 \times 1642}[1.642 - 1 - 1.08 \times 1.3025] \\ &= \frac{-ml^2}{2.504} \\ &= \frac{18.225 \times 300^2}{2.504} \\ &= 656.100\ \text{kip/ft}^2 \end{aligned} \tag{10.44}$$

Warping stresses are given by $\omega = B_0 \times \omega_s / I_\omega$.

At D,

$$\begin{aligned} \sigma_\omega &= \frac{656{,}100}{927{,}180} \times 189 \\ &= 133.74\ \text{ksf} \\ &= 0.928\ \text{ksi} \end{aligned} \tag{10.45}$$

At C,

$$\begin{aligned} \sigma_\omega &= \frac{656{,}100}{927{,}180} \times 117 \\ &= 82.79\ \text{ksf} \\ &= 0.575\ \text{ksi} \end{aligned} \tag{10.46}$$

Figure 10.20d shows a comparison of bending and warping stresses. The importance of warping torsion is obvious.

Example

The warping theory described for the twin-core example can also be used to determine the bending stresses in, and torsional rotations of buildings consisting of, randomly distributed shear walls. To demonstrate the method, a 15-story building is analyzed for torsional rotation using the warping theory, and then compared with computer results. The example also illustrates the method for calculating bending stresses in shear walls.

Consider the building shown in Figure 10.21a, consisting of three walls, W_1, W_2, and W_3, in the transverse direction and two walls W_4 and W_5 in the longitudinal direction. A uniform wind load of 25 psf (1.197 kN/m^2) assumed for the full height results in a horizontal load equal to 25 × 70 = 1.75 kip/ft (2.37 kN m), acting at the center of gravity of the plan.

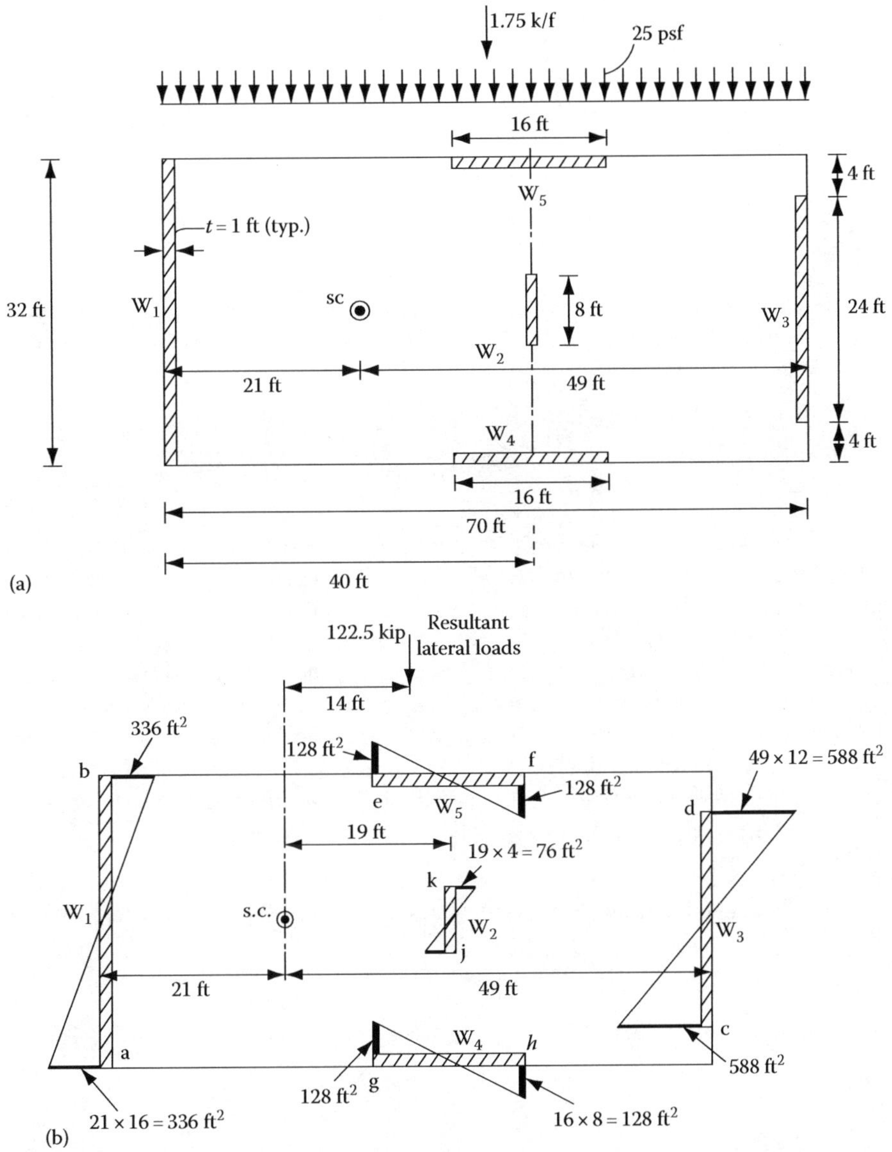

FIGURE 10.21 Torsion example; randomly distributed shear walls. (a) Plan, (b) warping coordinates. *(Continued)*

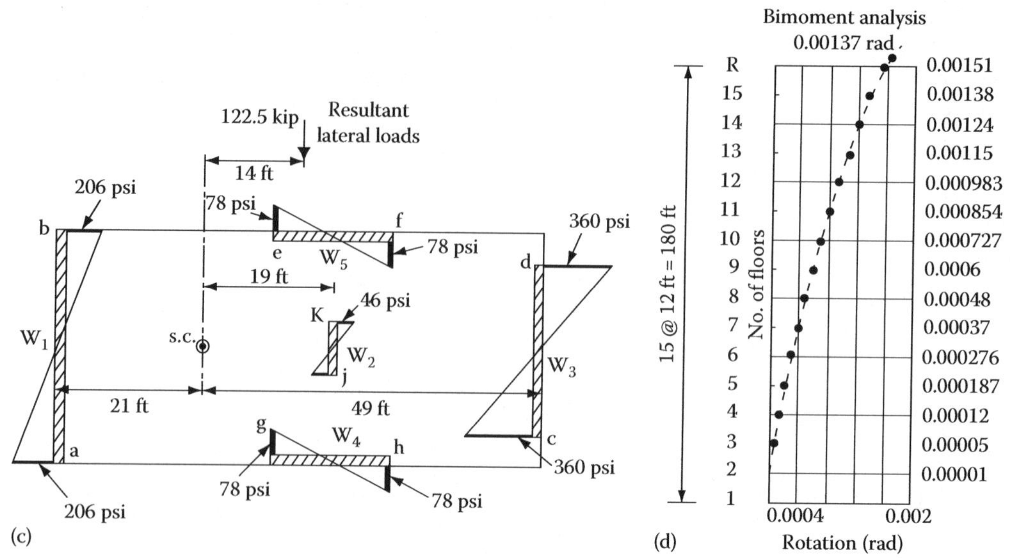

FIGURE 10.21 (*Continued*) Torsion example; randomly distributed shear walls. (c) axial stresses due to torsion, and (d) rotation comparison.

Analysis outline: By inspection, the location of the shear center (s.c.) of the building is judged to be midway between walls W_4 and W_5. The distance x of the shear center in the east–west direction from wall W_1 is obtained from the relation

$$X = \frac{\sum I_{xx}x}{\sum I_{xx}} \tag{10.47}$$

Next, the eccentricity, e, which is the distance from the line of action of the wind resultant to the shear center of the building, is determined to be 14 ft (4.27 m), as shown in the detailed calculations. The resulting torque, which is equal to the product of the wind load and eccentricity, is calculated at 1.75 × 14 = 24.5 kip/ft (10.12 kN m/m).

To keep the mathematics simple, we limit our analyses to the determination of bending stresses in the walls due to torsion only. We begin the analysis by considering the center of gravity of each wall as a point of zero warping deflection. These points are used as the origins for determining the values of the sectorial coordinate ω. Next, we calculate the warping moment of inertia, I_ω, of the shear wall assemblage and the bimoment, B_0, at the base. These are used to determine the bending stresses in each of the shear walls, as shown in the following calculations.

Location of shear center: By inspection, the location of shear center (s.c.) is determined to be on the common neutral axis x–x. Its distance from the center line of wall W_1 in the x-direction is given by

$$X = \frac{\sum I_{xx}x}{\sum I_x} \tag{10.48}$$

Observe the similarity between this and the following equation:

$$X = \frac{\sum Ay}{\sum A} \tag{10.49}$$

which is well known for determining the location of neutral axis of built-up sections. Just as we use the areas of individual parts of a built-up section to find the neutral axis, we use the moments of inertia of individual walls to determine the location of the shear center of the building. The procedure is the same; select a reference axis (y–y), and determine I_x for each shear wall (about its own neutral axis) and its distance x from the reference axis (y–y). The distance x of the shear center of the entire group of shear walls from the reference axis is given by

$$X = \frac{I_{x1}x_1 + I_{x2}x_2 + I_{x3}x_3 + I_{x4}x_4}{I_{x1} + I_{x2} + I_{x3} + I_{x4}} \tag{10.50}$$

$$X = \frac{\sum I_x x}{\sum I_x}$$

The summation of the moment of inertia, EI_x, of the walls W_1, W_2, and W_3 is given by

$$\sum I_x = 1\times\frac{32^3}{12} = 1\times\frac{8^3}{12} + 1\times\frac{24^3}{12}$$

$$= 2730.67 + 42.67 + 1152$$

$$= 3925.34 \text{ ft}^4 \tag{10.51}$$

$$X = \frac{\sum I_x x}{\sum I_x}$$

$$= \frac{(2736.67\times 0 + 42.67\times 40 + 1152\times 70)}{3925.24}$$

$$= \frac{82{,}346.8}{3925.24}$$

$$= 20.979 \text{ ft } (\text{use } 21 \text{ ft}) \tag{10.52}$$

Verify the location of s.c. from the center line of the east wall W_3:

$$X_2 = \frac{(1152\times 0 + 42.67\times 30 + 2730.67\times 70)}{3925.24}$$

$$= \frac{192{,}427}{3925.24} = 49 \text{ ft} \quad x + x_2 = 21 + 49 = 70 \text{ ft} \tag{10.53}$$

The eccentricity e of the line of action of wind resultant from the shear center

$$e = 35 - 21 = 14 \text{ ft.} \tag{10.54}$$

Torsional moment m per foot height of the building

$$m = 1.75 \times 14 = 24.5 \text{ kip-f/f} \tag{10.55}$$

Torsion properties: As a first step, we calculate the warping moment of inertia for the entire building assuming the floor slabs are rigid. Using the centers of each wall as the principal poles, the sectorial coordinate diagram for the composite building is drawn by sweeping the radius vector passing through the shear center of the building and the principal poles of each wall. The resulting ω_s diagram is shown in Figure 10.21b.

The warping moment of inertia is calculated as before by using the product integral table (see Table 10.3):

$$I_\omega = \frac{2}{3}\times 16\times 336^2 + \frac{2}{3}\times 4\times 76^2 + \frac{2}{3}\times 12\times 588^2 + \frac{2}{3}\times 8\times 128^2 + \frac{2}{3}\times 8\times 128^2$$

$$= 4{,}160{,}341 \text{ ft}^6 \tag{10.56}$$

St. Venant's torsion J is calculated from the relation

$$J = \Sigma bt^3$$

$$= 1^3(32 + 8 + 24 + 16 + 16)$$

$$= 96 \text{ ft}^4$$

$$GJ = 225{,}360 \times 96$$

$$= 21{,}634{,}560 \text{ kip/ft}^2 \tag{10.57}$$

$$k = l\sqrt{\frac{GJ}{EI_\omega}}, \quad \frac{G}{E} = \frac{1}{2.3}$$

$$= 180\sqrt{\frac{96}{4{,}160{,}341\times 2.3}} = 0.57$$

Torsional rotation: The rotation θ_l at the top due to a uniformly distributed torque of m units per unit height is given by

$$\theta_l = \frac{-ml^2}{GL\cosh k}\left[\frac{\cosh k - 1}{k^2} + \frac{\cosh k}{2} - \frac{\sinh k}{k}\right] \tag{10.58}$$

Substituting

$$k = 0.57 \quad \sinh k = \sinh 0.57 = 0.601$$

$$\cosh k = \cosh 0.57 = 1.167 \quad GJ = 21{,}634{,}560 \text{ kip/ft}^2$$

and

$$m = 1.75 \times 14 = 24.5 \text{ k-f/f}$$

$$\theta_l = 0.00137 \text{ rad}$$

Bending stresses due to torsion: The bimoment B_0 at the base is given by

$$B_0 = -\frac{ml^2}{k^2\cosh k}\left[\cosh k - 1 - k\sinh k\right] \tag{10.59}$$

Substituting

$k = 0.57 \quad \sinh k = \sinh 0.57 = 0.601$

$\cosh k = \cosh 0.57 = 1.167 \quad m = 1.75\ (35 - 21) = 24.5$ kip-ft/ft

$l = 15$ stories at 12 ft = 180 ft

$$B_0 = -\frac{24.5 \times 180^2}{0.57^2 \times 1.167}\left[1.167 - 1 - 0.57 + 0.601\right]$$

$$= 367{,}529 \text{ kip/ft}^2$$

The bending stresses, σ_ω, in the walls due to torsion are calculated by the equation $\sigma_\omega = B_0\omega_s/I_\omega$ as shown in the following.

Wall 1	σ_ω at a, b	= 367,529 × 336/4,160,341
		= 29.68 kip/ft²
		= 0.206 ksi
Wall 3	σ_ω at a, b	= 367,529 × 558/4,160,341
		= 51.941 kip/ft²
		= 0.360 ksi
Wall 5	σ_ω at a, b	= 367,529 × 128/4,160,341
		= 11.30 kip/ft²
		= 0.078 ksi
Wall 4	σ_ω at g, h-similar to wall 5 at e, f	
Wall 2	σ_ω at a, b	= 367,529 × 76/4,160,341
		= 671 kip/ft²
		= 0.046 ksi

The calculated stresses and a comparison of torsional rotations are shown in Figure 10.21.

10.9 WARPING TORSION CONSTANTS FOR OPEN SECTIONS

It is perhaps evident by now that although the concept of warping torsion is easy to assimilate, the calculations of sectorial properties are rather tedious. To alleviate this problem, formulas for the sectorial properties of open sections commonly used in shear wall structures are given in Table 10.6.

Let us verify the value of I_ω derived previously in the example, by using the formula given in the table (cross-section reference no. 7 in Table 10.6).

$$I_\omega = \frac{h^2 t_1 t_2 b_1^3 b_2^3}{12\left(t_1 b_1^3 + t_2 b_2^3\right)} \tag{10.60}$$

$h = 30 \text{ ft} \quad t_1 = t_2 = 2 \text{ ft} \quad b_1 = 20 \text{ ft} \quad b_2 = 10 \text{ ft}$

$$I_\omega = \frac{30^2 \times 2 \times 2 \times 10^3 \times 20^3}{12\left(2 \times 10^3 + 2 \times 20^3\right)} = 133{,}226 \text{ ft}^6$$

This confirms the accuracy of our previous calculations.

TABLE 10.6
Torsion Constants for Open Sections

Cross-Section Reference Number | **Constants**

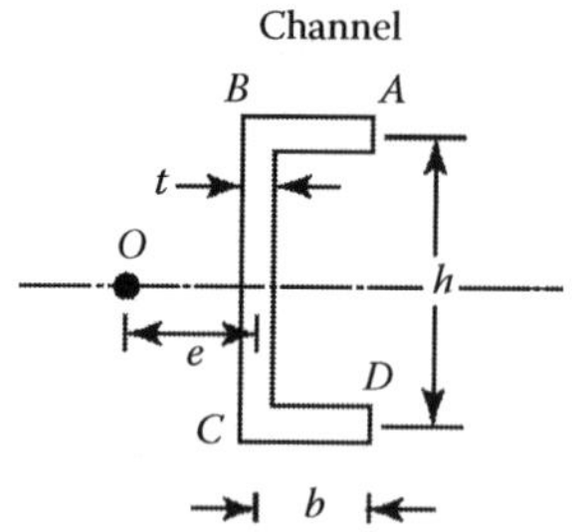

(1)

$$e = \frac{3b^2}{h+6b}$$

$$J = \frac{t^3}{3}(h+2b)$$

$$I_\omega = \frac{h^2b^2t}{12}\frac{2h+3b}{h+6b}$$

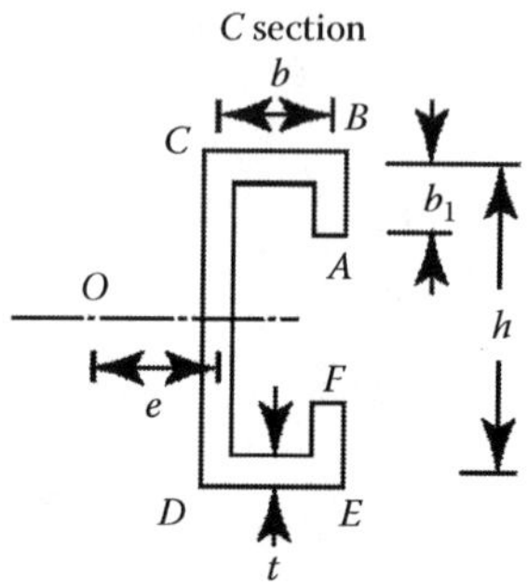

(2)

$$e = b\frac{3h^2b+6h^2b_1-8b_1^2}{h^2+6h^2b+6h^2b_1+8b_1^3-12hb_1^2}$$

$$J = \frac{t^3}{3}(h+2b+2b_1)$$

$$I_\omega = l\left[\frac{h^2b^2}{2}\left(b_1+\frac{b}{3}-e-\frac{2eb_1}{b}-\frac{2b_1^2}{h}\right) + \frac{h^2c^2}{2}\left(b+b_1+\frac{h}{6}+\frac{2b_1^2}{h}\right)+\frac{2b_1^2}{3}(b+e)^2\right]$$

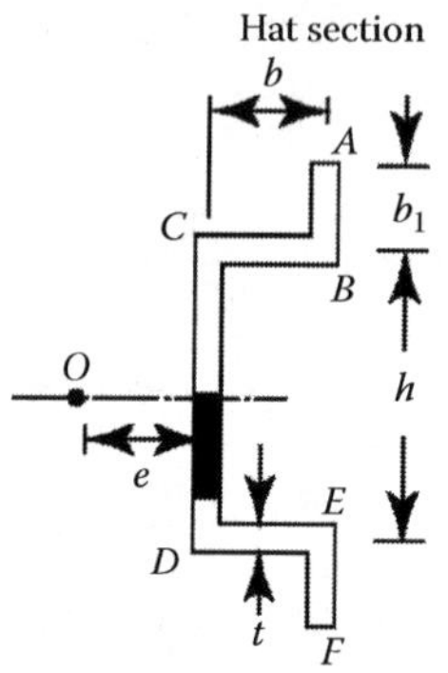

(3)

$$e = b\frac{3h^2b+6h^2b_1-8b_1^2}{h^2+6h^2b+6h^2b_1+8b_1^3+12hb_1^2}$$

$$J = \frac{t^2}{3}(h+2b+2b_1)$$

$$I_\omega = l\left[\frac{h^2b^2}{2}\left(b_1+\frac{b}{3}-e-\frac{2eb_1}{b}-\frac{2b_1^2}{h}\right) + \frac{h^2c^2}{2}\left(b+b_1+\frac{h}{6}+\frac{2b_1^2}{h}\right)+\frac{2b_1^2}{3}(b+e)^2\right]$$

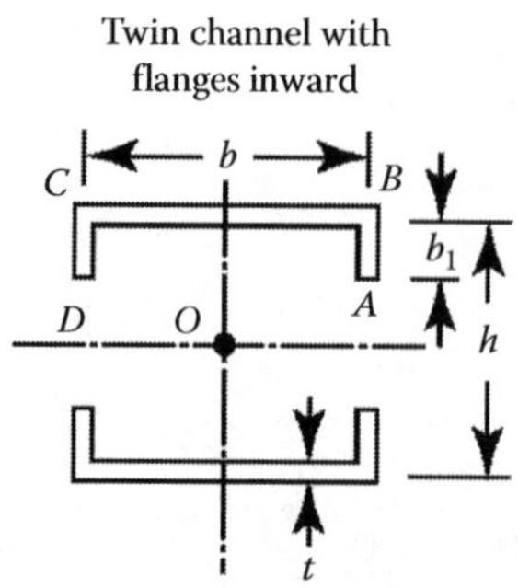

(4)

$$J = \frac{t^2}{3}(2b+4b_1)$$

$$I_\omega = \frac{tb^2}{24}\left(8b_1^2+6h^2b_1+h^2b+12b_1^2h\right)$$

(Continued)

TABLE 10.6 (*Continued*)
Torsion Constants for Open Sections

Cross-Section Reference Number	Constants

Twin channel with flanges inward

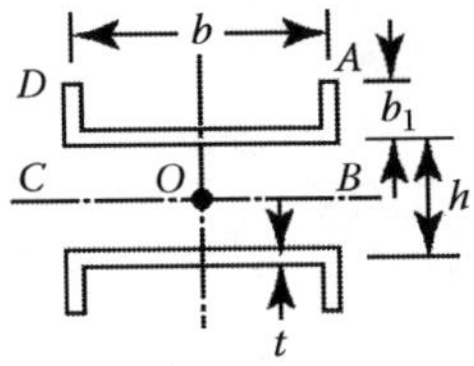

(5)

$$J = \frac{t^2}{3}(2h + 4b_1)$$

$$I_\omega = \frac{tb^2}{24}\left(8b_1^3 + 6h^2 b_1 + h^2 b - 12 b_1^2 h\right)$$

Wide flanged beam with equal flanges

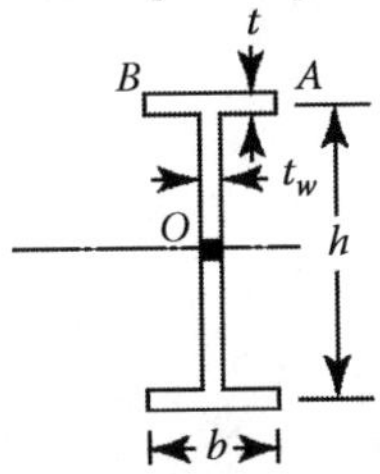

(6)

$$J = \frac{1}{3}\left(2t^3 b + t_\omega^3 h\right)$$

$$I_\omega = \frac{h^2 t b^3}{24}$$

Wide flanged beam with unequal flanges

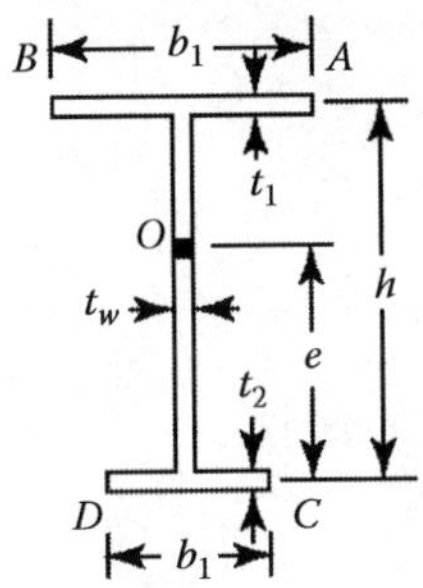

(7)

$$e = \frac{t_1 b_1^2 h}{t_1 b_1^3 + t_2 b_2^3}$$

$$J = \frac{1}{3}\left(t_1^3 b_1 + t_2^3 b_2 + t_w^3 h\right)$$

$$I_\omega = \frac{h^2 t_1 t_2 b_1^3 b_2^3}{12\left(t_1 b_1^3 + t_2 b_2^3\right)}$$

Z section

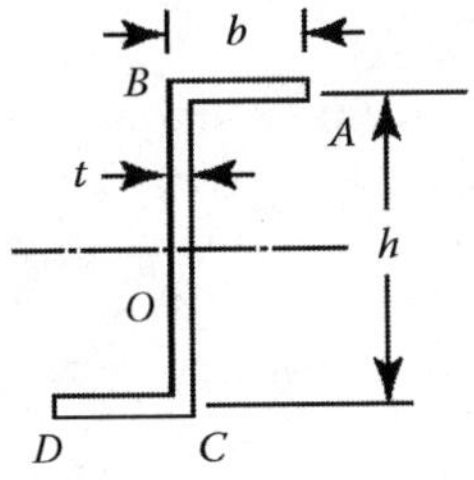

(8)

$$J = \frac{t^3}{3}(2b + h)$$

$$I_\omega = \frac{th^2 b^3}{12}\left(\frac{b + h}{2b + h}\right)$$

(*Continued*)

TABLE 10.6 (*Continued*)
Torsion Constants for Open Sections

Cross-Section Reference Number	Constants
Segment of a circular tube 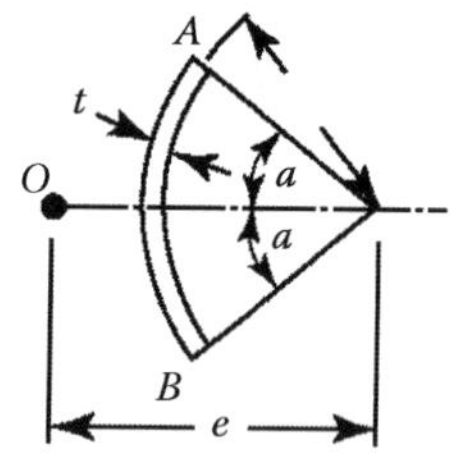 (9)	$e = 2r\dfrac{\sin\alpha - \alpha\cos\alpha}{\alpha - \sin\alpha\cos\alpha}$ $J = \dfrac{2}{3}t^3 r\alpha$ $I_\omega = \dfrac{2tr^5}{3}\left[\alpha^3 - 6\dfrac{(\sin\alpha - \alpha\cos\alpha)^2}{\alpha - \sin\alpha\cos\alpha}\right]$ For $\alpha = 45^\circ$ $e = 1.06r$ $= 90^\circ$ $e = 1.27r$ $= 180^\circ$ $e = 2r$
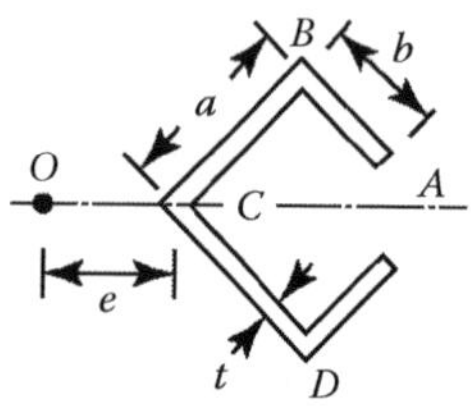	$e = 0.707ab^2\dfrac{3a-2b}{2a^3-(a-b)^3}$ $J = \dfrac{2}{3}t^3(a+b)$ $I_\omega = \dfrac{ta^4b^3}{6}\dfrac{3a+2b}{2a^3-(a-b)^2}$
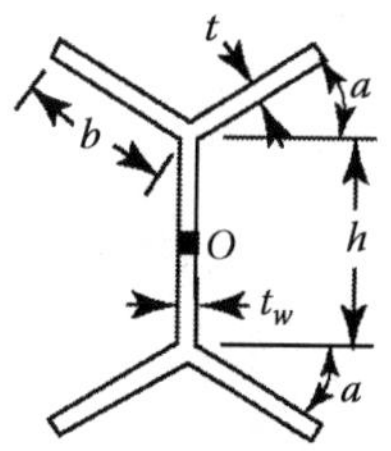 (10)	$J = \dfrac{1}{3}\left(4t^3b + t_w^3 a\right)$ $I_\omega = \dfrac{a^2b^2t}{3}\cos^2\alpha$

10.10 STIFFNESS METHOD USING WARPING-COLUMN MODEL

Building structures are generally analyzed as 3D frames, with the members oriented in any direction and subjected to axial force, shear, and moment in two orthogonal plans, and torsion about their linear axes. Therefore, a general beam or column element in the analysis of 3D frames must include forces in three directions and moments about three axes. Such a beam element with six displacements at each end is shown in Figure 10.22. The stiffness matrix, which is the relationship between the end forces and displacement, is a 12 × 12 matrix, corresponding to six degrees of freedom at each end. The stiffness coefficients depicting the force–displacement relation for the 3D beam element are found by combining the stiffness terms for axial deformation, bending about two axes, and torsion. The resulting 12 × 12 stiffness matrix is given in Figure 10.22.

A nonplanar shear wall such as an I- or C-shaped wall is modeled in a 3D analysis as an assemblage of floor-to-floor panel elements connected along their edges. The continuous

		u	v	w	θ_x	θ_y	θ_z	u	v	w	θ_x	θ_y	θ_z
		1	2	3	4	5	6	7	8	9	10	11	12
u	1	$\frac{12EI_y}{L^3}$	S										
v	2	0	$\frac{12EI_x}{L^3}$	Y									
w	3	0	0	$\frac{EA}{L}$	M								
θ_x	4	0	$\frac{6EI_x}{L^2}$	0	$\frac{4EI_x}{L}$	M							
θ_y	5	$\frac{-6EI_y}{L^2}$	0	0	0	$\frac{4EI_y}{L}$	E						
θ_z	6	0	0	0	0	0	$\frac{GJ}{L}$	T					
u	7	$\frac{-12EI_y}{L^3}$	0	0	0	$\frac{6EI_y}{L^2}$	0	$\frac{12EI_y}{L^3}$	R				
v	8	0	$\frac{-12EI_x}{L^3}$	0	$\frac{-6EI_x}{L^2}$	0	0	0	$\frac{12EI_x}{L^3}$	I			
w	9	0	0	$\frac{-EA}{L}$	0	0	0	0	0	$\frac{EA}{L}$	C		
θ_x	10	0	$\frac{-6EI_x}{L^2}$	0	$\frac{2EI_x}{L}$	0	0	0	$\frac{-6EI_x}{L^2}$	0	$\frac{4EI_x}{L}$	A	
θ_y	11	$\frac{-6EI_y}{L^2}$	0	0	0	$\frac{2EI_y}{L}$	0	$\frac{6EI_y}{L^2}$	0	0	0	$\frac{4EI_y}{L}$	L
θ_z	12	0	0	0	0	0	$\frac{-GJ}{L}$	0	0	0	0	0	$\frac{GJ}{L}$

(a)

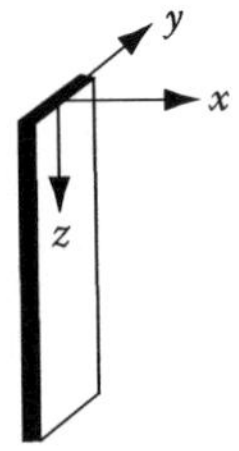

(b)

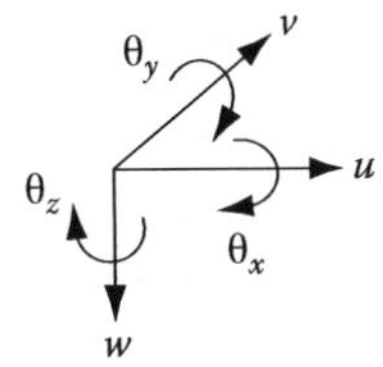

(c)

FIGURE 10.22 (a) 12 × 12 stiffness matrix for prismatic three-dimensional element, (b) coordinate axes, and (c) positive sign convention.

connection between the panels provides for the principal interaction and the vertical shear along their connecting edges.

As an alternative technique, a 3D wall may be represented in all its aspects of behavior including warping, by a warping-column element, with seven degrees of freedom at each floor level; its assigned properties would include the warping moment of inertia I_ω, in addition to the familiar area A, to represent its resistance to axial load, and inertias I_x and I_y to represent its St. Venant's resistance to torsion. Such a single-column model, with an extra seventh degree of freedom, is particularly suitable for open-section walls that are uniform over the height. The seventh, warping, degree of freedom is the parameter $d\phi/dz$, which expresses the magnitude of warping. It is used as the warping degree of freedom, while B, the bimoment, becomes the corresponding generalized force. Thus, with seven degrees of freedom per node, the column element (Figure 10.23) has a 14 × 14 stiffness matrix. A number of such story-height elements may be stacked vertically to represent a complete core. The interaction of slabs and beams at

		u	v	w	θ_x	θ_y	θ_z	θ'_z	u	v	w	θ_x	θ_y	θ_z	θ'_z
		1	2	3	4	5	6	7	8	9	10	11	12	13	14
u	1	$\frac{12EI_y}{L^3}$													
v	2	0	$\frac{12EI_x}{L^2}$	S											
w	3	0	0	$\frac{EA}{L}$	Y										
θ_x	4	0	$\frac{6EI_x}{L^3}$	0	$\frac{4EI_x}{L}$	M									
θ_y	5	$\frac{-6EI_y}{L^2}$	0	0	0	$\frac{4EI_y}{L}$	M								
θ_z	6	0	0	0	0	0	S_{11}	E							
θ'_z	7	0	0	0	0	0	S_{12}	S_{22}	T						
u	8	$\frac{-12EI_y}{L^3}$	0	0	0	$\frac{6EI_y}{L^2}$	0	0	$\frac{12EI_y}{L^3}$	R					
v	9	0	$\frac{-12EI_x}{L^3}$	0	$\frac{-6EI_x}{L^2}$	0	0	0	0	$\frac{12EI_y}{L^3}$	I				
w	10	0	0	$\frac{-EA}{L}$	0	0	0	0	0	0	$\frac{EA}{L}$	C			
θ_x	11	0	$\frac{6EI_x}{L^2}$	0	$\frac{2EI_x}{L}$	0	0	0	0	$\frac{-6EI_x}{L^2}$	0	0	A		
θ_y	12	$\frac{-6EI_y}{L^2}$	0	0	0	$\frac{2EI_y}{L}$	0	0	$\frac{6EI_y}{L^2}$	0	0	0	$\frac{4EI_y}{L}$	L	
θ_z	13	0	0	0	0	0	$-S_{11}$	$-S_{12}$	0	0	0	0	0	S_{11}	
θ'_z	14	0	0	0	0	0	$-S_{12}$	S_{24}	0	0	0	0	0	$-S_{12}$	S_{22}

(a)

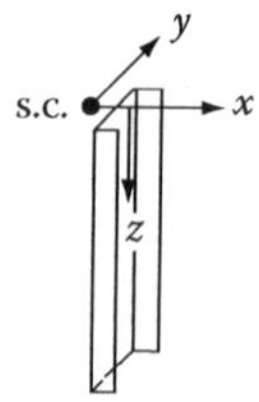

(b)

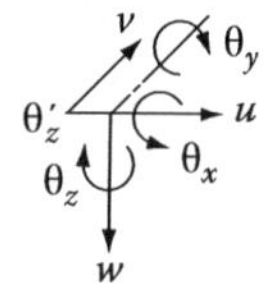

(c)

Torque bimoment relation as between P_x and θ_y

$$S_{11}=\alpha\frac{k}{\ell}\sinh k \quad S_{24}=\alpha\frac{\ell}{k}(\sinh k - k)$$

$$S_{12}=\alpha(1-\cosh k)$$

$$S_{22}=\alpha(k\cosh k-\sinh k)$$

$$\text{where } \alpha=\frac{GJ}{(2+k\sinh k-2\cosh k)}$$

FIGURE 10.23 (a) 14 × 14 stiffness matrix for thin walled open section, (b) coordinate axes, and (c) positive sign convention.

the floor levels may also be included in the stiffness matrix of the total structure by an appropriate combination of floor stiffness matrix with the stiffness matrix of the core (Taranath, 1986). Engineers engaged in developing special-purpose computer programs may find the reference useful for including the additional warping degree of freedom for open-section shear wall buildings.

11 Seismic Design
A Pictorial Review

PREVIEW

"One illustration is worth a thousand words"—so goes the adage, perhaps more so in structural engineering than in other scientific disciplines.

Accordingly, in this, and in subsequent chapters, an attempt is made to present design information in a concise format without an abundance of text; where necessary, a brief explanation of figures is included along with the illustrations.

11.1 FIGURES AND TABLES EXPLAINING THE FUNDAMENTALS OF SEISMIC DESIGN

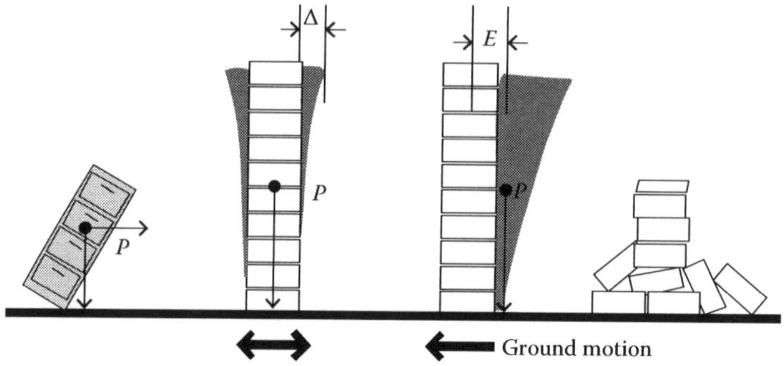

FIGURE 11.1 Why buildings generally fall down, not over.

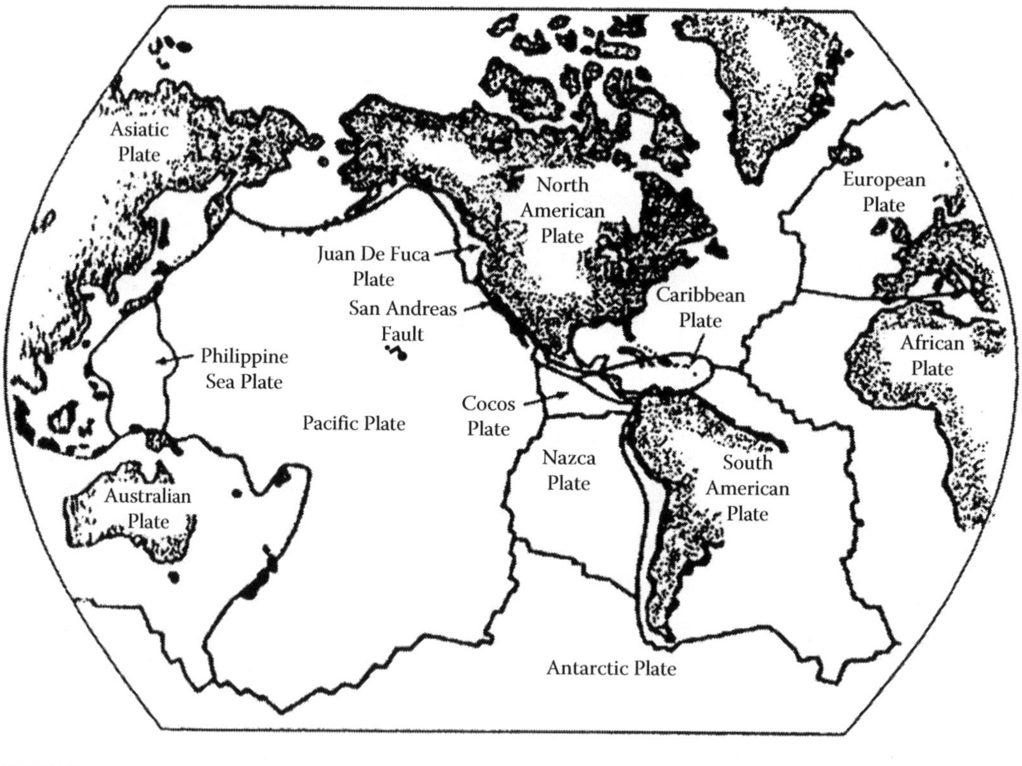

FIGURE 11.2 Tectonic plates.

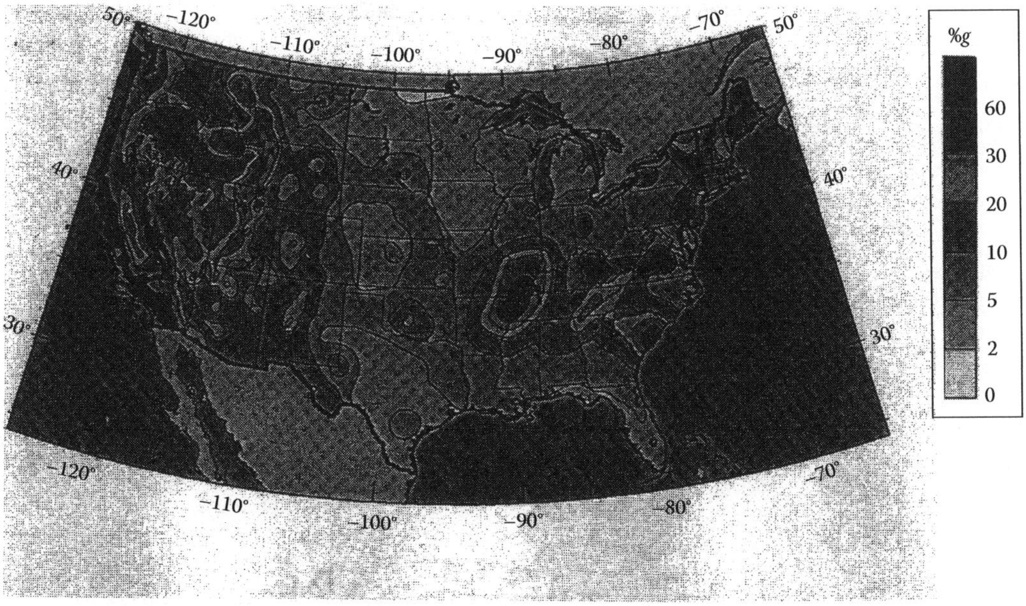

FIGURE 11.3 Horizontal ground acceleration (%g) with 2% probability of exceedance in 50 years. (From USGS National Seismic Hazard Mapping Project website: geohazards.cr.usgs.gov.)

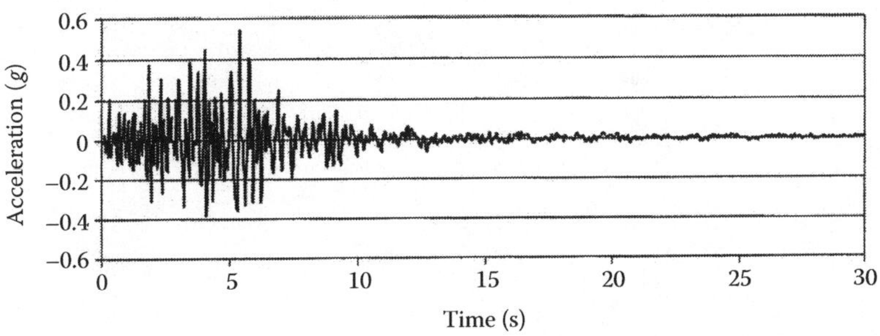

FIGURE 11.4 Typical acceleration time history of strong ground shaking.

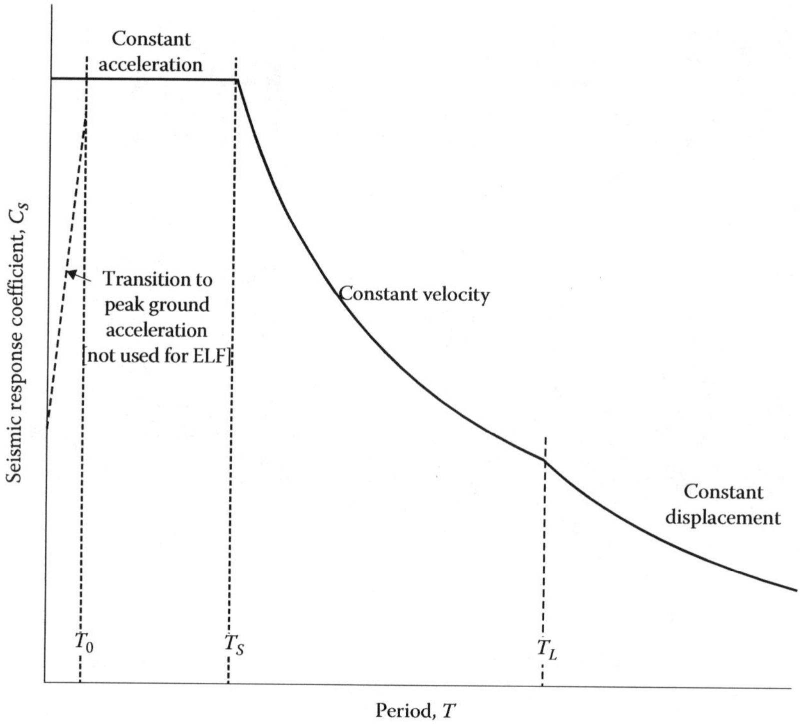

FIGURE 11.5 Seismic response coefficients versus period.

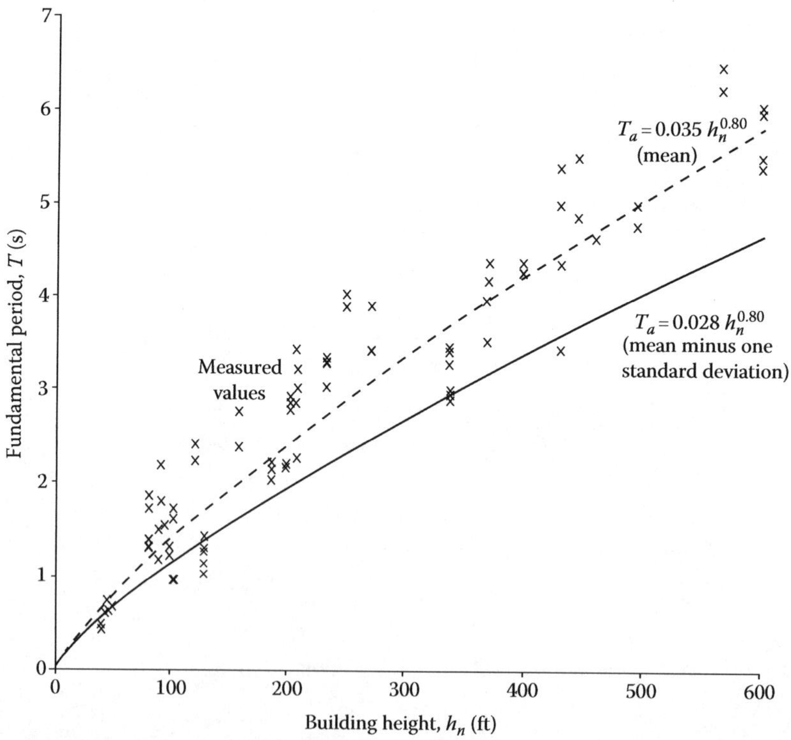

FIGURE 11.6 Variation of fundamental period versus building height.

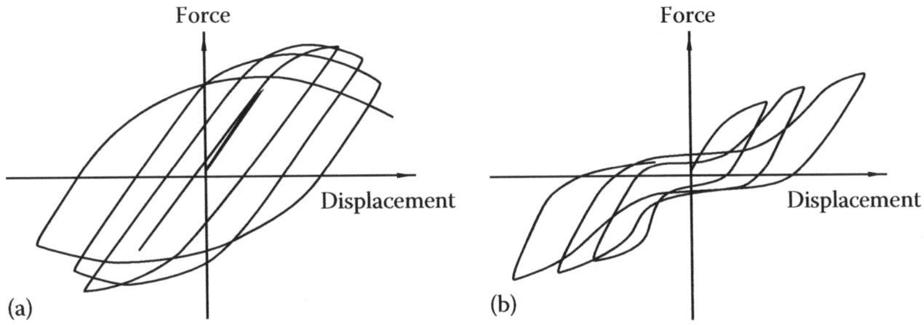

FIGURE 11.7 Typical hysteretic curves. (a) Ductile hysteretic loops. (b) Pinched hysteretic loops.

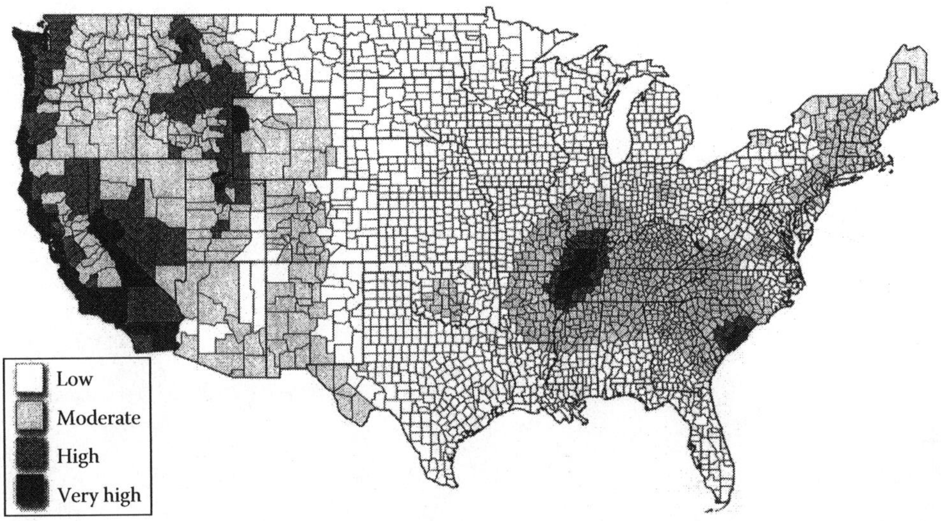

FIGURE 11.8 Map of continental United States showing countries and probabilities of earthquakes of varying magnitudes. (From U.S. Geological Survey, Reston, VA.)

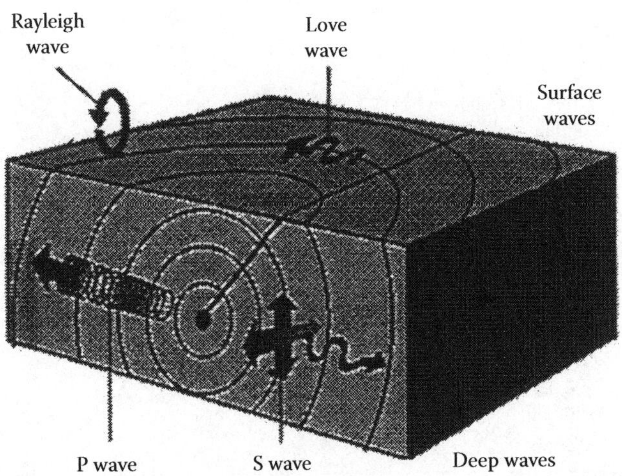

FIGURE 11.9 Types of seismic waves.

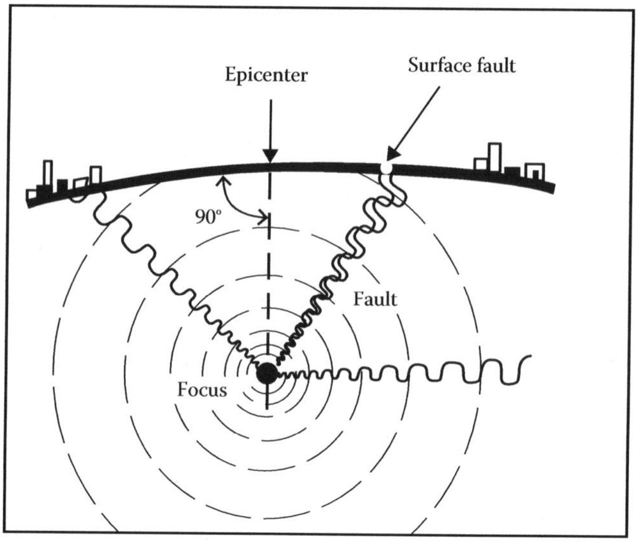

FIGURE 11.10 Common terms used in explaining earthquakes. Vibration waves radiate out from the fault break.

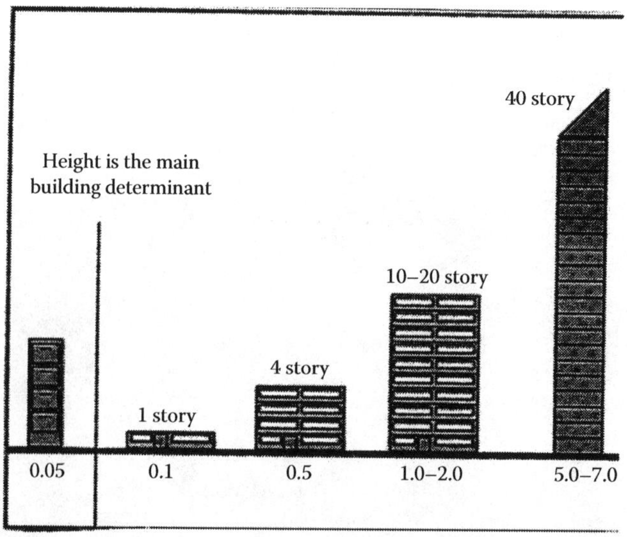

FIGURE 11.11 Fundamental period in seconds.

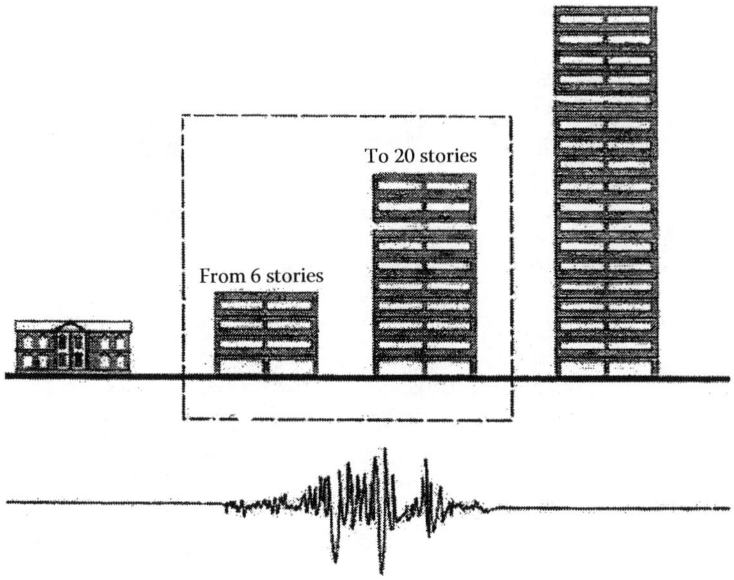

FIGURE 11.12 The vulnerable building group in Mexico City 1985 earthquake.

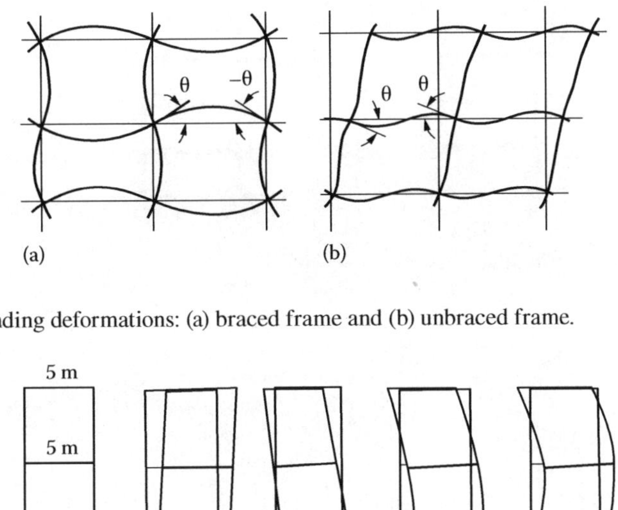

FIGURE 11.13 Bending deformations: (a) braced frame and (b) unbraced frame.

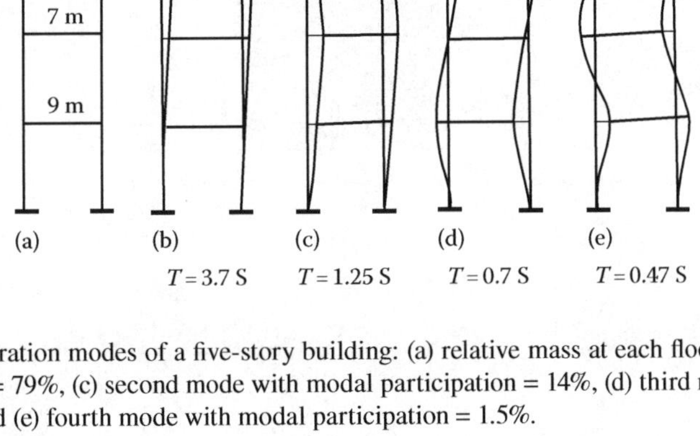

FIGURE 11.14 Vibration modes of a five-story building: (a) relative mass at each floor, (b) first mode with modal participation = 79%, (c) second mode with modal participation = 14%, (d) third mode with modal participation = 5.5%, and (e) fourth mode with modal participation = 1.5%.

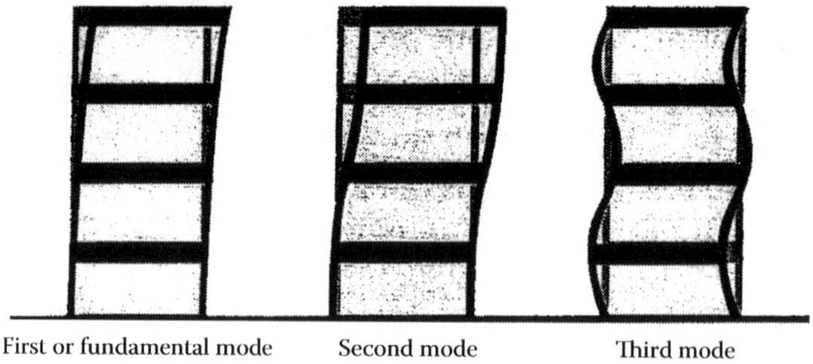

FIGURE 11.15 First three modes of vibration.

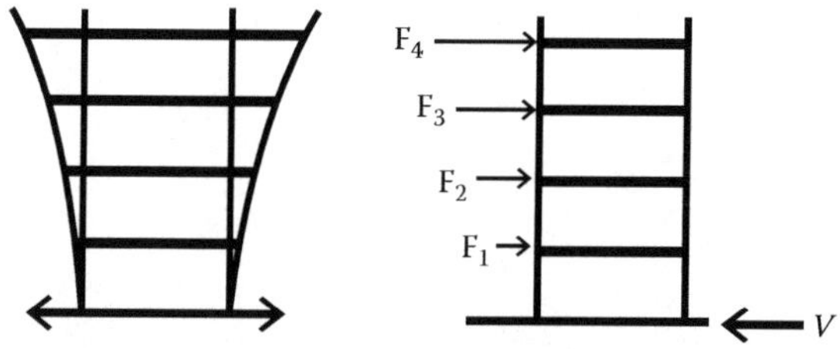

FIGURE 11.16 Typical distribution of base shear.

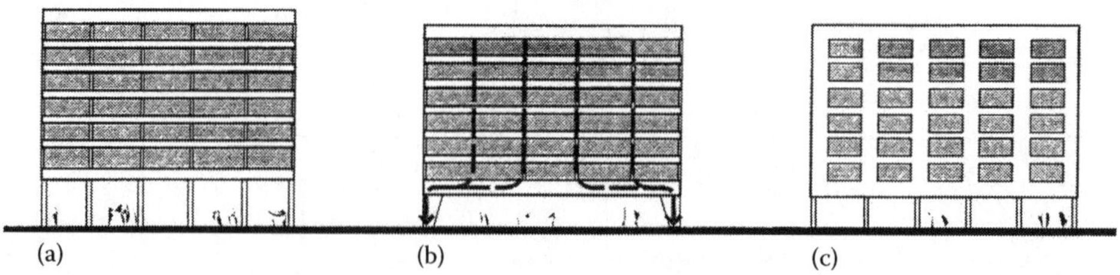

FIGURE 11.17 Three types of soft first story: (a) flexible first floor; (b) discontinuity, indirect load path; and (c) heavy superstructure.

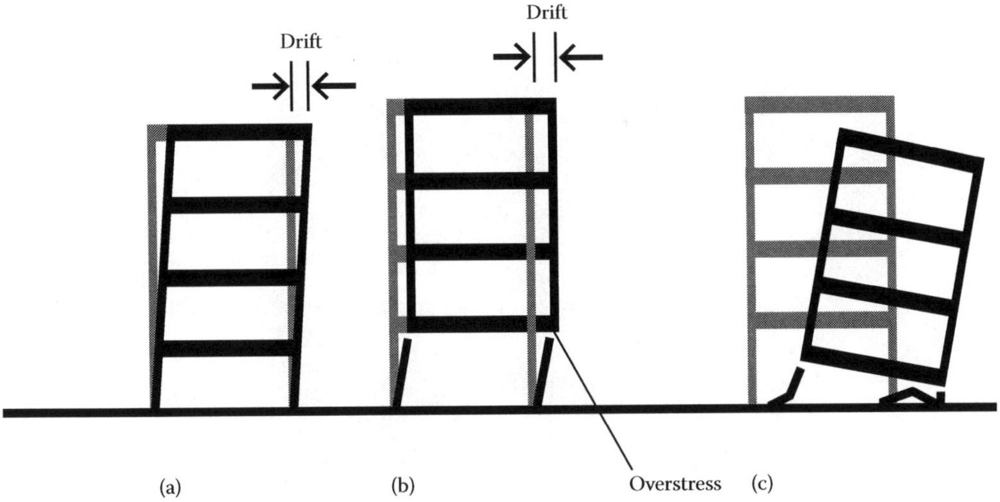

FIGURE 11.18 Soft first-story failure mechanism: (a) normal, (b) soft story, and (c) collapse.

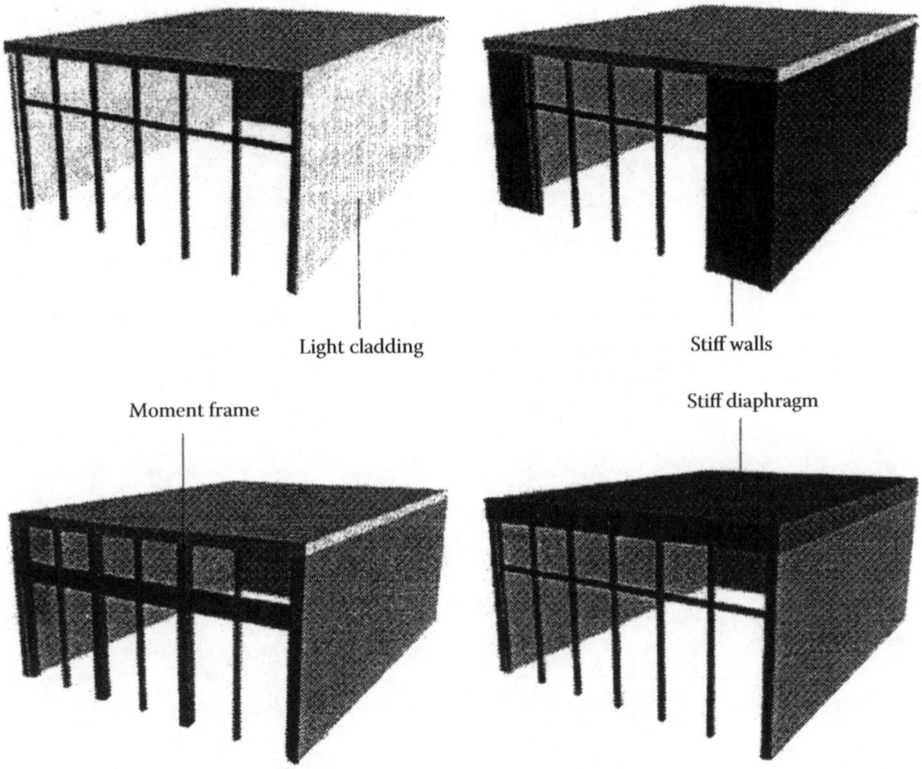

FIGURE 11.19 Example solutions to storefront-type unbalanced perimeter-resistance conditions.

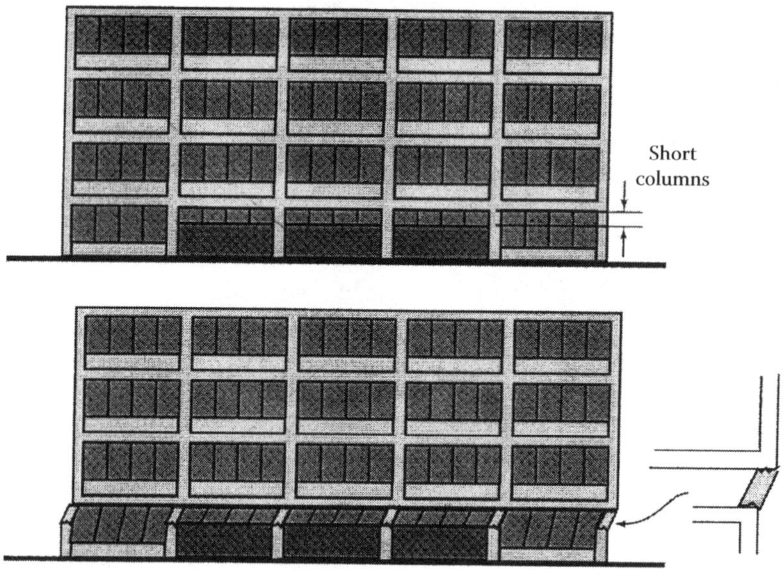

FIGURE 11.20 Short column effects.

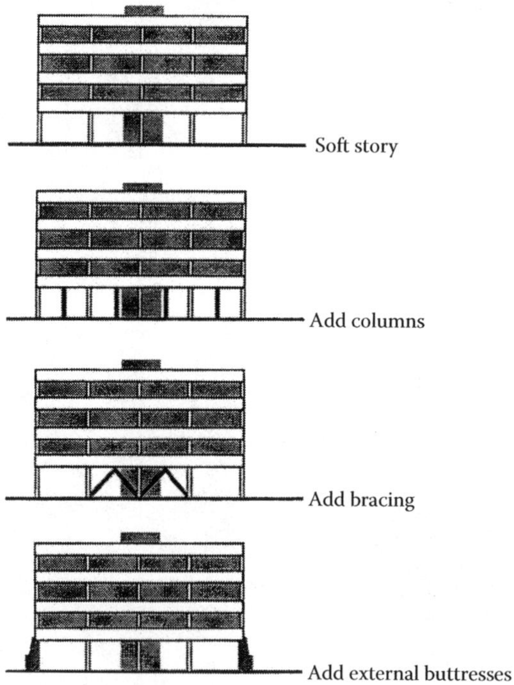

FIGURE 11.21 Example solutions to the soft first story.

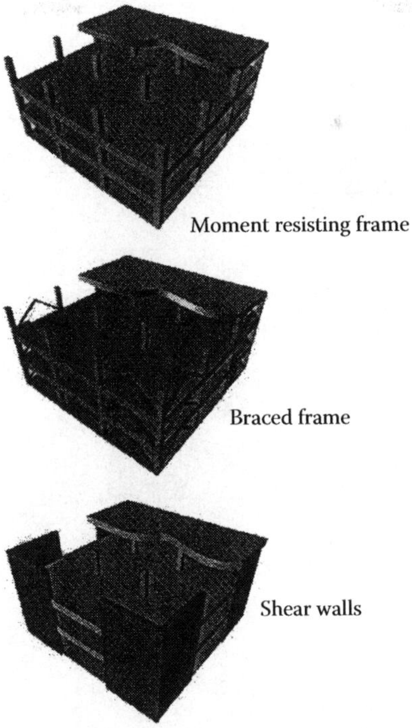

FIGURE 11.22 The optimal structural/architectural configurations.

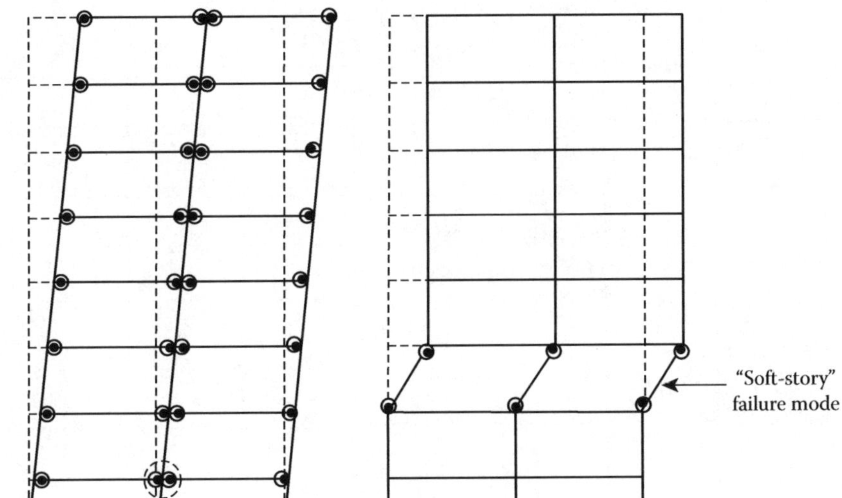

FIGURE 11.23 Comparison of collapse mechanism: (a) story-column/weak-girder mechanism and (b) soft-story mechanism.

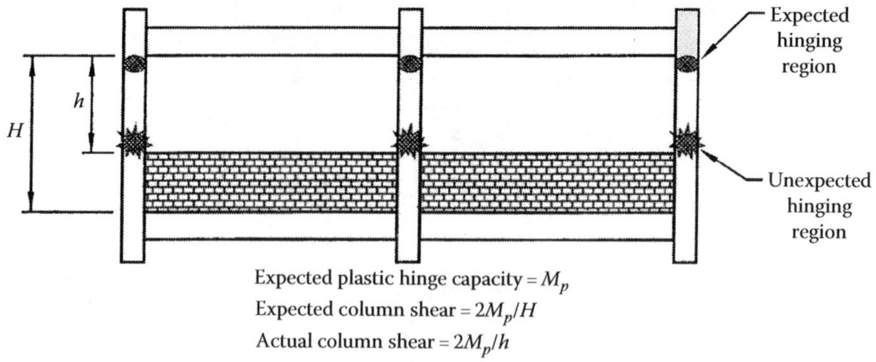

FIGURE 11.24 Undesired interaction effects.

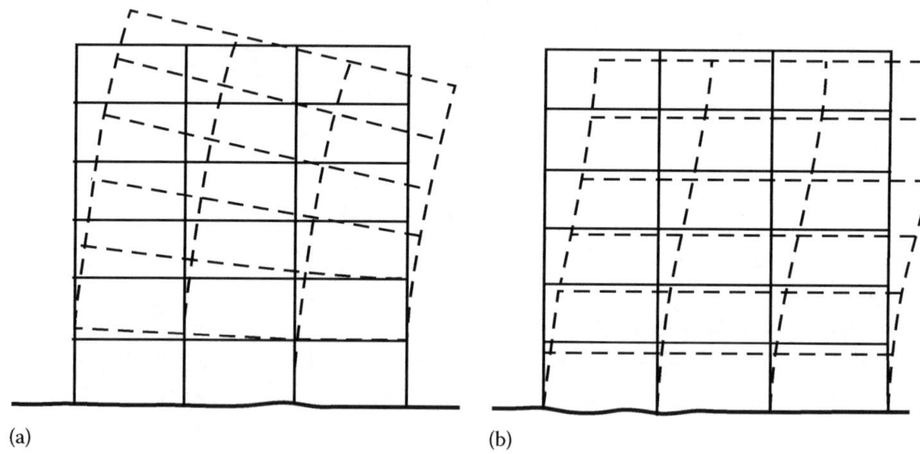

FIGURE 11.25 Building drift: caused by (a) bending and (b) shear.

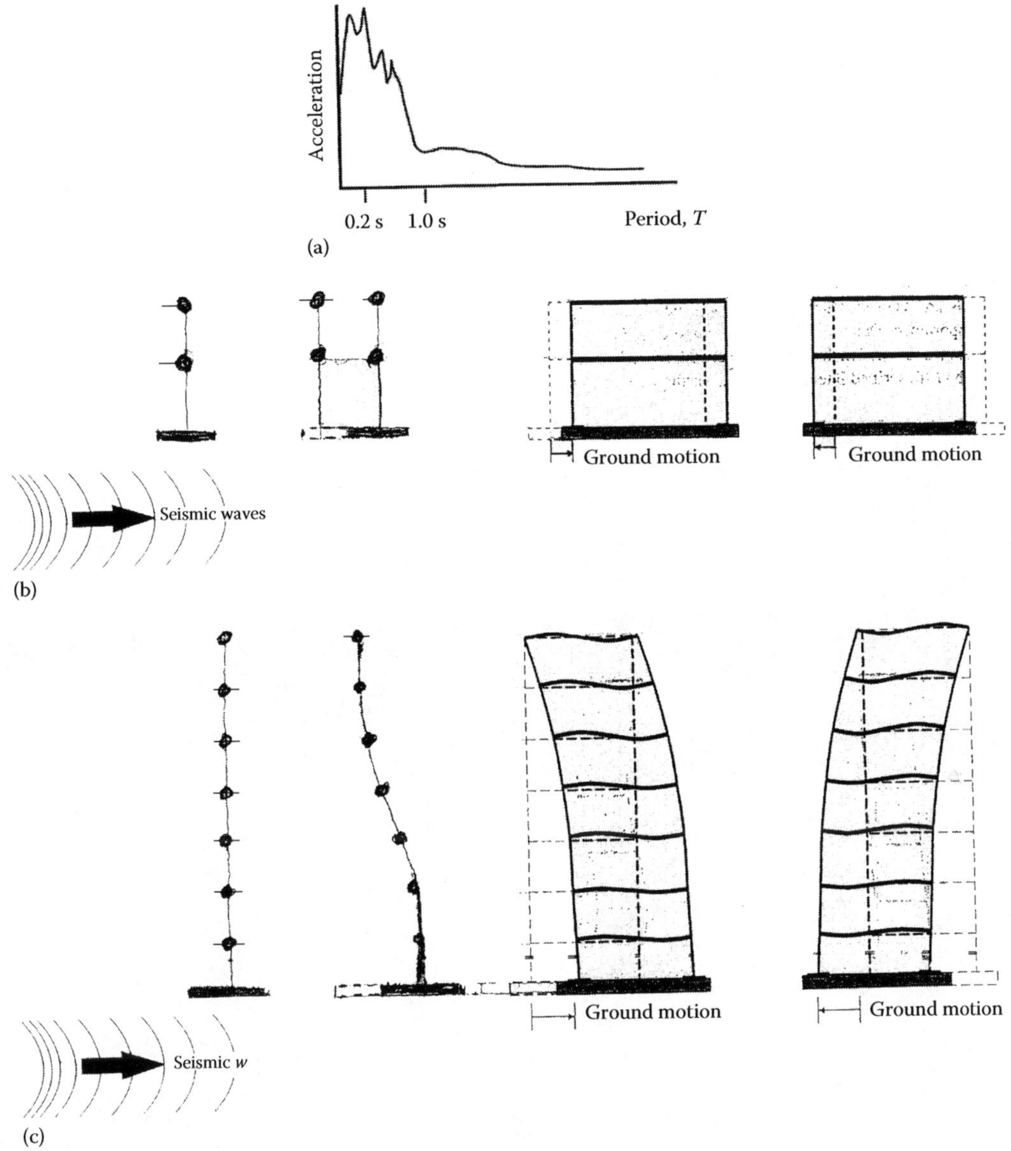

FIGURE 11.26 Behavior of stiff and flexible structures: (a) typical acceleration spectrum, (b) stiff structure ($T = 0.2$ s), and (c) flexible structure ($T = 5$–7 s).

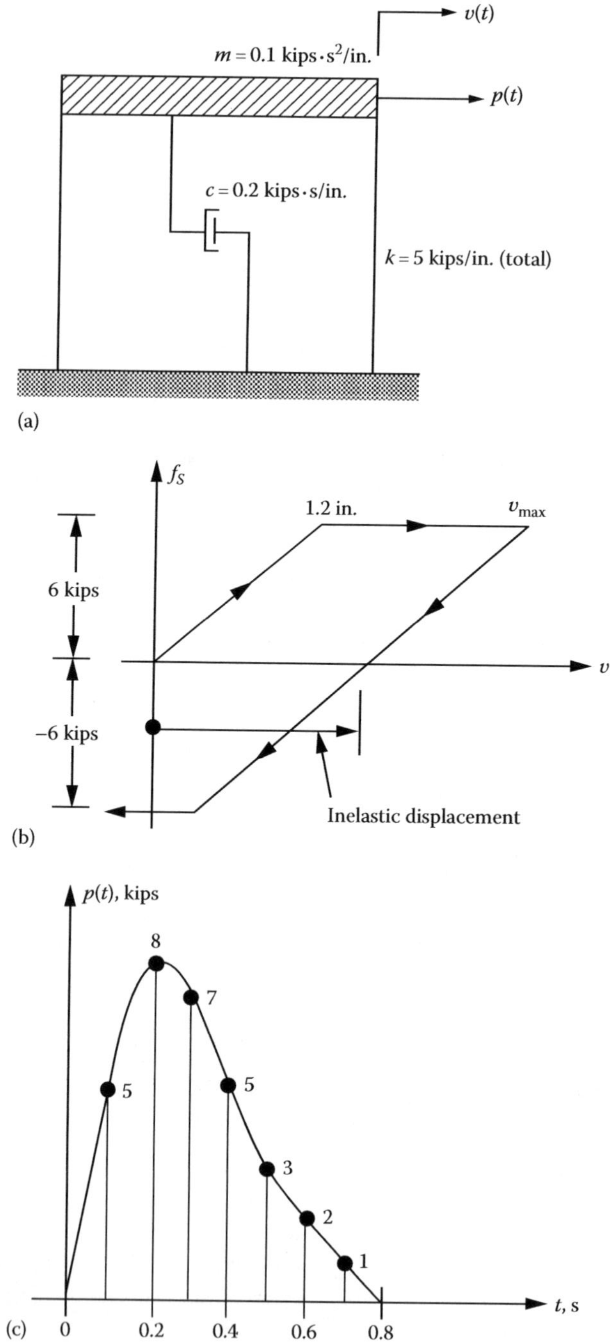

FIGURE 11.27 Elastoplastic frame subject to dynamic loading: (a) portal frame properties, (b) elastoplastic stiffness, and (c) dynamic loading history.

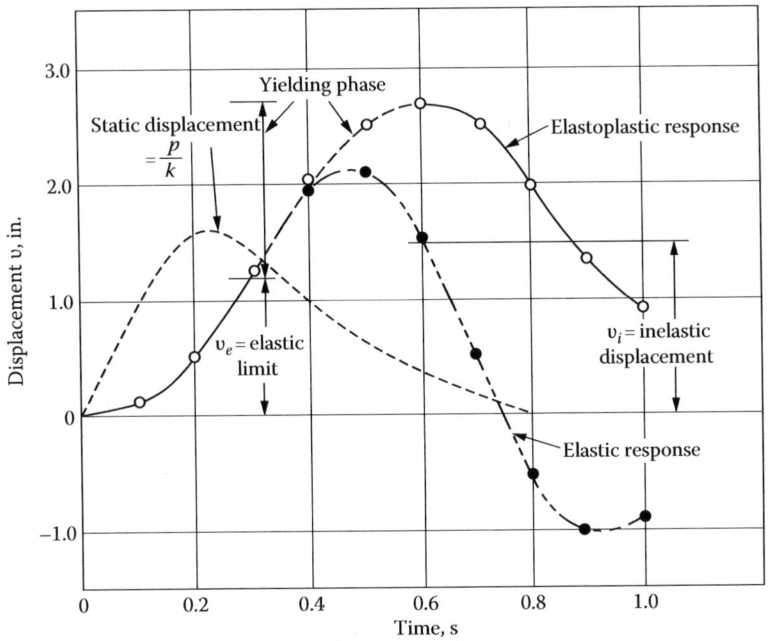

FIGURE 11.28 Comparison of elastic and elastoplastic response of frame shown in Figure 11.29.

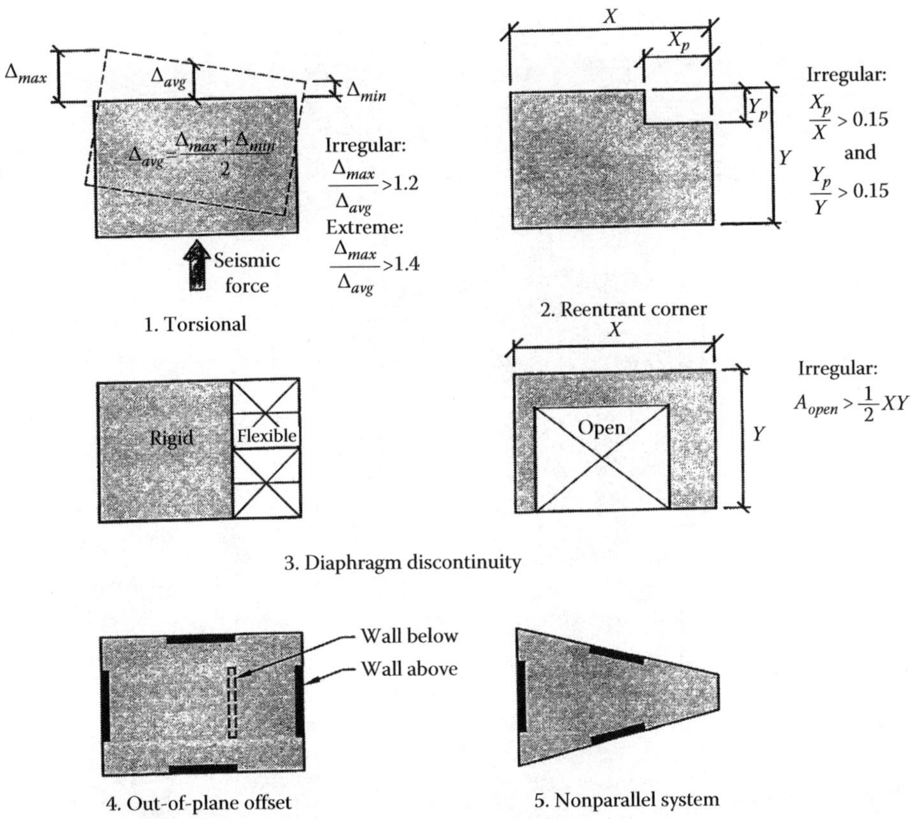

FIGURE 11.29 Building plan irregularities.

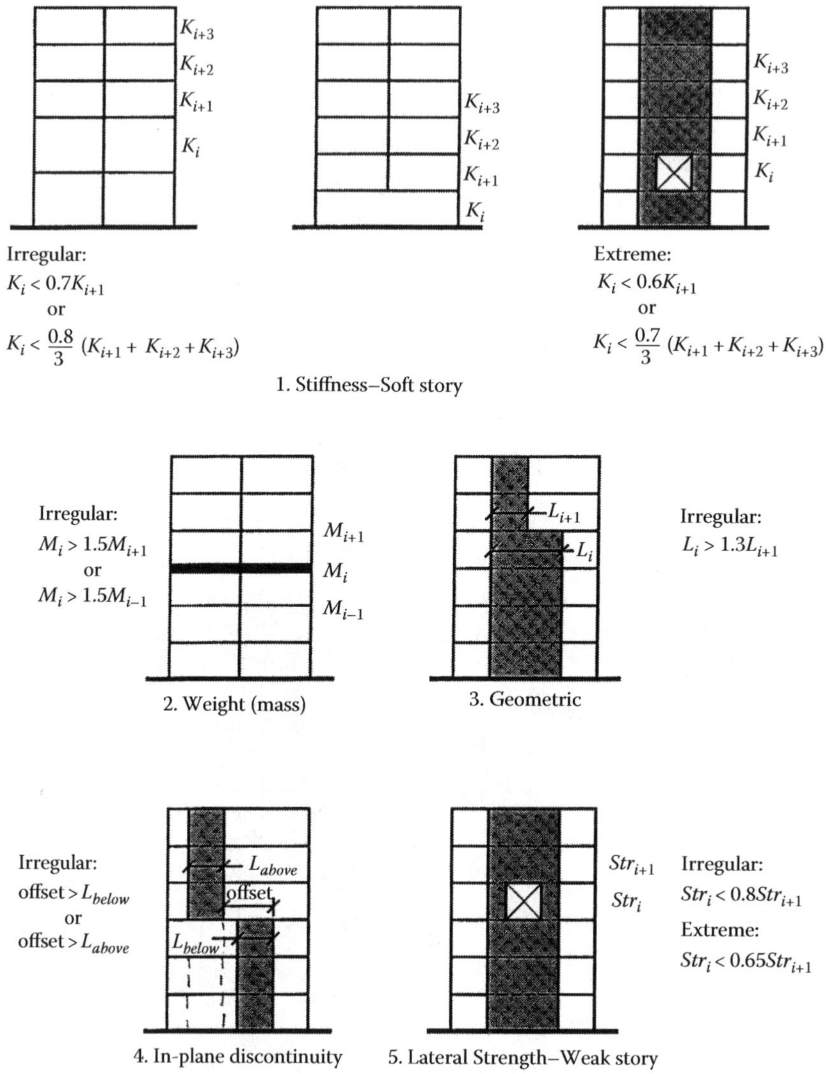

FIGURE 11.30 Building vertical irregularities.

Vertical Conditions	Resulting Failure Patterns	Performance	Code Remedies
		**V1 Stiffness irregularity: Soft story**	
		Common collapse mechanism. Death and much damage in Northridge earthquake.	Modal analysis, +65 ft high in SCD D,E,F. Extreme case not permitted in seismic use groups E and F.
		**V2 Weight/mass irregularity**	
		Collapse mechanism in extreme circumstances.	Modal analysis, +65 ft high in SDC D,E,F.
		**V3 Vertical geometric irregularity**	
		Localized structural damage.	Modal analysis, +65 ft high in SDC D,E,F.
	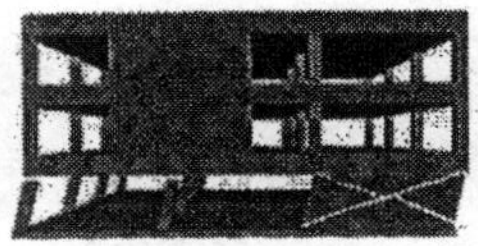	**V4 In-plane irregularity in vertical lateral force system**	
		Localized structural damage.	Modal analysis, +65 ft high is SDC D,E,F. 25% increase to diaphragm connection design force. Supporting members designed for increased forces.
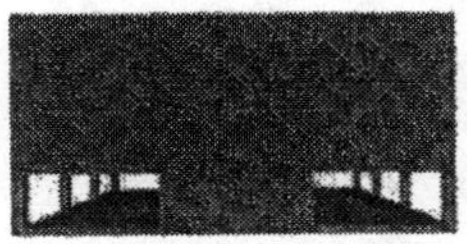		**V5 Capacity discontinuity: Weak story**	
		Collapse mechanism in extreme circumstances.	Modal analysis, +65 ft high in SDC D,E,F.

FIGURE 11.31 Vertical irregularities. (Based on IBC Section 1616.5.2.)

Plan Conditions	Resulting Failure Patterns	Performance	Code Remedies
		P1 Torsional irregularity: Unbalenced resistance	
		Localized damage. Collapse mechanism in extreme instances.	Modal analysis, +65 ft high in SDC D,E,F. 25% increase to diaphragm connection design forces. Amplified forces to max of X3.
		P2 Re-entrant corners	
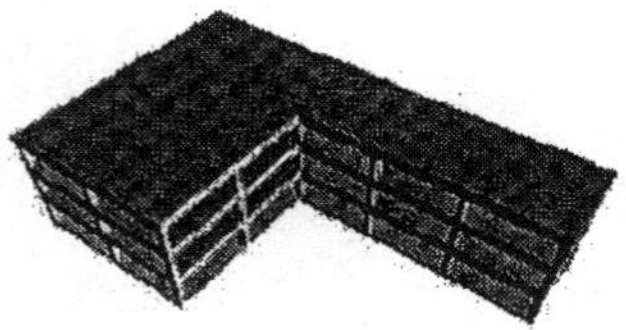		Local damage to diaphragm and attached elements. Collapse mechanism in extreme instances in large buildings.	25% increase in diaphragm connection design forces.
		P3 Diaphragm eccentricity and cutouts	
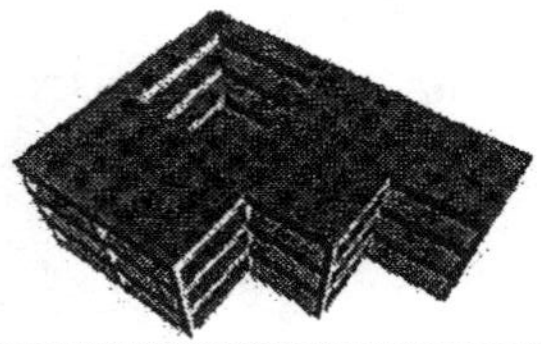	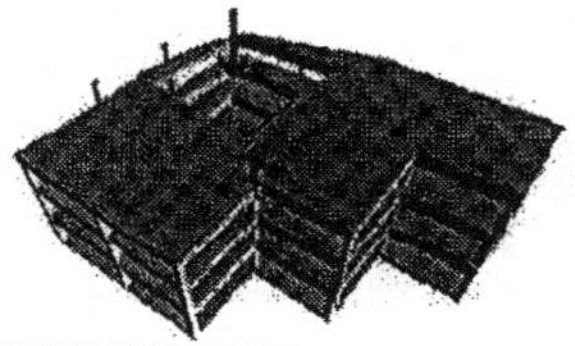	Localized structural damage.	25% increase in diaphragm connection design forces.
		P4 Nonparallel lateral force-resisting system	
		Leads to torsion and instability, localized damage.	Combine 100% and 30% of forces in two directions, use maximum.
		P5 Out-of-plane offsets: Discontinuous shearwalls	
	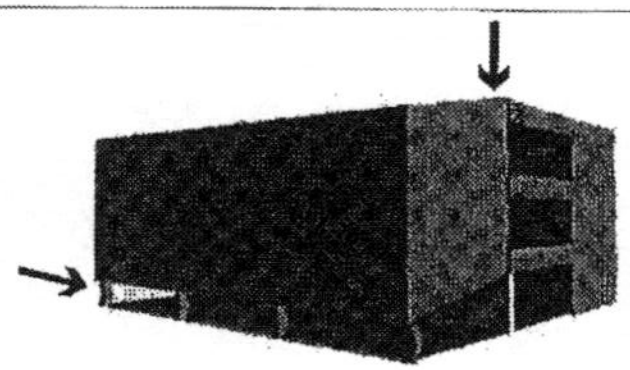	Collapse mechanism in extreme circumstances.	Modal analysis, +65 ft high in SDC D,E,F. 25% increase to diaphragm connection design forces.

FIGURE 11.32 Horizontal (plan) irregularities. (Based on IBC Section 1616.5.1.)

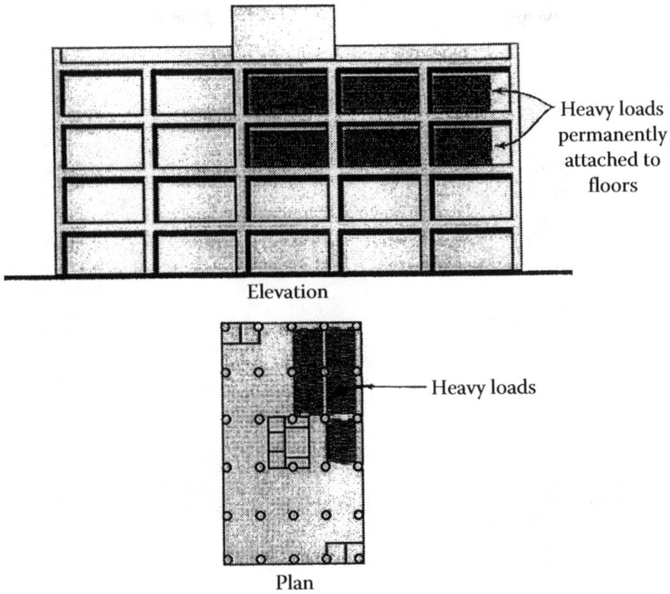

FIGURE 11.33 Nonsymmetric heavy loading.

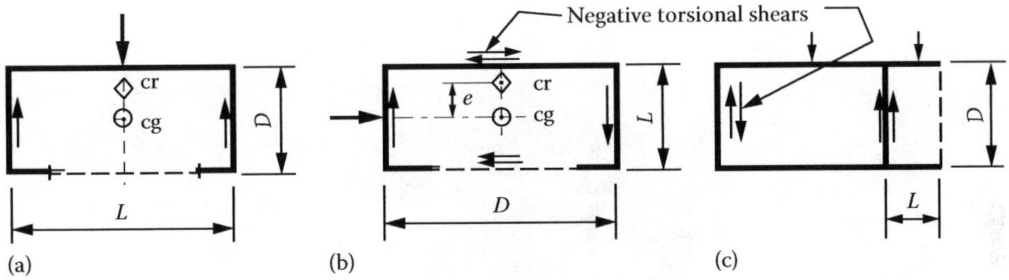

FIGURE 11.34 Torsional moment: (a) no diaphragm rotation, (b) rotation of diaphragm, and (c) cantilever diaphragm.

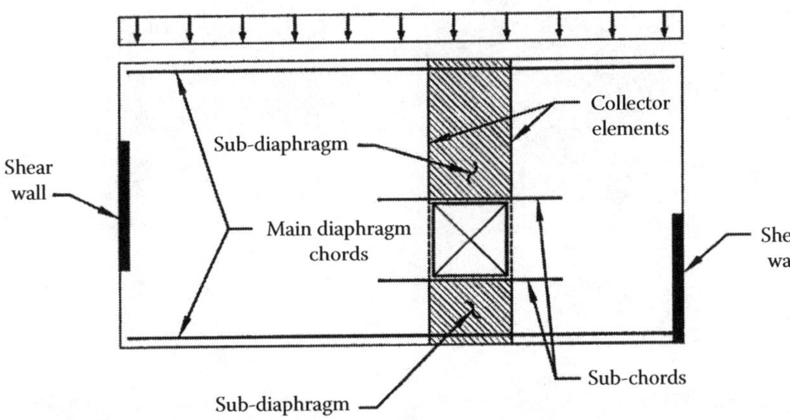

FIGURE 11.35 Diaphragm components.

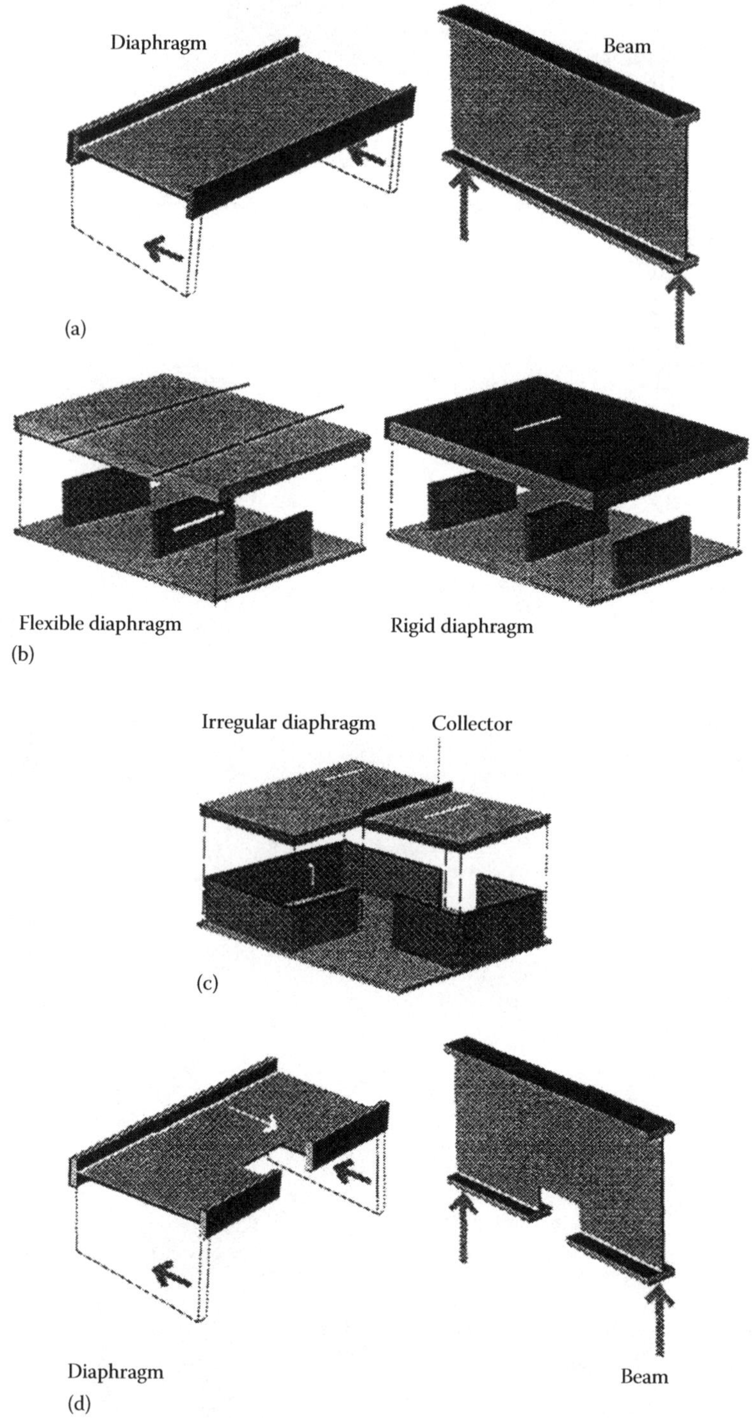

FIGURE 11.36 Deep beam action of diaphragm.

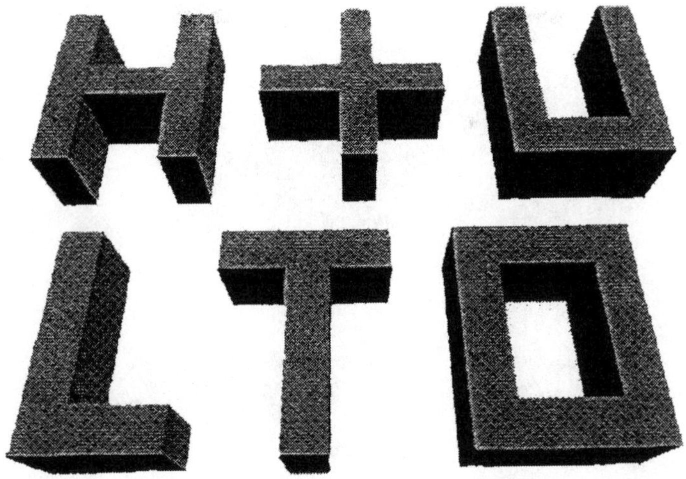

FIGURE 11.37 Reentrant corner plan forms.

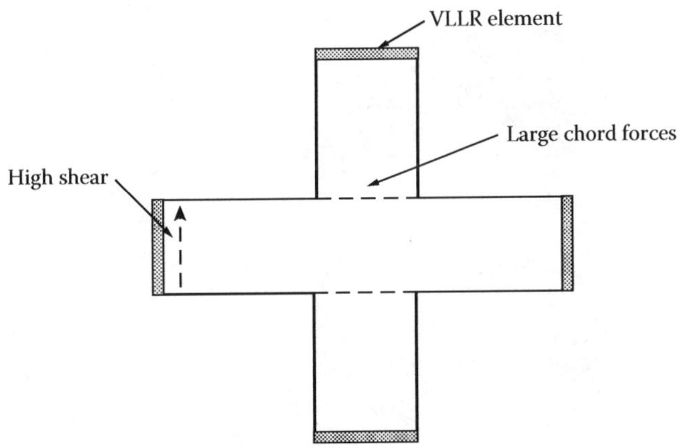

FIGURE 11.38 High shear and chord forces in a reentrant plan form.

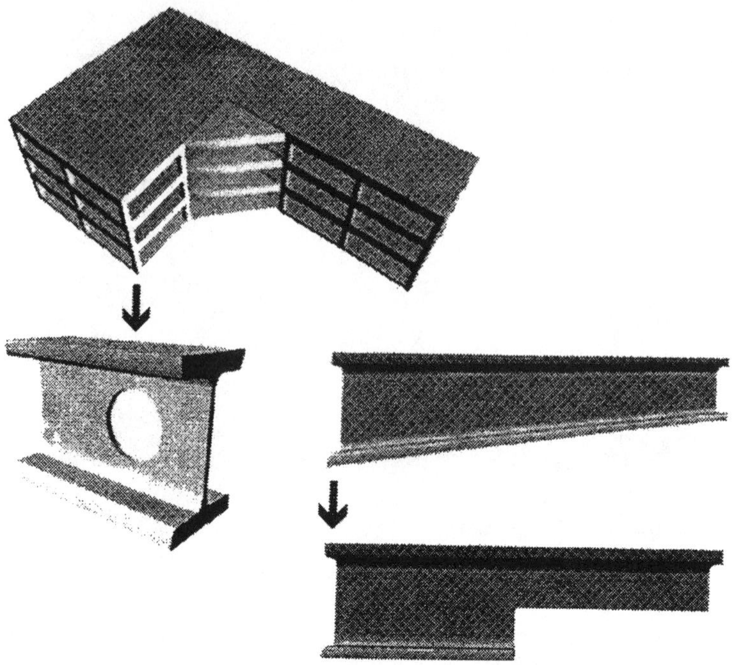

FIGURE 11.39 Relieving stress on a reentrant corner.

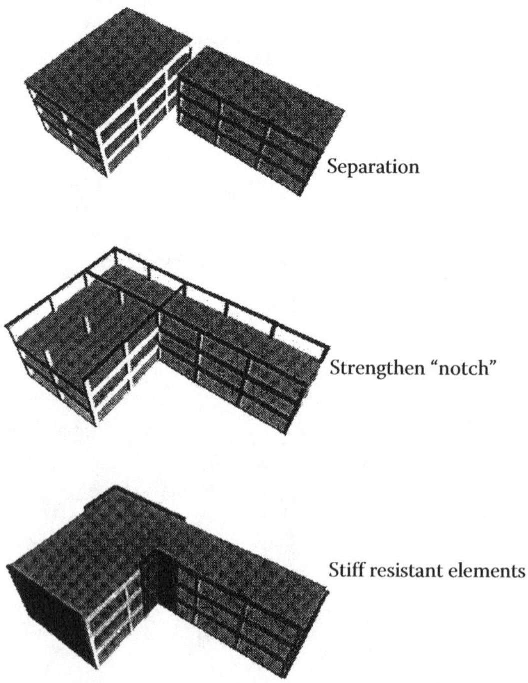

FIGURE 11.40 Solutions for reentrant corner condition.

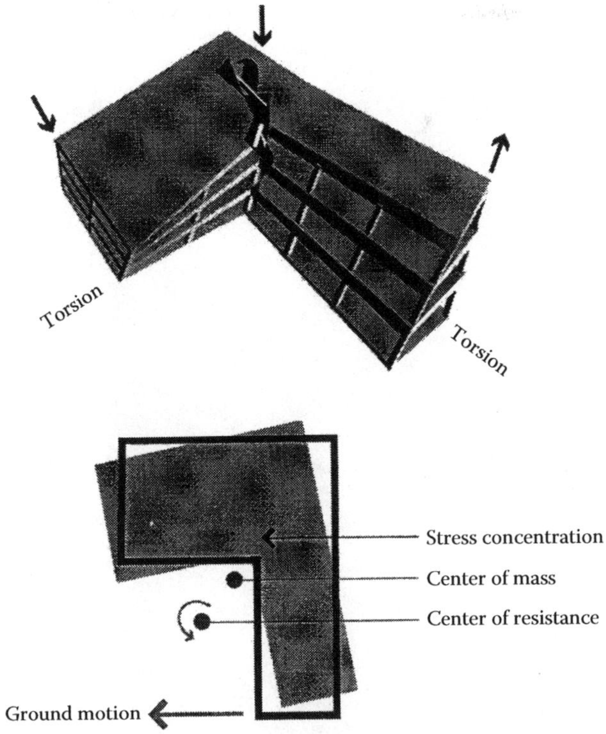

FIGURE 11.41 Reentrant corner conditions.

12 Steel Buildings
Bolted and Welded Connections, Gravity, and Lateral Load-Resisting Systems and Details

PREVIEW

This chapter starts with a discussion of joints in steel buildings using welds and/or bolts. The preparation and inspection required for both types of joints are discussed in some detail. Later sections explore odds and ends with particular reference to gravity and lateral load-resisting systems using self-explanatory tables and figures, and where required, additional *notes* are included along with the captions for the figures. Typical and special details are provided in the final section.

12.1 GENERAL CONSIDERATIONS FOR WELDS

Welded connections are used because of simplicity of design, fewer parts, less material, and decrease in shop handling and fabrication operations. Frequently, a combination of shop welding and field bolting is advantageous. With connection angle and/or plates shop welded to a beam, field connections can be made with high-strength bolts without the clearance problems that may arise in all-bolted connections.

Welded connections have a rigidity that can be advantageous if properly accounted for in design. Welded trusses, for example, deflect less than bolted trusses, because the end of a welded member at a joint cannot rotate relative to the other members there.

A disadvantage of welding, however, is that shrinkage of large welds must be considered. This is particularly important in large structures where there will be an accumulative effect.

Properly made, a weld is stronger than the base metal. Improperly made, even a good-looking weld may be worthless. Properly made, a weld has the required penetration and is not brittle.

In making a welded design, designers should specify only the amount and size of weld actually required. Generally, a 5/16 in. weld is considered the maximum size for a single pass. A 3/8 in. weld, while only 1/16 in. larger, requires three passes and results in an increase in cost.

The cost of fit-up for welding can range from about one-third to several times the cost of welding. In designing welded connections, therefore, designers should consider the work necessary for the fabricator and the erector in fitting members together so they can be welded.

The main types of welds used for structural steel are fillet, groove, plug, and slot. The most commonly used weld is the fillet. For light loads, it is the most economical, because little preparation of material is required. For heavy loads, groove welds are the most efficient, because the full strength of the base metal can be obtained easily. Use of plug and slot welds generally is limited to special conditions where fillet or groove welds are not practical.

More than one type of weld may be used in a connection. If so, the available strength of the connection is the sum of the available strengths of each type of weld used, separately computed with respect to the axis of the group. Refer to Figure 12.1 for stress–train curves.

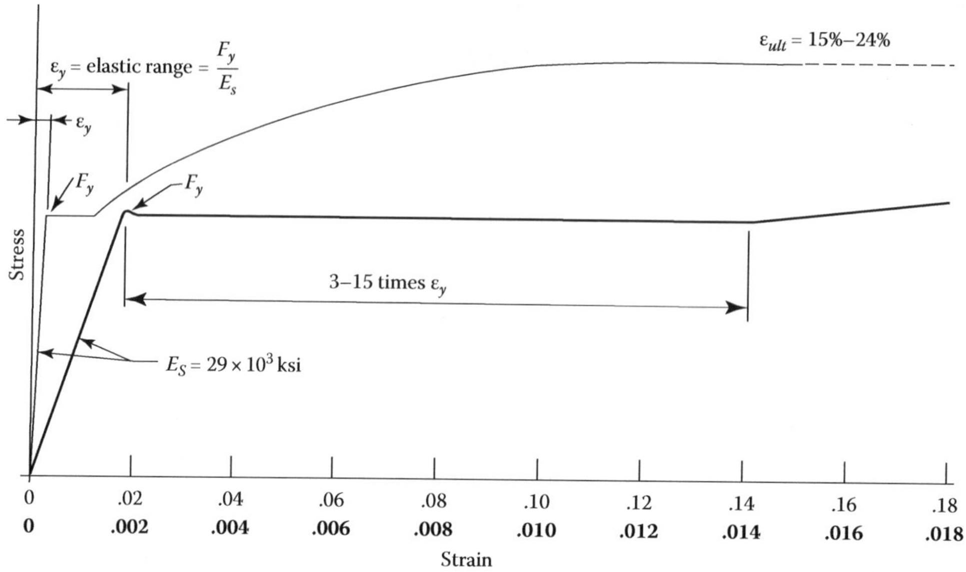

FIGURE 12.1 Idealized stress–strain curves over the entire range, structural steel. Bold values apply to magnified portion of curve (thicker line in graph).

Tack welds may be used for assembly or shipping. They are not assigned any stress-carrying capacity in the final structure. In some cases, these welds must be removed after final assembly or erection.

Fillet welds have the general shape of an isosceles right triangle (Figure 12.2). The size of the weld is given by the length of the leg. The strength is determined by the throat thickness, the shortest distance from the root (intersection of legs) to the face of the weld. If the two legs are unequal, the nominal size of the weld is given by the shorter of the legs. If welds are concave, the throat is diminished accordingly, and so is the strength.

Fillet welds are used to join two surfaces approximately at right angles to each other. The joints may be lap, tee or corner (Figure 12.3 a–c). Fillet welds also may be used with groove welds to reinforce corner joints. In a skewed tee joint, the included angle of weld deposit may vary up to a 30° from the perpendicular, and one corner of the edge to be connected may be raised up to 3/16 in. If the separation is greater than 1/16 in., the weld leg must be increased by the amount of the root opening.

Groove welds are made in a groove between the edges of two parts to be joined. These welds generally are used to connect two plates lying in the same plane (butt joint), but they also may be used for tee and corner joints.

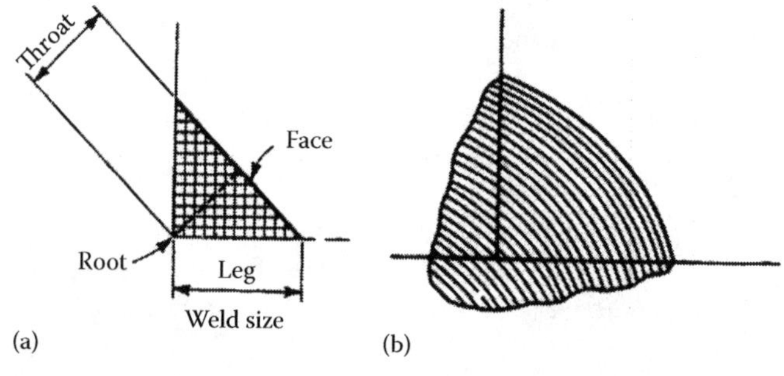

FIGURE 12.2 Fillet weld: (a) theoretical cross-section and (b) actual cross-section.

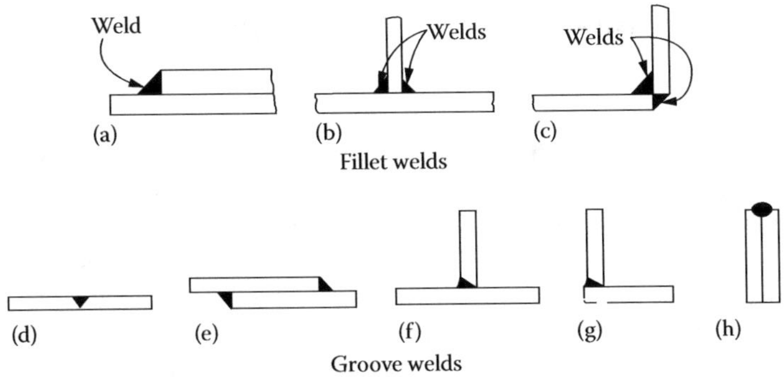

FIGURE 12.3 Types of welded joints: (a) lap joint, (b) tee joint, (c) corner joint, (d) butt-joint, (e) lap joint, (f) tee-joint, (g) corner joint, and (h) edge joint.

Standard types of groove welds are named in accordance with the shape given the edges to be welded: square, single V, double V, single bevel, double bevel, single U, double U, single J, and double J (Figure 12.3 d–h and Figure 12.4). Edges may be shaped by thermal cutting, arc-air gouging, or edge planning. Material up to 3/8 in. thick, however, may be groove welded with square-cut edges, depending on the welding process used.

Groove welds should extend the full width of the parts joined. Intermittent groove welds and butt joints not fully welded throughout the cross section are prohibited.

Groove welds also are classified as complete-penetration and partial-penetration welds.

In a *complete-joint-penetration weld*, the weld material and the base metal are fused throughout the depth of the joint. This type of weld is made by welding from both sides of the joint or from one

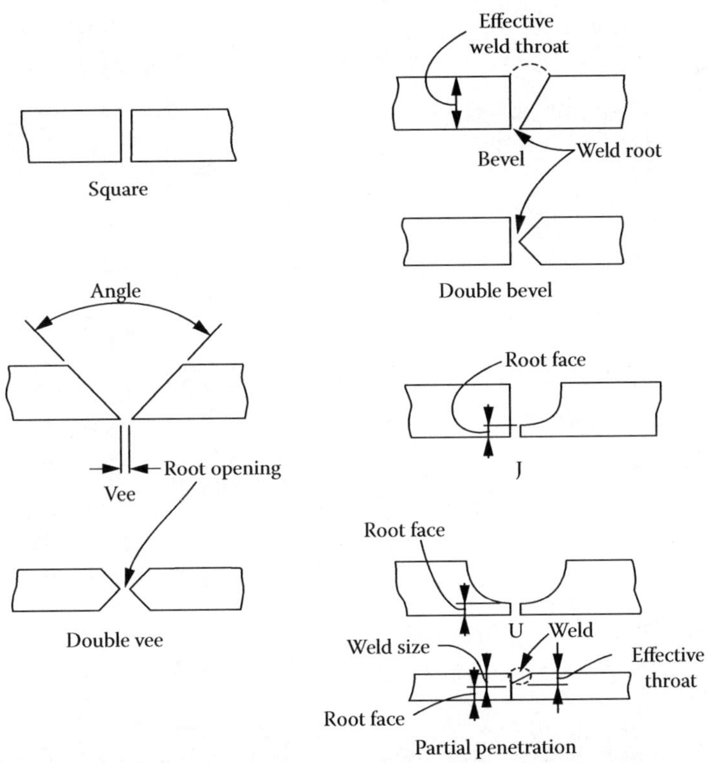

FIGURE 12.4 Groove weld.

side to a backing bar. When the joint is made by welding from both sides, the root of the first-pass weld is chipped or gouged to sound metal before the weld on the opposite side, or back pass, is made. The throat dimension of a complete-joint-penetration groove weld, for stress computations, is the full thickness of the thinner part joined, exclusive of weld reinforcement.

Partial-joint-penetration welds should be used when forces to be transferred are less than those requiring a complete-joint-penetration weld. The edges may not be shaped over the full joint thickness, and the depth of the weld may be less than the joint thickness (Figure 12.5). However, even if the edges are fully shaped, groove welds made from one side without a backing bar or made from both sides without back gouging are considered partial-joint-penetration welds. They are often used for splices in building columns carrying axial loads only.

Plug welds and *slot welds* are used to transmit shear in lap joints and to prevent buckling of lapped parts. In buildings, they also may be used to join components of built-up members. (Plug and slot welds, however, are not permitted on A514 steel.) The welds are made, with lapped parts in contact, by depositing weld metal in circular or slotted holes in one part. The openings may be partly or completely filled, depending on their depth. Load capacity of a plug or slot completely welded equals the product of hole area and allowable stress. Unless appearance is a main consideration, a fillet weld in holes or slots is preferable.

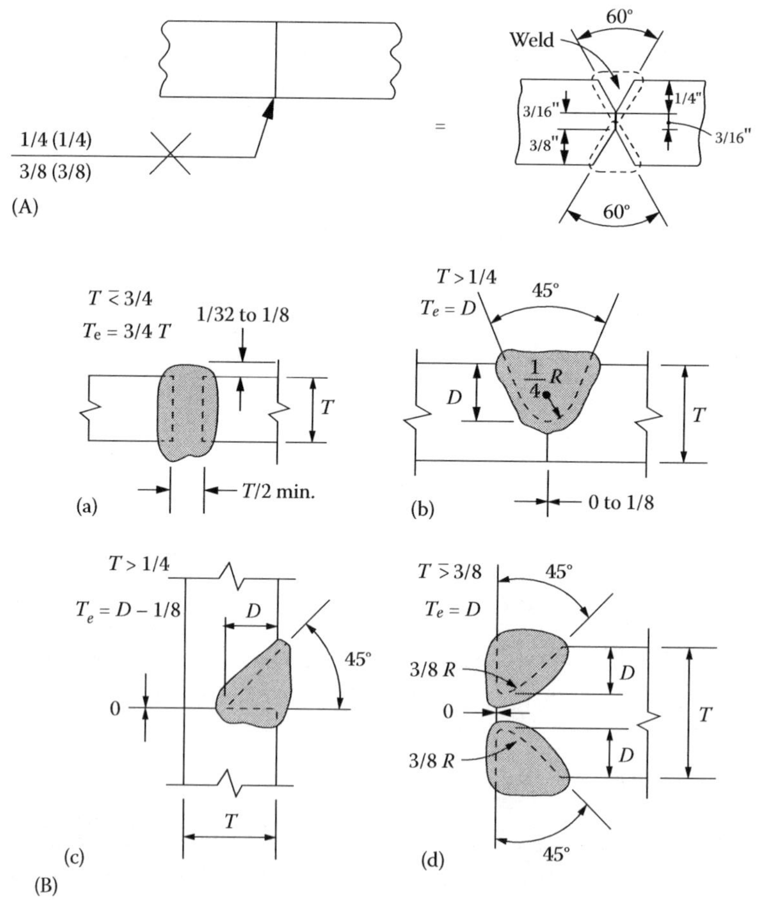

FIGURE 12.5 (A) Prequalified partial penetration joints (B) (a) Square-groove fillet butt, (b) single-U groove: butt or corner, (c) Single-bevel groove: butt, tee, or corner, (d) Double-J groove: butt, tee or corner.

12.2 METHODS OF WELDING INSPECTION

The five most commonly used methods for welding inspection discussed are

1. Visual testing (VT)
2. Penetrant testing (PT)
3. Magnetic partial testing (MT)
4. Ultrasonic testing (UT)
5. Radiographic testing (RT)

1. *Visual testing*

   Visual inspection provides the most economical way to check weld quality and is the most commonly used method. Joints are scrutinized prior to the commencement of welding to check fit-up, preparation bevels, gaps, alignment, and other variables. After the joint is welded, it is then visually inspected in accordance with AWS D1.1. If a discontinuity is suspected, the weld is either repaired or other inspection methods are used to validate the integrity of the weld. In most cases, timely visual inspection by an experienced inspector is sufficient and offers the most practical and effective inspection alternative to other, more costly methods.
2. *Penetrant testing*

   This test uses a red dye penetrant applied to the work from a pressure spray can. The dye penetrates any crack or crevice open to the surface. Excess dye is removed and a developer is sprayed on. Dye seeps out of the crack, producing a red image on the white developer.

   PT inspection can be used to detect tight cracks as long as they are open to the surface. However, only surface cracks are detectable. Furthermore, deep weld ripples and scratches may give a false indication when PT inspection is used.
3. *Magnetic-particle testing*

   A magnetizing current is introduced with contact prods into the weldment to be inspected. This induces a magnetic field in the work, which will be distorted by any cracks, seams, inclusions, etc., located on or near (within approximately 0.1 in. of) the surface. A dry magnetic powder, blown lightly on the surface by a rubber squirt bulb, will be picked up at such discontinuities, making a distinct mark. The magnetically held particles show the location, size, and shape of the discontinuity.

   Magnetic particle examination can be useful when a defect is suspected from visual inspection or when the absence of cracking in areas of high restraint must be confirmed. Relatively smooth surfaces are required for MT and, while cleanup is easy, demagnetization when necessary may not be.
4. *Ultrasonic testing*

   The ultrasonic (UT) inspection process is analogous to radar. A short pulse of high-frequency sound is broadcast from a crystal into a metal, after which the crystal waits to receive reflections from the far end of the metal member and from any voids encountered on the way through. The technique is called pulse echo. The sound beam is produced by a piezoelectric transducer energized by an electric current, which causes the crystal to vibrate and transmit through a liquid couplant into the metal. Any reflections are displayed as pips on a cathode-ray tube (CRT) grid whose horizontal scale represents distance through the metal. The vertical scale represents the strength (or area) of the reflecting surface.

   UT is a more versatile, rapid, and economical inspection method than radiography, but it does not provide a permanent record like the x-ray negative. The operator instead makes a written record of discontinuity indications appearing on his or her CRT. Certain joint geometry limits the use of the ultrasonic method.

Ultrasonic examinations have limited applicability. Relatively thin sections and variations in joint geometry can lead to difficulties interpreting the signals, although technicians with sufficient experience may be able to decipher UT readings in difficult geometries. Similarly, UT is usually not suitable for use with fillet welds and smaller PJP groove welds. CJP groove welds with and without backing bars also give readings that are subject to differing interpretations. Ultrasonic examination may be specified to validate the integrity of CJP groove welds that are subject to tension. Ultrasonic examination has largely replaced radiographic examination for the inspection of critical CJP groove welds.

5. *Radiographic testing*

   RT is basically an x-ray film process. To be detected by radiography, a crack must be oriented roughly parallel to the impinging radiation beam and occupy about $1^1/_2\%$ of the metal thickness along the beam. There are problems with radiography of fillets, tee, and corner joints, however, because the radiation beam must penetrate varying thicknesses.

   Precautions for avoiding radiation hazards interfere with shop work, and equipment and film costs make it the most expensive inspection method. Ultrasonic systems have gradually supplemented and even supplanted radiography.

Radiographic examination has very limited applicability in HSS fabrication, because of the irregular shape of common joints and the resulting variations in thickness of material as projected onto film. RT can be used successfully for butt splices but can only provide limited information about the condition of fusion at backing bars near the root corners. The general inability to place either the radiation source or the film inside the HSS means that exposures must usually be taken through both the front and back faces of the section with the film attached to the outside of the back face. Several such shots progressing around the member are needed to examine the complete joint.

12.3 FIELD TOLERANCES

Permissible variations from theoretical dimensions of an erected structure are specified in the AISC *Code of Standard Practice for Steel Buildings and Bridges*. It states that variations are within the limits of good practice or erected tolerance when they do not exceed the cumulative effect of permissible rolling and fabricating and erection tolerances. These tolerances are restricted in certain instances to total cumulative maximums.

The AISC *Code of Standard Practice* has a descriptive commentary that fully outlines and explains the application of the mill, fabrication, and erection tolerances for a building or bridge.

An example of tolerances that govern the plumbness of a multistory building is the tolerance for columns. In multistory buildings, columns are considered to be plumb if the error does not exceed 1:500, except for columns adjacent to elevator shafts and exterior columns, for which additional limits are imposed. The tolerances governing the variation of columns, as erected, from the theoretical centerline are sometimes wrongfully construed to be lateral-deflection (drift) limitations on the completed structure when, in fact, the two considerations are unrelated.

Structural steel is not static but moves due to varying ambient conditions and changing loads imposed during the construction process. Ambient conditions can be so extreme as to require final plumbing and span closing during nighttime hours.

Stability of the structure during construction and of each piece as it is lifted is considered by the erector. Pieces that are laterally supported and under a positive moment in service will frequently be unsupported and under a negative moment when they are raised, so precautions must be taken. Clearances for moving parts of lift equipment have to be monitored continually. Crane access and operating areas need to be capable of supporting superimposed loads.

Long slender members may have to be raised with a spreader beam. Others may have to be braced before the load line is released. Erection aids such as column lifting hitches must be designed and provided such that they will afford temporary support and allow easy access for assembly.

Full-penetration column splices are seldom necessary except on seismic moment frames but require special erection aids when encountered. Construction safety is regulated by the Federal Office of Safety and Health Administration (OSHA).

Multistory structures, or portions of multistory structures that lie within reach and capacity limitations of crawler cranes are usually erected with crawler cranes. For tall structures, a crawler crane places steel it can reach and then erects the guy derrick, which will continue erection. Alternatively, tower crawler cranes and climbing tower cranes are used extensively for multistory structures. Depending on height, these cranes can erect complete structures. They allow erection to proceed vertically, completing floors or levels for other trades to work on before the structure is topped out.

The sequence of placing the members of a multistory structure is, in general, columns, girders, bracing, and beams. The exact order depends on the erection equipment and type of framing.

In field-welded multistory buildings with continuous beam-to-column connections, the welded structure is not in its final alignment until beam-to-column connections are welded because of shrinkage caused by the welds. To accommodate the shrinkage, it is necessary to erect from the more restrained portion of the framing to the less restrained. If a structure has a braced center core, that area will be erected first to serve as a reference point, then steel will be erected toward the perimeter of the structure. If the structure is totally unbraced, an area in the center will be plumbed and temporarily braced for reference. Welding of column splices and beams is done after the structure is plumbed. The deck is attached for safety as it is installed, but final welding of deck and installation of studs and closures is completed after the tier is plumbed.

12.4 BRITTLE FRACTURE

Cracking of steel members or connections under moderate nominal stresses may be caused by brittle fracture. A brittle fracture may be defined as one that absorbs a relatively small amount of energy in propagation. Failure by cleavage of a large fraction of individual grains is the most common form in which brittle fracture occurs in structural steel. One way to prevent brittle fracture when using the allowable stress design method, is to reduce allowable stresses. This, however, may result in exceptionally low values for allowable stresses and expensive or impractical designs. The tendency to prevent brittle fracture by using special design details and types of steel that are sufficiently tough is usually a more fruitful approach to this problem.

12.4.1 Historical Review

As early as 1879, the problem of brittle fracture was recognized, and a discussion of the phenomenon took place at a meeting of the Iron and Steel Institute in the United States in 1886. The importance of brittle fracture in the design and subsequent service behavior of large structures such as ships and bridges, however, has been fully recognized only since the 1940s. The first recorded brittle fracture failure took place in October 1886. At Gravesend, Long Island, New York, during an acceptance test of a large riveted steel water standpipe, 1 in. thick plates in the lower section of the pipe cracked suddenly along a vertical crack of 10 ft length. In 1899, a large gas holder in New York City failed in brittle fracture. In January 1919, a molasses tank in Boston failed. Initial failures mainly took place in storage tanks or pipes, all of which are structures exposed to uniform high stresses. Until 1920, bridges were mostly riveted and comprised of several rather thick plates primarily subjected to uniaxial stress states.

In March 1938, the Hasselt Bridge in Belgium collapsed completely. The Duplessis Bridge in Canada collapsed in January 1951, 3 years after its construction in 30-degree-below-zero weather.

After the Zoo Brucke in Berlin collapsed in December 1939, it was concluded that high residual stresses caused by the welding of the extremely thick flange plates were the major cause. During World War II, brittle failures occurred in welded tankers and Liberty ships. In 1943, the S.S. Schenectady, an all-welded tanker, broke in two while being moored. The upper deck plate at that time was calculated to have a stress of only 10.8 ksi. During the next 10-year period, over 200 of these ships built during World War II collapsed. The majority of these ships were welded and most of the fractures started at points of high stress concentrations such as hatch corners.

12.4.2 Brittle Fracture Characteristics

The brittle fracture failures that took place in the aforementioned bridges and ships cannot be explained from overloading or discrepancies between actual and calculated stresses. From these low-level stresses, it appears that reducing the allowable stress is not the proper approach. Also, the use of more sophisticated and more detailed methods of analysis does not seem to provide a safer design against brittle fracture.

In most fractures, a lack of ductility of the fractured material after failure has been found to exist as compared with good initial ductility of the same material at the same temperature. This is one of the most significant characteristics of brittle fracture. However, many of the fractured materials did not possess an exceptionally high tensile strength or hardness, both of which are indicative of brittleness. Although in some instances fatigue created initial cracking, the kind of failure is different. Of all ships that have failed due to brittle fracture, only two authentic cases are known to be partially caused by fatigue.

When brittle fracture occurs in welded structures, the following common features have been noticed:

1. There existed residual stresses that in some portions of the welds caused triaxial tensions.
2. There were cracks in plate girders that occurred in areas where no plastic deformation had taken place, starting at the longitudinal welds (fillet or double-butt) between the web and the tensile flange.
3. The steels used were susceptible to aging and in some cases had large quantities of nonmetallic inclusions.
4. The plates were at least 3/4 in. thick.
5. Brittle fracture occurred not immediately after welding was completed but later under the action of a static external load approximately equal to that caused by the dead load of the structure and not at the maximum design load.
6. Brittle fracture in all cases occurred very suddenly.
7. Quite a number of failures occurred at low temperatures.
8. In most cases, no special heat treatment such as preheating was applied before welding.
9. The steels used had a limited capacity to absorb energy without crack initiation (notch sensitive).

For a given design, very seldom does only one failure criterion controls. Yielding deflections, dynamic response, instability, fatigue, and brittle fracture criteria could well apply to the design of a single structure. In most cases, a combination of these criteria has to be studied to achieve a safe design.

In spite of the fact that many different criteria could apply to any one design, the two major methods of structural steel design are the *allowable stress* and the *load resistance factor* design methods. Other methods need to be checked and modified to meet any additional criteria that are relevant.

The allowable stress method uses nominal local yielding as its strength criterion and makes use of allowable stresses by dividing the yield stress by a factor of safety. Although most building structures behave sufficiently linearly elastic under working loads, it should be recognized that this loading when multiplied with the same factor of safety quantitatively and qualitatively is different from the loading that would cause structural collapse. For reasons of economy and a better understanding of the real behavior of a steel structure, the plastic design method is superior to the allowable stress method. Its development during the past three decades has made great strides.

12.5 ASTM SPECIFICATIONS FOR STRUCTURAL SHAPES, PLATES AND BARS, AND FASTENERS

The materials and products used in building construction in the United States are almost universally designated by reference to an appropriate ASTM specification.

Figures 12.6 through 12.8 provide a summary of the common ASTM specifications used in steel building and construction, including structural shapes, plates, and fastening products. This information is based on similar and more extensive data in the of AISC's *Steel Construction Manual*, Fourteenth Edition.

12.6 THERMAL EFFECTS ON STRUCTURAL STEEL

Short-time elevated-temperature tensile tests indicate that the ratios of the elevated-temperature yield and tensile strengths to their respective room-temperature strength values are reasonably similar at any particular temperature for the various steels in the 300°F–700°F range, except for variations due to strain aging. (The tensile strength ratio may increase to a value greater than unity in the 300°F–700°F range when strain aging occurs.) Above this range, the ratio of elevated-temperature to room-temperature strength decreases as the temperature increases.

The composition of the steels is usually such that the carbon steels exhibit aging with attendant reduced notch toughness. The high-strength low-alloy and heat-treated constructional alloy steels exhibit less-pronounced or little strain aging.

As examples of the decreased ratio levels obtained at elevated temperature, the yield strength ratios for carbon and high-strength low-alloy steels are approximately 0.77 at 800°F, 0.63 at 1000°F, and 0.37 at 1200°F.

ASTM specification Standard Test Methods for Fire Tests of Building Construction and Materials outlines the procedures of fire testing of structural elements located inside a building and exposed to fire within the compartment or room in which they are located. The temperature criterion used requires that the average of the temperature readings not exceed 1000°F for columns and 1100°F for beams.

Steel buildings whose condition of exterior exposure and whose combustible contents under fire hazards will not produce a steel temperature greater than the foregoing criteria may therefore be considered fire resistive without the provisions of insulating protection for the steel.

A fire exposure of severity and duration sufficient to raise the temperature of the steel much above the fire test criteria temperature will seriously impair its ability to sustain loads permitted by the AISC specification. In such cases, the members upon which the stability of the structure depends should be insulated by fire-resistive materials or constructions capable of holding the average temperature of the steel to not more than that specified for the fire test standard.

The fire resistance rating is expressed as the time, in hours, that the assembly is able to withstand the fire exposure before the first critical point in its behavior is reached. These tests indicate the minimum period of time during which structural members, such as columns and beams, are capable of maintaining their strength and rigidity when subjected to the standard fire. They also establish the minimum period of time during which floors, roofs, walls, or partitions will prevent fire spread by protecting against the passage of flame, hot gases, and excessive heat.

To judge the effect of a fire on structural steel, it is necessary to consider what happens in such an exposure. Peculiarities of this exposure are (1) temperature attained by the steel can only be estimated, (2) time of exposure at any given temperature is unknown, (3) heating is uneven, (4) cooling rates vary and can only be estimated, and (5) the steel is usually under load and is sometimes restrained from normal expansion.

Steel that has been exposed to very high temperatures can be identified by very heavy scale, pitting, and surface erosion. Such temperatures may not only cause a loss of cross section but may also result in metallurgical changes. Normally these conditions will be accompanied by such severe deformation that the cost and difficulty of straightening such members, as compared to

Steel Type	ASTM Designation		F_y Min. Yield Stress (ksi)	F_u Tensile Stress[a] (ksi)	Applicable Shape Series									
												HSS		
					W	M	S	HP	C	MC	L	Rect.	Round	Pipe
Carbon	A36		36	58–80[b]	▨	■	■	▨	■	■	■			
	A53 Gr. B		35	60										■
	A500	Gr. B	42	58									■	
			46	58								■		
		Gr. C	46	62									▨	
			50	62								▨		
	A501	Gr. A	36	58								▨	▨	
		Gr. B	50	70								▨	▨	
	A529[c]	Gr. 50	50	65–100	▨	▨	▨	▨	▨	▨	▨			
		Gr. 55	55	70–100	▨	▨	▨	▨	▨	▨	▨			
High-strength low-alloy	A572	Gr. 42	42	60	▨	▨	▨	▨	▨	▨	▨			
		Gr. 50	50	65[d]	▨	▨	▨	■	▨	▨	▨			
		Gr. 55	55	70	▨	▨	▨	▨	▨	▨	▨			
		Gr. 60[e]	60	75	▨	▨	▨	▨	▨	▨	▨			
		Gr. 65[e]	65	80	▨	▨	▨	▨	▨	▨	▨			
	A618[f]	Gr. I & II	50[g]	70[g]								▨	▨	
		Gr. III	50	65								▨	▨	
	A913	50	50[h]	60[h]	▨	▨	▨	▨	▨	▨	▨			
		60	60	75	▨	▨	▨	▨	▨	▨	▨			
		65	65	80	▨	▨	▨	▨	▨	▨	▨			
		70	70	90	▨	▨	▨	▨	▨	▨	▨			
	A992		50–65[i]	65[i]	■	▨	▨	▨	▨	▨	▨			

FIGURE 12.6 Applicable ASTM specifications for various structural shapes.

(Continued)

Steel Type	ASTM Designation	F_y Min. Yield Stress (ksi)	F_u Tensile Stress[a] (ksi)	Applicable Shape Series: W	M	S	HP	C	MC	L	HSS Rect.	HSS Round	Pipe
Corrosion resistant high-strength low-alloy	A242	42[j]	63[j]	▒	□	□	□	□	□	□	□	□	□
		46[k]	67[k]	▒	□	□	▒	□	□	□	□	□	□
		50[l]	70[l]	▒	▒	▒	▒	▒	▒	▒	□	□	□
	A588	50	70	▒	▒	▒	▒	▒	▒	▒	□	□	□
	A847	50	70	□	□	□	□	□	□	□	▒	▒	□

■ Preferred material specification.

▒ Other applicable material specification, the availability of which should be confirmed prior to specification.

□ Material specification does not apply.

[a] Minimum unless a range is shown.

[b] For shapes over 426 lb/ft, only the minimum of 58 ksi applies.

[c] For shapes with a flange thickness less than or equal to 1.5 in. only. To improve weldability a maximum carbon equivalent can be specified (per ASTM Supplementary Requirement S78). If desired, maximum tensile stress of 90 ksi can be specified (per ASTM Supplementary Requirement S79).

[d] If desired, maximum tensile stress of 70 ksi can be specified (per ASTM Supplementary Requirement S91).

[e] For shapes with a flange thickness less than or equal to 2 in. only.

[f] ASTM A618 can also be specified as corrosion-resistant; see ASTM A618.

[g] Minimum applies for walls nominally ¾-in. thick and under. For wall thicknesses over ¾ in., $F_y = 46$ ksi and $F_u = 67$ ksi.

[h] If desired, maximum yield stress of 65 ksi and maximum yield-to-tensile strength ratio of 0.85 can be specified (per ASTM Supplementary Requirement S75).

[i] A maximum yield-to-tensile strength ratio of 0.85 and carbon equivalent formula are included as mandatory in ASTM A992.

[j] For shapes with a flange thickness greater than 2 in. only.

[k] For shapes with a flange thickness greater than 1.5 in. and less than or equal to 2 in. only.

[l] For shapes with a flange thickness less than or equal to 1.5 in. only.

FIGURE 12.6 (*Continued*) Applicable ASTM specifications for various structural shape.

Steel Type	ASTM Designation		F_y Min. Yield Stress (ksi)	F_u Tensile Stress[a] (ksi)	Plates and Bars: to 0.75 incl.	0.75 to 1.25 incl.	1.25 to 1.5 incl.	over 1.5 to 2 incl.	over 2 to 2.5 incl.	over 2.5 to 4 incl.	over 4 to 5 incl.	over 5 to 6 incl.	over 6 to 8 incl.	over 8
Carbon	A36		32	58–80										■
			36	58–80	■	■	■	■	■	■	■	■	■	
	A529	Gr. 50	50	70–100	▒	▒ b	▒ b	▒ b	▒ b					
		Gr. 55	55	70–100	▒	▒ b	▒ b							
High-strength low-alloy	A572	Gr. 42	42	60	▒	▒	▒	▒	▒	▒	▒	▒		
		Gr. 50	50	65	▒	▒	▒	▒	▒	▒				
		Gr. 55	55	70	▒	▒	▒	▒						
		Gr. 60	60	75	▒	▒								
		G4. 65	65	80	▒	▒								
Corrosion resistant high-strength low-alloy	A242		42	63				▒	▒	▒				
			46	67		▒	▒							
			50	70	▒									
	A588		42	63								▒	▒	
			46	67							▒			
			50	70	▒	▒	▒	▒	▒	▒				
Quenched and tempered alloy	A515[c]		90	100–130						▒	▒	▒		
			100	110–130	▒	▒	▒	▒	▒					
Quenched and tempered Low-alloy	A852[c]		70	90–110	▒	▒	▒	▒	▒	▒				

■ Preferred material specification.

▒ Other applicable material specification, the availability of which should be confirmed prior to specification.

□ Material specification does not apply.

[a] Minimum unless a range is shown.

[b] Applicable to bars only above 1-in. thickness.

[c] Available as plates only.

FIGURE 12.7 Applicable ASTM specifications for plates and bars.

ASTM Designation		F_y Min. Yield Stress (ksi)	F_u Tensile Stress[a] (ksi)	Diameter Range (in.)	High-Strength Bolts: Conventional	High-Strength Bolts: Twist-Off-Type Tension-Control	Common Bolts	Nuts	Washer	Direct-Tension-Indicators	Threaded Rods	Shear Stud Connectors	Anchor Rods: Hooked	Anchor Rods: Headed	Anchor Rods: Threaded & Nutted
A108		—	60	0.375–0.75 incl.								■			
A325[d]		—	105	over 1–1.5 incl.	■										
		—	120	0.5–1 incl.	■										
A490[d]		—	150	0.5–1.5	■										
F1852[d]		—	105	1.125		■									
			120	0.5–1 incl.		■									
F2280[d]		—	150	0.5–1.125 incl.		■									
A194 Gr. 2H		—	—	0.25–4				▒							
A563		—	—	0.25–4				■							
F436[b]		—	—	0.25–4					■						
F959		—	—	0.5–1.5						■					
A36		36	58–80	to 10							■		▒		▒
A 193 Gr. B7[e]		—	100	over 4–7							▒		▒		▒
		—	115	over 2.5–4							▒		▒		▒
		—	125	2.5 and under							▒		▒		▒
A307 Gr. A		—	60	0.25–4			■								
A354 Gr. BD		—	140	2.5–4 incl.							▒		▒	▒	▒
		—	150	0.25–2.5 incl.							▒		▒	▒	▒
A449		—	90	1.75–3 incl.	c						▒		▒	▒	▒
		—	105	1.125–1.5 incl.	c						▒		▒	▒	▒
		—	120	0.25–1 incl.	c						▒		▒	▒	▒
A572	Gr. 42	42	60	to 6							▒		▒		▒

FIGURE 12.8 Applicable ASTM specifications for various types of structural fasteners.

(Continued)

ASTM Designation		F_y Min. Yield Stress (ksi)	F_u Tensile Stress[a] (ksi)	Diameter Range (in.)	High-Strength Bolts: Conventional	High-Strength Bolts: Twist-Off-Type Tension-Control	Common Bolts	Nuts	Washer	Direct-Tension-Indicators	Threaded Rods	Shear Stud Connectors	Anchor Rods: Hooked	Anchor Rods: Headed	Anchor Rods: Threaded & Nutted
	Gr. 50	50	65	to 4							▒		▒		▒
	Gr. 55	55	70	to 2							▒		▒		▒
	Gr. 60	60	75	to 1.25							▒		▒		▒
	Gr. 65	65	80	to 1.25							▒		▒		▒
A588		42	63	over 5–8 incl.							▒		▒		▒
		46	67	over 4–5 incl.							▒		▒		▒
		50	70	4 and under							▒		▒		▒
A687		105	150 max.	0.625–3							▒		▒		▒
F1554	Gr. 36	36	58–80	0.25–4									■	■	■
	Gr. 55	55	75–95	0.25–4									■	■	■
	Gr. 105	105	125–150	0.25–3									▒	▒	▒

■ Preferred material specification.

▒ Other applicable material specification, the availability of which should be confirmed prior to specification.

☐ Material specification does not apply.

[a] Minimum unless a range is shown or maximum (max.) is indicated.

[b] Special washer requirements may apply per RCSC *Specification* Table 6.1 for some steel-to-steel bolting applications and per part 14 for anchor-rod applications.

[c] See AISC *Specification* Section J3.1 for limitations on use of ASTM A449 bolts.

[d] When atmospheric corrosion resistance is desired, Type 3 can be specified.

[e] For anchor rods with temperature and corrosion resistance characteristics.

— Indicates that a value is not specified in the material specification.

FIGURE 12.8 (*Continued*) Applicable ASTM specifications for various types of structural fasteners.

replacement, dictate that they be discarded. Steel members that have suffered rapid cooling will usually be so severely distorted that straightening for reuse will seldom be considered practicable.

In some cases, there may be some deformation in members whose normal thermal expansion is inhibited or prevented by the nature of the construction. Such members may usually be straightened and reused.

12.6.1 Effect of Heat due to Welding

Application of heat by welding produces residual stresses that are generally accompanied by distortion of various amounts. Both the stresses and distortions are minimized by controlled welding procedure and fabrications methods. Procedures normally followed include (1) proper positioning of the components of joints before welding, (2) selection of welding sequences determined by experience, (3) deposition of a minimum volume of weld metal with a minimum number of passes for the design condition, and (4) preheating as determined by experience (usually above the specified minimums).

12.6.2 Use of Heat to Straighten, Camber, or Curve Members

With modern fabrication techniques, a controlled application of heat can be effectively used to either straighten or to intentionally curve structural members. By this process, the member is rapidly heated in selected areas; the heated areas tend to expand but are restrained by adjacent cooler areas. This action causes a permanent plastic deformation or *upset* of the heated areas and, thus, a change of shape is developed in the cooled member.

Heat straightening is used in both normal shop fabrication operations and in the field to remove relatively severe accidental bends in members. Conversely, *heat cambering* and *heat curving* of either rolled beams or welded girders are examples of the use of heat to affect a desired curvature.

As with many other fabrication operations, the use of heat to straighten or curve will cause residual stresses in the member as a result of plastic deformations. These stresses are similar to those that develop in rolled structural shapes as they cool from the rolling temperature; in this case, the stresses arise because all parts of the shape do not cool at the same rate. In a like manner, welded members develop residual stresses from the localized heat of welding.

In general, the residual stresses from heating operations do not affect the ultimate strength of structural members. Figure 12.9 shows a variety of residual stresses.

The mechanical properties of steels are largely unaffected by heating operations provided that the maximum temperature does not exceed 1100°F for quenched and tempered alloy steels and 1300°F for other steels.

12.6.3 Coefficient of Expansion

The average coefficient of expansion for structural steel between 70°F and 100°F is 0.0000065 for each degree. For temperatures of 100°F to 1200°F, the coefficient is given by the approximate formula

$$\varepsilon = (6.1 + 0.00190t) \times 10^{-6}$$

in which ε is the coefficient of expansion for each degree Fahrenheit and t is the temperature in degree Fahrenheit.

The modulus of elasticity of structural steel is approximately 29,000 ksi at 70°F. It decreases linearly to about 25,000 ksi at 900° and then begins to drop at an increasing rate at higher temperatures.

The coefficient of linear expansion (e) is the change in length, per unit of length, for a change of 1° of temperature. The coefficient of surface expansion is approximately two times the linear coefficient, and the coefficient of volume expansion, for solids, is approximately three times the linear coefficient.

A bar, free to move, will increase in length with an increase in temperature and will decrease in length with a decrease in temperature. The change in length will be εtl, where ε is the coefficient of linear expansion, t the change in temperature, and l the length. If the ends of a bar are fixed, a

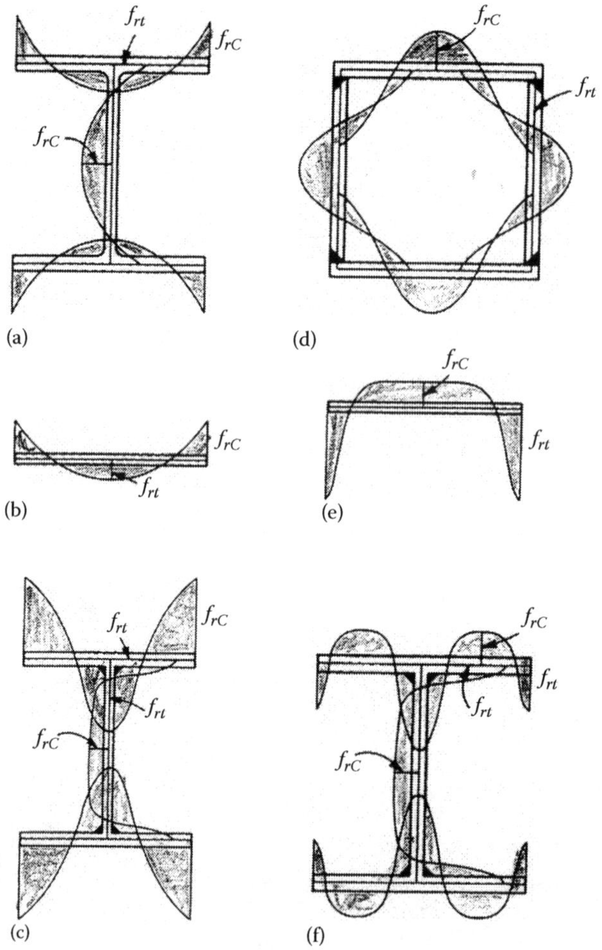

FIGURE 12.9 Residual stresses: (a) WF, (b) rolled plate, (c) H-section form UM plates, (d) welded box, (e) frame-cut plate, and (f) H-section from plane-cut plates.

change in temperature (t) will cause a change in the unit stress of $E\varepsilon t$, and in the total stress of $AE\varepsilon t$, where A is the cross-sectional area of the bar and E the modulus of elasticity.

Example 1

A piece of medium steel is exactly 40 ft long at 60°F. Find the length at 90°F assuming the ends are free to move.

Change of length—$\varepsilon tl = .00065 \times 30 \times 40/100 = .0078$ ft.

The length at 90°F is 40.0078 ft.

Example 2

A piece of medium steel is exactly 40 ft long and the ends are fixed. If the temperature increases 30°F, what is the resulting change in the unit stress?

Change in unit stress = $E\varepsilon t = 29{,}000{,}000 \times .00065 \times 30/100 = 5{,}655$ lb/in.^{2}

Note: The length of 40 ft is not used in calculating the stresses.

Structural steels lose strength and stiffness in a hurry when subjected to temperatures in excess of 800°F. From a design perspective, it is not necessary to quantify this strength loss, only guard against any significant likelihood of its occurrence. This is usually done by insulating the steel (fireproofing) or by reducing heat produced in fires (sprinklers). Heating or welding structural shapes that are supporting loads must be undertaken with care. The differential heating of steel shapes may cause permanent deformation.

Anticipated thermal variations must also be considered by the designers in developing the design for a long building. Long buildings are typically thermally relieved by the introduction of expansion joints.

Expansion joints have been successfully located at intervals of as much as 450 ft, though 200–250 ft, intervals are probably more common. Thermal changes will either cause a piece of steel to change dimension or, when change of dimension is inhibited, they will induce stresses in the member.

The spacing of expansion joints will then depend on how much variation in material temperature is expected and the extent to which dimensional changes or induced strains can be accommodated. The frequency of the thermal cycles will also be important. Exposed steel structures will experience daily fluctuations in temperature, which can be extreme in desert environments, while in severe climates, the seasonal change may be the primary concern. Steel frameworks are usually enclosed in climatized space and therefore once enclosed and climatized will experience only minor thermal variation and greater distances between expansion joints are reasonable.

The daily cyclic heating and cooling of steel framework during construction will often become a problem. If not anticipated, this daily thermal movement can impact the work of other trades as well as the field fit-up of the steel framework itself. Longitudinally flexible connections are often introduced into buildings to allow for a more localized movement of the steel framework during construction.

The thermal expansion coefficient for structural steel is 0.0000065°F. A 60 ft long beam subjected to a 100°F increase in temperature will expand 0.468 in.:

$$\Delta L = 0.0000065 \times 100 \times 60 \times 12 = 0.468 \text{ in.}$$

The unit stress induced in a member that is constrained and subjected to a thermal change of 100°F is almost 20,000 psi:

$$\begin{aligned} F &= \varepsilon E t \\ &= 0.0000065 \times 100 \times 29{,}000{,}000 \\ &= 18{,}850 \text{ psi} \end{aligned}$$

Observe that the induced stress is not a function of the length; therefore, one should avoid constraining even short steel members, which will be subjected to thermal variations.

The Federal Construction Council publication in 1974 provides guidance based on design temperature change for maximum spacing of structural expansion joints in beam-and-column-framed buildings with pinned column bases and heated interiors. The report includes data for numerous cities and gives five modification factors to be applied as appropriate:

1. If the building will be heated only and will have pinned column bases, use the maximum spacing as specified.
2. If the building will be air-conditioned as well as heated, increase the maximum spacing by 15%, provided the environmental control system will run continuously.

3. If the building will be unheated, decrease the maximum spacing by 33%.
4. If the building will have fixed column bases, decrease the maximum spacing by 15%.
5. If the building will have substantially greater stiffness against lateral displacement in one of the plan dimensions, decrease the maximum spacing by 25%.

When more than one of these design conditions prevails in a building, the percentile factor to be applied is the algebraic sum of the adjustment factors of all the various applicable conditions. Most building codes include restrictions on location and maximum spacing of fire walls, which often become default locations for expansion joints.

The most effective expansion joint is a double line of columns that provides a complete and positive separation. Alternatively, low-friction sliding elements can be used. Such systems, however, are seldom totally friction-free and will induce some level of inherent restraint to movement.

12.7 BOLTED CONNECTIONS

Bolts used in building construction are described by three basic ASTM designations, A307, A325, and A490. ASTM A307 bolts are commonly referred to as machine or black bolts and are made from carbon steel and have a minimum tensile strength of 60 ksi. The A307 bolt, which is available in diameters up to 4 in., has been the most commonly used bolt especially in small or low-rise buildings. Since 1951, however, high-strength bolts have become increasingly popular, especially in taller buildings. High-strength bolts are produced from two types of steel, quenched and tempered carbon steels, which possess a minimum tensile strength of between 105 and 120 ksi that is used to produce ASTM A325 bolts. An alloy steel that is quenched and tempered and capable of delivering a minimum tensile strength of 150 ksi is used to produce ASTM 490 bolts. High-strength bolts are available in diameters of ½–1½ in. inclusive. Standards for bolt holes are displayed in Figure 12.10.

Four basic load transfer mechanisms can be developed through bolted connections. The strength of the bolt itself in either *shear* or *tension* may serve to limit the load that may be transferred. The strength of the bolted assembly can also be a function of the *bearing* of the bolt on the members being joined. High-strength bolts can and often are pretensioned, thereby creating a *friction* load path between the two connected parts.

The four load transfer mechanisms are as follows:

1. Bolts in tension
2. Bolts in shear
3. Bolts in bearing
4. Slip-critical bolted connections

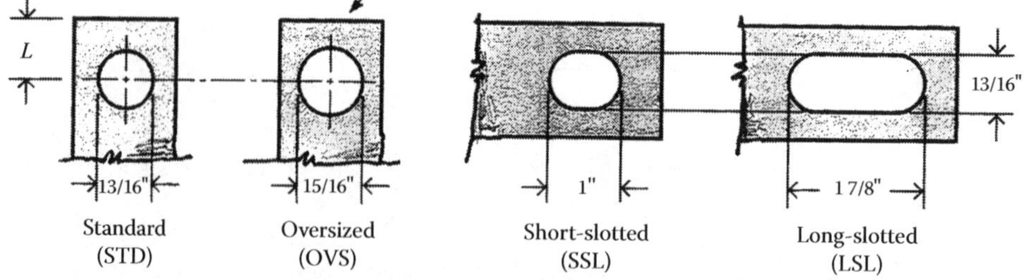

FIGURE 12.10 Bolt holes for ¾ in. diameter Bolt.

12.7.1 Bolts in Tension

The strength of a bolt in tension as described in AISC design specifications is the product of the nominal strength of the material (F_t) and the nominal area of the bolt (A_b):

$$T_n = A_b F_t$$

The factored or ultimate capacity of a bolt in tension is

$$T_u = \varphi A_b F_1$$

$$\varphi = 0.75$$

Upset threads (Figure 12.11) are occasionally used along tensile load paths to ensure that yielding occurs in the rod or bolt and not in the thread. The ultimate strength of an upset rod ($A_k > A_b$) is based on the actual tensile strength of the material at yield (F_y). Since yielding might occur, upset thread assemblies are usually constructed of more ductile material (i.e., A36 as opposed to A307). The capacity reduction factor for an upset road of A36 material is also based on tensile yield ($\varphi = 0.9$). Accordingly, the ultimate strength (φT_n) of a $^3/_4$ in. diameter A36 upset rod is

$$P_u = T_u = A_g F_y \varphi = 0.4418(36)0.9$$

$$= 14.4 \text{ kips}$$

12.7.2 Bolts in Shear

Bolt strength in shear is a function of the area sheared shown in Figure 12.12. As in tension bolts, the nominal shear strength and nominal area approach are used to develop capacities, and it is presumed that threads will exist on the shear plane of A307 bolts because most of the bolt is threaded. For example, the nominal strength in shear of an A307 bolt is 27 ksi (0.6 × 45) or 60% of its nominal capacity in tension. The bolt shown in Figure 12.12 must shear along two planes for the connector to fail; hence, it is referred to as being in double shear. When only one shear plane is present, the bolt is said to act in single shear. The resistance factor (φ) in shear is treated as a variable being 0.6 for an A307 and 0.65 for an A325 or A490 bolt. Accordingly, the ultimate shear

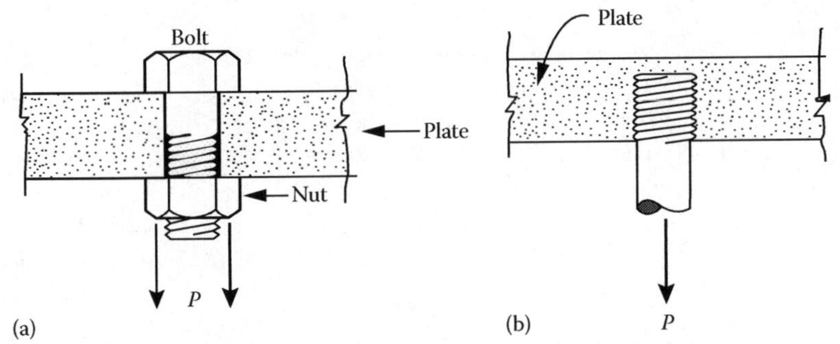

FIGURE 12.11 Bolt in tension: (a) standard bolt; (b) upset thread in tension rod.

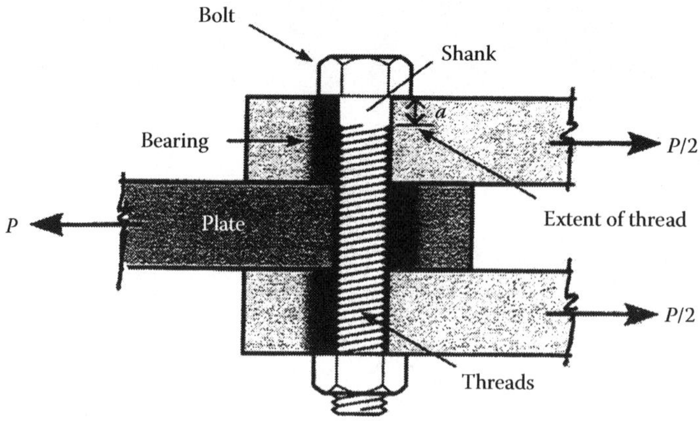

FIGURE 12.12 Bolt in double shear.

strength of a $^3/_4$ in. diameter bolted connector based on the capacity of the A307 bolt in double shear through the threads is

$$
\begin{aligned}
V_u &= 2\varphi A_b F_v \\
&= 2(0.60)\,(0.4418)27 \\
&= 14.3 \text{ kips}
\end{aligned}
$$

Bolts of a specified thread length can be purchased. The threading on the bolt shown in Figure 12.12 can be limited so as to ensure that neither shear plane passes through the thread. A307 bolts are not fabricated to special thread lengths, so when shear planes pass through unthreaded portions of an A307 bolt, special account is seldom taken.

A325 and A490 bolts are fabricated to a specified thread length. Therefore, in calculating the ultimate shear strength, it is permitted to recognize the actual sheared area. A designer may specify that the threads be excluded (X) from the shear plane or that it is acceptable for the thread to exist within the shear plane (N). For example, the shear transfer capacity (V_u) of $^3/_4$ in. diameter bolts in double shear would be

A325N = 31 kips A490N = 39 kips
A325X = 39 kips A490X = 48 kips

Observe that *N* bolts have 75% (A_e/A_b) of the capacity of *X* bolts.

12.7.3 Bolts in Bearing

The bolt shown in Figure 12.12 must bear on the connected plates. The permitted bearing stress (ASD) or bearing resistance (LRFD) is a function of several factors. First, the type of bolt hole must be considered. A standard bolt hole is $^1/_{16}$ in. larger than the specified diameter of the bolt. Alternative hole sizes are described as being OVS, short slotted (SSL), or long slotted (LSL). Connectors that rely on bearing to transfer a load must bear on standard length holes and this includes the long side of SSL and LSL holes. Shown in Figure 12.10 are various bolt-hole sizes for a $^3/_4$ in. diameter bolt. When the bolt bears on any of these holes except for the long side (1 7/8 in.) of an LSL hole, the nominal bearing resistance is

$$R_n = 2.4\, d\, t\, F_u$$

where

d is the nominal diameter of the bolt
t is the thickness of the plate

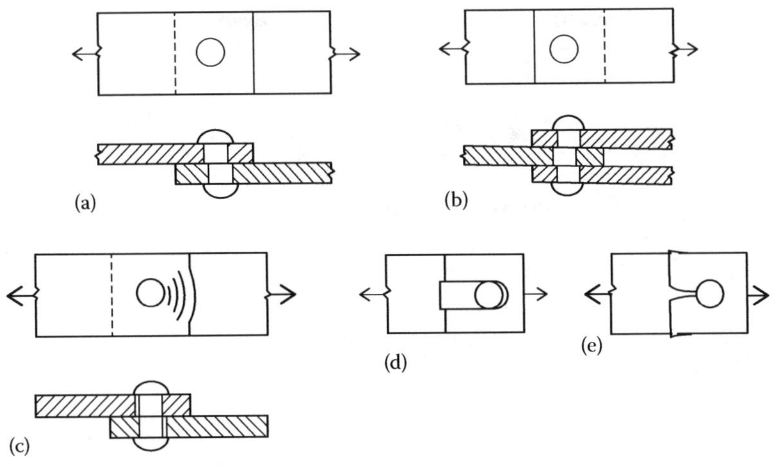

FIGURE 12.13 Failure mode, bolted connection: (a) bolt single-shear failure, (b) bolt double-shear failure, (c) bearing failure, (d) plate edge failure, and (e) transverse tension failure.

12.7.4 Slip-Critical Bolted Connection

In this type of connection, friction is used as a load path. It has several advantages. A friction-type connector need not slip to the edge of the bearing surface in order to develop a load transfer mechanism in the bolt. Hence, this type of connection is well suited where loads will fluctuate that may reverse the direction of shear. Since the connected plates do not move, a friction load path may be used in conjunction with welding to develop a connector.

A325 and A490 bolts may be used to develop a friction load path by pretensioning the bolt. Slip-critical connections are identified as an A325SC or A490SC bolt if the intent is to have the bolt pretensioned to specified minimums. The basic design approach is to develop the friction load path so that anticipated, unfactored, loads can be transferred by friction. Slip in a slip-resistant connector is expected to occur at about 1.4–1.5 times the nominal slip resistance. Nominal slip resistance is usually provide to at least 75% of the combined service load in joints, which are likely to be subjected to seismic or wind loads.

The development of a slip-critical connection requires a two-level approach. First, the connector must be developed so that it will not slip when nominal loads are applied. Second, the factored shear and bearing strength of the connector must exceed the factored load.

See Figure 12.13 for a summary of failure modes in bolted connections.

12.8 BEARING VERSUS SLIP-CRITICAL CONNECTIONS

Bolts in structural steel connections are commonly referred to as being either bearing bolts or slip-critical bolts. This is a misnomer, since the same high-strength bolts can be used for both bearing and slip-critical connections, though common bolts (A307) are restricted to use in bearing connections. In bearing connections, movement within the joint is prevented through contact between the shank of the bolt and the material. In slip-critical connections, movement of the joint is resisted through the friction between the faying surfaces caused by the tension in the bolt. Therefore, though either snug-tight or fully tensioned bolts may be installed in bearing-type connections, only fully tensioned bolts may be installed in slip-critical connections.

By definition, slip-critical connections are required where slip cannot be tolerated, which would seem to be a definitive statement, but in reality, there is a range of tolerance to slip. This range can be divided into two distinct levels, serviceability and strength.

Designing a slip-critical connection for serviceability can be viewed in two ways. It can be viewed as designing for a lower safety factor against slip than when designing as a strength limit state. This is analogous to what is done with sway due to wind loads in tall buildings. In such buildings, the limit states that govern the strength and stability of the structure are designed to perform satisfactorily for a 50- or 100-year storm, while the sway is limited based on an 8- to 10-year storm. Because of this approach, all limit states for bearing-type connections must also be checked when designing against slip as a serviceability limit state.

Slip criticality should be considered as a strength limit state where slip in the connection could be large enough to alter the usual analysis assumption that the undeformed structure can be used to calculate the internal forces. Examples might include braced frames where oversized (OVS) holes are used, which could potentially result in excessively large *P*-delta effects or long-span roof trusses with OVS holes, where slip could result in excessively large deflections.

Slip criticality should be considered as a serviceability limit state where slip in the connection would not violate the analysis assumptions of the structure. Examples might include structures that contain sensitive communication or testing equipment, where slip is undesirable but would not result in structural failure. Slip should also be viewed as a serviceability limit state when slip-critical connections are used for joints subjected to fatigue load with reversal of the loading direction.

Slip-critical connections are required for very few situations in building design. The specification requires the use of slip-critical connections for the following conditions:

1. Joints that are subject to fatigue load with reversal of the loading direction.
2. Joints that utilize OVS holes.
3. Joints that utilize slotted holes, except those with applied load approximately normal (within 80°–100°) to the direction of the long dimension of the slot.
4. Joints in which slip at the faying surface would be detrimental to the performance of the structure.

Items 1, 2, and 3 are quantitative. Item 4 is qualitative and requires judgment. Specifying slip-critical connections where bearing connections would suffice leads to uneconomical designs, usually with no accompanying increase in the overall safety of the structure.

It should be noted that wind and seismic loads do not produce fatigue loads that would require the use of slip-critical connections. The A1SC specification states: "Fatigue need not be considered for seismic effects or for the effects of wind loading on normal building lateral force-resisting systems and building enclosure components." This is because most such load changes occur only a small number of times or produce only minor stress fluctuations. The occurrence of full design wind or earthquake loads is too infrequent to warrant consideration in fatigue design. On the other hand, crane runways and supporting structures for machinery and equipment are often subjected to fatigue loading conditions.

12.8.1 Bolt Installation: Snug-Tight versus Fully Tensioned

High-strength bolts (A325 and A490) can be installed either snug-tight or fully tensioned. Common bolts (A307) can only be installed snug tight. Snug-tight installation is achieved when all plies are in contact. It can be attained by a few impacts of an impact wrench or the full effort of a person using an ordinary spud wrench. Fully tensioned installation is achieved when the bolt is stressed in tension to 70% of its tensile strength. The specification requires fully tensioned installation for the following conditions:

1. Column splices in all tier structures 125 ft or more in height.
2. Connections of all beams and girders to columns and any other beams and girders on which the bracing of columns is dependent, in structures over 125 ft in height.

3. In all structures carrying cranes of over 5-ton capacity, roof truss splices and connections of trusses to columns, column splices, column bracing, knee braces, and crane supports.
4. Connections for support of running machinery or of other live loads that produce impact or reversal of stress.
5. Joints that are subjected to significant load reversal.
6. Joints that are subjected to fatigue load with no reversal of the loading direction.
7. Joints with ASTM A325 or F1852 bolts that are subject to tensile fatigue.
8. Joints with ASTM A490 bolts that are subject to tension or combined shear and tension, with or without fatigue.

Fully tensioned bolts can be installed using four different methods: calibrated wrench, turn-of-nut, twist-off-type tension-control bolts, or with direct-tension indicators. In all installation methods, the plies are first brought together as in a snug-tight condition, before tensioning begins. Bolts are tensioned starting with the most rigid element and moving to the most flexible element, to minimize relaxation in the previously tensioned bolts. The calibrated-wrench method is a torque-controlled method, in which the wrench is calibrated to stop torquing after the required tension is achieved in the bolt. An ASTM F436 washer must be used under the turned element, and the unturned element must be prevented from turning. The wrench should be set to cut off at 5% above the required tension. Because the torque-controlled methods of installation rely on so many variables for proper performance, it is imperative that the wrench be calibrated at least daily, when changes occur in the bolting setup (including changes in bolt diameter and hose length), or when a number of wrenches run off the same air supply. It is also important that fasteners be kept protected from dirt and moisture to ensure that the proper tension is achieved.

In the turn-of-nut method, the specified tension is achieved by turning the nut a specified rotation while the unturned element is prevented from turning.

Twist-off-type tension-control bolts consist of a splined end that extends beyond the threaded portion of the bolt. The splined end is held in place by the wrench during installation, so that the nut turns relative to the bolt. When the specified tension is achieved, the splined end is severed and rotation stops. An ASTM F436 washer must be provided under the nut. Like the calibrated-wrench method, the twist-off-type tension-control bolts behave as a torque-controlled installation method. However, since the torque is controlled within the fastener, the variability of the wrench and power supply is eliminated. Nevertheless, it is still important that fasteners be kept protected from dirt and moisture to ensure the proper tension is achieved. If the splined end is severed during the first step of installation, when the plies are being brought into contact, the fastener must be removed and replaced.

Direct-tension indicators are hardened washer-shaped disks with arched protrusions that flatten when the specified tension is achieved. The protrusions must bear against the bolt head or nut or against a hardened flat washer. If the protrusions flatten to the job-inspection gap while the connection is being brought into the snug-tight condition, the direct-tension indicator must be removed and replaced.

12.9 BOLTS SUBJECTED TO SHEAR AND TENSION

Tests have shown that the strength of bolts subjected to combined shear and tension can be described by an elliptical expression. The linear approximation of this ellipse shown therein is the relationship used in the LRFD specification to limit the capacity of bolts subjected to both tension and shear. It is clear that unless both shear and externally induced tensile stresses are high, the interactive effects will not alter conclusions based on tension or shear acting alone.

13 Composite Buildings
Structural System and Details

PREVIEW

The first composite structure, believe it or not, was cast iron with masonry back up for stability. This cast iron facade became decorative art and was related to the robust cast iron columns on the interior. The practice of composite construction today includes using steel deck composite with studded beams as well as concrete-filled tubular sections and concrete-encased steel sections. Gravity composite systems may be broadly classified into composite floors and composite building columns. Composite floor systems typically consist of simply supported steel rolled sections and occasionally nonprismatic steel sections such as tapered, dapped, or castellated beams including haunch girders, trusses, or stub girders. Invariably, the steel member is attached to the concrete topping via shear connectors welded to the member through a formed composite steel deck. The entire floor system consisting of steel sections, steel deck, and concrete topping responds as a series of flexural T-beams under gravity loads. Shown in Figure 13.1 are the typical components of composite floor used nearly exclusively in steel-framed buildings in North America.

Composite construction is not new. It is at least 40 years older than the oldest person reading this book. It is characterized by interactive behavior between structural steel and concrete components designed to take advantage of the best load-resisting characteristics and economy of each material.

We begin this chapter with a discussion of formed, profiled steel deck sheeting used nearly ubiquitously in steel-framed buildings in North America. The deck itself acts compositely with the concrete slab as the tension reinforcement. Next, we discuss the most common composite floor system consisting of rolled or built-up steel beams connected to form steel deck and concrete slab. The steel deck and the concrete slab are most often mechanically connected to each other using headed steel studs to the almost complete exclusion of other types of shear connectors. Also discussed in this section are other composite floor systems such as tapered and dapped girders, joists, and trusses that attempt to use structural steel more efficiently.

The next section centers on the design of two basic types of composite columns: those with steel sections encased in concrete and those with the steel sections filled with concrete, the latter most often consisting of hollow sections fabricated from plates. In the first type, often referred to as encased composite column, the steel section can be designed to provide support for the erection of several floors ahead of concreting the steel columns for composite action. The procedure expedites construction schedule and thus has the potential for reducing overall construction cost.

The design of composite beams particularly when designed for partial composite action is not an easy task. Nor is the design of encased or filled composite columns. Unless design aids or a computer with a suitable design program is used, it is extremely cost prohibitive to perform manual calculations in a design office. Therefore, instead of presenting manual design examples in excruciating details, in this chapter, as in other chapters of this work, we present abbreviated versions emphasizing basic concepts, fundamental structural actions, and practical applications.

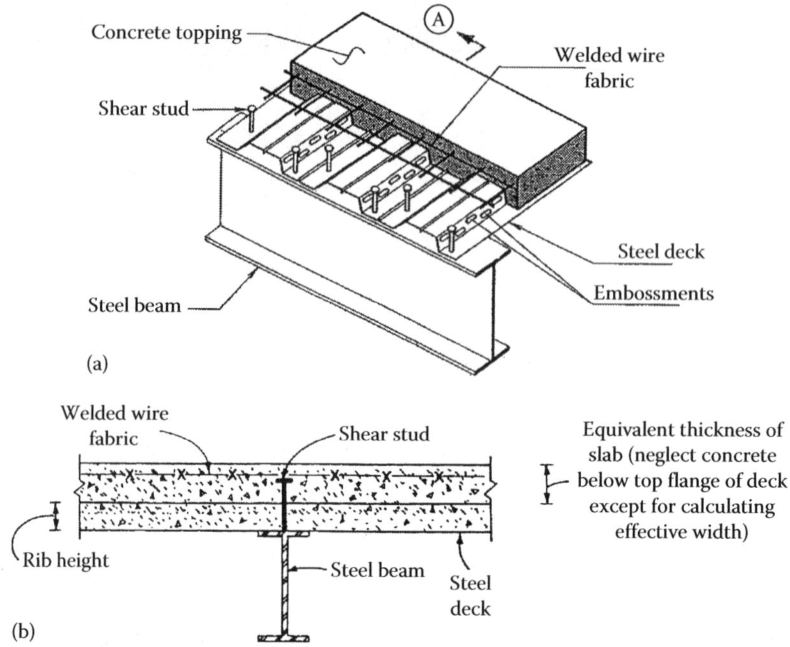

FIGURE 13.1 Components of composite floor system: (a) schematics showing steel deck perpendicular to beam; (b) section.

13.1 COMPOSITE STEEL DECK

Steel deck is manufactured from steel sheets by a fully mechanized, high-speed cold rolling process. Although it is possible to produce shapes up to ½ in. and even in ¾ in. thick by cold forming, cold-formed steel construction is generally restricted to plates and sheets weighing from 0.5 psf to a maximum of 9 psf.

Composite steel deck is manufactured with embossments formed in the steel sheet specifically designed to increase composite action between the steel deck and concrete. The shear connection between the two is provided through lugs, corrugations, ridges, or embossments formed in the profile of the sheet to increase the chemical bond between the two materials. The steel deck profile is typically trapezoidal with relatively wide flutes suitable for through-deck welding of shear studs. Steel deck may also include closed cells to accommodate floor electrification system. Noncellular deck panels may be blended with cellular panels as part of the total floor system. Steel deck is commonly available in $1^1/_2$, 2, and 3 in. depths with rib spacings of 6, $7^1/_2$, 8, 9, and 12 in.. See Figure 13.2 for some typical deck profiles.

A composite slab is usually designed as a simply supported reinforced concrete slab with steel deck acting as positive reinforcement. Typical mesh used for control of temperature cracking does not provide enough negative reinforcement for typical beam spacing of 8–15 ft. It is a good practice, however, to provide a nominal top reinforcement of say #4 at 24″ c-c in the negative moment regions

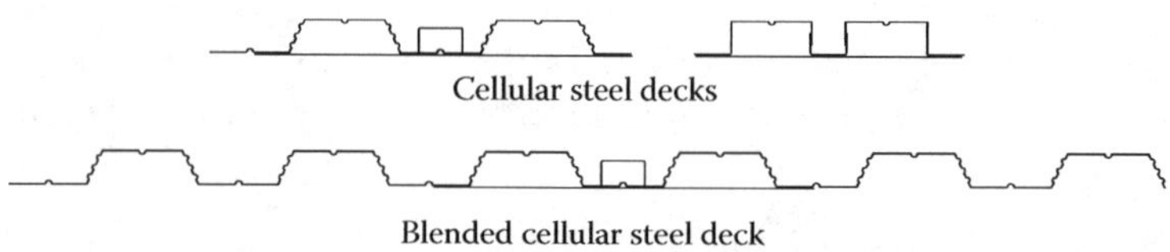

FIGURE 13.2 Typical composite steel deck profiles.

at the top to control excessive cracking of slab. It is generally believed that cracking of the slab in the negative regions does not materially impair the composite-beam strength.

The product of steel deck includes composite floor deck, deep roof deck, cellular roof deck, and cellular floor deck. Emphasis here is on composite steel decks with concrete topping, the most common form of floor construction in steel and composite buildings. Since hazards may be associated with the handling, installation and use of steel deck and its accessories, prudent construction practices should always be followed.

After installation and adequate fastening, composite steel decks (floor decks) serve several purposes. They act as working platforms, stabilize the frame, serve as concrete forms for slabs, and provide long-term positive bending reinforcement. Mechanical interlock and chemical bond resist horizontal shear and provide composite action between concrete and steel deck. All composite decks are made to mechanically interlock with the concrete through *rolled-in* embossments.

Deck should be selected to provide a working platform capacity of at least 50 psf. If temporary shoring is required to obtain this capacity, it should be available to support the deck as the deck is being installed. Generally, deck is selected to perform without temporary shores. The concrete volumes and the weights used to determine unshored spans are nominal and include no additional concrete due to frame deflection. As the deck is being erected, it is important to immediately attach it to the structural frame so a working platform is made. All OSHA rules for erection must be followed. The SDI Manual of Construction with Steel Deck is a recommended reference. When placing concrete, care must be taken to avoid high pileups of concrete and to avoid impacts caused by dropping or dumping. If buggies are used, runways should be planked and deck damage caused by careless placement practices should not be allowed. Field cutting that is not shown on the approved erection drawings or changes in fastener types or spacing should be authorized by the designer. Field cutting around openings or for temporary guys can create simple spans where multispans are intended.

13.1.1 Finishes

Composite deck is available, galvanized (G30, G60 or G90) and *phosphatized/painted*. When the deck is furnished phosphatized/painted only the side not in contact with the concrete is painted so chemical bond between steel and concrete can occur. (*Phosphatizing* is a cleaning process.) When spray fire proofing is required over *phosphatized/painted* deck, this must be clearly shown in the contract documents.

13.1.2 Venting

With the increased use of sealers on concrete slabs and more stringent regulations concerning solvent emissions at job sites, venting requirements are sometimes specified to allow curing and drying of structural concrete. Venting has long been specified with lightweight insulation fills. Venting is a function of vent area and spacing and is typically located in the bottom flats of flutes. Vents are not possible with cellular deck and integral hanger tabs are commonly used in composite deck. The nominal vent area provided by integral hangers in composite deck is ¼% but ½% is available in some circumstances. When venting is a concern, properly designed (rich) concrete mixes and increased curing time before sealing will significantly help.

13.1.3 Wire Mesh

Temperature reinforcing should be present in composite slabs. The wire mesh recommendations shall follow the SDI recommendation for steel area of 0.00075 times the area of concrete above the deck flutes and not less than 6 × 6-W1.4 × 1.4. The mesh is not proportioned to act as negative reinforcement but it does add some strength to the system. For best crack control, mesh should be kept near the top of the slab in negative bending regions (3/4″–1″ cover)—shrinkage and bending cracks are possible over supports. Mesh also helps to distribute loads, both during construction and during the service life of the slab. It can also be a secondary safety device if there is a collapse during concrete placement.

13.1.4 Parking Garages

Composite floor deck is not recommended for parking garages in the northern part of the United States; salt introduced by snow removal can deteriorate the deck. Deck can be used as a temporary form and reinforcing (mesh or bars) should be used for positive reinforcement. If sealing membranes are used, sufficient time after concrete placement must be allowed or other means for concrete curing must be provided. Negative steel may be required to control strain in the membrane.

13.1.5 Fork Lifts

Dynamic and repetitive loads can overcome bond and interlock. When very repetitive moving traffic loads can overcome bond and interlock. When very repetitive moving traffic is anticipated, it is recommended that the slab be designed as conventional reinforced slab or that the required composite slab bending strength be less than ½ the factored nominal capacity and that negative steel and distribution steel be designed.

The tables supplied by manufacturers of steel deck typically show the uniform live-load capacities for both lightweight and normal-weight concrete; both types assume a concrete strength of 3000 psi. The published uniform live load is the unfactored service load, typically obtained in codes. The tables are based on steel yield strength of 40 ksi; however, 50 ksi minimum yield steel is also available with sufficient lead time and quantity. SDI allows 50 ksi as the maximum design strength in the determination of slab capacities. Maximum unshored spans given in the manufacturers' tables may be taken as clear spans. The SDI construction loading is used to determine the values.

The research done on composite deck has shown that the presence of shear studs influences the resistance of the system. When a sufficient number of shear studs are present, the composite slab can achieve its predicted ultimate strength. When no shear studs are present, the factored moment is found by $M_{no} = \phi S_c F_y$, where ϕ is 0.85 and S_c is the cracked composite section modulus of the composite slab. If the number of studs present is between the amount required to produce the *fully* studded moment and zero, then a straight-line interpolation is valid. Generally, the load capacity of composite slabs is greater than required by the intended used, and the number of studs is not of importance. Studs are used primarily to make beams composite and the composite slab simply uses what is there.

13.2 SPECIFICATIONS FOR STEEL DECK: OVERVIEW

13.2.1 Material and Design

1. Composite floor deck shall be manufactured from steel conforming to ASTM A1008 or ASTM A653 with minimum yield strength (*fy*) of 40 ksi.
2. Floor deck shall extend over three or more spans if possible (the depth and gage of floor deck shall be selected to not exceed the unshored spans as calculated using LRFD methods under the construction loadings recommended by SDI). Deflection relative to support beams caused by the dead load of wet concrete and deck shall not exceed (deck span)/180 or ¾″.
3. Live-load capacities shall be calculated in accordance with the SDI Composite Deck Design Handbook. The type and gage of the metal floor deck shall be selected to carry, acting compositely with the concrete slab, the superimposed live loads shown on the project drawings without exceeding a deflection of 1/360 of the span.

13.2.2 Finishes

1. Galvanizing shall conform to the requirements of ASTM A653 coating class G30, G60, or G90 or Federal Specification QQ-S-775e, Class D or Class E.
2. Primer paint shall be shop applied over cleaned and phosphatized steel—paint applied only on the exposed side of the deck. The side of the deck that is to be in contact with the concrete is to be uncoated or galvanized.

13.2.3 Tolerances

Manufacturer's standard tolerances apply (deck sheet length is plus or minus $^1/_2$″). Base steel thickness shall be greater than or equal to 9% of the design thickness. (This is consistent with the SDI and AISI Standards.)

13.2.4 Installation

1. Installation of floor deck and accessories shall be done in accordance with the SDI Manual of Construction with Steel Deck. To form a working platform, immediately fasten sheets to the supports. Welds to supports shall be 5/8″ diameter puddle welds with an average weld spacing of at least 12″ on center. Side laps are to be welded at a maximum spacing of 36″ on center for spans over 5′0″. (Button punches or fasteners other than welds may be acceptable. Refer to the Underwriters Laboratories (UL) design for other restrictions.)
2. Deck shall be butted over supports (ending lapping or staggering is not recommended). As deck profile dictates, overlap at interlocking side laps or nest side laps without back lapping. Maintain end alignment. Minimum bearing of deck ends on supports shall be 1 1/2″, unless otherwise shown.
3. Floor openings located and detailed on the structural drawings shall be cut by the floor deck contractor. Holes for other trades plus any reinforcing for these holes shall be cut and reinforced by the other trades. (OSHA Regulations require that most openings be decked over during construction.)
4. All sheets from opened bundles must be fastened before the end of the working day; bundles must be left secured to prevent wind blowing the individual sheets.
5. Shoring shall be present at time of deck placement. Do not remove shoring until the concrete has attained 75% of its design compressive strength and in no case less than 7 days or as instructed by the engineer of record.
6. Maximum unshored spans, slab thickness, and concrete density shall be posted on the erection drawings.

13.2.5 Concrete

1. Placement of concrete shall conform to the applicable sections of the ACI Specifications and the SDI Manual of Construction with Steel Deck. If buggies are used, the deck shall be planned to prevent damage. Contractor shall not exceed the SDI construction loads during either concrete placement or finishing. Ponding of concrete and large screed machines or buggies should be avoided unless special project loading is defined in the contract documents and maximum unshored spans are determined prior to approval of the erection layout.
2. Calcium chloride (or any admixture containing chloride salts) shall not be used in concrete placed on steel deck.

13.2.6 Site Storage

Steel deck delivery should be scheduled to arrive at the jobsite as required for erection. If site storage is needed, the bundles of deck (either painted, uncoated, or galvanized) shall be stored off the ground with one end elevated to provide drainage and shall be protected against condensation with a ventilated waterproof covering.

13.3 ANSI/SDI (C1.0 STANDARD FOR COMPOSITE FLOOR DECK): A BRIEF REVIEW

1. *Fire resistance*: Many fire-related assemblies that use composite floor decks are available. In the UL Fire Resistance Directory, the composite deck constructions show hourly ratings for restrained and unrestrained assemblies. ASTM El 19 provides information in appendix X3 called *Guide for Determining Conditions of Restraint for Floor and Roof Assemblies and Individual Beams.*
2. *Products*: Most composite steel floor deck is manufactured from steel conforming to ASTM Designation A1008 (A1008M), grades 33 and 40, or from A653 (A653M). Structural Sheet Steel. When specifying alternative steels, certain restrictions apply (see the North American Specification for the Design of Cold-Formed Steel Structural Members Section A 2-3.2); 2.1 A refers to the use of galvanized deck, while 2.1B refers to the use of uncoated or phosphatized top/painted underside deck. In most cases, the designer in concert with the project owner and architect will choose one finish or the other. However, both types of finish may be used on a job, in which case the designer must indicate on the plans and project specifications in the areas in which each is used.
3. *Finish*: The finish on the steel composite deck shall be as specified by the designer and be suitable for the environment of the structure. Since the composite deck is the positive bending reinforcement for the slab, it must be designed to last the life of the structure. A galvanized finish equal to ASTM A653 (A653M)—G30 minimum—is recommended. When composite deck with a phosphatized top and painted bottom is used, the primer coat is intended to protect the steel for only a short period of exposure in ordinary atmospheric conditions and shall be considered an impermanent and provisional coating.
4. *Design*: The loading for steel deck design is representative of the sequential loading of wet concrete on the deck. The 150 lb load (per foot of width) is the result of distributing a 300 lb (1.33 kN) man over a 2 ft. (600 mm) width. Experience has shown this to be a conservative distribution. The metric equivalent of the 150 lb load is 2.2 kN per meter of width. For single span deck conditions, the ability to control the concrete placement may be restricted and an amplification factor of 1.5 is applied to the concrete load to address this condition; however, in order to keep this 50% load increase within a reasonable limit, the increase is not to exceed 30 psf (1.44 kPa). In LRFD, a load factor for construction of 1.4 is applied to this load. Whenever possible, the deck shall be multispan and not require shoring during concrete placement.
5. *Deck deflection*: The deflection calculations do not take into account construction loads because these are considered temporary loads. The deck is designed to always be in the elastic range so removal of temporary loads would allow the deck to recover. The structural steel also deflects under the loading of the wet concrete.

The designer is urged to check the deflection of the total system, especially if composite beams and girders are being used. If the designer wants to include additional concrete loading on the deck because of frame deflection, the additional load should be shown on the design drawings or stated in the deck section of the job specifications.

6. *Minimum bearing*: Experience has shown that 1–1/2 in. (38 mm) of bearing is sufficient for composite floor decks. If less than 1-1/4 in. (38 mm) of end bearing is available, or if high support reactions are expected, the design professional should check the deck web crippling capacity. The deck must be adequately attached to the structure to prevent slip off.
7. *Diaphragm shear capacity*: Calculations of diaphragm strength and stiffness should be made using the SDI Diaphragm Design Manual. If testing is used as the means for determining the diaphragm strength and stiffness, then it should follow the AIS1 TS 7-02 test protocol.
8. *Concentrated loads*: High concentrated loads, diaphragm loads, etc., required additional analysis. Horizontal load capacities can be determined by referring to the SDI Diaphragm Design Manual. Concentrated loads can be analyzed by the methods shown in the *SDI Composite Deck Design Handbook*. Most published live-load tables are based on a simple span analysis of the composite system; that is, the slab is assumed to crack over each support.

   By using standard reinforced concrete design procedures or test results, the deck manufacturer determines the live loads that can be applied to the composite deck slab combinations. The results are usually published as uniform load tables. For most applications, the deck thickness and profile is selected so that shoring is not required; the live-load capacity of the composite system is usually more than adequate for the superimposed live loads. In calculating the section properties of the deck, the AISI provisions may require that compression zones in the deck be reduced to *effective width*, but as tensile reinforcement, the total area of the cross section may be used.
9. *Concrete strength*: Load tables are generally calculated by using a concrete strength of 3 ksi (30 MPa). Composite slab capacities are not greatly affected by variations in concrete compressive strength; but, if the strength falls below 3 psi (20 MPa), it would be advisable to check shear stud strengths. Fire-rating requirements may dictate the minimum concrete strength. The use of admixtures containing chloride salts is not allowed because the salts will corrode the steel deck.
10. *Live-load deflection*: Live-load deflections are seldom a design factor. The deflection of the slab/deck combination can be predicted by using the average of the cracked and uncracked moments of inertia as determined by the transformed section method of analysis. Refer to Attachment C5 of the *SDI Composite Deck Design Handbook*.
11. *Suspended loads*: The designer must take into account the sequence of loading. Suspended loads may include ceilings, light fixtures, ducts, or other utilities. The designer must be informed of any loads applied after the composite slab has been installed.

    Care should be used during the placement of loads on all types of hanger tabs or other hanging devices for the support of ceilings so that an approximate uniform loading is maintained. The individual manufacturer should be consulted for allowable loading on single hanger tabs. Improper use of hanger tabs or other hanging devices could result in the overstressing of tabs and/or the overloading of the composite deck slab.
12. *Reinforcement*: Temperature and shrinkage reinforcement, consisting of welded wire fabric or reinforcing bars, shall have a minimum area of 0.00075 times the area of the concrete above the deck (per foot or meter of width), but shall not be less than the area provided by 6 × 6-W1.4 × W1.4 welded wire fabric.

    Fibers shall be permitted as a suitable alternative to the welded wire fabric specified for temperature and shrinkage reinforcement. Cold-drawn steel fibers meeting the criteria of ASTM A820, at a minimum addition rate of 25 lb/yd^3 (13.8 kg/m^3) or macrosynthetic fibers *coarse fibers* (per ASTM Subcommittee C09.42) made from virgin polyolefin, which shall have an equivalent diameter between 0.4 mm (0.016 in.) and 1.25 mm (0.05 in), having a minimum aspect ratio (length/equivalent diameter) of 50, at a minimum addition rate of 4 lb/yd^3. (2.4 kg/m^3) are suitable to be used as minimum temperature and shrinkage reinforcement.

Neither welded wire fabric nor fibers will prevent cracking; however, they have been shown to do a good job of crack control. The welded wire fabric must be placed near the top of the slab (3/4 in. to 1 in. cover (20–25 mm) at supports and draped toward the center of the deck span. If a welded wire fabric is used with a steel area given by the aforementioned formula, it will not be sufficient as the total negative reinforcement. If the minimum quantity of steel fibers, or macro synthetic fibers, is used for shrinkage and temperature reinforcement, they will not be sufficient as a total negative reinforcement.

Composite steel deck does not function as compression reinforcing steel in areas of negative moment. If the designer wants a continuous slab, then negative bending reinforcing should be designed using conventional reinforced concrete design techniques in compliance with the ACI Building Code Requirements for Reinforced Concrete. The welded wire fabric, chosen for temperature reinforcing, may not supply enough area for continuity. The deck is not considered to be compression reinforcement. Typically, negative reinforcement is required at all cantilevered slabs or if a continuous slab is desired.

When localized loads exceed the published uniform composite deck load tables, the designer shall proportion distribution reinforcement using conventional concrete design methods.

Distribution steel may be required in addition to the welded wire fabric or steel fibers. Concentrated loads, either during construction or in-service, are the most common example of this requirement. Concentrated loads may be analyzed by the methods in the latest *SDI Composite Deck Design Handbook*.

13. *Cantilever loads*: When cantilevered slabs are encountered, the deck acts only as a permanent form; top reinforcing steel shall be proportioned by the designer. For construction loads, the deck shall be designed for the more severe of (a) deck plus slab weight plus 20 psf (1 kPa) construction load on both cantilever and adjacent span or (b) deck plus slab weight on both cantilever and adjacent span plus a 150 lb (665 N) concentrated load per foot of width at end of cantilever. The load factors for bending, shear, and interior bearing shall be as required by ASCE 7. Resistance factors for bending, shear, and interior bearing shall be in accordance with the North American Specification for the Design of Cold-Formed Structural Members.

    The maximum cantilever deflection as a form, under deck plus slab weight, shall be $a/90$ where a is the cantilever length and shall not exceed 3/4 inches (19 mm).

    Side laps shall be attached at the end of the cantilever and a maximum spacing of 12 in. (300 mm) o.c. from the cantilever end. Each corrugation shall be fastened at both the perimeter support and the first interior support. The deck shall be completely attached to the supports and at the side laps before any load is applied to the cantilever. Concrete shall not be placed on the cantilever until after placement on the adjacent span.
14. *Accessories*
    a. Pour stops, column closures, end closures, cover plates, and girder fillers shall be the type suitable for the application. Pour stop minimum gages shall be in accordance with the Steel Deck Institute. (See Pour Stop Selection Table, Attachment C2.)
    b. Mechanical fasteners or welds shall be permitted for deck and accessory attachment.
15. *Execution*: Temporary shoring, if required, shall be installed before placing deck panels. Temporary shoring shall be designed to resist a minimum uniform load of 50 psf (2.4 kPa) and loading criteria indicated on Attachment C1. Shoring shall be securely in place before the floor deck erection begins. The shoring shall be designed and installed in accordance with the ACI Building Code Requirements for Reinforced Concrete and shall be left in place until the slab attains 75% of its specified design strength and a minimum of 7 days.

Deck panels shall be placed on structural supports and adjusted to final position ends aligned and attached securely to the supports immediately after placement in order to form a safe working platform. All deck sheets shall have adequate bearing and fastening to all supports to prevent slip-off during construction. Deck ends over supports shall be installed with a minimum end bearing of 1–1/2 in. (38 mm). Deck areas subject to heavy or repeated traffic, concentrated loads, impact loads, wheel loads, etc., shall be adequately protected by planking or other approved means to avoid overloading and/or damage.

Lapping composite deck ends can be difficult because shear lugs (web embossment) or profile shape can prevent a tight metal to metal fit. The space between lapped sheets can make welded attachments more difficult. Gaps are acceptable up to 1 in. (25 mm) at butted ends.

16. *Installation/anchorage*: Floor deck units shall be anchored to steel supporting members including perimeter support steel and/or bearing walls by arc spot puddle welds of the following diameter and spacing, fillet welds of equal strength, or mechanical fasteners.

    All welding of deck shall be in strict accordance with ANSI/AWS D1.3, Structural Welding Code-Sheet Steel. Each welder shall demonstrate an ability to produce satisfactory welds using a procedure such as shown in the SDI Manual of Construction with Steel Deck or as described in ANSI.AWS D1.3.

    A minimum visible 5/8 in. (15 mm) diameter arc puddle weld shall be used. Weld metal shall penetrate all layers of deck material and shall have good fusion to the supporting members.

    Edge ribs of panels shall be welded at each support. Space additional welds an average of 12 in. (300 mm) apart but not more than 18 in. (460 mm).

    When used, fillet welds shall be at least 1–1/2 in. (38 mm) long.

    Mechanical fasteners, either powder actuated, pneumatically driven, or screwed, shall be permitted in lieu of welding to fasten deck to supporting framing if fasteners meet all project service requirements. When the fasteners are powder actuated or pneumatically driven, the load value per fastener used to determine the maximum fastener spacing is based on a minimum structural support thickness of not less than 1/8 in. (3 mm) and on the fastener providing a minimum 5/16 in. (8 mm) diameter bearing surface (fastener head size). When the structural support thickness is less than 1/8 in. (3 mm), powder actuated or pneumatically driven fasteners shall not be used, but screws are acceptable.

    Mechanical fasteners (screws powder or pneumatically driven fasteners, etc.) are recognized as viable anchoring methods, provided the type and spacing of the fastener satisfies the design criteria. Documentation in the form of test data, design calculations, or design charts should be submitted by the fastener manufacturer as the basis for obtaining approval.

    For deck units with spans greater than 5 ft (1.5 m), side laps and perimeter edges of units between span supports shall be fastened at intervals not exceeding 36 in. (1 m) on center, using one of the following methods:

    a. #10 self-drilling screws
    b. Crimp or button punch
    c. Arch puddle welds 5/8 in. (15 mm) minimum visible diameter or minimum 1 in. (25 mm) long fillet weld

    The above side lap spacing is a minimum. Service loads or diaphragm design may require closer spacing or larger side lap welds. Good metal to metal contact is necessary for a good side lap weld. Bum holes are to be expected.

17. *Accessory attachment* (pour stop and girder fillers): Pour stops and girder fillers shall be fastened to supporting structure in accordance with the SDI Standard Practice Details.

    Column closures, cell closures, girder closures, and Z closures shall be fastened to provide tight fitting closures at open ends of ribs and sides of decking. Fasten cell closures at changes of direction of floor deck units unless otherwise directed.

13.4 COMPOSITE BEAMS

Three types of composite-beam construction shown in Figure 13.3 are recognized in the AISC specifications: (1) fully encased steel beams, (2) concrete-filled HSS, and (3) steel beams with mechanical anchorage to slab. In fully encased steel beams, the natural bond between concrete and steel interface is considered sufficient to provide the resistance to horizontal shear provided that (1) the concrete thickness is 2 in. or more on the beam sides and soffit with the top of the beam at least 1½ in. below the top and 2 in. above the bottom of the slab and that (2) the encasement is cast integrally with the slab and has adequate mesh or other reinforcing steel throughout the depth and across the soffit of the beam to prevent spalling of concrete.

The third type consisting of steel beams, steel deck, concrete topping, and shear connectors is by far the most popular in the construction of buildings in North America.

Invariably composite action is achieved by providing shear connectors between top flange of the steel beam and concrete topping. Fully encased steel beam construction is not used because encasing of beams with concrete requires expensive form work.

Composite sections have greater stiffness than the summation of the individual stiffness of slab and beam and, therefore, can carry larger loads or similar loads with appreciably smaller deflection and are less prone to transient vibrations. Composite action results in an overall reduction of floor depth, and for high-rise buildings, the cumulative savings in curtain walls, electrical wiring, mechanical ductwork, interior walls, plumbing, etc., can be considerable.

Composite beams can be designed either for shored or unshored construction. For shored construction, the cost of shoring should be evaluated in relation to the savings achieved by the use of lighter beams. For unshored construction, steel is designed to support by itself and the wet weight of concrete and construction loads. The steel section, therefore, is heavier than in shored construction.

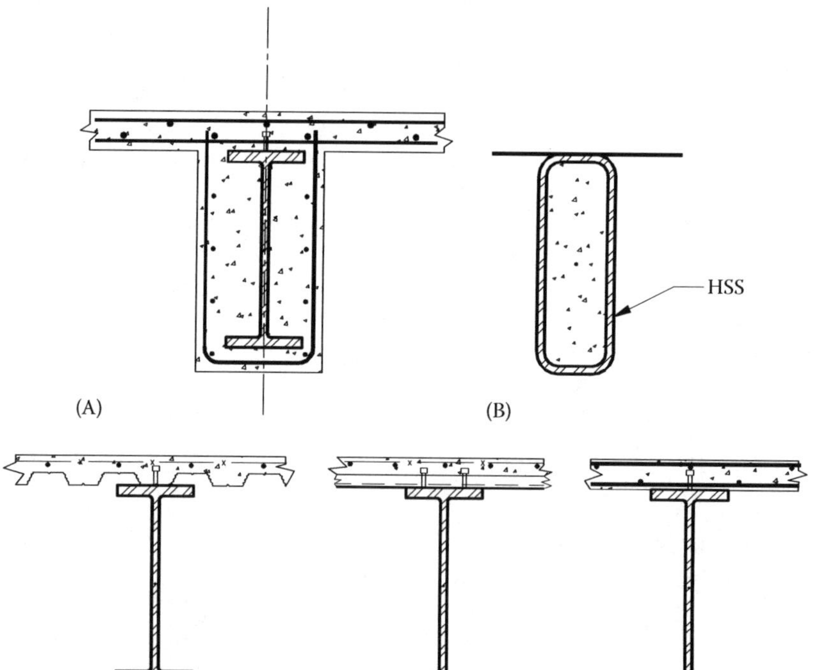

FIGURE 13.3 Three types of composite beams addressed in the AISC manual: (A) fully encased steel beam; (B) concrete filled HSS; (C) steel beams with mechanical anchorage to slab. (1) Steel deck parallel to beam; (2) steel deck perpendicular to beam; (3) cast-in-place slab without steel deck.

In composite floor construction, because top flange of steel beam is attached to concrete by the use of shear connectors, the slab becomes part of the compression flange. As a result, the neutral axis of the section shifts upward, making the bottom flange of the beam more effective in tension. Since concrete is required to serve as floor system, the only additional cost is that of the shear connectors. In addition to transmitting horizontal shear forces from the slab into the beam, the shear connector prevents any tendency for the slab to rotate independently of the beam. The stud shear connector is a short length of round steel bar welded to the steel beam at one end and having an anchorage provided in the form of a round head at the other end. The most common diameters are 1/2, 5/8, and 3/4 in. The length is dependent on the depth of steel deck and should extend at least 1½ in. above the top of the deck. The welding process typically reduces their length by about 3/16 in. The upset head thickness of the studs is usually 3/8 or 1/2 in., and the diameter 1/2 in. larger than the stud diameter. The studs are normally welded to the beam with an automatic welding gun, and when properly executed, the welds are stronger than the steel studs. Studs located on the side of the trough toward the beam support are more effective than studs located toward the beam centerline. The larger volume of concrete between the stud and the pushing side of the trough helps in the development of a larger failure cone in concrete, thus increasing its horizontal shear resistance.

Steel decks for composite construction are available in the United States in three depths, 1½, 2, and 3 in. The earlier types of steel deck did not have embossments, and the interlocking between concrete and steel deck was achieved by welding reinforcement transverse to the beam. Later developments of steel deck introduced embossments to engage concrete and steel deck and dispensed with the transverse welded reinforcement.

The spans utilizing composite steel deck are generally in the range of 8–15 ft.

In floor systems using 1½ in. decks, the electrical and telephone services are generally provided by the so-called poke-through system, which simply punches through the slab at given locations for passage of underfloor ducts. A deeper deck is required if the power distribution system is integrated as part of the structural slab; as a result, 2 and 3 in. steel decks were developed. Experiments have shown that there is very little loss of composite-beam stiffness due to the ribbed configuration of the steel deck in the depth range of 1½–3 in. As long as the ratio of width to depth of the steel deck is at least 1.75, the entire capacity of the shear stud can be developed similar to beams with solid slabs. However, with deeper deck, a substantial decrease in shear strength of the stud occurs, which is attributed to a different type of failure mechanism. Instead of the failure of shear stud, the mode of failure is initiated by cracking of the concrete in the rib corners. Eventual failure takes place by separation of concrete from the steel deck. When more than one stud is used in a steel deck flute, a failure cone can develop over the shear stud group, resulting in lesser shear capacity per each stud. The shear stud strength is therefore closely related to the steel deck configuration and factors related to the surface area of the shear cone.

Often, special details are required in composite design to achieve optimum results. Openings interrupt slab continuity, affecting capacity of the composite beam. For example, beams adjacent to elevator and stair openings may have full effective width for part of their length and perhaps half that value adjacent to the openings. Elevator sill details normally require a recess in the slab for door installations, rendering the slab ineffective for part of the beam length. A similar problem occurs in the case of trench header ducts, which require elimination of concrete, as opposed to the standard header duct, which is completely encased in concrete. When the trench is parallel to the composite beam, its effect can be incorporated in the design by suitably modifying the effective width of compression flange. The effect of the trench oriented perpendicular to the composite beam could range from negligible to severe depending upon its location. If the trench can be located in the region of minimum bending moment, such as near the supports in a simply supported beam, and if the required number of connectors could be placed between the trench and the point of maximum bending moment, its effect on the composite-beam design is minimum. If, on the other hand, the trench must be placed in an area of high bending moment, its effect may be so severe as to require that the beam be designed as a noncomposite beam.

The slab thickness normally employed in high-rise construction with composite steel deck is usually governed by fire-rating requirements rather than the thickness required by the bending capacity of the slab. In certain parts of the United States, it may be economical to use the minimum thickness required for strength and to use sprayed-on or some other methods of fireproofing to obtain the required ratings. Some major projects have used 2½ in. thick concrete on 3 in. deep steel deck spanning as much as 15 ft. A comparative study that takes into consideration the vibration characteristics of the floor is necessary to converge on the most economical floor system.

In continuous composite beams, the negative moment regions can be designed such that (1) the steel beam alone resists the negative moment or (2) it acts compositely with mild steel reinforcement placed in the slab parallel to the beam. In the latter case, shear connectors must be provided through the negative moment region.

Careful attention should be paid to the deflection characteristics of composite construction because the slender not-yet-composite shape deflects as wet concrete is placed on it. There are three ways to handle the deflection problem:

1. Use relatively heavy steel beams to limit the dead-load deflection and pour lens-shaped tapering slabs to obtain a nearly flat top. Although a reasonably flat surface results from this construction, the economic restraints of speculative office buildings do not usually permit the luxury of the added cost of additional concrete and heavier beams.
2. Camber the steel beam to compensate for the weight of steel beam and concrete. Place a constant thickness of slab by finishing the concrete to screeds set from the cambered steel. Continuous lateral bracing as provided by the steel deck is required to prevent the lateral torsion buckling of the beam. If steel deck is not used, this system requires a substantial temporary bracing system to stabilize the beam during construction.
3. Camber and shore the steel beam. The beam is fabricated with a camber calculated to compensate for the deflection of the final cured composite section. Shores are placed to hold the steel at its curved position while the concrete is being poured. As in method 2, slab is finished to screeds set from cambered steel. Although methods 1 and 3 are occasionally used, the trend is to use method 2 because it is the least expensive.

13.4.1 AISC Design Criteria: Composite Beams with Steel Deck and Concrete Topping

For purposes of design, it is useful to consider two types of composite beams:

1. Deck perpendicular to the beam (Figure 13.4)
   a. Concrete below the top of steel decking shall be neglected in computations of section properties and in calculating the number of shear studs, but the concrete below the top flange of deck may be included for calculating the effective width.
   b. The maximum spacing of shear connectors shall not exceed 32 in. along the beam length.
   c. The steel deck shall be anchored to the beam either by welding or by other means at a spacing not exceeding 16 in.
   d. Shear stud reduction factors, R_g and R_p, should be used for reducing the shear capacity of stud connectors.
   e. The recommended sequence for installing studs when deck is perpendicular to beams is as follows:
      i. Deck ribs at 6 in. on center. Start at beam ends and place a single stud at every fourth flute, working toward the center of the beam. If studs remain, fill in empty ribs, again starting at beam ends and working toward the center without exceeding 30 in. for stud spacing.

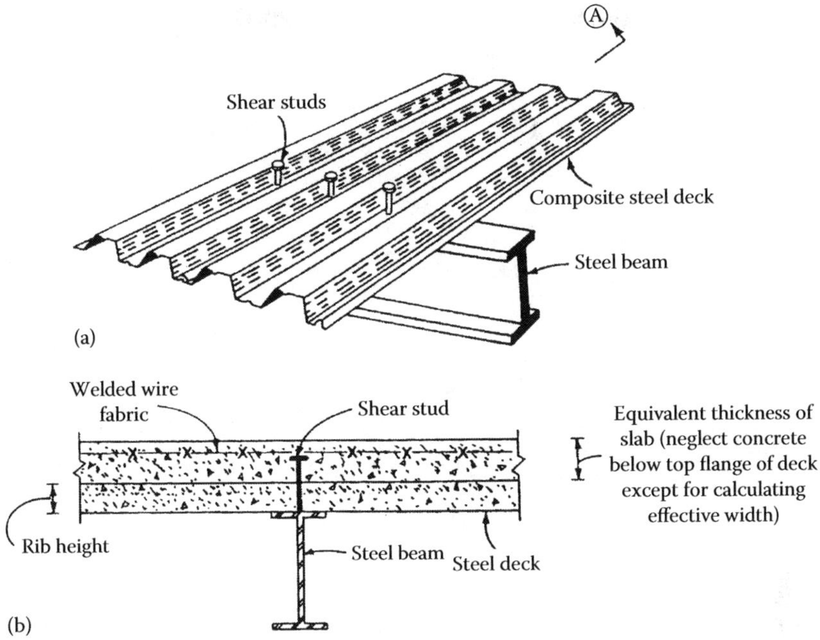

FIGURE 13.4 Composite beam with deck ribs perpendicular to beam: (a) schematic view; (b) section showing equivalent thickness of slab.

ii. Deck ribs at 12 in. on center. Start at beam ends and place a single stud in every other flute working toward the center of the beam. If studs remain, fill in empty ribs, again starting at beam ends and working toward the center of the beam without exceeding 24 in. (610 mm) for study spacing.

iii. If the number of studs is more than the number of ribs, place a double or triple row as needed, always starting from beam ends and working toward the beam center. In general, if studs cannot be uniformly spaced, the greatest number of studs should occur at the ends.

2. Deck ribs parallel to the beam (Figure 13.5)
   a. The major difference between the perpendicular and parallel orientation of deck ribs is that when deck is parallel to the beam, the concrete below the top of the decking can be included in the calculations of section properties and must be included when calculating the number of shear studs.
   b. If steel deck ribs occur on supporting beam flanges, it is permissible to cut high hat to form a concrete haunch.
   c. When nominal rib height is 1½ in. or greater, minimum average width of deck flute should not be less than 2 in. for the first stud in the transverse row plus four stud diameters for each additional stud. This gives minimum average widths of 2 in. for one stud, 2 in. plus $4d$ for two studs, 2 in. plus $8d$ for three studs, etc., where d is the diameter of stud. Note that if a steel deck cannot accommodate this width requirement, the deck can be split over the girder to form a haunch.
   d. Shear stud reduction factors, R_g and R_p as explained presently, shall be used for reducing the shear capacity of stud connectors.
   e. The recommended sequence for installing studs when deck is parallel to the girder is as follows. Start at the girder ends by placing the first stud at approximately 12 in. from the centerline of support and work toward the center of girder with uniform spaces between the studs. If a double row of studs is required, it is a good practice to place them in a staggered pattern rather than side by side.

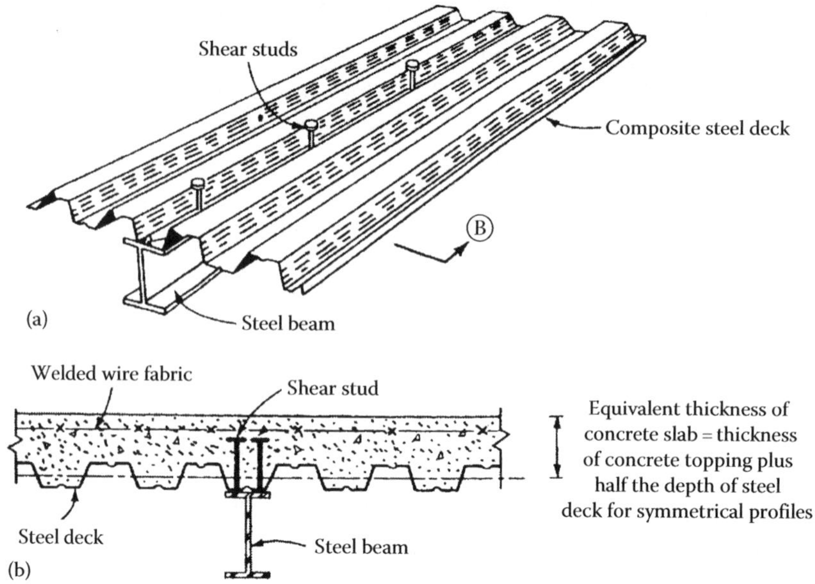

FIGURE 13.5 Composite beam with deck ribs parallel to beam: (a) schematic view; (b) section showing equivalent thickness of slab.

13.4.2 AISC Requirements: General Comments

a. The rib deck height shall not exceed 3 in.
b. Rib average width shall not be less than 2 in. If the deck profile is less than 2 in., this minimum clear width shall be used in the calculation.
c. The section properties do not change a great deal from deck running perpendicular or parallel to the beam, but the change in the required number of studs can be significant.
d. The reduction formula for stud length is based on rib geometry, number of studs per rib, and embedment length of the studs.
e. The provision for calculating the partial section modulus makes selection of heavier, stiffer beams with fewer studs economically more attractive.
f. High shear values can be used in longer shear studs. Concrete cover over the top of the stud is not limited by the AISC specifications, but for practical reasons, the author recommends a minimum of ½ in.
g. Studs can be placed as close to the web of deck as needed for installation and to maintain the necessary spacing.
h. Deck anchorages can be provided by the stud welds.
i. Maximum diameter of shear connectors is limited to ¾ in.
j. After installation, the studs should extend *a* minimum of 1½ in above the steel deck.
k. Total slab thickness including the ribs is used in determining the effective width without regard to the orientation of the deck with respect to the beam axis.
I. The slab thickness above the steel deck shall not be less than 2 in.
m. Shear studs may be welded directly through the deck or through prepunched or cut-in-place holes in the deck. The usual procedure is to install studs by welding directly through the deck; however, when the deck thickness is greater than 16 gage for single thickness, or 18 gage for each sheet of double thickness, or when the total thickness of galvanized coating is greater than 1.25 oz/sq ft., special precautions and procedures as recommended by the study manufacturer should be followed.

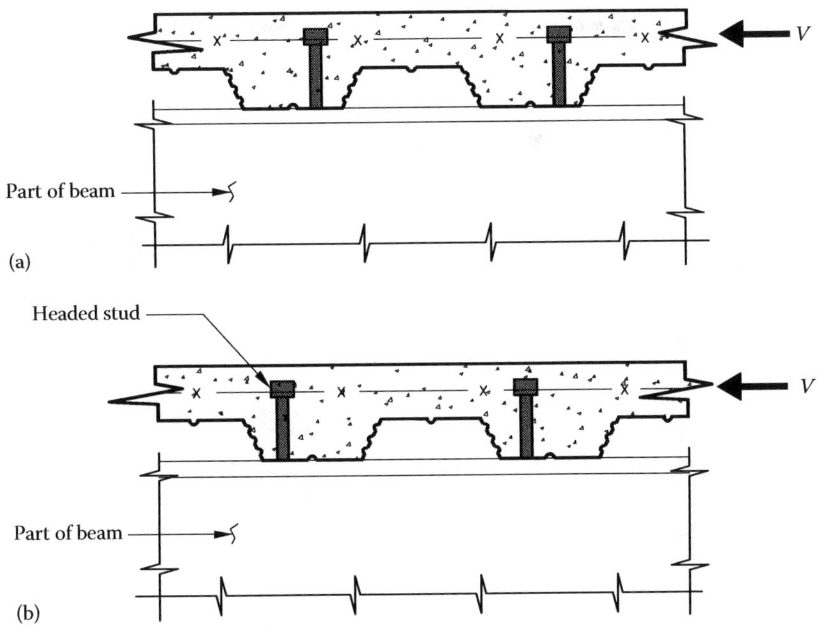

FIGURE 13.6 Possible stud positions: (a) weak position; (b) strong position. Note sets the default value for shear strength equal to that for the stud weak position.

n. The majority of composite steel floor decks used today have a stiffening rib in the middle of each deck flute. Because of the stiffener, studs must be welded off-center in the deck rib.
o. Although it is recommended that studs be detailed in the strong position, ensuring that studs are placed in the strong position is not necessarily an easy task because it is not always easy for the installer to determine where along the beam the particular rib is located, relative to the end, midspan, or point of zero shears. Therefore, the installer may not be clear on which is the strong and which is the weak position. However, it is reassuring to note that even a large change in shear strength does not result in a proportional decrease in the flexural strength of composite beams. (See Figures 13.6a and b for definition of strong and weak stud position.)
p. Uniform spacing of shear connectors is permitted, except in the presence of heavy concentrated loads.
q. Studs not located directly over the web of a beam tend to tear out of a thin flange before attaining full shear-resisting strength. To guard against this mechanism, the size of a stud not located over the beam web is limited to 2½ times the flange thickness. The practical application of this limitation is to select only beams with flanges thicker than the stud diameter divided by 2.5.
r. The minimum spacing of connectors along the length of the beam, in both flat soffit concrete slabs and in formed steel deck with ribs parallel to the beam, is six diameters. If the steel beam flange is narrow, this spacing requirement may be achieved by staggering the studs with a minimum transverse spacing of three diameters between the staggered rows of studs. When deck ribs are parallel to the beam and the design requires more studs than can be placed in the rib, the deck may be split so that adequate spacing is available for stud installation; Figure 13.7 shows possible connector arrangements.

See Figure 13.8 for a graphical display of some of the AISC requirements outlined.

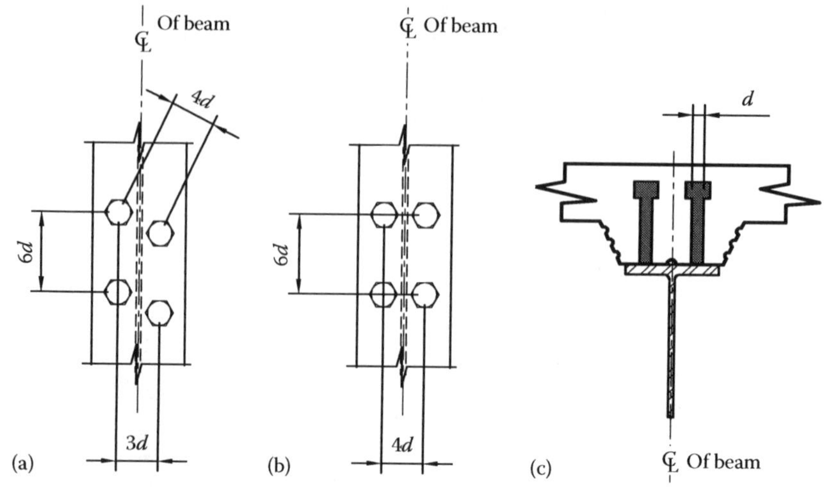

FIGURE 13.7 Shear connector arrangements: (a and b) plans; (c) section.

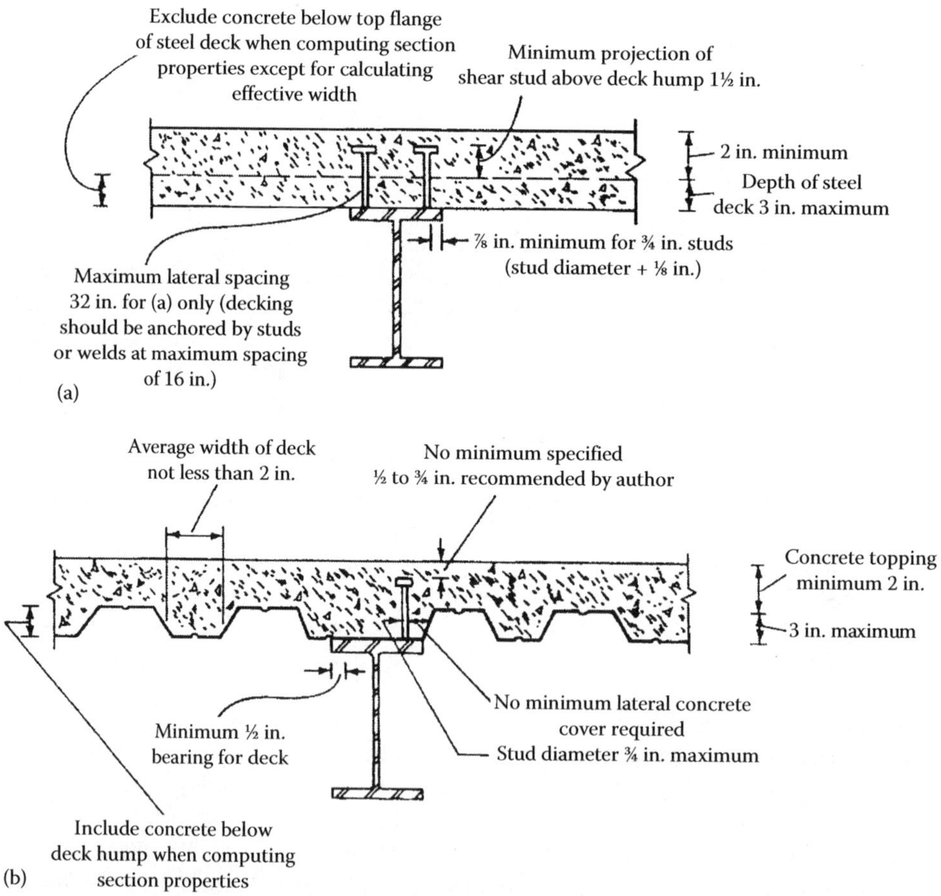

FIGURE 13.8 Composite beam, AISC requirements: (a) deck perpendicular to beam; (b) deck parallel to beam. *Note:* Dimension and clearance restrictions shown in either (a) or (b) apply to both unless noted.

13.4.3 Effective Width

The same effective width rules apply to composite beams with a slab on either one side or both sides of the beam. In cases where the effective stiffness of a beam with a one-sided slab is important, special care should be exercised since this model can substantially overestimate stiffness. To simplify design, effective width is based on the full span, center to center of supports, for both simple and continuous beams.

The *effective width* of the concrete slab shown in Figure 13.9 is the sum of the effective width for each side of the *beam* centerline, each of which shall not exceed

1. One-eighth of the beam span, center to center of supports
2. One-half the distance to the centerline of the adjacent beam
3. The distance to the edge of the slab

For design purposes, a composite floor system is assumed to consist of a series of T-beams, each made up of one steel beam and a portion of the concrete slab. The AISC limits on the width of slab that can be considered effective in the composite action are shown in Figure 13.9. When slab extends on one side of the beam only, as in beams adjacent to floor openings, the effective width naturally is less than when the slab extends on both sides of the beam.

The design of composite beams is achieved by the transformed area method, in which the concrete effective area of the composite beam is transformed into an equivalent steel area.

The method assumes transverse compatibility at the concrete and steel interface. The unit stress in each material is equal to the strain times its modulus of elasticity. Because of strain compatibility, the stress in steel is n times the stress in concrete, where n is the modular ratio E_s/E_c. A unit area of steel is, therefore, mathematically equivalent to n times the concrete area. Therefore, the effective area of concrete $A_c = bt$ can be replaced by an equivalent steel area A_c/n.

For strength calculations, the AISC specification uses the value of n for normal-weight concrete of the specified strength. For deflection computations, n depends not only on the specified strength but also the weight of concrete. Therefore, in computing deflections, especially for beams subjected to heavy sustained loads, it is necessary to account for the effects of creep. This is even more important in shored construction when dead load of the concrete is also resisted by composite action. Creep effect is accounted for in computing deflections by using a higher modular ratio, n.

A multiplication factor of 2 for creep effects appears to be adequate in building designs. Live loads are always resisted by the composite section. If they are of short durations, the deflections are computed using the short-term modular ratio.

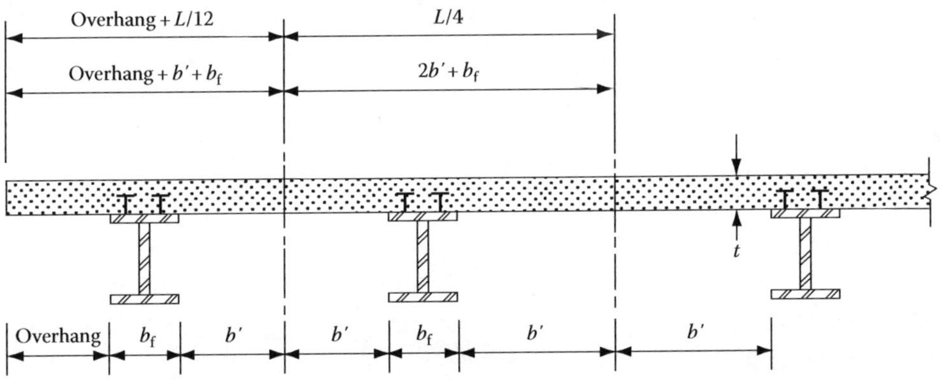

FIGURE 13.9 Effective width concept as defined in the AISC 360-05/10 specifications.

13.4.4 Positive Flexural Strength

The design positive flexural strength $\phi_b M_n$ is determined using ϕ_b, the resistance factor for flexure, equal to 0.9, and M_n, the nominal flexural strength, determined as follows:

a. For $h/t_w \leq 3.76\sqrt{E/F_y}$
   M_n is to be determined from the plastic *stress* distribution of the *composite* section for the limit state *of yielding* (plastic moment).
   It should be noted that all current ASTM A6, W, S, and HP shapes satisfy the aforementioned requirement for $F_y \leq 50$ ksi.
b. For $h/t_w \leq 3.76\sqrt{E/F_y}$
   M_n is determined from the superposition of elastic stresses, considering the effects of shoring for the limit state of yielding.

13.4.5 Negative Flexural Strength

The design negative flexural strength h, $\phi_b M_n$, is determined for the steel section alone using

$$\phi_b = 0.9$$

Alternatively, the available negative flexural strength may be determined from the plastic stress distribution of the *composite* section, provided that

1. The steel *beam* is *compact* and is adequately braced
2. *Shear connectors* connect the slab to the steel beam in the negative moment region
3. The slab reinforcement parallel to the steel beam, within the *effective width* of the slab, is *properly developed.*

13.4.6 Shear Connectors

1. Load transfer for positive moment
   The entire *horizontal shear* at the interface between the steel *beam* and the concrete slab shall be assumed to be transferred by shear connectors, except for *concrete-encased beams.* For *composite* action with concrete subject to flexural compression, the total horizontal shear force, V', between the point of maximum positive moment and the point of zero moment shall be taken as the lowest value according to *the limit states of concrete crushing, tensile yielding* of the steel section, or strength of the shear connectors:
   a. Concrete crushing

$$V' = 0.85 f_c' A_c$$

   b. Tensile yielding of the steel section

$$V' = F_y A_s$$

   c. Strength of shear connectors

$$V' = \Sigma Q_n$$

   where
   A_c is the area of concrete slab with effective width, in.^{2}
   A_s is the area of steel cross section, in.^{2}
   ΣQ_n is the sum of nominal strengths of shear connectors between the point of maximum positive moment and the point of zero moment, kips

2. Load transfer for negative moment

   In continuous *composite beams* where longitudinal reinforcing steel in the negative moment regions is considered to act compositely with the steel *beam*, the total *horizontal shear force* between the point of maximum negative moment and the point of zero moment shall be taken as the lower value of tensile yielding of the steel reinforcement in the slab or strength of the shear connectors:

   a. Tensile yielding of the slab reinforcement

$$V = A_r F_{yr}$$

   where

   A_r is the area of adequately developed longitudinal reinforcing steel within the *effective width* of the concrete slab, in.[2]

   F_{yr} is the specified *minimum yield stress* of the reinforcing steel, ksi

   b. Strength of shear connectors

$$V' = \Sigma Q_n$$

3. Strength of stud shear connectors

   The *nominal strength* of one stud shear connector embedded in solid concrete or in a *composite* slab is

$$Q_n = 0.54 A_{sc}\sqrt{f_c' E_c} \leq R_g R_p A_{sc} F_u$$

   where

   A_{sc} is the cross-sectional area of stud shear connector, in.[2]

   E_c is the modulus of elasticity of concrete $W_c^{1.5}\sqrt{f_c'}$ ksi

   F_u is the specified *minimum tensile* strength of a stud shear connector, ksi

   R_g and R_p are the stud shear capacity reduction factors as described in the following:

   R_g = 1.0 (a) for one stud welded in a steel deck rib with the deck oriented perpendicular to the steel shape, (b) for any number of studs welded in a row directly to the steel shape, and (c) for any number of studs welded in a row through steel deck with the deck oriented parallel to the steel shape and the ratio of the *average rib width* to rib depth >1.5 = 0.85 (a) for two studs welded in a steel deck rib with the deck oriented perpendicular to the steel shape and (b) for one stud welded through steel deck with the deck oriented parallel to the steel shape and the ratio of the average rib width to rib depth ≤1.5 = 0.7 for three or more studs welded in a steel deck rib with the deck oriented perpendicular to the steel shape

   R_p = 1.0 for studs welded directly to the steel shape (in other words, not through steel deck or sheet) and having a haunch detail with not more than 50% of the top flange covered by deck or sheet steel closures = 0.75 (a) for studs welded in a composite slab with the deck oriented perpendicular to the beam and $e_{mid-hi} \geq 2$ in. and (b) for studs welded through steel deck, or steel sheet used as girder filler material, and embedded in a composite slab with the deck oriented parallel to the beam = 0.6 for studs welded in a composite slab with deck oriented perpendicular to the beam and $e_{mid-hi} \geq 2$ in.

   $e_{mid-hi} \geq 2$ in. is the distance from the edge of stud shank to the steel deck web, measured at midheight of the deck rib, and in the *load*-bearing direction of the stud (in other words, in the direction of maximum moment for a simply supported beam)

   w_c is the weight of concrete per unit volume ($90 \leq w_c \leq 155$ lb/ft[3])

Values for R_g and R_p for typical cases are given in Figure 4.11. Note that the stud reduction factors R_g and R_p are equal to unity for composite steel beams with solid slab (Figure 13.3a).

4. Required number of shear connectors
The number of shear connectors required between the section of maximum bending moment, positive or negative, and the adjacent section of zero moment shall be equal to the *horizontal shear force* divided by the nominal strength of one shear connector.
5. Shear connector placement and spacing
Shear connectors required on each side of the point of maximum bending moment, positive or negative, shall be distributed uniformly between that point and the adjacent points of zero moment. However, the number of shear connectors placed between any concentrated *load and* the nearest point of zero moment shall be sufficient to develop the maximum moment required at the *concentrated load* point.

13.4.7 Deflection Considerations

When a selection of composite beam is controlled by deflection, the design should limit the behavior of the beam to the elastic range under serviceability load combinations. In other words, calculate deflections using elastic properties of composite beam and working stress design. It is often not practical to make accurate stiffness calculations of composite flexural members. Therefore, for realistic deflection calculations, one may use a lower bound moment of inertia, I_{lb}, as defined in the following:

$$I_{lb} = I_s + A_s(Y_{ENA} - d_3)^2 + \left(\frac{\sum Q_n}{F_y}\right)(2d_3 + d_1 - Y_{ENA})^2$$

where

A_s is the area of steel crosses section, in.^{2}
d_1 is the distance from the compression force in the concrete to the top of the steel section, in.
d_3 is the distance from the resultant steel tension force for full section tension yield to the top of the steel, in.
I_{lb} is the lower bound moment of inertia, in.^{4}

where

S_s is the section modulus for the structural steel section, referred to the tension flange, in.^{3}
S_{tr} is the section modulus for the fully composite uncracked transformed section, referred to as the tension flange of the steel section, in.^{3}

These equations should not be used for ratios, $\Sigma Q_n/C_f$, less than 0.25. This restriction is to prevent excessive slip, as well as substantial loss in beam stiffness.

I_s is the moment of inertia for the structural steel section, in.^{4}

ΣQ_n is the sum of the nominal strengths of shear connectors between the point of maximum positive moment and the point of zero moment to either side, kips:

$$Y_{ENA} = \left[\frac{\left(A_s d_3 + \left(\frac{\sum Q_n}{F_y}\right)(2d_3 + d_1)\right)}{\left(A_s + \left(\frac{\sum Q_n}{F_y}\right)\right)}\right]$$

The use of constant stiffness in elastic analyses of continuous beams is analogous to the practice in reinforced concrete design. The stiffness calculated using a weighted average of moments of inertia in the positive moment region and negative moment regions may take the following form:

$$I_1 = aI_{pos} + bI_{neg}$$

where

I_{pos} is the effective moment of inertia for positive moment, in.^{4}
I_{neg} is the effective moment of inertia for negative moment, in.^{4}

The effective moment of inertia is based on the cracked transformed section considering the degree of composite action. For continuous beams subjected to gravity loads only, the value of a may be taken as 0.6 and the value of b may be taken as 0.5 for calculations related to drift.

For cases where elastic properties of partially composite beams are needed, the elastic moment of inertia may be approximated by

$$I_{equiv} = I_s + \sqrt{\left(\frac{\sum Q_n}{C_f}\right)}\left(I_{tr} - I_s\right)$$

where

I_s is the moment of inertia for the structural steel section, in.^{4}
I_{tr} is the moment of inertia for the fully composite uncracked transformed section, in.^{4}
ΣQ_n is the strength of shear connectors between the points of maximum positive moment and the points of zero moment to either side, kips
C_f is the compression force in concrete slab for fully composite beam, smaller of A_sF_y and $0.85 f_c' A_c$, kips
A_c is the area of *concrete* slab within the effective width, in.^{2}

The effective section modulus S_{eff}, referred to the tension flange of the steel section for a partially composite beam, may be approximated by

$$S_{eff} = S_s + \sqrt{\left(\frac{\sum Q_n}{C_f}\right)}\left(S_{tr} - S_s\right)$$

13.4.8 Design Outline for Composite Beam

1. Material limitations
   a. Normal-weight concrete: 3 ksi ≤ f_c' ≤ 10 ksi
   b. Lightweight concrete: 3 ksi ≤ f_c' ≤ 6 ksi
   c. Structural steel and mild steel reinforcement: f_y ≤ 75 ksi
2. The design positive flexural strength, Q_bM_n, shall be determined from a plastic stress distribution using $Q_b = 0.90$.
3. The design negative flexural strength QM_n is determined from the steel section alone.
4. For positive moment region, the effective width of concrete flange in each side of the beam shall not exceed
   a. One-eighth of the beam span
   b. One-half the distance to the adjacent beam
   c. The distance to the edge of the slab

In this day and of computers ruling the world, it is beyond the bounds of possibility or reason to expect engineers to perform hand calculations for designing composite beam and for that matter any other structural component. However, in keeping up with the spirit of this book, in lieu of intense calculations, we will outline only the required design steps. The purpose of this indulgence is to enhance our understanding of what goes on in the black box, the computer.

The design steps are the following:

1. *Limit material properties as follows*:
   Normal-weight concrete: 3 ksi $\leq f_c' \leq$ 10 ksi
   Lightweight concrete: 3 ksi $\leq f_c' \leq$ 6 ksi
   Structural steel and reinforcing: $F_y \leq$ 75 ksi
2. *Loads.*
   Dead loads typically consist of slab and beam weight, miscellaneous of 10 psf for ceiling, etc., and an allowance of 15 psf for partitions (when live loads are less than 100 psf):
   Live loads shall be as specified in the applicable building codes.
   Use higher loads if specified by the building user.
   Reduce live loads if permitted by the applicable building code.
3. *Determine the required flexural strength.*
   In the most typical case of simply supported beams subject to uniformly distributed load, w_u, and span ℓ

$$M_u = \frac{w\ell^2}{8}$$

4. *Check the beam deflections and available strength.*
   Check the deflection of the steel beam under construction loads considering only the weight of steel beam and concrete topping and an allowance of 20 psf for construction live loads as contributing to the construction load.
   Limit deflection to a maximum of 2.5 in. to facilitate concrete placement. Camber the beam if the calculated deflection is greater than $^3/_4$ in. Provide camber equal to 75% of the calculated dead load deflection:

$$I_{req} = \frac{5}{384} \frac{w_{DL} l^4}{E\Delta}$$

   Revise trial section if I_{req} is greater than that of the trial section.
5. *Check selected member strength as an unshored beam under construction loads assuming adequate lateral bracing through the deck attachment to the beam flange.*
   Also determine ΣQ_u. Calculate the required strength of steel beam alone using the weight of the steel beam and concrete slab as dead loads, and an allowance of 20 psf construction load. Label this load as $w_{(const)}$.
6. *Determine* b_{eff}.
   The effective width of the concrete slab is the sum of the effective widths for each side of the beam centerline, which shall not exceed
   1. One-eighth of the beam span, center to center of supports
   2. One-half the distance to center line of the adjacent beam
   3. The distance to the edge of the slab
7. *Calculate the moment arm for the concrete force measured from the top of the steel shape, Y2.*
   Assume a = 1.0 in. (Some assumptions must be made to start the design process. An assumption of 1.0 in. has proven to be a reasonable starting point in many design problems.)

$$Y2 = t_{slab} - a/2$$

Enter AISC Steel Design Manual, Table 3-19 with the required strength and the calculated value of $Y2$. Select a beam and a plastic neutral axis location that indicates sufficient available strength. The trial beam is okay if $\phi_b M_n = 0.9M_n \geq M_u$:

$$M_{u(unshored)} = \frac{w(const) \times \ell^2}{8}$$

$$\phi_b M_{n(steel)} \geq M_{u(unshored)}$$

Using the assumed values of $Y2$ and the plastic neutral axis, PNA, determine the required horizontal shear ΣQ_n.

8. *Check a using the relation*

$$a = \frac{\Sigma Q_u}{0.55 f'_c b}$$

If the calculated value of a is less than the assumed value, the design is OK.

9. *Check live-load deflection.*

$$\Delta_{LL} = \frac{5}{384}\frac{w_{LL} l^4}{EI_{LB}}$$

10. *Determine if the beam has sufficient available shear strength.*
    $\phi V_n \geq V_u$ (This is not typically an issue for uniformly loaded beams.)
11. *Determine the required number of shear connectors = 2N.*
    With the known direction and profile of steel deck, determine shear capacity of a 1¾ in. diameter stud per rib for the given strength and weight of the concrete slab. Here, we identify ¾ in. diameter studs, because they are the most common type used in building construction, in North America:

$$N = \frac{\Sigma Q_n}{Q_n} \text{(on each side of the beam)}$$

12. *Check the spacing of shear connectors.*
    Use one stud every flute, starting at each support. If studs remain, double up studs on each end of the span. Check spacing requirement:

$$6d_{stud} < 12. < 8t_{slab}$$

Note that stud length shall extend a minimum of 1½ in. into slab after welding.

13.4.9 Composite-Beam Design Examples

Example 1

Given

Composite floor beams at 10 ft spacing. The beam is W21 × 44 and has 40 shear studs ¾ in. diameter by 5 in. long, equally spaced along the beam span. The span is 40 ft with an effective width of 10 ft.

DL = 70 psf including the weight of the beam, concrete, floor finishes, ceiling, Allowance for mechanical and other partition load = 20 psf

$$LL = 115 \text{ psf}$$

The floor construction consists of 3¼ in. normal-weight, 3000 psi concrete on 3 in. deep composite steel deck.

Required

Flexural capacity check of the composite beam using LRFD

Solution

1. Determine the required strength of beam using the factored load combination.

$$1.2D + 1.6L$$

$$D = 10 \text{ ft} \times 70 \text{ psf} = 700 \text{ plf} = 0.7 \text{ klf}$$

$$L = 10 \text{ ft } (20 \text{ psf} + 110 \text{ psf}) = 1300 \text{ plf} = 1.3 \text{ klf}$$

(**Note:** partition load is considered a live load)

$$1.2D + 1.6L = 1.2 \times 0.7 + 1.6 \times 1.3 = 2.92 \text{ klf}$$

$M_n = 2.92 \times 40^2/8 = 584$ ft-k (nominal strength)
Design strength = $\phi_b M_n = 0.85$
ϕ_b = the resistance factor = 0.85
M_n = nominal flexural strength obtained from the plastic distribution

2. Determine the nominal horizontal shear capacity, Q_n, of one stud. Assuming shear studs are placed two per rib, in the strong stud position, from Table 3-21 of *AISC Steel Construction Manual*, 13th Edition, we get

$$Q_n = 18.3 \text{ kips}$$

for $^3/_4$ in. diameter studs placed in normal-weight concrete (W_c = 145 pcf, f_c = 3 ksi)

3. Determine the governing plastic limit state, and the design flexural strength, ϕM_n.
When a composite beam reaches plastic limit state, the stresses will be distributed in one of the three ways shown in Figure 13.15B.
To determine which of the three cases governs, compute the compressive resultant as the smallest of
   a. $A_s F_y = 13 \times 50 = 650^K$
   b. $0.85 f_c' A_c = 0.85 \times 3 \times 10 \times 12\ (6.25-) = 994.5^K$
   c. $\Sigma Q_n = 44/2 \times 18 = 396^K$

Note: Because W21 × 44 is a beam with deck perpendicular to span, concrete below the steel deck is neglected in calculating the compressive resultant.

Check the lower limit for partial composite action

$$\Sigma Q_n \geq 0.25 A_s F_y$$

$$396 > 0.25 \times 650 - 162.5^K$$

Note: The limitation $\Sigma Q_n \geq 0.25 A_s F_y$ is not required by the AISC specification but is deemed to be a practical minimum value.

To use the AISC LRFD design tables, first find the portion of the table corresponding to the steel shape and grade and proceed as follows:

Note that the table for W21 × 44, grade 50 steel is on page 5–43 of the *AISC Steel Construction Manual*, 13th Edition.

a. Select ΣQ_n. This is the manual's notation for the compressive force C, the smallest of A_sF_y, $0.85f_CA_C$, and the total shear connector strength, ΣQ_n.
b. Determine the depth of compression block a, corresponding to $C = 396^K$. The depth of compression block is

$$a = C/0.85f_c'b = 396/0.85 \times 3 \times 120 = 1.294 \text{ in.} \approx 1.3 \text{ in.}$$

c. Calculate $Y2$, the distance from the top of steel to compressive force C

$$Y2 = t - a/2 = 6.25 - 1.3/2 = 5.6 \text{ in.}$$

d. Enter the tables with $\Sigma Q_n = 396^K$ and $Y2 = 5.6$ in. and read the corresponding QM_n.

The closest values we can read without interpolation are for $\Sigma Q_n = 358k$ and $Y2 = 5.5$ in., $\Sigma Q_n = 610$ kft. Therefore, W21 × 44, 50 ksi beams with 44 shear studs are okay in bending for the given loads.

Example 2: Girder Design

Given

Interior W24 × 76, 36 ksi existing steel girder with 16 – ¾ in. × 5 in. studs

Span $L = 28.5$ ft, effective width $b = 85.5$ in., $^3/_4$ in. lightweight concrete on 3 in. deep composite deck

Nominal stud shear strength = 18 kips

Find the moment capacity

Q_bM_n using LRFD

Compute the smallest compressive resultant:

1. $A_sF_y = 22.4 \times 36 = 806.4$ kips
2. $0.85f_C'A_C = 0.85 \times 3 \times 85.5\ (3.25 + 1.5) = 1035.6$ kips
3. $\Sigma Q_n = 16/2 \times 18 = 144$ kips

Note: $\Sigma Q_n = 144$ is less than $0.25A_sF_y = 0.25 \times 806.4 = 201.6$ kips

However, this limitation is not required by AISC specification.

Had this design been for a new beam perhaps we would have used a minimum of $16 \times 201.6/144 = 22.4$, say, 23 studs; because we are verifying the capacity of an existing beam, it is reasonable to not meet this nonmandated criteria.

The table for W24 × 76, 36 ksi steel is on page 5–25:

1. Select $SQ_n = 144$ kips as noted earlier.
2. Determine depth of compression a:

$$a = c/0.85 \times f_c'b = 144/0.85 \times 3 \times 85.5 = 0.66 \text{ in.}$$

3. Calculate $Y2$:

$$Y2 = t - a/2 = 6.25 - 0.66/2 = 5.91 \text{ in.}$$

Conservatively, use 5.5 in. in reading the table.

4. Enter the tables with $\Sigma Q_n = 144$ kips and $Y2 = 5.5$.
   Note: the lowest value of ΣQ_n in the tables is 202, which is equal to $0.25A_sF_y$. Therefore, we interpolate the value of QM_n:

   $Q_bM_n = 716$ kft for $\Sigma Q_n = 202$ kips

   $Q_bM_n = 773$ kft for $\Sigma Q_n = 284$ kips

   Q_bM_n for $\Sigma Q_n = 144$ kips $= 716 - (773–716)/(284–202) \times (202–144) = 716 - 40.03 = 675.7$ kft

13.5 COMPOSITE JOISTS AND TRUSSES

13.5.1 Composite Joists

Composite floors using joists and trusses typically involve simply supported members. The joists and trusses are linked by shear connectors to the concrete slab to form an effective T-beam to resist gravity loads. The versatility of the system results from the inherent strength of the concrete floor component in compression and the ability of the joists and trusses to span long distances. Composite floor systems are advantageous in reducing material cost, on-site labor, and construction time. They also result in reduced structural depth. The composite action of the framing element is by direct welding of shear studs through the steel deck while the composite action of the steel deck as flexural reinforcement for the concrete slab itself results from side embossments incorporated into the steel-sheet profile. The slab-and-beam arrangement typical in composite floor systems produces a horizontal diaphragm that provides stability to the overall building system while distributing wind and seismic shears to the lateral-load-resisting elements.

Pre-engineered and proprietary open-web floor joists, joist girders, and fabricated floor trusses when combined with a concrete floor slab are viable composite members. The advantages of the system are the increased span length and stiffness due to the greater structural depth and ease in accommodating electrical conduits, plumbing and heating, and air-conditioning ductwork. The now nonexistent WTC Towers, New York, used composite open-web joists.

13.5.2 Composite Trusses

Built-up fabricated composite floor trusses combine material efficiency in relatively long-span applications with maximum flexibility for incorporating building services ductwork and piping into the ceiling cavity. The triangular openings formed by web members of the truss allow the passage of large mechanical air ducts as well as other piping and electrical lines. The increased depth of the composite truss system over a standard rolled shape composite-beam system, with building services ductwork and piping passing below the beam, results in maximum material efficiency and high flexural stiffness. Generally, composite floor trusses are considered economical for floor spans in excess of 45 ft. A further requirement for floor truss systems is that the framing layout must be uniform and repetitive, resulting in relatively few types of trusses, which can be readily built in the fabrication shop. Otherwise, the high level of fabrication inherent in the floor truss assemblage tends to offset the relative material efficiency. For this reason, composite floor truss systems are particularly attractive in high-rise office building applications where large column-free areas are required and floor configurations are generally repetitive over the height of the building.

Shown in Figure 13.10 are schematics of floor framing with composite trusses. Observe that the top and bottom chords consist of T-sections to allow for direct welding of double-angle web members without using gusset plates. When the space between the diagonals is not sufficient to allow passage of mechanical and air-conditioning ducts, vertical webs may be welded between the truss chords to form a Vierendeel panel (see Figure 13.11).

A variety of chord and web member cross sections may be used in developing a truss system. The previous example showed the chords may be T-sections for top and bottom to allow direct connection of web members without gusset plates. Web members are most often tees, single-angle or

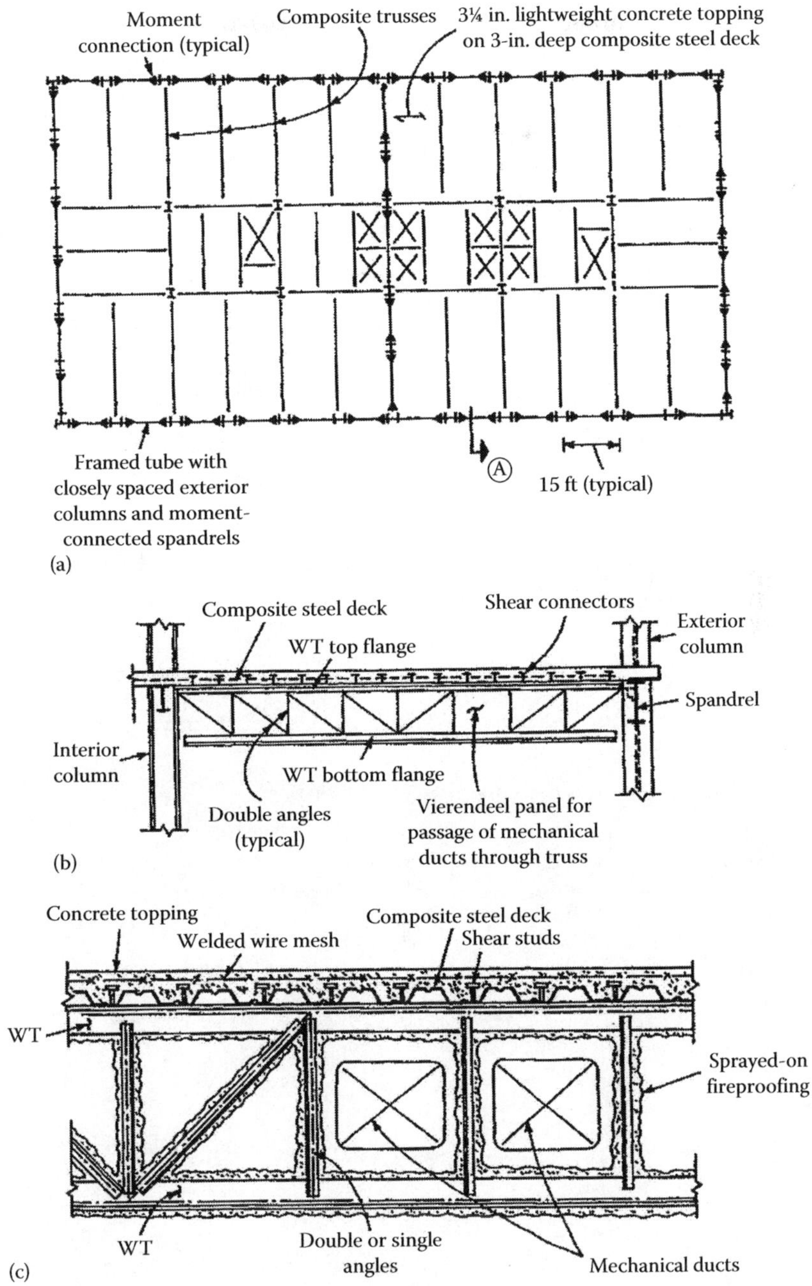

FIGURE 13.10 Composite truss: (a) floor framing plan; (b) section A; (c) Vierendeel panels.

double-angle sections welded directly to the chord tee. Some examples are shown for Figure 13.11a. The composite floor trusses are connected to the concrete floor slab by shear connectors welded to the top chord of the truss. The floor trusses are normally spaced such that the steel deck can span between the trusses without shoring. The space between the diagonals is used for the passage of mechanical and air-conditioning ducts. When the space between the diagonals is not sufficient, vertical members may be welded between the chords to form a Vierendeel panel.

Figure 13.11 shows some examples of composite truss elevations while in Figure 13.12 schematic cross sections illustrating truss components are shown.

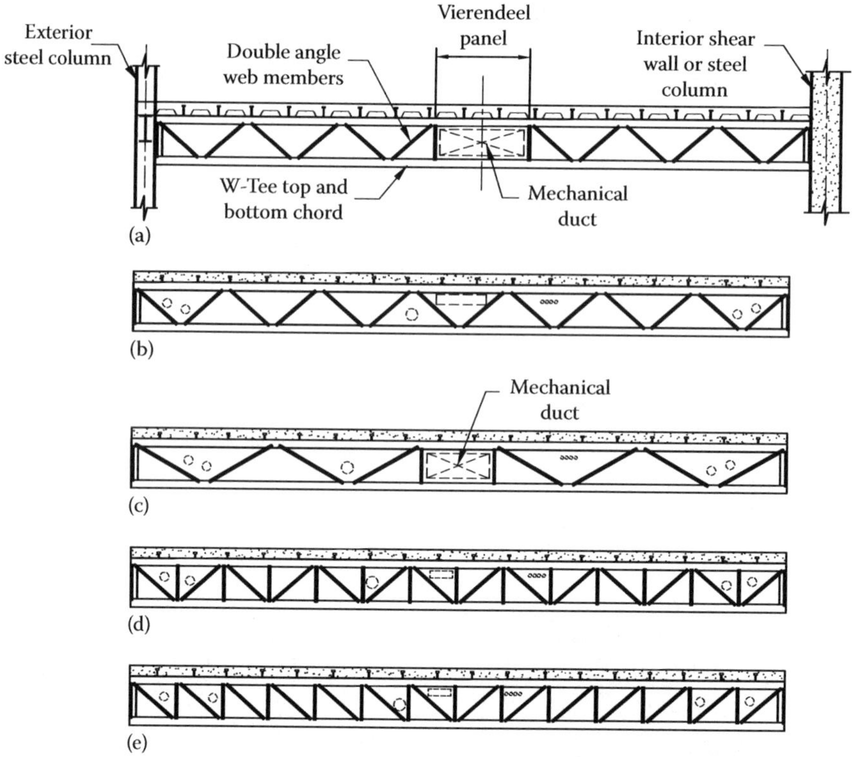

FIGURE 13.11 Examples of composite floor trusses: (a–e) elevations.

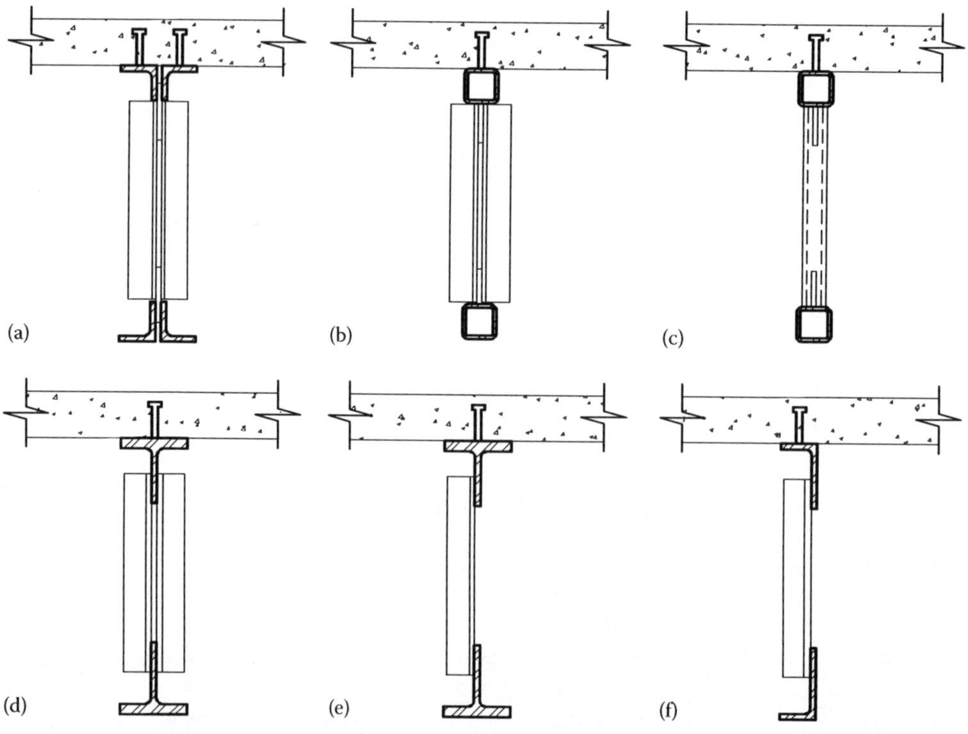

FIGURE 13.12 Composite truss schematic sections: (a–c) top and bottom chords with gusset plates; (d–f) top and bottom chords without gusset plates.

13.6 OTHER TYPES OF COMPOSITE FLOOR CONSTRUCTION

The slab elements of a composite floor system may take the form of a flat soffit reinforced concrete slab, precast-concrete planks, or floor panels with a cast-in-place topping. See Figures 13.13 and 13.14.

Early composite floor systems used concrete-encased steel beams to support a formed reinforced concrete slab spanning between the supporting beams. The concrete encasement of the steel beam was eliminated with the development of economical lightweight sprayed-on fireproofing. The reinforcement for the slab in the direction perpendicular to the beam span is determined through conventional continuous reinforced concrete design. Light slab reinforcement is placed parallel to the beam span to control shrinkage and thermal cracking.

To eliminate additional cost in temporary formwork for the slab, precast-concrete planks may be used effectively as permanent formwork, which also provides an effective working platform for the construction trades. Precast, prestressed hollow-core concrete planks spanning between the steel floor beams are well suited for jobs where a repetitive organization of beam spacings allow for effective prefabrication of the planks. Two to three inches of concrete is cast atop the planks to provide an effective diaphragm as well as a level, continuous floor surface. The concrete topping is reinforced, usually with welded wire fabric, to control concrete cracking. In this type of construction, the topping extends down to the surface of the steel beams. Note that the shear studs extend up into the topping above the hollow concrete planks. See Figure 13.14 for schematics.

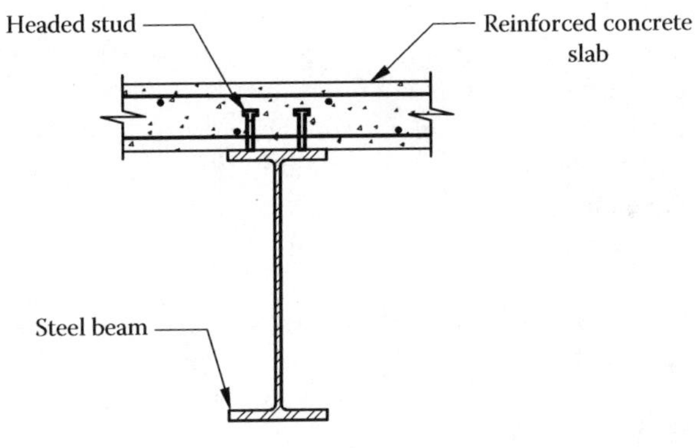

FIGURE 13.13 Composite beam with flat soffit reinforced concrete slab.

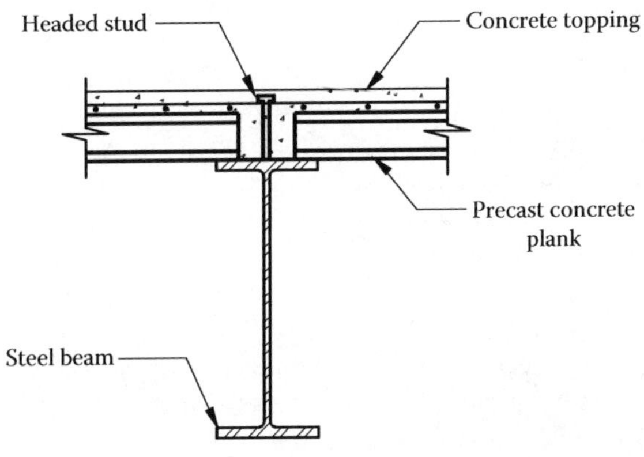

FIGURE 13.14 Composite beam with precast concrete plank and topping slab.

13.7 CONTINUOUS COMPOSITE BEAMS

Successful composite-beam design requires consideration of serviceability issues such as floor vibrations. Of particular concern is the issue of perceptibility of occupant-induced floor vibrations.

The high flexural stiffness of composite floor framing systems results in relatively low vibration amplitudes from transitory heel-drop excitations and therefore is effective in reducing perceptibility. Typically, composite floor framing systems perform quite well and rarely transmit annoying vibrations to the occupants. However, particular care for spans in the 30–40 ft range is recommended. Anticipated damping provided by partitions, ceiling construction, and the structure itself should be considered in conjunction with state-of-the-art prediction models to evaluate the potential for perceptible floor vibrations.

The serviceability characteristics of composite beams may be significantly improved through the use of semirigid connections. Such a typical semirigid joint is shown in Figure 13.13 in which partial end restraint is mobilized through the addition of a bottom angle and heavier reinforcement within the concrete slab in a band centered on the steel column. For unshored construction, the initial dead-load deflection is hardly altered, but live-load deflections, and quite possibly floor vibrations, are significantly reduced through the partial end restraint. Reduced positive moment results in a saving in steel tonnage. The partial end restraint may be used to shorten the effective unbraced length of the column as well. Full benefit of the system is obtained through shored construction where all loads are resisted by the semirigid composite connection. However, application of semirigid connections is limited because of the difficulty in determining the moment–rotation characteristics of the joint.

A composite beam subject to negative bending moment experiences tensile stresses in the concrete zone and therefore loses much of its advantage. However, when mild steel reinforcement is placed parallel to the beam within the effective width of slab, and is anchored adequately to develop the tensile forces, the advantage of continuous construction is restored. The reinforcing steel used in the tensile zone is included in computing the property of the composite section. Similarly, when the compressive stress in concrete subject to positive moment exceeds the allowable stress, it is permissible to use compressive steel in the effective width zone to reduce stresses.

Design negative flexural strength, $\phi_b M_n$ may be determined for the steel section alone. However, it is prudent to include the contribution of longitudinal reinforcement by considering plastic stress distribution in the composite section. Schematics of plastic stress distribution for negative and positive moments are shown in Figure 13.15.

13.8 NONPRISMATIC COMPOSITE BEAMS AND GIRDERS

A prismatic composite steel beam has two apparent disadvantages over other types of composite floor framing. One, the member must be designed for the maximum bending moment occurring near midspan and thus is understressed at all other sections along the span and, two, building services ductwork and piping must pass beneath the beam or the beam must be provided with penetrations (usually reinforced with plates leading to high fabrication costs). For this reason, a number of composite girder articulations that allow passage of mechanical ducts and related services through the depth of the girder have been developed. These include tapered and dapped girders, castellated beams, and stub-girder systems.

Use of tapered composite girders with a maximum depth at center span is limited, as the main mechanical duct loop normally runs through the center of the beam span rather than at each end. On the other hand, a castellated beam has a constant depth. It is formed from a single rolled wide-flange

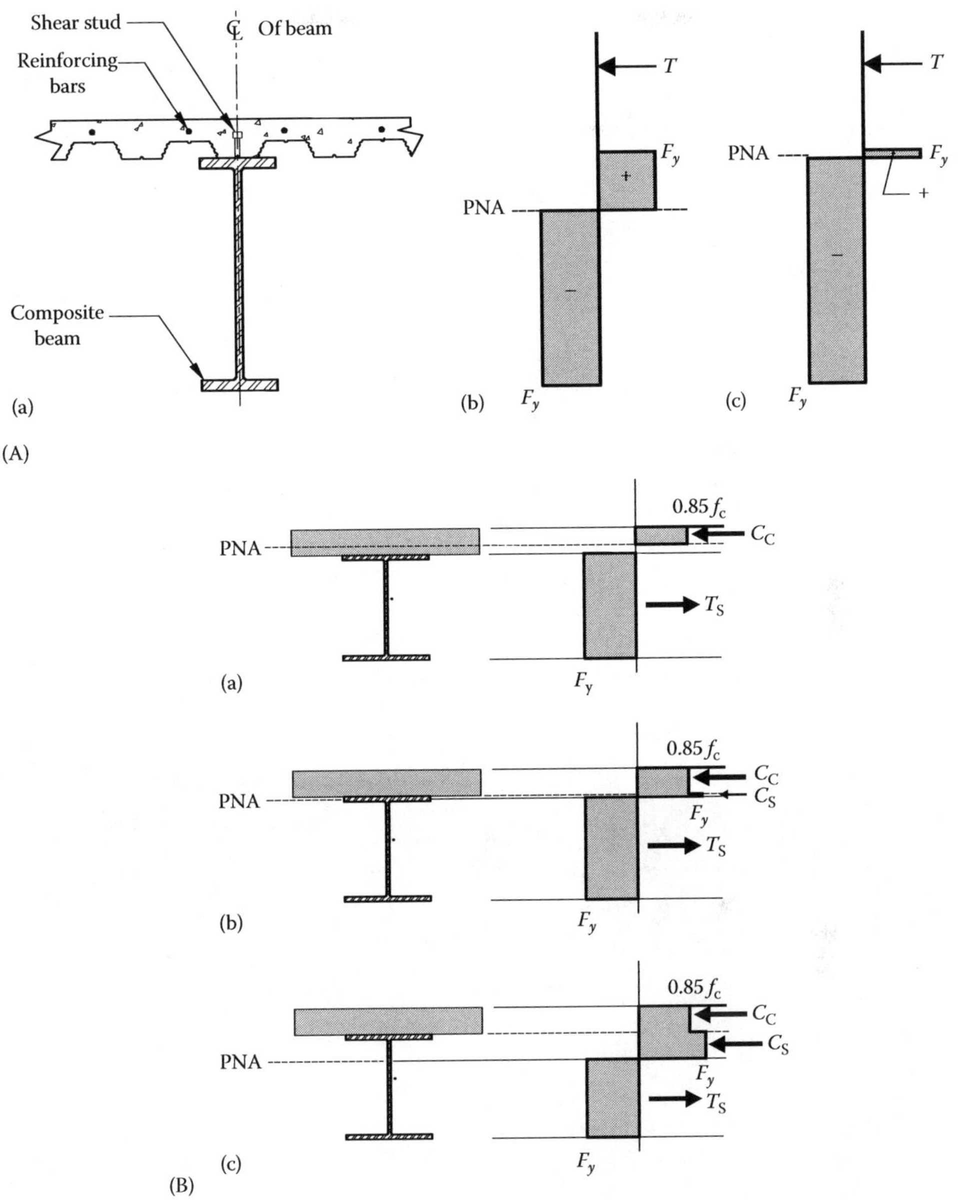

FIGURE 13.15 Composite beam with precast concrete plank and topping slab.

steel beam cut and reassembled by welding. The resulting beam has increased depth and provides for hexagonal openings over the entire span. However, its use in the United States is somewhat limited because of increased fabrication costs, and standard castellated openings may not be large enough to accommodate the required mechanical ductwork.

Continuous transverse secondary beams are positioned over the girder while the mechanical ducts pass through the openings formed between the beam stubs. This system, quite popular in the 1970s and 1980s, has been used in many building projects totaling well over 30 million square feet. The system, however, requires shoring that may offset some of the cost savings.

13.9 MOMENT-CONNECTED COMPOSITE HAUNCH GIRDERS

Composite haunch girders, although not often used as a floor framing system, merit mention because they minimize the floor-to-floor height without requiring complicated fabrication. As an example, shown in Figure 13.16 is a schematic floor plan in which composite haunch girders frame between exterior columns and interior core framing. The haunch girder typically consists of a shallow steel beam, 10–12 in. deep for span in the 35–45 ft range. At each end of the girder, a triangular haunch is formed by welding a diagonally cut wide-flange beam usually 24–27 in. deep. See Figure 13.16. The haunch is welded to the shallow beam and to the columns at each end of the girder. In this manner, the last 8 or 9 ft of the haunch girder at either end flares out toward the column with a depth varying from about 10 to 12 in. at the center to about 27 in. at the ends. The system uses less steel and provides greater flexibility for mechanical ducts, which can be placed anywhere under the shallow central span. The reduction in floor-to-floor height further cuts costs of exterior cladding and heating and cooling loads. The system, however, is not common because of higher fabrication costs.

A variation of the same concept shown in Figure 13.17 uses nontapered haunches at each end. The square haunch girder can be fabricated using a shallow-rolled section in the center and two relatively deep wide-flange sections, one at each end. Another method of fabricating the girder is to notch the bottom portion of the girder at midspan and reweld the flange to the web. The method requires more steel but comparatively less fabrication work.

In comparison to a shallow girder of constant depth, a haunch girder is significantly stiffer. Hence, its use as a frame–beam for resisting lateral loads warrants due considerations.

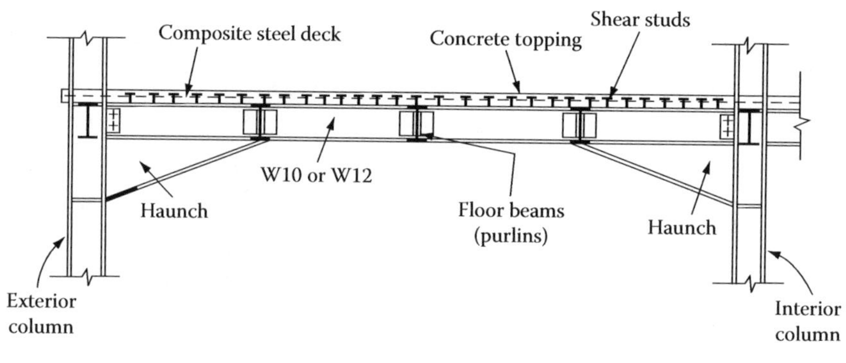

FIGURE 13.16 Composite girder with tapered haunch.

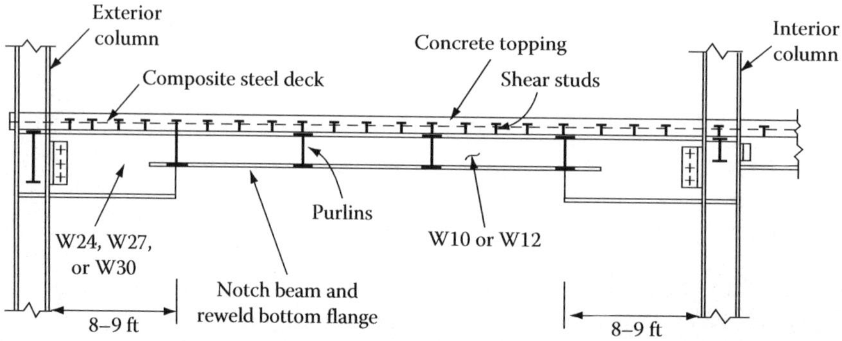

FIGURE 13.17 Composite girder with square haunch

13.10 COMPOSITE COLUMNS

A steel-concrete composite column is a compression member, comprising either a concrete-encased hot-rolled steel section or a concrete-filled tubular section of hot-rolled steel, and is generally used as a load-bearing member in a composite framed structure. Figures 13.18 and 13.19 show typical cross sections of concrete-filled tubular sections. Note that there is no requirement to provide additional reinforcing steel for composite concrete-filled tubular sections, except for requirements of fire resistance where appropriate.

In a composite column, both the steel and concrete resist the external loading by interacting together by bond and friction. Supplementary vertical reinforcement in the concrete encasement increases capacity but must be tied adequately to prevent excessive spalling of concrete.

In composite construction, the bare steel sections support the initial construction loads, including the weight of structure during construction. Concrete is later cast around the steel section or filled inside the

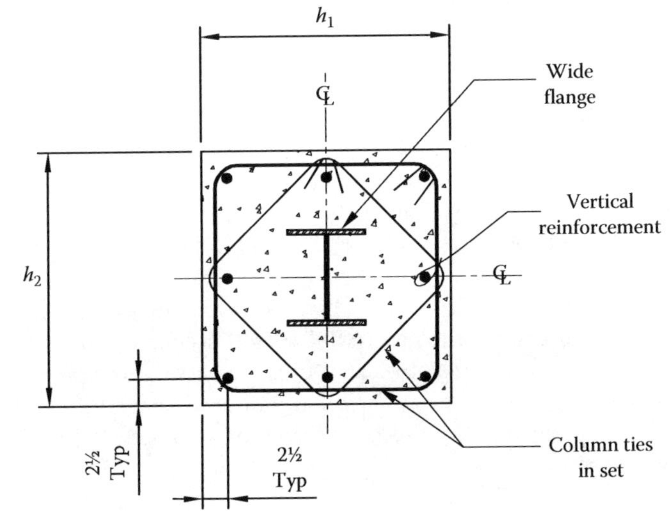

FIGURE 13.18 Encased composite column.

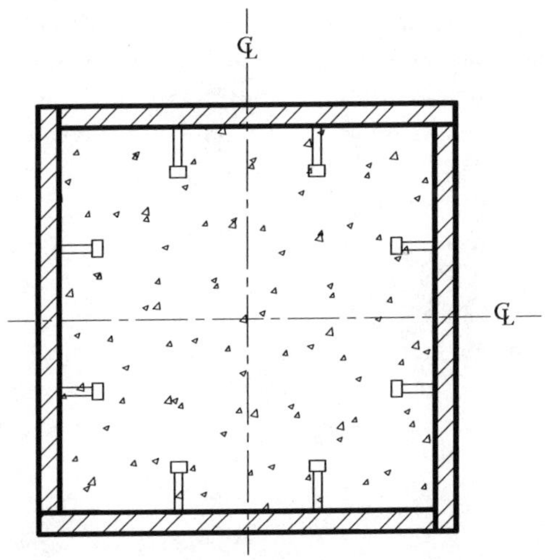

FIGURE 13.19 Filled composite column.

tubular sections. The concrete and steel are combined in such a fashion that the advantages of both the materials are utilized effectively in composite column. The subsequent concrete addition increases the strength and stiffness of columns, enabling the building frame to limit the sway and later deflections.

The use of composite columns along with composite decking is now a common design in high-rise structures. There is quite a vertical spread of construction activity carried out simultaneously at any one time, with numerous trades working simultaneously. The following are examples:

- One group of workers will be erecting the steel beams and columns for one or two stories at the top of frame.
- Two or three stories below another group of workers will be attaching the steel decking for the floors.
- A few stories below, another group will be placing concrete on steel deck concrete.
- Further down the building, another group will be tying the column reinforcing bars.
- Yet another group below this crew will be placing concrete into the column form work.

The *advantages* of composite columns are

- Increased strength for a given cross-sectional dimension
- Increased stiffness, leading to reduced slenderness and increased buckling resistance
- Good fire resistance for concrete-encased columns
- Corrosion protection for encased columns
- Possible economic advantages over either pure steel or reinforced concrete alternatives

The term *composite column* in building industry represents a unique form of construction in which structural steel interacts compositely with concrete. The steel section can be a tubular section filled with structural concrete or it could be used as a central core steel section encased in reinforced concrete.

Historically, composite columns evolved from concrete encasement of structural steel shapes primarily intended as fire protection. Although the increase in strength and stiffness of steel members due to concrete encasement was intuitively known, it was not until the 1940s that composite design methods were developed. In fact, in the earlier days, the design of the steel column was penalized by considering the weight of concrete as an additional dead load on the steel column. Later developments took into consideration the increased radius of gyration of columns because of concrete encasement, resulting in nominal reduction in structural steel weight. In some earlier high-rise designs, concrete encasement was ignored from strength considerations but the additional stiffness due to concrete was included in the lateral deflections calculations.

After the development of sprayed-on fireproofing in the 1950s and 1960s, use of concrete for fireproofing of structural steel was no longer an economical proposition. The high formwork cost of concrete could not be justified for fireproofing. However, this plight was overcome if one could include the stiffness and strength properties of concrete in the design of columns.

Therefore, over the last 40 years, the use of encased steel columns has found applications in buildings varying from as low as 10 stories to as high as 70 stories.

Composite systems have successfully captured the traditional advantages steel and concrete construction: the speed of steel construction and the increased stiffness and moldability of concrete. Concrete columns with relatively small steel cores used as erection columns were perhaps the earliest applications. Later, much heavier steel sections were used to limit the size of composite vertical elements.

13.10.1 Behavior

In a broad sense, the behavior of a composite column is a combination of the response of reinforced concrete and structural steel section. Generally, composite interaction is synergistic;

that is, it enhances the performance of the whole to something superior to the sum of the component parts.

The basic function of *columns* is the delivery of gravity loads to the base of the structure. However, columns can be more than compression members. For example, frame columns that are connected to beams with moment-resisting connections restrain the bending deflections of floor members as well as the lateral drift of the overall structure. Their design must be chosen to satisfy both axial and flexural demands.

Concrete is a material with reliable compressive strength and low cost per square inch of the cross section. However, an axially loaded structural member without lateral restraint requires not only strength but also flexural stiffness when axial forces must be delivered over significant lengths. In other words, the column needs to be strong *and* stable.

The stability of slender columns is a measure of material stiffness rather than material strength. Steel, which has five to eight times the stiffness and strength of concrete, is the more efficient structural material for slender columns. In building applications, structural columns are neither slender nor stocky. Instead, most are in the intermediate range between these two extremes.

Two basic types of composite columns are used in buildings: those with the steel section encased in concrete and those with the steel section filled with concrete.

Encased composite columns consist of structural shapes surrounded by structural concrete. The concrete encasement requires vertical and horizontal reinforcement to sustain the encasement of the steel core. Shear connectors may be needed as well to ensure interaction and force transfer between the steel shape and the concrete encasement. The shear connectors transfer forces between the steel and concrete through attachment by welds to the steel shape and by bearing against the surrounding concrete.

Filled composite columns are perhaps the most efficient application of the two materials for column cross sections. The steel shell can be a pipe, tubing or a hollow section fabricated from plates. The steel shell provides transverse confinement to the contained concrete, making the filled composite column ductile and tough to survive overloads. Since the concrete core is confined by the steel shell, interaction between the steel and concrete is assured. However, it may be desirable to provide additional bearing surfaces for shear transfer using welded studs or reinforcing steel particularly near the connections of the columns to the floor beams.

Design of composite columns for combined axial force and flexure is most often accomplished by the plastic distribution method. To assist in developing the load–moment interaction curve, a series of equations is provided in the AISC specifications for specific cases of reinforcement arrangement. When design cases deviate from these cases, the appropriate interaction equation can be derived from first principles. Slenderness effects are not directly considered in the derivation of interaction curves.

13.10.2 Encased Composite Columns: Design Overview

- Limitations
  1. Steel shape must be at least 1% of gross composite section; must have longitudinal reinforcing bars, at least 0.004 times the gross area; and must have lateral ties at least 0.009 in.^{2}/in. of tie spacing.
  2. Limitations on material strengths are the same as for the composite beams.
- The design compressive strength, $\phi_c P_n$, for the axially loaded columns is determined for the limit state of flexural buckling using $\phi_c = 0.75$.
- Maximum nominal compression strength, P_o, is determined as

$$P_o = A_s F_y + A_{sr} F_{yr} + 0.85 f'_C A_C$$

- Effective critical buckling strength, P_e, is given by

$$P_e = \pi^2 EI_{eff}/(KL)^2$$

where

$$EI_{eff} = E_s I_s + 0.5E_s I_{sr} + C_i E_c I_c$$

$$C_i = 0.1 + 2(A_s/A_c + A_s) \le 0.3$$

13.10.3 Filled Composite Columns: Design Overview

- **Limitations**
  1. Steel shape must be at least 1% of gross composite section.
  2. Maximum b/t ratio for rectangular HSS = $2.26\sqrt{E/F_y}$.
  3. Maximum D/t ratio for round HSS = $0.75\ (E/F_y)$.
  4. Limitations on steel and concrete strengths are the same as for encased composite columns.
- The design compressive strength, $\phi_c P_n$, for axially loaded columns is determined for the limit state of buckling using $\phi_c = 0.75$.
- Maximum nominal compressive strength is given by

$$P_o = A_s F_y + A_{sr} F_{yr} + C_2 f_c' A_c$$

where

$$P_e = \pi^2 EI_{eff}/(KL)^2$$

$$EI_{eff} = E_s I_S + E_s I_{sr} + C_3 E_c I_c$$

$$C_3 = 0.6 + 2(A_s/A_c + A_s) \le 0.9$$

13.10.3.1 Encased Composite Columns: AISC Design Criteria

See Figure 13.19 for schematics of AISC design criteria.

13.10.3.2 Limitations

To qualify as an encased *composite column*, the following limitations shall be met:

1. The cross-sectional area of the steel core shall comprise at least 1% of the total composite cross section.
2. Concrete encasement of the steel core shall be reinforced with continuous longitudinal bars and lateral ties or spirals. The minimum *transverse reinforcement* shall be at least 0.009 in.^{2}/in. of tie spacing.
3. The minimum reinforcement ratio for continuous longitudinal reinforcing, ρ_{sr}, shall be 0.004, where ρ_{sr} is given by

$$\rho_{sr} = \frac{A_{sr}}{A_g}$$

where

A_{sr} is the area of continuous reinforcing bars, in.^{2}
A_g is the gross area of composite member, in.^{2}

13.10.3.3 Compressive Strength

The *design compressive strength,* $\phi_c P_n$, for axially loaded *encased composite columns* shall be determined for the limit state of *flexural buckling* based on column slenderness as follows:

$$\phi_c = 0.75$$

a. When $P_e \geq 0.44P_o$,

$$P_n = P_o\left[0.658^{(P_o/P_e)}\right]$$

b. When $P_e < 0.44P_o$,

$$P_n = 0.877P_e$$

where

$$P_o = A_sF_y + A_{sr}F_{yr} + 0.85A_cf_c'$$

$$P_e = \pi^2(EI_{eff})/(KL)^2$$

and where

A_s is the area of steel section, in.^{2}
A_c is the area of concrete, in.^{2}
A_{sr} is the area of continuous *reinforcing* bars, in.^{2}
E_c is the modulus of elasticity of *concrete* = $w_c^{1.5}\sqrt{f_c'}$ ksi
E_s is the modulus of elasticity of steel = 29,000 ksi
f_c' is the *specified* compressive strength of concrete, ksi
F_y is the *specified minimum yield stress* of steel section, ksi
F_{yr} is the *specified* minimum yield stress of reinforcing bars, ksi
I_c is the *moment* of inertia of the concrete section, in.^{4}
I_s is the *moment* of inertia of steel shape, in.^{4}
I_{sr} is the moment of inertia of reinforcing bars, in.^{4}
K is the effective length factor
L is the laterally unbraced length of the member, in.
w_c is the weight of concrete per unit volume ($90 \leq w_c \leq 155$ lb/ft^3)

where

EI_{eff} is the *effective* stiffness of composite section, kip-in.^{2}
$EI_{eff} = E_sI_s + 0.5E_sI_{sr} + C_1E_cI_c$

where

$$C_1 = 0.1 + \left(\frac{A_s}{A_c + A_s}\right) \leq 0.3$$

13.10.3.4 Tensile Strength

The design tensile strength, P_n, is given by

$$P_t = \phi_t P_n = 0.90(A_xF_y + A_{sr}F_{yr})$$

13.10.3.5 Shear Strength

The *available shear strength* shall be calculated based on either the shear strength of the steel section alone, plus the shear strength provided by tie reinforcement, if present, or the shear strength of the reinforced concrete portion alone. The nominal shear strength of tie reinforcement may be determined as $A_{st}F_{yr}$ (d/s) where A_{st} is the area of tie reinforcement, d is the effective depth of the concrete section, and s is the spacing of the tie reinforcement. The shear capacity of reinforced concrete may be determined according to ACI 318-11.

13.10.3.6 Load Transfer

Loads applied to axially loaded encased composite columns shall be transferred between the steel and concrete in accordance with the following requirements:

a. When the external *force* is applied directly to the steel section, *shear connectors* shall be provided to transfer the required shear force, V', as follows:

$$V' = V\left(1 - \frac{A_s F_y}{P_o}\right)$$

where

V is the required shear force introduced to column, kips
A_s is the area of steel cross section, in.^{2}
P_o is the nominal axial compressive strength without consideration of *length effects*, kips

b. When the external force is applied directly to the concrete encasement, shear connectors shall be provided to transfer the required shear force, V', as follows:

$$V' = V\left(\frac{A_s F_y}{P_o}\right)$$

c. When load is applied to the concrete of an encased composite column by direct bearing, the *design bearing strength*, $\phi_B P_p$, *of the concrete shall be*

$$P_p = 1.7 f_c' A_B$$

$$\phi_B = 0.65$$

where A_B is the loaded area of concrete, in.^{2}

13.10.3.7 Detailing Requirements

At least four continuous longitudinal reinforcing bars shall be used in encased composite columns. *Transverse reinforcement* shall be spaced at the smallest of 16 longitudinal bar diameters, 48 tie bar diameters, or 0.5 times the least dimension of the composite section. The encasement shall provide at least 1.5 in. of clear cover to the reinforcing steel.

Shear connectors shall be provided to transfer the required shear force and shall be distributed along the length of the member at least a distance of 2.5 times the depth of the encased composite column above and below the load transfer region. The maximum connector spacing shall be 16 in. Connectors that transfer axial load shall be placed on at least two faces of the steel shape in a configuration symmetrical about the steel shape axes.

If the composite cross section is built up from two or more encased steel shapes, the shapes shall be interconnected with *lacing, tie plates, batten plates,* or similar components to prevent buckling of individual shapes due to loads applied prior to hardening of the concrete.

13.10.3.8 Strength of Stud Shear Connectors

The *nominal strength* of one stud shear connector embedded in solid concrete is

$$Q_m = 0.5A_{sc}\sqrt{f'_c E_c} \leq A_{sc}F_u$$

where

A_{sc} is the cross-sectional area of stud shear connector, in.^{2}
F_u is the *specified minimum tensile strength* of a stud shear connector, ksi

13.10.3.9 Filled Composite Columns: AISC Design Criteria

See Figure 13.20 for structural schematics.

13.10.3.10 Limitations

To qualify as a filled *composite column*, the following limitations shall be met:

1. The cross-sectional area of the steel *HSS* shall comprise at least 1% of the total composite cross section.
2. The maximum *b/t* ratio for a rectangular HSS used as a composite column shall be equal to $2.26\sqrt{E/F_y}$. Higher ratios are permitted when their use is justified by testing or analysis.
3. The maximum *D/t* ratio for a round HSS filled with concrete shall be $0.15E/F_y$.

13.10.3.11 Compressive Strength

The design compressive strength, $P_c = \phi_c P_n$, for axially loaded filled composite columns shall be determined for the *limit state* of *flexural buckling* based on Section I2, lb with the following modifications:

$$P_o = A_s F_y + A_{sr} F_{yr} + C_2 A_c f'_c$$

$C_2 = 0.85$ for rectangular sections and 0.95 for circular sections

$$EI_{eff} = E_s I_s + E_s I_{sr} + C_3 E_c I_c$$

$$C_3 = 0.6 + 2\left(\frac{A_s}{A_c + A_s}\right) \leq 9$$

13.10.3.12 Tensile Strength

The *design tensile strength, $P_t = \phi_t P_n$*, for filled composite columns shall be determined as

$$P_t = \phi_t P_n = 0.90(A_s F_y + A_{sr} F_{yr})$$

13.10.3.13 Shear Strength

The *available shear strength* shall be calculated based on either the shear strength of the steel section alone or the shear strength of the reinforced concrete portion alone. The shear strength of reinforced concrete may be determined by ACI 318-11.

13.10.3.14 Load Transfer

Loads applied to filled composite columns shall be transferred between the steel and concrete. When the external force is applied either to the steel section or to the concrete infill, the transfer of force from the steel section to the concrete core is required through *bond interaction*, shear *connection*, or direct bearing. The force transfer *mechanism* providing the largest *nominal strength* may be used. These force transfer mechanisms shall not be superimposed.

When load is applied to the concrete of an encased or filled composite column by direct bearing, the *design bearing strength*, $\phi_B P_p$, of the concrete shall be

$$P_p = 1.7 f_c' A_B$$

$$\phi_B = 0.65$$

where A_B is the loaded area, in.^{2}

13.10.4 Summary of AISC Design Criteria for Composite Columns

Design of composite sections requires consideration of both steel and concrete behavior. The AISC specification uses a strength approach for design consistent with that used in reinforced concrete design and assumes that the user is familiar with reinforced concrete design specifications needed for the concrete portion of the design, such as material specifications, anchorage and splice lengths, and shear and torsion provisions.

13.10.4.1 Nominal Strength of Composite Sections

The strength of composite sections may be computed based on the plastic distribution approach. The plastic stress distribution method is based on the assumption of linear strain across the cross section and elastic–plastic behavior. It assumes that the concrete has reached its crushing strength in compression at a strain of 0.003 and a corresponding stress (typically $0.85 f_c'$) on a rectangular stress block and that the steel has exceeded its yield strain, typically taken as F_y/E_s.

Based on these simple assumptions, the cross-sectional strength for different combinations of axial force and bending moment may be approximated, for typical composite column cross sections. The actual interaction diagram for moment and axial force for a composite section is thus similar to that of a reinforced concrete section. As a simplification, a conservative linear interaction between selected anchor points, depending on axis of bending, can be used.

The plastic stress approach for columns assumes that no slip has occurred between the steel and concrete portions and that the required width-to-thickness ratios prevent local buckling from occurring until extensive yielding has taken place.

As for reinforced concrete design, a limit of 10 ksi is imposed for strength calculations. A lower limit of 3 ksi is specified for both normal- and lightweight concrete and an upper limit of 6 ksi specified for lightweight concrete. The use of higher strengths in computing the modulus of elasticity is permitted.

The design of concrete-encased and concrete-filled composite columns is treated separately, although they have much in common. The design for length effects is consistent with that for steel columns. The equations used are the same as in steel design, albeit in a slightly different format, and as the percent of concrete in the section decreases, the design defaults to that of a steel section.

13.10.5 Encased Composite Column Limitations

1. Composite columns may be designed with as little as 1% steel ratio (area of steel shape divided by the gross area of column).
2. The specified minimum quantity for transverse reinforcement is intended to provide good confinement to the concrete.
3. A minimum amount of longitudinal reinforcing steel is prescribed so that at least four continuous corner bars are used. Other longitudinal bars may be needed to provide the required restraint to the crossties but that longitudinal steel cannot be counted toward the cross-sectional strength unless it is continuous and properly anchored.

13.10.5.1 Compressive Strength

The compressive strength of the cross section is given as the sum of the ultimate strengths of the components. The strength is not capped as in reinforced concrete column design for a combination of the following reasons:

1. The resistance factor has been lowered from 0.85 in previous editions to 0.75 in this specification.
2. The required transverse steel provides better performance than a typical reinforced column.
3. The presence of a steel section near the center of the section reduces the possibility of a sudden failure due to bulking of the longitudinal steel.

13.10.5.2 Shear Strength

The provisions require either the use of the steel section alone plus the contribution from any transverse shear reinforcement present in the form of ties or the shear strength calculated based on the reinforced concrete portion of the cross section alone (in other words, longitudinal and transverse reinforcing bars plus concrete). This implies the following shear strengths:

$$V_n = 0.6F_yA_w + A_{st}F_{yr}\frac{d}{s}$$

$$\phi = 0.9$$

or

$$V_n = 2\sqrt{f_c'}bd + A_{st}F_{yr}\frac{d}{s}$$

$$\phi = 0.75$$

13.10.5.3 Load Transfer

To avoid overstressing either the structural steel section or the concrete at connections in encased composite columns, a transfer of load by direct bearing, shear connection, or a combination of both is required. Although it is recognized that force transfer also occurs by direct bond interaction between the steel and concrete, this is typically ignored for encased composite columns.

When the shear connectors are used in encased composite columns, a uniform spacing is appropriate in most situations, but when large forces are applied, other connector arrangements may be needed to avoid overloading the component to which the load is applied directly.

When supporting concrete area is wider on all sides than the loaded area, the nominal bearing strength for concrete may be taken as

$$Nb = 0.85f_c'\sqrt{\frac{A_2}{A_1}}$$

where

A_1 is the loaded area

A_2 is the maximum area of the supporting surface that is geometrically similar and concentric with the loaded area

The value of $\sqrt{A_2/A_1}$ must be less than or equal to 2. This specification uses the maximum nominal bearing strength of $1.7f_c'A_B$. The resistance factor for bearing ϕ_B is 0.65 in accordance with ACI 318-11.

13.10.6 Filled Composite Column: Limitations

1. As discussed for encased columns, it is now permissible to design composite columns with a steel ratio as low as 1%.
2. The specified minimum wall slenderness now takes into account the restraining effect of the concrete on the local buckling of the section wall.

13.10.6.1 Compressive Strength

The compressive strength of the cross section is given as the sum of the ultimate strengths of the components. The beneficial confining effect of the steel shell can be taken into account by increasing the crushing strength of the concrete to $0.95 f_c'$.

13.10.6.2 Load Transfer

To avoid overstressing either the structural steel section or the concrete at connections in filled composite columns, a transfer of load by direct bearing, shear connection, or direct bond interaction is permitted, with the mechanism providing the largest resistance being permissible for use. However, superposition of these force transfer mechanisms is not permitted for filled composite columns.

Force transfer by direct bond is commonly used in filled composite columns as long as the connections are detailed to limit local deformations.

13.10.7 Combined Axial Force and Flexure

One of the three permitted methods in the AISC specifications is the plastic distribution method for determining the combined effect of axial force, P_u, and moment, M_u. This method is not unlike the method we use in everyday practice to generate the P_u–M_u interaction diagrams for reinforced concrete columns.

For an encased composite column, perhaps the most general and fundamental approach is to treat the steel section as an equivalent reinforcement placed at discrete locations within concrete section. Thus, the generation of the P_u–M_u interaction diagram for a composite column reverts back to that for a reinforced concrete column.

Shown in Figure 13.20 is an interaction diagram obtained by using such an approach. The interaction diagram is for a 36 × 36 in. column with 12 #18 reinforcing bars and a W14 × 150 structural steel shape. For comparison purposes, the interaction diagram for the same concrete column without the embedded structural steel shape is given.

However, to assist in developing the interaction curves without an elaborate discretization of the steel section, the AISC specification provides a series of simplified equations for calculating the P_u–M_u values on the interaction curve. The defining points on the interaction curve are shown as A, B, C, D, and E in Figure 13.21. The reader is referred to the AISC Commentary Sections for further details.

13.11 DESIGN TABLES AND DETAILS

The remainder of this chapter presents certain selected design tables and details that apply to composite construction of buildings (Figures 13.22 through 13.44; Tables 13.1 through 13.3).

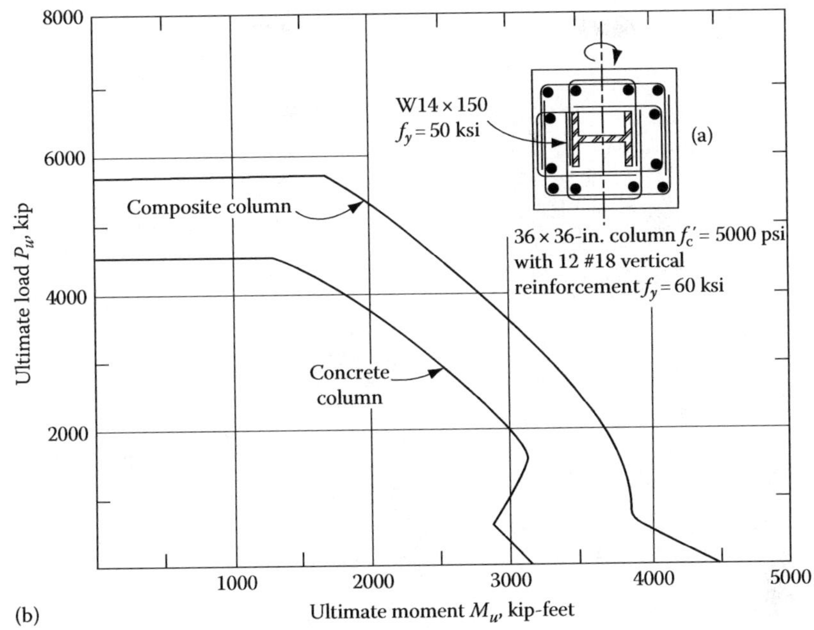

FIGURE 13.20 Composite column interaction diagram: (a) column detail; (b) interaction diagram.

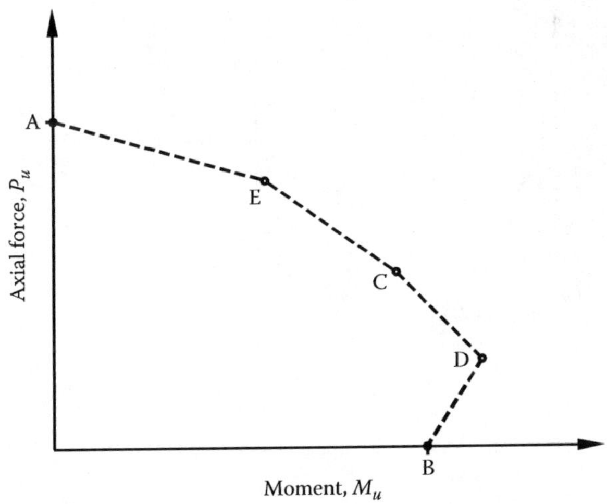

FIGURE 13.21 Interaction diagram for beam-column design.

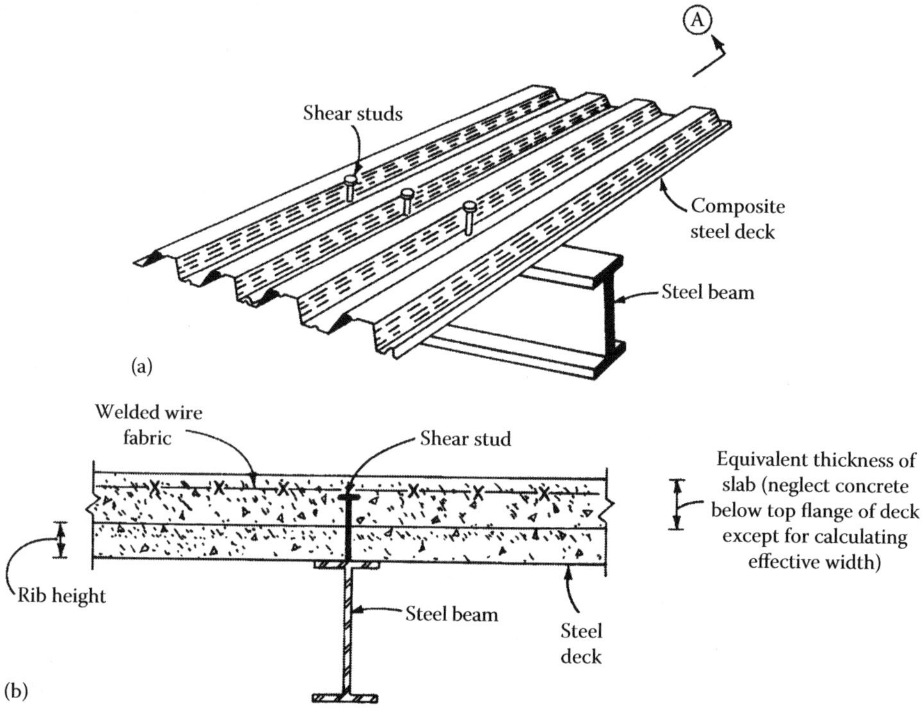

FIGURE 13.22 Composite beam with formed steel deck: (a) schematic view; (b) section A.

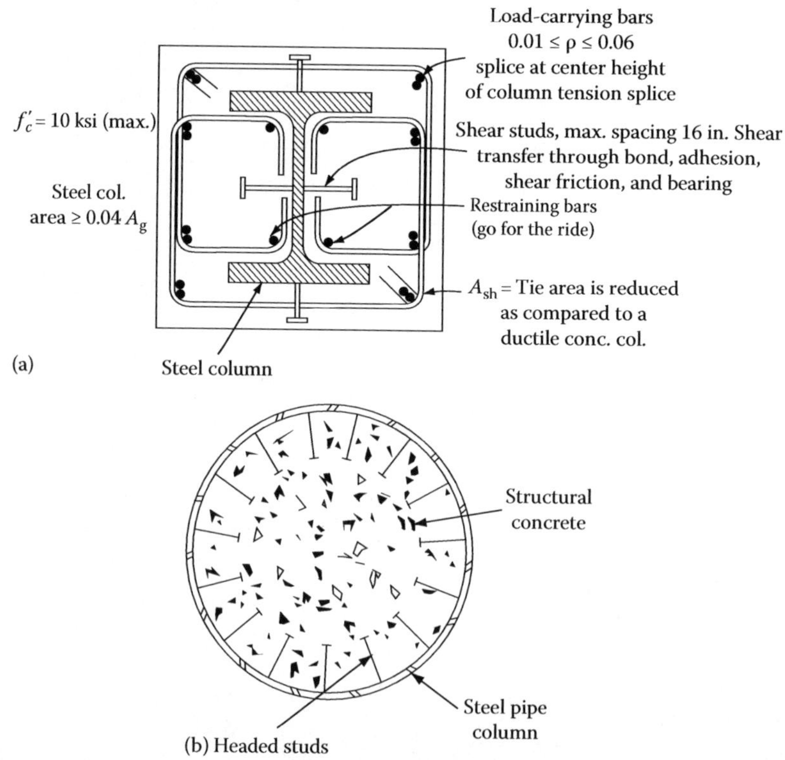

FIGURE 13.23 (a) Concrete encased composite column; design considerations. *Note*: Bond and adhesion must be ignored in calculating shear transfer. (b) Concrete-filled composite pipe column.

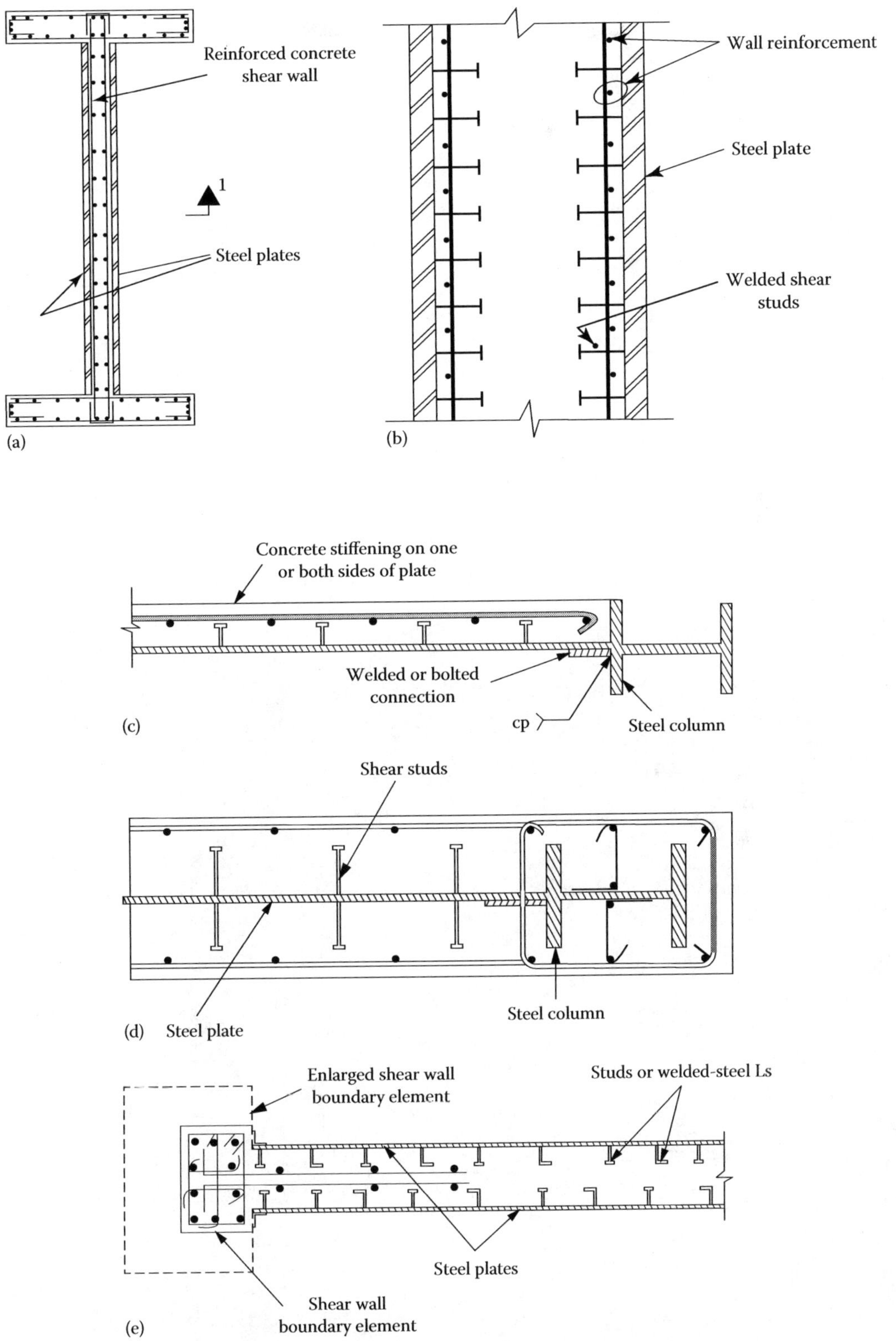

FIGURE 13.24 Composite steel plate shear walls: (a) plan; (b) section; (c) and (d) shear wall with single plate; steel plate on both faces of wall; (e) concrete wall with steel boundary elements. *(Continued)*

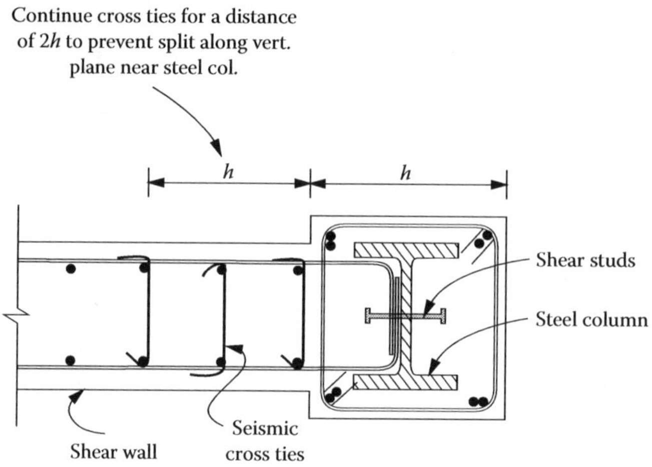

(f)

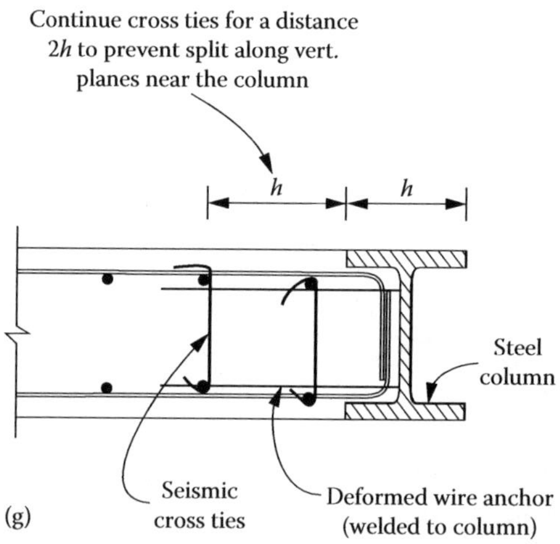

(g)

FIGURE 13.24 (*Continued*) Composite steel plate shear walls: (f) concrete wall with steel boundary elements; (g) concrete wall with steel boundary elements.

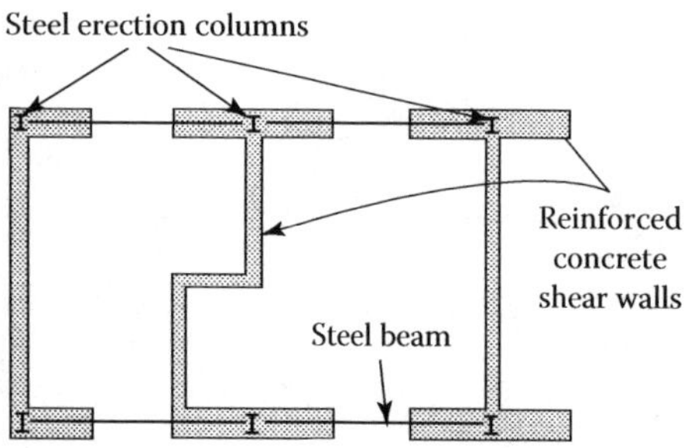

(a)

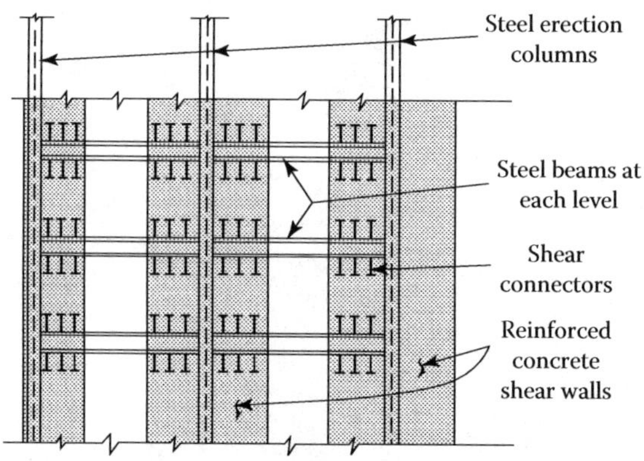

(b)

FIGURE 13.25 Composite shear wall with steel link beams: (a) plan; (b) elevation.

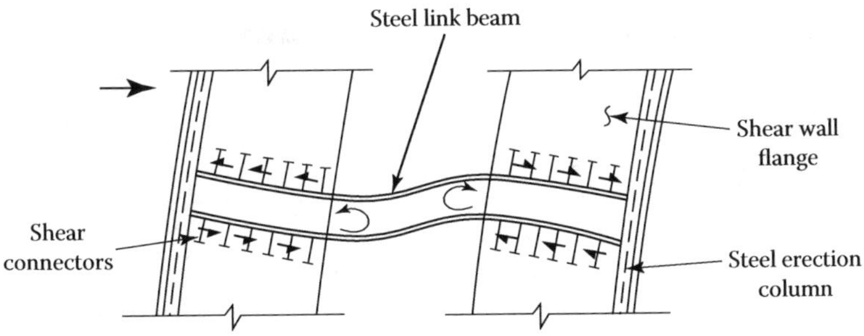

FIGURE 13.26 Moment transfer between steel link and beam and concrete wall.

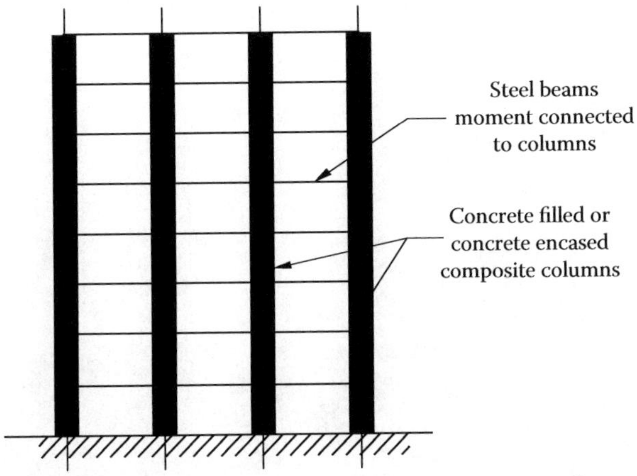

FIGURE 13.27 Composite moment frame.

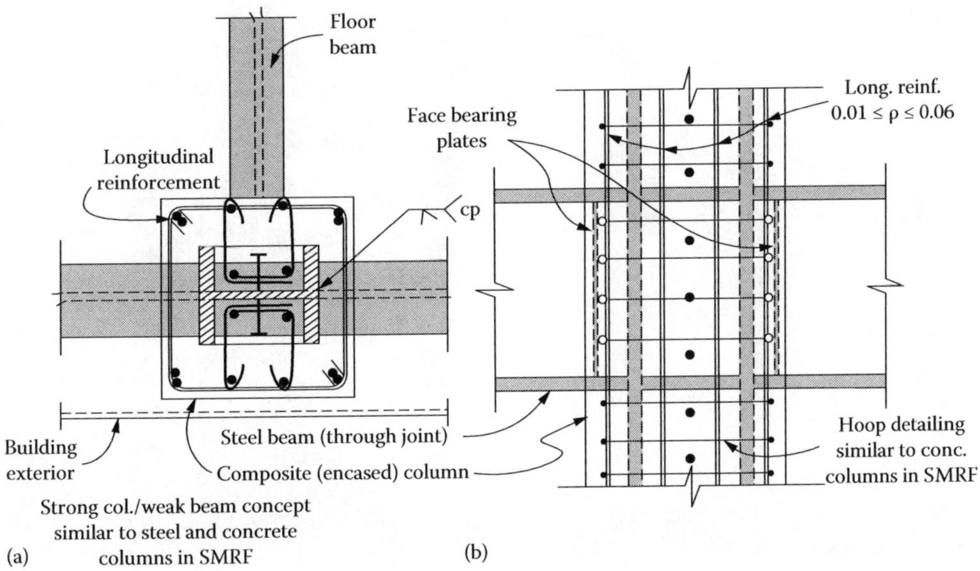

FIGURE 13.28 Spandrel beam-to-composite column connection: (a) plan; (b) elevation.

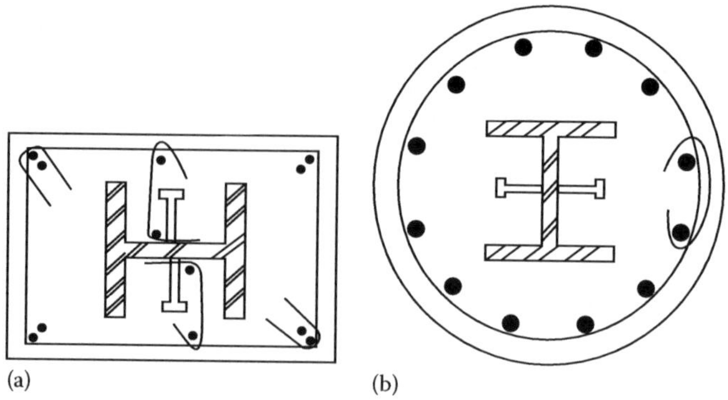

FIGURE 13.29 Seismic tie-arrangement in composite columns: (a) rectangular column; (b) circular column.

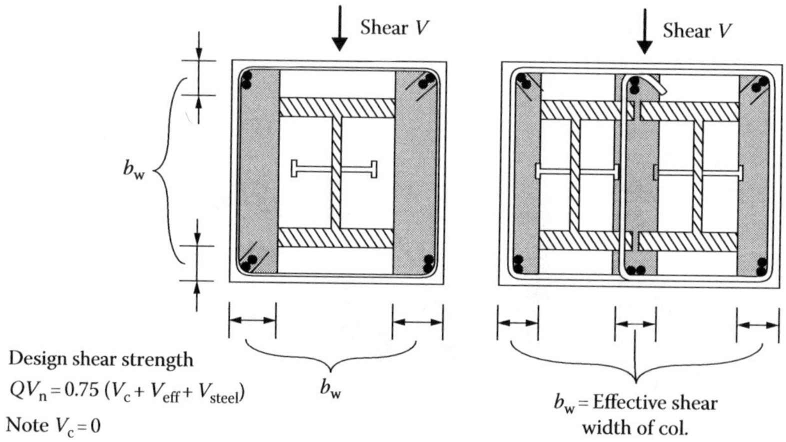

FIGURE 13.30 Encased composite column; shear design parameters.

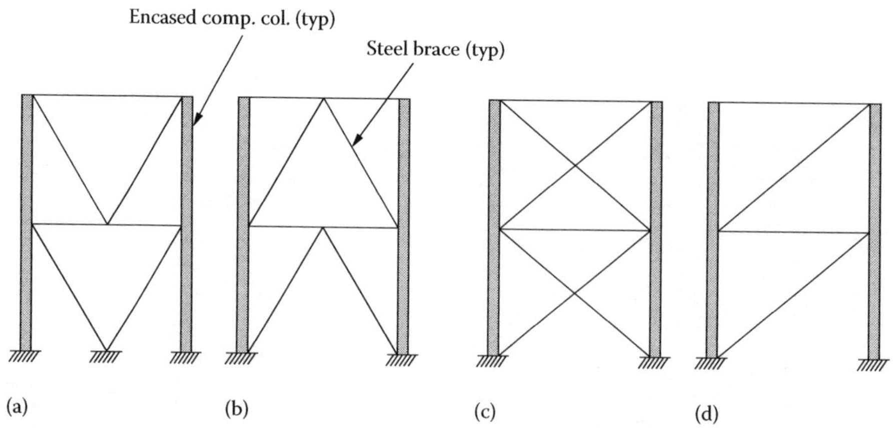

FIGURE 13.31 Composite concentrically braced frames: (a) V-bracing; (b) inverted V-bracing; (c) X-bracing (d) diagonal bracing.

(Continued)

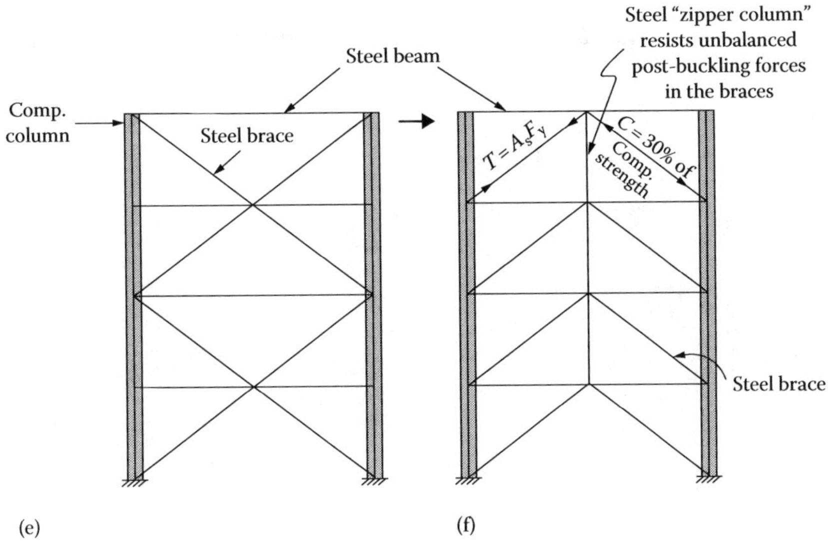

FIGURE 13.31 (*Continued*) Composite concentrically braced frames: (e) two-story X-bracing; (f) zipper column with inverted V-bracing.

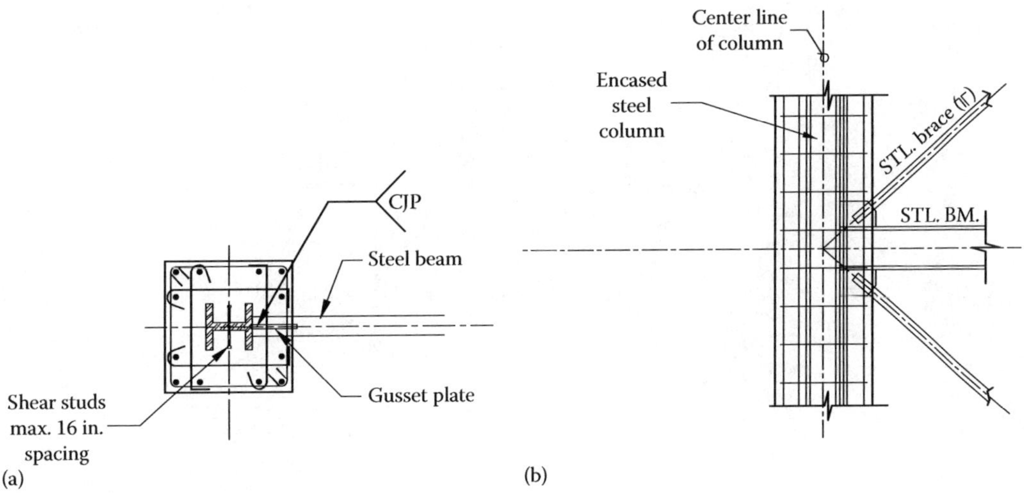

FIGURE 13.32 Composite concentric braced frame: Connection schematics; (a) plan; (b) elevation.

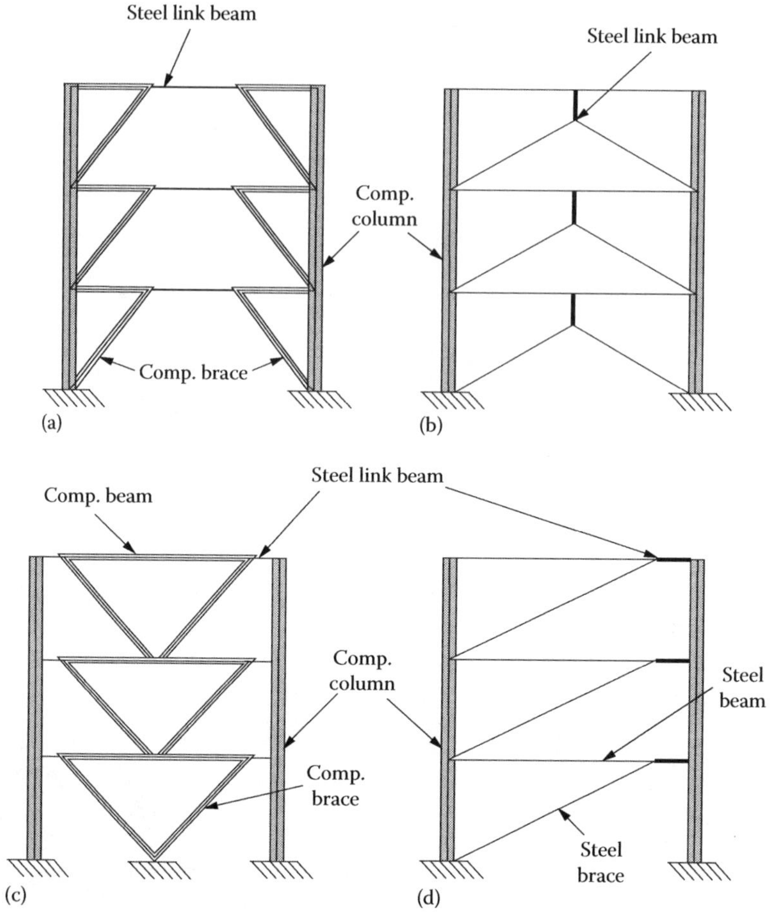

FIGURE 13.33 Examples of composite eccentrically braced frames.

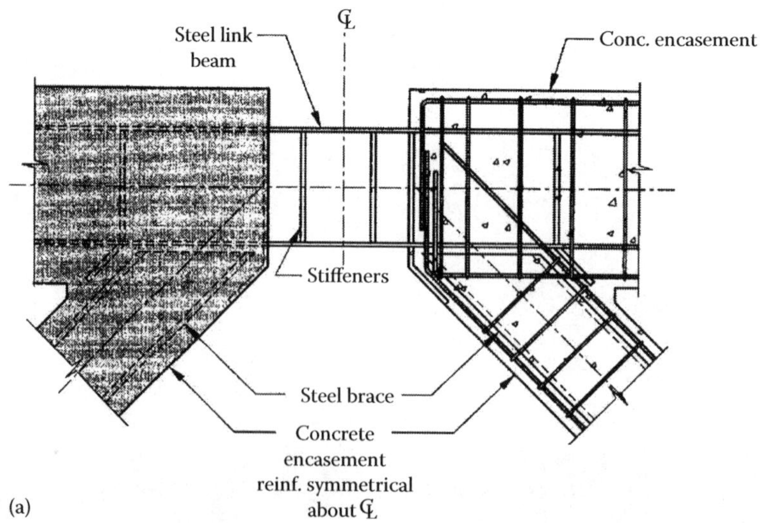

FIGURE 13.34 Schematic details of link beam: (a) link at center of beam. *(Continued)*

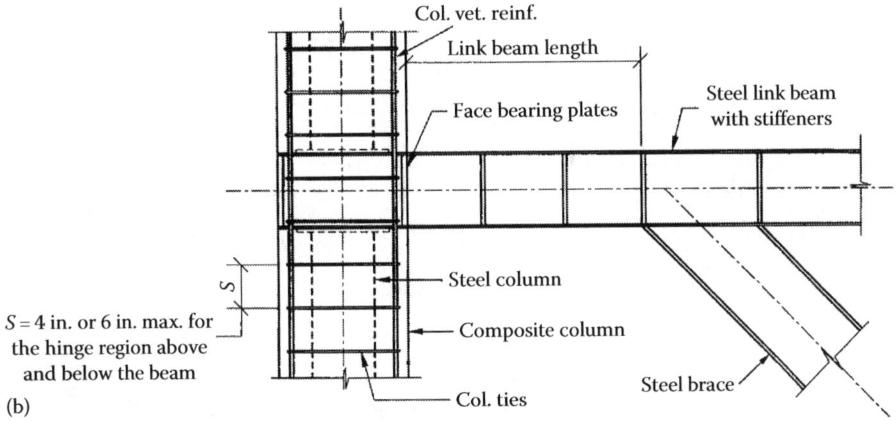

FIGURE 13.34 (*Continued*) Schematic details of link beam: (b) link adjacent to column.

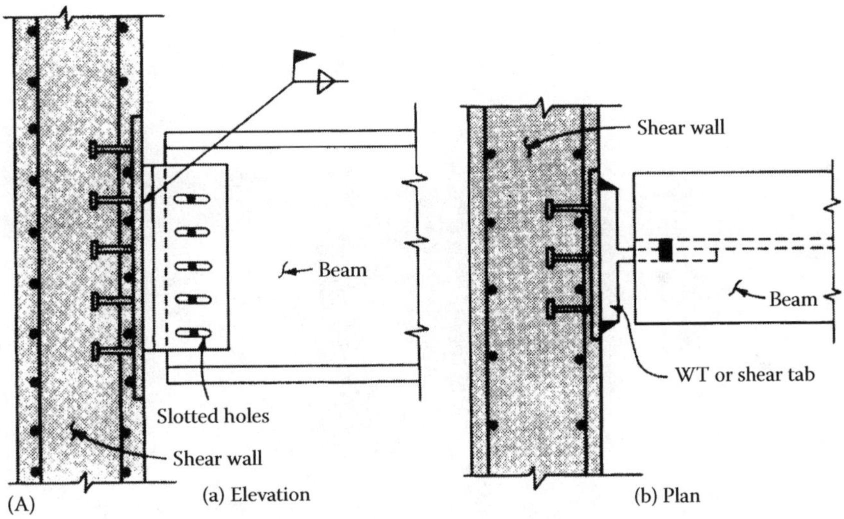

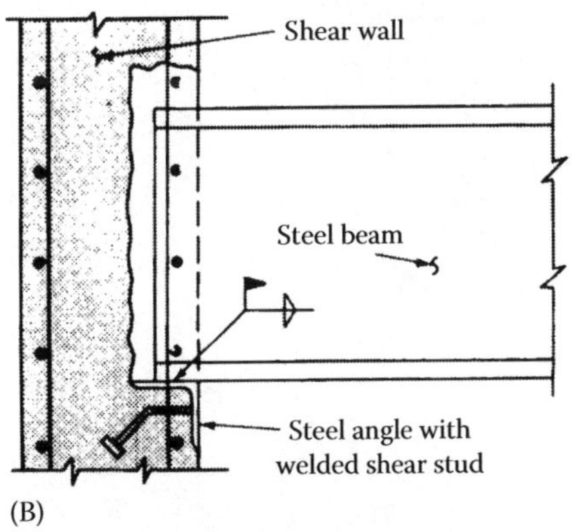

FIGURE 13.35 Beam-to-shear wall connection: (a) embedded plate detail; (A) elevation; (b) pocket detail. (B) plan.

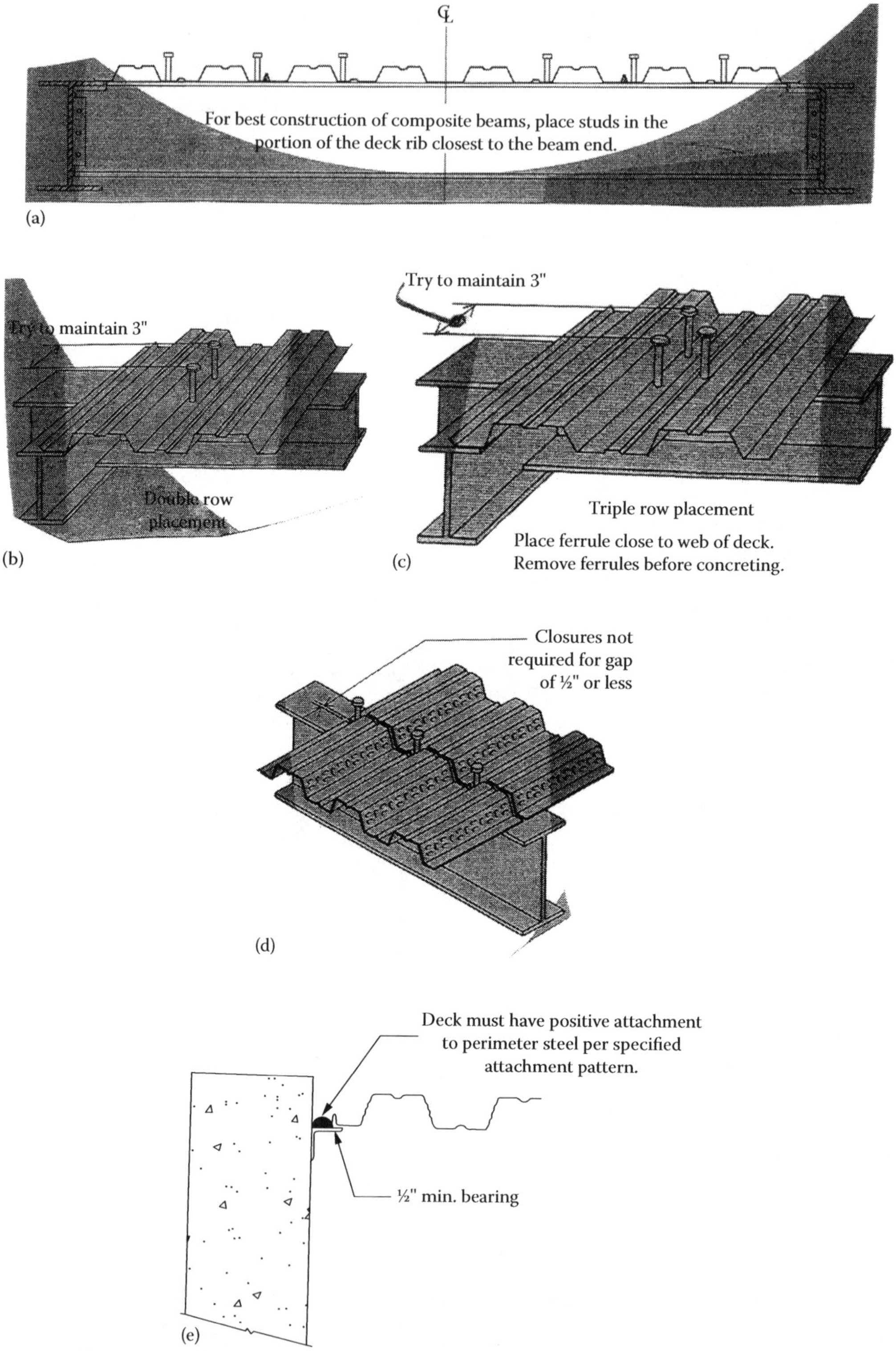

FIGURE 13.36 *(Continued)*

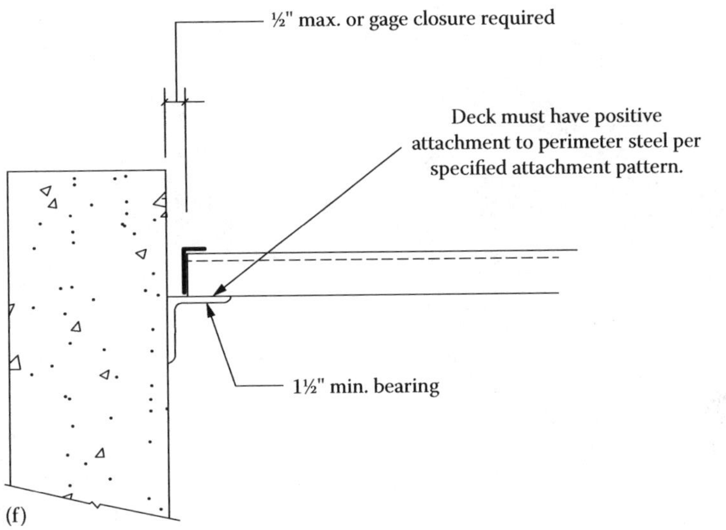

(f)

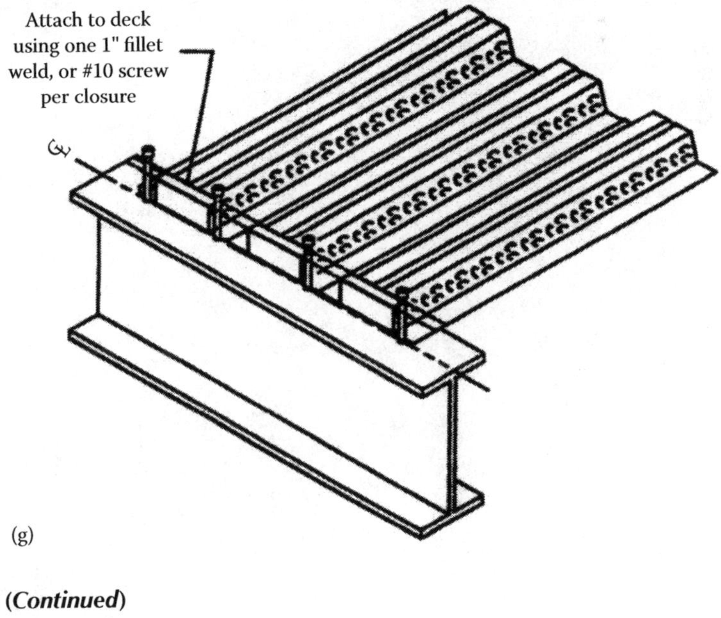

(g)

FIGURE 13.36 (*Continued*)

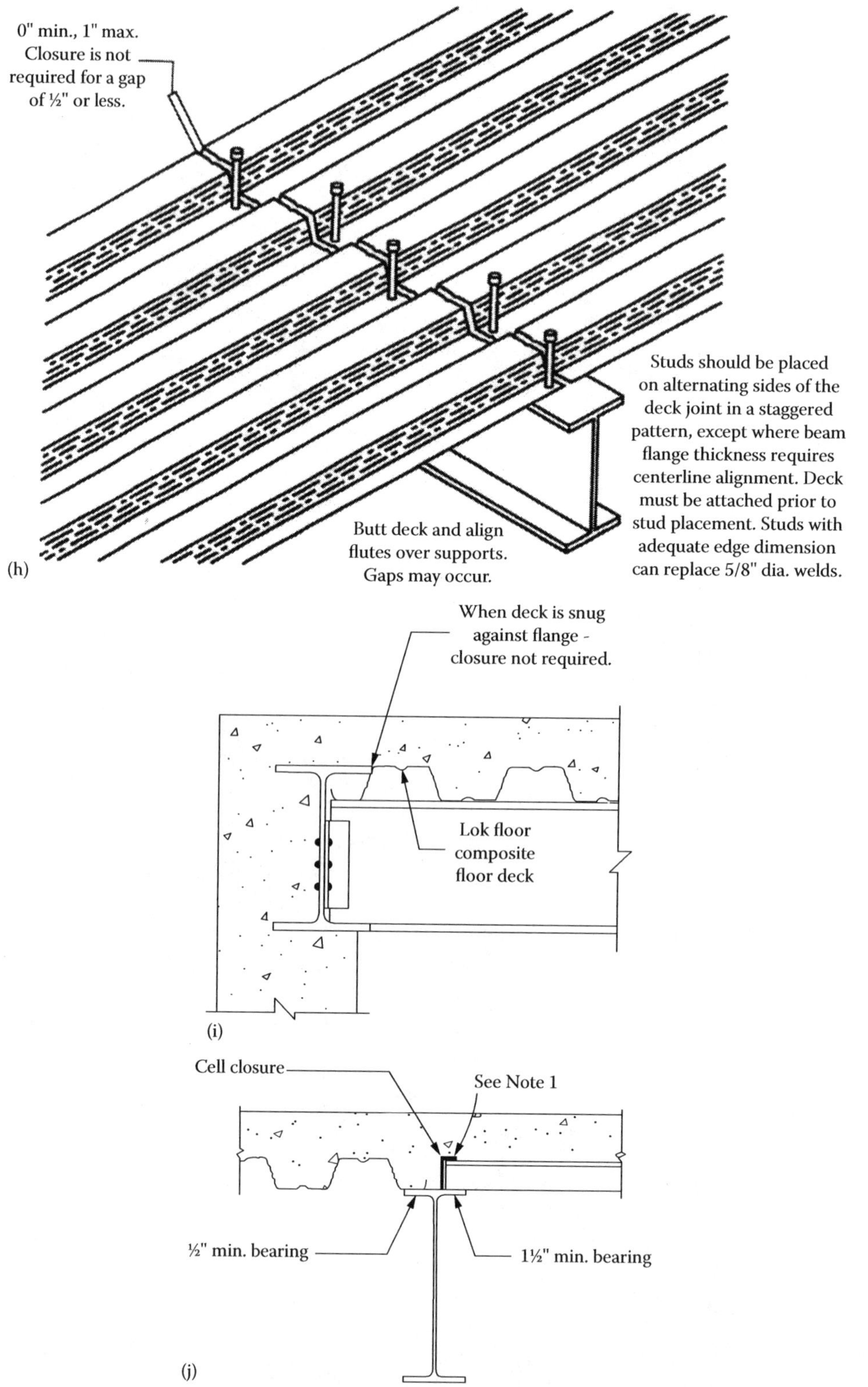

FIGURE 13.36 (*Continued*) (*Continued*)

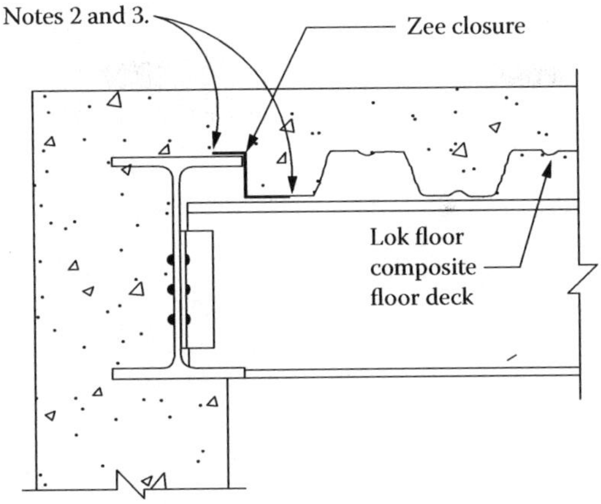

(k)

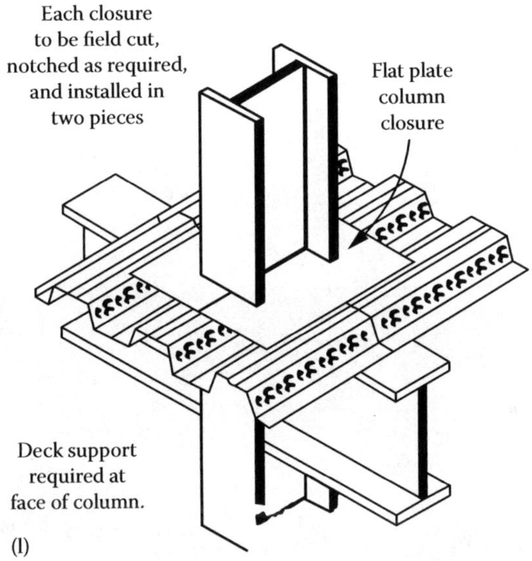

(l)

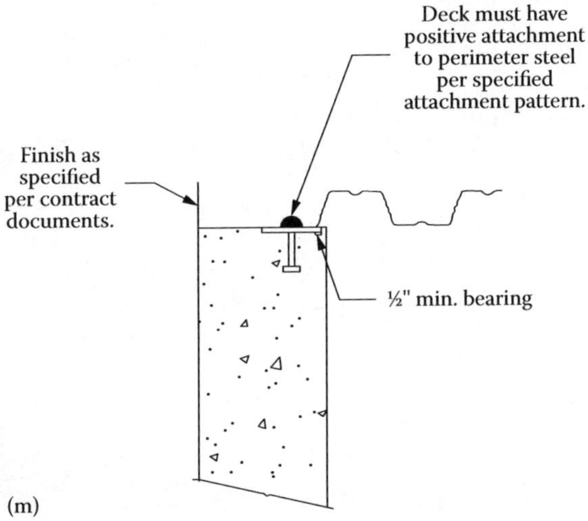

(m)

FIGURE 13.36 (*Continued*)

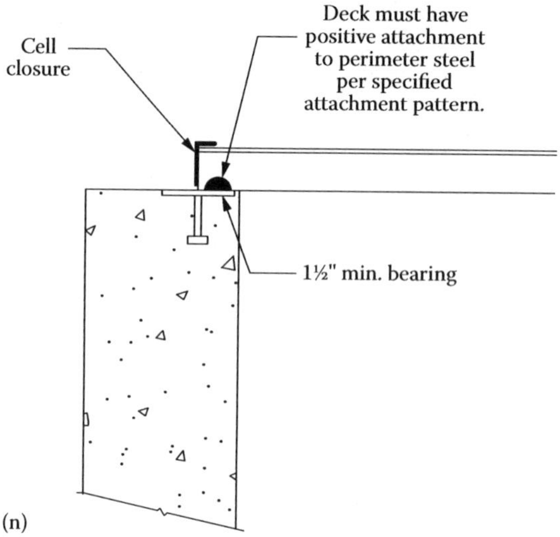

(n)

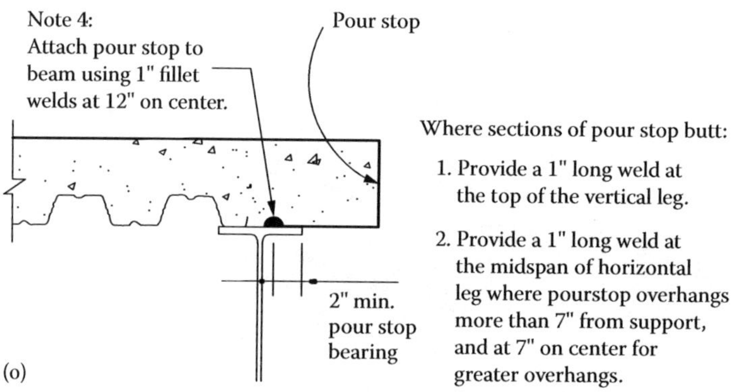

(o)

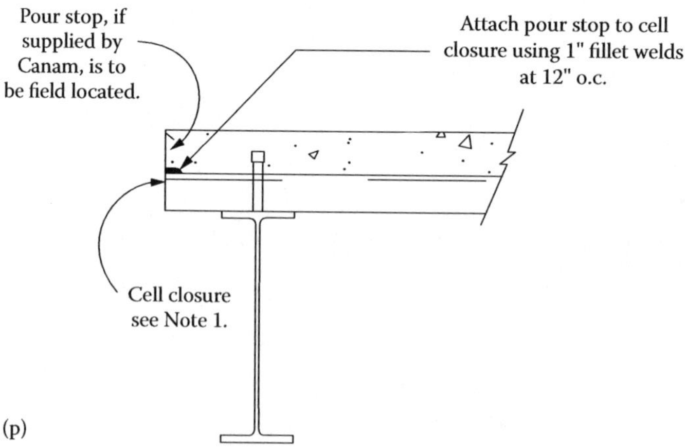

(p)

FIGURE 13.36 (*Continued*)

(Continued)

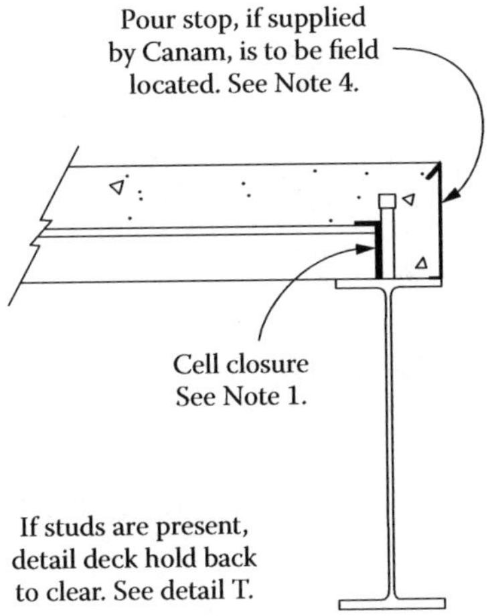

(q)

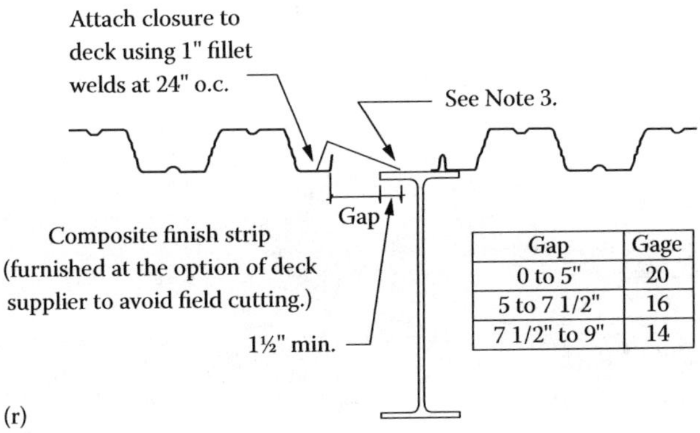

Gap	Gage
0 to 5"	20
5 to 7 1/2"	16
7 1/2" to 9"	14

(r)

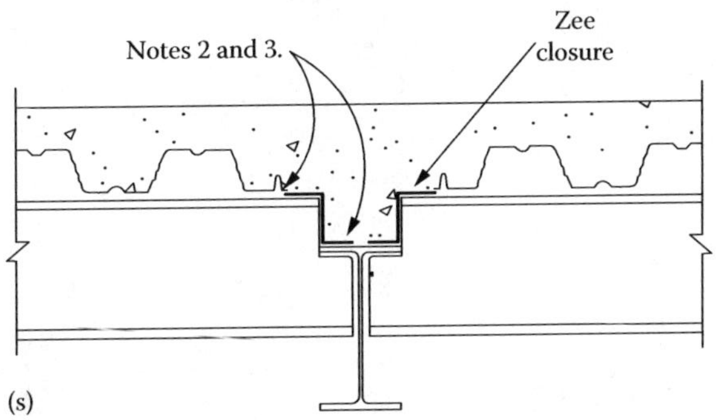

(s)

FIGURE 13.36 (*Continued*)

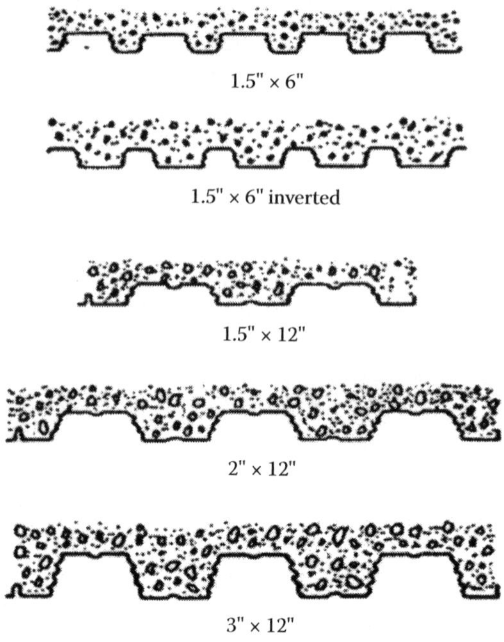

FIGURE 13.37

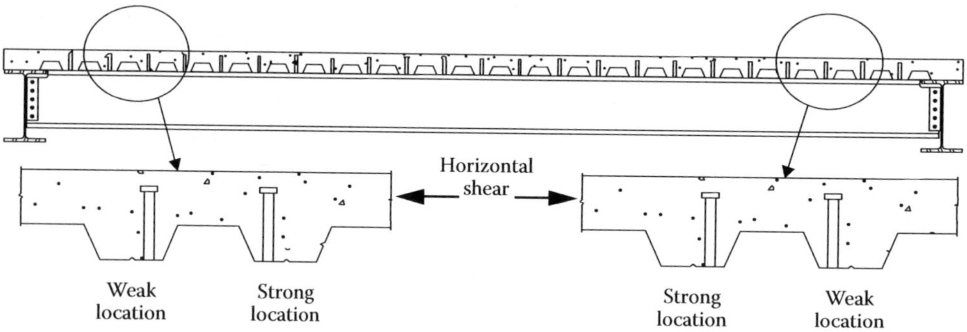

FIGURE 13.38

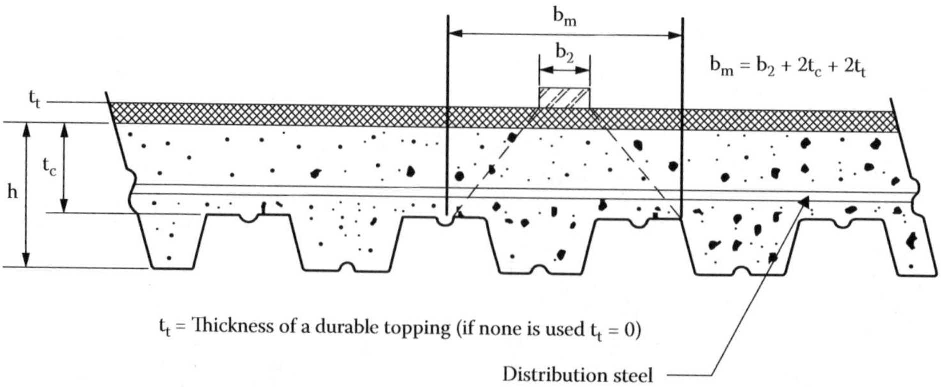

FIGURE 13.39

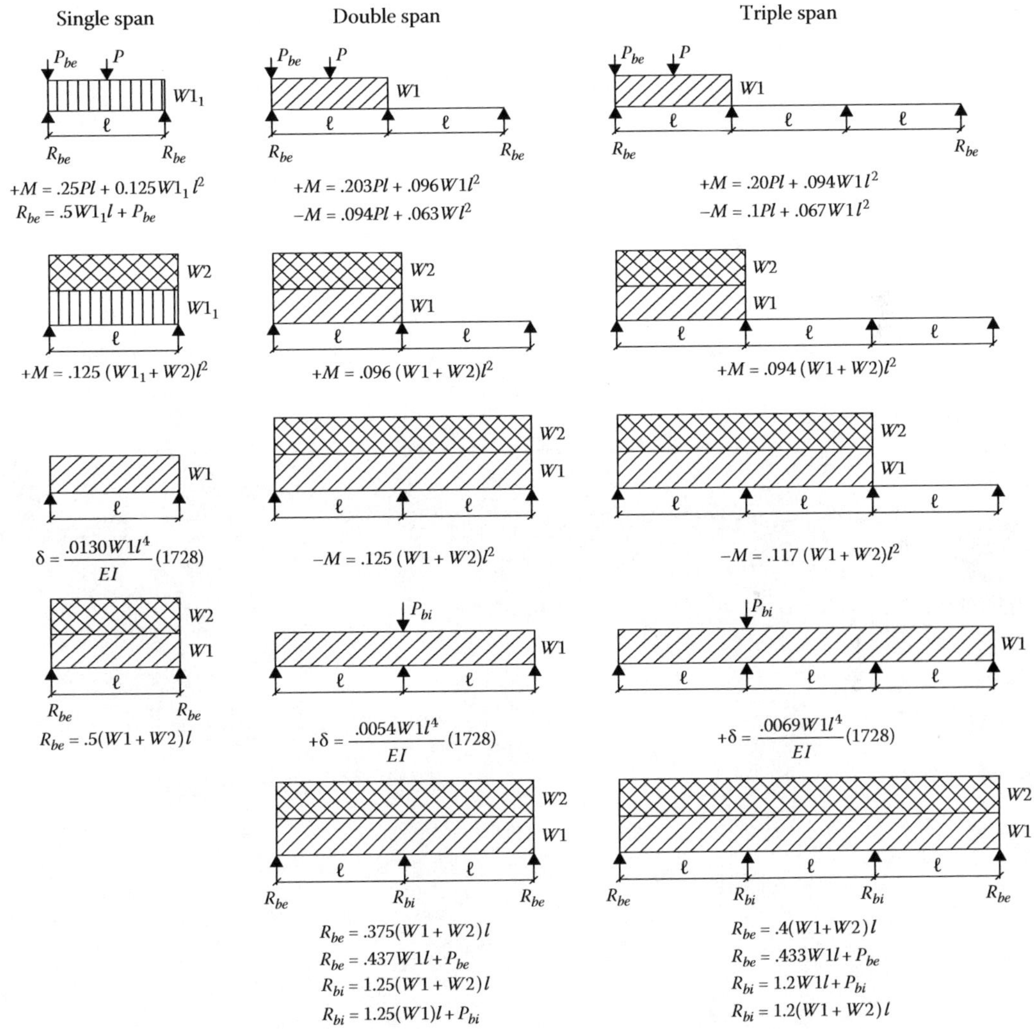

FIGURE 13.40 SDI formulas for construction loads.

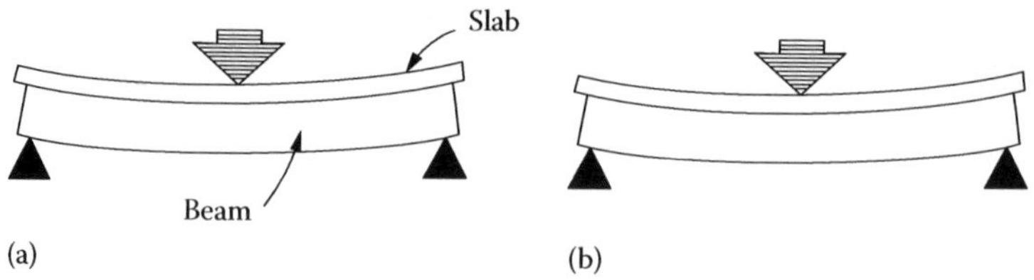

FIGURE 13.41

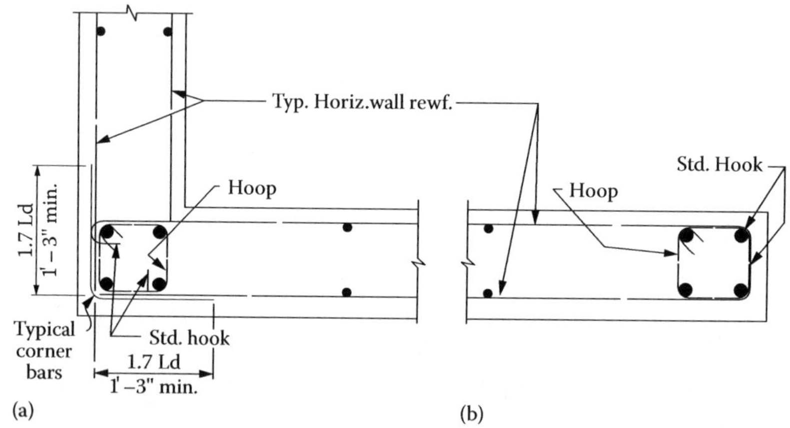

FIGURE 13.42 (a) Concrete wall intersection. (b) JAMB.

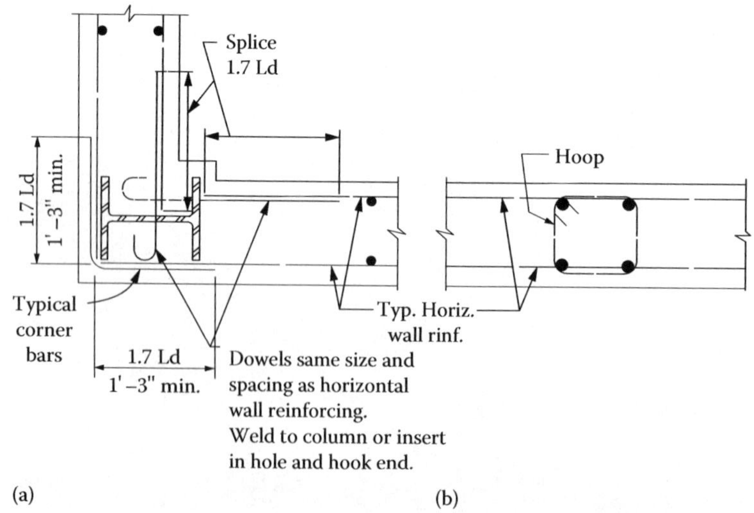

FIGURE 13.43 (a) Steel member in concrete wall. (b) Wall-column.

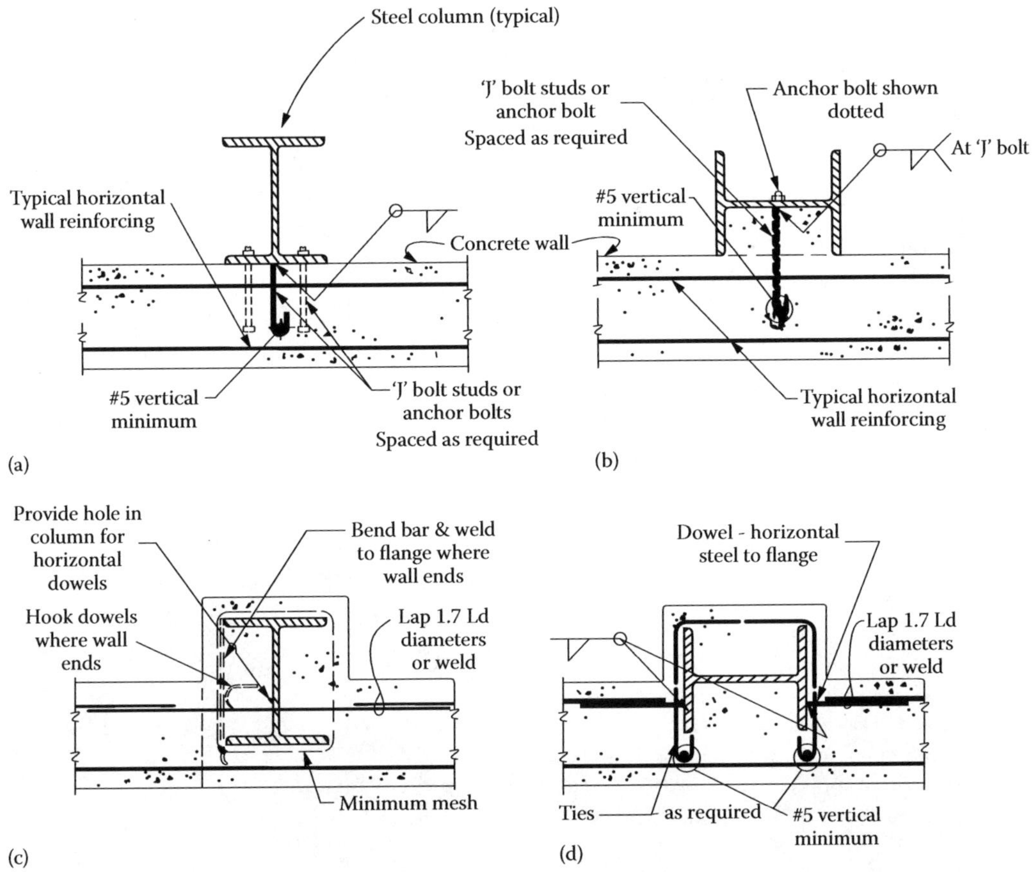

FIGURE 13.44 Concrete wall to steel column.

TABLE 13.1
Deck and Slab Selection for Required Fire Rating

Deck Depth (in.)	Fire Rating	Concrete Density	Minimum Concrete Thickness above Flutes (in.)	Stud Length (in.) Minimum After Welding	Stud Length (in.) Recommended* Before Welding	Stud Length (in.) Recommended* After Welding
2	2-h	LWC	3.5	3 1/2	4 3/16	4
3			3.5	4 1/2	5 3/16	5
2		NWC	4.5	3 1/2	5 3/16	5
3			4.5	4 1/2	6 3/16	6
2	1-h	LWC	2.75	3 1/2	3 7/8	3 11/16
3			2.75	4 1/2	4 7/8	4 11/16
2		NWC	3.5	3 1/2	4 3/16	4
3			3.5	4 1/2	5 3/16	5

Notes:

- Studs must be 1½ in. (min.) above steel deck.
- For sand-lightweight concrete, 3½ in. thickness over flutes required for 2-h fire rating.
- Thinner slabs require sprayed on fire protection.

TABLE 13.2
Steel Deck Maximum Spans

		Deck Gage						
2-h Fire Rating	3.5″ LWC above Flutes	Deck Gage						
	Deck Depth	22	21	20	19	18	17	16
	2	8′-5″	9′-0″	9′-6″	10′-6″	11′-4″	12′-1″	12′-8″
	3	9′-1″	10′-9″″	11′-7″	12′-10″	13′-8″	14′-7″	15′-3″″
	4.5″ NWC above Flutes	Deck Gage						
	Deck Depth	22	21	20	19	18	17	16
	2	6′-2″	7′-6″	8′-1″	8′-11″	9′-8″	10′-3″	10′-9″
	3	6′-6″	7′-9″	8′-10″	10′-10″	11′-7″	12′-4″	13′-0″
1-h Fire Rating	2.75″ LWC above Flutes	Deck Gage						
	Deck Depth	22	21	20	19	18	17	16
	2	8′-10″	9′-6″	10′-1″	11′-2″	12′-0″	12′-10″″	13′-5″
	3	10′-1″	11′-7″	12′-3″	13′-6″	14′-5″	15′-4″	16′-1″
	3.5″ NWC above Flutes	Deck Gage						
	Deck Depth	22	21	20	19	18	17	16
	2	7′-2″	8′-2″	8′-8″	9′-7″	10′-4″	11′-0″	11′-7″
	3	7′-5″	8′-10″	10′-1″	11′-7″	12′-5″	13′-3″	13′-11″

Notes:

1. No shoring is required for two or three continuous spans.
2. Deck depth and gage chosen to set beam spacing.
3. Use 18 gage and thinner deck for greatest economy in deck system.
4. Greater deck span allows economy in steel framing (reduction in piece count, connection, etc.).
5. Check bearing width and web crippling of deck during construction.
6. Check shear strength of deck.
7. Check for concentrated loads and verify design using SDI criteria.
8. Provide reinforcing in the form of welded wire reinforcing or rebar.

TABLE 13.3
Steel Deck Gage and Thickness

	Design Thickness		Minimum Thickness	
Gage No.	**in.**	**mm**	**in.**	**mm**
22	0.0295	0.75	0.028	0.71
21	0.0329	0.84	0.031	0.79
20	0.0358	0.91	0.034	0.86
19	0.0418	1.06	0.040	1.01
18	0.0474	1.20	0.045	1.14
17	0.0538	1.37	0.051	1.30
16	0.0598	1.52	0.057	1.44

Bibliography

2003 International Building Code. ICC Publications.

2003 International Building Code Commentary, vol. 2. ICC Publications.

2000 IBC Structural/Seismic Design Manual, vols. 1, 2, and 3. Structural Engineers Association of California, Sacramento, CA: ICC Publications.

ACI, *Building Code Requirements for Structural Concrete* (ACI 318-95) *and Commentary* (ACI 318R-95). Farmington Hills, MI: American Concrete Institute, 1996.

ACI, *Building Code Requirements for Structural Concrete* (ACI 318-08) and Commentary, Farmington Hills, MI: American Concrete Institute, 2008.

AISC, *Manual of Steel Construction: Allowable Stress Design*, 9th edn. Chicago, IL: American Institute of Steel Construction, 1989.

AISC, *Steel Construction Manual*, 13th edn. Chicago, IL: American Institute of Steel Construction, 2005.

AISC, *Seismic Design Manual.* Chicago, IL: American Institute of Steel Construction, 2006.

ANSI A58.1. *American National Standard Minimum Design Loads for Buildings and Other Structures*, Washington, DC: American National Standards Institute, 1982. (Need doc title)

ANSI/AISC 341-10, Seismic provisions for structural steel buildings, Public review draft dated April 17, 2009.

ANSI/AISC 358-05, Prequalified connections for special and intermediate steel moment frames for seismic applications.

ANSI/AISC 360-I0, Specification for structural steel buildings, June 22, 2010.

ASCE/SEI 7-10. *Minimum Design Loads for Buildings and Other Structures.* Reston, VA: American Society of Civil Engineers, 2010.

ASCE/SEI 31-03, *Seismic Evaluation of Existing Buildings*, American Society of Civil Engineers, 2003.

ASCE/SEI 41-06, *Seismic Rehabilitation of Existing Buildings*, American Society of Civil Engineers, 2006.

Biggs, J. M., *Introduction to Structural Dynamics.* New York: McGraw-Hill, 1964.

Bozorgnia, Y. and V. V. Bertero (eds.), *Earthquake Engineering: From Engineering Seismology to Performance-Based Engineering.* ICC Publications, 2004.

Cermak, J. E., Applications of fluid mechanics to wind engineering, *Journal of Fluids Engineering*, 97, 9, March 1975.

Chen, W.-F. and C. Scawthorn (eds.), *Earthquake Engineering Handbook.* ICC Publications.

Chopra, A. K., *Dynamic of Structures, Theory and Applications to Earthquake Engineering,* 2nd edn. Upper Saddle River, NJ: Pearson Education, Inc., 2001.

Clough, R. W. and J. Penzien, Dynamics of Structures, 2nd edn. New York: McGraw-Hill, Inc., 1993.

Coleman, R. A., *Structural System Designs.* Englewood Cliffs, NJ: Prentice Hall, 1983.

Cooper, S. E. and A. C. Chen, *Designing Steel Structures—Methods and Cases.* Englewood Cliffs, NJ: Prentice Hall, 1985.

FEMA 140, *Guide to Application of the NEHRP Recommended Provisions in Earthquake-Resistant Building Design*. Washington, DC: Federal Emergency Management Agency, 1990.

FEMA 172, *NEHRP Handbook for Seismic Rehabilitation of Existing Buildings*. Washington, DC: Federal Emergency Management Agency, 1992.

FEMA 178, *NEHRP Handbook for the Seismic Evaluation of Existing Buildings.* Washington, DC: Developed by the Building Seismic Safety Council for the Federal Emergency Management Agency, 1992.

FEMA 222 and 223, *NEHRP Recommended Provisions for the Development of Seismic Regulations for New Buildings, Part 1, Provisions, Part 2, Commentary*. Washington, DC: Federal Emergency Management Agency, 1991.

FEMA 267, *Interim Guidelines, Inspection, Evaluation, Repair, Upgrade and Design of Welded Moment-Resisting Steel Structures.* Washington, DC: Prepared by the SAC Joint Venture for the Federal Emergency Management Agency, 1995.

FEMA 267A, *Interim Guidelines Advisory No. 1.* Washington, DC: Prepared by the SAC Joint Venture for the Federal Emergency Management Agency, 1996.

FEMA 267B, *Interim Guidelines Advisory No. 2.* Washington, DC: Prepared by the SAC Joint Venture for the Federal Emergency Management Agency, 1999.

FEMA 273, *NEHRP Guidelines for the Seismic Rehabilitation of Buildings*. Washington, DC: Prepared by the Applied Technology Council for the Building Seismic Safety Council, published by the Federal Emergency Management Agency, 1997.

FEMA 274, *NEHRP Commentary on the Guidelines for the Seismic Rehabilitation of Buildings*. Washington, DC: prepared by the Applied Technology Council for the Building Seismic Safety Council, published by the Federal Emergency Management Agency, 1997.

FEMA 302, *NEHRP Recommended Provisions for Seismic Regulations for New Buildings and Other Structures, Part I—Provisions*. Washington, DC: Prepared by the Building Seismic Safety Council for the Federal Emergency Management Agency, 1997.

FEMA 303, *NEHRP Recommended Provisions for Seismic Regulations for New Buildings and Other Structures, Part 2—Commentary*. Washington, DC: Prepared by the Building Seismic Safety Council for the Federal Emergency Management Agency, 1997.

FEMA 310, *Handbook for the Seismic Evaluation of Buildings—A Prestandard*. Washington, DC: Prepared by the American Society of Civil Engineers for the Federal Emergency Management Agency, 1998.

FEMA 350, *Recommended Seismic Design Criteria for New Steel Moment-Frame Buildings*. Washington, DC: Prepared by the SAC Joint Venture for the Federal Emergency Management Agency, 2000.

FEMA 351, *Recommended Seismic Evaluation and Upgrade Criteria for Existing Welded Steel Moment—Frame Buildings*. Washington, DC: Prepared by the SAC Joint Venture for the Federal Emergency Management Agency, 2000.

FEMA 352, *Recommended Postearthquake Evaluation and Repair Criteria for Welded Steel Moment—Frame Buildings*. Washington, DC: Prepared by the SAC Joint Venture for the Federal Emergency Management Agency, 2000.

FEMA 353, *Recommended Specifications and Quality Assurance Guidelines for Steel Moment—Frame Construction for Seismic Applications*. Washington, DC: Prepared by the SAC Joint Venture for the Federal Emergency Management Agency, 2000.

FEMA 354, *A Policy Guide to Steel Moment-Frame Construction*. Washington, DC: Prepared by the SAC Joint Venture for the Federal Emergency Management Agency, 2000.

FEMA 355A, *State-of-the-Art Report on Base Metals and Fracture*. Washington, DC: Prepared by the SAC Joint Venture for the Federal Emergency Management Agency.

FEMA 355B, *State-of-the-Art Report on Welding and Inspection*. Washington, DC: Prepared by the SAC Joint Venture for the Federal Emergency Management Agency, 2000.

FEMA 355C, *State-of-the-Art Report on Systems Performance of Steel Moment-Frames Subject to Earthquake Ground Shaking*. Washington, DC: Prepared by the SAC Joint Venture for the Federal Emergency Management Agency, 2000.

FEMA 355D, *State-of-the-Art Report on Connection Performance*. Washington, DC: Prepared by the SAC Joint Venture for the Federal Emergency Management Agency, 2000.

FEMA 355E, *State-of-the-Art Report on Past Performance of Steel Moment-Frame Buildings in Earthquakes*. Washington, DC: Prepared by the SAC Joint Venture for the Federal Emergency Management Agency, 2000.

FEMA 355F, *State of the Art Report on Performance Prediction and Evaluation of Steel Moment—Frame Buildings*. Washington, DC: Prepared by the SAC Joint Venture for the Federal Emergency Management Agency, 2000.

FEMA 356, *Prestandard and Commentary for the Seismic Rehabilitation of Buildings*. Washington, DC: Prepared by the SAC Joint Venture for the Federal Emergency Management Agency, 2000.

FEMA 403, *World Trade Center Building Performance Study: Data Collection, Preliminaiy Observations, and Recommendations*. Washington, DC: Federal Insurance Mitigation Association, and New York: The Federal Emergency Management Agency, Region II, 2000.

FEMA 445, *Next-Generation Performance-Based Seismic Design Guidelines*. Washington, DC: Federal Emergency Management Agency, 2006.

FEMA 450, *NEHRP Recommended Provisions for Seismic Regulations for New Buildings and Other Structures*, Part 1, Provisions, Part 2, Commentary. Washington, DC: Federal Emergency Management Agency, 2003.

FEMA 547, *Techniques for the Seismic Rehabilitation of Existing Buildings*. Washington, DC: Federal Emergency Management Agency, 2006.

FEMA P-750, *NEHRP Recommended Seismic Provisions*. Washington, DC: Federal Emergency Management Agency, 2009.

Gaylord, Jr. E. H. and C. N. Gaylord, *Design of Steel Structures*, 2nd edn. New York: McGraw-Hill, 1972.

Gaylord, Jr. E. H. and C. N. Gaylord, *Structural Engineering Handbook*, 3rd edn. New York: McGraw-Hill, 1990.

Ghosh, S. K., Seismic design using structural dynamics, in *2000 International Building Code.* ICC Publications.

Ghosh, S. K., Seismic and wind design of concrete buildings, in *2000 International Building Code.* ICC Publications.

Halvorson, R. and N. Isyumov, Comparison of predicted and measured dynamic behavior of Allied Bank Plaza, in N. Isyumov and T. Tschanz (eds.), *Building Motion in Wind.* New York: ASCE, 1982.

Hamburger, R. O., A. B. Court, and J. R. Soulages, Vision 2000: A framework for performance-based engineering of buildings, in *Proceedings of Sixth SEAOC Convention*, pp. 127–146, 1995.

Kanchi, M. B., *Matrix Methods of Structural Analysis.* New York: Wiley Eastern Limited, 1994.

Ketchum, M. S. (ed.), *Structural Engineering Practice*, vol. 1 and 2, 1982; vol. 4, 1982–1983; vol. 1, 2, and 3, 1983; vol. 4, 1983–1984. New York: Marcel Dekker.

Lenzen, K. H., Vibrations of steel joist-concrete slab floors, *AISC Engineering Journal*, 3(3), 133–136, July 1966.

Lin, T. Y. and S. D. Stotesbury, *Structural Concepts and Systems for Architects and Engineers.* New York: Wiley, 1981.

Maeim, F. (ed.), *The Seismic Design Handbook*, 2nd edn. Boston, MA: Kluwer Academic Publishers, 2001.

Merritt, F. S., *Building Design and Construction Handbook*, 4th edn. New York: McGraw-Hill, 1982.

Mota, M. and B. S. Taranath, Structural systems for concrete buildings, in *Performance-Based Design of Concrete Building for Wind Loads*, ed. J. Bracci. Farmington Hills, MI: American Concrete Institute, 2006.

Murray, T. M., Design to prevent floor vibrations, *Engineering Journal American Institute of Steel Construction*, 12(3), 82, 1975.

Nester, M. R. and A. R. Porush, A rational system for earthquake risk management, in *SEAOC Conference Proceedings*, 1991.

Newmark, N. M. and E. Rosenblueth, *Fundamentals of Earthquake Engineering.* Englewood Cliffs, NJ: Prentice Hall, 1971.

Parmalee, R. A., *Wind Effects on High-Rise Buildings: Proceedings of the Chicago Design Symposium*, Northwestern University, Evanston, IL. March 1970.

Paz, M., *Structural Dynamics.* New York: Van Nostrand Reinhold, 1985.

PCI Design Handbook, 5th ed. PCI, Chicago, 1999.

Priestly, M. J. N., G. M. Calvi, and M. J. Kowalsky, *Displacement-Based Seismic Design of Structures.* Pavia, Italy: IUSS Press, 2007.

Ramaswamy, S. D. and C. T. Tam, *Tall Buildings*. Proceedings of the International Conference on Tall Buildings, 22–26 October 1984, Institution of Engineers, Singapore.

Reinhold, T. J. (ed.), *Wind Tunnel Modeling for Civil Engineering Applications.* Cambridge, U.K.: Cambridge University Press, 1982.

Rush, R., Technics: Structuring tall buildings, *Progressive Architecture*, 61(12), December 1980.

Schueller, W., *High-Rise Building Structures.* New York: John Wiley & Sons, 1977.

Seismic Design Guidelines for Upgrading Existing Buildings, PB 89-220453. Departments of the Army, the Navy, and the Air Force, Washington, DC, 1988.

Seismology Committee, Structural Engineers Association of California, *Recommended Lateral Force Requirements and Commentary* (Blue Book), 7th edn. Sacramento, CA: Seismology Committee, Structural Engineers Association of California, 1999.

Smith, B. S. and A. Coull, *Tall Building Structures, Analysis and Design.* New York: John Wiley & Sons, 1991.

Tall buildings, in *Conference Proceedings,* Singapore, 1984.

Taly, N., *Loads and Load Paths in Buildings: Principles of Structural Design.* Los Angeles, CA: ICC Publications.

Taranath, B. S., Torsional behavior of open-section shear wall structures, PhD thesis, University of Southampton, Southampton, U.K., 1968.

Taranath, B. S., Optimum belt truss locations for high-rise buildings, *AISC Engineering Journal*, 11(1), 18–21, 1974.

Taranath, B. S., Torsion analysis of braced multi-storey buildings, *The Structural Engineer*, 53(8), 345–347, 1975.

Taranath, B. S., The effect of warping on interconnected shear wall flat plate structures, *Proceedings of the Institution of Civil Engineers*, 1976.

Taranath, B. S., *Steel, Concrete, and Composite Design of Tall Buildings.* New York: McGraw-Hill, 1997.

Taranath, B. S., Differential shortening of columns in high-rise buildings, *Journal of Torsteel*, 1981.

Taranath, B. S., Composite design of First City Tower, *The Structural Engineer*, 60, 9, 271–281, 1982.

Taranath, B. S., A new look at composite high-rise construction, *Our World in Concrete and Structures*, Singapore, 1983.

Taranath, B. S., Analysis of interconnected open section wall structures, *ASCE Journal*, 1986.

Taranath, B. S., *Structural Analysis and Design of Tall Buildings*. New York: McGraw-Hill, 1988.

Taranath, B. S., *Wind and Earthquake Resistant Buildings, Structural Analysis and Design*. Boca Raton, FL: CRC Press, 2005.

Taranath, B. S., *Reinforced Concrete Design of Tall Buildings*. Boca Raton, FL: CRC Press, 2010.

Taranath, B. S. et al., A practical computer method of analysis for complex shear wall structure, in *Eighth ASCE Conference on Electronic Computation*, New York, 1983.

Tschanz, T., Measurement of total dynamic loads using elastic models with high natural frequencies, in T. J. Reinhold (ed.), *Wind Tunnel Modeling for Civil Engineering Applications*. Cambridge, U.K.: Cambridge University Press, 1982.

Wakabayashi, M., *Design of Earthquake Resistant Buildings*. New York: McGraw-Hill, 1986.

Williams, A., *Seismic and Wind Forces: Structural Design Examples*. Country Club Hills, IL: ICC Publications, 2007.

Wiss, J. F. and R. H. Parmalee, Human perception of transient vibrations, *Journal of Structural Division ASCE*, 100, 773–783, April 1974.

Index

T

Z